T0185564

Lecture Notes in Computer Science   11712

More information about this series at http://www.springer.com/series/7412

Frank Nielsen · Frédéric Barbaresco (Eds.)

Geometric Science of Information

4th International Conference, GSI 2019
Toulouse, France, August 27–29, 2019
Proceedings

 Springer

Editors
Frank Nielsen
Sony Computer Science Laboratories, Inc.
Tokyo, Japan

Frédéric Barbaresco
Thales
Limours, France

ISSN 0302-9743 ISSN 1611-3349 (electronic)
Lecture Notes in Computer Science
ISBN 978-3-030-26979-1 ISBN 978-3-030-26980-7 (eBook)
https://doi.org/10.1007/978-3-030-26980-7

LNCS Sublibrary: SL6 – Image Processing, Computer Vision, Pattern Recognition, and Graphics

Cover page painting: Woman teaching Geometry, from French medieval edition of Euclid's Elements (14th century) © The British Library, used with granted permission.

This Springer imprint is published by the registered company Springer Nature Switzerland AG
The registered company address is: Gewerbestrasse 11, 6330 Cham, Switzerland

Foreword

On behalf of both the Organizing and the Scientific Committees, it is our great pleasure to welcome you to the proceedings of the 4th International SEE Conference on Geometric Science of Information (GSI 2019), hosted at ENAC in Toulouse, during August 27–29, 2019.

GSI 2019 benefited from the following scientific and financial sponsors: SMF, SMAI, GDR CNRS ISIS & MIA, ENAC, THALES, Ecole Polytechnique, Mines ParisTech, Sony Computer Science Laboratories Inc (Sony CSL). GSI 2019 was the opening event of the CIMI labex trimester on "Statistics with Geometry and Topology": https://perso.math.univ-toulouse.fr/statistics-geometry-and-topology/.

The three-day conference was also organized in the framework of the relations set up between SEE and scientific institutions or academic laboratories: ENAC, Institut Mathématique de Bordeaux, Ecole Polytechnique, Ecole des Mines ParisTech, Inria, CentraleSupélec, Institut Mathématique de Bordeaux, Sony Computer Science Laboratories.

We would like to express all our thanks to the local organizers (ENAC, IMT and CIMI Labex) for hosting this event at the interface between geometry, probability, and information geometry.

The GSI conference cycle was initiated by the Brillouin Seminar Team as early as 2009. The GSI 2019 event was motivated in the continuity of the first initiatives launched in 2013 (https://www.see.asso.fr/gsi2013) at Mines ParisTech, consolidated in 2015 (https://www.see.asso.fr/gsi2015) at Ecole Polytechnique, and opened up to new communities in 2017 (https://www.see.asso.fr/gsi2017) at Mines ParisTech. We mention that in 2011, we organized an Indo-French workshop on "Matrix Information Geometry" that yielded an edited book in 2013, and in 2017, we collaborated with the CIRM seminar in Luminy TGSI 2017 "Topological and Geometrical Structures of Information" (http://forum.cs-dc.org/category/94/tgsi2017). The last GSI 2017 proceedings were published by Springer in their *Lecture Notes in Computer Science* series (https://www.springer.com/fr/book/9783319684444) and the most important contributions to the GSI 2017 conference are collected in the Springer volume *Geometric Structures of Information* (https://www.springer.com/us/book/9783030025199).

The technical program of GSI 2019 covered all the main topics and highlights in the domain of "geometric science of information" including information geometry manifolds of structured data/information and their advanced applications. These proceedings consist solely of original research papers that were carefully peer-reviewed by two or three experts and revised before acceptance.

The GSI 2019 program included a renowned honorary speaker, one guest honorary speaker, and three keynote distinguished speakers, as well as a history session with a talk on "Fermat, Pascal, and the Geometry of Chance" and a "Tribute to Jean-Louis Koszul" (who passed away in January 2018).

As with GSI 2013, GSI 2015, and GSI 2017, GSI 2019 addressed inter-relations between different mathematical domains like shape spaces (geometric statistics on manifolds and Lie groups, deformations in shape space, etc.), probability/optimization and algorithms on manifolds (structured matrix manifold, structured data/Information etc.), relational and discrete metric spaces (graph metrics, distance geometry, relational analysis, etc.), computational and Hessian information geometry, geometric structures in thermodynamics and statistical physics, algebraic/infinite dimensional/Banach information manifolds, divergence geometry, tensor-valued morphology, optimal transport theory, manifold and topology learning, and applications such as geometries of audio-processing, inverse problems, and signal/image processing. The GSI 2019 topics were enriched with contributions on Lie group machine learning, harmonic analysis on Lie groups, geometric deep learning, geometry of Hamiltonian Monte Carlo, geometric and (poly)symplectic integrators, contact geometry and Hamiltonian control, geometric and structure preserving discretizations, probability density estimation and sampling in high dimension, geometry of graphs and networks and geometry in neuroscience and cognitive sciences.

At the turn of the century, new and fruitful interactions were discovered between several branches of science: information science (information theory, digital communications, statistical signal processing), mathematics (group theory, geometry and topology, probability, statistics, sheaves theory) and physics (geometric mechanics, thermodynamics, statistical physics, quantum mechanics). The GSI conference cycle is an attempt to discover joint mathematical structures to all these disciplines by elaboration of a "general theory of information" embracing physics, information science, and cognitive science in a global scheme.

GSI 2019 addressed the following sessions with associated chairs:

- Probability on Riemannian Manifolds – Marc Arnaudon, Ana Bela Cruzeiro
- Optimization on Manifold – Salem Said, Rodolphe Sepulchre
- Shape Space – Nicolas Charon, Pietro Gori
- Statistics on Non-linear Data – Xavier Pennec, Stefan Sommer
- Lie Group Machine Learning – Elena Celledoni, Frédéric Barbaresco
- Statistical Manifold and Hessian Information Geometry – Michel Nguiffo Boyom, Hiroshi Matsuzoe
- Monotone Embedding and Affine Immersion of Probability Models – Jun Zhang, Atsumi Ohara
- Non-parametric Information Geometry – Lorenz Schwachhöfer, John Armstrong
- Divergence Geometry – Frank Nielsen, Wolfgang Stummer
- Computational Information Geometry – Frank Nielsen, Olivier Schwander
- Wasserstein Information Geometry/Optimal Transport – Guido Montufar, Wuchen Li
- Geometric Structures in Thermodynamics and Statistical Physics – Goffredo Chirco, François Gay-Balmaz
- Geometric and Structure-Preserving Discretizations – François Gay-Balmaz, Joël Bensoam
- Geometry of Quantum States – Florio Maria Ciaglia, Giuseppe Marmo
- Geometry of Tensor-Valued Data – Jesús Angulo, Geert Verdoolaege

- Geometric Mechanics – Géry de Saxcé, Jean Lerbet
- Geometric Science of Information Libraries – Nina Miolane, Alice Le Brigant
- Poster Session – Pierre Baudot

We were also honored to have the following keynote talks:

- Invited Honorary Speaker

  - Alain Chenciner (Université Paris 7, Observatoire de Paris), "N-body Relative Equilibria in Higher Dimensions"

- Guest Honorary Speaker

  - Karl Friston (Wellcome Trust Centre for Neuroimaging), "Markov Blankets and Bayesian Mechanics"

- Keynote Speakers

  - Elena Celledoni (Norwegian University of Science and Technology), "Structure preserving algorithms for geometric numerical integration"
  - Gabriel Peyré (CNRS, Ecole Normale Supérieure), "Optimal Transport for Machine Learning"
  - Jean-Baptiste Hiriart-Urruty (Université de Toulouse), "Fermat, Pascal: Geometry and Chance"

August 2019   Frank Nielsen
                                                          Frédéric Barbaresco

Organization

Program Chairs

Frédéric Barbaresco   Thales Air Systems, France
Frank Nielsen   Sony Computer Science Laboratories Inc., Japan

Patronage

SEE: Société de l'électricité, de l'électronique et des technologies de l'information et de la communication

Jean Vieille   SEE Webmaster, France
Valérie Alidor   SEE, France
Marianne Emorine   SEE, France

Scientific Committee

Alain Trouvé   ENS Paris-Saclay, France
Alessandro Barp   Imperial College London, UK
Alex Lin   University of California, Los Angeles, USA
Alexis Arnaudon   Imperial College London, UK
Ali Mohammad-Djafari   CNRS, France
Alice Barbara   Universite Lille, France
Alice Le Brigant   Thales Air Systems/Institut Mathématique de Bordeaux, France
Amor Keziou   Université de Reims, France
Ana Bela Cruzeiro   Universidade de Lisboa, Portugal
Anatole Chessel   École polytechnique, France
Anton Thalmaier   University of Luxembourg, Luxembourg
Antonio Mucherino   IRISA, University of Rennes 1, France
Arjan Van Der Schaft   University of Gröningen, The Netherlands
Atsumi Ohara   University of Fukui, Japan
Barbara Opozda   Jagiellonian University, Kraków, Poland
Barbara Trivellato   Politecnico di Torino, Italy
Benjamin Charlier   Université de Montpellier, France
Bernhard Maschke   Université Lyon 1, France
Casimiro Cavalcante   Federal University of Ceará, Brazil
Christian Leonard   Université Paris Nanterre, France
Cyrus Mostajeran   University of Cambridge, UK
Daniel Luft   University Trier, Germany
Darryl Holm   Imperial College London, UK

Dominique Spehner	University of Grenoble Alpes, France
Elena Celledoni	Norwegian University of Science and Technology, Norway
Emmanuel Chevallier	Weizmann Institute, Israel
Eric Grivel	Université de Bordeaux, France
Evan Gawlik	University of Hawaii at Manoa, USA
Fabio Di Cosmo	Università degli Studi di Napoli Federico II, Italy
Fabio Mele	University of Regensburg, Germany
Fabrice Gamboa	Institut de Mathématiques de Toulouse, France
Flavien Leger	University of California, Los Angeles, USA
Florence Nicol	ENAC, Toulouse, France
Florio Maria	Max Planck Institute for Mathematics in the Sciences, Leipzig, Germany
Frédéric Barbaresco	Thales Air Systems, France
Francesco Becattini	Università di Firenze, Italy
Francois Gay-Balmaz	Laboratoire de Météorologie Dynamique, École Normale Supérieure, France
Francois Xavier Vialard	Université Paris Est Marne La Vallée, France
Frank Nielsen	Sony Computer Science Laboratories, Tokyo, Japan
Geert Verdoolaege	Ghent University, Belgium
Gery de Saxce	Laboratoire de Mecanique de Lille, France
Giovanni Conforti	Ecole Polytechnique, Palaiseau, France
Giuseppe Marmo	University of Naples Federico II, Italy
Goffredo Chirco	Max Planck Institute for Gravitational Physics, Albert Einstein Institute, Germany
Guido Montufar	Max Planck institute for Mathematics in the Sciences, Germany
Haomin Zhou	Georgia Institute of Technology, USA
Hiroaki Yoshimura	Waseda University, Japan
Hiroshi Matsuzoe	Nagoya Institute of Technology, Japan
Hiroyasu Satoh	Nippon Institute of Technology, Japan
Hitoshi Furuhata	Hokkaido University, Japan
Hong van Le	Institute of Mathematics of ASCR, Czech Republic
Ishi Hideyuki	Nagoya University, Japan
Itoh Mitsuhiro	University of Tsukuba, Japan
Jésus Angulo	Centre de Morphologie Mathématique, École des Mines de Paris, France
Jan Naudts	Universiteit Antwerpen, Belgium
Jean Pierre Gazeau	Université Paris Diderot Paris 7, France
Joël Bensoam	IRCAM, France
Juhyun Par	Lancaster University, UK
Jun Zhang	University of Michigan, USA
Kathrin Welker	Helmut Schmidt University/University of the Federal Armed Forces Hamburg, Germany
Keisuke Haba	Nagoya Institute of Technology, Japan

Kinobu Shimizu	Tokyo University of Agriculture and Technology, Japan
Klas Modin	Chalmers University of Technology, Sweden
Koichi Tojo	The University of Tokyo, Japan
Leland McInnes	Tutte Institute for Mathematics and Computing, Canada
Lorenz Schwachhoefer	Technical University Dortmund, Germany
Luca Schiavone	Ostrava University, Italy
Luigi Malagò	Romanian Institute of Science and Technology, Romania
Marc Arnaudon	Université de Bordeaux, France
Marco Laudato	Università dell'Aquila, Italy
Marion Pilté	Mines ParisTech, France
Martin Bauer	Florida State University, USA
Maziar Esfahanian	University of Polytechnic of Turin, Italy
Michel Nguiffo Boyom	University Montpellier, France
Michel Broniatowski	LSTA Université Pierre et Marie Curie, Paris, France
Michele Pavon	Università di Padova, Italy
Miguel Tierz	Universidade de Lisboa, Portugal
Minh Ha Quang	RIKEN, Japan
Nicolas Brunel	ENSIIE, France
Nicolas Charon	Johns Hopkins University, USA
Nicolas Le Bihan	Gipsa-Lab, CNRS, France
Nihat Ay	Max Planck Institute for Mathematics in the Sciences, Germany
Nikolas Tapi	Norwegian University of Science and Technology, Norway
Nina Marinette Miolane	Stanford, USA
Olivier Peltre	Université Denis Didcrot Paris 7, France
Olivier Schwander	LIP6 Sorbonne University, Paris, France
Paola Siri	Politecnico di Torino, Italy
Paul Marriott	University of Waterloo, Canada
Philipp Harms	University of Freiburg, Germany
Pierre Baudot	Inserm U1072, France
Pierre Roussillon	Télécom ParisTech, Paris, France
Pierre-Antoine Absil	Université Catholique de Louvain, Belgium
Pietro Gori	Telecom ParisTech, France
Qi Feng	University of Southern California, USA
Radmila Pribic	Thales Nederland BV, The Netherlands
Remco Duits	Eindhoven University of Technology, The Netherlands
Rodolphe Sepulchre	University of Cambridge, UK
Salem Said	Laboratoire IMS, Bordeaux, France
Shiwei Lan	University of Illinois Urbana-Champaign, USA
Stefan Sommer	University of Copenhagen, Denmark
Stephane Puechmorel	Ecole Nationale de l'Aviation Civile, Toulouse, France
Tomonari Sei	The University of Tokyo, Japan

Wolfgang Stummer	University of Erlangen-Nurnberg, Germany
Wuchen Li	University of California, Los Angeles, USA
Xavier Pennec	Inria, France
Yifan Chen	California Institute of Technology, USA

Local Organizing Committee

Gilles Baroin	ENAC, Toulouse, France
Alice Le Brigant	ENAC, Toulouse, France
Murat Bronz	ENAC, Toulouse, France
Georges Mykoniatis	ENAC, Toulouse, France
Simon Le Sénéchal	ENAC, Toulouse, France
Florence Nicol	ENAC, Toulouse, France
Ludovic d'Estampes	ENAC, Toulouse, France
Stéphane Puechmorel	ENAC, Toulouse, France
Thierry Klein	ENAC, Toulouse, France
Tat Dat Tô	ENAC, Toulouse, France
Simon Le Senechal	ENAC, Toulouse, France

Sponsors and Organizers

ENAC, Toulouse, France
Institut Mathematique de Bordeaux, France
Ecole Polytechnique, France
Ecole des Mines ParisTech, France
Inria, France
Centrale Supélec, France
Institut Mathématique de Bordeaux, France
Sony Computer Science Laboratories, Tokyo, Japan

Contents

Computational Information Geometry

Statistical Manifold and Hessian Information Geometry

Non-parametric Information Geometry

Statistics on Non-linear Data

Geometric and Structure Preserving Discretizations

Optimization on Manifold

Geometry of Quantum States

Probability on Riemannian Manifolds

Shape Space

On Geometric Properties of the Textile Set and Strict Textile Set

Tomonari Sei[1(✉)] and Ushio Tanaka[2]

[1] The University of Tokyo, 7-3-1 Hongo, Bunkyo-ku, Tokyo 113-8656, Japan
sei@mist.i.u-tokyo.ac.jp
[2] Osaka Prefecture University, 1-1 Gakuen-cho, Naka-ku, Sakai-shi,
Osaka 599-8531, Japan
utanaka@mi.s.osakafu-u.ac.jp

Abstract. The textile plot is a tool for data visualisation proposed by Kumasaka and Shibata (2008). The textile set is a geometric object constructed to understand the textile plot outputs. In this study, we find additional facts on a proper subset called the strict textile set. Furthermore, we investigate differential and analytical geometric properties of the textile set.

Keywords: Textile plot · Textile set · Strict textile set · Submanifold · Canonical form

1 Introduction

The textile plot is a useful tool for data visualisation proposed by Kumasaka and Shibata [3]. The method transforms a given dataset consisting of continuous and/or categorical variables into a real matrix and draws a parallel coordinate plot based on it. The order of variables is determined using some variance criteria or clustering methods. We refer readers to [2] for a comprehensive study on parallel coordinate plots and [6] for geometric observation of parallel coordinate plots.

We first briefly review the textile plot (see [3] for details). We only consider continuous variables and fix the order of variables for simplicity. Suppose that a real matrix $\boldsymbol{X} = (x_{ti}) = (\boldsymbol{x}_1, \ldots, \boldsymbol{x}_p) \in \mathbb{R}^{n \times p}$ is given. Let $y_{ti} = a_i + b_i x_{ti}$ for each t and i, where each a_i and b_i are determined as follows. Let $\bar{y}_{t.} = p^{-1} \sum_{i=1}^{p} y_{ti}$ be the 'horizontal' mean. Then each a_i and b_i are determined in such a way that the deviation

$$\sum_{t=1}^{n} \sum_{i=1}^{p} (y_{ti} - \bar{y}_{t.})^2$$

is minimised under the restrictions of $y_{ti} = a_i + b_i x_{ti}$ and $\sum_t \sum_i y_{ti}^2 = 1$. The textile plot draws a line graph of $(y_{ti})_{i=1}^{p}$ for each t. Figure 1 explains the construction of the textile plot when $p = 5$.

© Springer Nature Switzerland AG 2019
F. Nielsen and F. Barbaresco (Eds.): GSI 2019, LNCS 11712, pp. 3–12, 2019.
https://doi.org/10.1007/978-3-030-26980-7_1

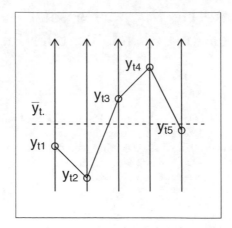

Fig. 1. Construction of the textile plot. The sum of deviations of y_{ti} against the horizontal average $\bar{y}_{t\cdot}$ is minimised under the restrictions $y_{ti} = a_i + b_i x_{ti}$ and $\sum_{t,i} y_{ti}^2 = 1$.

The obtained matrix $\boldsymbol{Y} = (y_{ti})$ satisfies a set of conditions. A set of such matrices is called the textile set (see also [5]). In [5], we have shown that a canonical part of the textile set can be written as a union of submanifolds.

A visualisation method of linkage disequilibrium in genetic studies, where multiple-single nucleotide polymorphism (SNP) genotype data are considered, has been developed as an application of the textile plot [4]. Since the data considered in [4] are categorical, they are first quantified by dummy variables and then the textile plot is applied. See [3] and [4] for details.

In this study, we derive some geometric properties of the textile set that have not been explored in [5]. Furthermore, we define a proper subset called the strict textile set in order to fill the gap between the textile set and the textile plot.

The rest of this paper is organised as follows. In Sect. 2, we provide the definition of the textile set and represent it as an inverse image of a differentiable map. In Sect. 3, we define the strict textile set and investigate its representations. In Sect. 4, we investigate the textile set from the viewpoint of differential and analytic geometry. Section 5 concludes the paper.

2   The Textile Set

The matrix $\boldsymbol{Y} \in \mathbb{R}^{n \times p}$ constructed for the textile plot as described in Sect. 1 satisfies the following two conditions:

$$\exists \lambda \in \mathbb{R}, \quad \forall i \in \{1, \ldots, p\}, \quad \sum_{j=1}^{p} \boldsymbol{y}_i^\top \boldsymbol{y}_j = \lambda \|\boldsymbol{y}_i\|^2 \tag{1}$$

and

$$\sum_{j=1}^{p} \|\boldsymbol{y}_j\|^2 = 1. \qquad (2)$$

The two conditions are necessary for $\boldsymbol{Y}$ to be an output of the textile plot but not sufficient. Indeed, two data matrices $\boldsymbol{Y} = (\boldsymbol{v}, \boldsymbol{v})$ and $\tilde{\boldsymbol{Y}} = (\boldsymbol{v}, -\boldsymbol{v})$ for any vector $\boldsymbol{v} \in \mathbb{R}^n$ with $\|\boldsymbol{v}\|^2 = 1/2$ satisfy the conditions (1) and (2) for $\lambda = 2$ and $\lambda = 0$, respectively, but only the former is the output of the textile plot.

The textile set is defined as follows (see also [5]).

Definition 1. *A set of all matrices $\boldsymbol{Y} \in \mathbb{R}^{n \times p}$ satisfying Eqs. (1) and (2) is called* the textile set *and denoted by $T_{n,p}$.*

We point out that the textile set is an inverse image of a differentiable map. Let $\mathcal{S}_+(p)$ be a set of all positive semi-definite matrices. For a data matrix $\boldsymbol{Y} \in \mathbb{R}^{n \times p}$, we denote the Gram matrix as $g(\boldsymbol{Y}) = \boldsymbol{Y}^\top \boldsymbol{Y} \in \mathcal{S}_+(p)$. Then, g is a function from $\mathbb{R}^{n \times p}$ to $\mathcal{S}_+(p)$. Let $T_+(p)$ be the image of $T_{n,p}$ by g. Explicitly, $T_+(p)$ consists of positive semi-definite matrices $\boldsymbol{S}$ satisfying the following two conditions:

$$\exists \lambda \in \mathbb{R}, \quad \forall i \in \{1, \ldots, p\}, \quad \sum_j S_{ij} = \lambda S_{ii},$$

and $\sum_j S_{jj} = 1.$

The following theorem is important for understanding the structure of $T_{n,p}$.

Theorem 1. *The textile set is given by $T_{n,p} = g^{-1}(T_+(p))$.*

Proof. The proof is straightforward.

Figure 2 shows the relation of these objects. Note that $T_+(p)$ does not depend on n.

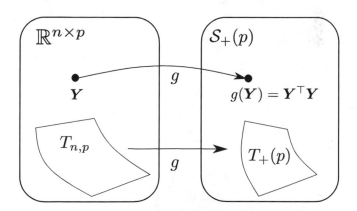

Fig. 2. The textile set $T_{n,p}$ as an inverse image.

3   The Strict Textile Set

In this section, we define the strict textile set as a proper subset of the textile set and characterise it.

The quantity λ in Eq. (1) is one of the eigenvalues of the correlation matrix $r_{ij} = \boldsymbol{y}_i^\top \boldsymbol{y}_j/(\|\boldsymbol{y}_i\|\|\boldsymbol{y}_j\|)$, where $\|\boldsymbol{y}_i\| \neq 0$ is assumed for simplicity. However, in the original definition of the textile plot (see [3]), λ is the maximum eigenvalue of the correlation matrix. The following lemma characterises the condition of maximality.

Lemma 1. *Let* $\boldsymbol{Y} = (\boldsymbol{y}_1, \ldots, \boldsymbol{y}_p)$ *be an element of* $T_{n,p}$ *and assume that* $\|\boldsymbol{y}_i\| \neq 0$ *for all* i. *Then* λ *in Eq. (1) is the maximal eigenvalue of* $r_{ij} = \boldsymbol{y}_i^\top \boldsymbol{y}_j/\|\boldsymbol{y}_i\|\|\boldsymbol{y}_j\|$ *if and only if a matrix*

$$Q_{ij} = \boldsymbol{y}_i^\top \left(\sum_{k=1}^p \boldsymbol{y}_k\right) \delta_{ij} - \boldsymbol{y}_i^\top \boldsymbol{y}_j \tag{3}$$

is positive semi-definite.

Proof. The maximality condition is equivalent to

$$\sum_i \sum_j a_i \frac{\boldsymbol{y}_i^\top \boldsymbol{y}_j}{\|\boldsymbol{y}_i\|\|\boldsymbol{y}_j\|} a_j - \lambda \sum_i a_i^2 \leq 0$$

for all $\boldsymbol{a} \in \mathbb{R}^p$. Let $b_i = a_i/\|\boldsymbol{y}_i\|$. Then we have

$$\sum_i \sum_j b_i (\boldsymbol{y}_i^\top \boldsymbol{y}_j) b_j - \lambda \sum_i b_i^2 \|\boldsymbol{y}_i\|^2 \leq 0.$$

Since Eq. (1) holds, this is further equivalent to

$$\sum_i \sum_j b_i \left(\boldsymbol{y}_i^\top \boldsymbol{y}_j - \boldsymbol{y}_i^\top \left(\sum_k \boldsymbol{y}_k\right) \delta_{ij}\right) b_j \leq 0,$$

and the proof is completed.

Now we define the strict textile set.

Definition 2. *The* strict textile set $T_{n,p}^1$ *consists of matrices* $\boldsymbol{Y} \in T_{n,p}$ *such that the matrix* $\boldsymbol{Q} = (Q_{ij})$ *defined by Eq. (3) is positive semi-definite.*

The matrix $\boldsymbol{Q}$ is a function of the Gram matrix $\boldsymbol{S} = g(\boldsymbol{Y}) = \boldsymbol{Y}^\top \boldsymbol{Y}$. For the dependence, we can write

$$Q_{ij}(\boldsymbol{S}) = \sum_k S_{ik}\delta_{ij} - S_{ij}$$

or

$$\boldsymbol{Q}(\boldsymbol{S}) = \mathrm{diag}(\boldsymbol{S}\boldsymbol{1}_p) - \boldsymbol{S},$$

where $\mathbf{1}_p$ is the all-ones vector and $\mathrm{diag}(\boldsymbol{v})$ is the diagonal matrix with the diagonal part $\boldsymbol{v}$. We also define

$$T_+^1(p) := T_+(p) \cap \{\boldsymbol{S} \in \mathcal{S}_+(p) \mid \boldsymbol{Q}(\boldsymbol{S}) \in \mathcal{S}_+(p)\}. \tag{4}$$

Recall that $\mathcal{S}_+(p)$ is the set of positive semi-definite matrices. From the definition, we obtain the following lemma.

Lemma 2. *The strict textile set is given by* $T_{n,p}^1 = g^{-1}(T_+^1(p))$.

In what follows, we study the set $T_+^1(p)$ instead of $T_{n,p}^1$.
For example, if $p = 2$, then

$$\boldsymbol{Q}(\boldsymbol{S}) = \begin{pmatrix} S_{12} & -S_{12} \\ -S_{12} & S_{12} \end{pmatrix} = S_{12} \begin{pmatrix} 1 & -1 \\ -1 & 1 \end{pmatrix}.$$

The condition $\boldsymbol{Q} \succeq \mathbf{0}$ is obviously equivalent to $S_{12} \geq 0$.
If $p = 3$, then

$$\boldsymbol{Q}(\boldsymbol{S}) = \begin{pmatrix} S_{12} + S_{13} & -S_{12} & -S_{13} \\ -S_{12} & S_{12} + S_{23} & -S_{23} \\ -S_{13} & -S_{23} & S_{13} + S_{23} \end{pmatrix}$$

$$= S_{12} \begin{pmatrix} 1 & -1 & 0 \\ -1 & 1 & 0 \\ 0 & 0 & 0 \end{pmatrix} + S_{13} \begin{pmatrix} 1 & 0 & -1 \\ 0 & 0 & 0 \\ -1 & 0 & 1 \end{pmatrix} + S_{23} \begin{pmatrix} 0 & 0 & 0 \\ 0 & 1 & -1 \\ 0 & -1 & 1 \end{pmatrix}.$$

A sufficient condition for positive semi-definiteness of $\boldsymbol{Q}$ is

$$S_{12} \geq 0, \quad S_{13} \geq 0, \quad S_{23} \geq 0.$$

This is not necessary: a counter-example is

$$\boldsymbol{S} = \begin{pmatrix} 100 & -1 & 10 \\ -1 & 100 & 10 \\ 10 & 10 & 100 \end{pmatrix}.$$

It is directly shown that $\boldsymbol{Q}$ is positive semi-definite if and only if $S_{12}+S_{13}+S_{23} \geq 0$ and $S_{12}S_{13} + S_{12}S_{23} + S_{13}S_{23} \geq 0$.

Now let us consider general p. We denote the set appearing in the definition (4) as

$$\mathcal{A} := \{\boldsymbol{S} \in \mathcal{S}_+(p) \mid \boldsymbol{Q}(\boldsymbol{S}) \in \mathcal{S}_+(p)\}.$$

Theorem 2. *The set $\mathcal{A}$ is a convex cone, which has an interior point.*

Proof. First, we show that $\mathcal{A}$ is a convex cone. Let $\boldsymbol{S}_1, \boldsymbol{S}_2 \in \mathcal{A}$ and $c_1, c_2 \geq 0$. Then $c_1\boldsymbol{S}_1 + c_2\boldsymbol{S}_2 \in \mathcal{S}_+(p)$ and

$$\boldsymbol{Q}(c_1\boldsymbol{S}_1 + c_2\boldsymbol{S}_2) = c_1\boldsymbol{Q}(\boldsymbol{S}_1) + c_2\boldsymbol{Q}(\boldsymbol{S}_2) \in \mathcal{S}_+(p).$$

Hence $c_1 S_1 + c_2 S_2 \in \mathcal{A}$. Next, we prove that $\mathcal{A}$ has an interior point. Observe that if $S_{ij} > 0$ for all pairs $i \neq j$, then $Q(S) \in \mathcal{S}_+(p)$. Hence, we obtain

$$\mathcal{S}_{++}(p) \cap \{S \mid S_{ij} > 0, \ i \neq j\} \subset \mathcal{A},$$

where $\mathcal{S}_{++}(p)$ denotes a set of all positive definite matrices. Since the two sets in the left-hand side are open, it is sufficient to show that their intersection is not empty. Indeed, a matrix

$$S = \begin{pmatrix} 2 & 1 & \cdots & 1 \\ 1 & 2 & & \vdots \\ \vdots & & \ddots & 1 \\ 1 & \cdots & 1 & 2 \end{pmatrix}$$

belongs to the two sets. This completes the proof.

We return to the space of Y. Define

$$\begin{aligned} \mathcal{B} &:= g^{-1}(\mathcal{A}) \\ &= \{Y \in \mathbb{R}^{n \times p} \mid Q(Y^\top Y) \in \mathcal{S}_+(p)\}. \end{aligned}$$

The strict textile set is given by $T_{n,p}^1 = T_{n,p} \cap \mathcal{B}$. Indeed, we have

$$\begin{aligned} T_{n,p}^1 &= g^{-1}(T_+^1) \\ &= g^{-1}(T_+ \cap \mathcal{A}) \\ &= g^{-1}(T_+) \cap g^{-1}(\mathcal{A}) \\ &= T_{n,p} \cap \mathcal{B}. \end{aligned}$$

Corollary 1. *If $n \geq p$, then $\mathcal{B}$ has an interior point.*

Proof. Note that $g(Y) = Y^\top Y$ is continuous. If $n \geq p$, then g is also surjective. Indeed, for a given $S \in \mathcal{S}_+(p)$, a matrix

$$Y = \begin{pmatrix} S^{1/2} \\ 0 \end{pmatrix}$$

with the matrix square root $S^{1/2}$ satisfies $g(Y) = S$. Since $\mathcal{A}$ has an interior point, $\mathcal{B}$ also has an interior point.

Figure 3 summarises the relations we obtained. Here we denote a set of all $p \times p$ symmetric matrices as $\mathcal{S}(p)$.

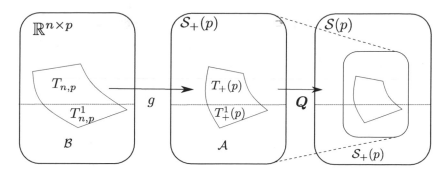

Fig. 3. The strict textile set $T_{n,p}^1$ and related objects.

4 Geometric Properties of the Textile Set from the Viewpoint of Differential and Analytic Geometry

In this section, we demonstrate that the textile set $T_{n,p}$ is a regular submanifold of $\mathbb{R}^{n \times p}$ with codimension $p + 1$. The result is independent from the preceding study presented in [5], where a canonical part of $T_{n,p}$ was studied. Furthermore, we obtain an envelope of the textile set and a canonical form of the envelope under the hypothesis that $n = p \geq 2$. Inselberg [1] has discussed the parallel coordinate from the analytical geometric point of view, which motivates our study.

Our observation starts with the quantity λ in Eq. (1).

Lemma 3. *Let* $\boldsymbol{Y} \in T_{n,p}$. *Then* λ *is bounded as follows:*

$$0 \leq \lambda \leq p.$$

Proof. For the lower bound, take the summation of Eq. (1) with respect to i and use Eq. (2) to obtain $\lambda = \| \sum_i \boldsymbol{y}_i \|^2 \geq 0$. For the upper bound, first consider the case $\|\boldsymbol{y}_i\| > 0$ for all i. Then Eq. (1) is equivalent to the condition that λ is an eigenvalue of the correlation matrix $r_{ij} = \boldsymbol{y}_i^\top \boldsymbol{y}_j / \|\boldsymbol{y}_i\| \|\boldsymbol{y}_j\|$ because $\sum_j r_{ij} \|\boldsymbol{y}_j\| = \lambda \|\boldsymbol{y}_i\|$. Since the trace of (r_{ij}) is p, we have $\lambda \leq p$. If $\|\boldsymbol{y}_i\|$ is zero for some i, consider a submatrix $(r_{ij})_{i,j \in I}$, where $I := \{i \mid \|\boldsymbol{y}_i\| > 0\}$. Note that I is not empty due to Eq. (2). It is shown that λ is an eigenvalue of the submatrix and therefore $\lambda \leq |I| \leq p$.

Remark 1. From the proof, we observe that $\lambda = 0$ if and only if $\sum_i \boldsymbol{y}_i = \boldsymbol{0}$, and that $\lambda = p$ if and only if $\boldsymbol{y}_1 = \cdots = \boldsymbol{y}_p$.

For each $\lambda \in [0, p]$, define a map $f_\lambda \colon \mathbb{R}^{n \times p} \to \mathbb{R}^{p+1}$ as

$$f_\lambda(\boldsymbol{y}_1, \ldots, \boldsymbol{y}_p) := \left(\sum_{j=1}^p \boldsymbol{y}_1^\top \boldsymbol{y}_j - \lambda \|\boldsymbol{y}_1\|^2, \ldots, \sum_{j=1}^p \boldsymbol{y}_p^\top \boldsymbol{y}_j - \lambda \|\boldsymbol{y}_p\|^2, \sum_{j=1}^p \|\boldsymbol{y}_j\|^2 - 1 \right).$$

Remark 2. $\{ f_\lambda^{-1}(\mathbf{0}) \mid 0 \leq \lambda \leq p \}$ yields a classification of the textile set, i.e.,

$$T_{n,p} = \bigsqcup_{0 \leq \lambda \leq p} f_\lambda^{-1}(\mathbf{0}). \tag{5}$$

The following theorem is a result of the textile set from the viewpoint of differential geometry. This theorem shows that each $f_\lambda^{-1}(\mathbf{0})$ is an $np - (p+1)$-dimensional differentiable manifold.

Theorem 3. *Suppose that*

$$0 < \lambda \, (\leq p), \quad y_{11} \neq 0, \tag{6}$$

$$y_{11} y_{jj} - y_{1j} y_{j1} \neq 0, \quad j = 2, \ldots, p, \tag{7}$$

$$\exists \ell \in \{ 2, \ldots, p \}; \sum_{j=2}^{p} y_{ij} + y_{i\ell}(1 - 2\lambda) \neq 0, \quad i = 1, \ldots, n. \tag{8}$$

We call the above equations the regularity condition, which implies that a natural inclusion map $\iota \colon f_\lambda^{-1}(\mathbf{0}) \hookrightarrow \mathbb{R}^{n \times p}$ is a homeomorphism onto its image. Then $f_\lambda^{-1}(\mathbf{0})$ is a regular submanifold of $\mathbb{R}^{n \times p}$ with codimension $p + 1$.

Proof. We outline the proof. We derive the sufficient condition for the Jacobi matrix of f_λ over $f_\lambda^{-1}(\mathbf{0})$ to be of full rank $(= p + 1)$. Each of Eqs. (6)–(8) establishes the desired conclusion.

The following theorem shows an application of Theorem 3.

Theorem 4. *Assume that in Eq. (5), $T_{n,p}$ is given by the finitely disjoint union of $f_\lambda^{-1}(\mathbf{0})$ in addition to the regularity condition. Then, $T_{n,p}$ is an $np - (p+1)$-dimensional compact differentiable manifold, where its differential structure is induced from the disjoint union of open sets of the differential manifold $f_\lambda^{-1}(\mathbf{0})$.*

Proof. We outline the proof. We have observed that $T_{n,p}$ is compact (see [5] for details). Combining Theorem 3 with the assumption of the finiteness leads to the conclusion.

The remainder of this section is devoted to the study of the textile set from the viewpoint of analytic geometry. Let $n = p \geq 2$. The following lemma shows an envelope of $T_{n,n}$.

Lemma 4. *Fix $\lambda \in [0, n]$, $n \geq 2$. Let $F_\lambda \colon \mathbb{R}^{n \times n} \to \mathbb{R}$ be a quadratic form defined as*

$$F_\lambda(\boldsymbol{y}_1, \ldots, \boldsymbol{y}_n) := \sum_{i=1}^{n} \sum_{j=1}^{n} y_{ij} \left(\sum_{k \neq j} y_{ik} \right) - (\lambda - 1).$$

Then $T_{n,n} \subset F_\lambda^{-1}(0)$.

Proof. We deduce that Eqs. (1) and (2) yield the following quadric: for all $\boldsymbol{y}_1, \ldots, \boldsymbol{y}_n \in T_{n,n}$,

$$F_\lambda(\boldsymbol{y}_1, \ldots, \boldsymbol{y}_n) = 0, \quad 0 \le \lambda \le n. \tag{9}$$

This completes the proof.

We proceed with the study on the canonical form of the quadric given by (9). The following theorem is a result of the textile set from the viewpoint of analytic geometry.

Theorem 5. *Let F_λ be defined as in Lemma 4. Then, the canonical form of the quadric defined from F_λ is given as follows:*

$$-\frac{1}{\lambda - 1}{z_1}^2 - \cdots - \frac{1}{\lambda - 1}{z_{n(n-1)}}^2$$
$$+\frac{n-1}{\lambda - 1}{z_{n(n-1)+1}}^2 + \cdots + \frac{n-1}{\lambda - 1}{z_{n^2}}^2 = 1, \quad \lambda \ne 1,$$
$$-{z_1}^2 - \cdots - {z_{n(n-1)}}^2 + (n-1){z_{n(n-1)+1}}^2 + \cdots + (n-1){z_{n^2}}^2 = 0, \quad \lambda = 1,$$

where each z_i, $i = 1, \ldots, n^2$, denotes a transformed coordinate to obtain the stated canonical form.

Proof. We outline the proof. For each $i, j = 1, \ldots, n$, identifying y_{ij} with $y_{(i-1)n+j}$ $(\in \mathbb{R}^{n^2})$, we can rewrite (9) as following:

$$F_\lambda(y_1, \ldots, y_{n^2}) = (y_1, \ldots, y_{n^2}) A (y_1, \ldots, y_{n^2})^\top - (\lambda - 1) = 0, \quad 0 \le \lambda \le n, \tag{10}$$

where

$$A := \begin{pmatrix} A_1 & & \mathbf{0} \\ & \ddots & \\ \mathbf{0} & & A_n \end{pmatrix}, \quad A_k := \begin{pmatrix} 0 & 1 & \ldots & 1 \\ 1 & 0 & \ldots & 1 \\ \vdots & \vdots & \ddots & \vdots \\ 1 & 1 & \ldots & 0 \end{pmatrix}, \quad k = 1, \ldots, n.$$

It can be noticed that the eigenvalues of A are given by -1 and $n - 1$ with their multiplicities $n(n-1)$ and n, respectively. Hence, we have $\det A = ((-1)^{n-1}(n-1))^n \ne 0$ because $n \ge 2$, from which it can be derived that F_λ given by Eq. (10) is a central quadric. Consequently, a proper coordinate transformation gives us the desired conclusion.

5   Conclusions

In this study, we have obtained geometric properties of the textile and strict textile sets as follows: The textile set can be characterised as an inverse image of the map g (Theorem 1). We have also defined the strict textile set and demonstrated its relation to a convex cone (Theorem 2). Furthermore, we have investigated the textile set from the viewpoint of differential geometry (Theorems 3 and 4) and analytic geometry (Lemma 4 and Theorem 5).

In the future, we plan to practically apply the results reported here and describe the intrinsically differential and analytical geometric structure of employed datasets. In fact, R. Shibata, who proposed the textile plot in [3], has suggested this direction to us. We are also concerned with defining a proper metric for the textile set $T_{n,p}$ and its class $f_\lambda^{-1}(\mathbf{0})$ as a differentiable manifold (stated in Theorems 3 and 4), and a quadric of the textile set $T_{n,n}$ itself as well as its envelop.

We could not investigate probabilistic properties of $\mathbf{Y}$ and $g(\mathbf{Y})$ defined in Sect. 2 when the data matrix $\mathbf{X}$ is distributed according to some multivariate distributions. The distribution of $\mathbf{Y}$ should be studied to understand the behaviour of the textile plot. For instance, the variable selection based on the norm $\|\mathbf{y}_i\|$ has to be justified in the framework of sampling distributions.

Acknowledgements. We would like to thank two referees for their helpful comments. This work was supported by JSPS KAKENHI Grant Numbers JP26108003, JP17K00044 and JP19K11865.

References

1. Inselberg, A.: The plane with parallel coordinates. Vis. Comput. **1**(2), 69–91 (1985). https://doi.org/10.1007/BF01898350
2. Inselberg, A.: Parallel Coordinates: Visual Multidimensional Geometry and Its Applications. Springer, New York (2009). https://doi.org/10.1007/978-0-387-68628-8
3. Kumasaka, N., Shibata, R.: High-dimensional data visualisation: the textile plot. Comput. Stat. Data Anal. **52**(7), 3616–3644 (2008). https://doi.org/10.1016/j.csda.2007.11.016
4. Kumasaka, N., Nakamura, Y., Kamatani, N.: The textile plot: a new linkage disequilibrium display of multiple-single nucleotide polymorphism genotype data. PLoS One **5**(4), e10207 (2010). https://doi.org/10.1371/journal.pone.0010207
5. Sei, T., Tanaka, U.: Geometric properties of textile plot. In: Nielsen, F., Barbaresco, F. (eds.) GSI 2015. LNCS, vol. 9389, pp. 732–739. Springer, Cham (2015). https://doi.org/10.1007/978-3-319-25040-3_78
6. Wegman, E.: Hyperdimensional data analysis using parallel coordinates. J. Am. Stat. Assoc. **85**(411), 664–675 (1990). https://doi.org/10.2307/2290001

Inexact Elastic Shape Matching in the Square Root Normal Field Framework

Martin Bauer[1] , Nicolas Charon[2]([⊠]) , and Philipp Harms[3]

[1] Florida State University, Tallahassee, FL 32304, USA
bauer@math.fsu.edu
[2] John Hopkins University, Baltimore, MD 21218, USA
charon@cis.jhu.edu
[3] Albert-Ludwig-University Freiburg, 79104 Freiburg, Germany
philipp.harms@stochastik.uni-freiburg.de

Abstract. This paper puts forth a new formulation and algorithm for the elastic matching problem on unparametrized curves and surfaces. Our approach combines the frameworks of square root normal fields and varifold fidelity metrics into a novel framework, which has several potential advantages over previous works. First, our variational formulation allows us to minimize over reparametrizations without discretizing the reparametrization group. Second, the objective function and gradient are easy to implement and efficient to evaluate numerically. Third, the initial and target surface may have different samplings and even different topologies. Fourth, texture can be incorporated as additional information in the matching term similarly to the `fshape` framework. We demonstrate the usefulness of this approach with several numerical examples of curves and surfaces.

Keywords: Square root normal field · Varifold metrics · Functional data analysis · Shape analysis

1 Introduction

Context. The statistical analysis of datasets of curves and surfaces is an active research field with many applications in e.g. computer vision, robotics, and medical imaging; see [4,20,24] and references therein. A recurring and fundamental task is finding optimal point correspondences between given shapes (i.e., the matching or registration problem), where optimality is typically expressed in terms of an elastic deformation energy. Solving the elastic matching problem in a numerically efficient way, which scales well to high-dimensional data encountered in real-world applications, remains a major challenge to date.

P. Harms is supported by the Freiburg Institute of Advanced Studies in the form of a Junior Fellowship. N. Charon is supported by NSF grant n° 1819131.

© Springer Nature Switzerland AG 2019
F. Nielsen and F. Barbaresco (Eds.): GSI 2019, LNCS 11712, pp. 13–20, 2019.
https://doi.org/10.1007/978-3-030-26980-7_2

Relation to Previous Work. This paper draws on two lines of work: square root normal fields (SRNFs) [11,12,14,19], which allow one to efficiently calculate elastic distances between parametrized shapes, and varifold distances [7,9,13, 18,22], which are distances between unparametrized shapes without any elastic interpretation. For each of these frameworks, efficient numerical implementations have been developed.

Contribution. We propose a new algorithm which combines SRNFs with varifold distances and inherits many advantages of both approaches. The key idea is to use varifold distances to relax the terminal constraint in the elastic matching problem. This bypasses the discretization of the reparametrization group, thereby eliminating the main computational burden in previous implementations of SRNF-based elastic shape matching. The resulting optimization problem is easy to implement and yields good results on some preliminary experiments on curves and surfaces. Moreover, the varifold distances allow one to match shapes with different meshes and even different topologies and to use texture information as in the `fshape` framework.

2   Shape Analysis of Curves and Surfaces

Elastic Shape Analysis. Elastic shape analysis operates in a Riemannian framework where infinitesimal shape deformations are measured by a Riemannian metric, which is often related to an elastic (or plastic) deformation energy; see the surveys [4,12]. We consider parameterized shapes as elements of the Fréchet manifold $\mathrm{Imm}(M, \mathbb{R}^d)$ of immersed hypersurfaces of a $(d-1)$-dimensional compact manifold M into $\mathbb{R}^d$. The corresponding space of unparameterized shapes is the quotient space $B_i(M, \mathbb{R}^d) = \mathrm{Imm}(M, \mathbb{R}^d)/\mathrm{Diff}(M)$, whose elements are denoted by $[f] = \{f \circ \varphi; \varphi \in \mathrm{Diff}(M)\}$. Given a $\mathrm{Diff}(M)$-invariant weak Riemannian metric G on $\mathrm{Imm}(M, \mathbb{R}^d)$, one defines a pseudo-distance between any two immersions $f_0, f_1 \in \mathrm{Imm}(M, \mathbb{R}^d)$ and their equivalence classes $[f_0], [f_1] \in B_i(M, \mathbb{R}^d)$ by

$$\mathrm{dist}_{\mathrm{Imm}}(f_0, f_1)^2 = \inf_{\substack{f \in C^\infty([0,1], \mathrm{Imm}(M, \mathbb{R}^d)) \\ f(0)=f_0, f(1)=f_1}} \int_0^1 G_f(\partial_t f, \partial_t f) dt, \qquad (1)$$

$$\mathrm{dist}_{B_i}([f_0], [f_1]) = \inf_{\varphi \in \mathrm{Diff}(M)} \mathrm{dist}_{\mathrm{Imm}}(f_0, f_1 \circ \varphi). \qquad (2)$$

Symmetry of the pseudo-distance on $B_i(M, \mathbb{R}^d)$ follows from the invariance of the metric G with respect to reparametrizations. Under suitable conditions on G the pseudo-distance is a distance, i.e., it separates points in the shape space of curves [16,17] or surfaces [5].

From a numerical perspective, the challenge is to calculate the above distances and the corresponding optimizers efficiently. This minimization can be solved numerically by path straightening and geodesic shooting methods (see e.g. [2,10]) or as in the next section by exploiting isometries to simpler spaces.

Square Root Normal Fields. Problem (1) simplifies considerably for certain first order Sobolev metrics [3,11,14,19,21,23,25]. One class of such metrics is defined using square root normal fields (SRNFs), which were introduced by Srivastava et al. [14,19] for planar curves and later generalized to surfaces by Jermyn et al. [11]. The SRNF of an oriented immersed hypersurface $f \in \mathrm{Imm}(M, \mathbb{R}^d)$ is defined as $\tilde{n}_f = n_f \mathrm{vol}_f^{1/2}$, where n_f is the unit normal field and $\mathrm{vol}_f^{1/2}$ the half density of f. For example, the SRNF of a planar curve $f \in \mathrm{Imm}(S^1, \mathbb{R}^2)$ is given in coordinates $\theta \in S^1$ by $\tilde{n}_f = i f_\theta \| f_\theta \|_{\mathbb{R}^2}^{-1/2}$, where i denotes rotation by $90°$ and coordinates in subscripts denote derivatives. Similarly, the SRNF of a surface $f \in \mathrm{Imm}(S^2, \mathbb{R}^3)$ is given in coordinates $(u, v) \in S^2$ as $\tilde{n}_f = (f_u \times f_v) \| f_u \times f_v \|_{\mathbb{R}^3}^{-1/2}$. In general, one obtains an elastic pseudo-Riemannian metric on $\mathrm{Imm}(M, \mathbb{R}^d)$ by setting

$$G_f(h, k) = \int_M \langle D_{(f,h)} \tilde{n}_f, D_{(f,k)} \tilde{n}_f \rangle_{\mathbb{R}^d},$$

where $D_{(f,h)} \tilde{n}_f$ denotes the directional derivative of $\tilde{n}_f$ at f in the direction h. This pseudo-Riemannian metric G is $\mathrm{Diff}(M)$-invariant, and by construction the map $f \mapsto \tilde{n}_f$ is a Riemannian isometry into the flat space of square integrable vector-valued half densities. For curves one obtains a Riemannian metric by modding out translations. For surfaces the situation is more complicated, as described in [11], and the kernel of the pseudo-metric may be larger than only translations. The metric belongs to the class of first order Sobolev metrics, which have been studied in great detail [5,16,17].

The advantage of this construction is that the Riemannian distance of G on $\mathrm{Imm}(M, \mathbb{R}^d)$ can be approximated efficiently as follows:

$$\mathrm{dist}_{\mathrm{Imm}}(f_0, f_1) \approx \| \tilde{n}_{f_0} - \tilde{n}_{f_1} \|_{L^2}. \tag{3}$$

Equality holds whenever the straight line between $\tilde{n}_{f_0}$ and $\tilde{n}_{f_1}$ is contained in the range of the map $f \mapsto \tilde{n}_f$. In general, equality holds up to first order for f_0 close to f_1 because the map $f \mapsto \tilde{n}_f$ is a Riemannian isometry.

The approximate distance (3) descends to the quotient space $B_i(M, \mathbb{R}^3)$ as described in (2). However, (2) involves a minimization over the reparametrization group, which is computationally costly. For curves this can be solved by dynamic programming [19] or using an explicit formula [15]. For surfaces in spherical coordinates, Jermyn et al. [11] proposed to discretize the diffeomorphism group of the two-dimensional sphere using spherical harmonics. This article puts forth an alternative method for minimization over the reparametrization group, which is based on varifold distances.

Varifold Distances. Geometric measure theory provides several embeddings of shape spaces into Banach spaces of distributions [7,9,13,18,22] with corresponding metrics. Varifold embeddings are one instance of this construction and are defined as follows (cf. [13] for details). Given a reproducing kernel Hilbert space W of real-valued functions on $\mathbb{R}^d \times S^{d-1}$, one associates to any immersion

$f \in \mathrm{Imm}(M, \mathbb{R}^d)$ the varifold $\mu_f \in W^*$ which satisfies

$$\forall w \in W: \qquad (\mu_f | w)_{W^*, W} = \int_M w(f(x), n(x)) \, \mathrm{vol}_f(dx).$$

The map $f \mapsto \mu_f$ is reparametrization-invariant and, under suitable assumptions on the kernel of W, injective [13]. Thus, one obtains a well-defined distance on the quotient space $B_i(M, \mathbb{R}^d)$ by defining for any two immersions $f_0, f_1 \in \mathrm{Imm}(M, \mathbb{R}^d)$:

$$\mathrm{dist}_{\mathrm{Var}}([f_0], [f_1]) = \|\mu_{f_0} - \mu_{f_1}\|_{W^*}.$$

From a computational point of view, these distances have explicit expressions in terms of the kernel function of W and are easy to implement for discrete curves or surfaces. We will use such distances to relax the terminal constraint in the boundary value problem for geodesics on shape space, as described next.

Combining SRNFs and Varifold Distances. Square root normal fields and varifold distances can be combined in an efficient matching algorithm for unparametrized shapes. This idea has been previously used in combination with large deformation models in [7,13] and with H^2 metrics on the space of curves in [1]. The boundary value problem (2) for geodesics on $B_i(M, \mathbb{R}^d)$ can be formulated as the program

$$\underset{f}{\mathrm{minimize}} \quad \mathrm{dist}(f_0, f) \qquad \text{subject to} \quad \mathrm{dist}_{\mathrm{Var}}([f], [f_1]) = 0. \qquad (4)$$

Relaxation using a (large) Lagrange multiplier λ and approximation of the elastic distance as in (3) yields

$$\underset{f}{\mathrm{minimize}} \; \|\tilde{n}_{f_0} - \tilde{n}_f\|_{L^2}^2 + \lambda \, \mathrm{dist}_{\mathrm{Var}}([f], [f_1])^2. \qquad (5)$$

This program has several advantages over previous alternative formulations of the SRNF matching problem [11,14,19]. First, the objective function and its gradient are easy to implement and can be computed efficiently. Second, the initial and target surface may have different discretizations and even different topologies. Third, texture information can be incorporated into the varifold matching term similarly to the fshape framework [6,8].

3 Numerical Implementation and Results

Algorithm. Given a pair (f_0, f_1) of curves or surfaces, the program (5) looks for a minimizer f with $\mathrm{dist}_{B_i}([f_0], [f_1]) = \mathrm{dist}_{\mathrm{Imm}}(f_0, f)$ and $[f] = [f_1]$. Thus, the algorithm solves the registration problem and calculates the distance between the unparametrized shapes $[f_0]$ and $[f_1]$. Note that it does, however, not provide a geodesic homotopy between these shapes. Such a homotopy can be obtained from the linear homotopy between $\tilde{n}_{f_0}$ and $\tilde{n}_{f_1}$ by (approximate) inversion of the SRNF map $f \mapsto \tilde{n}_f$. For open curves this inversion is exact and easy

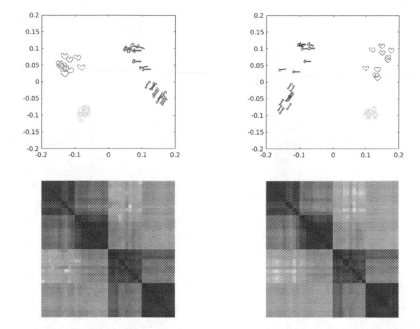

Fig. 1. Distances and clusters produced by our algorithm are comparable to state-of-the-art curve matching using dynamic programming [19] when tested on curves in the Kimia database. Left: our SRNF-varifold algorithm; right: dynamic programming; top: distance-based multi-dimensional scaling; bottom: symmetrized distance matrix.

to implement. For closed curves, the range of the SRNF map is not convex, and an approximate inverse has to be used [19]. For surfaces, this is a delicate issue [11], and to the best of our knowledge there exists no publicly available implementation for general triangulated surfaces.

Implementation. To implement the program (5) numerically, one has to discretize the space of parametrized shapes. An advantage over [11] is that the reparametrization group does not need to be discretized. Piecewise linear curves and triangular meshes are suitable discretizations in our context, the reason being that square root normal fields and kernel-based varifold distances extend naturally to these spaces. The minimization is performed using an L-BFGS method. The gradient of the discretized energy functional (5), which is needed by the L-BFGS method, has an explicit form and can be implemented efficiently.

Curves. For curves, our algorithm is comparable to state of the art methods. On the Kimia dataset[1] it produces distances and clusters which are similar to those based on dynamic programming, as shown in Fig. 1. A nice feature of our

---

[1] Computer Vision Group at LEMS at Brown University: Database of 99 binary shapes. https://vision.lems.brown.edu/content/available-software-and-databases.

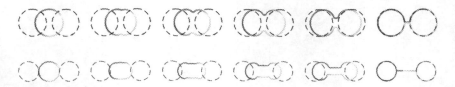

Fig. 2. Our algorithm can be used to compare shapes with different topologies. Left to right: geodesic interpolation (colored) between a single circle and a pair of circles (black, dashed); top: small distance between the pair of circles; bottom: large distance between the pair of circles. (Color figure online)

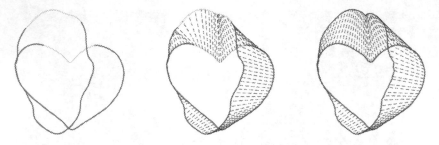

Fig. 3. Elastic matching of curves with functional data. Left: source and target curves with binary functional data in red/cyan. Middle: matching using functional data. Right: purely geometrical matching without functional data. (Color figure online)

algorithm, which stems from the use of varifold distances, is that the initial and target shapes are allowed to have different topologies. For example, one can match a single circle to a pair of circles, as demonstrated in Fig. 2. This is not possible using previous methods for shape matching using SRNFs. There are potential applications in cell division and removal of topological noise. Another feature of our algorithm is that it can account for functional data on the given shapes, as demonstrated in Fig. 3. To this aim, the varifold distance in (5) is replaced by a functional shape distance, as developed in [6,8]. This has several applications. The functional data may be dictated by the application at hand, as e.g. in the case of texture information. An interesting alternative to be explored in future work is to use shape descriptors as functional data to guide the matching algorithm.

Surfaces. For surfaces, we obtain some promising first results and see a high potential of improvement over alternative methods. An example is presented in Fig. 4, where the optimal point correspondences between two hand postures were calculated. As the two triangulated surfaces in this experiment had different mesh connectivities, and no point-to-point correspondences were initially available, we had to initialize the optimization procedure with the template surface. After optimization using an adaptive choice of Lagrange multiplier λ in (5), we obtained an excellent fit of the deformed template onto the target with anatomically correct point correspondences.

Fig. 4. Anatomically correct correspondences obtained by elastic matching of two surfaces. Left: template f_0 (blue, 2322 vertices) and target f_1 (red, 2829 vertices). Right: output f of the matching algorithm (green) and a linear homotopy between f_0 and f. (Color figure online)

References

1. Bauer, M., Bruveris, M., Charon, N., Møller-Andersen, J.: A relaxed approach for curve matching with elastic metrics. In: ESAIM COCV (2018, forthcoming)
2. Bauer, M., Bruveris, M., Harms, P., Møller-Andersen, J.: A numerical framework for Sobolev metrics on the space of curves. SIAM J. Imaging Sci. **10**(1), 47–73 (2017)
3. Bauer, M., Bruveris, M., Marsland, S., Michor, P.W.: Constructing reparameterization invariant metrics on spaces of plane curves. Differential Geom. Appl. **34**, 139–165 (2014)
4. Bauer, M., Bruveris, M., Michor, P.W.: Overview of the geometries of shape spaces and diffeomorphism groups. J. Math. Imaging Vis. **50**(1–2), 60–97 (2014)
5. Bauer, M., Harms, P., Michor, P.W.: Sobolev metrics on shape space of surfaces. J. Geom. Mech. **3**(4), 389–438 (2011)
6. Charlier, B., Charon, N., Trouvé, A.: The fshape framework for the variability analysis of functional shapes. Found. Comput. Math. **17**(2), 287–357 (2017)
7. Charon, N., Trouvé, A.: The varifold representation of nonoriented shapes for diffeomorphic registration. SIAM J. Imaging Sci. **6**(4), 2547–2580 (2013)
8. Charon, N., Trouvé, A.: Functional currents: a new mathematical tool to model and analyse functional shapes. J. Math. Imaging Vis. **48**(3), 413–431 (2014)
9. Glaunès, J., Qiu, A., Miller, M.I., Younes, L.: Large deformation diffeomorphic metric curve mapping. Int. J. Comput. Vis. **80**(3), 317 (2008)
10. Huang, W., Gallivan, K.A., Srivastava, A., Absil, P.A.: Riemannian optimization for registration of curves in elastic shape analysis. J. Math. Imaging Vis. **54**(3), 320–343 (2016)
11. Jermyn, I.H., Kurtek, S., Klassen, E., Srivastava, A.: Elastic shape matching of parameterized surfaces using square root normal fields. In: Fitzgibbon, A., Lazebnik, S., Perona, P., Sato, Y., Schmid, C. (eds.) ECCV 2012. LNCS, vol. 7576, pp. 804–817. Springer, Heidelberg (2012). https://doi.org/10.1007/978-3-642-33715-4_58

12. Jermyn, I.H., Kurtek, S., Laga, H., Srivastava, A.: Elastic shape analysis of three-dimensional objects. Synth. Lect. Comput. Vis. **12**(1), 1–185 (2017)
13. Kaltenmark, I., Charlier, B., Charon, N.: A general framework for curve and surface comparison and registration with oriented varifolds. In: Computer Vision and Pattern Recognition (CVPR) (2017)
14. Kurtek, S., Klassen, E., Gore, J.C., Ding, Z., Srivastava, A.: Elastic geodesic paths in shape space of parameterized surfaces. IEEE Trans. Pattern Anal. Mach. Intell. **34**(9), 1717–1730 (2012)
15. Lahiri, S., Robinson, D., Klassen, E.: Precise matching of PL curves in $\mathbb{R}^N$ in the square root velocity framework. Geom. Imaging Comput. **2**(3), 133–186 (2015)
16. Mennucci, A., Yezzi, A., Sundaramoorthi, G.: Properties of Sobolev-type metrics in the space of curves. Interfaces Free Boundaries **10**(4), 423–445 (2008)
17. Michor, P.W., Mumford, D.: An overview of the Riemannian metrics on spaces of curves using the Hamiltonian approach. Appl. Comput. Harmon. Anal. **23**(1), 74–113 (2007)
18. Roussillon, P., Glaunes, J.A.: Kernel metrics on normal cycles and application to curve matching. SIAM J. Imaging Sci. **9**(4), 1991–2038 (2016)
19. Srivastava, A., Klassen, E., Joshi, S.H., Jermyn, I.H.: Shape analysis of elastic curves in Euclidean spaces. IEEE Trans. Pattern Anal. Mach. Intell. **33**(7), 1415–1428 (2011)
20. Srivastava, A., Klassen, E.P.: Functional and Shape Data Analysis. Springer, New York (2016). https://doi.org/10.1007/978-1-4939-4020-2
21. Sundaramoorthi, G., Mennucci, A., Soatto, S., Yezzi, A.: A new geometric metric in the space of curves, and applications to tracking deforming objects by prediction and filtering. SIAM J. Imaging Sci. **4**(1), 109–145 (2011)
22. Vaillant, M., Glaunès, J.: Surface matching via currents. In: Christensen, G.E., Sonka, M. (eds.) IPMI 2005. LNCS, vol. 3565, pp. 381–392. Springer, Heidelberg (2005). https://doi.org/10.1007/11505730_32
23. Younes, L.: Computable elastic distances between shapes. SIAM J. Appl. Math. **58**(2), 565–586 (1998)
24. Younes, L.: Shapes and Diffeomorphisms, vol. 171. Springer, Heidelberg (2010). https://doi.org/10.1007/978-3-642-12055-8
25. Younes, L., Michor, P.W., Shah, J., Mumford, D.: A metric on shape space with explicit geodesics. Atti Accad. Naz. Lincei Rend. Lincei Mat. Appl. **19**(1), 25–57 (2008)

Signatures in Shape Analysis: An Efficient Approach to Motion Identification

Elena Celledoni⬤, Pål Erik Lystad⬤, and Nikolas Tapia$^{(\boxtimes)}$⬤

Department of Mathematical Sciences, NTNU Trondheim, Trondheim, Norway
nikolas.tapia@ntnu.no

Abstract. Signatures provide a succinct description of certain features of paths in a reparametrization invariant way. We propose a method for classifying shapes based on signatures, and compare it to current approaches based on the SRV transform and dynamic programming.

Keywords: Shape analysis · Signature · Motion identification

1 Introduction

Shape analysis is a broad and growing subject addressing the analysis of different types of data ranging from surfaces, landmarks, animation data etc. In this paper shapes are unparametrized curves. Mathematically a shape is an equivalence class of curves under reparameterization, that is, two curves $c_0, c_1 : [0, 1] \to M$ are equivalent and determine the same shape if there exists a strictly increasing smooth bijection $\varphi : [0, 1] \to [0, 1]$ such that $c_1 = c_0 \circ \varphi$. For a given curve c we denote by $[c]$ the corresponding shape.

The similarity between two shapes $[c_0], [c_1]$ is then defined by creating a distance function $d_{\mathcal{S}}$ on the space of shapes $\mathcal{S}$,

$$d_{\mathcal{S}}([c_0], [c_1]) := \inf_{\varphi} d_{\mathcal{P}}(c_0, c_1 \circ \varphi) \tag{1}$$

where $d_{\mathcal{P}}$ is a suitable reparameterization invariant Riemannian distance on the manifold of parametrized curves.

Finding the optimal reparameterization φ is however computationally demanding, and in many applications simply unnecessary. This is specifically the case of applications where the optimal parametrization is not explicitly used for further calculations, e.g. problems of identification and classification. Ways of circumventing this step are therefore of great interest.

In recent years, after extensive work by Terry Lyons and collaborators, the theory of rough paths has gained considerable importance as a toolbox for mathematical analysis and for mathematical modeling in applications. In this context, the signature map provides a faithful representation of paths, capturing their essential global properties. A fundamental property of the signature is its invariance under reparameterization, surmising its importance for shapes.

© Springer Nature Switzerland AG 2019
F. Nielsen and F. Barbaresco (Eds.): GSI 2019, LNCS 11712, pp. 21–30, 2019.
https://doi.org/10.1007/978-3-030-26980-7_3

In this paper, we define a measure of similarity between shapes in $\mathcal{S}$ by means of the signature. We define a distance directly on $\mathcal{S}$. We test the viability of this approach and use it to classify motion capture animations from the CMU motion capture database [7]. Indeed, this leads to an efficient technique that delivers results comparable to what is obtainable with methodologies based on the SRV transform, but at a much lower computational cost.

2   Shape Analysis on Lie Groups

In the following, G will denote a finite-dimensional Lie group under multiplication with identity element denoted by e. We let $\mathfrak{g}$ denote the corresponding right Lie algebra $\mathfrak{g} := \mathcal{L}_R(G)$. For a fixed $g \in G$, left and right translation by g will be denoted $L_g(h) = g \cdot h$ and $R_g(h) = h \cdot g$ respectively.

2.1   Shape Space

We consider the space $C^\infty([0,1], G)$ of parameterized smooth curves on G, i.e. smooth maps $c : [0,1] \to G$. To model the curves as unparameterized, or independent of parameterization, we define the *shape space* $\mathcal{S}$ as the quotient space

$$\mathcal{S} = C^\infty([0,1], G)/\mathrm{Diff}^+, \tag{2}$$

where Diff^+ is the group of orientation preserving diffeomorphisms of the parameter space $[0,1]$. The elements of $\mathcal{S}$ are equivalence classes of curves. The elements of the same class are curves which can be mapped to one another by changing their parameterization, that is, two curves $c_0, c_1 \in C^\infty(I, G)$ are equal in shape space if there exists $\varphi \in \mathrm{Diff}^+$ such that $c_1 = c_0 \circ \varphi$.

In the setting of our application, the search for optimal time parametrizations can be viewed as syncing up the animations, removing disturbances due to small pauses, different periodicity, or asynchronous starting and stopping, by shifting the movement of one character to match the other as closely as possible.

2.2   Geodesic Distances on Shape Space

Our goal is to introduce a meaningful and computable distance $d_\mathcal{S}$ on $\mathcal{S}$ to estimate the similarity between two shapes. This area of research started with the efforts of Younes [16]. We will restrict the space of curves to the space of immersions, i.e. curves with non-vanishing first derivative, which we denote by

$$\mathcal{P} = \mathrm{Imm}([0,1], G). \tag{3}$$

Let $d_\mathcal{P}$ be a pseudo-metric on $\mathcal{P}$. We define $d_\mathcal{S}$, for two elements $[c_0], [c_1] \in \mathcal{S}$, by

$$d_\mathcal{S}([c_0], [c_1]) := \inf_{\varphi \in \mathrm{Diff}^+} d_\mathcal{P}(c_0, c_1 \circ \varphi). \tag{4}$$

As shown in [3, Lemma 3.4], d_S will be a pseudo-metric on $\mathcal{S}$ if $d_{\mathcal{P}}$ is a *reparameterization invariant* or, in other words, if for any two $c_0, c_1 \in \mathcal{P}$ and any $\varphi \in \mathrm{Diff}^+$ we have that

$$d_{\mathcal{P}}(c_0 \circ \varphi, c_1 \circ \varphi) = d_{\mathcal{P}}(c_0, c_1). \tag{5}$$

An obvious choice of metric on $\mathcal{P}$ is the familiar L_2-metric. However, as shown by Michor and Mumford [13], this metric leads to vanishing geodesic distance which renders it useless. They further show in [14] that one solution to this problem is to consider metrics based on arc-length derivatives, creating a class of Sobolev-type metrics.

There are multiple possible metrics in this class. One option is based on what is usually referred to as the *Square Root Velocity Transform* (SRVT). This transform and accompanying metric was first introduced, in the context of shape analysis, by Srivastava et al. [15], who used the transformation when working with curves in Euclidian spaces. The transformation has later been adopted to more general shapes. Of particular interest is the formulation for shapes that are represented as Lie-group valued curves [3].

We define the SRVT $\mathcal{R} : \mathcal{P} \to C^{\infty}([0,1], \mathfrak{g} \setminus \{0\})$ by

$$\mathcal{R}(c)(t) := \frac{R_{c(t)*}^{-1}(\dot{c}(t))}{\sqrt{\|\dot{c}(t)\|}}. \tag{6}$$

This transformation has the following useful properties [3, Lemma 3.6]:

1. For every $c \in \mathcal{P}$ and $\varphi \in \mathrm{Diff}^+$, the following equivariant property holds:

$$\mathcal{R}(c \circ \varphi) = \mathcal{R}(c) \circ \varphi \cdot \sqrt{\dot{\varphi}}. \tag{7}$$

2. It is translation invariant: for all $c \in \mathcal{P}$ and $g \in G$

$$\mathcal{R}(R_g(c)) = \mathcal{R}(c).$$

A similar result is true for shapes with values in Euclidean spaces [15].

Further, one can obtain a Riemannian metric $d_{\mathcal{P}_*}$ that coincides with the geodesic distance on a submanifold $\mathcal{P}_* \subset \mathcal{P}$ by using the SRVT to pull back the L_2-metric on $C^{\infty}(I, \mathfrak{g} \setminus \{0\})$ [3]. Further restricting the immersion space to $\mathcal{P}_* = \{c \in \mathcal{P} : c(0) = e\}$, where e is the identity element in G, the distance $d_{\mathcal{P}_*}$ turns out to be reparameterization invariant.

This invariance implies, in particular, that it will also yield a geodesic distance on $\mathcal{S}_* := \mathcal{P}_*/\mathrm{Diff}^+$ [2]. The restriction to $\mathcal{P}_*$ isn't very troublesome as any curve can be transferred to this space by right translation by the inverse of its initial value, that is $R_{c(0)^{-1}}$ [3].

Using the equivariant property for the SRVT from Eq. (7) and defining $q_i = \mathcal{R}(c_i)$ for $i = 0, 1$, the problem of calculating the metric for the shape space $\mathcal{S}_*$ in Eq. (4) can be written as

$$d_{\mathcal{S}_*}(c_0, c_1) = \inf_{\varphi \in \mathrm{Diff}^+(I)} \sqrt{\int_I \|q_0(t) - q_1(\varphi(t)) \cdot \sqrt{\dot{\varphi}}\|^2 dt}. \tag{8}$$

Finding this infimum will generally be very difficult. The usual approach is therefore to discretize the curves and solve instead a finite dimensional optimization problem. The most common methods used to solve this problem in shape analysis [15] are based on either the gradient descent method or a dynamic programming algorithm (DP). In our experiments we use the DP approach described in [1].

3   Signatures

Signatures, introduced by Chen [4] for smooth paths and later generalized by Lyons [11] under the name of geometric rough paths, are an important tool for the study of the solutions of controlled differential equations, but have also proved useful for solving classification problems of time series, Machine Learning and Topological Data Analysis [6].

In the usual framework, signatures are defined for paths taking values in a Banach space. From a geometric point of view, and in light of our purposes, this setting has to be adapted. Luckily, Chen also considered signatures for curves taking values on a smooth manifold [4]. This definition is quite general and relies on the selection of a frame bundle. For Lie groups there is a canonical choice: the Maurer–Cartan form. This is the unique right-invariant one form ω such that $\omega_e = \mathrm{id}_{\mathfrak{g}}$, i.e. $\omega(v) = (R_g^{-1})_* v$ for $v \in T_g(G)$ [8, p. 311].

Below we denote, for a finite-dimensional vector space V of dimension $d = \dim V$, the tensor algebra over V,

$$T(V) := \bigoplus_{n \geq 0} V^{\otimes n}.$$

We observe that $T(V)$ is always infinite-dimensional. Its dual space is denoted by $T((V)) := T(V)^*$, and it may be identified with the ring of formal power series in d noncommuting variables $\{e_1, \ldots, e_d\}$.

Definition 1. *Let G be a d-dimensional Lie group and $\alpha \in C^\infty([0,1], G)$ be a smooth curve and ω the Maurer-Cartan form on G. The signature $S(\alpha)$ of α is the family of linear maps on $T(\mathbb{R}^d)$ recursively defined by $\langle S(\alpha)_{s,t}, 1 \rangle := 1$ and*

$$\langle S(\alpha)_{s,t}, e_{i_1 \cdots i_p} \rangle := \int_s^t \langle S(\alpha)_{s,u}, e_{i_1 \cdots i_{p-1}} \rangle \, \omega_{\alpha(u)}^{i_p}(\dot\alpha(u)) \, \mathrm{d}u.$$

In this definition, the notation $\omega_g^j(v)$ denotes the j-th component of the vector $\omega_g(v) \in \mathfrak{g}$ in a basis of the Lie algebra $\mathfrak{g}$ of G.

The signature provides a compact description of certain features of a path [5]. One of its main advantages in our context is its reparameterization invariance: for any orientation-preserving diffeomorphism φ on $[s,t]$ we have that

$$S(\alpha \circ \varphi)_{s,t} = S(\alpha)_{s,t}.$$

Other fundamental properties include:

1. For each $0 \leq s < t \leq 1$, the signature $S(x)_{s,t}$ belongs to the set of group-like elements of $T((\mathbb{R}^d))$, and for any $0 \leq s \leq 1$, $S(x)_{s,s} = 1$, the neutral element in the group.
2. **Chen's rule:** For any three $0 \leq s < u < t \leq 1$ we have

$$S(x)_{s,u} \otimes S(x)_{u,t} = S(x)_{s,t}.$$

Using these properties, signatures may be efficiently computed for some restricted classes of paths. For example, if x is a straight line in $\mathbb{R}^d$ with base point $a \in \mathbb{R}^d$ direction $b \in \mathbb{R}^d$, i.e. $x_t = a + tb$ for $t \in [0,1]$, then

$$
\begin{aligned}
S(x)_{s,t} &= \exp_{\otimes}((t-s)b) \\
&= 1 + (t-s)b + \frac{(t-s)^2}{2} b \otimes b + \frac{(t-s)^3}{6} b \otimes b \otimes b + \cdots .
\end{aligned}
\tag{9}
$$

A similar statement is true for geodesic curves on a finite-dimensional compact Lie group.

We may think of signatures as an infinite vector indexed by *words* over the alphabet $\{1, \ldots, d\}$. In particular, for a piecewise linear path the above formula means that if we want to know the component in (9) corresponding to the word $w = i_1 \cdots i_k$ then

$$\langle S(x)_{s,t}, e_w \rangle = \frac{(t-s)^k}{k!} \prod_{j=1}^{k} b_{i_j}$$

For a general piecewise linear path x, we may use the above formula and Chen's rule to deduce that

$$S(x)_{s,t} = \exp_{\otimes}(\Delta t_1 b_1) \otimes \exp_{\otimes}(\Delta t_2 b_2) \otimes \cdots \otimes \exp_{\otimes}(\Delta t_m b_m)$$

where $\Delta t_k = t_k - t_{k-1}$ are the length of the time intervals where the path is sampled and $b_1, \ldots, b_k$ are the slopes of the path in each of these intervals. The entries of this expression may be computed by using a Baker–Campbell–Hausdorff-type formula, for example.

Finally, we remark that the signature possesses another interesting property, namely it is an homomorphism from path space with concatenation to the tensor algebra $T((\mathbb{R}^d))$. This means that if we are given two paths $x\colon [0,1] \to G$ and $y\colon [0,1] \to G$, and we concatenate them to form a new path $x \cdot y$, then

$$S(x \cdot y)_{0,1} = S(x)_{0,1} \otimes S(y)_{0,1}.$$

Moreover, if we reverse the path x, i.e. we define $\overleftarrow{x}(t) := x(1-t)$ then

$$S(\overleftarrow{x})_{0,1} = S(x)_{0,1}^{-1}$$

where the inverse is taken in the group-like elements of the tensor algebra.

It can be shown that actually, as a function of time the signature satisfies the differential equation

$$\frac{\mathrm{d}}{\mathrm{d}t} S(x)_{s,t} = S(x)_{s,t} \otimes \dot{x}_t, \quad S(x)_{s,s} = 1$$

in the tensor algebra. From this point of view, the signature map corresponds to the flow map of the vector field given by the base path. Thus, the signature belongs to an infinite-dimensional Lie group whose Lie algebra is the free Lie algebra over $\mathbb{R}^d$ which we denote by $\mathfrak{L}(\mathbb{R}^d)$. It does not, however, constitute a one-parameter subgroup. Therefore, for each fixed time interval $[s, t]$ we can map the signature to the free Lie algebra via a logarithm map, and we define

$$\Lambda(x)_{s,t} = \log(S(x)_{s,t}) \in \mathfrak{L}(\mathbb{R}^d).$$

This element, called the log-signature in the literature, provides a minimal description of the path, which is equivalent to the full signature.

There are many ways in which signatures can be used to compare shapes, but the essential feature is that since the map S is reparameterization invariant, one obtains a way of directly comparing shapes instead of parameterized curves. For our experiments we chose a particular distance on $T((\mathbb{R}^d))$ (see next section for the precise formula), but this is by no means the only possible choice.

In making this choice one has to truncate the signature to obtain a finite-dimensional object. Due to the factorial decay of iterated integrals little information is lost in the process; still, some level has to be chosen and usually this done by running experiments. Once the truncation level is chosen, several choices of metric are available: the truncated tensor algebra becomes finite-dimensional so it has a nice linear structure and we are free to choose norms on it subject to some compatibility restrictions. There is also the notion of homogeneous norm on group-like elements, which takes into account the geometry of this group. Finally, the logarithm in this group maps signatures into a linear space (the free Lie algebra) in a bijective way, so no information is lost, but there is a substantial dimensional reduction.

According to our observations, is the last option which represents the most robust choice in terms of noise sensitivity, while also providing an accurate way of comparing signatures.

4   Experiments

Motion capture animations are usually recorded as the angle of every joint in a skeleton for every frame in an animation. A natural setting for the rotating joints is the Lie group of 3D rotations, $SO(3)$. Every frame consists of d independently rotating joints so the frame can be modeled as an element in $SO(3)^d$, where $SO(3)^d$ is the Cartesian product of d copies of $SO(3)$. Interpolating between the frames will then allow us to model the animation as a parameterized curve.

We use an interpolation scheme in which one uses the log map to linearly interpolate on the Lie algebra, and then pull back to the Lie group with the exponential map. Let $A, B \in SO(3)$, we define the interpolation $\kappa : [0,1] \to SO(3)$ between A and B as

$$\kappa(s) := \exp\big(s\log(B \cdot A^T)\big) \cdot A.$$

Notice that $\kappa(0) = A$ and $\kappa(1) = B$. Applying this interpolation component-wise to the frames in $SO(3)^d$ will enable us to construct a piece-wise interpolation between the frames of the animation. The Maurer–Cartan form along the interpolation is piece-wise constant, making it easy to compute SRV representations, $d_{\mathcal{P}_*}$-metrics, and signatures.

To test the effectiveness of the proposed frameworks we check whether they are able to identify different types of character motion. We have selected animations from the CMU motion capture database with descriptions "walk", "run/jog" and "forward jump". These are similar in length, and should produce results that conform with human intuition.

The test will calculate a distance matrix using the proposed similarity measures. From the distance matrix we produce a multidimensional scaling plot (MDS), depicting how similar, or dissimilar, the animations are. MDS tries to place the data points in 2-dimensional scatter plot while preserving the distances given by the distance matrix. See Kruskal [9] for more information on this method.

In Fig. 2a we calculate the distance matrix using the metric $d_{\mathcal{P}_*}$ on interpolation curves in $\mathcal{P}_*$, and in Fig. 2b we use the metric $d_{\mathcal{S}_*}$, Eq. (8), on the shapes generated by the curves in $\mathcal{S}_*$, where the optimal reparameterization is calculated with a DP algorithm. There are little to no patterns when projecting to the space $(\mathcal{P}_*, d_{\mathcal{P}_*})$, as seen in Fig. 2a. In Fig. 2b however, we observe that modelling

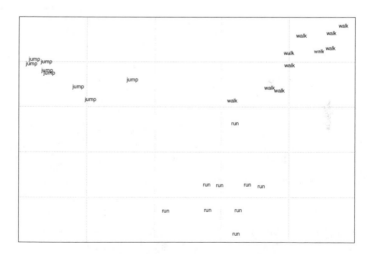

Fig. 1. Multi dimensional scaling plot of distance matrix calculated from by projecting animations to the space $\mathcal{S}_*$ equipped with the distance function d_{sig}. In this plot we have taken animation with descriptions "run/jog", "forward jump" and "walk" from the CMU Motion Capture Database [7].

the curves as being parameterization invariant yields three easily distinguishable clusters of animations. Compared to Fig. 2a we see a big benefit from this model assumption.

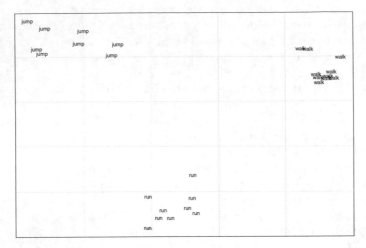

(a) Animations projected to $\mathcal{P}_*$ with distance matrix calculated with the metric $d_{\mathcal{P}_*}$.

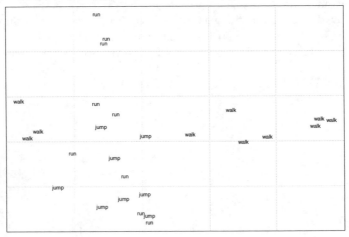

(b) Animations projected to $\mathcal{S}_*$ with distance matrix calculated with metric $d_{\mathcal{S}_*}$ using a DP algorithm.

Fig. 2. Multi dimensional scaling plots of distance matrix based on geodesic distances calculated in $\mathcal{P}_*$ and $\mathcal{S}_*$, figure (a) and (b) respectively. In this plot we have taken animation with descriptions "run/jog", "forward jump" and "walk" from the CMU Motion Capture Database [7].

In Fig. 1 the animations are projected to the shape space $\mathcal{S}$ equipped with the distance function $d_{\text{sig}}(c_0, c_1) = \left\| \frac{\log S(c_0)}{\|\log S(c_0)\|} - \frac{\log S(c_1)}{\|\log S(c_1)\|} \right\|$. While this figure does reveal the same structure as seen in Fig. 2b, the clusters exhibit both a higher internal and a lower external variability. An important take away from this experiment is that this distance function in fact does preserve some of the structure of the shape space.

5   Concluding Remarks

Our preliminary experiments, show that classifying animations using a distance function on $\mathcal{S}_*$ based on signatures produces very encouraging results. The proposed method is computationally very efficient, even though somewhat less accurate than known methods in shape analysis.

The Riemannian metric (4) requires calculating the optimal reparameterizations between every pair of animations. The proposed signature method instead only requires calculating the signature once for every animation, and then compares animations by computing inexpensive norms. The optimisation procedure is no longer necessary.[1]

In our experiments, the signature method outperformed the optimal reparameterization metric by a factor of $\sim$2000 when classifying animations. A more precise comparison with the SRVT approach and other methods, see e.g. [10] goes beyond the scope of this work and will be considered in future work. Still our preliminary experiments give an idea of the possible performance benefits gained with the signature approach.

Increasing the accuracy of the signature method might also be possible by defining a more precise similarity measure. Nonetheless, our results can be seen as proof of concept for using signatures as an efficient way of classifying shapes.

Acknowledgements. This paper contains work done as part of P.E.L.'s master thesis. The master thesis will be published separately as part of NTNU's Master of Science program in Applied Physics and Mathematics [12]. N.T. acknowledges that part of this work was carried out during his tenure of an ERCIM 'Alain Bensoussan' Fellowship Programme. This work was supported by the European Union's Horizon 2020 research and innovation programme under the Marie Sklodowska-Curie, grant agreement No.691070.

The data used in this project was obtained from http://mocap.cs.cmu.edu. The database was created with funding from NSF EIA-0196217.

---

[1] A more thorough analysis of the run time complexities associated with these algorithms has been left out due to space constraints.

References

1. Bauer, M., Eslitzbichler, M., Grasmair, M.: Landmark-guided elastic shape analysis of human character motions. Inverse Probl. Imaging **11**, 601–621 (2015). https://doi.org/10.3934/ipi.2017028
2. Bruveris, M., Michor, P., Mumford, D.: Geodesic completeness for Sobolev metrics on the space of immersed plane curves. Forum Math. Sigma [electronic only] **2** (2013). https://doi.org/10.1017/fms.2014.19
3. Celledoni, E., Eslitzbichler, M., Schmeding, A.: Shape analysis on lie groups with applications in computer animation. J. Geom. Mech. **8**(3), 273–304 (2016)
4. Chen, K.T.: Iterated integrals and exponential homomorphisms. Proc. Lond. Math. Soc. **s3-4**(1), 502–512 (1954). https://doi.org/10.1112/plms/s3-4.1.502
5. Chen, K.T.: Integration of paths-a faithful representation of paths by noncommutative formal power series. Trans. Am. Math. Soc. **89**(2), 395–407 (1958). http://www.jstor.org/stable/1993193
6. Chevyrev, I., Nanda, V., Oberhauser, H.: Persistence paths and signature features in topological data analysis. IEEE Trans. Pattern Anal. Mach. Intell. (2018, to appear). https://doi.org/10.1109/TPAMI.2018.2885516
7. CMU Graphics Lab: CMU Graphics Lab Motion Capture Database. http://mocap.cs.cmu.edu/. Accessed 10 Dec 2018
8. Hilgert, J., Neeb, K.H.: Structure and Geometry of Lie Groups. Springer, New York (2012). https://doi.org/10.1007/978-0-387-84794-8
9. Kruskal, J., Wish, M.: Multidimensional Scaling. Sage Publications, Newbury Park (1978)
10. Lahiri, S., Robinson, D., Klassen, E.: Precise matching of PL curves in $\mathbb{R}^N$ in the square root velocity framework. Geom. Imaging Comput. **2** (2015). https://doi.org/10.4310/GIC.2015.v2.n3.a1
11. Lyons, T.J.: Differential equations diriven by rough signals. Revista Matemática Iberoamericana **14**(2), 215–310 (1998). https://doi.org/10.4171/RMI/240
12. Lystad, P.E.: Signatures in shape analysis: an efficient approach to motion identification. Master's thesis, Department of Mathematical Sciences, NTNU, Trondheim, Norway (2019, manuscript submitted for publication)
13. Michor, P., Mumford, D.: Vanishing geodesic distance on spaces of submanifolds and diffeomorphisms. Doc. Math. **10**, 217–245 (2005)
14. Michor, P., Mumford, D.: An overview of the Riemannian metrics on spaces of curves using the Hamiltonian approach. Appl. Comput. Harmon. Anal. **23**, 74–113 (2006). https://doi.org/10.1016/j.acha.2006.07.004
15. Srivastava, A., Klassen, E., Joshi, S.H., Jermyn, I.H.: Shape analysis of elastic curves in Euclidean spaces. IEEE Trans. Pattern Anal. Mach. Intell. **33**(7), 1415–1428 (2011)
16. Younes, L.: Computable elastic distances between shapes. SIAM J. Appl. Math. **58**, 565–586 (1998)

Dilation Operator Approach and Square Root Velocity Transform for Time/Doppler Spectra Characterization on SU(n)

Guillaume Bouleux[1]([⊠]) [iD] and Frederic Barbaresco[2]

[1] Univ Lyon, INSA-Lyon, UJM-Saint-Etienne, DISP, EA 4570,
69621 Villeurbanne, France
`guillaume.bouleux@insa-lyon.fr`
[2] THALES LAS Representative at KTD Processing, Control & Cognition Board,
Advanced Radar Concepts (ARC), Thales Surface Radar (SRA),
Land & Air Systems, Hameau de Roussigny, 91470 Limours, France
`frederic.barbaresco@thalesgroup.com`

Abstract. We propose in this work the use of Dilation theory for non-stationary signals and their time/Doppler spectra to embed the underlying spectral measure on the Special Unitary group $SU(n)$. The Dilation theory gives access to rotation-like matrices built in with partial correlation coefficients. Due to the non-stationary condition, the time/Doppler spectra is associated with a path on $SU(n)$. We use next the Square root Velocity Transform which has been proven to be equivalent to a first order Sobolev metric on the space of shapes. Because the metric in the space of curves naturally extends to the space of shapes, this enables a comparison between curves' shapes and allows then the classification of time/Doppler spectra.

Keywords: Time/Doppler · $SU(n)$ · Dilation theory · Shape space · Elastic metric · SRV transform

1 Introduction

The analysis and/or representation of non-stationary processes has been approached for four or five decades now by time-scale/time-frequency analysis, or by the Fourier representation when processes belong to the class of periodically correlated processes (PC), or finally by the use of partial correlation coefficients (parcors). One of the advantages of the parcors utilization is their strong relationship with process measure through the point-to-point relationship with correlation coefficients [23]. They consequently appear explicitly in the Orthogonal Polynomial on the Real Line/Unit Circle decomposition of the measure [6,19], on the Matrices Orthogonal Polynomials on the Unit Circle [12]

© Springer Nature Switzerland AG 2019
F. Nielsen and F. Barbaresco (Eds.): GSI 2019, LNCS 11712, pp. 31–38, 2019.
https://doi.org/10.1007/978-3-030-26980-7_4

and its applications [4], are the elements for the construction of dilation matrices that appear in the Cantero Moral, and Velazquez (CMV)/Geronimus, Gragg, and Teplyaev (GGT) matrices [20], for the Schur flows problem with upper Hessenberg matrices [1] that are also seen in the literature as evolution operators [19] or shift operator [15], and finally appear in the state-space representation [9,10]. The Dilation theory is inspired by the theory of operators [21], which links the measure and unit operators of the process. In its simplest version, the dilation theory corresponds to the Naimark [21] dilation, and states that given a sequence of correlation coefficients $\{R_i\}_{i\in\mathbb{N}}$, there is a unit matrix W such that $R_i \triangleq (1\ 0\ 0\ldots)W^i(1\ 0\ 0\ldots)^T$ where $\cdot^T$ denotes transposition. When the process is not stationary, its associated correlation matrix is no longer structured Toeplitz, a set of matrices is then required [9] and the previous expression becomes $R_{i,j} \triangleq (1\ 0\ 0\ldots)W_{i+1}W_{i+2}\cdots W_j(1\ 0\ 0\ldots)^T$. The matrices W_i are theoretically understood as matrices of infinite rotation, which become finite when the sequence of the correlation coefficients is itself finite. In this particular case, the matrices W_i belong to $SO(n)$ or $SU(n)$, the orthogonal or unit special group, respectively, and the measure of the process is fully described by the set of W_i. As a result, the spectral measure of the process is characterized for the non-stationary case, by a sampled trajectory induced by the Dilation matrices on a Lie group.

In this work, we propose to analyse Time/Doppler spectra given by drones. This topic has been already tackled by the use of parcors but results were not as strong as expected [2,4]. The curve distance or their related shapes concerned only one parcor at a time and not the whole associated spectral measure. This is why in this work we analyse the curves of Dilation matrices associated with Time/Doppler spectra on the Special Unitary Group, $SU(n)$.

We will first recall how to obtain dilation matrices and then remind how to compare the associated curve on $SU(n)$, through the Square Root Velocity Transform. Finally, results will be given based on real Time/Doppler spectra and a conclusion will follow.

2   Building the Dilations

The idea, mainly explained in [9] consists in the parametrization of correlation matrices by one or more sequences of numbers, called choice sequence, Geronimo or Verblunsky coefficients, Schur parameters or partial reflection coefficients (parcors) according to the context. To simplify the study, we place ourselves within the framework of positive definite matrices. There is in fact a bijection between the set of positive definite matrices and the parcors, as Levinson, Burg, and more recently Barbaresco [2–4,23] have noted, in the case of stationary signals. This parameterization in its simplest form associates the partial correlation coefficients that we will call $\{\Gamma_{i,j}\}_{i<j}$, to the correlation matrix. In fact, the partial correlation coefficients appeared to be the elementary angles in the Gramm-Schmidt orthogonalization procedure which orthogonalizes the canonical basis of $\mathbb{R}^n$ in the basis in which the non-stationary kernel of the process is

represented. This is a consequence of the semi positive definite property of the correlation matrix.

With this idea in mind, (more explanations shall be obtained in [7, 9, 13]), the first orthogonalization step is therefore associated with an elementary rotation of the first canonical vector of $\mathbb{R}^n$ with angle $\arccos(\Gamma_{1,1})$, the second orthogonalization step is the composition of two rotations and so on. We briefly give the procedure as follows.

From the sequence of $\{\Gamma_{i,j}\}_{i<j}$ we build Givens rotation in such a way that:

$$G_{j-k}(\Gamma_{k,k+l}) = I \oplus \begin{pmatrix} \Gamma_{k,k+l} & D_{\Gamma_{k,k+l}^*} \\ D_{\Gamma_{k,k+l}} & -\Gamma_{k,k+l}^* \end{pmatrix} \oplus I \tag{1}$$

with I the identity matrix and $D_{\Gamma_{i,j}} = (I - \Gamma_{(i,j)}^* \Gamma_{i,j})$ the defect operator of the contraction $\Gamma_{i,j}$. By noting:

$$W_i = G_{j-i}(\Gamma_{i,i+1}) G_{j-i}(\Gamma_{i,i+2}) \ldots G_{j-i}(\Gamma_{i,j}) \tag{2}$$

we can therefore write:

$$R_{i,j} = P_1 W_i W_{i-1} \cdots W_{j-1} |_{H_1} \tag{3}$$

where $P_1 = (1\ 0 \cdots)^T$ that is, the orthogonal projection on the first line, and H_1 is the restriction to the first column. We then find the relation previously stated $R_{i,j} = (1\ 0 \cdots) W_{i+1} W_{i+2} \cdots W_j (1\ 0 \cdots)^T$, corresponding to the Kolmogorov decomposition of a non-stationary process [9]. It is the extension of the Naimark Dilation operator theorem. The structure of a Dilation matrix W_i is such that

$$W_i = \begin{pmatrix} \Gamma_1 & D_{1*}\Gamma_2 & D_{1*}D_{2*}\Gamma_3 & \cdots \\ D_1 & -\Gamma_1^*\Gamma_2 & -\Gamma_1^*D_{2*}\Gamma_3 & \cdots \\ 0 & D_2 & -\Gamma_2^*\Gamma_3 & \cdots \\ 0 & 0 & D_3 & \cdots \\ 0 & 0 & 0 & \cdots \\ 0 & 0 & 0 & \cdots \end{pmatrix}.$$

W_i belongs to the class of so-called higher Hessenberg matrices with positive subdiagonal coefficients. Of course, when the process is stationary, the calculations are greatly simplified. We have in that case $W = W_i$, $\forall i$, $\Gamma_{i,j} = \Gamma_{j-i}$.

3   Comparison of PC Processes

3.1   Shape Space

We are therefore interested in curves based on a variety, and more precisely on a Lie group. Let c be such a curve, $c : [0,1] \in SO(n)$. To study the geometric structure of these objects, consider the set of curves of $SO(n)$ defined by $\mathcal{M} = \{c \in \mathcal{C}^\infty([0,1], SU(n)) : c'(t) \neq 0\ \forall t\}$. When comparing two curves, it is natural to define a distance between these curves that do not depend on their parameterizations or their rotations. This amounts to reparametrizing the curves by an increasing diffeomorphism $\phi \to \mathcal{D} : [0,1] \to [0,1]$ and imposes that the metric on $\mathcal{M}$ be invariant by reparametrization. This property leads to the equivalence relation

$$c_o \sim c_1 \Leftrightarrow \exists\, \phi \in \mathcal{D} : c_0 = c_1 \circ \phi. \tag{4}$$

for two curves c_0, c_1 of $\mathcal{M}$. From this relation we naturally get a definition of a quotient space of $\mathcal{M}$, usually called the shape space and

$$\mathcal{S} = \mathcal{M}/\sim, \ or \ \mathcal{S} = \mathcal{M}/\mathcal{D}. \tag{5}$$

We deduce a distance on the space of forms from that defined on $\mathcal{M}$

$$d_{\mathcal{S}}\left([c_0],[c_1]\right) = \inf_{\phi \in \mathcal{D}} d_{\mathcal{M}}\left(c_0, c_1 \circ \phi\right), \tag{6}$$

where $[c_0], [c_1]$ are respectively the representatives of the equivalence classes of c_0 et c_1.

3.2   Metric and Distance on $\mathcal{S}$ then $\mathcal{M}$

We now propose to give some information on the choice of a relevant metric on $\mathcal{M}$ in order to compare the different closed curves. Since the base space is a Lie group, we use the Celledoni *et al.* [8] approach, but many other works have been proposed to take into account the homogeneous (or not) structure of the basic variety.

In [5,16,17] it is shown that unfortunately we can not use the simple metric L^2 on $\mathcal{M}$ because it leads to a null metric on the space of forms $\mathcal{S}$. In this case it is impossible to differentiate two different curve shapes. To overcome this difficulty, the family of elastic metrics, derived from the Sobolev metric, has been studied. In the case of an Euclidean space $\mathbb{R}^n$, it admits the expression:

$$g_c^{a,b}(u,v) = \int \left(a^2 \langle D_l u^N, D_l v^N \rangle + b^2 \langle D_l u^T, D_l v^T \rangle\right) ||c'|| dt,$$

where $D_l u = h' / \parallel c' \parallel$, $D_l u^T = \langle D_l u, w \rangle w$, avec $w = c' / \parallel c' \parallel$ and $D_l u^N = D_l u - D_l u^T$. In our case it is shown that $a, b = 1, 1/2$ allows to have an equivalence with a transformation at the same time simple interpretation and implementation, the TSRV. The TSRV, adapted to our case, sets the curve c by its velocity vector and its anchor point and then transports the new curves belonging to the successive tangent spaces by translation on the right, all at the same point of reference, which will be the identity element e of $SU(n)$. Its definition is

$$F_{Lie} : \mathcal{M} \longrightarrow SU(n) \times L^2([0,1], \mathfrak{g}) \tag{7}$$

$$F_{Lie}(c)(t) = (e, q(t)) = \left(e, \frac{T_c^{c(t) \to e}(c'(t))}{\sqrt{\parallel c'(t) \parallel}} \right) \tag{8}$$

where $\mathfrak{g}$ is the Lie algebra, $|| \cdot ||$ a norm induced by a right invariant metric on $SU(n)$, and where $T_c^{c(t) \to e}$ is the transport of $c(t)$ to the identity following the curve c. With this transformation, which corresponds to a first-order Sobolev metric, the desired distance in (6) leads to solving the optimization problem:

$$d_{\mathcal{S}}([c_0], [c_1]) = \inf_{\phi \in \mathcal{D}} \left(\int_0^1 ||q_0(t) - q_1(\phi(t)) \sqrt{\phi'(t)}||^2 \right)^{1/2},$$

which is solved by a traditional gradient descent algorithm or dynamic linear programming [8]. We also get a familiar expression for the geodetic interpolation between two curves c_0 and c_1, expressed in their TSRV domain:

$$\mathcal{Q}(s) = F_{Lie}^{-1} ((1 - s) F_{Lie}(c_0) + s F_{Lie}(c_1)) \text{ pour } s \in [0, 1].$$

We now have all the ingredients to describe the steps to characterize and compare non-stationary processes:

1. **Input:** A set of rotation matrices $\{W_i\}_i$ associated with the Time/Doppler spectra;
2. Inject the set of matrices $\{W_i\}$ into their respective Lie algebras $\{V_i\}$ via the inverse exponential application;
3. Use Spline to interpolate between the matrices V_i [14,18];
4. Come back to the base manifold $SU(n)$ via the exponential application;
5. Shift the curve in order to fulfil the condition $c(0) = e$ and then perform the SRV transform, given by (7);
6. Compute the distance defined by Sect. 3.2.
   The optimisation problem (non-crossing matching graph) is solved by dynamic programming;
7. **Output:** Distance between the curves on $\mathcal{M}$, and geodesic interpolation between the curves.

4   Application on Time/Doppler

This section is devoted to show how the approach proposed can help in deciding through Time Doppler whether a target is detected. For this scenario, we have used a Time/Doppler of a drone and computed the sequence of parcors for two different time instants. The periodically correlated property of the Time/Doppler signal is showed in Fig. 1, which is used for estimating the periodicity (with the lowest frequency) of the second order statistics. Here, the periodicity has been estimated to be roughly 50 samples. There will be consequently 50 sequences of parcors for each of the two signals analyzed. In order to illustrate the trajectory made of the dilation matrices associated with each sequence of parcors, we propose to use only two parcors. The dilation matrices will belong to $SU(4)$ so. The next step is to take a suitable parametrization in order to display the trajectories on $\mathbb{R}^3$.

4.1   Representation of $SU(4)$

We propose here to briefly explain the parametrization we have chosen. It is based on [22] which states that the vector of the Lie algebra eigenvalues for $SU(4)$ belongs to the 2-sphere. Actually, the eigenvalues of the Lie algebra are parametrized by two angles and a radius in the following way:

$$Tr(H) = \sum_{k=1}^{4} \lambda_k = 0, \quad Tr(H^2) = \sum_{k=1}^{4} \lambda_k^2 = r^2$$

$$Tr(H^3) = \sum_{k=1}^{4} \lambda_k^3 = \frac{3}{4} r^3 \sin(\theta) \sin(2\theta) \cos(\phi)$$

$$det(H) = \prod_{k=1}^{4} \lambda_k = \frac{1}{8} \left(\left(Tr(H^2) \right)^2 - 2Tr(H^4) \right)$$

with H the Lie algebra of $SU(4)$, $Tr()$ and det, the trace and the determinant respectively and λ_k the k−th eigenvalue of H. Once r, θ, ϕ are estimated, we are in position to represent the trajectory made of the dilation matrices.

4.2   Results

Curiously enough, the angle ϕ estimated has a value which is very closed to $\frac{\pi}{2}$ for all the $SU(4)$ matrices. Then, the 2-sphere representation of $SU(4)$ reduces here to a plan. This represented by Fig. 2. This figure allows us to clearly see a trajectory difference between a detected target or not. When no target is detected, Fig. 2-(a), the set of dilation matrices concentrates around a closed area which is obviously not the case for Fig. 2-(b) where the dilation matrices spread

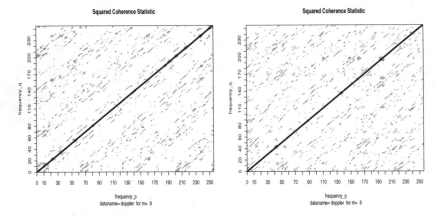

Fig. 1. Square Coherence Statistics for two different time instants, showing the periodically correlated property of the Time/doppler. (a) when there is no target detected by the Time/Doppler, (b) when a target is detected by the Time/Doppler

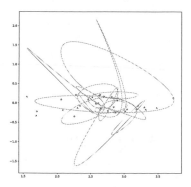 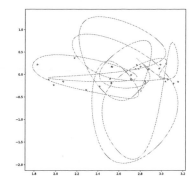

Fig. 2. Trajectory made of the Dilation matrices belonging to $SU(4)$ for two different time instants and represented here inside a ball (Sect. 4.1). (a) when there is no target detected by the Time/Doppler spectra, (b) when a target is detected by the Time/Doppler spectra

off much more, providing a closed curve with much more amplitude variations. The SRV next applied to these two trajectories gave a distance of 80.

5 Conclusion

In this paper, a new representation of Time/Doppler spectra associated with non-stationary processes has been proposed through the theory of Dilation. It allows to represent the spectral measure of the RADAR process on a Lie group by drawing a curve on $SU(n)$. The comparison of the processes, and then their associated spectra, can then be carried out directly by comparing these curves using tools coming from the differential geometry. In particular, an adapted metric is proposed in order to obtain a fast distance calculation and a simple geodesic interpolation. We note that the theory remains valid for real signals and that work in progress shows that it is possible to classify the processes by topological characteristics of the spectral measure of the processes. Finally, the formalism of the differential geometry makes it possible to interpolate between the rotation matrices and thus to interpolate the spectral measure of the process. Going a step further, the sequences of Time/Doppler cells might then be interpolated yielding to more accurate spectra.

References

1. Ammar, G., Gragg, W., Reichel, L.: Constructing a unitary Hessenberg matrix from spectral data. In: Golub, G.H., Van Dooren, P. (eds.) Numerical Linear Algebra, Digital Signal Processing and Parallel Algorithms, pp. 385–395. Springer, Heidelberg (1991). https://doi.org/10.1007/978-3-642-75536-1_18
2. Arnaudon, M., Barbaresco, F., Yang, L.: Riemannian medians and means with applications to radar signal processing. IEEE J. Sel. Top. Signal Proces. **7**, 595–604 (2013)

3. Barbaresco, F.: Interactions between symmetric cone and information geometries: Bruhat-Tits and Siegel spaces models for high resolution autoregressive Doppler imagery. In: Nielsen, F. (ed.) ETVC 2008. LNCS, vol. 5416, pp. 124–163. Springer, Heidelberg (2009). https://doi.org/10.1007/978-3-642-00826-9_6

4. Barbaresco, F.: Radar micro-Doppler signal encoding in Siegle unit poly-disk for machine learning in Fisher metric space. In: Proceedings of the 2018 19th International Radar Symposium (IRS), Bonn, Germany, 20–22 June 2018

5. Bauer, M., Bruveris, M., Michor, P.W.: Why use Sobolev metrics on the space of curves. In: Turaga, P., Srivastava, A. (eds.) Riemannian Computing in Computer Vision. Springer, Cham (2016). https://doi.org/10.1007/978-3-319-22957-7_11

6. Bingham, N.H.: Szego's theorem and its probabilistic descendants (2011). http://arxiv.org/abs/1108.0368

7. Bouleux, G., Dugast, M., Marcon, E.: Information topological characterization of periodically correlated processes by dilation operators. IEEE Trans. Inf. Theor. (2019, in press). https://doi.org/10.1109/TIT.2019.2923217

8. Celledoni, E., Eslitzbichler, M., Schmeding, A.: Shape analysis on Lie groups with applications in computer animation. J. Geom. Mech. **8**, 273–304 (2016)

9. Constantinescu, T.: Schur Parameters, Factorization and Dilation Problems. Birkhäuser, Basel (1995)

10. Desbouvries, F.: Unitary Hessenberg and state-space model based methods for the harmonic retrieval problem. IEE Proc. Radar Sonar Navig. **143**, 346–348 (1996)

11. Dégerine, S., Lambert-Lacroix, S.: Characterization of the partial autocorrelation function of a nonstationary time series. J. Multivariate Anal. **2**, 1296–1301 (2003)

12. Delsarte, P., Genin, Y.V., Kamp, Y.G.: Orthogonal polynomial matrices on the unit circle. IEEE Trans. Circ. Syst. **25**, 149–160 (1978)

13. Dugast, M., Bouleux, G., Marcon, E.: Representation and characterization of nonstationary processes by dilation operators and induced shape space manifolds. Entropy **20**(9), 717 (2018)

14. Hofer, M., Pottmann, H.: Energy-minimizing splines in manifolds. ACM Trans. Graph. **23**(3), 284–293 (2004)

15. Masani, P.: Dilations as propagators of Hilbertian varieties. SIAM J. Math. Anal. **9**, 414–456 (1978)

16. Michor, P., Mumford, D.: Vanishing geodesic distance on spaces of submanifolds and diffeomorphisms. Documenta Mathematica **10**, 217–245 (2004)

17. Michor, P.W., Mumford, D.: An overview of the Riemannian metrics on shape spaces of curves using the Hamiltonian approach. Appl. Comput. Harmon. Anal. **23**, 74–113 (2007)

18. Shingel, T.: Interpolation in special orthogonal groups. IMA J. Numer. Anal. **29**(3), 731–745 (2009)

19. Simon, B.: Orthogonal Polynomials on the Unit Circle Part 1 and Part 2, vol. 54. American Mathematical Society, Providence (2009)

20. Simon, B.: CMV matrices: five years after. J. Comput. Appl. Math. **208**, 120–154 (2007)

21. Sz.-Nagy, B., Foias, C., Bercovici, H., Kérchy, L.: Harmonic Analysis of Operators on Hilbert Space. Springer, New York (2010). https://doi.org/10.1007/978-1-4419-6094-8

22. Van Kortryk, T.S.: Matrix exponentials, $SU(N)$ group elements, and real polynomial roots. J. Math. Phys. **57**, 021701 (2016)

23. Yang, L., Arnaudon, M., Barbaresco, F.: Riemannian median, geometry of covariance matrices and radar target detection. In: 2010 European Radar Conference (EuRAD), pp. 415–418, September 2010

Selective Metamorphosis for Growth Modelling with Applications to Landmarks

Andreas Bock, Alexis Arnaudon[✉], and Colin Cotter

Department of Mathematics, Imperial College London, London SW7 2AZ, UK
alexis.arnaudon@imperial.ac.uk

Abstract. We present a framework for shape matching in computational anatomy allowing users control of the degree to which the matching is diffeomorphic. The control is a function defined over the domain describing where to violate the diffeomorphic constraint. The location can either be specified from prior knowledge of the growth location or learned from data. We consider landmark matching and infer the distribution of a finite dimensional parameterisation of the control via Markov chain Monte Carlo. Preliminary analytical and numerical results are shown and future paths of investigation are laid out.

Keywords: LDDMM · Computational anatomy · Metamorphosis · MCMC

1 Introduction

In computational anatomy [10,11] one of the most fundamental problems is to continuously deform an image or shape into another and thereby obtain a natural notion of distance between them as the energy required for such a deformation. Common methods to compute image deformations are based on diffeomorphic deformations which assume that the images are continuously deformed into one another with the additional property that the inverse deformation is also continuous. This is a strong requirement for images which implies that the 'mass' of any part of the image is conserved: we cannot create or close 'holes'. This is also a crucial property in fluid mechanics and in fact the theory of diffeomorphic matching carrying the moniker *Large Deformation Diffeomorphic Metric Mapping* (LDDMM) [5,23] has been inspired by fluid mechanics. Indeed, Arnold [4] made the central observation that the geodesic equations for the diffeomorphism group induced by divergence-free vector fields corresponded to that of incompressible flows. If a strictly diffeomorphic matching is not possible or necessary, an extension of LDDMM called metamorphosis [14,25] is available which introduces a parameter σ^2 parameterising the deviation from diffeomorphic matching

A. Arnaudon acknowledges EPSRC funding through award EP/N014529/1 via the EPSRC Centre for Mathematics of Precision Healthcare.

F. Nielsen and F. Barbaresco (Eds.): GSI 2019, LNCS 11712, pp. 39–48, 2019.
https://doi.org/10.1007/978-3-030-26980-7_5

allowing for topological variations e.g. growth via image intensity. In particular, if $\sigma^2 = 0$ the deformation is purely diffeomorphic as in LDDMM. See [18,22,24] for technical details pertaining to the construction of the metamorphosis problem. While diffeomorphic paths always exist for landmark problems [12] this theory allows one to match images of shapes with different topological features, which is ill-conditioned for standard LDDMM. Indeed, even inexact matching in LDDMM for such problems yields large energies and spurious geodesics that do not contribute to an intuitive matching, see Fig. 1. As observed here, introducing $\sigma^2 > 0$ regularises the problem and qualitative improves the matching.

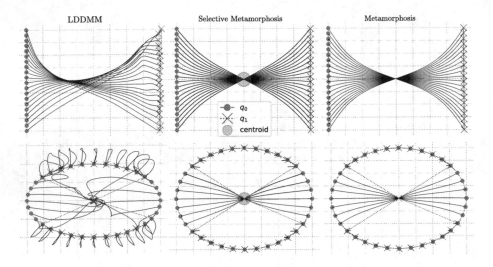

Fig. 1. This figure illustrates landmark matching with classical LDDMM (left column), metamorphosis (right column) and our selective metamorphosis approach (middle column). We perform a matching between two landmark configurations q_0 (circles) and q_1 (crosses), with the continuous lines between them describing trajectories. LDDMM fails to perform the matching and we observe unnatural landmark trajectories whereas metamorphosis achieves a more intuitive matching. Selective metamorphosis has the additional advantage of only breaking the diffeomorphic property where needed in along the matching, thus preserving more of the desired diffeomorphic property of the matching. These simulations where done for landmarks with Gaussian kernel of variance 0.5, 100 timesteps from $t = 0$ to $t = 1$, and a metamorphosis kernel of variance 0.2.

In this work, we modify metamorphosis to include a spatially dependent control parameter $x \mapsto \nu(x)$ in order to selectively allow non-diffeomorphic (*metamorphic*) matching in parts of the domain. For $\nu(\cdot) = \sigma^2$ our theory recovers the standard metamorphosis model. However, with a localised control (e.g. a Gaussian centred at a point in $\mathbb{R}^d$), we can selectively introduce metamorphosis in an image and model local topological effects such as growth phenomena. The difficulty of this problem is to infer the function ν without prior knowledge of the location of the topological effects. This problem is similar to the one treated in [3], where such functions were parameterising the randomness in LDDMM matching of shapes. We will use a Markov chain Monte Carlo (MCMC) approach to infer appropriate functions ν, such that the topological effects are well described and a large part of the matching remains diffeomorphic. In this paper we focus on landmark matching but aim to extend the theory to data images.

2 Metamorphosis for Landmarks

In this paper we are concerned with diffeomorphometric approaches to shape matching. To this end, we use time-dependent velocity fields $u_t \in V$, where the Hilbert space V is continuously embedded in $C_0^k(\mathbb{R}^d)$, $k \geq 1$. It induces a curve φ_t on a subgroup $\text{Diff}_V(\mathbb{R}^d)$ of diffeomorphisms [4, 26] via the equation

$$\dot{\varphi}_t = u_t \circ \varphi_t, \qquad \varphi_0 = \text{id}. \tag{1}$$

This is used in the matching problem of two images I_0 and I_1 with cost

$$S(u) = \int_0^1 \frac{1}{2}\|u_t\|_V^2 dt + \frac{1}{2\lambda^2} F(I_0 \circ \varphi_1^{-1}, I_1) \longrightarrow \text{min. subject to (1),} \tag{2}$$

where F denotes a similarity measure between the deformed initial image $I_0 \circ \varphi_1$ and the target image I_1 to allow inexact matching parameterised by λ^2. The LDDMM approach takes F as an L^2 norm of the difference between its arguments. In this work, we will consider singular solutions for M landmarks with positions $q_t^i \in \mathbb{R}^d$ and momenta $p_t^i \in \mathbb{R}^d$ for $i = 1..M$ so that the velocity is

$$u_t(x) = \sum_{i=1}^M p_t^i K(x - q_t^i), \tag{3}$$

where $K : \mathbb{R}^d \times \mathbb{R}^d \to \mathbb{R}$ is the kernel associated to the norm $\|\cdot\|_V$. For metamorphosis, we follow the notation of [14, Definition 1], and in addition to the deformation φ_t, we introduce a template variable $\boldsymbol{\eta}_t$ such that the positions $\mathbf{q}_t$ of a set of landmarks and the template velocity $\mathbf{z}_t$ are given by $\mathbf{z}$ as

$$\mathbf{q}_t = \varphi_t \boldsymbol{\eta}_t \qquad \text{and} \qquad \mathbf{z}_t = \varphi_t \dot{\boldsymbol{\eta}}. \tag{4}$$

We then extend the action functional (2) to account for the template variable as

$$S_m(\mathbf{q}_t, \mathbf{p}_t, \mathbf{z}_t) = \int_0^1 \frac{1}{2}\left(\|u_t\|_V^2 + \frac{1}{\sigma^2}\sum_{i=1}^M |z_t^i|^2\right) dt, \tag{5}$$

where now the reconstruction relation is

$$\dot{\mathbf{q}}_t = u_t(\mathbf{q}_t) + \mathbf{z}_t, \tag{6}$$

see [14, 25] for more details. By taking variations carefully, we obtain the equations of motion

$$\begin{aligned} \dot{\mathbf{p}}_t &= -\nabla u_t(\mathbf{q}_t)^T \mathbf{p}_t \\ \dot{\mathbf{q}}_t &= u_t(\mathbf{q}_t) + \sigma^2 \mathbf{p}_t, \end{aligned} \tag{7}$$

where $\mathbf{z}_t = \sigma^2 \mathbf{p}_t$ and u_t is defined in (3).

3   Selective Metamorphosis for Landmarks

We can now extend the metamorphosis setting to be able to locally control the amount of non-diffeomorphic evolution. For this, we introduce a function $\nu : \mathbb{R}^d \to \mathbb{R}$ replacing the parameter σ^2 such that $\nu(x) = \sigma^2$ corresponds to the classic landmark metamorphosis. The action for selective metamorphosis thus becomes

$$S_{sm}^{\nu}(\mathbf{q}_t, u_t, \mathbf{z}_t) = \int_0^1 \frac{1}{2}\left(\|u_t\|_V^2 + \sum_{i=1}^M \frac{1}{\nu(q_t^i)}|z_t^i|^2\right)\,\mathrm{d}t, \tag{8}$$

which we minimise subject to the reconstruction Eq. (4) and the boundary conditions $\mathbf{q}_0$ and $\mathbf{q}_1$ at time $t = 0, 1$. In the case of landmarks we have as before that $\mathbf{z}_t = \nu(\mathbf{q}_t)\mathbf{p}_t$ so we can eliminate the template variable $\mathbf{z}_t$ and write

$$S_{sm}^{\nu}(\mathbf{q}_t, u_t, \mathbf{p}_t) = \int_0^1 \frac{1}{2}\left(\|u_t\|_V^2 + \sum_{i=1}^M \nu(q_t^i)|p_t^i|^2\right)\,\mathrm{d}t. \tag{9}$$

The problem defined by (9) yields the following equations for selective metamorphosis for landmarks:

$$\begin{aligned}
\dot{\mathbf{p}}_t &= -\nabla u_t(\mathbf{q}_t)^T \mathbf{p}_t - \frac{1}{2}\nabla \nu(\mathbf{q}_t)|\mathbf{p}_t|^2 \\
\dot{\mathbf{q}}_t &= u_t(\mathbf{q}_t) + \nu(\mathbf{q}_t)\mathbf{p}_t,
\end{aligned} \tag{10}$$

with $\mathbf{q}_0$, $\mathbf{q}_1$ fixed. Again, the velocity is fully described by $\mathbf{p}$ and $\mathbf{q}$ via (3). As we see from these equations, our approach offers a granularity not attainable via classical inexact landmark matching or metamorphosis. Namely, with ν it is possible to specify where in the image growth is allowed. As an example, a medical expert may want to allow for metamorphic growth near a tumour-prone area of the brain whilst allowing for purely diffeomorphic growth of the skull of the patient.

A practical procedure for solving (10) with the velocity defined in (3) is called *shooting*, where we replace the end-point condition $\mathbf{q}_1$ with a guess for $\mathbf{p}_0$, and iteratively update $\mathbf{p}_0$ using automatically computed adjoint (or backward) equations until $\mathbf{q}_1$ compares to $\mathbf{q}(1)$ below a certain tolerance. We will perform this procedure directly with an automatic differentiation package Theano [21], see [16,17] for more details on the implementation.

Theorem 1. *Let ν be bounded from below away from zero by $\nu_{inf} \in \mathbb{R}$ and from above by $0 < \sigma^2 \in \mathbb{R}$. Then there exists a minimiser of (9) admissible to (6).*

Proof. The functional in (9) is not convex so we work with a reformulation to ensure the required lower semi-continuity. Define a variable $w_t^i = \sqrt{\nu(q_t^i)}p_t^i$ in the problem:

$$\inf_{\substack{u \in L^2([0,1],\,V) \\ \mathbf{q} \in H^1([0,1],\,\mathbb{R}^{d \times M}) \\ \mathbf{w} \in L^2([0,1],\,\mathbb{R}^{d \times M})}} \int_0^1 \frac{1}{2} \left(\|u_t\|_V^2 + \sum_{i=1}^M |w_t^i|^2 \right) dt$$

$$\dot{\mathbf{q}}_t^i = u_t(\mathbf{q}_t) + \sqrt{\nu(\mathbf{q}_t)} \mathbf{w}_t$$

$$\mathbf{q}_0,\ \mathbf{q}_1 \text{ fixed}$$

First, note that owing to the constraint effectively being a boundary value problem, we cannot always find a $\mathbf{q}$ for arbitrary pairs of $(u, \mathbf{w})$. We define a bounded operator $(\mathbf{q}, u_t) \mapsto \frac{\dot{\mathbf{q}}_t - u_t(\mathbf{q}_t)}{\sqrt{\nu(\mathbf{q}_t)}} \triangleq \mathbf{w}$:

$$\left(\sum_{i=1}^M |w_t^i|^2 \right)^{\frac{1}{2}} = \|\mathbf{w}\|_2 = \|\frac{\dot{\mathbf{q}}_t - u_t(\mathbf{q}_t)}{\sqrt{\nu(\mathbf{q}_t)}}\|_2 \lesssim \nu_{\inf}^{-1} \left(\|\dot{\mathbf{q}}_t\|_2 + \|u_t(\mathbf{q}_t)\|_V \right).$$

From this we generate a minimising sequence $(\mathbf{q}^n, u^n, \mathbf{w}^n)_{n \geq 0}$ admissible to (11). The rest of the proof is standard, see e.g. [26]. We show the constraint equation is continuous with respect to the weak topology on $X \triangleq H^1([0,1], \mathbb{R}^{d \times M}) \times L^2([0,1], V) \times L^2([0,1], \mathbb{R}^{d \times M})$ i.e. $e(\mathbf{q}_t^n, \mathbf{w}_t^n, u_t^n) \rightharpoonup e(\mathbf{q}_t, \mathbf{w}_t, u_t)$ where $e(q, w, u) \triangleq \dot{q} - u(q) - \sqrt{\nu(q)} w$. Then,

$$\langle \sqrt{\nu(\mathbf{q}_t)} \mathbf{w}_t - \sqrt{\nu(\mathbf{q}_t^n)} \mathbf{w}_t^n, \phi \rangle \lesssim \nu_{\inf} \langle \mathbf{w}_t - \mathbf{w}_t^n, \phi \rangle \to 0, \quad \forall \phi \in L^2([0,1], \mathbb{R}^{d \times M}).$$

Further, for $\phi \in L^2([0,1], V)$,

$$\langle u_t(\mathbf{q}_t) - u_t^n(\mathbf{q}_t^n), \phi \rangle = \langle u_t(\mathbf{q}_t) - u_t^n(\mathbf{q}_t), \phi \rangle + \langle u_t^n(\mathbf{q}_t) - u^n(\mathbf{q}_t^n), \phi \rangle.$$

The first term vanishes trivially, while for the second we see

$$\langle u_t^n(\mathbf{q}_t) - u_t^n(\mathbf{q}_t^n), \phi \rangle \leq \mathrm{Lip}(u_t^n) \langle \mathbf{q}_t - \mathbf{q}_t^n, \phi \rangle \to 0$$

Since linear operators are naturally compatible with the weak topology the required continuity follows. Passing to subsequences where necessary we can by classic results extract bounded subsequences converging to weak limits where necessary to obtain a minimiser. Convexity of S implies weak lower semicontinuity concluding the proof. □

Theorem 2. *Assume $\nu \in W^{2,\infty}(\mathbb{R}^d)$ and V is embedded in $C_0^k(\mathbb{R}^d)$, $k \geq 1$ (continuous functions with continuous derivatives to order k vanishing at infinity). Then, given $\mathbf{p}_0, \mathbf{q}_0, \in \mathbb{R}^{d \times M}$, (10) with (3) are integrable for all time.*

Proof. Establishing appropriate Lipschitz conditions implies integrability of the system akin to [6, Theorem 5]. We note that the kernel in (3) is Lipschitz in (p_t, q_t) by assumption, so the composition $(p, q) \mapsto u \circ q$ is also Lipschitz. $u(q) \mapsto \nabla u(q)^T$ consider $v, w \in V$ and $x, y \in \mathbb{R}^d$:

$$\|\nabla v(x) - \nabla w(y)\|_2 \lesssim \|v\|_V \|x - y\|_2 + \|v - w\|_V \|y\|_2 \tag{11}$$

so the mapping is Lipschitz in both the position and velocity. Given the conditions on ν the mappings

$$(q, p) \mapsto \nu(q)p$$
$$(q, p) \mapsto \nabla\nu(q)|p|^2$$

$$(12)$$

are locally Lipschitz. Consequently we verify that for any $(\mathbf{p}_0, \mathbf{q}_0) \in B(0, r) \subset \mathbb{R}^{d \times M} \times \mathbb{R}^{d \times M}$, the system (10) is locally Lipschitz with constant L_{r,t_0} for some $t_0 > 0$. By the conservation of the Hamiltonian we can extend the existence of solutions to arbitary $t > t_0$. □

4   Bayesian Framework

We now place a stochastic model on ν inspired by the approach taken in [6] to infer most probable such functions. See also [1,2,20] for similar Bayesian approaches in computational anatomy. The goal is to develop an algorithm to infer ν from a given set of localised functions. We refer to [7,8] for an exposition of function space MCMC but we will consider a simpler case here. We consider ν as a sum of time-independent Gaussian functions

$$\nu_h(x) = \sum_{k=1}^{K} e^{-\sigma_k^{-2}\|h_k - x\|^2}.$$

$$(13)$$

This means that metamorphosis permitted in the neighbourhood of a point x (determined by the radius σ_k and *centroids* $h_k \in \mathbb{R}^2$ selected on the template) is proportional to the value of $\nu_h(x)$. As described in (10), ν_h follows the trajectory of the landmarks in the dynamics of $\mathbf{q}$. Note the number of landmarks, M, differ from the number of centroids K. For instance, we selected $K = 1$ in Fig. 1 due to our a priori knowledge of the trajectories (e.g. there is only a single point where landmarks cross or intersect).

Defining a density $p_{sm} \propto e^{-S_{sm}^\nu}$ over the space of triples $(\nu, \mathbf{q}_\nu, \mathbf{p}_\nu)$ leads to the preconditioned Crank-Nicholson Algorithm 1, see e.g. [13]. The parameter β has to be set such that the samples are un-correlated, which corresponds to an acceptance rate in the range 0.5–0.8.

In general, ν should accommodate the granularity of the deformation between two shapes and be able to resolve the topological changes necessary. This constitutes an interesting problem in and of itself, as it is *a priori* difficult to say what constitutes a good ν simply by inspecting the template and targets. Here we use Gaussians for their smoothness and simplicity, but we comment on extensions in Sect. 6.

5   Numerical Examples

This section displays some numerical results for our method to infer a distribution for the growth location using the landmark configurations seen in Fig. 1.

Algorithm 1. MCMC for selective metamorphosis

---

procedure MCMCSM(N, K, $\mathbf{q}_0$, $\mathbf{q}_1$, $\beta \in (0,1]$)

    $j \leftarrow 1$

    $\nu^j \leftarrow$ initial guess in $\mathbb{R}^{d \times K}$

    Solve (10) with ν^j and $\mathbf{q}_0$, $\mathbf{q}_1$ to obtain $\omega^j = (\mathbf{q}^j, \mathbf{p}^j, u^j)$

    while $j < N$ **do**

        Sample a random point $\xi \in \mathcal{N}(0, \mathrm{Id}_{\mathbb{R}^d})^K$

        $\nu \leftarrow \beta\xi + \sqrt{1 - \beta^2}\nu^j$

        Solve (10) with ν and $\mathbf{q}_0$, $\mathbf{q}_1$ to obtain $\omega = (\mathbf{q}, \mathbf{p}, u)$

        if RANDOMUNIT() $< \min(1, e^{-S_{sm}^{\nu^j}(\omega^j)+S_{sm}^{\nu}(\omega)})$ **then**

            $\nu^{j+1} \leftarrow \nu$

            $\omega^{j+1} \leftarrow \omega$

        else

            $\nu^{j+1} \leftarrow \nu^j$

            $\omega^{j+1} \leftarrow \omega^j$

        $j \leftarrow j + 1$

    return $\{\nu^j, \omega^j\}_{j=1}^N$

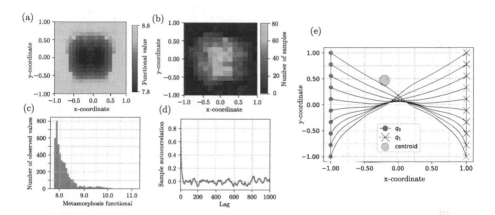

Fig. 2. We display the result of the MCMC Algorithm 1 applied to the inverted landmarks example of Fig. 1. (a) shows the analytical values for the functional (8) obtained for various positions of a single Gaussian ν. We observe a bimodal minimum near $(0,0)$, which depends on the choice of the model parameters, and in particular on the landmark interaction length corresponding to the Gaussian kernel K and σ_ν. (b) displays a heat map for the sampled positions of the centroid from the MCMC method, where the bimodality is not clearly visible. (c) is histogram of the sampled values of the functional which rapidly decays, indicating a good sampling of the minimum value of the functional. (d) shows the autocorrelation function of the Markov chain, which decays rapidly to reach an un-correlated state after 50 iterations. (e) shows one of the MAP estimators where the centroid is near on the edge of one of the wells of the top left panel. The simulations parameters are set to $\sigma_\nu = 0.2$, and 0.7 for the velocity kernel, $K = 1$ and $\beta = 0.2$ across 5000 samples.

The parameters and results for the first configuration is shown in Fig. 2. These preliminary results demonstrate that even for a small number of samples the density of accepted samples corresponds at least heuristically to the analytical density histogram obtained by computing the value of the metamorphosis functional in (8).

We arrive at the same conclusion for the second example, for which the results are shown in Fig. 3. Moreover, we note that the geodesic equations for p and q are time-reversible meaning that the configuration in Fig. 3 corresponds to both particle collapse as well as hole creation. It is numerically relatively simple to control the behaviour of ν by simple scaling or by adding regularisation terms to (8) to e.g. penalise having ν's far away from the support of the images. Such cost can easily be added to the MCMC algorithm, depending on the prior information one can have on the shape matching problem.

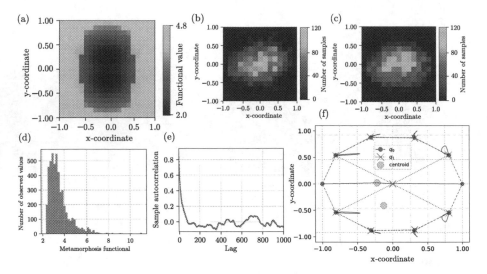

Fig. 3. Here we display the results for the second example (landmark collapse) of Fig. 1. Again, (a) shows the analytical values for a single ν field (8), which has also a bimodal structure, but in the other direction. For the MCMC we choose $K = 2$ Gaussian ν fields, and (b) and (c) displays two heat maps for the sampled positions of these centroids. (d) is a histogram of the sampled values of the functional, which has a peak at slightly higher values, possibly due to the redundant choice of two ν functions. (e) shows the autocorrelation function of the Markov chain which shows decorrelation after 100 steps. (f) shows the geodesics yielding one of the lowest functional values, where the two ν fields are close to each other, demonstrating the fact that only 1 would have been enough for this landmark configuration. The simulation parameters are the same as in Fig. 2 with the exception of $K = 2$.

6 Conclusion

We have presented a preliminary approach for selectively allowing photometric variation in a diffeomorphic image matching. We analysed the selective metamorphosis problem, the associated geodesic equations and demonstrated a proof of concept MCMC algorithm inferring a simple parameterisation of ν. This generalises LDDMM and metamorphosis and could provide a first-order exploratory tool for physicians to see if the development of a biological feature stems from a few violations of diffeomorphic evolution. This paper paves the way towards surgically investigating growth phenomena between topologically different images.

For future works we aim at extending the equations of Sect. 3 to images e.g. using the kernel framework in [19] or developing a space-time method.

In addition, there are many aspects of the probabilistic framework for the estimation of ν that need rigorous treatment and improvements. First, already in this simple setting, one would need to add additional penalties for the position of the centroids h_k to force them to remain for example near the centre of the image during the MCMC evolution. Second, natural extensions of our probabilistic approach by treating ν as a function could be considered and thus interpreting the resulting inverse problem through the appropriate measure-theoretical lens. Adding a time-dependency to ν can also be explored. Determining a truncated Fourier series of ν could lead to efficient numerical methods. Finally, we only used a simple MCMC algorithm, but a Metropolis-adjusted Langevin algorithm or Hamiltonian Monte-Carlo algorithm may be more appropriate to solve this problem.

To conclude, we hope that this framework could be used to model growth, in the spirit of the approaches of [15] or [9].

References

1. Allassonnière, S., Amit, Y., Trouvé, A.: Towards a coherent statistical framework for dense deformable template estimation. J. R. Stat. Soc. Ser. B (Stat. Methodol.) **69**(1), 3–29 (2007)
2. Allassonnière, S., Kuhn, E., Trouvé, A.: Map estimation of statistical deformable templates via nonlinear mixed effects models: deterministic and stochastic approaches. In: 2nd MICCAI Workshop on Mathematical Foundations of Computational Anatomy, pp. 80–91 (2008)
3. Arnaudon, A., Holm, D.D., Sommer, S.: A geometric framework for stochastic shape analysis. Found. Comput. Math. **19**, 653–701 (2018)
4. Arnold, V.I.: Sur la géométrie différentielle des groupes de lie de dimension infinie et ses applicationsa l'hydrodynamique des fluides parfaits. Ann. Inst. Fourier **16**(1), 319–361 (1966)
5. Beg, M.F., Miller, M.I., Trouvé, A., Younes, L.: Computing large deformation metric mappings via geodesic flows of diffeomorphisms. Int. J. Comput. Vis. **61**(2), 139–157 (2005)
6. Cotter, C.J., Cotter, S.L., Vialard, F.-X.: Bayesian data assimilation in shape registration. Inverse Probl. **29**(4), 045011 (2013)

7. Cotter, S.L., Roberts, G.O., Stuart, A.M., White, D.: MCMC methods for functions: modifying old algorithms to make them faster. Stat. Sci. **28**, 424–446 (2013)
8. Dashti, M., Stuart, A.M.: The Bayesian approach to inverse problems. In: Ghanem, R., Higdon, D., Owhadi, H. (eds.) Handbook of Uncertainty Quantification, pp. 311–428. Springer, Cham (2017). https://doi.org/10.1007/978-3-319-12385-1_7
9. Goriely, A.: The Mathematics and Mechanics of Biological Growth, vol. 45. Springer, New York (2017). https://doi.org/10.1007/978-0-387-87710-5
10. Grenander, U., Miller, M.I.: Representations of knowledge in complex systems. J. R. Stat. Society. Ser. B (Methodol.) **56**, 549–603 (1994)
11. Grenander, U., Miller, M.I.: Computational anatomy: an emerging discipline. Q. Appl. Math. **56**(4), 617–694 (1998)
12. Guo, H., Rangarajan, A., Joshi, S.: Diffeomorphic point matching. In: Paragios, N., Chen, Y., Faugeras, O. (eds.) Handbook of Mathematical Models in Computer Vision, pp. 205–219. Springer, Boston (2006). https://doi.org/10.1007/0-387-28831-7_13
13. Hairer, M., Stuart, A.M., Vollmer, S.J., et al.: Spectral gaps for a metropolis-hastings algorithm in infinite dimensions. Ann. Appl. Probab. **24**(6), 2455–2490 (2014)
14. Holm, D., Trouvé, A., Younes, L.: The Euler-Poincaré theory of metamorphosis. Q. Appl. Math. **67**(4), 661–685 (2009)
15. Kaltenmark, I.: Geometrical growth models for computational anatomy. Ph.D. thesis, Université Paris-Saclay (2016)
16. Kühnel, L., Arnaudon, A., Sommer, S.: Differential geometry and stochastic dynamics with deep learning numerics. arXiv preprint arXiv:1712.08364 (2017)
17. Kühnel, L., Sommer, S.: Computational anatomy in Theano. In: Cardoso, M.J., et al. (eds.) GRAIL/MFCA/MICGen 2017. LNCS, vol. 10551, pp. 164–176. Springer, Cham (2017). https://doi.org/10.1007/978-3-319-67675-3_15
18. Miller, M.I., Younes, L.: Group actions, homeomorphisms, and matching: a general framework. Int. J. Comput. Vis. **41**(1–2), 61–84 (2001)
19. Richardson, C.L., Younes, L.: Metamorphosis of images in reproducing kernel Hilbert spaces. Adv. Comput. Math. **42**(3), 573–603 (2016)
20. Schiratti, J.-B., Allassonnière, S., Colliot, O., Durrleman, S.: A Bayesian mixed-effects model to learn trajectories of changes from repeated manifold-valued observations. J. Mach. Learn. Res. **18**(1), 4840–4872 (2017)
21. Theano Development Team, Al-Rfou, R., et al.: Theano: a Python framework for fast computation of mathematical expressions. arXiv preprint arXiv:1605.02688 (2016)
22. Trouvé, A.: An infinite dimensional group approach for physics based models in pattern recognition. Preprint (1995)
23. Trouvé, A.: Diffeomorphisms groups and pattern matching in image analysis. Int. J. Comput. Vis. **28**(3), 213–221 (1998)
24. Trouvé, A., Younes, L.: Local geometry of deformable templates. SIAM J. Math. Anal. **37**(1), 17–59 (2005)
25. Trouvé, A., Younes, L.: Metamorphoses through Lie group action. Found. Comput. Math. **5**(2), 173–198 (2005)
26. Younes, L.: Shapes and Diffeomorphisms, vol. 171. Springer, Heidelberg (2010). https://doi.org/10.1007/978-3-642-12055-8

Geometric Mechanics

Intrinsic Incremental Mechanics

Jean Lerbet[1](⊠) [ID], Noël Challamel[2], François Nicot[3], and Félix Darve[4]

[1] LaMME, Univ Evry, Université Paris-Saclay, Evry, France
jlerbet@gmail.com
[2] IRDL-UBS, Université de Bretagne Sud, Lorient, France
[3] IRSTEA, ETNA, Université Grenoble-Alpes, Grenoble, France
[4] Lab 3SR, Grenoble INP, Université Grenoble-Alpes, Grenoble, France

Abstract. We produce a coordinate free presentation of some concepts usually involved in incremental mechanics (tangent linear stiffness matrix, stability, loading paths for example) but not always well founded. Thanks to the geometric language of vector bundles, a well defined geometrical object may be associated to each of these tools that allows us to understand some latent difficulties linked with these tools due to the absence of a natural connection and also to extend some of our recent results of linear stability to a non linear framework.

Keywords: Vector bundles · Transversality · Vertical derivative

1 Motivations

1.1 Kinematic Structural Stability

For the last ten years, we developed tools to tackle an old question regarding the conflict between two criteria of stability involved in rate-independent mechanical systems. We call these two criteria the divergence Lyapounov criterion (it is the usual one) and the Hill criterion also called the second order work criterion. These two criteria are identical for elastic conservative or for piece-wise rate-independent mechanical systems but they give different critical divergence stability values for elastic non conservative systems or for non linear rate-independent mechanical systems like non associate plastic materials. The usual language in mechanics characterizes these last systems by a non symmetric tangent stiffness matrix $K(p)$ whereas for the first class of system $K(p)$ is symmetric.

Thanks to the new concept of Kinematic Structural Stability (KISS) and an original variational formulation on all the possible kinematic constraints, we proved, in the discrete linear elastic nonconservative framework, that the two criteria become again equivalent ([4] for example). The most elegant proof of this result involves the geometric concept of compression of operator which can be extended to Hilbert spaces and which allowed us to extend the result to continuous linear elastic systems [6]: all the compressions of the operator are one-to-one if and only if the symmetric part of the operator is definite.

© Springer Nature Switzerland AG 2019
F. Nielsen and F. Barbaresco (Eds.): GSI 2019, LNCS 11712, pp. 51–54, 2019.
https://doi.org/10.1007/978-3-030-26980-7_6

1.2   Geometric Degree of Nonconservativity

In parallel to these stability issues, we also investigated the dual problem which questions the minimal number of kinematic constraints necessary to make conservative the elastic mechanical system. In a linear discrete framework, this number called the geometric degree of nonconservativity (GDNC) is the half of the rank of the skew symmetric part $K_a(p)$ of $K(p)$ [3]. The extension of the GDNC to infinite dimension Hilbert space involved for continuous systems is not obvious whereas the extension to the differentiable non linear framework is possible. Indeed, the skew-symmetric part $K_a(p)$ must be replaced by the exterior derivative $\mathbf{d}\omega_{\mathcal{F}}$ of the 1-form $\omega_{\mathcal{F}}$ defining the corresponding force system on the mechanical system. $\omega_{\mathcal{F}}$ is a section of the cotangent bundle $T^*\mathbb{M}$ of the configuration manifold $\mathbb{M}$ and the GDNC is then the half of the class of the 2-form $\mathbf{d}\omega_{\mathcal{F}}$ [5].

1.3   Main Issue

The problem investigated in this paper is to provide such a non linear extension but for the original KISS issue. Whereas the exterior derivative $\mathbf{d}\omega_{\mathcal{F}}$ provides a "natural" non linear extension of the skew symmetric part $K_a(p)$ of $K(p)$, it appears that there is no such natural extension for $K(p)$ nor for its symmetric part $K_s(p)$.

Indeed the incremental point of view necessitates to make a derivative of $\omega_{\mathcal{F}}$. However there is no natural connection on $\mathbb{M}$ to do it. To solve this problem, we will use the fact that the incremental quasi-static evolution of the mechanical system lies on the nil section of $T^*\mathbb{M}$ (which represents the equilibrium manifold) and we will use this canonical and global section of $T^*\mathbb{M}$ as a horizontal space for the derivative of $\omega_{\mathcal{F}}$. It allows for example to provide an intrinsic meaning of the common concept of tangent stiffness matrix of a system. We have to stress that we stay here within the differentiable framework which means that only hypoelasticity and not plasticity is investigated even if it is the long-term goal of these investigations. We also have to stress that the tools used in these investigations are usual (see [2] or [1] for example) in classical mechanics for so-called Lagrangian or Hamiltonian mechanics or even multisymplectic mechanics. However, here, by principle we do not suppose an Hamiltonian or Lagrangian functions to describe the evolution of the mechanical systems.

2   Some Results

We now present three intrinsic objects or results that geometrically extend more or less usual concepts of the linear framework to the not linear case. A large part of these developments are in [7]. In all these developments, the mechanical system is called Σ and is described by a finite number n of parameters which means that the configuration space is a n dimension manifold $\mathbb{M}$. With this language, any system of forces is represented by a section of the cotangent bundle

$T^*\mathbb{M}$. As usual, we sometimes identify the nil section of $T^*\mathbb{M}$ with $\mathbb{M}$ itself $(0_{T^*\mathbb{M}}(\mathbb{M}) \simeq \mathbb{M})$. We also note $\pi : T^*\mathbb{M} \to \mathbb{M}$ the natural projection so that $T_m^*\mathbb{M} = \pi^{-1}\{m\} \;\; \forall m \in \mathbb{M}$.

2.1   Tangent Stiffness Operator

Let $\mathcal{F}$ be a force system described by a section $\omega_{\mathcal{F}}$ of $T^*\mathbb{M}$ and $m_e \in \mathbb{M}$ an equilibrium configuration of Σ subjected to $\mathcal{F}$. Then, considering the derivative $d\omega_{\mathcal{F}}$ of $\omega_{\mathcal{F}}$ (and NOT the exterior derivative $\mathbf{d}\omega_{\mathcal{F}}$ as above), we have

$$d\omega_{\mathcal{F}}(m_e) : T_{m_e}\mathbb{M} \to T_{(m_e,0)}T^*\mathbb{M} = T_{x_e}\mathbb{M} \oplus \pi_{m_e}^{-1}$$
$$u \mapsto u + d\omega_{\mathcal{F}}(m_e)^{ver}(m_e)(u) \tag{1}$$

where $d\omega_{\mathcal{F}}(m_e)^{ver}(m_e)$ is a linear map and then belongs to $\mathcal{L}(T_{m_e}\mathbb{M}, T_{m_e}^*\mathbb{M})$. Then, we are led to put the

Definition 1. *The above linear map $d\omega_{\mathcal{F}}^{ver}(m_e) \in \mathcal{L}(T_{m_e}\mathbb{M}, T_{m_e}^*\mathbb{M})$ is called the tangent stiffness operator or the tangent stiffness tensor of Σ at m_e. Because of the involved spaces, it is a covariant 2-tensor on the vector space $T_{m_e}\mathbb{M}$. It obviously depends on m_e and on the force system $\mathcal{F}$.*

In local coordinates on the manifold $\mathbb{M}$, this tensor is represented by a square matrix of $\mathcal{M}_n(\mathbb{R})$: it is the "usual" tangent stiffness matrix K at the equilibrium m_e and for the force system $\mathcal{F}$.

2.2   T-Stability

We adopt the following

Definition 2. *Let $\mathcal{F}$ be a force system described by a section $\omega_{\mathcal{F}}$ of $T^*\mathbb{M}$ and $m_e \in \mathbb{M}$ an equilibrium configuration of Σ subjected to $\mathcal{F}$. Thus $\omega_{\mathcal{F}}(m_e) = 0 = 0_{T^*\mathbb{M}}(m_e)$. m_e is then called Transversality-stable or T-stable if $\omega_{\mathcal{F}}$ intersects or cuts transversally the nil section $0_{T^*\mathbb{M}}$.*

This definition is then purely geometric and does not involves the tangent stiffness. The infinitesimal characterization of the transversality of the intersection of manifolds leads to the following property:

Proposition 1. *m_e is T-stable if and only if $d\omega_{\mathcal{F}}^{ver}(m_e)$ is an invertible map.*

The T-transversality leads then to the usual characterization of the divergence Lyapounov stability.

2.3   KISS Issue

The KISS issue necessitates to consider all the submanifolds $\mathbb{V}$ of (embedded in) $\mathbb{M}$ and also the definition of loading paths $\mathcal{L} = (\mathbb{M}, \omega_{\mathcal{L}})$ on $\mathbb{M}$. The above

T-stability can be extended to loading paths which are then called regular loading paths (see [7] for more precisions). Then, for any embedded submanifold $j : \mathbb{V} \to \mathbb{M}$ of $\mathbb{M}$ we may define by a pullback j^* the induced loading path $\mathcal{L}_{\mathbb{V}}$ on $\mathbb{V}$ and is called the subloading path $\mathcal{L}_{\mathbb{V}} = (\mathbb{V}, \omega_{\mathcal{L}_{\mathbb{V}}})$ of $\mathcal{L}$. We then have the following extension of the KISS result to the non linear framework:

Theorem 1. *Let $\mathcal{L}$ be a regular loading path and $m_{e,\sigma} = \pi(\omega_{\mathcal{L}}(\sigma)) \in \mathbb{M}$. If the symmetric part $d\omega_{\mathcal{L}}(\sigma)^{ver,s}(m_{e,\sigma})$ of $d\omega_{\mathcal{L}}(\sigma)^{ver}(m_{e,\sigma})$ is a degenerated $(0,2)$ symmetric tensor then there is a submanifold $\mathbb{V} \ni m_{e,\sigma}$ of $\mathbb{M}$ such that $\mathcal{L}_{\mathbb{V}} = (\mathbb{V}, \omega_{\mathcal{L}_{\mathbb{V}}})$ is singular at σ.*

3   Some Open Questions

For future works, the fundamental open problem is to establish such a geometric framework to tackle similar questions for plastic evolutions. Two main issues are then to describe the internal irreversibility and to take into account the non differentiability along the incremental evolutions.

References

1. Aldaya, V., De Azcarraga, J.A.: Geometric formulation of classical mechanics and field theory. Rivista Del Nuovo Cimento **3**(10), 1–66 (1980)
2. Godbillon, C.: Géométrie différentielle et Mécanique Analytique. Hermann, Paris (1969)
3. Lerbet, J., Aldowaji, M., Challamel, N., Nicot, F., Kirillov, O., Darve, F.: Geometric degree of nonconservativity. Memocs **2**(2), 123–139 (2014)
4. Lerbet, J., Challamel, N., Nicot, F., Darve, F.: Variational formulation of divergence stability for constrained systems Appl. Math. Model. **39**(2324), 7469–7482 (2015)
5. Lerbet, J., Challamel, N., Nicot, F., Darve, F.: Geometric degree of nonconservativity: set of solutions for the linear case and extension to the differentiable non-linear case. Appl. Math. Model. **40**(11–12), 5930–5941 (2016)
6. Lerbet, J., Challamel, N., Nicot, F., Darve, F.: On the stability of nonconservative continuous systems under kinematic constraints. ZAMM **97**(9), 1100–1119 (2017)
7. Lerbet, J., Challamel, N., Nicot, F., Darve, F.: Intrinsic nonlinear incremental discrete mechanics. ZAMM **98**(10), 1813–1833 (2018)

Multi-symplectic Lie Group Thermodynamics for Covariant Field Theories

Goffredo Chirco[1]($\boxtimes$)(iD), Marco Laudato[2]($\boxtimes$)(iD), and Fabio M. Mele[3]($\boxtimes$)(iD)

[1] Max Planck Institute for Gravitational Physics, Albert Einstein Institute,
Am Mühlenberg 1, 14476 Potsdam, Germany
goffredo.chirco@aei.mpg.de
[2] Dipartimento di Ingegneria e Scienze dell'Informazione e Matematica,
Università degli Studi dell' Aquila, Via Vetoio (Coppito 1),
67100, Coppito L'Aquila, Italy
marco.laudato@graduate.univaq.it
[3] Institute for Theoretical Physics, University of Regensburg, Universitätsstraße 31,
93040 Regensburg, Germany
fabio.mele@physik.uni-regensburg.de

Abstract. We propose a multi-symplectic generalisation of Souriau's Lie group thermodynamics for first order parametrised classical field theories. A new notion of general covariant Gibbs state functional is defined in terms of the multi-momentum map associated to the lifted action of the diffeomorphisms group on the fields extended phase space. We elaborate on the use of such functional toward a covariant statistical mechanic description of fully constrained field theories, at the crossroad between geometrical methods and information theory.

Keywords: Lie group thermodynamics · Covariant moment map · Parametrised field theories

1 Introduction

In the covariant, or multi-symplectic framework, for first order parametrised field theories, canonical initial value constraints have been shown to coincide with the vanishing of the *instantaneous* reduction of the covariant momentum map [1–3], associated to the action of the gauge group of the theory on its extended phase-space. Remarkably, such an induced energy-momentum map appears to encode all the dynamical information carried by a given classical field theory [1].

Besides classical field theory, the notion of momentum map provides a natural higher-dimensional generalization of the Hamiltonian function, comprising all conserved charges associated to the symplectic action of some dynamical group on a given phase space. In this terms, the momentum map can be used to generalize the standard Maxwell-Boltzmann-Gibbs approach to thermodynamics [4], to the case where the energy function is vector-valued. In the 70's Souriau

© Springer Nature Switzerland AG 2019
F. Nielsen and F. Barbaresco (Eds.): GSI 2019, LNCS 11712, pp. 55–65, 2019.
https://doi.org/10.1007/978-3-030-26980-7_7

was the first to explore such an epistemic perspective by proposing a symplectic reformulation of statistical mechanics [5–7]. In short, consider a connected $2n$-dimensional symplectic manifold $(\mathcal{M}, \omega)$ and a connected Lie group $\mathcal{G}$ acting on $\mathcal{M}$ by a Hamiltonian action Φ. Let $\mathfrak{g}$ be the Lie algebra of $\mathcal{G}$, $\mathfrak{g}^*$ be its dual space and $J : \mathcal{M} \rightarrow \mathfrak{g}^*$ be a *momentum map* of the $\mathcal{G}$-action [8]. A *statistical state* on $(\mathcal{M}, \omega)$ is a probability measure μ on $\mathcal{M}$ defined by the product of the Liouville density of $\mathcal{M}$ with a classical distribution function $\rho(x) \in \mathcal{M}$ [5,6]. For a given constant mean value of the equivariant momentum map J, thermodynamic equilibria are states

$$\rho_{eq}(b) = \frac{1}{Z(b)} e^{-\langle J(x), b \rangle}, \quad \text{with} \quad Z(b) = \int_{\mathcal{M}} e^{-\langle J(x), b \rangle} \omega^n(x), \quad b \in \mathfrak{g} \quad (1.1)$$

which are invariant under the action of any one-parameter subgroup of $\mathcal{G}$ on $\mathcal{M}$, which maximize the Shannon entropy $S(\rho)$ [4–7], and such that $S(\rho_{eq})$ is stationary w.r.t. all infinitesimal smooth variations of the probability density.

The (co)momentum map $\langle J(x), b \rangle$, with $b \in \mathfrak{g}$, provides a natural vector-valued generalisation of the Hamiltonian function, retaining the operational information of all conserved charges associated to the symplectic action of a dynamical group on a given system's phase space. In particular, as first shown by Souriau [5,6], this leads to a remarkable *covariant* generalization of Gibbs's equilibrium as soon as we consider the action of the symmetry group of the system (e.g. Galileo, Poincaré) on its manifold of motions.

In this short contribution, we explore a radical conceptual extension of Souriau Lie group thermodynamics to the case where the covariant symmetry group of the system is *gauge* and the symplectic phase space consists of the full extended phase space of the system. In particular, building on the notion of covariant (multi)momentum map in the multi-symplectic framework for first order parametrised field theories, we aim at an *off-shell*, spacetime (diffeomorphism) covariant notion of equilibrium state for fully constrained field theories.

2   Multi-symplectic Formulation of Generally Covariant Field Theories

The first step in our construction consists in recalling a suitable covariant Hamiltonian framework for our approach, where the key momentum map construction can be generalized in spacetime covariant terms, for the case of first order parametrised field theories. Seminal works on this topic are [1] and references within to which we refer for a more detailed exposition.

2.1   Parametrization and Covariant Multiphase Space

Let $\mathcal{X}$ be an oriented $(n + 1)$-dimensional manifold, which in many examples is spacetime, and let $\mathcal{Y} \xrightarrow{\pi_{\mathcal{X}\mathcal{Y}}} \mathcal{X}$ be a finite-dimensional fiber bundle over $\mathcal{X}$ whose fibers $\mathcal{Y}_x$ over $x \in \mathcal{X}$ have dimension N. This is called the *configuration bundle* and is the field theoretic analogue of the configuration space in classical mechanics. Physical fields correspond to sections of this bundle. A set of local coordinates (x^μ, y^A) on $\mathcal{Y}$ is provided by the $n+1$ local coordinates x^μ, $\mu = 0, \dots, n$, on $\mathcal{X}$ and the N fiber coordinates y^A, $A = 1, \dots, N$, which represent the field components at a given point $x \in \mathcal{X}$. Let then $\mathscr{L} : J^1(\mathcal{Y}) \to \Lambda^{n+1}(\mathcal{X})$ be the Lagrangian density for a first order classical field theory, where $J^1(\mathcal{Y})$ is the first jet bundle of $\mathcal{Y}$ and $\Lambda^{n+1}(\mathcal{X})$ is the space of $(n+1)$-forms on $\mathcal{X}$. The first jet bundle $J^1(\mathcal{Y})$ of $\mathcal{Y}$ here plays the role of the field-theoretic analogue of the tangent bundle of classical mechanics[1]. Local coordinates (x^μ, y^A) on $\mathcal{Y}$ induce coordinates v_μ^A on the fibers of $J^1(\mathcal{Y})$ so that the first jet prolongation $j^1\phi$ of a section ϕ of the bundle $\mathcal{Y} \xrightarrow{\pi_{\mathcal{X}\mathcal{Y}}} \mathcal{X}$ is given by $j^1\phi : x^\mu \longmapsto (x^\mu, y^A, v_\mu^A) = (x^\mu, y^A(x), y_{,\mu}^A(x))$, where $y_{,\mu}^A = \partial_\mu y^A$ and $\partial_\mu = \partial/\partial x^\mu$. The Lagrangian then reads $\mathscr{L}(j^1\phi) = L\left(x^\mu, y^A(x), y_{,\mu}^A(x)\right) \mathrm{d}^{n+1}x$, where $\mathrm{d}^{n+1}x = \mathrm{d}x^0 \wedge \cdots \wedge \mathrm{d}x^n$ is the volume form on $\mathcal{X}$.

As it is well known from the pioneering work of Dirac [9], further developed by Kuchař and Isham [10, 11], field theories with a fixed background metric can be made generally covariant, i.e., with the spacetime diffeomorphism group as symmetry group, by means of the so-called *parametrisation* procedure. A precise geometric reformulation of such a procedure within the context of multi-symplectic field theories was developed by Castrillón López, Gotay and Marsden in [12, 13]. The main steps of the construction are the following: (1) (oriented) diffeomorphisms of $\mathcal{X}$, reinterpreted as sections $\eta : \mathcal{X} \to \tilde{\mathcal{X}}$ of the bundle $\tilde{\mathcal{X}} \times \mathcal{X} \xrightarrow{\tilde{\pi}} \mathcal{X}$ over $\mathcal{X}$, with $(\tilde{\mathcal{X}}, g)$ a copy the base manifold, are introduced as new dynamical fields called *covariance fields*; (2) the configuration bundle $\mathcal{Y}$ is then replaced by the fibered product $\tilde{\mathcal{Y}} = \mathcal{Y} \times_\mathcal{X} (\tilde{\mathcal{X}} \times \mathcal{X})$ whose sections are thought of as pairs (ϕ, η); (3) the Lagrangian density of the starting theory is modified by introducing the new Lagrangian density $\tilde{\mathscr{L}}(j^1\phi, j^1\eta) := \mathscr{L}(j^1\phi, \eta^*g)$ which, denoting coordinates on $J^1(\tilde{\mathcal{Y}})$ by $(x^\mu, y^A, v_\mu^A, u^a, u_\mu^a)$ with u_μ^a the jet coordinates associated to u^a on $\tilde{\mathcal{X}}$, reads

$$\tilde{\mathscr{L}}(x^\mu, y^A, v_\mu^A, u^a, u_\mu^a) = \mathscr{L}(x^\mu, y^A, v_\mu^A ; G_{\mu\nu}), \tag{2.2}$$

where $G_{\mu\nu} \equiv (\eta^*g)_{\mu\nu} = \eta_{,\mu}^a \eta_{,\nu}^b g_{ab} \circ \eta = u_\mu^a u_\nu^b g_{ab} \circ \eta$. Let then $\alpha_\mathcal{X} \in \mathsf{Diff}(\mathcal{X})$, we denote by $\alpha_\mathcal{Y} \in \mathsf{Aut}(\mathcal{Y})$ its lift to $\mathcal{Y}$. This can be extended to an action by bundle automorphisms on $\tilde{\mathcal{Y}}$ by requiring that $\mathsf{Diff}(\mathcal{X})$ acts trivially on $\tilde{\mathcal{X}}$, i.e.

$$\alpha_{\tilde{\mathcal{X}}} : \tilde{\mathcal{X}} \times \mathcal{X} \longrightarrow \tilde{\mathcal{X}} \times \mathcal{X} \qquad \text{by} \qquad (u, x) \longmapsto (u, \alpha_\mathcal{X}(x)). \tag{2.3}$$

The induced action on the space $\tilde{\mathscr{Y}} \equiv \Gamma(\mathcal{X}, \tilde{\mathcal{Y}})$ of sections of $\tilde{\mathcal{Y}}$ is then given by

$$\alpha_{\tilde{\mathscr{Y}}}(\phi, \eta) = (\alpha_\mathscr{Y}(\phi), \alpha_{\tilde{\mathscr{X}}}(\eta)), \tag{2.4}$$

where

$$\alpha_\mathscr{Y}(\phi) = \alpha_\mathcal{Y} \circ \phi \circ \alpha_\mathcal{X}^{-1} \qquad , \qquad \phi \in \mathscr{Y} \equiv \Gamma(\mathcal{X}, \mathcal{Y}) \tag{2.5}$$

---

[1] In this case $\mathcal{Y} = \mathbb{R} \times \mathcal{Q}$ is the extended configuration space regarded as an $\mathbb{R}$-bundle over $\mathcal{Q}$, and $J^1(\mathcal{Q} \times \mathbb{R})$ is isomorphic to the bundle $T\mathcal{Q} \times T\mathbb{R}$.

generalizes the usual push-forward action on tensor fields, and

$$\alpha_{\tilde{\mathcal{X}}}(\eta) = \eta \circ \alpha_{\mathcal{X}}^{-1}, \tag{2.6}$$

is the (left) action by composition on sections of the trivial bundle $\tilde{\mathcal{X}} \times \mathcal{X}$. The modified field theory on $J^1(\tilde{\mathcal{Y}})$ is $\mathsf{Diff}(\mathcal{X})$-*covariant*, i.e., the Lagrangian density (2.2) is $\mathsf{Diff}(\mathcal{X})$-*equivariant* [12,13]:

$$\tilde{\mathscr{L}}\left(j^1(\alpha_{\mathcal{Y}}(\phi)), j^1(\alpha_{\tilde{\mathcal{X}}}(\eta))\right) = (\alpha_{\mathcal{X}}^{-1})^* \left[\tilde{\mathscr{L}}(j^1\phi, j^1\eta)\right]. \tag{2.7}$$

The fixed background metric g is then no longer thought of as living on $\mathcal{X}$, but rather just as a geometric object on the copy $\tilde{\mathcal{X}}$ in the fiber of the extended configuration bundle $\tilde{\mathcal{Y}}$. On the other hand, the metric variable $G = \eta^* g$ on $\mathcal{X}$ inherits a dynamical character via the covariance field η. The true dynamical fields of the parametrised theory are thus provided by ϕ and η, the latter not modifying the physical content of the original theory thus providing an efficient way of parametrizing it [12,13].

By introducing the covariant Hamiltonian $\tilde{p}$ and the multimomenta p_A^μ, ϱ_a^μ (respectively conjugate to the multivelocities v_μ^A and u_μ^a) defined via Legendre transformation as

$$\tilde{p} = \tilde{L} - \frac{\partial \tilde{L}}{\partial v_\mu^A} v_\mu^A - \frac{\partial \tilde{L}}{\partial u_\mu^a} u_\mu^a, \qquad p_A^\mu = \frac{\partial \tilde{L}}{\partial v_\mu^A} = \frac{\partial L}{\partial v_\mu^A}, \qquad \varrho_a^\mu = \frac{\partial \tilde{L}}{\partial u_\mu^a} = \mathcal{T}^{\mu\nu} u_\nu^b g_{ab}, \tag{2.8}$$

with $\mathcal{T}^{\mu\nu} = 2\frac{\partial L}{\partial G_{\mu\nu}}$ the so-called *Piola-Kirchhoff stress-energy-momentum tensor density* [12,13], a covariant Hamiltonian formalism for (first order) parametrised field theories can be now developed. The field-theoretic analogue of the phase space of classical mechanics is provided by the so-called *covariant* or *parametrised multiphase space* $\tilde{\mathcal{Z}} \cong J^1(\tilde{\mathcal{Y}})^*$ equipped with a canonical Poincaré-Cartan $(n+1)$-form

$$\tilde{\Theta} = \tilde{p}\, \mathrm{d}^{n+1}x + p_A^\mu \mathrm{d}y^A \wedge \mathrm{d}^n x_\mu + \varrho_a^\mu \mathrm{d}u^a \wedge \mathrm{d}^n x_\mu, \tag{2.9}$$

and the *multisymplectic* $(n+2)$-form

$$\tilde{\Omega} = \mathrm{d}y^A \wedge \mathrm{d}p_A^\mu \wedge \mathrm{d}^n x_\mu + \mathrm{d}u^a \wedge \mathrm{d}\varrho_a^\mu \wedge \mathrm{d}^n x_\mu - \mathrm{d}\tilde{p} \wedge \mathrm{d}^{n+1}x. \tag{2.10}$$

Let now $\mathcal{G}$ be a Lie group (perhaps infinite-dimensional) realizing the gauge group of the theory and denote by $\mathfrak{g}$ its Lie algebra. In the case of generally covariant field theories, $\mathcal{G}$ is a subgroup of $\mathsf{Aut}(\tilde{\mathcal{Y}})$ covering diffeomorphisms on $\mathcal{X}$. Given an element $\xi \in \mathfrak{g}$, we denote by $\xi_{\mathcal{X}}, \xi_{\mathcal{Y}}, \xi_{\tilde{\mathcal{Y}}}$, and $\xi_{\tilde{\mathcal{Z}}}$ the infinitesimal generators of the corresponding transformations on $\mathcal{X}, \mathcal{Y}, \tilde{\mathcal{Y}}$, and $\tilde{\mathcal{Z}}$, i.e., the infinitesimal generators on $\mathcal{X}, \mathcal{Y}, \tilde{\mathcal{Y}}$, and $\tilde{\mathcal{Z}}$ of the one-parameter group generated by ξ. The group $\mathcal{G}$ is said to act on $\tilde{\mathcal{Z}}$ by *covariant canonical transformation* if the $\mathcal{G}$-action corresponds to an infinitesimal multi-symplectomorphism, i.e. $\mathcal{L}_{\xi_{\tilde{\mathcal{Z}}}} \tilde{\Omega} = 0$, while it is said to act by *special covariant canonical transformations* if $\tilde{\Theta}$ is $\mathcal{G}$-invariant, that is $\mathcal{L}_{\xi_{\tilde{\mathcal{Z}}}} \tilde{\Theta} = 0$. This is the Hamiltonian counterpart of the $\mathcal{G}$-equivariance property (2.7) of the Lagrangian.

In analogy to the definition of momentum maps in symplectic geometry [8], a *covariant momentum map* (or a *multimomentum map*) associated to the $\mathcal{G}$-action on $\tilde{\mathcal{Z}}$ by covariant canonical transformations is given by

$$\tilde{\mathcal{J}} : \tilde{\mathcal{Z}} \longrightarrow \mathfrak{g}^* \otimes \Lambda^n(\tilde{\mathcal{Z}}) \quad , \quad \mathrm{d}\tilde{\mathcal{J}}(\xi) = \mathrm{i}_{\xi_{\tilde{z}}} \tilde{\Omega}, \tag{2.11}$$

where $\tilde{\mathcal{J}}(\xi)$ is the n-form on $\tilde{\mathcal{Z}}$ whose value at $\tilde{z} \in \tilde{\mathcal{Z}}$ is $\langle \tilde{\mathcal{J}}(\tilde{z}), \xi \rangle$ with $\langle \cdot, \cdot \rangle$ being the pairing between the Lie algebra $\mathfrak{g}$ and its dual $\mathfrak{g}^*$. Let then $\alpha \in \mathcal{G}$ be the transformation associated to $\xi \in \mathfrak{g}$, the covariant momentum map $\tilde{\mathcal{J}}$ is said to be Ad*-*equivariant* if

$$\tilde{\mathcal{J}}(\mathrm{Ad}_\alpha^{-1}\xi) = \alpha_{\tilde{\mathcal{Z}}}^*[\tilde{\mathcal{J}}(\xi)]. \tag{2.12}$$

For special covariant canonical transformations, the momentum map (2.11) admits an explicit expression given by $\tilde{\mathcal{J}}(\xi) = \mathrm{i}_{\xi_{\tilde{z}}} \tilde{\Theta}$ that, when the $\mathcal{G}$-action on $\tilde{\mathcal{Z}}$ is the lift of an action of $\mathcal{G}$ on $\tilde{\mathcal{Y}}$, reads

$$\langle \tilde{\mathcal{J}}(\tilde{z}), \xi \rangle = \left(\tilde{p} \, \xi^\mu + p_A^\mu \xi^A \right) \mathrm{d}^n x_\mu - p_A^\mu \xi^\nu \mathrm{d}y^A \wedge \mathrm{d}^{n-1} x_{\mu\nu} - \varrho_a^\mu \xi^\nu \mathrm{d}u^a \wedge \mathrm{d}^{n-1} x_{\mu\nu}, \tag{2.13}$$

where $\mathrm{d}^{n-1} x_{\mu\nu} = \mathrm{i}_{\partial_\nu} \mathrm{i}_{\partial_\mu} \mathrm{d}^{n+1} x$, $\mathrm{d}^n x_\mu = \mathrm{i}_{\partial_\mu} \mathrm{d}^{n+1} x$ accordingly.

2.2   Canonical Phase Space and Energy-Momentum Map

To construct the canonical formulation of a field theory we need to introduce a foliation of spacetime and consequently of the bundles over it[2]. Let then Σ be a compact, oriented, connected, boundaryless 3-manifold and let $\mathsf{Emb}_G(\Sigma, \mathcal{X})$ be the set of all space-like embeddings of Σ in $\mathcal{X}$. A foliation $\mathfrak{s}_{\mathcal{X}} : \Sigma \times \mathbb{R} \to \mathcal{X}$ of $\mathcal{X}$ then corresponds to a 1-parameter family of space-like embeddings $\tau \equiv \tau_\lambda : \Sigma \to \mathcal{X}$ by $\tau(\boldsymbol{x}) \equiv \tau_\lambda(\boldsymbol{x}) := \mathfrak{s}_{\mathcal{X}}(\boldsymbol{x}, \lambda)$, where $\boldsymbol{x}$ is a shorthand notation for the spatial coordinates on the space-like hypersurface $\Sigma_\tau = \tau(\Sigma)$. A foliation $\mathfrak{s}_{\mathcal{X}}$ of $\mathcal{X}$ induces a *compatible slicing* of bundles over it whose generating vector fields project onto the generating vector field $\zeta_{\mathcal{X}} = \frac{\partial}{\partial \lambda} \mathfrak{s}_{\mathcal{X}}(\boldsymbol{x}, \lambda)$ of $\mathfrak{s}_{\mathcal{X}}$. The flow of such a generating vector field defines a one-parameter group of bundle automorphisms.

For parametrised field theories, we are interested in a so-called $\mathcal{G}$-*slicing* in which case the one-parameter group of automorphisms of the extended configuration bundle is induced by a one-parameter subgroup of the gauge group $\mathcal{G}$, i.e. $\zeta_{\tilde{\mathcal{Y}}} = \xi_{\tilde{\mathcal{Y}}}$ for some $\xi \in \mathfrak{g}$. The corresponding slicing $\mathfrak{s}_{\tilde{\mathcal{Z}}}$ of $\tilde{\mathcal{Z}}$ is then generated by the canonical lift $\zeta_{\tilde{\mathcal{Z}}} = \xi_{\tilde{\mathcal{Z}}}$ of $\xi_{\tilde{\mathcal{Y}}}$ to $\tilde{\mathcal{Z}}$ whose flow defines a one-parameter group of special canonical transformations on $\tilde{\mathcal{Z}}$. Spatial fields will then be identified with smooth sections of the pull-back bundle $\mathcal{Y}_\tau \to \Sigma_\tau$ over a Cauchy surface given by $\varphi := \phi_\tau = \tau^*\phi$. According to the parametrization procedure discussed in Sect. 2.1, the space-like embedding $\tau \in \mathsf{Emb}_G(\Sigma, \mathcal{X})$ acquires a dynamical character through the covariance fields. Indeed, we have $\tau = \eta^{-1} \circ \tilde{\tau}$ for a given space-like embedding $\tilde{\tau} \in \mathsf{Emb}_g(\Sigma, \tilde{\mathcal{X}})$ of Σ into $\tilde{\mathcal{X}}$ associated to the slicing of $\tilde{\mathcal{X}}$ w.r.t. the fixed metric g.

---

[2] In what follows, we assume spacetime to be globally hyperbolic, i.e. $\mathcal{X} \cong \Sigma \times \mathbb{R}$.

The canonical parametrised configuration space then consists of the pairs (φ, τ) of spatial fields defined over a Cauchy slice and the space-like embedding identifying a $\mathcal{G}$-slicing of spacetime w.r.t. one-parameter subgroups of diffeomorphisms. Let then $(x^0, x^1, \ldots, x^n)$ be a chart on $\mathcal{X}$ adapted to τ, i.e. such that Σ_τ is locally a level set of x^0. Denoting by (φ, Π, τ, P) a point in the canonical parametrised phase space $T^*\tilde{\mathscr{Y}}_\tau = T^*\mathscr{Y}_\tau \times T^*\mathsf{Emb}_G(\Sigma, \mathcal{X})$, the canonical symplectic structure $\tilde{\omega}_\tau$ on $T^*\tilde{\mathscr{Y}}_\tau$ reads as [14]

$$\tilde{\omega}_\tau(\varphi, \Pi, \tau, P) = \int_{\Sigma_\tau} \left(\mathrm{d}\varphi^A \wedge \mathrm{d}\Pi_A + \mathrm{d}\tau^\mu \wedge \mathrm{d}P_\mu \right) \otimes \mathrm{d}^n x_0. \qquad (2.14)$$

Following the construction of [1] (cfr. Ch. 5), the multisymplectic structure on $\tilde{\mathcal{Z}}$ induces a presymplectic structure on the space $\tilde{\mathscr{Z}}_\tau$ of sections of the bundle $\tilde{\mathcal{Z}}_\tau \to \Sigma_\tau$ given by

$$\tilde{\Omega}_\tau(\sigma)(V, W) = \int_{\Sigma_\tau} \sigma^*(\mathrm{i}_W \mathrm{i}_V \tilde{\Omega}), \qquad \sigma \in \tilde{\mathscr{Z}}_\tau, \ V, W \in T_\sigma \tilde{\mathscr{Z}}_\tau \qquad (2.15)$$

which in turn is related to $\tilde{\omega}_\tau$ via $\tilde{\Omega}_\tau = R_\tau^* \tilde{\omega}_\tau$, where R_τ is the bundle map $R_\tau : \tilde{\mathscr{Z}}_\tau \to T^*\tilde{\mathscr{Y}}_\tau$ relating in adopted coordinates the momenta Π_A and P_a respectively to the temporal components of the multimomenta p_A^μ and ϱ_a^μ as

$$\Pi_A = p_A^0 \circ \sigma, \qquad P_a = \varrho_a^0 \circ \sigma. \qquad (2.16)$$

Let now $\sigma \in \tilde{\mathscr{Z}} \equiv \Gamma(\mathcal{X}, \tilde{\mathcal{Z}})$ be a section of the bundle $\tilde{\mathcal{Z}}$ over $\mathcal{X}$, and let $\alpha_{\tilde{\mathcal{Z}}} : \tilde{\mathcal{Z}} \to \tilde{\mathcal{Z}}$ be a covariant canonical transformation covering a diffeomorphism $\alpha_{\mathcal{X}} : \mathcal{X} \to \mathcal{X}$ whose induced action on sections is given by $\alpha_{\tilde{\mathscr{Z}}}(\sigma) = \alpha_{\tilde{\mathcal{Z}}} \circ \sigma \circ \alpha_{\mathcal{X}}^{-1}$ (cfr. Eq. (2.5)). The corresponding transformation on $\tilde{\mathscr{Z}}_\tau \equiv \Gamma(\Sigma_\tau, \tilde{\mathscr{Z}})$ given by

$$\alpha_{\tilde{\mathscr{Z}}_\tau} : \tilde{\mathscr{Z}}_{\eta^{-1} \circ \tilde{\tau}} \to \tilde{\mathscr{Z}}_{\alpha_{\tilde{\mathscr{Z}}}(\eta)^{-1} \circ \tilde{\tau}}, \quad \sigma \mapsto \alpha_{\tilde{\mathscr{Z}}_\tau}(\sigma) = \alpha_{\tilde{\mathcal{Z}}} \circ \sigma \circ \alpha_\tau^{-1} \qquad (2.17)$$

with $\alpha_{\tilde{\mathscr{Z}}}(\eta)$ defined in (2.6) and $\alpha_\tau := \alpha_{\mathcal{X}}|_{\Sigma_\tau}$ is a (special) covariant canonical transformation relative to the presymplectic 2-form (2.15) if $\alpha_{\tilde{\mathcal{Z}}}$ is a (special) covariant canonical transformation [1]. The covariant multimomentum map (2.11) associated to the $\mathcal{G}$-action on $\tilde{\mathcal{Z}}$ will then induce a so-called (*parametrised*) *energy-momentum map* on $\tilde{\mathscr{Z}}_\tau$ defined by

$$\tilde{\mathcal{E}}_\tau : \tilde{\mathscr{Z}}_\tau \longrightarrow \mathfrak{g}^*, \qquad \tilde{\mathcal{E}}_\tau(\sigma, \eta) = \tilde{\mathcal{E}}_{\eta^{-1} \circ \tilde{\tau}}(\sigma) := \int_{\Sigma_\tau} \sigma^* \langle \tilde{\mathcal{J}}, \xi \rangle, \qquad (2.18)$$

which is Ad^*-equivariant w.r.t. the action (2.17), namely $\langle \tilde{\mathcal{E}}_\tau(\sigma, \eta), \mathrm{Ad}_\alpha^{-1} \xi \rangle = \langle \alpha_{\tilde{\mathscr{Z}}_\tau}^* [\tilde{\mathcal{E}}_\tau(\sigma, \eta)], \xi \rangle$ [18]. Now, let $\tilde{\mathscr{P}}_\tau = R_\tau(\tilde{\mathcal{N}}_\tau) \subset T^*\tilde{\mathscr{Y}}_\tau$ be the primary constraint submanifold in $T^*\tilde{\mathscr{Y}}_\tau$ with $\tilde{\mathcal{N}}_\tau = \mathbb{FL}((j^1\tilde{\mathscr{Y}})_\tau) \subset \tilde{\mathscr{Z}}_\tau$, $\mathbb{FL}$ being the Legendre transform. For any σ *holonomic lift* of (φ, Π, τ, P) to $\tilde{\mathcal{N}}_\tau$, that is $\sigma \in R_\tau^{-1}\{(\phi, \Pi, \tau, P)\} \cap \tilde{\mathcal{N}}_\tau$, we have

$$\sigma^*(\mathrm{i}_{\zeta_{\tilde{\mathcal{Z}}}} \tilde{\Theta}) = - \left(\Pi_A \dot{\varphi}^A + P_a \zeta^\mu \eta_{,\mu}^a - \tilde{L}(\sigma)\zeta^0 \right) \mathrm{d}^n x_0, \qquad (2.19)$$

with $\tilde{L}(\sigma)\zeta^0 d^n x_0 = \tau^* i_{\zeta_{\mathcal{X}}} \mathscr{L}(j^1\phi, j^1\eta) = i_{\zeta_{\mathcal{X}}} \mathscr{L}(j^1\varphi, \dot{\varphi}, j^1\eta_\tau, \dot{\eta}_\tau)$, and we used the expression (2.13) for lifted actions [18]. Therefore, for lifted actions (and this is the case for a $\mathcal{G}$-slicing), the parametrised energy-momentum map (2.18) induces a functional $\tilde{\mathcal{J}}_H : \tilde{\mathscr{P}}_\tau \to \mathfrak{g}^*$ on $\tilde{\mathscr{P}}_\tau$ given by

$$\langle \tilde{\mathcal{J}}_H(\varphi, \Pi, \tau, P), \zeta \rangle = - \int_{\Sigma_\tau} d^n x_0 (\zeta^\mu \mathcal{H}_\mu^{(\varphi)} + \zeta^\mu P_\mu), \qquad (2.20)$$

where $P_\mu = \eta_{,\mu}^a P_a$ is the pull-back of P_a to Σ along η. The functional (2.20) is nothing but the total Hamiltonian whose components in the tangential and transversal direction to the spatial slice yields the super-momenta and Hamiltonian constraints.

2.3 Representation of Spacetime Diffeomorphisms

Let $\mathcal{G} = \text{Diff}(\mathcal{X})$ be the group of diffeomorphisms of $\mathcal{X}$. The Lie algebra $\mathfrak{g} = \text{diff}(\mathcal{X})$ can be realized as the set of all (complete) vector fields on $\mathcal{X}$. To any element $\xi \in \text{diff}(\mathcal{X})$, we associate a vector field $\xi_{\mathcal{X}} \in \mathfrak{X}(\mathcal{X})$ generating a one-parameter group of diffeomorphisms. In the instantaneous canonical formalism, the generating vector field is decomposed into "lapse" and "shift" components

$$\xi_{\mathcal{X}}^\mu(\boldsymbol{x}) = N(\boldsymbol{x})n^\mu(\tau(\boldsymbol{x})) + N^k(\boldsymbol{x})\tau_{,k}^\mu(\boldsymbol{x}) \qquad (2.21)$$

where $n^\mu = G^{\mu\nu}n_\nu$ is the future-pointing normal such that $\tau^* n = 0$ for any $\boldsymbol{x} \in \Sigma_\tau$ and $G^{\mu\nu}n_\nu n_\nu = -1$. The so called *lapse function* $N \in C^\infty(\Sigma, \mathcal{X})$ and *shift vector* $\boldsymbol{N} \in T\Sigma$ of the foliation [2] respectively specify the magnitude of the normal and tangential deformation at every point on a spatial hypersurface and play the role of arbitrary Lagrange multipliers in the action implementing the first-class constraints of the theory. The "space+time" decomposition however deforms the algebra of spacetime diffeomorphisms with the result that only the subalgebra $\text{diff}(\Sigma)$ of spatial diffeomorphisms can be represented in the canonical formalism [9, 15]. This reflects into the non-Diff($\mathcal{X}$)-equivariance of the *(instantaneous) energy-momentum map* $\mathcal{E}_\tau : T^*\text{Emb}_G(\Sigma, \mathcal{X}) \times T^*\mathscr{Y}_\tau \to \Lambda_d^0 \times \Lambda_d^1$

$$\mathcal{E}_\tau[N, \boldsymbol{N}] = \int_\Sigma d^n x_0 \left(N\mathcal{H} + N^k \mathcal{H}_k \right), \qquad (2.22)$$

Λ_d^0 and Λ_d^1 spaces of function densities and 1-form densities on Σ.

The spacetime equivariance can be recovered by considering the Diff-action on the embeddings induced by the action (2.6) on the covariance fields η. Specifically, the left action of Diff($\mathcal{X}$) on $\mathcal{X}$ induces a natural left action on the space $\text{Emb}(\Sigma, \mathcal{X})$ of all embeddings of Σ in $\mathcal{X}$

$$\Psi : \text{Emb}(\Sigma, \mathcal{X}) \times \text{Diff}(\mathcal{X}) \longrightarrow \text{Emb}(\Sigma, \mathcal{X}) \qquad \text{by} \qquad (\tau, \alpha_{\mathcal{X}}) \longmapsto \alpha_{\mathcal{X}} \circ \tau. \quad (2.23)$$

The corresponding generating vector field

$$\xi_\tau(\boldsymbol{x}) = \xi_{\mathcal{X}}(\tau(\boldsymbol{x})) = \xi_{\mathcal{X}}^\mu(\tau(\boldsymbol{x})) \left. \frac{\partial}{\partial x^\mu} \right|_{\tau(\boldsymbol{x}, \lambda)} \qquad (2.24)$$

yields a representation of the algebra $\mathsf{diff}(\mathcal{X})$ by vector fields on $\mathsf{Emb}_G(\Sigma, \mathcal{X})$. Such a vector field restricted to the embeddings can be again decomposed into the corresponding lapse and shift components which now are not freely specifiable but are some definite functionals of τ [11]. By taking this dependence into account, for any $\xi \in \mathsf{diff}(\mathcal{X})$, it is possible to define a new Hamiltonian functional on the parametrised phase space related to the equivariant momentum map (2.20) via

$$H(\xi)(\varphi, \Pi, \tau, P) = -\langle \xi, \tilde{\mathcal{J}}_H(\varphi, \Pi, \tau, P) \rangle = \int_\Sigma \mathrm{d}^n x_0 \, \xi_{\mathcal{X}}^\mu(\tau(\boldsymbol{x})) \Big(\mathcal{H}_\mu^{(\varphi)} + P_\mu \Big). \tag{2.25}$$

The complete derivation of the equivariance property of (2.25) is given in [18].

Eventually, the total Hamiltonian (2.25) is constructed in such a way that the constraints are preserved along the flow generated by $H(\xi)$. Moreover, as any functional of the embedding commutes with $H^{(\varphi)}$, we have $\dot{\tau}^\mu(\boldsymbol{x}) = \xi_{\mathcal{X}}^\mu(\tau(\boldsymbol{x}))$, i.e., $\xi_{\mathcal{X}}(\tau(\boldsymbol{x}))$ is the deformation vector of the foliation which can be decomposed into its transversal $\boldsymbol{\xi}_{\parallel}(\tau(\boldsymbol{x}))$ and normal $\xi_{\perp}(\tau(\boldsymbol{x}))$ components which, unlike the Lagrange multipliers N and $\boldsymbol{N}$ entering the parametrised action, are now specific functionals of the embedding. On the other hand, since $P(\xi)$ commutes with the field variables, the rates of change of the field φ and its conjugate momentum Π yield the Hamiltonian field equations with deformation vector $\xi_{\mathcal{X}}(\tau(\boldsymbol{x}))$. In other words, the canonical action of $\mathsf{Diff}(\mathcal{X})$ represented by $H(\xi)$ generates a displacement of the spatial hypersurface embedded in spacetime and also set the correctly evolved Cauchy data for fields on the deformed hypersurface.

3 Multisymplectic Lie Group Thermodynamics

We can finally proceed to extend the generalized notion of thermodynamic equilibrium states á la Souriau to parametrised field theories in which the Hamiltonian action we are interested in is that of the spacetime diffeomorphism group.

Denoting the canonical parametrised phase space by $\Upsilon \equiv T^*\mathscr{Y}_\tau \times T^*\mathsf{Emb}_G(\Sigma, \mathcal{X})$, a statistical state $\rho : \Upsilon \to \mathbb{R}([0, +\infty[)$ on Υ is a smooth probability density on Υ such that, for any Borel subset $\mathscr{A}$ of Υ, the integral

$$\mu(\mathscr{A}) = \int_{\mathscr{A}} \mathcal{D}[\varphi, \Pi, \tau, P] \, \rho(\varphi, \Pi, \tau, P) \tag{3.26}$$

defines a probability measure on Υ with the normalization condition

$$Z(\rho) = \int_\Upsilon \mathcal{D}[\varphi, \Pi, \tau, P] \, \rho(\varphi, \Pi, \tau, P) = 1, \tag{3.27}$$

where $\mathcal{D}[\varphi, \Pi, \tau, P]$ formally denotes the integration measure on Υ. In complete analogy to Sect. 1, we consider the mean value $\mathbb{E}_\rho(\tilde{\mathcal{J}}_H)$ of the momentum map (2.20) such that, for $\xi \in \mathfrak{g} = \mathsf{diff}(\mathcal{X})$, we have

$$\langle \xi, \mathbb{E}_\rho(\tilde{\mathcal{J}}_H) \rangle = \int_\Upsilon \mathcal{D}[\varphi, \Pi, \tau, P] \rho(\varphi, \Pi, \tau, P) \, \langle \xi, \tilde{\mathcal{J}}_H(\varphi, \Pi, \tau, P) \rangle. \tag{3.28}$$

Stationarity of the entropy functional

$$S(\rho) = -\int_{\Upsilon} \mathcal{D}[\varphi, \Pi, \tau, P] \, \rho(\varphi, \Pi, \tau, P) \log \rho(\varphi, \Pi, \tau, P), \qquad (3.29)$$

under an infinitesimal smooth variation $\rho_s(\varphi, \Pi, \tau, P)$ with $s \in]-\varepsilon, \varepsilon[$, $\varepsilon > 0$ of the statistical state ρ with fixed mean value of $\tilde{\mathcal{J}}_H$ can be therefore implemented by introducing two Lagrange multipliers $b \in \mathfrak{g} = \operatorname{diff}(\mathcal{X}), a \in \mathbb{R}$ respectively associated to the constraint $\mathbb{E}_\rho(\tilde{\mathcal{J}}_H(b)) = const.$ and the normalization condition (3.27) via

$$\mathcal{S}(\rho_s) = S(\rho_s) - \langle b, \mathbb{E}_{\rho_s}(\tilde{\mathcal{J}}_H) \rangle - aZ(\rho_s) \quad \text{s.t.} \quad \left.\frac{\delta \mathcal{S}(\rho_s)}{\delta s}\right|_{s=0} = 0 \ \forall \rho_s \qquad (3.30)$$

from which it follows that

$$\begin{aligned}
\rho_b^{(\mathrm{eq})}(\varphi, \Pi, \tau, P) &= \frac{1}{Z(b)} \exp\left(-\langle b, \tilde{\mathcal{J}}_H(\varphi, \Pi, \tau, P) \rangle\right) \\
&= \frac{1}{Z(b)} \exp\left(-\int_\Sigma \mathrm{d}^n x_0 \, \xi_{(b)}^\mu(\tau(\boldsymbol{x})) \left(\mathcal{H}_\mu^{(\varphi)}(\boldsymbol{x}) + P_\mu(\boldsymbol{x})\right)\right), \qquad (3.31)
\end{aligned}$$

where $\xi_{(b)}$ denotes the vector field on $\mathcal{X}$ associated to $b \in \operatorname{diff}(\mathcal{X})$ generating a one-parameter family of spacetime diffeomorphisms. The statistical state (3.31) is now a functional of (φ, Π, τ, P) through the comomentum map $\langle b, \tilde{\mathcal{J}}_H \rangle$. In particular, the dependence from the spacetime coordinates occurs only through the dynamical variables thus respecting the coordinate-independence of relativistic theories. Moreover, being a functional of the embeddings, the statistical state (3.31) is generally covariant in the sense that the momentum map is evaluated on any space-like hyper-surface without fixing the slicing a priori. The one parameter group of automorphisms of the extended configuration space generated by $b \in \operatorname{diff}(\mathcal{X})$ identifies a generalized concept of "time evolution" w.r.t. which the Gibbs state is of equilibrium.

4 Conclusions

The application of Souriau's Lie group thermodynamic formalism to the multi-symplectic framework of parametrised field theory leads to a new notion of spacetime covariant Gibbs-like state. Such a state is of equilibrium w.r.t. the one-parameter group of diffeomorphisms generated by the first-class constraints vector field $\xi_\mathcal{X}$ associated to $\xi \in \mathfrak{g} = \operatorname{diff}(\mathcal{X})$ and, in this sense, it defines a consistent spacetime covariant notion of thermodynamical equilibrium. Remarkably, the replacing of a dynamical symmetry with a gauge one moves our analysis from the fully reduced symplectic space of motions (on-shell) to the unconstrained extended phase space of the constrained theory (off-shell). While being defined off-shell, the covariant Gibbs state is by construction an observable of the theory, and it encodes, via the covariant momentum map functional, all the dynamical information carried by the given parametrised field theory: its canonical Hamiltonian, its initial value constraints, its gauge freedom and its stress energy-momentum tensor.

The proposed result points to a deep connection between geometrical methods, information theory and field theories. We expect our approach to open the road for a spacetime covariant formulation of statistical mechanics, possibly capable of describing the fluctuations of the gravitational field in a general relativistic context (see e.g. [16,17]). From an information-geometric viewpoint, we further expect the derived covariant Gibbs state functional to provide a useful tool for exploring a *statistical* generalisation of symplectic reduction in field theory [18], as well as a further support to the use of momentum map and Lie group formalism in the study of covariant generative models in machine learning.

References

1. Gotay, M.J., Isenberg, J., Marsden, J.E., Montgomery, R.: Momentum maps and classical relativistic fields. Part I, II, III. arXiv:physics/9801019 [math-ph]
2. Giulini, D.: Dynamical and Hamiltonian formulation of general relativity. arXiv:1505.01403 [gr-qc] (2015)
3. Fischer, A.E., Marsden, J.E.: The initial value problem and the dynamical formulation of general relativity. In: Hawking, S., Israel, W. (eds.) General Relativity: An Einstein's Centenary Survey, pp. 138–211. Cambridge University Press, London (1979)
4. Jaynes, E.T.: Information theory and statistical mechanics-I, II. Phys. Rev. **106**, 620 (1957)
5. Souriau, J.-M.: Structure des Systèmes Dynamiques. Dunod, Malakoff (1969)
6. Souriau, J.-M.: Definition covariante des équilibres thermodynamiques. Supplemento al Nuovo cimento **IV**(1), 203–216 (1966)
7. Marle, C.-M.: From tools in Symplectic and Poisson geometry to J.-M. Souriau's theories of statistical mechanics and thermodynamics. Entropy **18**(10), 370 (2016)
8. Marsden, J.E., Ratiu, T.: Introduction to Mechanics and Symmetry, A Basic Exposition of Classical Mechanical Systems. Texts in Applied Mathematics, vol. 17. Springer, New York (1999). https://doi.org/10.1007/978-0-387-21792-5
9. Dirac, P.A.M.: The Hamiltonian form of field dynamics. Can. J. Math. **3**, 1–23 (1951)
10. Kuchař, K.V.: Canonical quantization of gravity. In: Israel, W. (ed.) Relativity. Astrophysics and Cosmology, pp. 237–288. Reidel, Dordrecht (1973)
11. Isham, C.J., Kuchař, K.V.: Representations of spacetime diffeomorphisms. I. Canonical parametrized field theories. Ann. Phys. **164**, 288–315 (1985)
12. Castrillón Lopez, M., Gotay, M.J., Marsden, J.E.: Parametrization and stress-energy-momentum tensors in metric field theories. J. Phys. A. **41**, 34–400 (2008)
13. Castrillón Lopez, M., Gotay, M.J.: Covariantizing classical field theories. J. Geom. Mech. **3**(4), 487–506 (2010)
14. Hajicek, P., Isham, C.J.: The symplectic geometry of a parametrized scalar field on a curved background. J. Math. Phys. **37**, 3505–3521 (1996)
15. Dirac, P.A.M.: The theory of gravitation in Hamiltonian form. Proc. Roy. Soc. **A246**, 333–343 (1958)

16. Chirco, G., Josset, T., Rovelli, C.: Class. Quant. Grav. **33** (2016). arXiv:1503.08725; Chirco, G., Josset, T.: arXiv:1606.04444
17. Iglesias, P., Souriau, J.: Heat, cold and geometry. In: Cahen, M., DeWilde, M., Lemaire, L., Vanhecke, L. (eds.) Differential Geometry and Mathematical Physics. Mathematical Physics Studies, vol. 3, pp. 37–68. Reidel, Dordrecht (1983)
18. Chirco, G., Laudato, M., Mele, F.M. (in preparation)

Euler-Poincaré Equation for Lie Groups with Non Null Symplectic Cohomology. Application to the Mechanics

Géry de Saxcé[✉][iD]

Univ. Lille, CNRS, Centrale Lille, FRE 2016 – LaMcube – Laboratoire de mécanique multiphysique multiéchelle, Villeneuve-d'Ascq, France
gery.de-saxce@univ-lille.fr

Abstract. Let G be a symplectic group on a symplectic manifold $\mathcal{N}$. To any momentum map $\psi : \mathcal{N} \to \mathfrak{g}^*$, one can associate a class of symplectic cohomology *cocs*. It does not depend on the choice of the momentum map but only on the structure of the Lie group G. It forwards an affine left action $\mu = a \cdot \mu' = Ad^*(a)\,\mu' + cocs(a)$ of G on $\mathfrak{g}^*$.

If G is a Lie subgroup of the affine group, one can define an associated affine connection as a field of $\mathfrak{g}$-valued 1-forms $\tilde{\Gamma}$ on a G-principal bundle of affine frames $\pi : \mathcal{F} \to \mathcal{M}$. Let ω be a smooth field of 2-form on $\mathfrak{g}^* \times \mathcal{F}$ defined by: $(\mu, f) \mapsto \frac{1}{2}\,d\mu \wedge \tilde{\Gamma}$.

On each orbit $\boldsymbol{\mu}$, ω is the pull-back of Kirillov-Kostant-Souriau symplectic form by the projection $\psi_{\boldsymbol{\mu}} : (\mu, f) \mapsto \mu$. The G-principal bundle $\mathfrak{g}^* \times \mathcal{F}$ is a presymplectic bundle of symplectic form ω and $\psi_{\boldsymbol{\mu}}$ is a momentum map.

The equation of motion $d(\mu, f) \in Ker(\omega)$ expresses the fact that the momentum is parallel-transported. It generalizes Euler-Poincaré equation when the class of symplectic cohomology of the group is not null, especially for the important case of Galileo's group.

Keywords: Symplectic geometry · Lie group ·
Connection on a manifold

1   Symplectic Cohomology

Let $(\mathcal{N}, \omega)$ be a symplectic manifold. A Lie group G smoothly left acting on $\mathcal{N}$ and preserving the symplectic form ω is said to be symplectic. The interior product of a vector $\overrightarrow{V}$ and a p-form ω is denoted $\iota(\overrightarrow{V})\,\omega$. A map $\psi : \mathcal{N} \to \mathfrak{g}^*$ such that:

$$\forall \eta \in \mathcal{N}, \quad \forall Z \in \mathfrak{g}, \qquad \iota(Z \cdot \eta)\,\omega = -d(\psi(\eta)Z),$$

is called a **momentum map** of G. It is the quantity involved in Noether's theorem that claims ψ is constant on each leaf of $\mathcal{N}$.

© Springer Nature Switzerland AG 2019
F. Nielsen and F. Barbaresco (Eds.): GSI 2019, LNCS 11712, pp. 66–74, 2019.
https://doi.org/10.1007/978-3-030-26980-7_8

We call **symplectic cocycle** of a Lie group G a smooth map $cocs$ from G into the dual $\mathfrak{g}^*$ of its Lie algebra such that:

(i) $cocs(a'a) = cocs(a') + Ad^*(a')\, cocs(a)$,
(ii) $dcocs = D\, cocs\,(e)$ is skew-symmetric.

Formula (i) is called the **symplectic cocycle identity**.

Let $\mu_0 \in \mathfrak{g}^*$. We call **symplectic coboundary** of G a smooth map $cobs_{\mu_0}$ of G into $\mathfrak{g}^*$ such that:

$$cobs_{\mu_0}(a) = Ad^*(a)\,\mu_0 - \mu_0.$$

As maps valued in the linear space $\mathfrak{g}^*$, the symplectic cocycles form a linear space and the symplectic coboundaries form a linear subspace thereof. The relation "$cocs_1$ and $cocs_2$ differs by a coboundary" is an equivalence relation and the set of equivalence classes is a linear space called the space of **classes of symplectic cohomology**. The class of symplectic cohomology does not depend on the choice of the momentum map but only on the structure of the Lie group G.

The reason why one introduces this definition is that Souriau proved in ([8] (Theorem (11.17), page 109, or its English translation [9]) that the smooth map $cocs$ from G into $\mathfrak{g}^*$:

$$cocs(a) = \psi(a \cdot \eta) - Ad^*(a)\,\psi(\eta), \tag{1}$$

is a symplectic cocycle.

Replacing η by $a^{-1} \cdot \eta$ in (1), this formula reads:

$$\psi(\eta) = Ad^*(a)\,\psi'(\eta) + cocs(a),$$

where $\psi \mapsto \psi' = a \cdot \psi$ is the induced action of the one of G on $\mathcal{N}$. It is worth observing it is just the action on $\mathfrak{g}^*$:

$$\mu = a \cdot \mu' = Ad^*(a)\,\mu' + cocs(a), \tag{2}$$

with $\mu = \psi(\eta)$ and $\mu' = \psi'(\eta)$. Because $cocs$ is a symplectic cocycle, (2) is an affine representation of G. Hence G may be seen as a Lie subgroup of the affine group of $\mathbb{R}^n$. Let $\mathfrak{g}$ be the Lie algebra of G, that is the set of infinitesimal generators $Z = da = (dC, dP)$ with $a \in G$. Let us identify the space of the momentum components $\mu = (F, L)$ to the dual $\mathfrak{g}^*$ of the Lie algebra thanks to the dual pairing:

$$\mu\, Z = \mu\, da = (F, L)\,(dC, dP) = F\, dC + Tr(L\, dP) \tag{3}$$

For convenience, we represent the affine transformations $X = P\,X' + C$ of $\mathbb{R}^n$ by linear relationships in $\mathbb{R}^{n+1}$:

$$\tilde{X} = \begin{pmatrix} 1 \\ X \end{pmatrix} = \begin{pmatrix} 1 & 0 \\ C & P \end{pmatrix}\begin{pmatrix} 1 \\ X' \end{pmatrix} = \tilde{P}\,\tilde{X}' \tag{4}$$

Galileo's group $\mathbb{GAL}$ is the set of affine transformations of the space-time $\mathbb{R}^4$ of the form:

$$\tilde{X} = \begin{pmatrix} 1 \\ t \\ x \end{pmatrix}, \qquad \tilde{P} = \begin{pmatrix} 1 & 0 & 0 \\ \tau_0 & 1 & 0 \\ k & u & R \end{pmatrix}.$$

where t is the time, $x \in \mathbb{R}^3$ is the position, $u \in \mathbb{R}^3$ is a Galilean boost, $R \in \mathbb{SO}(3)$ is a rotation, $k \in \mathbb{R}^3$ is a spatial translation and $\tau_0 \in \mathbb{R}$ is a clock change. Hence Galileo's group is a Lie group of dimension 10. In [3] (Theorem 17.4, page 374), we proved:

Theorem 1. *The most general symplectic cohomology of Galileo's group is defined by:*

$$cocs(a)\, Z' = l(a) \cdot d\varpi' - q(a) \cdot du' + p(a) \cdot dk' - e(a)\, d\tau_0', \tag{5}$$

where the components are:

$$p(a) = m\, u, \qquad e(a) = \frac{1}{2} m \parallel u \parallel^2, \tag{6}$$

$$q(a) = m\,(k - \tau_0 u), \qquad l(a) = m\, k \times u. \tag{7}$$

The space of symplectic cohomology of Galileo's group is of dimension 1. Moreover one has:

$$dcocs\,(Z, Z') = m\,(du \cdot dk' - dk \cdot du') \tag{8}$$

Proof. The condition (i) of the definition of the symplectic cocycles means that the action (2) is an affine representation of G in $\mathfrak{g}^*$. In [3] (Theorem 16.3, page 329), it is proved that, for Galileo's group, the most general affine representation of this form is, *modulo* a symplectic coboundary:

$$p(a) = m\, u, \qquad e(a) = \frac{1}{2} m \parallel u \parallel^2 + e_1\, \tau_0,$$

$$q(a) = m\,(k - \tau_0 u), \qquad l(a) = m\, k \times u + s\, u.$$

where $s, e_1 \in \mathbb{R}$. By differentiation, the components of $D\,coc\,(e)$ are:

$$dp = m\, du, \qquad de = \frac{1}{2} m\, u \cdot du + e_1\, d\tau_0,$$

$$dq = m\,(dk - \tau_0\, du - d\tau_0\, u), \qquad dl = m\,(dk \times u + k \times du) + s\, du.$$

that leads to:

$$dcocs\,(Z, Z') = m\,(du \cdot dk' - dk \cdot du') + s\, du \cdot d\varpi' - e_1 d\tau_0\, d\tau_0',$$

As the 2-form is skew-symmetric, $s = e_1 = 0$, that achieves the proof. ∎

In the sequel, we need an important result, **Kirillov-Kostant-Souriau theorem**, revealing the **orbit symplectic structure**:

Theorem 2. *Let G be a Lie group and an orbit of the coadjoint representation $orb(\mu) \subset \mathfrak{g}^*$. Then:*

(i) *The inclusion map $orb(\mu) \to \mathfrak{g}^*$ is a regular imbedding. A vector $d\mu \in T_\mu \mathfrak{g}^*$ is tangent to the orbit if there exists $Z_d \in \mathfrak{g}$ such that:*

$$d\mu = \mu \circ ad(Z_d) + dcocs(Z_d) = -ad^*(Z_d)\mu + dcocs(Z_d).$$

(ii) *The orbit $orb(\mu)$ is a symplectic manifold of which the symplectic form is defined by:*

$$\omega_{KKS}(d\mu, \delta\mu) = \mu[Z_d, Z_\delta] + dcocs(Z_d, Z_\delta),$$

The dimension of the orbit is even.

(iii) *G is a symplectic group and any $\mu \in \mathfrak{g}^*$ is its own momentum.*

2 Affine Connections

Let $\pi : \mathcal{F} \to \mathcal{M}$ be a G-principal bundle of affine frames with the free action $(a, f) \mapsto f' = a \cdot f$ on each fiber. We built the associated G-principal bundle:

$$\hat{\pi} : \mathfrak{g}^* \times \mathcal{F} \to (\mathfrak{g}^* \times \mathcal{F})/G : (\mu, f) \mapsto \boldsymbol{\mu} = orb(\mu, f),$$

for the free action:

$$(a, (\mu, f)) \mapsto (\mu', f') = a \cdot (\mu, f) = (a \cdot \mu, a \cdot f),$$

where the action on $\mathfrak{g}^*$ is (2). Clearly, (2) is the transformation law of the components $\boldsymbol{\mu}$ of a momentum tensor in the frame f. The tensor may be identified to the orbit $\boldsymbol{\mu} = orb(\mu, f)$. The orbit space $(\mathfrak{g}^* \times \mathcal{F})/G$ is sometimes denoted $\mathfrak{g}^* \times_G \mathcal{F}$.

Let $ver_f = Ker(D\pi)$ the vertical space at f. An **Ehresmann connection** on the G-principle bundle $\mathcal{F}$ is a field of supplementary subspaces hor_f in $T_f\mathcal{F}$:

$$T_f\mathcal{F} = ver_f \oplus hor_f.$$

The decomposition $df = df_v + df_h$ is unique and the map $hor : T_f\mathcal{F} \to hor_f : df \mapsto df_h$ is called the horizontal projection.

Alternatively, a connection can be defined by a field of $\mathfrak{g}$-valued 1-forms $\tilde{\Gamma}$ on $\mathcal{F}$ such that $hor_f = Ker\,\tilde{\Gamma}$ and:

- $\tilde{\Gamma}$ is **vertical**: $\forall df_h \in hor_f$, $\qquad \tilde{\Gamma}(df_h) = 0$,
- $\tilde{\Gamma}(Z \cdot f) = Z$,
- $\tilde{\Gamma}$ is Ad-**equivariant**: $L_a\tilde{\Gamma} = Ad(a)\,\tilde{\Gamma}$ where $Ad(a)$ is the adjoint representation.

The covariant derivative $\tilde{\nabla}_{\overrightarrow{dX}} \mu$ of a momentum field $X \mapsto \mu(X)$ in a moving frame $X \mapsto f(X)$ is defined by:

$$\tilde{\nabla}_{dX} \mu = d\mu - (\tilde{\Gamma}(df)) \cdot \mu, \tag{9}$$

that can reads:

$$\tilde{\nabla}_{\overrightarrow{dX}} \mu = orb(\tilde{\nabla}_{dX}\mu, f) = orb(d\mu - (\tilde{\Gamma}(df)) \cdot \mu, f). \tag{10}$$

As G is a subgroup of $\mathbb{A}ff(n)$, the connection is decomposed as $\tilde{\Gamma} = (\Gamma_A, \Gamma)$ where:

- The $\mathfrak{gl}(n)$-valued 1-form Γ is a classical linear connection describing the infinitesimal motion of the basis of the affine frame f. In general relativity, it represents the gravitation.
- the $\mathbb{R}^n$-valued 1-form Γ_A is the affine part of the connection describing the infinitesimal motion of the origin of the affine frame f. Its physical meaning is not so strong as the gravitation but it represents the observer.

It is worth to remark that usual connections are defined for a right action of G on the components: $\mu \cdot a = a^{-1} \cdot \mu$ and, differentiating around the identity, there is a sign change in the infinitesimal action of $\mathfrak{g}$: $\mu \cdot Z = -Z \cdot \mu$. Hence the rule to swap the usual connections $\tilde{\Gamma}'$ for the corresponding ones $\tilde{\Gamma}$ considered here is:

$$\tilde{\Gamma}' = -\tilde{\Gamma}. \tag{11}$$

For more details on connections and in particular on affine connections, the reader is referred for instance to Kobayashi's book [5].

We call **Galilean connections** the symmetric connections associated to Galileo's group [6]. In a Galilean chart, they are given by the 4×4 connection matrix:

$$\Gamma(dX) = \begin{pmatrix} 0 & 0 \\ j(\Omega)\,dx - g\,dt & j(\Omega)\,dt \end{pmatrix} \tag{12}$$

where $g \in \mathbb{R}^3$ is identified to the gravity [4], while $j(\Omega)$ is the unique skew-symmetric matrix associated to $\Omega \in \mathbb{R}^3$ that can be interpreted as representing Coriolis' effects.

In ([1,3]), we proved that the affine connexion is given by:

$$\Gamma_A(dX) = dX - \nabla_{dX}C, \tag{13}$$

which gives for a galilean connexion ([2,3]):

$$\Gamma_A(dX) = d\begin{pmatrix} t \\ x \end{pmatrix} - d\begin{pmatrix} 0 \\ x \end{pmatrix} - \begin{pmatrix} 0 & 0 \\ j(\Omega)\,dx - g\,dt & j(\Omega)\,dt \end{pmatrix}\begin{pmatrix} 0 \\ x \end{pmatrix} = \begin{pmatrix} dt \\ -\Omega \times x\,dt \end{pmatrix}. \tag{14}$$

3   Factorized Symplectic Form

According to the dual pairing (3), we can put momenta and connections into duality:
$$\mu\,\tilde{\Gamma} = F\,\Gamma_A + Tr\,(L\,\Gamma).$$

This suggests to introduce the factorized 2-form:

$$\boxed{\omega = \tfrac{1}{2}\,d\mu \wedge \tilde{\Gamma}} \tag{15}$$

exterior product of the $\mathfrak{g}^*$-valued 1-form $d\mu$ and the $\mathfrak{g}$-valued 1-form $\tilde{\Gamma}$, hence a scalar valued 2-form.

We begin with a preliminary result.

Theorem 3. *Let $f \mapsto \tilde{\Gamma}$ be a field of connection 1-form on a G-principal bundle $\pi : \mathcal{F} \to \mathcal{M}$. The group G left acts on the dual $\mathfrak{g}^*$ of the Lie algebra of G and on the associated G-principal bundle: $\hat{\pi} : \mathfrak{g}^* \times \mathcal{F} \to (\mathfrak{g}^* \times \mathcal{F})/G$ by: $a \cdot \eta = a \cdot (\mu, f) = (a \cdot \mu, a \cdot f)$. Then:*

- *The tangent space to $\mathfrak{g}^* \times \mathcal{F}$ at η is a direct sum:*

$$T_\eta(\mathfrak{g}^* \times \mathcal{F}) = ver_\eta \oplus hor_\eta,$$

  where $ver_\eta = Ker\,(D\hat{\pi})$ is the space of vertical vectors and $hor_\eta = \mathfrak{g}^* \times hor_f$ is the space of horizontal vectors.
- *Any tangent vector $d\eta = (d\mu, df)$ can be decomposed in an unique way as $d\eta = d\eta_v + d\eta_h$, where:*

$$d\eta_v = (\tilde{\Gamma}(df)) \cdot \eta,$$

$$d\eta_h = (\nabla_{dX}\,\mu, hor\,(df)), \quad \text{with:} \quad D\pi(df) = \overrightarrow{d\mathbf{X}}.$$

We can now state the main result:

Theorem 4. *Let $f \mapsto \tilde{\Gamma}$ be a field of connection 1-form on a G-principal bundle $\pi : \mathcal{F} \to \mathcal{M}$. The group G left acts on $\mathfrak{g}^*$ by the affine representation (2) and on the G-principal bundle $\hat{\pi} : \mathfrak{g}^* \times \mathcal{F} \to (\mathfrak{g}^* \times \mathcal{F})/G$ by: $a \cdot \eta = a \cdot (\mu, f) = (a \cdot \mu, a \cdot f)$. Let $\eta \to \omega$ be a smooth field of 2-form on $\mathfrak{g}^* \times \mathcal{F}$ defined by:*

$$\omega = \frac{1}{2}\,d\mu \wedge \tilde{\Gamma}.$$

Then:

(i) *Let $\boldsymbol{\mu} = orb(\mu, f)$ be an orbit in $\mathfrak{g}^* \times \mathcal{F}$, and $\psi_{\boldsymbol{\mu}}$ be the projection on the first component $\eta = (\mu, f) \mapsto \mu$, restricted to the orbit. Then the smooth map $\psi_{\boldsymbol{\mu}}$ is a submersion.*

(ii) One has:

$$\omega(\delta\eta, \delta\eta) = \frac{1}{2}(d\,\mu\,Z_\delta - \delta\,\mu\,Z_d) \qquad with: \quad Z_d = \tilde{\Gamma}(df), \qquad Z_\delta = \tilde{\Gamma}(\delta f).$$

On each orbit, ω is the pull-back of Kirillov-Kostant-Souriau symplectic form ω_{KKS} by ψ_μ:

$$\omega = \psi_\mu^*(\omega_{KKS}),$$

and is invariant by left action:

$$L_a^*\,\omega = \omega.$$

The G-principal bundle $\mathfrak{g}^* \times \mathcal{F}$ is a presymplectic bundle of symplectic form ω.

(iii) ψ_μ is a momentum map and:

$$\psi_\mu \circ L_a = Ad^*(a)\,\psi_\mu + cocs(a).$$

(iv) The equation of motion is:

$$d\eta \in Ker(\omega) \quad \Leftrightarrow \quad \tilde{\nabla}_{dX}\mu = d\,\mu + ad^*(\tilde{\Gamma})\mu - dcocs\,(\tilde{\Gamma}) = 0, \qquad (16)$$

and the momentum $\boldsymbol{\mu} = orb(\mu, f)$ is parallel-transported:

$$\tilde{\nabla}_{\overrightarrow{dX}}\boldsymbol{\mu} = \mathbf{0}.$$

Using the swap rule (11), Eq. (16) reads:

$$\boxed{\nabla_{dX}\,\mu = d\,\mu - ad^*(\tilde{\Gamma}')\mu + dcocs\,(\tilde{\Gamma}') = 0} \qquad (17)$$

It is worth to remark that, if $dcocs = 0$, it is nothing else **Euler-Poincaré equation** [7]. In fact, (17) generalizes this equation when the class of symplectic cohomology of the group is not null, especially for the important case of Galileo's group.

4   Application to Classical Mechanics

Considering the representation of a Galilean transformation a by a 5×5 matrix (4), we have by differentiation around the identity:

$$Z_d = d\tilde{P} = \begin{pmatrix} 0 & 0 & 0 \\ d\tau_0 & 0 & 0 \\ dk & du & j(d\varpi) \end{pmatrix}.$$

The Lie bracket of two infinitesimal Galilean transformations is:

$$[Z_d, Z_\delta] = Z_\delta Z_d - Z_d Z_\delta =$$

$$\begin{pmatrix} 0 & 0 & 0 \\ 0 & 0 & 0 \\ d\tau_0\,\delta u - \delta\tau_0\,du + \delta\varpi \times dk - d\varpi \times \delta k & \delta\varpi \times du - d\varpi \times \delta u & j(\delta\varpi \times d\varpi) \end{pmatrix}.$$

Owing to:

$$(ad^*(Y)\mu)(Z) = -\mu\,(Ad\,(Y)\,Z) = -\mu[Y, Z],$$

we obtain the infinitesimal coadjoint representation of Galileo's group $\mu' = ad^*(Z_d)\mu$ represented by:

$$e' = -du \cdot p, \qquad p' = -d\varpi \times p,$$

$$q' = -d\varpi \times q + d\tau_0 p, \qquad l' = -d\varpi \times l + du \times q - dk \times p.$$

On the other hand, $dcocs$ is given for Galileo's group by (8). Hence, applying (17), and combining with the expressions (12) and (14) of the Galilean connection, the equation of motion (16) reads:

$$\tilde{\nabla}e = de + p \cdot (\Omega \times dx - g\,dt) = de - p \cdot g\,dt = 0$$

$$\tilde{\nabla}p = dp + \Omega \times p\,dt + \underline{m\,(\Omega \times dx - g\,dt)} = dp - m\,(g - 2\,\Omega \times v)\,dt = 0.$$

$$\tilde{\nabla}q = dq + \Omega \times (q - \underline{m\,x})\,dt - p\,dt = 0$$

$$\tilde{\nabla}l = dl + \Omega \times l\,dt - q \times (g\,dt - \Omega \times dx) + p \times (\Omega \times x)\,dt = 0.$$

The terms resulting from the non null class of symplectic cohomology of Galileo's group are underlined and absolutely necessary to find the expected equation fitting the experiments of classical dynamics. After dividing by dt and some simplifications, we can recast them as:

- balance of energy: $\dot{e} = g \cdot p$
- balance of linear momentum: $\dot{p} = m\,(g - 2\,\Omega \times v)$
- balance of passage: $\dot{q} = p$
- balance of angular momentum: $\dot{l} + \Omega \times l_0 = x \times\ m\,(g - 2\,\Omega \times v)$

These equations are full covariant. The right hand member of the second equation takes into account both Newton's gravity and Coriolis' force, allowing to understand the experience of Foucault's pendulum. The last equation allows for instance to explain the motion of a satellite or Lagrange's top.

References

1. de Saxcé, G., Vallée, C.: Affine tensors in shell theory. J. Theor. Appl. Mech. **41**(3), 593–621 (2003)
2. de Saxcé, G., Vallée, C.: Affine tensors in mechanics of freely falling particles and rigid bodies. Math. Mech. Solid J. **17**(4), 413–430 (2011)
3. de Saxcé, G., Vallée, C.: Galilean Mechanics and Thermodynamics of Continua. Wiley-ISTE, London (2016)
4. Cartan, É.: Sur les variétés à connexion affine et la théorie de la relativité généralisée (première partie). Annales de l'École Normale Supérieure **40**, 325–412 (1923)

5. Kobayashi, S., Nomizu, K.: Foundations of Differential Geometry, vol. 1. Wiley, New York (1963)
6. Künzle, H.P.: Galilei and Lorentz structures on space-time: comparison of the corresponding geometry and physics. Annales de l'Institut Henri Poincaré, section A **17**(5), 337–362 (1972)
7. Poincaré, H.: Sur une forme nouvelle des équations de la Mécanique. C.R. Acad. Sci. Paris, Tome CXXXII **7**, 369–371 (1901)
8. Souriau, J.-M.: Structure des systèmes dynamiques. Dunod (out of print), Paris (1970)
9. Souriau, J.-M.: Structure of Dynamical Systems: A Symplectic View of Physics. Birkhäuser Verlag, New York (1997)

Geometric Numerical Methods with Lie Groups

Joël Bensoam$^{(\boxtimes)}$ and Pierre Carré

Ircam, CNRS, Sorbonne Université, Ministàre de la Culture, Science et Technologies
de la Musique et du son, STMS, 75004 Paris, France
joel.bensoam@ircam.fr

Abstract. Due to the increasing demands for modeling large-scale and
complex systems, designing optimal controls, and conducting optimization tasks, many real-world applications require sophisticated models.
Geometric methods are designed to capture the underlying structure of
the system at hand and to preserve the global qualitative or geometric properties of the flow, such as symplecticity, volume preservation
and symmetry. A survey on three of such structure preserving numerical
methods is proposed in the present article. Testing the validity of such
simulations is achieved by exhibiting analytically solvable models and
comparing the result of simulations with their exact behavior.

Keywords: Geometric integrators ·
Structure preserving numerical methods · Variational methods

1 Introduction

It is today well established that geometrical methods connected to powerful
numerical tools (i.e. Runge-Kutta, Butcher series) can be applied to equations
on the Lie algebra to design high order methods and determine their numerical
convergence. Beyond these structure preservations, approaches for integration
algorithms based on variational principles give a unified treatment of many symplectic numerical schemes. In this context, the Noether theorem [10] allows for
a numerical formulation that preserves symmetries and conservation laws.

In the case of homogeneous spaces (smooth manifold on which a Lie group
acts transitively), the so-called Lie group integrators, comprising Runge-Kutta-
Munthe-Kaas [12] methods is presented shortly in Sect. 2. The main preoccupa-
tion is to ensure that discrete solutions are guaranteed to stay on the given manifold. However in this case no particular preservation of symmetries is obtained
without further constraints. This is why variational methods are revisited in
Sect. 3 to be compared to Lie-Poisson Hamilton-Jacobi algorithm based on generating function for which higher order designs are also available (Sect. 4).

© Springer Nature Switzerland AG 2019
F. Nielsen and F. Barbaresco (Eds.): GSI 2019, LNCS 11712, pp. 75–84, 2019.
https://doi.org/10.1007/978-3-030-26980-7_9

2 Lie Group Integrators, Runge-Kutta-Munthe-Kaas Methods

The Runge-Kutta-Munhe-Kaas methods (RKMK) developed in a serie of articles [12], are an example of Lie group methods. Let $Y(t)$ be a curve in a matrix Lie group G verifying

$$\dot{Y} = A(t, Y)\, Y, \qquad Y(0) = Y_0 \tag{1}$$

where $A(t, Y) \in \mathfrak{g}$ for all $t, Y \in \mathbb{R} \times G$. The starting point is to describe the solution of (1) as $Y(t) = \exp(\Omega(t))Y_0$ and to deduce an ODE on Ω. Computing the derivative of Y we get

$$\dot{Y}(t) = \frac{d}{dt}\exp(\Omega(t))Y_0 = \mathrm{dexp}_{\Omega(t)}(\dot{\Omega}(t))\, Y_0 = \mathrm{d}^{\mathrm{R}}\exp_{\Omega(t)}(\dot{\Omega}(t))\, Y(t),$$

where the right trivialized derivative $\mathrm{d}^{\mathrm{R}}\exp_{\Omega} := \mathrm{d}R_{\exp(\Omega)^{-1}} \circ \mathrm{dexp}_{\Omega}$ is introduced. Using this expression in (1) and inverting[1] $\mathrm{d}^{\mathrm{R}}\exp_{\Omega}$ a differential equation is obtained for Ω lying on the Lie algebra $\mathfrak{g}$

$$\dot{\Omega}(t) = \mathrm{d}^{\mathrm{R}}\exp_{\Omega(t)}^{-1}\left(A(t, Y(t))\right), \qquad \Omega(0) = 0. \tag{2}$$

The advantage is that the non linear invariants defining the Lie group become linear invariants on the Lie algebra, and will be preserved by any numerical method [7]. This ensures that the solution stays on the Lie group.

The idea behind RKMK methods is to approximate the solution Y of Eq. (1) with a discrete solution (Y_n) by approximately solving Eq. (2) with a general Runge-Kutta method $\dot{\Omega} = f(\Omega)$ with $f = \mathrm{d}^{\mathrm{R}}\exp_{\Omega}^{-1}$ and updating the position via the exponential map. Knowing that $\mathrm{d}^{\mathrm{R}}\exp_{\Omega}^{-1}(\Theta) = \Theta + \sum_{k=1}^{\infty}\frac{B_k}{k!}\,\mathrm{ad}_{\Omega}^k(\Theta)$ where B_k are Bernouilli numbers, a truncated sum up to order q is used in Eq. (2). If the Runke-Kutta method is order p and the truncature order is such that $q \geq p - 2$, then the associated RKMK method is order p [12].

Application to the Rigid Body Problem. We consider here the free rigid body problem. Let $\pi \in \mathfrak{so}(3)^* \approx \mathbb{R}^3$ be the angular momentum in the body frame and $\mathbb{J} = \mathrm{diag}(J_1, J_2, J_3)$ the inertia tensor, it verifies the Euler-Poincaré equation $\dot{\pi} = \pi \wedge \xi$, $\pi(0) = \pi_0$ where $\xi = \mathbb{J}^{-1}\pi \in \mathfrak{so}(3)$ and π_0 is the initial angular momentum. In terms of matrix product, this yields

$$\dot{\pi} = \begin{bmatrix} 0 & \frac{\pi_3}{J_3} & -\frac{\pi_2}{J_2} \\ -\frac{\pi_3}{J_3} & 0 & \frac{\pi_1}{J_1} \\ \frac{\pi_2}{J_2} & -\frac{\pi_1}{J_1} & 0 \end{bmatrix} \pi, \qquad \pi(0) = \pi_0. \tag{3}$$

This is in the form of Eq. (1), hence π can be approximately solved using a RKMK method where $SO(3)$ is the acting group. The Lie group $SO(3)$ leaves

---

[1] Here we made the assumption that $\mathrm{d}^{\mathrm{R}}\exp_{\Omega} : \mathfrak{g} \to \mathfrak{g}$ is invertible, which is the case for SO(3) whenever $\|\Omega\| < \pi$.

the vector space $\mathfrak{so}(3)^*$ invariant, reflecting the conservation of $\|\pi(t)\|$ in time. Applying a Lie group method guarantees the preservation of that constraint, ensuring that the angular momentum $\pi(t)$ lies on the sphere of radius $\|\pi_0\|$ for all t.

Defining Ω with $\pi(t) = \exp(\Omega(t))\pi_0$, the following expression is obtained for Eq. (2)

$$\dot{\Omega} = \sum_{k=0}^{\infty} \frac{B_k}{k!} \operatorname{ad}_{\Omega}^k (\mathbb{J}^{-1} \exp(\Omega)\pi_0), \quad \Omega(0) = 0. \tag{4}$$

We build an order 4 RKMK method by truncating the sum (4) up to order 2 and a applying a classical order 4 RK method. The results shown in Fig. 1, outputting the expected behaviour, have been computed for the following parameters:

$$\mathbb{J} = \operatorname{diag}\left(2/3, 1, 2\right), \qquad \pi_0 = \left[\cos\left(\tfrac{\pi}{3}\right)\, 0\, \sin\left(\tfrac{\pi}{3}\right)\right]^T, \qquad h = 0.5s, \qquad N = 200.$$

3   Covariant Variational Methods

Here we build a covariant variational method based on the Hamilton principle associated to a discrete Lagrangian following a similar approach to [5]. We take the case where the configuration space of the system is a Lie group G together with a reduced Lagrangian $\ell : \mathfrak{g} \to \mathbb{R}$.

Let a time step h divide equally the time interval, the set of discrete paths is defined by $\mathcal{C}_d(G) = \left\{ g_d : \{t_k\}_{0 \leq k \leq N} \to G \right\}$ where $\forall k,\ t_k = kh$. To determine an approximate trajectory $g_d \in \mathcal{C}_d(G)$ such that $g_k := g_d(k) \approx g(t_k)$, we define a discrete reduced Lagrangian ℓ_d approximating the action

$$\ell_d(\xi_0) \approx \int_{t_0}^{t_1} \ell(\xi)\,\mathrm{d}t$$

where $\xi(0) = \xi_0$ and $\xi = g^{-1}\dot{g}$ such that g is an action extremum on $[t_0, t_1]$. To discretize the relation $\xi = g^{-1}\dot{g}$ we introduce a local diffeomorphism $\tau : \mathfrak{g} \to G$ defined on an open set containing the identity and such that $\tau(0) = e_G$ (the exponential map is an example of such a diffeomorphism). Starting from the reconstruction formula

$$g_{i+1} = g_i \tau(h\,\xi_i), \tag{5}$$

we define $\xi_i := \frac{1}{h}\tau^{-1}(g_i^{-1}g_{i+1})$

The discrete action is approximated from the classical action by the sum $S_d(g_d) = \sum_{i=0}^{N-1} \ell_d(\xi_i)$. Applying the Hamilton principle on S_d evaluated on a discrete path g_d yields $\delta S_d(g_d) = \sum_{i=0}^{N-1} \left\langle \frac{\partial \ell_d}{\partial \xi}(\xi_i), \delta\xi_i \right\rangle$. The variation $\delta\xi_i$ is expressed using (5) as

$$\delta\xi_i = \frac{1}{h} \operatorname{d}\tau^{-1}_{g_i^{-1}g_{i+1}} \left(-g_i^{-1}\delta g_i g_i^{-1} g_{i+1} + g_i^{-1}\delta g_{i+1} \right)$$

$$= \frac{1}{h} \operatorname{d}\tau^{-1}_{\tau(h\xi_i)} \left(\left(-\zeta_i + \operatorname{Ad}_{\tau(h\xi_i)} \zeta_{i+1} \right) \tau(h\xi_i) \right)$$

where $\zeta_i = g_i^{-1} \delta g_i$. Here the right trivialized differential $\mathrm{d}^{\mathrm{R}} \tau^{-1} : \mathfrak{g} \to \mathfrak{g}$ defined by $\mathrm{d}^{\mathrm{R}} \tau_\xi^{-1} := T_{\tau(\xi)} \tau^{-1} \circ T R_{\tau(\xi)}$ is introduced, allowing us to write

$$\delta \xi_i = \frac{1}{h} \mathrm{d}^{\mathrm{R}} \tau_{h\xi_i}^{-1} \left(-\zeta_i + \mathrm{Ad}_{\tau(h\xi_i)} \zeta_{i+1} \right)$$

Using the definition of the adjoint $\langle \pi, A\xi \rangle = \langle A^*\pi, \xi \rangle$ where $\pi \in \mathfrak{g}^*$ and $\xi \in \mathfrak{g}$, the variation of the action functional now reads

$$\delta S_d(g_d) = \sum_{i=0}^{N-1} \left\langle \frac{1}{h} \left(\mathrm{d}^{\mathrm{R}} \tau_{h\xi_i}^{-1} \right)^* \frac{\partial \ell_d}{\partial \xi}(\xi_i), \mathrm{Ad}_{\tau(h\xi_i)} \zeta_{i+1} - \zeta_i \right\rangle.$$

Introducting the momentum μ_i associated to ξ_i via the formula

$$\mu_i := \left(\mathrm{d}^{\mathrm{R}} \tau_{h\xi_i}^{-1} \right)^* \frac{\partial \ell_d}{\partial \xi}(\xi_i) \tag{6}$$

and changing the indexes in the sum (discrete integration by part), we finally get, by the independence of ζ_i for all $i \in \{1, \ldots, N-1\}$, the discrete Euler-Poincaré equations

$$\mu_i - \mathrm{Ad}^*_{\tau(h\xi_{i-1})} \mu_{i-1} = 0. \tag{7}$$

This allows us to define the general formulation of a covariant method in Algorithm 1 for given boundary conditions g_0 et ξ_0. The momentum μ_i is computed from (7), and the associated $\xi_i \in \mathfrak{g}$ is then deduced from (6). This equation being implicit, it is typically solved using a numerical solver such as a Newton method. Finally, the position is updated via the reconstruction formula (5).

Algorithm 1. General implementation of the covariant variational method.

Data: g_0, ξ_0

$g_1 = g_0 \tau(h\xi_0), \quad \mu_0 = h \left(\mathrm{d}^{\mathrm{R}} \tau_{h\xi_0}^{-1} \right)^* \frac{\partial \ell_d}{\partial \xi}(\xi_0)$

for $i = 1$ **to** $N - 1$ **do**

    Compute $\mu_i = \mathrm{Ad}^*_{\tau(h\xi_{i-1})} \mu_{i-1}$ (equation (7))

    Find ξ_i **solution of** $\left(\mathrm{d}^{\mathrm{R}} \tau_{h\xi_i}^{-1} \right)^* \frac{\partial \ell_d}{\partial \xi}(\xi_i) - h\mu_i = 0$ (equation (6))

    Update $g_{i+1} = g_i \tau(h\xi_i)$ (equation (5))

end

Application to the Rigid Body Problem. A rigid body is represented by an element of the rotation group $SO(3)$. The reduced Lagrangian for this system is defined for $\xi \in \mathfrak{so}(3)$ as the rotation kinetic energy $\ell(\xi) := 1/2 \langle \mathbb{J}\xi, \xi \rangle$. We chose to approximate this Lagrangian with ℓ_d defined by $\ell_d(\xi_0) := h\ell(\xi_0) = \frac{h}{2} \langle \mathbb{J}\xi_0, \xi_0 \rangle$ and choose the local diffeomorphism τ to be defined as the Cayley map $\tau :=$ cay $: \mathfrak{so}(3) \to SO(3)$ (details can be found in [4]).

The results of the application of Algorithm 1 for the parameters given in Sect. 2 are also plotted on Fig. 1 for comparison.

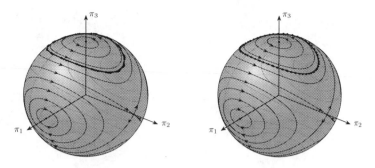

Fig. 1. Numerical angular momentum of the rigid body problem computed for RKMK4 (left) and covariant method (right); both exactly lie one the sphere. Exact solutions have been plotted for comparison.

4   Methods Based on Generating Functions, Hamilton-Jacobi Equation

4.1   Classical Case

The Hamilton Jacobi equation plays an important role in the development of numerical integrators that preserve the symplectic structure. In this section Hamilton-Jacobi theory is approached from the point of view of extended phase space as it is presented by Marsden [11] (p 206) and Arnold [1] (chapter 9). A link between Hamilton-Jacobi integrators and variational integrators could also be find in ([10]).

By definition, canonical transformations preserve the (pre)-symplectic 2-form, which can be deduced from the differential of the Poincaré-Cartan form. Let us consider a canonical transformation in the extended phase space $(t, q, p) \mapsto (T, Q, P)$ depicted in Fig. 2. Let (t, q, p) be coordinate functions in some chart of extended phase space considered as a manifold M. The Poincaré-Cartan form $\theta = p\,dq - H\,dt$ is a differential 1-form on M for which $H(t, q, p)$ is a Hamiltonian function. The coordinates (t, Q, P) can be considered as giving another chart on M associated to the 1-form $\Theta = P\,dQ - K\,dt$ with a corresponding Hamiltonian function $K(T, Q, P)$.

As it is well-know, it is possible to find four[2] generating functions depending of all mixes of old and new variables: (q, Q), (q, P), (p, Q), or (p, P). It appears that the second kind (q, P) of generating function is easily used to generate an infinitesimal transformation closed to the identity. And in turn, defines, by construction, a structure preserving numerical method. The mixed coordinates system (t, q, P) may be related to the previous ones through two mappings h and f: such that

$$h : (t, q, P) \mapsto p(t, q, P) \quad \text{and} \quad f : (t, q, P) \mapsto Q(t, q, P)$$

---

[2] At least four since many generating functions can be constructed.

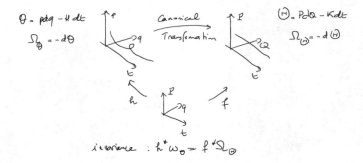

Fig. 2. Canonical transformation $(t, q, p) \mapsto (T, Q, P)$. Independent variables (q, P) are used to construct the second kind of generating function $G(t, q, P)$.

If each the (pre-)symplectic forms $\omega_\theta = -\,d\theta$ and $\Omega_\Theta = -\,d\Theta$ are invariantly associated to one another, their pull-back should agree: $h^*\omega_\theta = f^*\Omega_\Theta$. Since the operator (d) and (*) commute, that means $d(h^*\theta) = d(f^*\Theta)$. Consequently, $h^*\theta$ and $f^*\Theta$ differ from a closed form $dS = h^*\theta - f^*\Theta$ which is

$$dS(t, q, P) = h^* \left(p\,dq - H\,dt \right) - f^* \left(P\,dQ - K\,dT \right).$$

Replacing $P\,dQ = d(QP) - Q\,dP$ and introducing $G = (f^*Q)P + S$, one computes

$$\frac{\partial G}{\partial t}\,dt + \frac{\partial G}{\partial q}\,dq + \frac{\partial G}{\partial P}\,dP = h^* \left(p\,dq - H\,dt \right) - f^* \left(Q\,dP - K\,dT \right)$$

and obtains

$$\begin{cases} f^*K = h^*H + \frac{\partial G}{\partial t} \\ f^*Q = \frac{\partial G}{\partial P} \\ h^*p = \frac{\partial G}{\partial q} \end{cases} \mapsto \begin{cases} K(t, Q(t, q, P), P) = H(t, q, p(t, q, P)) + \frac{\partial G}{\partial t} \\ Q(t, q, P) = \frac{\partial G}{\partial P} \\ p(t, q, P) = \frac{\partial G}{\partial q} \end{cases}$$

$$(8)$$

Now suppose that $G(t, q, P)$ satisfies the so-called Hamilton-Jacobi equation,

$$H(t, q, \frac{\partial G}{\partial q}), + \frac{\partial G}{\partial t} = 0 \qquad (9)$$

for a given time dependent Hamiltonian H. This equation is obtained by taking $K \equiv 0$ in (8-a). The generating function G generates a time dependent canonical transformation ψ that transforms the Hamiltonian vector fields X_H to equilibrium: $\psi_* X_H = X_{K=0}$. That means that the integral curves of X_K are represented by straight lines in the image space. The vector field has been "integrated" by the transformation (see Fig. 2).

The choice of the second kind of generating function is convenient to easily generate the identity transformation. Choosing $G = qP$ in (8b) and (8c) reads $Q = \frac{\partial G}{\partial P} = q$ and $p = \frac{\partial G}{\partial q} = P$. So, a canonical (infinitesimal) transformation is obtained by plugging the ansatz

$$G(t, q, P) = qP + \sum_{m=1}^{\infty} \frac{t^m}{m!} G_m(q, P) = qP + tG_1(q, P) + \frac{t^2}{2} G_2(q, P) + \dots \quad (10)$$

into the Hamilton-Jacobi Eq. (9). Equating coefficients of equal powers of t gives

$$G_1 = -H(t, q, P), \quad G_2 = -\frac{\partial H}{\partial p} \frac{\partial G_1}{\partial q}, \quad G_3 = -\frac{\partial H}{\partial p} \frac{\partial G_2}{\partial q} - \frac{\partial^2 H}{\partial p^2} \frac{\partial G_1}{\partial q} \quad G_4 = \dots$$

A numerical method of the order k is obtained by truncating the serie (10) to a certain order k (see also [2]). The remaining variables (p, Q) are computed using the generating function G in (8b) and (8c): $Q = \frac{\partial G}{\partial P}$ and $p = \frac{\partial G}{\partial q}$. Putting (q, p) in the left-hand size, the numerical algorithm is finally

$$\begin{cases} q = Q - \sum_{m=1}^{k} \frac{t^m}{m!} \frac{\partial G_m}{\partial P}(q, P) \\ p = P + \sum_{m=1}^{k} \frac{t^m}{m!} \frac{\partial G_m}{\partial q}(q, P) \end{cases}$$

As it can be seen, the first step may be implicit for the variable q. But when it is solved, the second step is explicit for p. The symplectic Euler method is an example of such methods of order 1 with $G_1 = -H(q, P)$ given by

$$\begin{cases} q = Q + t\frac{\partial H}{\partial P}(q, P) \\ p = P - t\frac{\partial H}{\partial q}(q, P) \end{cases}$$

for which the "discrete Hamiltonian structure" is easily recognizable.

4.2   Lie-Poisson Hamilton-Jacobi Integrators

Following the same approach as the preceding section, the Hamilton-Jacobi theory is reduced from T^*G to $\mathfrak{g}^*$, the dual Lie algebra. Let (t, q_0, π_0) be coordinate functions in some chart of extended phase space considered as a manifold $M = \mathbb{R} \times G \times \mathfrak{g}^*$ (see Fig. 3). The 1-form

$$\theta = \pi_0 \lambda_{q_0} - H \, dt$$

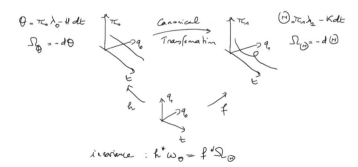

Fig. 3. Canonical transformation using the dual Lie algebra $(t, q_0, \pi_0) \mapsto (t, q_1, \pi_1)$. Independent variables (q_1, π_1) are used to construct the first kind of generating function $S(t, q_1, \pi_1)$.

is the reduced Poincaré-Cartan for which the Maurer-Cartan form is defined by $\lambda_{q_0}(v) = (L_{q_0^{-1}})_*(v)$. The coordinates (t, q_1, π_1) can be considered as giving another chart in M associated to the 1-form $\Theta = \pi_1 \lambda_{q_1} - K \, dt$ with $\lambda_{q_1}(v) = (L_{q_1^{-1}})_*(v)$. The mixed coordinates system (t, q_0, q_1) may be related to the previous ones through two mappings $h : (t, q_0, q_1) \mapsto \pi_0(t, q_0, q_1)$ and $f(t, q_0, q_1) \mapsto \pi_1(t, q_0, q_1)$.

For the left invariant system, the Hamiltonian function is left invariant. It is then natural to seek for left invariant generating functions satisfying $S_t(q_0, q_1) = S_t(h q_0, h q_1), \forall h \in G$. Choosing $h = q_0^{-1}$ we can construct a left invariant function $\bar{S}_t$ given by

$$S_t(q_0, q) = S_t(e, q_0^{-1} q) = S_t(e, g) = \bar{S}_t(g), \quad g = q_0^{-1} q_1.$$

The invariance of the (pre-)symplectic forms $\omega_\theta = -d\theta$ and $\Omega_\Theta = -d\Theta$ gives now rise to a function $\bar{S}_t(g)$ such that

$$d\bar{S}_t = f^* \Theta - h^* \theta = f^* \left(\pi \lambda_{q_1} - K \, dt \right) - h^* \left(\pi_0 \lambda_{q_0} - H \, dt \right) \tag{11}$$

So computing $d\bar{S}_t = \frac{\partial \bar{S}_t}{\partial t} \, dt + \frac{\partial \bar{S}_t}{\partial g} \, dg$, it appears that dg must also be computed in term of λ_{q_0} and λ_q,

$$dg = d(q_0^{-1} q_1) = dq_0^{-1} q_1 + q_0^{-1} \, dq_1 = -q_0^{-1} \, dq_0 q_0^{-1} q_1 + q_0^{-1} q_1 q_1^{-1} \, dq_1$$
$$= -\lambda_{q_0} g + g \lambda_{q_1} = -(R_g)_* \lambda_{q_0} + (L_g)_* \lambda_{q_1}.$$

So, comparing the expression $d\bar{S}_t = \frac{\partial \bar{S}_t}{\partial t} \, dt - \frac{\partial \bar{S}_t}{\partial g}(R_g)_* \lambda_{q_0} + \frac{\partial \bar{S}_t}{\partial g}(L_g)_* \lambda_{q_1}$ with (11), one obtains

$$\begin{cases} h^* H = f^* K + \frac{\partial \bar{S}_t}{\partial t} \\ f^* \pi_1 = (L_g)^* \frac{\partial \bar{S}_t}{\partial g} \\ h^* \pi_0 = (R_g)^* \frac{\partial \bar{S}_t}{\partial g} \end{cases} \mapsto \begin{cases} H(t, \pi_0(t, g)) = K(t, \pi_1(t, g)) + \frac{\partial \bar{S}_t}{\partial t} \\ \pi_1(t, g) = (L_g)^* \frac{\partial \bar{S}_t}{\partial g} \\ \pi_0(t, g) = (R_g)^* \frac{\partial \bar{S}_t}{\partial g} \end{cases} \tag{12}$$

For $H \equiv 0$, this yields the Lie-Poisson Hamilton-Jacobi equation

$$K \left(t, (L_g)^* \frac{\partial \bar{S}_t}{\partial g} \right) + \frac{\partial \bar{S}_t}{\partial t} = 0, \quad g = q_0^{-1} q_1 \tag{13}$$

So Eq. (12c)

$$\pi_0(t, g) = (R_g)^* \frac{\partial \bar{S}_t}{\partial g} \tag{14}$$

plugged into Eq. (12b) gives

$$\pi_1(t, g) = Ad_g^* \pi_0(t, g) \tag{15}$$

Marsden [6,9], Li [8] and de Degio [3] obtained a slightly different result using the convention $g = q_1^{-1} q_0$. Nevertheless, one can obtain a Lie-Poisson integrator by approximately solving the Lie-Poisson Hamilton-Jacobi Eq. (13) and then using (14) and (15) to generate the algorithm. This last Eq. (15) manifestly preserves the co-adjoint orbit $\mathcal{O}_{\pi_0} = \{\pi \in \mathfrak{g}^* | \pi = Ad_g^* \pi_0, \forall g \in G\}$. As in the classical case, one can generate algorithms of arbitrary accuracy by approximating the generative function by an ansatz such as the one given by (10), i.e $\bar{S}_t(g) = S_0(g) + \sum_{m=1}^{\infty} \frac{t^m}{m!} S_m(g)$ The main difficulty is to determine S_0 that can generate the identity map. Marsden propose to use in [6] the function $S_0 = trace(Ad_g^*)$ and astoundingly, de Diego [3], approximating the solution by taking the Taylor series in t of S up to order k, mention $S_0 = 0$.

Li [8] propose to reformulate the above theory of a generating function on $TG*$ by the exponential mapping in terms of algebra variable. For $g \in G$, choose $\xi \in \mathfrak{g}$ so that $g = \exp \xi$. He use Channel and Scovel's [2] results for which $S_0 = (\xi, \xi)/2$.

4.3   Conclusions and Future Research

In our case, our perspective is to relate the Lie-Poisson Hamilton-Jacobi algorithm to the Euler-Poincaré algorithm developed in Sect. 3 based on the Cayley map. In particular, since Eqs. (15) and (7) are the same in both algorithm, it will be instructive to compare the approximation of the Lie-Poisson Hamilton-Jacobi Eq. (13) to the relationship between μ and ξ given by Eq. (6).

References

1. Arnol'd, V.I.: Mathematical Methods of Classical Mechanics, vol. 60. Springer, New York (2013)
2. Channell, P.J., Scovel, C.: Symplectic integration of hamiltonian systems. Nonlinearity **3**(2), 231 (1990)
3. de Diego, D.M.: Lie-Poisson integrators. arXiv e-prints arXiv:1803.01427, March 2018
4. Demoures, F., Gay-Balmaz, F., Kobilarov, M., Ratiu, T.S.: Multisymplectic lie group variational integrator for a geometrically exact beam in R3. Commun. Nonlinear Sci. Numer. Simul. **19**(10), 3492–3512 (2014)
5. Demoures, F., Gay-Balmaz, F., Ratiu, T.S.: Multisymplectic variational integrators and space/time symplecticity. Anal. Appl. **14**(03), 341–391 (2016)
6. Ge, Z., Marsden, J.E.: Lie-Poisson Hamilton-Jacobi theory and Lie-Poisson integrators. Phys. Lett. A **133**(3), 134–139 (1988)
7. Hairer, E., Lubich, C., Wanner, G.: Geometric Numerical Integration: Structure-Preserving Algorithms for Ordinary Differential Equations, vol. 31. Springer, Heidelberg (2006). https://doi.org/10.1007/3-540-30666-8
8. Li, S.T., Qin, M.: A note for Lie-Poisson Hamilton-Jacobi equation and Lie-Poisson integrator. Comput. Math. Appl. **30**(7), 67–74 (1995)
9. Marsden, J.E., Pekarsky, S., Shkoller, S.: Discrete Euler-Poincaré and Lie-Poisson equations. Nonlinearity **12**(6), 1647–1662 (1999)
10. Marsden, J.E., West, M.: Discrete mechanics and variational integrators. Acta Numerica **10**, 357–514 (2001)

11. Marsen, J., Ratiu, T.: Introduction to Mechanics and Symmetry: A Basic Exposition of Classical Mechanical Systems. Texts in Applied Mathematics, vol. 17. Springer, New York (1994). https://doi.org/10.1007/978-0-387-21792-5
12. Munthe-Kaas, H.: High order Runge-Kutta methods on manifolds. Appl. Numer. Math. **29**(1), 115–127 (1999)

Souriau Exponential Map Algorithm for Machine Learning on Matrix Lie Groups

Frédéric Barbaresco[✉]

Thales Land & Air Systems, 91470 Limours, France
frederic.barbaresco@thalesgroup.com

Abstract. Jean-Marie Souriau extended Urbain Jean Joseph Leverrier algorithm to compute characteristic polynomial of a matrix in 1948. This Souriau algorithm could be used to compute exponential map of a matrix that is a challenge in Lie Group Machine Learning. Main property of Souriau Exponential Map numerical scheme is its scalability with highly parallelization.

Keywords: Souriau algorithm · Matrix characteristic polynomial · Exponential map · Lie group machine learning

1 From Le Verrier to Souriau Algorithm

The algorithm to compute characteristic polynomial of a matrix was discovered by Urbain Jean Joseph Leverrier in 1840, and was rediscovered in 1948 by Jean-Marie Souriau and modified to its present form, but published only in French. Other authors, P. Horst, D. K. Faddejew and Sominski, J. S. Frame, U. Wegner and L. Csanky, were credited with rediscovering the technique. As soon as 1955, Souriau algorithm was tested and benchmarked by the National Bureau of Standards, Los Angeles, under the sponsorship of the Wright Air Development Center, U. S. Air Force, and the Office of Naval Research, and was concluded at the University of California, by the Office of Naval Research. As observed and illustrated by Souriau, for $n = 10$, his algorithm uses only 8 thousands of additions and multiplications, compared to 37 million of additions and 62 million of multiplications for classical approach (Gaussian elimination). Main drawback of most efficient classical algorithm based on Krylov iterates cannot be parallelized. Souriau algorithm has a complexity $O(n^4)$ or $O(n^{\omega+1})$ in sequential computation, and so cannot compete with Krylov-based algorithm, but Souriau algorithm has been parallelized by L. Csanky, proving that characteristic polynomial computation could be solved in parallel time $log^2 n$ with a polynomial number of processors. Souriau algorithm parallelization by Czansky has been improved more recently by Preparata and Sarwate using fast matrix product, and by Keller-Gehrig using matrix reduction. Reduction to complexity $O(n^{\omega})$ is given for generic matrices, but for non-generic ones, only $O(n^{\omega} logn)$ complexity could be achieved. A disadvantage of both algorithms (Le verrier and Gaussian elimination) is the presence of divisions.

The computation of the matrix exponential is a classical problem in numerical mathematics as explained in 1978 paper of Moler and van Loan and many efficient

© Springer Nature Switzerland AG 2019
F. Nielsen and F. Barbaresco (Eds.): GSI 2019, LNCS 11712, pp. 85–95, 2019.
https://doi.org/10.1007/978-3-030-26980-7_10

algorithms described in 1998 paper Hochbruck, Lubich and Selhofer 1998. But this problem is very far from being fully solved, especially to approximate an exponential of a matrix which resides in a Lie algebra, a central problem in geometric integration as studied by Iserles, Munthe-Kaas, Nørsett, Zanna and Celledoni.

2 Souriau Matrix Characteristic Polynomial Computation

Jean-Marie Souriau introduced his algorithm in the framework of his lecture on Multilinear Algebra by consideration on *volume form*. In a vector space E of dimension n, we can prove that vector space of n-forms (*n-form* as an anti-symmetric n-linear operator with scalar value). After Selecting a frame $(e_1, e_2, \ldots, e_n)$ of E, we can define an *n-form* called "*volume form*" with:

$$vol(e_1)(e_2)\ldots(e_n) = 1 \tag{1}$$

Volume of parallelepiped generated by vectors $(x_1, x_2, \ldots, x_n)$ is given by:

$$|vol(x_1)(x_2)\ldots(x_n)| \tag{2}$$

Souriau called "espace jaugé (jauged space)", all vector space E, of finite size, where we have selected a "unit-jauge" defined by vol. If we define a linear operator $A : E \to E$, considered as "affiner" in a jauged space, we can then give definition of:

- *Determinant of A,* $\det(A)$ by:

$$\det(A)vol(v_1)(v_2)\ldots(v_n) = vol(Av_1)(Av_2)\ldots(Av_n) \tag{3}$$

- *Adjoint linear operator of A*, $adj(A)$, by:

$$vol(adj(A)v_1)(v_2)\ldots(v_n) = vol(v_1)(Av_2)\ldots(Av_n) \tag{4}$$

- *Trace number of A*, $tr(A)$, by:

$$\begin{aligned} tr(A)vol(v_1)(v_2)\ldots(v_n) &= vol(Av_1)(v_2)\ldots(v_n) + vol(v_1)(Av_2)\ldots(v_n) \\ &\ldots + vol(v_1)(v_2)\ldots(Av_n) \end{aligned} \tag{5}$$

By using the following relation deduced from previous equations:

$$vol(adj(A)Av_1)(v_2)\ldots(v_n) = vol(Av_1)(Av_2)..(Av_n) = \det(A)vol(v_1)(v_2)\ldots(v_n) \tag{6}$$

If A is invertible, we recover classical equations:

$$adj(A)A = \det(A)I \text{ and } A^{-1} = [\det(A)]^{-1}adj(A) \tag{7}$$

Using these formulas, we can try to invert $[\lambda I - A]$ assuming that $\det(\lambda I - A) \neq 0$. If we use previous determinant definition, we have:

$$
\begin{aligned}
\det(\lambda I - A)vol(v_1)(v_2)\ldots(v_n) &= vol(\lambda v_1 - Av_1)(\lambda v_2 - Av_2)\ldots(\lambda v_n - Av_n) \\
&= \lambda^n vol(v_1)(v_2)\ldots(v_n) + \ldots
\end{aligned}
\tag{8}
$$

where $\det(\lambda I - A)$ is the characteristic polynomial of A, a polynomial in λ of degree n, with:

$$
adj(\lambda I - A)[\lambda I - A] = \det(\lambda I - A)I \Leftrightarrow A.Q(\lambda) = \lambda Q(\lambda) - P(\lambda)I
\tag{9}
$$

(if λ is an eigenvalue of A, the nonzero columns of $Q(\lambda)$ are corresponding eigenvectors). We can then observe that $adj(\lambda I - A)$ is a polynomial of degree $n - 1$. We can then define both $P(\lambda)$ and $Q(\lambda)$ by polynomials:

$$
P(\lambda) = \det(\lambda I - A) = \sum_{i=0}^{n} k_i \lambda^{n-i} \quad \text{and} \quad Q(\lambda) = adj(\lambda I - A) = \sum_{i=0}^{n-1} \lambda^{n-i-1} B_i
\tag{10}
$$

with

$$
k_0 = 1, k_n = (-1)^n \det(A), B_0 = I \text{ and } B_{n-1} = (-1)^{n-1} adj(A)
\tag{11}
$$

By developing equation $adj(\lambda I - A)[\lambda I - A] = \det(\lambda I - A)I$, we can write:

$$
\sum_{i=0}^{n} k_i \lambda^{n-i} I = \sum_{i=0}^{n-1} \lambda^{n-i-1} B_i [\lambda I - A] = \lambda B_{n-1} + \sum_{i=1}^{n-1} \lambda^{n-i} [B_i - B_{i-1}A] - B_{n-1}A
\tag{12}
$$

By identification term by term, we find the expression of matrices B_i:

$$
\begin{cases}
B_0 = I \\
B_i = B_{i-1}A + k_i I, \quad i = 1, \ldots, n-1 \\
B_{n-1}A + k_n I = 0
\end{cases}
\tag{13}
$$

We can observe that $A^{-1} = -\frac{B_{n-1}}{k_n}$ and also the Cayley-Hamilton theorem:

$$
k_0 A^n + k_1 A^{n-1} + \ldots + k_{n-1}A + k_n I = 0
\tag{14}
$$

To go further, we have to use this classical result from analysis on differentiation given by $\delta[\det(G)] = tr(adj(G)\delta G)$. If we set $G = (\lambda I - A)$ and $\delta = \frac{d}{d\lambda}$, we then obtain $tr(adj(\lambda I - A)) = \frac{d}{d\lambda}\det(\lambda I - A)$ providing:

$$
\sum_{i=0}^{n-1} \lambda^{n-i-1} tr(B_i) = \frac{d}{d\lambda}\left(\sum_{i=0}^{n} k_i \lambda^{n-i}\right) = \sum_{i=0}^{n-1} (n-i)k_i \lambda^{n-i-1}
\tag{15}
$$

We can then deduce that $tr(B_i) = (n - i)k_i$, $i = 0, \ldots, n - 1$.
As

$$B_i = B_{i-1}A + k_iI, tr(B_i) = tr(B_{i-1}A) + n.k_i, \text{ and then } k_i = -\frac{tr(B_{i-1}A)}{i} \quad (16)$$

We finally obtain the Souriau Algorithm:

$$
\begin{cases}
k_0 = 1 \quad \text{and} \quad B_0 = I \\
A_i = B_{i-1}A, \quad k_i = -\frac{1}{i}tr(A_i), i = 1, \ldots, n - 1 \\
B_i = A_i + k_iI \quad \text{or} \quad B_i = B_{i-1}A - \frac{1}{i}tr(B_{i-1}A)I \\
A_n = B_{n-1}A \quad \text{and} \quad k_n = -\frac{1}{n}tr(A_n)
\end{cases}
\quad (17)
$$

3   Souriau Algorithm to Compute Exponential Map of Matrix

Souriau approach of Exponential computation is based on algebraic analogy:

$$[\lambda I - A]^{-1} = \frac{Q(\lambda)}{P(\lambda)} \Leftrightarrow [\lambda I - A]Q(\lambda) = P(\lambda)I \quad (18)$$

and the differential property (with $\lambda = \frac{d}{dt}$):

$$\left[I\frac{d}{dt} - A\right]Q\left(\frac{d}{dt}\right) = P\left(\frac{d}{dt}\right)I \quad (19)$$

If a numeric function γ verifies $P(\frac{d}{dt})\gamma = 0$, then:

$$P\left(\frac{d}{dt}\right)\gamma = \sum_{i=0}^{n} k_i\gamma^{(n-i)} = k_0\gamma^{(n)} + k_1\gamma^{(n-1)} + \ldots + k_{n-1}\gamma^{(1)} + k_n\gamma = 0 \quad (20)$$

with $\gamma^{(n)} = \frac{d^n\gamma(t)}{dt^n}$ n-*th* derivative of function γ, with initial conditions:

$$\gamma(0) = \gamma^{(1)}(0) = \ldots = \gamma^{(n-2)} = 0 \text{ and } \gamma^{(n-1)}(0) = 1 \quad (21)$$

In this case, the matrix function $\Phi = Q(\frac{d}{dt})\gamma$ is solution of the differential equation $\frac{d\Phi(t)}{dt} = A\Phi(t)$, with initial condition $\Phi(0) = I$:

$$\Phi = Q\left(\frac{d}{dt}\right)\gamma = \sum_{i=0}^{n-1} \gamma^{(n-i-1)}B_i = \gamma^{(n-1)}B_0 + \gamma^{(n-2)}B_1 + \ldots + \gamma B_{n-1} \quad (22)$$

We can then observe that the exponential map of matrix tA is given by:

$$\begin{cases} e^{tA} = \sum_{i=0}^{n-1} \gamma^{(n-i-1)} B_i = \gamma^{(n-1)} B_0 + \gamma^{(n-2)} B_1 + \ldots + \gamma B_{n-1} \\ \text{with } B_0 = I \text{ and } B_i = B_{i-1}A - \frac{tr(B_{i-1}A)}{i} I \end{cases} \tag{23}$$

$$\begin{cases} \gamma \text{ such that } k_0 \gamma^{(n)} + k_1 \gamma^{(n-1)} + \ldots + k_{n-1} \gamma^{(1)} + k_n \gamma = 0 \\ \text{with } k_i = -\frac{tr(B_{i-1}A)}{i}, \ \gamma(0) = \ldots = \gamma^{(n-2)} = 0 \text{ and } \gamma^{(n-1)}(0) = 1 \end{cases} \tag{24}$$

The solution $\gamma(t)$ of characteristic ordinary differential equation is obtained in $[0, h]$ of the spectral interval of integration. In the remaining part, the exponential function $\Phi(t)$ is computed by:

$$\Phi(ph) = (\Phi(h))^p \tag{25}$$

Souriau Algorithm for Exponential Map of Matrix is given by:

$$\begin{aligned} &(1) \begin{cases} B_0 = I \text{ and } B_i = B_{i-1}A - \frac{tr(B_{i-1}A)}{i} I \\ k_0 = 1, \ k_i = -\frac{tr(B_{i-1}A)}{i} \ \ i = 1, \ldots, n \end{cases} \\ &(2) \begin{cases} \gamma \text{ integrated on } [0, h] \text{ such that} \\ k_0 \gamma^{(n)} + k_1 \gamma^{(n-1)} + \ldots + k_{n-1} \gamma^{(1)} + k_n \gamma = 0 \\ \text{with } \gamma(0) = \ldots = \gamma^{(n-2)} = 0 \text{ and } \gamma^{(n-1)}(0) = 1 \end{cases} \\ &(3) \text{ Computation of } \Phi(t) = e^{tA} = \sum_{i=0}^{n-1} \gamma^{(n-i-1)}(t) B_i \text{ on } [0, h] \\ &(4) \text{ Extension of Computation on } [0, ph] \text{ by } \Phi(pt) = (\Phi(t))^p \\ &(5) \ X(t) = \Phi(t) X_0 \text{ with } X_0 = X(0) \end{aligned} \tag{26}$$

If we observe that $\ln(A) = \int_{-\infty}^{0} [sI - A]^{-1} - [sI - I]^{-1} ds$, this algorithm could be used also to compute $A^s = e^{s \ln(A)}$ such as $A^{1/2}$. This Souriau algorithm to solve $\frac{d\Phi(t)}{dt} = A\Phi(t)$ by computation of exponential $\Phi(t) = e^{tA}$ could be extended to solve $L\frac{d^2\Phi(t)}{dt^2} + M\frac{d\Phi(t)}{dt} + N\Phi(t) = 0$ by substituting $\left[\lambda^2 L + \lambda M + N \right] Q(\lambda) = P(\lambda)I$ to $[\lambda I - A]Q(\lambda) = P(\lambda)I$ through the following algebraic relations:

$$\left(\lambda^2 L + \lambda M + N \right) adj \left(\lambda^2 L + \lambda M + N \right) = \det \left(\lambda^2 L + \lambda M + N \right) I \tag{27}$$

$$\begin{aligned} P(\lambda) &= \det \left(\lambda^2 L + \lambda M + N \right) = \sum_{i=0}^{2n} k_i \lambda^{2n-i}, Q(\lambda) = adj \left(\lambda^2 L + \lambda M + N \right) \\ &= \sum_{i=0}^{2n-2} \lambda^{2n-i-2} B_i \end{aligned}$$

4 Examples of Souriau Exponential Map Algorithm

We can illustrate Souriau algorithm with:

$$J = \begin{bmatrix} 0 & -1 \\ 1 & 0 \end{bmatrix} \text{ and } e^{tJ} = \cos(t)I + \sin(t)J = \begin{bmatrix} \cos(t) & -\sin(t) \\ \sin(t) & \cos(t) \end{bmatrix} \in SO(2) \qquad (28)$$

$$\begin{vmatrix} B_0 = I \text{ and } k_0 = 1 \\ B_1 = B_0 J - tr(B_0 J)I = J \text{ and } k_1 = -tr(B_0 J) = 0 \\ B_2 = B_1 J - \frac{tr(B_1 J)}{2}I = J^2 - \frac{tr(J^2)}{2}I = \begin{bmatrix} -1 & 0 \\ 0 & -1 \end{bmatrix} + \begin{bmatrix} 1 & 0 \\ 0 & 1 \end{bmatrix} = \begin{bmatrix} 0 & 0 \\ 0 & 0 \end{bmatrix} \\ k_2 = -\frac{tr(J^2)}{2} = 1 \end{vmatrix} \qquad (29)$$

$$\begin{cases} \gamma \text{ on } [0, h] \text{ such that } \frac{d^2\gamma(t)}{dt^2} + \gamma = 0 \\ \text{with } \gamma(0) = 0 \text{ and } \gamma^{(1)}(0) = 1 \end{cases} \Rightarrow \gamma(t) = \sin(t) \qquad (30)$$

$$\begin{cases} \Phi(t) = \frac{d\gamma(t)}{dt}B_0 + \gamma(t)B_1 = \cos(t)I + \sin(t)J \text{ on } [0, h] \\ \frac{d\Phi}{dt} = J\Phi(t) \end{cases} \qquad (31)$$

Another example is given by harmonic oscillator:

$$\frac{d}{dt}\begin{pmatrix} p \\ q \end{pmatrix} = \begin{pmatrix} -q \\ p \end{pmatrix} = \begin{pmatrix} 0 & -1 \\ 1 & 0 \end{pmatrix}\begin{pmatrix} p \\ q \end{pmatrix} = J\begin{pmatrix} p \\ q \end{pmatrix} \text{ with } J^2 = -I \qquad (32)$$

$$\text{then } e^{tJ}\begin{pmatrix} p \\ q \end{pmatrix} = \begin{pmatrix} \cos t & -\sin t \\ \sin t & \cos t \end{pmatrix}\begin{pmatrix} p \\ q \end{pmatrix}, \text{ rotation in } \begin{pmatrix} p \\ q \end{pmatrix}\text{-plane.} \qquad (33)$$

Next example, is given for skew-symmetric matrix, corresponding to exponential map for $so(3)$, the Lie Algebra of Lie group $SO(3) = \{R/R^{-1} = R^T\}$:

$$\omega_\times = \begin{pmatrix} 0 & -\omega_3 & \omega_2 \\ \omega_3 & 0 & -\omega_1 \\ -\omega_2 & \omega_1 & 0 \end{pmatrix} = \omega_1 L_1 + \omega_2 L_2 + \omega_3 L_3 \in so(3) \text{ and } \omega$$
$$= (\omega_1, \omega_2, \omega_3) \in \mathbb{R}^3 \qquad (34)$$

The generators of $so(3)$ correspond to the derivatives of rotation around the each of the standard axes, evaluated at identity. The exponential map that takes skew symmetric matrices to rotation matrices is simply the matrix exponential over a linear combination of the generators. We compute this exponential map by Souriau algorithm:

$$e^{\omega_\times} = \gamma^{(2)}B_0 + \gamma^{(1)}B_1 + \gamma B_2 \qquad (35)$$

Souriau algorithm provides:

$$B_0 = I \text{ and } k_0 = 1 \tag{36}$$

$$B_1 = I.\omega_\times - \frac{Tr(I.\omega_\times)}{1}I = \omega_\times \text{ and } k_1 = -\frac{Tr(I\omega_\times)}{1} = 0 \tag{37}$$

$$B_2 = B_1.\omega_\times - \frac{Tr(\omega_\times.\omega_\times)}{2}I = \omega_\times.\omega_\times + \|\omega\|^2 I \text{ and } k_2 = -\frac{Tr(\omega_\times.\omega_\times)}{2} = \|\omega\|^2 \tag{38}$$

We can observe that $B_2 = \omega_\times.\omega_\times + \|\omega\|^2 I = \omega \otimes \omega^T$ and $k_3 = 0$, and we obtain:

$$e^{\omega_\times} = \gamma^{(2)}I + \gamma^{(1)}\omega_\times + \gamma\omega \otimes \omega^T \tag{39}$$

The function $\gamma(t)$ should verify:

$$k_0\gamma^{(3)}(t) + k_1\gamma^{(2)}(t) + k_2\gamma^{(1)}(t) + k_3\gamma(t) = 0 \text{ with } k_0 = 1, k_1 = 0, k_2 = \|\omega\|^2, k_3 = 0 \tag{40}$$

$$\gamma^{(3)}(t) + \|\omega\|^2\gamma^{(1)}(t) = 0 \text{ with } \gamma^{(2)}(0) = 1, \gamma^{(1)}(0) = 0, \gamma(0) = 0 \tag{41}$$

We can then deduce that:

$$\gamma^{(1)}(t) = \frac{1}{\|\omega\|}\sin(\|\omega\|t) \text{ and } \gamma(t) = \frac{1}{\|\omega\|^2}(1 - \cos(\|\omega\|t)) \tag{42}$$

We can then deduce the exponential map of $\mathsf{so}(3)$:

$$e^{t.\omega_\times} = \cos(\|\omega\|t)I + \frac{1}{\|\omega\|}\sin(\|\omega\|t)\omega_\times + \frac{1 - \cos(\|\omega\|t)}{\|\omega\|^2}\omega \otimes \omega^T \tag{43}$$

But using the relation $\omega \otimes \omega^T = \omega_\times.\omega_\times + \|\omega\|^2 I$, we recover Rodrigues formula:

$$e^{t.\omega_\times} = I + \frac{1}{\|\omega\|}\sin(\|\omega\|t)\omega_\times + \frac{1 - \cos(\|\omega\|t)}{\|\omega\|^2}\omega_\times^2 \tag{44}$$

The exponential map from $\mathsf{se}(3)$ to $SE(3) = \left\{ C/C = \begin{bmatrix} R & t \\ 0 & 1 \end{bmatrix}, R \in SO(3), t \in \right.$
$\left. \mathbb{R}^3 \right\}$ is the matrix exponential on a linear combination of the generators:

$$\delta = \begin{pmatrix} \omega_\times & u \\ 0 & 0 \end{pmatrix} = u_1 G_1 + u_2 G_2 + u_3 G_3 + \omega_1 G_4 + \omega_2 G_5 + \omega_3 G_6 \tag{45}$$
$$\delta = (u \quad \omega) \in \mathsf{se}(3) \text{ and } (u \quad \omega)^T \in \mathbb{R}^6$$

$$e^\delta = \exp\begin{pmatrix} \omega_\times & u \\ 0 & 0 \end{pmatrix} = \begin{pmatrix} e^{\omega_\times} & Vu \\ 0 & 1 \end{pmatrix} \text{ with } V = I + \frac{1}{2!}\omega_\times + \frac{1}{3!}(\omega_\times)^2 + \dots \tag{46}$$

By using the identity, $(\omega_\times)^3 = -(\omega^T\omega).\omega_\times = -\|\omega\|^2\omega_\times$:

$$V = I + \sum_{i=0}^{\infty}\left[\frac{\omega_\times^{2i+1}}{(2i+2)!} + \frac{\omega_\times^{2i+2}}{(2i+3)!}\right] = I + \left(\sum_{i=0}^{\infty}\frac{(-1)^i\theta^{2i}}{(2i+2)!}\right)\omega_\times + \left(\sum_{i=0}^{\infty}\frac{(-1)^i\theta^{2i}}{(2i+3)!}\right)\omega_\times^2 \tag{47}$$

$$V = I + \left(\frac{1 - \cos(\|\omega\|)}{\|\omega\|^2}\right)\omega_\times + \left(\frac{\|\omega\| - \sin(\|\omega\|)}{\|\omega\|^3}\right)\omega_\times^2 \tag{48}$$

We can apply Souriau formula for exponential map of $\mathsf{su}(2)$, the Lie Algebra of Lie group $SU(2)$ through a linear combination of the generators given by the Pauli spin matrices:

$$a.I + i(c.\sigma_x + b.\sigma_y + d.\sigma_z) = \begin{pmatrix} a + id & b + ic \\ -b + ic & a - di \end{pmatrix} \text{ with } (a, b, c, d) \in \mathbb{R}^4 \tag{49}$$

Last example deals with "Geodesic Shooting" for multivariate Gaussian densities $\aleph(m, R)$. Information Geometry provides an invariant Koszul-Fisher metric and geodesic by Euler-Lagrange equations:

$$\begin{cases} \ddot{R} + \dot{m}\dot{m}^T - \dot{R}R^{-1}\dot{R} = 0 \\ \ddot{m} - \dot{R}R^{-1}\dot{m} = 0 \end{cases} \tag{50}$$

Using Souriau theorem of moment map (geometrization of Noether theorem):

$$\Rightarrow \begin{cases} R^{-1}\dot{R} + R^{-1}\dot{m}m^T = B = cste \\ R^{-1}\dot{m} = b = cste \end{cases} \tag{51}$$

This moment map could be computed if we consider the following Lie group action in case of Gaussian densities:

$$\begin{bmatrix} Y \\ 1 \end{bmatrix} = \begin{bmatrix} R^{1/2} & m \\ 0 & 1 \end{bmatrix} \begin{bmatrix} X \\ 1 \end{bmatrix} = \begin{bmatrix} R^{1/2}X = m \\ 1 \end{bmatrix}, \quad \begin{cases} (m, R) \in R^n \times Sym^+(n) \\ M = \begin{bmatrix} R^{1/2} & m \\ 0 & 1 \end{bmatrix} \in G_{aff} \end{cases} \quad (52)$$

$$X \approx \aleph(0, I) \rightarrow Y \approx \aleph(m, R)$$

With $R^{1/2}$, square root of R, is given by Cholesky decomposition of R. $R^{1/2}$ is the Lie group of triangular matrix with positive elements on the diagonal. Euler-Poincaré equations, reduced equations from Euler-Lagrange equations, are then given by:

$$\begin{cases} \dot{m} = Rb \\ \dot{R} = R(B - bm^T) \end{cases} \quad (53)$$

Geodesic shooting is obtained by using equations established by Eriksen for "*exponential map*" using the following change of variables:

$$\begin{cases} \Delta(t) = R^{-1}(t) \\ \delta(t) = R^{-1}(t)m(t) \end{cases} \Rightarrow \begin{cases} \dot{\Delta} = -B\Delta + bm^T \\ \dot{\delta} = -B\delta + (1 + \delta^T \Delta^{-1}\delta)b \quad \text{with} \\ \Delta(0) = I_p, \delta(0) = 0 \end{cases} \begin{cases} \dot{\Delta}(0) = -B \\ \dot{\delta}(0) = b \end{cases} \quad (54)$$

The method based on geodesic shooting consists in iteratively approaching the solution by geodesic shooting in direction $\left(\dot{\delta}(0), \dot{\Delta}(0) \right)$, using Souriau exponential map:

$$\Lambda(t) = \exp(tA) = \sum_{n=0}^{\infty} \frac{(tA)^n}{n!} = \begin{pmatrix} \Delta & \delta & \Phi \\ \delta^T & \varepsilon & \gamma^T \\ \Phi^T & \gamma & I' \end{pmatrix}$$

$$\text{with } A = \begin{pmatrix} -B & b & 0 \\ b^T & 0 & -b^T \\ 0 & -b & B \end{pmatrix} \quad (55)$$

$$A^2 = \begin{pmatrix} -B & b & 0 \\ b^T & 0 & -b^T \\ 0 & -b & B \end{pmatrix}^2 = \begin{pmatrix} B^2 + bb^T & -Bb & -bb^T \\ -b^T B & 2b^T b & -b^T B \\ -bb^T & -Bb & B^2 + bb^T \end{pmatrix} \quad (56)$$

$$k_0 = 1, B_0 = I \text{ and } k_1 = 0, B_1 = A \text{ because } tr(A) = 0$$
$$B_2 = A^2 - \frac{tr(A^2)}{2}I, k_2 = -\frac{tr(A^2)}{2}, B_i = B_{i-1}A - \frac{tr(B_{i-1}A)}{i}I, k_i = -\frac{tr(B_{i-1}A)}{i} \quad (57)$$

$$k_0\gamma^{(n)} + k_1\gamma^{(n-1)} + \ldots + k_{n-1}\gamma^{(1)} + k_n\gamma = 0 \text{ with } \gamma(0) = \ldots = \gamma^{(n-2)} = 0, \gamma^{(n-1)}(0) = 1$$
$$e^{tA} = \sum_{i=0}^{n-1} \gamma^{(n-i-1)}(t)B_i$$

$$(58)$$

Acknowledgement. I would like to show my gratitude Ms. Danielle Fortuné from University of Poitiers, for giving me some documents from Claude Vallée and Jean-Marie Souriau archives.

References

1. Souriau, J.-M.: Une méthode pour la décomposition spectrale et l'inversion des matrices. CRAS **227**(2), 1010–1011 (1948)
2. Souriau, J.-M.: Calcul Linéaire. EUCLIDE, Introduction aux études Scientifiques, vol. 1. Presses Universitaires de France, Paris (1959)
3. Souriau, J.-M.; Vallée, C.; Réaud, K.; Fortuné, D.: Méthode de Le Verrier–Souriau et équations différentielles linéaires. CRAS s. IIB Mech. **328**(10), 773–778 (2000)
4. Souriau, J.-M.: Grammaire de la Nature. Private publication (2007)
5. Thomas, F.: Nouvelle méthode de résolution des équations du mouvement de systèmes vibratoires linéaire, discrets, DEA Mécanique, université de Poitiers (1998)
6. Réaud, K., Fortuné, D., Prudhorffne, S. Vallée, C.: Méthode d'étude des vibrations d'un système mécanique non basée sur le calcul de ses modes propres. XVème Congrès français de Mécanique, Nancy, (2001)
7. Champion-Réaud, K.: Méthode d'étude des vibrations d'un système mécanique non basée sur le calcul de ses modes propres. SupAéro PhD (2002)
8. Réaud, K., Vallée, Cl., Fortuné, D.: Détermination des vecteurs propres d'un système vibratoire par exploitation du concept de matrice adjuguée. 6ème Colloque national en calcul des structures, Giens (2003)
9. Champion-Réaud, K.; Vallée, C.; Fortuné, D. Champion-Réaud, J.L.: Extraction des pulsations et formes propres de la réponse d'un système vibratoire. 16ème Congrès Français de Mécanique, Nice, (2003)
10. Vallée, C., Fortuné, D., Champion-Réaud, K.: A general solution of a linear dissipative oscillatory system avoiding decomposition into eigenvectors. J. Appl. Math. Mech. **69**, 837–843 (2005)
11. Le Verrier, U.: Sur les variations séculaires des éléments des orbites pour les sept planètes principales. J. de Math. (1) **5**, 230 (1840)
12. Le Verrier, U.: Variations séculaires des éléments elliptiques des sept planètes principales. I Math. Pures Appli. **4**, 220–254 (1840)
13. Juhel, A.: Le Verrier et la première détermination des valeurs propres d'une matrice, Bibnum, Physique (2011)
14. Tong, M.D., Chen, W.K.: A novel proof of the Souriau-Frame-Faddeev algorithm. IEEE Trans. Autom. Control **38**, 1447–1448 (1993)
15. Faddeev, D. K.; Sominsky, I. S.: Problems in Higher Algebra, Problem 979. Mir Publishers, Moskow-Leningrad (1949)
16. Frame, J.S.: A simple recursion formula for inverting a matrix. Bull. Amer. Math. Soc. **56**, 1045 (1949)
17. Forsythe, G.E., Straus, L.W.: The Souriau-Frame characteristic equation algorithm on a digital computer. J. Math. Phys. Stud. Appl. Math. **34**(1–4), 152–156 (1955)
18. Fadeev, D.K. Fadeeva, V.N.: Computational Methods of Linear Algebra (translated from Russian by R. C. Williams). W. H. Freeman and Co., San Francisco (1963)
19. Greville, T.N.E.: The Souriau-Frame algorithm and the Drazin pseudoinverse. Linear Algebr. Its Appl. **6**, 205 (1973)
20. Downs, T.: Some properties of the Souriau-frame algorithm with application to the inversion of rational matrices. SIAM J. on Applied Mathematics **28**(2), 237–251 (1975)

21. Csanky, L.: Almost parallel matrix inversion algorithms. SIAM 618–623 (1976)
22. Hartwig, R.E.: More on the Souriau-Frame algorithm and the Drazin inverse. SIAM J. Appl. Math. **31**(1), 42–46 (1976)
23. Hou, S.-H.: A simple proof of the Leverrier-Faddeev characteristic polynomial algorithm. SIAM Rev. **40**(3), 706–709 (1998)
24. Helmberg, G., Wagner, P., Veltkamp, G.: On Faddeev-Leverrier's method fort the computation of the characteristic polynomial of a matrix and of eigenvectors. Linear Algebra Its Appl. **185**, 219–233 (1993)
25. Barnett, S.: Leverrier's algorithm: a new proof and extensions. SIAM J. Matrix Anal. Appl. **10**, 551–556 (1989)
26. Keller-Gehrig, W.: Fast algorithms for the characteristic polynomial. Theor. Comput. Sci. **36**, 309–317 (1985)
27. Preparata, F., Et Sarwate, D.: An improved parallel processor bound in fast matrix inversion. Inf. Process. Lett. **7**(3), 148–150 (1978)
28. Pernet, C.: Algèbre linéaire exacte efficace: le calcul du polynôme caractéristique, PhD Université Joseph Fourier, 27 (2006)
29. Eriksen, P.S.: Geodesics connected with the fisher metric on the multivariate normal manifold. Technical report, 86-13; Inst. of Elec. Sys., Aalborg University (1986)
30. Eriksen, P.S.: Geodesics connected with the Fisher metric on the multivariate normal manifold. In Proceedings of the GST Workshop, Lancaster, UK, 28–31 October 1987
31. Moler, C.B., van Loan, C.F.: Nineteen dubious ways to compute the exponential of a matrix. SIAM Rev. **20**, 801–836 (2003)
32. Hochbruck, M., Lubich, C., Selhofer, H.: Exponential integrators for large systems of differential equations. SIAM J. Sci. Comput. **19**(5), 1552–1574 (1998)
33. Iserles, A., Zanna, A.: Efficient computation of the matrix exponential by generalized polar decompositions. SIAM J. Numer. Anal. **42**(5), 2218–2256 (2005)
34. Celledoni, E., Iserles, A.: Approximating the exponential from a Lie algebra to a Lie group. Math. Comput. **69**, 1457–1480 (2000)
35. Celledoni, E., Iserles, A.: Methods for the approximation of the matrix exponential in a Lie-algebraic setting. IMA J. Numer. Anal. **21**, 463–488 (2001)
36. Leite, F.S., Crouch, P.: Closed forms for the exponential mapping on matrix Lie groups based on Putzer's method. J. Math. Phys. **40**(7), 3561–3568 (1999)
37. Lewis, D., Olver, P.J.: Geometric integration algorithms on homogeneous manifolds'. Found. Comput. Math. **2**, 363–392 (2002)
38. Munthe-Kaas, H., Quispel, R.G.W., Zanna, A.: Generalized polar decompositions on Lie groups with involutive automorphisms. Found. Comput. Math. **1**(3), 297–324 (2001)
39. Saad, Y.: Analysis of some Krylov subspace approximations to the matrix exponential operator. SIAM J. Numer. Anal. **29**, 209–228 (1992)
40. Zanna, A.: Recurrence relation for the factors in the polar decomposition on Lie groups. Technical report, Report no. 192, Dep. of Infor., Univ. of Bergen, Math. Comp. (2000)
41. Zanna, A., Munthe-Kaas, H.Z.: Generalized polar decompositions for the approximation of the matrix exponential'. SIAM J. Matrix Anal. **23**(3), 840–862 (2002)
42. Nobari, E., Hosseini, S.M.: A method for approximation of the exponential map in semidirect product of matrix Lie groups and some applications. J. Comput. Appl. Math. **234** (1), 305–315 (2010)

Geometry of Tensor-Valued Data

ℝ-Complex Finsler Information Geometry Applied to Manifolds of Systems

Christophe Corbier[(✉)]

Université de Lyon, UJM-Saint-Etienne, LASPI, IUT de Roanne,
42334 Roanne, France
christophe.corbier@univ-st-etienne.fr

Abstract. In this article information geometry is applied to the models of linear discrete time-invariant systems $\mathcal{M}_S^{LTI}$ with external input and zeros outside the unit disc of the z-plane. A new information geometry of manifolds of systems based ℝ-Complex Finsler spaces with three metric tensors is presented. First, two metric tensors in Hermitian spaces $\mathcal{H}_\delta$ where $\mathcal{H}_{-1}$ corresponds to the zeros outside the unit disc (*exogenous zeros*) and $\mathcal{H}_{+1}$ to the zeros and poles inside the unit disc (*endogenous zeros-poles*). Then, one metric tensor as a mixing of *exogenous zeros* and *endogenous zeros-poles* in a non-Hermitian space $\overline{\mathcal{H}}$. Experimental results are presented from a semi-finite acoustic waves guide.

Keywords: Manifolds of systems · ℝ-Complex Finsler spaces · System-model structure · Exogenous zeros

1 Introduction

Information geometry has links with many engineering domains such as robotic, acoustic, mechanical, signal processing or automatic control [14,16,17,19]. These authors applied information geometries to AR, MA, ARMA or ARFIMA models with success to their analyzes and different points of view. But Amari in [5] introduced the differential geometry of a parametric family of invertible linear systems with Riemannian metric, dual affine connections, divergences and the differential-geometrical methods to statistics in [6]. Chapter 5 in [7] developed the geometry of time series and linear systems. Information properties of parameter estimation in spectral analysis of stationary time series based geometrical framework have been studied in ARMA models [12]. Using linear and nonlinear observation models, in [2] the authors developed fundamental sample complexity bounds in order to recover sparse and structured signals. In the same vein, based on high-dimensional data applied to generalized linear models, the author in [1] proposed a wide class of model selection criteria based on penalized maximum likelihood. Among manifolds in information geometry, there exist Kahlerian manifolds as an interesting topic in many domains. On a Kahlerian manifold, the tensor metric and the Levi-Civita connection are determined from a Kahlerian potential [20] and the Ricci tensor is yielded from the determinant

© Springer Nature Switzerland AG 2019
F. Nielsen and F. Barbaresco (Eds.): GSI 2019, LNCS 11712, pp. 99–106, 2019.
https://doi.org/10.1007/978-3-030-26980-7_11

of the metric tensor. It is well known that every Kahlerian metric tensor can be written as $a_{\mu\bar\nu}(\boldsymbol{Z}, \overline{\boldsymbol{Z}}) = \frac{\partial^2 \mathcal{K}(\boldsymbol{Z}, \overline{\boldsymbol{Z}})}{\partial Z^\mu \partial \bar{Z}^\nu}$ with respect to local complex coordinates $\boldsymbol{Z} = (Z^\mu)$, $(\mu, \nu = 1, ..., n)$, where $\mathcal{K}(\boldsymbol{Z}, \overline{\boldsymbol{Z}})$ is a real valued function as Kahler potential. Kahlerian manifolds are present in many domains such as mathematics and theoretical physics, and its applications to information geometry for time series models have been introduced by Barbaresco [8]. A fundamental work in [9] has been carried out related to Kahlerian information geometry for signal processing. The authors proved that information geometry of a signal filter with a finite complex cepstrum norm was a Kahlerian manifold but only focused on a signal filter with *minimum-phase* (mP), that is all zeros/poles of AR and ARMA discrete transfer functions inside the unit disc.

In this paper, the study is extended and focused on the models of linear discrete time-invariant systems $\mathcal{M}_S^{LTI}$ with external input and zeros outside the unit disc of the z-plane in discrete transfer functions. Indeed after identification, many estimated system-models present a non null probabality to have some zeros outside the unit disc. Such a system is considered as a *maximum-phase* (MP) system [11]. Only one zero outside the unit disc changes Kahler information geometry into $\mathbb{R}$-Complex Finsler spaces denoted $\mathbb{R}^{\mathbb{C}(\mathcal{F})} = \mathcal{H}_\delta \oplus \overline{\mathcal{H}}$, where $\mathcal{H}_\delta$ are Hermitian spaces ($\delta = \pm 1$) with square metric tensors and $\overline{\mathcal{H}}$ a non-Hermitian space with a non-square metric tensor. Let M be a complex manifold with $\dim_{\mathbb{C}} M = n$, (Z^μ) be a local complex coordinates in a chart (U, φ) and T'M its holomorphic tangent bundle. Let $\mathcal{E}_o = \{1_o, ..., n_o\}$ and $\mathcal{E}_{ip} = \{1_i, ..., m_i, 1_p, ..., s_p\}$ be two sets such that in $\mathcal{H}_\delta$ square metric tensors are

$$c_{\mu\bar\nu}^{(\delta)}(\boldsymbol{Z}, \overline{\boldsymbol{Z}}) = \frac{\alpha_{\mu\bar\nu}^{(\delta)}(\mathcal{S}_\mathcal{M})}{1 - (Z^\mu \bar{Z}^\nu)^\delta}, \quad |Z^t| < 1, \ \delta = \pm 1 \tag{1}$$

where (μ, ν) run over $\mathcal{E}_o$ for $\delta = -1$ and $\mathcal{E}_{ip}$ for $\delta = +1$. Likewise the non-square metric tensor in $\overline{\mathcal{H}}$ is

$$a_{\mu\nu}(\boldsymbol{Z}, \overline{\boldsymbol{Z}}) = \frac{\beta_{\mu\nu}(\mathcal{S}_\mathcal{M})}{Z^\mu (Z^\mu - Z^\nu)}, \quad |Z^\mu| > 1, |Z^\nu| < 1 \tag{2}$$

where (μ, ν) run over $\mathcal{E}_o \cup \mathcal{E}_{ip}$. Here $\mathcal{S}_\mathcal{M}$ is a system-model structure with $\alpha_{\mu\bar\nu}^{(\delta)}$ and $\beta_{\mu\nu}$ depending on $\mathcal{S}_\mathcal{M}$.

The remainder of this article is structured as follows. Section 2 describes the $\mathbb{R}$-Complex Finsler spaces. Manifolds of systems in these spaces are presented in Sect. 3. Section 4 focuses on experimental results on a real acoustic system. Conclusions and perspectives are drawn in Sect. 5.

2 $\mathbb{R}$-Complex Finsler spaces

In order to extend the general relativity to electromagnetic field, Finsler was inspired to define a metric form as $F(\boldsymbol{X}, \boldsymbol{Y}) = \sqrt{g_{\mu\nu}(\boldsymbol{X}, \boldsymbol{Y}) \frac{dX^\mu}{dt} \frac{dX^\nu}{dt}}$ with $\boldsymbol{Y} = (Y^\mu)$, $Y^\mu = \frac{dX^\mu}{dt}$, $\mu = 1...n$ where $\dim_{\mathbb{R}} M = n$. A direct consequence is the dependence of $g_{\mu\nu}(\boldsymbol{X}, \boldsymbol{Y})$ of the parameter $t \in \mathbb{R}$. An additional condition was required on the homogeneity of F in $\boldsymbol{Y}$, meaning that $F(\boldsymbol{X}, \lambda\boldsymbol{Y}) = \lambda F(\boldsymbol{X}, \boldsymbol{Y})$ for any $\lambda \in \mathbb{R}$. Rizza in [18] extended to the complex case where F :T'M$\to \mathbb{R}_+$ with the homogeneity property $F(\boldsymbol{Z}, \lambda\boldsymbol{\Omega}) = |\lambda| F(\boldsymbol{Z}, \boldsymbol{\Omega})$ for any $\lambda \in \mathbb{C}$. In this approach $(\boldsymbol{Z}, \boldsymbol{\Omega})$ are the coordinates in holomorphic tangent bundle T'M. In complex Finsler geometry the arc length is determinated for the curves which depend on a real parameter $c : t \to (Z^\mu)$ and $\Omega^\mu = \frac{dZ^\mu}{dt}$ and the invariance of the integral to the change of parameters is ensured only for the real parameters. Subsequently there exists a homogeneity condition F :T'M$\to \mathbb{R}_+$ to the real scalars where $F(\boldsymbol{Z}, \lambda\boldsymbol{\Omega}) = |\lambda| F(\boldsymbol{Z}, \boldsymbol{\Omega})$ for any $\lambda \in \mathbb{R}_+$ and $L(\boldsymbol{Z}, \boldsymbol{\Omega}) = F^2(\boldsymbol{Z}, \boldsymbol{\Omega})$ satisties the so called $\mathbb{R}$-homogeneity condition by producing two metric tensors $g_{\mu\nu}(\boldsymbol{Z}) = \frac{\partial^2 L(Z, \Omega)}{\partial\Omega^\mu\partial\Omega^\nu}$ and $g_{\mu\bar{\nu}}(\boldsymbol{Z}) = \frac{\partial^2 L(Z, \Omega)}{\partial\Omega^\mu\partial\bar{\Omega}^\nu}$ [4, 15]. Let M be a complex manifold with dim$_\mathbb{C}$M= n, (Z^μ) be a local complex coordinates in a chart (U, φ) and T'M its holomorphic tangent bundle. There exists a natural structure of complex manifold, dim$_\mathbb{C}$M= $2n$ and the induced coordinates in a local chart on $u \in$ T'M are given by $u = (Z^\mu, \Omega^\mu)$. Here will be considered a continuous function F :T'M$\to \mathbb{R}_+$ depending on $(\boldsymbol{Z}, \boldsymbol{\Omega})$ and $(\bar{\boldsymbol{Z}}, \bar{\boldsymbol{\Omega}})$ as the holomorphic and antiholomorphic coordinates on tangent bundle T'M, respectively. A $\mathbb{R}$-Complex Finsler metric on M is a continuous function F :T'$M \to \mathbb{R}_+$ satisfying (i) $L(\boldsymbol{Z}, \boldsymbol{\Omega}, \bar{\boldsymbol{Z}}, \bar{\boldsymbol{\Omega}}) = F^2(\boldsymbol{Z}, \boldsymbol{\Omega}, \bar{\boldsymbol{Z}}, \bar{\boldsymbol{\Omega}})$ (smooth on T'M), (ii) $F(\boldsymbol{Z}, \boldsymbol{\Omega}, \bar{\boldsymbol{Z}}, \bar{\boldsymbol{\Omega}}) \geq 0$ and (iii) $F(\boldsymbol{Z}, \lambda\boldsymbol{\Omega}, \bar{\boldsymbol{Z}}, \lambda\bar{\boldsymbol{\Omega}}) = \lambda F(\boldsymbol{Z}, \boldsymbol{\Omega}, \bar{\boldsymbol{Z}}, \bar{\boldsymbol{\Omega}})$, $\lambda \in \mathbb{R}_+$. It follows that $L(\boldsymbol{Z}, \boldsymbol{\Omega}, \bar{\boldsymbol{Z}}, \bar{\boldsymbol{\Omega}})$ is $(2, 0)$ homogeneous with respect to λ. On the other hand the function L satisfies Hessian and Levy forms introducing metric tensors $g_{\mu\nu} = \frac{\partial^2 L}{\partial\Omega^\mu\partial\Omega^\nu}; g_{\mu\bar{\nu}} = \frac{\partial^2 L}{\partial\Omega^\mu\partial\bar{\Omega}^\nu}; g_{\bar{\mu}\bar{\nu}} = \frac{\partial^2 L}{\partial\bar{\Omega}^\mu\partial\bar{\Omega}^\nu}$. Consider now $\boldsymbol{Z} \in$ M and $\boldsymbol{\Omega} \in$ T'$_Z$M such that $\boldsymbol{\Omega} = \Omega^\mu \frac{\partial}{\partial\Omega^\mu}$ a section in a holomorphic tangent space where $\mathbb{R}$-Complex Finsler spaces (M, F) are named (α, β)-metrics if F is homogeneous by means of the functions (α, β) investigated in [3] from Randers $(F = \alpha + \beta)$ and Kropina metrics $(F = \alpha^2/\beta)$. Therefore metric tensors are defined as $a_{\mu\nu} = \frac{\partial^2 \alpha^2}{\partial\Omega^\mu\partial\Omega^\nu}$, $c_{\mu\bar{\nu}}^{(-1)} = \frac{\partial^2 \alpha^2}{\partial\Omega_*^\mu\partial\Omega_*^\nu}$ and $c_{\mu\bar{\nu}}^{(+1)} = \frac{\partial^2 \alpha^2}{\partial\Omega^\mu\partial\bar{\Omega}^\nu}$.

3 Manifolds of Systems with External Input and Exogenous Zeros

Consider that a linear discrete time-invariant system $\mathcal{M}_S^{LTI}$ with external input $u(t)$ and additive disturbance $e(t)$ ($\mathbb{E}e(t) = 0, \mathbb{E}e^2(t) = \lambda$) can be described as a linear time-invariant parameter model

$$s(t) = G(q, \boldsymbol{\theta})u(t) + H(q, \boldsymbol{\theta})e(t) \tag{3}$$

where q is the lag operator such that $q^{-1}u(t) = u(t-1)$ and $\boldsymbol{\theta}$ the parameter vector belongs to a compact $D_\mathcal{M}$ as a set of values over which $\boldsymbol{\theta}$ ranges in a linear system model structure $\mathcal{M}_S^{LTI}$. Using z-transformation with $z = e^{i\omega T_s}$ (T_s sampling period), transfer functions G (process model) and H (noise model) become discrete transfer functions $G(z, \boldsymbol{\theta})$ and $H(z, \boldsymbol{\theta})$, respectively. Let $\mathbb{V}_o$ and $\mathbb{W}_{ip}$ be two compacts such that $\mathbb{V}_o = \{Z^\mu/|Z^\mu| > 1, \mu = 1_o, 2_o, ..., n_o\}$ and

$\mathbb{W}_{ip} = \{Z^\nu/|Z^\nu| < 1, \nu = 1_i, 2_i, ..., m_i, 1_p, 2_p, ..., s_p\}$. Let $G_{\mathbb{V}_o,b}(z, \boldsymbol{Z})$ and $G_{\mathbb{W}_{ip},b}(z, \boldsymbol{Z})$ be two discrete transfer functions defined as a backward (MP property) and forward (mP property) system-model filters [11]. For a system with a MP property, transfer function $G(z, \boldsymbol{Z})$ can be written as $G(z, \boldsymbol{Z}) = G_{\mathbb{V}_o,b}(z, \boldsymbol{Z})G_{\mathbb{W}_{ip},f}(z, \boldsymbol{Z})$ with a general form

$$G(z, \boldsymbol{Z}) = |b_1|z^{-1} \prod_{\mu=1}^{n} \left(1 - Z^{\mu_o}z^{-1}\right) \prod_{\mu=1}^{m} \left(1 - Z^{\mu_i}z^{-1}\right) \prod_{\mu=1}^{s} \left(1 - Z^{\mu_p}z^{-1}\right)^{-1} \quad (4)$$

For ARX(Auto Regressive with eXternal input) and OE (Output Error) system-model structures $H(z, \boldsymbol{Z}) = \prod_{\mu=1}^{s}\left(1 - Z^{\mu_p}z^{-1}\right)^{-1}$ and $H(z, \boldsymbol{Z}) = 1$, respectively. See [13]. Now the main order is to transform $G_{\mathbb{V}_o,b}(z, \boldsymbol{Z})$ into $G_{\mathbb{V}_o,f}(z, \boldsymbol{Z})$ with the same frequency response by $G_{\mathbb{V}_o,b}(z, \boldsymbol{Z}) = z^{-n}\bar{G}_{\mathbb{V}_o,f}(\frac{1}{\bar{z}}, \boldsymbol{Z})$ to keep the mP property. After straightforward calculations

$$G_{\mathbb{V}_o,f}(z, \boldsymbol{Z}) = |b_1|z^{1-n} \prod_{\mu=1}^{n} \left(1 - \bar{Z}^{\mu_o}z\right) \quad (5)$$

The previous expression is fundamental to explain and justify $\mathbb{R}$-Complex Finsler spaces information geometry. Transfer function $G(z, \boldsymbol{Z})$ is then a function of $\boldsymbol{Z}$ and $\bar{\boldsymbol{Z}}$ and becomes $G(z, \boldsymbol{Z}, \bar{\boldsymbol{Z}})$. Its associated natural logarithm transfer function is

$$lnG = ln|b_1| + (1-n)lnz + nln(-z) + \sum_{\mu=1}^{n} ln(\bar{Z}^{\mu_o}) + \sum_{\mu=1}^{n} ln(1 - \left(\bar{Z}^{\mu_o}\right)^{-1}z^{-1})$$

$$+ \sum_{\mu=1}^{m} ln(1 - Z^{\mu_i}z^{-1}) - \sum_{\mu=1}^{s} ln(1 - Z^{\mu_p}z^{-1}) \quad (6)$$

Therefore $lnG = f\left(\bar{Z}^{\mu_o}, Z^{\mu_i}, Z^{\mu_p}\right)$, $ln\bar{G} = f\left(Z^{\mu_o}, \bar{Z}^{\mu_i}, \bar{Z}^{\mu_p}\right)$, $lnH = f\left(Z^{\mu_p}\right)$ and $ln\bar{H} = f\left(\bar{Z}^{\mu_p}\right)$. For short notations $\frac{\partial}{\partial Z^\mu} = \partial_\mu$ and $\frac{\partial}{\partial \bar{Z}^\mu} = \partial_{\bar{\mu}}$. Now define the Ψ-function by $\Psi_\mu = \partial_\mu ln\Phi_j^X$ where Φ_j^X is the *joint-power spectral density* (j-psd) and μ run over $\mathcal{E}_o \cup \mathcal{E}_{ip}$ (resp. $\bar{\mathcal{E}}_o \cup \bar{\mathcal{E}}_{ip}$). Here $X = P$ (process) and $X = N$ (noise). Indeed for a signal-model $G \equiv 0$ and the psd is $\Phi^{sig} = \Phi^N = \lambda H\bar{H}$ where λ is the variance of $e(t)$. For a system-model given by (3) $\Phi^{sys} = \Phi^P + \Phi^N = G\bar{G}\Phi_u + \lambda H\bar{H}$ where Φ_u is the psd of the external $u(t)$ and independent of zeros-poles. Consider a general form of the psd as $\Phi^X = \alpha_X \left(\frac{\Phi^A}{2} + \frac{\Phi^B}{2}\right)$ and the j-psd as $\Phi_j^X = \sqrt{\Phi^A \Phi^B}$. For a signal-model $A = N, B = N, \alpha_X = 1$ leading to $\Phi^{sig} = \Phi^N$ and $\Phi_j^{sig} = \sqrt{\Phi^N \Phi^N} = \Phi^N$. Likewise for a system-model $A = P, B = N, \alpha_X = 2$ leading to $\Phi^{sys} = \Phi^P + \Phi^N$ and $\Phi_j^{sys} = \sqrt{\Phi^P \Phi^N}$. From $\Phi^P = G\bar{G}\Phi_u$ and $\Phi^N = \lambda H\bar{H}$ the Ψ-functions for process and noise in holomorphic and anti-holomorphic coordinates are

$$\Psi_\mu^P = \partial_\mu lnG + \partial_{\bar{\mu}}lnG + \partial_\mu ln\bar{G} + \partial_{\bar{\mu}}ln\bar{G}; \Psi_\mu^N = \partial_\mu lnH + \partial_{\bar{\mu}}lnH + \partial_\mu ln\bar{H} + \partial_{\bar{\mu}}ln\bar{H} \quad (7)$$

and the Ψ-function is $\Psi_\mu^{sys} = \frac{1}{2}\Psi_\mu^P + \frac{1}{2}\Psi_\mu^N$. Three conditions on the transfer function of a system-filter are necessary for defining the information geometry of a linear system, that is stability, minimum phase and $\frac{1}{2i\pi}\oint_{\mathbb{D}} |\ln F|^2 \frac{dz}{z} < \infty$ where $F = (G, H)$ and $\mathbb{D}$ the unit disc. Now define the fundamental metric tensor as $h_{\mu\nu} = \frac{1}{2i\pi}\oint_{\mathbb{D}} \Psi_\mu^{sys} \Psi_\nu^{sys} \frac{dz}{z}$ with (μ, ν) run over $\mathcal{E}_o \cup \mathcal{E}_{ip}$ (resp. $\bar{\mathcal{E}}_o \cup \bar{\mathcal{E}}_{ip}$) and given by

$$h_{\mu\nu} = \frac{1}{8i\pi} \oint_{\mathbb{D}} \left(\Psi_\mu^P \Psi_\nu^P + \Psi_\mu^P \Psi_\nu^N + \Psi_\mu^N \Psi_\nu^P + \Psi_\mu^N \Psi_\nu^N \right) \frac{dz}{z} \tag{8}$$

For ARX model the metric tensor is equal to (8) and for OE model $h_{\mu\nu} = \frac{1}{8i\pi}\oint_{\mathbb{D}} \Psi_\mu^P \Psi_\nu^P \frac{dz}{z}$. Equation (7) leads to thirty six expressions gathering nine non-zero terms with their respective conjugates and eighteen zero terms. Five Hermitian metric tensors in $\mathcal{H}_{-1}$ and $\mathcal{H}_{+1}$ and four non-Hermitian in $\bar{\mathcal{H}}$ are given respectively by

$$c_{\mu_o\bar{\nu}_o}^{(-1)} = \frac{\alpha_{\mu_o\bar{\nu}_o}^{(-1)}(\mathcal{S_M})}{1 - (Z_*^{\mu_o}\bar{Z}_*^{\nu_o})^{-1}}; c_{\mu_i\bar{\nu}_i}^{(+1)} = \frac{\alpha_{\mu_i\bar{\nu}_i}^{(+1)}(\mathcal{S_M})}{1 - Z^{\mu_i}\bar{Z}^{\nu_i}}; c_{\mu_p\bar{\nu}_p}^{(+1)} = \frac{\alpha_{\mu_p\bar{\nu}_p}^{(+1)}(\mathcal{S_M})}{1 - Z^{\mu_p}\bar{Z}^{\nu_p}}$$

$$c_{\mu_i\bar{\nu}_p}^{(+1)} = \frac{\alpha_{\mu_i\bar{\nu}_p}^{(+1)}(\mathcal{S_M})}{1 - Z^{\mu_i}\bar{Z}^{\nu_p}}; c_{\mu_p\bar{\nu}_i}^{(+1)} = \frac{\alpha_{\mu_p\bar{\nu}_i}^{(+1)}(\mathcal{S_M})}{1 - Z^{\mu_p}\bar{Z}^{\nu_i}}, \quad Z_*^x = (Z^x)^{-1}, |Z_*^x| < 1 \tag{9}$$

and

$$a_{\mu_o\nu_i} = \frac{\beta_{\mu_o\nu_i}(\mathcal{S_M})}{Z^{\mu_o}(Z^{\mu_o} - Z^{\nu_i})}; a_{\mu_i\nu_o} = \frac{\beta_{\mu_i\nu_o}(\mathcal{S_M})}{Z^{\nu_o}(Z^{\nu_o} - Z^{\mu_i})}$$

$$a_{\mu_o\nu_p} = \frac{\beta_{\mu_o\nu_p}(\mathcal{S_M})}{Z^{\mu_o}(Z^{\mu_o} - Z^{\nu_p})}; a_{\mu_p\nu_o} = \frac{\beta_{\mu_p\nu_o}(\mathcal{S_M})}{Z^{\nu_o}(Z^{\nu_o} - Z^{\mu_p})} \tag{10}$$

Equation (9) leads to (1) in $\mathcal{H}_\delta$ where square matrices C_{-1} and C_{+1} are

$$C_{-1} = \begin{pmatrix} c_{1_o\bar{1}_o}^{(-1)} & c_{1_o\bar{2}_o}^{(-1)} & \cdots & c_{1_o\bar{n}_o}^{(-1)} \\ c_{2_o\bar{1}_o}^{(-1)} & c_{2_o\bar{2}_o}^{(-1)} & \cdots & c_{2_o\bar{n}_o}^{(-1)} \\ \vdots & \vdots & \ddots & \vdots \\ c_{n_o\bar{1}_o}^{(-1)} & c_{n_o\bar{2}_o}^{(-1)} & \cdots & c_{n_o\bar{n}_o}^{(-1)} \end{pmatrix}; C_{+1} = \begin{pmatrix} c_{1_i\bar{1}_i}^{(+1)} & \cdots & c_{1_i\bar{m}_i}^{(+1)} & c_{1_i\bar{1}_p}^{(+1)} & \cdots & c_{1_i\bar{s}_p}^{(+1)} \\ \vdots & \ddots & \vdots & \vdots & \ddots & \vdots \\ c_{m_i\bar{1}_i}^{(+1)} & \cdots & c_{m_i\bar{m}_i}^{(+1)} & c_{m_i\bar{1}_p}^{(+1)} & \cdots & c_{m_i\bar{s}_p}^{(+1)} \\ c_{1_p\bar{1}_i}^{(+1)} & \cdots & c_{1_p\bar{m}_i}^{(+1)} & c_{1_p\bar{1}_p}^{(+1)} & \cdots & c_{1_p\bar{s}_p}^{(+1)} \\ \vdots & \ddots & \vdots & \vdots & \ddots & \vdots \\ c_{s_p\bar{1}_i}^{(+1)} & \cdots & c_{s_p\bar{m}_i}^{(+1)} & c_{s_p\bar{1}_p}^{(+1)} & \cdots & c_{s_p\bar{s}_p}^{(+1)} \end{pmatrix} \tag{11}$$

Likewise Eq. (10) leads to (2) in $\bar{\mathcal{H}}$ where the non-square matrix $\bar{A}$ is

$$\bar{A} = \begin{pmatrix} a_{1_o1_i} & \cdots & a_{1_om_i} & a_{1_o1_p} & \cdots & a_{1_os_p} \\ \vdots & \ddots & \vdots & \vdots & \ddots & \vdots \\ a_{n_o1_i} & \cdots & a_{n_om_i} & a_{n_o1_p} & \cdots & a_{n_os_p} \end{pmatrix} \tag{12}$$

For ARX model $\alpha_{\mu\bar{\nu}}^{(-1)}(\mathcal{S}_{\mathcal{M}}^{ARX}) = -1/4$ for $(\mu\nu) = (\mu_o\nu_o)$, $\alpha_{\mu\bar{\nu}}^{(+1)}(\mathcal{S}_{\mathcal{M}}^{ARX}) = (1/4; 1; -1/2; -1/2)$ for $(\mu\nu) = (\mu_i\nu_i; \mu_p\nu_p; \mu_i\nu_p; \mu_p\nu_i)$ and $\beta_{\mu\nu}(\mathcal{S}_{\mathcal{M}}^{ARX}) = (-1/4; -1/4; 1/2; 1/2)$ for $(\mu\nu) = (\mu_o\nu_i; \mu_i\nu_o; \mu_o\nu_p; \mu_p\nu_o)$, respectively. For OE model $\alpha_{\mu\bar{\nu}}^{(-1)}(\mathcal{S}_{\mathcal{M}}^{OE}) = -1/4$ for $(\mu\nu) = (\mu_o\nu_o)$, $\alpha_{\mu\bar{\nu}}^{(+1)}(\mathcal{S}_{\mathcal{M}}^{OE}) = (1/4; 1/4; -1/4; -1/4)$ for $(\mu\nu) = (\mu_i\nu_i; \mu_p\nu_p; \mu_i\nu_p; \mu_p\nu_i)$ and $\beta_{\mu\nu}(\mathcal{S}_{\mathcal{M}}^{OE}) = (-1/4; -1/4; 1/4; 1/4)$ for $(\mu\nu) = (\mu_o\nu_i; \mu_i\nu_o; \mu_o\nu_p; \mu_p\nu_o)$, respectively.

Fig. 1(left) shows the real part of the square metric tensor $c_{\mu_o\bar{\nu}_o}^{(-1)}$ for $Z^{\mu_o} = x^{\mu_o} + iy^{\mu_o}$ with x^{μ_o} and y^{μ_o} varying in the interval range $[-3, +3] \times [-4, +4]$ when $Z^{\bar{\nu}_o}$ is fixed. In the same vein Fig. 1(right) shows the real part of the square metric tensor $c_{\mu\bar{\nu}}^{(+1)}$ for $Z^{\mu} = x^{\mu} + iy^{\mu}$ with x^{μ} and y^{μ} varying in the interval range $[-1, +1] \times [-1, +1]$ when $Z^{\bar{\nu}}$ is fixed. These tensorial manifolds are fundamendal to see all zeros-poles dynamical and trajectories in time for a complex dynamical system. Indeed all system in time has a zeros-poles motion. Their trajectories on *tensorial metric manifolds* will provide informations of the behavior of the $\mathbb{R}$-Complex Finsler metric $L = F^2$, for example the *distance* in time between zeros, poles or between them.

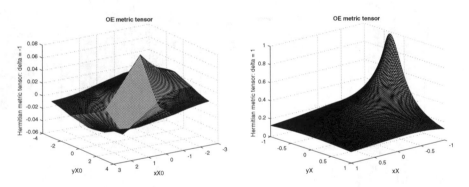

Fig. 1. (left): Tensorial metric manifold of $c_{\mu_o\bar{\nu}_o}^{(-1)}$ for OE model in $[-3, +3] \times [-4, +4]$. (right): Tensorial metric manifold of $c_{\mu\bar{\nu}}^{(+1)}$ for OE model in $[-1, +1] \times [-1, +1]$.

4   Experimental Results

$\mathbb{R}$-Complex Finsler information geometry is applied on a semi-finite acoustic waves guide from system identification using mixed L_p estimators. See [10] for more detail. Figure 2 shows experimental setup including secondary loudspeaker, amplifier, acoustic secondary path, measurement microphone and its amplifier/anti-aliasing filter. This identification has been carried out from a PRBS (Pseudo Random Binary Sequence) as external input signal. Best system-models have been obtained for an OE model structure $(G = B/F)$ where parameter vector is $\boldsymbol{\theta} = (\boldsymbol{\theta}_B \boldsymbol{\theta}_F)^T$ with $\boldsymbol{\theta}_B = (-0.0927, -0.1415, 0.1829, 0.1553, 0.0447, 0.0412, -0.1857)$ and $\boldsymbol{\theta}_F = $

$(-0.0699, -0.7319, 0.0578, 0.0387, -0.0021, 0.3249, -0.2332)$. A zeros-poles analysis shows that there are three exogenous zeros, three endogenous zeros and seven poles leading to $n_o = 3$, $m_i = 2$ and $s_p = 4$. Such a complex dynamical system presents a $\mathbb{R}$-Complex Finsler information geometry where $\mathbb{V}_o = \{Z^{1o} = -2.06, Z^{2o} = -1.18, Z^{3o} = 1.08\}$, $\mathbb{W}_{ip} = \{Z^{1i} = -0.16 + i0.87, Z^{2i} = 0.96, Z^{1p} = -0.9 + i0.34, Z^{2p} = -0.09 + i0.75, Z^{3p} = 0.65 + i0.36, Z^{4p} = 0.76\}$. Subsequently the matrix C_{-1} is

$$C_{-1} = \begin{pmatrix} 0,0765 & 0,1741 & -0,0772 \\ 0,1741 & 0,6416 & -0,1098 \\ -0,0772 & -0,1098 & 1,4617 \end{pmatrix} \tag{13}$$

Each component of C_{-1} is a point of the tensorial metric manifold as depicted in Fig. 1 (left). For example $c_{1_o \bar{1}_o}^{(-1)} = 0,0765$ is on this tensorial manifold with $Z^{1o} = x^{1o} + iy^{1o} = -2.06$ and $\bar{Z}^{1o} = x^{1o} - iy^{1o} = -2.06$.

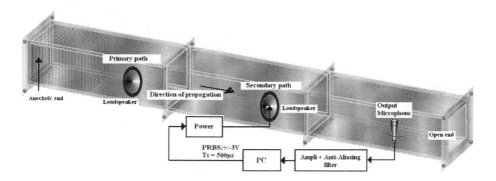

Fig. 2. Experimental setup of acoustic duct for identification.

5 Conclusions and Perspectives

In this paper a new information geometry based $\mathbb{R}$-Complex Finsler spaces on manifolds of systems has been shown from a specific characteristic of discrete transfer functions when some zeros are outside the unit disc. Square and non-square metric tensors have been established and this theory has been applied on a real complex dynamical system. Tensorial metric manifolds have been presented in order to provide informations on trajectories in time t of the zeros-poles. Indeed in the future, their motions in time will give space informations on manifolds from a time (α, β)-metric such that

$$\alpha^2(t) = \frac{1}{2}\left(a_{\mu\nu}(t)\Omega^x(t)\Omega^y(t) + a_{\bar{\mu}\bar{\nu}}(t)\bar{\Omega}^{\mu}(t)\bar{\Omega}^{\nu}(t) + 2c_{\mu\bar{\nu}}^{(\delta)}(t)\Omega^{\mu}(t)\bar{\Omega}^{\nu}(t)\right)$$

and

$$\beta(t) = \frac{1}{2}\left(S_{\mu}(t)\Omega^{\mu}(t) + S_{\bar{\mu}}(t)\bar{\Omega}^{\mu}(t)\right)$$

where $S_\mu(t)$ is the differential entropy in time of each zeros-poles and $\Omega^\mu(t) = \frac{dZ^\mu(t)}{dt}$. Each time metric tensor will be inserted in a Randers metric to calculate $g_{\mu\nu}(t)$, their inverse and associated α-connections and curvatures.

References

1. Abramovich, F., Grinshtein, V.: Model selection and minimax estimation in generalized linear models. IEEE Trans. Inf. Theory **62**(6), 3721–3730 (2016)
2. AksoylarAksoylar, C., Atia, K., Saligrama, V.: Sparse signal processing with linear and nonlinear observations: a unified shannon-theoretic approach. IEEE Trans. Inf. Theory **63**(2), 749–776 (2017)
3. Aldea, N., Purcaru, M.: ℝ-complex finsler spaces with (α, β)-metric. Novi Sad J. Math. **38**(1), 1–9 (2008)
4. Aldea, N., Campean, G.: On some classes of ℝ-complex Hermitian Finsler spaces. J. Korean Math. Soc. **52**(3), 587–601 (2015)
5. Amari, S.: Differential geometry of a parametric family of invertible linear systems-Riemannian metric and dual affine connections and divergences. Math. Syst. Theory **20**, 53–82 (1987)
6. Amari, S.: Differential-Geometrical Methods in Statistics. Springer, Heidelberg (1990)
7. Amari, S., Nagaoka, H.: Methods of Information Geometry. Oxford University Press, Oxford (2000)
8. Barbaresco, F.: Information intrinsic geometric flows. AIP Conf. Proc. **872**, 211–218 (2006)
9. Choi, J., Mullhaupt, A.P.: Kahlerian information geometry for signal processing. Entropy **17**, 1581–1605 (2015)
10. Corbier, C., Carmona, J.-C.: Mixed L_p estimators variety for model order reduction in control oriented system identification. Hindawi Publ. Corp. Math. Probl. Eng. 1–18 (2014). Article ID 349070
11. Haykin, S.: Adaptive Filter Theory, 4th edn. Prentice Hall, Upper Saddle River (2002)
12. Jiang, W., Cheng, S., Xi, Y., Wang, S.: Information properties in spectral analysis of stationary times series. Statistica Sinica **10**, 191–201 (2000)
13. Ljung, L.: System Identification: Theory for the User. Prentice Hall PTR, New York (1999)
14. Martin, R.J.: A metric for ARMA processes. IEEE Trans. Signal Process **48**, 1164–1170 (2000)
15. Munteanu, G., Purcaru, M.: On ℝ-complex Finsler spaces. Balkan J. Geom. Appl. **14**(1), 52–59 (2009)
16. Ravishanker, N., Melnik, E.L., Tsai, C.-L.: Differential geometry of ARMA models. J. Times Ser. Anal. **11**(3), 259–274 (1989)
17. Ravishanker, N.: Differential geometry of ARFIMA processes. Commun. Stati. Theory Methods **30**, 1889–1902 (2001)
18. Rizza, G.B.: Strutture di Finsler di tipo quasi Hermitiato Riv. Mat. Univ. Parma **4**, 83–106 (1963)
19. Van Garderen, K.J.: Exact geometry of autoregressive models. J. Times Ser. Anal. **20**(1), 2–21 (1996)
20. Watanabe, Y.: Kahlerian metrics given by certain smooth potential functions. Kodai Math. J. **5**, 329–338 (1982)

Minkowski Sum of Ellipsoids and Means of Covariance Matrices

Jesús Angulo[✉]

MINES ParisTech, PSL-Research University,
CMM-Centre de Morphologie Mathématique, Fontainebleau, France
jesus.angulo@mines-paristech.fr

Abstract. The Minkowski sum and difference of two ellipsoidal sets are in general not ellipsoidal. However, in many applications, it is required to compute the ellipsoidal set which approximates the Minkowski operations in a certain sense. In this study, an approach based on the so-called ellipsoidal calculus, which provides parameterized families of external and internal ellipsoids that tightly approximate the Minkowski sum and difference of ellipsoids, is considered. Approximations are tight along a direction l in the sense that the support functions on l of the ellipsoids are equal to the support function on l of the sum and difference. External (resp. internal) support function-based approximation can be then selected according to minimal (resp. maximal) measures of volume or trace of the corresponding ellipsoid. The connection between the volume-based approximations to the Minkowski sum and difference of two positive definite matrices and their mean using their Euclidean or Riemannian geometries is developed, which is also related to their Bures-Wasserstein mean.

Keywords: Minkowski sum · Ellipsoid calculus · Positive definite matrices

1 Introduction

For any pair of sets X and Y, $X, Y \subset \mathbb{R}^n$, their Minkowski sum (or addition) $\oplus$ and Minkowski difference (or subtraction) $\ominus$ are defined as follows:

$$X \oplus Y = \bigcup_{y \in Y} X_y = \{x + y \ : \ x \in X, y \in Y\} = \{p \in E \ : \ X \cap \check{Y}_p \neq \emptyset\}, \quad (1)$$

$$X \ominus Y = \bigcap_{y \in Y} X_{-y} = \{p \in E \ : \ Y_p \subset X\} = \{x \ : \ \forall p \in \check{Y}, x \in X_p\}. \quad (2)$$

These set operations are fundamental in mathematical morphology [5], since set dilation and erosion of set X by structuring element B are just defined respectively as $\delta_B(X) = X \oplus B$ and $\varepsilon_B(X) = X \ominus B$. The space of convex sets is closed under Minkowski sum and difference. In this paper, we deal with the

© Springer Nature Switzerland AG 2019
F. Nielsen and F. Barbaresco (Eds.): GSI 2019, LNCS 11712, pp. 107–115, 2019.
https://doi.org/10.1007/978-3-030-26980-7_12

particular covex case of ellipsoids and the Minkowski sum and difference of two ellipsoidal sets are in general not ellipsoidal. However, in many applications, we are interested in computing the ellipsoidal set which approximates in a certain sense the Minkowski operations between them since the ellipsoids represent a positive (semi-)definite symmetric, a matrix covariance matrix, a Riemannian metric, etc. Indeed, ellipsoidal sets appear nowadays in different imaging techniques, e.g., structure tensor images or DTI. In data analysis, the dispersion of a scatter set of points can be described by a multivariate Gaussian distribution where the covariance matrix may be seen as an ellipsoidal shape centered at the mean position. Ellipsoids are usually taken as canonical sets because they: (i) can be concisely described using matrices interpretable as covariance matrices; (ii) provide a satisfactory approximation of convex sets in most applications; (iii) are invariant under affine transformations.

A classical way to solve the problem will be to, firstly, to compute convex set S corresponding to the Minkowski sum (resp. difference) of two ellipsoids; secondly, to compute the minimum volume ellipsoid that contains S, also called the Löwner-John ellipsoid (resp. maximum volume ellipsoid that lies inside a bounded convex set). Both constrained sets are convex semidefinite programming problems which therefore can be solved using classical techniques from convex optimization. Using this approach, little can be said about the set properties of such an approximation to Minkowski sum and difference. In this study, a different approach based on the so-called *ellipsoidal calculus* [4] is adopted, which is a method for solving problems in control and estimation theory, having unknown but bounded errors in terms of sets of approximating ellipsoidal-value functions. From ellipsoidal calculus (explicit) parameterized families of external and internal ellipsoids that tightly approximate the Minkowski sum and difference of ellipsoids are well formulated. It is also possible to select optimal approximations according to a given criterium. Here we focus in particular on those optimal ellipsoids according to volume.

There are classical results on the topological equivalence between the space of ellipsoids endowed with the Hausdorff metric and the space of their shape matrices endowed with the spectral metric. The goal of this paper is to state another more explicit connection between some particular approximations to the Minkowski sum and difference of ellipsoids and some means between their shape matrices.

2 Basic Notions on Elipsoidal Space

Let us assume that everything takes place in the Euclidean space $\mathbb{R}^n$. Let $\mathbb{P}(n)$ be the set of positive semidefinite (psd) matrices of size $n \times n$. An ellipsoid, noted by $\mathcal{E}(c, Q)$, in $\mathbb{R}^n$, with center $c \in \mathbb{R}^n$ and shape matrix $Q \in \mathbb{P}(n)$ is the set

$$\mathcal{E}(c, Q) = \left\{ x \in \mathbb{R}^n \ : \ \langle (x - c), Q^{-1}(x - c) \rangle \leq 1 \right\}.$$

Geometrically, an ellipsoid can also be defined as a translated and deformed version of the unit sphere B_1 of $\mathbb{R}^n$, i.e., $\mathcal{E}(c, Q) = c + Q^{1/2}B_1$. By this parametrization, it is obvious that there is a one-to-one correspondence between ellipsoids and points of the product space $(c, Q) \in \mathbb{R}^n \times \mathbb{P}(n)$.

Hausdorff Distance and Support Function of Ellipsoids. The set of subsets of $\mathbb{R}^n$ can be metrized by the Hausdorff distance. More precisely, given two non-empty sets $X, Y \subset \mathbb{R}^n$, their Hausdorff distance $d_{\mathrm{H}}(X, Y)$ can be defined by means of the Minkowski sum as

$$d_{\mathrm{H}}(X, Y) = \inf\{\lambda \geq 0 \,:\, X \subseteq Y \oplus B_\lambda \text{ and } Y \subseteq X \oplus B_\lambda\},$$

where B_r is the ball of radius r of $\mathbb{R}^n$.

The support function h_A is a tool for a dual representation of the set as the intersection of half-spaces. The support function $h_A \colon \mathbb{R}^n \to \mathbb{R}$ of a non-empty closed convex set $A \in \mathbb{R}^n$ is given by

$$h_A(x) = \sup\{\langle x, a\rangle,\ a \in A\}, \quad x \in \mathbb{R}^n,$$

and it is a real valued, continuous and convex function, satisfiying many relevant properties. In particular, one has:

$$h_{\alpha A + b}(x) = \alpha h_A(x) + \langle x, b\rangle, \qquad \alpha \geq 0; x,\ b \in \mathbb{R}^n.$$

The Hausdorff distance $d_{\mathrm{H}}(A, B)$ of two nonempty compact convex sets A and B can be expressed in terms of their support functions:

$$d_{\mathrm{H}}(A, B) = \sup\left\{|h_A(x) - h_B(x)| \,:\, \|x\| = 1\right\},$$

which uses the uniform norm on the unit sphere.

For our particular case, the support function of an ellipsoid $\mathcal{E}(c, Q)$ is just given by

$$h_{\mathcal{E}(c,Q)}(x) = \langle x, c\rangle + \langle x, Qx\rangle^{1/2}.$$

Therefore, given two ellipsoids, $\mathcal{E}(c_1, Q_1)$ and $\mathcal{E}(c_2, Q_2)$, the Hausdorff distance between them is

$$d_{\mathrm{H}}(\mathcal{E}(c_1, Q_1), \mathcal{E}(c_2, Q_2)) = \sup_{\|x\|=1}\left\{|h_{\mathcal{E}(c_1,Q_1)}(x) - h_{\mathcal{E}(c_2,Q_2)}(x)|\right\}$$

$$= \sup_{\|x\|=1}\left\{|\langle x, c_1\rangle - \langle x, c_2\rangle + \langle x, Q_1x\rangle^{1/2} - \langle x, Q_2x\rangle^{1/2}|\right\}.$$

It seems clear that for metric purposes, it will be sufficient to study ellipsoids centered at the origin.

Remark on the Topology of the Space of Ellipsoids. Let us review the main result by Goffin and Hoffman [3] on the relationship between the Hausdorff distance and the matrix distance of ellipsoids. Firstly, in order to simplify the notation and avoding the term $1/2$ in later expressions, we introduce the following change of variable: $Q \mapsto P = Q^{1/2}$, $P \in \mathbb{P}(n)$. Let $\mathcal{E}(c_1, P_1)$ and $\mathcal{E}(c_2, P_2)$ be two ellipsoids in $\mathbb{P}^n$. Then, define the so-called spectral distance as follows

$$d_{Spectral}(\mathcal{E}(\mathbf{0}, P_1), \mathcal{E}(\mathbf{0}, P_2)) = \|P_1 - P_2\|_S,$$

where $\| \cdot \|_S$ is the matrix spectral norm, i.e.,

$$\|A\|_S = \sup\{\|Ax\| : x \in \mathbb{R}^n \text{ with } \|x\| = 1\} = \sqrt{\lambda_{\max}(A^T A)}.$$

As discussed above on Hausdorff distance, it is sufficient to study ellipsoids centered at the origin. In that case, one has

$$d_{Spectral}(\mathcal{E}(\mathbf{0}, P_1), \mathcal{E}(\mathbf{0}, P_2)) = \sup_{\|x\|=1} \{\|(P_1 - P_2)x\|\},$$

$$d_{Hausdorff}(\mathcal{E}(\mathbf{0}, P_1), \mathcal{E}(\mathbf{0}, P_2)) = \sup_{\|x\|=1} \{|\, \|P_1 x\| - \|P_2 x\| \,|\}.$$

Now, the fundamental result is as follows,

Theorem 1 (Goffin and Hoffman, 1983 [3]). *Let $\mathcal{E}(\mathbf{0}, P_1)$ and $\mathcal{E}(\mathbf{0}, P_2)$ be two centred ellipsoids in $\mathbb{R}^n$, with $P_1, P_2 \in \mathbb{P}(n)$. Then*

$$k_n^{-1} d_S(\mathcal{E}(\mathbf{0}, P_1), \mathcal{E}(\mathbf{0}, P_2)) \le d_H(\mathcal{E}(\mathbf{0}, P_1), \mathcal{E}(\mathbf{0}, P_2)) \le d_S(\mathcal{E}(\mathbf{0}, P_1), \mathcal{E}(\mathbf{0}, P_2))$$

$$d_H(\mathcal{E}(\mathbf{0}, P_1), \mathcal{E}(\mathbf{0}, P_2)) \le d_S(\mathcal{E}(\mathbf{0}, P_1), \mathcal{E}(\mathbf{0}, P_2)) \le k_n d_H(\mathcal{E}(\mathbf{0}, P_1), \mathcal{E}(\mathbf{0}, P_2))$$

where $k_n = 2\sqrt{(2)} n (n + 2)$.

Inequalities from Theorem 1 imply that the two metrics define the same topology on the space of ellipsoids, but, more strongly, the rates of convergence of a sequence of ellipsoids may be studied within a space of sets, or within a space of matrices. In fact, both rates are identical. Additionally, that means that the Hausdorff distance for ellipsoids is essentially a spectral matrix distance.

3   Ellipsoidal Approximations to Minkowski Sum and Difference of Ellipsoids

Let parameter l be a direction in $\mathbb{R}^n$, $l \in \mathbb{S}^{n-1}$. Given two ellipsoids $\mathcal{E}(c_1, Q_1)$ and $\mathcal{E}(c_2, Q_2)$, the external and internal ellipsoidal approximation to their Minkowski sum according to direction l, noted respectively by $\mathcal{E}(c^\oplus, Q_l^{\oplus,+})$ and $\mathcal{E}(c^\oplus, Q_l^{\oplus,-})$, are tight along the direction l in the sense that the value of support functions at l are equal:

$$h_{\mathcal{E}(c^\oplus, Q_l^{\oplus,-})}(\pm l) = h_{\mathcal{E}(c_1,Q_1)\oplus\mathcal{E}(c_2,Q_2)}(\pm l) = h_{\mathcal{E}(c^\oplus, Q_l^{\oplus,+})}(\pm l).$$

The center of both approximations is just the vector sum, i.e., $c^\oplus = c_1 + c_2$. In the case of the internal ellipsoid, the shape matrix is given by [4]:

$$Q_l^{\oplus,-} = \left(Q_1^{1/2} + S Q_2^{1/2} \right)^T \left(Q_1^{1/2} + S Q_2^{1/2} \right),$$

with matrix S being orthogonal and vectors $Q_1^{1/2}l$ and $SQ_2^{1/2}l$ are parallel. The shape matrix of the external ellipsoid is given by [4]

$$Q_l^{\oplus,+} = \left(1 + p^{-1}\right) Q_1 + (1 + p)\, Q_2, \quad p > 0,$$

where $p = \langle l, Q_1 l\rangle^{1/2} / \langle l, Q_2 l\rangle^{1/2}$.

Unlike the Minkowski sum, ellipsoidal approximations for the Minkowski difference do not exist for every direction l. Similar internal and external approximation for valid directions can be defined in the context of ellipsoidal calculus. See [4] for the expressions.

3.1   Volume-Based Optimal Approximations

From these expressions, it is possible to find the direction l such as the corresponding ellipsoids will be optimal according to a given criterion, typically the trace or the volume (i.e., related to the determinant), minimal for the external or maximal for the internal approximations [4]. Let us focus in particular on the approximations of optimal volume. There is a unique ellipsoid of maximal volume contained in the Minkowski sum and its shape matrix is given by [4]

$$Q_{\max vol}^{\oplus,-} = Q_1 + Q_2 + 2Q_2^{1/2} \left[Q_2^{-1/2} Q_1 Q_2^{-1/2}\right]^{1/2} Q_2^{1/2}. \tag{3}$$

Similarly, there is a unique ellipsoid of minimal volume contained in the Minkowski difference and its shape matrix is given by

$$Q_{\min vol}^{\ominus,+} = Q_1 + Q_2 - 2Q_2^{1/2} \left[Q_2^{-1/2} Q_1 Q_2^{-1/2}\right]^{1/2} Q_2^{1/2}. \tag{4}$$

4   Means on Space of $\mathbb{P}(n)$ and Minkowski Sum and Difference

We discuss in this section an interpretation of the approximations to Minkowski sum and difference in terms of the means of the corresponding shape matrices.

4.1   Means in Two Riemannian Geometries on $\mathbb{P}(n)$

The standard Riemannian metric distance for any A, B in $\mathbb{P}(n)$ is given by [1]

$$d_{Riemannian}(A, B) = \left\| \log\left(A^{-1/2} B A^{-1/2}\right) \right\|_2.$$

Associated to this distance, the space $\left(\mathbb{P}(n), ds_{Riem}^2\right)$ is a Riemannian manifold where the local metric is the natural metric in the cone:

$$ds_{Riem}^2 = \mathrm{Tr}\left(Q^{-1} dQ Q^{-1} dQ\right).$$

Any two points $A, B \in \mathbb{P}(n)$ can be joined by a unique geodesic with respect to this metric:

$$\gamma_{A,B}^{Riem}(t) = A^{1/2} \left(A^{-1/2} B A^{-1/2} \right)^t A^{1/2}, \ 0 \le t \le 1.$$

The geometric mean $M_{Riemannian}(A, B)$ between matrices A and B is evidently the midpoint of this geodesic, i.e.,

$$M_{Riemannian}(A, B) = \gamma_{A,B}^{Riem}(0.5) = A^{1/2} \left(A^{-1/2} B A^{-1/2} \right)^{1/2} A^{1/2}. \quad (5)$$

This geometry and the mean for the case of two covariance matrices are well known in information geometry. It corresponds to that of the Fisher metric for the case of Gaussian densities of zero-mean and covariance given by the psd matrix. This mean is symmetric in A and B. In fact, it is a kind of symmetrization of the equivalent geometric mean $(ab)^{1/2}$ for matrices, since in general $AB \ne BA$. In $\mathbb{P}(n)$, the matrix AB has positive eigenvalues and it has a unique square root $(AB)^{1/2}$ that has positive eigenvalues. The eigenvalues of AB are the same as those of BA One has $M_{Riemannian}(A, B) = A \left(A^{-1} B \right)^{1/2} = \left(AB^{-1} \right)^{1/2} B$. Thus, one also has [2]

$$(AB)^{1/2} = A M_{Riemannian}(A^{-1}, B) = A^{1/2} \left(A^{1/2} B A^{1/2} \right)^{1/2} A^{-1/2}. \quad (6)$$

Given A, B in $\mathbb{P}(n)$, the Bures metric distance (in quantum information) and the Wasserstein metric distance (in optimal transport) is

$$d_{Bures-Wasser}(A, B) = \left[\operatorname{Tr} A + \operatorname{Tr} B - 2 \operatorname{Tr} \left(A^{1/2} B A^{1/2} \right)^{1/2} \right]^{1/2}.$$

If A and B are diagonal matrices (vectors), then $d_{Bures-Wasser}(A, B)$ reduces to the Hellinger distance between probability distributions. In quantum theory, a density matrix (or state) is a psd matrix A with $\operatorname{Tr} A = 1$. Bures distance for density matrices is the particular case of $d_{Bures-Wasser}(A, B)$. It corresponds to the $2-$Wasserstein distance between two Borel probability measures μ and ν in $\mathbb{R}^n$, when μ and ν are zero-mean Gaussian measures with covariance matrices A and B.

Bures–Wasserstein distance and the underlying Rieamannian geometry has been recently studied in a deep and illuminating perspective in [2]. The geodesic joining A and B in the Bures-Wassertein metric space is:

$$\gamma_{A,B}^{B-W}(t) = (1-t)^2 A + t^2 B + t(1-t) \left[(AB)^{1/2} + (BA)^{1/2} \right]$$

$$= A^{-1/2} \left[(1-t)A + t(A^{1/2} B A^{1/2})^{1/2} \right]^2 A^{-1/2}, \ 0 \le t \le 1.$$

Therefore, using $t = 0.5$ in this geodesic and the equality (6), the Bures–Wassertein mean of A and B is

$$
\begin{aligned}
M_{Bures-Wasser}(A, B) &= \frac{1}{4}\left[A + B + (AB)^{1/2} + (BA)^{1/2}\right] \\
&= \frac{1}{2}\left[M_{Euclidean}(A, B) + \frac{1}{2}\left[A M_{Riemannian}(A^{-1}, B) \right.\right. \\
&\quad \left.\left. + B M_{Riemannian}(A, B^{-1})\right]\right]
\end{aligned}
$$

where $M_{Euclidean}(A, B) = \frac{A+B}{2}$ is just the Euclidean (Frobenious norm-based) mean of two matrice in the flat space.

4.2 Optimal Approximations to Minkowski Sum and Difference in Terms of Means

Using the notation of the Euclidean and Riemannian means, it obvious that the internal approximation to the Minkowski sum of maximal volume (3) can be just rewritten as:

$$
Q^{\oplus,-}_{\max vol} = 2\left[M_{Euclidean}(Q_1, Q_2) + M_{Riemannian}(Q_1, Q_2)\right],
$$

and similarly for the external approximation to the Minkowski difference of minimal volume (4):

$$
Q^{\ominus,+}_{\min vol} = 2\left[M_{Euclidean}(Q_1, Q_2) - M_{Riemannian}(Q_1, Q_2)\right].
$$

Therefore, one has

$$
M_{Euclidean}(Q_1, Q_2) = \frac{1}{4}\left[Q^{\oplus,-}_{\max vol} + Q^{\ominus,+}_{\min vol}\right], \tag{7}
$$

$$
M_{Riemannian}(Q_1, Q_2) = \frac{1}{4}\left[Q^{\oplus,-}_{\max vol} - Q^{\ominus,+}_{\min vol}\right]. \tag{8}
$$

Euclidean and Riemannian means of covariance matrices are consequently related to the Minkowski sum and difference of the corresponding ellipsoids. This result is not surprising since as we have discussed, the topology of both spaces are equivalent. However, we can observe that the relationship is straightforward in this very particular case.

Furthermore, we can notice that in the case where the matrix product commute, i.e., $AB = BA$, which involves $(AB)^{1/2} = M_{Riemannian}(A, B)$, one just has

$$
Q^{\oplus,-}_{\max vol} = 4 M_{Bures-Wasser}(A, B).
$$

A sufficient condition for product commutation is that two matrices are simultaneously diagonalizable. In the case of ellipsoids, it corresponds to the case when they are aligned, i.e., they have the same orientation axis.

4.3 A Riemannian Product Space

For the sake of understanding, let us precise that $Q^{\oplus,-}_{\max\,vol}$ does not correspond to the midpoint on a geodesic space product of two copies of $\mathbb{P}(n)$ with the Euclidean and Riemannian metrics. Let us consider the Riemannian manifolds $\left(\mathbb{P}(n), ds^2_{Euclid}\right)$ and $\left(\mathbb{P}(n), ds^2_{Riem}\right)$, where the flat metric is just $ds^2_{Euclid} = dQ^2$.

Let us consider now the space $\mathbb{P}(2n)$, where, on the one hand, for each matrix $Q \in \mathbb{P}(n)$, a map associates it to the matrix $Q^{\times} \in \mathbb{P}(2n)$ and, on the other hand, the Riemannian metric $g_{\times} = \alpha g_{Euclid} \otimes \beta g_{Rieman}$, $\alpha, \beta > 0$, which are respectively given by

$$Q \mapsto Q^{\times} = \begin{pmatrix} Q & \mathbf{0}_{n \times n} \\ \mathbf{0}_{n \times n} & Q \end{pmatrix} \quad g_{\times} = \begin{pmatrix} \alpha g_{Euclid} & \mathbf{0} \\ \mathbf{0} & \beta g_{Rieman} \end{pmatrix}.$$

Note that one has $ds^2_{\times} = \alpha ds^2_{Euclid} + \beta ds^2_{Riem}$. Let $A^{\times}$ and $B^{\times}$ be two different points in this product manifold $\left(\mathbb{P}(2n), ds^2_{\times}\right)$. In this manifold, the Riemannian distance between two points $A^{\times}$ and $B^{\times}$ is given by

$$d_{\times}(A^{\times}, B^{\times})^2 = \alpha d_{Euclidean}(A, B)^2 + \beta d_{Riemannian}(A, B)^2,$$

where $d_{Euclidean}(A, B)^2 = \|A - B\|^2_2$. In the product manifold, the geodesic from $A^{\times}$ and $B^{\times}$ is given by

$$\gamma^{\times}_{A,B}(t) = \mathrm{diag}\left(\gamma^{Euclid}_{A,B}\left(\frac{\beta d_{Riemannian}(A, B)}{d_{\times}(A, B)} t\right), \gamma^{Riem}_{A,B}\left(\frac{\alpha d_{Euclidean}(A, B)}{d_{\times}(A, B)} t\right)\right),$$

the scaling of the arc lenght is evident since the lenght of both geodesics is different. In conclusion, the geometry of the space associated to $Q^{\oplus,-}_{\max\,vol}$ as the midpoint of a geodesic is not the trivial product of Euclidean and Riemannian geometry. In any case, since the tangent space of the product manifold $T_{Q^{\times}}\mathbb{P}(2n) = T_{Q,Euclid}\mathbb{P}(2) \otimes T_{Q,Rieman}\mathbb{P}(2)$, the exponential maps of the corresponding spaces can be used to deal with the tangent spaces.

5 Conclusions and Perspectives

Ellipsoidal approximations to Minkowski sum and difference are based on the approximation in terms of the support function, which is merely related to approximation in terms of Hausdorff distance. The corresponding metric space of ellipsoids is equivalent to the spectral space of their shape matrices. For a very particular case of optimal approximated ellipsoids in terms of their volume, this equivalence leads to an explicit interpretation based on the mean of the two ellipsoids in two different geometries. Some questions about the underlying Riemannian geometry are still open and deserves additional work. The interest of these approximations to Minkowski sum and difference of ellipsoids in tensor-valued image processing tasks, typically regularization and interpolation, will be explored in ongoing work.

References

1. Bhatia, R.: Positive Definite Matrices. Princeton University Press, Princeton (2007)
2. Bhatia, R., Jain, T., Lim, Y.: On the Bures-Wassertein distance between positive definite matrices. Expositiones Mathematicae (2019, to appear)
3. Goffin, J.-L., Hoffman, A.J.: On the relationship between the hausdorff distance and matrix distances of ellipsoids. Linear Algebra Appl. **52–53**, 301–313 (1983)
4. Kurzhanski, A.B., Vályi, I.: Ellipsoidal Calculus for Estimation and Control. Birkhäuser, Boston (1996)
5. Matheron, G.: Random Sets and Integral Geometry. Wiley, New York (1975)

Hyperquaternions: An Efficient Mathematical Formalism for Geometry

Patrick R. Girard[1]([✉])[iD], Patrick Clarysse[1][iD], Romaric Pujol[2], Robert Goutte[1],
and Philippe Delachartre[1][iD]

[1] Univ Lyon, INSA-LYON, Université Claude Bernard Lyon 1, UJM-Saint Etienne,
CNRS, Inserm, CREATIS UMR 5220, U1206, 69621 Lyon, France
`patrick.girard@creatis.insa-lyon.fr`
[2] Pôle de Mathématiques, INSA-Lyon, Bât. Léonard de Vinci,
21 avenue Jean Capelle, 69621 Villeurbanne, France

Abstract. Hyperquaternions being defined as a tensor product of
quaternion algebras (or a subalgebra thereof), they constitute Clifford
algebras endowed with an associative exterior product providing an
efficient mathematical formalism for differential geometry. The paper
presents a hyperquaternion formulation of pseudo-euclidean rotations
and the Poincaré groups in n dimensions (via dual hyperquaternions).
A canonical decomposition of these groups is developed as an extension
of an euclidean formalism and illustrated by a $5D$ example. Potential
applications include in particular, moving reference frames and machine
learning.

Keywords: Quaternions · Hyperquaternions ·
Pseudo-euclidean rotations · Poincaré groups ·
Canonical decomposition

1 Introduction

Clifford algebras allow an excellent representation of pseudo-euclidean rotations
which are important symmetry groups of physics [1–4]. A decomposition of these
groups into orthogonal, commuting planar rotations is called a canonical decomposition. Various canonical decompositions have been developed which deal with
either specific rotations or dimensions and are often expressed in terms of matrices [5,6]. In a recent paper, we have introduced a hyperquaternion formulation of
Clifford algebras and applied them to the unitary and unitary symplectic groups
[7]. Here, we consider pseudo-euclidean rotations and the Poincaré groups in
n dimensions (via dual hyperquaternions). A canonical decomposition of these
groups is developed within that framework as an extension of an euclidean formalism introduced by Moore [8,9]. After a short presentation of hyperquaternions and multivectors, we derive the pseudo-euclidean rotations and the canonical decomposition. Then we go on to the Poincaré groups and a $5D$ example.
Potential applications are moving reference frames and machine learning [10].

© Springer Nature Switzerland AG 2019
F. Nielsen and F. Barbaresco (Eds.): GSI 2019, LNCS 11712, pp. 116–125, 2019.
https://doi.org/10.1007/978-3-030-26980-7_13

Table 1. Biquaternion multivector structure

1		$i = e_3e_2$	$j = e_1e_3$	$k = e_2e_1$
$I = e_1e_2e_3$	$Ii = e_1$	$Ij = e_2$	$Ik = e_3$

2 Background: Quaternions, Hyperquaternions and Multivectors

In this section, we briefly introduce quaternions, hyperquaternions and multivectors [7,11–15]. The quaternion algebra $\mathbb{H}$ which contains $\mathbb{R}$ and $\mathbb{C}$ as particular cases is constituted by quaternions

$$a = a_1 + a_2 i + a_3 j + a_4 k \qquad (a_i \in \mathbb{R}) \tag{1}$$

where i, j, k multiply according to

$$i^2 = j^2 = k^2 = ijk = -1, ij = -ji = k, etc. \tag{2}$$

The product of two quaternions a, b is given by

$$ab = (a_1b_1 - a_2b_2 - a_3b_3 - a_4b_4) + (a_1b_2 + a_2b_1 + a_3b_4 - a_4b_3)\, i \tag{3}$$
$$+ (a_1b_3 + a_3b_1 + a_4b_2 - a_2b_4)\, j + (a_1b_4 + a_4b_1 + a_2b_3 - a_3b_2)\, k. \tag{4}$$

The conjugate of a quaternion is $a_c = a_1 - a_2 i - a_3 j - a_4 k$ with

$$aa_c = a_1^2 + a_2^2 + a_3^2 + a_4^2, (ab)_c = b_c a_c \tag{5}$$

The hyperquaternion algebra (over $\mathbb{R}$) is defined as the tensor product of quaternion algebras (or a subalgebra thereof). Examples of hyperquaternion algebras are the quaternions $\mathbb{H}$, tetraquaternions $\mathbb{H} \otimes \mathbb{H}$ and so on $\mathbb{H} \otimes \mathbb{H} \otimes ... \otimes \mathbb{H}$; subalgebras are the complex numbers $\mathbb{C}$, biquaternions $\mathbb{H} \otimes \mathbb{C}$, Dirac algebra $\mathbb{H} \otimes \mathbb{H} \otimes \mathbb{C}$, etc.

Calling (i, j, k) the first quaternionic system, (I, J, K) the second one and (l, m, n) the third one, all systems commuting with each other, one has

$$i \otimes i \otimes i = iIl, \quad i \otimes j \otimes k = iJn, etc. \tag{6}$$

which uniquely defines the multiplication.

Hyperquaternions having n generators e_i such that $e_i e_j + e_j e_i = 0$ $(i \neq j)$, $e_i^2 = \pm 1$ constitute Clifford algebras C_n. The choice of the generators entails a multivector structure as shown, in the case of biquaternions, in Table 1. The 2^n elements of the algebra are composed of scalars, vectors e_i , bivectors $e_i e_j$, trivectors $e_i e_j e_k$ etc. yielding respectively the multivector spaces $V_0, V_1, V_2, V_3, ... V_n$. C^+ is the subalgebra constituted by products of an even number of e_i, C^- is the rest of the algebra. The multivector structure allows to define basic operations like conjugation, duality and the interior and exterior products.

Considering a general element A of the algebra, the conjugate A_c is obtained by replacing the e_i by their opposite $-e_i$ and reversing the order of the elements

$$(A_c)_c = A, (AB)_c = (B_c)(A_c). \qquad (7)$$

The dual of A is $A^* = i_d A$ where $i_d = e_1 \wedge e_2 ... \wedge e_n$ (to be defined below) and the commutator of two hyperquaternions is

$$[A, B] = \frac{1}{2}(AB - BA). \qquad (8)$$

The interior and exterior products of two vectors a, b are obtained as follows. From the identity

$$2ab = \lambda \lambda^{-1}[(ab + ba) + (ab - ba)] \qquad (9)$$

where $\lambda = \pm 1$ is a given coefficient (allowing to eventually change the sign of the metric), one defines

$$2a.b = \lambda^{-1}(ab + ba), 2a \wedge b = \lambda^{-1}(ab - ba) \qquad (10)$$

which are respectively a scalar and a bivector. A multivector $A_p = a_1 \wedge a_2 \wedge ... \wedge a_p$ $(2 \leq p < n)$ where a_p are vectors, is then defined by recurrence

$$2a.A_p = \lambda^{-p}[aA_p - (-1)^p A_p a] \in V_{p-1} \qquad (11)$$
$$2a \wedge A_p = \lambda^{-p}[aA_2 + (-1)^p A_2 a] \in V_{p+1} \qquad (12)$$

By definition, we take

$$A_p.a \equiv (-1)^{p-1} a.A_p, A_p \wedge a \equiv (-1)^p a \wedge A_p. \qquad (13)$$

An important property of the exterior product is its associativity.

Interior and exterior products between multivectors are defined by

$$A_p \wedge B_q = a_1 \wedge (a_2 \wedge ... \wedge a_p \wedge B_q) \qquad (14)$$
$$A_p.B_q = (a_1 \wedge ... \wedge a_{p-1}).(a_p.B_q), \quad (p \leq q) \qquad (15)$$

with $A_p.B_q = (-1)^{p(q+1)} B_q.A_p$ [16]. In particular, we have the following useful formulas where Bi are bivectors and $V_p[A]$ the multivector part V_p of A

$$B_1 B_2 = B_1.B_2 + B_1 \wedge B_2 + [B_1, B_2] \qquad (16)$$
$$B_1 \wedge B_2 = V_4[B_1 B_2] \qquad (17)$$
$$B_1 \wedge B_2 \wedge B_3 = V_6[B_1(B_2 \wedge B_3)] \qquad (18)$$
$$B_1.(B_2 \wedge B_3) = V_2[B_1(B_2 \wedge B_3)] \qquad (19)$$
$$(B_1 \wedge B_2).(B_3 \wedge B_4 \wedge B_5) = V_2[(B_1 \wedge B_2)(B_3 \wedge B_4 \wedge B_5)]. \qquad (20)$$

Hyperquaternions yield all real, complex and quaternionic square matrices as well as the transposition, adjunction and transpose quaternion conjugate via a hyperconjugation defined as $\mathbb{H}_c \otimes \mathbb{H}_c \otimes ... \otimes \mathbb{H}_c$ as indicated in Table 2.

Table 2. Hyperquaternions and matrices

$$
\begin{array}{ll}
\mathbb{H} \otimes \mathbb{H} \simeq m(4, \mathbb{R}) & \mathbb{H}_c \otimes \mathbb{H}_c \simeq [m(4, \mathbb{R})]^t \\
\hline
\mathbb{H} \otimes \mathbb{H} \otimes \mathbb{C} \simeq m(4, \mathbb{C}) & \mathbb{H}_c \otimes \mathbb{H}_c \otimes \mathbb{C}_c \simeq [m(4, \mathbb{C})]^\dagger \\
\mathbb{H} \otimes \mathbb{H} \otimes \mathbb{H} \simeq m(4, \mathbb{H}) & \mathbb{H}_c \otimes \mathbb{H}_c \otimes \mathbb{H}_c \simeq [m(4, \mathbb{H})]^t_c .
\end{array}
\tag{21}
$$

3 Pseudo-Orthogonal Rotations

Here, we derive a hyperquaternion formulation of pseudo-euclidean rotations and develop a canonical decomposition. Historically, the formula of n dimensional euclidean rotations $x' = axa^{-1}$ ($a \in C_n^+$) was given by Lipschitz [17] and Moore developed a canonical decomposition thereof [8,9]. We introduce, as an extension of Moore's method, within the hyperquaternion Clifford algebra framework, a canonical decomposition of pseudo-euclidean rotations and the Poincaré groups. After a brief review of the basic definitions and the Cartan theorem, we develop the canonical decomposition.

3.1 Definitions and Theorem

Let $C_{p,q}$ be a hyperquaternion algebra having $n = p + q$ generators e_i and the quadratic form

$$
x.y = x_1 y_1 + ... + x_p y_p - (x_{p+1} y_{p+1} ... - x_{p+q} y_{p+q})
\tag{22}
$$
$$
= \lambda^{-1} (xy + yx) / 2
\tag{23}
$$

where x, y are vectors ($x = x_i e_i$). A vector x is timelike if $x.x > 0$, spacelike if $x.x < 0$ and isotropic if $x.x = 0$.

An orthogonal symmetry with respect to a plane going through the origin and perpendicular to a unit vector a ($a^2 = \pm 1$) is given by [12,13]

$$
x' = \pm axa
\tag{24}
$$

with $x'x' = (\pm axa)(\pm axa) = xx$.

Definition 1. *The pseudo-orthogonal group $O(p, q)$ is the group of linear operators which leave invariant the form $x \cdot y$.*

Theorem 1. *Every rotation of $O(p, q)$ is the product of an even number $2m \leq n$ of symmetries.*

Definition 2. *The special orthogonal group $SO^+(p, q)$ is constituted by rotations which preserve the orientation of the space of positive norm vectors and the space of negative norm vectors.*

A rotation of $SO^+(p, q)$ can thus be expressed as

$$
x' = axa_c \quad (aa_c = 1)
\tag{25}
$$

with $a = a_1 a_2 ... a_{2m}, \in C^+$, where a_i are unit vectors (with an even number of timelike and spacelike vectors). Developing the product (with $\lambda = 1$)

$$a_i a_j = a_i.a_j + a_i \wedge a_j \tag{26}$$

one sees that it contains a simple plane $B = a_i \wedge a_j$ such that $B^2 = B.B + B \wedge B$ is a scalar since $B \wedge B = 0$. Hence, a rotation involves at most $m \leq n/2$ simple planes. A canonical decomposition of rotations is obtained by choosing these simple planes to be orthogonal.

3.2   Canonical Decomposition

A rotation of $SO^+(p,q)$ can be decomposed as

$$a = e^{\frac{\Phi_1}{2} B_1} e^{\frac{\Phi_2}{2} B_2} ... e^{\frac{\Phi_m}{2} B_m} \quad (aa_c = 1) \tag{27}$$

where B_i are m simple orthogonal commuting planes such that $B_i^2 = \pm 1$ together for $i \neq j$

$$B_i.B_j = 0, B_i B_j = B_j B_i, B_i B_j = B_i \wedge B_j; \tag{28}$$

Φ_i are the angles of rotation within the planes B_i. According to whether $B_i^2 = -1$ or $B_i^2 = 1$, one has respectively

$$e^{\frac{\Phi_i}{2} B_i} = \cos \frac{\Phi_i}{2} + \sin \frac{\Phi_i}{2} B_i, e^{\frac{\Phi_i}{2} B_i} = \cosh \frac{\Phi_i}{2} + \sinh \frac{\Phi_i}{2} B_i. \tag{29}$$

The rotation can be developed as

$$a = S (1 + b_1 B_1) (1 + b_2 B_2) ... (1 + b_m B_m) \tag{30}$$

with $b_i = \tan \frac{\Phi_i}{2}$ (or $\tanh \frac{\Phi_i}{2}$). Since $aa_c = 1$ one has

$$S^2 (1 + b_1^2) (1 - b_2^2) (1 - b_3^2) = 1 \tag{31}$$

$$S = \frac{1}{\sqrt{(1 \pm b_1^2) ... (1 \pm b_m^2)}} \tag{32}$$

which shows that S is determined by the b_i. Writing

$$B = b_1 B_1 + b_2 B_2 + b_3 B_3 \tag{33}$$

one can express a as

$$a = S \left(1 + B + \frac{B \wedge B}{2! S^2} + ... \frac{B \wedge B \wedge B \wedge ... (m \text{ terms})}{m! S^m} \right) \tag{34}$$

which shows that the bivector B determines completely the rotation.

If the scalar is nil, for example if ($\Phi_1 = \pm\pi$, $B_1^2 = -1$), then a is proportional to B_1

$$a = B_1 e^{\frac{\Phi_2}{2} B_2} e^{\frac{\Phi_3}{2} B_3}; \tag{35}$$

one then computes $B_1^{-1} a$ and comes back to the general expression to evaluate the remaining b_i and B_i.

To determine the b_i and B_i, one makes a change of variable $X_i = b_i B_i$, $x_i = X_i^2 = \pm b_i^2$ and considers the linear system of equations in X_i [9]

$$P_1 = B = \sum_{i=1}^{m} X_i \tag{36}$$

$$P_2 = (B \wedge B) \,.\, B = 2 \sum_{i,j=1}^{m} X_i x_j \quad (i \neq j) \tag{37}$$

$$P_3 = (B \wedge B \wedge B) \,.\, (B \wedge B) = 3!2! \sum_{i,j,k=1}^{m} X_i x_j \, x_k \quad (i \neq j, j < k) \tag{38}$$

$$\cdots\cdots \tag{39}$$

$$P_m = (B \wedge B \wedge \ldots m \text{ factors}) \,.\, (B \wedge B \ldots (m-1) \text{ factors}) \tag{40}$$

$$= m! \, (m-1)! \sum_{i=1}^{m} x_1 x_2 \ldots x_{i-1} x_{i+1} \ldots x_m X_i. \tag{41}$$

The determinant Δ is the product

$$\Delta = \left\{ m! \left[(m-1)! \right]^2 \left[(m-2)! \right]^2 \ldots 1 \right\} \prod_{i,j=1}^{m} (x_i - x_j) \quad (i \neq j, i < j). \tag{42}$$

If $\Delta \neq 0$, one obtains the bivectors X_i as a function of P_m and x_i. To determine the x_i, one writes the equations

$$S_1 = P_1 . P_1 = \sum_{i=1}^{m} x_i \tag{43}$$

$$S_2 = P_2 . P_1 = 2! \sum_{i,j=1}^{m} x_i x_j \quad (i \neq j) \tag{44}$$

$$S_3 = P_3 . P_1 = (3!)^2 \sum_{i,j,k=1}^{m} x_i x_j x_k \quad (i \neq j, j < k) \tag{45}$$

$$\cdots\cdots \tag{46}$$

$$S_m = P_m . P_1 = (m!)^2 \, (x_1 x_2 \ldots x_m). \tag{47}$$

The solutions yield $x_i = \pm b_i^2$, thus one obtains b_i and B_i

$$b_i = \sqrt{|x_i|}, \, B_i = \frac{X_i}{b_i}. \tag{48}$$

If $\Delta = 0$, the Eqs. (36–41) are not independent, the B bivector can nevertheless be decomposed in m mutually orthogonal simple planes but this decomposition is not unique.

4 Poincaré Group in n Dimensions (via Dual Hyperquaternions)

Much of physics being covariant with respect to the $4D$ Poincaré group, we provide here a hyperquaternion representation of the nD Poincaré groups in terms of dual hyperquaternions. Thereby one comes back to a $(n+1)D$ rotation which one can be decomposed canonically. The procedure is illustrated by a $5D$ case (for example a color image with 2 spatial and 3 color dimensions) which might be of interest in machine learning [10].

4.1 General Formalism

The Poincaré group of the pseudo-euclidean space associated with the Clifford algebra $C_{p,q}$ $(n = p + q)$ is constituted by the isometries of the metric

$$ds^2 = \left(dx_1^2 + ... + dx_p^2\right) - \left(dx_{p+1}^2 + ... + dx_{p+q}^2\right). \tag{49}$$

It includes the rotations $SO^+(p,q)$, translations and reflections (time or spacelike). The reflections having already been dealt with above, we shall focus on the rotations and translations.

Consider a hyperquaternion algebra $\mathbb{H} \otimes \mathbb{H}...\otimes \mathbb{H}$ (or a subalgebra thereof) with $n + 1$ generators $e_1, e_2, ...e_n, e_{n+1}$ and let X be a dual vector such that

$$X = e_{n+1} + \varepsilon x \tag{50}$$

where x belongs to the vector space V_1 with $x = \sum_{i=1}^{n} e_i x_i$ $(x_i \in \mathbb{R})$ and $\varepsilon^2 = 0$ (ε commuting with e_i). An nD hyperbolic rotation in V_1 leaves the last variable unchanged. Hence,

$$X' = aXa_c = e_{n+1} + \varepsilon x' \tag{51}$$

with $x' = axa_c, x'x'_c = xx_c, aa_c = 1$. A translation in V_1 can be expressed as

$$X' = bXb_c \tag{52}$$

with

$$b = e^{\varepsilon e_{n+1}\frac{t}{2}} = 1 + \varepsilon e_{n+1}\frac{t}{2}, (t = \sum_{i=1}^{n} e_i t_i , t_i \in \mathbb{R}) \tag{53}$$

and $bb_c = 1$. Developing Eq. (52), one obtains, assuming $e_{n+1}^2 = -1$

$$X' = \left(1 + \varepsilon e_{n+1}\frac{t}{2}\right)(e_{n+1} + \varepsilon x)\left(1 - \varepsilon e_{n+1}\frac{t}{2}\right) \tag{54}$$

$$= e_{n+1} + \varepsilon x - \varepsilon e_{n+1}e_{n+1}\frac{t}{2} - \varepsilon e_{n+1}e_{n+1}\frac{t}{2} \tag{55}$$

$$= e_{n+1} + \varepsilon(x + t) \tag{56}$$

which is a translation on the variables $1...n$ (if $e_{n+1}^2 = 1$, one simply takes $b = e^{\varepsilon \frac{t}{2} e_{n+1}}$). A combination of an nD rotation and translation gives with $f = ab$ (or ba)

$$X' = fXf_c \quad (ff_c = 1, \ f \in C^+) \tag{57}$$

which can be viewed as a a particular $(n+1)D$ rotation. One thus obtains a hyperquaternion representation of the Poincaré groups, distinct from the matrix one. A canonical decomposition leads to simple dual planes as will be illustrated in the following example.

4.2   Example: 5D Poincaré Group

As application consider a $5D$-space (for example a $2D$ color image) imbedded in the $6D$ hyperquaternion algebra $\mathbb{H} \otimes \mathbb{H} \otimes \mathbb{H}$ having six generators (see Appendix)

$$e_1 = kI, e_2 = kJ, e_3 = kKl, e_4 = kKm, e_5 = kKn, e_6 = j \tag{58}$$

with the generic vector $X = e_6 + \varepsilon x$ $(x = \sum_{i=1}^{5} e_i x_i)$. The transformation $X' = fXf_c$ with

$$f = e^{\frac{\Phi_2}{2} Jl} e^{\varepsilon i (2I + Kn)} e^{\frac{\Phi_1}{2} I(m+n)} \tag{59}$$

$$= \left(2 + \sqrt{3}Jl\right) \left[1 + \varepsilon i \left(2I + Kn\right)\right] \left[\sqrt{3} + \sqrt{2}I \left(\frac{m}{\sqrt{2}} + \frac{n}{\sqrt{2}}\right)\right] \tag{60}$$

and $\tanh \frac{\Phi_1}{2} = \sqrt{\frac{2}{3}} (= b_1)$, $\tanh \frac{\Phi_2}{2} = \frac{\sqrt{3}}{2} (= b_2)$ is a $5D$ -Poincaré transform. Applying the canonical decomposition presented above, one obtains

$$f = e^{\frac{\Phi_2}{2} B_2} e^{X_3} e^{\frac{\Phi_1}{2} B_1} \tag{61}$$

with the same values of Φ_1, Φ_2 as above and the following simple commuting orthogonal dual planes B_1, B_2, X_3

$$B_1 = \frac{1}{\sqrt{2}} I(m+n) + \varepsilon \frac{1}{\sqrt{2}} \left[\frac{\sqrt{3}}{2} K(m+n) - iJ\right] \tag{62}$$

$$B_2 = Jl + 2\varepsilon i \left(\frac{2}{\sqrt{3}} I - Kl\right) \tag{63}$$

$$X_3 = \frac{\varepsilon}{2} iK(-m+n). \tag{64}$$

with $(B_1)^2 = (B_2)^2 = 1, (X_3)^2 = 0$.

5   Conclusion

The paper has given a hyperquaternion representation of pseudo-euclidean rotations and the Poincaré groups in n dimensions, distinct from the matrix one.

A canonical decomposition of these groups was introduced, as an extension of an euclidean formalism, within a hyperquaternion Clifford algebra framework and illustrated by a $5D$ example. Potential geometric applications include in particular, moving reference frames and machine learning.

Acknowledgements. This work was supported by the LABEX PRIMES (ANR-11-LABX-0063) and was performed within the framework of the LABEX CELYA (ANR-10-LABX-0060) of Université de Lyon, within the program "Investissements d'Avenir" (ANR-11-IDEX-0007) operated by the French National Research Agency (ANR).

A   Multivector Structure of $\mathbb{H} \otimes \mathbb{H} \otimes \mathbb{H}$

$$
\begin{bmatrix}
1 & l = e_4 e_5 & m = e_5 e_3 & n = e_3 e_4 \\
I = e_2 e_3 e_4 e_5 & I\,l = e_3 e_2 & I\,m = e_4 e_2 & I\,n = e_5 e_2 \\
J = e_3 e_1 e_4 e_5 & J\,l = e_1 e_3 & J\,m = e_1 e_4 & J\,n = e_1 e_5 \\
K = e_2 e_1 & Kl = e_2 e_1 e_4 e_5 & Km = e_1 e_2 e_3 e_5 & Kn = e_2 e_1 e_3 e_4
\end{bmatrix}
$$

$$
+i
\begin{bmatrix}
1 = e_1 e_2 e_3 e_4 e_5 e_6 & l = e_2 e_1 e_3 e_6 & m = e_2 e_1 e_4 e_6 & n = e_2 e_1 e_5 e_6 \\
I = e_6 e_1 & I\,l = e_4 e_1 e_5 e_6 & I\,m = e_5 e_1 e_3 e_6 & I\,n = e_3 e_1 e_4 e_6 \\
J = e_6 e_2 & J\,l = e_4 e_2 e_5 e_6 & J\,m = e_5 e_2 e_3 e_6 & J\,n = e_3 e_2 e_4 e_6 \\
K = e_3 e_4 e_5 e_6 & Kl = e_6 e_3 & Km = e_6 e_4 & Kn = e_6 e_5
\end{bmatrix}
$$

$$
+j
\begin{bmatrix}
1 = e_6 & l = e_4 e_5 e_6 & m = e_6 e_5 e_3 & n = e_3 e_4 e_6 \\
I = e_2 e_3 e_4 e_5 e_6 & I\,l = e_3 e_2 e_6 & I\,m = e_6 e_4 e_2 & I\,n = e_6 e_5 e_2 \\
J = e_4 e_3 e_5 e_6 e_1 & J\,l = e_1 e_3 e_6 & J\,m = e_1 e_4 e_6 & J\,n = e_1 e_5 e_6 \\
K = e_2 e_1 e_6 & Kl = e_2 e_1 e_4 e_5 e_6 & Km = e_1 e_2 e_3 e_5 e_6 & Kn = e_2 e_1 e_3 e_4 e_6
\end{bmatrix}
$$

$$
+k
\begin{bmatrix}
1 = e_2 e_1 e_3 e_4 e_5 & l = e_1 e_2 e_3 & m = e_1 e_2 e_4 & n = e_1 e_2 e_5 \\
I = e_1 & I\,l = e_1 e_4 e_5 & I\,m = e_3 e_1 e_5 & I\,n = e_1 e_3 e_4 \\
J = e_2 & J\,l = e_2 e_4 e_5 & J\,m = e_3 e_2 e_5 & J\,n = e_2 e_3 e_4 \\
K = e_4 e_3 e_5 & Kl = e_3 & Km = e_4 & Kn = e_5
\end{bmatrix}
$$

References

1. Ungar, A.: Beyond Pseudo-Rotations in Pseudo-Euclidean Spaces. Academic Press, London (2018)
2. Ferreira, M., Sommen, F.: Complex boosts: a Hermitian Clifford algebra approach. Adv. Appl. Clifford Algebras **23**, 339–362 (2013)
3. Lounesto, P.: Clifford Algebras and Spinors. Cambridge University Press, Cambridge (2001)
4. Crumeyrolle, A.: Orthogonal and Symplectic Clifford Algebras: Spinor Structures. Kluwer Academic Publishers, Dordrecht (1990)
5. Perez-Gracia, A., Thomas, F.: On Cayley's factorization of 4D rotations and applications. Adv. Appl. Clifford Algebras **27**, 523–538 (2017)

6. Richard, A., Fuchs, L., Andres, E., Largeteau-Skapin, G.: Decomposition of nD-rotations: classification, properties and algorithm. Graph. Models **73**(6), 346–353 (2011). Elsevier
7. Girard, P.R., Clarysse, P., Pujol, R., Goutte, R., Delachartre, P.: Hyperquaternions: a new tool for physics. Adv. Appl. Clifford Algebras **28**, 68 (2018)
8. Moore, C.L.E.: Hyperquaternions. J. Math. Phys. **1**, 63–77 (1922)
9. Moore, C.L.E.: Rotations in hyperspace. In: Proceedings of the American Academy of Arts and Sciences, vol. 53, no. 8, pp. 651–694 (1918)
10. Zhu, X., Xu, Y., Xu, H., Chen, C.: Quaternion convolutional neural networks. In: Ferrari, V., Hebert, M., Sminchisescu, C., Weiss, Y. (eds.) ECCV 2018. LNCS, vol. 11212, pp. 645–661. Springer, Cham (2018). https://doi.org/10.1007/978-3-030-01237-3_39
11. Girard, P.R., Clarysse, P., Pujol, R., Wang, L., Delachartre, P.: Differential geometry revisited by biquaternion clifford algebra. In: Boissonnat, J.-D., et al. (eds.) Curves and Surfaces 2014. LNCS, vol. 9213, pp. 216–242. Springer, Cham (2015). https://doi.org/10.1007/978-3-319-22804-4_17
12. Girard, P.R.: Quaternions Clifford Algebras and Relativistic Physics. Birkhäuser, Basel (2007)
13. Girard, P.R.: Algèbre de Clifford et Physique relativiste. PPUR, Lausanne (2004)
14. Girard, P.R.: Einstein's equations and Clifford algebra. Adv. Appl. Clifford Algebras **9**(2), 225–230 (1999)
15. Girard, P.R.: The quaternion group and modern physics. Eur. J. Phys. **5**, 25–32 (1984)
16. Casanova, G.: L'algèbre vectorielle. PUF, Paris (1976)
17. Lipschitz, R.: Principes d'un calcul algébrique qui contient comme espèces particulières le calcul des quantités imaginaires et des quaternions. C.R. Acad. Sci. Paris, vol. 91, pp. 619–621, 660–664 (1880)

α-power Sums on Symmetric Cones

Keiko Uohashi$^{(\boxtimes)}$

Tohoku Gakuin University, Tagajo, Miyagi 985-8537, Japan
uohashi@mail.tohoku-gakuin.ac.jp

Abstract. In this paper, we define α-power sums of two or more elements on symmetric cones. For two elements, α-power sums, which are generalized parallel sums, are defined on our previous paper. We mention interpolation for α-power sums, which is not defined on our previous paper. It is shown that the synthesized resistances of α-series parallel circuits naturally correspond to α-power sums. We also mention relations with power sums and arithmetic, geometric, harmonic and α-power means, where α is a parameter of dualistic structure on information geometry.

Keywords: Parallel sum · Power sum · Mean · Operator monotone function · Symmetric cone · Series parallel circuit

1 Introduction

Arithmetic, geometric and harmonic mean are well known means on positive operators [1–3]. α-power mean (or power mean) is a generalized geometric mean, and corresponds to arithmetic, geometric and harmonic mean for $\alpha = 1, 0$ and -1, respectively [4–7]. On a symmetric cone, the α-power mean is the midpoint on the α-geodesic connecting two points, where α is a parameter of dualistic structure on information geometry [8,9].

Parallel sum is the half of harmonic mean [10,11]. However, it seems that few literatures treat sums related to geometric and α-mean for reasons of difficulty of convergence. Then, we define α-power means which are continuous for α and are arithmetic sum, parallel sum for $\alpha = 1, -1$, respectively.

First, we recall definitions and properties on symmetric cones. In Sect. 3, means and monotone functions are mentioned. In Sect. 4, we show definitions of α-power sums and the operator monotone function generating α-power sums. In Sect. 5, we define interpolation for α-power sums. Finally, we show a continuous deformation of the series circuit into the parallel circuit in which resistance elements have fixed resistivity and fixed volumes. The circuits realize arithmetic sum and parallel sum for $\alpha = 1, -1$, respectively.

Applications of α-power mean appear in fields of functional analysis, quantum mechanics, nonextensive statistical mechanics, optimization and information geometry. We expect to find applications of α-power sum as α-power mean.

© Springer Nature Switzerland AG 2019
F. Nielsen and F. Barbaresco (Eds.): GSI 2019, LNCS 11712, pp. 126–134, 2019.
https://doi.org/10.1007/978-3-030-26980-7_14

2   Symmetric Cones

A vector space V is called a Jordan algebra if a product $*$ defined on V satisfies

$$x * y = y * x, \quad x * (x^2 * y) = x^2 * (x * y) \tag{1}$$

for all $x, y \in V$ by setting $x^2 = x * x$. Let V be an n-dimensional Jordan algebra over $\mathbf{R}$ with an identity element e, i.e., $x * e = e * x = x$. An element $x \in V$ is said to be invertible if there exists $y \in \mathbf{R}[x]$ such that $x * y = e$, where $\mathbf{R}[X]$ is polynomials of X over $\mathbf{R}$. Since $\mathbf{R}[x]$ is an associative algebra, y is unique, called the inverse of x and denoted by $x^{-1} = y$ [8, 12, 13].

For x in V, let $L(x)$ and $P(x)$ be endomorphisms of V defined by

$$L(x)y = x * y, \quad y \in V \tag{2}$$

$$P(x) = 2L(x)^2 - L(x^2). \tag{3}$$

The following results, about P the quadratic representation of V, are known.

Proposition 1. *([12]) (i) An element x is invertible if and only if $P(x)$ is invertible, and*

$$P(x)x^{-1} = x, \quad P(x)^{-1} = P(x^{-1}). \tag{4}$$

(ii) If x and y are invertible, so is $P(x)y$ and

$$(P(x)y)^{-1} = P(x^{-1})y^{-1}. \tag{5}$$

(iii) For all x and y,

$$P(P(y)x) = P(y)P(x)P(y). \tag{6}$$

Let Ω be an open convex cone on a vector space V. We denote by G the identity component of the linear automorphism group of Ω. If G acts on Ω transitively, Ω is said to be homogeneous. The dual cone of Ω is defined by

$$\Omega^* = \{y \in V \mid (x, y) > 0, \forall x \in \bar{\Omega} \backslash \{0\}\}, \tag{7}$$

where $(,)$ is an inner product on V, $\bar{\Omega}$ the closure of Ω. If $\Omega = \Omega^*$, a cone Ω is said to be self-dual. A cone Ω is called symmetric if it is homogeneous and self-dual.

3   Means and Operator Monotone Functions

We consider a symmetric cone Ω a set of positive operators.

Let $x = \sum_{i=1}^r \lambda_i p_i$ be a spectral decomposition of $x \in V$, where r and $\{p_1, \ldots, p_r\}$ are the rank and a Jordan frame of V, respectively, and $\lambda_1, \ldots, \lambda_r$ are eigenvalues of x [12]. For a function $f(t)$ on an interval $\mathbf{I} \subseteq \mathbf{R}$, $f(x)$ is defined by

$$f(x) = \sum_{i=1}^r f(\lambda_i)p_i \tag{8}$$

if $\lambda_1,\ldots,\lambda_r \in \mathbf{I}$. A function $f(t)$ on an interval $\mathbf{I} \subseteq \mathbf{R}$ satisfying Inequation () is called an operator monotone function on $\mathbf{I}$.

$$a \leq b \Rightarrow f(a) \leq f(b), \tag{9}$$

where a and $b \in \Omega$ have eigenvalues on $\mathbf{I}$, respectively.

A binary operation $\sigma : (a,b) \in \bar{\Omega} \times \bar{\Omega} \mapsto a\sigma b \in \bar{\Omega}$ is called an operator connection if the following requirements are fulfilled.

(i) Monotonicity; $a \leq c$ and $b \leq d$ imply $a\sigma b \leq c\sigma d$,
(ii) Transformer inequality; $P(c)(a\sigma b) \leq (P(c)(a))\sigma(P(c)(b))$,
(iii) Semi-continuity; $a_n \downarrow a$ and $b_n \downarrow b$ imply $(a_n\sigma b_n) \downarrow a\sigma b$,

where $a \leq b$ (resp. $a < b$) is $b - a \in \bar{\Omega}$ (resp. in Ω) [1,8].

On transformer inequality, it holds that $P(c)(a\sigma b) = (P(c)(a))\sigma(P(c)(b))$ for Ω. If satisfying normalization $e\sigma e = e$, an operator connection σ is called an operator mean (or a mean).

It is known that α-power mean on Ω is generated by

$$a\sigma^{(\alpha)}b = P(a^{\frac{1}{2}})f^{(\alpha)}(P(a^{-\frac{1}{2}})b), \quad -1 \leq \alpha \leq 1, \tag{10}$$

where f is an operator monotone function defined by

$$f^{(\alpha)}(t) = \left(\frac{1+t^\alpha}{2}\right)^{\frac{1}{\alpha}} \quad (\alpha \neq 0), \quad f^{(0)}(t) = \sqrt{t} \tag{11}$$

[6,8]. Arithmetic, geometric and harmonic mean are described by $a\sigma^{(1)}b$, $a\sigma^{(0)}b$ and $a\sigma^{(-1)}b$, respectively. In particular, for positive definite matrices A and B, they are

(i) arithmetic mean; $A\sigma^{(1)}B = (A+B)/2$,
(ii) geometric mean; $A\sigma^{(0)}B = A\#B = A^{\frac{1}{2}}(A^{-\frac{1}{2}}BA^{-\frac{1}{2}})^{\frac{1}{2}}A^{\frac{1}{2}}$,
(iii) harmonic mean; $A\sigma^{(-1)}B = ((A^{-1}+B^{-1})/2)^{-1}$,
(iv) α-power mean; $A\sigma^{(\alpha)}B = ((A^\alpha+B^\alpha)/2)^{1/\alpha}$.

For scalar A and B, the geometric mean is $A\#B = \sqrt{AB}$.

4 α-power Sums and Operator Monotone Functions

In our previous paper, we defined α-power sum via an operator monotone function, which interpolates generalized sum between arithmetic sum and parallel sum [14].

For $-1 \leq \alpha \leq 1$, a function

$$f^{(\alpha)}(t) = \frac{(1+t)^{1+\alpha}}{1+t^\alpha}, \quad t > 0 \tag{12}$$

is an operator monotone function on $\{t|t^\alpha - \alpha t^{\alpha-1} + \alpha + 1 > 0\}$. Function (12) is obviously monotone increasing and operator monotone with α.

Definition 1. *([14]) Let $f^{(\alpha)}(t)$ be a function defined by Eq. (12). For $-1 \leq \alpha \leq 1$, we define the α-power sum $:^{(\alpha)}$ of a and $b \in \Omega$ by*

$$a :^{(\alpha)} b = P(a^{\frac{1}{2}})f^{(\alpha)}(P(a^{-\frac{1}{2}})b). \tag{13}$$

Theorem 1. *([14]) The α-power sum $:^{(\alpha)}$ of a and $b \in \Omega$ corresponds to arithmetic sum $a+b$ for $\alpha = 1$, and to parallel sum $a : b = (a^{-1}+b^{-1})^{-1}$ for $\alpha = -1$. The 0-power sum $a :^{(0)} b$ is arithmetic mean $(a + b)/2$.*

Corollary 1. *([14]) For positive definite matrices A and B, α-power sums are*

(i) arithmetic sum; $A :^{(1)} B = A + B$,
(ii) 0-power sum; $A :^{(0)} B = (A + B)/2$ (arithmetic mean),
(iii) parallel sum; $A :^{(-1)} B = (A^{-1} + B^{-1})^{-1}$ (the half of harmonic mean),
(iv) α-power sum; $A :^{(\alpha)} B = (A^\alpha + B^\alpha)^{-1/2}(A + B)^{1+\alpha}(A^\alpha + B^\alpha)^{-1/2}$
    $= (A + B)^{(1+\alpha)/2}(A^\alpha + B^\alpha)^{-1}(A + B)^{(1+\alpha)/2}$.

For scalar A and B, the α-power sum is

$$A :^{(\alpha)} B = \frac{(A + B)^{1+\alpha}}{A^\alpha + B^\alpha}. \tag{14}$$

If defined by $(A^\alpha + B^\alpha)^{1/\alpha}$ which is α-power mean without normalization property, generalized sum diverges to $\pm\infty$ as $\alpha = 0$. The α-power sum by Eqs. (12), (13) possesses continuity at $\alpha = 0$. It satisfies (i) Monotonicity for elements with eigenvalues on an interval $\{t|t^\alpha - \alpha t^{\alpha-1} + \alpha + 1 > 0\}$. It satisfies (ii) Transformer inequality and (iii) Semi-continuity on $\bar{\Omega}$ (resp. Ω).

The α-power sum of $a_1, \dots, a_n \in \Omega$ for $n \geq 2$ is defined as follows.

Definition 2. *For $-1 \leq \alpha \leq 1$, we define the α-power sum of $a_1, \dots, a_n \in \Omega$ for $n \geq 2$ by*

$$a_1 :^{(\alpha)} \dots :^{(\alpha)} a_n = P(a_1^{\frac{1}{2}})P((e + \sum_{i=2}^{n} P(a_1^{-\frac{1}{2}})a_i)^{\frac{1+\alpha}{2}})(e + \sum_{i=2}^{n}(P(a_1^{-\frac{1}{2}})a_i)^\alpha)^{-1}. \tag{15}$$

If $n = 2$, the α-power sum $a_1 :^{(\alpha)} a_2$ defined by Definition 2 coincides with $a_1 :^{(\alpha)} a_2$ defined by Definition 1 for a_1 and $a_2 \in \Omega$. In general, it holds that $(a_1 :^{(\alpha)} a_2) :^{(\alpha)} a_3 \neq a_1 :^{(\alpha)} a_2 :^{(\alpha)} a_3$ for a_1, a_2 and $a_3 \in \Omega$.

We obtain the next theorem similar to Corollary 1.

Theorem 2. *For positive definite matrices $A_1, \dots, A_n$, $n \geq 2$, α-power sums are*

(i) arithmetic sum; $A_1 :^{(1)} \dots :^{(1)} A_n = A_1 + \dots + A_n$,
(ii) 0-power sum; $A_1 :^{(0)} \dots :^{(0)} A_n = (A_1 + \dots + A_n)/n$,
(iii) parallel sum; $A_1 :^{(-1)} \dots :^{(-1)} A_n = (A_1^{-1} + \dots + A_n^{-1})^{-1}$,
(iv) α-power sum; $A_1 :^{(\alpha)} \dots :^{(\alpha)} A_n$
    $= (A_1^\alpha + \dots + A_n^\alpha)^{-1/2}(A_1 + \dots + A_n)^{1+\alpha}(A_1^\alpha + \dots + A_n^\alpha)^{-1/2}$
    $= (A_1 + \dots + A_n)^{(1+\alpha)/2}(A_1^\alpha + \dots + A_n^\alpha)^{-1}(A_1 + \dots + A_n)^{(1+\alpha)/2}$.

For scalar $A_1, \ldots, A_n$, $n \geq 2$, the α-power sum is

$$A_1 :^{(\alpha)} \cdots :^{(\alpha)} A_n = \frac{(A_1 + \cdots + A_n)^{1+\alpha}}{A_1^\alpha + \cdots + A_n^\alpha}. \tag{16}$$

Proof. The theorem is proved by calculations similar to techniques on the proof of Corollary 1.

Remark 1. For scalar $A_1, \ldots, A_n$, $n \geq 2$, the α-power sum (16) is the arithmetic sum $A_1 + \cdots + A_n$ multiplied by the ratio of the α-coordinate for $A_1 + \cdots + A_n$ and the arithmetic sum for the α-coordinates A_i^α, $i = 1, \ldots, n$.

5   Interpolation for α-power Sums

Uhlmann's interpolation for an α-power mean $\sigma^{(\alpha)}$ $(-1 \leq \alpha \leq 1)$ is defined by an operator monotone function

$$f_s^{(\alpha)}(t) = (1 - s + st^\alpha)^{\frac{1}{\alpha}} \ (\alpha \neq 0), \quad f_s^{(0)}(t) = t^s, \quad 0 \leq s \leq 1 \tag{17}$$

[4,5]. We define interpolation for α-power sums as follows.

Definition 3. *For $-1 \leq \alpha \leq 1$, we define interpolation $:_s^{(\alpha)}$ for an α-power sum $:^{(\alpha)}$ on a symmetric cone Ω by $a :_s^{(\alpha)} b = P(a^{\frac{1}{2}}) f^{(\alpha)}(P(a^{-\frac{1}{2}})b)$, $a, b \in \Omega$, where*

$$f_s^{(\alpha)}(t) = \frac{(2(1 - s) + 2st)^{1+\alpha}}{2(1 - s) + 2st^\alpha} = \frac{2^\alpha (1 - s + st)^{1+\alpha}}{1 - s + st^\alpha} \tag{18}$$

We have the next theorem via simple calculations.

Theorem 3. *For $-1 \leq \alpha \leq 1$ and $a, b \in \Omega$, they hold that*

$$a :_0^{(\alpha)} b = 2^\alpha a, \quad a :_{\frac{1}{2}}^{(\alpha)} b = a :^{(\alpha)} b, \quad a :_1^{(\alpha)} b = 2^\alpha b. \tag{19}$$

Proof. For a function (18), they hold that

$$f_0^{(\alpha)}(t) = 2^\alpha, \quad f_{\frac{1}{2}}^{(\alpha)}(t) = \frac{(1 + t)^{1+\alpha}}{1 + t^\alpha}, \quad f_1^{(\alpha)}(t) = 2^\alpha t. \tag{20}$$

Thus, we obtain Eq. (19).

Corollary 2. *For $\alpha = 1, 0$ and -1, interpolation $:_s^{(\alpha)}$ between a and $b \in \Omega$ is described as follows, respectively.*

(i) $a :_s^{(1)} b = 2((1 - s)a + sb)$ *(interpolation for arithmetic sum)*

$$a :_0^{(1)} = 2a, \quad a :_{\frac{1}{2}}^{(1)} = a + b, \quad a :_1^{(1)} = 2b \tag{21}$$

(ii) $a :_s^{(0)} b = (1 - s)a + sb$ *(interpolation for arithmetic mean)*

$$a :_0^{(0)} = a, \quad a :_{\frac{1}{2}}^{(0)} = \frac{1}{2}(a + b), \quad a :_1^{(0)} = b \tag{22}$$

(iii) $a :_s^{(-1)} b = (2((1 - s)a^{-1} + sb^{-1}))^{-1}$ *(interpolation for the half of harmonic mean)*

$$a :_0^{(-1)} = \frac{1}{2}a, \quad a :_{\frac{1}{2}}^{(-1)} = (a^{-1} + b^{-1})^{-1}, \quad a :_1^{(-1)} = \frac{1}{2}b \tag{23}$$

Corollary 3. *For* $-1 \le \alpha \le 1$ *and for scalar A and B, it holds that*

$$A :_s^{(\alpha)} B = \frac{2^\alpha ((1 - s)A + sB)^{1+\alpha}}{(1 - s)A^\alpha + sB^\alpha}. \tag{24}$$

6 Series Parallel Circuits Realizing α-power Sums

In our previous paper, we show series parallel circuits realizing α-power sums of two positive numbers [14]. In this section, we show series parallel circuits realizing α-power sums of two or more positive numbers.

Let the symbol of a parallel sum $A_1 : \cdots : A_n$ be also one of the circuit connecting resistances $A_1, \ldots, A_n$ in parallel. We suppose that electric resistances R_j, $j = 1, \ldots, n$ consist of element with fixed resistivity 1 and fixed cross-sectional areas 1, and that lengths of resistances R_j, $j = 1, \ldots, n$ are $R_j > 0$, respectively. Then, the synthetic resistance of the parallel circuit connecting n resistances with resistivity 1 and length R_j and with cross-sectional areas $R_1/(R_1 + \cdots + R_n), \ldots, R_n/(R_1 + \cdots + R_n)$ is R_j for each j. We give a continuous deformation of $R_1 + \cdots + R_n$ into $R_1 : \cdots : R_n$, using resistances R_{ij} with cross-sectional areas $(R_i/(R_1 + \cdots + R_n))^{(1+\alpha)/2}$, lengths $(R_i/(R_1 + \cdots + R_n))^{(1-\alpha)/2} R_j$ and volumes $R_i R_j/(R_1 + \cdots + R_n)$, $i, j = 1, \ldots, n$, respectively (Fig. 1). Note that, for each i, j, the volume $R_i R_j/(R_1 + \cdots + R_n)$ is constant for all $-1 \le \alpha \le 1$.

Theorem 4. *Let* $R_j > 0$, $j = 1, \ldots, n$ *be constant real numbers, and for* $-1 \le \alpha \le 1$,

$$R_{ij} = \left(\frac{R_1 + \cdots + R_n}{R_i} \right)^\alpha R_j, \quad i, j = 1, \ldots, n \tag{25}$$

be resistances in an electric circuit. Then, the synthetic resistance of the series circuit connecting parallel circuits $R_{1j} : \cdots : R_{nj}$, $j = 1, \ldots, n$, *which we call the* α-series parallel circuit, *is the* α-power sum of $R_1, \ldots, R_n$, *i. e.*,

$$R_1 :^{(\alpha)} \cdots :^{(\alpha)} R_n = \frac{(R_1 + \cdots + R_n)^{1+\alpha}}{R_1^\alpha + \cdots + R_n^\alpha}. \tag{26}$$

Proof. It follows from Eq. (25) that the synthetic resistance of the series circuit connecting $R_{1j} : \cdots : R_{nj}, j = 1, \ldots, n$ is

$$(R_{11} : \cdots : R_{n1}) + \cdots + (R_{1n} : \cdots : R_{nn})$$
$$= (R_{11}^{-1} + \cdots + R_{n1}^{-1})^{-1} + \cdots + (R_{1n}^{-1} + \cdots + R_{nn}^{-1})^{-1}$$
$$= ((R_1 + \cdots + R_n)^{-\alpha} R_1^{\alpha} R_1^{-1} + \cdots + (R_1 + \cdots + R_n)^{-\alpha} R_n^{\alpha} R_1^{-1})^{-1} + \cdots$$
$$+ ((R_1 + \cdots + R_n)^{-\alpha} R_1^{\alpha} R_n^{-1} + \cdots + (R_1 + \cdots + R_n)^{-\alpha} R_n^{\alpha} R_n^{-1})^{-1}$$
$$= (R_1 + \cdots + R_n)^{\alpha} (R_1^{\alpha} + \cdots + R_n^{\alpha})^{-1} (R_1 + \cdots + R_n)$$
$$= (R_1 + \cdots + R_n)^{1+\alpha} (R_1^{\alpha} + \cdots + R_n^{\alpha})^{-1} = R_1 :^{(\alpha)} \cdots :^{(\alpha)} R_n .$$

Remark 2. For $\alpha = 1$, it holds that

$$R_{1j} : \cdots : R_{nj} = (R_{1j}^{-1} + \cdots + R_{nj}^{-1})^{-1} = R_j , \quad j = 1, \ldots, n .$$

Then, the 1-series parallel circuit is equivalent to series circuit $R_1 + \cdots + R_n$ (Fig. 2).

Remark 3. For $\alpha = 0$, it holds that

$$R_{ij} = R_j, \quad i, j = 1, \ldots, n$$

(Fig. 3).

If $n = 2$, the 0-series parallel circuit is equivalent to the balanced Wheatstone bridge connecting two R_1 in parallel and two R_2 in parallel [15].

Remark 4. For $\alpha = -1$, it holds that

$$R_{i1} + \cdots + R_{in} = (R_1 + \cdots + R_n)^{-1} R_i R_1 + \cdots + (R_1 + \cdots + R_n)^{-1} R_i R_n = R_i ,$$

$i = 1, \ldots, n$. Then, the (-1)-series parallel circuit is equivalent to parallel circuit $R_1 : \cdots : R_n$ (Fig. 4).

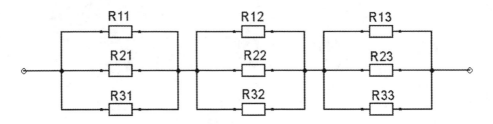

Fig. 1. The α-series parallel circuit ($n = 3$).

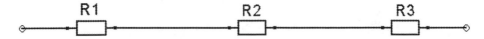

Fig. 2. The series circuit ($\alpha = 1$) ($n = 3$).

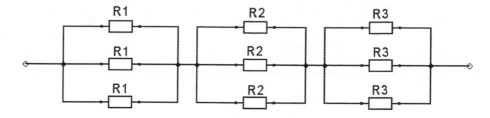

Fig. 3. The 0-series parallel circuit $(n = 3)$.

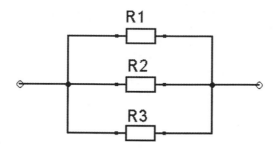

Fig. 4. The parallel circuit $(\alpha = -1)$ $(n = 3)$.

7   Conclusions

In this paper, we defined α-power sums of two or more elements on symmetric cones. They are generalized sums for arithmetic and parallel sums. We compared monotone functions of α-power sums and means. We also mentioned interpolation describing weighted sums for each α-power sum.

It was shown that the synthesized resistances of α-series parallel circuits naturally correspond to α-power sums. An α-series parallel circuit is the series circuit and the parallel circuit for $\alpha = 1, -1$, respectively.

The assumed medium of the resistances is free to deform. The results may be applicable to the comparison of the electrical properties of metal elements. In addition, characteristics such as fluid and blood flow may be compared with characteristics of the electrical circuit. Applications to fluid in tubes that combine in series and parallel in complexity are also conceivable.

It is a future subject to investigate these through α-power sum and information geometry.

References

1. Kubo, F., Ando, T.: Means of positive linear operators. Math. Ann. **246**, 205–224 (1980)
2. Bernstein, D.S.: Matrix Mathematics; Theory, Facts, and Formulas. Princeton University Press, New Jersey (2009)

3. Bhatia, R.: Positive Definite Matrices. Princeton University Press, New Jersey (2007)
4. Uhlmann, A.: Relative entropy and the Wigner-Yanase-Dyson-Lieb concavity in an interpolation theory. Commun. Math. Phys. **80**, 21–32 (1977)
5. Fujii, J., Kamei, E.: Uhlmann's interpolational method for operator means. Math. Jap. **34**, 541–547 (1989)
6. Kamei, E.: Paths of operators parametrized by operator means. Math. Jap. **39**, 395–400 (1994)
7. Hardy, G.H., Littlewood, J.E., Polya, G.: Inequalities, 2nd edn. Cambridge University Press, Cambridge (1952)
8. Ohara, A.: Geodesics for dual connections and means on symmetric cones. Integr. Eq. Oper. Theory **50**, 537–548 (2004)
9. Amari, S.: Information Geometry and Its Applications. Springer, Tokyo, Japan (2016)
10. Morley, T.D.: An alternative approach to the parallel sum. Adv. Appl. Math. **10**, 358–369 (1989)
11. Berkics, P.: On parallel sum of matrices. Linear Multilinear A. **65**, 2114–2123 (2017)
12. Faraut, J., Korányi, A.: Analysis on Symmetric Cones. Clarendon Press, Oxford (1994)
13. Uohashi, K., Ohara, A.: Jordan algebras and dual affine connections on symmetric cones. Positivity **8**, 369–378 (2004)
14. Uohashi, K.: Generalized parallel sums on symmetric cones and series parallel circuits (2019, submitted)
15. Hambley, A.R.: Electrical Engineering: Principles and Applications, 4th edn. Prentice Hall, New Jersey (2007)

Packing Bounds for Outer Products
with Applications to Compressive Sensing

Sebastian Semper$^{(\boxtimes)}$ and Thomas Hotz

Institut für Mathematik, Technische Universität Ilmenau, 98684 Ilmenau, Germany
{sebastian.semper,thomas.hotz}@tu-ilmenau.de

Abstract. In order to obtain good reconstruction guarantees for typical compressive sensing scenarios, we translate the search for good compression matrices into a ball packing problem in a suitable projective space. We then derive such reconstruction guarantees for two relevant scenarios, one where the matrices are unstructured and one where they have to be Khatri-Rao products. Finally, we demonstrate how the proposed method can be implemented with a physically motivated numerical optimization scheme, and how it compares to a conventional scheme of random compression matrices.

Keywords: Compressive sensing · Packing bounds ·
Khatri-Rao products · Outer products · Projective spaces

1 Motivation

In compressive sensing one starts from some input vector $y \in \mathbf{R}^N$ which can be represented as $y = B \cdot x$ for a *sparse* $x \in \mathbf{R}^N$ where $B \in \mathbf{R}^{N \times N}$ is a full rank matrix. One then observes m some so-called linear samples $\phi_i(y)$ of the vector y which can be written concisely as

$$z = \Phi \cdot y = \Phi \cdot B \cdot x, \tag{1}$$

where $z \in \mathbf{R}^m$, $\Phi \in \mathbf{R}^{m \times N}$ and $m < N$. In the model above, one generally assumes that x, and thus also y, are unknown, i.e. we only have access to z, whereas B and Φ are known. Theoretical results on compressive sensing now specify conditions on x, B and Φ under which one is able to recover x (and thus y) from z. Normally, we assume B to be given and fixed, and we allow Φ to be chosen freely. Generally, if the above system of linear equations has at least one solution, there are infinitely many, since $\Phi \cdot B$ has non-trivial kernel. The main condition in compressive sensing is that the vector x is *sparse*. To make this notion precise, we consider the so-called ℓ_0-*norm* of x which is given by $x \mapsto \|x\|_0 = |\{i \in \mathbf{N} \mid x_i \neq 0\}|$ (though this is not a norm in the usual sense as

This work was funded by the Deutsche Forschungsgemeinschaft (DFG) project CoSMoS (grant GA 2062/2-1).

F. Nielsen and F. Barbaresco (Eds.): GSI 2019, LNCS 11712, pp. 135–143, 2019.
https://doi.org/10.1007/978-3-030-26980-7_15

it is not homogeneous). Now, x is *sparse*, or in other words y possesses a sparse representation in the basis given by B, if $\|x\|_0 \ll N$. This additional assumption on x may suffice such that the solution of

$$\min_{x \in \mathbf{R}^N} \|x\|_0 \quad \text{s.t.} \quad \Phi \cdot B \cdot x = z \tag{2}$$

allows us to recover x from z. The combinatorial problem above essentially searches for the sparsest solution in the affine subspace of the solutions to the linear system $\Phi \cdot B \cdot x = z$. For simplicity, we will from now on assume that $B = I_N$; this is no restriction as B may be incorporated into Φ.

It can be shown that (2) is an NP hard problem (see e.g. [3, Theorem 2.17]), which motivates the search for a proxy problem that allows us to efficiently recover x in at least some situations. One popular approach is *basis pursuit* which may be viewed as the convex relaxation of (2) given by

$$\min_{x \in \mathbf{R}^N} \|x\|_1 \quad \text{s.t.} \quad \Phi \cdot x = z, \tag{3}$$

where $\|x\|_p^p = \sum_{i=1}^N |x_i|^p$ denotes the ℓ_p-norm of x for $p \geqslant 1$. Indeed, it can be shown that (3) is equivalent to a linear program [2, Sect. 1.5] and as such can be tackled by efficient algorithms like interior point methods. Now the natural question is under which circumstances a solution x^* of (3) is also a solution to (2), and in fact equal to x – the original problem of interest. To this end we consider the following a quantity associated to the matrix Φ; note that for the matrix $\Phi \in \mathbf{R}^{m \times N}$ we denote its i-th column of by $\phi_i \in \mathbf{R}^N$, $1 \leqslant i \leqslant m$, and analogously for other matrices.

Definition 1 (Coherence). *For $\Phi \in \mathbf{R}^{m \times N}$ its coherence is given by*

$$\Phi \mapsto \mu(\Phi) = \max_{1 \leqslant i < j \leqslant N} \frac{|\phi_i^t \cdot \phi_j|}{\|\phi_i\|_2 \|\phi_j\|_2}.$$

From a geometric point of view, the coherence $\mu(\Phi)$ is given by the largest correlation between pairs of columns. The coherence $0 \leq \mu(\Phi) \leq 1$ now allows to pose a sufficient condition such that basis pursuit recovers the solution we desire.

Theorem 1 (see e.g. [2, Theorem 4.5]). *If the columns of Φ are ℓ_2-normalized, i.e. $\|\phi_i\|_2 = 1$ for $i = 1, \ldots, N$, further let $x \neq 0$ and*

$$\mu(\Phi) < \frac{1}{2\|x\|_0 - 1},$$

then the unique solution x^* of (3) is equal to x which simultaneously is the unique solution of (2).

Theorem 1 tells us how many non-zeros x^* is allowed to possess before we no longer can guarantee that (3) recovers x. But since we are able to design Φ

freely there is room for optimization. Since in practice each row of Φ generates one element of z we can think of m as the measurement effort one has to take. One question immediately arising asks for the maximal s given m such that basis pursuit recovers the correct solution for every x with $\|x\|_0 \leqslant s$, or conversely, how large m – the measurement effort – has to be in order to recover all x with $\|x\|_0 \leq s$ for some pre-specified s.

In a probabilistic setting, this question has already been answered. If the entries of Φ are sampled i.i.d. from a centered subgaussian distribution with appropriate variance factor γ, then Φ has a sufficiently low coherence with high probability for all x with $\|x\|_0 \leqslant s$, if $m \gtrsim s \ln(s/N)$, see e.g. [3, Theorem 9.2]. However, in practice randomly sampled matrices behave unpredictably for varying x and as such a more deterministic approach should be followed in order to avoid this irregular behavior.

Taking a closer look at $\mu(\Phi)$ for some column normalized Φ, we see that it is determined by the pair of columns, now being points on the unit sphere $S^{m-1} = \{x \in \mathbf{R}^m \mid \|x\|_2 = 1\}$, which enclose the smallest angle between each other. To make things more specific let us define the real projective space and equip it with a metric as follows.

Definition 2 (Projective space). *Call x and y equivalent, $x \sim y$, iff there exists a $\lambda \in \{-1, 1\}$ such that $x = \lambda y$ for $x, y \in S^{m-1}$. Now we define the $(m-1)$-dimensional projective space via $\mathbb{P}^{m-1} = S^{m-1}/\sim$. Further, for the geodesic distance d_S on S^{m-1} where $(x, y) \mapsto d_S(x, y) = \arccos(x^T y)$ we define the corresponding metric on $\mathbb{P}^{m-1}$ as*

$$(x, y) \mapsto d_{\mathbb{P}}(x, y) = \min_{x' \in [x], y' \in [y]} d_S(x', y').$$

With this definition at hand, we can rephrase the coherence of a column-normalized matrix Φ as

$$\mu(\Phi) = \min_{1 \leq i < j \leq N} \cos(d_{\mathbb{P}}([\phi_i], [\phi_j])).$$

This embeds the notion of the coherence $\mu(\Phi)$ into a setting where one has packed a fixed number of points, i.e. the equivalence classes of columns of Φ, into the right projective space, where one considers these points' distances in that space. Thus the question whether there is a matrix Φ with a coherence lower than a certain specified $\hat{\mu}$ transforms into the problem of showing the existence of a packing in $\mathbb{P}^{m-1}$ where the points have distance at least $\arccos(\hat{\mu})$, i.e. open balls of radii $\frac{1}{2} \arccos(\hat{\mu})$ centered at these points are disjoint. The following result originally stated in [1] can trivially be extended to state packing bounds on $\mathbb{P}^{m-1}$ instead of S^{m-1}. As such it relates all the quantities of interest with each other and allows to address the mentioned questions.

Theorem 2. *For $\nu < \pi/4$ the maximum number $N(m, \nu)$ of open balls with radii at least 2ν that can be packed in $\mathbb{P}^{m-1}$ equipped with the geodesic distance satisfies*

$$\sqrt{2\pi} \sin(2\nu)^{1-m} < 2N(m, \nu) < 23(m-1)^{3/2} \left(\sqrt{2} \sin(\nu)\right)^{1-m}.$$

Note that this result merely states the existence of such a packing, and its proof is not constructive, so the actual question how to construct such a packing has to be postponed.

But, we can estimate an upper bound on the minimum number of measurements m one has to acquire in z such that basis pursuit recovers the correct solution of (2).

Theorem 3. *If $x \neq 0$ is s-sparse, then for*

$$
m > \frac{\ln\left(\sqrt{\frac{\pi}{2N^2}}\right)}{\ln\left(\sin\left(\arccos\left(\frac{1}{2s-1}\right)\right)\right)} + 1
$$

there is a matrix $\Phi \in \mathbf{R}^{m \times N}$ such that (3) has x as its unique solution.

Proof. Solving for ν in the lower bound in Theorem 2, we find that

$$
\nu > \frac{1}{2}\arcsin\left(\left(\frac{\pi}{2N^2}\right)^{\frac{1}{2m-2}}\right)
$$

Now, requiring that $\cos(2\nu) < 1/(2s-1)$ as demanded by Theorem 1 we find the requirement on m as stated.

The derivation of the above theorem can be viewed as a general procedure to derive the minimum number of measurements required. This is of great use when imposing structural requirements on the measurement matrix Φ which generally can be stated as $\Phi \in T \subsetneq S^{m-1}$ for some set T describing the structural constraints.

2   Packings of Outer Products and Coherence Bounds

We now shift our focus to a more complicated measurement process of the form

$$
H = \Psi \cdot B_1 \cdot \Gamma \cdot B_2 \cdot \Sigma \tag{4}
$$

for known $B_1 \in \mathbf{R}^{M_1 \times N_1}$, $B_2 \in \mathbf{R}^{N_2 \times M_2}$, sparse and unknown $\Gamma \in \mathbf{R}^{N_1 \times N_2}$ and arbitrary but fixed matrices $\Psi \in \mathbf{R}^{m_1 \times M_1}$, $\Sigma \in \mathbf{R}^{M_2 \times m_1}$ such that $H \in \mathbf{R}^{m_1 \times m_2}$. To formulate this as a compressive sensing problem we vectorize the matrix $H \in \mathbf{R}^{m_1 \times m_2}$, i.e. we consider

$$
\mathrm{vec}(H) = [H_1^{\mathrm{T}}, \ldots, H_{m_2}^{\mathrm{T}}]^{\mathrm{T}} = h = (\Psi \diamond \Sigma^{\mathrm{T}}) \cdot (B_1 \diamond B_2^{\mathrm{T}}) \cdot \gamma \in \mathbf{R}^{m_1 m_2} \tag{5}
$$

where $\gamma = \mathrm{vec}(\Gamma) \in \mathbf{R}^{N_1 N_2}$ is now a sparse vector and $\diamond$ denotes the *Khatri-Rao product* or column-wise Kronecker product defined by

$$
\Theta \diamond \Lambda = [\mathrm{vec}(\theta_1 \lambda_1^{\mathrm{T}}), \ldots, \mathrm{vec}(\theta_N \lambda_N^{\mathrm{T}})]
$$

for $\Theta \in \mathbf{R}^{m_1 \times N}$ and $\Lambda \in \mathbf{R}^{m_2 \times N}$.

In fact, it has recently been suggested to apply the compressive sensing paradigm to the task of bi-static channel sounding [5] where one aims to estimate the propagation paths taken by a few planar waves through an environment of interest. In this case, the sparsity of Γ stems from the physical properties of high frequency electro-magnetic planar waves. In another application one is interested in estimating the sparsity order $\|x\|_0 = s$ directly from z in (1) without estimating x first; for this it has been shown in [9] that this is possible if Φ is a Khatri-Rao product. In both cases one ends up with a model of the form (5), where $\Phi = (\Psi \diamond \Sigma^{\mathrm{T}})$ still has to be a good compression matrix in the sense that it allows an efficient recovery of γ. So, one is again interested in estimating a lower bound for the minimal number of measurements necessary in order to recover Γ from H.

To this end, we proceed as before and derive this bound by first estimating packing sizes in the right projective space. In this case it is the projective space of matrices of the form $a \cdot b^{\mathrm{T}}$ for $a \in \mathbf{R}^{m_1}$ and $b \in \mathbf{R}^{m_2}$ which we collect in the set T_{m_1,m_2}. The following result sheds some light on the structure of T_{m_1,m_2}. Since the proof consists of straightforward calculations, we omit it here.

Lemma 1. *Let $a_1, a_2 \in \mathbf{R}^{m_1}$ and $b_1, b_2 \in \mathbf{R}^{m_2}$ be arbitrary vectors. Furthermore, let $\langle \cdot, \cdot \rangle$ be the inner product on $\mathbf{R}^{m_1 \times m_2}$ defined as $\langle A, B \rangle = \mathrm{tr}(A^{\mathrm{T}} \cdot B)$ and $\|\cdot\|$ the induced (Frobenius) norm. Let further $t_1 = a_1 b_1^{\mathrm{T}}$ and $t_2 = a_2 b_2^{\mathrm{T}}$. Then:*

1. $\langle t_1, t_2 \rangle = \langle a_1, a_2 \rangle \cdot \langle b_1, b_2 \rangle$.
2. *For $t \in T_{m_1,m_2}$ with $\|t\| = 1$, there exist $a \in S^{m_1-1}$ and $b \in S^{m_2-1}$ such that $t = ab^{\mathrm{T}}$.*
3. *For $\|a_1\| = \|a_2\| = \|b_1\| = \|b_2\|$ we have*
   (a) $\|t_1 - t_2\|^2 = \|a_1 - a_2\|^2 + \|b_1 - b_2\|^2 + \frac{1}{2}\|a_1 - a_2\|^2 \|b_1 - b_2\|^2$.
   (b) $\|t_1 - t_2\| \leqslant \|a_1 - a_2\| + \|b_1 - b_2\|$.
   (c) $\max\{\|a_1 - a_2\|^2, \|b_1 - b_2\|^2\} \leqslant \|t_1 - t_2\|^2 \leqslant \|a_1 - a_2\|^2 + \|b_1 - b_2\|^2$

We now use this to derive bounds on the packing numbers for $T = \mathrm{vec}(T_{m_1,m_2}) \cap S^{m_1 m_2 - 1}$ which is the set of unit norm columns of a Khatri-Rao product. Since we ultimately want to derive coherence bounds, we construct the projective space $\mathbb{P}^{m_1 m_2 - 1}$ as before and consider the equivalence classes of the elements in T.

Let P_A and P_B be packings of open balls with radii ν in $\mathbb{P}^{m_1-1}$ and $\mathbb{P}^{m_2-1}$ respectively. Then for arbitrary but fixed $a_1, a_2 \in P_A$ and $b_1, b_2 \in P_B$ we have by (3c) from Lemma 1 and the properties of the geodesic distance that $\|a_1 b_1^{\mathrm{T}} - a_2 b_2^{\mathrm{T}}\| \geqslant \|a_1 - a_2\| \geqslant 2\sin(\nu)$ and $\|a_1 b_1^{\mathrm{T}} - a_2 b_2^{\mathrm{T}}\| \geqslant \|b_1 - b_2\| \geqslant 2\sin(\nu)$. Now, if we use each pair $(a_i, b_j) \in P_1 \times P_2$ as an element $t = \mathrm{vec}(a_i b_j^{\mathrm{T}})$, this results in a packing with geodesic distance 2ν on T that contains $|P_A| \cdot |P_B|$ elements. So, given two matrices Ψ and Σ, each obeying the bound in Theorem 2, we consider the matrix $\Phi = \Psi \otimes \Sigma$ with columns $\mathrm{vec}(\psi_i \sigma_j^{\mathrm{T}})$ where $\otimes$ denotes the *Kronecker product* obtained by vectorizing the outer products of all possible combinations of columns of its factors, i.e. Φ is in fact a (special) Khatri-Rao product. This immediately leads to the following theorem.

Theorem 4. *For $\nu < \pi/4$ the maximum number $N(m_1, m_2, \nu)$ of points that can be packed into $[T] \cap \mathbb{P}^{m_1 m_2 - 1}$ satisfies*

$$\pi \sin(2\nu)^{2-m_1-m_2} < N(m_1, m_2, \nu).$$

Now we are in the position to state the main result of this paper which estimates the number of measurements needed in order to recover x by solving (3) for $\Phi = \Psi \otimes \Sigma$.

Theorem 5. *There is a matrix $\Phi = \Psi \diamond \Sigma \in \mathbf{R}^{m \times N}$ such that for any s-sparse $x \neq 0$ the unique solution to (3) is in fact x, if*

$$m > \left\lceil \frac{\ln\left(\frac{\pi}{N}\right)}{2 \cdot \ln\left(\sin(\arccos(1/(2s-1)))\right)} + 1 \right\rceil^2$$

where $\lceil \cdot \rceil$ denotes the ceiling function.

Proof. From Theorem 4 we know that

$$m_1 + m_2 > \frac{\ln\left(\frac{\pi}{N}\right)}{\ln\left(\sin(\arccos(1/(2s-1)))\right)} + 2 = c,$$

which suggests to choose $m_1, m_2 > \lceil c/2 \rceil$, and since this implies $m = m_1 \cdot m_2 > \lceil c/2 \rceil^2$ we get the statement.

Clearly the above procedure of first deriving packing bounds in a suitable projective space is a general way of attaining reconstruction guarantees in compressive sensing, because the coherence of the measurement matrix is tightly related to the packing distance.

3   Simulations

To remedy the shortcoming that there is no explicit construction of a packing with the desired density, we present a physically motivated algorithm to approximate a sufficiently separated configuration in the projective space $\mathbb{P}^{m-1}$, namely a variant of a spring embedding algorithm [7] adapted to the problem at hand. In light of Theorem 4 we only need an algorithm constructing unstructured matrices of low coherence. To this end, we consider for a set of points $x_1, \ldots, x_N \in \mathbf{R}^m$ the function $V : \mathbf{R}^{m \times N} \to \mathbf{R}$ where

$$(x_1, \ldots, x_N) \mapsto V(x_1, \ldots, x_N) = \sum_{i \neq j} \frac{1}{\|x_j - x_i\|_2^2} + \frac{1}{\|x_j + x_i\|_2^2}$$

The function V essentially defines a potential at each point of the x_i given by the current configuration in such a way that points and antipodal points repel each other. Then we simply run a gradient descent algorithm to find a local minimum of V with respect to the $(x_1, \ldots, x_N)$ with step size $\Delta t = 10^{-5}$ for a

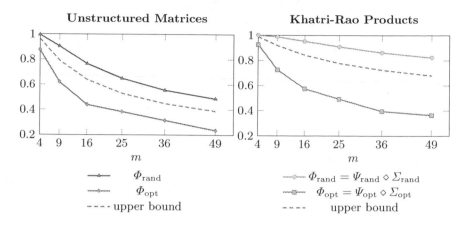

Fig. 1. Achieved median coherence of random matrices of 2500 trials and optimized matrices trials compared to the upper bounds derived from Theorems 2 and 4.

maximum number of 10^4 steps, where we sampled the initial x_i i.i.d. from the uniform distribution on the appropriate sphere S^{m-1}.

The first simulation (Fig. 1) demonstrates that the optimization outlined above actually yields good compression matrices in two ways. Firstly, they substantially improve upon the coherence obtained by simply generating them randomly from a centered and properly scaled element-wise i.i.d. Gaussian distribution. Secondly, the numerical results indicate that the iterative optimization above also produces packings which are actually better than the upper bounds derived in Theorems 2 and 4. In Fig. 1 we vary the number of measurements m and calculate the achieved coherence after optimization, the median coherence of 2500 trials for the random generation and the corresponding upper bounds. We do this for both the unstructured and Khatri-Rao structured case. Note that we select the number of measurements according to $m = k^2$ for $k \in \{2, \ldots, 7\}$ such that each factor in the Khatri-Rao product has k rows.

Finally, we present numerical results in Fig. 2 to show how the optimized coherences influence the actual reconstruction performance. This was estimated by calculating the relative frequency for the solution of (3) to match x (up to numerical inaccuracies) for several trials, i.e. instances of x, while Φ was kept fixed for each m. (3) was solved by reformulating it as an equivalent linear program which in turn was optimized with the standard solver for linear programs provided by SciPy [6], a standard Python [4] package for scientific computing. The underlying ground truth x of sparsity s for the 2500 trials was generated by first drawing the indices J where x is non-zero from the uniform distribution over all subsets of the set $\{1, \ldots, N\}$ of magnitude $|J| = s$, then drawing the corresponding entries of x from the standard normal distribution. As we can see, the structured matrices perform significantly worse compared to the unstructured ones, which is expected since the imposed structure reduces the degrees of freedom and as such the achievable coherence. Moreover, the optimization

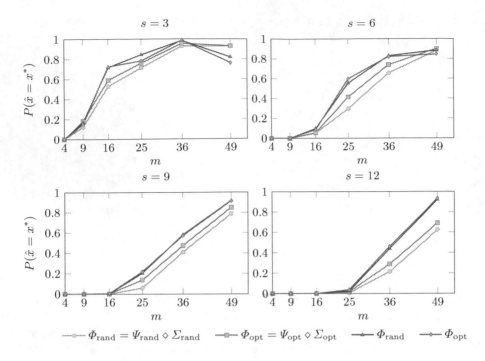

Fig. 2. Empirical probability of successful recovery depending on the number of measurements m over 2500 trials for various sparsity orders s.

routine results in an overall slightly better reconstruction performance. In the case $s = 3$ we observe a slight decrease in performance for $m = 49$, which is due to the fact that the numerical optimization routine got unstable from too many side constraints. Further experiments (not shown) suggest that this difference is even larger in a setting where one observes only noisy measurements z or when one uses a faster approximate algorithm, like Orthogonal Matching Pursuit [8], instead of solving (3) exactly.

4   Conclusion

Summarizing, we proposed a scheme to obtain coherence bounds for structured matrices by deriving packing bounds in the corresponding projective space. We applied this to the the standard unstructured case as well as the case where the packed points are columns of a Khatri-Rao product. Finally we showed that random compression matrices which serve as a theoretical benchmark in compressive sensing are far from the derived packing bounds whereas our proposed numerical optimization produces even better point configurations. Moreover, the compression matrices from this packing procedure result in a slightly superior reconstruction performance compared to the random compression ensemble.

References

1. Böröczky, K.: Finite packing and covering by congruent convex domains. Discret. Comput. Geom. **30**(2), 185–193 (2003)
2. Elad, M.: Sparse and Redundant Representations - from Theory to Applications in Signal and Image Processing. Springer, New York (2010)
3. Foucart, S.: A Mathematical Introduction to Compressive Sensing. Birkhäuser, Basel (2013)
4. Foundation, P.S.: Python. http://www.python.org
5. Ibrahim, M., et al.: Compressive spatial channel sounding. In: Proceedings of the 12th European Conference on Antennas and Propagation (EuCAP 2018), London, UK, Apr 2018
6. Jones, E., Oliphant, T., Peterson, P., et al.: Scipy: Open source scientific tools for python (2001). http://www.scipy.org/
7. Kobourov, S.G.: Spring embedders and force directed graph drawing algorithms. CoRR abs/1201.3011 (2012). http://arxiv.org/abs/1201.3011
8. Pati, Y.C., Rezaiifar, R., Krishnaprasad, P.S.: Orthogonal Matching Pursuit: Recursive Function Approximation with Applications to Wavelet Decomposition, pp. 40–44, vol. 1 (1993)
9. Semper, S., Römer, F., Hotz, T., DelGaldo, G.: Sparsity order estimation from a single compressed observation vector. IEEE Trans. Signal Process. **66**(15), 3958–3971 (2018). https://doi.org/10.1109/TSP.2018.2841867

Lie Group Machine Learning

On a Method to Construct Exponential Families by Representation Theory

Koichi Tojo[1,2]([✉]) and Taro Yoshino[3]

[1] RIKEN Center for Advanced Intelligence Project, Tokyo, Japan
koichi.tojo@riken.jp
[2] Department of Mathematics, Faculty of Science and Technology, Keio University,
3-14-1 Hiyoshi, Kohoku-ku, Yokohama 223-8522, Japan
[3] Graduate School of Mathematical Science, The University of Tokyo,
3-8-1 Komaba, Meguro-ku, Tokyo 153-8914, Japan
yoshino@ms.u-tokyo.ac.jp

Abstract. Exponential family plays an important role in information geometry. In [TY18], we introduced a method to construct an exponential family $\mathcal{P} = \{p_\theta\}_{\theta \in \Theta}$ on a homogeneous space G/H from a pair (V, v_0). Here V is a representation of G and v_0 is an H-fixed vector in V. Then the following questions naturally arise: (Q1) when is the correspondence $\theta \mapsto p_\theta$ injective? (Q2) when do distinct pairs (V, v_0) and (V', v_0') generate the same family? In this paper, we answer these two questions (Theorems 1 and 2). Moreover, in Sect. 3, we consider the case $(G, H) = (\mathbb{R}_{>0}, \{1\})$ with a certain representation on $\mathbb{R}^2$. Then we see the family obtained by our method is essentially generalized inverse Gaussian distribution (GIG).

Keywords: Exponential family · Representation theory ·
Homogeneous space · Generalized inverse Gaussian distribution

1 Introduction

Let G be a Lie group and H its closed subgroup. In [TY18], we introduced a method to construct an exponential family $\mathcal{P} = \{p_\theta\}_{\theta \in \Theta}$ on the homogeneous space $X := G/H$ from (V, v_0). In this paper, we answer two natural questions on our method.

1.1 Correspondence Parameters and Probability Measures

In the theory of exponential family, "minimal representation" is important [BN70]. If an exponential family is realized by "minimal representation", then we obtain one-to-one correspondence between the parameter space and the family of probability measures, which enable us to make use of the family. Moreover, from the perspective of information geometry, the correspondence is used as a coordinate. Then we would like to consider the following:

© Springer Nature Switzerland AG 2019
F. Nielsen and F. Barbaresco (Eds.): GSI 2019, LNCS 11712, pp. 147–156, 2019.
https://doi.org/10.1007/978-3-030-26980-7_16

Question 1. When is the following correspondence injective?

$$\Theta \ni \theta \mapsto p_\theta \in \mathcal{P}. \tag{1.1}$$

We want to answer this question for families obtained by our method. We give a necessary and sufficient condition for the injectivity of (1.1) in Theorem 1. It is, however, a little bit difficult to check. So, we will see the following easier equivalent conditions (A) and (B) are necessary.

(A) The orbit Gv_0 is not contained in any proper affine subspace of V.
(B) (1) v_0 is cyclic,
    (2) $V^\vee$ has no nonzero G-fixed vector.

In the case where G is compact or connected semisimple, they are also sufficient (see Remark 2).

1.2 Equivalence Relation

Our method in [TY18] constructs an exponential family from a pair (V, v_0). In some cases, the same exponential family comes from distinct pairs (V, v_0) and (V', v_0'). To reduce the choice of (V, v_0), it is useful to give an answer to the following question.

Question 2. When do distinct pairs (V, v_0) and (V', v_0') generate the same family?

We give an answer to this question in Theorem 2. More precisely, we introduce an equivalence relation on the set of pairs $\{(V, v_0)\}$ and show that two families obtained by (V, v_0), (V', v_0') coincide if $(V, v_0) \sim (V', v_0')$.

2 Main Theorems

2.1 Method Introduced in [TY18]

Before stating our main results, we recall the method introduced in [TY18]. Let G be a Lie group and H its closed subgroup. Then the quotient space $X := G/H$ naturally equips manifold structure, which is called the *homogeneous space* of G.

Let V be a finite dimensional real vector space, and $\rho: G \to GL(V)$ a Lie group homomorphism. Then the pair $V := (\rho, V)$ is called a *representation* of G. We often use simpler notation $gv := \rho(g)v$ for $g \in G$ and $v \in V$.

A vector $v_0 \in V$ is said to be *H-fixed* if $hv_0 = v_0$ for any $h \in H$. We denote by V^H the linear subspace consisting of all H-fixed vectors. Let (V, v_0) be a pair of representation of G and an H-fixed vector.

We put

$$\Omega_0(G, H) := \{\chi : G \to \mathbb{R}_{>0} \mid \chi \text{ is a continuous group homomorphism}, \chi|_H = 1\}, \tag{2.1}$$

$$\log \Omega_0(G, H) := \{\log \chi : G \to \mathbb{R} \mid \chi \in \Omega_0(G, H)\}. \tag{2.2}$$

Take a relatively G-invariant measure μ on X. Then we define a measure $\tilde{p}_\theta$ on X parameterized by $V^\vee \times \Omega_0(G, H)$ as follows:

$$d\tilde{p}_\theta(x) = d\tilde{p}_{\xi,\chi}(x) := \exp(-\langle \xi, xv_0 \rangle)\chi(x)d\mu(x) \quad (x \in X), \qquad (2.3)$$

where $\theta = (\xi, \chi) \in V^\vee \times \Omega_0(G, H)$.

Remark 1. Since v_0 is H-fixed, the notion xv_0 in (2.3) is well-defined. Owing to $\chi|_H = 1$, the notion $\chi(x)$ is also well-defined for $\chi \in \Omega_0(G, H)$.

Then we consider the normalization of the measures above. Put

$$\Theta := \{\theta = (\xi, \chi) \in V^\vee \times \Omega_0(G, H) \mid \int_X d\tilde{p}_\theta < \infty\}, \qquad (2.4)$$

$$\varphi(\theta) := \log \int_X d\tilde{p}_\theta \quad (\theta \in \Theta), \qquad (2.5)$$

$$dp_\theta := e^{-\varphi(\theta)}d\tilde{p}_\theta. \qquad (2.6)$$

Then we obtain a family of distributions on X as follows:

$$\mathcal{P} := \{p_\theta\}_{\theta \in \Theta}. \qquad (2.7)$$

This is an exponential family if $\Theta \neq \emptyset$ [TY18].

2.2   Correspondence

In this section, we give an answer to Question 1. Namely, we state a criterion of the injectivity of the correspondence (1.1). Moreover, we also give necessary conditions, which one can easily check (Proposition 1)

Theorem 1. *In the setting as in Sect. 2.1, the following three conditions are equivalent:*

(i) *The correspondence $\Theta \ni \theta \mapsto p_\theta \in \mathcal{P}$ is injective.*
(ii) *There does not exist $\xi \in V^\vee \setminus \{0\}$ such that $f_\xi \in \log \Omega_0(G, H)$.*
(iii) *There does not exist a triple $(\xi, \chi, c) \in (V^\vee \setminus \{0\}) \times \Omega_0(G, H) \times \mathbb{R}$ satisfying $\langle \xi, gv_0 \rangle = \log \chi(g) + c$ for any $g \in G$.*

Here, $f_\xi(g) := \langle \xi, gv_0 - v_0 \rangle$ for $g \in G$.

We prove this theorem in Sect. 4.2.

Moreover, we also give necessary conditions for the injectivity of (1.1). To state them, we prepare the notion of cyclic.

Definition 1 (cyclic). *We say a vector $v \in V$ is cyclic if $\mathrm{span}\{gv \mid g \in G\} = V$.*

Proposition 1. *If the correspondence (1.1) is injective, then the following equivalent conditions (A) and (B) are satisfied. Namely, ((1.1) is injective) $\Rightarrow$ (A) $\Leftrightarrow$ (B).*

(A) The orbit Gv_0 is not contained in any proper affine subspace of V.
(B) (1) $v_0 \in V$ is cyclic,
      (2) $\rho^\vee : G \to GL(V^\vee)$ has no nonzero G-fixed vector.

Here $\rho^\vee$ is the contragredient representation of G. Moreover, in the case where $\Omega_0(G, H) = \{1\}$, the converse implication also holds.

We prove this proposition in Sect. 4.3

Remark 2. In the case where G is compact or connected semisimple, we have $\Omega_0(G, H) = \{1\}$. See [TY18] for the details.

2.3 Equivalence

We use the same notation as in Sect. 2.1. In this subsection, we give an answer to Question 2. To state it, we introduce the notations $\tilde{\mathcal{V}}(G)$ and $\tilde{\mathcal{V}}(G, H)$.

Definition 2. *We put*
$$\tilde{\mathcal{V}}(G) := \{(V, v_0) \mid V \text{ is a finite dimensional real representation of } G, v_0 \in V \text{ is cyclic}\},$$
$$\tilde{\mathcal{V}}(G, H) := \{(V, v_0) \in \tilde{\mathcal{V}}(G) \mid v_0 \in V^H\}.$$

We say elements (V, v_0) and (V', v_0') in $\tilde{\mathcal{V}}(G)$ are equivalent if there exists a G-equivariant linear isomorphism $\psi : V \to V'$ such that $\psi(v_0) = v_0'$ and denote it by $(V, v_0) \sim (V', v_0')$. This is an equivalence relation on $\tilde{\mathcal{V}}(G)$. By definition, this is also an equivalence relation on $\tilde{\mathcal{V}}(G, H)$.

Theorem 2. *Equivalent elements in $\tilde{\mathcal{V}}(G, H)$ generate the same family by our method.*

We prove this theorem in Sect. 4.4.

Remark 3. From Theorem 2, in the special case $\dim V^H = 1$, the choice of v_0 is essentially unique. In the next section, we also see an example in which the choice of v_0 is essentially unique even if $\dim V^H > 1$.

3   Generalized Inverse Gaussian Distribution

Throughout this section, we put $G = \mathbb{R}_{>0}$, $H = \{1\}$ and $V = \mathbb{R}^2$, and consider a representation $\rho \colon G \to GL(V)$ given by $\rho(g) = \begin{pmatrix} g & \\ & g^{-1} \end{pmatrix}$ for $g \in G$. We answer Questions 1 and 2 for this case.

We consider the following two cases.

(Case 1) In the case where $\begin{pmatrix} r \\ s \end{pmatrix} \in V^H = V$ with $r = 0$ or $s = 0$:

Vectors $\begin{pmatrix} r \\ 0 \end{pmatrix}$, $\begin{pmatrix} 0 \\ s \end{pmatrix}$ are not cyclic. Therefore the obtained families have "unessential parameters".

(Case 2) In the case where $\begin{pmatrix} r \\ s \end{pmatrix} \in V^H$ with $r \neq 0$ and $s \neq 0$:

Proposition 2. *The pairs* $(V, \binom{r}{s})$ *with* $r \neq 0$ *and* $s \neq 0$ *are equivalent each other. Moreover, we obtain the family* $\{dp_{a,b,\lambda}\}_{(a,b,\lambda)\in\Theta}$ *of GIG (3.1) by applying our method to* $(V, \binom{r}{s})$, *where* $\Theta = \{(a, b, \lambda) \in \mathbb{R}^3 \mid (a, b, \lambda) \text{ satisfies } (3.2)\}$.

Definition 3 (Generalized inverse Gaussian distribution. See [J82] for the details). *The following distribution on* $\mathbb{R}_{>0}$ *is called generalized inverse Gaussian distribution.*

$$c_{a,b,\lambda} x^{\lambda-1} e^{-(ax+b/x)/2} dx \quad (x \in \mathbb{R}_{>0}), \tag{3.1}$$

where dx *denotes Lebesgue measure on* $\mathbb{R}_{>0}$, *and* (a, b, λ) *satisfies one of the following three conditions:*

$$(i)\ a > 0, b > 0,\ (ii)\ a > 0, b = 0, \lambda > 0,\ (iii)\ a = 0, b > 0, \lambda < 0. \tag{3.2}$$

Here $c_{a,b,\lambda}$ *is the normalizing constant given as follows, respectively.*

$$(i)\ \frac{(a/b)^{\frac{\lambda}{2}}}{2K_\lambda(\sqrt{ab})},\ (ii)\ \frac{1}{\Gamma(\lambda)}\left(\frac{a}{2}\right)^\lambda,\ (iii)\ \frac{1}{\Gamma(-\lambda)}\left(\frac{b}{2}\right)^{-\lambda}, \tag{3.3}$$

where K_λ *is the modified Bessel function of the second kind with index* λ.

Proof (Proposition 2). Put $v_0 := \frac{1}{2}\binom{1}{1}$. For $r, s \neq 0$, a G-linear isomorphism $\begin{pmatrix} 2r & 0 \\ 0 & 2s \end{pmatrix} \in GL(V)$ gives $(V, v_0) \sim (V, \binom{r}{s})$, which implies the former part.

For the latter part, it is enough to show the case (V, v_0) by Theorem 2. It is easily checked that $\Omega_0(G, H) = \{x \mapsto x^\lambda \mid \lambda \in \mathbb{R}\}$. Take a relatively invariant measure $\frac{dx}{x}$ on $\mathbb{R}_{>0}$. We identify $(\mathbb{R}^2)^\vee$ with $\mathbb{R}^2$ by taking the standard inner product. Then we have

$$d\tilde{p}_{a,b,\lambda}(x) := \exp(-\langle \binom{a}{b}, \binom{x}{x^{-1}} v_0 \rangle) x^\lambda \frac{dx}{x} \quad (\binom{a}{b} \in \mathbb{R}^2)$$
$$= \exp(-(ax + bx^{-1})/2) x^{\lambda-1} dx.$$

We get $\Theta = \{\theta = (a, b, \lambda) \in \mathbb{R}^3 \mid (a, b, \lambda) \text{ satisfies } (3.2)\}$. By normalizing these distributions, we obtain the desired family of GIG (3.1).

Finally, let us check the injectivity of the correspondence (1.1). For $(a, b, c, \lambda) \in \mathbb{R}^4$,

$$axg + byg^{-1} = \lambda \log g + c \quad \text{for any } g \in G$$

holds only if $(a, b, c, \lambda) = 0$. Thus, the condition (iii) of Theorem 1 is satisfied. $\square$

4   Proof of Main Theorems

In this section, we give proofs to Theorems 1 and 2 and Proposition 1.

4.1   Preliminary

In this subsection, we prepare some notations for proofs in the following sections. Let G be a Lie group, H a closed subgroup of G and V a finite dimensional real vector space.

Notation 3. *We denote by $C(G)$ the vector space consisting of all $\mathbb{R}$-valued continuous functions on G. The constant function 1 is an element of $C(G)$. The space $C(G)$ admits the left and right regular representations $L, R : G \to GL(C(G))$, respectively. We put $C(G)^H := \{f \in C(G) \mid R_h f = f$ for any $h \in H\}$.*

Remark 4. The set $\log \Omega_0(G, H)$ is a subspace of $C(G)$ (see (2.2)). For $f \in C(G)$, the condition $f \in \log \Omega_0(G, H)$ is equivalent to the pair of the following conditions:

(a) $f(h) = 0$ for any $h \in H$,
(b) $f(gg') = f(g) + f(g')$ for any $g, g' \in G$.

Notation 4. *We denote by* ev *the evaluation map. We identify V with $(V^\vee)^\vee$ canonically as follows:*

$$V \to (V^\vee)^\vee, \quad x \mapsto \text{ev}_x. \tag{4.1}$$

Let W be a subspace of V. Then we put

$$W^\perp := \{f \in V^\vee \mid \langle f, w \rangle = 0 \text{ for any } w \in W\}. \tag{4.2}$$

Notation 5. *For a representation $\rho : G \to GL(V)$, we denote the contragredient representation by $\rho^\vee : G \to GL(V^\vee)$. We often use simpler notation $g^\vee \xi := \rho^\vee(g)\xi$ for $g \in G$ and $\xi \in V^\vee$. Then, the following equality holds:*

$$\langle g^\vee \xi, v \rangle = \langle \xi, g^{-1} v \rangle \quad (g \in G, \ v \in V, \ \xi \in V^\vee). \tag{4.3}$$

4.2   Proof of Theorem 1

Proof (Theorem 1). We are enough to show $\neg$(ii)$\Rightarrow \neg$(iii)$\Rightarrow \neg$(i)$\Rightarrow \neg$(ii).

First, we see $\neg$(ii)$\Rightarrow \neg$(iii). Take $\xi \in V^\vee \setminus \{0\}$ such that $f_\xi \in \log \Omega_0(G, H)$. Then there exists $\chi \in \Omega_0(G, H)$ such that $\langle \xi, gv_0 - v_0 \rangle = \langle \xi, gv_0 \rangle - \langle \xi, v_0 \rangle = \log \chi(g)$ for any $g \in G$, so $\neg$(iii) is proved.

Next, we see $\neg$(iii)$\Rightarrow \neg$(i). Assume there exist $\xi \in V^\vee \setminus \{0\}$, $c \in \mathbb{R}$ and $\chi \in \Omega_0(G, H)$ satisfying $\langle \xi, gv_0 \rangle = \log \chi(g) + c$ for any $g \in G$. Take any $\theta_1 = (\xi_1, \chi_1) \in \Theta$ and put $\theta_2 := (\xi_1 + \xi, \chi_1 \chi) \in V^\vee \times \Omega_0(G, H)$. It is enough to show that $\theta_2 \in \Theta$ and $p_{\theta_1} = p_{\theta_2}$. This comes from $d\tilde{p}_{\theta_2}(x) = e^{-\langle \xi_1 + \xi, x v_0 \rangle} \chi_1(x) \chi(x) d\mu(x) = e^{-\langle \xi, x v_0 \rangle + \log \chi(x)} e^{-\langle \xi_1, x v_0 \rangle} \chi_1(x) d\mu(x) = e^{-c} d\tilde{p}_{\theta_1}(x)$.

Finally, we see $\neg$(i)$\Rightarrow \neg$(ii). Assume two distinct elements $\theta_1 = (\xi_1, \chi_1)$ and $\theta_2 = (\xi_2, \chi_2) \in \Theta$ satisfy $p_{\theta_1} = p_{\theta_2}$. Put $\xi := \xi_2 - \xi_1$. It is enough to show the following:

Claim. $\xi \neq 0$ and $f_\xi \in \log \Omega_0(G, H)$.

From $p_{\theta_1} = p_{\theta_2}$, we have for almost every $x \in X$,

$$\exp(-\langle \xi_1, xv_0 \rangle + \log \chi_1(x) - \varphi(\theta_1) + \langle \xi_2, xv_0 \rangle - \log \chi_2(x) + \varphi(\theta_2)) = \frac{dp_{\theta_1}}{dp_{\theta_2}}(x) = 1.$$

Therefore we have

$$\langle \xi, gv_0 \rangle + \varphi(\theta_2) - \varphi(\theta_1) = \log \chi_2(g) - \log \chi_1(g) \in \log \Omega_0(G, H). \tag{4.4}$$

From Remark 4(a), we have $\varphi(\theta_2) - \varphi(\theta_1) = -\langle \xi, v_0 \rangle$, that is, $f_\xi \in \log \Omega_0(G, H)$. Moreover, from (4.4) and $\theta_1 \neq \theta_2$, we obtain $\xi \neq 0$.

4.3   Proof of Proposition 1

In this subsection, we prove Proposition 1 by using Lemma 1 below.

Lemma 1. *For $\xi \in V^\vee \setminus \{0\}$, we consider the following three conditions:*

(i) $g^\vee \xi = \xi$ for any $g \in G$,
(ii) $f_\xi = 0$ (see Theorem 1 for the definition of f_ξ),
(iii) there exists $c \in \mathbb{R}$ satisfying $Gv_0 \subset \{v \in V \mid \langle \xi, v \rangle = c\}$.

Then, we have (i)$\Rightarrow$(ii)$\Leftrightarrow$(iii). Moreover, under the assumption that v_0 is cyclic, the implication (iii)$\Rightarrow$(i) also holds.

Proof. Since the implications (i)$\Rightarrow$(ii)$\Leftrightarrow$(iii) are easy, we prove only the implication (iii)$\Rightarrow$(i) under the assumption that v_0 is cyclic. Take any $g \in G$. It is enough to show that $\langle g^\vee \xi, g'v_0 \rangle = \langle \xi, g'v_0 \rangle$ for any $g' \in G$. From (4.3), we have

$$\langle g^\vee \xi, g'v_0 \rangle = \langle \xi, g^{-1}g'v_0 \rangle = c = \langle \xi, g'v_0 \rangle.$$

Proof (Proposition 1). First, note that we have the following three easy implications (a), (b) and (c):

(a) $\neg$(A) $\Longleftrightarrow$ there exists $\xi \in V^\vee \setminus \{0\}$ satisfying Lemma 1(iii),
(b) $\neg$(B)(2) $\Longleftrightarrow$ there exists $\xi \in V^\vee \setminus \{0\}$ satisfying Lemma 1(i),
(c) (A) $\Longrightarrow$ v_0 is cyclic.

Therefore, the equivalence (A)$\Leftrightarrow$(B) comes from Lemma 1.

Next, the implication ((1.1) is injective)$\Rightarrow$(A) follows from (a). In fact, the condition Theorem 1(ii) fails if there exists $\xi \in V^\vee \setminus \{0\}$ satisfying Lemma 1(ii).

Finally, assume $\Omega_0(G, H) = \{1\}$. The converse implication above also holds. So, (A) implies the injectivity of (1.1).

4.4   Proof of Theorem 2

We show Theorem 2 by using Lemmas 2 and 3 below. We prove Lemma 2 in the next subsection.

Proof (Theorem 2). It is enough to show that $\{g \mapsto \langle \xi, gv_0 \rangle \mid \xi \in V^\vee\} = \{g \mapsto \langle \xi', gv_0' \rangle \mid \xi' \in V'^\vee\}$ as a subspace of $C(G)^H$ if $(V, v_0), (V', v_0') \in \tilde{\mathcal{V}}(G, H)$ are equivalent. This follows from Lemmas 2 and 3 below.

Lemma 2. *Put*

$$\mathcal{V}(G) := \tilde{\mathcal{V}}(G)/\sim,$$

$$\mathcal{W}(G) := \{W \subset C(G) \mid W \text{ is a finite dimensional } L_G\text{-invariant subspace}\}.$$

The following map gives a one-to-one correspondence.

$$\mathcal{V}(G) \to \mathcal{W}(G), \ (V, v_0) \mapsto \eta(V^\vee), \tag{4.5}$$

where

$$\eta := \eta_{V, v_0} : V^\vee \to C(G), \ \xi \mapsto (g \mapsto \langle \xi, gv_0 \rangle). \tag{4.6}$$

Lemma 3. *Let H be a closed subgroup of G. Suppose $(V, v_0) \in \mathcal{V}(G)$ corresponds to $W \in \mathcal{W}(G)$ in Lemma 2. Then v_0 is H-fixed if and only if any element $w \in W$ is R_H-fixed.*

Proof. We have

$$\text{the function } \eta(\xi) : G \to \mathbb{R} \text{ is } R_H\text{-fixed for any } \xi \in V^\vee,$$
$$\Longleftrightarrow \langle \xi, ghv_0 \rangle = \langle \xi, gv_0 \rangle \ \text{ for any } g \in G, h \in H \text{ and } \xi \in V^\vee,$$
$$\Longleftrightarrow ghv_0 = gv_0 \text{ for any } g \in G \text{ and } h \in H,$$
$$\Longleftrightarrow v_0 \text{ is } H\text{-fixed.}$$

4.5   Proof of Lemma 2

In this subsection, we prove Lemma 2. To show this lemma, we use Lemmas 4 and 5 below.

Lemma 4 (property of η). *The map $\eta : V^\vee \to C(G)$ defined in (4.6) satisfies the following:*

(1) η is a G-equivariant linear map,
(2) v_0 is cyclic if and only if η is injective,
(3) $(V, v_0) \sim (V', v_0') \Rightarrow \eta(V^\vee) = \eta'(V'^\vee)$, where $\eta = \eta_{V, v_0}$ and $\eta' = \eta_{V', v_0'}$.

We give a proof of this lemma at the end of this subsection.

Lemma 5. *Let $W \subset C(G)$ be a finite dimensional L_G-invariant subspace. Then $v_0 := \mathrm{ev}_e \mid_W \in W^\vee$ is $L_G^\vee$-cyclic in $W^\vee$.*

Proof. Put $E := \mathrm{span}\{L_g^\vee v_0 \mid g \in G\} \subset W^\vee$. It is enough to show $E^\perp = \{0\}$. Take any function $f \in E^\perp$, then we have $f(g) = (L_{g^{-1}} f)(e) = \langle v_0, L_{g^{-1}} f \rangle = \langle L_g^\vee v_0, f \rangle = 0$. Therefore, we obtain $f = 0$.

Proof (Lemma 2). From Lemmas 4(1) and 5, the following maps are well-defined:

$$\Phi : \tilde{\mathcal{V}}(G) \to \mathcal{W}(G), \qquad (V, v_0) \mapsto \eta(V^\vee), \qquad (4.7)$$

$$\Psi : \mathcal{W}(G) \to \tilde{\mathcal{V}}(G), \qquad W \mapsto (W^\vee, \mathrm{ev}_e \mid_W). \qquad (4.8)$$

Then it is enough to show the following:

(a) $(V, v_0) \sim (V', v_0')$ in $\tilde{\mathcal{V}}(G) \Rightarrow \Phi(V, v_0) = \Phi(V', v_0')$,
(b) $\Phi \circ \Psi = \mathrm{id}_{\mathcal{W}(G)}$,
(c) $\Psi \circ \Phi(V, v_0) \sim (V, v_0)$ in $\tilde{\mathcal{V}}(G)$ for $(V, v_0) \in \tilde{\mathcal{V}}(G)$.

First, the condition (a) follows from Lemma 4(3).

Next, we show the condition (b). Let W be an element of $\mathcal{W}(G)$. Since we have $\Psi(W) = (W^\vee, \mathrm{ev}_e \mid_W)$, we get $\Phi \circ \Psi(W) = \{g \mapsto \langle \xi, L_g^\vee(\mathrm{ev}_e \mid_W) \rangle \mid \xi \in (W^\vee)^\vee\}$. Then, we have

$$\langle \xi, L_g^\vee(\mathrm{ev}_e \mid_W) \rangle = (L_g^\vee(\mathrm{ev}_e \mid_W))(\xi) = (\mathrm{ev}_e \mid_W)(L_{g^{-1}}\xi) = (L_{g^{-1}}\xi)(e) = \xi(g).$$

Therefore, we obtain $\Phi \circ \Psi(W) = W$.

Finally, we show the condition (c). Let (V, v_0) be an element of $\tilde{\mathcal{V}}(G)$. Put $W := \eta(V^\vee)$ and $(V', v_0') := \Psi \circ \Phi(V, v_0) = \Psi(W) = (W^\vee, \mathrm{ev}_e \mid_W)$. Since $\eta^\vee : W^\vee \to (V^\vee)^\vee$ is a G-linear isomorphism by Lemma 4(1) and (2), it is enough to show that $\eta^\vee(\mathrm{ev}_e \mid_W) = v_0$. For any $\xi \in V^\vee$, we have

$$\langle \xi, \eta^\vee(\mathrm{ev}_e \mid_W) \rangle = \langle \eta(\xi), \mathrm{ev}_e \mid_W \rangle = \eta(\xi)(e) = \langle \xi, v_0 \rangle. \qquad (4.9)$$

Therefore, we obtain $\eta^\vee(\mathrm{ev}_e \mid_W) = v_0$.

Proof (Lemma 4)

(1) Clearly, η is a linear map. The G-equivariance of η follows from the definition of the contragredient representation.
(2) Since η is linear, it is enough to show that v_0 is cyclic if and only if $\ker \eta = \{0\}$. The condition $\ker \eta = \{0\}$ means that for $\xi \in V^\vee$, $\langle \xi, gv_0 \rangle = 0$ for any $g \in G$ implies $\xi = 0$. Therefore this is equivalent to the condition v_0 is cyclic.
(3) Take a G-equivariant linear isomorphism $\psi : V \to V'$ with $\psi(v_0) = v_0'$. Then it is enough to show $\eta' = \eta \circ \psi^\vee : V'^\vee \to C(G)$. For any $\xi' \in V'^\vee$ and $g \in G$,

$$\eta \circ \psi^\vee(\xi')(g) = \langle \psi^\vee \xi', gv_0 \rangle = \langle \xi', \psi(gv_0) \rangle = \langle \xi', g\psi(v_0) \rangle = \langle \xi', gv_0' \rangle = \eta'(\xi')(g).$$

Acknowledgements. The authors would like to thank Dr. Frédéric Barbaresco for recommending us to submit a paper to the conference Geometric Science of Information 2019. The authors wish to thank referees for several helpful comments, particularly the comment concerning the condition (A) in Proposition 1.

References

[BN70] Barndorff-Nielsen, O.E.: Exponential Families: Exact Theory. Various Publication Series, vol. 19. Matematisk Institut, Aarhus Universitet, Aarhus (1970)

[TY18] Tojo, K., Yoshino, T.: A method to construct exponential families by representation theory. arXiv:1811.01394v2

[J82] Jørgensen, B.: Statistical Properties of the Generalized Inverse Gaussian Distribution. Lecture Notes in Statistics, vol. 9. Springer, New York (1982). https://doi.org/10.1007/978-1-4612-5698-4

Lie Group Machine Learning and Gibbs Density on Poincaré Unit Disk from Souriau Lie Groups Thermodynamics and SU(1,1) Coadjoint Orbits

Frédéric Barbaresco[✉]

Thales Land & Air Systems, Limours, France
frederic.barbaresco@thalesgroup.com

Abstract. In 1969, Jean-Marie Souriau has introduced a "Lie Groups Thermodynamics" in Statistical Mechanics in the framework of Geometric Mechanics. This Souriau's model considers the statistical mechanics of dynamic systems in their "space of evolution" associated to a homogeneous symplectic manifold by a Lagrange 2-form, and defines thanks to cohomology (non equivariance of the coadjoint action on the moment map with appearance of an additional cocyle) a Gibbs density (of maximum entropy) that is covariant under the action of dynamic groups of physics (e.g., Galileo's group in classical physics). Souriau model is more general if we consider another Souriau theorem, that we can associate to a Lie group, an homogeneous symplectic manifold with a KKS 2-form on their coadjoint orbits. Souriau method could then be applied on Lie Groups to define a covariant maximum entropy density by Kirillov representation theory. We will illustrate this method for homogeneous Siegel domains and more especially for Poincaré unit disk by considering SU(1,1) group coadjoint orbit and by using its Souriau's moment map. For this case, the coadjoint action on moment map is equivariant.

Keywords: Lie groups thermodynamics · Lie group machine learning · Kirillov representation theory · Coadjoint orbits · Moment map · Covariant Gibbs density · Maximum entropy density · Souriau-Fisher metric

1 Lie Groups Thermodynamics and Covariant Gibbs Density

We identify the Riemanian metric introduced by Souriau based on cohomology, in the framework of "Lie groups thermodynamics" as an extension of classical Fisher metric introduced in information geometry. We have observed that Souriau metric preserves Fisher metric structure as the Hessian of the minus logarithm of a partition function, where the partition function is defined as a generalized Laplace transform on a sharp convex cone. Souriau's definition of Fisher metric extends the classical one in case of Lie groups or homogeneous manifolds. Souriau has developed this "Lie groups thermodynamics" theory in the framework of homogeneous symplectic manifolds in geometric statistical mechanics for dynamical systems, but as observed by Souriau, these model equations are no longer linked to the symplectic manifold but equations

© Springer Nature Switzerland AG 2019
F. Nielsen and F. Barbaresco (Eds.): GSI 2019, LNCS 11712, pp. 157–170, 2019.
https://doi.org/10.1007/978-3-030-26980-7_17

only depend on the Lie group and the associated cocycle. This analogy with Fisher metric opens potential applications in machine learning, where the Fisher metric is used in the framework of information geometry, to define the "natural gradient" tool for improving ordinary stochastic gradient descent sensitivity to rescaling or changes of variable in parameter space. In machine learning revised by natural gradient of information geometry, the ordinary gradient is designed to integrate the Fisher matrix. Amari has theoretically proved the asymptotic optimality of the natural gradient compared to classical gradient. With the Souriau approach, the Fisher metric could be extended, by Souriau-Fisher metric, to design natural gradients for data on homogeneous manifolds. Information geometry has been derived from invariant geometrical structure involved in statistical inference. The Fisher metric defines a Riemannian metric as the Hessian of two dual potential functions, linked to dually coupled affine connections in a manifold of probability distributions. With the Souriau model, this structure is extended preserving the Legendre transform between two dual potential function parametrized in Lie algebra of the group acting transentively on the homogeneous manifold. Classically, to optimize the parameter θ of a probabilistic model, based on a sequence of observations y_t, is an online gradient descent:

$$\theta_t \leftarrow \theta_{t-1} - \eta_t \frac{\partial l_t(y_t)^T}{\partial \theta} \tag{1}$$

with learning rate η_t, and the loss function $l_t = -\log p(y_t/\hat{y}_t)$. This simple gradient descent has a first drawback of using the same non-adaptive learning rate for all parameter components, and a second drawback of non invariance with respect to parameter re-encoding inducing different learning rates. Amari has introduced the natural gradient to preserve this invariance to be insensitive to the characteristic scale of each parameter direction. The gradient descent could be corrected by $I(\theta)^{-1}$ where I is the Fisher information matrix with respect to parameter θ, given by:

$$I(\theta) = [g_{ij}] \text{ with } g_{ij} = \left[-E_{y \sim p(y/\theta)} \left[\frac{\partial^2 \log p(y/\theta)}{\partial \theta_i \partial \theta_j} \right] \right]_{ij} \tag{2}$$

with natural gradient:

$$\theta_t \leftarrow \theta_{t-1} - \eta_t I(\theta)^{-1} \frac{\partial l_t(y_t)^T}{\partial \theta} \tag{3}$$

Amari has proved that the Riemannian metric in an exponential family is the Fisher information matrix defined by:

$$g_{ij} = -\left[\frac{\partial^2 \Phi}{\partial \theta_i \partial \theta_j} \right]_{ij} \text{ with } \Phi(\theta) = -\log \int_{\mathbb{R}} e^{-\langle \theta, y \rangle} dy \tag{4}$$

and the dual potential, the Shannon entropy, is given by the Legendre transform:

$$S(\eta) = \langle \theta, \eta \rangle - \Phi(\theta) \text{ with } \eta_i = \frac{\partial \Phi(\theta)}{\partial \theta_i} \text{ and } \theta_i = \frac{\partial S(\eta)}{\partial \eta_i} \tag{5}$$

In geometric statistical mechanics, Souriau has developed a "Lie groups thermodynamics" of dynamical systems where the (maximum entropy) Gibbs density is covariant with respect to the action of the Lie group. In the Souriau model, previous structures of information geometry are preserved:

$$I(\beta) = -\frac{\partial^2 \Phi}{\partial \beta^2} \text{ with } \Phi(\beta) = -\log \int_M e^{-\langle \beta, U(\xi) \rangle} d\omega \text{ and } U : M \to \mathfrak{g}^* \tag{6}$$

$$S(Q) = \langle \beta, Q \rangle - \Phi(\beta) \text{ with } Q = \frac{\partial \Phi(\beta)}{\partial \beta} \in \mathfrak{g}^* \text{ and } \beta = \frac{\partial S(Q)}{\partial Q} \in \mathfrak{g} \tag{7}$$

In the Souriau Lie groups thermodynamics model, β is a "geometric" (Planck) temperature, element of Lie algebra $\mathfrak{g}$ of the group, and Q is a "geometric" heat, element of dual Lie algebra $\mathfrak{g}^*$ of the group. Souriau has proposed a Riemannian metric that we have identified as a generalization of the Fisher metric:

$$I(\beta) = [g_\beta] \text{ with } g_\beta([\beta, Z_1], [\beta, Z_2]) = \tilde{\Theta}_\beta(Z_1, [\beta, Z_2]) \tag{8}$$

$$\text{with } \tilde{\Theta}_\beta(Z_1, Z_2) = \tilde{\Theta}(Z_1, Z_2) + \langle Q, ad_{Z_1}(Z_2) \rangle \text{ where } ad_{Z_1}(Z_2) = [Z_1, Z_2] \tag{9}$$

Souriau has proved that all co-adjoint orbit of a Lie Group given by $O_F = \left\{ Ad_g^* F = g^{-1} F g, g \in G \right\}$ subset of $\mathfrak{g}^*, F \in \mathfrak{g}^*$ carries a natural homogeneous symplectic structure by a closed G-invariant 2-form. If we define $K = Ad_g^* = \left(Ad_{g^{-1}} \right)^*$ $K_*(X) = -(ad_X)^*$ with $\left\langle Ad_g^* F, Y \right\rangle = \langle F, Ad_{g^{-1}} Y \rangle, \forall g \in G, Y \in \mathfrak{g}, F \in \mathfrak{g}^*$ where if $X \in \mathfrak{g}$, $Ad_g(X) = gXg^{-1} \in \mathfrak{g}$, the G-invariant 2-form is given by the following expression $\sigma_\Omega(K_{*X} F, K_{*Y} F) = B_F(X, Y) = \langle F, [X, Y] \rangle, X, Y \in \mathfrak{g}$. Souriau Fundamental Theorem is that « every symplectic manifold is a coadjoint orbit ». We can observe that for Souriau model (8), Fisher metric is an extension of this 2-form in non-equivariant case $g_\beta([\beta, Z_1], [\beta, Z_2]) = \tilde{\Theta}(Z_1, [\beta, Z_2]) + \langle Q, [Z_1, [\beta, Z_2]] \rangle$.

The Souriau additional term $\tilde{\Theta}(Z_1, [\beta, Z_2])$ is generated by non-equivariance through Symplectic cocycle. The tensor $\tilde{\Theta}$ used to define this extended Fisher metric is defined by the moment map $J(x)$, application from M (homogeneous symplectic manifold) to the dual Lie algebra $\mathfrak{g}^*$, given by:

$$\tilde{\Theta}(X, Y) = J_{[X,Y]} - \{J_X, J_Y\} \tag{10}$$

$$\text{with } J(x) : M \to \mathfrak{g}^* \text{ such that } J_X(x) = \langle J(x), X \rangle, X \in \mathfrak{g} \tag{11}$$

This tensor $\tilde{\Theta}$ is also defined in tangent space of the cocycle $\theta(g) \in \mathfrak{g}^*$ (this cocycle appears due to the non-equivariance of the coadjoint operator Ad_g^*, action of the group on the dual lie algebra):

$$Q(Ad_g(\beta)) = Ad_g^*(Q) + \theta(g) \tag{12}$$

$$\tilde{\Theta}(X,Y) : \mathfrak{g} \times \mathfrak{g} \to \Re \qquad \text{with } \Theta(X) = T_e\theta(X(e)) \tag{13}$$
$$X,Y \mapsto \langle \Theta(X), Y \rangle$$

In Souriau's Lie groups thermodynamics, the invariance by re-parameterization in information geometry has been replaced by invariance with respect to the action of the group. When an element of the group g acts on the element $\beta \in \mathfrak{g}$ of the Lie algebra, given by adjoint operator Ad_g. Under the action of the group $Ad_g(\beta)$, the entropy $S(Q)$ and the Fisher metric $I(\beta)$ are invariant:

$$\beta \in \mathfrak{g} \to Ad_g(\beta) \Rightarrow \begin{cases} S[Q(Ad_g(\beta))] = S(Q) \\ I[Ad_g(\beta)] = I(\beta) \end{cases} \tag{14}$$

In the framework of Lie group action on a symplectic manifold, equivariance of moment could be studied to prove that there is a unique action $a(.,.)$ of the Lie group G on the dual $\mathfrak{g}^*$ of its Lie algebra for which the moment map J is equivariant, that means for each $x \in M$:

$$J(\Phi_g(x)) = a(g, J(x)) = Ad_g^*(J(x)) + \theta(g) \tag{15}$$

When coadjoint action is not equivariant, the symmetry is broken, and new "cohomological" relations should be verified in Lie algebra of the group. A natural equilibrium state will thus be characterized by an element of the Lie algebra of the Lie group, determining the equilibrium temperature β. The entropy $s(Q)$, parametrized by Q the geometric heat (mean of energy U, element of the dual Lie algebra) is defined by the Legendre transform of the Massieu potential $\Phi(\beta)$ parametrized by β ($\Phi(\beta)$ is the minus logarithm of the partition function $\psi_\Omega(\beta)$). Souriau has then defined a Gibbs density that is covariant under the action of the group:

$$p_{Gibbs}(\xi) = e^{\Phi(\beta) - \langle \beta, U(\xi) \rangle} = \frac{e^{-\langle \beta, U(\xi) \rangle}}{\int_M e^{-\langle \beta, U(\xi) \rangle} d\omega} \ , \ \text{with } \Phi(\beta) = -\log \int_M e^{-\langle \beta, U(\xi) \rangle} d\omega$$

$$Q = \frac{\partial \Phi(\beta)}{\partial \beta} = \frac{\int_M U(\xi) e^{-\langle \beta, U(\xi) \rangle} d\omega}{\int_M e^{-\langle \beta, U(\xi) \rangle} d\omega} = \int_M U(\xi) p(\xi) d\omega \tag{16}$$

We will illustrate computation of this covariant Souriau-Gibbs density for the Lie group $SU(1,1)$ and the unit disk considered as an homogeneous symplectic manifold.

2 Souriau Moment Map

$i_V\omega$ is the (p − 1)-form on M obtained by inserting $V(x)$ as the first argument of ω:

Interior product

$$i_V\omega(v_2, \cdots v_p) = \omega(V(x), v_2, \cdots, v_p) \tag{17}$$

$\theta \wedge \omega$ is the (p + 1)-form on X where ω is a p-form and θ is a 1-form on M:

Exterior product $\theta \wedge \omega(v_0, \cdots, v_p) = \sum\limits_{i=0}^{p} (-1)^i \theta(v_i)\omega(v_0, \cdots, \hat{v}_i, \cdots, v_p)$ (where the hat indicates a term to be omitted).

$L_V\omega$ is a p-form on M, and $L_V\omega = 0$ if the flow of V consists of symmetries of ω:

Lie derivative

$$L_V\omega(v_1, \cdots, v_p) = \frac{d}{dt} e^{tV^*}\omega(v_1, \cdots, v_p)\Big|_{t=0} \tag{18}$$

$d\omega$ is the (p + 1)-form on M defined by taking the ordinary derivative of ω and then antisymmetrizing:

Exterior derivative

$$d\omega(v_0, \cdots, v_p) = \sum\limits_{i=0}^{p} (-1)^i \frac{\partial \omega}{\partial x}(v_i)(v_0, \cdots, \hat{v}_i, \cdots, v_p) \tag{19}$$

$p = 0, [d\omega]_i = \partial_i\omega$; $p = 1, [d\omega]_{ij} = \partial_i\omega_j - \partial_j\omega_i$; $p = 2, [d\omega]_{ijk} = \partial_i\omega_{jk} + \partial_j\omega_{ki} + \partial_k\omega_{ij}$. The properties of the exterior and Lie Derivative are the following:

$$L_V\omega = di_V\omega + i_V d\omega \, (\textbf{\textit{E. Cartan}}), \, i_{[U,V]}\omega = i_V L_U\omega - L_U i_V\omega \, (\textbf{\textit{H. Cartan}}) \tag{20}$$

$$L_{[U,V]}\omega = L_V L_U\omega - L_U L_V\omega \, (\textbf{\textit{S. Lie}}) \tag{21}$$

Let (M, σ) be a connected symplectic manifold. A vector field η on M is called symplectic if its flow preserves the 2-form: $L_\eta\sigma = 0$. If we use Elie Cartan's formula, we can deduce that $L_\eta\sigma = di_\eta\sigma + i_\eta d\sigma = 0$ but as $d\sigma = 0$ then $di_\eta\sigma = 0$. We observe that the 1-form $i_\eta\sigma$ is closed. When this 1-form is exact, there is a smooth function $x \mapsto H$ on M with: $i_\eta\sigma = -dH$. This vector field η is called Hamiltonian and could be defined as symplectic gradient $\eta = \nabla_{Symp}H$.

Let a Lie group G that acts on M and that also preserve σ. A moment map exists if these infinitesimal generators are actually hamiltonian, so that a map $J : M \rightarrow \mathbf{g}^*$ exists with $i_{Z_X}\sigma = -dH_Z$ where

$$H_Z = \langle J(x), Z \rangle \tag{22}$$

We define also the Poisson bracket of two functions H, H' by:

$$\{H, H'\} = \sigma(\eta, \eta') = \sigma\left(\nabla_{Symp} H', \nabla_{Symp} H\right) \text{ with } i_\eta \sigma = -dH \text{ and } i_{\eta'} \sigma = -dH' \tag{23}$$

3 Coadjoint Orbits and Moment Map for SU(1,1)

3.1 Poincaré Unit Disk and SU(1,1) Lie Group

The group of complex unimodular pseudo-unitary matrices $SU(1, 1)$, is the set of elements u such that: $uMu^+ = M$ with

$$M = \begin{pmatrix} +1 & 0 \\ 0 & -1 \end{pmatrix} \tag{24}$$

We can show that the most general matrix u belongs to the Lie group given by:

$$G = SU(1, 1) = \left\{ \begin{pmatrix} a & b \\ b^* & a^* \end{pmatrix} / |a|^2 - |b|^2 = 1, \ a, b \in \mathbb{C} \right\} \tag{25}$$

Its Cartan decomposition is given by:

$$\begin{pmatrix} a & b \\ b^* & a^* \end{pmatrix} = |a| \begin{pmatrix} 1 & z \\ z^* & 1 \end{pmatrix} \begin{pmatrix} a/|a| & 0 \\ 0 & a^*/|a| \end{pmatrix} \text{ with } z = b(a^*)^{-1}, |a| = \left(1 - |z|^2\right)^{-1/2} \tag{26}$$

$$\begin{pmatrix} a & b \\ b^* & a^* \end{pmatrix} \begin{pmatrix} 1 & z \\ z^* & 1 \end{pmatrix} = |a'| \begin{pmatrix} 1 & z' \\ z'^* & 1 \end{pmatrix} \begin{pmatrix} a'/|a'| & 0 \\ 0 & a'^*/|a'| \end{pmatrix} \text{ with } \begin{cases} d' = bz^* + a \\ z' = \dfrac{az + b}{b^*z + a^*} \end{cases} \tag{27}$$

$SU(1, 1)$ is associated to group of holomorphic automorphisms of the Poincaré unit disk $D = \{z = x + iy \in \mathbb{C}/|z| < 1\}$ in the complex plane, by considering its action on the disk as $g(z) = (az + b)/(b^*z + a^*)$. The following measure on Unit disk:

$$d\mu_0(z, z^*) = \frac{1}{2\pi i} \frac{dz \wedge dz^*}{\left(1 - |z|^2\right)^2} \tag{28}$$

is invariant under the action of $SU(1,1)$ captured by the fractional holomorphic transformation:

$$\frac{dz' \wedge dz'^*}{\left(1 - |z'|^2\right)^2} = \frac{dz \wedge dz^*}{\left(1 - |z|^2\right)^2} \tag{29}$$

The complex unit disk admits a Kähler structure determined by potential function:

$$\Phi(z', z^*) = -\log(1 - z'z^*) \tag{30}$$

The invariant 2-form is:

$$\Omega = \frac{1}{i} \frac{\partial^2 \Phi(z, z^*)}{\partial z \partial z^*} dz \wedge dz^* = \frac{1}{i} \frac{dz \wedge dz^*}{\left(1 - |z|^2\right)^2} \tag{31}$$

which is closed $d\Omega = 0$. This group $SU(1,1)$ is isomorphic to the group $SL(2, \mathbb{R})$ as a real Lie group, and the Lie algebra $\mathbf{g} = \mathsf{su}(1,1)$ is given by:

$$g = \left\{ \begin{pmatrix} -ir & \eta \\ \eta^* & ir \end{pmatrix} / r \in \mathbb{R}, \eta \in \mathbb{C} \right\} \tag{32}$$

with the bases $(u_1, u_2, u_3) \in \mathbf{g}$: $u_1 = \frac{1}{2}\begin{pmatrix} 0 & -i \\ i & 0 \end{pmatrix}$, $u_2 = \frac{1}{2}\begin{pmatrix} 0 & 1 \\ 1 & 0 \end{pmatrix}$, $u_3 = \frac{1}{2}\begin{pmatrix} -i & 0 \\ 0 & i \end{pmatrix}$ with the commutation relation:

$$[u_3, u_2] = u_1, [u_3, u_1] = -u_2, [u_2, u_1] = -u_3 \tag{33}$$

Dual base on dual Lie algebra is named $(u_1^*, u_2^*, u_3^*) \in \mathbf{g}^*$. The dual vector space $\mathbf{g}^* = \mathsf{su}^*(1,1)$ can be identified with the subspace of $\mathsf{sl}(2, \mathbb{C})$ of the form:

$$g^* = \left\{ \begin{pmatrix} z & x+iy \\ -x+iy & -z \end{pmatrix} = x\begin{pmatrix} 0 & 1 \\ -1 & 0 \end{pmatrix} + y\begin{pmatrix} 0 & i \\ i & 0 \end{pmatrix} + z\begin{pmatrix} 1 & 0 \\ 0 & -1 \end{pmatrix} / x, y, z \in \mathbb{R} \right\} \tag{34}$$

Coadjoint action of $g \in G$ on dual Lie algebra $\xi \in \mathbf{g}^*$ is written $g.\xi$.

3.2 Coadjoint Orbit of SU(1,1) and Souriau Moment Map

We will use results of Cishahayo and de Bièvre [7] and Cahen [8, 9] for computation of moment map of $SU(1,1)$. Let $r \in \mathbb{R}^{*+}$, orbit $O(ru_3^*)$ of ru_3^* for the coadjoint action of $g \in G$ could be identified with the upper half sheet $x_3 > 0$ of $\{\xi = x_1 u_1^* + x_2 u_2^* + x_3 u_3^* / -x_1^2 - x_2^2 + x_3^2 = r^2\}$, the two-sheet hyperboloid. The stabilizer of ru_3^* for the coadjoint action of G is torus $K = \left\{ \begin{pmatrix} e^{i\theta} & 0 \\ 0 & e^{-i\theta} \end{pmatrix}, \theta \in \mathbb{R} \right\}$. K induces rotations of the unit disk, and leaves 0 invariant. The stabilizer for the origin 0 of unit disk is maximal compact subgroup K of $SU(1,1)$. We can observe [8] that $O(ru_3^*) \simeq G/K$. On the other hand $O(ru_3^*) \simeq G/K$ is diffeomorphic to the unit disk $D = \{z \in \mathbb{C}/|z| < 1\}$, then by composition, the moment map is given by:

$$J : D \to O(ru_3^*)$$

$$z \mapsto J(z) = r\left(\frac{z+z^*}{\left(1-|z|^2\right)}u_1^* + \frac{z-z^*}{i\left(1-|z|^2\right)}u_2^* + \frac{1+|z|^2}{\left(1-|z|^2\right)}u_3^*\right) \tag{35}$$

J is linked to the natural action of G on D (by fractional linear transforms) but also the coadjoint action of G on $O(ru_3^*) \simeq G/K$. J^{-1} could be interpreted as the stereographic projection from the two-sphere S^2 onto $\mathbb{C} \cup \infty$. In case $r = \frac{n}{2}$ where $n \in \mathbb{N}^+, n \geq 2$ then the coadjoint orbit is given by $O_n = O(\xi_n)$ with $\xi_n = \frac{n}{2}u_3^* \in \mathfrak{g}^*$, with stabilizer of ξ_n for coadjoint action the torus $K = \left\{\begin{pmatrix} e^{i\theta} & 0 \\ 0 & e^{-i\theta} \end{pmatrix}, \theta \in \mathbb{R}\right\}$ with Lie algebra $\mathbb{R}u_3$. $O_n = O(\xi_n)$ is associated with a holomorphic discrete series representation π_n of G by the KKS (Kirillov-Kostant-Souriau) method of orbits.

$$J : D \to O_n$$

$$z \mapsto J(z) = \frac{n}{2}\left(\frac{z+z^*}{\left(1-|z|^2\right)}u_1^* + \frac{z-z^*}{i\left(1-|z|^2\right)}u_2^* + \frac{1+|z|^2}{\left(1-|z|^2\right)}u_3^*\right) \tag{36}$$

Group G act on D by homography $g.z = \begin{pmatrix} a & b \\ b^* & a^* \end{pmatrix}.z = \frac{az+b}{b^*z+a^*}$. *This action corresponds with coadjoint action of G on O_n.* The Kirillov-Kostant-Souriau 2-form of O_n is given by:

$$\Omega_n(\zeta)(X(\zeta), Y(\zeta)) = \langle \zeta, [X, Y] \rangle, X, Y \in \mathfrak{g} \text{ and } \zeta \in O_n \tag{37}$$

and is associated in the frame by J with:

$$\omega_n = \frac{in}{\left(1-|z|^2\right)^2}dz \wedge dz^* \tag{38}$$

with the corresponding Poisson Bracket:

$$\{f, g\} = i\left(1-|z|^2\right)^2\left(\frac{\partial f}{\partial z}\frac{\partial g}{\partial z^*} - \frac{\partial f}{\partial z^*}\frac{\partial g}{\partial z}\right) \tag{39}$$

It has been also observed that there are 3 basic observables generating the $SU(1,1)$ symmetry on classical level:

$$\begin{cases} D \to \mathbb{R} \\ z \mapsto k_3(z) = \frac{1+|z|^2}{1-|z|^2} \end{cases}, \begin{cases} D \to \mathbb{R} \\ z \mapsto k_1(z) = \frac{1}{i}\frac{z-z^*}{1-|z|^2} \end{cases}, \begin{cases} D \to \mathbb{R} \\ z \mapsto k_2(z) = \frac{z+z^*}{1-|z|^2} \end{cases} \tag{40}$$

With the Poisson commutation rule:

$$\{k_3, k_1\} = k_2, \{k_3, k_2\} = -k_1, \{k_1, k_2\} = -k_3 \tag{41}$$

(k_1, k_2, k_3) vector points to the upper sheet of the two-sheeted hyperboloid in $\mathbb{R}^3$ given by $k_3^2 - k_1^2 - k_2^2 = 1$, whose the stereographic projection onto the open unit disk is:

$$\begin{cases} (k_1, k_2, k_3) \in H^+ \to D \\ z = \frac{k_2 + ik_1}{1 + k_3} = \sqrt{\frac{k_3-1}{k_3+1}} e^{i\arg z} \end{cases} \tag{42}$$

Under the action of $g \in G = SU(1,1) = \left\{ \begin{pmatrix} a & b \\ b^* & a^* \end{pmatrix} / |a|^2 - |b|^2 = 1, \, a, b \in \mathbb{C} \right\}$:

$$\begin{pmatrix} k_- & k_3 \\ k_3 & k_+ \end{pmatrix} = \begin{pmatrix} k_2 + ik_1 & k_3 \\ k_3 & k_2 - ik_1 \end{pmatrix} = \frac{1}{1-|z|^2} \begin{pmatrix} 2z & 1+|z|^2 \\ 1+|z|^2 & 2z^* \end{pmatrix} \quad \text{is trans-}$$

form in:

$$\begin{pmatrix} k'_- & k'_3 \\ k'_3 & k'_+ \end{pmatrix} = \begin{pmatrix} k_-(g^{-1}.z) & k_3(g^{-1}.z) \\ k_3(g^{-1}.z) & k_+(g^{-1}.z) \end{pmatrix} = g^{-1} \begin{pmatrix} k_- & k_3 \\ k_3 & k_+ \end{pmatrix} (g^{-1})^t \tag{43}$$

This transform can be viewed as the co-adjoint action of $SU(1,1)$ on the coadjoint orbit identified with $k_3^2 - k_1^2 - k_2^2 = 1$.

4 Covariant Gibbs Density by Souriau Thermodynamics

Representation theory studies abstract algebraic structures by representing their elements as linear transformations of vector spaces, and algebraic objects (Lie groups, Lie algebras) by describing its elements by matrices and the algebraic operations in terms of matrix addition and matrix multiplication, reducing problems of abstract algebra to problems in linear algebra. Representation theory generalizes Fourier analysis via harmonic analysis. The modern development of Fourier analysis during XXth century has explored the generalization of Fourier and Fourier-Plancherel formula for non-commutative harmonic analysis, applied to locally compact non-Abelian groups. This has been solved by geometric approaches based on "orbits methods" (Fourier-Plancherel formula for G is given by coadjoint representation of G in dual vector space of its Lie algebra) with many contributors (Dixmier, Kirillov, Bernat, Arnold, Berezin, Kostant, Souriau, Duflo, Guichardet, Torasso, Vergne, Paradan, etc.).

For classical commutative harmonic analysis, we consider the following groups:

$G = \mathrm{T}^n = \mathbb{R}^n / \mathbb{Z}^n$ for Fourier series, $G = \mathbb{R}^n$ for Fourier Transform

G group character (linked to e^{ikx}) : $\chi : G \to U$ with $U = \{ z \in \mathbb{C} / |z| = 1 \}$

$\hat{G} = \{ \chi / \chi_1 \cdot \chi_2(g) = \chi_1(g) \chi_2(g) \}$ and Fourier transform is given by:

$$\begin{aligned} \varphi : G &\to \mathbb{C} & \hat{\varphi} : \hat{G} &\to \mathbb{C} \\ g \mapsto \varphi(g) &= \int_{\hat{G}} \hat{\varphi}(\chi) \chi(g)^{-1} d\chi \quad \text{and} & \chi \mapsto \hat{\varphi}(\chi) &= \int_{G} \varphi(g) \chi(g) dg \end{aligned} \tag{44}$$

For non-commutative harmonic analysis, Group unitary irreductible representation is $\mathrm{U} : G \to U(\mathrm{H})$ with H Hilbert space and character by $\chi_{\mathrm{U}}(g) = tr \mathrm{U}_g$. Fourier transform for non-commutative group is $\mathrm{U}_\varphi = \int_G \varphi(g) \mathrm{U}_g dg$ with character $\chi_{\mathrm{U}}(g) = tr \mathrm{U}_\varphi$. If we describe group element with exponential map $\mathrm{U}_\psi = \int_{\mathfrak{g}} \psi(X) \mathrm{U}_{\exp(X)} dX$, we have:

$$tr \mathrm{U}_\psi = \dim \tau . \mu_{G.f} \left(\overset{\wedge}{\psi . j^{-1}} \right) \quad \text{with} \quad \begin{cases} \mu_{G.f} : \text{Liouville meas. on } \mathrm{O} = G.f, f \in \mathfrak{g}^* \\ \mu_{G.f} \left(\overset{\wedge}{\psi . j^{-1}} \right) : \text{Integral of } \overset{\wedge}{\psi . j^{-1}} \text{ wrt } \mu_{G.f} \end{cases}$$

$$\overset{\wedge}{\psi . j^{-1}} : \mathfrak{g} \to \mathfrak{g}^*, \text{ Four. Transf.} \tag{45}$$

where

$$j(X) = (\det s(ad_X))^{1/2} \text{ with } s(x) = \sum_{n=0}^{\infty} \frac{1}{(2n+1)!} \left(\frac{x}{2} \right)^{2n} = sh\left(\frac{x}{2} \right) / \left(\frac{x}{2} \right) \tag{46}$$

Kirillov Character formula is:

$$\chi_U(\exp(X)) = tr \mathrm{U}_{\exp(X)} = j(X)^{-1} \int_{\mathrm{O}} e^{i\langle f, X \rangle} d\mu_{\mathrm{O}}(f) \tag{47}$$

$$\int_{\mathrm{O}} e^{i\langle f, X \rangle} d\mu_{\mathrm{O}}(f) = j(X) tr \mathrm{U}_{\exp(X)} \text{ with } j(X) = \left(\det \left(\frac{e^{ad_X/2} - e^{-ad_X/2}}{ad_X/2} \right) \right)^{1/2} \tag{48}$$

We will use Kirillov representation theory and his character formula [10–19] to compute Souriau covariant Gibbs density in the unit Poincaré disk. For any Lie group G, a coadjoint orbit $\mathrm{O} \subset \mathfrak{g}^*$ has a canonical symplectic form ω_{O} given by KKS 2-form. As seen, if G is finite dimensional, the corresponding volume element defines a

G-invariant measure supported on O, which can be interpreted as a tempered distribution. The Fourier transform (where d is the half of the dimension of the orbit O):

$$\Im(x) = \int_{O \subset g^*} e^{-i\langle x, \lambda \rangle} \frac{1}{d!} d\omega_{O^d} \text{ with } \lambda \in g^* \text{ and } x \in g \tag{49}$$

is Ad G-invariant. When O $\subset g^*$ is an integral coadjoint orbit, Kirillov formula, given previously, expresses Fourier transform $\Im(x)$ by Kirillov character χ_O:

$$\Im(x) = j(x)\chi_O(e^x) \text{ where } j(x) = \det^{1/2} \left(\frac{\sinh(ad(x/2))}{ad(x/2)} \right) \tag{50}$$

χ_O is, as defined previously, the "*Kirillov character*" of a unitary representation associated to the orbit. We will consider the universal covering of $PSU(1, 1)$, the Lie algebra is:

$$g^* = su(1, 1)^* = \left\{ \begin{pmatrix} iE & p^* \\ p & -iE \end{pmatrix} / E \in \mathbb{R}, p \in \mathbb{C} \right\} \tag{51}$$

As observed in [8], the Ad-invariant form $m^2 = E^2 - |p|^2$ allows to identify the following operator Ad and Ad^*, m could be considered analogously as rest mass, E as energy, and $p = p_1 + ip_2$ as the momentum vector. The coadjoint orbits are the rest mass shells. Let $D = \{w \in \mathbb{C}/|w| < 1\}$ Poincaré unit disk, for any $m > 0$, there is a corresponding action of the universal covering of $PSU(1, 1)$ on $\kappa^{m/2}$ (with κ the holomorphic cotangent bundle of unit disk), with the invariant symplectic form

$$\omega = curv(\kappa) = -i\partial\bar{\partial}^* \log|dw|^2 = 2i \frac{dw \wedge dw^*}{\left(1 - |w|^2\right)^2} \tag{52}$$

The moment map is an equivariant isomorphism (O_m^+ coadjoint orbit for $m^2 > 0$ and $E > 0$):

$$J : w \in \left(D, curv\left(\kappa^{m/2}\right)\right) \mapsto (p, E) = \frac{m}{\left(1 - |w|^2\right)} \left(2iw, 1 + |w|^2\right) \in O_m^+ \tag{53}$$

In case $m > 1$, the Kirillov character formula is given by:

$$\chi_m \left(\exp \left(\begin{pmatrix} x & . \\ . & -x \end{pmatrix} \right) \right) = j(x)^{-1} \int_{O_{m-1}^+} e^{-i \left\langle \begin{pmatrix} x & . \\ . & -x \end{pmatrix}, \begin{pmatrix} iE & p^* \\ p & -iE \end{pmatrix} \right\rangle} \omega_{O_{m-1}^+} \tag{54}$$

where

$$j(x) = \det^{1/2} \left[\sinh \left(ad \begin{pmatrix} x/2 \\ & -x/2 \end{pmatrix} \right) \middle/ ad \begin{pmatrix} x/2 \\ & -x/2 \end{pmatrix} \right] = \frac{\sinh(x)}{x} \qquad (55)$$

which reduces to:

$$\frac{e^{mx}}{1 - e^{2x}} j(x) = \int_D e^{(m-1)x \frac{1+|w|^2}{1-|w|^2}} \frac{1}{\left(1 - |w|^2\right)^2} dw \wedge dw^* \qquad (56)$$

Finally, the Souriau-Gibbs density is given by:

$$p_{Gibbs}(w) = \frac{e^{-\left\langle \begin{pmatrix} ix & -\eta \\ -\eta^* & -ix \end{pmatrix}, \begin{pmatrix} im\frac{1+|w|^2}{1-|w|^2} & 2m\frac{w}{1-|w|^2} \\ 2m\frac{w}{1-|w|^2} & -im\frac{1+|w|^2}{1-|w|^2} \end{pmatrix} \right\rangle}}{j(x)\chi_m \left(e^{\begin{pmatrix} x & i\eta \\ i\eta^* & -x \end{pmatrix}} \right)}$$

$$= \frac{e^{2m\left(x\frac{1+|w|^2}{1-|w|^2} + \frac{w(\eta+\eta^*)}{1-|w|^2} \right)}}{j(x)\chi_m \left(e^{\begin{pmatrix} x & i\eta \\ i\eta^* & -x \end{pmatrix}} \right)} \qquad (57)$$

5 Extension from Poincaré to Siegel Homogeneous Domains

V. Bargmann has proposed the covering of the general symplectic group $Sp(2N, \mathbb{R})$:

$$Sp(2N, \mathbb{R}) = \left\{ g = \begin{pmatrix} A & B \\ C & D \end{pmatrix} / gJ_{2N}g^T = J_{2N}, J_{2N}^T = -J_{2N}, J_{2N} = \begin{pmatrix} 0 & I_N \\ -I_N & 0 \end{pmatrix} \right\} \qquad (58)$$

$$AB^T = BA^T, AC^T = CA^T, BD^T = DB^T, CD^T = DC^T, AD^T - BC^T = I_N \qquad (59)$$

Bargmann has observed that although $Sp(2N, \mathbb{R})$ is not isomorphic to any pseudo-unitary group, its inclusion in $U(N, N)$ will display the connectivity properties through its unitary $U(N)$ maximal compact subgroup, generalizing the role of $U(1) = SO(2)$ in $Sp(2, \mathbb{R})$: $W_N = W \otimes I_N, 2N \times 2N$ matrix

where

$$W = W_1 = \frac{1}{\sqrt{2}} \begin{pmatrix} \omega_{\pi/4}^{-1} & \omega_{\pi/4}^{-1} \\ -\omega_{\pi/4} & \omega_{\pi/4} \end{pmatrix} \text{ with } \omega = e^{i\pi/4} = \frac{1}{\sqrt{2}}(1+i) \qquad (60)$$

$$u(g) = W_N^{-1} g W_N = \frac{1}{2} \begin{pmatrix} [A+D] - i[B-C] & [A-D] + i[B+C] \\ [A-D] - i[B+C] & [A+D] + i[B-C] \end{pmatrix} = \begin{pmatrix} \alpha & \beta \\ \beta^* & \alpha^* \end{pmatrix} \qquad (61)$$

with

$$\alpha\alpha^+ - \beta\beta^+ = I_N, \alpha^+\alpha - \beta^T\beta^* = I_N \text{ and } \alpha\beta^T - \beta\alpha^T = 0, \alpha^T\beta^* - \beta^+\alpha = 0 \qquad (62)$$

The symplecticity property of g becomes:

$$uM_{2N}u^+ = M_{2N}, M_{2N} = iW_N^{-1}J_{2N}W_N = \begin{pmatrix} I_N & 0 \\ 0 & -I_N \end{pmatrix} \qquad (63)$$

$$\begin{pmatrix} A & B \\ C & D \end{pmatrix} = g(u) = W_N u W_N^{-1} = \begin{pmatrix} \text{Re}(\alpha+\beta) & -\text{Im}(\alpha-\beta) \\ \text{Im}(\alpha+\beta) & \text{Re}(\alpha-\beta) \end{pmatrix} \qquad (64)$$

References

1. Bargmann, V.: Irreducible unitary representations of the Lorentz group. Ann. Math. **48**, 588–640 (1947)
2. Souriau, J.-M.: Mécanique statistique, groupes de Lie et cosmologie, Colloques int. du CNRS numéro 237. Aix-en-Provence, France, 24–28, pp. 59–113 (1974)
3. Souriau, J.-M.: Structure des systèmes dynamiques, Dunod (1969)
4. Kirillov, A.A.: Elements of the Theory of Representations. Springer, Berlin (1976). https://doi.org/10.1007/978-3-642-66243-0
5. Marle, C.-M.: From tools in symplectic and poisson geometry to J.-M. Souriau's theories of statistical mechanics and thermodynamics. Entropy **18**, 370 (2016)
6. Barbaresco, F.: Higher order geometric theory of information and heat based on poly-symplectic geometry of Souriau lie groups thermodynamics and their contextures: the bedrock for lie group machine learning. Entropy **20**, 840 (2018)
7. Cishahayo, C., de Bièvre, S.: On the contraction of the discrete series of SU(1;1). Annales de l'institut Fourier **43**(2), 551–567 (1993)
8. Cahen B.: Contraction de SU(1,1) vers le groupe de Heisenberg, Travaux mathématiques, Fascicule XV, pp. 19–43 (2004)
9. Cahen, M., Gutt, S., Rawnsley, J.: Quantization on Kähler manifolds I, Geometric interpretation of Berezin quantization. J. Geom. Phys. **7**, 45–62 (1990)
10. Dai, J.: Conjugacy classes, characters and coadjoint orbits of DiffS¹, Ph.D. dissertation, The University of Arizona, Tucson, AZ, 85721, USA (2000)

11. Dai, J., Pickrell, D.: The orbit method and the Virasoro extension of Diff+(S1): I. Orbital integrals. J. Geom. Phys. **44**, 623–653 (2003)
12. Knapp, A.: Representation Theory of Semisimple Groups: An Overview Based on Examples. Princeton University Press, Princeton (1986)
13. Frenkel, I.: Orbital theory for affine Lie algebras. Invent. Math. **77**, 301–354 (1984)
14. Libine, M.: Introduction to Representations of Real Semisimple Lie Groups, arXiv:1212.2578v2 (2014)
15. Guichardet, A.: La methode des orbites: historiques, principes, résultats. Leçons de mathématiques d'aujourd'hui, vol. 4, Cassini, pp. 33–59 (2010)
16. Vergne, M.: Representations of Lie groups and the orbit method. In: Srinivasan, B., Sally, J. D. (eds.) Actes Coll. Bryn Mawr, pp. 59–101. Springer, New York (1983). https://doi.org/10.1007/978-1-4612-5547-5_5
17. Duflo, M., Heckman, G., Vergne, M.: Projection d'orbites, formule de Kirillov et formule de Blattner, Mémoires de la SMF, Série 2, no. 15, pp. 65–128 (1984)
18. Witten, E.: Coadjoint orbits of the Virasoro group. Com. Math. Phys. **114**, 1–53 (1988)
19. Pukanszky, L.: The Plancherel formula for the universal covering group of SL(2, R). Math. Ann. **156**, 96–143 (1964)
20. Clerc, J.L., Orsted, B.: The Maslov index revisited. Transform. Groups **6**(4), 303–320 (2001)
21. Foth, P., Lamb M.: The poisson geometry of SU(1,1). J. Math. Phys. **51** (2010)
22. Perelomov, A.M.: Coherent states for arbitrary lie group. Commun. Math. Phys. **26**, 222–236 (1972)
23. Ishi, H.: Kolodziejek, B.: Characterization of the Riesz Exponential Family on Homogeneous Cones. arXiv:1605.03896 (2018)
24. Tojo, K., Yoshino, T.: A Method to Construct Exponential Families by Representation Theory. arXiv:1811.01394 (2018)
25. Tojo, K., Yoshino, T.: On a method to construct exponential families by representation theory. In: Nielsen, F., Barbaresco, F. (eds.) GSI 2019. LNCS, vol. 11712, pp. 147–156. Springer, Cham (2019)
26. Pukanszky, L.: The Plancherel formula for the universal covering group of SL(2,R). Math Annalen **156**, 96–143 (1964)
27. Pukanszky, L.: Leçons sur les représentations des groupes. Monographies de la Société Mathématique de France, Dunod, Paris (1967)
28. Bernat, P., et al.: Représentations des groupes de Lie. Monographie de la Société Mathématique de France, Dunod, Paris (1972)
29. Dixmier, J.: Les algèbres enveloppantes. Gauthier-Villars, Paris (1974)
30. Duflo, M.: Construction des représentations unitaires d'un groupe de Lie, C.I.M.E. (1980)
31. Guichardet, A.: Théorie de Mackey et méthode des orbites selon M. Duflo. Expo. Math. **3**, 303–346 (1985)
32. Mnemné, R., Testard, F.: Groupes de Lie classiques, Hermann (1985)
33. Yahyai, M.: Représentations étoile du revêtement universel du groupe hyperbolique et formule de Plancherel, Thèse Université de Metz, 23 Juin 1995
34. Rais, M.: Orbites coadjointes et représentations des groupes, cours C.I.M.P.A. (1980)
35. Rais, M.: La représentation coadjointe du groupe affine. Annales de l'Institut Fourier **28**(1), 207–237 (1978)
36. Barbaresco, F.: Souriau exponential map algorithm for machine learning on matrix lie groups. In: Nielsen, F., Barbaresco, F. (eds.) GSI 2019. LNCS, vol. 11712, pp. 85–95. Springer, Cham (2019)
37. Barbaresco, F.: Geometric theory of heat from Souriau lie groups thermodynamics and Koszul Hessian geometry: applications in information geometry for exponential families. Entropy **18**, 386 (2016)

Irreversible Langevin MCMC
on Lie Groups

Alexis Arnaudon[1], Alessandro Barp[1,2]($\boxtimes$), and So Takao[1]

[1] Department of Mathematics, Imperial College London, London SW7 2AZ, UK
a.barp16@imperial.ac.uk
[2] Alan Turing Institute, British Library, 96 Euston Rd, London NW1 2DB, UK

Abstract. It is well-known that irreversible MCMC algorithms converge faster to their stationary distributions than reversible ones. Using the special geometric structure of Lie groups $\mathcal{G}$ and dissipation fields compatible with the symplectic structure, we construct an irreversible HMC-like MCMC algorithm on $\mathcal{G}$, where we first update the momentum by solving an OU process on the corresponding Lie algebra $\mathfrak{g}$, and then approximate the Hamiltonian system on $\mathcal{G} \times \mathfrak{g}$ with a reversible symplectic integrator followed by a Metropolis-Hastings correction step. In particular, when the OU process is simulated over sufficiently long times, we recover HMC as a special case. We illustrate this algorithm numerically using the example $\mathcal{G} = SO(3)$.

Keywords: Hamiltonian Monte Carlo · MCMC ·
Irreversible diffusions · Lie Groups · Geometric mechanics ·
Langevin dynamics · Sampling

1 Introduction

In this work, we construct an irreversible MCMC algorithm on Lie groups, which generalises the standard Hamiltonian Monte Carlo (HMC) algorithm on $\mathbb{R}^n$. The HMC method [13] generates samples from a probability density (with respect to an appropriate reference measure) known up to a constant factor by generating proposals using Hamiltonian mechanics, which is approximated by a reversible symplectic numerical integrator and followed by a Metropolis-Hastings step to correct for the bias introduced during the numerical approximations. The resulting time-homogeneous Markov chain is thus reversible, and allow distant proposals to be accepted with high probability, which decreases the correlations between samples (for a basic reference on HMC see [18], and for a geometric description see [4,6]). However, it is well-known that ergodic irreversible diffusions converge faster to their target distributions [14,21], and several irreversible MCMC algorithms based on Langevin dynamics have been proposed [19,20].

From a mechanical point of view, diffusions on Lie groups are important since they form the configuration space of many interesting systems, such as the free rigid body. For example in [9] Euler-Poincaré reduction of group invariant

© Springer Nature Switzerland AG 2019
F. Nielsen and F. Barbaresco (Eds.): GSI 2019, LNCS 11712, pp. 171–179, 2019.
https://doi.org/10.1007/978-3-030-26980-7_18

symplectic diffusions on Lie groups are considered in view of deriving dissipative equations from a variational principle, and in [1] Langevin systems on coadjoint orbits are constructed by adding noise and dissipation to Hamiltonian systems on Lie groups. The phase transitions of this system were analysed using a sampling method [2]. In lattice gauge theory one typically uses the HMC algorithm for semi-simple compact Lie groups which was originally presented in [15] and extended to arbitrary Lie groups in [3], see also [10,11,16]. In [5], it was shown how to construct HMC on homogeneous manifolds using symplectic reduction, which includes sampling on Lie groups as a special case.

To construct an irreversible algorithm on Lie groups, we first extend Langevin dynamics to general symplectic manifolds $\mathcal{M}$ based on Bismut's symplectic diffusion process [7]. Our generalised Langevin dynamics with multiplicative noise and nonlinear dissipation has the Gibbs measure as the invariant measure, which allows us to design MCMC algorithms that sample from a Lie group $\mathcal{G}$ when we take $\mathcal{M} = T^*\mathcal{G}$. In our Langevin system the irreversible component is determined by Hamiltonian vector fields which are compatible with the symplectic structure, thus avoiding the appearance of divergence terms associated to the volume distortion. We are then free to choose the noise-generating Hamiltonians to best suit the target distribution. Choosing Hamiltonians that only depend on position allows us to proceed with a Strang splitting of the dynamics into a position-dependent OU process in the fibres which can be solved exactly, and a Hamiltonian part which is approximated using a leapfrog scheme, followed by a Metropolis-Hastings acceptance/rejection step in a similar fashion to [8,19,20]. Ideally one wants to choose these Hamiltonians to achieve the fastest convergence to stationarity.

On a general manifold, it would be necessary to introduce local coordinates in order to solve the OU process on the fibres, making it difficult to implement. However, since our base manifold is a Lie group, the Maurer-Cartan form defines an isomorphism between the cotangent bundle $T^*\mathcal{G}$ and the trivial bundle $\mathcal{G} \times \mathfrak{g}^*$, which, given an inner-product on $\mathfrak{g}$, may further be identified with $\mathcal{G} \times \mathfrak{g}$. As a result, one may pull back the OU process on $T_g^*\mathcal{G}$ to an OU process on $\mathfrak{g}$ for any $g \in \mathcal{G}$, thus avoiding the problem of having to choose appropriate charts. Hence on Lie groups, we obtain a practical irreversible MCMC algorithm which generalises the $\mathbb{R}^n$-version of the irreversible algorithm considered in [19,20].

Finally, we simulate this algorithm in the special case $\mathcal{G} = SO(3)$ and perform a Maximum Mean Discrepancy (MMD) test to show that on average, the irreversible algorithm converges faster to the stationary measure than the corresponding reversible HMC on $SO(3)$.

2   Diffusions on Symplectic Manifolds

We consider diffusion processes on symplectic manifolds $(\mathcal{M}, \omega)$, where we have a natural volume form ω^n, and define the canonical Poisson bracket $\{g, f\} := \omega(X_f, X_g) = X_g f$, where X_g is the Hamiltonian vector field associated to the

Hamiltonian $g : \mathcal{M} \to \mathbb{R}$, i.e. $dg = \iota_{X_g}\omega$. Given arbitrary functions H and H_i for $i = 1, \ldots, m$ on $\mathcal{M}$, we consider the SDE

$$dZ_t = \left(X_H(Z_t) - \frac{\beta}{2}\sum_{i=1}^{m}\{H_i, H\}X_{H_i}(Z_t)\right)dt + \sum_{i=1}^{m}X_{H_i}(Z_t) \circ dW_t^i, \quad (1)$$

which has the generator, or forward Kolmogorov operator

$$\mathcal{L}f = -\{f, H\} - \frac{\beta}{2}\sum_{i=1}^{m}\{H, H_i\}\{f, H_i\} + \frac{1}{2}\sum_{i=1}^{m}\{\{f, H_i\}, H_i\}. \quad (2)$$

To show that the Gibbs measure is an invariant measure for (1), we will need the following lemma:

Lemma 1. *For a symplectic manifold $(\mathcal{M}, \omega)$ and two functions $f, g \in C^{\infty}(\mathcal{M})$ such that either $\partial\mathcal{M} = \varnothing$ or $g|_{\partial\mathcal{M}} = 0$, we have the following identity*

$$\int_{\mathcal{M}}\{f, g\}\omega^n = 0.$$

For the proof, see [12], Sect. 4.3. Hence (1) enables us to build MCMC algorithms on any symplectic manifold, and in particular the cotangent bundle of Lie groups, that converge to the Gibbs measure:

Theorem 1. *Given a symplectic manifold $(\mathcal{M}, \omega)$ without boundary, equation (1) on $\mathcal{M}$ has the Gibbs measure*

$$\mathbb{P}_{\infty}(z) = p_{\infty}\omega^n := \frac{1}{Z}e^{-\beta H(z)}\omega^n, \quad Z = \int_{\mathcal{M}}e^{-\beta H(z)}\omega^n,$$

as its stationary measure for any choice of $H_i : \mathcal{M} \to \mathbb{R}$ where $i = 1, \ldots, m$.

Proof. Using the Leibniz rule $\{fg, h\} = f\{g, h\} + g\{f, h\}$, we have

$$g\{f, H_i\}\{H, H_i\} = \{gf, H_i\}\{H, H_i\} - f\{g, H_i\}\{H, H_i\}$$
$$= \cdots = \{gf\{H, H_i\}, H_i\} - f\{g\{H, H_i\}, H_i\},$$

and similarly

$$g\{\{f, H_i\}, H_i\} = \{g\{f, H_i\}, H_i\} - \{f\{g, H_i\}, H_i\} + f\{\{g, H_i\}, H_i\}.$$

Hence one can compute the $L^2(\mathcal{M}, \omega^n)$-adjoint of the operator $\mathcal{L}$ as follows

$$\int_{\mathcal{M}}g(\mathcal{L}f)\,\omega^n = \int_{\mathcal{M}}g\left(-\{f, H\} - \frac{\beta}{2}\sum_{i=1}^{m}\{f, H_i\}\{H, H_i\} + \frac{1}{2}\sum_{i=1}^{m}\{\{f, H_i\}, H_i\}\right)\omega^n$$

$$= \int_{\mathcal{M}}\left(-\{fg, H\} - \frac{\beta}{2}\sum_{i=1}^{m}\{fg\{H, H_i\}, H_i\} + \frac{1}{2}\sum_{i=1}^{m}(\{g\{f, H_i\}, H_i\} - \{f\{g, H_i\}, H_i\})\right)\omega^n$$

$$+ \int_{\mathcal{M}}f\left(\{g, H\} + \frac{\beta}{2}\sum_{i=1}^{m}\{g\{H, H_i\}, H_i\} + \frac{1}{2}\sum_{i=1}^{m}\{\{g, H_i\}, H_i\}\right)\omega^n$$

$$= \int_{\mathcal{M}}f\left(\{g, H\} + \frac{\beta}{2}\sum_{i=1}^{m}\{g\{H, H_i\}, H_i\} + \frac{1}{2}\sum_{i=1}^{m}\{\{g, H_i\}, H_i\}\right)\omega^n = \int_{\mathcal{M}}f(\mathcal{L}^*g)\,\omega^n,$$

where we have used Lemma 1 to integrate the Poisson brackets to 0. Hence, we obtain the Fokker-Planck operator

$$\mathcal{L}^* g = \{g, H\} + \frac{\beta}{2} \sum_{i=1}^{m} \{g\{H, H_i\}, H_i\} + \frac{1}{2} \sum_{i=1}^{m} \{\{g, H_i\}, H_i\} \, .$$

Now, by the derivation property of the Poisson bracket, $\{f \circ g, h\} = f' \circ g \{g, h\}$ and noting that $p'_\infty(H) = -Z^{-1}\beta e^{-\beta H} = -\beta p_\infty(H)$, one can check that

$$\mathcal{L}^* p_\infty = p'_\infty(H)\{H, H\} + \frac{\beta}{2} \sum_{i=1}^{m} \{p_\infty(H)\{H, H_i\}, H_i\} - \frac{\beta}{2} \sum_{i=1}^{m} \{p_\infty(H)\{H, H_i\}, H_i\} = 0 \, .$$

Therefore $\mathbb{P}_\infty(z) = p_\infty(z)\omega^n$ is indeed an invariant measure for (1).

If $\mathcal{M} = T^*\mathcal{Q}$ is the cotangent bundle of a manifold without boundary $\mathcal{Q}$, we define the marginal measure $\mathbb{P}^1_\infty$ on $\mathcal{Q}$ by

$$\int_A \mathbb{P}^1_\infty = \int_{T^*A} \iota^* \mathbb{P}_\infty \, , \qquad (3)$$

for any measurable set $A \subset \mathcal{Q}$, where $\iota : T^*A \to T^*\mathcal{Q}$ is the inclusion map. In addition, if $(\mathcal{Q}, \gamma)$ is a Riemannian manifold, we can consider the Hamiltonian function $H(q, p) = \frac{1}{2}\gamma_q(p, p) + V(q)$, for $(q, p) \in T^*\mathcal{Q}$, and the marginal invariant measure $\mathbb{P}^1_\infty(dq)$ of the process (1) is simply

$$\mathbb{P}^1_\infty(dq) = \frac{1}{Z_1} e^{-V(q)} \sqrt{|g|} dq, \quad Z_1 = \int_{\mathcal{Q}} e^{-V(q)} \sqrt{|g|} dq \, ,$$

where $\sqrt{|g|} dq$ is the Riemannian volume form.

The MCMC algorithm which we will derive in Sect. 3 is based on a Strang splitting of the dynamics (1) into a Hamiltonian part and a Langevin part. Hereafter, we identify $T^*\mathcal{Q}$ with $T\mathcal{Q}$ through the metric and just consider the dynamics on $T\mathcal{Q}$ instead of $T^*\mathcal{Q}$.

3 Irreversible Langevin MCMC on Lie Groups

Consider a n dimensional Lie group $\mathcal{G}$ and let $e_i, \theta^i, i = 1, \dots, n$ be an orthonormal basis of left-invariant vector fields and dual one-forms respectively. We consider $H = V \circ \pi + T : T\mathcal{G} \to \mathbb{R}$, where T is the kinetic energy associated to a bi-invariant metric on $\mathcal{G}$ and $V \propto \log \chi : \mathcal{G} \to \mathbb{R}$ is the potential energy (with the projection $\pi : T\mathcal{G} \to \mathcal{G}$), where χ is the distribution we want to sample from on $\mathcal{G}$. We let $v^i : T\mathcal{G} \to \mathbb{R}$ be the fibre coordinate functions with respect to the left-invariant vector fields, $v^i(g, u_g) := \theta^i_g(u_g)$.

Vector fields tangent to $T\mathcal{G}$ (i.e., elements of $\Gamma(TT\mathcal{G})$) can be expanded in terms of left-invariant vector fields e_i and the fibre-coordinate vector fields ∂_{v^i}, (i.e., $\Gamma(TT\mathcal{G}) \cong \Gamma(T\mathcal{G} \oplus T\mathfrak{g})$). We consider noise Hamiltonians that depend only on position, $H_i = U_i \circ \pi$ where $U_i : \mathcal{G} \to \mathbb{R}$, so the corresponding Hamiltonian vector fields can be written as $X_{H_i} = -e_j(U_i)\partial_{v^j}$, (see [3]). Hence the stochastic process (1) on $T\mathcal{G}$ can be split up into a Langevin part

$$\mathrm{d}Z_t = \frac{\beta}{2}\sum_{i=1}^m X_H(H_i)X_{H_i}(Z_t)\mathrm{d}t + \sum_{i=1}^m X_{H_i}(Z_t)\circ\mathrm{d}W_t^i\,,$$

$$= -\frac{\beta}{2}\sum_{i=1}^m v^k e_k(U_i)e_j(U_i)\partial_{v^j}(Z_t)\mathrm{d}t - \sum_{i=1}^m e_j(U_i)\partial_{v^j}(Z_t)\circ\mathrm{d}W_t^i\,. \qquad (4)$$

and a Hamiltonian part

$$\frac{\mathrm{d}Z_t}{\mathrm{d}t} = X_H(Z_t)\,. \qquad (5)$$

Note the geodesics are given by the one-parameter subgroups, with Hamiltonian vector field $X_T = v^k e_k$. Since X_{H_i} only has components in the fibre direction ∂_{v^i} (i.e., it has no e_i components along $\mathcal{G}$) the diffusion starting at any point $(g, v) \in T\mathcal{G}$ remains in $T_g\mathcal{G}$, i.e., with the same base point g. When $\mathcal{G} = \mathbb{R}^n$ and we use the standard kinetic energy $T = \frac{1}{2}\|v\|_{\mathbb{R}^n}^2$, then vector fields become gradients, i.e. $e_j = \partial_{q^j}$, and equation (4) becomes the space dependent Langevin equation for $(q, v) \in T\mathbb{R}^n$,

$$\dot{q} = 0, \qquad \mathrm{d}v_t = -\frac{\beta}{2}\nabla_q U_i(q)\nabla_q U_i(q)^T v_t\mathrm{d}t - \nabla_q U_i(q)\circ\mathrm{d}W_t^i\,. \qquad (6)$$

This equation has been considered for instance in [17] to construct MCMC algorithm with space dependent noise.

Now let $\xi_i := e_i(1)$ be a basis of the Lie algebra $\mathfrak{g}$, where 1 is the identity. Then $e_i(g) = \partial_1 L_g\xi$ and we can identify $T\mathcal{G}$ with $\mathcal{G}\times\mathfrak{g}$ through the relation $(g, v^i e_i(g)) \sim (g, v^i\xi_i)$. In other words, we may now think of v^i as the Lie algebra coordinate functions $v^i : \mathcal{G}\times\mathfrak{g} \to \mathbb{R}$ with $v^i(g, u) = \theta_1^i(u)$, and since $\mathfrak{g}$ is a vector space, we can identify $\partial_{v^i} \sim \xi_i$ and write

$$X_{H_i}(g, v) = -e_j|_g(U_i)\xi_j =: \sigma_{ji}(g)\xi_j\,. \qquad (7)$$

for $i = 1, \ldots, m$ and $j = 1, \ldots, n$. For matrix Lie groups, this becomes

$$X_{H_i}(g, v) = -\mathrm{Tr}\big(\nabla U_i^T g\xi_j\big)\xi_j = \sigma_{ji}(g)\xi_j\,, \qquad (8)$$

where $(\nabla U_i)_{ab} := \partial_{x_{ab}}U_i$, where x_{ab} are the matrix coordinates of $g \in \mathcal{G}$. The Langevin equation (4) can then be written as

$$\dot{g} = 0, \qquad \mathrm{d}v_t = -\frac{\beta}{2}(\sigma(g)\sigma(g)^T)_{jk}v_t^k\xi_j\mathrm{d}t + \sigma_{ji}(g)\xi_j\mathrm{d}W_t^i\,, \qquad (9)$$

and if we identify $\mathfrak{g} \sim \mathbb{R}^n$, $v^i\xi_i \sim \mathbf{v} \in \mathbb{R}^n$, we get a standard OU process on $\mathbb{R}^n$

$$\mathrm{d}\mathbf{v}_t = -\frac{\beta}{2}\sigma\sigma^T\mathbf{v}_t\mathrm{d}t + \sigma\mathrm{d}\mathbf{W}_t\,, \qquad (10)$$

for the vector-valued Wiener process $\mathbf{W}_t = (W_t^1, \ldots, W_t^m) \in \mathbb{R}^m$. Note that the noise term can be interpreted as an Itô integral since the diffusion coefficient σ does not depend on $\mathbf{v}$. We can solve this Langevin equation explicitly if the matrix $D = \sigma\sigma^T$ is invertible. This is the case if the vectors ∇U_i form an orthonormal basis of $\mathbb{R}^n$, or more generally if they satisfy the Hörmander condition. The explicit solution is then given by

$$v_{t+h} = e^{-\frac{\beta}{2}Dh} v_t + \sigma \int_t^{t+h} e^{-\frac{\beta}{2}D(t+h-s)} \mathrm{d}\mathbf{W}_s \,, \tag{11}$$

which has the transition probability

$$p(v_0, v) = \frac{1}{(2\pi)^{n/2}|\mathrm{det}\Sigma_h|} \exp\left(-\frac{1}{2}\left\|v - e^{-\frac{\beta}{2}Dh}v_0\right\|_{\Sigma_h^{-1}}^2\right) \,, \tag{12}$$

where $\quad \Sigma_h = \frac{1}{\beta}\left(\mathrm{Id} - e^{-\beta Dh}\right) \,.$

3.1   MCMC Algorithm on Lie Groups

From the Langevin system considered in the previous section, we can construct the following MCMC algorithm to sample from the distribution $\chi := e^{-\frac{\beta}{2}V}$ on $\mathcal{G}$. Starting from $(g_0, v_0) \in T\mathcal{G}$, we iterate the following

1. Solve equation (9) exactly until time h by sampling

$$v^* \simeq \mathcal{N}\left(e^{-\frac{\beta}{2}Dh}v_0, \Sigma_h\right) \,, \tag{13}$$

   to obtain $(\bar{g}_0, \bar{v}_0) = (g_0, v^*)$;
2. Approximate the Hamiltonian system (5) using N Leapfrog trajectories with step size $\delta > 0$. Starting at $(\bar{g}_0, \bar{v}_0) = (g_0, v^*)$:

For $k = 0, \ldots, N - 1$:[1]

$$\bar{v}_{k+\frac{1}{2}} = \bar{v}_k - \frac{\delta}{2}\mathrm{Tr}\left(\partial_x V^T \bar{g}_k \xi_i\right)\xi_i$$

$$\bar{g}_{k+1} = \bar{g}_k \exp\left(\delta\, \bar{v}_{k+\frac{1}{2}}\right)$$

$$\bar{v}_{k+1} = \bar{v}_{k+\frac{1}{2}} - \frac{\delta}{2}\mathrm{Tr}\left(\partial_x V^T \bar{g}_{k+1}\xi_i\right)\xi_i$$

   to obtain $(\bar{g}_N, \bar{v}_N)$. The time step δ and number of steps N are to be tuned appropriately by the users.
3. Accept or reject the proposal by a Metropolis-Hastings step. We accept the proposal $(\bar{g}_N, \bar{v}_N)$ with probability

$$\alpha = \min\left\{1, \exp\left(-H(\bar{g}_N, \bar{v}_N) + H(\bar{g}_0, \bar{v}_0)\right)\right\} \,,$$

and set $(g_1, v_1) = (\bar{g}_N, \bar{v}_N)$. On the other hand, if the proposal is rejected, we set $(g_1, v_1) = (\bar{g}_0, -\bar{v}_0)$.

---

[1] For a non-matrix group, simply replace $\mathrm{Tr}\left(\partial_x V^T \bar{g}_k \xi_i\right)$ with $e_i|_g(V)$.

Notice also that in the limit $h \to \infty$, this algorithm becomes the standard HMC algorithm, which is reversible, but for finite h, the algorithm is irreversible (see [19] for a detailed discussion).

4 Example on $SO(3)$

As an example we will pick the rotation group $\mathcal{G} = SO(3)$, where the Lie algebra $\mathfrak{so}(3)$ consists of 3×3 anti-symmetric matrices so that the kinetic energy on the Lie algebra is $T(v) = \frac{1}{2}\mathrm{Tr}(v^T v)$, for $v \in \mathfrak{so}(3)$. The potentials on $SO(3)$ will be defined by functions $V, U_i : \mathbb{R}^{3 \times 3} \to \mathbb{R}$ on matrices, which we choose to be of the form $V(g) = e^{\alpha Tr(g)}$, and isotropic noise with

$$U_i(g) = \epsilon \mathrm{Tr}(e^{-\xi_i} g) \qquad \text{for} \qquad \mathrm{span}(\xi_i) = \mathfrak{so}(3). \tag{14}$$

We then obtain samples g_t on $SO(3)$, which we can project onto the sphere by simply letting the group act on a vector $z = (0, 0, 1)$, to get $x_t = g_t z$, see panel (a) of Fig. 1. From these samples g_t, we can also estimate the convergence rate of the MCMC algorithm by computing the maximum mean discrepancy (MMD) between the set of first N samples and the whole sequence, using the values on the diagonal of the matrices g_t. We see that in Fig. 1, small values for h give MCMC algorithms with a faster convergence rate than the HMC limit, i.e., $h \to \infty$. This is a direct consequence from the irreversibility of the algorithm, as explained in [14, 21]. Even if a faster convergence is desirable, one has to ensure that the correct distribution is sampled, and if h is taken too small, the algorithm will be close to pure Hamiltonian dynamics, with additional reversal steps $v \to -v$ when the proposed state is rejected. We observe this effect already for $h = 0.01$ where the convergence is as fast as $h = 0.1$ for the first steps of the chain, but then later slows down, as the distribution is not sampled correctly.

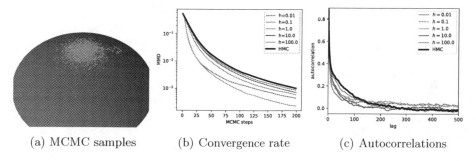

(a) MCMC samples   (b) Convergence rate   (c) Autocorrelations

Fig. 1. This figure illustrates the irreversible MCMC algorithm on Lie group with $\mathcal{G} = SO(3)$. For each run, we sampled 5000 samples, shown in panel (a), along with the first one it red and the next 50 in green. We ran several chains with several parameters h, corresponding to the integration time of the Langevin dynamics. For large h, the MCMC algorithm converges to the HMC algorithm, also displayed in black. Each line in panel (b) and (c) are averages over 20 chains with the same parameters. For the leapfrog integrator we use 5 timesteps with a total time $\delta = 0.5$. The MMD computation on $\mathcal{G}$ has been run with Gaussian kernel of variance $\sigma = 1$.

Acknowledgements. The authors thank A. Duncan for the useful insights that helped improve this work.AA acknowledges EPSRC funding through award EP/N014529/1 via the EPSRCCentre for Mathematics of Precision Healthcare. ST acknowledges the Schrödinger scholarship scheme for funding during this work. AB was supported by a Roth Scholarship funded by the Department of Mathematics, Imperial College London, by EPSRC Fellowship (EP/J016934/3) and by The Alan Turing Institute under the EPSRC grant [EP/N510129/1].

References

1. Arnaudon, Alexis, De Castro, A.L., Holm, D.D.: Noise and dissipation on coadjoint orbits. J. Nonlinear Sci. **28**(1), 91–145 (2018)
2. Arnaudon, A., Takao, S., Networks of coadjoint orbits: from geometric to statistical mechanics, arXiv preprint arXiv:1804.11139 (2018)
3. Barp, A.: Hamiltonian Monte Carlo On Lie Groups and Constrained Mechanics on Homogeneous Manifolds, arXiv preprint arXiv:1903.04662 (2019)
4. Barp, Alessandro, Briol, François-Xavier, Kennedy, A.D., Girolami, M.: Geometry and dynamics for markov chain monte carlo. Annu. Rev. Stat. Appl. **5**, 451–471 (2018)
5. Barp, A., Kennedy, A.D., Girolami, M.: Hamiltonian Monte Carlo on Symmetric and Homogeneous Spaces via Symplectic Reduction, arXiv preprint arXiv:1903.02699 (2019)
6. Betancourt, Michael, Byrne, Simon, Livingstone, Sam, Girolami, Mark, et al.: The geometric foundations of hamiltonian monte carlo. Bernoulli **23**(4A), 2257–2298 (2017)
7. Bismut, J.M.: Méecanique aléeatoire lect. Notes Maths **966**, 19881 (1981)
8. Bou-Rabee, N., Vanden-Eijnden, E.: A patch that imparts unconditional stability to certain explicit integrators for SDEs, arXiv preprint arXiv:1010.4278 (2010)
9. Chen, X., Cruzeiro, A.B., Ratiu, T.S.: Constrained and stochastic variational principles for dissipative equations with advected quantities, arXiv preprint arXiv:1506.05024 (2015)
10. Clark, M.A., Kennedy, A.D.: Accelerating staggered-fermion dynamics with the rational hybrid Monte Carlo algorithm. Phys. Rev. D **75**(1), 011502 (2007)
11. Clark, M.A., Kennedy, A.D., Silva, P.J.: Tuning HMC using Poisson brackets. arXiv preprint arXiv:0810.1315 (2008)
12. Cruzeiro, A.B., Holm, D.D., Ana Bela Cruzeiro: Momentum maps and stochastic clebsch action principles. Commun. Math. Phys. **357**(2), 873–912 (2018)
13. Duane, Simon, Kennedy, A.D., Pendleton, B.J., Roweth, D.: Hybrid monte carlo. Phys. lett. B **195**(2), 216–222 (1987)
14. Duncan, A.B., Lelievre, T., Pavliotis, G.A.: Variance reduction using nonreversible Langevin samplers. J. Stat. Phys. **163**(3), 457–491 (2016)
15. Kennedy, A.D., Rossi, P.: Classical mechanics on group manifolds. Nucl. Phys. B **327**, 782–790 (1989)
16. Kennedy, A.D., Silva, P.J., Clark, M.A.: Shadow Hamiltonians, poisson brackets, and gauge theories. Phys. Rev. D **87**(3), 034511 (2013)
17. Mathias, Rousset, Gabriel, Stoltz, Tony, L.: Free Energy Computations: A Mathematical Perspective. World Scientific, Singapore (2010)
18. Neal, R.M., et al.: MCMC using Hamiltonian dynamics. Handbook of Markov Chain Monte Carlo **2**(11), 2 (2011)

19. Ottobre, M.: Markov Chain Monte Carlo and irreversibility. Rep. Math. Phys. **77**(3), 267–292 (2016)
20. Ottobre, M., Pillai, N.S., Pinski, F.J., Stuart, A.M., et al.: A function space HMC algorithm with second order Langevin diffusion limit. Bernoulli **22**(1), 60–106 (2016)
21. Rey-Bellet, L., Spiliopoulos, K.: Irreversible langevin samplers and variance reduction: a large deviations approach. Nonlinearity **28**(7), 2081 (2015)

Predicting Bending Moments
with Machine Learning

Elena Celledoni⬛, Halvor S. Gustad, Nikita Kopylov$^{(\boxtimes)}$⬛,
and Henrik S. Sundklakk

NTNU, Trondheim, Norway
{elena.celledoni,nikita.kopylov,henrik.s.sundklakk}@ntnu.no,
halvorsg@stud.ntnu.no
http://www.ntnu.edu/imf/research/dna/numerics

Abstract. We investigate the possibility of predicting the bending moment of slender structures based on a limited number of deflection measurements. These predictions can help to estimate the wear and tear of the structures. We compare linear regression and a recurrent neural network on numerically simulated Euler–Bernoulli beam and drilling riser.

Keywords: Euler–Bernoulli beam · Drilling riser ·
Slender structures · Material fatigue · Machine learning

1 Introduction

Slender flexible structures like beams, pipes, and drilling risers are subjected to repeated loads due to the environment, thus the material fatigue should be assessed to guarantee performance and safety. Frequently, data which describe the state of the structure are scarce because the possibility to install sensors is limited. Moreover, the measurements are often noisy as it is not possible to ensure that the sensors stay well-oriented with respect to the structure. For these reasons, the problem of learning the wear from a large amount of data from few sensors attracted considerable attention [1,8,9,13].

We consider two applications: estimating the maximal bending moment of an Euler–Bernoulli beam and the fatigue of a marine drilling riser given a small number of sensors.

2 Slender Structures

2.1 Euler–Bernoulli Beam

The first structure we consider is a one-dimensional Euler–Bernoulli beam which has a unit length and a unit stiffness, is clamped at the ends, and is subjected

This work was carried out during the tenure of an ERCIM "Alain Bensoussan" Fellowship Programme.

F. Nielsen and F. Barbaresco (Eds.): GSI 2019, LNCS 11712, pp. 180–187, 2019.
https://doi.org/10.1007/978-3-030-26980-7_19

to an external load:

$$\frac{d^4 w(x)}{dx^4} = q(x)$$
$$w(0) = w(1) = 0,$$
$$w'(0) = w'(1) = 0.$$
(1)

Here $w \colon (0,1) \to \mathbb{R}$ is the transverse deflection, and $q \colon (0,1) \to \mathbb{R}$ is the load applied normally to the beam. As a measure of tear and wear we use the bending moment M, which is proportional to the second derivative of the deflection:

$$M = -\frac{d^2 w(x)}{dx^2}.$$
(2)

2.2   Drilling Riser

Our second, more advanced model is the drilling riser. It connects an offshore oil platform to the wellhead which is on the seabed and, therefore, allows for little movement. One of the reasons for the fatigue at the connection point is the riser's oscillations due to waves and currents. We aim to predict bending at the wellhead to assess the remaining lifetime of the riser based on its stress cycles.

Riser Equations. The set of equations governing a one-dimensional riser arise from the conservation of energy [6]. Assuming the oil rig to be directly above the wellhead, we get the following partial differential equation [6,14]:

$$EI\frac{\partial^4 u(x,t)}{\partial x^4} - T\frac{\partial^2 u(x,t)}{\partial x^2} + \rho\frac{\partial^2 u(x,t)}{\partial t^2} + c\frac{\partial u(x,t)}{\partial t} = f(x,t),$$
(3)

where EI is the bending stiffness, T is the tension, ρ is the linear density of the riser, and c is a damping constant, and $f(x,t)$ represents external forces acting on the riser. The riser is assumed to hang freely from the motionless oil rig positioned at $x = 0$. At the seabed the riser is clamped to the wellhead, hence the boundary conditions:

$$u(0,t) = 0, \ \ u(L,t) = 0$$
$$u''(0,t) = 0, \ \ u'(L,t) = 0$$
(4)

where L is the length of the riser.

External Forces. We assume that the only forces acting on the riser come from ocean currents and waves. This is a simplifying assumption since vortex-induced vibrations (VIV) also occur in a real situation. VIV can be included in a three-dimensional model with a significant increase in simulation complexity. We use Morison's equation [11] that describes the forces acting on a cylinder submerged in water. Then, the resulting force f on a body in an unsteady viscous flow is the sum of a drag term f_D and an inertial term f_I [14]:

$$f_D = \frac{1}{2}\rho_w C_D R(v - \dot{u})|v - \dot{u}|, \tag{5}$$

$$f_I = \rho_w C_I A_e \dot{v} - \rho_w (C_I - 1)A_e \ddot{u}, \tag{6}$$

where ρ_w is the density of water, C_D is the drag coefficient, R is the radius of the riser, v is the flow speed, and $\dot{u}$ is the speed of the riser in the direction orthogonal to its axis; C_I is the inertia coefficient, A_e is the external cross-section area, and $\ddot{u}$ is it the derivative of the orthogonal velocity component of the riser.

In Eq. (5), we consider v that obey Airy theory, which describes gravity-dominated surface waves [11,14]. It gives a sufficiently good approximation for waves in a homogeneous fluid over a uniform seabed. In our model, the depth is large enough to neglect the seabed variation after scaling. The fluid velocity $v(x,t)$ is given as

$$v(x,t) = \frac{H}{2}\frac{\cosh(k(L-x))}{\sinh(kL)}\omega\cos(\omega t) \tag{7}$$

with $k = \frac{2\pi}{\lambda}$, $\omega = \frac{2\pi}{T_\lambda}$, where H is the wave amplitude, λ is the wave length, T_λ is the wave period, and L is the length of the riser [14, p. 266].

Our riser has the following parameters: $EI = 3.186 \times 10^5\,\mathrm{Nm^2}$, $T = 7.554 \times 10^6\,\mathrm{N}$, $c = 2$, $\rho = 1200\,\mathrm{kg/m^3}$; $\rho_w = 1000\,\mathrm{kg/m^3}$, $C_D = 2$, $C_I = 2$, $A_e = \pi/100\,\mathrm{m^2}$, $R = 1/10\,\mathrm{m}$; $L = 100\,\mathrm{m}$. To calculate the forcing term, we use a combination of waves with amplitudes 4.1 m, 8.5 m , and 14.8 m , which are typical for the Norwegian continental shelf [15].

3   Measurements Simulation

Machine learning deals with large amounts of data. As we do not possess real-world measurements of our structures, we use the mathematical models described above to generate training and test sets. The approach is similar for both structures: given an external force, we solve numerically the corresponding equations using a finite element method. Having obtained a solution, we "measure" displacements at certain points, and in this way we get training examples. Then, we calculate the bending moments to obtain corresponding targets for prediction.

3.1   Finite Element Method Solution

Numerical solutions are obtained with the finite elements method with Hermite polynomials. To avoid the interference between discretization and predictions, a sufficiently large number of finite elements is used.

Beam. In case of our first structure, the discretization is rather simple. Four Hermite interpolation polynomials ensure C^1-continuity. As the basis functions are 3rd-order polynomials, the second derivative is linear and attains maximum at an endpoint. Since the solution is only C^1, the second derivative is not necessarily continuous at the nodes. Consequently, it is evaluated twice on each element, and the largest absolute value is used as the target bending moment M in the training set. On the spatial interval $[0, 1]$ we use 21 FEs.

Riser. Riser (3) to (5) are discretized in the spatial domain using a finite element method. Compared to the beam, here we use 5th-order Hermite elements because it makes the boundary conditions easier to implement and the second derivative of the solution in the nodes is instantly available:

$$u_h(x, t) = \sum_{j=0}^{N} u_j(t)\phi_{3j}(x) + \theta_j(t)\phi_{3j+1}(x) + c_j(t)\phi_{3j+2}(x). \tag{8}$$

where $u_j(t)$, $\theta_j(t)$ and $c_j(t)$ are the value, the inclination and the curvature in x_j, respectively. The function $\phi_{3j+k}(x)$ for $k = 0, 1, 2$ takes the value 1 in either its 0th, 1st or 2nd derivative and 0 in the remaining two. This gives three active basis functions in the jth node hence the subscript $3j + k$. For the riser of length $L = 100\,\text{m}$ we use 100 FEs.

Eventually, we get a system of non-linear ODEs:

$$\boldsymbol{D}\boldsymbol{u}(t) + \boldsymbol{A}\boldsymbol{u}(t) + \boldsymbol{M_p}\ddot{\boldsymbol{u}}(t) + \boldsymbol{M_c}\dot{\boldsymbol{u}}(t) = \boldsymbol{F}(t, \boldsymbol{u}), \quad \boldsymbol{u}(0) = 0, \quad \dot{\boldsymbol{u}}(0) = 0, \tag{9}$$

where the ith element on the right-hand side of (9) is written as

$$F_i = \rho_w \int_0^L \frac{1}{2}C_D R \left(v - \sum_{j=0}^{N} \dot{u}_j\phi_j \right) \left| v - \sum_{j=0}^{N} \dot{u}_j\phi_j \right| + C_M A_e \dot{v}\phi_i \, dx.$$

Transforming the equation into a system of first order ODEs by setting $\boldsymbol{u}_0 = \boldsymbol{u}$ and $\boldsymbol{u}_1 = \dot{\boldsymbol{u}}$ gives the following system

$$\begin{bmatrix} \dot{\boldsymbol{u}}_0(t) \\ \dot{\boldsymbol{u}}_1(t) \end{bmatrix} + \begin{bmatrix} 0 & -\boldsymbol{I} \\ \boldsymbol{M_p}^{-1}(\boldsymbol{A} + \boldsymbol{D}) & \boldsymbol{M_p}^{-1}\boldsymbol{M_c} \end{bmatrix} \begin{bmatrix} \boldsymbol{u}_0(t) \\ \boldsymbol{u}_1(t) \end{bmatrix} = \begin{bmatrix} 0 \\ \boldsymbol{M_p}^{-1}\boldsymbol{F}(t, u) \end{bmatrix}$$

which may be written as

$$\dot{\boldsymbol{u}}(t) + \boldsymbol{B}\boldsymbol{u} = \boldsymbol{F}(t, \boldsymbol{u}(t)), \quad \boldsymbol{u}(0) = 0. \tag{10}$$

The problem is stiff, so we use the exponential Euler method [7] with a sufficiently small time step 0.01 to solve it for $t \in [0, 100]$.

3.2   Beam Dataset

For the beam, we use the following procedure. First, we draw n values x_i from a uniform distribution on the interval $[0, 1]$ and explicitly add the beam endpoints.

At these $n+2$ gridpoints a random load is applied: we get $n+2$ samples from the standard normal distribution which represent magnitudes $q(x_i)$, $i = 1, \ldots, n+2$. The load $q(x)$ on the all the beam is then taken as the linear interpolation these points. In the experiments we have used $n = 2$.

Once we have the numerical solutions for $w(x)$, we calculate the maximum bending moment M. We consider two setups: putting one deflection sensor at equidistant points on the riser, and putting two symmetrically around $x = 0.5$.

3.3   Riser Dataset

We aim to predict the bending moment (and, consequently, the fatigue) at the wellhead. Similarly to the beam, we use one and two sensors.

For installation, maintenance, and reliability of the signals, it would be ideal to have only one sensor on the top of the riser. However, it is reasonable to expect that the top position is not the optimal one. We simulate measurements with the sensor placed in 9 distinct points at the riser (10m between them) and compare how predictions change.

In the second case, we also take into account the inclination of the riser which is relatively easy to measure. From the engineering point of view, the corresponding equipment is cheaper and easier to maintain because the sensor is not submerged. One may expect that the additional information will improve prediction accuracy.

After placing a sensor at a certain point, we construct datasets in the following way. The vector of displacement measurements for $t \in [0, 100]$ is sliced with a moving window of size $w = 50$ with stride equal to one, and each of these subvectors is treated as a training example. Its corresponding target is the bending moment at the final time of each sliced subinterval. That is, we predict the bending using the $w = 50$ preceding measurements of the displacement.

The model is tested with two sets: the first one corresponds to waves with smaller amplitudes and the second set contains waves with larger amplitudes.

4   Machine Learning Prediction

4.1   Linear Regression for the Beam

With the procedure described above, 5000 simulations have been performed with different loads. Linear regression from Scikit-learn [12] is used to predict the maximal bending moment M. The coefficient of determination r^2, describing the amount of the variance in the data explained by the model, is chosen to asses the goodness of fit. Results for one and two symmetrically put sensors are summarized in Table 1. Although two sensors give a slightly better result, one sensor seems to be sufficient. Additional sensors do not improve prediction yielding $r^2 = 0.948$ (3 symmetric points) and $r^2 = 0.964$ (6 symmetric points), however, in these cases we note that points which are closer to the ends are given more weight. One may conclude that linear regression represents well the linear Euler–Bernoulli beam.

4.2 Recurrent Neural Network for the Riser

Since the riser is described by a time-dependent non-linear equation, we expect that a more advanced machine learning should have advantages over linear regression. As the measurements form a time series, we use a recurrent neural network to capture the temporal dynamics [2,4].

We have chosen the antisymmetric RNN [3], in which forward propagation through l layers is defined as

$$a^{(0)} = \Delta t\sigma(V^{(0)}x^{(0)} + b^{(0)})$$
$$a^{(k)} = a^{(k-1)} + \Delta t\sigma(W^{(k-1)}a^{(k-1)} + V^{(k)}x^{(k)} + b^{(k)}) \quad \text{for } k = 1,\dots,l$$
$$a^{(l+1)} = g(W^{(l)}a^{(l)} + b^{(l+1)}).$$

The vector $a^{(k)}$ contains the value of the nodes in the hidden layer, $b^{(k)}$ is a vector of the biases, $x^{(k)}$ is the features being passed to the kth layer, $V^{(k)}$ and $W^{(k)}$ are weight matrices. The functions $\sigma(x)$ and (output) $g(x)$ are activation functions applied elementwise. The latter is forced to be antisymmetric to enhance the accuracy and the training properties of the NN [3,5].

Results for one accelerometer are presented in Table 2. We observe that the RNN has a higher accuracy when the sensor is far form the wellhead, where the bending is expected to be the greatest, while regression provides nearly the same results with the sensor close to the connection point. A similar difference between two models can be observed when an inclination sensor is added to the top of the riser (Table 3). We attribute the improvement in prediction to the ability of RNNs to capture non-linear behaviour of time-dependent systems.

Table 1. Coefficient of determination with one and two deflection sensors on the beam

Position of 1st sensor	0.1	0.2	0.3	0.4	0.5	0.6	0.7	0.8	0.9
One sensor	0.956	0.948	0.982	0.931	0.967	0.986	0.975	0.946	0.983
Two sensors	0.968	0.968	0.965	0.969	0.969	0.971	0.973	0.966	0.971

Table 2. Coefficient of determination with a displacement sensor in different positions on the riser

Distance from the top, m		10	20	30	40	50	60	70	80	90
Training set	LR	0.208	0.292	0.366	0.44	0.565	0.734	0.816	0.887	0.977
	RNN	0.819	0.845	0.872	0.911	0.953	0.980	0.993	0.997	0.998
Test set 1	LR	0.197	0.271	0.357	0.434	0.584	0.742	0.822	0.912	0.986
	RNN	0.294	0.335	0.392	0.532	0.713	0.825	0.925	0.962	0.982
Test set 2	LR	0.32	0.416	0.494	0.566	0.709	0.802	0.89	0.957	0.995
	RNN	0.627	0.656	0.726	0.739	0.883	0.954	0.987	0.995	0.995

Table 3. Coefficient of determination with an inclination sensor on the top and a displacement sensor at the riser

Distance from the top, m		10	20	30	40	50	60	70	80	90
Training set	LR	0.351	0.414	0.481	0.544	0.605	0.761	0.819	0.889	0.977
	RNN	0.912	0.950	0.959	0.969	0.980	0.988	0.992	0.997	0.998
Test set 1	LR	0.296	0.374	0.449	0.535	0.63	0.767	0.822	0.914	0.986
	RNN	0.436	0.481	0.563	0.650	0.727	0.876	0.895	0.963	0.983
Test set 2	LR	0.208	0.300	0.424	0.554	0.754	0.822	0.891	0.958	0.995
	RNN	0.601	0.713	0.800	0.854	0.932	0.974	0.979	0.995	0.995

5   Discussion and Conclusion

In this work, using simulated data, we have showed how two machine learning models can be used for fatigue estimation from sensors: predicting the bending moment of a clamped Euler–Bernoulli beam and of a drilling riser. We have demonstrated how sensors position affect the accuracy: the closer a sensor is to the point with high stress the better is the model prediction. We note that, although in certain setups linear regression can make predictions for non-linear dynamics, the more advanced RNN is more accurate. Therefore, we expect that including stronger non-linearities into the model may require more advanced machine learning techniques than the RNN used in this work.

We see the presented experiments as a starting point for work on various slender structures and consider several possible directions for developing the results. A more complex model of a riser (e.g. with moving rig and vortex-induced vibrations) should be considered in the absence of real measurements. Concerning the beam, it is more favourable to utilize a model formulated with Lie group $SO(3)$, like Kirchhoff–Love or Simo–Reissner (for a review see, e.g. [10]). When real data is available from industry, it is necessary to validate the modelled results versus the actual behaviour of the structures considered in this work.

References

1. Authén, K.: Learning from riser analyses and predicting results with artificial neural networks. In: Volume 3B: Structures, Safety and Reliability. ASME, V03BT02A056, June 2017. https://doi.org/10.1115/OMAE2017-61775
2. Burkov, A.: The Hundred-Page Machine Learning Book, Quebec (2019)
3. Chang, B.: et al.: AntisymmetricRNN: a dynamical system view on recurrent neural networks. In: International Conference on Learning Representations (2019)
4. Chollet, F.: Deep Learning with Python, vol. 28, p. 384. Manning Publications, New York (2017)
5. Haber, E., Ruthotto, L.: Stable architectures for deep neural networks. Inverse Prob. **34**(1), 014004 (2017). https://doi.org/10.1088/1361-6420/aa9a90
6. He, W., et al.: Dynamics and Control of Mechanical Systems in Off Shore Engineering. Springer-Verlag, London (2014). https://doi.org/10.1007/978-1-4471-5337-5

7. Hochbruck, M., Ostermann, A.: Exponential integrators. Acta Numerica **19**, 209–286 (2010). https://doi.org/10.1017/S0962492910000048

8. Liu, K., et al.: Nonlinear dynamic analysis and fatigue damage assessment for a deepwater test string subjected to random loads. Petrol. Sci. **13**(1), 126–134 (2016). https://doi.org/10.1007/s12182-015-0063-4

9. Liu, X., et al.: Analysis on the operation fatigue of deepwater drilling riser system. Open Petrol. Eng. J. **9**(1), 279–287 (2016). https://doi.org/10.2174/1874834101609010279

10. Meier, C., Popp, A., Wall, W.A.: Geometrically exact finite element formulations for slender beams: Kirchhoff–Love theory versus Simo–Reissner theory. Arch. Comput. Methods Eng. **26**(1), 163–243 (2019). https://doi.org/10.1007/s11831-017-9232-5

11. Morison, J.R., et al.: The force exerted by surface waves on piles. Petrol. Trans., AIME **189**(4), 149–154 (1950)

12. Pedregosa, F., et al.: Scikit-learn: machine learning in Python. J. Mach. Learn. Res. **12**, 2825–2830 (2011)

13. Raissi, M., et al.: Deep learning of vortex-induced vibrations. J. Fluid Mech. **861**, 119–137 (2019). https://doi.org/10.1017/jfm.2018.872

14. Sparks, C.: Fundamentals of Marine Riser Mechanics. PennWell Corporation, Tulsa (2007)

15. Vikebø, F., et al.: Wave height variations in the North Sea and on the Norwegian Continental Shelf, 1881–1999. Cont. Shelf Res. **23**(3–4), 251–263 (2003). https://doi.org/10.1016/s0278-4343(02)00210-8

The Exponential of Nilpotent Supergroups in the Theory of Harish-Chandra Representations

Claudio Carmeli[1] , Rita Fioresi[2]([⊠]) , and V. S. Varadarajan[3]

[1] DIME, Università di Genova, Genova, Italy
carmeli@dime.unige.it
[2] Dipartimento di Matematica, University of Bologna,
piazza Porta San Donato 5, 40126 Bologna, Italy
rita.fioresi@unibo.it
[3] Department of Mathematics, UCLA, Los Angeles, CA 90095-1555, USA

Abstract. In this paper we discuss the exponential map in the case of nilpotent superalgebras. This provides global coordinates for nilpotent analytic supergroups, which are useful in the applications.

Keywords: Supergeometry · Lie theory · Representation theory

1 Introduction

In supersymmetry, originally introduced by Berezin (see [3] and also [4,21]), the concept of a Lie supergroup is central and it is indeed the search for extra symmetries of physical systems that let to the discovery of supersymmetry and later on of supergeometry (see [25,26] and the comprehensive treatments [5,16, 17,23,24]).

First, the notion of Lie superalgebra was introduced and only later on, there was a formalization of the notion of Lie and algebraic supergroup. The exponential map plays an eminent role and was originally introduced and studied by Koszul in [22] and later on the theory was further developed in [18].

In the algebraic setting, the recipe presented in [13–15] to construct an algebraic supergroup starting from a Lie superalgebra, uses in a implicit way the notion of exponential (see also [6,22]).

In this paper we want to restrict our attention to a special case, which is however important in the applications, namely the case of a nilpotent analytic subsupergroup of an analytic complex matrix supergroup. We shall employ freely the language of Super Harish-Chandra pair (SCHP) introduced in [21] and developed by Koszul in his fundamental work [22].

In Sect. 2, we present the construction of the exponential, while in the subsequent sections we give applications, important in the study of the Harish-Chandra representations of supergroups (see [7,8,11,19]).

Supported by funded project GHAIA (GA 777822).

F. Nielsen and F. Barbaresco (Eds.): GSI 2019, LNCS 11712, pp. 188–195, 2019.
https://doi.org/10.1007/978-3-030-26980-7_20

2   The Exponential Map

Let $\mathfrak{g}$ be a complex contragredient Lie superalgebra, $\mathfrak{g} \neq A(n,n)$, $\mathfrak{g}_1 \neq 0$; hence $\mathfrak{g}$ will be one in the following list of Lie superalgebras (see [20] Prop. 1.1):

$$A(m,n) \text{ with } m \neq n, \, B(m,n), \, C(n), \, D(m,n), \, D(2,1;\alpha), \, F(4), \, G(3) \quad (1)$$

$\mathfrak{g}_0$ is either semisimple or with a one-dimensional center. Hence, by the ordinary theory, we know that the simply connected Lie group $\widetilde{G}$, $\mathfrak{g}_0 = \mathrm{Lie}(\widetilde{G})$ is a matrix complex analytic and algebraic group. Then, the super Harish-Chandra pair SHCP $G = (\widetilde{G}, \mathfrak{g})$ (see [5] Ch. 11 and [12]) can be viewed either as a complex analytic or algebraic supergroup, via the theory of SHCP that establishes an equivalence of categories between the categories of analytic supergroups and SHCP (see [6]). We shall take the first point of view and regard G as a complex analytic matrix supergroup, but later on we will also view G as a pair $G = (\widetilde{G}, \mathcal{O}_G)$, $\mathcal{O}_G$ a sheaf of superalgebras, (see [6]).

Fix $\mathfrak{h}$ a CSA of $\mathfrak{g}$ and fix P a positive system. Let us define $\mathfrak{b}^{\pm}$ and $\mathfrak{n}^{\pm}$ the *Borel and nilpotent subsuperalgebras*:

$$\mathfrak{g} = \mathfrak{h} \oplus \bigoplus_{\alpha \in \Delta} \mathfrak{g}_\alpha, \qquad \mathfrak{b}^{\pm} := \mathfrak{h} \oplus \sum_{\alpha \in \pm P} \mathfrak{g}_\alpha, \qquad \mathfrak{n}^{\pm} := \sum_{\alpha \in \pm P} \mathfrak{g}_\alpha. \quad (2)$$

We will call $B^{\pm}$ *Borel subsupergroup* and $N^{\pm}$ *unipotent subsupergroup*, their corresponding analytic Lie supergroups in G. In particular, $B^{\pm}$ and $N^{\pm}$ are connected and are algebraic subsupergroups of G. Let A be the torus with $\mathrm{Lie}(A) = \mathfrak{h}$.

We want to define the exponential diffeomorphism: $exp : \mathfrak{n}^- \longrightarrow N^-$ for the analytic supergroup N^-. To ease the notation we shall drop the index "$-$".

Our purpose in the construction of the exponential diffeomorphism is to obtain *global coordinates* on the nilpotent supergroup N; such coordinates are going to be essential for some important applications, (see [7,8]).

We start with some general remarks on the functor of the Λ-points, we invite the reader to consult [1] and [2] for the complete details.

Let M be a supermanifold. Instead of looking at the whole functor of points $M(\cdot) \colon (\mathrm{smflds}) \to (\mathrm{sets})$, it is sometimes convenient to restrict the functor of points from the category (smflds) to the subcategory (spts) consisting of just the *superpoints*: $k^{0|n}$. These are the supermanifolds $(\{*\}, \Lambda^n)$, where Λ^n denotes the Grassmann algebra in n generators over k. In this approach the set $M(k^{0|n})$ can be endowed with the structure of an ordinary manifold, but with some peculiarities. The tangent space at a point is a Λ_0^n-module and the change of coordinates induced by a change of coordinates in M must have Λ_0^n-linear differential. These are called Λ_0-manifolds and we denote with $(\Lambda_0\mathrm{mflds})$ the corresponding category. The functor

$$(\mathrm{spts}) \to (\Lambda_0\mathrm{mflds}) \qquad k^{0|n} \mapsto M(k^{0|n}) \quad (3)$$

is a full and faithful embedding (see [1] Sec. 4, Theorem 4.5). We notice that, if V is a vector superspace, we have the identification $V(k^{0|n}) \simeq (V \otimes \Lambda^n)_0$ and the previous result is known as the *even rules principle* (see also [10]).

Proposition 1. *1. If G is a complex matrix supergroup as above, the Λ_0-manifold $G(k^{0|n})$ is a group object in the category $(\Lambda_0 \text{mflds})$ and in particular it is an analytic Lie group. Similarly $\mathfrak{g}(k^{0|n})$ is an ordinary Lie algebra.*
2. The ordinary exponential $\exp_{k^{0|n}}\colon \mathfrak{g}(k^{0|n}) \to G(k^{0|n})$ is a morphism of Λ_0 manifolds.

Proof. (1) is a simple check. As for (2), one can readily see that the differential of this map is Λ_0-linear and the correspondence $\mathfrak{g}(k^{0|n}) \to G(k^{0|n})$ is functorial.

Since the functoriality property of exp in Proposition 1 (refer also to [1,2] for a thorough treatment of Λ-points), we can immediately define the exponential morphism for an analytic supergroup G.

Definition 1. *Let G and $\mathfrak{g}$ as above. We define the exponential map as the morphism of analytic supermanifolds given on the Λ-points as the ordinary exponential as in Proposition 1 (2).*

Proposition 2. *Let N be a nilpotent supergroup as above. Then the exponential morphism $\exp\colon \mathfrak{n} \to N$ is a global superdiffeomorphism.*

Proof. In case N is a unipotent Lie supergroup as in (2), each $G(\Lambda)$ is also a unipotent Lie group and, by a classical result, each $\exp_{k^{0|n}}$ is a diffeomorphism. Hence exp is a superdiffeomorphism.

3   The Nilpotent Subsupergroup N^-

In this section we give some applications of the global coordinates we have built in the previous section. Let $\Gamma = N^- A N^+$ denote the *big cell*; it is the open analytic subsupermanifold $\Gamma = (\widetilde{N^- A N^+}, \mathcal{O}_G|_{\widetilde{N^- A N^+}})$ of the analytic supergroup $G = (\widetilde{G}, \mathcal{O}_G)$. We need some preliminary propositions.

Proposition 3. *N^- is a section for $\Gamma \to \Gamma/B^+$, the left action of A reads:*

$$A \times \Gamma/B^+ \longrightarrow \Gamma/B^+, \quad (h, nB^+(A)) \mapsto hnh^{-1}B^+(A),$$

where $n \in N^{\pm(T)}$, $h \in A(T)$, $T \in (\text{smflds})_{\mathbb{C}}$ ($(\text{smflds})_{\mathbb{C}}$ denoting the category of analytic supermanifolds).

Proof. Since the big cell $\Gamma \subset G$ is right B^+-invariant and open, and the canonical projection $p\colon G \to G/B^+$ is a submersion, we can define the open subsupermanifold of G/B^+:

$$\Gamma/B^+ := (\widetilde{\Gamma/B^+}, \mathcal{O}_{G/B^+}|_{\widetilde{\Gamma/B^+}})$$

We have a N^- equivariant diffeomorphism $N^- \longrightarrow \Gamma/B^+$, $n^- \mapsto n^- B^+(T)$, $n^- \in N^-(T)$, $T \in (\text{smflds})_{\mathbb{C}}$. In fact, by the ordinary theory we have a diffeomorphisms of the underlying differentiable manifolds and the differential at the identity is an isomorphism: $\mathfrak{n}^- \cong \mathfrak{g}/\mathfrak{b}^+$.

Clearly $p^{-1}(\Gamma/B^+) = \Gamma$. We are going to construct a section $s \colon \Gamma/B^+ \to \Gamma$. The local splitting $\gamma \colon N^- \times B^+ \to \Gamma$ is an holomorphic morphism such that $\gamma^* \mathcal{O}_{\Gamma/B^+} = \mathcal{O}_{N^-} \otimes 1$. Hence we have an isomorphism $N^- \to \Gamma/B^+$ given by the composition of the "canonical" embedding $i \colon N^- \hookrightarrow N^- \times B^+$ with γ and p (which is essentially the same as considering $p \circ \gamma|_{N^- \times \{e\}}$). Its inverse is the required section.

Proposition 4. *The (ordinary) torus A normalizes $N^\pm$.*

Proof. We give the proof for $N^+ = N$. We want to prove that the conjugation

$$\mathrm{conj}(a) \colon G \to G \qquad \mathrm{conj}(a) = \ell_{a^{-1}} \circ r_a, \qquad a \in \tilde{A}$$

stabilizes N. Since N is connected and the exponential map $\exp \colon \mathfrak{n} \to N$ is surjective it is enough to prove that $(d\mathrm{conj}(a))_1(\mathfrak{n}) \subseteq \mathfrak{n}$ We know from the infinitesimal theory that $\mathrm{ad}(\mathfrak{h})(\mathfrak{n}) = \mathfrak{n}$. Hence, we have

$$\mathrm{Ad}(e^{tX})Y = e^{t\,\mathrm{ad}X}(Y) \quad \forall X \in \mathfrak{h},\, Y \in \mathfrak{n}$$

so that $\mathrm{Ad}(e^{tX})\mathfrak{n} = \mathfrak{n}$. Since the exponential map of an abelian connected Lie group is surjective we have that $\mathrm{Ad}(A)\mathfrak{n} = \mathfrak{n}$.

By the simply connectedness of $\tilde{N}$, we get a map $\widetilde{\mathrm{conj}(a)} \colon \tilde{N} \to \tilde{N}$. It is easy to check that the pair

$$\mathrm{Ad}(a) \colon \mathfrak{n} \to \mathfrak{n} \qquad \widetilde{\mathrm{conj}(a)} \colon \tilde{N} \to \tilde{N}$$

is a SHCP morphism: $(\widetilde{\mathrm{conj}(a)}, \mathrm{Ad}(a)) \colon (\tilde{N}, \mathfrak{n}) \longrightarrow (\tilde{G}, \mathfrak{g})$, so that, by the equivalence of categories between analytic SHCP and analytic supergroups, we have a morphism of super Lie groups $N \to G$. Since its differential coincides with the differential of $\mathrm{conj}(a) \colon N \longrightarrow G$ and the reduced maps are the same, the two morphisms coincide, hence $\mathrm{conj}(a)N = N$.

Let us fix a character $\chi \colon A \longrightarrow \mathbb{C}^\times$ of the ordinary torus, that we can trivially extend to a character (still denoted by χ) of the supergroup B^+. Define:

$$L^\chi(\Gamma) := \{f \in \mathcal{O}_G(\Gamma) \mid f(gb) = \chi(b)^{-1}f(g)\}$$

We can geometrically view this superspace as the superspace of sections of the line bundle uniquely associated with χ. This superspace is the key for the construction of the infinite dimensional representations of real forms of the analytic supergroup G (see [7,8] for more details). The actions that we are going to describe are absolutely essential for the realization of such representations.

Since A acts on N^- by conjugation (see Proposition 4), we have a global action of A on Γ defined as:

$$a \cdot (n^- b^+) = (an^- a^{-1})b^+, \qquad a \in \tilde{A},\, n^- \in N^-(T),\, b^+ \in B^+(T).$$

Since A also acts on B^+ by left translation, we can define the left action of A on Γ as:

$$a \cdot (n^- b^+) = (an^- a^{-1})a \cdot b^+.$$

Both actions commute with right translations by B^+ and hence define representations of A on $L^\chi(\Gamma)$

$$i, \ell \colon A \times L^\chi(\Gamma) \to L^\chi(\Gamma)$$

where:

$$i_a(f)(n^- b^+) = f((a^{-1}n^- a)b^+), \qquad \ell_a(f)(n^- b^+) = f((a^{-1}n^- a)a^{-1}b^+) \quad (4)$$

and $a \in \tilde{A}$, $n^- \in N^-(T)$, $b^+ \in B^+(T)$, $f \in L^\chi(\Gamma)$.

Let t_α denote the global homogeneous exponential coordinates on N^- obtained by Proposition 2.

Lemma 1. *Let the notation be as above. Then*

1. $\ell_a f = \chi(a)(i_a f)$
2. $i_a t_\alpha = \chi_\alpha(a) t_\alpha \qquad \forall a \in \tilde{A}$

where χ_α is the character of the maximal torus A obtained by exponentiating the root $\alpha \in \mathfrak{h}^*$.

Proof. (1) follows immediately from the definitions. For (2) let $n = \exp(\sum_{\beta \in P} y_\beta X_{-\beta})$ be an element in N^-, then the result comes from the following formal calculation in the exponential global coordinates:

$$t_\alpha(a^{-1}na) = t_\alpha\big(\exp(\sum_{\beta \in P} y_\beta Ad(a)X_{-\beta})\big) = t_\alpha(\exp(\sum_{\beta \in P} y_\beta \chi_\beta(a)X_{-\beta})$$

$$= \chi_\alpha(a) t_\alpha(n), \quad a \in \tilde{A}, \, y_\beta \in \mathbb{C}$$

4   The Action of $\mathcal{U}(\mathfrak{g})$ and G on $L^\chi(\Gamma)$

Now we want to use the theory developed so far and extend the action of the maximal torus $A \subset G$ to an action of the whole group on $L^\chi(\Gamma)$. We start by defining the natural action of $\mathcal{U}(\mathfrak{g})$ on the holomorphic functions on any neighbourhood W of the identity of the supergroup G.

Definition 2. *Let $W \subset G$ be an open neighbourhood of the identity 1_G in G. There are two well defined actions of $\mathfrak{g}$, hence of $\mathcal{U}(\mathfrak{g})$, on $\mathcal{O}_G(W)$, that read as follows:*

$$\ell(X)f = (-X \otimes 1)\mu^*(f), \qquad \partial(X)f = (1 \otimes X)\mu^*(f), \qquad X \in \mathfrak{g}$$

The actions ℓ and ∂ commute with each other. Moreover, if $\widetilde{U}$ is open in $\widetilde{G/B^+}$, then ℓ is a well defined action on $L^\chi(\widetilde{U})$.

We now want to show that the natural action ℓ of $\mathcal{U}(\mathfrak{g})$ on $L^\chi(N^-B^+)$ preserves the polynomial sections on $\widetilde{N^-}$. For this we need some preliminary notation. Since $\mathfrak{g} = \mathfrak{n}^- \oplus \mathfrak{b}^+$, if we fix bases of $\mathfrak{n}^-$ and $\mathfrak{b}^+$, by the PBW (Poincaré Birkhoff Witt) theorem any $X \in \mathcal{U}(\mathfrak{g})$ can be written as

$$X = \sum_{I,J} c_{IJ}(X)\mathrm{B}_I\mathrm{N}_I, \qquad \mathrm{B}_I \in \mathcal{U}(\mathfrak{b}^+), \mathrm{N}_I \in \mathcal{U}(\mathfrak{n}^-) \qquad (5)$$

Lemma 2. *Let $\phi \in \mathcal{O}_G(N^-B^+)$. In the SHCP notation, ϕ is in $L^\chi(N^-B^+)$ if and only if*

$$\phi(X)(nb) = \widetilde{\chi}(b)^{-1}\sum_{IJ} c_{IJ}(b.X)\lambda(\overline{\mathrm{B}}_I)\phi(\mathrm{N}_J)(n), \quad X \in \mathcal{U}(\mathfrak{g}), \quad \lambda = d\chi$$

where $b.X$ is the adjoint action of $b \in \widetilde{B^+}$ on $\mathcal{U}(\mathfrak{g})$ and as usual $\overline{\mathrm{B}}_I$ denotes the antipode of B_I in the Hopf superalgebra $\mathcal{U}(\mathfrak{g})$.

Proof. By the very definition we have $\phi \in L^\chi(N^-B^+)$ if

1. $r_b^*\phi = \widetilde{\chi}(b)^{-1}\phi, \; b \in \widetilde{B^+}$
2. $D_Y^L(\phi) = \lambda(\overline{Y})\phi, \; \lambda|_{\mathfrak{g}_0} = d\widetilde{\chi}$.

where as usual $\widetilde{\chi}$ denotes the reduced morphism. The result comes with a calculation. $\qquad\square$

Notice that once the lemma is established, if p is a polynomial in the global coordinates of N^-, we can define $p^\sim \in L^\chi(N^-B^+)$ as:

$$p^\sim(X)(nb) = \widetilde{\chi}(b)^{-1}\sum_{IJ} c_{IJ}(b.X)\lambda(\overline{\mathrm{B}^+}_I)p(\mathrm{N}_J)(n)$$

Vice-versa we can recover p from $p^\sim$ restricting to N^-. In the language of SHCP this amounts to two restrictions: we impose $b = 1$ and $X \in \mathcal{U}(\mathfrak{n}^-)$. We shall denote the set of such $p^\sim$ with $\mathcal{P}^\sim$.

Proposition 5. *The actions ℓ of $\mathcal{U}(\mathfrak{g})$ on $L^\chi(\widetilde{U})$, $p^{-1}(\widetilde{U}) \subset \Gamma$ leave $\mathcal{P}^\sim$ invariant.*

Proof. We need to show that, given $Z \in \mathcal{U}(\mathfrak{g})$ and $X \in \mathcal{U}(\mathfrak{n}^-)$, $(D_Z^R p^\sim|_{N^-})(X)$ is a polynomial section. We have (see [5] Sec. 7.4):

$$(D_Z^R p^\sim)(X)(g) = (-1)^{|Z||p|}[p^\sim((g^{-1}.Z)X)](g)$$

Hence if $n \in N^-$, we have:

$$(D_Z^R p^\sim)(X)(n) = (-1)^{|Z||p|}[p^\sim((n^{-1}.Z)X)](n)$$
$$= (-1)^{|Z||p|}\sum_{IJ} c_{IJ}((n^{-1}.Z)X)[\lambda(\overline{\mathrm{B}}_I)p^\sim(\mathrm{N}_J)](n)$$

where B_I and N_J are obtained as in (5) applied to $(n^{-1}.Z)X$. The last equality is true by Lemma 2. $\qquad\square$

Once this is established, we have the following result.

Theorem 1. *There is a non-singular $\mathcal{U}(\mathfrak{g})$ invariant pairing between $\mathcal{P}^\sim$ and the Verma module V_λ:*

$$\langle,\rangle : \mathcal{P}^\sim \times \mathcal{U}(\mathfrak{g}) \longrightarrow \mathbb{C}, \qquad \langle f, u \rangle := (-1)^{|u||f|}(\partial(u)f)(1_G)$$

Proof. In order for $\langle,\rangle$ to be a $\mathcal{U}(\mathfrak{g})$ invariant pairing, we need to verify:

$$\langle \ell(c)f, u \rangle = \langle f, (-1)^{|f||c|}c^T u \rangle, \qquad c, u \in \mathcal{U}(\mathfrak{g}), f \in \mathcal{P}^\sim$$

where $(\cdot)^T$ denotes the antiautomorphism of $\mathcal{U}(\mathfrak{g})$ induced by $X \mapsto -X$ with $X \in \mathfrak{g}$. This is just a check.

Now let $\mathfrak{g}_r$ be a real form of $\mathfrak{g}$ and define the real supergroup $G_r = (\widetilde{G_r}, \mathfrak{g}_r)$, where $\widetilde{G_r}$ is a real form of $\widetilde{G}$, $\mathrm{Lie}(\widetilde{G_r}) = \mathfrak{g}_{r,0}$. Since $\mathfrak{g}_r + \mathfrak{b}^+ = \mathfrak{g}$ as real superalgebras (see [9], Iwasawa decomposition), we have that $S := G_r B^+$ is an open subsupermanifold of G.

Theorem 2. *Assume $L^\chi(S) \neq 0$ modulo J the submodule generated by the odd part. Then $L^\chi(S)$ contains an element ψ which is an analytic continuation of $1^\sim$ and*

$$\ell(\mathcal{U}(\mathfrak{g}))\psi = \mathcal{P}^\sim \cong \pi_{-\lambda}$$

where $\pi_{-\lambda}$ the irreducible representation with lowest weight $-\lambda$. Furthermore $L^\chi(S)$ carries a G_r representation defined as:

$$\begin{cases} (g \cdot f) = l^*_{g^{-1}}f & g \in \widetilde{G_r} \\ X.f = D^R_{\overline{X}}f & X \in \mathfrak{g}_\mathbb{C} \end{cases}$$

where, as usual, $\overline{X}$ is the antipode of $X \in \mathcal{U}(\mathfrak{g})$.

Proof. Direct check.

The closure of $\ell(\mathcal{U}(\mathfrak{g}))\psi$ in $L^\chi(S)$, with a Fréchet superspace structure is a Harish-Chandra representation, with $\ell(\mathcal{U}(\mathfrak{g}))\psi$ as its K_r finite part, where K_r is the supergroup corresponding to the subalgebra $\mathfrak{k}_r$ in the Cartan decomposition of $\mathfrak{g}_r$ (see [9]). The proof of these facts is non trivial, we invite the reader to see [7,8].

Acknowledgements. We are indebted to Prof. F. Gavarini for remarks and suggestions.

References

1. Balduzzi, L., Carmeli, C., Fioresi, R.: The local functors of points of supermanifolds. Exp. Math. **28**(3), 201–217 (2010)

2. Balduzzi, L., Carmeli, C., Fioresi, R.: A comparison of the functors of points of supermanifolds. J. Algebra Appl. **12**(3), 125–152 (2013). 41 p
3. Berezin, F.A.: Introduction to Superanalysis. D. Reidel Pub., Holland (1987)
4. Berezin, F.A., Leites, D.: Supermanifolds. Dokl. Akad. Nauk SSSR **224**(3), 505–508 (1975)
5. Carmeli, C., Caston, L., Fioresi, R.: Mathematical Foundation of Supersymmetry. EMS Ser. Lect. Math., European Math. Soc., Zurich (2011)
6. Carmeli, C., Fioresi, R.: Super distributions, analytic and algebraic super Harish-Chandra pairs. Pac. J. Math. **263**, 29–51 (2013)
7. Carmeli, C., Fioresi, R., Varadarajan, V.S.: Super bundles. Universe **4**(3), 46 (2018)
8. Carmeli, C., Fioresi, R., Varadarajan, V.S.: Highest weight Harish-Chandra super-modules and their geometric realizations. Transform. Groups (2019). Preprint
9. Chuah, M.K., Fioresi, R.: Hermitian real forms of contragredient Lie superalgebras. J. Algebra **437**, 161–176 (2015)
10. Deligne, P., Morgan, J.: Notes on supersymmetry (following J. Bernstein). In: Quantum Fields and Strings. A Course for Mathematicians, vol. 1. AMS (1999)
11. Fioresi, R.: Compact forms of complex Lie supergroups. J. Pure Appl. Algebra **218**, 228–236 (2014)
12. Fioresi, R.: Smoothness of algebraic supervarieties and supergroups. Pac. J. Math. **234**, 295–310 (2008)
13. Fioresi, R., Gavarini, F.: Chevalley Supergroups. AMS Memoirs, vol. 215, pp. 1–64 (2012)
14. Fioresi, R., Gavarini, F.: On algebraic supergroups with Lie superalgebras of classical type. J. Lie Theory **23**, 143–158 (2013)
15. Fioresi, R., Gavarini, F.: On the construction of Chevalley supergroups. In: Ferrara, S., Fioresi, R., Varadarajan, V. (eds.) Supersymmetry in Mathematics and Physics. Lecture Notes in Mathematics, vol. 2027, pp. 101–123. Springer, Heidelberg (2011)
16. Fioresi, R., Lledo, M.A.: The Minkowski and Conformal Superspaces: The Classical and Quantum Descriptions. World Scientific Publishing, Singapore (2015)
17. Fioresi, R., Lledo, M.A., Varadarajan, V.S.: The Minkowski and conformal super-spaces. J. Math. Phys. **48**, 113505 (2007)
18. Garnier, S., Wurzbacher, T.: Integration of vector fields on smooth and holomorphic supermanifolds. Documenta Mathematica **18**, 519–545 (2013)
19. Harish-Chandra: Representations of semi-simple Lie groups IV, V, VI. Amer. J. Math. (77), 743–777 (1955). no. 78, 1–41 and 564–628
20. Kac, V.G.: Lie superalgebras. Adv. Math. **26**, 8–26 (1977)
21. Kostant, B.: Graded manifolds, graded Lie theory, and prequantization. In: Bleuler, K., Reetz, A. (eds.) Differential Geometrical Methods in Mathematical Physics. LNM, vol. 570, pp. 177–306. Springer, Heidelberg (1977). https://doi.org/10.1007/BFb0087788
22. Koszul, J.-L.: Graded manifolds and graded Lie algebras. In: Proceedings of the International Meeting on Geometry and Physics (Florence, 1982), pp. 71–84, Pitagora, Bologna (1982)
23. Leites, D.A.: Introduction to the theory of supermanifolds. Russ. Math. Surv. **35**(1), 1–64 (1980)
24. Varadarajan, V.S.: Supersymmetry for mathematicians: an introduction. Courant Lecture Notes **1**. AMS (2004)
25. Volkov, D.V., Akulov, V.P.: Is the neutrino a Goldstone particle? Phys. Lett. B **46**, 109 (1973)
26. Wess, J., Zumino, B.: Supergauge transformationsin four dimensions. Nucl. Phys. B **70**, 39–50 (1974)

Geometric Structures in Thermodynamics and Statistical Physics

Dirac Structures in Open Thermodynamics

Hiroaki Yoshimura[1]([⊠]) and François Gay-Balmaz[2]

[1] Waseda University, 3-4-1, Okubo, Shinjuku, Tokyo 169-8555, Japan
yoshimura@waseda.jp
[2] Ecole Normale Supérieure, Paris, France
francois.gay-balmaz@lmd.ens.fr

Abstract. Dirac structures are geometric objects that generalize Poisson structures and presymplectic structures on manifolds. They naturally appear in the formulation of constrained mechanical systems and play an essential role in the understanding of the interrelations between system elements in implicit dynamical systems. In this paper, we show how nonequilibrium thermodynamic systems can be naturally understood in the context of Dirac structures, by mainly focusing on the case of open systems, i.e., thermodynamic systems exchanging heat and matter with the exterior.

Keywords: Dirac structures · Open systems ·
Dirac dynamical systems · Nonequilibrium thermodynamics

1 Introduction

Nonequilibrium thermodynamics is a phenomenological theory which aims to identify and describe the relations among the observed macroscopic properties of a physical system and to determine the macroscopic dynamics of this system with the help of fundamental laws, see [15]. A novel Lagrangian variational approach for nonequilibrium thermodynamic has been proposed by the authors [4,5] for both finite dimensional (discrete) and infinite dimensional (continuum) systems. This variational formulation was extended to the case of open systems as in [6]. The authors have also shown that, in the case of adiabatically closed systems, the variational formulation leads to an associated geometric formulation in terms of Dirac structures [7]. Recall that Dirac structures are geometric objects that extend both Poisson structures and presymplectic structures on manifolds [2]. Such structures play an essential role in formulating constrained systems such as electric circuits and nonholonomic mechanical systems (e.g., [16,17]). On the

H. Yoshimura is partially supported by JSPS Grant-in-Aid for Scientific Research (S) 24224004, the MEXT Top Global University Project and Waseda University (SR 2019C-176 and SR 2019Q-020); F. Gay-Balmaz is partially supported by the ANR project GEOMFLUID, ANR-14-CE23-0002-01.

F. Nielsen and F. Barbaresco (Eds.): GSI 2019, LNCS 11712, pp. 199–208, 2019.
https://doi.org/10.1007/978-3-030-26980-7_21

other hand, for *equilibrium thermodynamics*, the geometric formulations have been mainly given by using contact geometry, see [1,8,9,11–13]. In this geometric setting, the thermodynamic properties are encoded by Legendre submanifolds of the thermodynamic phase space. It was shown by [3] that a geometric formulation of irreversible processes can be made by lifting port-Hamiltonian systems to the thermodynamic phase space. The underlying geometric structure is again given in the context of contact geometry.

In this paper, we show that the equations of evolutions for an *open system* exchanging matter with the exterior can be geometrically formulated by using Dirac structures. This geometric formulation is associated to the variational formulation given in [6]. To achieve this goal, we first recall below the first and second laws as they apply to an open system. Then, we develop a general Dirac formulation for a class of systems with time-dependent nonlinear nonholonomic constraints. In particular, we introduce a time-dependent Dirac structure on the covariant Pontryagin bundle over a thermodynamic configuration manifold. Finally, we apply our Dirac formulation of systems with nonlinear time-dependent constraints to the case of open thermodynamic systems and we show that the system of evolution equations of the open system can be directly formulated as a Dirac dynamical system.

2   A Fundamental Setting for Open Systems

2.1   The First Law for Open Thermodynamic Systems

The first law of thermodynamics, following [15], asserts that for every system there exists an extensive state function, the energy, which satisfies

$$\frac{d}{dt}E = P_W^{\text{ext}} + P_H^{\text{ext}} + P_M^{\text{ext}},$$

where t denotes time, P_W^{ext} is the power associated to the work done on the system, P_H^{ext} is the power associated to the transfer of heat into the system, and P_M^{ext} is the power associated to the transfer of matter into the system. In particular, a system in which $P_M^{\text{ext}} \neq 0$ is called *open*. For such an open system, matter can flow into or out of the system through several ports, $a = 1, ..., A$. We suppose, for simplicity, that the system involves only one chemical species and denote by N the number of moles of this species. In this case, the mole balance equation is

$$\frac{d}{dt}N = \sum_{a=1}^{A} \mathcal{J}^a,$$

where $\mathcal{J}^a$ is the molar flow rate *into* the system through the a-th port, so that $\mathcal{J}^a > 0$ for flow into the system and $\mathcal{J}^a < 0$ for flow out of the system.

As matter enters or leaves the system, it carries its internal, potential, and kinetic energy. This energy flow rate at the a-th port is the product $\mathsf{E}^a \mathcal{J}^a$ of the energy per mole (or molar energy) E^a and the molar flow rate $\mathcal{J}^a$ at the a-th port. In addition, as matter enters or leaves the system it also exerts work on the system that is associated with pushing the species into or out of the system. The associated energy flow rate is given at the a-th port by $p^a \mathsf{V}^a \mathcal{J}^a$, where p^a and V^a are the pressure and the molar volume of the substance flowing through the a-th port. In this case, the power exchange due to the mass transfer is

$$P_M^{\text{ext}} = \sum_{a=1}^{A} \mathcal{J}^a (\mathsf{E}^a + p^a \mathsf{V}^a).$$

A system is called adiabatically closed if $P_H^{\text{ext}} = P_M^{\text{ext}} = 0$.

2.2   The Second Law for Open Thermodynamic Systems

Following [15], the evolution part of the second law of thermodynamics asserts that for every adiabatically closed system, there exists an extensive state function, the entropy, which satisfies

$$\frac{d}{dt} S = I \geq 0,$$

where I is the entropy production of the system. Let us deduce the expression of the entropy production in an open system of one chemical component, with constant volume and an internal energy given by $U = U(S, N)$. The balance of mole and the balance energy, i.e., the first law, are respectively given by

$$\frac{d}{dt} N = \sum_{a=1}^{A} \mathcal{J}^a, \quad \frac{d}{dt} U = \sum_{a=1}^{A} \mathcal{J}^a (\mathsf{U}^a + p^a \mathsf{V}^a) = \sum_{a=1}^{A} \mathcal{J}^a \mathsf{H}^a,$$

where $\mathsf{H}^a = \mathsf{U}^a + p^a \mathsf{V}^a$ is the molar enthalpy at the a-th port and where U^a, p^a, and V^a are respectively the molar internal energy, the pressure and the molar volume at the a-th port, see [10,14]. From these equations and the second law, one obtains the equations for the rate of change of the entropy of the system as

$$\frac{d}{dt} S = I + \sum_{a=1}^{A} \mathsf{S}^a \mathcal{J}^a,$$

where S^a is the molar entropy at the a-th port and I is the rate of internal entropy production of the system given by

$$I = \frac{1}{T} \sum_{a=1}^{A} \left[\mathcal{J}_S^a (T^a - T) + \mathcal{J}^a (\mu^a - \mu) \right],$$

where $T = \frac{\partial U}{\partial S}$ denotes the temperature and $\mu = \frac{\partial U}{\partial N}$ the chemical potential. The entropy flow rate is given by $\mathcal{J}_S^a := \mathsf{S}^a \mathcal{J}^a$ and we also have the relation $\mathsf{H}^a = \mathsf{U}^a + p^a \mathsf{V}^a = \mu^a + T^a \mathsf{S}^a$. The thermodynamic quantities known at the ports are usually the pressure and the temperature p^a, T^a, from which the other thermodynamic quantities, such as $\mu^a = \mu^a(p^a, T^a)$ or $\mathsf{S}^a = \mathsf{S}^a(p^a, T^a)$ are deduced from the state equations of the gas.

3 Dirac Formulation of Time-Dependent Nonlinear Nonholonomic Systems

3.1 Variational and Kinematic Time Dependent Constraints

In order to formulate an open thermodynamic system in the context of Dirac structures, we first introduce two different constraints C_V and C_K which depend explicitly on time t. For a thermodynamic configuration manifold Q which is the space of the thermodynamic variables as well as the mechanical variables, we define the *extended configuration manifold* as $Y := \mathbb{R} \times Q \ni (t, x)$, which can be seen as a trivial vector bundle $Y = \mathbb{R} \times Q \to \mathbb{R}$, $(t, x) \mapsto t$, over the space of time $\mathbb{R}$. Consider the vector bundle $(\mathbb{R} \times TQ) \times_Y TY \to Y$ over Y, whose vector fiber at $y = (t, x)$ is given by $T_x Q \times T_{(t,x)} Y = T_x Q \times (\mathbb{R} \times T_x Q)$. An element in the fiber at (t, x) is denoted $(v, \delta t, \delta x)$. In general, by definition a *variational constraint* is a subset $C_V \subset (\mathbb{R} \times TQ) \times_Y TY$, such that $C_V(t, x, v)$, defined by $C_V(t, x, v) := C_V \cap (\{(t, x, v)\} \times T_{(t,x)} Y)$, is a vector subspace of $T_{(t,x)} Y$, for all $(t, x, v) \in \mathbb{R} \times TQ$. In general, a *kinematic constraint* is a submanifold $C_K \subset TY$. More concretely, given functions $A^r : \mathbb{R} \times TQ \to T^* Q$ and $B^r : \mathbb{R} \times TQ \to \mathbb{R}$, $r = 1, ..., m$, the variational constraint C_V is given by

$$C_V = \big\{ (t, x, v, \delta t, \delta x) \in (\mathbb{R} \times TQ) \times_Y TY \mid \\ A_i^r(t, x, v) \delta x^i + B^r(t, x, v) \delta t = 0, \ r = 1, ..., m \big\} \tag{1}$$

and the associated kinematic constraint C_K of thermodynamic type is

$$C_K = \big\{ (t, x, \dot{t}, \dot{x}) \in TY \mid A_i^r(t, x, \dot{x}) \dot{x}^i + B^r(t, x, \dot{x}) \dot{t} = 0, \ r = 1, ..., m \big\}. \tag{2}$$

We will see later how C_V and C_K are concretely given in thermodynamics.

3.2 Covariant Pontryagin Bundles and the Generalized Energy

Associated to the extended configuration manifold $Y = \mathbb{R} \times Q$ for the time-dependent system, we define the *covariant Pontryagin bundle* by

$$\pi_{(\mathcal{P}, Y)} : \mathcal{P} = (\mathbb{R} \times TQ) \times_Y T^* Y = (\mathbb{R} \times TQ) \times_{\mathbb{R} \times Q} T^*(\mathbb{R} \times Q) \to Y = \mathbb{R} \times Q.$$

An element in the fiber at (t, x) is denoted (v, p, p). Given the Lagrangian $\mathcal{L} : \mathbb{R} \times TQ \to \mathbb{R}$, the *covariant generalized energy* is defined on $\mathcal{P}$ as

$$\mathcal{E} : \mathcal{P} \to \mathbb{R}, \quad \mathcal{E}(t, x, v, \mathsf{p}, p) = \mathsf{p} + \langle p, v \rangle - \mathcal{L}(t, x, v). \tag{3}$$

3.3 Dirac Structures on the Covariant Pontryagin Bundle

Given a variational constraint $C_V \subset (\mathbb{R} \times TQ) \times_Y TY$, we consider the distribution $\Delta_{\mathcal{P}}$ on the covariant Pontryagin bundle defined by

$$\Delta_{\mathcal{P}}(t, x, v, \mathsf{p}, p) := \big(T_{(t,x,v,\mathsf{p},p)} \pi_{(\mathcal{P}, Y)} \big)^{-1} (C_V(t, x, v)) \subset T_{(t,x,v,\mathsf{p},p)} \mathcal{P}.$$

From the expression of C_V in (1), we get

$$
\Delta_{\mathcal{P}}(t, x, v, \mathsf{p}, p) = \big\{ (\delta t, \delta x, \delta v, \delta \mathsf{p}, \delta p) \in T\mathcal{P} \mid \\
A_i^r(t, x, v)\delta x^i + B^r(t, x, v)\delta t = 0, \ r = 1, ..., m \big\}. \tag{4}
$$

Consider the canonical symplectic form on T^*Y given by $\Omega_{T^*Y} = -\mathbf{d}\Theta_{T^*Y}$, where Θ_{T^*Y} is the canonical one-form on T^*Y. In local coordinates, we have $\Theta_{T^*Y} = p_i dx^i + \mathsf{p}dt$ and $\Omega_{T^*Y} = dx^i \wedge dp_i + dt \wedge d\mathsf{p}$. Using the projection $\pi_{(\mathcal{P},T^*Y)} : \mathcal{P} \to T^*Y$, $(t, x, v, \mathsf{p}, p) \mapsto (t, x, \mathsf{p}, p)$ onto T^*Y, we get the presymplectic form on the covariant Pontryagin bundle given by $\Omega_{\mathcal{P}} = \pi_{(\mathcal{P},T^*Y)}^* \Omega_{T^*Y}$. The local expression is given by $\Omega_{\mathcal{P}} = dx^i \wedge dp_i + dt \wedge d\mathsf{p}$.

Given the distribution $\Delta_{\mathcal{P}}$ in (4) and the presymplectic form $\Omega_{\mathcal{P}}$, the Dirac structure $D_{\Delta_{\mathcal{P}}}$ on $\mathcal{P}$ is given by

$$
D_{\Delta_{\mathcal{P}}}(\mathsf{x}) = \big\{ (\mathsf{u}_\mathsf{x}, \mathfrak{a}_\mathsf{x}) \in T_\mathsf{x}\mathcal{P} \times T_\mathsf{x}^*\mathcal{P} \mid \mathsf{u}_x \in \Delta_{\mathcal{P}}(\mathsf{x}), \\
\langle \mathfrak{a}_\mathsf{x}, \mathfrak{v}_\mathsf{x} \rangle = \Omega_{\mathcal{P}}(\mathsf{x})(\mathsf{u}_\mathsf{x}, \mathfrak{v}_\mathsf{x}), \ \forall \ \mathfrak{v}_\mathsf{x} \in \Delta_{\mathcal{P}}(\mathsf{x}) \big\}, \tag{5}
$$

for all $\mathsf{x} \in \mathcal{P}$.

3.4   Dirac Dynamical Systems

Using the Dirac structure $D_{\Delta_{\mathcal{P}}}$ on $\mathcal{P}$ in (5), we can define a Dirac dynamical system for a curve $x(t)$ in $\mathcal{P}$ as follows:

$$
\big(\dot{\mathsf{x}}, \mathbf{d}\mathcal{E}(\mathsf{x}) \big) \in D_{\Delta_{\mathcal{P}}}(\mathsf{x}). \tag{6}
$$

Equivalently, condition (6) gives the equations of motion

$$
\mathbf{i}_{\dot{\mathsf{x}}}\Omega_{\mathcal{P}} - \mathbf{d}\mathscr{E}(\mathsf{x}) \in \Delta_{\mathcal{P}}(\mathsf{x})^\circ, \ \ \dot{\mathsf{x}} \in \Delta_{\mathcal{P}}(\mathsf{x}). \tag{7}
$$

Using coordinates, we can now explicitly express these equations as follows. The differential of $\mathcal{E}$ is given by

$$
\mathbf{d}\mathcal{E}(t, x, v, \mathsf{p}, p) = \left(-\frac{\partial \mathcal{L}}{\partial t}, -\frac{\partial \mathcal{L}}{\partial x}, p - \frac{\partial \mathcal{L}}{\partial v}, 1, v \right)
$$

and the tangent vector $\dot{\mathsf{x}}$ to $T_{\mathsf{x}(t)}\mathcal{P}$ is given by $(t, x, \dot{t}, \dot{x})$. We deduce that the Dirac dynamical system (7) gives the following conditions on the curve $\mathsf{x}(t) \in \mathcal{P}$,

$$
\dot{x} = v, \qquad \dot{t} = 1, \qquad p = \frac{\partial \mathcal{L}}{\partial v},
$$
$$
(t, x, \dot{t}, \dot{x}) \in C_V(t, x, v), \qquad \left(\dot{\mathsf{p}} - \frac{\partial \mathcal{L}}{\partial t}, \dot{p} - \frac{\partial \mathcal{L}}{\partial x} \right) \in C_V(t, x, v)^\circ. \tag{8}
$$

In local expressions, these evolution Eq. (8) read

$$
\begin{cases}
\dot{x}^i = v^i, \qquad \dot{t} = 1, \qquad p_i - \dfrac{\partial \mathcal{L}}{\partial v^i} = 0, \qquad A_i^r(t, x, v)\dot{q}^i + B^r(t, x, v) = 0, \\[2ex]
\dot{p}_i - \dfrac{\partial \mathcal{L}}{\partial x^i} = \lambda_r A_i^r(t, x, v), \qquad \dot{\mathsf{p}} - \dfrac{\partial \mathcal{L}}{\partial t} = \lambda_r B^r(t, x, v).
\end{cases}
$$
$$\tag{9}$$

3.5 Energy Balance Equations

One immediately notices that the covariant generalized energy $\mathcal{E}(t, x, v, \mathsf{p}, p)$ defined in (3) is preserved along the solution curve $\mathsf{x}(t) = (t, x(t), v(t), \mathsf{p}(t), p(t))$ of the Dirac dynamical system (9),

$$\frac{d}{dt} \mathcal{E}(t, x, v, \mathsf{p}, p) = 0. \tag{10}$$

Note that $\mathcal{E}$ does not represent the total energy of the system. The total energy is represented by the generalized energy $E : \mathbb{R} \times T\mathcal{Q} \times T^*\mathcal{Q} \to \mathbb{R}$ defined as

$$E(t, x, v, p) = \langle p, v \rangle - \mathcal{L}(t, x, v)$$

and Eq. (10) yields

$$\frac{d}{dt} E(t, x, v, p) = -\frac{d}{dt}\mathsf{p} = \frac{\partial \mathcal{L}}{\partial t}(t, x, v) - \lambda_r B^r(t, x, v).$$

This is the *balance of energy* for the Dirac system. Note that $\frac{d}{dt}\mathsf{p}$ is interpreted as the power flowing out of the system. The first term on the right hand side is essentially due to the explicit dependence of the Lagrangian on time. The second term is due to the affine character of the constraint and will be interpreted later as the energy flowing in or out of the systems through its ports.

4 Dirac Formulation of Open Thermodynamics

4.1 Geometric Setting

Consider a simple open finite dimensional system with a single entropy S and a single chemical species with number of moles N in a single compartment. The system has a constant volume $V = V_0$, it has A external ports, through which matter can flow into or out of the system as well as B ports, through which heat can flow in or out of the system. Let $U(S, N)$ be the internal energy of the system. Let $\mathcal{J}^a(t)$, $\mathsf{S}^a(t)$, $T^a(t)$, $\mu^a(t)$ be given functions of time associated to the external flow of matter into the system through the a-port and define $\mathcal{J}^a_S(t) = \mathcal{J}^a(t)\mathsf{S}^a(t)$. We assume that there exist external heat sources at the b-port with entropy flow rate $\mathcal{J}^b(t)$, molar entropy $\mathsf{S}^b(t)$ and temperature $T^b(t)$.

The thermodynamic configuration space is $\mathcal{Q} = \mathbb{R}^5 \ni x = (S, N, \Gamma, W, \Sigma)$, with Γ, W, Σ the thermodynamic displacements, see [6]. As in Sect. 3.1, let $Y = \mathbb{R} \times \mathcal{Q}$ be the trivial bundle over $\mathbb{R}$ and consider the thermodynamic phase space $\mathbb{R} \times T\mathcal{Q}$ over Y with coordinates $(t, x, v) \in \mathbb{R} \times T\mathcal{Q}$, where $v = (v_S, v_N, v_\Gamma, v_W, v_\Sigma) \in T_q\mathcal{Q}$. Let us employ the local coordinates for $(t, x, \delta t, \delta x) \in TY$ and $(t, x, \mathsf{p}, p) \in T^*Y$, where $\delta x = (\delta S, \delta N, \delta \Gamma, \delta W, \delta \Sigma) \in T_q\mathcal{Q}$ and $p = (p_S, p_N, p_\Gamma, p_W, p_\Sigma) \in T^*_x\mathcal{Q}$.

4.2   Nonholonomic Constraints in Thermodynamics

For open thermodynamic systems, the constraint (1) reads

$$C_V = \Big\{ (t,x,v,\delta t, \delta x) \in (\mathbb{R} \times T\mathcal{Q}) \times_Y TY \; \Big| $$

$$-\frac{\partial U}{\partial S}\delta \Sigma = \sum_{a=1}^{A} \big[\mathcal{J}^a \delta W + \mathcal{J}_S^a \delta \Gamma - (\mu^a \mathcal{J}^a + T^a \mathcal{J}_S^a)\, \delta t \big] + \sum_{b=1}^{B} \mathcal{J}_S^b (\delta \Gamma - T^b \delta t) \Big\}.$$

Hence the nonlinear kinematic constraint (2) becomes

$$C_K = \Big\{ (t,x,\dot{t},\dot{x}) \in TY \; \Big| $$

$$-\frac{\partial U}{\partial S}\dot{\Sigma} = \sum_{a=1}^{A} \big[\mathcal{J}^a \dot{W} + \mathcal{J}_S^a \dot{\Gamma} - (\mu^a \mathcal{J}^a + T^a \mathcal{J}_S^a)\, \dot{t} \big] + \sum_{b=1}^{B} \mathcal{J}_S^b (\dot{\Gamma} - T^b \dot{t}) \Big\},$$

where we have denoted $\dot{x} = (\dot{S}, \dot{N}, \dot{\Gamma}, \dot{W}, \dot{\Sigma})$.

4.3   Dirac Structures on $\mathcal{P}$ for Open Thermodynamic Systems

As in Sect. 3.2, let $\mathcal{P} = (\mathbb{R} \times T\mathcal{Q}) \times_Y T^*Y$ be the covariant Pontryagin bundle over Y, whose coordinates are given by $\mathrm{x} = (t,x,v,\mathsf{p},p) \in \mathcal{P}$. The canonical one form Θ_{T^*Y} and the canonical symplectic form $\Omega_{T^*Y} = -\mathrm{d}\Theta_{T^*Y}$ are expressed as

$$\begin{aligned}\Theta_{T^*Y} &= p_i dx^i + \mathsf{p}\, dt \\ &= p_S dS + p_N dN + p_\Gamma d\Gamma + p_W dW + p_\Sigma d\Sigma + p_\Sigma d\Sigma + \mathsf{p}\, dt,\end{aligned}$$

$$\begin{aligned}\Omega_{T^*Y} &= dx^i \wedge dp_i + dt \wedge d\mathsf{p} \\ &= dS \wedge dp_S + dW \wedge dp_W + dN \wedge dp_N + d\Gamma \wedge dp_\Gamma + d\Sigma \wedge dp_\Sigma + dt \wedge d\mathsf{p}.\end{aligned}$$

Recall that the presymplectic form on $\mathcal{P}$ is defined by $\Omega_\mathcal{P} = \pi^*_{(\mathcal{P},T^*Y)}\Omega_{T^*Y}$.

Associated with $\mathcal{P}$, we have the natural projection $\pi_{(\mathcal{P},Y)} : \mathcal{P} \to Y$, given by $(t,x,v,\mathsf{p},p) \mapsto (t,x)$, and we can lift the constraint subspace $C_V(t,x,v) \subset TY$ to get the constraint distribution $\Delta_\mathcal{P}$ on $\mathcal{P}$ defined as

$$\Delta_\mathcal{P} = (T\pi_{(\mathcal{P},Y)})^{-1}(C_V(t,x,v)) \subset T\mathcal{P}.$$

As shown in (5), from the distribution $\Delta_\mathcal{P}$ and the presymplectic form $\Omega_\mathcal{P}$, we can define the induced Dirac structure $D_{\Delta_\mathcal{P}} \subset T\mathcal{P} \oplus T^*\mathcal{P}$ on $\mathcal{P}$.

4.4   Dirac Thermodynamic Systems on $\mathcal{P} = (\mathbb{R} \times T\mathcal{Q}) \times_Y T^*Y$

For each $\mathrm{x} = (t,x,v,\mathsf{p},p) \in \mathcal{P}$, we write the vector and the covector in (5) as

$$\mathfrak{u}_\mathrm{x} = (\dot{t}, \dot{x}, \dot{v}, \dot{\mathsf{p}}, \dot{p}) \in T_\mathrm{x}\mathcal{P} \quad \text{and} \quad \mathfrak{a}_\mathrm{x} = (\pi, \alpha, \beta, u, w) \in T^*_\mathrm{x}\mathcal{P},$$

where $\dot{v} = (\dot{v}_S, \dot{v}_N, \dot{v}_\Gamma, \dot{v}_W, \dot{v}_\Sigma), \dot{p} = (\dot{p}_S, \dot{p}_N, \dot{p}_\Gamma, \dot{p}_W, \dot{p}_\Sigma), \alpha = (\alpha_S, \alpha_N, \alpha_\Gamma, \alpha_W, \alpha_\Sigma), \beta = (\beta_S, \beta_N, \beta_\Gamma, \beta_W, \beta_\Sigma),$ and $w = (w_S, w_N, w_\Gamma, w_W, w_\Sigma)$. From (6) the Dirac system reads

$$((\dot{t}, \dot{x}, \dot{v}, \dot{\mathsf{p}}, \dot{p}), (\pi, \alpha, \beta, u, w)) \in D_{\Delta_{\mathscr{P}}}(t, x, v, \mathsf{p}, p).$$

Using the definition of the Dirac structure in terms of C_V, we get

$$\dot{x} = w, \quad \dot{t} = u, \quad \beta = 0, \quad (t, x, \dot{t}, \dot{x}) \in C_V(t, x, v), \quad (\dot{\mathsf{p}} + \pi, \dot{p} + \alpha) \in C_V(t, x, v)^\circ.$$

Following [6], the Lagrangian is given by $\mathcal{L}(t, x, v) = -U(S, N) + v_W N + v_\Gamma (S - \Sigma)$. The covariant generalized energy is here given by

$$
\begin{aligned}
\mathcal{E}(t, x, v, \mathsf{p}, p) &= \mathsf{p} + \langle p, v \rangle - \mathcal{L}(t, x, v) \\
&= \mathsf{p} + p_S v_S + p_N v_N + (p_\Gamma + \Sigma - S) v_\Gamma + (p_W - N) v_W + p_\Sigma v_\Sigma + U(S, N).
\end{aligned}
$$

The differential of $\mathbf{d}\mathcal{E}$ is obtained as

$$\mathbf{d}\mathcal{E}(t, x, v, \mathsf{p}, p) = \left(-\frac{\partial \mathcal{L}}{\partial t}, -\frac{\partial \mathcal{L}}{\partial x}, p - \frac{\partial \mathcal{L}}{\partial v}, 1, v \right) = (\pi, \alpha, \beta, \gamma, w),$$

where $\pi = -\frac{\partial L}{\partial t} = 0, \; \alpha = -\frac{\partial L}{\partial x} = \left(-v_\Gamma + \frac{\partial U}{\partial S}, -v_W + \frac{\partial U}{\partial N}, 0, 0, v_\Gamma \right), \; \beta = p - \frac{\partial L}{\partial v} = (p_S, p_N, p_\Gamma + \Sigma - S, p_W - N, p_\Sigma), \; w = v = (v_S, v_N, v_\Gamma, v_W, v_\Sigma).$

By using this, the Dirac dynamical system

$$((\dot{t}, \dot{q}, \dot{v}, \dot{\mathsf{p}}, \dot{p}), \mathbf{d}\mathcal{E}(t, q, v, \mathsf{p}, p)) \in D_{\Delta_{\mathscr{P}}}(t, q, v, \mathsf{p}, p)$$

is equivalent to the following evolution equations:

$$
\begin{cases}
\dot{\mathsf{p}} = -\sum_{a=1}^{A}(\mu^a \mathcal{J}^a + T^a \mathcal{J}_S^a) - \sum_{b=1}^{B} \mathcal{J}_S^b T^b, \quad \dot{\Gamma} = \dfrac{\partial U}{\partial S}, \quad \dot{W} = \dfrac{\partial U}{\partial N}, \\[2mm]
\dot{t} = 1, \quad \dot{N} = \sum_{a=1}^{A} \mathcal{J}^a, \quad \dot{S} = \dot{\Sigma} + \sum_{a=1}^{A} \mathcal{J}_S^a + \sum_{b=1}^{B} \mathcal{J}_S^b, \\[2mm]
-\dfrac{\partial U}{\partial S}\dot{\Sigma} = \sum_{a=1}^{A} \left[\mathcal{J}^a(\dot{W} - \mu^a) + \mathcal{J}_S^a(\dot{\Gamma} - T^a) \right] + \sum_{b=1}^{B} \mathcal{J}_S^b(\dot{\Gamma} - T^b).
\end{cases}
$$

Making arrangements, this system yields the equations of evolution as

$$
\begin{cases}
\dot{N} = \sum_{a=1}^{A} \mathcal{J}^a, \\[2mm]
\dot{S} = I + \sum_{a=1}^{A} \mathcal{J}_S^a + \sum_{b=1}^{B} \mathcal{J}_S^b = \dfrac{d_i S}{dt} + \dfrac{d_e S}{dt},
\end{cases}
\tag{11}
$$

where $I = \frac{1}{T} \sum_{a=1}^{A} \left[\mathcal{J}^a(\mu^a - \mu) + \mathcal{J}_S^a(T^a - T) \right] + \frac{1}{T} \sum_{b=1}^{B} \mathcal{J}_S^b(T^b - T)$ and $\frac{d_i S}{dt} := I \geq 0$ denotes the internal entropy production due to the mixing of matter flowing into the system and $\frac{d_eS}{dt}$ is the entropy flow. The system of equation also gives the definition of the thermodynamic displacement as $\dot{\Gamma} = T$, $\dot{W} = \mu$, and $\dot{\Sigma} = I$. The momentum p represents the part of the energy associated to the interaction of the system with exterior through its ports. In fact, it follows

$$-\frac{d}{dt}E = \frac{d}{dt}\mathsf{p} = -\sum_{a=1}^{A}(\mathcal{J}^a\mu^a + \mathcal{J}^a_S T^a) - \sum_{b=1}^{B}\mathcal{J}^b_S T^b = -P_M^{\text{ext}} - P_H^{\text{ext}},$$

where E is the total energy of the system, defined by

$$E(t, x, v) = \frac{\partial \mathcal{L}}{\partial v^i}v^i - \mathcal{L}(t, x, v).$$

Example: A Piston Device with Ports and Heat Sources. As illustrated in Fig. 1, we consider an open chamber containing a species with internal energy $U(S, N)$, where we assume that the cylinder has two external heat sources with entropy flow rates $\mathcal{J}^{b_i}$, $i = 1, 2$, the volume of the chamber is constant V_0 and two ports through which the species is injected into or flows out of the cylinder with molar flow rates $\mathcal{J}^{a_i}$, $i = 1, 2$. The entropy flow rates at the ports are given by $\mathcal{J}^{a_i}_S = \mathcal{J}^{a_i}\mathsf{S}^{a_i}$.

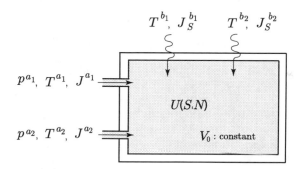

Fig. 1. An open chamber with ports and heat sources.

Recall that the Lagrangian is given by

$$\mathcal{L}(t, x, v) = -U(S, N) + v_W N + v_\Gamma(S - \Sigma).$$

The equations of evolution of the Dirac open thermodynamic system (11) read

$$\dot{N} = \sum_{a=1}^{A}\mathcal{J}^a, \qquad \dot{S} = I + \sum_{i=1}^{2}\mathcal{J}^{a_i}_S + \sum_{j=1}^{2}\mathcal{J}^{b_j}_S,$$

where $I = \dot{\Sigma}$ is the internal entropy production given by

$$I = \frac{1}{T}\sum_{i=1}^{2}\left[(\mu^{a_i} - \mu) + \mathsf{S}^{a_i}(T^{a_i} - T)\right]\mathcal{J}^{a_i} + \frac{1}{T}\sum_{j=1}^{2}\mathcal{J}^{b_j}_S(T^{b_j} - T).$$

The first term represents the entropy production associated to the mixing of gas flowing into the cylinder at the two ports a_1, a_2, and the third term denotes the entropy production due to the external heating. The second law requires that each of these terms is positive. The first law, i.e., the energy balance holds as

$$\frac{d}{dt}E = \underbrace{\sum_{j=1}^{2} \partial_S^{b_j} T^{b_j}}_{=P_H^{\text{ext}}} + \underbrace{\sum_{i=1}^{2}(\partial^{a_i}\mu^{a_i} + \partial_S^{a_i}T^{a_i})}_{=P_M^{\text{ext}}}.$$

References

1. Carathéodory, C.: Untersuchungen über die Grundlagen der Thermodynamik. Math. Ann. **67**, 355–386 (1909)
2. Courant, T.J.: Dirac manifolds. Trans. Am. Math. Soc. **319**, 631–661 (1990)
3. Eberard, D., Maschke, B.M., van der Schaft, A.J.: An extension of Hamiltonian systems to the thermodynamic phase space: towards a geometry of nonreversible processes. Rep. Math. Phys. **60**(2), 175–198 (2007)
4. Gay-Balmaz, F., Yoshimura, H.: A Lagrangian variational formulation for nonequilibrium thermodynamics. Part I: discrete systems. J. Geom. Phys. **111**, 169–193 (2017a)
5. Gay-Balmaz, F., Yoshimura, H.: A Lagrangian variational formulation for nonequilibrium thermodynamics. Part II: continuum systems. J. Geom. Phys. **111**, 194–212 (2017b)
6. Gay-Balmaz, F., Yoshimura, H.: A variational formulation of nonequilibrium thermodynamics for discrete open systems with mass and heat transfer. Entropy **20**(3), 1–26 (2018a). https://doi.org/10.3390/e20030163
7. Gay-Balmaz, F., Yoshimura, H.: Dirac structures in nonequilibrium thermodynamics. J. Math. Phys. **59**, 012701–29 (2018b)
8. Gibbs, J.W.: Graphical methods in the thermodynamics of fluids. Trans. Connecticus Acad. **2**, 309–342 (1873)
9. Hermann, R.: Geometry, Physics and Systems. Dekker, New York (1973)
10. Klein, S., Nellis, G.: Thermodynamics. Cambridge University Press, Cambridge (2011)
11. Mrugala, R.: Geometrical formulation of equilibrium phenomenological thermodynamics. Rep. Math. Phys. **14**, 419–427 (1978)
12. Mrugala, R.: A new representation of thermodynamic phase space. Bull. Polish Acad. Sci. **28**, 13–18 (1980)
13. Mrugala, R., Nulton, J.D., Schon, J.C., Salamon, P.: Contact structure in thermodynamic theory. Rep. Math. Phys. **29**, 109–121 (1991)
14. Sandler, S.I.: Chemical, Biochemical, and Engineering Thermodynamics. Wiley, Hoboken (2006)
15. Stueckelberg, E.C.G., Scheurer, P.B.: Thermocinétique Phénoménologique Galiléenne, Birkhäuser (1974)
16. Yoshimura, H., Marsden, J.E.: Dirac structures in Lagrangian mechanics. Part I: Implicit Lagrangian systems. J. Geom. Phys. **57**, 133–156 (2006a)
17. Yoshimura, H., Marsden, J.E.: Dirac structures in Lagrangian mechanics. Part II: Variational structures. J. Geom. Phys. **57**, 209–250 (2006b)

From Variational to Bracket Formulations in Nonequilibrium Thermodynamics of Simple Systems

François Gay-Balmaz[1]([envelope]) and Hiroaki Yoshimura[2]

[1] Ecole Normale Supérieure de Paris, Paris, France
francois.gay-balmaz@lmd.ens.fr
[2] Waseda University, Okubo, Shinjuku, Tokyo 169-8555, Japan
yoshimura@waseda.jp

Abstract. A variational formulation for nonequilibrium thermodynamics was recently proposed in [7,8] for both discrete and continuum systems. This formulation extends the Hamilton principle of classical mechanics to include irreversible processes. In this paper, we show that this variational formulation yields a constructive and systematic way to derive from a unified perspective several bracket formulations for nonequilibrium thermodynamics proposed earlier in the literature, such as the single generator bracket and the double generator bracket. In the case of a linear relation between the thermodynamic fluxes and the thermodynamic forces, the metriplectic or GENERIC brackets are recovered. A similar development has been presented for continuum systems in [6] and applied to multicomponent fluids.

Keywords: Nonequilibrium thermodynamics ·
Variational formulation · Bracket formulation

1 Variational Formulation of Nonequilibrium Thermodynamics

A Lagrangian variational approach for nonequilibrium thermodynamic has been proposed in [7,8] for finite dimensional and continuum closed systems and for open systems in [9]. This variational formulation extends the Hamilton principle of classical mechanics to include irreversible processes such as friction, heat or mass transfer in the equations of motion. It is a type of Lagrange-d'Alembert principle with nonlinear constraints and it follows a very systematic construction from the given thermodynamic fluxes and forces of the irreversible processes. This formulation is based on the concept of thermodynamic displacements, which

---

F. Gay-Balmaz is partially supported by the ANR project GEOMFLUID, ANR-14-CE23-0002-01; H. Yoshimura is partially supported by JSPS Grant-in-Aid for Scientific Research (S) 24224004, the MEXT Top Global University Project and Waseda University (SR 2019C-176).

F. Nielsen and F. Barbaresco (Eds.): GSI 2019, LNCS 11712, pp. 209–217, 2019.
https://doi.org/10.1007/978-3-030-26980-7_22

are primitive in time of the thermodynamic forces. This variational formulation naturally has an associated geometric description given in terms of Dirac structures [10].

Historically, the proposed general formalisms of nonequilibrium thermodynamics have been mainly constructed via appropriate modifications of Poisson brackets, as initiated by [11,13,14]. Since then, this approach has been developed for a large list of systems, see, e.g. [12]. Other classes of bracket have been proposed, e.g. [3–5]. Unlike the variational formalism, most of these bracket formalisms do not follow from a systematic construction and have often been derived via a case-by-case approach, with slightly different axioms used in different situations.

In this paper, we shall show that the variational formalism directly yields the two main bracket formalisms, namely, the single and double generator brackets. Moreover, in the case of a linear relation between the thermodynamic fluxes and the thermodynamic forces, the metriplectic ([16]) or GENERIC ([12,17]) brackets are recovered. We focus on the case of simple thermodynamic systems, in which only one entropy variable is needed. The general case will be studied elsewhere. The derivation of such brackets from the variational formulation for continuum system has been illustrated in [6] in the context of multicomponent fluids. We refer for instance to [1,2] for an introduction to the nonequilibrium thermodynamics of continuum systems.

1.1   Variational Formulation for Mechanical Systems with Friction

Consider a thermodynamic system described only by a mechanical variable $q \in Q$ and an entropy variable $S \in \mathbb{R}$. The Lagrangian of this thermodynamic system is a function

$$L : TQ \times \mathbb{R} \to \mathbb{R}, \quad (q, v, S) \mapsto L(q, v, S),$$

where TQ denotes the tangent bundle of the mechanical configuration manifold Q. We assume that the system involves external and friction forces given by fiber preserving maps $F^{\text{ext}}, F^{\text{fr}} : TQ \times \mathbb{R} \to T^*Q$, i.e., such that $F^{\text{fr}}(q, v, S) \in T_q^*Q$, similarly for F^{ext}. As stated in [7], the variational formulation for this system is given as follows:

Find the curves $q(t)$, $S(t)$ which are critical for the *variational condition*

$$\delta \int_{t_1}^{t_2} L(q, \dot{q}, S)\mathrm{d}t + \int_{t_1}^{t_2} \left\langle F^{\text{ext}}(q, \dot{q}, S), \delta q \right\rangle \mathrm{d}t = 0, \tag{1}$$

subject to the *phenomenological constraint*

$$\frac{\partial L}{\partial S}(q, \dot{q}, S)\dot{S} = \left\langle F^{\text{fr}}(q, \dot{q}, S), \dot{q} \right\rangle, \tag{2}$$

---

and for variations subject to the *variational constraint*

$$\frac{\partial L}{\partial S}(q, \dot{q}, S)\delta S = \langle F^{\text{fr}}(q, \dot{q}, S), \delta q \rangle, \tag{3}$$

with $\delta q(t_1) = \delta q(t_2) = 0$.

This variational formulation yields the system of equations

$$\frac{d}{dt}\frac{\partial L}{\partial \dot{q}} - \frac{\partial L}{\partial q} = F^{\text{fr}}(q, \dot{q}, S), \qquad \frac{\partial L}{\partial S}\dot{S} = \langle F^{\text{fr}}(q, \dot{q}, S), \dot{q} \rangle. \tag{4}$$

The first equation is the balance of mechanical momentum, while the second one gives the rate of entropy production of the system

$$\dot{S} = -\frac{1}{T}\langle F^{\text{fr}}(q, \dot{q}, S), \dot{q} \rangle,$$

with $T = -\frac{\partial L}{\partial S}(q, \dot{q}, S)$ the temperature of the system. From the second law the friction force F^{fr} must satisfy $\langle F^{\text{fr}}(q, \dot{q}, S), \dot{q} \rangle \leq 0$ for all $(q, \dot{q}, S)$. For instance, for a friction force linear in velocities, we have

$$F_i^{\text{fr}} = -\lambda_{ij}\dot{q}^j,$$

where λ_{ij}, $i, j = 1, ..., n$ are functions of the state variables with the symmetric part of the matrix λ_{ij} positive semi-definite.

1.2 Variational Formulation for Systems with Internal Mass Transfer

The previous variational formulation can be extended to systems experiencing internal diffusion processes. Diffusion is particularly important in biology, where many processes depend on the transport of chemical species through bodies. Consider a thermodynamic system consisting of K compartments that can exchange matter by diffusion across walls (or membranes) on their common boundaries. We assume that the system has a single species and denote by N_k the number of moles of the species in the k-th compartment, $k = 1, ..., K$. We assume that the thermodynamic system is simple; i.e., a uniform entropy S, the entropy of the system, is attributed to all the compartments. The Lagrangian of this thermodynamic system is thus a function

$$L : TQ \times \mathbb{R}^{K+1} \to \mathbb{R}, \quad (q, v, S, N_1, .., N_K) \mapsto L(q, v, S, N_1, .., N_K). \tag{5}$$

We denote $\mathcal{J}^{\ell \to k} = -\mathcal{J}^{k \to \ell}$ the molar flow rate from compartment ℓ to compartment k due to diffusion of the species. In general, we have the dependence

$$\mathcal{J}^{\ell \to k} = \mathcal{J}^{\ell \to k}\left(S, N_k, N_\ell, \frac{\partial L}{\partial N_k}, \frac{\partial L}{\partial N_\ell}\right). \tag{6}$$

The variational formulation involves the new variables W^k, $k = 1, ..., K$, which are examples of *thermodynamic displacements* and play a central role in our formulation. In general, we define the *thermodynamic displacement associated to a irreversible process* as the primitive in time of the thermodynamic force (or affinity) of the process. This force (or affinity) thus becomes the rate of change of the thermodynamic displacement. In the case of matter transfer, $\dot{W}^k$ corresponds to the chemical potential of N_k. The variational formulation for a simple system with internal diffusion process is stated as follows.

Find the curves $q(t)$, $S(t)$, $W^k(t)$, $N_k(t)$, which are critical for the *variational condition*

$$\delta \int_{t_1}^{t_2} \left[L(q, \dot{q}, S, N_1, ..., N_K) + \dot{W}^k N_k \right] dt + \int_{t_1}^{t_2} \langle F^{\text{ext}}, \delta q \rangle \, dt = 0, \qquad (7)$$

subject to the *phenomenological constraint*

$$\frac{\partial L}{\partial S} \dot{S} = \langle F^{\text{fr}}, \dot{q} \rangle + \sum_{k,\ell=1}^{K} J^{\ell \to k} \dot{W}^k, \qquad (8)$$

and for variations subject to the *variational constraint*

$$\frac{\partial L}{\partial S} \delta S = \langle F^{\text{fr}}, \delta q \rangle + \sum_{k,\ell=1}^{K} J^{\ell \to k} \delta W^k, \qquad (9)$$

with $\delta q(t_1) = \delta q(t_2) = 0$ and $\delta W^k(t_1) = \delta W^k(t_2) = 0$, $k = 1, ..., K$.

These conditions, combined with the phenomenological constraint (8), yield the system of evolution equations for the curves $q(t)$, $S(t)$, and $N^k(t)$:

$$\begin{cases} \dfrac{d}{dt} \dfrac{\partial L}{\partial \dot{q}} - \dfrac{\partial L}{\partial q} = F^{\text{fr}} + F^{\text{ext}}, \qquad \dfrac{d}{dt} N_k = \sum_{\ell=1}^{K} J^{\ell \to k}, \quad k = 1, ..., K, \\ \dfrac{\partial L}{\partial S} \dot{S} = \langle F^{\text{fr}}, \dot{q} \rangle - \sum_{k<\ell} J^{\ell \to k} \left(\dfrac{\partial L}{\partial N_k} - \dfrac{\partial L}{\partial N_\ell} \right). \end{cases} \qquad (10)$$

The last equation in (10) yields the rate of entropy production of the system as

$$\dot{S} = -\frac{1}{T} \langle F^{\text{fr}}, \dot{q} \rangle - \frac{1}{T} \sum_{k<\ell} J^{\ell \to k} (\mu^k - \mu^\ell), \qquad (11)$$

with $\mu^k = -\frac{\partial L}{\partial N_k}$ the chemical potentials. The two terms in (11) correspond, respectively, to the rate of entropy production due to mechanical friction and to matter transfer. From the second law, F^{fr} and $J^{k \to \ell}$ must satisfy

$$\langle F^{\text{fr}}, \dot{q} \rangle \leq 0 \qquad \text{and} \qquad J^{\ell \to k}(\mu^k - \mu^\ell) \leq 0. \qquad (12)$$

When a linear relation is assumed between the forces and fluxes, we have relations

$$F_i^{\mathrm{fr}} = -\lambda_{ij}\dot{q}^j \qquad \text{and} \qquad \mathcal{J}^{\ell \to k} = -G^{k\ell}(\mu^k - \mu^\ell), \qquad (13)$$

where λ_{ij}, $i,j = 1,...,n$ and $G^{k\ell}$, $k,\ell = 1,...,K$ are functions of the state variables, with the symmetric part of the matrix λ_{ij} positive semi-definite and with $G^{k\ell} \geq 0$, for all k,ℓ.

Note that in both variational formulations (1)–(3) and (7)–(9), the two constraints are related in a very systematic way, suggested by the relation

$$\sum_\alpha J_\alpha \dot{\Lambda}^\alpha \quad \rightsquigarrow \quad \sum_\alpha J_\alpha \delta \Lambda^\alpha \qquad (14)$$

with J_α the thermodynamic flux and Λ^α the thermodynamic displacement of the process α. This systematic correspondence holds for finite dimensional and continuum closed systems, and is at the core of the formulation in terms of Dirac structures, [7,8,10].

For simplicity, from now on we set the external forces F^{ext} to zero. They can be easily included in our developments below, and yield an additional term in the bracket formalisms.

2   Single and Double Generator Brackets

In this section, we shall show that the variational formulation induces and unifies several bracket formulations for nonequilibrium thermodynamics proposed earlier in the literature, such as the single generator bracket, the double generator bracket, and the metriplectic (or GENERIC) bracket.

2.1   Bracket Formulations in Nonequilbirium Thermodynamics

There are two main approaches to the bracket formulation for irreversible processes in the literature: the *single generator* and *double generator* formulations. In this paragraph, we quickly review the structure of these two brackets. Let M be a Poisson manifold, with Poisson bracket $\{\,,\}$. We denote by $H \in C^\infty(M)$ the Hamiltonian and $S \in C^\infty(M)$ the entropy. We assume that $\{H, S\} = 0$.

In the single generator formalism, [3–5], the evolution of an arbitrary functional $F \in C^\infty(M)$ is governed by

$$\frac{d}{dt}F = \{F, H\} + [F, H], \qquad (15)$$

where the dissipation bracket $[F, H]$ is linear and a derivation in F, it can be nonlinear in H, and satisfies $[H, H] = 0$ and $[S, H] \geq 0$. These last two requirements are consistent with the first and second laws of thermodynamics, respectively. Since both the reversible (Poisson) and dissipation brackets use the same generator H, this is referred to as the *single generator formalism*. The bracket formulation (15) yields the dynamical system $\dot{m}(t) = X_H(m(t)) + D_H(m(t))$, where $X_H = \mathbf{J}\mathbf{d}H$ is the Hamiltonian vector field associated to H, with $\mathbf{J} : T^*M \to TM$ the Poisson tensor, and the vector field D_H is determined from $[F, H] = \mathbf{d}F \cdot D_H$, for all F, which follows since $F \mapsto [F, H]$ is a derivation.

In the double generator formalism, the evolution of an arbitrary functional $F \in C^\infty(M)$ is governed by

$$\frac{d}{dt}F = \{F, H\} + (F, S), \tag{16}$$

where the dissipation bracket (F, G) is symmetric, bilinear and satisfies the Leibniz rule, as well as $(H, S) = 0$ and $(S, S) \geq 0$. These are precisely the axioms given in [13]. Since the Poisson and dissipation brackets use different generators (H for Poisson and S for dissipation), this is referred to as the *double generator formalism*. The bracket formulation (16) yields the dynamical system $\dot{m}(t) = \mathbf{J}\mathbf{d}H(m(t)) + \mathbf{K}\mathbf{d}S(m(t))$, where as before $\mathbf{J}\mathbf{d}H = X_H$ is the Hamiltonian vector field associated to H, and the symmetric vector bundle linear map $K : T^*M \to TM$, $K^* = K$, is such that $(F, G) = \langle \mathbf{d}F, K\mathbf{d}G \rangle$, which follows from the fact that (F, G) is symmetric and a derivation in each factor.

Sometimes, the stronger requirements that $\{G, S\} = 0$, $(H, G) = 0$, $(G, G) \geq 0$, for arbitrary $G \in C^\infty(M)$ are imposed, in which case the system (16) is termed *metriplectic*, [16]. For example, this is what is used in the GENERIC formalism [12, 17]. When considering macroscopic systems, typically only bilinearity, $(H, S) = 0$, and $(S, S) \geq 0$ seem to be required on physical grounds.

2.2 Derivation of the Single Generator Bracket

Consider the system (10), assume that the Lagrangian L in (5) is hyperregular with respect to the mechanical part and define the associated Hamiltonian $H :$ $T^*Q \times \mathbb{R}^{K+1} \to \mathbb{R}$. In terms of H, system (10) can be equivalently written as

$$\begin{cases} \dot{q} = \dfrac{\partial H}{\partial p}, \quad \dot{p} = -\dfrac{\partial H}{\partial q} + F^{\mathrm{fr}}\Big(q, \dfrac{\partial H}{\partial p}, S\Big), \quad \dfrac{d}{dt}N_k = \displaystyle\sum_{\ell=1}^{K} \jmath^{\ell \to k}, \\[3mm] -\dfrac{\partial H}{\partial S}\dot{S} = \Big\langle F^{\mathrm{fr}}\Big(q, \dfrac{\partial H}{\partial p}, S\Big), \dfrac{\partial H}{\partial p}\Big\rangle + \displaystyle\sum_{k<\ell} \jmath^{\ell \to k}\Big(\dfrac{\partial H}{\partial N_k} - \dfrac{\partial H}{\partial N_\ell}\Big). \end{cases} \tag{17}$$

In this system, the dependence of the fluxes in (6) is written in terms of H by using $\frac{\partial L}{\partial N_k} = -\frac{\partial H}{\partial N_k}$. From (17), we directly deduce that the evolution of an arbitrary function $F \in C^\infty(T^*Q \times \mathbb{R}^{K+1})$ is of the form (15) with $\{,\}$ the direct sum of the canonical Poisson bracket on T^*Q and the zero bracket on $\mathbb{R}^{K+1}$, where the dissipation bracket is computed as

$$\begin{aligned} [F, H] = \Big\langle F^{\mathrm{fr}}, \dfrac{\partial F}{\partial p}\Big\rangle &+ \sum_{k<\ell} \jmath^{\ell \to k}\Big(\dfrac{\partial F}{\partial N_k} - \dfrac{\partial F}{\partial N_\ell}\Big) \\[2mm] &- \dfrac{\frac{\partial F}{\partial S}}{\frac{\partial H}{\partial S}}\Big[\Big\langle F^{\mathrm{fr}}, \dfrac{\partial H}{\partial p}\Big\rangle + \sum_{k<\ell} \jmath^{\ell \to k}\Big(\dfrac{\partial H}{\partial N_k} - \dfrac{\partial H}{\partial N_\ell}\Big)\Big]. \end{aligned} \tag{18}$$

In this expression, we recall that both F^{fr} and $\mathcal{J}^{\ell\to k}$ may depend on H via their dependence on $(q, \frac{\partial H}{\partial p}, S)$ and $(S, N_k, \frac{\partial H}{\partial N_k}, N_\ell, \frac{\partial H}{\partial N_\ell})$, respectively. One directly checks that the conditions $\{H, S\} = 0$, $[H, H] = 0$ are satisfied. The condition $[S, H] \geq 0$ is satisfied if and only if (12) holds.

We have thus recovered the single generator formalism. This formulation does not impose a specific dependence (such as a linear dependence) of the thermodynamic fluxes F^{fr} and $\mathcal{J}^{\ell\to k}$ on the thermodynamic forces.

2.3 Derivation of the Double Generator Bracket

Starting again from the system (10) obtained from the variational formulation, we compute the evolution of an arbitrary function F. The expression (18) has now to be interpreted as the bracket (F, S). Hence it suffices to multiply this expression by $1 = \frac{\partial S}{\partial S}$, to symmetrize in F and S the resulting expression, and finally to replace S by an arbitrary function G to finally get the symmetric bracket

$$
\begin{aligned}
(F, G) =\ & \left\langle F^{\mathrm{fr}}, \frac{\partial F}{\partial p} \right\rangle \frac{\partial G}{\partial S} + \left\langle F^{\mathrm{fr}}, \frac{\partial G}{\partial p} \right\rangle \frac{\partial F}{\partial S} \\
& + \sum_{k<\ell} \mathcal{J}^{\ell\to k} \left(\frac{\partial F}{\partial N_k} - \frac{\partial F}{\partial N_\ell} \right) \frac{\partial G}{\partial S} + \sum_{k<\ell} \mathcal{J}^{\ell\to k} \left(\frac{\partial G}{\partial N_k} - \frac{\partial G}{\partial N_\ell} \right) \frac{\partial F}{\partial S} \quad (19) \\
& - \frac{1}{\frac{\partial H}{\partial S}} \left[\left\langle F^{\mathrm{fr}}, \frac{\partial H}{\partial p} \right\rangle + \sum_{k<\ell} \mathcal{J}^{\ell\to k} \left(\frac{\partial H}{\partial N_k} - \frac{\partial H}{\partial N_\ell} \right) \right] \frac{\partial F}{\partial S} \frac{\partial G}{\partial S}.
\end{aligned}
$$

One directly checks that the bracket (F, G) is symmetric, bilinear and satisfies the Leibniz rule, as well as $(H, S) = 0$. The condition $(S, S) \geq 0$ is satisfied if and only if (12) holds.

2.4 Derivation of the Metriplectic Bracket

In general the bracket (18) is not metriplectic, since one has

$$
(F, H) = \left\langle F^{\mathrm{fr}}, \frac{\partial F}{\partial p} \right\rangle \frac{\partial H}{\partial S} + \sum_{k<\ell} \mathcal{J}^{\ell\to k} \left(\frac{\partial F}{\partial N_k} - \frac{\partial F}{\partial N_\ell} \right) \frac{\partial H}{\partial S} \neq 0. \qquad (20)
$$

Let us assume as in (13) that the thermodynamic fluxes F^{fr} and $\mathcal{J}^{\ell\to k}$ depend linearly on their corresponding thermodynamic forces as

$$
F^{\mathrm{fr}}\left(q, \frac{\partial H}{\partial p}, S, N\right) = -\lambda(q, S) \cdot \frac{\partial H}{\partial p} \quad \text{and} \quad \mathcal{J}^{\ell\to k} = -G^{k\ell}(S, N_k, N_\ell)\left(\frac{\partial H}{\partial N_k} - \frac{\partial H}{\partial N_\ell} \right),
$$

where $\lambda(q, S) : T_q Q \to T_q^* Q$ is symmetric positive semi-definite and where $G^{k\ell}(S, N_k, N_l) \geq 0$ for all k, ℓ. Using these relations in the expression (20) by writing them in terms of an arbitrary function G, and subtracting it from (F, G), we get the symmetric bracket

$$
(F, G)_{\mathrm{met}} = (F, G) + \left\langle \lambda \cdot \frac{\partial G}{\partial p}, \frac{\partial F}{\partial p} \right\rangle \frac{\partial H}{\partial S} + \sum_{k<\ell} G^{k\ell}\left(\frac{\partial G}{\partial N_k} - \frac{\partial G}{\partial N_\ell} \right)\left(\frac{\partial F}{\partial N_k} - \frac{\partial F}{\partial N_\ell} \right) \frac{\partial H}{\partial S}.
$$

A direct computation using (19) and rearranging the terms finally yields the expression

$$
\begin{aligned}
(F,G)_{\mathrm{met}} =& \frac{1}{\frac{\partial H}{\partial S}} \left\langle \frac{\partial F}{\partial p}\frac{\partial H}{\partial S} - \frac{\partial H}{\partial p}\frac{\partial F}{\partial S}, \lambda \cdot \left(\frac{\partial G}{\partial p}\frac{\partial H}{\partial S} - \frac{\partial H}{\partial p}\frac{\partial G}{\partial S} \right) \right\rangle \\
&+ \frac{1}{\frac{\partial H}{\partial S}} \sum_{k<\ell} G^{kl} \left[\left(\frac{\partial F}{\partial N_k} - \frac{\partial F}{\partial N_\ell} \right)\frac{\partial H}{\partial S} - \left(\frac{\partial H}{\partial N_k} - \frac{\partial H}{\partial N_\ell} \right)\frac{\partial F}{\partial S} \right] \\
&\times \left[\left(\frac{\partial G}{\partial N_k} - \frac{\partial G}{\partial N_\ell} \right)\frac{\partial H}{\partial S} - \left(\frac{\partial H}{\partial N_k} - \frac{\partial H}{\partial N_\ell} \right)\frac{\partial G}{\partial S} \right].
\end{aligned}
\tag{21}
$$

From this, one directly checks that $(H,G)_{\mathrm{met}} = 0$, and $(G,G)_{\mathrm{met}} \geq 0$, for arbitrary $G \in C^\infty(T^*Q \times \mathbb{R}^{K+1})$ by (13), therefore $(F,G)_{\mathrm{met}}$ is a metriplectic (or GENERIC) bracket. We note that $\dot{S} = (S,S) = (S,S)_{\mathrm{met}}$. The structure of the bracket (21) is a finite dimensional analogue of that of the metriplectic bracket for viscous heat conducting fluid presented in [15].

References

1. Balian, R.: De la mécanique statistique hors équilibre aux équations de transport. École thématique. École Joliot Curie "Noyaux en collisions", Maubuisson, France, du 11–16 septembre 1995: 14ème session (1995)
2. Balian, R.: Introduction à la thermodynamique hors équilibre (2003). http://e2phy.in2p3.fr/2003/actesBalian.pdf
3. Beris, A.N., Edwards, B.J.: Thermodynamics of Flowing Systems with Internal Microstructure. Oxford University Press, New York (1994)
4. Edwards, B.J., Beris, A.N.: Noncanonical poisson bracket for nonlinear elasticity with extensions to viscoelasticity. Phys. A Math. Gen. **24**, 2461–2480 (1991a)
5. Edwards, B.J., Beris, A.N.: Unified view of transport phenomena based on the generalized bracket formulation. Ind. Eng. Chem. Res. **30**, 873–881 (1991b)
6. Eldred, C., Gay-Balmaz, F.: Single and double generator bracket formulations of geophysical fluids with irreversible processes (2019). https://arxiv.org/pdf/1811.11609.pdf
7. Gay-Balmaz, F., Yoshimura, H.: A Lagrangian variational formulation for nonequilibrium thermodynamics. Part I: discrete systems. J. Geom. Phys. **111**, 169–193 (2017a)
8. Gay-Balmaz, F., Yoshimura, H.: A Lagrangian variational formulation for nonequilibrium thermodynamics. Part II: continuum systems. J. Geom. Phys. **111**, 194–212 (2017b)
9. Gay-Balmaz, F., Yoshimura, H.: A variational formulation of nonequilibrium thermodynamics for discrete open systems with mass and heat transfer. Entropy **163**, 1–26 (2018a). https://doi.org/10.3390/e20030163
10. Gay-Balmaz, F., Yoshimura, H.: Dirac structures in nonequilibrium thermodynamics. J. Math. Phys. **59**, 012701–29 (2018b)
11. Grmela, M.: Bracket formulation of dissipative fluid mechanics equations. Phys. Lett. A **102**, 355–358 (1984)
12. Grmela, M., Öttinger, H.-C.: Dynamics and thermodynamics of complex fluids. I. Development of a general formalism. Phys. Rev. E **56**(3), 6620–6632 (1997)

13. Kaufman, A.: Dissipative Hamiltonian systems: a unifying principle. Phys. Lett. A **100**, 419–422 (1984)
14. Morrison, P.: Bracket formulation for irreversible classical fields. Phys. Lett. A **100**, 423–427 (1984a)
15. Morrison, P.: Some observations regarding brackets and dissipation. Technical report, University of California, Berkeley (1984b)
16. Morrison, P.: A paradigm for joined hamiltonian and dissipative systems. Physica D **18**, 410–419 (1986)
17. Öttinger, H.-C., Grmela, M.: Dynamics and thermodynamics of complex fluids. II. Illustrations of a general formalism. Phys. Rev. E **56**, 6633–6655 (1997)
18. Stueckelberg, E.C.G., Scheurer, P.B.: Thermocinétique phénoménologique galiléenne, Birkhäuser (1974)

A Homological Approach to Belief Propagation and Bethe Approximations

Olivier Peltre$^{(\boxtimes)}$

Université Paris 7 Denis Diderot, équipe Géométrie et Dynamique, Paris, France
`olivier@peltre.xyz`

Abstract. We introduce a differential complex of local observables given a set of random variables covered by subsets. Its boundary operator ∂ allows us to define a transport equation $\dot{u} = \partial\Phi(u)$ equivalent to Belief Propagation. This definition reveals a maximal set of conserved quantities under Belief Propagation and gives new geometric insight on the relationship of its equilibria with the critical points of Bethe free energy.

1 Introduction

A common feature of statistical physics and statistical learning is to consider a very large number of random variables, each of them mostly interacting with only a small subset of neighbours. Both lead to the effort of extracting relevant information about collective phenomena in spite of intractable global computations, hence motivating the development of local techniques where only small enough subsets of variables are simultaneously considered.

In the present note, we work with a collection of local algebras of observables, on which a boundary operator describes relations between intersecting subsystems. The construction of this differential complex is exposed in Sect. 2. It allows for a homological interpretation of the equivalence established by Yedidia *et al.* between critical points of the Bethe free energy approximation and fixed points of the Belief Propagation algorithm. We review this beautiful theorem bridging statistical learning and thermodynamics[1] in Sect. 3.

This is part of a PhD work done under the kind supervision of Daniel Bennequin. I wish to thank him as well as Grégoire Sergeant-Perthuis and Juan-Pablo Vigneaux, for their sustained collaboration and many fruitful discussions.

---

[1] See the longer version of this note for more context and historical background.

© Springer Nature Switzerland AG 2019
F. Nielsen and F. Barbaresco (Eds.): GSI 2019, LNCS 11712, pp. 218–227, 2019.
https://doi.org/10.1007/978-3-030-26980-7_23

2 Differential and Combinatorial Structures

2.1 Statistical System

Regions. We call system a finite set Ω equipped with a covering $X \subseteq \mathcal{P}(\Omega)$ by subsets such that:

- the empty set $\varnothing$ is in X,
- if $\alpha \in X$ and $\beta \in X$, then $\alpha \cap \beta$ is also in X.

We view X as a subcategory of the partial order $\mathcal{P}(\Omega)$ having an arrow $\alpha \to \beta$ whenever $\beta \subseteq \alpha$. We call $\alpha \in X$ a region[2] of Ω and denote by $\Lambda^\alpha \subseteq X$ the subsystem of those regions contained in α.

Chains and Nerve. A p-chain $\bar{\alpha}$ is a totally ordered sequence $\alpha_0 \to \ldots \to \alpha_p$ in X, it is said non-degenerate when all inclusions are strict. A p-chain $\bar{\alpha}$ may be viewed as a p-simplex, whose $p+1$ faces are the chains $\bar{\alpha}^{(k)}$ obtained by removing α_k, for $0 \leq k \leq p$. The nerve of X is the simplicial complex $NX = \bigsqcup_p N^p X$ formed by all non-degenerate chains.

Microscopic States. For each $i \in \Omega$, suppose given a finite set E_i. A microscopic state of a region $\alpha \subseteq \Omega$ is an element of the cartesian product[3]:

$$E_\alpha = \prod_{i \in \alpha} E_i$$

We denote by $\pi^{\beta\alpha} : E_\alpha \to E_\beta$ the canonical projection of E_α onto E_β whenever β is a subregion of α.

2.2 Scalar Fields

Differentials. We call scalar field a collection $\lambda \in \mathbb{R}(X)$ of scalars indexed by the nerve of X. We denote by $\mathbb{R}_p(X) = \mathbb{R}^{N^p X}$ the space of p-fields, vanishing everywhere but on the p-simplices of NX.

Through the canonical scalar product of $\mathbb{R}(X)$, scalar fields can be identified with chains or cochains with real coefficients in NX eitherwise. We denote by ∂ the boundary operator of $\mathbb{R}(X)$ and by d its adjoint differential:

$$\partial : \mathbb{R}_0(X) \leftarrow \mathbb{R}_1(X) \leftarrow \ldots$$
$$d : \mathbb{R}_0(X) \to \mathbb{R}_1(X) \to \ldots$$

---

[2] The term refers to the notion of region-graphs in Yedidia et al.

[3] The configuration space $E_\varnothing$ is thus a point, unit for the cartesian product and terminal element in **Set**.

Convolution. Let $\tilde{\mathbb{R}}_1(X) = \mathbb{R}_0(X) \oplus \mathbb{R}_1(X)$. Identifying the degenerate 1-chain $\alpha \to \alpha$ with α, an element of $\tilde{\mathbb{R}}_1(X)$ is indexed by general 1-chains in X. Equipped with Dirichlet convolution, $\tilde{\mathbb{R}}_1(X)$ is the incidence algebra[4] of X:

$$(\varphi * \psi)_{\alpha\gamma} = \sum_{\alpha \to \beta' \to \gamma} \varphi_{\alpha\beta'} \cdot \psi_{\beta'\gamma}$$

The unit of $*$ is $1 \in \mathbb{R}_0(X)$, sometimes viewed as a Kronecker symbol in $\tilde{\mathbb{R}}_1(X)$.

The space of 0-fields $\mathbb{R}_0(X)$ also has a $\tilde{\mathbb{R}}_1(X)$-bimodule structure, where the left action of $\varphi \in \tilde{\mathbb{R}}_1(X)$ on $\lambda \in \mathbb{R}_0(X)$ is given by:

$$(\varphi \cdot \lambda)_\alpha = \sum_{\alpha \to \beta'} \varphi_{\alpha\beta'} \lambda_{\beta'}$$

Möbius Inversion. The Dirichlet zeta function $\zeta \in \tilde{\mathbb{R}}_1(X)$ is defined by $\zeta_{\alpha\beta} = 1$ for every $\alpha \to \beta$ in X. When X is locally finite[5], ζ is invertible. Its inverse μ, known as the Möbius function, satisfies:

$$\mu_{\alpha\beta} = \sum_{k \geq 0} (-1)^k (\zeta - 1)_{\alpha\beta}^{*k}$$

where $(\zeta - 1)_{\alpha\beta}^{*k}$ counts the number of non-degenerate k-chains from α to β.

The coefficients $c = (1 \cdot \mu) \in \mathbb{R}_0(X)$ contain all the combinatorics of Bethe approximations. They satisfy the following «inclusion-exclusion» formula:

$$(c \cdot \zeta)_\beta = \sum_{\alpha' \to \beta} c_{\alpha'} = 1$$

2.3   Observables, Densities and Statistical States

Observable Fields. Denote by $\mathfrak{a}_\alpha = \mathbb{R}^{E_\alpha}$ the commutative algebra of observables on $\alpha \subseteq \Omega$. For every subregion $\beta \subseteq \alpha$, an observable $u_\beta \in \mathfrak{a}_\beta$, as a real function on E_β, admits a cylindrical extension $j_{\alpha\beta}(u_\beta)$ on E_α.

For every $\bar{\alpha} \in N_pX$, let $\mathfrak{a}_{\bar{\alpha}}$ denote a copy of the algebra $\mathfrak{a}_{\alpha_p}$ of observables on its smallest region. There is an injection $j_{\bar{\beta}\bar{\alpha}} : \mathfrak{a}_{\bar{\alpha}} \to \mathfrak{a}_{\bar{\beta}}$ whenever $\bar{\beta}$ is a subchain of $\bar{\alpha}$.[6] We define the graded vector space $\mathfrak{a}(X)$ of observable fields by:

$$\mathfrak{a}_p(X) = \bigoplus_{\bar{\alpha} \in N^p X} \mathfrak{a}_{\bar{\alpha}}$$

---

[4] See Rota [12] for a deeper treatment of these combinatorial structures.

[5] X is locally finite if for any $\alpha, \beta \in X$ there is only a finite number of non-degenerate chains from α to β.

[6] Observable fields form a simplicial algebra $\mathfrak{a}(X) : \mathbf{Ord}^{op} \to \mathbf{Alg}$. To relate to this more general theory, see Segal's note on classifying spaces [14] for instance.

It is equipped with a boundary[7] operator $\partial : \mathfrak{a}_{p+1}(X) \to \mathfrak{a}_p(X)$. When $p = 0$, we have for instance[8]:

$$\partial_\beta \varphi = \sum_{\alpha' \to \beta} \varphi_{\alpha'\beta} - \sum_{\beta \to \gamma'} \varphi_{\beta\gamma'}$$

Belief Propagation is essentially a dynamic up to a boundary term $\partial\varphi$ in $\mathfrak{a}_0(X)$, although it is usually viewed in the multiplicative group $G_0(X) = \prod_{\alpha \in X} G_\alpha$ with $G_\alpha = (\mathbb{R}_+^*)^{E_\alpha}$.

Density Fields. We call density on $\alpha \subseteq \Omega$ a linear form on observables $\omega_\alpha \in \mathfrak{a}_\alpha^*$. Denote by $\Sigma^{\beta\alpha}(\omega_\alpha) \in \mathfrak{a}_\beta^*$ the partial integration of ω_α along the fibers of $\pi^{\beta\alpha}$:

$$\Sigma^{\beta\alpha}(\omega_\alpha)(x_\beta) = \sum_{x' \in E_{\alpha \setminus \beta}} \omega_\alpha(x_\beta, x')$$

It satisfies $\langle \Sigma^{\beta\alpha}(\omega_\alpha) \,|\, u_\beta \rangle = \langle \omega_\alpha \,|\, j_{\alpha\beta}(u_\beta) \rangle$ for every $u_\beta \in \mathfrak{a}_\beta$.

The complex $\mathfrak{a}^*(X)$ is equipped with a differential $d : \mathfrak{a}_p^*(X) \to \mathfrak{a}_{p+1}^*(X)$, adjoint of ∂. For $p = 0$, we have for instance:

$$(d\omega)_{\alpha\beta} = \omega_\beta - \Sigma^{\beta\alpha}(\omega_\alpha)$$

A field $\omega \in \mathfrak{a}_0^*(X)$ is said consistent if $d\omega = 0$. The notion of consistent densities will replace that of a global measure on E_Ω.

Statistical Fields. Denote by $\Delta_\alpha \subseteq \mathfrak{a}_\alpha^*$ the convex subset of probability measures. It consists of all the positive densities ω_α satisfying $\omega_\alpha(1_\alpha) = 1$, and any non-trivial positive density $\omega_\alpha \in \mathfrak{a}_\alpha^*$ defines a normalised density $[\omega_\alpha] \in \Delta_\alpha$.

Its interior $\overset{\circ}{\Delta}_\alpha$ admits a natural Lie group structure, as it is diffeomorphic to the quotient of G_α by scalings of $\mathbb{R}_+^*$, itself isomorphic to the quotient of $\mathfrak{a}_\alpha$ by the action of additive constants. We denote by $[e^{-U_\alpha}] \in \overset{\circ}{\Delta}_\alpha$ the Gibbs state associated to $U_\alpha \in \mathfrak{a}_\alpha$ and by $[e^-]_\alpha : \mathfrak{a}_\alpha \to \overset{\circ}{\Delta}_\alpha$ this surjective group morphism.

We denote by $\Delta(X) \subseteq \mathfrak{a}^*(X)$ the convex subset of statistical fields, by $\overset{\circ}{\Delta}(X)$ its interior, and by $\overset{\circ}{\Delta}{}^d \subseteq \overset{\circ}{\Delta}_0(X)$ the subset of consistent ones.

2.4   Homology

Gauss Formulas. For every region $\alpha \in X$, let us define the coboundary of the subsystem Λ^α as the subset of arrows $\delta\Lambda^\alpha \subseteq N^1 X$ that meet Λ^α but are not contained in Λ^α:

$$\delta\Lambda^\alpha = \{\alpha' \to \beta' \,|\, \alpha' \notin \Lambda^\alpha \quad \text{and} \quad \beta' \in \Lambda^\alpha\}$$

---

[7] A boundary ∂ satisfies $\partial^2 = \partial \circ \partial = 0$.
[8] We will generally drop the injection in our notation.

The following proposition may then be thought of as a Gauss formula on Λ^α:

Proposition 1. *For every $\varphi \in \mathfrak{a}_1(X)$ and $\alpha \in X$ we have:*

$$\sum_{\beta' \in \Lambda^\alpha} \partial_{\beta'} \varphi = \sum_{\alpha'\beta' \in \delta\Lambda^\alpha} \varphi_{\alpha'\beta'}$$

In particular, the above vanishes if φ is supported in Λ^α.

A similar formula holds on the cone V_β over β in X, formed by all the regions containing β with coboundary the set δV_β of arrows leaving V_β. The sums however need to be embedded in the space of global observables.

Proposition 2. *For every $\varphi \in \mathfrak{a}_1(X)$ and $\beta \in X$ we have:*

$$\sum_{\alpha' \in V_\beta} \partial_{\alpha'} \varphi = - \sum_{\alpha'\beta' \in \delta V_\beta} \varphi_{\alpha'\beta'}$$

as global observables of $\mathfrak{a}_\Omega$.

Interaction Decomposition. We call boundary observable on a region $\alpha \in X$ any observable generated by observables on strict subregions of α in X. Suppose chosen for every α a supplement $\mathfrak{z}_\alpha$ of boundary observables, so that:

$$\mathfrak{a}_\alpha = \mathfrak{z}_\alpha \oplus \left(\sum_{\alpha > \beta'} \mathfrak{a}_{\beta'} \right)$$

We may inductively continue this procedure, as illustrated by the following well known[9] theorem.

Theorem 1 (Interaction Decomposition). *Given supplements $(\mathfrak{z}_\alpha)$ of boundary observables for every $\alpha \in X$, we have the decompositions:*

$$\mathfrak{a}_\alpha = \bigoplus_{\alpha \to \beta'} \mathfrak{z}_{\beta'}$$

They induce a projection P of $\mathfrak{a}_0(X)$ onto $\mathfrak{z}_0(X) = \bigoplus_\alpha \mathfrak{z}_\alpha$ defined by:

$$P^\beta(u) = \sum_{\alpha' \to \beta} P^{\beta\alpha'}(u_{\alpha'})$$

where $P^{\beta\alpha}$ denotes the projection of $\mathfrak{a}_\alpha$ onto $\mathfrak{z}_\beta$ for all $\alpha \to \beta$ in X.

Given a field $u \in \mathfrak{a}_0(X)$, define the global observable $\zeta_\Omega(u) \in \mathfrak{a}_\Omega$ by:

$$\zeta_\Omega(u) = \sum_{\alpha \in X} u_\alpha$$

Corollary 1. *For any $u \in \mathfrak{a}_0(X)$, we have the equivalence:*

$$P(u) = 0 \quad \Leftrightarrow \quad \zeta_\Omega(u) = 0$$

In particular, $\mathfrak{z}_0(X)$ is isomorphic to the image of ζ_Ω in $\mathfrak{a}_\Omega{}^{10}$.

---

[9] The first appearance of this now very common result in statistics seems to be in Kellerer [4]. See also [7] for a proof via harmonic analysis.

[10] They both represent the inductive limit of $\mathfrak{a}$ over X.

Homology Groups. The complex of observable fields $\mathfrak{a}(X)$ is acyclic[11] and we only focus on the first homology group.

Theorem 2. *The interaction decomposition P induces an isomorphism on the first homology group of observable fields:*

$$\mathfrak{a}_0(X)/\partial\mathfrak{a}_1(X) \sim \mathfrak{z}_0(X)$$

Proof. The Gauss formula on the cone V_β above β in X first ensures that P vanishes on boundaries:

$$P^\beta(\partial\varphi) = \sum_{\alpha'\to\beta} P^\beta(\partial_{\alpha'}\varphi) = \sum_{\alpha'\to\beta}\sum_{\beta'\not\to\beta} P^\beta(\varphi_{\alpha'\beta'}) = 0$$

as $P^\beta(\mathfrak{a}_{\beta'})$ is non-zero if and only if β' contains β. Let us denote by $[P]$ the quotient map induced by P. Given $u \in \mathfrak{a}_0(X)$, consider the flux φ defined by $\varphi_{\alpha\beta} = P^{\beta\alpha}(u_\alpha)$:

$$\partial_\beta\varphi = \sum_{\alpha'\to\beta}\varphi_{\alpha'\beta} - \sum_{\beta\to\gamma'}\varphi_{\beta\gamma'} = P^\beta(u) - u_\beta$$

When $P(u) = 0$ this gives $u = -\partial\varphi$, hence $[P]$ is injective.

Corollary 2. *Let $V = \zeta \cdot v$ in $\mathfrak{a}_0(X)$. We have the equivalence:*

$$cV \in \mathrm{Im}(\partial) \quad \Leftrightarrow \quad v \in \mathrm{Im}(\partial)$$

Proof. According to the theorem, it suffices to show that $P(v) = P(cV)$ and:

$$P^\gamma(v) = P^\gamma(\mu \cdot V) = \sum_{\alpha'\to\beta'\to\gamma} P^{\gamma\beta'}(\mu_{\alpha'\beta'}V_{\beta'}) = \sum_{\beta'\to\gamma} P^{\gamma\beta'}(c_{\beta'}V_{\beta'}) = P^\gamma(cV)$$

3   First Applications

3.1   Critical Points of Bethe Free Energy

Gibbs Free Energy. For every $\alpha \subseteq \Omega$, denote by $\mathcal{F}_\alpha$ its local Gibbs free energy, viewed as the functional on $\Delta_\alpha \times \mathfrak{a}_\alpha$ defined by:

$$\mathcal{F}_\alpha(p_\alpha, H_\alpha) = \mathbb{E}_{p_\alpha}[H_\alpha] - S(p_\alpha)$$

where $S(p_\alpha) = -\sum p_\alpha \ln(p_\alpha)$ denotes Shannon entropy.

Given a global hamiltonian $H_\Omega \in \mathfrak{a}_\Omega$, the global Gibbs state $[\mathrm{e}^{-H_\Omega}] \in \Delta_\Omega$ is the global minimum of $\mathcal{F}_\Omega(\,\cdot\,, H_\Omega)$. This definition being hardly computable in practice, we shall seek to estimate its marginals $\Sigma^{\alpha\Omega}(p_\Omega)$ by an approximation on the global Gibbs free energy $\mathcal{F}_\Omega$.

---

[11] We do not provide a proof here, a treatment of higher degrees shall be given in later work.

Bethe Approximation. The Bethe-Peierls approach and its refinements[12] essentially consist in writing an approximate decomposition of $\mathcal{F}_\Omega$ as a sum of local free energy summands f_β, for $\beta \in X$. This localisation procedure can be made exact on any $\alpha \in X$ by Möbius inversion:

$$\mathcal{F}_\alpha = \sum_{\alpha \to \beta'} f_{\beta'} \quad \Leftrightarrow \quad f_\beta = \sum_{\beta \to \gamma'} \mu_{\beta\gamma'} \, \mathcal{F}_{\gamma'}$$

The approximation only comes when Ω is not in X and we may then write the error $\mathcal{F}_\Omega - \check{\mathcal{F}}$ as a global free energy summand f_Ω. One should expect f_Ω to be small when sufficiently large regions are taken in X, by extensivity of the global Gibbs free energy[13].

The Bethe free energy $\check{\mathcal{F}}$ is thus defined for $p \in \Delta_0(X)$ and $H \in \mathfrak{a}_0(X)$ by:

$$\check{\mathcal{F}}(p, H) = \sum_{\beta \in X} c_\beta \cdot \mathcal{F}_\beta(p_\beta, H_\beta)$$

Given $H \in \mathfrak{a}_0(X)$, we denote by $\check{\mathcal{F}}^H$ the induced functional on $\Delta_0(X)$.

Critical Points. Because of the Möbius numbers c_β appearing in its definition, the Bethe free energy $\check{\mathcal{F}}$ is no longer convex in general, and $\check{\mathcal{F}}^H$ might have a great multiplicity[14] of consistent critical points in $\mathring{\Delta}^d$.

Theorem 3. *A non-vanishing consistent statistical field $p \in \mathring{\Delta}^d$ is a critical point of the Bethe free energy $\check{\mathcal{F}}^H$ constrained to $\mathring{\Delta}^d$ if and only if there exists a flux $\varphi \in \mathfrak{a}_1(X)$ such that:*

$$-\ln(p) \simeq H + \zeta \cdot \partial\varphi \quad \mathrm{mod} \; \mathbb{R}_0(X)$$

Proof. To describe the normalisation constraints, we may look at the quotient $\mathfrak{a}_0(X)/\mathbb{R}_0(X)$ as the cotangent space of $\mathring{\Delta}_0(X)$ at p, and write the differential of $\check{\mathcal{F}}^H$ as:

$$\frac{\partial \check{\mathcal{F}}}{\partial p} \simeq \sum_{\beta \in X} c_\beta \big(H_\beta + \ln(p_\beta) \big) \quad \mathrm{mod} \; \mathbb{R}_0(X)$$

The flux term comes as a collection of Lagrange multipliers for the consistency constraints. Whenever p is a critical point, the differential of $\check{\mathcal{F}}^H$ vanishes on $\mathrm{Ker}(d) = \mathrm{Im}(\partial)^\perp$ and we have:

$$c\big(H + \ln(p)\big) \in \mathrm{Im}(\partial) + \mathbb{R}_0(X)$$

The corollary of Theorem 2 is crucial[15] to state that this implies:

$$H + \ln(p) \in \zeta \cdot \mathrm{Im}(\partial) + \mathbb{R}_0(X)$$

---

[12] For reference see [2,5,8,11].

[13] Schlijper [13] proved this procedure convergent to the true free energy per lattice point for the infinite Ising 2D-model.

[14] For numerical studies see [6,9,15], A first mathematical proof of multiplicity is given by Bennequin in [1].

[15] The proof given in [17] is problematic when there exists β such that $c_\beta = 0$.

3.2   Belief Propagation as a Transport Equation

Effective Energy. For every $\alpha \to \beta$ in X, call effective energy the smooth submersion $\mathbb{F}^{\beta\alpha}$ of $\mathfrak{a}_\alpha$ onto $\mathfrak{a}_\beta$ defined by:

$$\mathbb{F}^{\beta\alpha}(U_\alpha) = -\ln\left(\Sigma^{\beta\alpha}(e^{-U_\alpha})\right)$$

Physically $\mathbb{F}^{\beta\alpha}(U_\alpha)(x_\beta)$ can be thought of as the conditional free energy of Λ^α given x_β. It is functorial in the category of smooth manifolds and we have the commutative diagram: $\Sigma^{\beta\alpha} \circ [e^-]_\alpha = [e^-]_\beta \circ \mathbb{F}^{\beta\alpha}$.

Let us call effective gradient the smooth functional $\nabla^{\mathbb{F}}$ from $\mathfrak{a}_0(X)$ to $\mathfrak{a}_1(X)$ defined by:

$$\nabla^{\mathbb{F}}(H)_{\alpha\beta} = H_\beta - \mathbb{F}^{\beta\alpha}(H_\alpha)$$

The hamiltonian H is related to a field of local potentials h by $H = \zeta \cdot h$. Letting $\Phi = -\nabla^{\mathbb{F}} \circ \zeta$, we have:

$$\Phi_{\alpha\beta}(h) = \mathbb{F}^{\beta\alpha}\left(\sum_{\beta' \in \Lambda^\alpha \setminus \Lambda^\beta} h_{\beta'}\right)$$

which is the effective contribution of $\Lambda^\alpha \setminus \Lambda^\beta$ to the energy of Λ^β.

Belief Propagation. Consider the following transport equation:

$$\dot{u} = \partial\Phi(u)$$

and denote by $\Xi = \partial\Phi$ the induced vector field on $\mathfrak{a}_0(X)$. In absence of normalisation, Belief Propagation[16] is equivalent to the naive Euler scheme[17] approximating the flow of Ξ by:

$$e^{n\tau\Xi} \simeq (1 + \tau\Xi)^n$$

The beliefs are given by $q = e^{-U}$ with $U = \zeta \cdot u$.

This new perspective reveals the strong homological character of Belief Propagation. Denote by $T_h(\varphi)$ the transport of h by a flux $\varphi \in \mathfrak{a}_1(X)$:

$$T_h(\varphi) = h + \partial\varphi$$

With initial condition h, the potentials u remain in the image of T_h. This yields a maximal set of conserved quantities in light of Theorem 2.

Theorem 4. *Let $q \in G_0(X)^{\mathbb{N}}$ denote a sequence of belief fields obtained by iterating BP. The following quantity remains constant:*

$$q_\Omega = \prod_{\alpha \in X} (q_\alpha)^{c_\alpha}$$

---

[16] For reference and the algorithm formula see [3,9,10,15–17].

[17] BP is actually for $\tau = 1$, a different time scale would appear as exponent in the multiplicative formulation.

Proof. The fact that $u \in \mathrm{Im}(T_h)$ is equivalent to $P(u) = P(h)$. According to Corollary 1, this is also equivalent to $\zeta_\Omega(u) = \zeta_\Omega(h)$ where:

$$\zeta_\Omega(u) = \sum_{\alpha \in X} u_\alpha = \sum_{\beta \in X} c_\beta U_\beta$$

Letting $u = h + \partial\varphi$, BP can also be viewed as a dynamic over messages:

$$\dot{\varphi} = \Phi\big(T_h(\varphi)\big)$$

Although it converges on trees, this algorithm is generally divergent in presence of loops, and beliefs need to be normalised in order to attain projective equilibria.

Normalisation. Because the effective gradient $\nabla^{\mathbb{F}}$ is additive along constants and both ζ and ∂ preserve scalar fields, Ξ induces a vector field on the quotient $\mathfrak{a}_0(X)/\mathbb{R}_0(X)$. Normalised belief are given by $q = [\mathrm{e}^{-U}]$ with $U = \zeta \cdot u$.

Given an initial hamiltonian $H = \zeta \cdot h$, a belief field $q \in \mathring{\Delta}_0(X)$ obtained by iterating BP satisfies:

$$-\ln(q) \simeq H + \zeta \cdot \partial\varphi \quad \mathrm{mod}\ \mathbb{R}_0(X)$$

In virtue of Theorem 3, this implies that q is a critical point of the Bethe free energy $\check{\mathcal{F}}^H$ constrained to $\mathring{\Delta}^d$ if and only if q is a consistent statistical field. Considering all beliefs that may be obtained by such a choice of messages, let:

$$\mathring{\Delta}_H = \big\{ [\mathrm{e}^{-U}] \mid U \in H + \zeta \cdot \mathrm{Im}(\partial) \big\} \subseteq \mathring{\Delta}_0(X)$$

Following Yedidia *et al.*, call any consistent $q \in \mathring{\Delta}_H \cap \mathring{\Delta}^d$ a fixed point of Belief Propagation[18]. With this terminology, we can rephrase their initial claim [17]:

Theorem 5. *Fixing a reference hamiltonian field H, fixed points of Belief Propagation are in one to one correspondence with critical points of the Bethe free energy.*

References

1. Bennequin, D., Peltre, O., Sergeant-Perthuis, G., Vigneaux, J.P.: Informations, Energies and Messages. Preprint (2019)
2. Bethe, H.A., Bragg, W.L.: Statistical theory of superlattices. Proc. R. Soc. Lond. Ser. Math. Phys. Sci. **150**(871), 552–575 (1935)
3. Gallager, R.G.: Low-Density Parity-Check Codes. MIT Press, New York (1963)
4. Kellerer, H.G.: Maßtheoretische Marginalprobleme. Mathematische Annalen **153**(3), 168–198 (1964)
5. Kikuchi, R.: A theory of cooperative phenomena. Phys. Rev. **81**, 988–1003 (1951)

---

[18] This terminology is somewhat ambiguous as it does not mean that q may be obtained by iterating BP from $[\mathrm{e}^{-H}]$.

6. Knoll, C., Pernkopf, F.: On loopy belief propagation - local stability analysis for non-vanishing fields. In: Uncertainty in Artificial Intelligence (2017)
7. Matúš, F.: Discrete marginal problem for complex measures. Kybernetika **24**, 36–46 (1988)
8. Morita, T.: Cluster variation method of cooperative phenomena and its generalization I. J. Phys. Soc. Jpn. **12**(7), 753–755 (1957)
9. Murphy, K.P., Weiss, Y., Jordan, M.I.: Loopy belief propagation for approximate inference: an empirical study. In: UAI (1999)
10. Pearl, J.: Networks of plausible inference. In: Probabilistic Reasoning in Intelligent Systems (1988)
11. Peierls, R.: On Ising's model of ferromagnetism. Math. Proc. Camb. Philos. Soc. **32**(3), 477–481 (1936)
12. Rota, G.-C.: On the foundations of combinatorial theory - I. Theory of Möbius functions. Z. Warscheinlichkeitstheorie **2**, 340–368 (1964)
13. Schlijper, A.G.: Convergence of the cluster-variation method in the thermodynamic limit. Phys. Rev. B **27**, 6841–6848 (1983)
14. Segal, G.: Classifying spaces and spectral sequences. Publications Mathématiques de l'IHÉS **34**, 105–112 (1968)
15. Weiss, Y.: Belief propagation and revision in networks with loops. Technical report, MIT (1997)
16. Yedidia, J.S., Freeman, W.T., Weiss, Y.: Bethe free energy, Kikuchi approximations, and belief propagation algorithms. Technical Report TR2001-16, MERL - Mitsubishi Electric Research Laboratories, Cambridge, MA 02139, May 2001
17. Yedidia, J.S., Freeman, W.T., Weiss, Y.: Constructing free energy approximations and generalized belief propagation algorithms. IEEE Trans. Inf. Theory **51**(7), 2282–2312 (2005)

About Some System-Theoretic Properties of Port-Thermodynamic Systems

Arjan van der Schaft[1(✉)] and Bernhard Maschke[2]

[1] Bernoulli Institute for Mathematics, Computer Science and AI,
University of Groningen, Groningen, The Netherlands
a.j.van.der.schaft@rug.nl
[2] Université Claude Bernard Lyon 1, UMR CNRS 5007 LAGEP,
Villeurbanne, France
bernhard.maschke@univ-lyon1.fr

Abstract. Recently a class of Hamiltonian control systems was introduced for geometric modeling of open irreversible thermodynamic processes. These systems are defined as ordinary Hamiltonian input-output systems on a symplectic manifold, with the special property that the Hamiltonian is homogeneous in the generalized momentum variables, and that there is an invariant homogeneous Lagrangian submanifold characterizing the state properties of the thermodynamic system. After recalling the basic framework we study the passivity, controllability and observability properties of such systems.

Keywords: Thermodynamic systems · Symplectic geometry · Nonlinear control

1 Port-Thermodynamic Systems

It was argued in [2] that the phase space of thermodynamic systems can be defined as the projectivization of a symplectic manifold, and that reversible and irreversible thermodynamic processes can be expressed as Hamiltonian dynamics with respect to a Hamiltonian that is homogeneous of degree one in the generalized momentum variables. In this section, we will recall the recently proposed generalization of this framework to non-isolated thermodynamic systems, summarizing the definition of homogeneous Hamiltonian control systems and port-thermodynamic systems as given in [13, 22, 23].

1.1 Homogeneous Hamiltonian Control Systems

Consider an $(n+1)$-dimensional manifold Q^e, with elements denoted by q^e (the vector of *extensive variables* in thermodynamics). Consider its cotangent bundle, denoted by T^*Q^e, with generalized momentum variables $p^e \in T^*_{q^e}Q^e$, equipped with the canonically defined *Liouville one-form* α and *symplectic form* $\omega = d\alpha$.

© Springer Nature Switzerland AG 2019
F. Nielsen and F. Barbaresco (Eds.): GSI 2019, LNCS 11712, pp. 228–238, 2019.
https://doi.org/10.1007/978-3-030-26980-7_24

In natural cotangent bundle coordinates $(q^e, p^e) = (q_0^e, \ldots, q_n^e, p_0^e, \ldots, p_n^e)$ for T^*Q^e the Liouville form α has the expression

$$\alpha = \sum_{i=0}^{n} p_i^e \, dq_i^e, \tag{1}$$

while the symplectic form ω is given as $\omega = \sum_{i=0}^{n} dp_i^e \wedge dq_i^e$. On the symplectic manifold T^*Q^e we consider Hamiltonian vector fields which are generated by Hamiltonian functions that are *homogeneous of degree* 1 in the momentum variables p^e. Such vector fields are characterized in the following proposition.

Proposition 1. *[22, Prop. A.1] If the function $h : T^*Q^e \to \mathbb{R}$ is homogeneous of degree 1 in p^e, then the Hamiltonian vector field $X = X_h$ satisfies*

$$\mathbb{L}_X \alpha = 0, \tag{2}$$

where $\mathbb{L}_X$ denotes the Lie derivative with respect to the vector field X. Conversely, if a vector field X satisfies (2) then $X = X_h$ for some locally defined Hamiltonian h that is homogeneous of degree 1 in p^e. [1]

Hamiltonian vector fields X_h with h homogeneous of degree 1 in p^e will be referred to as *homogeneous Hamiltonian vector fields*.

By Gibbs' relation the *state properties* of any thermodynamic system are specified by a Lagrangian submanifold of the cotangent bundle T^*Q^e with the following additional homogeneity property [22].

Definition 1. *A homogeneous Lagrangian submanifold $\mathcal{L} \subset T^*Q^e$ is a Lagrangian submanifold $\mathcal{L} \subset T^*Q^e$ (i.e., $\omega|_{\mathcal{L}} = 0$ and $\mathcal{L}$ is maximal with respect to this property) satisfying $(q^e, p^e) \in \mathcal{L} \Rightarrow (q^e, \lambda p^e) \in \mathcal{L}$, for every $\lambda \in \mathbb{R}^*$.*

Equivalently, in [22] homogeneous Lagrangian submanifolds are geometrically characterized as maximal submanifolds satisfying $\alpha|_{\mathcal{L}} = 0$.

Motivated by thermodynamics we require that the *dynamics* specified by a homogeneous Hamiltonian vector field is compatible with the *state properties* defined by a homogeneous Lagrangian submanifold, in the sense of leaving this submanifold invariant. This is characterized as follows.

Proposition 2. *[11, 23] A homogeneous Hamiltonian vector field X_h leaves invariant the homogeneous Lagrangian submanifold $\mathcal{L} \subset T^*Q^e$ if and only if $h|_{\mathcal{L}} = 0$.*

This leads to the following definition of a class of Hamiltonian control systems.

---

[1] Note that (2) is stronger than the standard condition that the vector field X is (locally) Hamiltonian, i.e., $\mathbb{L}_X \omega = 0$ [11, p. 97].

Definition 2. *[23] Consider an $(n+1)$-dimensional manifold Q^e. A homogeneous Hamiltonian control system on T^*Q^e is defined as a pair $(\mathcal{L}, K)$, composed of a homogeneous Lagrangian submanifold $\mathcal{L} \subset T^*Q^e$ and a nonlinear control system*

$$\dot{x} = X_{K^a}(x) + X_{K^c}(x)\, u, \quad x = \begin{pmatrix} q^e \\ p^e \end{pmatrix}, \tag{3}$$

generated by the control dependent Hamiltonian function $K := K^a + K^c u :$ $T^*Q^e \to \mathbb{R}$, $u \in \mathbb{R}^m$, where K^a and the elements of the m-dimensional row vector K^c are functions which are homogeneous of degree 1 in p^e, and satisfy the invariance conditions $K^a|_{\mathcal{L}} = K^c|_{\mathcal{L}} = 0$.

Hence homogeneous Hamiltonian control systems leave invariant the homogeneous Lagrangian submanifold $\mathcal{L}$ characterizing the state properties of the system; in particular energy-storage. In the control-theoretic sense, this homogeneous Lagrangian submanifold is the actual state space of the system and only the *restriction* of the homogeneous Hamiltonian control system to this submanifold is relevant. Thus the state space is defined as a submanifold of a covering manifold, similarly to approaches to differential-algebraic equation systems such as described in [3,4,21].

1.2 Relation with Irreversible Thermodynamic Systems

The Thermodynamic Phase Space. In the thermodynamic case the $(n+1)$-dimensional manifold Q^e consists of the space of all *extensive variables*. For simple thermodynamic systems, the extensive variables are volume, number of moles of chemical species, as well as entropy and internal energy. Following [2], the *thermodynamic phase space* is the $(2n+1)$-dimensional manifold $\mathbb{P}(T^*Q^e)$, called the projectivization of T^*Q^e, the $(2n+2)$-dimensional cotangent bundle T^*Q^e without its zero-section. The projectivization $\mathbb{P}(T^*Q^e)$ is defined as the fiber bundle over Q^e with fiber at any point $q^e \in Q^e$ given by the projective space $\mathbb{P}\left(T^*_{q^e}Q^e\right)$, with projection map $\pi : T^*Q^e \to \mathbb{P}\left(T^*_{q^e}Q^e\right)$.

It is a classical result, see e.g. [11, chap. V], [1, Appendix 4], that the $(2n+1)$-dimensional manifold $\mathbb{P}(T^*Q^e)$ is a *contact manifold*, endowed with a locally defined canonical contact form, denoted by θ. In fact, in a neighborhood where $p_0 \neq 0$, the contact form θ is given as

$$\theta = dq_0^e - \sum_{i=1}^{n} \gamma_i dq_i^e,$$

where $\gamma_i = -\frac{p_i}{p_0}$, $i = 1, \ldots, n$, are the homogeneous coordinates corresponding to the condition $p_0 \neq 0$. For a thermodynamic system, the n coordinates γ_i, $i = 1, \ldots, n$, are called the *intensive variables*. Whenever $p_1 \neq 0$ we may define *different* homogeneous coordinates $\hat{\gamma}_i = -\frac{p_i}{p_1}$, $i = 0, 2, \ldots, n$, corresponding to the contact form $\hat{\theta} = dq_1^e - \hat{\gamma}_0 dq^0 - \sum_{i=2}^{n} \hat{\gamma}_i dq_i^e$, and so on for

$p_2 \neq 0, p_3 \neq 0, \ldots$. In thermodynamics this reflects the choice of different intensive variables corresponding to, e.g., the energy or entropy representation of the thermodynamic system.

The formulation of the thermodynamic phase space as a contact manifold is well-known [1,9,15]. The covering symplectization of the contact manifold $\mathbb{P}(T^*Q^e)$ by the symplectic manifold T^*Q^e, together with the resulting different choices of intensive variables, unifies the different representations of the thermodynamic phase space; in particular, the energy and entropy representations [2,22,23]. In the same vein, the description of the state properties of the thermodynamic system by a Legendre submanifold of the thermodynamic phase space, see e.g. [15], is extended to a covering homogeneous Lagrangian submanifold. In this way, the formulation of reversible and irreversible processes of thermodynamic systems using contact vector fields as given before in [7,8,16,17], and for open thermodynamic systems as control contact systems in [5,6,14,19], is now replaced by ordinary, but homogeneous, Hamiltonian dynamics on the symplectization T^*Q^e of the thermodynamic phase space $\mathbb{P}(T^*Q^e)$. Apart from unifying different representations as mentioned above, this covering has other advantages as well. From a computational point of view, Hamiltonian dynamics is more easy than contact dynamics. More importantly, as we will see in the next subsection, the homogeneous Hamiltonian formulation admits to define in a natural way the *outputs* of the thermodynamic system. For further details on the mathematical relation between the symplectic, but homogeneous, and the contact representations we refer to [22].

Port-Thermodynamic Systems. In order to ensure the compatibility with the First and Second Law of thermodynamics, and to define natural outputs which are conjugate to the inputs, we first recall Euler's theorem for homogeneous functions. Since the Hamiltonian functions K^a and the elements of the row vector K^c are homogeneous of degree 1 in the momenta p^e, Euler's theorem yields the identities

$$K^a = p^{e^T} \frac{\partial K^a}{\partial p^e} \quad \text{and} \quad K^c = p^{e^T} \frac{\partial K^c}{\partial p^e}, \tag{4}$$

where $\frac{\partial K^a}{\partial p^e}$ and $\frac{\partial K^c}{\partial p^e}$ are homogeneous of degree 0 in the p^e variables, and thus project to well-defined functions on the thermodynamic phase space $\mathbb{P}(T^*Q^e)$. Furthermore, by definition of Hamiltonian vector fields, $\frac{\partial K^a}{\partial p^e}$ equals the autonomous drift dynamics, and $\frac{\partial K^c}{\partial p^e}$ the input-dependent dynamics, of the extensive variables $q^e \in T^*Q^e$. In view of the First and Second Law of thermodynamics this leads to the following additional requirements on the autonomous Hamiltonian function K^a, and to the following definition of natural outputs, which combined with the previous definition of homogeneous Hamiltonian control systems culminates in the definition of *port-thermodynamic systems*.

Definition 3. *[22] Port-thermodynamic systems are homogeneous Hamiltonian control systems $(\mathcal{L}, K)$ (as in Definition 2) for which the set of extensive variables contains a coordinate q_0^e corresponding to the total energy of the system* and q_1^e corresponding to the total entropy of the system, and the autonomous Hamiltonian K^a satisfies the following additional conditions in relation to the homogeneous Lagrangian submanifold

$$\left.\frac{\partial K^a}{\partial p_0^e}\right|_{\mathcal{L}} = 0, \qquad \left.\frac{\partial K^a}{\partial p_1^e}\right|_{\mathcal{L}} \geq 0 \tag{5}$$

Furthermore, the power-conjugate outputs of the thermodynamic system are defined by the row vector

$$y_p = \left.\frac{\partial K^c}{\partial p_0^e}\right|_{\mathcal{L}} \tag{6}$$

while the entropy flow-conjugate outputs are defined as the row vector

$$y_e = \left.\frac{\partial K^c}{\partial p_1^e}\right|_{\mathcal{L}} \tag{7}$$

The above definition of the outputs y_p and y_e of a port-thermodynamic system, together with the conditions (5) , imply the following *balance laws* for the dynamics restricted to the invariant manifold $\mathcal{L}$:

$$\begin{aligned}\frac{d}{dt}E &= y_p u \\ \frac{d}{dt}S &\geq y_e u\end{aligned} \tag{8}$$

As an illustrative example we discuss the gas-piston-damper system. (Compare with the treatment of this example in a contact geometry setting in [5].)

Example 1 (Actuated gas-piston-damper system). Consider extensive variables V (volume of the gas), π (momentum of the piston with mass m), entropy S and total energy E. The state properties of the system are described by the homogeneous Lagrangian submanifold $\mathcal{L}$ with generating function (in energy representation) $-p_E\left(U(V,S) + \frac{\pi^2}{2m}\right)$, where $U(V,S)$ is the internal energy of the gas (expressed as a function of volume and entropy):

$$\begin{aligned}\mathcal{L} = \{(V,\pi,S,E,p_V,p_\pi,p_S,p_E) \mid E &= U(V,S) + \frac{\pi^2}{2m}, \\ p_V = -p_E\frac{\partial U}{\partial V}, p_\pi = -p_E\frac{\pi}{m}, p_S &= -p_E\frac{\partial U}{\partial S}\}\end{aligned} \tag{9}$$

The dynamics of the system is defined by the homogeneous Hamiltonian

$$\begin{aligned}K &= K^a + K^c u \\ &= \left[p_V A\frac{\pi}{m} + p_\pi\left(-\frac{\partial U}{\partial V} - d\frac{\pi}{m}\right) + p_S\frac{d(\frac{\pi}{m})^2}{\frac{\partial U}{\partial S}}\right] + \left(p_\pi + p_E\frac{\pi}{m}\right)u,\end{aligned} \tag{10}$$

where A is the area of the piston, and u the external force applied to the piston. The power-conjugate output $y_p = \frac{\pi}{m}$ is the velocity of the piston. The entropy-conjugate output y_e is identically zero, implying $\frac{d}{dt}S \geq 0$ for every u.

2   System-Theoretic Properties of Port-Thermodynamic Systems

In the first part of this section we make some observations regarding the *passivity* properties of port-thermodynamic systems, while in the second part we study their *observability* in relation with the *controllability* properties as treated before in [22].

2.1   Passivity Properties of Port-Thermodynamic systems

Throughout this subsection we will denote $q^e = (q, S, E)$, with E the energy, S the entropy (previously denoted by, respectively, q_0^e and q_1^e), and q the remaining extensive variables.

Following *passivity theory* [20, 24] the equations (8) imply that port-thermo-dynamic systems restricted to their invariant Lagrangian submanifold $\mathcal{L}$ are *cyclo-lossless* with respect to the supply rate $y_p u$ and the storage function E (expressed as a function of the extensive variables (q, S)), and *cyclo-passive* with respect to the supply rate $-y_e u$ and the storage function $-S$ (expressed as a function of the extensive variables (q, E)). Here, 'cyclo' [20, 24] refers to the fact that in general the storage function E need not be bounded from below, nor S is bounded from above.

Let us in particular concentrate on the dissipation inequality $\frac{d}{dt}(-S) \leq -y_e u$ corresponding to the Second Law of thermodynamics. If the entropy S happens to be *bounded from above*, and thus $-S$ is bounded from below, then the port-thermodynamic system is truly *passive* with respect to the supply rate $-y_e u$. Furthermore, see [20, 24], in this case the *available storage* given as

$$V(q, E) := \sup_{\tau \geq 0, u:[0,\tau] \to \mathbb{R}^m} \int_0^\tau y_e(t)u(t)dt, \tag{11}$$

where $y_e(t)$ is the output time-function corresponding to an input time-function $u(t)$ and initial condition (q, E) at time 0, is well-defined (i.e., finite for all (q, E)), while obviously $V \geq 0$. Furthermore, V is itself a storage function, and actually the *minimal* one among all other non-negative storage functions.

Interpreting minus the entropy as 'information', it follows that V as defined in (11) can be interpreted as the *maximally extractable information* of the system, where $y_e(t)u(t)$ is the rate of extracted information from the system at time t.

Alternatively, $-V$ is *maximal* among all functions $S \leq 0$ satisfying the inequality

$$\frac{d}{dt}S \geq y_e u \tag{12}$$

In this sense, $-V$ can be interpreted as the *maximal entropy* function for the thermodynamic system. (Obviously, this relates to the question of identifying the entropy function from the external behavior of the system.)

However, as mentioned before, in general the entropy function S need not be bounded from above. A possible approach to resolve this problem, as already discussed in [25], is to consider an *exergy function*

$$\mathcal{E}(q, S) := E(q, S) - T_0 S, \tag{13}$$

with T_0 a constant temperature. Indeed, by combining the energy and entropy balance laws (8) one obtains

$$\frac{d}{dt}(E - T_0 S) \le y_p u - T_0 y_e u = (y_p - T_0 y_e) u, \tag{14}$$

showing that for every $T_0 \ge 0$ the system is cyclo-passive with respect to the output $y_p - T_0 y_e$, with storage function $\mathcal{E}(q, S)$ given by (13).

Furthermore, in monophase thermodynamic systems E is a *convex* function of S. This implies that under additional conditions the exergy function is actually bounded from below; in this case yielding true passivity. For example, if $E(q, S)$ only depends on S then convexity yields that for any constant S_0 the function

$$A(S) := E(S) - E'(S_0)(S - S_0) - E(S_0) \tag{15}$$

is non-negative. This function is known as the Bregman divergence in convex analysis, or availability function [10] in thermodynamics. Denoting the temperature $T_0 := E'(S_0) \ge 0$ this function equals the exergy function (13) modulo a constant, implying that the exergy is bounded from below.

2.2   Controllability and Observability Properties

First we recall from [22], with some extensions, how the *controllability* properties of a port-thermodynamic system $(\mathcal{L}, K)$ can be directly studied in terms of the homogeneous Hamiltonians K^a and K_j^c, $j = 1, \cdots, m$, and their *Poisson brackets*. First note the following property proved in [22].

Proposition 3. *Consider the Poisson bracket $\{h_1, h_2\}$ of functions h_1, h_2 on $T^* Q^e$ defined with respect to the symplectic form $\omega = d\alpha$. Then*

(a) *If h_1, h_2 are both homogeneous of degree 1 in p^e, then also $\{h_1, h_2\}$ is homogeneous of degree 1 in p^e.*
(b) *If h_1 is homogeneous of degree 1 in p^e, and h_2 is homogeneous of degree 0 in p^e, then $\{h_1, h_2\}$ is homogeneous of degree 0 in p^e.*
(c) *If h_1, h_2 are both homogeneous of degree 0 in p^e, then $\{h_1, h_2\}$ is zero.*

In particular, Poisson brackets of the homogeneous (degree 1 in p^e) Hamiltonians K^a and K_j^c, $j = 1, \cdots, m$, are again homogeneous of degree 1 in p^e. Secondly, we recall the well-known correspondence [11] between Poisson brackets of Hamiltonians h_1, h_2, and Lie brackets of their corresponding Hamiltonian vector fields

$$[X_{h_1}, X_{h_2}] = X_{\{h_1, h_2\}} \tag{16}$$

In particular, this property implies that if the homogeneous Hamiltonians h_1, h_2 are zero on the homogeneous Lagrangian submanifold $\mathcal{L}$, and thus the homogeneous Hamiltonian vector fields X_{h_1}, X_{h_2} are tangent to $\mathcal{L}$, then also $[X_{h_1}, X_{h_2}]$ is tangent to $\mathcal{L}$, and therefore the Poisson bracket $\{h_1, h_2\}$ is also zero on $\mathcal{L}$. Furthermore, with respect to the projection to the corresponding Legendre submanifold L, we note the following property of homogeneous Hamiltonians

$$\widehat{\{h_1, h_2\}} = \{\widehat{h_1}, \widehat{h_2}\} \tag{17}$$

where $\widehat{h}$ is the contact Hamiltonian of the contact vector field obtained by projection of the Hamiltonian vector field X_h corresponding to a homogeneous Hamiltonian h. This leads to the following characterization of the *accessibility algebra* of a port-thermodynamic system characterizing controllability, cf. [22].

Proposition 4. *Consider a port-thermodynamic system $(\mathcal{L}, K)$ on $\mathbb{P}(T^*Q^e)$ with homogeneous $K := K^a + \sum_{j=1}^{m} K_j^c u_j : T^*Q^e \rightarrow \mathbb{R}$, zero on $\mathcal{L}$. Consider the algebra $\mathcal{P}$ (with respect to the Poisson bracket) generated by K^a, K_j^c, $j = 1, \cdots, m$, consisting of homogeneous functions that are zero on $\mathcal{L}$, and the corresponding algebra $\widehat{\mathcal{P}}$ generated by $\widehat{K}^a, \widehat{K}_j^c$, $j = 1, \cdots, m$, on the corresponding Legendre submanifold L. The accessibility algebra [18] of the port-thermodynamic system is spanned by all contact vector fields $X_{\widehat{h}}$ on L, with $\widehat{h}$ in the algebra $\widehat{\mathcal{P}}$.*

It follows that the thermodynamic system $(\mathcal{L}, K)$ is locally accessible [18] if the dimension of the co-distribution $d\widehat{\mathcal{P}}$ on L defined by the differentials of $\widehat{h}$, with h in the Poisson algebra $\mathcal{P}$, is equal to the dimension of L. Conversely, if the system is locally accessible then the co-distribution $d\widehat{\mathcal{P}}$ on L has dimension equal to the dimension of L on an open and dense subset of L.

Similar statements can be made with respect to local *strong* accessibility [18] of port-thermodynamic systems. In this case the same conditions need to be satisfied for the algebra $\mathcal{P}_s$, which is equal to $\mathcal{P}$ *minus* the drift Hamiltonian K^a. (Thus, possibly repeated, Poisson brackets with K^a are taken into account, but not the function K^a itself.)

With regard to *observability* we proceed as follows. First, let us consider the observability properties with respect to the *power-conjugate* output $y_p = \left.\frac{\partial K^c}{\partial p_0^e}\right|_{\mathcal{L}}$ as in (6), with q_0^e and p_0^e corresponding to the energy variable E. Note that

$$\frac{\partial K_j^c}{\partial p_0^e} = \{K_j^c, q_0^e\} \tag{18}$$

Recall (cf. [18] for further information) the definition of the *observation space* $\mathcal{O}$ as given by the linear span of functions of the form

$$\mathbb{L}_{X_1} \mathbb{L}_{X_2} \cdots \mathbb{L}_{X_k} \{K_j^c, q_0^e\}, \quad j = 1, \cdots, m, \tag{19}$$

with $X_i, i = 1, \cdots, k$, taken from the set $\{X_{K^a}, X_{K_j^c}, j = 1, \cdots, m\}$. Using the equality $\mathbb{L}_{X_h} K = \{h, K\}$ it follows that the observation space $\mathcal{O}$ of the port-thermodynamic system with power-conjugate outputs y_p is given by the linear span of all expressions

$$\{h_1, \{h_2, \{\cdots, \{h_k, \{K_j^c, q_0^e\}\} \cdots \}\}\}, \quad j = 1, \cdots, m, \tag{20}$$

with $h_i, i = 1, \cdots, k$, taken from the set $\{K^a, K_j^c, j = 1, \cdots, m\}$.

Furthermore, since by (5) $\{K^a, q_0^e\} = 0$ the following results.

Proposition 5. *The observation space $\mathcal{O}$ of a port-thermodynamic system $(\mathcal{L}, K)$ with power-conjugate outputs y_p is equal to the linear span of all functions*

$$\{h_1, \{h_2, \{\cdots, \{h_k, \{h_{k+1}, q_0^e\}\} \cdots \}\}\}, \quad j = 1, \cdots, m, \tag{21}$$

with $h_i, i = 1, \cdots, k+1$, taken from the set $\{K^a, K_j^c, j = 1, \cdots, m\}$.

Furthermore, analogously to [18, Proposition 3.8], $\mathcal{O}$ is equal to the linear span of all functions (21) with $h_i, i = 1, \cdots, k+1$, taken from the accessibility algebra $\mathcal{P}$.

Since the functions h_i in (21) are all homogeneous of degree 1 in p^e, and clearly the function q_0^e is homogeneous of degree 0 in p^e, it follows by Proposition 3 that all functions in $\mathcal{O}$ are homogeneous of degree 0, and therefore project to functions $\widehat{h}$ on the thermodynamic phase space $\mathbb{P}(T^*Q^e)$. As a result, we obtain the following proposition.

Proposition 6. *Consider the thermodynamic system with power-conjugate outputs y_p. It is locally observable if $\dim d\widehat{\mathcal{O}} = \dim L (= n+1)$, where $\widehat{\mathcal{O}}$ is the set of all functions on $L \subset \mathbb{P}(T^*Q^e)$ obtained by projection of the functions in $\mathcal{O}$. Conversely, if the system is locally observable then $\dim d\widehat{\mathcal{O}} = \dim L (= n+1)$ on an open and dense subset of L.*

Comparing Proposition 6 with Proposition 4 we notice a close relation between the two conditions. In fact, the situation is similar to the relation between controllability and observability of lossless port-Hamiltonian systems as discussed in [12].

More or less the same story holds for the *entropy flow*-conjugate output $y_e = \left. \frac{\partial K^c}{\partial p_1^e} \right|_{\mathcal{L}}$ as in (7), with the difference that in this case $\{K^a, q_1^e\} \neq 0$, and hence in this case the observation space is slightly *smaller* than the linear span of all functions (21), with q_0^e replaced by q_1^e.

3   Conclusion

In this paper we have considered homogenous Hamiltonian control systems, which are generated by Hamiltonian drift and control Hamiltonian functions that are all homogeneous of degree one in the momentum variables, and leave invariant a homogeneous Lagrangian submanifold representing the actual state space of the system. They represent open thermodynamic systems once additional conditions are satisfied corresponding to the First and Second Law of thermodynamics, leading to the definition of *port-thermodynamic systems*. The 'symplectization' point of view also enables the definition of outputs which are conjugate in the sense of external energy or entropy flow, and leads to elegant results concerning controllability and observability. Furthermore, we have made some initial observations regarding the passivity properties of port-thermodynamic systems, and on the use of exergy functions; thus suggesting further research on passivity-based control of port-thermodynamic systems.

References

1. Arnold, V.: Mathematical Methods of Classical Mechanics, 2nd edn. Springer, New York (1989). https://doi.org/10.1007/978-1-4757-2063-1
2. Balian, R., Valentin, P.: Hamiltonian structure of thermodynamics with gauge. Eur. Phys. J. B-Condens. Matter Complex Syst. **21**(2), 269–282 (2001)
3. Barbero-Linàn, M., Cendra, H., Andrés, E.G.T., de Diego, D.M.: New insights in the geometry and interconnection of port-Hamiltonian systems. J. Phys. A Math. Theor. **51**(37), 201–375 (2018)
4. Beattie, C., Mehrmann, V., Xu, H., Zwart, H.: Port-Hamiltonian descriptor systems. Math. Control, Signals, Syst. **30**(4), 1–27 (2018)
5. Eberard, D., Maschke, B., van der Schaft, A.: An extension of Hamiltonian systems to the thermodynamic space: towards a geometry of non-equilibrium Thermodynamics. Rep. Math. Phys. **60**(2), 175–198 (2007)
6. Favache, A., Dochain, D., Maschke, B.M.: An entropy-based formulation of irreversible processes based on contact structures. Chem. Eng. Sci. **65**, 5204–5216 (2010)
7. Grmela, M.: Reciprocity relations in thermodynamics. Phys. A **309**, 304–328 (2002)
8. Grmela, M.: Contact geometry of mesoscopic thermodynamics and dynamics. Entropy **16**(3), 1652 (2014)
9. Hermann, R.: Geometry, Physics and Systems. Dekker, New-York (1973)
10. Keenan, J.H.: Availability and irreversibility in thermodynamics. Br. J. Appl. Phys. **2**, 183 (1952)
11. Libermann, P., Marle, C.M.: Symplectic Geometry and Analytical Mechanics. D. Reidel Publishing Company, Dordrecht (1987)
12. Maschke, B., van der Schaft, A.: Port-controlled Hamiltonian systems: modelling origins and system theoretic properties. In: Proceedings 2nd IFAC Symposium on Nonlinear Control Systems (NOLCOS92), Fliess, M. (Ed.) pp. 282–288. Bordeaux, France (1992)
13. Maschke, B., van der Schaft, A.: Homogeneous Hamiltonian control systems, Part II: Applications to thermodynamic systems. In: 6th IFAC Workshop on Lagrangian and Hamiltonian Methods for Nonlinear Control (LHMNC), IFACPapersOnLine, **51**(3), pp. 7–12 (2018)
14. Merker, J., Krüger, M.: On a variational principle in thermodynamics. Continuum Mech. Thermodyn. **25**(6), 779–793 (2013)
15. Mrugała, R.: Geometrical formulation of equilibrium phenomenological thermodynamics. Rep. Math. Phys. **14**(3), 419–427 (1978)
16. Mrugała, R.: Continuous contact transformations in thermodynamics. Rep. Math. Phys. **33**(1/2), 149–154 (1993)

17. Mrugała, R.: On a special family of thermodynamic processes and their invariants. Rep. Math. Phys. **46**(3), 461–468 (2000)
18. Nijmeijer, H., van der Schaft, A.J.: Nonlinear Dynamical Control Systems. Springer-Verlag, New York (1990). https://doi.org/10.1007/978-1-4757-2101-0. Corrected printing 2016
19. Ramirez, H., Maschke, B., Sbarbaro, D.: Irreversible port-Hamiltonian systems: a general formulation of irreversible processes with application to the CSTR. Chem. Eng. Sci. **89**, 223–234 (2013)
20. van der Schaft, A.J.: L2-Gain and Passivity Techniques in Nonlinear Control. Springer-Verlag, Berlin (2017). https://doi.org/10.1007/978-3-319-49992-5
21. van der Schaft, A., Maschke, B.: Generalized port-Hamiltonian DAE systems. Syst. Control Lett. **121**, 31–37 (2018)
22. van der Schaft, A., Maschke, B.: Geometry of thermodynamic processes. Entropy **20**(12), 925 (2018)
23. van der Schaft, A., Maschke, B.: Homogeneous Hamiltonian control systems Part I: Geometric formulation. In: 6th IFAC Workshop on Lagrangian and Hamiltonian Methods for Nonlinear Control (LHMNC), IFACPapersOnLine, **51**(3), pp. 1–6 (2018)
24. Willems, J.C.: Dissipative dynamical systems Part I: general theory. Arch. Rat. Mech. Anal. **45**(5), 321–351 (1972)
25. Ydstie, E., Alonso, A.: Process systems and passivity via the Clausius-Planck inequality. Syst. Control Lett. **30**, 253–264 (1997)

Expectation Variables on a Para-Contact Metric Manifold Exactly Derived from Master Equations

Shin-itiro Goto$^{(\boxtimes)}$ (iD) and Hideitsu Hino (iD)

The Institute of Statistical Mathematics,
10-3 Midori-cho, Tachikawa, Tokyo 190-8562, Japan
{sgoto,hino}@ism.ac.jp

Abstract. Based on information and para-contact metric geometries, in this paper a class of dynamical systems is formulated for describing time-development of expectation variables. Here such systems for expectation variables are exactly derived from continuous-time master equations describing nonequilibrium processes.

Keywords: Master equations · Para-contact manifolds · Nonequilibrium statistical mechanics · Information geometry

1 Introduction

Information geometry is a geometrization of mathematical statistics [1,2], and its differential geometric aspects and applications in statistics have been investigated. Examples of applications of information geometry include thermodynamics, and some links between equilibrium thermodynamics and information geometry have been clarified. In addition, links between information geometry and contact geometry have been argued [3,4]. In this context, it was found that para-Sasakian geometry is suitable for describing thermodynamics [5,6], where para-Sasakian manifolds are para-contact metric manifolds satisfying some additional condition. We then ask how para-contact metric manifolds describe thermodynamics.

In this paper a class of nonequilibrium thermodynamic processes are formulated on a para-contact metric manifold. Most of discussions in this paper have been in [7], and those involving an almost para-contact structure are given in this contribution.

2 Preliminaries

In this paper manifolds are assumed smooth and connected. In addition tensor fields are assumed smooth and real. The set of vector fields on a manifold $\mathcal{M}$ is denoted by $\mathcal{S}T\mathcal{M}$, and the Lie derivative along $X \in \mathcal{S}T\mathcal{M}$ by $\mathcal{L}_X$.

© Springer Nature Switzerland AG 2019
F. Nielsen and F. Barbaresco (Eds.): GSI 2019, LNCS 11712, pp. 239–247, 2019.
https://doi.org/10.1007/978-3-030-26980-7_25

In this section definitions and some existing statements are summarized. (see [5,8]).

Let $\mathcal{M}$ be a $(2n+1)$-dimensional manifold ($n \geq 1$). An *almost para-contact structure* on $\mathcal{M}$ is a triplet (ϕ, ξ, λ), where ξ is a vector field, λ a one-form, $\phi : \mathcal{STM} \to \mathcal{STM}$ a $(1,1)$-tensor field such that

(i) : $\phi^2 = \mathrm{Id} - \lambda \otimes \xi$, (ii) : $\lambda(\xi) = 1$, and (iii) : $\ker(\lambda) = \mathrm{Im}(\phi) = \mathcal{D}^+ + \mathcal{D}^-$,

where $\ker(\lambda) := \{X \in \mathcal{STM} \mid \lambda(X) = 0\}$, $\mathrm{Im}(\phi) := \{\phi(X, -) \in \mathcal{STM} \mid X \in \mathcal{STM}\}$, $\mathcal{D}^{\pm}$ are eigen-spaces whose eigenvalues are ± 1, and Id is an identity operator. A pseudo Riemannian metric tensor field g satisfying

$$g(\phi X, \phi Y) = -g(X, Y) + \lambda(X)\lambda(Y), \qquad \forall X, Y \in \mathcal{STM}$$

is referred to as a *metric tensor compatible with an almost para-contact structure*. It is verified for non-compact manifolds that any almost para-contact structure admits a metric tensor field compatible with an almost para-contact structure. Then $(\mathcal{M}, \phi, \xi, \lambda, g)$ is referred to as an *almost para-contact metric manifold*.

On almost para-contact metric manifolds, one can show that

$$\lambda \phi = 0, \ \phi \xi = 0, \ \lambda(X) = g(X, \xi), \ g(\xi, \xi) = 1, \ \text{and } g(\phi X, Y) + g(X, \phi Y) = 0,$$

for $\forall X, Y \in \mathcal{STM}$. If g of an almost para-contact metric manifold satisfies

$$g(X, \phi Y) = \frac{1}{2}\mathrm{d}\lambda(X, Y), \qquad \forall X, Y \in \mathcal{STM} \tag{1}$$

then, $(\mathcal{M}, \phi, \xi, \lambda, g)$ is referred to as a *para-contact metric manifold*, where the convention of the numerical factor $\mathrm{d}\lambda(X, Y) = X\lambda(Y) - Y\lambda(X) - \lambda([X, Y])$, $([X, Y] := XY - YX)$ has been adopted. Para-Sasakian manifolds are para-contact metric manifolds satisfying the so-called normality condition.

Coordinate expressions were given for a para-contact metric manifold (and a para-Sasakian manifold) $(\mathcal{M}, \phi, \xi, \lambda, g)$ in [5]. They are summarized here. Let (x, y, z) be coordinates for $\mathcal{M}$ with $x = \{x^1, \ldots, x^n\}$ and $y = \{y_1, \ldots, y_n\}$ such that $\lambda = \mathrm{d}z - y_a \mathrm{d}x^a$ where the Einstein convention has been used. Introduce the pseudo-Riemannian metric tensor field, referred to as the *Mruagala metric tensor field* [11],

$$g^{\mathrm{M}} = \frac{1}{2}\mathrm{d}x^a \otimes \mathrm{d}y_a + \frac{1}{2}\mathrm{d}y_a \otimes \mathrm{d}x^a + \lambda \otimes \lambda, \tag{2}$$

which is shown to induce a para-contact metric manifold. In what follows we consider the case where $y_a > 0$ for all $a \in \{1, \ldots, n\}$. Introduce the co-frame $\{\widehat{\theta}^0, \widehat{\theta}^1_-, \widehat{\theta}^1_+, \ldots, \widehat{\theta}^n_-, \widehat{\theta}^n_+\}$ and frame $\{e_0, e^-_1, e^+_1, \ldots, e^-_n, e^+_n\}$ with

$$\widehat{\theta}^0 := \lambda, \qquad \widehat{\theta}^a_{\pm} := \frac{1}{2\sqrt{y_a}}\left[y_a \mathrm{d}x^a \pm \mathrm{d}y_a\right], \qquad \text{(no sum over } a\text{)},$$

$$e_0 := \xi, \qquad e^{\pm}_a := \sqrt{y_a}\left[\frac{1}{y_a}\left(\frac{\partial}{\partial x^a} + y_a \frac{\partial}{\partial z}\right) \pm \frac{\partial}{\partial y_a}\right], \qquad \text{(no sum over } a\text{)},$$

so that

$$\widehat{\theta}^0(e_0) = 1, \qquad \widehat{\theta}^a_+(e^+_b) = \widehat{\theta}^a_-(e^-_b) = \delta^a_b, \qquad \text{others vanish,}$$

where δ^a_b is the Kronecker delta, giving unity for $a = b$ and zero otherwise. One can then show that

$$g^M = \widehat{\theta}^0 \otimes \widehat{\theta}^0 + \sum_{a=1}^n \widehat{\theta}^a_+ \otimes \widehat{\theta}^a_+ - \sum_{a=1}^n \widehat{\theta}^a_- \otimes \widehat{\theta}^a_-, \quad \xi = \frac{\partial}{\partial z}, \quad \phi = -\widehat{\theta}^a_- \otimes e^+_a - \widehat{\theta}^a_+ \otimes e^-_a.$$

Then, introducing the abbreviation $\phi(X) := \phi(X, -) \in \mathcal{STM}$ for $X \in \mathcal{STM}$, one has $\phi(e^+_a) = -e^-_a$, $\phi(e^-_a) = -e^+_a$, and $\phi(e_0) = 0$.

In the context of geometry of thermodynamics, contact manifold is identified with the so-called thermodynamic phase space [12]. This manifold is defined as follows (see [9] for details). Let $\mathcal{C}$ be a $(2n + 1)$-dimensional manifold ($n = 1, 2, \ldots$), and λ a one-form. If λ satisfies

$$\lambda \wedge \underbrace{d\lambda \wedge \cdots \wedge d\lambda}_{n} \neq 0,$$

then the pair $(\mathcal{C}, \lambda)$ is referred to as a *contact manifold*, and λ a *contact one-form*. It has been known as the Darboux theorem that there exists a special set of coordinates (x, y, z) with $x = \{x^1, \ldots, x^n\}$ and $y = \{y_1, \ldots, y_n\}$ such that $\lambda = dz - y_a dx^a$. It follows from (1) that para-contact metric manifolds are contact manifolds.

The *Legendre submanifold* $\mathcal{A} \subset \mathcal{C}$ is an n-dimensional submanifold where $\lambda|_{\mathcal{A}} = 0$ holds. One can verify that

$$\mathcal{A}_\varpi = \left\{ (x, y, z) \,\middle|\, y_a = \frac{\partial \varpi}{\partial x^a}, \quad \text{and} \quad z = \varpi(x) \right\}, \tag{3}$$

is a Legendre submanifold, where $\varpi : \mathcal{C} \to \mathbb{R}$ is a function of x on $\mathcal{C}$. The submanifold $\mathcal{A}_\varpi$ is referred to as the *Legendre submanifold generated by ϖ*, and is used for describing equilibrium thermodynamic systems [12].

As shown in [3] and [6], a class of relaxation processes, initial states approach to the equilibrium state as time develops, can be formulated as contact Hamiltonian vector fields on contact manifolds. This statement on a class of contact Hamiltonian vector fields can be summarized as follows.

Proposition 1 (*Legendre submanifold as an attractor, [3]*). *Let $(\mathcal{C}, \lambda)$ be a $(2n + 1)$-dimensional contact manifold with λ being a contact form, (x, y, z) its coordinates so that $\lambda = dz - y_a dx^a$, and ϖ a function depending only on x. Then, one has*

1. *The contact Hamiltonian vector field associated with the contact Hamiltonian $h : \mathcal{C} \to \mathbb{R}$ such that $h(x, y, z) = \varpi(x) - z$, gives*

$$\frac{d}{dt}x^a = 0, \qquad \frac{d}{dt}y_a = \frac{\partial \varpi}{\partial x^a} - y_a, \qquad \frac{d}{dt}z = \varpi(x) - z. \tag{4}$$

2. *The Legendre submanifold generated by ϖ, given by (3), is an invariant manifold for the contact Hamiltonian vector field.*
3. *Every point on $\mathcal{C}\backslash\mathcal{A}_{\varpi}$ approaches to $\mathcal{A}_{\varpi}$ along an integral curve as time develops. Equivalently $\mathcal{A}_{\varpi}$ is an attractor in $\mathcal{C}$.*
4. *Let $\{x(0), y(0), z(0)\}$ be a point on $\mathcal{C}\backslash\mathcal{A}_{\varpi}$. Then for any $t \in \mathbb{R}$,*

$$h(x(t), y(t), z(t)) = \exp(-t)\, h(x(0), y(0), z(0)).$$

3   Solvable Master Equations

In this section a set of master equations with particular Markov kernels is introduced, and then its solvability is shown.

Let Γ be a set of finite discrete states, $t \in \mathbb{R}$ time, and $p(j, t)\, \mathrm{d}t$ a probability that a state $j \in \Gamma$ is found in between t and $t + \mathrm{d}t$. The first objective is to realize a given distribution function p_θ^{eq} that can be written as

$$p_\theta^{\mathrm{eq}}(j) = \frac{\pi_\theta(j)}{Z(\theta)}, \qquad Z(\theta) := \sum_{j \in \Gamma} \pi_\theta(j)$$

where $\theta \in \Theta \subset \mathbb{R}^n$ is a parameter set with $\theta = \{\theta^1, \ldots, \theta^n\}$, and $Z : \Theta \to \mathbb{R}$ the so-called partition function so that p_θ^{eq} is normalized: $\sum_{j \in \Gamma} p_\theta^{\mathrm{eq}}(j) = 1$.

In what follows, attention is focused on a class of master equations. Let $p : \Gamma \times \mathbb{R} \to \mathbb{R}_{\geq 0}$ be a time-dependent probability function. Then, consider the set of master equations

$$\frac{\partial}{\partial t} p(j, t) = \sum_{j'(\neq j)} \left[\, w(j|j')\, p(j', t) - w(j'|j)\, p(j, t)\,\right], \tag{5}$$

where $w : \Gamma \times \Gamma \to I,\ (I := [0, 1] \subset \mathbb{R})$ is such that $w(j|j')$ denotes a probability that a state jumps from j' to j. With (5) and the assumptions

$$w_\theta(j|j') = p_\theta^{\mathrm{eq}}(j), \qquad \text{and} \qquad p_\theta^{\mathrm{eq}}(j) \neq 0, \quad \forall j \in \Gamma,$$

one derives the *solvable master equations*:

$$\frac{\partial}{\partial t} p(j, t) = p_\theta^{\mathrm{eq}}(j) - p(j, t). \tag{6}$$

An explicit form of $p(j, t)$ is obtained by solving (6). Then the following proposition can easily be shown.

Proposition 2 *(Solutions of the master equations, [7]). The solution of (6) is*

$$p(j, t) = \mathrm{e}^{-t} p(j, 0) + (1 - \mathrm{e}^{-t}) p_\theta^{\mathrm{eq}}(j), \quad \text{from which} \quad \lim_{t \to \infty} p(j, t) = p_\theta^{\mathrm{eq}}(j).$$

With this proposition, one notices that every solution p depends on θ. Taking into account this, $p(j, t)$ is denoted $p(j, t; \theta)$. Also notice that the equilibrium state is realized with (6) as the time-asymptotic limit.

4   Time-Development of Observables

In this section differential equations describing time-development of observables are derived with the solvable master equations under some assumptions. Then, the time-asymptotic limit of such observables is stated. Here *observable* in this paper is defined as a function that does not depend on a random variable or a state. Thus expectation values with respect to a probability distribution function are observables.

Let $\mathcal{O}_a : \Gamma \to \mathbb{R}$ be a function with $a \in \{1, \ldots, n\}$, and $p : \Gamma \times \mathbb{R} \to \mathbb{R}_{\geq 0}$ a distribution function that follows (6). Then

$$\langle \mathcal{O}_a \rangle_\theta (t) := \sum_{j \in \Gamma} \mathcal{O}_a(j) \, p(j, t; \theta), \qquad \text{and} \qquad \langle \mathcal{O}_a \rangle_\theta^{\mathrm{eq}} := \sum_{j \in \Gamma} \mathcal{O}_a(j) \, p_\theta^{\mathrm{eq}}(j),$$

are referred to as the *expectation variable* of $\mathcal{O}_a$ with respect to p, and that with respect to p_θ^{eq}, respectively.

If an equilibrium distribution function belongs to the exponential family, then the function $\Psi^{\mathrm{eq}} : \Theta \to \mathbb{R}$ with

$$\Psi^{\mathrm{eq}}(\theta) := \ln \left(\sum_{j \in \Gamma} e^{\theta^b \mathcal{O}_b(j)} \right), \tag{7}$$

plays various roles. Here and in what follows, (7) is assumed to exist. In the context of information geometry, this function is referred to as a θ-*potential*. Discrete distribution functions are considered in this paper and it has been known that such distribution functions belong to the exponential family, then Ψ^{eq} in (7) also plays a role throughout this paper. The value $\Psi^{\mathrm{eq}}(\theta)$ can be interpreted as the negative dimension-less free-energy. It follows from (7) that

$$\langle \mathcal{O}_a \rangle_\theta^{\mathrm{eq}} = \frac{\partial \Psi^{\mathrm{eq}}}{\partial \theta^a}.$$

One then can generalize Ψ^{eq} defined at equilibrium state to a function defined in nonequilibrium states as $\Psi : \Theta \times \mathbb{R} \to \mathbb{R}$,

$$\Psi(\theta, t) := \left(\frac{1}{J^0} \sum_{j \in \Gamma} \frac{p(j, t; \theta)}{p_\theta^{\mathrm{eq}}(j)} \right) \Psi^{\mathrm{eq}}(\theta), \qquad \text{where} \qquad J^0 := \sum_{j' \in \Gamma} 1.$$

Since $p_\theta^{\mathrm{eq}}(j) \neq 0$ and $\Psi^{\mathrm{eq}}(\theta) < \infty$ by assumptions, the function Ψ exists. Generalizing the idea for the equilibrium case, the function Ψ may be interpreted as a nonequilibrium negative dimension-less free-energy.

A set of differential equations for $\{\langle \mathcal{O}_a \rangle_\theta\}$ and Ψ can be derived as follows.

Proposition 3 *(Dynamical system obtained from the master equations, [7]). Let θ be a time-independent parameter set characterizing a discrete distribution function p_θ^{eq}. Then $\{\langle \mathcal{O}_a \rangle_\theta\}$ and Ψ are solutions to the differential equations on $\mathbb{R}^{2n+1}$*

$$\frac{\mathrm{d}}{\mathrm{d}t} \theta^a = 0, \qquad \frac{\mathrm{d}}{\mathrm{d}t} \langle \mathcal{O}_a \rangle_\theta = - \langle \mathcal{O}_a \rangle_\theta + \frac{\partial \Psi^{\mathrm{eq}}}{\partial \theta^a}, \qquad \text{and} \qquad \frac{\mathrm{d}}{\mathrm{d}t} \Psi = - \Psi + \Psi^{\mathrm{eq}}.$$

Remark 1. The explicit time-dependence for this system is obtained as $\theta^a(t) = \theta^a(0)$, and $\Psi(\theta, t) = e^{-t}\left[\Psi(0) - \Psi^{eq}(\theta)\right] + \Psi^{eq}(\theta)$, and

$$\langle \mathcal{O}_a \rangle_\theta (t) = e^{-t}\left[\langle \mathcal{O}_a \rangle_\theta (0) - \frac{\partial \Psi^{eq}}{\partial \theta^a}\right] + \frac{\partial \Psi^{eq}}{\partial \theta^a}.$$

From these, one can verify that the time-asymptotic limit of these variables are those defined at equilibrium. In this paper this dynamical system is referred to as the *moment dynamical system.*

5   Geometric Description of Dynamical Systems

Several geometrization of nonequilibrium states for some models and methods have been proposed. Yet, suffice to say that there remains no general consensus on how best to extend a geometry of equilibrium states to a geometry of nonequilibrium states. In this section, a geometrization of nonequilibrium states is proposed for the moment dynamical system.

5.1   Geometry of Equilibrium States

Equilibrium states are identified with the Legendre submanifolds generated by functions in the context of geometric thermodynamics [10, 12]. Besides, in the context of information geometry, equilibrium states are identified with dually flat spaces [1]. Combining these identifications, one has the following.

Proposition 4 *(A contact manifold and a strictly convex function induce a dually flat space, [3]).* Let $(\mathcal{C}, \lambda)$ *be a contact manifold,* (x, y, z) *a set of coordinates such that* $\lambda = dz - y_a dx^a$ *with* $x = \{x^1, \dots, x^n\}$ *and* $y = \{y_1, \dots, y_n\}$, *and* ϖ *a strictly convex function depending only on* x. *Then,* $((\mathcal{C}, \lambda), \varpi)$ *induces an* n-*dimensional dually flat space*

To apply the proposition above to physical systems, the coordinate sets x and y are chosen such that x^a and y_a form a thermodynamic conjugate pair for each a. Here it is assumed that such thermodynamic variables can be defined even for nonequilibrium states, and that they are consistent with those variables defined at equilibrium. In addition to this, the physical dimension of ϖ should be equal to that of $y_a dx^a$. Also Ψ and its Legendre transform are chosen as ϖ.

5.2   Geometry of Nonequilibrium States

So far geometry of equilibrium states have been discussed. One remaining issue is how to give the physical meaning of the set outside $\mathcal{A}_\varpi$, $\mathcal{C} \setminus \mathcal{A}_\varpi$. A natural interpretation of $\mathcal{C} \setminus \mathcal{A}_\varpi$ would be some set of nonequilibrium states. We make this interpretation in this paper.

As shown in Proposition 2, initial states approach to the equilibrium state as time develops. This can be reformulated on contact manifolds and para-contact metric manifolds. In the contact geometric framework of nonequilibrium thermo-dynamics, the equilibrium state is identified with a Legendre submanifold. Then, as found in [3] and [6], some dynamical systems expressing nonequilibrium process can be identified with a class of contact Hamiltonian vector fields on a contact manifold. The above claim also holds on para-contact metric manifolds.

Geometry of Moment Dynamical System. Proposition 3 is written in a contact geometric language here. In what follows phase space is identified with a $(2n+1)$-dimensional para-contact metric manifold $(\mathcal{C}, \phi, \xi, \lambda, g^{\mathrm{M}})$.

As shown below, the moment dynamical system is a contact Hamiltonian system.

Proposition 5 *(Moment dynamical system as a contact Hamiltonian system, [7]). The dynamical system in Proposition 3 can be written as a contact Hamiltonian system.*

One is interested in how a $(1,1)$-tensor field ϕ plays a role for geometric nonequilibrium thermodynamics. To give an answer, one needs the following.

Lemma 1. *Let $\{\dot{x}_a\}, \{\dot{y}_a\}, \dot{z}$ be some functions, and X_0 the vector field*

$$X_0 = \dot{x}^a \frac{\partial}{\partial x^a} + \dot{y}_a \frac{\partial}{\partial y_a} + \dot{z} \frac{\partial}{\partial z}.$$

Then, $\phi(X_0)$ and $\phi^2(X_0)$ are calculated as

$$\phi^\mu(X_0) = (-1)^\mu \dot{x}^a \left(\frac{\partial}{\partial x^a} + y_a \frac{\partial}{\partial z} \right) + \dot{y}_a \frac{\partial}{\partial y_a}, \quad \mu = 1, 2.$$

Proof. Throughout this proof, the Einstein convention is not used. With the local expressions shown in Sect. 2, one has

$$\widehat{\theta}^a_\pm(X_0) = \frac{\sqrt{y_a}}{2} \dot{x}^a \pm \frac{\dot{y}_a}{2\sqrt{y_a}},$$

$$e^+_a + e^-_a = \frac{2}{\sqrt{y_a}} \left(\frac{\partial}{\partial x^a} + y_a \frac{\partial}{\partial z} \right), \quad \text{and} \quad e^+_a - e^-_a = 2\sqrt{y_a} \frac{\partial}{\partial y_a}.$$

Combining these, one has

$$\phi(X_0) = \sum_a \left[-\widehat{\theta}^a_-(X_0) e^+_a + \widehat{\theta}^a_+(X_0) e^-_a \right]$$

$$= \sum_a \left[-\frac{\sqrt{y_a}}{2} \dot{x}^a (e^+_a + e^-_a) + \frac{\dot{y}_a}{2\sqrt{y_a}} (e^+_a - e^-_a) \right]$$

$$= \sum_a \left[-\dot{x}^a \left(\frac{\partial}{\partial x^a} + y_a \frac{\partial}{\partial z} \right) + \dot{y}_a \frac{\partial}{\partial y_a} \right].$$

For $\phi^2(X_0)$, substituting $\lambda(X_0) = \dot{z} - \sum_a y_a \dot{x}^a$ into $\phi^2(X_0) = X_0 - \lambda(X_0)\xi$, one has the desired expression. $\qquad\square$

Applying this Lemma, one has the following.

Theorem 1 (*Roles of ϕ of X_h for the moment dynamical system*). *Let X_h be the contact Hamiltonian vector field in Proposition 1. Then*

$$\mathcal{L}_{\phi(X_h)}h = \mathcal{L}_{\phi^2(X_h)}h = 0.$$

Proof. Substituting $\dot{x}^a = 0$ into $\phi^\mu(X_0)$ in Lemma 1, one has

$$\phi^\mu(X_h) = \dot{y}_a\frac{\partial}{\partial y_a}, \qquad \text{where} \quad \dot{y}_a = \frac{\partial\varpi}{\partial x^a} - y_a, \qquad \mu = 1, 2.$$

Then, with $\partial h/\partial y_a = 0$, one has

$$\mathcal{L}_{\phi^\mu(X_h)}h = [\phi^\mu(X_h)]\,h = 0, \qquad \mu = 1, 2.$$

$\square$

This states that the h is preserved along $\phi^\mu(X_h) \in \mathcal{STC}$, which should be compared with the case of $\mathcal{L}_{X_h}h$:

$$\mathcal{L}_{X_h}h = -\dot{z} = -(\varpi(x) - z) = -h.$$

Curve Length from the Equilibrium State. In nonequilibrium statistical physics, attention is often concentrated on how far a state is close to the equilibrium state. In general, to define and measure such a distance in terms of geometric language, length of a curve can be used. In Riemannian geometry, length is a measure for expressing how far given two points are away, where these points are connected with an integral curve of a vector field on a manifold.

The following can easily be proven.

Lemma 2 (*[7]*). *The length between a state and the equilibrium state for the moment dynamical system calculated with (2) is*

$$l[X_\Psi]_\infty^t := \int_\infty^t \sqrt{g^{\mathrm{M}}(X_\Psi, X_\Psi)}\,dt = |\,h(\theta, \langle\mathcal{O}\rangle_\theta, \Psi)\,|, \qquad (8)$$

where $\langle\mathcal{O}\rangle_\theta = \{\langle\mathcal{O}_1\rangle_\theta, \ldots, \langle\mathcal{O}_n\rangle_\theta\}$, h is such that $h(\theta, \langle\mathcal{O}\rangle_\theta, \Psi) = \Psi^{\mathrm{eq}}(\theta) - \Psi$ (see Proposition 1), and X_Ψ its corresponding contact Hamiltonian vector field. Then the convergence rate for (8) is exponential.

Combining Lemma 2 and discussions in the previous sections, one arrives at the main theorem in this paper.

Theorem 2 (*Geometric description of the expectation variables and its convergence*). *The moment dynamical system derived from solvable master equations are described on a para-contact metric manifold, and its convergence rate associated with the metric tensor field (2) is exponential.*

6   Conclusions

This paper has offered a viewpoint that expectation variables of the moment dynamical system derived from master equations can be described on a para-contact metric manifold. To give a geometric description of these variables a contact Hamiltonian vector field has been introduced on a para-contact metric manifold. Also, roles of the $(1,1)$-tensor field ϕ have been clarified in this paper (Theorem 1). Then, with the Mrugala metric tensor field, the convergence rate has been shown to be exponential on this para-contact metric manifold (Theorem 2).

Acknowledgments. The author S.G. is partially supported by JSPS (KAKENHI) grant number 19K03635. The other author H.H. is partially supported by JSPS (KAKENHI) grant number 17H01793. In addition, both of the authors are partially supported by JST CREST JPMJCR1761.

References

1. Amari, S., Nagaoka, H.: Methods of Information Geometry. AMS. Oxford University Press, Oxford (2000)
2. Ay, N., et al.: Information Geometry. Springer, Cham (2017). https://doi.org/10.1007/978-3-319-56478-4
3. Goto, S.: Legendre submanifolds in contact manifolds as attractors and geometric nonequilibrium thermodynamics. J. Math. Phys. **56**, 073301 (2015). [30 pages]
4. Goto, S.: Contact geometric descriptions of vector fields on dually flat spaces and their applications in electric circuit models and nonequilibrium thermodynamics. J. Math. Phys. **57**, 102702 (2016). [40 pages]
5. Bravetti, A., Lopez-Monsalvo, C.S.: Para-Sasakian geometry in thermodynamic fluctuation theory. J. Phys. A: Math. Theor. **48**, 125206 (2015). [21 pages]
6. Bravetti, A., Lopez-Monsalve, C.S., Nettel, F.: Contact symmetries and Hamiltonian thermodynamics. Ann. Phys. **361**, 377–400 (2015)
7. Goto, S., Hino, H.: Information and contact geometric description of expectation variables exactly derived from master equations. arXiv:1805.10592v2
8. Zamkovoy, S.: Canonical connections on paracontact manifolds. Ann. Glob. Geom. **36**, 37–60 (2009)
9. da Silva, A.C.: Lectures on Symplectic Geometry, 2nd edn. Springer, Heidelberg (2008). https://doi.org/10.1007/978-3-540-45330-7
10. Mrugala, R.: Geometrical formulation of equilibrium phenomenological thermodynamics. Rep. Math. Phys. **14**, 419–427 (1978)
11. Mrugala, R.: Statistical approach to the geometric structure of thermodynamics. Phys. Rev. A **41**, 3156–3160 (1990)
12. Mrugala, R.: On contact and metric structures on thermodynamic spaces. Suken kokyuroku **1142**, 167–181 (2000)

Monotone Embedding and Affine Immersion of Probability Models

Doubly Autoparallel Structure
on Positive Definite Matrices
and Its Applications

Atsumi Ohara[⊠]

University of Fukui, 3-9-1 Bunkyo Fukui 910-8507, Fukui, Japan
`ohara@fuee.u-fukui.ac.jp`

Abstract. In a statistical manifold, we can naturally define submanifolds that are simultaneously autoparallel with respect to both the primal and the dual affine connections of the statistical manifold. We call them *doubly autoparallel* submanifolds. The aim of this paper is to mainly introduce doubly autoparallelism on positive definite matrices in linear algebraic way and show its applicability to two related topics.

Keywords: Semidefinite program · Structured covariance estimation · Doubly autoparallelism · Jordan subalgebra

1 Introduction

Let us consider an information geometric structure [3] (g, ∇, ∇^*) on a manifold $\mathcal{M}$, where g, ∇, ∇^* are, respectively, a Riemannian metric and a pair of torsion-free affine connections satisfying

$$Xg(Y, Z) = g(\nabla_X Y, Z) + g(Y, \nabla_X^* Z), \quad \forall X, Y, Z \in \mathcal{X}(\mathcal{M}).$$

Here, $\mathcal{X}(\mathcal{M})$ denotes the set of all vector fields on $\mathcal{M}$. Such a manifold with the structure (g, ∇, ∇^*) is called a *statistical manifold* and we say ∇ and ∇^* are *mutually dual* with respect to g.

In a statistical manifold, we can naturally define a submanifold $\mathcal{N}$ that is simultaneously autoparallel with respect to both ∇ and ∇^*.

Definition 1. *Let $(\mathcal{M}, g, \nabla, \nabla^*)$ be a statistical manifold and $\mathcal{N}$ be its submanifold. We call $\mathcal{N}$ doubly autoparallel in $\mathcal{M}$ when the followings hold:*

$$\nabla_X Y \in \mathcal{X}(\mathcal{N}), \ \nabla_X^* Y \in \mathcal{X}(\mathcal{N}), \quad \forall X, Y \in \mathcal{X}(\mathcal{N}).$$

The concept of doubly autoparallelism has been investigated for symmetric cones [19] and probability simplex [16] in algebraic ways, where Jordan or Hadamard subalgebras play crucial roles. In addition, the concept plays important roles in several applications of information geometry. The first purpose of

Supported by JSPS Grant278 in-Aid (C) 15K04997.

F. Nielsen and F. Barbaresco (Eds.): GSI 2019, LNCS 11712, pp. 251–260, 2019.
https://doi.org/10.1007/978-3-030-26980-7_26

the paper shows a tractable characterization of doubly autoparallel submanifolds in particular for positive definite matrices using basic knowledges of linear algebra.

Next, we consider two applications: the one is a statistical problem called the maximum likelihood estimation of structured covariance matrix [1,5], and the other is a convex optimization problem called semidefinite program [20]. We show that solvabilities or difficulties of both problems are, in some sense, related with the doubly autoparallelism.

2   Dually Flat Structure on Positive Definite Matrices

We summarize the facts given in [17].

Let $Sym(n)$ denote the set of n by n real symmetric matrices, which is a vector space on $\mathbf{R}$ of dimension $N := n(n+1)/2$. We use the standard inner product

$$\langle X, Y \rangle := \operatorname{tr}(XY), \quad X, Y \in Sym(n). \tag{1}$$

In the following we show $PD(n)$ is dually flat. Denote by $T_P PD(n)$ the tangent vector space at $P \in PD(n)$. Since each $T_P PD(n)$ is isomorphic to $Sym(n)$, we identify the tangent vector in $T_P PD(n)$ with the element in $Sym(n)$. In other words, for $P(x) := \sum_{i=1}^{N} x^i E_i \in PD(n)$ with arbitrary fixed basis matrices $\{E_i\}_{i=1}^{N}$ of $Sym(n)$, we consider $(\partial/\partial x^i)_P \equiv E_i$.

Define a *Riemannian metric* at P for each tangent vector space $T_P PD(n)$ by

$$g_P(X, Y) := \operatorname{tr}(P^{-1} X P^{-1} Y), \ X, Y \in T_P PD(n).$$

Next, consider two linear isomorphisms between $T_{P_1} PD(n)$ and $T_{P_2} PD(n)$:

$$\Pi : \ X \in T_{P_1} PD(n) \mapsto X \in T_{P_2} PD(n),$$
$$\Pi^* : \ X \in T_{P_1} PD(n) \mapsto P_2 P_1^{-1} X P_1^{-1} P_2 \in T_{P_2} PD(n).$$

Regarding these isomorphisms as parallel shifts, we see that the corresponding affine connections, respectively denoted by ∇ or ∇^*, have the following properties:

(d1) For any P_1, P_2, it holds $g_{P_1}(X, Y) = g_{P_2}(\Pi(X), \Pi^*(Y))$.
(d2) Both torsions and curvatures derived from ∇ and ∇^* vanish.

Hence, the structure $(PD(n), g, \nabla, \nabla^*)$ is *dually flat*. They are represented by

$$(\nabla_{\partial_i} \partial_j)_P \equiv 0, \quad \partial_i := \partial/\partial x^i$$
$$(\nabla^*_{\partial_i} \partial_j)_P \equiv -E_i P^{-1} E_j - E_j P^{-1} E_i. \tag{2}$$

Let $P(t), t \in \mathcal{I} \subset \mathbf{R}$ be a smooth curve in $PD(n)$ defined on a interval $\mathcal{I}$, and $X(t)$ be its tangent vector at $P(t)$. The curve satisfying

$$\Pi(X(t_1)) = X(t_2) \quad \Pi^*(X(t_1)) = X(t_2), \qquad \forall t_1, t_2 \in \mathcal{I}$$

are respectively called ∇- and ∇^*-geodesics.

The ∇-*geodesic* is nothing but the ordinary straight line, i.e,

$$P(t) = P + tX \in PD(n), P \in PD(n), X \in Sym(n).$$

The tangent vector is always X. On the other hand, the ∇^*-*geodesic* is represented by

$$P(t) = (P + tY)^{-1} \in PD(n), P \in PD(n), Y \in Sym(n).$$

Actually, the tangent vectors of this curve at $P(t_1)$ and $P(t_2)$ are, respectively,

$$X_1 = -P(t_1)^{-1}YP(t_1)^{-1}, X_2 = -P(t_2)^{-1}YP(t_2)^{-1}.$$

It would be easy to see $\Pi^*(X_1) = X_2$. Consequently, ∇^*-geodesic is the straight line in the space of inverse matrices.

Finally we state the relation with Jordan algebras [7]. The Jordan product on $Sym(n)$ is defined by

$$X * Y := \frac{1}{2}(XY + YX). \tag{3}$$

The associated quadratic representation is

$$\mathcal{P}(X)Y := XYX, \quad X, Y \in Sym(n). \tag{4}$$

Then we see that the logarithmic characteristic function on $PD(n)$

$$\psi(P) = -\log \det P \tag{5}$$

is the potential function of the dually flat structure $(PD(n), g, \nabla, \nabla^*)$ as follows:
For $P \in PD(n)$, there exists $Q \in Sym(n)$ such that $Q * Q = P$. We have

$$g_P(X, Y) = D^2\psi(P)[X, Y] = \langle \mathcal{P}(Q^{-1})X, \mathcal{P}(Q^{-1})Y \rangle, \tag{6}$$

where $D^k\psi(P)[\bullet, \ldots, \bullet]$ denote k-th directional derivative of $\psi(P)$. Further, ∇^*-connection meets the following relation:

$$g_P(\nabla^*_{\partial_i}\partial_j, \partial_k) = D^3\psi(P)[E_i, E_j, E_k] = -2\langle \mathcal{P}(Q^{-1})E_i * \mathcal{P}(Q^{-1})E_j, \mathcal{P}(Q^{-1})E_k \rangle$$

In particular, the connection ∇^* in (2) satisfies the definition of Jordan product c1) and c2) when it is regarded as a binary operation of E_i and E_j. The right-hand side of (2) is sometimes called *mutation* of the Jordan product. In particular the connection ∇^* at $P = I$ corresponds to the standard Jordan product (3) on $Sym(n)$. These relations between information geometry and Jordan algebra can be extended on general symmetric cones [19].

3 Related Topics in Applications

In this section, as application problems on $PD(n)$, we present a brief introduction to semidefinite programming (SDP) [20] and maximum likelihood estimation of structured covariance matrix [1,5]. A certain common class of structured matrices in $PD(n)$ is featured in relation with the solvability or difficulties of these mathematical problems.

3.1   Semidefinite Programming

Let $c \in \mathbf{R}^m$ and $E_i \in Sym(n), i = 0, \ldots, m$ be given vector and matrices, where $\{E_i\}_{i=1}^m$ are linearly independent. Semidefinite program is defined (in a dual form) as the optimization problem:

$$\min_x c^T x, \text{ s.t. } P(x) = E_0 + \sum_{i=1}^m x^i E_i \succeq O. \tag{7}$$

Since the correspondence between x and $P(x)$ is one-to-one, we call the set of $P(x)$ satisfying (7) the *feasible region*. Define

$$\mathcal{V} := \mathrm{span}\{E_i\}_{i=1}^m, \quad E_0 + \mathcal{V} := \{X | X - E_0 \in \mathcal{V}\},$$

then the feasible region of semidefinite program is the closure of $\mathcal{L}$, which is defined by

$$\mathcal{L} := PD(n) \cap (E_0 + \mathcal{V}). \tag{8}$$

From a computational viewpoint, semidefinite program can be solved via a certain type of interior point algorithms [14]. In that case, computational complexity depends on the choice of barrier functions for $\mathcal{L}$, essentially for the convex cone $PD(n)$. It is proved that if a barrier function is *self-concordant* [14], the interior point method works efficiently. The logarithmic characteristic function ψ in (5) is an example of such functions on $PD(n)$.

Several researchers [4,6,10,18] have introduced the differential geometric point of view to analyze the mathematical structures behind the interior point methodology. In [8,9,15] the authors discussed the relation between *the second fundamental form (or the embedding curvature)* of $\mathcal{L}$ and computational complexity (the number of Newton steps) by investigating the affine scaling trajectories. In particular the following result was shown:

Proposition 1. *Suppose the following two assumptions:*

(i) *the ∇^*-embedding curvature of $\mathcal{L}$ in $PD(n)$ vanishes everywhere on the feasible region $\mathcal{L}$ of semidefinite program,*
(ii) *one of any feasible points is known.*

Then, the SDP with the arbitrary linear objective function can be explicitly solved *via the formula of the optimal solutions.*

The specific geometric structure of $\mathcal{L}$ given in (i) would be of interest. The condition (i) can be equivalently rewritten in terms of $\{E_i\}_{i=1}^m$, which characterize the linear subspace $\mathcal{V}$, as follows:

$$\text{(i)} \iff \nabla^*_{\partial_i}\partial_j \equiv -E_i P^{-1} E_j - E_j P^{-1} E_i \in \mathcal{V}, i, j = 1, \cdots, m, \quad \forall P \in \mathcal{L}. \tag{9}$$

However, note that it is practically difficult to check the condition (9) at *all* $P \in \mathcal{L}$. Another algebraic characterization of this class of matrices will be given in the Sect. 4.1.

The above result suggests the close relation with geometric property such as embedding curvature of the feasible region and the computational complexity. It not only proposes the interesting theoretical problem but also give a clue to consider efficiently-solved structures for large-scale SDP problems.

3.2 Maximum Likelihood Estimation of Structured Covariance Matrices

Suppose the finite observed data $\{z_1, z_2, ..., z_n\}$ obey to the p-dimensional multivariate normal distribution with zero-mean and covariance matrix P

$$p(z) = (2\pi)^{-p/2} (\det P)^{-1/2} \exp\{-\frac{1}{2}z^T P^{-1} z\}$$

and consider the problem to estimate P from the observed data. The structured covariance estimation problem assumes that true P should lie in a certain subset, denoted by $\mathcal{L}$, in $PD(n)$ depending on the mechanisms of data generations.

For example, several elements of P may be known beforehand. Or in signal processing the data observed from the stationary time series or images involve (block) Toeplitz or Hankel structure for covariance matrices. These are examples of linear constraints of P.

On the other hand, in the field of statistical estimation called graphical Gaussian modelling or covariance selection [11], the structure in which several elements of P^{-1} are specified to be zero, plays an important role. In this case the constraints are inversely-linear, i.e., linear with respect to the elements of P^{-1}. Further, in the area of factor analysis, more complicated structures may be possibly imposed on covariance matrices.

Thus, attention has been paid to structured covariance estimation as an important problem in practice for a long time [1,5].

Now consider the maximum likelihood estimation for the case of linear constraints on covariance matrices, which is basic and important in practice, e.g, factor analysis or Toeplitz structure in signal processing. In other words, we consider the following general linear structure that is same as in SDP:

$$\mathcal{L} = \{P | P = E_0 + \sum_{i=1}^{m} x^i E_i \in PD(n)\}.$$

Since the likelihood function $L(P)$ for the observed data $\{z_1, z_2, ..., z_n\}$ is

$$L(P) = \prod_{i=1}^{n} f(z_i) = (2\pi)^{-np/2} (\det P)^{-n/2} \prod_{i=1}^{n} \exp\{-\frac{1}{2}z_i^T P^{-1} z_i\},$$

we have

$$\log L(P) = -\frac{np}{2} \log 2\pi + \frac{n}{2}\{-\log \det P - \frac{1}{n}\sum_{i=1}^{n} z_i^T P^{-1} z_i\}$$

$$= -\frac{np}{2} \log 2\pi + \frac{n}{2}\{-\log \det P - \operatorname{tr}(P^{-1}S)\},$$

where S is the sample covariance matrix calculated by

$$S = \frac{1}{n}\sum_{i=1}^{n} z_i z_i^T.$$

and it is not generally in $\mathcal{L}$.

Hence, solving the maximizer of $\log L(P)$ on $\mathcal{L}$, i.e, the maximum likelihood estimation for covariance structure $\mathcal{L}$ is equivalent with the optimization:

$$\min_{P \in \mathcal{L}} h(P) \text{ s.t. } h(P) = -\log \det P^{-1} + \text{tr}(P^{-1}S).$$

Unfortunately, the function $h(P)$ is not convex with respect to P. Hence, in general we can only expect to solve local minimizers. However, note that the first and second terms of $h(P)$ are, respectively, self-concordant and linear with respect to P^{-1} as was shown previously, and thus, $h(P)$ is also self-concordant with respect to not P but P^{-1}. We can summarize as follows:

Proposition 2. *If the inverses of all matrices in $\mathcal{L}$ are also equipped with linear structure, then the MLE for $\mathcal{L}$ can be reduced to convex program for a self-concordant function $h(P)$ and solved efficiently by interior point methodology.*

The theoretical problem would be the characterization of such both linear and inversely linear structure of positive definite matrices. This will be discussed in the Sect. 4.

Unfortunately, it is shown that not so many general or useful linear structures are in this nice class. For example, Toeplitz structure of order greater than three is generally not. However, to tackle the problem the so-called EM (Expectation Maximization) algorithm are often employed with combining the technique to embed the general linearly structured matrix as a submatrix of the larger order one in the nice class [12,13]. Hence, the characterization problem is of importance in this sense, too (cf. the Sect. 4).

4   Doubly Autoparallel Submanifolds in Positive Definite Matrices

In the previous sections we show that the class of semidefinite program whose feasible region satisfies the condition (9) can be solved explicitly. In addition, it is shown maximum likelihood estimation for structured covariance matrices can be reduced to the self-concordant convex programming if the structure is inversely linear.

Motivated by the special structure, we characterize, in this section, *doubly autoparallelism* on $PD(n)$ and discuss the relation with the Jordan algebra in $Sym(n)$. Further, we show examples of the doubly autoparallel submanifolds in $PD(n)$.

4.1   Doubly Autoparallelism

From the Definition 1 and the arguments in the Sect. 2, we see that the following results hold:

Theorem 1. *For $m < N :=$ dim $Sym(n)$ an m dimensional submanifold $\mathcal{L}$ in $PD(n)$ is doubly autoparallel if and only if it satisfies the following two conditions:*

(1) $\mathcal{L}$ is the intersection of an affine subspace of $Sym(n)$ and the positive define cone $PD(n)$, i.e., there exist $E_i \in Sym(n), i = 0, \cdots, m$ where $\{E_i\}_{i=1}^m$ are linearly independent and satisfy

$$\mathcal{L} := PD(n) \cap (E_0 + \mathcal{V}), \quad \text{where } \mathcal{V} := \mathrm{span}\{E_i\}_{i=1}^m, \quad E_0 + \mathcal{V} := \{X | X - E_0 \in \mathcal{V}\}.$$

(2) For all $\{E_i\}_{i=1}^m$ in the condition (1), it holds

$$E_i P^{-1} E_j + E_j P^{-1} E_i \in \mathcal{V}, i, j = 1, \ldots, m, \quad \forall P \in \mathcal{L}, \tag{10}$$

i.e., ∇^*-embedding curvature (the second fundamental form for ∇^*) vanishes on $\mathcal{L}$.

Remark. The definition of the doubly autoparallelism is equivalent with autoparallelism in both ∇- and ∇^*-connections and the concept can be extended to general statistical manifold. In the above definition (1) and (2) imply ∇- and ∇^*-autoparallelism, respectively.

Theorem 2. *The following statements are equivalent:*

1. *$\mathcal{L} \subset PD(n)$ is doubly autoparallel,*
2. *there exist $E_i, F^i \in Sym(n), i = 0, \ldots, m$, where $\{E_i\}_{i=1}^m$ and $\{F^i\}_{i=1}^m$ are linearly independent sets of matrices and $\mathcal{L}$ is simultaneously represented by*

$$\mathcal{L} = \{P | P = E_0 + \sum_{i=1}^m x^i E_i \succ O, \exists x = (x^i) \in \mathbf{R}^m\}$$

$$= \{Q | Q^{-1} = F^0 + \sum_{i=1}^m y_i F^i \succ O, \exists y = (y_i) \in \mathbf{R}^m\}.$$

Thus, from Proposition 2 we immediately see that MLE for structured covariance matrices in $\mathcal{L}$ is cast to convex optimization with a self concordant function if and only if $\mathcal{L}$ is doubly autoparallel in $PD(n)$.

Next, we discuss on the structure and concrete examples of doubly autoparallel submanifolds in $PD(n)$. We have a simpler condition in the following special case:

Proposition 3. *Let E_0 be in $\mathcal{V} = \mathrm{span}\{E\}_{i=1}^m$ ($\mathcal{L} = \mathcal{V} \cap PD(n)$) and the identity matrix I be in $\mathcal{L}$. Then, $\mathcal{L}$ is doubly autoparallel if and only if $\mathcal{V}$ is a Jordan subalgebra of $Sym(n)$, i.e., the following two conditions holds:*

1. *$\mathcal{V}$ is a subspace of $Sym(n)$,*
2. *$\mathcal{V}$ is closed under the Jordan product, i.e., $\mathcal{V}$ is a Jordan subalgebra:*

$$E_i * E_j := (E_i E_j + E_j E_i)/2 \in \mathcal{V}, \quad \forall E_i, E_j, \quad i, j = 1, \ldots, m. \tag{11}$$

Remark. As examples of Jordan subalgebra in $Sym(n)$, we have

(i) *doubly symmetric matrices* defined by

$$\{X | X = (x_{ij}), x_{ij} = x_{ji}, \ x_{ij} = x_{n+1-j \ n+1-i}\}, \tag{12}$$

i.e., the set of matrices that are symmetric with respect to both main and anti-main diagonal elements,

(ii) matrices that have the prescribed eigenvectors

and so on. For more further information on Jordan subalgebras in $Sym(n)$, consult with [12].

The doubly autoparallel submanifolds characterized by Jordan subalgebras in the above proposition are the intersections of $PD(n)$ and linear but not affine subspaces, i.e., they consist of subcones in $PD(n)$ and contain the origin. Hence, we give an example of doubly autoparallel submanifolds that are not subcones.

We denote by $\mathcal{JS}(k)$ one of the arbitrary Jordan subalgebras of $Sym(k)$ that contain the k-th order identity matrix I_k. When it is unnecessary to specify the order k, we denote it simply by $\mathcal{JS}$.

Example 1. Define the affine subspace as follows:

$$\mathcal{A}_1 := \left\{ \begin{pmatrix} X & A \\ A^T & B \end{pmatrix} \middle| X \in \mathcal{JS}, \ \det B \neq 0, AB^{-1}A^T \in \mathcal{JS} \right\},$$

where A and B are constant matrices. Then if $\mathcal{L}_1 := \mathcal{A}_1 \cap PD(n)$ is not empty, it is doubly autoparallel.

4.2 Explicit Formula of the Optimal Solution for SDP

We only show an explicit formula of one of the optimal solutions $P(x^*)$ of SDP in the case when its feasible region $\mathcal{L}$ given by (8) is doubly autoparallel. Full proofs can be found in [15].

Step 1: Let $P(x_0)$ be a given feasible point in $\mathcal{L}$. Define

$$F^i = -P(x_0)^{-1} E_i P(x_0)^{-1}$$

and $F^0 = P(x_0)^{-1}$. Using these $\{F^i\}_{i=0}^m$ and that $\mathcal{L}$ is doubly autoparallel, any $P \in \mathcal{L}$ can be represented by

$$P^{-1} = F^0 + \sum_{i=1}^m y_i F^i.$$

Step 2: Solve the matrix C that satisfies

$$c^T x = \mathrm{tr} C P(x) + \text{constant}, \quad C \in \mathrm{span}\{F^i\}_{i=1}^m.$$

For this, solve the following linear equation

$$H\tilde{c} = c, \quad H = (h_{ij}) := \left(\mathrm{tr}(E_i F^j)\right),$$

where H is ensured to be always nonsingular because the Hesse matrix of $\varphi(P)$ is positive definite. Using the solution $\tilde{c} = (\tilde{c}_i)$, the matrix C can be constructed by

$$C = \sum_{i=1}^{m} \tilde{c}_i F^i.$$

Step 3: Compute the singular decomposition of C:

$$C = \begin{pmatrix} V_1 & V_2 \end{pmatrix} \begin{pmatrix} \Sigma_1 & O \\ O & O \end{pmatrix} \begin{pmatrix} V_1^T \\ V_2^T \end{pmatrix} = V_1 \Sigma_1 V_1^T$$

.

Step 4: The formula of the optimal solution $P(x^*)$ is

$$P(x^*) = P(x_0) - P(x_0) V_1 \{ V_1^T P(x_0) V_1 \}^{-1} V_1^T P(x_0).$$

Each element of x^* is obtained, by using E^i's that satisfy $\mathrm{tr}(E^i E_j) = \delta_j^i$ (Kronecker's delta), as follows:

$$x^{*i} = \mathrm{tr}(E^i P(x^*)) - \mathrm{tr}(E^i E_0).$$

Thus, when the feasible region of SDP is doubly autoparallel, it can be solved by only solving a linear equation and the singular decomposition.

5   Conclusions

We have discussed the structured positive definite matrices with both linear and inversely linear constraints. We have shown such a structure is characterized by information geometric concept, which we call doubly autoparallelism. Further, we have also shown it is characterized algebraically by Jordan subalgebra under a certain condition. Finally, it is demonstrated that this class of structure has close relation with certain kinds of solvability of semidefinite program and the maximum likelihood estimation of structured covariance matrices.

References

1. Anderson, T.W.: Asymptotically efficient estimation of covariance matrices with linear structure. Ann. Math. **1**(1), 135–141 (1971)
2. Amari, S.: Differential-Geometrical Methods in Statistics. Springer, Heidelberg (1985)
3. Amari, S., Nagaoka, H.: Methods of Information Geometry. AMS and Oxford, New York (2000)

4. Bayer, D.A., Lagarias, J.C.: The nonlinear geometry of linear programming I. Trans. Amer. Math. Soc. **314**, 499–526 (1989). II, Trans. Amer. Math. Soc. **314**, 527–581 (1989)
5. Burg, J.P., Luengberger, D.G., Wenger, D.L.: Estimation of structured covariance matrices. Proc. IEEE **70**(9), 963–974 (1982)
6. Faybusovich, L.: A hamilton structure for generalized affine-scaling vector fields. J. Nonlinear Sci. **5**, 11–28 (1995)
7. Faraut, J., Korányi, A.: Analysis on Symmetric Cones. Oxford Press, New York (1994)
8. Kakihara, S., Ohara, A., Tsuchiya, T.: Information geometry and interior-point algorithms in semidefinite programs and symmetric cone programs. J. Optim. Theory Appl. **157**, 749–780 (2013)
9. Kakihara, S., Ohara, A., Tsuchiya, T.: Curvature integrals and iteration complexities in SDP and symmetric cone programs. Comput. Optim. Appl. **57**, 623–665 (2014)
10. Karmarkar, N.: Riemannian geometry underlying interior-point methods for linear programming. Contemp. Math. **114**, 51–75 (1990)
11. Lauritzen, S.L.: Graphical Models, Oxford (1996)
12. Malley, J.D.: Statistical Applications of Jordan Algebras. Springer, Heidelberg (1994)
13. Morgera, S.D.: The role of abstract algebra in structured estimation theory. IEEE IT **38**(3), 1053–1065 (1992)
14. Nesterov, Y., Nemirovskii, A.: Interior-Point Polynomial Algorithms in Convex Programming. SIAM, Philadelphia (1994)
15. Ohara, A.: Information geometric analysis of semidefinite programming problems. Proc. Inst. Stat. Math. **46**(2), 317–334 (1998). in Japanese
16. Ohara, A., Ishi, H.: Doubly autoparallel structure on the probability simplex. In: Ay, N., Gibilisco, P., Matúš, F. (eds.) Information Geometry and Its Applications, vol. 252, pp. 323–334. Springer, Cham (2018). https://doi.org/10.1007/978-3-319-97798-0_12
17. Ohara, A., Suda, N., Amari, S.: Dualistic differential geometry of positive definite matrices and its applications to related problems. Linear Algorithm Appl. **247**, 31–53 (1996)
18. Tanabe, K., Tsuchiya, T.: New geometry for linear programming. Math. Sci. **303**, 32–37 (1988). in Japanese
19. Uohashi, K., Ohara, A.: Jordan algebra and dual affine connections on symmetric cones. Positivity **8**, 369–378 (2004)
20. Wolkowicz, H., Saigal, R., Vandenberghe, L. (eds.): Handbook of Semidefinite Programming. Theory, Algorithms, and Applications. Kluwer, Boston (2000)

Toeplitz Hermitian Positive Definite Matrix Machine Learning Based on Fisher Metric

Yann Cabanes[1,2]([⊠]), Frédéric Barbaresco[1], Marc Arnaudon[2], and Jérémie Bigot[2]

[1] Thales Surface Radar, Advanced Radar Concepts, Limours, France
yann.cabanes@gmail.com, frederic.barbaresco@thalesgroup.com
[2] Institut de Mathématiques de Bordeaux, Bordeaux, France
{marc.arnaudon,jeremie.bigot}@math.u-bordeaux.fr

Abstract. Here we propose a method to classify radar clutter from radar data using an unsupervised classification algorithm. The data will be represented by Positive Definite Hermitian Toeplitz matrices and clustered using the Fisher metric. Once the clustering algorithm dispose of a large radar database, new radars will be able to use the experience of other radars, which will improve their performances: learning radar clutter can be used to fix some false alarm rate created by strong echoes coming from hail, rain, waves, mountains, cities; it will also improve the detectability of slow moving targets, like drones, which can be hidden in the clutter, flying close to the landform.

Keywords: Radar clutter · Machine learning ·
Unsupervised classification · k-means · Autocorrelation matrix ·
Burg algorithm · Reflection coefficients · Kähler metric

1 Introduction

Our aim is to classify the radar clutter cell by cell. The idea is to classify each cell according to its autocorrelation matrix. In [1] this autocorrelation matrix is said to be equivalent to coefficients of an autoregressive model, called reflection coefficients, which will be estimated thanks to Burg algorithms. We will then classify the cells according to these reflection coefficients. Finally we will present a classification algorithm called k-means, and test it on simulated data. The unsupervised classification of radar data is dealt in [2] with a mean-shift algorithm. Here we will present another classification algorithm called k-means, and test it on simulated data, showing promising results.

2 Introduction to Signal Processing Theory

2.1 From Radar Data to Complex Matrices

In this study, the input data will be taken on a single burst, for a single elevation corresponding to the horizontal beam.

© Springer Nature Switzerland AG 2019
F. Nielsen and F. Barbaresco (Eds.): GSI 2019, LNCS 11712, pp. 261–270, 2019.
https://doi.org/10.1007/978-3-030-26980-7_27

Therefore, the radar provides us a 2D complex matrix of size $(\#impulses) \times (\#cells)$:

$$U = \begin{bmatrix} \begin{bmatrix} u_{0,0} \\ u_{1,0} \\ \vdots \\ u_{n-1,0} \end{bmatrix} \begin{bmatrix} u_{0,1} \\ u_{1,1} \\ \vdots \\ u_{n-1,1} \end{bmatrix} \begin{matrix} u_{0,2} & \cdots & u_{0,p-1} \\ u_{1,2} & \ddots & u_{1,p-1} \\ \vdots & \ddots & \vdots \\ u_{n-1,2} & \cdots & u_{n-1,p-1} \end{matrix} \end{bmatrix} \tag{1}$$

where n denotes the number of pulses of the burst, p the number of cells.

The complex coefficient u_{ij} represents the amplitude and phase after pulse compression of the echo beam at distance index i from the radar, at time index j (jth impulse).

The data to classify are the cells, each cell being represented by a column of the matrix U.

2.2   Model and Hypothesis

In this section, we will focus on a single column of the matrix U defined in Eq. 1. We will define its autocorrelation matrix and explain how to estimate an equivalent formulation of this autocorrelation matrix.

We denote by $\cdot^T$ the matrix transposition, $\cdot^H$ the complex matrix conjugate transpose and $\cdot^*$ the complex scalar conjugate.

We denote:

$$\mathbf{u} = [u(0), u(1), ..., u(n-1)]^T \tag{2}$$

the one dimensional complex signal registered in a cell.

We assume this signal to be stationary with zero mean:

$$\mathbb{E}[u(n)] = 0 \text{ for all } n \tag{3}$$

We also assume that this signal can be modeled as an autoregressive Gaussian process.

Interested readers may refer to [3] for a comprehensive course on complex signal processing theory.

2.3   From Input Vector to Autocorrelation Matrix

We define the autocorrelation matrix:

$$\mathbf{R} = \mathbb{E}[\mathbf{u} \, \mathbf{u}^H] \tag{4}$$

$$r_{i,j} = \mathbb{E}[u(k+i)u(k+j)^*] \tag{5}$$

We define the lag: $t = i - j$.

Proposition 1 (autocorrelation and stationarity). *The signal is supposed to be stationary, so $r_{i,j}$ depends only of the lag t.*

$$\begin{aligned}
r_{i,j} &= \mathbb{E}[u(k+i)u(k+j)^*] \\
&= \mathbb{E}[u(k+i-j)u(k)^*] \\
&= \mathbb{E}[u(k+t)u(k)^*] \\
&= r_t
\end{aligned} \tag{6}$$

Proposition 2 (autocorrelation and conjugation).

$$\begin{aligned}
r_{-t} &= \mathbb{E}[u(k-t)u(k)^*] \\
&= \mathbb{E}[u(k)u(k+t)^*] \\
&= \mathbb{E}[u(k+t)u(k)^*]^* \\
&= r_t^*
\end{aligned} \tag{7}$$

Consequence $\mathbf{R}$ is a Toeplitz Hermitian Positive Definite matrix.

$$\mathbf{R} = \begin{bmatrix}
r_0 & r_1^* & r_2^* & \cdots & r_{n-1}^* \\
r_1 & r_0 & r_1^* & \cdots & r_{n-2}^* \\
r_2 & r_1 & r_0 & \cdots & r_{n-3}^* \\
\vdots & \vdots & \vdots & \ddots & \vdots \\
r_{n-1} & r_{n-2} & r_{n-3} & \cdots & r_0
\end{bmatrix} \tag{8}$$

Note that the assumptions made in Sect. 2.2 that the signal can be modeled as a complex stationary autoregessive Gaussian process with zero mean has the following equivalent formulation: $\mathbf{u} = \mathbf{R}^{1/2}x$ with $\mathbf{R}$ a Toeplitz Hermitian Positive Definite matrix and x a standard complex Gaussian random vector which dimension is equal to the number of pulses.

2.4   Autocorrelation Matrix Estimation

In our classification problem, the autocorrelation matrix $\mathbf{R}_i$ will be estimated independently for each cell $\mathbf{u}_i$:

$$U = \begin{bmatrix} \begin{bmatrix} u_{0,0} \\ u_{1,0} \\ \vdots \\ u_{n-1,0} \end{bmatrix} & \begin{bmatrix} u_{0,1} \\ u_{1,1} \\ \vdots \\ u_{n-1,1} \end{bmatrix} & \begin{bmatrix} u_{0,2} \\ u_{1,2} \\ \vdots \\ u_{n-1,2} \end{bmatrix} & \cdots & \begin{bmatrix} u_{0,p-1} \\ u_{1,p-1} \\ \vdots \\ u_{n-1,p-1} \end{bmatrix} \end{bmatrix}$$

$$\begin{array}{ccccc}
\downarrow & \downarrow & \downarrow & & \downarrow \\
\widehat{\mathbf{R}_0} & \widehat{\mathbf{R}_1} & \widehat{\mathbf{R}_2} & & \widehat{\mathbf{R}_{n-1}}
\end{array} \tag{9}$$

Empirical Covariance Matrix. To estimate the Toeplitz autocorrelation matrix $\mathbf{R}$ from the data vector $\mathbf{u}$, we can estimate each coefficient r_t by the following empirical mean:

$$\widehat{r_t} = \frac{1}{n-t} \sum_{k=0}^{n-1-t} u(k+t)u(k)^* \quad t = 0, ..., n-1 \tag{10}$$

Note that this method is unprecise when the vector length n is small, especially when the lag t is close to $n-1$. We now propose a more robust method to estimate the autocorrelation matrix with few data, based on an autoregessive model.

Burg Algorithm. The Burg algorithm principle is to minimize the forward and the backward prediction errors. The regularised Burg algorithm of order M and regularization coefficient γ is described in Algorithm 1 and detailed in [4,5].

The regularized Burg algorithm allows us to transform the original data into a power factor in $\mathbb{R}_+^*$ and reflection coefficients in $\mathbb{D}^{n-1}$, where $\mathbb{D}$ represents the complex unit disk.

According to [1], the following transformation is a bijection:

$$\begin{aligned} \mathcal{T}_n^+ &\to \mathbb{R}_+^* \times \mathbb{D}^{n-1} \\ R_n &\mapsto (p_0, \mu_1, ..., \mu_{n-1}) \end{aligned} \tag{18}$$

where $\mathcal{T}_n^+$ denotes the set of Toeplitz Hermitian Positive Definite matrices of size n.

It is therefore equivalent to estimate the coefficients $(p_0, \mu_1, ..., \mu_{n-1})$ and the autocorrelation matrix R_n.

2.5   The Kähler Metric

Each data vector $\mathbf{u}_i$ is now represented by an estimation of its autocorrelation matrix $\widehat{\mathbf{R}_i}$ which is a Toeplitz Hermitian Positive Definite matrix. We define the metric on the set $\mathcal{T}_n^+$ of Toeplitz Hermitian Positive Definite matrices as coming from the Fisher metric on the manifold of complex Gaussian distributions with zero means, Toeplitz Hermitian Positive Definite covariance matrices and null relation matrices.

According to the previous bijection, we will represent a Toeplitz Hermitian Positive Definite matrix T_i by the corresponding coefficients $(p_{0,i}, \mu_{1,i}, ..., \mu_{n-1,i})$. The following distance has been introduced by Barbaresco in [6] on the set $\mathbb{R}_+^* \times \mathbb{D}^{n-1}$ to make this bijection an isometry. In the Encyclopedia of Distance by Deza [7], this distance is called Barbaresco distance:

$$\begin{aligned} d_{\mathcal{T}_n^+}^2(T_1, T_2) &= d_{\mathcal{T}_n^+}^2\left((p_{0,1}, \mu_{1,1}, ..., \mu_{n-1,1}), (p_{0,2}, \mu_{1,2}, ..., \mu_{n-1,2})\right) \\ &= n \log^2\left(\frac{p_{0,2}}{p_{0,1}}\right) + \sum_{l=1}^{n-1} \frac{n-l}{4} \log^2\left(\frac{1 + \frac{\mu_{l,1} - \mu_{l,2}}{1 - \mu_{l,1}\mu_{l,2}^*}}{1 - \frac{\mu_{l,1} - \mu_{l,2}}{1 - \mu_{l,1}\mu_{l,2}^*}}\right) \end{aligned} \tag{19}$$

Algorithm 1. regularised Burg algorithm

Initialization:

$$f_{0,k} = b_{0,k} = u_k \quad k = 0, ..., n-1 \tag{11}$$

$$a_{0,k} = 1 \quad k = 0, ..., n-1 \tag{12}$$

$$p_0 = \frac{1}{n} \sum_{k=0}^{n-1} |u_k|^2 \tag{13}$$

for $i = 1, ..., M$: **do**

$$\mu_i = -\frac{\left(\frac{2}{n-i} \sum_{k=i}^{n-1} f_{i-1,k} \bar{b}_{i-1,k-1} + 2 \sum_{k=1}^{i-1} \beta_{k,i} a_{k,i-1} a_{i-k,i-1} \right)}{\left(\frac{1}{n-i} \sum_{k=i}^{n-1} |f_{i-1,k}|^2 + |b_{i-1,k-1}|^2 + 2 \sum_{k=0}^{i-1} \beta_{k,i} |a_{k,i-1}|^2 \right)} \tag{14}$$

where:

$$\beta_{k,i} = \gamma (2\pi)^2 (k-i)^2 \tag{15}$$

$$\begin{cases} a_{k,i} = a_{k,i-1} + \mu_i \bar{a}_{i-k,i-1} & k = 1, ..., i-1 \\ a_{i,i} = \mu_i \end{cases} \tag{16}$$

and

$$\begin{cases} f_{i,k} = f_{i-1,k} + \mu_i b_{i-1,k-1} & k = i, ..., n-1 \\ b_{i,k} = b_{i-1,k-1} + \bar{\mu}_i f_{i-1,k} & k = i, ..., n-1 \end{cases} \tag{17}$$

end for

return $(p_0, \mu_1, ..., \mu_{n-1})$

The equations of the geodesics of the set $\mathbb{R}_+^* \times \mathbb{D}^{n-1}$ endowed with the Kähler metric are described in [4].

2.6 The Kähler Mean

The Kähler mean of $(T_0, ..., T_{m-1})$ is defined as the point T_{mean} such that the following function $f(T) = \sum_{i=0}^{m-1} d^2(T, T_i)$, sum of the squared distances from T to T_i, reaches its unique minimum.

The Kähler mean algorithm is performed in [4,8] as a gradient descent on the function f. The gradient expression of f is:

$$\vec{\nabla} f(T) = \sum_{i=0}^{m-1} 2\vec{\nabla} d(T,T_i)\, d(T,T_i) = 2\sum_{i=0}^{m-1} -\frac{\overrightarrow{T\,T_i}}{d(T,T_i)}\, d(T,T_i) = -2\sum_{i=0}^{m-1} \overrightarrow{T\,T_i}$$

(20)

where $\vec{\nabla}$ denotes the gradient operator and $\overrightarrow{T\,T_i}$, also written $exp_T^{-1}(T_i)$, denotes the element of the tangent space of the manifold $\mathbb{R}_+^* \times \mathbb{D}^{n-1}$ at T such that the geodesic starting at T at time 0 with inital tangent vector $\overrightarrow{T\,T_i}$ arrives at T_i at time 1.

Note that the squared distance between two matrices T_1 and T_2 is a linear combination of squared distances between the coordinates $(p_{0,1}, \mu_{1,1}, ..., \mu_{n-1,1})$ and $(p_{0,2}, \mu_{1,2}, ..., \mu_{n-1,2})$. Hence the coordinates can be averaged independently:

$$
\begin{array}{ccccc}
T_0 & \mapsto (& \boxed{p_{0,0},} & \boxed{\mu_{1,0},} & \cdots , & \boxed{\mu_{n-1,0}} &) \\
\vdots & & \vdots & \vdots & & \vdots & \\
T_{m-1} & \mapsto (& \boxed{p_{0,m-1},} & \boxed{\mu_{1,m-1},} & \cdots , & \boxed{\mu_{n-k,m-1}} &) \\
& & \downarrow & \downarrow & & \downarrow & \\
T & \leftarrow (& p_0, & \mu_1, & \cdots , & \mu_{n-1} &)
\end{array}
$$

(21)

The gradient descent on the function f is therefore equivalent to a gradient descent on each coordinate. At each step of the algorithm, once the gradient is computed, we move on $\mathbb{R}_+^* \times \mathbb{D}^{n-1}$ following its geodesics.

3   Simulation Model

Each cell is simulated independently from the others. For each cell, we simulate a complex vector using a SIRV (Spherically Invariant Random Vectors) model:

$$Z = \underbrace{\sqrt{\tau} R^{1/2} x}_{\text{information coming from the environment}} + \underbrace{b_{radar}}_{\text{noise coming from the radar itself}}$$

(22)

with:

τ: clutter texture coefficient (positive real random variable).

R: scaled autocorrelation matrix (Toeplitz Hermitian Positive Definite).

x, b_{radar}: independent standard complex Gaussian random vectors which dimension is equal to the number of pulses.

The radar noise b_{radar} is assumed to be small enough in comparaison with the information coming from the environment $\sqrt{\tau} R^{1/2} x$ for estimating the autocorrelation matrix τR using the methods described in Sect. 2.4.

To choose the matrix R, we learn experimentally from radar measures the spectrum shape of the clutter we want to simulate. The scaled autocorrelation coefficients of the matrix R can then be computed from the spectrum using the inverse Fourier transform.

See [9,10] for more details about the clutter modeling.

4   Classification Problem

4.1   Methodology

Using the previous model, we simulate 100 vectors with the model parameters (τ_1, R_1) and 100 vectors with the model parameters (τ_2, R_2). Then for each vector we try to recover the parameters used to simulate it thanks to Burg algorithm. In this paper, we classify the data only on the scaled autocorrelation matrix R, represented by the reflection coefficients $(\mu_1, ..., \mu_{n-1})$. Future work might also use the texture parameter τ, influencing the power coefficient p_0, to classify the data.

Each vector is now represented by its reflection coefficients in the metric space $\mathbb{D}^{n-1}$ endowed with the Kähler metric. We classify these vectors using a k-means algorithm described in the next section. The k-means algorithm is a classical clustering algorithm in Euclidean spaces, the main difficulty was to adapt it to the Riemannian manifold $\mathbb{D}^{n-1}$ endowed with the Kähler metric. In Fig. 1, we plot the FFT of each simulated vector on the left graphic, each FFT being drawn horizontally; the vertical axis represents the different cells along the distance axis. On the graphic in the middle of Fig. 1, we plot the result of the corresponding k-means clustering. We present in Fig. 2 a visualization of the clustering on the first coefficients of reflection.

Once the clustering is done, we compute the F1 score of the classification. The F1 score is a way to measure the performance of a supervised classification algorithm. We adapted it to our unsupervised classification algorithm by doing all possible permutations in the classification results labels in order to find the best matching with the expected results. Finally we plot on Fig. 3 the normalized confusion matrix using the labels corresponding to this best matching.

4.2   k-means on $\mathbb{D}^{n-1}$ with the Kähler Metric

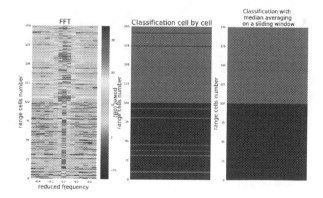

Fig. 1. FFT and classification results, k-means on $\mathbb{D}^{n-1}$, Kähler metric

First Coefficients of Reflection

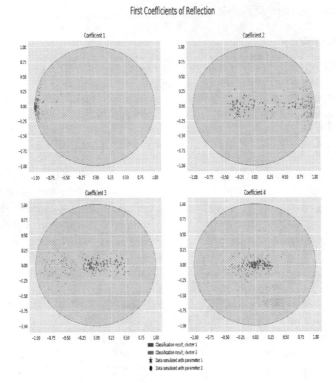

Fig. 2. First coefficients of reflection, k-means on $\mathbb{D}^{n-1}$, Kähler metric

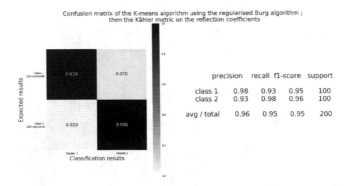

Fig. 3. Confusion matrix and F1 score, k-means on $\mathbb{D}^{n-1}$, Kähler metric

The Algorithm. The k-means algorithm is described in Algorithm 2.

Algorithm 2. k-means algorithm for N clusters

Initialization:
Pick randomly N points in the dataset. They now represent the barycenters of each class.
for i = 1 to loop number **do**
    Assign each point of the dataset to the closest barycenter.
    Compute the new barycenter of each class.
end for
return Each point is labeled according to the closest barycenter.

Predictions. Once an effective k-means algorithm is developed, we can easily predict the class of the new radar data: they will be assigned to the cluster having the closest barycenter.

4.3   Median Averaging

During all this study, we classified the data cell by cell, regardless of the spatial positioning of the data, each cell being considered independently from its neighbours. If we assume that each cell is correlated to the neighbouring cells, we can avoid missclassification due to outliers by associating to each data an average of its neigbouring cells, and performing the classification on the averaged data.

In Fig. 1, the graphic on the right represents the classification result given by a sliding window of size 9 (the classification result was perfect). In each window, we compute a median of the data in $\mathbb{D}^{n-1}$. The median of a set of points $(x_1, x_2, ..., x_n)$ in a metric space (E, d) is defined as follows:

$$median(x_1, x_2, ..., x_n) = argmin_{x \in E} \sum_{i=1}^{n} d(x, x_i) \qquad (23)$$

The median is more robust to outliers than the mean, the mean being the point minimizing the sum of squared distances. We then select the closest points of the barycenter to get rid of outliers, keeping half of the points, and compute the new median of these selected points. The center cell of the sliding window is now represented by this last median. Interested reader will find in [11] an algorithm to compute the median of several points in $\mathbb{D}^{n-1}$.

5   Conclusion

We developed a k-means algorithm to classify the radar clutter. This algorithm has been adapted to the Kähler metric and has given promising results. Future work may also take into account the texture coefficient τ; the normalized Burg algorithm presented in [12] might help to take this texture coefficient τ into consideration. More clustering algorithms will be adapted to the Kähler metric to deal with clusters of unusual interlaced shapes, like the mean-shift algorithm presented in [2,13]. These clustering algorithms will also be used to cluster groups of neighbouring cells: we will use a multidimensional spatial autoregressive model to represent the data (the autocorrelation matrices will be Positive Definite Block-Toeplitz matrices) and adapt our clustering algorithms to this higher dimensional space [1,14].

Acknowledgments. We thank the French MoD DGA MRIS for funding (convention CIFRE $N°2017.0008$).

References

1. Jeuris, B., Vandrebril, R.: The Kähler mean of Block-Toeplitz matrices with Toeplitz structured blocks (2016)
2. Chevallier, E., Forget, T., Barbaresco, F., Angulo, J.: Kernel Density Estimation on the Siegel Space with an Application to Radar Processing. Entropy (2016)
3. Haykin, S.: Adaptive Filter Theory. Pearson (2014)
4. Arnaudon, M., Barbaresco, F., Yang, L.: Riemannian medians and means with applications to radar signal processing. IEEE J. **7**(4), 595–604 (2013)
5. Barbaresco, F.: Super resolution spectrum analysis regularization: burg, capon and AGO-antagonistic algorithms. In: EUSIPCO 1996, Trieste, Italy, pp. 2005–2008 (1996)
6. Barbaresco, F.: Information geometry of covariance matrix: cartan-siegel homogeneous bounded domains, mostow/berger fibration and fréchet median. In: Nielsen, F., Bhatia, R. (eds.) Matrix Information Geometry, pp. 199–256. Springer, Heidelberg (2012). https://doi.org/10.1007/978-3-642-30232-9_9
7. Deza, M.M., Deza, E.: Encyclopedia of Distances. Springer, Heidelberg (2016). https://doi.org/10.1007/978-3-662-52844-0. ISBN 978-3-662-52844-0. http://www.springer.com/us/book/9783662528433
8. Bini, D., Iannazzo, B., Jeuris, B., Vandebril, R.: Geometric means of structured matrices. BIT **54**(1), 55–83 (2014)
9. Barrie Billingsley, J.: Low-Angle Radar Land Clutter, Measurements and Empirical Models. William Andrew Publishing, Norwich (2002)
10. Greco, M.S., Gini, F.: Radar Clutter Modeling
11. Arnaudon, M., Barbaresco, F., Yang, L.: Riemannian medians and means with applications to radar signal processing. IEEE Trans. Sig. Proc.
12. Decurninge, A., Barbaresco, F.: Robust burg estimation of radar scatter matrix for mixtures of gaussian stationary autoregressive vectors. IET Radar Sonar Navig. **11**(1), 78–89 (2016)
13. Barbaresco, F., Forget, T., Chevallier, E., Angulo, J.: Doppler spectrum segmentation of radar sea clutter by mean-shift and information geometry metric (2017)
14. Barbaresco, F.: Radar micro-doppler signal encoding in siegel unit poly-disk for machine learning in fisher metric space. In: IRS 2018, Bonn, June 2018

Deformed Exponential and the Behavior of the Normalizing Function

Francisca Leidmar Josue Vieira[1]([✉]), Rui Facundo Vigelis[2],
Luiza Helena Felix de Andrade[3], and Charles Casimiro Cavalcante[4]

[1] Department of Mathematics, Regional University of Cariri, Crato-CE, Brazil
leidmar.vieira@urca.br
[2] Computer Engineering, Campus Sobral, Federal University of Ceará,
Sobral-CE, Brazil
rfvigelis@ufc.br
[3] Center of Exact and Natural Sciences,
Federal Rural University of Semi-Arid Region, Mossoró-RN, Brazil
luizafelix@ufersa.edu.br
[4] Department of Teleinformatics Engineering, Federal University of Ceará,
Fortaleza-CE, Brazil
charles@ufc.br

Abstract. In this paper we consider the statistical manifold defined in terms of a deformed exponential φ. For the non-atomic case we establish a relation between the behavior of the deformed exponential function and the Δ_2-condition and analyze the comportment of the normalizing function near to the boundary of its domain. In the purely atomic case we find an equivalent condition to the behavior that characterizes the deformed exponential discussed in this work. Moreover, we prove a consequence from the fact the Musielak-Orlicz function does not satisfy the δ_2-condition.

Keywords: Deformed exponential ·
φ-Families of probability distribuitions · Musielak-Orlicz space ·
Normalizing function

1 Introduction

Consider (T, Σ, μ), a σ-finite, non-atomic measure space and $\mathcal{P}_\mu$ the set of μ-equivalent strictly positive probability densities. Let $\varphi : \mathbb{R} \to (0, \infty)$ be a deformed exponential, which is a convex function such that $\lim_{u \to -\infty} \varphi(u) = 0$ and $\lim_{u \to \infty} \varphi(u) = \infty$. The φ-families were constructed based on the replacement of the classical exponential function by a deformed exponential function [6,8], which satisfies the property that there exists a measurable function $u_0 \colon T \to (0, \infty)$ such that [11]

$$\int_T \varphi(c + \lambda u_0) d\mu < \infty, \qquad \text{for all } \lambda > 0, \tag{1}$$

F. Nielsen and F. Barbaresco (Eds.): GSI 2019, LNCS 11712, pp. 271–278, 2019.
https://doi.org/10.1007/978-3-030-26980-7_28

for each measurable function $c \colon T \to \mathbb{R}$ satisfying $\int_T \varphi(c)d\mu < \infty$. The construction of a φ-families is based on Musielak-Orlicz spaces [5].

Another statement with respect to deformed exponential that satisfies condition (1), is that two probability densities in $\mathcal{P}_\mu$ can be connected by an open arc [2,3,9,13]. This allows us to define the generalization of Rényi divergence and consequently a family of α-connections [4,7].

In the non-atomic case, it was proved in [10] that if the Musielak-Orlicz function does not satisfy the Δ_2-condition, then the boundary of the parametrization domain is non-empty. Moreover, the authors considered that condition (1) occurs and studied the behavior of the normalizing function near to the boundary of their domain. In [1], it was considered that condition (1) does not occur and analyzed the behavior of the normalizing function near to the boundary of its domain in points that are not in the Musielak-Orlicz class.

In this work, we establish a relationship between condition (1) and the Δ_2-condition, which allows us to study points on the boundary of the parametrization domain. Furthermore, continuing the discussion initiated in [1], we analyse the behavior of the normalizing function in the points that are in the Musielak-Orlicz class. More precisely, we observe that regardless of the condition (1) occurs, given a function in the Musielak-Orlicz class, we have that the normalizing function converges near the boundary of its domain. In the purely atomic case, where μ is a counting measure on the set $T = \mathbb{N}$, as in [5], the analogous of the Δ_2-condition will be denoted by δ_2-condition. In this case, we find an equivalent condition to (1) and a result related to a Musielak-Orlicz function Φ_c which does not satisfy the δ_2-condition.

This paper is organized as follows. In Sect. 2 we recall some important results about φ-families of probability distributions. In Sects. 3 and 4 we develop our main results regarding the condition described in Eq. (1) and for the purely atomic case. Finally, in Sect. 5 we state our conclusions and future perspectives for later works.

2 φ-Families of Probability Distribution: Revisiting Some Results

In this section we recall some results and we fix notations that will be important for the understanding of the text. The Musielak-Orlicz space L^{Φ_c} and Musielak-Orlicz class $\tilde{L}^{\Phi_c}$, when the Musielak-Orlicz function is $\Phi_c(t, u) = \varphi(t, c(t) + u) - \varphi(t, c(t))$ are denoted by L_c^φ and $\tilde{L}_c^\varphi$, respectively, and defined as [11]

$$L_c^\varphi = \left\{ u \in L^0 : \int_T \varphi(c + \lambda u)d\mu < \infty \text{ for each } \lambda \in (-\varepsilon, \varepsilon), \exists \ \varepsilon > 0 \right\},$$

$$\tilde{L}_c^\varphi = \left\{ u \in L^0 : \int_T \varphi(c + u)d\mu < \infty \right\},$$

are used to build the sets $B_c^\varphi = \left\{ u \in L_c^\varphi; \int_T u \varphi_+'(c)d\mu = 0 \right\}$,

$$\mathcal{K}_c^\varphi = \left\{ u \in L_c^\varphi; \int_T \varphi(c + \lambda u) < \infty \text{ for each } \lambda \in (-\varepsilon, 1 + \varepsilon), \ \exists \ \varepsilon > 0 \right\}$$

and the parametrization $\varphi_c : \mathcal{B}_c^\varphi \to \mathcal{F}_c^\varphi$, where $\varphi_c(u) = \varphi(c + u - \psi(u)u_0)$, for each $u \in \mathcal{B}_c^\varphi = B_c^\varphi \cap \mathcal{K}_c^\varphi$. The mapping $\psi : \mathcal{B}_c^\varphi \to [0, \infty)$ is called *normalizing function* and is defined in such way that $\varphi_c(u) = \varphi(c + u - \psi(u)u_0) \in \mathcal{P}_\mu$.

We say that the function Φ_c satisfies the Δ_2-condition ($\Phi_c \in \Delta_2$) if it can be found a constant $K > 0$ and a non-negative function f belonging to Musielak-Orlicz class $\tilde{L}_c^\varphi$, such that $\Phi_c(t, u) \leq K\Phi_c(t, u)$, for all $u > f(t)$ and μ-a.e. $t \in T$. In [10] was proved that if Φ_c does not satisfy the Δ_2-condition then the boundary of $\mathcal{B}_c^\varphi$ is non-empty. A function $u \in B_c^\varphi$ belongs to the boundary of $\mathcal{B}_c^\varphi$ (denoted by $\partial\mathcal{B}_c^\varphi$) if, and only if, $\int_T \varphi(c + \lambda u)d\mu < \infty$ for all $\lambda \in (0, 1)$ and $\int_T \varphi(c + \lambda u)d\mu = \infty$, for each $\lambda > 1$.

In [10], supposing that (1) is satisfied and given u, $w \in \partial\mathcal{B}_c^\varphi$ such that $\int_T \varphi(c + w)d\mu < \infty$ and $\int_T \varphi(c + u)d\mu = \infty$, then $\psi(\alpha w) \to \beta$, with $\beta \in (0, \infty)$ as $\alpha \uparrow 1$ and $\psi'_+(\alpha u) \to \infty$ as $\alpha \uparrow 1$. This last result was complemented by [1] when it was proved that given $u \in \partial\mathcal{B}_c^\varphi$ such that $\int_T \varphi(c + u)d\mu = \infty$ we have that $\lim_{\alpha\uparrow 1} \psi(\alpha u) = \infty$, as $\alpha \uparrow 1$.

Now, supposing that (1) does not occur, in [1] was shown that there exists $u \in \partial\mathcal{B}_c^\varphi$ such that $\int_T \varphi(c+u)d\mu = \infty$ but $\psi(\alpha u) \to \beta$, with $\beta \in (0, \infty)$ as $\alpha \uparrow 1$. In the next section we analyse the behavior of the normalizing function in the points of the boundary of $\mathcal{B}_c^\varphi$ such that $\int_T \varphi(c + u)d\mu < \infty$.

3   The Condition (1) on the deformed exponential and Its consequences

In this section, we notice that regardless of the occurence condition (1) if $u \in \partial\mathcal{B}_c^\varphi$ is such that $\int_T \varphi(c+u)d\mu < \infty$, then the *normalizing function* converges near to the boundary of its domain. Moreover, we relate condition (1) and condition Δ_2.

Remark 1. *Consider the deformed exponential function φ. Given $u \in \partial\mathcal{B}_c^\varphi$ such that $\int_T \varphi(c + u)d\mu < \infty$, we have that $\psi(\alpha u) \to \beta$, with $\beta \in (0, \infty)$ as $\alpha \uparrow 1$. In fact, since $\int_T \varphi(c + u)d\mu < \infty$, we have $\int_T \varphi(c + u - \lambda u_0)d\mu < \infty$ for all $\lambda > 0$. Suppose that $\psi(\alpha u) \uparrow \infty$, as $\alpha \uparrow 1$. Then, for all $A > 0$, there exists $\delta > 0$, such that $0 < 1 - \alpha < \delta \Rightarrow \psi(\alpha u) > A$. Since $\int_T \varphi(c + u - \lambda u_0)d\mu < \infty$ for all $\lambda > 0$, we have that there exists $\gamma > \lambda$, such that $\int_T \varphi(c+u-\gamma u_0)d\mu < 1$. In particular, take $A = \gamma$. Then, from of the Dominated Convergence Theorem it follows that*

$$1 = \lim_{\alpha\uparrow 1} \int_T \varphi(c + \alpha u - \psi(\alpha u)u_0)d\mu$$

$$\leq \lim_{\alpha\uparrow 1} \int_T \varphi(c + \alpha u - \gamma u_0)d\mu$$

$$= \int_T \varphi(c + u - \gamma u_0)d\mu$$

$$< 1,$$

which is an absurd. Therefore, we obtain the desired result.

From the previous remark, supposing that the deformed exponential function φ does not satisfy the condition (1), we can find $u \in \partial \mathcal{B}_c^\varphi$, such that $\int_T \varphi(c + u)d\mu < \infty$ and $\psi(\alpha u)$ converges as $\alpha \uparrow 1$.

We have the following condition that is equivalent the condition (1).

Proposition 1. *[1, Proposition 2] We say that a deformed exponential function φ and a measurable function $u_0 : T \to (0, \infty)$ satisfy the condition (1) if and only if, for some measurable function $c : T \to \mathbb{R}$ such that $\varphi(c)$ is μ-integrable, we can find constants $\overline{\lambda}, \alpha > 0$ and a non-negative function $f \in \widetilde{L}^{\Phi_c}$ such that*

$$\alpha \Phi_c(t, u) \leq \Phi_{c - \overline{\lambda} u_0}(t, u), \text{ for all } u > f(t), \tag{2}$$

where $\Phi_c(t, u) = \varphi(t, c(t) + u(t)) - \varphi(c(t))$ is a Musielak-Orlicz function.

For what follows we will define $I_{\Phi_c}(u(t)) = \int_T \Phi_c(t, |u(t)|)d\mu$ for any $u \in L^0$.

Lemma 1. *[1, Lemma 3] Consider $c : T \to [0, \infty)$ a measurable function such that $\int_T \varphi(c)d\mu < \infty$. Suppose that, for each $\lambda > 0$, we cannot find $\alpha > 0$ and $f \in \widetilde{L}^{\Phi_c}$ such that*

$$\alpha \Phi_c(t, u) \leq \Phi_{c - \lambda u_0}(t, u), \text{ for all } u > f(t). \tag{3}$$

Then we can find a strictly decreasing sequence $0 < \lambda_n \downarrow 0$, a sequence $\{u_n\}$ of measurable functions finite-value and a sequence $\{A_n\}$ of measurables sets pairwise disjoint such that

$$I_{\Phi_c}(u_n \chi_{A_n}) = 1 \text{ and } I_{\Phi_{c - \lambda_n u_0}}(u_n \chi_{A_n}) \leq 2^{-n}, \text{ for all } n \geq 1. \tag{4}$$

In the next corollary, we prove that there exists a relationship between the condition (1) and the Δ_2-condition. For this, we use the fact of that if $\Phi_c \in \Delta_2$ then $L_c^\varphi = \widetilde{L}_c^\varphi$, and consequently $I_{\Phi_c}(u) < \infty$, for every $u \in L_c^\varphi$.

Proposition 2. *If the deformed exponential function φ does not satisfiy the condition (1), then $\Phi_c \notin \Delta_2$.*

Proof. Suppose that (1) does not occur. Take $\lambda > 0$. Then, there exists a $n_0 \in \mathbb{N}$, such that $\lambda > \lambda_n$, for all $n \geq n_0$. By [1, Proposition 2] and Lemma 1, we can take $u = \sum_{n=n_0}^{\infty} u_n \chi_{A_n}$. Since $u \in L_c^\varphi$ and $I_{\Phi_c}(u) = \infty$ it follows the result.

The reciprocal of the Proposition 2 is not valid, since the exponential function satisfies the condition (1) but does not satisfy the Δ_2-condition.

Proposition 3 *Suposing that the deformed exponential φ does not satisfy (1) and that $u_0 \in E_c^\varphi$, then there exists $w \in \partial \mathcal{B}_c^\varphi$ such that $\int_T \varphi(c + w)d\mu < \infty$, $\int_T \varphi(c + w + \overline{\lambda} u_0)d\mu = \infty$, for all $\overline{\lambda} > 0$, and $\psi(\alpha w) \to \beta$, with $\beta \in (0, \infty)$, as $\alpha \uparrow 1$.*

Proof. Let $\{\lambda_n\}$, $\{u_n\}$ and $\{A_n\}$ be defined as in Lemma 1. Given $\alpha > 1$, there exists $n_1 \in \mathbb{N}$, such that $\alpha(u_n - \lambda_n u_0) > u_n$ for all $n \geq n_1$. Given any $\bar{\lambda} > 0$, there exists $n_2 \in \mathbb{N}$ such that $\bar{\lambda} > \lambda_n$ for all $n \geq n_2$. Take $n_0 = \max\{n_1, n_2\}$. Consider $B = \cup_{n=n_0}^{\infty} A_n$ and $u = \sum_{n=n_0}^{\infty}(u_n - \lambda_n u_0)\chi_{A_n}$. Then,

$$
\begin{aligned}
\int_T \varphi(c+u)d\mu &= \int_T \varphi\left(c + \sum_{n=n_0}^{\infty}(u_n - \lambda_n u_0)\chi_{A_n}\right)d\mu \\
&= \int_{T\setminus B} \varphi(c)d\mu + \sum_{n=n_0}^{\infty}\left\{\int_{A_n}\varphi(c)d\mu + I_{\Phi_c}(u_n - \lambda_n u_0)\chi_{A_n}\right\} \\
&\leq \int_T \varphi(c)d\mu + \sum_{n=n_0}^{\infty}\left\{I_{\Phi_{c-\lambda_n u_0}}(u_n\chi_{A_n})\right\} \\
&< \infty.
\end{aligned}
$$
(5)

For $\alpha \in (0,1)$ we have

$$
\int_T \varphi(c+\alpha u)d\mu \leq \alpha\int_T \varphi(c+u-\lambda u_0)d\mu + (1-\alpha)\int_T \varphi\left(c + \frac{\alpha\lambda}{1-\alpha}u_0\right)d\mu \\
< \infty
$$

and considering $\alpha > 1$ we have

$$
\begin{aligned}
\int_T \varphi(c+\alpha u)d\mu &= \int_{T\setminus B}\varphi(c)d\mu + \sum_{n=n_0}^{\infty}\int_{A_n}\varphi(c+\alpha(u_n - \lambda_n u_0))d\mu \\
&\geq \int_T \varphi(c)d\mu + \sum_{n=n_0}^{\infty}I_{\Phi_c}(u_n\chi_{A_n})d\mu \\
&= \infty.
\end{aligned}
$$

Then, since (5) occurs, it follows from Remark 1 that $\psi(\alpha u) \to \beta$, with $\beta \in (0,\infty)$, as $\alpha \uparrow 1$. Take $\lambda > 0$, such that $w = \sum_{n=n_0}^{\infty}\alpha(u_n - \lambda_n u_0) - \lambda u_0 \chi_{T\setminus B} \in B_c^{\varphi}$. Clearly, $\int_T \varphi(c+w)d\mu = \infty$, $\int_T \varphi(c+\alpha w)d\mu < \infty$ for $\alpha \in (0,1)$ and $\int_T \varphi(c+\alpha w)d\mu = \infty$ for $\alpha > 1$. Therefore, $w \in \partial B_c^{\varphi}$ and we obtain

$$
\begin{aligned}
\int_T \varphi(c+w+\bar{\lambda}u_0)d\mu &> \int_{T\setminus B}\varphi(c-\lambda u_0)d\mu + \sum_{n=n_0}^{\infty}\int_{A_n}\varphi(c+u_n - \lambda_n u_0 + \bar{\lambda}u_0)d\mu \\
&> \int_{T\setminus B}\varphi(c-\lambda u_0)d\mu + \sum_{n=n_0}^{\infty}\left(I_{\Phi_c}(u_n) - \int_{A_n}\varphi(c)d\mu\right) \\
&= \infty
\end{aligned}
$$

4   Purely Atomic Case

This section is devoted to present new results involving the condition (1) and the δ_2-condition in the purely atomic case. In this case, the integrals present in the definition of condition (1) are replaced by sums and the functions be replaced by sequences of functions.

Given $\Phi = \{\Phi_i\}$ a Musielak-Orlicz function, we define the Musielak-Orlicz class in the purely atomic case as being $\widetilde{\ell}^{\Phi} = \{u \in L^0; \sum_{i=1}^{\infty} \Phi_i(c) < \infty\}$.

In the next proposition, we find a condition equivalent to the sequence of functions $u_0 = \{u_{0,i}\}$ and a deformed exponential function φ satisfying the condition (1). In order to provide such result we will need of the following lemma.

Lemma 2. *Let Ψ and Φ be finite-valued Musielak-Orlicz functions. Then, the inclusion $\widetilde{\ell}^{\Phi} \subseteq \widetilde{\ell}^{\Psi}$ holds if, and only if, there exist ε, $\alpha > 0$ and a sequence of non-negative real numbers $f = \{f_i\} \in \widetilde{\ell}^{\Phi}$ such that*

$$\alpha\Psi_i(u) \leq \Phi_i(u), \text{ for all } u > f_i \text{ with } \Phi_i(u) < \varepsilon. \tag{6}$$

The proof of the Lemma 2 is analogous to the proof provided in [5, Theorem 8.4].

Proposition 4. *A sequence $u_0 = \{u_{0,i}\}$ and a deformed exponential function φ satisfy the condition (1) if, and only if, for some sequence $c = \{c_i\}$ of numbers such that $\sum_{i=1}^{\infty} \varphi(c_i) = 1$, we can find constants ε, λ, $\alpha > 0$ and a sequence $f = \{f_i\} \in \widetilde{\ell}^{\Phi_c}$ of non-negative real numbers such that*

$$\alpha\Phi_{c,i}(u) \leq \Phi_{c-\lambda u_{0,i}}(u), \text{ for all } u > f_i \text{ with } \Phi_{c-\lambda u_{0,i}}(u) < \varepsilon. \tag{7}$$

The proof of the Proposition 4 is analogous to the proof provided in [1, Proposition 2].

The authors in [12] have proven that if a Musielak-Orlicz function Φ does not satisfy the Δ_2-condition then we can find sequences satisfying some conditions and, from this, we obtain functions admitting the properties in (10).

The sequence $\Phi = \{\Phi_i\}$ does satisfy the δ_2-condition if, and only if, for every $\lambda \in (0,1)$, there exist constants $\varepsilon > 0$, $\alpha \in (0,1)$, and a non-negative sequence $f = \{f_i\}$ with $I_\Phi(f) < \infty$ such that

$$\alpha\Phi_i(u) \leq \Phi_i(\lambda u), \text{ for all } u > f_i \text{ with } \Phi_i(u) < \varepsilon. \tag{8}$$

Lemma 3. *Let $\Phi = \{\Phi_i\}$ be a finite-valued Musielak-Orlicz function which does not satisfy the δ_2-condition. Then we can find a strictly increasing sequence $\{\lambda_n\}$ in $(0,1)$ converging upward to 1, and sequences $\{u_n\}$ and $\{A_n\}$ of finite-valued real numbers, and pairwise disjoint sets in $\mathbb{N}$, respectively, such that*

$$1 - 2^{-n} \leq I_\Phi(u_n\chi_{A_n}) \leq 1 \text{ and } I_\Phi(\lambda_n u_n \chi_{A_n}) \leq 2^{-n}, \tag{9}$$

for all $n \geq 1$.

Proof. Suppose that the Musielak-Orlicz function Φ does not satisfy the δ_2-condition. Let $\{\lambda_n\}$ be a strictly increasing sequence in $(0,1)$ such that $\lambda_n \uparrow 1$. For each $n \geq 1$, we define the non-negative sequence $u_n = \{u_{n,i}\}$ by

$$u_{n,i} = \sup\{u > 0; 2^{-n}\Phi_i(u) \leq \Phi_i(\lambda_n u) \text{ and } \Phi_i(u) < 2^{-n}\},$$

where we adopt the convention $\sup \emptyset = 0$. Since (8) is not satisfied, we have that $I_\Phi(u_n) = \infty$ for each $n \geq 1$. Because $\Phi_i(u_{n,i}) \leq 2^{-n}$, we can find an increasing sequence $\{k_n\} \subset \mathbb{N}$ such that

$$1 - 2^{-n} \leq \sum_{i=k_{n-1}}^{k_n} \Phi_i(u_{n,i}) \leq 1.$$

The second inequality above in conjunction with $2^{-n}\Phi_i(u_{n,i}) \geq \Phi_i(\lambda_n u_{n,i})$ implies that

$$\sum_{i=k_{n-1}}^{k_n} \Phi_i(\lambda_n u_{n,i}) \leq 2^{-n}.$$

Thus, expression (9) follows with $A_n = [k_{n-1}, k_n - 1] \cap \mathbb{N}$.

Similar to what was done in [12, Remark 3.12], using Lemma 3 let $\Phi = \{\Phi_i\}$ be a finite-valued Musielak-Orlicz function does not satisfying the δ_2-condition. Then we can find functions $u_* = \sum_{n=1}^{\infty} \lambda_n u_n \chi_{A_n}$ and $u^* = \sum_{n=1}^{\infty} u_n \chi_{A_n}$ in L^Φ such that

$$\begin{cases} I_\Phi(\lambda u_*) < \infty, \text{ for } 0 \leq \lambda \leq 1, \\ I_\Phi(\lambda u_*) = \infty, \text{ for } 1 < \lambda, \end{cases} \quad \begin{cases} I_\Phi(\lambda u^*) < \infty, \text{ for } 0 \leq \lambda < 1, \\ I_\Phi(\lambda u^*) = \infty, \text{ for } 1 \leq \lambda. \end{cases} \tag{10}$$

5   Conclusion

We conclude that in the case non-atomic regardless of the condition (1) occurs, the normalizing function converges to a finite value near the boundary of its domain in the case that the functions u belong to the Musielak-Orlicz class. We prove that if the condition (1) does not occur, then the Musielak-Orlicz function does not satisfy the Δ_2-condition. Another important fact is that in the purely atomic case we find an equivalence for the occurence of condition (1) and given a Musielak-Orlicz function not satisfying the δ_2-condition, we find functions in Musielak-Orlicz space satisfying the equations in (10). The perspective for future works is to study, in the case purely atomic, the possibility of relating the condition (1) with the δ_2-condition and the behavior of the normalizing function near the boundary of $\mathcal{B}_c^\varphi$. We also want to investigate the behavior of normalizing function, considering that the deformed exponential function is not injective in all its domain.

Funding. The authors would like to thank Coordenação de Aperfeiçoamento de Pessoal de Nível Superior - Brasil (CAPES) - Finance Code 001, Conselho Nacional de Desenvolvimento Científico e Tecnológico (CNPq) (Procs. 309472/2017-2 and 408609/2016-8) and FUNCAP (Proc. IR7-00126-00037.01.00/17).

References

1. de Andrade, L.H.F., Vigelis, R.F., Vieira, F.L.J., Cavalcante, C.C.: Normalization and φ-function: definition and consequences. In: Nielsen, F., Barbaresco, F. (eds.) GSI 2017. LNCS, vol. 10589, pp. 231–238. Springer, Cham (2017). https://doi.org/10.1007/978-3-319-68445-1_27

2. Andrade, L.H.F., Vieira, F.L.J., Vigelis, R.F., Cavalcante, C.C.: Mixture and exponential arcs on generalized statistical manifold. Entropy **20**(3), 147 (2018)

3. Cena, A., Pistone, G.: Exponential statistical manifold. Ann. Inst. Statist. Math. **59**(1), 27–56 (2007)

4. de Souza, D.C., Vigelis, R.F., Cavalcante, C.C.: Geometry induced by a generalization of Rényi divergence. Entropy **18**(11), 407 (2016)

5. Musielak, J.: Modular spaces. Orlicz Spaces and Modular Spaces. LNM, vol. 1034, pp. 1–32. Springer, Heidelberg (1983). https://doi.org/10.1007/BFb0072211

6. Naudts, J.: Generalised Thermostatistics. Springer, London (2011). https://doi.org/10.1007/978-0-85729-355-8

7. Naudts, J., Zhang, J.: Rho-tau embedding and gauge freedom in information geometry. Inf. Geom. **1**(1), 79–115 (2018)

8. Naudts, J., Zhang, J.: Rho-tau embedding of statistical models. In: Nielsen, F. (ed.) Geometric Structures of Information. SCT, pp. 1–13. Springer, Cham (2019). https://doi.org/10.1007/978-3-030-02520-5_1

9. Santacroce, M., Siri, P., Trivellato, B.: On mixture and exponential connection by open arcs. In: Nielsen, F., Barbaresco, F. (eds.) GSI 2017. LNCS, vol. 10589, pp. 577–584. Springer, Cham (2017). https://doi.org/10.1007/978-3-319-68445-1_67

10. Vigelis, R.F., Cavalcante, C.C.: The Δ_2-condition and ϕ-families of probability distributions. In: Nielsen, F., Barbaresco, F. (eds.) GSI 2013. LNCS, vol. 8085, pp. 729–736. Springer, Heidelberg (2013). https://doi.org/10.1007/978-3-642-40020-9_81

11. Vigelis, R.F., Cavalcante, C.C.: On ϕ-families of probability distributions. J. Theoret. Probab. **26**(3), 870–884 (2013)

12. Vigelis, R.F., Cavalcante, C.C.: Smoothness of the Orlicz norm in Musielak-Orlicz function spaces. Math. Nachr. **287**(8–9), 1025–1041 (2014)

13. Vigelis, R.F., de Andrade, L.H.F., Cavalcante, C.C.: On the existence of paths connecting probability distributions. In: Nielsen, F., Barbaresco, F. (eds.) GSI 2017. LNCS, vol. 10589, pp. 801–808. Springer, Cham (2017). https://doi.org/10.1007/978-3-319-68445-1_92

Normalization Problems for Deformed Exponential Families

Hiroshi Matsuzoe[1(✉)], Antonio M. Scarfone[2], and Tatsuaki Wada[3]

[1] Department of Computer Science and Engineering, Graduate School of Engineering, Nagoya Institute of Technology, Gokiso-cho, Showa-ku, Nagoya 466-8555, Japan
matsuzoe@nitech.ac.jp
[2] Istituto dei Sistemi Complessi (ISC-CNR) c/o, Politecnico di Torino, Corso Duca degli Abruzzi 24, Torino 10129, Italy
[3] Region of Electrical and Electric Systems Engineering, Ibaraki University, Nakanarusawa-cho, Hitachi 316-8511, Japan

Abstract. In this study, after reviewing fundamental properties of ordinary exponential families, we consider geometry of deformed exponential families. We redefine a deformed logarithm function and a deformed exponential function. In fact, by adjusting the initial condition of deformed logarithm, we find that the α-representations in information geometry and the rectified linear unit (ReLU) belong the class of q-exponentials. Under the new definition of deformed exponential function, we discuss dually flat structures for deformed exponential families. We also consider normalization problems for those families using the ReLU.

Keywords: Deformed exponential family · Deformed logarithm · Information geometry · α-representation · Rectified linear function

1 Introduction

In anomalous statistics, observations have strong non-linear correlations, and often contain outliers. To overcome these problems, a non-standard statistical model, which is called a deformed exponential family, plays important roles. Since such a statistical model may contain heavily tailed probability distributions, it is useful for robust statistics. However, such a model is not an ordinary exponential family, standard statistical methods may not work efficiently.

In this study, after we review fundamental properties of ordinary exponential families, we redefine a deformed logarithm function and a deformed exponential function. For these definitions, a choice of initial condition is important.

H. Matsuzoe—This research was partially supported by JSPS (Japan Society for the Promotion of Science), KAKENHI (Grants-in-Aid for Scientific Research) Grant Numbers JP15K04842, JP16KT0132 and 19K03489.
T. Wada—This research was partially supported by JSPS (Japan Society for the Promotion of Science), KAKENHI (Grants-in-Aid for Scientific Research) Grant Number JP17K05341.

© Springer Nature Switzerland AG 2019
F. Nielsen and F. Barbaresco (Eds.): GSI 2019, LNCS 11712, pp. 279–287, 2019.
https://doi.org/10.1007/978-3-030-26980-7_29

In fact, by adusting the initial conditions, the α-representations in information geometry and the ReLU can be regarded as q-deformed exponential functions. Under the new definition of deformed exponential function, we discuss dually flat structures for deformed exponential families. We also find that a choice of initial condition affects the normalization of probability distribution. Therefore, we consider normalization problems for deformed exponential families.

2   Exponential Families

We start with summarizing the geometry of exponential families. Throughout this paper, we assume that the Lebesgue measure as a dominating measure since we consider normalization properties of statistical models.

Let Ω be a total sample space. Suppose that $\bar{\theta} \in \overline{\Theta} \subset \mathbf{R}^n$ is a parameter of probability distributions. We define statistical models by the following equations:

$$S_{ee} := \left\{ p(x; \bar{\theta}) \ \middle| \ p(x; \bar{\theta}) = \exp \left[C(x) + \sum_{i=1}^{n} \bar{F}_i(x) \bar{\theta}^i - \bar{\psi}(\bar{\theta}) \right] \right\}, \qquad (1)$$

$$S_{em} := \left\{ p(x; \bar{\theta}) \ \middle| \ p(x; \bar{\theta}) = \frac{1}{Z(x)} \exp \left[\sum_{i=1}^{n} \bar{F}_i(x) \bar{\theta}^i - \bar{\psi}(\bar{\theta}) \right] \right\}, \qquad (2)$$

where $C, Z, F_1, \ldots, F_n$ are functions on Ω and $\bar{\psi}$ is a function on $\overline{\Theta}$. The functions C, Z and $\bar{\psi}$ are normalizations of probability density. We say that S_{ee} is an *exponential family of e-form* or an *exponential family of exponential form*. On the other hand, we say that S_{em} is an *exponential family of m-form* or an *exponential family of mixture form*. In statistical physics, S_{ee} and S_{em} are called a *S-form* and a *Z-form*, respectively [3]. Obviously, two statistical models are equivalent if $C(x) = -\ln Z(x)$. In particular, when $C(x) = 0$ or $Z(x) = 1$, we have

$$S_e := \left\{ p(x; \theta) \ \middle| p(x; \theta) = \exp \left[\sum_{i=1}^{n} F_i(x) \theta^i - \psi(\theta) \right] \right\}. \qquad (3)$$

This statistical model is called an *exponential family of purely exponential form*.

Proposition 1. *Suppose that S_{ee} is an exponential family of e-form defined (1). Then it has a pure exponential form (3) if the normalization $C(x)$ is linearly dependent from $\bar{F}_1(x), \ldots, \bar{F}_n(x)$.*

Proof. Suppose that $C(x)$ is given by $C(x) = \sum_{i=1}^{n} c^i \bar{F}_i(x)$ for some constants $c^1, \ldots, c^n \in \mathbf{R}$. By setting $F_i(x) = (c^i + 1)\bar{F}_i(x)$ $(1 \le i \le n)$(here we do not use Einstein's summation rule), S_{ee} can be written as a pure exponential form. □

We remark the ambiguity of affine transformations for exponential families. For $p(x; \bar{\theta}) \in S_{ee}$, we change an affine coordinate system $\bar{\theta}$ to $\tilde{\theta} = A\bar{\theta} + b$ where $\bar{\theta}$ is a natural parameter in e-form (1), $A \in GL(n; \mathbf{R})$ and $b \in \mathbf{R}^n$, then the functions should be changed as follows

$$(\tilde{F}_1(x), \ldots, \tilde{F}_n(x)) = (\bar{F}_1(x), \ldots, \bar{F}_n(x))A,$$
$$\tilde{C}(x) = \bar{C}(x) + (\bar{F}_1(x), \ldots, \bar{F}_n(x))b,$$

and the probability distribution $p(x; \tilde{\theta})$ coincides with $p(x; \bar{\theta})$.

Let us consider the case of log-normal family. Suppose that S is the set of all log-normal distributions, that is, the set of probability distributions on $\Omega = \mathbf{R}_{++}$ such that

$$S = \left\{ p(x; \mu, \sigma) \,\middle|\, p(x; \mu, \sigma) = \frac{1}{\sqrt{2\pi}\sigma x} \exp\left[-\frac{(\ln x - \mu)^2}{2\sigma^2} \right] \right\},$$

where μ and σ are parameters defined on $\mu \in \mathbf{R}$ and $\sigma \in \mathbf{R}_{++}$, respectively. By setting $t = \ln x$, a log-normal distribution is reduced to an ordinary normal distribution. However, we now consider normalization functions.

By setting $\bar{C}(x) = -\ln x$, $\bar{F}_1(x) = \ln x$, $\bar{F}_2(x) = (\ln x)^2$, $\bar{\theta}^1 = \mu/\sigma^2$, $\bar{\theta}^2 = -1/2\sigma^2$ and

$$\bar{\psi}(\bar{\theta}) = \frac{\mu^2}{2\sigma^2} + \frac{1}{2}\ln(2\pi\sigma^2) = -\frac{(\bar{\theta}^1)^2}{4\bar{\theta}^2} + \frac{1}{2}\ln\left(-\frac{\pi}{\bar{\theta}^2}\right),$$

a log-normal distribution can be written by

$$p(x; \bar{\theta}) = \exp\left[-\ln x + \frac{\mu}{\sigma^2}\ln x - \frac{1}{2\sigma^2}(\ln x)^2 - \frac{\mu^2}{2\sigma^2} - \frac{1}{2}\ln(2\pi\sigma^2) \right]$$
$$= \exp\left[\bar{C}(x) + \bar{F}_1(x)\bar{\theta}^1 + \bar{F}_2(x)\bar{\theta}^2 - \bar{\psi}(\bar{\theta}) \right].$$

This implies that the log-normal family is an exponential family of e-form. The Fisher metric with respect to the natural coordinate $\{\bar{\theta}^1, \bar{\theta}^2\}$ is given by

$$g^F(\bar{\theta}) = -\frac{1}{2\bar{\theta}^2} \begin{pmatrix} 1 & -\dfrac{\bar{\theta}^1}{\bar{\theta}^2} \\ -\dfrac{\bar{\theta}^1}{\bar{\theta}^2} & \dfrac{(\bar{\theta}^1)^2 - \bar{\theta}^2}{(\bar{\theta}^2)^2} \end{pmatrix}. \tag{4}$$

Recall that a log-normal distribution is obtained from a normal distribution with a change of random variable $x \mapsto \ln x$. The Fisher metric $g^F(\bar{\theta})$ coincides with the one of normal distributions [4], since they are invariant under the choice of sufficient statistics,

On the other hand, set $F_1(x) = \ln x$, $F_2(x) = (\ln x)^2$, $\theta^1 = \mu/\sigma^2 - 1$, $\theta^2 = -1/2\sigma^2$ (i.e., $\theta^1 = \bar{\theta}^1 - 1$, $\theta^2 = \bar{\theta}^2$) and

$$\psi(\theta) = -\frac{(\theta^1 + 1)^2}{4\theta^2} + \frac{1}{2}\ln\left(-\frac{\pi}{\theta^2}\right).$$

Then we have a purely exponential form of log-normal distribution as follows

$$p(x; \theta) = \exp\left[F_1(x)\theta^1 + F_2(x)\theta^2 - \psi(\theta) \right].$$

The Fisher metric with respect to the natural coordinate $\{\theta^1, \theta^2\}$ is given by

$$g^F(\theta) = -\frac{1}{2\theta^2}\begin{pmatrix} 1 & -\dfrac{\theta^1+1}{\theta^2} \\[2mm] -\dfrac{\theta^1+1}{\theta^2} & \dfrac{(\theta^1+1)^2 - \theta^2}{(\theta^2)^2} \end{pmatrix}.$$

Of course two Fisher metrics $g^F(\bar{\theta})$ and $g^F(\theta)$ are same Riemannian metrics. The coordinate expressions are only different.

3   Deformed Exponential Families

In this section, we redefine deformed exponential functions and deformed exponential families in order to discuss normalizations of statistical models. For foundations of deformed exponential families, see [7] and [8].

Let χ be a non-decreasing function from $\mathbf{R}_{++}$ to $\mathbf{R}_{++}$. We call χ a *deformation function*. For a deformation function χ, we define a χ^c-*logarithm function* by

$$\ln_\chi^c s := \int_c^s \frac{1}{\chi(t)}\,dt,$$

where c is a constant in $\mathbf{R}_{++}$. We assume that the integral of the RHS converges. The function $\ln_\chi^c$ is also called a *shifted deformed logarithm function* or just a *deformed logarithm function* for short. The inverse of $\ln_\chi s$ is called a χ^c-*exponential function*, a *shifted deformed exponential function*, or a *deformed exponential function*, which satisfies the following differential equation:

$$\frac{d}{dx}\chi(\exp_\chi^c x) = \chi(\exp_\chi^c x)$$

with the initial condition $\exp_\chi^c(0) = c$. When $c = 1$, we omit the initial condition and we denote the deformed logarithm by $\ln_\chi s$ and the deformed exponential by $\exp_\chi x$.

If the deformation function is a power function $\chi(t) = t^q$ $(q > 0, q \neq 1)$, we obtain a q^c-*logarithm function* and a q^c-*exponential function* by

$$\ln_q^c s := \frac{s^{1-q} - c^{1-q}}{1-q}, \qquad\qquad (s > 0),$$

$$\exp_q^c t := \left(c^{1-q} + (1-q)t\right)^{\frac{1}{1-q}}, \qquad\qquad (c^{1-q} + (1-q)t > 0),$$

respectively. When $c \neq 0$, taking a limit $q \to 1$, the ordinary logarithm function $\ln(s/c)$ and the ordinary exponential function $c\exp t$ are recovered.

In particular, if $c = 0$, then the q^0-logarithm function and the q^0-exponential function are given by

$$\ln_q^0 s := \frac{1}{1-q}s^{1-q}, \qquad\qquad (s > 0),$$

$$\exp_q^0 t := ((1-q)t)^{\frac{1}{1-q}}, \qquad\qquad ((1-q)t > 0),$$

respectively. The q^0-logarithm $\ln_q^0 s$ is known as an α-*representation* or an α-*embedding* ($\alpha = 2q - 1$) in information geometry [1,9].

Since the q^c-exponential function may not be defined entire $\mathbf{R}$, we also define a *rectified q^c-exponential function* by

$$\mathrm{Rexp}_q^c x := \left[c^{1-q} + (1-q)x \right]_+^{\frac{1}{1-q}},$$

where $[*]_+$ is the cutoff function $[x]_+ := \max\{0, x\}$. In particular, if $q = c = 0$, then the rectified deformed exponential function coincides with the cutoff function itself, that is, $\mathrm{Rexp}_0^0 x = [x]_+ = \max\{0, x\}$. This function is also called the *rectified linear unit* (ReLU).

Let us define a statistical model using a sifted deformed exponential function. Suppose that $C, F_1, \ldots, F_n$ are functions on Ω. For a deformation function χ and a constant c, we define a χ^c-*exponential family* by

$$S_\chi^c = \left\{ p(x, \theta) \,\middle|\, p(x; \theta) = \exp_\chi^c \left[C(x) + \sum_{i=1}^n \theta^i F_i(x) - \psi(\theta) \right], \ \theta \in \Theta \subset \mathbf{R}^n \right\},$$

where $\theta = {}^t(\theta^1, \ldots, \theta^n)$ is a parameter, and $\psi(\theta)$ is the normalization with respect to the parameter θ. The parameter $\{\theta^1, \ldots, \theta^n\}$ is called a *natural coordinate* of S_χ. We say that a statistical model is a q^c-*exponential family* if the deformed exponential function is a q^c-exponential function.

Theorem 1 (discrete distributions). *Suppose that Ω is the finite sample space $\Omega = \{x_0, x_1, \ldots, x_n\}$. Then a statistical model on Ω is defined by*

$$S_n = \left\{ p(x, \eta) \,\middle|\, \eta_i > 0, \sum_{i=0}^n \eta_i = 1, \ p(x; \eta) = \sum_{i=0}^n \eta_i \delta_i(x) \right\}, \tag{5}$$

where $\delta_i(x)$ equals one if $x = i$ and zero otherwise. Then S_n is a q^c-exponential family for arbitrary q and c.

Proof. The proof is quite same as in the ordinary exponential case [1]. By setting $\theta^i = \ln_q^c \eta_i - \ln_q^c \eta_0 = ((\eta_i)^{1-q} - (\eta_0)^{1-q})/(1-q)$, and $\psi(\theta) = -\ln_q^c \eta_0$, we have

$$\ln_q^c p(x; \theta(\eta)) = \frac{1}{1-q} \left\{ p^{1-q}(x; \eta) - c^{1-q} \right\} = \frac{1}{1-q} \left\{ \sum_{i=0}^n (\eta_i)^{1-q} \delta_i(x) - c^{1-q} \right\}$$

$$= \frac{1}{1-q} \left\{ \sum_{i=0}^n \left((\eta_i)^{1-q} - (\eta_0)^{1-q} \right) \delta_i(x) + (\eta_0)^{1-q} - c^{1-q} \right\}$$

$$= \sum_{i=1}^n \theta^i \delta_i(x) - \psi(\theta).$$

This implies that the set of discrete distributions is a q^c-exponential family. □

In the q-exponential case, a q^c-exponential distribution can be written as a q-exponential distribution. That is, the following relation holds

$$
p(x;\theta) = \exp_q^c\left[C(x) + \sum_{i=1}^n F_i(x)\theta^i - \psi(\theta)\right]
$$
$$
= \exp_q\left[\ln_q c + C(x) + \sum_{i=1}^n F_i(x)\theta^i - \psi(\theta)\right].
$$

This property also holds for another deformed exponential distribution.

In order to discuss geometric structures on deformed exponential family, let us recall escort expectations.

Let χ be a deformation function. For $p(x;\theta) \in S_\chi^c$ and a function $f(x)$ on Ω, we define the *normalized escort expectation* of $f(x)$ with respect to $p(x;\theta)$ by

$$
E_{\chi,p}^{esc}[f(x)] := \int_\Omega f(x)\frac{\chi\{p(x;\theta)\}}{Z_\chi(p)}dx,
$$

where $Z_\chi(p)$ is the normalization defined by

$$
Z_\chi(p) := \int_\Omega \chi\{p(x;\theta)\}dx.
$$

Assuming that an integration and a differentiation are interchangeable, we obtain

$$
E_{\chi,p}^{esc}\left[\frac{\partial}{\partial\theta^i}\ln_\chi^c p(x;\theta)\right] = 0. \tag{6}
$$

Hence $\partial/\partial\theta^i \ln_\chi^c p(x;\theta)$ is regarded as a generalization of a score function on S_χ^c (cf. [5] and [7]). In this paper, we only discuss the normalized first escort expectation. For further discussions of escort expectations, see [6].

Now we consider geometric structures on S_χ^c. Suppose that the normalization ψ is strictly convex. Then we can define a χ-*Fisher metric* g^χ and a χ-*cubic form* C^χ by

$$
g_{ij}^\chi(\theta) := \partial_i\partial_j\psi(\theta),
$$
$$
C_{ijk}^\chi(\theta) := \partial_i\partial_j\partial_k\psi(\theta),
$$

respectively. In Hessian geometry [10], the normalization ψ is called the *potential* of g^χ and C^χ with respect to $\{\theta^i\}$.

Denote by $\nabla^{\chi(0)}$ be the Levi-Civita connection with respect to g^χ. For a fixed $\alpha \in \mathbf{R}$, we define an affine connection $\nabla^{\chi(\alpha)}$ by

$$
g^\chi(\nabla_X^{\chi(\alpha)}Y, Z) := g^\chi(\nabla_X^{\chi(0)}Y, Z) - \frac{\alpha}{2}C^\chi(X,Y,Z).
$$

In particular, we denote $\nabla^{\chi(e)} = \nabla^{\chi(1)}$ and $\nabla^{\chi(m)} = \nabla^{\chi(-1)}$. Then we have the following theorem, which generalizes the theorems in [2] and [7].

Theorem 2. *For a deformed exponential family S_χ^c, the following hold:*

1. $(S_\chi^c, \nabla^{\chi(e)}, g^\chi)$ *and* $(S_\chi^c, \nabla^{\chi(m)}, g^\chi)$ *are mutually dual Hessian manifolds, that is,* $(S_\chi^c, g^\chi, \nabla^{\chi(e)}, \nabla^{\chi(m)})$ *is a dually flat space.*
2. $\{\theta^i\}$ *is a* $\nabla^{\chi(e)}$*-affine coordinate system on* S_χ^c.
3. $\psi(\theta)$ *is the potential of* g^χ *and* C^χ *with respect to* $\{\theta^i\}$.
4. *Set* $\eta_i := E_{\chi,p}^{\mathrm{esc}}[F_i(x)]$. *Then* $\{\eta_i\}$ *is a* $\nabla^{\chi(m)}$*-affine coordinate system on* S_χ^c *and the dual of* $\{\theta^i\}$ *with respect to* g^χ, *that is,* $g^\chi(\partial/\partial\theta^i, \partial/\partial\eta_j) = \delta_j^i$ *holds.*
5. *Set* $\phi(\eta) := E_{\chi,p}^{\mathrm{esc}}[\log_\chi^c p(x;\theta) - C(x)]$. *Then* $\phi(\eta)$ *is the potential of* g^χ *with respect to* $\{\eta_i\}$.

Proof. The proof is quite same as [7]. Statements 1, 2 and 3 are obtained from a well-known arguments of Hessian geometry [10] (See also [7]). Since $\eta_i = E_{\chi,p}^{\mathrm{esc}}[F_i(x)]$, we obtain

$$E_{\chi,p}^{\mathrm{esc}}[\log_\chi^c p(x;\theta) - C(x)] = E_{\chi,p}^{\mathrm{esc}}\left[\sum_{i=1}^n \theta^i F_i(x) - \psi(\theta)\right] = \sum_{i=1}^n \theta^i \eta_i - \psi(\theta).$$

From the unbiasedness of generalized score function (6), using the Legendre transformation on a dually flat space [1], we obtain the Statements 4 and 5. □

4   Normalization Problems

In this section, we discuss how to normalize a given function to a probability distribution. Such a normalization is not a trivial problem.

Let Ω be a sample space, and Θ be a parameter space. Suppose that F is a function on $\Omega \times \Theta$ and f is a function on the image $F(\Omega \times \Theta)$. We say that $f(F_\theta(x))$ is *e-normalizable* or *exponential normalizable* if there exists a function ψ on Θ such that $f(F_\theta(x) - \psi(\theta)) \geq 0$ and

$$\int_\Omega f(F_\theta(x) - \psi(\theta))dx = 1. \tag{7}$$

On the other hand, we say that $f(F_\theta(x))$ is *m-normalizable* or *mixture normalizable* if $f(F_\theta(x)) \geq 0$ and

$$0 < \int_\Omega f(F_\theta(x))dx < \infty. \tag{8}$$

From (7) and (8), the functions

$$f(F_\theta(x) - \psi(\theta)), \quad \text{and} \quad \frac{f(F_\theta(x))}{\int_\Omega f(F_\theta(x))dx}$$

can be regarded as probability density functions on Ω, respectively. If f is the ordinary exponential function, then two normalizations are equivalent. However, they are different in general.

Let us consider the difference to these two normalizations using the rectified linear function $\mathrm{Rexp}_0^0 x$ as an extreme case.

Suppose that $\Omega = \{x_0, \ldots, x_n\}$ is a finite sample space, and $S_n = \{p(x; \eta)\}$ is the set of all probability distributions on Ω defined by (5). From Theorem 1, S_n is a deformed exponential family.

By setting $\theta^i = \ln_0^0 p(x_i) - \ln_0^0 p(x_0) = \eta_i - \eta_0$, we obtain

$$\ln_0^0 p(x; \theta(\eta)) = \sum_{i=1}^{n} (\eta_i - \eta_0) \delta_i(x) + \eta_0 \;=\; \begin{cases} \eta_i & (x = x_i), \\ 1 - \displaystyle\sum_{i=1}^{n} \eta_i & (x = x_0). \end{cases}$$

Hence the e-normalization of $p(x; \eta)$ with the rectified linear function determines a natural coordinate, and it is a parallel transport of a local coordinate from $\{\eta_1, \ldots, \eta_n\}$ to $\{\theta^1, \ldots, \theta^n\} = \{\eta_1 - \eta_0, \ldots, \eta_n - \eta_0\}$.

On the other hand, we set $\Omega' = \{x \in \Omega \mid p(x_i; \eta) > p(x_0; \eta)\}$ and

$$f_\eta(x) = \mathrm{Rexp}_0^0 \left(\sum_{i=1}^{n} \theta^i \delta_i(x) \right) \;=\; \begin{cases} \eta_i - \eta_0 & (\, x_i \in \Omega' \,), \\ 0 & (\, x_i \notin \Omega' \,). \end{cases}$$

Therefore the m-normalization of $p(x; \eta)$ is given by

$$\bar{p}_m(x; \eta) = \frac{\sum_{i=1}^{n} \mathrm{Rexp}_0^0(\theta^i) \delta_i(x)}{\sum_{w \in \Omega'} \theta^i \delta_i(w)}.$$

This normalization is that all the probabilities smaller than $p(x_0; \eta) = \eta_0$ make zero, that is, $\bar{p}_m(x_i; \eta) = 0$ if $x_i \in \Omega \backslash \Omega'$. This normalization is meaningful if we reduce parameters of statistical models.

5   Discussions

In any case, the procedure of normalization of a non-probability distribution to a probability distribution is not a trivial problem. It affects the geometric structure of the statistical model. It is necessary to discuss what is a good normalization depending on the given problem.

References

1. Amari, S., Nagaoka, H.: Method of Information Geometry. Amer. Math. Soc. Providence, Oxford University Press, Oxford (2000)
2. Amari, S., Ohara, A., Matsuzoe, H.: Geometry of deformed exponential families: invariant, dually-flat and conformal geometry. Phys. A **391**, 4308–4319 (2012)
3. Bashkirov, A.G.: On maximum entropy principle, superstatistics, power-law distribution and Renyi parameter. Phys. A **340**, 153–162 (2004)
4. Lauritzen, S.L.: Statistical manifolds. In: Differential Geometry in Statistical Inferences, IMS Lecture Notes Monograph Series 10, Institute of Mathematical Statistics. Hayward California, pp. 96–163 (1987)

5. Matsuzoe, H.: Statistical manifolds and geometry of estimating functions. In: Prospects of Differential Geometry and its Related Fields, pp. 187–202. World Sci. Publ. (2013)
6. Matsuzoe, H.: A sequence of escort distributions and generalizations of expectations on q-exponential family. Entropy **19**(1), 7 (2017)
7. Matsuzoe, H., Henmi, M.: Hessian structures and divergence functions on deformed exponential families. In: Nielsen, F. (ed.) Geometric Theory of Information. SCT, pp. 57–80. Springer, Cham (2014). https://doi.org/10.1007/978-3-319-05317-2_3
8. Naudts, J.: Generalised Thermostatistics. Springer-Verlag, London (2011)
9. Naudts, J., Zhang, J.: Rho-tau embedding and gauge freedom in information geometry. Inf. Geom. **1**, 79–115 (2018)
10. Shima, H.: The Geometry of Hessian Structures. World Scientific, Singapore (2007)

New Geometry of Parametric Statistical Models

Jun Zhang$^{(\boxtimes)}$ and Gabriel Khan

University of Michigan, Ann Arbor, MI 48109, USA
junz@umich.edu

Abstract. We provide an alternative differential geometric framework of the manifold $\mathbb{M}$ of parametric statistical models. While adopting the Fisher-Rao metric as the Riemannian metric g on $\mathbb{M}$, we treat the original parameterization of the statistical model as affine coordinate chart on the manifold endowed with a flat connection, instead of using a pair of torsion-free affine connections with generally non-vanishing curvature. We then construct its g-conjugate connection which, while necessarily curvature-free, carries torsion in general. So instead of associating a statistical structure to $\mathbb{M}$, we construct a statistical manifold admitting torsion (SMAT). We show that $\mathbb{M}$ is dually flat if and only if torsion of the conjugate connection vanishes.

Keywords: Torsion · Weitzenböck connection · Hessian manifold

1   Introduction

Recall that in the now-classic information geometry, a parametric family of density functions, $p(\cdot|x)$, called a *parametric statistical model*, is the association $x \mapsto p(\cdot|x)$ of a point $x = [x^1, \cdots, x^n]$ in a connected open subset of $\mathbb{R}^n$ to p, such that x serves as a local coordinate chart of $p \in \mathbb{M}$ [Ama85, AN00]. The Fisher-Rao metric and the α-connections are given by

$$g_{ij}(x) = \int_{\Omega} d\omega \left\{ p(\omega|x) \frac{\partial \log p(\omega|x)}{\partial x^i} \frac{\partial \log p(\omega|x)}{\partial x^j} \right\};$$

$$\Gamma_{ij,k}^{(\alpha)}(x) = \int_{\Omega} d\omega \frac{\partial p(\omega|x)}{\partial x^k} \left(\frac{1-\alpha}{2} \frac{\partial \log p(\omega|x)}{\partial x^i} \frac{\partial \log p(\omega|x)}{\partial x^j} + \frac{\partial^2 \log p(\omega|x)}{\partial x^i \partial x^j} \right).$$

The α- and $(-\alpha)$-connection are conjugate to each other with respect to the Fisher-Rao metric g. Note that all α-connections are torsion-free; yet generally they have non-zero curvatures, with curvature of $(\pm\alpha)$-connections equal but opposite sign of each other. When the curvatures of (± 1)-connections vanish, g takes the form of a Hessian metric. It is important to keep in mind that each member of the α-connection is Codazzi-coupled to the Fisher-Rao metric g.

The project is supported by DARPA/ARO Grant W911NF-16-1-0383 ("Information Geometry: Geometrization of Science of Information", PI: Zhang).

© Springer Nature Switzerland AG 2019
F. Nielsen and F. Barbaresco (Eds.): GSI 2019, LNCS 11712, pp. 288–296, 2019.
https://doi.org/10.1007/978-3-030-26980-7_30

In this paper, we take a different perspective about the manifold $\mathbb{M}$ of parametric statistical models $p(\cdot|x)$. We take the parameter x to be the local coordinates of a parallelizable manifold after trivialization of its tangent bundle $T\mathbb{M}$, i.e, x is taken to be the affine coordinates of a flat connection ∇^* on $\mathbb{M}$. We continue to take Fisher-Rao metric g as the Riemannian metric on $\mathbb{M}$. Denote ∇ to be the g-conjugate of this flat connection ∇^*. Though ∇ is necessarily curvature-free, in general, ∇ will not be torsion-free. This connection is adapted to the g-conjugate frame, and we call it "pseudo-Weitzenböck connection." In the literature, a manifold $(\mathbb{M}, g, \nabla, \nabla^*)$ for which ∇^* is flat is called a "statistical manifold admitting torsion" or SMAT [Kur07,HM11], and ∇ and g are coupled by

$$(\nabla_Z g)(X,Y) - (\nabla_X g)(Z,Y) = g(T^\nabla(Z,X),Y).$$

Below, we actually describe parametric statistical model as SMAT by constructing *the* biorthogonal frame B based on g being the Fisher-Rao metric. ∇ is torsion-free, and hence becomes "flat", if and only if g is Hessian. For more details including proofs, see [ZK19].

2 Theoretical Foundation

2.1 g-Conjugate Connection

We recall that given any connection ∇ and an arbitrary Riemannian metric g, the g-conjugate connection ∇^* is defined as the (unique) connection that jointly preserves g with ∇:

$$Z(g(X,Y)) = g(\nabla_Z X, Y) + g(X, \nabla_Z^* Y), \tag{1}$$

where X, Y, Z are all vector fields on $\mathbb{M}$.

The curvature and torsion of conjugate connections ∇ and ∇^* are related:

(i) their curvature tensors R^∇, R^{∇^*} satisfy

$$g(R^\nabla(Z,W)X,Y) + g(R^{\nabla^*}(Z,W)Y,X) \equiv 0; \tag{2}$$

(ii) their torsion tensors T^∇, T^{∇^*} satisfy

$$g(T^{\nabla^*}(Z,X) - T^\nabla(Z,X),Y) \equiv (\nabla_Z g)(X,Y) - (\nabla_X g)(Z,Y). \tag{3}$$

A consequence of (2) is that if ∇ is curvature-free, then so is ∇^*. The consequence of (3) is that ∇ and ∇^* carry the same amount of torsion if and only if

$$(\nabla_Z g)(X,Y) = (\nabla_X g)(Z,Y),$$

which is known as the "Codazzi coupling" of (g, ∇). It is easily verified that (g, ∇) is Codazzi-coupled if and only if (g, ∇^*) is Codazzi-coupled. Both (2) and (3) are well-known facts in information geometry. A connection is called *flat* when it is both curvature-free and torsion-free. A manifold is called *dually flat* when it carries two flat connections ∇ and ∇^* that form a conjugate pair with respect to the (necessarily) Hessian metric constructed from either ∇ or ∇^*.

2.2 Connection Adapted to a Frame

Let us start by defining a local frame on a parallelizable manifold $\mathbb{M}$. A frame $\mathfrak{B} = \{\mathbf{b}_1, \cdots, \mathbf{b}_n\}$ with $n = \dim(\mathbb{M})$ is a collection of n locally linearly independent vector fields $\{\mathbf{b}_i\}_{i=1}^n$ on $\mathbb{M}$. Under local coordinate system $x = \{x^i\}_{i=1}^n$, the expression of a frame $\mathfrak{B}$ is $\mathbf{b}_i = B_i^j \partial_{x^j}$, where ∂_{x^j} is the shorthand for $\partial/\partial x^j$, and B_i^j is an $n \times n$ matrix, assumed to be of full rank and hence invertible:

$$(B^{-1})_l^i B_j^l = \delta_j^i = B_l^i (B^{-1})_j^l.$$

Here, and in the rest of the paper, B^{-1} denotes the matrix inverse of B_j^i, and Einstein summation notation is in effect.

When the B-matrix is taken to be the Jacobian matrix of coordinate transform: $x \longrightarrow y$

$$(B^{-1})_j^\alpha = \frac{\partial y^\alpha}{\partial x^j} \longleftrightarrow B_\alpha^j = \frac{\partial x^j}{\partial y^\alpha}, \tag{4}$$

then the frame $\{\mathbf{b}_i\}_{i=1}^n$ forms a coordinate frame:

$$\mathbf{b}_i = \frac{\partial x^\alpha}{\partial y^i} \frac{\partial}{\partial x^\alpha} = \frac{\partial}{\partial y^i} := \partial_{y^i}.$$

The necessary and sufficient condition for (4) is

$$\partial_{x^i}(B^{-1})_j^\alpha = \partial_{x^j}(B^{-1})_i^\alpha. \tag{5}$$

Necessity is obvious. As for sufficiency, note that when Eq. 5 is satisfied, then for each α there exists a function $y^\alpha = y^\alpha(x)$ such that

$$(B^{-1})_j^\alpha = \frac{\partial y^\alpha}{\partial x^j}.$$

Definition 1 (Adapted connection). *Given any frame $\mathfrak{B}$, the adapted connection $\nabla^{\mathfrak{B}}$ is defined by $\nabla^{\mathfrak{B}} = B\,\partial(B^{-1})$ or in component forms:*

$$\Gamma_{k\alpha}^\beta = B_j^\beta (\partial_{x^\alpha}(B^{-1})_k^j) = -(B^{-1})_k^j (\partial_{x^\alpha} B_j^\beta). \tag{6}$$

$\nabla^{\mathfrak{B}}$ as constructed is known as the "connection of parallelization" [BG80], since they always exist on a parallelizable manifold after trivialization of its tangent bundle with a global frame $\mathfrak{B}$. The following is well-known [BG80, p.223].

Proposition 1. *Given a frame $\mathfrak{B} = \{\mathbf{b}_1, \cdots, \mathbf{b}_n\}$, then*

(i) $\nabla_{\mathbf{b}_i}^{\mathfrak{B}} \mathbf{b}_j \equiv 0, \quad \forall i, j;$
(ii) $R^{\nabla^{\mathfrak{B}}} = 0 ;$
(iii) $T^{\nabla^{\mathfrak{B}}} = 0$ *iff $\mathfrak{B}$ is a coordinate frame, i.e., $[\mathbf{b}_i, \mathbf{b}_j] = 0$.*

Definition 2 (g-Biorthogonal frame). *Given any frame $\mathfrak{B} = \{\mathbf{b}_i\}_{i=1}^n$, the g-biorthogonal frame is defined as the (unique) frame $\mathfrak{B}^\star = \{\mathbf{b}_i^\star\}_{i=1}^n$ that is biorthogonal with respect to the given g:*

$$g(\mathbf{b}_i, \mathbf{b}_j^\star) \equiv \delta_{ij}.$$

We have the following nice property

Theorem 3 *[ZK19, Theorem 10]. With respect to any Riemannian metric g, the g-conjugation of a connection induced by a frame $\mathfrak{B}$ equals the connection indued by the g-biorthogonal frame $\mathfrak{B}^\star$:*

$$\left(\nabla^{\mathfrak{B}}\right)^* = \nabla^{(\mathfrak{B}^\star)}.$$

Historically, an affine connection adapted to an orthonormal frame is called the *Weitzenböck connection*, and has been used in theoretical physics to describe an alternative theory to Einstein's general relativity. Here this construction is extended to an arbitrary frame, and hence the terminology "pseudo-Weitzenböck connections." Theorem 3 shows that the notion of biorthogonal frames is compatible with the notion of conjugate connections when the pair of connections are both adapted connections.

2.3 Dually Flat Versus Partially-Flat Manifolds

Recall that the Hessian operator (second derivative) on a function Φ on a manifold is a bilinear form sometimes denoted as $(\nabla d\Phi)(X, Y)$. Operating on the coordinate base $(X = \partial_{x^i}, Y = \partial_{x^j})$ it takes the form

$$Hess_\nabla(\Phi)(\partial_{x^i}, \partial_{x^j}) = \frac{\partial^2 \Phi}{\partial x^i \partial x^j} - \Gamma_{ij}^k \frac{\partial \Phi}{\partial x^k}.$$

Torsion-freeness of ∇ is reflected as $\Gamma_{ij}^k = \Gamma_{ji}^k$. When ∇ is further curvature-free (and hence ∇ is flat), $\Gamma_{ij}^k = 0$ using x as affine coordinates, so that

$$Hess_\nabla(\Phi)(\partial_{x^i}, \partial_{x^j}) = \frac{\partial^2 \Phi}{\partial x^i \partial x^j}.$$

It is established in Zhang and Khan [ZK19] that

Proposition 2 *[ZK19, Theorem 3]. Given a torsion-free connection ∇ and a smooth function Φ on a manifold, then $(\nabla, Hess_\nabla(\Phi))$ is Codazzi coupled iff $d\Phi(R^\nabla) = 0$.*

A consequence is that any flat connection ∇ is always Codazzi coupled to $Hess_\nabla(\Phi)$, as [Shi07] observed. Denote ∇^* the conjugate connection with respect to the symmetric bilinear form $Hess_\nabla(\Phi)$ induced from a flat ∇. Then ∇^* is also flat (both curvature- and torsion-free). Assuming Φ is convex, then we have the standard Hessian manifold with

$$g = Hess_\nabla(\Phi) = Hess_\nabla^*(\Phi^*),$$

where Φ^* is the convex conjugate function of Φ.

The above analysis also tells us that given $(\mathbb{M}, g, \nabla)$ with a flat connection ∇, then whenever $g \neq Hess_\nabla(\Phi)$, then (∇, g) is in general *not* a Codazzi pair, as [Shi07] pointed out, so the g-conjugate connection ∇^* is *not* torsion-free. This is the situation of the so-called "partially-flat" manifold [Hen17]. Next, we apply this concept to the manifold of parametric statistical models.

3   Parametric Statistical Models as Partially-Flat Geometry

3.1   Riemannian Manifold of Parametric Statistical Models

Take $x = [x_1, \cdots, x_n]$, the parameter of a parametric statistical model $p(\cdot|x)$, to be the affine coordinates on a parallelizable manifold with flat connection ∇^*, i.e., the Christoffel symbol Γ_{jk}^{*i} vanishes. Writing out the equation of conjugate connections ∇, ∇^* under this coordinate chart

$$\frac{\partial g_{ij}}{\partial x^k} = g_{lj}\Gamma_{ki}^l + g_{il}\Gamma_{kj}^{*l} = g_{lj}\Gamma_{ki}^l \,.$$

Therefore, the pseudo-Weitzenböck connection ∇ of the parametric statistical model is (written as its Christoffel symbol Γ_{jk}^i)

$$\Gamma_{ki}^j = g^{jl}\frac{\partial g_{il}}{\partial x^k} \,, \tag{7}$$

with g^{ij} denoting the elements of the matrix inverse of g, the Fisher-Rao metric. It is well-known [BG80] that such a connection is always curvature-free., but carries torsion

$$T_{ik}^j = g^{jl}\left(\frac{\partial g_{kl}}{\partial x^i} - \frac{\partial g_{il}}{\partial x^k}\right).$$

In general, $T \neq 0$, unless $\frac{\partial g_{il}}{\partial x^k}$ is totally symmetric, i.e., g is Hessian. Otherwise, from any connection ∇ with torsion T^∇, we can construct a torsion-free connection $\nabla - \frac{1}{2}T^\nabla$; in the present case,

$$\Gamma_{ki}^j - \frac{1}{2}T_{ki}^j = \frac{g^{jl}}{2}\left(\frac{\partial g_{kl}}{\partial x^i} + \frac{\partial g_{il}}{\partial x^k}\right)$$

is always torson-free, and differs from the Levi-Civita connection of g by $\frac{1}{2}g^{jl}\partial_{x^l}g_{ik}$.

Even though Γ_{ki}^j given by (7) may carry torsion, its geodesic equation

$$\frac{d^2x^j}{ds^2} + g^{jl}\frac{\partial g_{il}}{\partial x^k}\frac{dx^i}{ds}\frac{dx^k}{ds} = 0$$

or equivalently

$$\frac{d}{ds}\left(g_{ij}\frac{dx^j}{ds}\right) = 0$$

still yields the same solution as given by

$$g_{ij}\frac{dx^j}{ds} = const, \qquad i = 1, 2, \cdots, n.$$

Torsion of Γ_{ki}^j is *not* captured in the geodesic curves themselves; it describes the "screw" component of the motion with axis of rotation precisely the tangent direction of the curve. When two connections differ only by torsion, then their associated geodesic equations are the same, since the anti-symmetric part of Γ is canceled after summation with $\frac{dx^i}{ds}\frac{dx^k}{ds}$.

The associated frame, which we call "canonical frame" of $\mathbb{M}$ and denote by upper script $\{\mathbf{b}^i\}_{i=1}^n$, is

$$\mathfrak{B} = \{\mathbf{b}^i\}_{i=1}^n = \{g^{ij}\partial_{x^j}, \ i = 1, 2, \cdots, n\}.$$

This frame is nothing but the "natural gradient" vector popularly known to the machine learning community after Amari [Ama98].

3.2   Pre-contrast Function and α-Connections

Just as a statistical structure may be induced by a contrast function, a SMAT may be induced by a pre-contrast function ρ [HM11] which, in the partially-flat case, has a canonical expression [Hen17]. We show that

Proposition 3. *The canonical pre-contrast function* $\mathbb{M} \times T\mathbb{M} \to \mathbb{R}$ *is*

$$\rho(\partial_{x^i}, x, x') = -g(\partial_{x^i}, (x'^j - x^j)\partial_{x^j}) = (x^j - x'^j)\, g_{ij}(x).$$

This can be seen from

$$-\left.\frac{\partial \rho}{\partial x'^j}\right|_{x'=x} = g_{ij}(x),$$

$$-\left.\frac{\partial^2 \rho}{\partial x'^k \partial x'^j}\right|_{x'=x} = 0,$$

$$-\left.\frac{\partial \rho}{\partial x^k \partial x'^j}\right|_{x'=x} = \frac{\partial g_{ij}}{\partial x^k} = \Gamma_{ki,j}.$$

where the canonical connection Γ carries torsion, $\Gamma_{ki,j} \neq \Gamma_{ik,j}$, in general.

The family of α-connections, $\widetilde{\nabla}^{(\alpha)} = \frac{1+\alpha}{2}\nabla + \frac{1-\alpha}{2}\nabla^* = \frac{1+\alpha}{2}\nabla$ all carry torsion (except $\alpha = -1$)

$$\widetilde{\Gamma}_{ki,j}^{(\alpha)}(x) = \frac{1+\alpha}{2}\left(\int_\Omega d\omega \left\{\frac{\partial^2 \log p(\omega|x)}{\partial x^k \partial x^i}\frac{\partial p(\omega|x)}{\partial x^j} + \frac{\partial^2 \log p(\omega|x)}{\partial x^k \partial x^j}\frac{\partial p(\omega|x)}{\partial x^i}\right\}\right.$$
$$\left. + \int_\Omega d\omega\, p(\omega|x) \frac{\partial \log p(\omega|x)}{\partial x^i}\frac{\partial \log p(\omega|x)}{\partial x^j}\frac{\partial \log p(\omega|x)}{\partial x^k}\right),$$

with torsion given by

$$\widetilde{T}_{ik}^{(\alpha)j} = \frac{1+\alpha}{2}\, g^{jl}\int_\Omega d\omega \left\{\frac{\partial^2 \log p(\omega|x)}{\partial x^i \partial x^l}\frac{\partial p(\omega|x)}{\partial x^k} - \frac{\partial^2 \log p(\omega|x)}{\partial x^k \partial x^l}\frac{\partial p(\omega|x)}{\partial x^i}\right\}.$$

3.3 Univariate Normal Distribution: An Example

We consider the univariate normal family on the real line $(-\infty < \omega < \infty)$

$$\mathcal{N}(\omega|\mu,\sigma) = \frac{1}{\sqrt{2\pi\sigma^2}} \exp\left(-\frac{(\omega-\mu)^2}{2\sigma^2}\right)$$

with parameters $m = \sqrt{2}\,\mu$ and σ (the factor of $\sqrt{2}$ is for later convenience). Reparametrizing, it is possible to consider $\mathcal{N}(\omega|\mu,\sigma)$ as an exponential family with natural coordinates $x = (x^1, x^2)$ and expectation coordinates $u = (u_1, u_2)$:

$$x^1 = \frac{\mu}{\sigma^2}, \qquad x^2 = -\frac{1}{2\sigma^2};$$
$$u_1 = \mu, \qquad u_2 = \mu^2 + \sigma^2.$$

When treating x (or u) as affine coordinates for the dually flat connections, the Fisher-Rao metric g becomes the Hessian metric with potential Φ

$$\Phi(x) = -\frac{x^1 \cdot x^1}{4x^2} + \frac{1}{2}\log\left(-\frac{\pi}{x^2}\right).$$

As the mean μ and variance σ parameters of the univariate normal model are intrinsically meaningful in statistics, it is desirable to treat (μ,σ) as affine coordinates for some flat connection. As such, if we consider the coordinate frame

$$\left\{\frac{\partial}{\partial m}, \frac{\partial}{\partial \sigma}\right\},$$

its biorthogonal frame with respect to the Fisher-Rao metric

$$g = \frac{2}{\sigma^2}(dm^2 + d\sigma^2)$$

is

$$\left\{\frac{\sigma^2}{2}\frac{\partial}{\partial m}, \frac{\sigma^2}{2}\frac{\partial}{\partial \sigma}\right\}.$$

By computing the Lie bracket of the biorthogonal frame, we find that

$$\left[\frac{\sigma^2}{2}\frac{\partial}{\partial m}, \frac{\sigma^2}{2}\frac{\partial}{\partial \sigma}\right] = -\frac{\sigma^3}{2}\frac{\partial}{\partial m}.$$

This is the torsion of the pseudo-Weitzenböck connection adapted to the g-biorthogonal frame. It is not a coordinate frame and the torsion of the g-conjugate connection is non-zero.

Note that the Fisher-Rao metric, when expressed in the (m,σ)-coordinates, is *not* Hessian. The pseudo-Weitzenböck connection derived above has geodesics which are reparametrizations of straight lines in the upper half-plane. This fact does not hold in general, but turns out in the present case because the Fisher-Rao metric, though not Hessian, is so simple for our choice of parametrization.

To summarize, we have constructed a presentation of the univariate normal family (as a manifold of upper half-plane), not as a manifold of dual flatness (Hessian manifold) in the conventionally-adopted natural and expectation coordinates, but as a partially-flat statistical manifold admitting torsion (SMAT) in the original (m,σ)-coordinates.

4   Discussions

Classical information geometry involves statistical manifolds, with two equivalent definitions as follows:

(i) Lauritzen's [Lau87] viewpoint: $(\mathbb{M}, g, \nabla, \nabla^*)$ where the pair of g-conjugated connections ∇ and ∇^* are both torsion-free;

(ii) Kurose's [Kur90] viewpoint: $(\mathbb{M}, g, \nabla)$ where ∇ is torsion-free and Codazzi-coupled to g.

With application to parametric statistical models, the Riemannian metric is the Fisher-Rao metric and the pair of conjugate connections are the (± 1)-connections, generated by divergence (contrast) functions. These are "canonical" objects once the parametric statistical model $p(\cdot|x)$ is specified, canonical because they are unique second- and third-order invariants for parametric statistical models (see [Dow18] and [AJVLS15]). Here we provide another "canonical" construction of a parametric statistical model as a parallelizable manifold with a "partially-flat" geometry [Hen17] under which both conjugate connections are curvature-free. A partially-flat structure (of a parallelizable manifold) is a slightest relaxation to the dually flat Hessian structure, by allowing one of the connections (say, ∇^*) to be torsion-free. The metric g need not be Hessian, nor is the flat connection required to be Codazzi coupled to g. In other words, our construction of this manifold $(\mathbb{M}, g, \nabla, \nabla^*)$ is such that ∇^* is flat and ∇ is curvature-free but usually carries torsion, while g is still the Fisher-Rao metric. This is a special case of a statistical manifold admitting torsion (SMAT, [Kur07]) that can be generated by "pre-contrast functions" [HM11]. Compared to statistical manifold $(\mathbb{M}, g, \nabla, \nabla^*)$ à la Lauritzen, our alternative approach selects a pair of connections both of which are, instead of torsion-free, curvature-free. Compared to statistical manifold $(\mathbb{M}, g, \nabla)$ à la Kurose, our alternative approach selects a connection that is, instead of Codazzi-coupled, SMAT-coupled to g. The switch of emphasis from curvature to torsion may lead to interesting reformulation of information geometry.

References

[AJVLS15] Ay, N., Jost, J., Lê, H.V., Schwachhöfer, L.: Information geometry and sufficient statistics. Probab. Theory Relat. Fields **162**(1–2), 327–364 (2015)

[Ama85] Amari, S.: Differential-Geometrical Methods in Statistics. Lecture Notes in Statistics. Springer, New York (1985). https://doi.org/10.1007/978-1-4612-5056-2

[Ama98] Amari, S.-I.: Natural gradient works efficiently in learning. Neural Comput. **10**(2), 251–276 (1998)

[AN00] Amari, S., Nagaoka, H.: Methods of Information Geometry, vol. 191. American Mathematical Soc. (2000)

[BG80] Bishop, R.L., Samuel, I.: Tensor Analysis on Manifolds. Courier Corporation, Goldberg (1980)

[Dow18] Dowty, J.G.: Chentsov's theorem for exponential families. Inf. Geom. **1**(1), 117–135 (2018)

[Hen17] Henmi, M.: Statistical manifolds admitting torsion, pre-contrast functions and estimating functions. In: Nielsen, F., Barbaresco, F. (eds.) GSI 2017. LNCS, vol. 10589, pp. 153–161. Springer, Cham (2017). https://doi.org/10.1007/978-3-319-68445-1_18

[HM11] Henmi, M., Matsuzoe, H.: Geometry of pre-contrast functions and non-conservative estimating functions. In: AIP Conference Proceedings, vol. 1340, pp. 32–41. AIP (2011)

[Kur90] Kurose, T.: Dual connections and affine geometry. Math. Z. **203**(1), 115–121 (1990)

[Kur07] Kurose, T.: Statistical manifolds admitting torsion. Geometry and Something (2007)

[Lau87] Lauritzen, S.L.: Statistical manifolds. Differ. Geom. Stat. Infer. **10**, 163–216 (1987)

[Shi07] Shima, H.: The Geometry of Hessian Structures. World Scientific, Singapore (2007)

[ZK19] Zhang, J., Khan, G.: From Hessian to Weitzenböck: manifolds with torsion-carrying connections. Inf. Geom. (2019)

Divergence Geometry

The Bregman Chord Divergence

Frank Nielsen[1]([✉])[iD] and Richard Nock[2,3,4]

[1] Sony Computer Science Laboratories, Inc., Tokyo, Japan
Frank.Nielsen@acm.org
[2] Data61, Sydney, Australia
Richard.Nock@data61.csiro.au
[3] The Australian National University, Canberra, Australia
[4] The University of Sydney, Sydney, Australia

Abstract. Distances are fundamental primitives whose choice significantly impacts the performances of algorithms in applications. However selecting the most appropriate distance for a given task is an endeavor. Instead of testing one by one the entries of an ever-expanding dictionary of *ad hoc* distances, one rather prefers to consider parametric classes of distances that are exhaustively characterized by axioms derived from first principles. Bregman divergences are such a class. However fine-tuning a Bregman divergence is delicate since it requires to smoothly adjust a functional generator. In this work, we propose an extension of Bregman divergences called the Bregman chord divergences. This new class of distances bypasses the gradient calculations, uses two scalar parameters that can be easily tailored in applications, and generalizes asymptotically Bregman divergences.

Keywords: Csiszár's f-divergence · Bregman divergence ·
Jensen divergence · Skewed divergence

1 Introduction

Distances are at the heart of many signal processing tasks [6,14], and the performance of algorithms solving those tasks heavily depends on the chosen distances. Historically, many *ad hoc* distances have been proposed and empirically benchmarked on different tasks in order to improve the state-of-the-art performances. However, getting the most appropriate distance for a given task is often an endeavour. Thus principled *classes* of distances[1] have been proposed

---

[1] Here, we use the word distance to mean a *dissimilarity* (or a distortion, a deviance, a discrepancy, etc.), not necessarily a metric distance [14]. A distance between arguments θ_1 and θ_2 satisfies $D(\theta_1, \theta_2) \geq 0$ with equality if and only if $\theta_1 = \theta_2$.

© Springer Nature Switzerland AG 2019
F. Nielsen and F. Barbaresco (Eds.): GSI 2019, LNCS 11712, pp. 299–308, 2019.
https://doi.org/10.1007/978-3-030-26980-7_31

and studied. Among those generic classes of distances, three main generic classes
have emerged:

- The *Bregman divergences* [5,7,22] defined for a strictly convex and differentiable generator $F \in \mathcal{B} : \Theta \to \mathbb{R}$ (where $\mathcal{B}$ denotes the class of strictly convex and differentiable functions defined modulo affine terms):

$$B_F(\theta_1 : \theta_2) := F(\theta_1) - F(\theta_2) - (\theta_1 - \theta_2)^\top \nabla F(\theta_2), \tag{1}$$

  measure the dissimilarity between *parameters* $\theta_1, \theta_2 \in \Theta$, where $\Theta \subset \mathbb{R}^d$ is a d-dimensional convex set. Bregman divergences have also been generalized to other types of objects like matrices [26].
- The Csiszár f-divergences [1,11,12] defined for a convex generator $f \in \mathcal{C}$ satisfying $f(1) = 0$ and strictly convex at 1:

$$I_f[p_1 : p_2] := \int_{\mathcal{X}} p_1(x) f\left(\frac{p_2(x)}{p_1(x)}\right) \mathrm{d}\mu(x) \geq f(1) = 0, \tag{2}$$

  measure the dissimilarity between *probability densities* p_1 and p_2 that are absolutely continuous with respect to a base measure μ (defined on a support $\mathcal{X}$).
- The Burbea-Rao divergences [9] also called Jensen differences or Jensen divergences because they rely on the Jensen's inequality [16] for a strictly convex function $F \in \mathcal{J} : \Theta \to \mathbb{R}$:

$$J_F(\theta_1, \theta_2) := \frac{F(\theta_1) + F(\theta_2)}{2} - F\left(\frac{\theta_1 + \theta_2}{2}\right) \geq 0, \tag{3}$$

  where θ_1 and θ_2 belong to a parameter space Θ.

These three fundamental classes of distances are *not* mutually exclusive, and their pairwise intersections (e.g., $\mathcal{B} \cap \mathcal{C}$ or $\mathcal{J} \cap \mathcal{C}$) have been studied in [2,17,27]. The ':' notation between arguments of distances emphasizes the potential asymmetry of distances (oriented distances with $D(\theta_1 : \theta_2) \neq D(\theta_2 : \theta_1)$), and the brackets surrounding distance arguments indicate that it is a *statistical distance* between probability densities, and not a distance between parameters. Using these notations, we express the Kullback-Leibler distance [10] (KL) as

$$\mathrm{KL}[p_1 : p_2] := \int p_1(x) \log \frac{p_1(x)}{p_2(x)} \mathrm{d}\mu(x). \tag{4}$$

The KL distance/divergence between two members p_{θ_1} and p_{θ_2} of a parametric family $\mathcal{F}$ of distributions amount to a parameter divergence

$$\mathrm{KL}_{\mathcal{F}}(\theta_1 : \theta_2) := \mathrm{KL}[p_{\theta_1} : p_{\theta_2}]. \tag{5}$$

For example, the KL statistical distance between two probability densities belonging to the same exponential family or the same mixture family amounts to a (parameter) Bregman divergence [3,23]. When p_1 and p_2 are finite discrete distributions of the d-dimensional probability simplex Δ_d, we have $\mathrm{KL}_{\Delta_d}(p_1 :$ $p_2) = \mathrm{KL}[p_1 : p_2]$. This explains why sometimes we can handle loosely distances between discrete distributions as both a parameter distance and a statistical distance. For example, the KL distance between two discrete distributions is a Bregman divergence $B_{F_{\mathrm{KL}}}$ for $F_{\mathrm{KL}}(x) = \sum_{i=1}^{d} x_i \log x_i$ (Shannon negentropy) for $x \in \Theta = \Delta_d$. Extending $\Theta = \Delta_d$ to positive measures $\Theta = \mathbb{R}_+^d$, this Bregman divergence $B_{F_{\mathrm{KL}}}$ yields the extended KL distance:

$$\mathrm{eKL}[p : q] = \sum_{i=1}^{d} p_i \log \frac{p_i}{q_i} + q_i - p_i. \tag{6}$$

Notice that the KL divergence of 4 between non-probability positive distributions may yield potential negativity of the measure (e.g., Example 2.1 of [28] and [8]). This case also happens when doing Monte Carlo stochastic integrations of the KL divergence integral.

Whenever using a functionally parameterized distance in applications, we need to choose the most appropriate functional generator, ideally from first principles [3,4,13]. For example, Non-negative Matrix Factorization (NMF) for audio source separation or music transcription from the signal power spectrogram can be done by selecting the Itakura-Saito divergence [15][2] that satisfies the requirement of being *scale invariant*:

$$B_{F_{\mathrm{IS}}}(\lambda \theta : \lambda \theta') = B_{F_{\mathrm{IS}}}(\theta : \theta') = \sum_{i} \left(\frac{\theta_i}{\theta_i'} - \log \frac{\theta_i}{\theta_i'} - 1 \right), \tag{7}$$

for any $\lambda > 0$. When no such first principles can be easily stated for a task [13], we are left by choosing manually or by cross-validation a generator. Notice that the convex combinations of Csiszár generators is a Csiszár generator (idem for Bregman divergences): $\sum_{i=1}^{d} \lambda_i I_{f_i} = I_{\sum_i \; i=1^d \lambda_i f_i}$ for λ belonging to the standard $(d-1)$-dimensional standard simplex Δ_d.

In this work, we propose a novel class of distances, termed *Bregman chord divergences*. A Bregman chord divergence is parameterized by a Bregman generator and two scalar parameters which make it easy to fine-tune in applications, and matches asymptotically the ordinary Bregman divergence.

The paper is organized as follows: In Sect. 2, we describe the skewed Jensen divergence, show how to bi-skew any distance by using two scalars, and report on the Jensen chord divergence [20]. In Sect. 3, we first introduce the univariate Bregman chord divergence, and then extend its definition to the multivariate case, in Sect. 4. Finally, we conclude in Sect. 5.

2   Geometric Design of Skewed Divergences

We can geometrically *design* divergences from convexity gap properties of the graph plot of the generator. For example, the Jensen divergence $J_F(\theta_1 : \theta_2)$ of

---

[2] A Bregman divergence for the Burg negentropy $F_{\mathrm{IS}}(x) = -\sum_i \log x_i$.

Eq. 3 is visualized as the ordinate (vertical) gap between the midpoint of the line segment $[(\theta_1, F(\theta_1)); (\theta_2, F(\theta_2))]$ and the point $(\frac{\theta_1+\theta_2}{2}, F(\frac{\theta_1+\theta_2}{2}))$. The non-negativity property of the Jensen divergence follows from the Jensen's midpoint convex inequality [16]. Instead of taking the midpoint $\bar{\theta} = \frac{\theta_1+\theta_2}{2}$, we can take *any* interior point $(\theta_1\theta_2)_\alpha := (1-\alpha)\theta_1 + \alpha\theta_2$, and get the skewed α-Jensen divergence (for any $\alpha \in (0,1)$):

$$J_F^\alpha(\theta_1 : \theta_2) := (F(\theta_1)F(\theta_2))_\alpha - F((\theta_1\theta_2)_\alpha) \geq 0. \tag{8}$$

A remarkable fact is that the scaled α-Jensen divergence $\frac{1}{\alpha}J_F^\alpha(\theta_1 : \theta_2)$ tends asymptotically to the reverse Bregman divergence $B_F(\theta_2 : \theta_1)$ when $\alpha \to 0$, see [21,30].

By measuring the ordinate gap between two non-crossing upper and lower chords anchored at the generator graph plot, we can extend the α-Jensen divergences to a tri-parametric family of Jensen chord divergences [20]:

$$J_F^{\alpha,\beta,\gamma}(\theta : \theta') := (F(\theta)F(\theta'))_\gamma - (F((\theta\theta')_\alpha)F((\theta\theta')_\beta))_{\frac{\gamma-\alpha}{\beta-\alpha}}, \tag{9}$$

with $\alpha, \beta \in [0,1]$ and $\gamma \in [\alpha, \beta]$. The α-Jensen divergence is recovered when $\alpha = \beta = \gamma$ (Fig. 1).

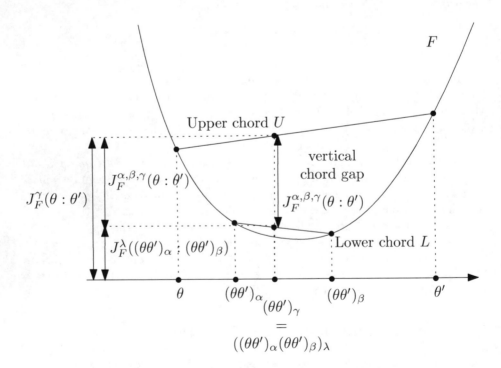

Fig. 1. The Jensen chord gap divergence.

For any given distance $D : \Theta \times \Theta \to \mathbb{R}_+$ (with convex parameter space Θ), we can bi-skew the distance by considering two scalars $\gamma, \delta \in \mathbb{R}$ (with $\delta \neq \gamma$) as:

$$D_{\gamma,\delta}(\theta_1 : \theta_2) := D((\theta_1\theta_2)_\gamma : (\theta_1\theta_2)_\delta). \tag{10}$$

Clearly, $(\theta_1\theta_2)_\gamma = (\theta_1\theta_2)_\delta$ if and only if $(\delta - \gamma)(\theta_1 - \theta_2) = 0$. That is, if (i) $\theta_1 = \theta_2$ or if (ii) $\delta = \gamma$. Since by definition $\delta \neq \gamma$, we have $D_{\gamma,\delta}(\theta_1 : \theta_2) = 0$ if and only if $\theta_1 = \theta_2$. Notice that both $(\theta_1\theta_2)_\gamma = (1-\gamma)\theta_1 + \gamma\theta_2$ and $(\theta_1\theta_2)_\delta = (1-\delta)\theta_1 + \delta\theta_2$ should belong to the parameter space Θ. A sufficient condition is to ensure that $\gamma, \delta \in [0,1]$ so that both $(\theta_1\theta_2)_\gamma \in \Theta$ and $(\theta_1\theta_2)_\delta \in \Theta$. When $\Theta = \mathbb{R}^d$, we may further consider any $\gamma, \delta \in \mathbb{R}$.

3   The Scalar Bregman Chord Divergence

Let $F : \Theta \subset \mathbb{R} \to \mathbb{R}$ be a univariate Bregman generator with open convex domain Θ, and denote by $\mathcal{F} = \{(\theta, F(\theta))\}_\theta$ its graph. Let us rewrite the ordinary univariate Bregman divergence [7] of Eq. 1 as follows:

$$B_F(\theta_1 : \theta_2) = F(\theta_1) - T_{\theta_2}(\theta_1), \tag{11}$$

where $y = T_\theta(\omega)$ denotes the equation of the tangent line of F at θ:

$$T_\theta(\omega) := F(\theta) + (\omega - \theta)F'(\theta), \tag{12}$$

Let $\mathcal{T}_\theta = \{(\theta, T_\theta(\omega)) \ : \ \theta \in \Theta\}$ denote the graph of that tangent line. Line $\mathcal{T}_\theta$ is tangent to curve $\mathcal{F}$ at point $P_\theta := (\theta, F(\theta))$. Graphically speaking, the Bregman divergence is interpreted as the *ordinate gap* (gap vertical) between the point $P_{\theta_1} = (\theta_1, F(\theta_1)) \in \mathcal{F}$ and the point of $(\theta_1, T_{\theta_2}(\theta_1)) \in \mathcal{T}_\theta$, as depicted in Fig. 2.

Now let us observe that we may *relax* the tangent line $\mathcal{T}_{\theta_2}$ to a *chord line* (or secant) $\mathcal{C}^{\alpha,\beta}_{\theta_1,\theta_2} = \mathcal{C}_{(\theta_1\theta_2)_\alpha,(\theta_1\theta_2)_\beta}$ passing through the points $((\theta_1\theta_2)_\alpha, F((\theta_1\theta_2)_\alpha))$ and $((\theta_1\theta_2)_\beta, F((\theta_1\theta_2)_\beta))$ for $\alpha, \beta \in (0,1)$ with $\alpha \neq \beta$ (with corresponding Cartesian equation $C_{(\theta_1\theta_2)_\alpha,(\theta_1\theta_2)_\beta}$), and still get a non-negative vertical gap between $(\theta_1, F(\theta_1))$ and $(\theta_1, C_{(\theta_1\theta_2)_\alpha,(\theta_1\theta_2)_\beta}(\theta_1))$ (because any line intersects a convex body in at most two points). By construction, this vertical gap is smaller than the gap measured by the ordinary Bregman divergence. This yields the Bregman chord divergence ($\alpha, \beta \in (0,1]$, $\alpha \neq \beta$):

$$B_F^{\alpha,\beta}(\theta_1 : \theta_2) := F(\theta_1) - C_F^{(\theta_1\theta_2)_\alpha,(\theta_1\theta_2)_\beta}(\theta_1) \leq B_F(\theta_1 : \theta_2), \tag{13}$$

illustrated in Fig. 3. By expanding the chord equation and massaging the equation, we get the following formula:

$$B_F^{\alpha,\beta}(\theta_1 : \theta_2) := F(\theta_1) - \Delta_F^{\alpha,\beta}(\theta_1, \theta_2)(\theta_1 - (\theta_1\theta_2)_\alpha) - F((\theta_1\theta_2)_\alpha), \tag{14}$$

$$= F(\theta_1) - F((\theta_1\theta_2)_\alpha) + \frac{\alpha\{F((\theta_1\theta_2)_\alpha) - F((\theta_1\theta_2)_\beta)\}}{\beta - \alpha},$$

where

$$\Delta_F^{\alpha,\beta}(\theta_1, \theta_2) := \frac{F((\theta_1\theta_2)_\alpha) - F((\theta_1\theta_2)_\beta)}{(\theta_1\theta_2)_\alpha - (\theta_1\theta_2)_\beta} \tag{15}$$

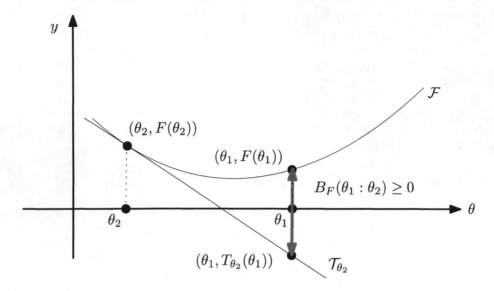

Fig. 2. Bregman divergence as the vertical gap between the generator graph $\mathcal{F}$ and the tangent line $\mathcal{T}_{\theta_2}$ at θ_2.

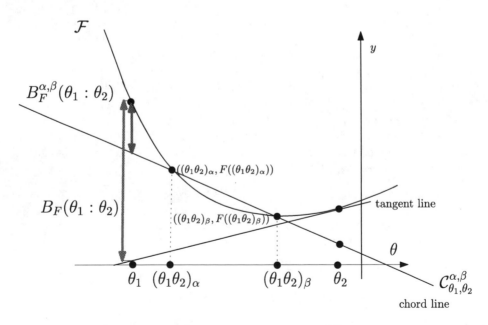

Fig. 3. The Bregman chord divergence $B_F^{\alpha,\beta}(\theta_1 : \theta_2)$.

is the slope of the chord, and since $(\theta_1\theta_2)_\alpha - (\theta_1\theta_2)_\beta = (\beta - \alpha)(\theta_1 - \theta_2)$ and $\theta_1 - (\theta_1\theta_2)_\alpha = \alpha(\theta_1 - \theta_2)$.

Notice the symmetry $B_F^{\alpha,\beta}(\theta_1 : \theta_2) = B_F^{\beta,\alpha}(\theta_1 : \theta_2)$. We have

$$\lim_{\alpha \to 1, \beta \to 1} B_F^{\alpha,\beta}(\theta_1 : \theta_2) = B_F(\theta_1 : \theta_2). \tag{16}$$

When $\alpha \to \beta$, the Bregman chord divergences yields a subfamily of *Bregman tangent divergences*:

$$B_F^\alpha(\theta_1 : \theta_2) = \lim_{\beta \to \alpha} B_F^{\alpha,\beta}(\theta_1 : \theta_2) \leq B_F(\theta_1 : \theta_2). \tag{17}$$

We consider the tangent line $\mathcal{T}_{(\theta_1\theta_2)_\alpha}$ at $(\theta_1\theta_2)_\alpha$ and measure the ordinate gap at θ_1 between the function plot and this tangent line:

$$\begin{aligned} B_F^\alpha(\theta_1 : \theta_2) &:= F(\theta_1) - F((\theta_1\theta_2)_\alpha) - (\theta_1 - (\theta_1\theta_2)_\alpha)^\top \nabla F((\theta_1\theta_2)_\alpha), \\ &= F(\theta_1) - F((\theta_1\theta_2)_\alpha) - \alpha(\theta_1 - \theta_2)^\top \nabla F((\theta_1\theta_2)_\alpha), \end{aligned} \tag{18}$$

for $\alpha \in (0, 1]$. The ordinary Bregman divergence is recovered when $\alpha = 1$. Notice that the *mean value theorem* yields $\Delta_F^{\alpha,\beta}(\theta_1, \theta_2) = F'(\xi)$ for $\xi \in (\theta_1, \theta_2)$. Thus $B_F^{\alpha,\beta}(\theta_1 : \theta_2) = B_F^\xi(\theta_1 : \theta_2)$ for $\xi \in (\theta_1, \theta_2)$. Letting $\beta = 1$ and $\alpha = 1 - \epsilon$ (for small values of $1 > \epsilon > 0$), we can approximate the ordinary Bregman divergence by the Bregman chord divergence without requiring to compute the gradient:

$$B_F(\theta_1 : \theta_2) \simeq_{\epsilon \to 0} B_F^{1-\epsilon,1}(\theta_1 : \theta_2). \tag{19}$$

4   The Multivariate Bregman Chord Divergence

When the generator is separable [3], i.e., $F(x) = \sum_i F_i(x_i)$ for univariate generators F_i, we extend easily the Bregman chord divergence as:

$$B_F^{\alpha,\beta}(\theta : \theta') = \sum_i B_{F_i}^{\alpha,\beta}(\theta_i : \theta'_i). \tag{20}$$

Otherwise, we have to carefully define the notion of "slope" for the multivariate case. An example of such a non-separable multivariate generator is the Legendre dual of the Shannon negentropy: The log-sum-exp function [24,25]:

$$F(\theta) = \log(1 + \sum_i e^{\theta_i}). \tag{21}$$

Given a multivariate (non-separable) Bregman generator $F(\theta)$ with $\Theta \subseteq \mathbb{R}^D$ and two prescribed distinct parameters θ_1 and θ_2, consider the following univariate function, for $\lambda \in \mathbb{R}$:

$$F_{\theta_1,\theta_2}(\lambda) := F((1 - \lambda)\theta_1 + \lambda\theta_2) = F(\theta_1 + \lambda(\theta_2 - \theta_1)), \tag{22}$$

with $F_{\theta_1,\theta_2}(0) = F(\theta_1)$ and $F_{\theta_1,\theta_2}(1) = F(\theta_2)$.

The functions $\{F_{\theta_1,\theta_2}\}_{\theta_1 \neq \theta_2}$ are strictly convex and differentiable univariate Bregman generators.

Proof. To prove the strict convexity of a univariate function G, we need to show that for any $\alpha \in (0, 1)$, we have $G((1 - \alpha)x + \alpha y) < (1 - \alpha)G(x) + \alpha G(y)$.

$$
\begin{aligned}
F_{\theta_1,\theta_2}((1 - \alpha)\lambda_1 + \alpha\lambda_2) &= F(\theta_1 + ((1 - \alpha)\lambda_1 + \alpha\lambda_2)(\theta_2 - \theta_1)), \\
&= F((1 - \alpha)(\lambda_1(\theta_2 - \theta_1) + \theta_1) + \alpha((\lambda_2(\theta_2 - \theta_1) + \theta_1))), \\
&< (1 - \alpha)F(\lambda_1(\theta_2 - \theta_1) + \theta_1) + \alpha F((\lambda_2(\theta_2 - \theta_1) + \theta_1)), \\
&< (1 - \alpha)F_{\theta_1,\theta_2}(\lambda_1) + \alpha F_{\theta_1,\theta_2}(\lambda_2).
\end{aligned}
$$

Then we define the multivariate Bregman chord divergence by applying the definition of the univariate Bregman chord divergence on these families of univariate Bregman generators:

$$
B_F^{\alpha,\beta}(\theta_1 : \theta_2) := B_{F_{\theta_1,\theta_2}}^{\alpha,\beta}(0 : 1), \tag{23}
$$

Since $(01)_\alpha = \alpha$ and $(01)_\beta = \beta$, we get:

$$
\begin{aligned}
B_F^{\alpha,\beta}(\theta_1 : \theta_2) &= F_{\theta_1,\theta_2}(0) + \frac{\alpha(F_{\theta_1,\theta_2}(\alpha) - F_{\theta_1,\theta_2}(\beta))}{\beta - \alpha} - F_{\theta_1,\theta_2}(\alpha), \\
&= F(\theta_1) - F((\theta_1\theta_2)_\alpha) - \frac{\alpha(F((\theta_1\theta_2)_\beta) - F((\theta_1\theta_2)_\alpha))}{\beta - \alpha},
\end{aligned}
$$

in accordance with the univariate case. Since $(\theta_1\theta_2)_\beta = (\theta_1\theta_2)_\alpha - (\beta - \alpha)(\theta_2 - \theta_1)$, we have the first-order Taylor expansion

$$
F((\theta_1\theta_2)_\beta) \simeq_{\beta \simeq \alpha} F((\theta_1\theta_2)_\alpha) - (\beta - \alpha)(\theta_2 - \theta_1)^\top \nabla F((\theta_1\theta_2)_\alpha). \tag{24}
$$

Therefore, we have:

$$
\frac{\alpha(F((\theta_1\theta_2)_\beta) - F((\theta_1\theta_2)_\alpha))}{\beta - \alpha} \simeq -\alpha(\theta_2 - \theta_1)^\top \nabla F((\theta_1\theta_2)_\alpha). \tag{25}
$$

This proves that

$$
\lim_{\beta \to \alpha} B_F^{\alpha,\beta}(\theta_1 : \theta_2) = B_F^{\alpha}(\theta_1 : \theta_2). \tag{26}
$$

Notice that the Bregman chord divergence does *not* require to compute the gradient ∇F The "slope term" in the definition is reminiscent to the q-derivative [18] (quantum/discrete derivatives). However the (p, q)-derivatives [18] are defined with respect to a *single* reference point while the chord definition requires *two* reference points.

5   Conclusion

In this paper, we geometrically designed a new class of distances using a Bregman generator and two additional scalar parameters, termed the *Bregman chord divergence*, and its one-parametric subfamily, the *Bregman tangent divergences* that includes the ordinary Bregman divergence. This generalization allows one to easily fine-tune Bregman divergences in applications by adjusting smoothly one or two (scalar) knobs. Moreover, by choosing $\alpha = 1 - \epsilon$ and $\beta = 1$ for small $\epsilon > 0$, the Bregman chord divergence $B_F^{1-\epsilon,1}(\theta_1 : \theta_2)$ lower bounds closely the Bregman divergence $B_F(\theta_1 : \theta_2)$ without requiring to compute the gradient (a different approximation without gradient is $\frac{1}{\epsilon}J_F^\epsilon(\theta_2 : \theta_1)$). We expect that this new class of distances brings further improvements in signal processing and information fusion applications [29] (e.g., by tuning $B_{F_{\mathrm{KL}}}^{\alpha,\beta}$ or $B_{F_{\mathrm{IS}}}^{\alpha,\beta}$). While the Bregman chord divergence defines an ordinate gap on the exterior of the epigraph, the Jensen chord divergence [20] defines the gap inside the epigraph of the generator. In future work, the dualistic information-geometric structure induced by the Bregman chord divergences shall be investigated from the viewpoint of gauge theory [19] and in contrast with the dually flat structures of Bregman manifolds [3].

Source code in Java^{TM} is available for reproducible research.[3]

References

1. Ali, S.M., Silvey, S.D.: A general class of coefficients of divergence of one distribution from another. J. Roy. Stat. Soc.: Ser. B (Methodol.) **28**(1), 131–142 (1966)
2. Amari, S.I.: α-divergence is unique, belonging to both f-divergence and Bregman divergence classes. IEEE Trans. Inf. Theory **55**(11), 4925–4931 (2009)
3. Amari, S.: Information Geometry and Its Applications. AMS, vol. 194. Springer, Tokyo (2016). https://doi.org/10.1007/978-4-431-55978-8
4. Banerjee, A., Guo, X., Wang, H.: On the optimality of conditional expectation as a Bregman predictor. IEEE Trans. Inf. Theory **51**(7), 2664–2669 (2005)
5. Banerjee, A., Merugu, S., Dhillon, I.S., Ghosh, J.: Clustering with Bregman divergences. J. Mach. Learn. Res. **6**(Oct), 1705–1749 (2005)
6. Basseville, M.: Divergence measures for statistical data processing: an annotated bibliography. Sig. Process. **93**(4), 621–633 (2013)
7. Bregman, L.M.: The relaxation method of finding the common point of convex sets and its application to the solution of problems in convex programming. USSR Comput. Math. Math. Phys. **7**(3), 200–217 (1967)
8. Broniatowski, M., Stummer, W.: Some universal insights on divergences for statistics, machine learning and artificial intelligence. In: Nielsen, F. (ed.) Geometric Structures of Information. SCT, pp. 149–211. Springer, Cham (2019). https://doi.org/10.1007/978-3-030-02520-5_8
9. Burbea, J., Rao, C.: On the convexity of some divergence measures based on entropy functions. IEEE Trans. Inf. Theory **28**(3), 489–495 (1982)
10. Cover, T.M., Thomas, J.A.: Elements of Information Theory. Wiley, New York (2012)
11. Csiszár, I.: Eine infonnationstheoretische Ungleichung und ihre Anwendung auf den Beweis der Ergodizitlit von Markoffschen Ketten. Magyar Tudományos Akadémia - MAT **8**, 85–108 (1963)
12. Csiszár, I.: Information-type measures of difference of probability distributions and indirect observation. Studia Scientiarum Mathematicarum Hungarica **2**, 229–318 (1967)

---

[3] https://franknielsen.github.io/~nielsen/BregmanChordDivergence/.

13. Csiszár, I.: Why least squares and maximum entropy? An axiomatic approach to inference for linear inverse problems. Ann. Stat. **19**(4), 2032–2066 (1991). https://doi.org/10.1007/978-1-4613-0071-7
14. Deza, M.M., Deza, E.: Encyclopedia of distances. In: Deza, M.M., Deza, E. (eds.) Encyclopedia of Distances, pp. 1–583. Springer, Heidelberg (2009). https://doi.org/10.1007/978-3-642-00234-2_1
15. Févotte, C.: Majorization-minimization algorithm for smooth Itakura-Saito non-negative matrix factorization. In: 2011 IEEE International Conference on Acoustics, Speech and Signal Processing (ICASSP), pp. 1980–1983. IEEE (2011)
16. Jensen, J.L.W.V.: Sur les fonctions convexes et les inégalités entre les valeurs moyennes. Acta Math. **30**(1), 175–193 (1906)
17. Jiao, J., Courtade, T.A., No, A., Venkat, K., Weissman, T.: Information measures: the curious case of the binary alphabet. IEEE Trans. Inf. Theory **60**(12), 7616–7626 (2014)
18. Kac, V., Cheung, P.: Quantum Calculus. Springer, New York (2001)
19. Naudts, J., Zhang, J.: Rho-tau embedding and gauge freedom in information geometry. Inf. Geom. **1**(1), 79–115 (2018)
20. Nielsen, F.: The chord gap divergence and a generalization of the Bhattacharyya distance. In: IEEE International Conference on Acoustics, Speech and Signal Processing (ICASSP), pp. 2276–2280, April 2018. https://doi.org/10.1109/ICASSP.2018.8462244
21. Nielsen, F., Boltz, S.: The Burbea-Rao and Bhattacharyya centroids. IEEE Trans. Inf. Theory **57**(8), 5455–5466 (2011)
22. Nielsen, F., Nock, R.: Sided and symmetrized Bregman centroids. IEEE Trans. Inf. Theory **55**(6), 2882–2904 (2009)
23. Nielsen, F., Nock, R.: On the geometry of mixtures of prescribed distributions. In: IEEE International Conference on Acoustics, Speech and Signal Processing (ICASSP), pp. 2861–2865. IEEE (2018)
24. Nielsen, F., Sun, K.: Guaranteed bounds on information-theoretic measures of univariate mixtures using piecewise log-sum-exp inequalities. Entropy **18**(12), 442 (2016)
25. Nielsen, F., Sun, K.: Guaranteed bounds on the Kullback-Leibler divergence of univariate mixtures. IEEE Signal Process. Lett. **23**(11), 1543–1546 (2016)
26. Nock, R., Magdalou, B., Briys, E., Nielsen, F.: Mining matrix data with Bregman matrix divergences for portfolio selection. In: Nielsen, F., Bhatia, R. (eds.) Matrix Information Geometry, pp. 373–402. Springer, Heidelberg (2013). https://doi.org/10.1007/978-3-642-30232-9_15
27. Pardo, M.C., Vajda, I.: About distances of discrete distributions satisfying the data processing theorem of information theory. IEEE Trans. Inf. Theory **43**(4), 1288–1293 (1997)
28. Stummer, W., Vajda, I.: On divergences of finite measures and their applicability in statistics and information theory. Statistics **44**(2), 169–187 (2010)
29. Üney, M., Houssineau, J., Delande, E., Julier, S.J., Clark, D.E.: Fusion of finite set distributions: pointwise consistency and global cardinality. CoRR abs/1802.06220 (2018). http://arxiv.org/abs/1802.06220
30. Zhang, J.: Divergence function, duality, and convex analysis. Neural Comput. **16**(1), 159–195 (2004)

Testing the Number and the Nature of the Components in a Mixture Distribution

Michel Broniatowski[1(✉)], Emilie Miranda[1], and Wolfgang Stummer[2]

[1] Sorbonne Universite, LPSM, 4 place Jussieu, 75252 Paris, France
`michel.broniatowski@sorbonne-universite.fr`
[2] Department of Mathematics, University of Erlangen–Nürnberg,
Cauerstrasse 11, 91058 Erlangen, Germany

Abstract. In this paper, we propose a test procedure for the number of components of mixture distributions in a parametric setting. The test statistic is based on divergence estimators derived through the dual form of the divergence in parametric models. We provide a standard limit distribution for the test statistic under the null hypothesis that holds for mixtures of any number of components $k \geq 2$.

Keywords: Test procedure · Mixture model · Number of components · Divergence estimation

1 Introduction

Consider a k-component parametric mixture model P_θ ($k \geq 2$) defined as follows:

$$P_\theta := \sum_{i=1}^{k} w_i P_{a_i}^{(i)} \tag{1}$$

where $\left\{ P_{a_1}^{(1)}; a_1 \in A_1 \right\}, \ldots, \left\{ P_{a_k}^{(k)}; a_k \in A_k \right\}$ are k parametric models and $A_1, \ldots, A_k$ are k sets in $\mathbb{R}^{d_1}, \ldots, \mathbb{R}^{d_k}$ with $d_1, \ldots, d_k \in \mathbb{N}^*$ and $0 \leq w_i \leq 1$, $\sum w_i = 1$. Note that we consider a nonstandard framework in which the weights w_i are allowed to be equal to 0. Note Θ the parameter space:

$$\theta \in \Theta := \left\{ (w_1, \ldots, w_k, a_1, \ldots, a_k)^T \in [0,1]^k \times A_1 \times \cdots \times A_k \text{ such that } \sum_{i=1}^{k} w_i = 1 \right\}, \tag{2}$$

and assume that the model is identifiable. Let $k_0 \in \{1, \ldots, k-1\}$.

We are willing to test if $(k - k_0)$ components in (1) have null coefficients. We assume that their labels are $k_0 + 1, \ldots, k$. Denote Θ_0 the subset of Θ defined by

$$\Theta_0 := \left\{ \theta \in \Theta \text{ such that } w_{k_0+1} = \cdots = w_k = 0 \right\}.$$

On the basis of an i.i.d sample $X_1, \ldots, X_n$ with distribution P_{θ_T}, $\theta_T \in \Theta$, we intend to perform tests of the hypothesis

$$\mathcal{H}_0 : \theta_T \in \Theta_0 \text{ against the alternative } \mathcal{H}_1 : \theta_T \in \Theta \setminus \Theta_0. \tag{3}$$

F. Nielsen and F. Barbaresco (Eds.): GSI 2019, LNCS 11712, pp. 309–318, 2019.
https://doi.org/10.1007/978-3-030-26980-7_32

When considering the test (3), it is known that the generalized likelihood ratio test, based on the statistic

$$2 \log \lambda := 2 \log \frac{\sup_{\theta \in \Theta} \prod_{i=1}^{n} p_\theta(X_i)}{\sup_{\theta \in \Theta_0} \prod_{i=1}^{n} p_\theta(X_i)}, \tag{4}$$

is not valid, since the asymptotic approximation by χ^2 distribution does not hold in this case; the problem is due to the fact that the null value of θ_T is not in the interior of the parameter space Θ. We clarify now this problem.

For simplicity, consider a mixture of two known densities p_0 and p_1 with $p_0 \neq p_1$:

$$p_\theta = (1 - \theta)p_0 + \theta p_1 \text{ where } \theta \in \Theta := [0,1]. \tag{5}$$

Given data $X_1, \ldots, X_n$ with distribution P_{θ_T} and density p_{θ_T}, $\theta_T \in [0,1]$, consider the test problem

$$\mathcal{H}_0 : \theta_T = 0 \text{ against the alternative } \mathcal{H}_1 : \theta_T > 0. \tag{6}$$

The generalized likelihood ratio statistic for this test problem is

$$W_n(0) := 2 \log \frac{L(\widehat{\theta})}{L(0)}, \tag{7}$$

where $\widehat{\theta}$ is the maximum likelihood estimator of θ_T.

Under suitable regularity conditions we can prove that the limit distribution of the statistic W_n in (7) is $0.5\delta_0 + 0.5\chi_1^2$, a mixture of the χ^2-distribution and the Dirac measure at zero; see e.g. Titterington et al. [16], Self and Liang [15] and Ciuperca [9].

Moreover, in the case of more than two components and $k - k_0 \geq 2$, the limit distribution of the GLR statistic (4) under $\mathcal{H}_0$ is complicate, and not standard (not a χ^2 distribution) which poses some difficulty in determining the critical value that will give correct asymptotic size; see Self and Liang [15]. Azais et al. [1] proposes for instance a likelihood ratio approach for mixtures and give the asymptotic properties of the test, but its numerical application is extremely complicated, especially under non-Gaussian mixtures. On the other hand, the likelihood ratio statistic can not be used to construct an asymptotic confidence region for the parameter θ_T since its limit law is not the same when $\theta_T = 0$ and $\theta_T > 0$.

The case where some parameter of the model belongs to the frontier of the domain is a special case of power models, see for instance Castillo et al. [8] for related statistical issues.

In the sequel, we propose a simple solution for testing the number of components of a parametric mixture model. This method consists in constructing a test statistic based on φ−divergences and their asymptotic properties. In the following section, we provide the general framework that will be used to construct the test procedure, i.e. the definitions, representation and properties of φ−divergences.

2 Some Definition and Notation in Relation with Minimum Divergence Inference

Let $\mathscr{P} := \{P_\theta, \theta \in \Theta\}$ be an identifiable parametric model on $\mathbb{R}^s$ where Θ is a subset of $\mathbb{R}^d$. All measures in $\mathscr{P}$ will be assumed to be measure equivalent sharing therefore the same support. The parameter space Θ does not need to be open in the present setting. It may even happen that the model includes measures which would not be probability distributions; cases of interest cover the present setting, namely models including unnormalized mixtures of probability distributions; see Broniatowski and Keziou [3].

The f-divergences were introduced by Csiszar [11] as convex non-negative dissimilarities between two probability distributions. Let f be a convex function on $\mathbb{R}_+$, that possibly takes infinite values at 0 and such that $f(1) = 0$. Denote by F the f-divergence between two probability distributions P and Q:

$$F(Q,P) := \int_{\mathbb{R}^s} f\left(\frac{dQ}{dP}(x)\right) \, dP(x).$$

Extensions to cases where Q is a finite signed measure and P a probability measure are called φ-divergences.

Let φ be a proper closed convex function from $]-\infty,+\infty[$ to $[0,+\infty]$ with $\varphi(1) = 0$ and such that its domain $\mathrm{dom}\varphi := \{x \in \mathbb{R} \text{ such that } \varphi(x) < \infty\}$ is an interval with endpoints $a_\varphi < 1 < b_\varphi$ (which may be finite or infinite). For two measures P_α and P_θ in $\mathscr{P}$ the φ-divergence between the two is defined by

$$\phi(\alpha,\theta) := \int_{\mathbb{R}^s} \varphi\left(\frac{dP_\alpha}{dP_\theta}(x)\right) \, dP_\theta(x).$$

The basic property of $\varphi-$ divergences states that when φ is strictly convex on a neighborhood of $x = 1$, then

$$\phi(\alpha,\theta) = 0 \text{ if and only if } \alpha = \theta.$$

We refer to Liese and Vajda [12] Chap. 1 for a complete study of those properties. See also Pardo [13].

2.1 Examples of φ-divergences

The Kullback-Leibler (KL), modified Kullback-Leibler (KL_m), χ^2, modified χ^2 (χ_m^2), Hellinger (H), and L_1 divergences are respectively associated to the convex functions $\varphi(x) = x\log x - x + 1$, $\varphi(x) = -\log x + x - 1$, $\varphi(x) = \frac{1}{2}(x-1)^2$, $\varphi(x) = \frac{1}{2}(x-1)^2/x$, $\varphi(x) = 2(\sqrt{x}-1)^2$ and $\varphi(x) = |x-1|$. All these divergences except the L_1 one, belong to the class of the so called "power divergences" introduced in Cressie and Read [10] (see also Liese and Vajda [12] Chap. 2), a class which takes its origin from Rényi [14]. They are defined through the class of convex functions

$$x \in]0,+\infty[\mapsto \varphi_\gamma(x) := \frac{x^\gamma - \gamma x + \gamma - 1}{\gamma(\gamma-1)} \tag{8}$$

if $\gamma \in \mathbb{R} \setminus \{0,1\}$, $\varphi_0(x) := -\log x + x - 1$ and $\varphi_1(x) := x\log x - x + 1$. So, the KL - divergence is associated to φ_1, the KL_m to φ_0, the χ^2 to φ_2, the χ_m^2 to φ_{-1} and the Hellinger distance to $\varphi_{1/2}$.

Consider any φ-divergence except the likelihood divergence, with φ being a differentiable function. When θ_T in $int\Theta$ is defined as the true parameter of the distribution of the i.i.d. sample $(X_1,..,X_n)$, it is convenient to assume that

> There exists a neighborhood $\mathcal{U}$ of θ_T for which   (A)
> $\phi(\theta,\theta')$ is finite whatever θ and θ' in $\mathcal{U}$.

We will only consider divergences defined through differentiable functions φ, which we assume to satisfy

> There exists a positive δ such that for all c in $[1-\delta,1+\delta]$,
> **(RC)**   we can find numbers c_1, c_2, c_3 such that
> $\varphi(cx) \le c_1 \varphi(x) + c_2 |x| + c_3$, for all real x.

Condition **(RC)** holds for all power divergences including KL and KL_m divergences.

For all divergences considered in this paper it will be assumed that for any α and θ in $\mathcal{U}$

$$\int \left| \varphi' \left(\frac{dP_\theta}{dP_\alpha} \right) \right| dP_\theta < \infty. \tag{9}$$

We state the following lemma covering nearly all classical divergences (see Liese and Vajda (1987) [12] and Broniatowski and Kéziou (2006) [2], Lemma 3.2).

Lemma 1. *Assume that RC holds and $\phi(\theta,\alpha)$ is finite. Then (9) holds.*

2.2 Dual Form of the Divergence and Dual Estimators in Parametric Models

The following representation is the cornerstone of parametric inference through divergence based methods.

Theorem 1. *Let θ belong to Θ and let $\phi(\theta,\theta_T)$ be finite. Assume that RC holds together with Condition (A). Then*

$$\phi(\theta,\theta_T) = \sup_{\alpha \in \mathcal{U}} \int \varphi' \left(\frac{dP_\theta}{dP_\alpha} \right) dP_\theta - \int \varphi^\# \left(\frac{dP_\theta}{dP_\alpha} \right) dP_{\theta_T}$$
$$= \sup_{\alpha \in \mathcal{U}} \int h(\theta,\alpha,x) dP_{\theta_T} \tag{10}$$

Furthermore the sup is reached at θ_T and uniqueness holds.

From (10), simple estimators for $\phi(\theta,\theta_T)$ and θ_T can be defined, plugging any convergent empirical measure in place of P_{θ_T} and taking the infimum in θ in the resulting estimator of $\phi(\theta,\theta_T)$.

In the context of simple i.i.d. sampling, introducing the empirical measure $P_n := \frac{1}{n}\sum_{i=1}^n \delta_{X_i}$ where the X_i's are i.i.d. r.v's with common unknown distribution P_{θ_T} in $\mathscr{P}$, the natural estimator of $\phi(\theta,\theta_T)$ is

$$\phi_n(\theta,\theta_T) := \sup_{\alpha \in \mathcal{U}} \left\{ \int h(\theta,\alpha,x) \, dP_n(x) \right\}$$
$$= \sup_{\alpha \in \mathcal{U}} \int \varphi' \left(\frac{dP_\theta}{dP_\alpha} \right) dP_\theta - \frac{1}{n} \sum_{i=1}^n \varphi^\# \left(\frac{dP_\theta}{dP_\alpha}(X_i) \right) \text{ when } (A) \text{ holds.}$$

As stated in Theorem 3.2 in Broniatowski and Keziou [3]:

Theorem 2. *Under some derivability assumptions on* $\varphi\left(\frac{dP_\theta}{dP_\alpha}\right)$ *(conditions A.0 to A.2 in Broniatowski and Keziou [3]),*

$$\text{If } \theta = \theta_T, \quad \text{then} \quad \frac{2n}{\varphi''(1)}\phi_n(\theta, \theta_T) \xrightarrow{d} \chi^2_{(d)} \quad \text{for } d = \dim(\Theta). \tag{11}$$

This last result of convergence of the estimated φ-divergence is of great interest in the problem we are taking on and serves as the basis for the test procedure that we propose.

3 A Simple Solution for Testing Finite Mixture Models

3.1 Testing Between Mixtures of Fully Characterized Components

Let us consider a set of signed measures defined by

$$p_\theta = (1 - \theta)p_0 + \theta p_1, \quad \theta \in \mathbb{R}, \tag{12}$$

where p_0 and p_1 are two known densities (belonging or not to the same parametric family).

In relation with (5), the case $\theta_T = 0$ is now an interior point of the parameter space.

We observe a random sample $X_1, \ldots, X_n$ of distribution p_T. We are willing to test:

$$H_0 : p_T = p_0 \text{ vs } H_1 : p_T = p_\theta \neq p_0 \tag{13}$$

which can be reduced to

$$H_0 : \theta = 0 \text{ vs } H_1 : \theta \neq 0 \tag{14}$$

whenever $p_0 \neq p_1$ is met. The latter condition ensures the identifiability of the model and enables to consider different parametric families for p_0 and p_1. Conversely, Chen et al. [7], for instance, assumes that $0 < \theta < 1$, and tests the equality of the parameters of p_0 and p_1 inside a unique family $\mathscr{F}$.

In the following, we thus assume that $p_0 \neq p_1$.

3.2 Test Statistics

The choice of the test statistic is driven by the result given in Theorem 2. Accordingly, let ϕ be any divergence associated with convex finite functions and such that 0 is an interior point of the space parameter defined by:

$$\Theta := \left\{ \alpha \in \mathbb{R} : \int | \varphi'\left(\frac{dP_0}{dP_\alpha}\right) | \, dP_0 < \infty \right\} \tag{15}$$

Then the statistic $2n\phi_n(0, \theta_T)$ can be used as a test statistic for (14) and

$$2n\phi_n(0, \theta_T) \longrightarrow \chi^2_{(1)} \text{ when } H_0 \text{ holds.} \tag{16}$$

Also, (16) holds when testing whether the true distribution is a k_0 component mixture or a k component mixture as in (3). In this case, the test statistic $2n\phi_n(\Theta_0, \theta_T)$ converges to a $\chi^2_{(k-k_0)}$ distribution when H_0 holds.

We restrict in the sequel to two generators.

Chi-Square Divergence

The first divergence that we consider is the χ^2-divergence. The corresponding φ function $\varphi_2(x) := \frac{1}{2}(x-1)^2$ is defined and convex on whole $\mathbb{R}$; an example when $\mathscr{P}$ may contain signed finite measures and not be restricted to probability measures is considered in Broniatowski and Keziou [2] in relation with a two components mixture model defined in (12) and where θ is allowed to assume values in an open neighborhood Θ of 0, in order to provide a test for (14), with θ an interior point of Θ.

Extended Kullback-Leibler Divergence

The second divergence that we retain is generated by a function described below, namely

$$\varphi_c(x) := (x+e^c-1)\cdot\log(x+e^c-1)+1-(x+e^c-1)+(1-c)\cdot(e^c-1)-c\cdot x \geq 0$$
$$x \in]1-e^c,\infty[, \quad c \in \mathbb{R}, \tag{17}$$

which has been derived within the recent general framework of Broniatowski and Stummer [6]. It is straightforward to see that φ_c is strictly convex and satisfies $\varphi_c(1) = 0 = \varphi_c'(1)$. For the special choice $c = 0$, (17) reduces to the omnipresent Kullback-Leibler divergence generator

$$\varphi_0(x) := x\log x - x + 1 \geq 0, \quad x \in]0,\infty[.$$

According to (17), in case of $c > 0$ the domain $]1-e^c,\infty[$ of φ_c covers also negative numbers (see Broniatowski and Stummer [5] for insights on divergence-generators with general real-valued domain); thus, the same facts holds for the new generator than for the χ^2 and this opens the gate to considerable comfort in testing mixture-type hypotheses against corresponding marginal-type alternatives, as we derive in the following. We denote KL_c the corresponding divergence functional for which $KL_c(Q,P)$ is well defined whenever P is a probability measure and Q is a signed measure.

It can be noted that the convergence of $I = \int |\phi'\left(\frac{p_0}{p_\theta}\right)| dP_0$ is not always guaranteed. This kind of considerations may guide the choice of the test statistic. For instance, in some cases, including scaling models, conditions that are required for the χ^2–divergence, do not apply to the KL_c–divergence.

For instance, consider a Gaussian mixture model with different variances:

$$p_0 \sim \mathcal{N}(\mu,\sigma_0^2), \quad p_1 \sim \mathcal{N}(\mu,\sigma_1^2).$$

The convergence of I with the χ^2 requires either $\sigma_1^2 > \sigma_0^2$ or $\sigma_0^2 > \sigma_1^2 > \frac{1}{2}\sigma_0^2$. On the other hand, the convergence is always ensured with the KL_c–divergence.

3.3 Generalization to Parametric Distributions with Unknown Parameters

In the previous section, the densities of each component were supposed to be known. We now generalize to the case where the components belong to parametric families with unknown parameter. We therefore deal with a way more complicate, issue and consider a generalized test procedure which aggregates tests of simple hypotheses over the components densities parameter spaces.

We present the generalization for a two component mixture, but it is valid as well for k component mixtures with $k \geq 2$. Let us assume $p_0 \in \mathscr{F}_0 = \{p_0(. \mid \lambda_0) : \lambda_0 \in \Lambda_0\}$ and $p_1 \in \mathscr{F}_1 = \{p_1(. \mid \lambda_1) : \lambda_1 \in \Lambda_1\}$, with Λ_0 and Λ_1 being compact subsets of $\mathbb{R}^d$, $d \geq 1$.

We consider aggregated tests of composite hypotheses.

For λ_0 fixed, $H_0(\lambda_0)$ is accepted if $\forall \lambda_1 \in \Lambda_1$, $H_0(\lambda_0)$ is accepted against $H_1(\lambda_0, \lambda_1)$. The aggregated hypothesis $H_0(\Lambda_0)$ is accepted if $\forall \lambda_0 \in \Lambda_0, \lambda_1 \in \Lambda_1$, $H_0(\lambda_0)$ is accepted against $H_1(\lambda_0, \lambda_1)$.

Thus the null hypothesis of homogeneity of the population is rejected when there exists at least one couple of parameters $(\lambda_0^*, \lambda_1^*) \in \Lambda_0 \times \Lambda_1$ with $\lambda_1^* \neq \lambda_0^*$ for which the simple hypothesis $H_0(\lambda_0^*)$ is rejected in favor of $H_1(\lambda_0^*, \lambda_1^*)$. In other words, the possibility that the underlying distribution is a mixture is enough for us to reject that there is a unique component.

Another perspective would be to consider that the null hypothesis $H_0(\Lambda_0)$ is rejected when there is no $\lambda_0 \in \Lambda_0$ such that $\forall \lambda_1 \in \Lambda_1$, $H_0(\lambda_0)$ is accepted against $H_1(\lambda_0, \lambda_1)$.

Note the condition $\{\lambda_1^* \neq \lambda_0^*\}$ is only required when p_0 and p_1 belong to the same parametric family.

Let $2n\phi_n(0, \theta_T \mid \lambda_0, \lambda_1)$ be the test statistic of the test (13) of the simple hypotheses $H_0(\lambda_0)$ vs $H_1(\lambda_0, \lambda_1)$ when λ_0 and λ_1 are fixed. Recall that $\phi_n(0, \theta_T \mid \lambda_0, \lambda_1)$ is the estimated divergence between $p_0(. \mid \lambda_0)$ and $p_{\theta_T}(. \mid \lambda_0, \lambda_1)$. The test statistic for (14) is derived from:

$$\Phi_n(0, \theta_T) = \sup_{\lambda_0 \in \Lambda_0} \sup_{\lambda_1 \in \Lambda_1 \setminus \lambda_0} \phi_n(0, \lambda_T \mid \lambda_0, \lambda_1) \tag{18}$$

where the parameter spaces Λ_0 and Λ_1 can be discretized in $\Lambda_{0,n}$ and $\Lambda_{1,n}$ for the sake of computational complexity.

In order to facilitate the computation of the test statistic, the successive optimizations have been rearranged as follows:

$$\Phi_n(0, \theta_T) = \sup_{\alpha \in \Theta} \left\{ \sup_{\lambda_0 \in \Lambda_0} \sup_{\lambda_1 \in \Lambda_1 \setminus \lambda_0} \int \varphi' \left(\frac{dP_{0, \lambda_0}}{dP_{\alpha, (\lambda_0, \lambda_1)}} \right) dP_{0, \lambda_0} - \frac{1}{n} \sum_{i=1}^n \varphi^\# \left(\frac{dP_{0, \lambda_0}}{dP_{\alpha, (\lambda_0, \lambda_1)}} (X_i) \right) \right\} \tag{19}$$

The critical region R_{Λ_0, Λ_1} associated with the aggregated test can be defined as follows:

$$\left\{ \Phi_n(0, \theta_T) \in R_{\Lambda_0, \Lambda_1} \right\} = \cup_{\lambda_0 \in \Lambda_0} \cup_{\lambda_1 \in \Lambda_1} \left\{ \phi_n(0, \theta_T \mid \lambda_0, \lambda_1) \in R_{\lambda_0, \lambda_1}(\alpha) \right\} \tag{20}$$

where $R_{\lambda_0, \lambda_1}(\alpha)$ is the critical region of risk α for the test of the simple hypotheses $H_0(\lambda_0)$ vs $H_0(\lambda_0, \lambda_1)$. α can then be tuned to ensure that the probability of (20) is of the wanted first kind level of risk $\alpha^\star$ for the global test.

Note that in this case, we do not have an equivalence to Theorem 2. Indeed, we do not directly estimate the true parameters of the densities p_0 and p_1, but rather aggregate the test over the parameter spaces Λ_0 and Λ_1. Thus, there is still no convergence result on the test statistic Φ_n. In the following, we evaluate the performances of the proposed test procedure through numerical simulations.

4 Numerical Simulations

4.1 Mixture of Fully Characterized Components

We here consider the simple case of a mixture between a Lognormal and a Weibull distribution whose parameters are supposed to be known. Results in Table 1 show that the test procedure of simple hypotheses performs well when the two components are fully characterized.

Table 1. Power of the test for a Lognormal and Weibull mixture with fully characterized components

	Lognormal and Weibull Mixture $lN(\lambda_0, 0.2)$ vs $0.8lN(\lambda_0, 0.2) + 0.2\mathcal{W}(\lambda_1, 2)$			
n = 250 observations				
	χ^2 **test statistic**		**KL$_c$ test statistic**	
First kind risk	0.05	0.10	0.05	0.10
Power	0.98	1	0.99	1

4.2 Mixture of Unknown Components Within a Parametric Family

The performances of the test procedure are evaluated numerically on three two-component mixtures. In the first two examples, both components belong to the same parametric family, while in the third, p_0 and p_1 are from different models. In each case, the distributions of the components are such that the resulting mixture is not bimodal. The following results are used on an illustrative basis. The critical regions are determined as to guarantee the value of the first kind risk $\alpha^\star$ at 0.05 and 0.10.

Lognormal Mixture
We first consider a Lognormal mixture. The two components belonging to the same parametric family, we can compare the performances of the divergence based test with the modified likelihood ratio test proposed by Chen et al. [7].
The alternate hypotheses are the following:

$$H_0 : p_T = p_0 \sim lN(\lambda_0, 1) \quad vs \quad H_1 : p_T = p_\theta \sim (1 - \theta)lN(\lambda_0, 1) + \theta lN(\lambda_1, 1),$$

where $\lambda_0 \in \Lambda_0 = [0.4, 1.6]$ and $\lambda_1 \in \Lambda_1 = [1.4, 2.6]$.
The critical region is computed numerically through Monte Carlo simulations under H_0. The power of the test is also computed numerically when the realizations are drawn

Table 2. Power of the tests for three types of mixtures whose components belong to a parametric family

	Lognormal Mixture $lN(1,1)$ vs $0.8lN(1,1)+0.2lN(2,1)$					
n = 250 observations						
	χ^2 **test statistic**		**KL$_c$ test statistic**		**Chen's modified lik ratio**	
First kind risk	0.05	0.10	0.05	0.10	0.05	0.10
Power	0.22	0.41	0.50	0.65	0.12	0.18
	Gamma Mixture $\mathscr{G}(2,1)$ vs $0.8\mathscr{G}(2,1)+0.2\mathscr{G}(5,2)$					
n = 250 observations						
	χ^2 **test statistic**		**KL$_c$ test statistic**		**Chen's modified lik ratio**	
First kind risk	0.05	0.10	0.05	0.10	0.05	0.10
Power	0.31	0.46	0.35	0.45	0.13	0.22
	Lognormal and Weibull Mixture $lN(\lambda_0,0.2)$ vs $0.8lN(\lambda_0,0.2)+0.2\mathscr{W}(\lambda_1,2)$					
n=250 observations						
	χ^2 **test statistic**		**KL$_c$ test statistic**			
First kind risk	0.05	0.10	0.05	0.10		
Power	0.28	0.47	0.34	0.57		

from the mixture model with $\theta = 0.2, \lambda_0 = 1$ and $\lambda_1 = 2$ for the χ^2 and KL_c test statistics and Chen's modified likelihood ratio.

The results in Table 2 show that both χ^2 and KL_c outperform the modified likelihood ratio test and the test based on the KL_c divergence achieves in this case the greatest power.

Gamma Mixture

We test the following hypothesis

$$H_0 : p_T = p_0 \sim \mathscr{G}(\lambda_0, 1) \quad vs \quad H_1 : p_T = p_\theta \sim (1 - \theta)\mathscr{G}(\lambda_0, 1) + \theta\mathscr{G}(\lambda_1, 2),$$

where $\lambda_0 \in \Lambda_0 = [1.4, 2.6]$ and $\lambda_1 \in \Lambda_1 = [4.4, 5.6]$. The realizations are drawn from the mixture model with $\theta = 0.2, \lambda_0 = 2$ and $\lambda_1 = 5$.

Here again, both divergence based statistics achieves higher power than the modified likelihood ratio test (cf. Table 2).

Weibull and Lognormal Mixture

We here consider the case where the two components are from different parametric families. We want to test

$$H_0 : p_T = p_0 \sim lN(\lambda_0, 0.2) \quad vs \quad H_1 : p_T = p_\theta \sim (1 - \theta)lN(\lambda_0, 0.2) + \theta\mathscr{W}(\lambda_1, 2),$$

where $\lambda_0 \in \Lambda_0 = [0.4, 1.6]$ and $\lambda_1 \in \Lambda_1 = [2.4, 3.6]$. The realizations are drawn from mixture model with $\theta = 1, \lambda_0 = 1$ and $\lambda_1 = 3$.

The results in Table 2 show that the test based on the KL_c divergence performs better than the χ^2 statistic.

Concerning the choice of the test statistic, we might note that the KL_c performs better when the two alternate distributions differ mainly in their central tendency, while the χ^2 might be prefered when the difference lays in the tails.

References

1. Azais, J.M., Gassiat, E., Mercadier, C.: Asymptotic distribution and local power of the log-likelihood ratio test for mixtures: bounded and unbounded cases. Bernoulli **12**(5), 775–799 (2006)
2. Broniatowski, M., Keziou, A.: Minimization of ϕ-divergences on sets of signed measures. Studia Sci. Math. Hungar. **43**(4), 403–442 (2006)
3. Broniatowski, M., Keziou, A.: Parametric estimation and tests through divergences and the duality technique. J. Multivar. Anal. **100**, 16–36 (2009)
4. Broniatowski, M., Jureckova, J., Kumar, A.M., Miranda, E.: Composite tests under corrupted data. Entropy **21**(1), 63 (2009)
5. Broniatowski, M., Stummer, W.: Some universal insights on divergences for statistics, machine learning and artificial intelligence. In: Nielsen, F. (ed.) Geometric Structures of Information, pp. 149–211. Springer, Switzerland (2019). https://doi.org/10.1007/978-3-030-02520-5_8
6. Broniatowski, M., Stummer, W.: A bare simulation approach to finding minimum distances. Preprint (2019)
7. Chen, H., Chen, J., Kalbfleisch, J.D.: A modified likelihood ratio test for homogeneity in finite mixture models. J. R. Statist. Soc. B **63**(1), 19–29 (2001)
8. Castillo, N.O., Gallardo, D.I., Bolfarine, H., Gomez, H.W.: Truncated power-normal distribution with application to non-negative measurements. Entropy **20**(6), 433 (2018)
9. Ciuperca, G.: Likelihood ratio statistic for exponential mixtures. Ann. Inst. Stat. Math. **54**(3), 585–594 (2002)
10. Cressie, N., Read, T.R.C.: Multinomial goodness-of-fit tests. J. Roy. Statist. Soc. Ser. B **46**(3), 440–464 (1984)
11. Csiszár, I.: Eine informationstheoretische Ungleichung und ihre Anwendung auf den Beweis der Ergodizität von Markoffschen Ketten. Magyar Tud. Akad. Mat. Kutató Int. Közl **8**, 85–108 (1963)
12. Liese, F., Vajda, I.: Convex Statistical Distances, vol. 95. BSB B. G. Teubner Verlagsgesellschaft, Leipzig (1987)
13. Pardo, L.: Statistical Inference Based on Divergence Measures. Statistics: Textbooks and Monographs, vol. 185. Chapman & Hall/CRC, Boca Raton (2006)
14. Rényi, A.: On measures of entropy and information. In: Proceedings of the 4th Berkeley Symposium Mathematical Statistics and Probability, vol. 1, pp. 547–561. Univ. California Press, Berkeley(1961)
15. Self, S.G., Liang, K.-Y.: Asymptotic properties of maximum likelihood estimators and likelihood ratio tests under nonstandard conditions. J. Am. Statist. Assoc. **82**(398), 605–610 (1987)
16. Titterington, D. M., Smith, A. F. M., and Makov, U. E.: Statistical Analysis of Finite Mixture Distributions. Wiley Series in Probability and Mathematical Statistics: Applied Probability and Statistics. John Wiley & Sons Ltd., Chichester (1985)

Robust Estimation by Means of Scaled Bregman Power Distances. Part I. Non-homogeneous Data

Birgit Roensch and Wolfgang Stummer[(✉)]

Department of Mathematics, University of Erlangen–Nürnberg,
Cauerstrasse 11, 91058 Erlangen, Germany
stummer@math.fau.de

Abstract. In contemporary data analytics, one often models uncertainty-prone data as samples stemming from a sequence of independent random variables whose distributions are non-identical but linked by a common (scalar or multidimensional) parameter. For such a context, we present in the current Part I a new robustness-featured parameter-estimation framework, in terms of minimization of the scaled Bregman power distances of Stummer and Vajda [23] (see also [21]); this leads to a wide range of outlier-robust alternatives to the omnipresent (non-robust) method of maximum-likelihood-examination, and extends the corresponding method of Ghosh and Basu [7]. In Part II (see [20]), we provide some applications of our framework to data from potentially rare but dangerous events described by approximate extreme value distributions.

1 Introduction

One of the most widely used procedures to obtain parameter estimates from data is the celebrated maximum likelihood method, which is known to be non-robust against outliers. In order to circumvent this deficiency, one can start by noticing that within ("well-behaved") frameworks of identically distributed (i.i.d.) observations, the maximum likelihood estimate MLE can be equivalently derived as the parameter which minimizes the Kullback-Leibler divergence KL (relative entropy) between the data-derived empirical distribution and a pregiven class of parametric candidate distributions of interest. Correspondingly, robustness can be achieved by replacing the KL by a more adequate density-based distribution-distance (divergence, disparity, (dis)similarity measure, proximity measure). Prominent choices are, amongst others, the "classical" Bregman distances CBD (see e.g. Pardo and Vajda [18]); this includes e.g. the density power divergences DPD of Basu et al. [4] (see also e.g. Basu et al. [5]), with the abovementioned KL and the more robust but less efficient squared L_2-norm as special cases. For independent and i.i.d. data, the general robustness properties of parameter estimation by means of minimizing DPDs (rather than KL) were studied by Basu et al. [4].

© Springer Nature Switzerland AG 2019
F. Nielsen and F. Barbaresco (Eds.): GSI 2019, LNCS 11712, pp. 319–330, 2019.
https://doi.org/10.1007/978-3-030-26980-7_33

Those results were extended by Ghosh and Basu [7] to the context of independent but non-identically distributed data, and applied by Ghosh [9] to generate a robustified generalization of the tail index (extreme value index) estimate of Matthys and Beirlant [15].

Concerning some recent progress of divergences, Stummer [21] and Stummer and Vajda [23] introduced the concept of *scaled Bregman distances* SBD, which enlarges and flexibilizes the above-mentioned CBD divergence class; further insights on them, including the explicit behaviour at density-zeros and several connections to information geometry, can be found in the recent comprehensive paper of Broniatowski and Stummer [6], within an even much wider framework. A prominent SBD subclass are the scaled Bregman power distances SBPD of Stummer and Vajda [23], which can also be interpreted as scaled extensions of the DPDs. For the SBPD subcase of *adaptive* scalings, results on robustness of minimum-distance parameter estimation, testing as well as applications can be found in Kißlinger and Stummer [11–14], Roensch and Stummer [19].

In the light of the above-mentioned explanations, the main goals of this paper are:

(i) To explicitly compute the SBPD for exponential families of distributions, where the non-adaptive scaling measure is of the same type but with arbitrary (e.g., infinite) total mass. This extends some results of Stummer and Vajda [23], Kißlinger and Stummer [11].

(ii) To develop a general parameter estimation method based on minimum SBPD estimation with non-adaptive scalings, for the setup of independent data observations from non-identical distributions linked by a common (scalar or multidimensional) parameter. In particular, we extend some corresponding method of Ghosh and Basu [7].

To achieve this, in the current Part I we first present some basic facts about scaled Bregman power distances, see Sect. 2. The goal (i) respectively (ii) is achieved in Sect. 3 respectively Sect. 4. In the separate Part II (see [20]) we shall employ (i), (ii) in order to derive a robustification of the tail index estimate of Matthys and Beirlant [15].

2   Scaled Bregman Power Divergences

Let us assume that the modeled respectively observed (random) data take values in a state space $\mathscr{X}$ (with at least two distinct values), equipped with a system $\mathscr{A}$ of admissible events (σ-algebra) and a fixed – maybe *nonprobability* – distribution[1] λ. On this we want to measure the dissimilarity between two *probability* distributions F, G which are described by their probability-λ-densities $x \mapsto f(x) \geq 0$, $x \mapsto g(x) \geq 0$ via $F[A] = \int_A f(x) \, d\lambda(x)$, $G[A] = \int_A g(x) \, d\lambda(x)$ ($A \in \mathscr{A}$), with normalizations $\int_{\mathscr{X}} f(x) \, d\lambda(x) = \int_{\mathscr{X}} g(x) \, d\lambda(x) = 1$. The set of all such *probability* distributions will be denoted by $\mathscr{M}_\lambda^1$. We also employ the

---

[1] Sigma-finite measure.

set $\mathscr{M}_\lambda$ of all general – maybe *nonprobability* – distributions M of the form $M[A] = \int_A m(x) \, d\lambda(x) \; (A \in \mathscr{A})$ with λ–density $x \mapsto m(x) > 0$. For instance, in the *discrete setup* where $\mathscr{X} = \mathscr{X}_{count}$ has countably many elements and $\lambda := \lambda_{count}$ is the counting measure (i.e., $\lambda_{count}[\{x\}] = 1$ for all $x \in \mathscr{X}_{count}$) then $f(\cdot), g(\cdot)$ are (e.g. binomial) probability mass functions and $m(\cdot)$ is a (e.g. unnormalized-histogram-related) general mass function. If λ is the Lebesgue measure on $\mathscr{X} = \mathbb{R}$ (and hence, except for rare cases, the integrals turn into Riemann integrals), then $f(\cdot), g(\cdot)$ are classical (e.g. Gaussian) probability density functions and $m(\cdot)$ is a general classical (possibly unnormalized) density function.

In such a setup, Stummer [21], Stummer and Vajda [23] (see also Broniatowski and Stummer [6]) introduced the following family $\{D_{\phi,M}(G,F) : \phi \in \Phi, M \in \mathscr{M}_\lambda\}$ of distances (divergences, measures of dissimilarities) between two probability distributions G and F, where Φ is the class of functions $\phi : (0, \infty) \mapsto \mathbb{R}$ which are continuously differentiable with derivative ϕ', strictly convex and continuously extended to the boundary points $v = 0$ and $v = \infty$:

Definition 1. *Let $\phi \in \Phi$. Then the Bregman distance (divergence) of $G, F \in \mathscr{M}_\lambda^1$ scaled by $M \in \mathscr{M}_\lambda$ is defined by*

$$0 \leq D_{\phi,M}(G,F) := B_\phi(G, F \| M)$$
$$:= \overline{\int}_{\mathscr{X}} \left[\phi\left(\frac{g(x)}{m(x)}\right) - \phi\left(\frac{f(x)}{m(x)}\right) - \phi'\left(\frac{f(x)}{m(x)}\right) \cdot \left(\frac{g(x)}{m(x)} - \frac{f(x)}{m(x)}\right) \right] m(x) \, d\lambda(x). \quad (1)$$

To guarantee the existence of the integral in (1), the zeros of f and g have to be incorporated by appropriate limit-taking and conventions; this is indicated by using $\overline{\int}$ instead of $\int$, see [6] (who even deal with a much more general framework) for details.

Generalizations thereof can be found in [22] and [6]. From Definition 1, the family $\{D_{\phi_\alpha,M}(G,F) : \alpha \in \mathbb{R}, M \in \mathscr{M}_\lambda\}$ of *scaled Bregman power divergences SBPD* (cf. Stummer and Vajda [23], see also [6,21]) with tuning parameter α and (tuning) scale-distribution M is obtained by

$$0 \leq B_{\phi_\alpha}(G, F \| M)$$
$$= \int_{\mathscr{X}} \frac{m(x)^{-\alpha}}{\alpha \cdot (\alpha+1)} \cdot \left[g(x)^{\alpha+1} + \alpha \cdot f(x)^{\alpha+1} - (\alpha+1) \cdot g(x) \cdot f(x)^\alpha \right] d\lambda(x), \quad \text{for } \alpha > 0,$$

$$0 \leq B_{\phi_\alpha}(G, F \| M)$$
$$= \int_{\mathscr{X}} \frac{m(x)^{-\alpha}}{(\alpha+1) \cdot \alpha} \cdot \left[g(x)^{\alpha+1} + \alpha \cdot f(x)^{\alpha+1} - (\alpha+1) \cdot g(x) \cdot f(x)^\alpha \right] \cdot \mathbf{I}_{]0,\infty[}\big(g(x) \cdot f(x)\big) \, d\lambda(x)$$

$$+ \int_{\mathscr{X}} \infty \cdot \mathbf{I}_{]0,\infty[}\big(g(x)\big) \cdot \mathbf{I}_{\{0\}}\big(f(x)\big) \, d\lambda(x)$$

$$+ \int_{\mathscr{X}} m(x)^{-\alpha} \cdot \left[\frac{f(x)^{\alpha+1}}{\alpha+1} \cdot \mathbf{I}_{]-1,0[}(\alpha) + \infty \cdot \mathbf{I}_{]-\infty,-1[}(\alpha) \right] \cdot \mathbf{I}_{]0,\infty[}\big(f(x)\big) \cdot \mathbf{I}_{\{0\}}\big(g(x)\big) \, d\lambda(x),$$

$$\text{for } \alpha \in \;] - \infty, -1[\cup] - 1, 0[, \quad (2)$$

$$0 \leq B_{\phi_0}(G, F \| M)$$

$$= \int_{\mathscr{X}} \left[g(x) \cdot \log\left(\frac{g(x)}{f(x)}\right) + f(x) - g(x) \right] \cdot \mathbf{I}_{]0,\infty[}(g(x) \cdot f(x)) \, \mathrm{d}\lambda(x)$$

$$+ \int_{\mathscr{X}} \infty \cdot \mathbf{I}_{]0,\infty[}(g(x)) \cdot \mathbf{I}_{\{0\}}(f(x)) \, \mathrm{d}\lambda(x) + \int_{\mathscr{X}} f(x) \cdot \mathbf{I}_{]0,\infty[}(f(x)) \cdot \mathbf{I}_{\{0\}}(g(x)) \, \mathrm{d}\lambda(x),$$

$$0 \leq B_{\phi_{-1}}(G, F \| M)$$

$$= \int_{\mathscr{X}} m(x) \cdot \left[-\log\left(\frac{g(x)}{f(x)}\right) + \frac{g(x)}{f(x)} - 1 \right] \cdot \mathbf{I}_{]0,\infty[}(g(x) \cdot f(x)) \, \mathrm{d}\lambda(x)$$

$$+ \int_{\mathscr{X}} \infty \cdot \mathbf{I}_{]0,\infty[}(g(x)) \cdot \mathbf{I}_{\{0\}}(f(x)) \, \mathrm{d}\lambda(x) + \int_{\mathscr{X}} \infty \cdot \mathbf{I}_{]0,\infty[}(f(x)) \cdot \mathbf{I}_{\{0\}}(g(x)) \, \mathrm{d}\lambda(x), \quad (3)$$

where we have employed the indicator function $\mathbf{I}_A(\cdot)$ of the set A and the generators

$$\phi_\alpha(v) := \frac{v^{\alpha+1} - (\alpha+1) \cdot v + \alpha}{\alpha \cdot (\alpha+1)} \geq 0, \quad v \in]0, \infty[, \quad \alpha \in \mathbb{R} \backslash \{-1, 0\},$$

$$\phi_0(v) := \lim_{\alpha \to 0} \phi_\alpha(v) = v \cdot \log v + 1 - v \geq 0, \quad v \in]0, \infty[,$$

$$\phi_{-1}(v) := \lim_{\alpha \to -1} \phi_\alpha(v) = -\log v + v - 1 \geq 0, \quad v \in]0, \infty[,$$

and appropriately incorporated the zeros of f and g. It is straightforward to see by construction, that for all $\alpha \in \mathbb{R}$ there holds $B_{\phi_\alpha}(G, F \| M) \geq 0$, as well as, $B_{\phi_\alpha}(G, F \| M) = 0$ if and only if $G = F$. Notice that $B_{\phi_0}(G, F \| M)$ does not depend on the scaling M. If $f(x) > 0$ for (λ−almost) all $x \in \mathscr{X}$, then one gets the following:

(ai) $B_{\phi_0}(G, F \| M) = \int_{\mathscr{X}} g(x) \cdot \log\left(\frac{g(x)}{f(x)}\right) \mathrm{d}\lambda(x)$ which is nothing but the well-known Kullback-Leibler divergence (relative entropy);

(aii) all the integrals in (2) to (3), which involve $\mathbf{I}_{\{0\}}(f(x))$, become zero;

(aiii) if g is allowed to have zeros, one should only use $\alpha > -1$.

In the special subcase for which $\alpha > 0$ and $m(x) = 1$, the SBPD $B_{\phi_\alpha}(G, F \| M)$ reduces to a multiple of the α−order density power divergence (DPD) of Basu et al. [4].

3   Scaled Bregman Power Divergences for Exponential Families

Let us consider the special sub-setup of exponential families which is important e.g. for information geometry (see e.g. Amari and Nagaoka [2], Amari [1], Ay et al. [3] and Nielsen and Garcia [16] for comprehensive overviews, and Nielsen and Hadjeres [17] for an exemplary very recent application). To begin with, suppose that the data come from a subspace of the d−dimensional Euclidean space, i.e. $(\mathscr{X}, \mathscr{A}) \subseteq (\mathbb{R}^d, \mathscr{B}^d)$, $d \in \mathbb{N}$. Moreover, for fixed integer $p \in \mathbb{N}$ denote by $\mathscr{E}^1(p,\boldsymbol{a},a_0,b,\boldsymbol{c},\boldsymbol{\Theta},\lambda)$ the family of parametric probability distributions $F_{\boldsymbol{\theta}}$ with λ−densities given by

$$f_{\boldsymbol{\theta}}(x) = \exp\left\{\boldsymbol{a}(x)^{\mathrm{t}}\boldsymbol{c}(\boldsymbol{\theta}) + a_0(x) - b(\boldsymbol{c}(\boldsymbol{\theta}))\right\} > 0, \quad x \in \mathscr{X},\ \boldsymbol{\theta} \in \boldsymbol{\Theta},$$

where $a_0,a_1,\ldots,a_p$ are real-valued functions on $\mathscr{X}$, $c_1,\ldots,c_p$ are real-valued functions on $\boldsymbol{\Theta}$, $\boldsymbol{a}(x) := (a_1(x),\ldots,a_p(x))^{\mathrm{t}}$ respectively $\boldsymbol{c}(\boldsymbol{\theta}) := (c_1(\boldsymbol{\theta}),\ldots,$ $c_p(\boldsymbol{\theta}))^{\mathrm{t}}$ are corresponding column vectors, and $\boldsymbol{a}(x)^{\mathrm{t}}\boldsymbol{c}(\boldsymbol{\theta}) = \sum_{i=1}^p a_i(x)\cdot c_i(\boldsymbol{\theta})$ their scalar product; furthermore, $\boldsymbol{\Theta} \subset \mathbb{R}^\ell$ for some $\ell \in \mathbb{N}$ such that

$$\boldsymbol{c}(\boldsymbol{\Theta}) \subset \mathscr{Y} := \left\{\boldsymbol{y} \in \mathbb{R}^p : \int_{\mathscr{X}} \exp\left\{\boldsymbol{a}(x)^{\mathrm{t}}\boldsymbol{y} + a_0(x)\right\}\,\mathrm{d}\lambda(x) \in\,]0,\infty[\right\},$$

and $b(\boldsymbol{y}) := \log\int_{\mathscr{X}} \exp\left\{\boldsymbol{a}(x)^{\mathrm{t}}\boldsymbol{y} + a_0(x)\right\}\,\mathrm{d}\lambda(x)$ for $\boldsymbol{y} \in \mathscr{Y}$. In other words, $\mathscr{E}^1(p,\boldsymbol{a},b,\boldsymbol{c},\boldsymbol{\Theta},\lambda)$ is a p−parametric exponential family of *probability* distributions from which we chose F and G. For the scaling M we use a connected exponential family $\mathscr{E}(p,\boldsymbol{a},\widetilde{b},\widetilde{\boldsymbol{c}},\boldsymbol{\Xi},\lambda)$ of general – maybe *nonprobability* – distributions $M_{\boldsymbol{\eta}}$ with λ−densities given by

$$m_{\boldsymbol{\eta}}(x) = \exp\left\{\boldsymbol{a}(x)^{\mathrm{t}}\widetilde{\boldsymbol{c}}(\boldsymbol{\eta}) + a_0(x) - \widetilde{b}(\widetilde{\boldsymbol{c}}(\boldsymbol{\eta}))\right\} \in\,]0,\infty[, \quad x \in \mathscr{X},\ \boldsymbol{\eta} \in \boldsymbol{\Xi},$$

where $\boldsymbol{\Xi} \subset \mathbb{R}^\ell$ and $\widetilde{b}$ is an arbitrary (measurable) real-valued function on $\mathbb{R}^p$. For such a setup, we obtain the following explicit SBPD representation:

Theorem 1. *Let* $F_{\boldsymbol{\theta}_1} \neq F_{\boldsymbol{\theta}_2}$ *be from* $\mathscr{E}^1(p,\boldsymbol{a},b,\boldsymbol{c},\boldsymbol{\Theta},\lambda)$ *with some* $\boldsymbol{\theta}_1 \neq \boldsymbol{\theta}_2$, *and* $M_{\boldsymbol{\eta}}$ *be from* $\mathscr{E}(p,\boldsymbol{a},\widetilde{b},\widetilde{\boldsymbol{c}},\boldsymbol{\Xi},\lambda)$ *with some* $\boldsymbol{\eta} \in \boldsymbol{\Xi}$. *Furthermore, suppose that all four quantities* $|\boldsymbol{c}(\boldsymbol{\theta}_i)|$ $(i=1,2)$, $|\widetilde{\boldsymbol{c}}(\boldsymbol{\eta})|$, $|\widetilde{b}(\widetilde{\boldsymbol{c}}(\boldsymbol{\eta}))|$ *are finite.*

(a) If $\alpha\cdot(\alpha+1) \neq 0$ *and* $(\alpha+1)\cdot\boldsymbol{c}(\boldsymbol{\theta}_i) - \alpha\cdot\widetilde{\boldsymbol{c}}(\boldsymbol{\eta}) \in \mathscr{Y}$ $(i=1,2)$, *then one gets*

$$B_{\phi_\alpha}(F_{\boldsymbol{\theta}_1},F_{\boldsymbol{\theta}_2}\,|\,M_{\boldsymbol{\eta}}) = \frac{e^{\rho_\alpha(\boldsymbol{c}(\boldsymbol{\theta}_1),\widetilde{\boldsymbol{c}}(\boldsymbol{\eta}))}}{\alpha(\alpha+1)} + \frac{e^{\rho_\alpha(\boldsymbol{c}(\boldsymbol{\theta}_2),\widetilde{\boldsymbol{c}}(\boldsymbol{\eta}))}}{\alpha+1} - \frac{e^{\sigma_\alpha(\boldsymbol{c}(\boldsymbol{\theta}_1),\boldsymbol{c}(\boldsymbol{\theta}_2),\widetilde{\boldsymbol{c}}(\boldsymbol{\eta}))}}{\alpha}, \quad (4)$$

with (for $i=1,2$)

$$\rho_\alpha(\boldsymbol{c}(\boldsymbol{\theta}_i),\widetilde{\boldsymbol{c}}(\boldsymbol{\eta})) := b((\alpha+1)\cdot\boldsymbol{c}(\boldsymbol{\theta}_i) - \alpha\cdot\widetilde{\boldsymbol{c}}(\boldsymbol{\eta})) - (\alpha+1)\cdot b(\boldsymbol{c}(\boldsymbol{\theta}_i)) + \alpha\cdot\widetilde{b}(\widetilde{\boldsymbol{c}}(\boldsymbol{\eta})),$$
$$\sigma_\alpha(\boldsymbol{c}(\boldsymbol{\theta}_1),\boldsymbol{c}(\boldsymbol{\theta}_2),\widetilde{\boldsymbol{c}}(\boldsymbol{\eta})) := b(\boldsymbol{c}(\boldsymbol{\theta}_1) + \alpha\cdot[\boldsymbol{c}(\boldsymbol{\theta}_2) - \widetilde{\boldsymbol{c}}(\boldsymbol{\eta})])$$
$$-b(\boldsymbol{c}(\boldsymbol{\theta}_1)) + \alpha\cdot[\widetilde{b}(\widetilde{\boldsymbol{c}}(\boldsymbol{\eta})) - b(\boldsymbol{c}(\boldsymbol{\theta}_2))].$$

If additionally $\boldsymbol{c}(\boldsymbol{\theta}_1) + \alpha\cdot[\boldsymbol{c}(\boldsymbol{\theta}_2) - \widetilde{\boldsymbol{c}}(\boldsymbol{\eta})] \in \mathscr{Y}$ *holds, then the SBPD in (4) is finite.*

(b) If $\alpha = 0$ *and* $\boldsymbol{c}(\boldsymbol{\theta}_1) \in \overset{\circ}{\mathscr{Y}}$, *then one obtains for the KL*

$$B_{\phi_0}(F_{\boldsymbol{\theta}_1},F_{\boldsymbol{\theta}_2}\,|\,M_{\boldsymbol{\eta}}) = b(\boldsymbol{c}(\boldsymbol{\theta}_2)) - b(\boldsymbol{c}(\boldsymbol{\theta}_1)) - [\boldsymbol{c}(\boldsymbol{\theta}_2) - \boldsymbol{c}(\boldsymbol{\theta}_2)]^{\mathrm{t}}\nabla b(\boldsymbol{c}(\boldsymbol{\theta}_1)) < \infty,$$

with $\nabla b\left(\boldsymbol{y}\right) = \int_{\mathscr{X}} \exp\left\{\boldsymbol{a}(x)^{\mathrm{t}}\mathbf{y} + a_0(x) - b\left(\boldsymbol{y}\right)\right\} \cdot \boldsymbol{a}(x) \, \mathrm{d}\lambda(x), \quad \boldsymbol{y} \in \overset{\circ}{\mathscr{Y}}.$

The special subcase of Theorem 1 with $\boldsymbol{a}(x) = x$, $\boldsymbol{\varXi} = \boldsymbol{\Theta}$, $\widetilde{c}(\boldsymbol{\theta}) = \boldsymbol{\theta} = c(\boldsymbol{\theta})$, $\widetilde{b}(\cdot) = b(\cdot)$, $\boldsymbol{\eta} = \boldsymbol{\theta}_0$ (i.e. $M_{\boldsymbol{\eta}}$ is chosen as probability distribution being a natural exponential family member $F_{\boldsymbol{\theta}_0}$) for some $\boldsymbol{\theta}_0 \in \boldsymbol{\Theta}$, has been first treated in Stummer and Vajda [23]; later on, Kißlinger and Stummer [11] have dealt with the wider subcase of general $\boldsymbol{a}(\cdot)$, $\boldsymbol{\varXi} = \boldsymbol{\Theta}$, $\widetilde{c}(\cdot) = c(\cdot)$, $\widetilde{b}(\cdot) = b(\cdot)$, $\boldsymbol{\eta} = \boldsymbol{\theta}_0$ (i.e. $M_{\boldsymbol{\eta}}$ is chosen as probability distribution $F_{\boldsymbol{\theta}_0}$ from the underlying non-natural exponential family) for some $\boldsymbol{\theta}_0 \in \boldsymbol{\Theta}$.

We shall employ Theorem 1 in the separate Part II (see [20]), within an extreme-values-modeling exponential regression setup for tail index estimation. Another important potential field of application of Theorem 1 is the *nonhomogeneous*-data concerning framework of the omnipresent generalized linear models (GLM), which – e.g. for $d = p = 1$ – employs independent response (outcome) variables X_i whose distribution stem from a natural exponential family with $a(x) = x$, $c(\theta_i) = \theta_i$, arbitrary $a_0(x)$, and cumulant function $b(c(\theta_i)) = b(\theta_i)$ $(i = 1, \ldots, n)$, where for the sake of brevity we have set the typically involved nuisance parameter to be one. Moreover, the natural parameters θ_i are supposed to depend on a (for the sake of brevity) non-stochastic vector $\boldsymbol{z}_i$ of m covariates (features) in the following connected way: $\theta_i = g(h^{-1}(\beta_0 + \boldsymbol{\beta}^{\mathrm{t}}\boldsymbol{z}_i))$, where $g^{-1}(\theta_i) = b'(\theta_i) = \mu_i$ linking the mean μ_i of X_i to θ_i, and $\mu_i = h^{-1}(\beta_0 + \boldsymbol{\beta}^{\mathrm{t}}\boldsymbol{z}_i)$ linking the mean μ_i to the linear predictor $\beta_0 + \boldsymbol{\beta}^{\mathrm{t}}\boldsymbol{z}_i$, with the parameters β_0, $\boldsymbol{\beta} := (\beta_1, \ldots, \beta_m)^{\mathrm{t}}$ to be estimated from the observed data. This has been used – for the unscaled case $m_{\boldsymbol{\eta}} \equiv 1$ (called "density power divergences"), $\beta_0 = 0$ and for $\alpha \geq 0$ – by Ghosh and Basu [8] for the estimation of $\boldsymbol{\beta}$, and by Ghosh and Basu [10] for associated testing.

4   Robust Minimum SBPD Estimation

Let us now fix a general data space $(\mathscr{X}, \mathscr{A})$ and a general distribution λ on it. Furthermore, for arbitrarily fixed sample size $n \in \mathbb{N}$, suppose that our data observations $Y_1, \ldots, Y_n$ are independent and have – not necessarily identical but somehow linked – "true" probability distributions $Y_i \sim G_i \in \mathscr{M}_\lambda^1$ with (for the sake of brevity) strictly positive λ-density g_i $(i = 1, \ldots, n)$. Such a non-homogeneous setup we want to model by the parametric probability-distribution family

$$\mathscr{F}_{\boldsymbol{\Theta}}^{(n)} := \{F_{\boldsymbol{\theta}}^{(n)} := F_{1,\boldsymbol{\theta}} \otimes \cdots \otimes F_{n,\boldsymbol{\theta}} \mid F_{i,\boldsymbol{\theta}} \in \mathscr{M}_\lambda^1 \text{ with strictly positive } \lambda\text{-density } f_{i,\boldsymbol{\theta}}$$
$$\text{for each } i = 1, \ldots, n \text{ and each } \boldsymbol{\theta} \in \boldsymbol{\Theta}\},$$

where $\boldsymbol{\Theta} \subset \mathbb{R}^p$ for some fixed $p \in \mathbb{N}$; notice the slight abuse of notation since $\mathscr{F}_{\boldsymbol{\Theta}}^{(n)}$ need not be an exponential family. Accordingly, the model distributions may be non-identical (i–dependent) but share a common parameter $\boldsymbol{\theta}$. We also suppose that each $G_i \in \mathcal{M}_\lambda^1$ respectively each $M_i \in \mathcal{M}_\lambda$ has strictly positive λ−density m_i respectively g_i, as well as that for each $i = 1, \ldots, n$, $\boldsymbol{\theta} \in \boldsymbol{\Theta}$ and each (for the sake of brevity) $\alpha \in]-1, \infty[$ one can separate the corresponding SBPD into the finite integrals

$$B_{\phi_\alpha}(G_i, F_{i,\boldsymbol{\theta}} \| M_i) = \int_{\mathcal{X}} \frac{m_i(x)^{-\alpha}}{(\alpha+1) \cdot \alpha} \cdot g_i(x)^{\alpha+1} \, d\lambda(x)$$

$$+ \int_{\mathcal{X}} \frac{m_i(x)^{-\alpha}}{\alpha+1} \cdot f_{i,\boldsymbol{\theta}}(x)^{\alpha+1} \, d\lambda(x) + \int_{\mathcal{X}} \frac{m_i(x)^{-\alpha}}{\alpha} \cdot f_{i,\boldsymbol{\theta}}(x)^\alpha \, dG_i(x),$$

for $\alpha \in]-1, 0[\cup]0, \infty[$, \hfill (5)

$$B_{\phi_0}(G_i, F_{i,\boldsymbol{\theta}} \| M_i) = \int_{\mathcal{X}} g_i(x) \cdot \log(g_i(x)) \, d\lambda(x) - \int_{\mathcal{X}} \log(f_{i,\boldsymbol{\theta}}(x)) \, dG_i(x). \quad (6)$$

Furthermore, we assume that the scaling densities $m_i(\cdot)$ in (5), (6) do not depend on $f_{i,\boldsymbol{\theta}}(\cdot)$ and $g_i(\cdot)$. Moreover, for $\boldsymbol{G} := (G_1, \ldots, G_n)$, $\boldsymbol{M} := (M_1, \ldots, M_n)$ we define the average-SBPD-minimizing functional $\boldsymbol{T}_{\alpha, \boldsymbol{M}}(\boldsymbol{G})$ via

$$\frac{1}{n} \sum_{i=1}^n B_{\phi_\alpha}(G_i, F_{i,\boldsymbol{T}_{\alpha,\boldsymbol{M}}(\boldsymbol{G})} | M_i) := \inf_{\boldsymbol{\theta} \in \boldsymbol{\Theta}} \frac{1}{n} \sum_{i=1}^n B_{\phi_\alpha}(G_i, F_{i,\boldsymbol{\theta}} | M_i), \quad (7)$$

and require that "the best-fitting parameter" $\boldsymbol{\theta}^G := \boldsymbol{T}_{\alpha, \boldsymbol{M}}(\boldsymbol{G})$ is contained in an non-empty open subset $\widetilde{\boldsymbol{\Theta}}$ of $\boldsymbol{\Theta}$ such that for almost all $x \in \mathcal{X}$, all $\boldsymbol{\theta} \in \widetilde{\boldsymbol{\Theta}}$ and all $i \in \{1, \ldots, n\}$ the λ−densities $f_{i,\boldsymbol{\theta}}(x)$ are thrice continuously differentiable in $\boldsymbol{\theta}$; all $\boldsymbol{\theta}$−dependent integrals in (5), (6) are assumed to be thrice $\boldsymbol{\theta}$−differentiable such that the derivatives can be taken under the integral sign. Notice that $\boldsymbol{\theta}^G = \boldsymbol{\theta}^{true}$ in case that $G_i = F_{i,\boldsymbol{\theta}^{true}}$ for some $\boldsymbol{\theta}^{true} \in \boldsymbol{\Theta}$ $(i = 1, \ldots, n)$.

By first removing from $B_{\phi_\alpha}(G_i, F_{i,\boldsymbol{\theta}} | M_i)$ all $\boldsymbol{\theta}$−independent terms (being obsolete for the right-hand side of (7)), and afterwards in (7) plugging in $\boldsymbol{G}^{(n)} := (G_1^{(n)}, \ldots, G_n^{(n)}) := (\delta_{Y_1}, \ldots, \delta_{Y_n})$ instead of $\boldsymbol{G}$ (where δ_{Y_i} is the one-point (Dirac) distribution putting mass 1 to the observation Y_i), we arrive at the *minimum SBPD estimator* $\widehat{\boldsymbol{\theta}}_{n,\alpha,\boldsymbol{M}} := \boldsymbol{T}_{\alpha, \boldsymbol{M}}(\boldsymbol{G}^{(n)})$ to be the $\boldsymbol{\theta}$−minimizer of

$$H_{n,\alpha,\boldsymbol{M}}(\boldsymbol{\theta}) := \begin{cases} \frac{1}{n} \sum_{i=1}^n \left[\int_{\mathcal{X}} \frac{f_{i,\boldsymbol{\theta}}(x)^{1+\alpha}}{(\alpha+1) \cdot m_i(x)^\alpha} \, d\lambda(x) - \frac{f_{i,\boldsymbol{\theta}}(Y_i)^\alpha}{\alpha \cdot m_i(Y_i)^\alpha} \right], & \text{if } \alpha \in]-1, 0[\cup]0, \infty[, \\[4ex] -\frac{1}{n} \sum_{i=1}^n \log f_{i,\boldsymbol{\theta}}(Y_i), & \text{if } \alpha = 0. \end{cases}$$
(8)

Notice that $H_{n,\alpha,\boldsymbol{M}}(\boldsymbol{\theta})$ may become negative and hence does generally not constitute a distance/divergence anymore. Clearly, $\widehat{\boldsymbol{\theta}}_{n,0,\boldsymbol{M}}$ does not depend on the scaling $\boldsymbol{M}$ and is nothing but the celebrated maximum-likelihood estimator, and the corresponding estimating equation is

$$\nabla_{\boldsymbol{\theta}} H_{n,0,\boldsymbol{M}}(\boldsymbol{\theta}) = \sum_{i=1}^n \boldsymbol{u}_{i,\boldsymbol{\theta}}(Y_i) = \boldsymbol{0}, \quad (9)$$

where $\boldsymbol{u}_{i,\theta}(\cdot) := \nabla_\theta \log f_{i,\theta}(\cdot)$ is the likelihood score function of the $i-$th density. For the other tuning-parameter case $\alpha \in]-1, 0[\cup]0, \infty[$ we obtain the estimating equation

$$\nabla_\theta \, H_{n,\alpha,M}\,(\boldsymbol{\theta}) = \frac{1}{n} \sum_{i=1}^n \left[\int_{\mathscr{X}} \frac{f_{i,\theta}(x)^{1+\alpha} \cdot \boldsymbol{u}_{i,\theta}(x)}{m_i(x)^\alpha} \, \mathrm{d}\lambda(x) - \frac{f_{i,\theta}(Y_i)^\alpha \cdot \boldsymbol{u}_{i,\theta}(Y_i)}{m_i(Y_i)^\alpha} \right] = \boldsymbol{0}, \quad (10)$$

which reduces to (9) for $\alpha \to 0$. In the special subsetup of unscaled SBPD $m_i(x) = 1$ (called DPD), (10) reduces to the estimating equations of Ghosh and Basu [7]. Notice that another important subcase is the i.i.d. case $f_{i,\theta} = f_\theta$ together with $i-$independent scalings $M_i = M$ (for which (8) and thus (10) can be rewritten in terms of the usual empirical distribution).

In order to establish robustness measures, let us closer investigate the above-mentioned functional $\boldsymbol{\theta}^G = T_{\alpha,M}(\boldsymbol{G}) = T_{\alpha,M}(G_1, \ldots, G_n)$ for $\alpha > -1$. For one-point contaminations at $t_0 \in \mathscr{X}$ with degree $\epsilon \in [0,1]$ in the i_0-th data observation ($i_0 \in \{1, \ldots, n\}$) we achieve by straightforward calculations the corresponding influence function

$$t_0 \longrightarrow IF(t_0; i_0, T_{\alpha,M}, \boldsymbol{G}) := \frac{\partial}{\partial \epsilon} T\Big(G_1, \ldots, (1-\epsilon)\cdot G_{i_0} + \epsilon \cdot \delta_{t_0}, \ldots, G_n\Big)\Big|_{\epsilon=0}$$
$$= \left(\sum_{i=1}^n \mathbb{J}_{i,\alpha}^M(\boldsymbol{\theta})\right)^{-1} \left(\frac{f_{i_0,\theta}(t_0)^\alpha}{m_{i_0}(t_0)^\alpha} \cdot \boldsymbol{u}_{i_0,\theta}(t_0) - \boldsymbol{\xi}_{i_0,\alpha}^M(\boldsymbol{\theta})\right), \quad (11)$$

with matrix $\quad \mathbb{J}_{i,\alpha}^M(\boldsymbol{\theta}) := \displaystyle\int_{\mathscr{X}} \frac{f_{i,\theta}(x)^{\alpha+1}}{m_i(x)^\alpha} \cdot \boldsymbol{u}_{i,\theta}(x)\,\boldsymbol{u}_{i,\theta}(x)^{\mathrm{t}}\,\mathrm{d}\lambda(x)$

$$- \int_{\mathscr{X}} \frac{f_{i,\theta}(x)^\alpha}{m_i(x)^\alpha} \cdot (g_i(x) - f_{i,\theta}(x)) \cdot \left[\nabla \boldsymbol{u}_{i,\theta}(x) + \alpha \cdot \boldsymbol{u}_{i,\theta}(x)\,\boldsymbol{u}_{i,\theta}(x)^{\mathrm{t}}\right] \mathrm{d}\lambda(x)$$

and $\quad \boldsymbol{\xi}_{i,\alpha}^M(\boldsymbol{\theta}) := \displaystyle\int_{\mathscr{X}} \frac{f_{i,\theta}(x)^\alpha g_i(x)}{m_i(x)^\alpha} \cdot \boldsymbol{u}_{i,\theta}(x)\,\mathrm{d}\lambda(x)$.

Analogously, for contamination in all data observations we can deduce

$$\mathbf{t} \longrightarrow IF(\mathbf{t}; T_{\alpha,M}, \boldsymbol{G}) := \frac{\partial}{\partial \epsilon} T\Big((1-\epsilon)\cdot G_1 + \epsilon \cdot \delta_1, \ldots, (1-\epsilon)\cdot G_n + \epsilon \cdot \delta_n\Big)\Big|_{\epsilon=0}$$
$$= \left(\sum_{i=1}^n \mathbb{J}_{i,\alpha}^M(\boldsymbol{\theta})\right)^{-1} \sum_{i=0}^n \left(\frac{f_{i,\theta}(t_i)^\alpha}{m_i(t_i)^\alpha} \cdot \boldsymbol{u}_{i,\theta}(t_i) - \boldsymbol{\xi}_{i,\alpha}^M(\boldsymbol{\theta})\right) \quad (12)$$

with $\mathbf{t} := (t_1, ..., t_n) \in \mathscr{X}^n$. From (11) and (12), one can also derive other connected robustness measures such as e.g. the gross-error sensitivity and the self-standardized sensitivity. Breakdown-point analyses can be performed as well. For the sake of brevity, these investigations will appear elsewhere, including many $(\alpha, \boldsymbol{M})$–subcases which perform much more robust than the MLE estimator (which corresponds to $\alpha = 0$ with obsolete $\boldsymbol{M}$ and which has unbounded influence function). In the subsetup of unscaled SBPD $m_i(x) = 1$ (called DPD), (11) and (12) reduce to the influence functions of Basu and Ghosh [7].

In order to derive asymptotic properties, for $\alpha > -1$ we use the following notations $H_{n,\alpha,M}(\boldsymbol{\theta}) =: \frac{1}{n} \sum_{i=1}^{n} V_{\boldsymbol{\theta}}^{(i)}(Y_i)$ (cf. (8)), $\mathbb{A}_n := \frac{1}{n} \sum_{i=1}^{n} \mathbb{J}_{i,\alpha}^M(\boldsymbol{\theta}^G)$,

$$\Omega_n := \frac{1}{n} \sum_{i=1}^{n} \left[\int_{\mathscr{X}} \frac{g_i(x) \cdot f_{i,\theta^G}(x)^{2\alpha}}{m_i(x)^{2\alpha}} \, \boldsymbol{u}_{i,\theta^G}(x) \, \boldsymbol{u}_{i,\theta^G}(x)^{\mathsf{t}} \, \mathrm{d}\lambda(x) \; - \; \xi_{i,\alpha}^M(\boldsymbol{\theta}^G) \, \xi_{i,\alpha}^M(\boldsymbol{\theta}^G)^{\mathsf{t}} \right],$$

and employ the following additional assumptions for unlimited sample size $n \in \mathbb{N}$:

(AdA1) $\mathbb{J}_{i,\alpha}^M(\boldsymbol{\theta}^G)$ is positive definite for each $i \in \mathbb{N}$ and $\inf_{n \in \mathbb{N}} [\min$ eigenvalue of $\mathbb{A}_n] > 0$.

(AdA2) For each $j, l, r \in \{1, ..., p\}$ there exists a function $Z_{jlr}^{(i)}(\cdot)$ with

$$\frac{1}{n} \sum_{i=1}^{n} \mathbb{E}_{G_i}[Z_{jlr}^{(i)}(Y_i)] = O(1), \text{ such that } |\nabla_{\boldsymbol{\theta},jlr}^3 V_{\boldsymbol{\theta}}^{(i)}(x)| \leq Z_{jlr}^{(i)}(x) \text{ for all}$$

$\boldsymbol{\theta} \in \boldsymbol{\Theta}$, $i \in \{1, \dots, n\}$ and alm. all $x \in \mathscr{X}$.

(AdA3) For all $j, l \in \{1, ..., p\}$ we have for $\boldsymbol{\theta} = \boldsymbol{\theta}^G$

$$\lim_{N \to \infty} \sup_{n > 1} \left\{ \frac{1}{n} \sum_{i=1}^{n} \mathbb{E}_{G_i}\left[|\nabla_{\boldsymbol{\theta},j} V_{\boldsymbol{\theta}}^{(i)}(Y_i)| \cdot \mathsf{J}_{]N,\infty[}(|\nabla_{\boldsymbol{\theta},j} V_{\boldsymbol{\theta}}^{(i)}(Y_i)|) \right] \right\} = 0, \quad (13)$$

$$\lim_{N \to \infty} \sup_{n > 1} \left\{ \frac{1}{n} \sum_{i=1}^{n} \mathbb{E}_{G_i}\left[\, |\nabla_{\boldsymbol{\theta},jl}^2 V_{\boldsymbol{\theta}}^{(i)}(Y_i) - \mathbb{E}_{G_i}[\nabla_{\boldsymbol{\theta},jl}^2 V_{\boldsymbol{\theta}}^{(i)}(Y_i)]| \right.\right.$$

$$\left.\left. \cdot \mathsf{J}_{]N,\infty[}(|\nabla_{\boldsymbol{\theta},jl}^2 V_{\boldsymbol{\theta}}^{(i)}(Y_i) - \mathbb{E}_{G_i}[\nabla_{\boldsymbol{\theta},jl}^2 V_{\boldsymbol{\theta}}^{(i)}(Y_i)]|) \, \right] \right\} = 0. \quad (14)$$

(AdA4) For all $\epsilon > 0$ we have for $\boldsymbol{\theta} - \boldsymbol{\theta}^G$

$$\lim_{n \to \infty} \left\{ \frac{1}{n} \sum_{i=1}^{n} \mathbb{E}_{G_i}\left[|| \Omega_n^{-1/2} \nabla_{\boldsymbol{\theta}} V_{\boldsymbol{\theta}}^{(i)}(Y_i) ||^2 \mathsf{J}_{]\epsilon\sqrt{n},\infty[}(|| \Omega_n^{-1/2} \nabla_{\boldsymbol{\theta}} V_{\boldsymbol{\theta}}^{(i)}(Y_i) ||) \right] \right\} = 0.$$

Theorem 2. *Under the above-mentioned assumptions, the following assertions hold:*

i. *There exists a consistent sequence $\boldsymbol{\theta}_n$ of roots to the estimating Eq. (10).*
ii. *The asymptotic distribution (for $n \to \infty$) of $\sqrt{n} \cdot \Omega_n^{-1/2} \mathbb{A}_n (\boldsymbol{\theta}_n - \boldsymbol{\theta}^G)$ is p-dimensional normal with zero mean vector $\mathbf{0}_p$ and covariance matrix $\mathbb{I}_p$ (identity matrix).*

In the subsetup of unscaled SBPD $m_i(x) = 1$ (called DPD), Theorem 13 reduces to Theorem 3.1 of Ghosh and Basu [7]. The lengthy proof of Theorem 13, which extends the verification lines of the latter, will appear elsewehere; for the sake of brevity, we only give a short sketch thereof, in the following. To begin with part i, at parameter points $\boldsymbol{\theta}$ in a closed ball $B(\boldsymbol{\theta}^G, R) \subset \tilde{\boldsymbol{\Theta}}$ with center

$\boldsymbol{\theta}^G$ and sufficiently small radius $R > 0$ we perform a Taylor-series expansion of $H_{n,\alpha,M}(\boldsymbol{\theta})$ around $\boldsymbol{\theta}^G$ to achieve

$$H_{n,\alpha,M}(\boldsymbol{\theta}^G) - H_{n,\alpha,M}(\boldsymbol{\theta}) = \underbrace{\sum_{j=1}^{p} \left(\nabla_{\boldsymbol{\theta},j} H_{n,\alpha,M}(\boldsymbol{\theta}^G) \right) \cdot (\theta_j^G - \theta_j)}_{=:C_1}$$

$$+ \underbrace{\frac{1}{2} \sum_{j=1}^{p} \sum_{l=1}^{p} \left(\nabla_{\boldsymbol{\theta},jl}^2 H_{n,\alpha,M}(\boldsymbol{\theta}^G) \right) \cdot (\theta_j^G - \theta_j) \cdot (\theta_l - \theta_l^G)}_{=:C_2}$$

$$+ \underbrace{\frac{1}{6} \sum_{j=1}^{p} \sum_{l=1}^{p} \sum_{r=1}^{p} (\theta_j^G - \theta_j) \cdot (\theta_l - \theta_l^G) \cdot (\theta_r - \theta_r^G) \cdot \frac{1}{n} \sum_{i=1}^{n} \kappa_{jlr}^{(i)}(y_i) \cdot Z_{jlr}^{(i)}(y_i)}_{=:C_3} ,$$

where $0 \leq |\kappa_{jlr}^{(i)}(Y_i)| \leq 1$ for all $j, l, r \in \{1, ..., p\}$ and all $i \in \{1, ..., n\}$. Because of $\mathbb{E}_{G_i}\left[\nabla_{\boldsymbol{\theta},j} V_{\boldsymbol{\theta}}^{(i)}(Y_i) \right] = 0$ and formula (13), one can apply a generalization of the weak law of large numbers (GWLLN) to end up with $\lim_{n \to \infty} \mathbb{P}_G \left[|C_1| < p \cdot R^3 \right] = 1$. Similarly, from (14), the GWLLN, some appropriate diagonalization as well as assumption (AdA1) one can deduce the existence of a constant $b > 0$ such that $\lim_{n \to \infty} \mathbb{P}_G \left[C_2 < -b \cdot R \right] = 1$. Furthermore, by assumption (AdA2) and the GWLLN there follows $\lim_{n \to \infty} \mathbb{P}_G \left[|C_3| < c \cdot R^3 \right] = 1$ for some $c > 0$. Since $(p+c) \cdot R^3 - b \cdot R < 0$ for sufficiently small R, one can derive that $H_{n,\alpha,M}(\cdot)$ has a local minimum in the interior of $B(\boldsymbol{\theta}^G, R)$ which satisfies (10). From this, the assertion of part i follows in a straightforward manner. To derive part ii, we perform a Taylor-series expansion of $\widetilde{\boldsymbol{\theta}} \mapsto \nabla_{\boldsymbol{\theta},j} H_{n,\alpha,M}(\widetilde{\boldsymbol{\theta}})$ around $\boldsymbol{\theta}^G$ to obtain for the root-case $\widetilde{\boldsymbol{\theta}} = \boldsymbol{\theta}_n$ via (10) the formula

$$h_{n,j} := -\sqrt{n} \cdot \nabla_{\boldsymbol{\theta},j} H_{n,\alpha,M}(\boldsymbol{\theta}^G)$$

$$= \sum_{l=1}^{p} \underbrace{\sqrt{n} \cdot (\theta_{n,l} - \theta_l^G)}_{=: d_{n,l}} \cdot \underbrace{\left[\left(\nabla_{\boldsymbol{\theta},jl}^2 H_{n,\alpha,M}(\boldsymbol{\theta}^G) \right) + \frac{1}{2} \sum_{r=1}^{p} (\theta_{n,r} - \theta_r^G) \cdot \nabla_{\boldsymbol{\theta},jlr}^3 H_{n,\alpha,M}(\boldsymbol{\theta}_n^*) \right]}_{=: A_{n,jl}} ,$$

where $\boldsymbol{\theta}_n^* := \boldsymbol{\theta}_n + \tilde{\kappa} \cdot (\boldsymbol{\theta}^G - \boldsymbol{\theta}_n)$ with $\tilde{\kappa} \in [0, 1]$. Writing this in vector/matrix-form as $\mathbf{h}_n = \mathbb{A}_n \mathbf{d}_n$, one can deduce via (cf. (7))

$$\nabla_{\boldsymbol{\theta}} \frac{1}{n} \cdot \sum_{i=1}^{n} B_{\phi_\alpha}(G_i, F_{i,\boldsymbol{\theta}} | M_i) \bigg|_{\boldsymbol{\theta} = \boldsymbol{\theta}^G} = \mathbf{0}$$

that $\mathbb{E}_G[\mathbf{h}_n] = \mathbf{0}$. By lengthy but straightforward calculations, we can also compute the corresponding covariance matrix as $\mathbb{C}\text{ov}_G[\mathbf{h}_n] = \mathfrak{N}_n$. From Assumption (AdA4) together with the appropriate version of the central limit theorem, as well as some limit relation between $\mathbb{A}_n$ and $\mathbb{A}_n$, we end up with the assertion of part ii.

Acknowledgement. We are grateful to the five referees for their comments and very useful suggestions.

References

1. Amari, S.-I.: Information Geometry and Its Applications. Springer, Tokyo (2016)
2. Amari, S.-I., Nagaoka, H.: Methods of Information Geometry. Oxford University Press, Oxford (2000)
3. Ay, N., Jost, J., Lê, H.V., Schwachhöfer, L.: Information Geometry. EMG-FASMSM, vol. 64. Springer, Cham (2017). https://doi.org/10.1007/978-3-319-56478-4
4. Basu, A., Harris, I.R., Hjort, N.L., Jones, M.C.: Robust and efficient estimation by minimizing a density power divergence. Biometrika **85**(3), 549–559 (1998)
5. Basu, A., Shioya, H., Park, C.: Statistical Inference: The Minimum Distance Approach. CRC Press, Boca Raton (2011)
6. Broniatowski, M., Stummer, W.: Some universal insights on divergences for statistics, machine learning and artificial intelligence. In: Nielsen, F. (ed.) Geometric Structures of Information. SCT, pp. 149–211. Springer, Cham (2019). https://doi.org/10.1007/978-3-030-02520-5_8
7. Ghosh, A., Basu, A.: Robust estimation for independent non-homogeneous observations using density power divergence with applications to linear regression. Electron. J. Stat. **7**, 2420–2456 (2013)
8. Ghosh, A., Basu, A.: Robust estimation in generalized linear models: the density power divergence approach. TEST **25**, 269–290 (2016)
9. Ghosh, A.: Divergence based robust estimation of the tail index through an exponential regression model. Stat. Methods Appl. **26**, 181–213 (2017)
10. Ghosh, A., Basu, A.: Robust bounded influence tests for independent non-homogeneous observations. Statistica Sinica **28**, 1133–1155 (2018)
11. Kißlinger, A.-L., Stummer, W.: Some decision procedures based on scaled Bregman distance surfaces. In: Nielsen, F., Barbaresco, F. (eds.) GSI 2013. LNCS, vol. 8085, pp. 479–486. Springer, Heidelberg (2013). https://doi.org/10.1007/978-3-642-40020-9_52
12. Kißlinger, A.-L., Stummer, W.: New model search for nonlinear recursive models, regressions and autoregressions. In: Nielsen, F., Barbaresco, F. (eds.) GSI 2015. LNCS, vol. 9389, pp. 693–701. Springer, Cham (2015). https://doi.org/10.1007/978-3-319-25040-3_74
13. Kißlinger, A.-L., Stummer, W.: Robust statistical engineering by means of scaled Bregman distances. In: Agostinelli, C., Basu, A., Filzmoser, P., Mukherjee, D. (eds.) Recent Advances in Robust Statistics - Theory and Applications, pp. 81–113. Springer, India (2016). https://doi.org/10.1007/978-81-322-3643-6_5
14. Kißlinger, A.-L., Stummer, W.: A new toolkit for robust distributional change detection. Appl. Stoch. Models Bus. Ind. **34**, 682–699 (2018)
15. Matthys, G., Beirlant, J.: Estimating the extreme value index and high quantiles with exponential regression models. Statistica Sinica **13**, 853–880 (2003)
16. Nielsen, F., Garcia, V.: Statistical exponential families: a digest with flash cards. Preprint, arXiv:0911.4863v2 (2011)
17. Nielsen, F., Hadjeres, G.: Monte Carlo information-geometric structures. In: Nielsen, F. (ed.) Geometric Structures of Information. SCT, pp. 69–103. Springer, Cham (2019). https://doi.org/10.1007/978-3-030-02520-5_5

18. Pardo, M.C., Vajda, I.: On asymptotic properties of information-theoretic divergences. IEEE Trans. Inf. Theor. **49**(7), 1860–1868 (2003)
19. Roensch, B., Stummer, W.: 3D insights to some divergences for robust statistics and machine learning. In: Nielsen, F., Barbaresco, F. (eds.) GSI 2017. LNCS, vol. 10589, pp. 460–469. Springer, Cham (2017). https://doi.org/10.1007/978-3-319-68445-1_54
20. Roensch, B., Stummer, W.: Robust estimation by means of scaled Bregman power distances. Part II. Extreme values. In: Nielsen, F., Barbaresco, F. (eds.) GSI 2019. LNCS, vol. 11712, pp. 331–340. Springer, Cham (2019). https://doi.org/10.1007/978-3-030-26980-7_34
21. Stummer, W.: Some Bregman distances between financial diffusion processes. Proc. Appl. Math. Mech. **7**(1), 1050503–1050504 (2007)
22. Stummer, W., Kißlinger, A.-L.: Some new flexibilizations of Bregman divergences and their asymptotics. In: Nielsen, F., Barbaresco, F. (eds.) GSI 2017. LNCS, vol. 10589, pp. 514–522. Springer, Cham (2017). https://doi.org/10.1007/978-3-319-68445-1_60
23. Stummer, W., Vajda, I.: On Bregman distances and divergences of probability measures. IEEE Trans. Inform. Theory **58**(3), 1277–1288 (2012)

Robust Estimation by Means of Scaled Bregman Power Distances. Part II. Extreme Values

Birgit Roensch[1(✉)] and Wolfgang Stummer[1,2]

[1] Department of Mathematics, University of Erlangen–Nürnberg,
Cauerstrasse 11, 91058 Erlangen, Germany
roensch@math.fau.de

[2] Faculty Member of the School of Business and Economics, University of Erlangen–Nürnberg,
Lange Gasse 20, 90403 Nürnberg, Germany

Abstract. In the separate Part I (see [23]), we have derived a new robustness-featured parameter-estimation framework, in terms of minimization of the scaled Bregman power distances of Stummer and Vajda [25] (see also [24]); this leads to a wide range of outlier-robust alternatives to the omnipresent non-robust method of maximum-likelihood-examination. In the current Part II, we provide some applications of our framework to data from potentially rare but dangerous events (modeled with approximate extreme value distributions), by estimating the correspondingly characterizing extreme value index (reciprocal of tail index); as a special subcase, we recover the method of Ghosh [9] which is essentially a robustification of the procedure of Matthys and Beirlant [19]. Some simulation studies demonstrate the potential partial superiority of our method.

1 Introduction

In the last three decades, statistical and machine-learning issues in modelling (here, univariate) extremes of random data have been widely applied in many research fields, such as e.g. environmetrics (natural desasters), biometrics, insurance, finance, and corresponding triggered-loss-concerning risk management tasks. The basic idea is to quantify the potential danger that (losses from) rare events can have enormous – and even catastrophic – consequences on natural, human-made (including economic) and hybrid systems. For this, a large variety of quantification methods have been developed, see e.g. the books of Beirlant et al. [3], Castillo et al. [5], Coles et al. [6], Embrechts et al. [7], de Haan and Ferreira [8], Reiss and Thomas [20], Resnick [21], the recent survey paper of Gomes and Guillou [14], as well as the references therein. One fundamental result in the extreme value theory of independent random variables $(X_i)_{i \in \mathbb{N}}$ with identical distribution $\mathbb{P}^X[\cdot] := \mathbb{P}[X_i \in \cdot]$ ($i \in \mathbb{N}$) stems from the Fisher-Tippett-Gnedenko theorem (cf. [13]) which states that the only possible non-degenerate (i.e. not of one-point mass type) limit distributions (as $n \to \infty$) for normalized block-maxima

© Springer Nature Switzerland AG 2019
F. Nielsen and F. Barbaresco (Eds.): GSI 2019, LNCS 11712, pp. 331–340, 2019.
https://doi.org/10.1007/978-3-030-26980-7_34

$(\max\{X_1, \ldots, X_n\} - a_n)/b_n$ are members $H_{\theta,\mu,\sigma}$ of the so-called *generalized-extreme-value (GEV) distribution family* described by the probability density functions

$$h_{\theta,\mu,\sigma}(x) := \frac{\exp\left(-(1+\theta \cdot \frac{x-\mu}{\sigma})^{-1/\theta}\right)}{\sigma \cdot (1+\theta \cdot \frac{x-\mu}{\sigma})^{(\theta+1)/\theta}} \cdot \mathbb{1}_{]-1,\infty[}\left(\theta \cdot \frac{x-\mu}{\sigma}\right), \quad x \in \mathbb{R}, \ \theta \in \mathbb{R}\backslash\{0\},$$

$$h_{0,\mu,\sigma}(x) := \frac{1}{\sigma} \cdot \exp(-\frac{x-\mu}{\sigma}) \cdot \exp\left(-\exp(-\frac{x-\mu}{\sigma})\right), \quad x \in \mathbb{R}, \tag{1}$$

for some location parameter $\mu \in \mathbb{R}$ and some scale parameter $\sigma \in]0,\infty[$, where we have employed the indicator function $\mathbb{1}_A(\cdot)$ of the set A. Accordingly, $\mathbb{P}^X$ is said to be in the *maximum domain of attraction of* $H_{\theta,0,0}$, denoted by $\mathbb{P}^X \in MDA(H_{\theta,0,0})$. The shape parameter θ is called *extreme value index* EVI (and its reciprocal $1/\theta$ is called tail index, although some authors such as e.g. Ghosh [9] call θ itself the tail value index) and plays a key role in extreme value analytics. It is well known that essentially all the common continuous distributions $\mathbb{P}^X$ (e.g. Gaussian, etc.) used in statistics, machine learning and artificial intelligence are covered. Hence, for fixed large enough block-size $n \in \mathbb{N}$ and $n \cdot N$ data point observations, one can approximate the log-likelihood function of the $N \in \mathbb{N}$ block-maxima $Z_1 := \max\{X_1, \ldots, X_n\}$, $Z_2 := \max\{X_{n+1}, \ldots, X_{2n}\}$, $\ldots$, $Z_N := \max\{X_{n(N-1)+1}, \ldots, X_{Nn}\}$ of the corresponding N blocks in a straightforward manner from (1); accordingly, the maximum likelihood estimator MLE of (θ,μ,σ) can be derived. However, the MLE-procedure is known to be non-robust against outliers. One way to achieve a robust estimator $\widehat{\theta}$ of θ is given by Ghosh [9] who starts from Matthys and Beirlant's [19] nonhomogeneous-exponential-regression-type approximation for log-ratios of spacings of successive order statistics of the X's, and afterwards employs the corresponding special case of the general minimum density-power-divergence parameter-estimation method of Ghosh and Basu [10] (see also [11,12]). The latter has been extended and flexibilized in Roensch and Stummer [23] (cf. the separate Part I), by means of minimizing an appropriate scaled Bregman power divergence in the sense of Stummer and Vajda [25] (see also Stummer [24] and Sect. 2.2 of Broniatowski and Stummer [4]).

In the light of the above-mentioned explanations, the main goals of this paper are:

(i) To extend and flexibilize the Ghosh-method of extreme-value-index estimation by employing some of the results of Roensch and Stummer [23].

(ii) To present some significant simulation studies for which our method (i) performs superiorly to the Ghosh-method.

To achieve this, we first explain some basic facts about the fundamentally underlying Matthys-Beirlant method and about scaled Bregman power distances, see Sect. 2. The goal (i) respectively (ii) is realized in Sect. 3 respectively Sect. 4.

2   Matthys-Beirlant Model and Extreme-Value-Index Estimation

In their paper [19], Matthys & Beirlant derived the following approximation: for independent data observations $(X_i)_{i\in\mathbb{N}}$ of identical distributions $\mathbb{P}^X$ (satisfying the above-mentioned non-degeneracy assumption for the Fisher-Tippett-Gnedenko theorem), data size $n \in \mathbb{N}$, and corresponding order statistics $\min\{X_1,\ldots,X_n\} = X_{n:1} \leq X_{n:2} \leq \ldots \leq X_{n:n} = \max\{X_1,\ldots,X_n\}$, the log-ratios $(Y_j)_{j=1,\ldots,k-1}$ of spacings of the $k \in \{1,\ldots,n-1\}$ largest successive observations defined by

$$Y_j := j \cdot \log \left(\frac{X_{n:n-j+1} - X_{n:n-k}}{X_{n:n-j} - X_{n:n-k}} \right)$$

are approximately independent and exponentially distributed with non-homogeneous

means $\mathbb{E}[Y_j] = \dfrac{\theta}{1-(\frac{j}{k+1})^\theta} =: \gamma_j(\theta)^{-1} > 0,$ denoted by $\mathbb{P}^{Y_j} \overset{ind}{\approx} \mathbb{Exp}(\gamma_j(\theta))$,   (2)

depending on the extreme value index $\theta \in \mathbb{R}$. Accordingly, fitting θ to the data observations amounts to parameter estimation within an (approximately valid) exponential regression model. As a side remark, notice that

$$Y_j = j \cdot \log \left(\frac{\frac{X_{n:n-j+1}-a_n}{b_n} - \frac{X_{n:n-k}-a_n}{b_n}}{\frac{X_{n:n-j}-a_n}{b_n} - \frac{X_{n:n-k}-a_n}{b_n}} \right), \quad j = 1,\ldots,k, \ \ k \leq n-1,$$

for any sequences $(a_n)_{n\in\mathbb{N}}$ of real numbers and $(b_n)_{n\in\mathbb{N}}$ of strictly positive real numbers; consequently, the location parameter μ and scale parameter σ play no role here.

Let us imbed the current context as a special case of the general parameter-estimation framework of Roensch and Stummer [23] for independent non-homogeneous data observations, in terms of minimizing scaled Bregman power distances. To start with, we assume that the underlying "true" (approximate) joint distribution is of the (independence-reflecting product-)form $P^{(Y_1,\ldots,Y_{k-1})} = G_1 \otimes \cdots \otimes G_{k-1}$ with density $\prod_{j=1}^{k-1} g_j(x_j) > 0$ $(x_j \in [0,\infty[)$, and that it is modeled with the distribution family

$$\mathscr{F}_\Theta^{(k-1)} := \{F_\theta^{(k-1)} := F_{1,\theta} \otimes \cdots \otimes F_{k-1,\theta} \,|\, F_{j,\theta} = \mathbb{Exp}(\gamma_j(\theta)) \text{ having density}$$
$$[0,\infty) \ni x \mapsto f_{j,\theta}(x) = \gamma_j(\theta) \cdot \exp(-x \cdot \gamma_j(\theta)) > 0 \text{ for each } j = 1,\ldots,k-1 \text{ and } \theta \in \Theta\}$$

with $\Theta := \mathbb{R}$. Furthermore, let M be a fixed – maybe *nonprobability* – distribution (σ-finite measure) having – maybe nonnormalized – density $[0,\infty[\ni x \mapsto m(x) > 0$. Moreover, we fix an arbitrary tuning parameter $\alpha \in]-1,\infty[$. In terms of the α-order scaled Bregman power distance SBPD (cf. Stummer and Vajda [25] and [24]; see also [4,15–18,22] for further insights and applications)

$$B_{\phi_\alpha}(G_j, F_{j,\theta} \,||\, M) := \int_{\mathscr{X}} \frac{m(x)^{-\alpha}}{(\alpha+1)\cdot\alpha} \cdot g_j(x)^{\alpha+1} \, dx$$
$$+ \int_{\mathscr{X}} \frac{m(x)^{-\alpha}}{\alpha+1} \cdot f_{j,\theta}(x)^{\alpha+1} \, dx + \int_{\mathscr{X}} \frac{m(x)^{-\alpha}}{\alpha} \cdot f_{j,\theta}(x)^\alpha \, dG_j(x),$$
$$\text{for } \alpha \in]-1,0[\cup]0,\infty[, \quad (3)$$

$$B_{\phi_0}(G_j, F_{j,\theta} \,||\, M) = \int_{\mathscr{X}} g_j(x) \cdot \log(g_j(x)) \, dx - \int_{\mathscr{X}} \log(f_{j,\theta}(x)) \, dG_j(x), \quad (4)$$

as well as the short-hand notation $G := (G_1, \ldots, G_{k-1})$, we define the average-SBPD-minimizing functional $T_{\alpha,M}(G)$ via

$$\frac{1}{k-1} \sum_{j=1}^{k-1} B_{\phi_\alpha}(G_j, F_{j,T_{\alpha,M}(G)} \,||\, M) := \inf_{\theta \in \Theta} \frac{1}{k-1} \sum_{j=1}^{k-1} B_{\phi_\alpha}(G_j, F_{j,\theta} \,||\, M), \qquad (5)$$

with "best-fitting parameter" $\theta^G := T_{\alpha,M}(G)$. Notice that in case of $G_j = F_{j,\theta^{true}}$ for some $\theta^{true} \in \Theta$ $(j = 1, \ldots, k-1)$, one can apply the general Theorem 1 of Roensch and Stummer [23] to the special choices $d = 1 = p$, $\mathcal{X} =]0,\infty[= \mathcal{Y}$, $\Theta = \mathbb{R}$, $a_0(x) = 0$, $a(x) = -x$, $c(\overline{\theta}) = \gamma_j(\overline{\theta})$, $b(y) = -\log(y)$, $\Xi = \mathbb{R}$, $\tilde{c}(\eta) = c^M$, $\tilde{b}(\tilde{c}(\eta)) = b^M$ (and thus $m(x) = \exp(-x \cdot c^M - b^M)$), $\theta_1 = \theta^{true}$, $\theta_2 = \theta$, and derive from (3), (4) the SBPDs

$$\frac{1}{k-1} \sum_{j=1}^{k-1} B_{\phi_\alpha}(G_j, F_{j,\theta} \,||\, M) = \frac{1}{k-1} \sum_{j=1}^{k-1} \left\{ \frac{\gamma_j(\theta^{true})^{\alpha+1} \cdot e^{\alpha \cdot b^M}}{\alpha \cdot (\alpha+1)^2 \cdot \gamma_j(\theta^{true}) - \alpha^2 \cdot (\alpha+1) \cdot c^M} \right.$$

$$\left. + \frac{\gamma_j(\theta)^{\alpha+1} \cdot e^{\alpha \cdot b^M}}{(\alpha+1)^2 \cdot \gamma_j(\theta) - \alpha \cdot (\alpha+1) \cdot c^M} + \frac{\gamma_j(\theta^{true}) \cdot \gamma_j(\theta)^\alpha \cdot e^{\alpha \cdot b^M}}{\alpha \cdot \gamma_j(\theta^{true}) + \alpha^2 \cdot [\gamma_j(\theta) - c^M]} \right\} < \infty,$$

$$\text{if } \alpha \in]-1,0[\cup]0,\infty[,$$

$$\frac{1}{k-1} \sum_{j=1}^{k-1} B_{\phi_0}(G_j, F_{j,\theta} \,||\, M) = \frac{1}{k-1} \sum_{j=1}^{k-1} \left\{ \log(\gamma_j(\theta^{true})) - \log(\gamma_j(\theta)) + \frac{\gamma_j(\theta)}{\gamma_j(\theta^{true})} \right\} - 1 < \infty,$$

$$\text{if } \alpha = 0,$$

provided that for $\alpha \in]-1,0[\cup]0,\infty[$ the three constraints $(\alpha+1) \cdot \gamma_j(\theta^{true}) - \alpha \cdot c^M > 0$, $(\alpha+1) \cdot \gamma_j(\theta) - \alpha \cdot c^M > 0$, and $\gamma_j(\theta^{true}) + \alpha \cdot [\gamma_j(\theta) - c^M] > 0$ are satisfied; then, according to (5), one has the best-fitting parameter $\theta^G = T_{\alpha,M}(G) = \theta^{true}$ with $\frac{1}{k-1} \sum_{j=1}^{k-1} B_{\phi_\alpha}(G_j, F_{j,T_{\alpha,M}(G)} \,||\, M) = 0$ for all $\alpha > -1$. Notice that the above-mentioned first two constraints are always satisfied for $\alpha \in]-1,0[$ and $c^M \geq 0$, whereas all three constraints are fulfilled for $\alpha > 0$ and $c^M \leq 0$; the special choice $c^M = 0$ corresponds (up to the multiple constant b^M) to the unscaled case $m(x) \equiv 1$ which can be interpreted as a (set-up adapted) density power divergence in the sense of Basu et al. [1] (see also Basu et al. [2]). For the choice $c^M < 0$, the scaling M is not a finite distribution anymore.

Let us now return to the case where the transformed data (log-ratios) $(Y_j)_{j=1,\ldots,k-1}$ are generated from general (not necessarily $\mathscr{F}_\Theta^{(k-1)}$-family member) probability distributions G_j $(j = 1, \ldots, k-1)$. Following the lines of Roensch and Stummer [23], by first removing from $B_{\phi_\alpha}(G_j, F_{j,\theta} \,||\, M_j)$ $(\alpha > -1)$ in (3), (4) all θ-independent terms (being obsolete for the right-hand side of (5)), and afterwards plugging in $G^{(k-1)} := (G_1^{(k-1)}, \ldots, G_{k-1}^{(k-1)}) := (\delta_{Y_1}, \ldots, \delta_{Y_{k-1}})$ instead of G (where δ_{Y_j} is the one-point distribution putting mass 1 to the observation Y_j), we arrive at the *minimum SBPD estimator* $\hat{\theta}_{k-1,\alpha,M} := T_{\alpha,M}(G^{(k-1)})$ to be the θ-minimizer of

$$H_{k-1,\alpha,M}(\theta) := \frac{1}{k-1} \sum_{j=1}^{k-1} \left[\int_{\mathcal{X}} \frac{f_{j,\theta}(x)^{1+\alpha}}{(\alpha+1) \cdot m_j(x)^\alpha} \, dx - \frac{f_{j,\theta}(Y_j)^\alpha}{\alpha \cdot m_j(Y_j)^\alpha} \right]$$

$$= \frac{e^{\alpha \cdot b^M}}{k-1} \sum_{j=1}^{k-1} \left[\frac{\gamma_j(\theta)^{1+\alpha}}{(1+\alpha)^2 \cdot \gamma_j(\theta) - \alpha \cdot (\alpha+1) \cdot c^M} - \frac{\gamma_j(\theta)^\alpha}{\alpha} \cdot e^{-Y_j \cdot \alpha \cdot (\gamma_j(\theta) - c^M)} \right]$$

$$\text{if } \alpha \in]-1,0[\cup]0,\infty[, \qquad (6)$$

$$H_{k-1,0,M}(\theta) := -\frac{1}{k-1}\sum_{j=1}^{k-1}\log(f_{j,\theta}(Y_j)) = -\frac{1}{k-1}\sum_{j=1}^{k-1}[\log(\gamma_j(\theta)) - Y_j \cdot \gamma_j(\theta)]$$
$$\text{if } \alpha = 0, \tag{7}$$

provided that $(1+\alpha)\cdot\gamma_j(\theta) - \alpha\cdot c^M > 0$ for $\alpha \in]-1,0[\cup]0,\infty[$ (e.g. $\alpha \in]-1,0[$ and $c^M \geq 0$, or $\alpha > 0$ and $c^M \leq 0$). By recalling $\gamma_j(\theta) = (1-(\frac{j}{k+1})^\theta)/\theta > 0$ (cf. (2)) having derivative $\gamma_j'(\theta) = ((\theta\cdot\gamma_j(\theta) - 1)\cdot\log(\frac{j}{k+1}) - \gamma_j(\theta))/\theta$, we obtain from (6) respectively (7) the corresponding estimating equations

$$\frac{d}{d\theta}H_{k-1,\alpha,M}(\theta) = \frac{e^{\alpha\cdot b^M}}{\theta\cdot(k-1)}\sum_{j=1}^{k-1}\left[\left((\theta\cdot\gamma_j(\theta) - 1)\cdot\log(\frac{j}{k+1}) - \gamma_j(\theta)\right)\cdot\gamma_j(\theta)^\alpha\right.$$
$$\left. \cdot\frac{\alpha\cdot(\gamma_j(\theta) - c^M)}{[(\alpha+1)\cdot\gamma_j(\theta) - \alpha\cdot c^M]^2} - \left(\frac{1}{\gamma_j(\theta)} - Y_j\right)\cdot e^{-Y_j\cdot\alpha\cdot(\gamma_j(\theta) - c^M)}\right] = 0,$$
$$\text{if } \alpha \in]-1,0[\cup]0,\infty[, \tag{8}$$

$$\frac{d}{d\theta}H_{k-1,0,M}(\theta) = -\frac{1}{\theta\cdot(k-1)}\sum_{j=1}^{k-1}\left[\left((\theta\cdot\gamma_j(\theta) - 1)\cdot\log(\frac{j}{k+1}) - \gamma_j(\theta)\right)\cdot\left(\frac{1}{\gamma_j(\theta)} - Y_j\right)\right]$$
$$:= -\frac{1}{k-1}\sum_{j=1}^{k-1}u_{j,\theta}(Y_j) = 0, \qquad\qquad \text{if } \alpha = 0, \tag{9}$$

to be solved for θ. Notice that (9) corresponds (up to the non-effective outer minus sign) to the estimating equation derived from the maximization of the log-likelihood, which is exactly the MLE method for estimating the extreme-value-index θ of Matthys and Beirlant [19]. Moreover, in the special subsetup of unscaled SBPD where $c^M = b^M = 0$ (and hence, $m(x) \equiv 1$), (8) reduces to the estimating equation of Ghosh [9].

3   Robustness Against Degenerate Contaminations

It is well known that the degree of robustness with respect to degenerate (i.e., one-point) contaminations of the data, can be measured for instance in terms of appropriate influence functions; in the current setup, the latter can be derived from the above-mentioned functional $\theta^G = T_{\alpha,M}(G) = T_{\alpha,M}(G_1,\dots,G_{k-1})$ for $\alpha > -1$. Indeed, by employing the corresponding results of Roensch and Stummer [23], for one-point contaminations at $t_0 \in \mathscr{X} = \mathbb{R}$ with degree $\varepsilon \in [0,1[$ in the j_0–th (transformed, log-ratio) data observation ($j_0 \in \{1,\dots,k-1\}$) we achieve by straightforward calculations the corresponding influence function

$$\mathbb{R} \ni t_0 \longrightarrow IF(t_0; j_0, T_{\alpha,M}, G) := \frac{\partial}{\partial\varepsilon}T\left(G_1,\dots,(1-\varepsilon)\cdot G_{j_0} + \varepsilon\cdot\delta_{t_0},\dots,G_{k-1}\right)\Big|_{\varepsilon=0}$$
$$= \left(\sum_{j=1}^{k-1}J_{j,\alpha}^M(\theta)\right)^{-1}\cdot\left(\frac{f_{j_0,\theta}(t_0)^\alpha}{m(t_0)^\alpha}\cdot u_{j_0,\theta}(t_0) - \xi_{j_0,\alpha}^M(\theta)\right),$$

with $\quad J_{j,\alpha}^M(\theta) := \int\limits_{\mathscr{X}} \frac{f_{j,\theta}(x)^{\alpha+1}}{m(x)^\alpha} \cdot \left(u_{j,\theta}(x)\right)^2 dx$

$$- \int\limits_{\mathscr{X}} \frac{f_{j,\theta}(x)^\alpha}{m(x)^\alpha} \cdot (g_j(x) - f_{j,\theta}(x)) \cdot \left[\frac{\partial}{\partial\theta} u_{j,\theta}(x) + \alpha \cdot \left(u_{j,\theta}(x)\right)^2\right] dx$$

and $\quad \xi_{j,\alpha}^M(\theta) := \int\limits_{\mathscr{X}} \frac{f_{j,\theta}(x)^\alpha g_j(x)}{m(x)^\alpha} \cdot u_{j,\theta}(x) \, dx.$

By assuming that the true data-generating probability distributions come from the $\mathscr{F}_\Theta^{(k-1)}$–family, i.e. $\boldsymbol{G} = \boldsymbol{F}_\theta = (F_{1,\theta}, \dots, F_{k-1,\theta})$ and thus $g_j(\cdot) = f_{j,\theta}(\cdot)$ for all $j \in \{1, \dots, k-1\}$ and some $\theta \in \Theta$ (to be found out), and by plugging in the explicit forms of $f_{j,\theta}(\cdot)$ and $m(\cdot)$, we arrive by straightforward but lengthy calculations at

$$\mathbb{R} \ni t_0 \longrightarrow IF(t_0; j_0, T_{\alpha,M}, \boldsymbol{F}_\theta)$$

$$= \frac{-\gamma_{j_0}(\theta)^\alpha \cdot \gamma_{j_0}'(\theta) \cdot \left[\frac{\alpha \cdot (\gamma_{j_0}(\theta) - c^M)}{[(\alpha+1)\cdot\gamma_{j_0}(\theta) - \alpha\cdot c^M]^2} + \left(t_0 - \frac{1}{\gamma_{j_0}(\theta)}\right) \cdot e^{-t_0 \cdot \alpha \cdot (\gamma_{j_0}(\theta) - c^M)}\right]}{\sum\limits_{j=1}^{k-1}\left[(\gamma_j'(\theta))^2 \cdot \gamma_j(\theta)^{\alpha-1} \cdot \frac{\gamma_j(\theta)^2 + \alpha^2 \cdot (\gamma_j(\theta) - c^M)^2}{[(\alpha+1)\cdot\gamma_j(\theta) - \alpha\cdot c^M]^3}\right]}, \quad (10)$$

where the explicit form of $\gamma_{j_0}'(\theta)$ has been given above. Applying general theory, for good robustness performance it is desirable to choose the tuning parameter $\alpha > -1$ and the scaling constant c^M such that $t_0 \to IF(t_0; j_0, T_{\alpha,M}, \boldsymbol{F}_\theta)$ is bounded for each $j_0 \in \{1, \dots, k-1\}$ and each $\theta \in \Theta = \mathbb{R}$. For the MLE-case of Matthys and Beirlant [19], where $\alpha = 0$ and thus c^M becomes obsolete, boundedness is not possible since then (10) degenerates to a t_0–line on $\mathbb{R}$. By demanding $\alpha \cdot (\gamma_{j_0}(\theta) - c^M) > 0$ for the crucial exponent, and by inspecting the range of $\gamma_{j_0}(\cdot)$, it is recommendable to ideally employ either (i) $\alpha > 0$ and $c^M < 0$, or (ii) $\alpha > 0$ and $c^M = 0$ (which for the non-effective choice $b^M = 0$ becomes the unscaled framework of Ghosh [9] and which has the theoretical disadvantage that robustness fails asymptotically), or (iii) $\alpha \in\,]-1, 0[$ and

$$c^M > (1 - n^{-\theta_{min}})/\theta_{min} \geq \gamma_1(\theta_{min}) \geq \gamma_{j_0}(\theta_{min}) = \frac{1 - (\frac{j_0}{k+1})^{\theta_{min}}}{\theta_{min}} > 0$$ where θ_{min} is a (e.g. prior-knowledge or constraint-desire reflecting) lower bound for the unknown extreme value index θ (for $\theta_{min} > 0$ one has $1/\theta_{min} > (1 - n^{-\theta_{min}})/\theta_{min}$ leading to the more restrictive but n–independent lower bound $c^M > 1/\theta_{min}$). In Fig. 1(d), we exemplarily display some outcoming influence functions which demonstrate the partial superiority of our method.

Analogously to (10), for contamination in all data observations we can deduce from the corresponding general result of Roensch and Stummer [23] an explicit expression of

$$\mathbb{R}^{k-1} \ni \mathbf{t} \mapsto IF(\mathbf{t}; T_{\alpha,M}, \boldsymbol{G}) := \frac{\partial}{\partial\varepsilon} T\left((1-\varepsilon)\cdot G_1 + \varepsilon\cdot\delta_1, \dots, (1-\varepsilon)\cdot G_n + \varepsilon\cdot\delta_n\right)\Big|_{\varepsilon=0}$$

$$= \left(\sum_{j=1}^{k-1} J_{i,\alpha}^M(\theta)\right)^{-1} \cdot \sum_{j=1}^{k-1}\left(\frac{f_{j,\theta}(t_j)^\alpha}{m(t_j)^\alpha} \cdot u_{j,\theta}(t_j) - \xi_{j,\alpha}^M(\theta)\right) \quad (11)$$

with $\mathbf{t} := (t_1, \dots, t_{k-1})$ and $\boldsymbol{G} = \boldsymbol{F}_\theta$. From (10) and (11) one can also derive other connected robustness measures such as e.g. the gross-error sensitivity and the self-standardized sensitivity; breakdown-point analyses can be performed as well. Moreover, following the lines of Roensch and Stummer [23] one can also show for the (say) subcase $\boldsymbol{G} = \boldsymbol{F}_\theta$ ($\theta \in \mathbb{R}$), $\alpha > 0$, $c^M < 0$, that there exists a consistent sequence θ_k of roots to the estimating Eqs. (8), (9), and that the asymptotic distribution (for $k = k_n \to \infty$ as $n \to \infty$) of $\sqrt{k_n - 1} \cdot \Omega_{k_n}^{-1/2} \cdot \Lambda_{k_n} \cdot \left(\theta_{k_n} - \theta \right)$ is normal with zero mean and unit variance. Here, we use the notations $\Lambda_k := \frac{1}{k-1} \sum_{j=1}^{k-1} J_{j,\alpha}^M(\theta)$ and

$$\Omega_k := \frac{1}{k-1} \sum_{j=1}^{k-1} \left[\int_{\mathscr{X}} \frac{f_{j,\theta}(x)^{2\alpha+1}}{m(x)^{2\alpha}} \cdot (u_{j,\theta}(x))^2 \, \mathrm{d}x \; - \; \left(\xi_{j,\alpha}^M(\theta) \right)^2 \right],$$

which (as above) for our current context of exponential distributions can be calculated explicitly. For the sake of brevity, these robustness and limit investigations will appear elsewhere, including many (α, M)–subcases which perform much more robust than the MLE estimator.

4 Robustness Against Non-degenerate Contaminations

Let us now deal with the case of non-degenerate contaminations, in the sense that the original data are described by independent random variables $(X_i)_{i=1,\ldots,n}$ with identical probability distribution of the form $\mathbb{P}^X := (1-c) \cdot \mathbb{P} + c \cdot \mathbb{P}_{cont}$ where $\mathbb{P} \in MDA(H_{\theta,0,0})$, $\mathbb{P}_{cont} \in MDA(H_{\theta_{cont},0,0})$ for some $\mathbb{P} \neq \mathbb{P}_{cont}$, and $c \in [0,1[$ represents the contamination degree. In terms of the sign function, we call the case $sgn(\theta) \neq sgn(\theta_{cont})$ an *out-of-type contamination* and $sgn(\theta) = sgn(\theta_{cont})$ an *in-type contamination*; the latter we further subdivide into *in-family contaminations* (where $\mathbb{P}$, $\mathbb{P}_{cont}$ stem from the same distribution family, e.g. lognormal distributions) and *in-type out-of-family contaminations*. Within each of these three contamination setups, Matthys and Beirlant (MB) [19] respectively Ghosh (G) [9] have presented exemplary simulation studies, by employing the exponential regression model (2) with – notationally embedded in our wider framework – tuning-parameter constellations $\alpha = 0$ (which corresponds to MB-estimation being of MLE-type where c^M is obsolete) respectively $\alpha > 0$, $c^M = 0$ (unscaled case, which corresponds to G-estimation). Both [19] and [9] use sample size $n = 500$ and $rep = 100$ number of replications. We have extended these simulation studies to our component-widened set of tuning parameters $\alpha > -1$, $c^M \in \mathbb{R}$ (scaled case with enlargened divergence-parameter domain), by keeping $n = 500$ but using $rep = 10000$ which leads to more significant results but which is much more computer-runtime expensive. Expressed in terms of the standard performance measures

$$Bias(k; \alpha, c^M, c) := \frac{1}{rep} \sum_{r=1}^{rep} \widehat{\theta}_{k-1,\alpha,M}^{(r,c)} - \theta, \quad AE(k; \alpha, c^M, c) := \frac{1}{rep} \sum_{r=1}^{rep} \left| \widehat{\theta}_{k-1,\alpha,M}^{(r,c)} - \theta \right|,$$

$$MSE(k; \alpha, c^M, c) := \frac{1}{rep} \sum_{r=1}^{rep} \left(\widehat{\theta}_{k-1,\alpha,M}^{(r,c)} - \theta \right)^2, \quad k = 1,\ldots,n, \; \alpha \in \mathbb{R}, \; c^M \in \mathbb{R}, c \in [0,1[,$$

$$(12)$$

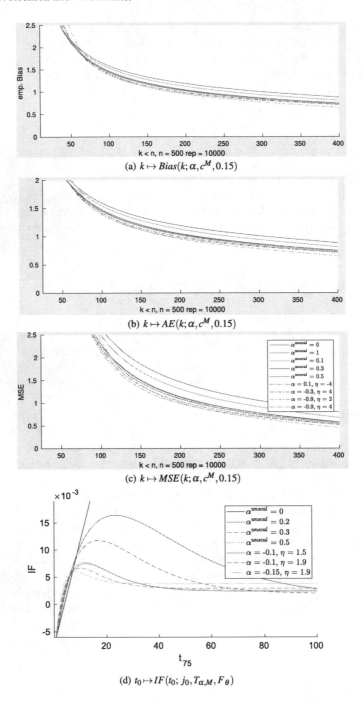

(a) $k \mapsto Bias(k; \alpha, c^M, 0.15)$

(b) $k \mapsto AE(k; \alpha, c^M, 0.15)$

(c) $k \mapsto MSE(k; \alpha, c^M, 0.15)$

(d) $t_0 \mapsto IF(t_0; j_0, T_{\alpha,M}, F_\theta)$

Fig. 1. (a)–(c): Performance measures (12) for $c = 15\%$ and various different tuning parameters α, $\eta := c^M$, with line types and colors explained in the legend of (c) (where, say, $\alpha^{unscaled} = 1$ means $\alpha = 1$ together with $c^M = 0$). (d): Influence functions (10) for $k = 100$, $\theta = 1$, contamination at the $j_0 = 75$-th observation, and various different tuning parameters α, $\eta := c^M$ (explained in an analogous legend).

(where the superscript (r, c) denotes the data-derived estimate obtained in the r-th replication with contamination-degree c), the main essences of our simulation findings are:

(1) for the *uncontaminated case* $c = 0$, the MB estimator $\widehat{\theta}_{k-1,0,obsolete}$ (cf. (9)) has the best performance and $\widehat{\theta}_{k-1,\alpha,M}$ produces slightly higher bias and MSE, for appropriate tuning-parameter choices;

(2) for the *contaminated case* $c \neq 0$ and for fixed $\alpha \in]0, 0.3[$, the estimator $\widehat{\theta}_{k-1,\alpha,M}$ with tuning-parameter choice $c^M < 0$ is often advantageous over the corresponding G-estimator (where $c^M = 0$);

(3) for $c \neq 0$, $\alpha \in]-1, 0[$ and large enough c^M (say, case-dependently, 2 or 4 or greater than 10), the corresponding estimator $\widehat{\theta}_{k-1,\alpha,M}$ is often very advantageous over that of any other tuning-parameter constellation $\alpha \in [0, 1]$, $c^M \leq 0$ (including the G-estimators). Moreover, the cases $\alpha \in]-1, 0[$, $c^M = 0$ (i.e., no scaling on the enlargened divergence-parameter-range), often lead to MSE's being unstable with changing excerpt-size k.

For the sake of brevity, we only present a representative simulation outcome here; the entire comprehensive simulation studies will appear elsewhere. Consider the out-of-type contamination context where $\mathbb{P}$ is a standard lognormal distribution (leading to $\theta = 0$ and being very important e.g. in finance, insurance), $\mathbb{P}_{cont}$ is a Student t-distribution with $df = 1/3$ degrees of freedom (leading to $\theta_{cont} = 3$), and contamination-degree $c = 15\%$. For this, we have run simulations of $r = 10000$ replicates with sample size $n = 500$ each. The resulting plots of the correspondingly evaluated performance measures (12), as functions of the excerpt-size k, are given in Fig. 1; they exemplarily reflect the above-mentioned empirical findings (for $k \gtrsim 80 \approx n/6$).

Acknowledgement. We are grateful to the three referees for their very useful comments.

References

1. Basu, A., Harris, I.R., Hjort, N.L., Jones, M.C.: Robust and efficient estimation by minimizing a density power divergence. Biometrika **85**(3), 549–559 (1998)
2. Basu, A., Shioya, H., Park, C.: Statistical Inference: The Minimum Distance Approach. CRC Press, Boca Raton (2011)
3. Beirlant, J., Goegebeur, Y., Teugels, J., Segers, J.: Statistics of Extremes: Theory and Applications. Wiley, Chichester (2004)
4. Broniatowski, M., Stummer, W.: Some universal insights on divergences for statistics, machine learning and artificial intelligence. In: Nielsen, F. (ed.) Geometric Structures of Information. SCT, pp. 149–211. Springer, Cham (2019). https://doi.org/10.1007/978-3-030-02520-5_8
5. Castillo, E., Hadi, A.S., Balakrishnan, N., Sarabia, J.M.: Extreme Value and Related Models with Applications in Engineering and Science. Wiley, Hoboken (2005)
6. Coles, S.: An Introduction to Statistical Modeling of Extreme Values. Springer, London (2001)
7. Embrechts, P., Klüppelberg, C., Mikosch, T.: Modelling Extremal Events for Insurance and Finance. Springer, Berlin (1997). https://doi.org/10.1007/978-3-642-33483-2

8. de Haan, L., Ferreira, A.: Extreme Value Theory: An Introduction. Springer, New York (2006). https://doi.org/10.1007/0-387-34471-3
9. Ghosh, A.: Divergence based robust estimation of the tail index through an exponential regression model. Stat. Methods Appl. **26**, 181–213 (2017)
10. Ghosh, A., Basu, A.: Robust estimation for independent non-homogeneous observations using density power divergence with applications to linear regression. Electron. J. Stat. **7**, 2420–2456 (2013)
11. Ghosh, A., Basu, A.: Robust estimation in generalized linear models: the density power divergence approach. TEST **25**, 269–290 (2016)
12. Ghosh, A., Basu, A.: Robust bounded influence tests for independent non-homogeneous observations. Statistica Sinica **28**, 1133–1155 (2018)
13. Gnedenko, B.: Sur la distribution limite du terme maximum dune série aléatoire. Ann. Math. **44**(3), 423–453 (1943)
14. Gomes, M.I., Guillou, A.: Extreme value theory and statistics of univariate extremes: a review. Intern. Stat. Rev. **83**(2), 263–292 (2015)
15. Kißlinger, A.-L., Stummer, W.: Some decision procedures based on scaled Bregman distance surfaces. In: Nielsen, F., Barbaresco, F. (eds.) GSI 2013. LNCS, vol. 8085, pp. 479–486. Springer, Heidelberg (2013). https://doi.org/10.1007/978-3-642-40020-9_52
16. Kißlinger, A.-L., Stummer, W.: New model search for nonlinear recursive models, regressions and autoregressions. In: Nielsen, F., Barbaresco, F. (eds.) GSI 2015. LNCS, vol. 9389, pp. 693–701. Springer, Cham (2015). https://doi.org/10.1007/978-3-319-25040-3_74
17. Kißlinger, A.-L., Stummer, W.: Robust statistical engineering by means of scaled Bregman distances. In: Agostinelli, C., Basu, A., Filzmoser, P., Mukherjee, D. (eds.) Recent Advances in Robust Statistics: Theory and Applications, pp. 81–113. Springer, New Delhi (2016). https://doi.org/10.1007/978-81-322-3643-6_5
18. Kißlinger, A.-L., Stummer, W.: A new toolkit for robust distributional change detection. Appl. Stochastic Models Bus. Ind. **34**, 682–699 (2018)
19. Matthys, G., Beirlant, J.: Estimating the extreme value index and high quantiles with exponential regression models. Statistica Sinica **13**, 853–880 (2003)
20. Reiss, R.-D., Thomas, M.: Statistical Analysis of Extreme Values, with Application to Insurance, Finance, Hydrology and Other Fields. Birkhäuser, Basel (2007)
21. Resnick, S.I.: Heavy-tail Phenomena: Probabilistic and Statistical Modeling. Springer, New York (2007)
22. Roensch, B., Stummer, W.: 3D insights to some divergences for robust statistics and machine learning. In: Nielsen, F., Barbaresco, F. (eds.) GSI 2017. LNCS, vol. 10589, pp. 460–469. Springer, Cham (2017). https://doi.org/10.1007/978-3-319-68445-1_54
23. Roensch, B., Stummer, W.: Robust estimation by means of scaled Bregman power distances. Part I. Non-homogeneous data. In: Nielsen, F., Barbaresco, F. (eds.) GSI 2019. LNCS, vol. 11712, pp. 319–330. Springer, Cham (2019). https://doi.org/10.1007/978-3-030-26980-7_33
24. Stummer, W.: Some Bregman distances between financial diffusion processes. Proc. Appl. Math. Mech. **7**(1), 1050503–1050504 (2007)
25. Stummer, W., Vajda, I.: On Bregman distances and divergences of probability measures. IEEE Trans. Inform. Theory **58**(3), 1277–1288 (2012)

Computational Information Geometry

Topological Methods for Unsupervised Learning

Leland McInnes[✉]

Tutte Institute for Mathematics and Computing, Ottawa, Canada
leland.mcinnes@gmail.com

Abstract. Unsupervised learning is a broad topic in machine learning with many diverse sub-disciplines. Within the field of unsupervised learning we will consider three major topics: dimension reduction; clustering; and anomaly detection. We seek to use the languages of topology and category theory to provide a unified mathematical approach to these three major problems in unsupervised learning.

Keywords: Unsupervised learning · Manifold learning · Clustering

1 Mathematical Foundations

Our goal is to develop a flexible theory that can cope with data sampled non-uniformly from an underlying manifold embedded in a high dimension ambient space. The underlying approach is based on combining together multiple local approximations of an underlying manifold on which the input data $X = \{x_1, x_2, \ldots, x_N\}$ is assumed to lie. Since most theory, such as the Nerve theorem, implicitly assumes uniform distribution with respect to the ambient metric, other approaches are called for. For example we can adapt to local density by locally approximating the Riemannian metric of the manifold for each point, yielding N incompatible local approximations of inter-point distance. This incompatibility is in turn tackled by transferring the problem to fuzzy topological representations where natural means of combination are available. Given a single fuzzy topological representation of the data the various subtopics of unsupervised learning fall out as natural further operations.

The foundation of this approach is based on the work of David Spivak on metric realization for fuzzy simplicial sets [10], and categorical versions of fuzzy sets by Michael Barr [1] – it is this work that allows us to transfer the problem to one of fuzzy topological representations. A sketch of the approach is as follows: Let I be the unit interval $(0, 1] \subseteq \mathbb{R}$ with topology given by intervals of the form $(0, a)$ for $a \in (0, 1]$. The category of open sets (with morphisms given by inclusions) can be imbued with a Grothendieck topology in the natural way for any poset category.

Definition 1. *A presheaf $\mathscr{P}$ on I is a functor from I^{op} to **Sets**. A fuzzy set is a presheaf on I such that all maps $\mathscr{P}(a \leq b)$ are injections.*

© Springer Nature Switzerland AG 2019
F. Nielsen and F. Barbaresco (Eds.): GSI 2019, LNCS 11712, pp. 343–350, 2019.
https://doi.org/10.1007/978-3-030-26980-7_35

Presheaves on I form a category with morphisms given by natural transformations. We can thus form a category of fuzzy sets by simply restricting to those presheaves that are fuzzy sets. We note that such presheaves are trivially sheaves under the Grothendieck topology on I. A section $\mathscr{P}([0, a))$ can be thought of as the set of all elements with membership strength at least a. We can now define the category of fuzzy sets.

Definition 2. *The category **Fuzz** of fuzzy sets is the full subcategory of sheaves on I spanned by fuzzy sets.*

Defining fuzzy simplicial sets is simply a matter of considering presheaves of Δ valued in the category of fuzzy sets rather than the category of sets.

Definition 3. *The category of fuzzy simplicial sets **sFuzz** is the category with objects given by functors from Δ^{op} to **Fuzz**, and morphisms given by natural transformations.*

Alternatively, a fuzzy simplicial set can be viewed as a sheaf over $\Delta \times I$, where Δ is given the trivial topology and $\Delta \times I$ has the product topology. We will use $\Delta^n_{<a}$ to denote the sheaf given by the representable functor of the object $([n], (0, a))$. The importance of this fuzzy (sheafified) version of simplicial sets is their relationship to metric spaces. We begin by considering the larger category of extended-pseudo-metric spaces.

Definition 4. *An extended-pseudo-metric space (X, d) is a set X and a map $d : X \times X \to \mathbb{R}_{\geq 0} \cup \{\infty\}$ such that*

1. *$d(x, y) \geqslant 0$, and $x = y$ implies $d(x, y) = 0$;*
2. *$d(x, y) = d(y, x)$; and*
3. *Either $d(x, z) \leqslant d(x, y) + d(y, z)$ or $d(x, z) = \infty$.*

*The category of extended-pseudo-metric spaces **EPMet** has as objects extended-pseudo-metric spaces and non-expansive maps as morphisms. We denote the subcategory of finite extended-pseudo-metric spaces **FinEPMet**.*

The choice of non-expansive maps in Definition 4 is due to Spivak, but we note that it closely mirrors the work of Carlsson and Memoli in [3] on topological methods for clustering as applied to finite metric spaces. This choice is significant since pure isometries are too strict and do not provide large enough Hom-sets.

In [10] Spivak constructs a pair of adjoint functors, Real and Sing between the categories **sFuzz** and **EPMet**. These functors are the natural extension of the classical realization and singular set functors from algebraic topology (see [5]). We are only interested in finite metric spaces, and thus use the analogous adjoint pair FinReal and FinSing. Formally we define the finite realization functor as follows:

Definition 5. *Define the functor* FinReal : **sFuzz** $\to$ **FinEPMet** *by setting*

$$\mathsf{FinReal}(\Delta^n_{<a}) \triangleq (\{x_1, x_2, \ldots, x_n\}, d_a),$$

where

$$d_a(x_i, x_j) = \begin{cases} -\log(a) & \text{if } i \neq j, \\ 0 & \text{otherwise} \end{cases}.$$

and then defining

$$\mathsf{FinReal}(X) \triangleq \operatorname*{colim}_{\Delta^n_{<a} \to X} \mathsf{FinReal}(\Delta^n_{<a}).$$

A morphism $(\sigma, \leq) : ([n], ([0, a)) \to ([m], ([0, b))$ only exists for $a \leq b$, and in that case we can define

$$\mathsf{FinReal}((\sigma, \leq)) : \mathsf{FinReal}(\Delta^n_{<a}) \to \mathsf{FinReal}(\Delta^m_{<b})$$

to be the map

$$(\{x_1, x_2, \ldots, x_n\}, d_a) \mapsto (\{x_{\sigma(1)}, x_{\sigma(2)}, \ldots, x_{\sigma(n)}\}, d_b),$$

which is non-expansive since $a \leq b$ implies $d_a \geq d_b$.

Since $\mathsf{FinReal}$ preserves colimits it admits a right adjoint, the fuzzy singular set functor $\mathsf{FinSing}$. To define the fuzzy singular set functor we require some further notation. Given a fuzzy simplicial set X let $X^n_{<a}$ be the set $X([n], (0, a))$. We can then define the fuzzy singular set functor in terms of the action of its image on $\Delta \times I$.

Definition 6. *Define the functor* $\mathsf{FinSing} : \mathbf{FinEPMet} \to \mathbf{sFuzz}$ *by*

$$\mathsf{FinSing}(Y)^n_{<a} \triangleq \hom_{\mathbf{FinEPMet}}(\mathsf{FinReal}(\Delta^n_{<a}), Y).$$

With the necessary theoretical background in place, the means to handle the family of incompatible metric spaces described earlier becomes clear. Each metric space in the family can be translated into a fuzzy simplicial set via the fuzzy singular set functor, distilling the topological information while still retaining metric information in the fuzzy structure. Resolving the incompatibilities of the resulting family of fuzzy simplicial sets can be done by taking a (fuzzy) union or intersection across the entire family. The result is a single fuzzy simplicial set which captures the relevant topological and underlying metric structure of the manifold $\mathcal{M}$. Thus if d_i is the induced distance metric approximated for the point x_i then the fuzzy topological representation of the manifold is given by either

$$\bigcup_{i=1}^{N} \mathsf{FinSing}((X, d_i)) \quad \text{or} \quad \bigcap_{i=1}^{N} \mathsf{FinSing}((X, d_i)).$$

depending on the kind of combination required. This can be phrased directly in terms of pushouts (or colimits) and pullbacks (or limits) of simplicial sheaves for those less comfortable with fuzzy set semantics.

2 Dimension Reduction

We can make use of the general theory outlined in Sect. 1 for the purposes of dimension reduction. The resulting algorithm, called Uniform Manifold Approximation and Projection (UMAP), is sketched here. For a fuller presentation see [9].

For the purposes of dimension reduction we assume the manifold hypothesis – that the data is (approximately) sampled from some manifold $\mathcal{M}$ embedded in the ambient feature space. If we assume the data to be uniformly distributed on $\mathcal{M}$ (with respect to a Riemannian metric g) then any ball of fixed volume should contain approximately the same number of points of X regardless of where on the manifold it is centered. Conversely, a ball centered at X_i that contains exactly the k-nearest-neighbors of X_i should have fixed volume regardless of the choice of $X_i \in X$. We can approximate geodesic distance from X_i to its neighbors by normalising distances with respect to the distance to the kth nearest neighbor of X_i.

For real data it is safe to assume that the manifold $\mathcal{M}$ is locally connected. In practice this can be realized by measuring distance in the extended-pseudo-metric space local to X_i as geodesic distance *beyond* the nearest neighbor of X_i. Since this sets the distance to the nearest neighbor to be equal to 0, this is only possible in the more relaxed setting of extended-pseudo-metric spaces. It ensures, however, that each 0-simplex is the face of some 1-simplex with fuzzy membership strength 1, meaning that the resulting topological structure derived from the manifold is locally connected. We note that this has a similar practical effect to the truncated similarity approach of Lee and Verleysen [7], but derives naturally from the assumption of local connectivity of the manifold. We may then use the results of the previous section to combine together the local fuzzy simplicial sets and arrive at a single global fuzzy topological representation.

Having obtained a representation of the manifold it only remains to find a low dimensional representation of the data that suitably approximates the manifold.

Let $Y = \{Y_1, \ldots, Y_N\} \subseteq \mathbb{R}^d$ be a low dimensional ($d \ll n$) representation of X such that Y_i represents the source data point X_i. In contrast to the source data where we want to estimate a manifold on which the data is uniformly distributed, we know the manifold for Y is $\mathbb{R}^d$ itself. Therefore we know the manifold and manifold metric apriori, and can compute the fuzzy topological representation directly. Of note, we still want to incorporate the distance to the nearest neighbor as per the local connectedness requirement. This can be achieved by supplying a parameter that defines the expected distance between nearest neighbors in the embedded space.

Given fuzzy simplicial set representations of X and Y, a means of comparison is required. If we consider only the 1-skeleton of the fuzzy simplicial sets we can describe each as a fuzzy graph, or, more specifically, a fuzzy set of edges. To compare two fuzzy sets we will make use of fuzzy set cross entropy. For these purposes we will revert to classical fuzzy set notation. That is, a fuzzy set is given by a reference set A and a membership strength function $\mu : A \to [0,1]$. Comparable fuzzy sets have the same reference set. Given a sheaf representation

$\mathscr{P}$ we can translate to classical fuzzy sets by setting $A = \bigcup_{a \in (0,1]} \mathscr{P}([0,a))$ and $\mu(x) = \sup\{a \in (0,1] \mid x \in \mathscr{P}([0,a))\}$.

Definition 7. *The cross entropy C of two fuzzy sets (A, μ) and (A, ν) is defined as*

$$C((A,\mu),(A,\nu)) \triangleq \sum_{a \in A} \mu(a) \log\left(\frac{\mu(a)}{\nu(a)}\right) + (1 - \mu(a)) \log\left(\frac{1 - \mu(a)}{1 - \nu(a)}\right).$$

Similar to t-SNE we can optimize the embedding Y with respect to fuzzy set cross entropy C by using stochastic gradient descent. However, this requires a differentiable fuzzy singular set functor. If the expected minimum distance between points is zero the fuzzy singular set functor is differentiable for these purposes, however for any non-zero value we need to make a differentiable approximation (chosen from a suitable family of differentiable functions).

This completes the algorithm: by using manifold approximation and patching together local fuzzy simplicial set representations we construct a topological representation of the high dimensional data. We then optimize the layout of data in a low dimensional space to minimize the error between the two topological representations.

We believe this algorithm [9] to be state-of-the-art for dimension reduction techniques.

3   Clustering

Clustering data requires a clear definition of what constitutes a cluster. Here we follow Hartigan [6], Stuetzle [11], Chaudhuri et al. [4], and others in viewing the problem as that of approximating the 'cluster tree' formed by level sets of the probability density function (pdf) from which the data was drawn.

Specifically, a statistically oriented view of density clustering begins with the assumption that there exists some unknown density function from which the observed data is drawn. From the density function f, defined on a metric space $(\mathcal{X}, d)$, one can construct a hierarchical cluster structure, where a cluster is a connected subset of an f-level set $\{x \in (\mathcal{X}, d) \mid f(x) \geq \lambda\}$. As $\lambda \geq 0$ varies these f-level sets nest in such a way as to construct an infinite tree, which is referred to as the *cluster tree*. Each cluster is a branch of this tree, extending over the range of λ values for which it is distinct. The goal of a clustering algorithm is to suitably approximate the cluster tree, converging to it in the limit of infinite observed data points.

For our purposes we assume that the data is drawn from a pdf defined on some manifold. In contrast to the dimension reduction case where it was beneficial to adjust the metric so as have the data uniformly distributed on the manifold, we specifically want to *preserve* the distribution. Indeed we may wish to go a step further – since we are ultimately interested in the high density regions of the pdf on the manifold it may be beneficial to *emphasise* these regions. This is due to the finite sampling producing "noise" points that can confound density approximations built on limited data. Thus while the dimension reduction case normalised distances based on the uniform distribution assumption we will instead do the opposite.

Specifically we have two issues when considering distances between points: first we need some level of global normalisation to bring distances into a common range regardless of the scaling of the source data. This can be achieved, similar to the dimension reduction case, by normalising by a k-nearest-neighbour distance – in this case, however we select the global average of knn distances since we do not want to locally normalise the data. Next we may wish to emphasise the pdf by extending distances in sparse regions of the space, while contracting them in dense regions. This would be similar to Wishart's mode analysis [12]. In practice we can simply do the *opposite* of the dimension reduction case; instead of normalizing by the knn-distance of each point we can expand distances by a factor of some power α of the knn-distance. In some practical cases α may be zero, but in high noise datasets values in the range $(0, 1]$ are desirable. Thus we can define a clustering distance local to x_i as

$$d_{C_i}(x_i, x_j) = \frac{d_{\mathbb{R}^n}(x_i, x_j)}{\left(\frac{1}{N}\sum_{k=1}^{N} r_k\right)} \cdot \left(\frac{r_i}{\left(\frac{1}{N}\sum_{k=1}^{N} r_k\right)}\right)^{\alpha}$$

where r_i is the distance to the kth nearest neighbor of x_i.

We now have the familiar state of having multiple incompatible local metric spaces, which we can convert and combine into a single fuzzy topological representation. Again we wish to do the opposite of the dimension reduction approach by using logical conjunction to join the fuzzy simplicial sets – i.e. we wish to use intersection in this case. In categorical terms this is simply a matter of taking the limit of a family of simplicial sheaves rather than the colimit.

In practical term this represents the requirement that a given simplex should have membership strength at most the lowest membership strength of any of the local representations. The result is a fuzzy topological representation of the manifold that preserves density information.

Our next concern is converting this topological representation S into actual density estimates which we can find level sets for. From an algebraic topology point of view the way is clear: we want to consider $\pi_0(S)$, the connected components of the fuzzy simplicial set S, assuming that such an object can be defined. We first note that S itself is a functor $S : \mathbf{\Delta}^{\text{op}} \to \mathbf{sFuzz}$, where, in turn, $\mathbf{sFuzz}$ is a sub-category of presheaves $\mathbf{Sets}^{I^{\text{op}}}$. It follows, via some trivial symbol pushing, that S also defines a functor $S : I^{\text{op}} \to \mathbf{sSet}$. Since the connected component functor π_0 is defined on $\mathbf{sSet}$, we can post compose with it to define a functor $\pi_0 \circ S$

$$I^{\text{op}} \xrightarrow{\quad S \quad} \mathbf{sSet} \xrightarrow{\quad \pi_0 \quad} \mathbf{Sets}$$

which defines a fuzzy set.

Expressing this fuzzy set in classical fuzzy set notation (A, μ) we have

$$A = \bigcup_{a \in I} \pi_0 \circ S([0, a))$$

and

$$\mu(a) = \sup\{i \in (0,1] \mid a \in \pi_0 \circ S(i)\}.$$

This is a very large set. Similar to HDBSCAN* [2], or the pruning of Stuetzle and Nugent [11], we can dismiss connected components with size less then some parameterized bound m. Thus we arrive at the fuzzy subset (A, ν) where

$$\nu(a) = \begin{cases} \sup\{i \in (0,1] \mid a \in \pi_0 \circ S(i)\} & \text{if } |a| \geq m \\ 0 & \text{otherwise} \end{cases}.$$

Finally we can further simplify the set by placing an equivalence relation on the carrier set A. First define a binary relation $\smile$ on elements of A by

$$a \smile b \iff |a \triangle b| \leq m,$$

where $\triangle$ is set symmetric difference. Clearly this relation is symmetric and reflexive, and thus the transitive closure $\sim$ of $\smile$ is an equivalence relation on A. The resulting fuzzy quotient set provides effectively a simplified tree of clusters, and similar methods to those of HDBSCAN* [2] can be used to extract a flat clustering if desired.

We note that all of this theory goes through much more cleanly than, for example, the topological interpretations of HDBSCAN* (section 2.3 of [8]). Furthermore the resulting clustering algorithm, while producing very similar results to HDBSCAN*, is more robust to hyper-parameter selection. We therefore contend, based on the limited experiments so far conducted, that this represents a new state-of-the-art clustering algorithm.

4   Anomaly Detection

Anomaly detection is the task of finding samples that are "inconsistent with the rest of the data". Ultimately such an approach requires a model of the expected distribution of the data through which to determine which samples are "surprising". While parametric techniques such as Gaussian mixture models can be used, they still impose significant parametric assumptions. Ideally an anomaly detection technique would make use of a non-parametric approximation of the distribution of the data.

Following the work of the previous sections such a non-parametric density estimate can be easily constructed. Specifically we can define a fuzzy set (X, φ) with carrier set the source data X and fuzzy membership given by

$$\varphi(x) = \sup\{\nu(a) \mid a \in A \text{ and } x \in a\},$$

where (A, ν) is defined as per the procedure in the previous section. This provides a $[0, 1]$-valued membership strength for each data point. By simply taking the fuzzy complement of this set we arrive at an outlier score with 1 representing a maximal outlier and 0 representing a maximal inlier.

In effect we are taking the fuzzy set (X, φ) to be a (scaled) approximation of the pdf from which the data was drawn. By making use of the noise-offsetting density exaggeration in the early steps of the clustering construction the effects of outlying points have limited impact on this approximation, and hence are readily identified as outliers.

We believe this represents a powerful new approach to anomaly detection that demands further research.

5 Summary

We have presented a mathematical formulation in terms of topology and category theory that allows for simple extensions to yield algorithms for a wide variety of unsupervised learning tasks. These algorithms demonstrate state-of-the-art performance in their tasks, despite being drawn from a single unified framework. Thus the topological approaches to unsupervised learning outlined here not only provide several new algorithms, but a rich field for further research into unsupervised learning in general.

References

1. Barr, M.: Fuzzy set theory and topos theory. Canad. Math. Bull **29**(4), 501–508 (1986)
2. Campello, R.J.G.B., Moulavi, D., Sander, J.: Density-based clustering based on hierarchical density estimates. In: Pei, J., Tseng, V.S., Cao, L., Motoda, H., Xu, G. (eds.) PAKDD 2013. LNCS (LNAI), vol. 7819, pp. 160–172. Springer, Heidelberg (2013). https://doi.org/10.1007/978-3-642-37456-2_14
3. Carlsson, G., Mémoli, F.: Classifying clustering schemes. Found. Comput. Math. **13**(2), 221–252 (2013)
4. Chaudhuri, K., Dasgupta, S.: Rates of convergence for the cluster tree. In: Proceedings of the 23rd International Conference on Neural Information Processing Systems. NIPS 2010, pp. 343–351. Curran Associates Inc., USA (2010)
5. Goerss, P.G., Jardine, J.F.: Simplicial Homotopy Theory. Springer, Basel (2009). https://doi.org/10.1007/978-3-0346-0189-4
6. Hartigan, J.A.: Consistency of single linkage for high-density clusters. J. Am. Stat. Assoc. **76**(374), 388–394 (1981)
7. Lee, J.A., Verleysen, M.: Shift-invariant similarities circumvent distance concentration in stochastic neighbor embedding and variants. Procedia Comput. Sci. **4**, 538–547 (2011)
8. McInnes, L., Healy, J.: Accelerated hierarchical density clustering. arXiv preprint arXiv:1705.07321 (2017)
9. McInnes, L., Healy, J.: Umap: uniform manifold approximation and projection for dimension reduction. arXiv preprint arXiv:1802.03426 (2018)
10. Spivak, D.I.: Metric realization of fuzzy simplicial sets. Self published notes http://math.mit.edu/~dspivak/files/metric_realization.pdf
11. Stuetzle, W.: Estimating the cluster tree of a density by analyzing the minimal spanning tree of a sample. J. Classif. **20**(1), 025–047 (2003)
12. Wishart, D.: Mode analysis: a generalization of nearest neighbor which reduces chaining effects. Numer. Taxonomy **76**(282–311), 17 (1969)

Geometry and Fixed-Rate Quantization in Riemannian Metric Spaces Induced by Separable Bregman Divergences

Erika Gomes-Gonçalves[1] , Henryk Gzyl[2] , and Frank Nielsen[3(✉)]

[1] Independent Consultant, Madrid, Spain
erikapat@gmail.com
[2] Centro de Finanzas IESA, Caracas, Venezuela
henryk.gzyl@iesa.edu.ve
[3] Sony Computer Science Laboratories, Inc., Tokyo, Japan
Frank.Nielsen@acm.org

Abstract. Dual separable Bregman divergences induce dual Riemannian metric spaces which are isometric to the Euclidean space after nonlinear monotone embeddings. We investigate fixed-rate quantization and the induced Voronoi diagrams in those metric spaces.

Keywords: Bregman divergence · Riemannian metric ·
Legendre transformation · Voronoi diagram · Quantization

1 Introduction

1.1 Riemannian Geometry from a Separable Bregman Divergence

Consider a Bregman divergence

$$\delta_\Phi(\boldsymbol{x}, \boldsymbol{x}') = \Phi(\boldsymbol{x}) - \Phi(\boldsymbol{x}') - (\boldsymbol{x} - \boldsymbol{x}')^\top \nabla\Phi(\boldsymbol{x}'), \tag{1.1}$$

with real-valued generator function $\Phi : \mathcal{M} \to \mathbb{R}$ and open convex subset $\mathcal{M}$ of $\mathbb{R}^K$. Here, $\mathcal{M} = \mathcal{J}^K$ where $\mathcal{J}$ is an interval (bounded or unbounded) in the real line, and the *separable* Bregman generator

$$\Phi(\boldsymbol{x}) = \sum_{j=1}^K \phi(x_j), \tag{1.2}$$

with $\phi : \mathcal{J} \to \mathbb{R}$ is a three times continuously differentiable function so that we have

$$\delta_\Phi(x, x') = \sum_{j=1}^K \delta_\phi(x_j, x_j'). \tag{1.3}$$

The Riemannian metric on $\mathcal{M}$ is given by

$$g_{i,j}(\boldsymbol{x}) = \phi''(x_i)\delta_{i,j} \tag{1.4}$$

F. Nielsen and F. Barbaresco (Eds.): GSI 2019, LNCS 11712, pp. 351–358, 2019.
https://doi.org/10.1007/978-3-030-26980-7_36

(where δ is the Krönecker'symbol), and the corresponding Riemannian distance [6] is

$$d_\phi(\boldsymbol{x}, \boldsymbol{x}')^2 = \sum_{j=1}^{K} \left(h(x_j) - h(x_j') \right)^2, \qquad (1.5)$$

where

$$h(x) = \int \sqrt{\phi''(x)} \mathrm{d}x \qquad (1.6)$$

is a an increasing continuously differentiable function defined as the antiderivative of $(\phi'')^{1/2}$. Its compositional inverse h^{-1} shall be denoted by H (with $H \circ h^{-1} = h^{-1} \circ H = \mathrm{id}$). Let $h(\boldsymbol{x})$ to denote the transpose of $(h(x_1), ..., h(x_K))$, and the same for H. We term distance d_ϕ the *Riemann-Bregman distance*.

The equation $\gamma(t) = (\gamma_1(t), \ldots, \gamma_K(t))$ of the Riemannian geodesic [6] between $\boldsymbol{x}$ and $\boldsymbol{x}'$ is

$$\gamma_i(t) = h^{-1}(h(x_i) + k_i t), \qquad (1.7)$$

where the $k_i = h(x_i') - h(x_i)$ are the constants of integration such that $\gamma(0) = \boldsymbol{x}$ and $\gamma(1) = \boldsymbol{x}'$. That is, the geodesics are expressed as

$$\gamma_i(t) = h^{-1}\left((1-t)h(x_i) + th(x_i') \right), \quad t \in [0,1]. \qquad (1.8)$$

The coordinates of the geodesics are interpreted as a quasi-arithmetic mean.

1.2   Dual Riemann-Bregman Distances

By introducing the Legendre convex conjugate

$$\phi^*(y) = \sup_{x \in \mathcal{J}} \{yx - \phi(x)\} = y{\phi'}^{-1}(y) - \phi({\phi'}^{-1}(y)),$$

with $y_i = \phi'(x_i)$ (and $x_i = {\phi'}^{-1}(y_i) = \phi^{*\prime}(y_i)$), we get the dual Riemannian metric as $g^{*ij}(\boldsymbol{y}) = \phi^{*\prime\prime}(y_i)\delta_{i,j}$, and we check the following Crouzeix identity: $g(\boldsymbol{x})g^*(\boldsymbol{y}) = \mathrm{id}$. It follows that we can express the Bregman-Riemannian distance either by using the primal affine **x**-coordinate system, or by using the dual affine **y**-coordinate system (with $y = \nabla\Phi(x)$ and $x = \nabla\Phi^*(y)$):

Theorem 1. *We have* $\delta_\Phi(\boldsymbol{x}, \boldsymbol{x}') = \delta_{\Phi^*}(\boldsymbol{y}', \boldsymbol{y})$, *and*

$$d_\phi(\boldsymbol{x}, \boldsymbol{x}') = d_{\phi^*}(\boldsymbol{y}, \boldsymbol{y}') = d_{\phi^*}(\nabla\Phi(\boldsymbol{x}), \nabla\Phi(\boldsymbol{x}')). \qquad (1.9)$$

Figure 1 illustrates the Legendre duality for defining the Riemann-Bregman distances in the dual coordinate systems induced by the Legendre transformation.

Example 1. Consider $\phi(x) = x \log x - x$ (extended Shannon negentropy). We have $\phi'(x) = \log x$ and $\phi''(x) = \frac{1}{x}$. It follows that $h(x) = 2\sqrt{x} + c_x$, where c_x is a constant. The Legendre conjugate is $\phi^*(y) = e^y$ and $h^*(y) = \int \sqrt{\phi^{*\prime\prime}} = 2e^{\frac{y}{2}} + c_y$, where c_y is a constant. We have $y(x) = \log x$, $d_\phi(x, x') = 2|\sqrt{x} - \sqrt{x'}|$ and $d_{\phi^*}(y, y') = 2|e^{\frac{y}{2}} - e^{\frac{y'}{2}}|$. We check that $d_\phi(x, x') = d_{\phi^*}(y, y')$ since $e^{\frac{y}{2}} = \sqrt{x}$.

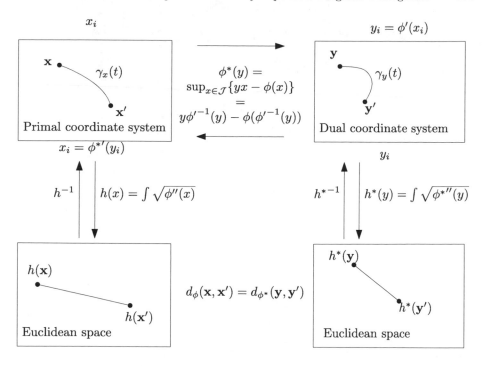

Fig. 1. The dual Riemann-Bregman distances.

In information geometry [1,5], a smooth dissimilarity measure D on a manifold $\mathcal{M}$ induces a dual structure $(\mathcal{M}, g_D, \nabla_D, \nabla_D^*)$ defined by a pair of affine connection (∇_D, ∇_D^*) coupled to the metric tensor g_D. The Riemannian geometry $(\mathcal{M}, g_D)$ is a self-dual information-geometric structure with the dual connections coinciding with the Levi-Civita connection ∇^{LC}. Bregman divergences δ_Φ yield dually-flat information-geometric spaces (meaning ∇_D-flat and ∇_D^*-flat) although the Riemann-Bregman geometry is usual curved [9] (meaning ∇_D^{LC}-curved).

Let us report a few examples of Riemann-Bregman distances:

- $\phi(x) = x^2/2$, $\phi''(x) = 1$ and $\phi'''(x) = 0$. The induced distance is the standard squared Euclidean distance

$$d_\phi(\boldsymbol{x}, \boldsymbol{y})^2 = \sum_{i=1}^{K} (x_i - y_i)^2. \tag{1.10}$$

- $\phi(x) = e^x$, $\phi''(x) = \phi'''(x) = e^x$. The geodesic distance between $\boldsymbol{x}$ and $\boldsymbol{y}$ is given by

$$d_\phi(\boldsymbol{x}, \boldsymbol{y})^2 = \sum_{i=1}^{K} (e^{y_i/2} - e^{x_i/2})^2. \tag{1.11}$$

– $\phi(x) = e^{-x}$: The geodesic distance between $\boldsymbol{x}$ and $\boldsymbol{y}$ is given by

$$d_\phi(\boldsymbol{x}, \boldsymbol{y})^2 = \sum_{i=1}^K (e^{-y_i/2} - e^{-x_i/2})^2. \tag{1.12}$$

– $\phi(x) = x \ln x$ (Shannon negentropy). The domain is $\mathcal{M} = (0, \infty)^K$ and $\phi'(x) = (\ln x) - 1$, $\phi''(x) = \frac{1}{x}$ and $h(x) = 2\sqrt{(x)}$. The geodesic distance between $\boldsymbol{x}$ and $\boldsymbol{y}$ is

$$d_\phi(\boldsymbol{x}, \boldsymbol{y})^2 = 2 \sum_{i=1}^K (\sqrt{y_i} - \sqrt{x_i})^2. \tag{1.13}$$

This is the squared Hellinger distance used in probability theory [10].

– $\phi(x) = -\ln x$ (Burg negentropy). To finish, we shall consider another example on $\mathcal{M} = (0, \infty)^K$. We have $\phi'(x) = -\frac{1}{x}$, $\phi''(x) = \frac{1}{x^2}$ and $h(x) = \log x + c$. Now, the distance between $\boldsymbol{x}$ and $\boldsymbol{y}$ is now given by

$$d_\phi(\boldsymbol{x}, \boldsymbol{y})^2 = \sum_{i=1}^K (\ln y_i - \ln x_i)^2. \tag{1.14}$$

1.3   Riemann-Bregman Centroids

Consider the best predictor of a random variable $\boldsymbol{X}$ taking values in $\mathcal{M}$ when the prediction error is measured in the d_ϕ-distance. The special relation of the distance (1.5) to the Euclidean distance makes the following result clear:

Theorem 2 (Separable Riemann-Bregman centroid). *Given a collection of points $\{\boldsymbol{x}_1, ..., \boldsymbol{x}_N\}$ in $\mathcal{M}$, the point $\hat{\boldsymbol{x}}$ the realizes the minimum of*

$$\sum_{n=1}^N d_\phi(\boldsymbol{x}_n, \boldsymbol{\xi})^2 \quad over \ \boldsymbol{\xi} \in \mathcal{M}$$

is unique and given by

$$\hat{\boldsymbol{x}} = H\Big(\frac{1}{N} \sum_{n=1}^N h(\boldsymbol{x}_n)\Big). \tag{1.15}$$

Observe that formula Eq. 1.15 coincides with the left-sided Bregman centroid [8]. However, left-sided Bregman k-means and Riemann-Bregman k-means will differ in the assignment steps because they rely on different dissimilarity measures.

1.4   Balls and Voronoi Cells

The examples listed show that $h : (\mathcal{M}, d_\phi) \to (\mathbb{R}^K, \|\cdot\|)$ is a non-linear isometry. This allows us to describe balls in $(\mathcal{M}, d_\phi)$ as

$$B_\phi(\boldsymbol{x}, r) = \{\boldsymbol{y} \in \mathcal{M} : d_\phi(\boldsymbol{y}, \boldsymbol{x}) \le r\} = h^{-1}\Big(B(h(\boldsymbol{x}), r)\Big). \tag{1.16}$$

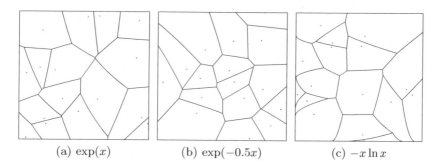

(a) $\exp(x)$ (b) $\exp(-0.5x)$ (c) $-x\ln x$

Fig. 2. Riemann-Bregman Voronoi diagrams.

Here, $B(h(\boldsymbol{x}), r)$ denotes the ball in $\mathbb{R}^K$, of radius r centered at $h(\boldsymbol{x})$.

Let $\mathcal{S} = \{\boldsymbol{y}_1, ..., \boldsymbol{y}_n\}$ be some subset $(\mathcal{M}, d_\phi)$. The Voronoi cell in $\mathcal{M}$ of $\boldsymbol{y}_i$ is defined by

$$\text{vor}_\phi(\boldsymbol{y}_i) = \{\boldsymbol{x} \in \mathcal{M} \;:\; d_\phi(\boldsymbol{x}, \boldsymbol{y}_i) \le d_\phi(\boldsymbol{x}, \boldsymbol{y}_j), \; \forall \boldsymbol{y}_j \in \mathcal{M}\}. \tag{1.17}$$

And as above, if we denote by $\text{vor}(h(\boldsymbol{y}_i))$ the Voronoi cell of $h(\boldsymbol{y}_i)$ in the Euclidean metric, it is clear that

$$\text{vor}_\phi(\boldsymbol{y}_i) = h^{-1}\big(\text{vor}(h(\boldsymbol{y}_i))\big). \tag{1.18}$$

A few graphical examples are shown in Fig. 2.

Figure 3 displays four types of Voronoi diagrams [2–4] induced by a Bregman generator; The pictures illustrate the fact that the symmetrized Bregman Voronoi diagram is different from the (symmetric metric) Riemann-Bregman Voronoi diagram.

2   Fixed Rate Quantization

Let $\mathcal{M} \subseteq \mathbb{R}^K$ be provided with a metric d_ϕ derived from a divergence. Let $N \ge 1$ be a given integer and $\mathcal{C} = \{y_1, ..., y_N\} \subset \mathcal{M}$ be a given set, called the *codebook* and its elements are called the *code vectors*. A fixed rate quantizer (with codebook $\mathcal{C}$) is a measurable mapping $q : \mathcal{M} \to \mathcal{C}$. The cardinality of $\mathcal{C}$ is called the quantization rate. Two quantizers have the same rate if their codebooks have the same cardinality

For a $\mathcal{M}$−valued random variable $\boldsymbol{X}$ defined on a probability space $(\Omega, \mathcal{F}, P)$, the ϕ−distortion of q in quantizing $\boldsymbol{X}$ is given by the expected reconstruction square error:

$$D_\phi(\boldsymbol{X}, q(\boldsymbol{X}))^2 = E[d_\phi(\boldsymbol{X}, q(\boldsymbol{X}))^2]. \tag{2.1}$$

If we are interested in only one $\mathcal{M}$−valued random variable X, we may replace Ω by $\mathcal{M}$, think of $\mathcal{F}$ as the Borel sets of $\mathcal{M}$, think of $\boldsymbol{X}$ as the identity mapping, and think of P as a probability measure on $(\mathcal{M}, \mathcal{F})$. Also, our notation is slightly different from that of Linder [7] in that we put a square into the definition of

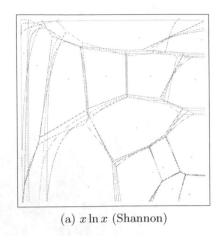 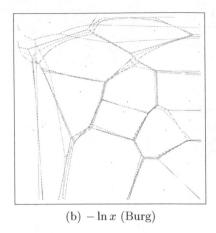

(a) $x \ln x$ (Shannon)   (b) $-\ln x$ (Burg)

Fig. 3. Four types of Voronoi diagrams induced by a Bregman generator: The left-sided Bregman Voronoi diagram (blue), the right-sided Bregman Voronoi diagram (red), the symmetrized Bregman Voronoi diagram (green), and the Riemann-Bregman Voronoi diagram (purple). (Color figure online)

the square error. Since $q(\boldsymbol{X})$ is constant on the "level" sets $S_i := \{q(X) = \boldsymbol{y}_i\}$ for $i = 1, ..., N$, the former expectation can be written as

$$D_\phi(\boldsymbol{X}, q(\boldsymbol{X}))^2 = E[E[d_\phi(\boldsymbol{X}, q(\boldsymbol{X}))^2 | \sigma(q)]].$$

Here we denote by $\sigma(q)$ the sub$-\sigma-$algebra of $\mathcal{F}$ generated by the partition $\{S_1, ..., S_N\}$ of Ω induced by the quantizer q. When $P(S_i) > 0$ holds for all blocks of the partition, the conditional expectation takes values

$$E[d_\phi(\boldsymbol{X}, q(\boldsymbol{X}))^2 | \sigma(q)] = \frac{1}{P(S_i)} \int_{S_i} d_\phi(\boldsymbol{X}, \boldsymbol{y}_i)^2 dP \quad \text{on the set} \quad S_i, i = 1, ..., N.$$

$$(2.2)$$

Notice that given $\sigma(q)$, there is a quantizer having the same rate and realizing an error smaller than $D_\phi(\boldsymbol{X}, q(\boldsymbol{X}))^2$. The existence and uniqueness of such quantizer is given by the following result in [6].

Theorem 3. *Let $\mathcal{G}$ be a sub$-\sigma-$algebra of $\mathcal{F}$. Let $\boldsymbol{X}$ be such that $h(\boldsymbol{X}) \in \mathcal{L}_2$ (Lebesgue space of functions that are square integrable). Then there exists a unique $\mathcal{G}$ measurable random variable, denoted by $E_\phi[X|\mathcal{G}]$ satisfying*

$$E_\phi[X|\mathcal{G}] = argmin\{D_\phi(\boldsymbol{X}, \boldsymbol{Y})^2 | \boldsymbol{Y} \in \mathcal{G}, \ h(\boldsymbol{Y}) \in \mathcal{L}_2\}$$

The result mentioned a few lines above can be stated as

Theorem 4. *Let q be a quantizer of rate N. Suppose that the blocks $\{S_i : i = 1, ..., N\}$ of the partition of Ω determined by q are such that $P(S_i) > 0$ for all $i = 1, ..., N$. Then*

$$q^*(\boldsymbol{X}) = E_\phi[X|\mathcal{G}] = \sum_{i=1}^{N} \frac{1}{P(S_i)} E_\phi[\boldsymbol{X}; S_i] I_{S_i}(\boldsymbol{X}) := \sum_{i=1}^{N} \boldsymbol{y}_i I_{S_i}(\boldsymbol{X}). \qquad (2.3)$$

is the quantizer with codebook C^*, *whose elements are* $\boldsymbol{y}_i^* = \frac{1}{P(S_i)} E_\phi[\boldsymbol{X}; S_i]$, *which makes the error* $D_\phi(\boldsymbol{X}, q(\boldsymbol{X}))^2$ *smaller over all codebooks which determine the same partition of* Ω.

An interesting way to associate sets to a codebook $C = \{\boldsymbol{y}_1, ..., \boldsymbol{y}_N\}$ of rate N is by means of Voronoi cells. $\{S_1, ..., S_N\}$ are the Voronoi cells determined by C, their intersections are polyhedra of dimension lower than K (faces), and the probability P does not change them, that is if $P(S_i \cap S_j) = 0$ whenever $i \neq j$. If we put $B_i = \text{int}(S_i)$ for the interior of the Voronoi cells, they generate a sub-σ-algebra of $\mathcal{F}$ such that $\cup_i B_i$ differs from $\mathcal{M}$ by a set of P-probability 0.

In the notation of Theorem 4 we have

Corollary 1. *If the* σ-*algebra* $\mathcal{G}$ *is generated buy the interiors of the Voronoi cells determined by a codebook* C, *then*

$$\boldsymbol{y}_i^* = \frac{1}{P(S_i)} E_\phi[\boldsymbol{X}; S_i] = \frac{1}{P(B_i)} E_\phi[\boldsymbol{X}; B_i] \in B_i \subset S_i, \quad \forall i = 1, ..., N.$$

3   Conclusion

We considered the dual Riemannian metric distances induced by the dual separable Bregman divergences, and studied the fixed rate quantization problem. We described the Riemann-Bregman Voronoi diagrams that can be obtained from an ordinary Euclidean Voronoi diagram after monotone isometric embeddings.

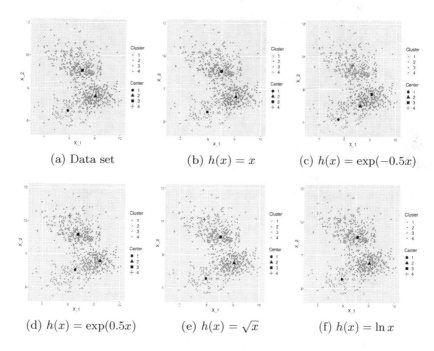

(a) Data set   (b) $h(x) = x$ (c) $h(x) = \exp(-0.5x)$

(d) $h(x) = \exp(0.5x)$ (e) $h(x) = \sqrt{x}$ (f) $h(x) = \ln x$

Fig. 4. Clustering by k-means

We can cluster datasets in those Riemann-Bregman metric spaces easily using either off-the-shelf partition-based (k-means), soft (Expectation-Maximization) or hierarchical clusterings: We map the data $\{x_1, \ldots, x_n\}$ using the monotone embedding $h(x)$, then apply the chosen clustering on $\{h(x_1), \ldots, h(x_n)\}$, and map back the clusters into the metric space using the inverse monotone function $h^{-1}(x)$ (see Fig. 4).

References

1. Amari, S.-i., Nagaoka, H.: Methods of Information Geometry. Translations of Mathematical Monographs, vol. 191. Oxford University Press, New York (2000)
2. Nielsen, F, Boissonnat, J-D., Nock, F.: On Bregman Voronoi diagrams. In: ACM-SIAM Symposium on Discrete Algorithms (SODA), pp. 746–755 (2007)
3. Boissonnat, J.-D., Nielsen, F., Nock, F.: Bregman Voronoi diagrams. Discret. Comput. Geom. **44**, 281–307 (2010)
4. Nielsen, F., Nock, R.: Skew Jensen-Bregman Voronoi diagrams. In: Gavrilova, M.L., Tan, C.J.K., Mostafavi, M.A. (eds.) Transactions on Computational Science XIV. LNCS, vol. 6970, pp. 102–128. Springer, Heidelberg (2011). https://doi.org/10.1007/978-3-642-25249-5_4
5. Calin, O., Urdiste, C.: Geometric Modeling in Probability and Statistics. Springer Internl. Pub, Switzerland (2010)
6. Gzyl, H.: Prediction in Riemannian metrics derived from divergence functions (2018). http://arxiv.org/abs/1808.01638
7. Linder, T.: Learning-theoretic methods in vector quantization. In: Györfi, L. (ed.) Principles of Nonparametric Learning. CISM, vol. 434, pp. 163–210. Springer, Vienna (2002). https://doi.org/10.1007/978-3-7091-2568-7_4
8. Nielsen, F., Nock, R.: Sided and symmetrized Bregman centroids. IEEE Trans. Inf. Theory **55**(6), 2882–2904 (2009)
9. Nielsen, F.: An elementary introduction to information geometry (2018). https://arxiv.org/abs/1808.08271
10. Pollard, D.: A User's Guide to Measure Theoretic Probability. Cambridge University Press, Cambridge (2002)

The Statistical Minkowski Distances: Closed-Form Formula for Gaussian Mixture Models

Frank Nielsen$^{(\boxtimes)}$ (iD)

Sony Computer Science Laboratories, Inc., Tokyo, Japan
Frank.Nielsen@acm.org

Abstract. The traditional Minkowski distances are induced by the corresponding Minkowski norms in real-valued vector spaces. In this work, we propose statistical symmetric distances based on the Minkowski's inequality for probability densities belonging to Lebesgue spaces. These statistical Minkowski distances admit closed-form formula for Gaussian mixture models when parameterized by integer exponents.

Keywords: Minkowski ℓ_p metrics · L_p spaces ·
Minkowski's inequality · Statistical mixtures of exponential families ·
Multinomial theorem · Gaussian Mixture Model

1 Introduction and Motivation

1.1 Statistical Distances Between Mixtures

Gaussian Mixture Models (GMMs) are flexible statistical models often used in machine learning, signal processing and computer vision [15,28] since they can arbitrarily closely approximate any smooth density. To measure the dissimilarity between probability distributions, one often relies on the principled information-theoretic Kullback-Leibler (KL) divergence [8]. However the lack of closed-form formula for the KL divergence between GMMs has motivated various KL lower and upper bounds [12,13,23,24] for GMMs or approximation techniques [10], and further spurred the *design* of novel distances that admit closed-form formula between GMMs [19].

A *distance* $D : \mathcal{X} \times \mathcal{X} \to \mathbb{R}$ is a non-negative real-valued function D on the *product space* $\mathcal{X} \times \mathcal{X}$ such that $D(p, q) = D((p, q)) = 0$ iff. $p = q$. Here, a distance may not be symmetric nor satisfy the triangular inequality. Two usual symmetrizations of the KL divergence are the Jeffreys' divergence and the Jensen-Shannon divergence [18].

The Cauchy-Schwarz (CS [14]) and Hölder [25] distances $D(p : q)$ are said to be *projective* because $D(\lambda p : \lambda' q) = D(p : q)$ for any $\lambda, \lambda' > 0$. An important family of projective divergences for robust statistical inference are the γ-divergences [11,22]. The Hölder projective divergences do not admit closed-form formula for GMMs, except for the very special case of the CS divergence.

© Springer Nature Switzerland AG 2019
F. Nielsen and F. Barbaresco (Eds.): GSI 2019, LNCS 11712, pp. 359–367, 2019.
https://doi.org/10.1007/978-3-030-26980-7_37

1.2 Minkowski Distances and Lebesgue Spaces

The renown Minkowski distances are norm-induced metrics [9] measuring distances between d-dimensional vectors $x, y \in \mathbb{R}^d$ defined for $\alpha \geq 1$ by:

$$M_\alpha(x, y) := \|x - y\|_\alpha = \left(\sum_{i=1}^{d} |x_i - y_i|^\alpha \right)^{\frac{1}{\alpha}}, \tag{1}$$

where the Minkowski norms are given by $\|x\|_\alpha = \left(\sum_{i=1}^{d} |x_i|^\alpha \right)^{\frac{1}{\alpha}}$. The Minkowski norms can be extended to countably infinite-dimensional ℓ_α spaces of sequences (see [1], p. 68).

Let $(\mathcal{X}, \mathcal{F})$ be a measurable space where $\mathcal{F}$ denotes the σ-algebra of $\mathcal{X}$, and let μ be a probability measure (with $\mu(\mathcal{X}) = 1$) with full support $\mathrm{supp}(\mu) = \mathcal{X}$ (where $\mathrm{supp}(\mu) := \mathrm{cl}(\{F \in \mathcal{F} : \mu(F) > 0\})$ and cl denotes the set closure). Let $\mathbb{F}$ be the set of all real-valued measurable functions defined on $\mathcal{X}$. We define the *Lebesgue space* [1,4] $L_\alpha(\mu) := \{f \in \mathbb{F} : \int_{\mathcal{X}} |f(x)|^\alpha \mathrm{d}\mu(x) < \infty\}$ for $\alpha \geq 1$. The Minkowski distance [16] of Eq. 1 can be generalized to probability densities belonging to Lebesgue $L_\alpha(\mu)$ spaces, to get the *statistical Minkowski distance* for $\alpha \geq 1$:

$$M_\alpha(p, q) := \left(\int_{\mathcal{X}} |p(x) - q(x)|^\alpha \mathrm{d}\mu(x) \right)^{\frac{1}{\alpha}}. \tag{2}$$

When $\alpha = 1$, we recover twice the *Total Variation* (TV) metric

$$\mathrm{TV}(p, q) := \frac{1}{2} \int |p(x) - q(x)| \mathrm{d}\mu(x) = \frac{1}{2} \|p - q\|_{L_1(\mu)} = \frac{1}{2} M_1(p, q). \tag{3}$$

In this work, we design novel distances based on the Minkowski's inequality (triangle inequality for $L_\alpha(\mu)$, which proves that $\|p\|_{L_\alpha(\mu)}$ is a norm (i.e., the L_α-norm), so that the statistical Minkowski's distance between functions of a Lebesgue space can be written as $M_\alpha(p, q) = \|p - q\|_{L_\alpha(\mu)}$). The space $L_\alpha(\mu)$ is a Banach space (i.e., complete normed linear space).

2 Distances from the Minkowski's Inequality

Let us state Minkowski's inequality:

Theorem 1 (Minkowski's inequality). *For $p(x), q(x) \in L_\alpha(\mu)$ with $\alpha \in [1, \infty)$, we have the following Minkowski's inequality:*

$$\left(\int |p(x) + q(x)|^\alpha \mathrm{d}\mu(x) \right)^{\frac{1}{\alpha}} \leq \left(\int |p(x)|^\alpha \mathrm{d}\mu(x) \right)^{\frac{1}{\alpha}} + \left(\int |q(x)|^\alpha \mathrm{d}\mu(x) \right)^{\frac{1}{\alpha}}, \tag{4}$$

with equality holding only when $q(x) = 0$ (almost everywhere, a.e.), or when $p(x) = \lambda q(x)$ a.e. for $\lambda > 0$ for $\alpha > 1$.

The usual proof of Minkowski's inequality relies on Hölder's inequality [25,27]. Following [25], we define distances by measuring in several ways the tightness of the Minkowski's inequality. When clear from context, we shall write $\| \cdot \|_\alpha$ for short instead of $\| \cdot \|_{L_\alpha(\mu)}$. Thus Minkowski's inequality writes as $\|p + q\|_\alpha \leq \|p\|_\alpha + \|q\|_\alpha$. Minkowski's inequality proves that the L_α-spaces are normed vector spaces. Notice that when $p(x)$ and $q(x)$ are probability densities (i.e., $\int p(x)\mathrm{d}\mu(x) = \int q(x)\mathrm{d}\mu(x) = 1$), Minkowski's inequality becomes an equality iff. $p(x) = q(x)$ almost everywhere, for $\alpha > 1$. Thus we can define the following novel Minkowski's distances between probability densities satisfying the identity of indiscernibles:

Definition 1 (Minkowski difference distance). *For probability densities* $p, q \in L_\alpha(\mu)$, *we define the Minkowski difference* $D_\alpha(\cdot, \cdot)$ *distance for* $\alpha \in (1, \infty)$ *as:*

$$D_\alpha(p, q) := \|p\|_\alpha + \|q\|_\alpha - \|p + q\|_\alpha \geq 0. \tag{5}$$

Definition 2 (Minkowski log-ratio distance). *For probability densities* $p, q \in L_\alpha(\mu)$, *we define the Minkowski log-ratio distance* $L_\alpha(\cdot, \cdot)$ *for* $\alpha \in (1, \infty)$ *as:*

$$L_\alpha(p, q) := - \log \frac{\|p + q\|_\alpha}{\|p\|_\alpha + \|q\|_\alpha} = \log \frac{\|p\|_\alpha + \|q\|_\alpha}{\|p + q\|_\alpha} \geq 0. \tag{6}$$

By construction, all these Minkowski distances are symmetric distances: Namely, $M_\alpha(p, q) = M_\alpha(q, p)$, $D_\alpha(p, q) = D_\alpha(q, p)$ and $L_\alpha(p, q) = L_\alpha(q, p)$. Notice that $L_\alpha(p, q)$ is *scale-invariant*[1]: $L_\alpha(\lambda p, \lambda q) = L_\alpha(p, q)$ for any $\lambda > 0$. Scale-invariance is a useful property in many signal processing applications. Distance $D_\alpha(p, q)$ is *homogeneous* since

$$D_\alpha(\lambda p, \lambda q) = |\lambda| D_\alpha(p, q), \tag{7}$$

for any $\lambda \in \mathbb{R}$ (and so is distance $M_\alpha(p, q)$).

3   Closed-Form Formula for Statistical Mixtures

In this section, we shall prove that D_α and L_α between statistical mixtures are in closed-form for all integer exponents (and M_α for all even exponents) for mixtures of exponential families with conic natural parameter spaces.

Notice that by considering a parametric family $\mathcal{E}$ of probability densities, those integral-based statistical Minkowski distances amount to equivalent parameter distances. For example, we can define the parametric distance

$$M'_\alpha(\theta, \theta') := M_\alpha(p_\theta, p_{\theta'}), \tag{8}$$

for $p_\theta, p_{\theta'} \in \mathcal{E}$. From the viewpoint of symbolic calculation, whether the definite integrals admit closed-form expressions or not when considering parametric distributions (using computer algebra systems), can be decided using Risch *semi-algorithm* [6,26]. The symbolic integration method is called a semi-algorithm because Risch algorithm requires an *oracle* to decide whether certain expressions

---

[1] Like any distance based on the log ratio of triangle inequality gap induced by a homogeneous norm.

are equivalent to zero or not. We shall derive manually below a closed-form expression for densities belonging to the same exponential family.

Let us first define the positively *weighed geometric integral* I of a set $\{p_1, \ldots, p_k\}$ of k probability densities of $L_\alpha(\mu)$ as:

$$I(p_1, \ldots, p_k; \alpha_1, \ldots, \alpha_k) := \int_{\mathcal{X}} p_1(x)^{\alpha_1} \ldots p_k(x)^{\alpha_k} d\mu(x), \quad \alpha \in \mathbb{R}_+^k. \tag{9}$$

An *exponential family* [7,21] $\mathcal{E}_{t,\mu}$ is a set $\{p_\theta(x)\}_\theta$ of probability densities wrt. μ which densities can be expressed proportionally canonically as $p_\theta(x) \propto \exp(t(x)^\top \theta)$, where $t(x)$ is a D-dimensional vector of sufficient statistics [7]. The term $t(x)^\top \theta$ can be written equivalently as $\langle t(x), \theta \rangle$, where $\langle \cdot, \cdot \rangle$ denotes the scalar product on $\mathbb{R}^D$. Thus the normalized probability densities of $\mathcal{E}_{t,\mu}$ can be written as

$$p_\theta(x) = \exp\left(t(x)^\top \theta - F(\theta)\right), \tag{10}$$

where

$$F(\theta) := \log \int_{\mathcal{X}} \exp(t(x)^\top \theta) d\mu(x), \tag{11}$$

is called the *log-partition function* (also called cumulant function [7] or log-normalizer [21]). The natural parameter space is

$$\Theta := \left\{ \theta \in \mathbb{R}^D : \int_{\mathcal{X}} \exp(t(x)^\top \theta) d\mu(x) < \infty \right\}. \tag{12}$$

Many common distributions (Gaussians, Poisson, Beta, etc.) belong to exponential families in disguise [7,21].

Lemma 1. *For probability densities $p_{\theta_1}, \ldots, p_{\theta_k}$ belonging to the same exponential family $\mathcal{E}_{t,\mu}$, we have:*

$$I(p_{\theta_1}, \ldots, p_{\theta_k}; \alpha_1, \ldots, \alpha_k) = \exp\left(F\left(\sum_{i=1}^k \alpha_i \theta_i \right) - \sum_{i=1}^k \alpha_i F(\theta_i) \right) < \infty, \tag{13}$$

provided that $\sum_{i=1}^k \alpha_i \theta_i \in \Theta$.

Proof.

$$I(p_{\theta_1}, \ldots, p_{\theta_k}; \alpha_1, \ldots, \alpha_k) = \int \prod_{i=1}^k \left(\exp\left(\left(t(x)^\top \theta_i - F(\theta_i) \right) \right) \right)^{\alpha_i} d\mu(x), \tag{14}$$

$$= \int \exp\left(t(x)^\top (\sum_i \alpha_i \theta_i) - \sum_i \alpha_i F(\theta_i) + \underbrace{F\left(\sum_i \alpha_i \theta_i \right) - F\left(\sum_i \alpha_i \theta_i \right)}_{=0} \right) d\mu(x), \tag{15}$$

$$= \exp\left(F\left(\sum_i \alpha_i \theta_i \right) - \sum_i \alpha_i F(\theta_i) \right) \underbrace{\int_{\mathcal{X}} \exp\left(t(x)^\top \left(\sum_i \alpha_i \theta_i \right) - F\left(\sum_i \alpha_i \theta_i \right) \right) d\mu(x)}_{=1},$$

$$\tag{16}$$

$$= \exp\left(F(\sum_i \alpha_i \theta_i) - \sum_i \alpha_i F(\theta_i) \right), \tag{17}$$

since

$$\int_{\mathcal{X}} \exp \left(t(x)^{\top} (\sum_i \alpha_i \theta_i) - F(\sum_i \alpha_i \theta_i) \right) \mathrm{d}\mu(x) = \int_{\mathcal{X}} p_{\sum_i \alpha_i \theta_i}(x) \mathrm{d}\mu(x) = 1, \tag{18}$$

provided that $\bar{\theta} := \sum_i \alpha_i \theta_i \in \Theta$ (and $p_{\bar{\theta}} \in \mathcal{E}_{t,\mu}$).

In particular, the condition $\sum_i \alpha_i \theta_i \in \Theta$ always holds when the natural parameter space Θ is a *cone*. In the remainder, we shall call those exponential families with natural parameter space being a cone, *Conic Exponential Families* (CEFs) for short. Note that when $\sum_i \alpha_i \theta_i \notin \Theta$, the integral $I(p_{\theta_1}, \ldots, p_{\theta_k}; \alpha_1, \ldots, \alpha_k)$ diverges (that is, $I(p_{\theta_1}, \ldots, p_{\theta_k}; \alpha_1, \ldots, \alpha_k) = \infty$). Observe that for a CEF density $p_\theta(x)$, we have $p_\theta(x)^\alpha$ in $L_\alpha(\mu)$ for *any* $\alpha \in [1, \infty)$.

Corollary 1. *We have* $I(p_{\theta_1}, \ldots, p_{\theta_k}; \alpha_1, \ldots, \alpha_k) = \exp(F(\sum_i \alpha_i \theta_i) - \sum_i \alpha_i F(\theta_i)) < \infty$ *for probability densities belonging to the same exponential family with natural parameter space Θ being a cone.*

Let us define:

$$J_F(\theta_1, \ldots, \theta_k; \alpha_1, \ldots, \alpha_k) := \sum_i \alpha_i F(\theta_i) - F\left(\sum_i \alpha_i \theta_i\right). \tag{19}$$

This quantity is called the *Jensen diversity* [20] when $\alpha \in \Delta_k$ (the $(k-1)$-dimensional standard simplex), or Bregman information[2] in [3]. Although the Jensen diversity is non-negative when $\alpha \in \Delta_k$, this Jensen diversity of Eq. 19 maybe negative when $\alpha \in \mathbb{R}_+^k$. When $\alpha \in \mathbb{R}_+^k$, we thus call the Jensen diversity the *generalized Jensen diversity*. Whenever we want to emphasize that $\alpha \in \mathbb{R}_+^k$, we denote this generalized Jensen diversity by J_F^+. It follows that we have

$$I(p_{\theta_1}, \ldots, p_{\theta_k}; \alpha_1, \ldots, \alpha_k) = \exp(-J_F(\theta_1, \ldots, \theta_k; \alpha_1, \ldots, \alpha_k)). \tag{20}$$

The CEFs include the Gaussian family, the Wishart family, the Binomial/multinomial family, etc. [7,19,21].

Let us consider a finite positive mixture $\tilde{m}(x) = \sum_{i=1}^k w_i p_i(x)$ of k probability densities, where the weight vector $w \in \mathbb{R}_+^k$ are not necessarily normalized to one.

Lemma 2. *For a finite positive mixture $\tilde{m}(x)$ with components belonging to the same CEF, $\|\tilde{m}\|_{L_\alpha(\mu)}$ is finite and in closed-form, for any integer $\alpha \geq 2$.*

Proof. Consider the multinomial expansion[3] $\tilde{m}(x)^\alpha$ obtained by applying the multinomial theorem [5]:

$$\tilde{m}(x)^\alpha = \sum_{\substack{\sum_{i=1}^k \alpha_i = \alpha \\ \alpha_i \in \mathbb{N}}} \binom{\alpha}{\alpha_1, \ldots, \alpha_k} \prod_{j=1}^k (w_j p_j(x))^{\alpha_j}, \tag{21}$$

---

[2] Because $\sum_i \alpha_i B_F(\theta_i : \bar{\theta}) = J_F(\theta_1, \ldots, \theta_k; \alpha_1, \ldots, \alpha_k)$ for the barycenter $\bar{\theta} = \sum_i \alpha_i \theta_i$, where $B_F(\theta : \theta') = F(\theta) - F(\theta') - (\theta - \theta')^{\top} \nabla F(\theta')$ is a Bregman divergence.

[3] To apply the multinomial expansion, we need elements to commute wrt. the product. Thus it does not apply to the matrix cases.

where

$$\binom{\alpha}{\alpha_1,\ldots,\alpha_k} := \frac{\alpha!}{\alpha_1! \times \ldots \times \alpha_k!}, \tag{22}$$

is the *multinomial coefficient* [2]. It follows that:

$$\int \tilde{m}(x)^\alpha d\mu(x) = \sum_{\substack{\sum_i \alpha_i = \alpha \\ \alpha_i \in \mathbb{N}}} \binom{\alpha}{\alpha_1,\ldots,\alpha_k} \left(\prod_{j=1}^k w_j^{\alpha_j}\right) I(p_1,\ldots,p_k;\alpha_1,\ldots,\alpha_k). \tag{23}$$

Thus the term $\int \tilde{m}(x)^\alpha d\mu(x)$ amounts to a positively weighted sum of integrals of monomials that are positively weighted geometric means of mixture components. When $p_i = p_{\theta_i}$, since $I(p_{\theta_1},\ldots,p_{\theta_k};\alpha_1,\ldots,\alpha_k) < \infty$ using Eq. 1, we conclude that $\tilde{m} \in L_\alpha(\mu)$ for $\alpha \in \mathbb{N}$, and we get the formula:

$$\|\tilde{m}\|_{L_\alpha(\mu)} = \left(\sum_{\substack{\sum_i \alpha_i = \alpha \\ \alpha_i \in \mathbb{N}}} \binom{\alpha}{\alpha_1,\ldots,\alpha_k} \left(\prod_{j=1}^k w_j^{\alpha_j}\right) \exp\left(-J_F(\theta_1,\ldots,\theta_k;\alpha_1,\ldots,\alpha_k)\right)\right)^{\frac{1}{\alpha}}, \tag{24}$$

for $\alpha \in \mathbb{N}$.

A naive multinomial expansion of $\tilde{m}(x)^\alpha$ yields k^α terms that can then be simplified. Using the multinomial theorem, there are $\binom{k+\alpha-1}{\alpha}$ integral terms in the formula of $\int (\sum_{i=1}^k w_i p_i(x))^\alpha d\mu(x)$. This number corresponds to the number of sequences of k disjoint subsets whose union is $\{1,\ldots,\alpha\}$ (also called the number of ordered partitions but beware that some sets may be empty). Notice that by setting $w_j p_j(x) = 1$ in Eq. 21 (with $\tilde{m}(x) = k$), we get the following identity of the sum of all multinomial coefficients:

$$\sum_{\substack{\sum_{i=1}^k \alpha_i = \alpha \\ \alpha_i \in \mathbb{N}}} \binom{\alpha}{\alpha_1,\ldots,\alpha_k} = k^\alpha. \tag{25}$$

The multinomial expansion can be calculated efficiently using a generalization of Pascal's triangle, called *Pascal's simplex* [17], thus avoiding to compute from scratch all the multinomial coefficients. An efficient way to implement the multinomial expansion using nested iterative loops follows from this identity:

$$\left(\sum_{i=1}^k x_i\right)^\alpha = \sum_{\alpha_1=0}^\alpha \sum_{\alpha_2=0}^{\alpha_1} \cdots \sum_{\alpha_{k-1}=0}^{\alpha_{k-2}} \binom{\alpha}{\alpha_1}\binom{\alpha_1}{\alpha_2}\cdots\binom{\alpha_{k-1}}{\alpha_{k-2}} x_1^{\alpha-\alpha_1} x_2^{\alpha_1-\alpha_2} \cdots x_{k-1}^{\alpha_{k-2}-\alpha_{k-1}} x_k^{\alpha_{k-1}}. \tag{26}$$

We are now ready to show when the statistical Minkowski's distances M_α, D_α and L_α are in closed-form for mixtures of CEFs using Lemma 2.

Theorem 2 (Closed-form formula for Minkowski's distances). *For mixtures* $m = \sum_{i=1}^{k} w_i p_{\theta_i}$ *and* $m' = \sum_{j=1}^{k'} w'_j p_{\theta'_j}$ *of CEFs* $\mathcal{E}_{\mu,t}$, D_α *and* L_α *admits closed-form formula for integers* $\alpha \geq 2$, *and* M_α *is in closed-form when* $\alpha \geq 2$ *is an even positive integer.*

Proof. For D_α and L_α, it is enough to show that $\|m\|_{L_\alpha(\mu)}, \|m'\|_{L_\alpha(\mu)}$ and $\|m + m'\|_{L_\alpha(\mu)}$ are all in closed-form. This follows from Lemma 2 by setting $\tilde{m}$ to be m, m' and $m + m'$, respectively. The overall number of generalized Jensen diversity terms in the formula of D_α or L_α is $O\left(\binom{k+k'+\alpha-1}{\alpha}\right)$.

Now, consider distance M_α. To get rid of the absolute value in M_α for even integers α, we rewrite M_α as follows:

$$M_\alpha(m, m') = \|m - m'\|_{L_\alpha(\mu)} = \left(\int |m(x) - m'(x)|^\alpha d\mu(x) \right)^{\frac{1}{\alpha}},$$

$$= \left(\int \left((m(x) - m'(x))^2 \right)^{\frac{\alpha}{2}} d\mu(x) \right)^{\frac{1}{\alpha}}.$$

Let $\tilde{m}(x) = (m(x) - m'(x))^2$. We have:

$$\tilde{m}(x) = (m(x) - m'(x))^2,$$
$$= m(x)^2 + m'(x)^2 - 2m(x)m'(x),$$
$$= \left(\sum_{i=1}^{k} w_i p_{\theta_i}(x) \right)^2 + \left(\sum_{j=1}^{k'} w'_j p_{\theta'_j}(x) \right)^2 - 2 \sum_{i=1}^{k} \sum_{j=1}^{k'} w_i w'_j p_{\theta_i}(x) p_{\theta'_j}(x). \quad (27)$$

We have the density products $p_{\theta,\theta'} := p_\theta p_{\theta'} = I(p_\theta, p_{\theta'}; 1, 1) \in L_{\frac{\alpha}{2}}(\mu)$ (using Lemma 2) for any $\theta, \theta' \in \Theta$ and $\alpha \geq 2$. When $\alpha = 2$, $\frac{\alpha}{2} = 1$, and we easily reach a closed-form formula for $M_2(m, m')$. Otherwise, let us expand all the terms in Eq. 27, and rewrite $\tilde{m}(x) = \sum_{l=1}^{K} w''_l p_{\theta_l, \theta'_l}$. Now, a key difference is that $w''_l \in \mathbb{R}$, and not necessarily positive. Nevertheless, since $\frac{\alpha}{2} \in \mathbb{N}$, we can still use the multinomial theorem to expand $\tilde{m}(x)^{\frac{\alpha}{2}}$, distribute the integral over all terms, and compute elementary integrals $I(p_{\theta_1,\theta'_1}, \ldots, p_{\theta_K,\theta'_K}; \alpha'_1, \ldots, \alpha'_K)$ with $\sum_{l=1}^{K} \alpha'_i = \frac{\alpha}{2}$ in closed-form. Thus M_α is available in closed-form for mixtures of CEFs for all even positive integers $\alpha \geq 2$. The number of terms in the M_α formula is $O\left(\binom{\max(k^2, k'^2)+\alpha-1}{\alpha}\right)$.

Notice that $D_\alpha^\lambda(m_1, m_2) = 0$ or $L_\alpha^\lambda(m_1, m_2) = 0$ if and only if $(1-\lambda)m_1(x) = \lambda m_2(x)$ almost everywhere. By integrating over the support, we find that a necessary condition is $\lambda = \frac{1}{2}$. But when $\lambda = \frac{1}{2}$, we have $L_\alpha^{\frac{1}{2}}(m_1 : m_2) = L_\alpha(m_1, m_2)$ and $D_\alpha^{\frac{1}{2}}(m_1 : m_2) = \frac{1}{2} D_\alpha(m_1, m_2)$. It follows that $D_\alpha^\lambda(m_1, m_2) = 0$ or $L_\alpha^\lambda(m_1, m_2) = 0$ if and only if $m_1 = m_2$ almost everywhere.

References

1. Alabiso, C., Weiss, I.: A Primer on Hilbert Space Theory: Linear Spaces, Topological Spaces, Metric Spaces, Normed Spaces, and Topological Groups. UNITEXT for Physics. Springer, Cham (2015). https://doi.org/10.1007/978-3-319-03713-4
2. Balakrishnan, V.K.: Introductory Discrete Mathematics. Courier Corporation, North Chelmsford (2012)
3. Banerjee, A., Merugu, S., Dhillon, I.S., Ghosh, J.: Clustering with Bregman divergences. J. Mach. Learn. Res. **6**, 1705–1749 (2005)
4. Bogachev, V.I.: Measure Theory, vol. 1. Springer, Heidelberg (2007). https://doi.org/10.1007/978-3-540-34514-5
5. Bolton, D.W.: The multinomial theorem. Math. Gaz. **52**(382), 336–342 (1968)
6. Bronstein, M.: Symbolic Integration I: Transcendental Functions, vol. 1. Springer, New York (2013)
7. Brown, L.D.: Fundamentals of Statistical Exponential Families: With Applications in Statistical Decision Theory. IMS (1986)
8. Cover, T.M., Thomas, J.A.: Elements of Information Theory. Wiley, Hoboken (2012)
9. Deza, M.M., Deza, E.: Encyclopedia of distances. In: Deza, E., Deza, M.M. (eds.) Encyclopedia of Distances, pp. 1–583. Springer, Heidelberg (2009). https://doi.org/10.1007/978-3-642-00234-2_1
10. Durrieu, J.L., Thiran, J.P., Kelly, F.: Lower and upper bounds for approximation of the Kullback-Leibler divergence between Gaussian mixture models. In: IEEE International Conference on Acoustics, Speech and Signal Processing (ICASSP), pp. 4833–4836. IEEE (2012)
11. Fujisawa, H., Eguchi, S.: Robust parameter estimation with a small bias against heavy contamination. J. Multivar. Anal. **99**(9), 2053–2081 (2008)
12. Goldberger, J., Aronowitz, H.: A distance measure between GMMs based on the unscented transform and its application to speaker recognition. In: European Conference on Speech Communication and Technology (INTERSPEECH), pp. 1985–1988 (2005)
13. Goldberger, J., Gordon, S., Greenspan, H.: An efficient image similarity measure based on approximations of KL-divergence between two Gaussian mixtures, p. 487 (2003)
14. Jenssen, R., Principe, J.C., Erdogmus, D., Eltoft, T.: The Cauchy-Schwarz divergence and Parzen windowing: connections to graph theory and Mercer kernels. J. Frankl. Inst. **343**(6), 614–629 (2006)
15. Jian, B., Vemuri, B.C.: Robust point set registration using Gaussian mixture models. IEEE Trans. Pattern Anal. Mach. Intell. (PAMI) **33**(8), 1633–1645 (2011)
16. Minkowski, H.: Geometrie der Zahlen, vol. 40 (1910)
17. Mousley, S., Schley, N., Shoemaker, A.: Planar rook algebra with colors and Pascal's simplex. arXiv preprint arXiv:1211.0663 (2012)
18. Nielsen, F.: A family of statistical symmetric divergences based on Jensen's inequality. arXiv preprint arXiv:1009.4004 (2010)
19. Nielsen, F.: Closed-form information-theoretic divergences for statistical mixtures. In: 21st International Conference on Pattern Recognition (ICPR), pp. 1723–1726. IEEE (2012)
20. Nielsen, F., Boltz, S.: The Burbea-Rao and Bhattacharyya centroids. IEEE Trans. Inf. Theory **57**(8), 5455–5466 (2011)

21. Nielsen, F., Garcia, V.: Statistical exponential families: a digest with flash cards. preprint arXiv:0911.4863 (2009)
22. Nielsen, F., Nock, R.: Patch matching with polynomial exponential families and projective divergences. In: Amsaleg, L., Houle, M.E., Schubert, E. (eds.) SISAP 2016. LNCS, vol. 9939, pp. 109–116. Springer, Cham (2016). https://doi.org/10.1007/978-3-319-46759-7_8
23. Nielsen, F., Sun, K.: Guaranteed bounds on information-theoretic measures of univariate mixtures using piecewise log-sum-exp inequalities. Entropy 18(12), 442 (2016)
24. Nielsen, F., Sun, K.: On the chain rule optimal transport distance. arXiv preprint arXiv:1812.08113 (2018)
25. Nielsen, F., Sun, K., Marchand-Maillet, S.: On Hölder projective divergences. Entropy 19(3), 122 (2017)
26. Risch, R.H.: The solution of the problem of integration in finite terms. Bull. Am. Math. Soc. 76(3), 605–608 (1970)
27. Tolsted, E.: An elementary derivation of the Cauchy, Hölder, and Minkowski inequalities from Young's inequality. Math. Mag. 37(1), 2–12 (1964)
28. Wang, F., Syeda-Mahmood, T., Vemuri, B.C., Beymer, D., Rangarajan, A.: Closed-form jensen-renyi divergence for mixture of gaussians and applications to group-wise shape registration. In: Yang, G.-Z., Hawkes, D., Rueckert, D., Noble, A., Taylor, C. (eds.) MICCAI 2009. LNCS, vol. 5761, pp. 648–655. Springer, Heidelberg (2009). https://doi.org/10.1007/978-3-642-04268-3_80

Parameter Estimation with Generalized Empirical Localization

Takashi Takenouchi[(✉)]

Future University Hakodate, RIKEN Center for Advanced Intelligence Project (AIP),
Hakodate, Japan
ttakashi@fun.ac.jp

Abstract. It is often difficult to estimate parameters of discrete models because of the computational cost for calculation of normalization constant, which enforces the model to be probability. In this paper, we consider a computationally feasible estimator for discrete probabilistic models using a concept of generalized empirical localization, which corresponds to the generalized mean of distributions and homogeneous γ-divergence. The proposed estimator does not require the calculation of the normalization constant and is asymptotically efficient.

Keywords: Unnormalized model · Asymptotic efficiency · γ-divergence

1 Introduction

In this paper, we focus on a problem of parameter estimation of discrete probabilistic models. A typical way for the estimation is the Maximum Likelihood Estimation (MLE) and the MLE is a "good" estimator which asymptotically satisfies the Cramér-Rao bound and is asymptotically efficient. In general, explicit solutions for the MLE cannot be obtained and then gradient-based optimization methods is usually required. But the calculation of the gradient includes the calculation of the normalization constant which makes the model to be in the probability space, and the calculation of the normalization constant is sometimes computationally intractable when the model is in a high-dimensional space. A typical example is the Boltzmann machine on $\mathcal{X} = \{+1, -1\}^p$,

$$\frac{\exp\left(\boldsymbol{\theta}_1 \boldsymbol{x} + \frac{1}{2}\boldsymbol{x}^T \boldsymbol{\theta}_2 \boldsymbol{x}\right)}{\sum_{\boldsymbol{x} \in \mathcal{X}} \exp\left(\boldsymbol{\theta}_1 \boldsymbol{x} + \frac{1}{2}\boldsymbol{x}^T \boldsymbol{\theta}_2 \boldsymbol{x}\right)} \tag{1}$$

and a calculation of the normalization constant of requires 2^p summation, which is hard to calculate as p is large. Other estimators derived from minimization of divergence measures [2] also suffer the computational problem. To tackle with the

This work was supported by JSPS KAKENHI Grant Numbers 16K00051 from MEXT, Japan.

© Springer Nature Switzerland AG 2019
F. Nielsen and F. Barbaresco (Eds.): GSI 2019, LNCS 11712, pp. 368–376, 2019.
https://doi.org/10.1007/978-3-030-26980-7_38

problem associated with the normalization constant, various kinds of approaches have been researched. Some methods are based on the Markov Chain Monte Carlo (MCMC) sampling and the contrastive divergence [7] is a well-known example. Another approach approximate the targeted probabilistic model by a tractable model by the mean-field approximation assuming independence of variables [11]. In this paper, we focus on an approach which considers an unnormalized model rather than the (normalized) probabilistic model. [8] defines information of "neighbor" by contrasting probability with that of a flipped variable and makes it possible to omit the calculation of normalization constant. [4] proposed a generalized local scoring rules on discrete sample spaces and [6] avoids the calculation of the normalization constant using a trick with auxiliary examples. [13] proposes an asymptotically efficient estimator without the calculation of the normalization constant, which consists of a concept of empirical localization and a homogeneous γ-divergence [5,10]. In this paper, we extend the concept of the empirical localization and propose a novel estimator which does not require the calculation of normalization constant. We investigate statistical properties of the proposed estimator and verify its validity with small experiments.

2   Settings

Let $\boldsymbol{x}$ be a d-dimensional vector in discrete space $\mathcal{X}$ such as $\{+1, -1\}^d$ or $\{1, 2, \ldots\}^d$, and a bracket $\langle f \rangle$ for a function f on $\mathcal{X}$ denotes a sum of f over $\mathcal{X}$, $\langle f \rangle = \sum_{x \in \mathcal{X}} f(\boldsymbol{x})$. For a given dataset $\mathcal{D} = \{\boldsymbol{x}_i\}_{i=1}^n$, the empirical distribution $\tilde{p}(\boldsymbol{x})$ is defined as

$$\tilde{p}(\boldsymbol{x}) = \begin{cases} \frac{n_x}{n} & \boldsymbol{x} \text{ is observed,} \\ 0 & \text{otherwise,} \end{cases} \qquad (2)$$

where n_x is number of examples $\boldsymbol{x}$ is observed. We consider a probabilistic model

$$\bar{q}_\theta(\boldsymbol{x}) = \frac{q_\theta(\boldsymbol{x})}{Z_\theta} \qquad (3)$$

where $q_\theta(\boldsymbol{x})$ is a unnormalized model expressed as

$$q_\theta(\boldsymbol{x}) = \exp(\psi_\theta(\boldsymbol{x})) \qquad (4)$$

with a function $\psi_\theta(\boldsymbol{x})$ parameterized by $\boldsymbol{\theta}$ and Z_θ is a normalization constant defined as $Z_\theta = \langle q_\theta \rangle$ which enforces the (3) to be a probability function. Note that the unnormalized model (4) is not a probability function and $\langle q_\theta \rangle = 1$ does not hold in general, and calculation of the normalization constant Z_θ often requires a high computational cost. Then calculation of the Maximum Likelihood Estimator (MLE)

$$\hat{\boldsymbol{\theta}}_{MLE} = \operatorname*{argmax}_\theta \sum_{i=1}^n \log \bar{q}_\theta(\boldsymbol{x}_i) \qquad (5)$$

or maximization process of the log-likelihood using its gradient

$$\sum_{i=1}^{n} \{\psi_\theta(\boldsymbol{x}_i) - \langle \bar{q}_\theta \psi_\theta \rangle\} \tag{6}$$

involves difficulty of computational cost derived from Z_θ. To overcome the difficulty of the computation of Z_θ, we consider combination of the γ-divergence and generalized mixture model.

2.1 γ-divergence

For two positive measure f, g, the γ-divergence [5] is defined as follows.

$$D_\gamma(f, g) = \frac{1}{1+\gamma} \log \langle f^{\gamma+1} \rangle + \frac{\gamma}{1+\gamma} \log \langle g^{\gamma+1} \rangle - \log \langle f g^\gamma \rangle \tag{7}$$

where γ is a positive constant. Note that $D_\gamma(f, g)$ is non-negative and is said to be homogeneous divergence because $D_\gamma(f, g) = 0$ holds if and only if $f \propto g$, rather than $f = g$. In the limit of $\gamma \to 0$, the γ-divergence reduces to the usual KL-divergence,

$$\left\langle f \log \frac{f}{g} - f + g \right\rangle. \tag{8}$$

Note that a combination of the γ-divergence and the unnormalized model does not solve the problem of computational cost because a term $D_\gamma(\tilde{p}, q_\theta)$ includes $\left\langle q_\theta^{\gamma+1} \right\rangle$ whose computation also requires the same order with the normalization constant Z_θ.

2.2 Empirical Localization

Firstly, we briefly introduce a concept of empirical localization of the (unnormalized) model $\bar{q}_\theta(\boldsymbol{x})$(or $q_\theta(\boldsymbol{x})$) with the empirical distribution $\tilde{p}(\boldsymbol{x})$ [13]. The empirical localization is interpreted as a generalized mean of q_θ and $\tilde{p}$, and lies in e-flat subspace [1] as

$$\tilde{r}_{\alpha,\theta}(\boldsymbol{x}) = \frac{\tilde{p}(\boldsymbol{x})^\alpha \bar{q}_\theta(\boldsymbol{x})^{1-\alpha}}{\langle \tilde{p}^\alpha \bar{q}_\theta^{1-\alpha} \rangle} = \frac{\tilde{p}(\boldsymbol{x})^\alpha q_\theta(\boldsymbol{x})^{1-\alpha}}{\langle \tilde{p}^\alpha q_\theta^{1-\alpha} \rangle}. \tag{9}$$

Note that the normalization constant Z_θ in $\bar{q}_\theta$ is canceled out and $\tilde{r}_{\alpha,\theta}(\boldsymbol{x})$ does not depend on Z_θ, except for $\alpha = 0$. Also note that the denominator $\langle \tilde{p}^\alpha \bar{q}^{1-\alpha} \rangle$ (or $\tilde{r}_{\alpha,\theta}(\boldsymbol{x})$ itself) can be easily calculated because the empirical distribution $\tilde{p}(\boldsymbol{x})$ has some values only on observed $\boldsymbol{x}$ in the dataset and is always 0 on the unobserved subset of $\mathcal{X}$. This implies the model $q_\theta(\boldsymbol{x})$ is empirically localized to the observed subset of domain $\mathcal{X}$ of dataset $\mathcal{D}$ and we can ignore the unobserved subset of $\mathcal{X}$, which leads to a drastic reduction of computational cost. We observe $\tilde{r}_{0,\theta}(\boldsymbol{x}) = \bar{q}(\boldsymbol{x})$ and $\tilde{r}_{1,\theta}(\boldsymbol{x}) = \tilde{p}(\boldsymbol{x})$, and (9) connects the empirical distribution $\tilde{p}$ and the normalized model $\bar{q}$ with the parameter α.

2.3 Estimator by Homogeneous Divergence and Empirical Localization

In [13], an estimator which does not require calculation of the normalization constant, was proposed by combining (7) and (9). The estimator is defined as

$$\hat{\boldsymbol{\theta}} = \operatorname*{argmin}_{\boldsymbol{\theta}} D_\gamma(\tilde{r}_{\alpha,\theta}^{1/(1+\gamma)}, \tilde{r}_{\alpha',\theta}^{1/(1+\gamma)}) \tag{10}$$

$$D_\gamma(\tilde{r}_{\alpha,\theta}^{1/(1+\gamma)}, \tilde{r}_{\alpha',\theta}^{1/(1+\gamma)}) = \frac{1}{1+\gamma} \log \left\langle \tilde{p}^\alpha q_\theta^{1-\alpha} \right\rangle + \frac{\gamma}{1+\gamma} \log \left\langle \tilde{p}^{\alpha'} q_\theta^{1-\alpha'} \right\rangle - \log \left\langle \tilde{p}^\beta q_\theta^{1-\beta} \right\rangle \tag{11}$$

where $\alpha \neq \alpha'$ and $\beta = (\alpha + \gamma\alpha')/(1+\gamma)$. We observe that a setting with $\alpha = 1, \alpha' = 0$ and $\gamma \to 0$ corresponds to the conventional MLE. Note that the empirical risk (11) does not include the calculation of the normalization constant Z_θ and can be easily calculated.

The estimator (10) has the following good statistical properties.

Proposition 1 ([13]). *Let us assume that $\psi_\theta(\boldsymbol{x})$ is written as $\boldsymbol{\theta}^T \boldsymbol{\phi}(\boldsymbol{x})$ with a fixed vector function $\boldsymbol{\psi}(\boldsymbol{x})$. Then the risk function (10) is convex with respect to $\boldsymbol{\theta}$ when $\beta = 1$ holds.*

Proposition 2 ([13]). *The estimator (10) is Fisher consistent and asymptotically efficient.*

3 Proposed Estimator

The empirical localization (9) can be interpreted as a generalized mean of a constant 1 and a distribution ratio $q_\theta/\tilde{p}$, and is rewritten as

$$\tilde{r}_{\alpha,\theta}(\boldsymbol{x}) \propto \tilde{p}(\boldsymbol{x})^\alpha q_\theta(\boldsymbol{x})^{1-\alpha} = \tilde{p}(\boldsymbol{x}) \left(\frac{q_\theta(\boldsymbol{x})}{\tilde{p}(\boldsymbol{x})} \right)^{1-\alpha}$$

$$= \tilde{p}(\boldsymbol{x}) \exp \left(\alpha \log 1 + (1-\alpha) \log \frac{q_\theta(\boldsymbol{x})}{\tilde{p}(\boldsymbol{x})} \right). \tag{12}$$

We can extend the concept of (12) to the quasi-arithmetic mean, with a monotonically increasing function u and its inverse function ξ, as follows.

$$\tilde{r}_{u,\alpha,\theta}(\boldsymbol{x}) = \tilde{p}(\boldsymbol{x}) u \left(\alpha\xi(1) + (1-\alpha)\xi \left(\frac{q_\theta(\boldsymbol{x})}{\tilde{p}(\boldsymbol{x})} \right) \right). \tag{13}$$

By transforming the function $u(z)$ to $u(z-a)$, we can set $\xi(1) = 0$ without loss of generality. The generalized version of empirical localization (13) is rewritten as

$$\tilde{r}_{u,\alpha,\theta}(\boldsymbol{x}) \propto \begin{cases} n_x u \left((1-\alpha)\xi \left(\frac{nq_\theta(\boldsymbol{x})}{n_x} \right) \right) & \boldsymbol{x} \text{ is observed} \\ 0 & \text{otherwise,} \end{cases} \tag{14}$$

and the model can be easily calculated because we can omit the unobserved domain. We show two examples associated with β-divergence and η-divergence [3] which are employed for the purpose of robust estimation [9,12].

Example 1. For $u(z) = (1 + \beta z)^{1/\beta}$ and $\xi(z) = \frac{z^\beta - 1}{\beta}$, we have

$$\tilde{r}_{u,\alpha,\theta}(\boldsymbol{x}) = \tilde{p}(\boldsymbol{x}) \left((1 - \alpha) \left(\frac{q_\theta(\boldsymbol{x})}{\tilde{p}(\boldsymbol{x})} \right)^\beta + \alpha \right)^{1/\beta} \tag{15}$$

Example 2. For $u(z) = (1 + \eta)e^z - \eta$ and $\xi(z) = \log \frac{z+\eta}{1+\eta}$, we have

$$\tilde{r}_{u,\alpha,\theta}(\boldsymbol{x}) = \tilde{p}(\boldsymbol{x}) \left\{ (1 + \eta) \left(\frac{\frac{q_\theta(\boldsymbol{x})}{\tilde{p}(\boldsymbol{x})} + \eta}{1 + \eta} \right)^{1-\alpha} - \eta \right\} \tag{16}$$

Example 3. For $u(z) = -\frac{1}{z}$ and $\xi(z) = -\frac{1}{z}$, we have

$$\tilde{r}_{u,\alpha,\theta}(\boldsymbol{x}) = \frac{\tilde{p}(\boldsymbol{x})q_\theta(\boldsymbol{x})}{\alpha q_\theta(\boldsymbol{x}) + (1 - \alpha)\tilde{p}(\boldsymbol{x})} \tag{17}$$

We propose a novel estimator for discrete probabilistic model, which can be constructed without calculation of the normalization constant Z_θ. The proposed estimator is defined by combining the (13) and γ-divergence with two hyperparameters $\alpha, \alpha' (\alpha \neq \alpha')$, as follows.

$$\hat{\boldsymbol{\theta}} = \operatorname*{argmin}_{\boldsymbol{\theta}} D_\gamma ((\tilde{r}_{u,\alpha,\theta})^{1/(1+\gamma)}, (\tilde{r}_{u,\alpha',\theta})^{1/(1+\gamma)}) \tag{18}$$

Note that when $q_\theta(\boldsymbol{x}) \propto \tilde{p}(\boldsymbol{x})$ holds, we observe that $\tilde{r}_{u,\alpha,\theta}(\boldsymbol{x}) \propto \tilde{r}_{u,\alpha',\theta}(\boldsymbol{x})$ and $D_\gamma ((\tilde{r}_{u,\alpha,\theta})^{1/(1+\gamma)}, (\tilde{r}_{u,\alpha',\theta})^{1/(1+\gamma)}) = 0$ holds.

4   Statistical Property

In this section, we investigate statistical property of the proposed estimator. Firstly, we show the Fisher consistency of the proposed estimator.

Proposition 3. *Let $\boldsymbol{\theta}_0$ be a true parameter of the underlying distribution, i.e., $p(\boldsymbol{x}) = \bar{q}_{\theta_0}(\boldsymbol{x})$. Then*

$$\boldsymbol{\theta}_0 = \operatorname*{argmin}_{\boldsymbol{\theta}} D_\gamma(r_{u,\alpha,\theta}, r_{u,\alpha',\theta}) \tag{19}$$

holds for arbitrary γ, α, $\alpha' (\alpha \neq \alpha')$ and $\boldsymbol{\theta}_0$.

Proof. The proposed estimator satisfies the equilibrium equation

$$0 = \left. \frac{\partial D_\gamma ((\tilde{r}_{u,\alpha,\theta})^{1/(1+\gamma)}, (\tilde{r}_{u,\alpha',\theta})^{1/(1+\gamma)})}{\partial \boldsymbol{\theta}} \right|_{\theta=\hat{\theta}} \tag{20}$$

implying the Fisher consistency.

Secondly, we investigate the asymptotic distribution of the proposed estimator.

Proposition 4. *Let $\boldsymbol{\theta}_0$ be a true parameter of the underlying distribution. Then, under mild regularity condition, the proposed estimator asymptotically follows*

$$\sqrt{n}(\hat{\boldsymbol{\theta}} - \boldsymbol{\theta}_0) \sim \mathcal{N}(\mathbf{0}, I(\boldsymbol{\theta}_0)^{-1}) \tag{21}$$

where $\mathcal{N}$ is the Normal distribution and $I(\boldsymbol{\theta}_0) = V_{\bar{q}_{\boldsymbol{\theta}_0}}[\psi'_{\boldsymbol{\theta}_0}]$ is the Fisher information matrix.

Proof. Let us assume that the empirical distribution is written as $\tilde{p}(\boldsymbol{x}) = \bar{q}_{\boldsymbol{\theta}_0}(\boldsymbol{x}) + \epsilon(\boldsymbol{x})$. By expanding the equilibrium condition (20) around $\boldsymbol{\theta} = \boldsymbol{\theta}_0$ and $\epsilon(\boldsymbol{x}) = 0$, we have

$$
\begin{aligned}
0 \simeq {} & \frac{\partial D_\gamma((\tilde{r}_{u,\alpha,\theta})^{1/(1+\gamma)}, (\tilde{r}_{u,\alpha',\theta})^{1/(1+\gamma)})}{\partial \boldsymbol{\theta}}\bigg|_{\boldsymbol{\theta}=\boldsymbol{\theta}_0} \\
& + \frac{\partial^2 D_\gamma((\tilde{r}_{u,\alpha,\theta})^{1/(1+\gamma)}, (\tilde{r}_{u,\alpha',\theta})^{1/(1+\gamma)})}{\partial \boldsymbol{\theta} \partial \boldsymbol{\theta}^T}\bigg|_{\boldsymbol{\theta}=\boldsymbol{\theta}_0} (\hat{\boldsymbol{\theta}} - \boldsymbol{\theta}_0).
\end{aligned} \tag{22}
$$

Using the delta method [14], we have

$$
\frac{\partial D_\gamma((\tilde{r}_{u,\alpha,\theta})^{1/(1+\gamma)}, (\tilde{r}_{u,\alpha',\theta})^{1/(1+\gamma)})}{\partial \boldsymbol{\theta}}\bigg|_{\boldsymbol{\theta}=\boldsymbol{\theta}_0} - \frac{\partial D_\gamma((r_{u,\alpha,\theta})^{1/(1+\gamma)}, (r_{u,\alpha',\theta})^{1/(1+\gamma)})}{\partial \boldsymbol{\theta}}\bigg|_{\boldsymbol{\theta}=\boldsymbol{\theta}_0} \tag{23}
$$

$$\simeq C \left\langle \psi'_{\boldsymbol{\theta}_0} \epsilon \right\rangle \tag{24}$$

where C is a constant, and from the central limit theorem, we observe $\sqrt{n} \left\langle \psi'_{\boldsymbol{\theta}_0} \epsilon \right\rangle$ asymptotically follows the normal distribution with mean 0 and variance $I(\boldsymbol{\theta}_0) = V_{\bar{q}_{\boldsymbol{\theta}_0}}[\psi'_{\boldsymbol{\theta}_0}]$. From the law of large number, we observe that the second term in the rhs of (22) converges to $-CI(\boldsymbol{\theta}_0)$ in the limit of $n \to \infty$, which concludes the proposition.

The asymptotic variance in (21) implies that the proposed estimator is asymptotically efficient and has the same efficiency with the MLE, which asymptotically attains the Cramér-Rao bound. Also note that the asymptotic variance of the proposed estimator does not depend on choice of α, α', γ.

5 Experiments

We numerically investigated properties of the proposed estimator with a small synthetic dataset. Let $\bar{q}_{\boldsymbol{\theta}}(\boldsymbol{x})$ be a 5-dimensional Boltzmann machine

$$\bar{q}_{\boldsymbol{\theta}}(\boldsymbol{x}) = \frac{\exp\left(\frac{1}{2}\boldsymbol{x}^T \boldsymbol{\theta} \boldsymbol{x}\right)}{Z_{\boldsymbol{\theta}}} \tag{25}$$

whose parameter $\boldsymbol{\theta}$ follows the normal distribution with mean 0 and variance 1. We generated 20 sets of datasets including 4000 examples and compared the following method.

1. MLE: Maximum likelihood estimator
2. gamma: The proposed estimator with $u(z) = \exp(z)$ [13]
3. IS: The proposed estimator with $u(z) = -\frac{1}{z}$
4. eta: The proposed estimator with $u(z) = (1 + \eta)\exp(z) - \eta$
5. beta: The proposed estimator with $u(z) = (1 + \beta z)^{1/\beta}$

Figure 1(a) shows a box plot of MSEs of parameters, $||\hat{\boldsymbol{\theta}} - \boldsymbol{\theta}||^2$ in a logarithmic scale, with various deformation function u in (13). Figure 1(b) shows a box plot of computational times for each estimator in a logarithmic scale. We observe that some of the proposed estimator is comparable with the MLE, while the computational time of the proposed estimator is drastically reduced compared with that of the MLE.

A reason of why the proposed estimator with some functions u are inferior to the MLE is a shortage of examples. The theoretical result shown in Sect. 4 is based on assumptions of asymptotics and requires a lot of examples to assure

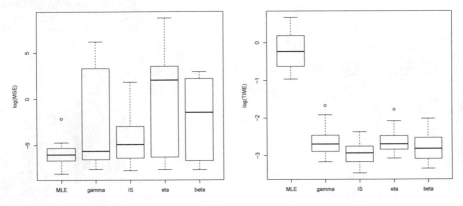

Fig. 1. $n = 4000$. (a) Box plot of estimation errors, $||\hat{\boldsymbol{\theta}} - \boldsymbol{\theta}||^2$ of each method. (b) Box plot of computational time of each method.

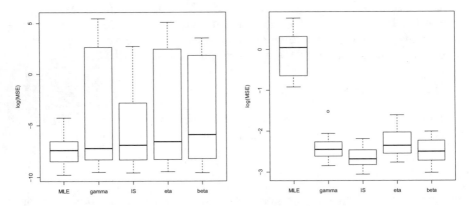

Fig. 2. $n = 16000$. (a) Box plot of estimation errors, $||\hat{\boldsymbol{\theta}} - \boldsymbol{\theta}||^2$ of each method. (b) Box plot of computational time of each method.

the asymptotic efficiency. We executed an another experiment with the same setting except for the number n of examples. Figure 2(a), (b) show results for $n = 16000$ and we observe that performance of the proposed estimator (IS, eta, beta) is improved at the same level as the MLE while required computational cost is still drastically fewer.

6   Conclusion

We proposed the novel estimator for discrete probabilistic model, which does not require calculation of the normalization constant. The proposed estimator is constructed by a combination of the γ-divergence and generalized empirical localization, which can be interpreted as the generalized mean of distributions. We investigated statistical properties of the proposed estimator and showed that the proposed estimator asymptotically has the same efficiency with the MLE and demonstrated the asymptotic efficiency with the small experiment.

References

1. Amari, S., Nagaoka, H.: Methods of Information Geometry, Translations of Mathematical Monographs, vol. 191. Oxford University Press, Oxford (2000)
2. Basu, A., Shioya, H., Park, C.: Statistical Inference: The Minimum Distance Approach. Chapman and Hall/CRC, Boca Raton (2011)
3. Cichocki, A., Cruces, S., Amari, S.i.: Generalized alpha-beta divergences and their application to robust nonnegative matrix factorization. Entropy **13**(1), 134–170 (2011)
4. Dawid, A.P., Lauritzen, S., Parry, M.: Proper local scoring rules on discrete sample spaces. Ann. Stat. **40**(1), 593–608 (2012)
5. Fujisawa, H., Eguchi, S.: Robust parameter estimation with a small bias against heavy contamination. J. Multivar. Anal. **99**(9), 2053–2081 (2008)
6. Gutmann, M., Hyvärinen, A.: Noise-contrastive estimation: a new estimation principle for unnormalized statistical models. In: Proceedings of the Thirteenth International Conference on Artificial Intelligence and Statistics, pp. 297–304 (2010)
7. Hinton, G.E.: Training products of experts by minimizing contrastive divergence. Neural Comput. **14**(8), 1771–1800 (2002)
8. Hyvärinen, A.: Some extensions of score matching. Comput. Stat. Data Anal. **51**(5), 2499–2512 (2007)
9. Mihoko, M., Eguchi, S.: Robust blind source separation by beta divergence. Neural Comput. **14**(8), 1859–1886 (2002)
10. Nielsen, F., Nock, R.: Patch matching with polynomial exponential families and projective divergences. In: Amsaleg, L., Houle, M.E., Schubert, E. (eds.) SISAP 2016. LNCS, vol. 9939, pp. 109–116. Springer, Cham (2016). https://doi.org/10.1007/978-3-319-46759-7_8
11. Opper, M., Saad, D. (eds.): Advanced Mean Field Methods: Theory and Practice. MIT Press, Cambridge (2001)
12. Takenouchi, T., Eguchi, S.: Robustifying AdaBoost by adding the naive error rate. Neural Comput. **16**(4), 767–787 (2004)

13. Takenouchi, T., Kanamori, T.: Statistical inference with unnormalized discrete models and localized homogeneous divergences. J. Mach. Learn. Res. **18**(56), 1–26 (2017). http://jmlr.org/papers/v18/15-596.html
14. Van der Vaart, A.W.: Asymptotic Statistics. Cambridge University Press, Cambridge (1998)

Properties of the Cross Entropy Between ARMA Processes

Eric Grivel[(✉)]

Bordeaux University - INP Bordeaux ENSEIRB-MATMECA - IMS - UMR
CNRS 5218, Talence, France
`eric.grivel@ims-bordeaux.fr`

Abstract. In this paper, we propose to analyze the properties of the cross entropy between the probability density functions of k consecutive samples of wide-sense stationary Gaussian ARMA processes. It is shown that, when k increases, the cross entropy has a transient behavior before tending to an affine function of k whose slope is defined by the so-called asymptotic increment. The latter depends on the parameters of the processes and can be a reasonable choice to characterize the cross entropy. Some illustrations are then given.

Keywords: Cross entropy · Jeffreys divergence ·
Kullback-Leibler divergence · ARMA processes

1 Introduction

Cross entropy between two probability density functions has been used in a wide range of applications. Thus, in the field of information theory, it can be viewed as the expected message-length per datum when a distribution has been used whereas the data are characterized by another one. In addition, it is often used to define the cost function useful for the backpropagation step when designing convolutional neural networks (CNNs) in the field of deep learning. Analyzing the behaviour of the cross entropy between Gaussian autoregressive with moving average (ARMA) processes can be of interest to compare short-memory processes and to detect a statistical change in time series. Such issues are of interest for instance in biomedical applications to detect a pathology. To address these issues, it is true that divergences can be used [1,2]. Several divergences have been proposed in the literature from [3] to the f-divergences. The reader may refer to [4,5] for more details and for information about recent works. All make it possible to compare the probability density functions of k consecutive samples of time series. Assuming that they correspond to autoregressive fractionally integrated with moving average (ARFIMA) processes[1], several works have been recently led to analyze the Kullback-Leibler (KL) divergence [7], its symmetric version

---

[1] Note that an ARMA process with orders (p,q) corresponds to an ARFIMA process with orders (p,q) and a differencing order d equal to 0.

ⓒ Springer Nature Switzerland AG 2019
F. Nielsen and F. Barbaresco (Eds.): GSI 2019, LNCS 11712, pp. 377–386, 2019.
https://doi.org/10.1007/978-3-030-26980-7_39

known as the Jeffreys divergence (JD), but also the Rényi divergence of order α [8], which can be seen as a generalization of the KL [9–16]. The authors studied the way the increment of the divergence, *i.e.* the difference between two divergences computed for $k + 1$ and k successive variates, evolves when k increases. Concerning the JD, the conclusions are the following: after a transient behavior, the JD increment tends to a finite value called asymptotic JD increment, except when the ARFIMA processes have different unit zeros and/or when the difference between the differencing orders of the ARFIMA processes is larger than 0.5. In these particular cases, the limit of the increment tends to infinity since it has been shown that the asymptotic JD increment amounts to summing two signal powers: the power of the first process filtered by the so-called inverse filter associated with the second one and the power of the second process filtered by the inverse filter associated with the first one. It should be noted this asymptotic increment can be also related to the so-called divergence rate, usually defined in information theory.

In this paper, we propose a complementary study to the above analysis. Our contribution is the following: we suggest analyzing the cross entropy between the joint distributions of k consecutive samples of two Gaussian ARMA processes. Among the questions that can be considered, one can wonder how the cross entropy is influenced by the number k of variates that are considered. We show that when k increases, the increment of the cross entropy tends to a limit called the asymptotic increment. The latter can hence be of interest to characterize the cross entropy. To derive its expression, the partial correlation coefficients (PACF) of the processes are used in this paper.

The remainder of this paper is organized as follows. In Sect. 2, properties of Gaussian ARMA processes are first recalled in terms of filtering interpretation, correlation and power spectral density. Then, in Sect. 3, after giving the definition of the Kullback-Leibler divergence, the Jeffreys divergence, the Shannon entropy and the corresponding cross entropy, their expressions for Gaussian processes are provided. Then, the evolution of the cross entropy is analyzed when two Gaussian ARMA processes are compared. The expression of the asymptotic cross entropy is also given. Finally, in Sect. 4, illustrations are provided and confirm the theoretical analysis.

2   About Gaussian ARMA(p, q) Processes

The Gaussian ARMA process of order p and q is used in a wide range of applications from speech processing to biomedical applications. Its kth sample satisfies:

$$x_k = -\sum_{l=1}^{p} a_l x_{k-l} + \sum_{l=0}^{q} b_l u_{k-l} \tag{1}$$

where u_k is the driving process, assumed to be white, Gaussian, zero-mean with variance σ_u^2. Given $a_0 = 1$ and $b_0 = 1$, $\{a_l\}_{l=0,...,p}$ and $\{b_l\}_{l=0,...,q}$ are the AR and MA parameters respectively.

The ARMA process x_k can be seen as the output of a linear filter whose input is the zero-mean white Gaussian noise u_k. Taking the z-transform of (1), relation interpreted as a difference equation, one has:

$$X(z) = H(z)U(z) = \frac{\sum_{l=0}^{q} b_l z^{-l}}{\sum_{l=0}^{p} a_l z^{-l}} U(z) = \frac{\prod_{l=1}^{q}(1 - z_l z^{-1})}{\prod_{l=1}^{p}(1 - p_l z^{-1})} U(z) \qquad (2)$$

with $X(z)$, $U(z)$ and $H(z)$ the z-transform of x_k, u_k and the impulse response of the filter respectively. In addition, $\{p_l\}_{l=1,\ldots,p}$ are the poles while $\{z_l\}_{l=1,\ldots,q}$ are the zeros.

The power spectral density (PSD), denoted as $S(\theta)$, with θ the normalized angular frequency, can be expressed as follows:

$$S(\theta) = \sigma_u^2 \frac{|\sum_{l=0}^{q} b_l e^{-jl\theta}|^2}{|\sum_{l=0}^{p} a_l e^{-jl\theta}|^2} = \sigma_u^2 \frac{\prod_{l=1}^{q}|1 - z_l e^{-j\theta}|^2}{\prod_{l=1}^{p}|1 - p_l e^{-j\theta}|^2} \qquad (3)$$

In the following, as the ARMA process x_k is zero-mean, the correlation function and the covariance function are equal. In addition, the values of the correlation function r_τ, with τ the lag, tend to decay to zero geometrically. $\sum_\tau r_\tau$ is absolutely summable. Therefore, this wide-sense stationary (w.s.s.) process is known to be a short-memory process. As a corollary, the Toeplitz correlation matrix of finite size $k \times k$, denoted as Q_k and defined from the correlation-function coefficients r_τ with $\tau = 1 - k, \ldots, k - 1$, belongs to the Wiener class Toeplitz matrices. As a consequence, according to [17], Q_k is non singular even if the PSD of the process is equal to zero at some frequencies. Q_k is hence invertible. Nevertheless, the infinite-size Toeplitz correlation matrix is no longer invertible when the corresponding transfer function of the ARMA process has unit roots.

Let us now focus our attention on the identification issue of ARMA processes. When dealing with a finite-order AR model, the Yule-Walker (YW) equations and the correlation method can be considered to estimate both the AR parameters and the variance of the driving process [18]. When dealing with a finite-order MA model, the most popular method is the Durbin algorithm, but other approaches based on spectral factorization for instance exist. The reader may refer to [6] for an exhaustive state of the art on the estimations of the ARMA parameters when the data are noise free or disturbed by an additive white noise.

Remark: different ARMA processes are characterized by the same correlation matrix or the same PSD. Indeed, let us take the toy example of a 1st-order MA process defined by the driving-process variance equal to σ_u^2 and the transfer function $H_{z_l}(z)$ where the zero z_l has its modulus larger than 1. In this case, the following transformation can be considered:

$$H_{z_l}(z) = (1 - z_l z^{-1}) = -z_l^* H_{bla,z_l}^{-1}(z)\left(1 - \frac{1}{z_l^*}z^{-1}\right) \qquad (4)$$

where z_l^* is the conjugate of z_l, $H_{bla,z_l}(z) = \frac{z^{-1}-z_l^*}{1-z_l z^{-1}}$ is a Blaschke product [19] up to a multiplicative value of the form $\pm e^{j\phi_l}$ with ϕ_l the argument of z_l. (4) amounts to saying that $H_{z_l}(z)$ corresponds to the cascade of two all-pass filters with gains respectively equal to $|z_l|$ and 1 and a minimum-phase filter with transfer function $1 - \frac{1}{z_l^*}z^{-1}$. Therefore, the MA process with a variance of the driving process equal to $K_l\sigma_u^2 = |z_l|^2\sigma_u^2$ and characterized by the minimum-phase transfer function $1 - \frac{1}{z_l^*}z^{-1}$ has the same PSD as the MA process defined by the variance of driving process equal to σ_u^2 and the transfer function $1 - z_l z^{-1}$. This can be easily generalized to any ARMA process by taking advantage of (2).

Given the above remark, let us now define the inverse filter that will be used in this communication for a given correlation matrix. When all the zeros are inside the unit-circle in the z-plane, $H(z)$ is minimum-phase and directly invertible. When one zero is outside the unit-circle in the z-plane, we suggest defining the inverse filter by taking advantage of the above remark and (4). Therefore, the BIBO-stable inverse filter is defined as follows:

$$H^{-1}(z) = \frac{1}{\sigma_u} \prod_{l=1}^{p} (1 - p_l z^{-1}) \prod_{l=1}^{q} H_{z_l}^{-1}(z) \tag{5}$$

with

$$H_{z_l}^{-1}(z) = \begin{cases} \frac{1}{1-z_l z^{-1}} & \text{if } |z_l| < 1 \\ \frac{1}{-z_l*} \frac{1}{1-\frac{1}{z_l^*}z^{-1}} & \text{if } |z_l| > 1 \end{cases} \tag{6}$$

3   Kullback-Leibler Divergence, Jeffreys Divergence, Entropy and Cross Entropy

3.1   Definitions and Applications to Gaussian Processes

In the following, two processes will be considered. By introducing the joint distributions of k successive values of the ith random process, denoted as $p_i(x_{1:k})$ for $i = 1, 2$, the Shannon entropy is given by:

$$H_k^{(i)} = - \int_{-\infty}^{+\infty} p_i(x_{1:k}) \ln p_i(x_{1:k}) dx_{1:k} \tag{7}$$

To study the dissimilarities between two processes defined by $p_1(x_{1:k})$ and $p_2(x_{1:k})$, the KL divergence can be evaluated. It satisfies [7]:

$$KL_k^{(1,2)} = \int_{x_{1:k}} p_1(x_{1:k}) \ln\left(\frac{p_1(x_{1:k})}{p_2(x_{1:k})}\right) dx_{1:k} \tag{8}$$

Remark: as the KL is not symmetric, Jeffreys divergence can be used[2]:

$$JD_k^{(1,2)} = KL_k^{(1,2)} + KL_k^{(2,1)} \tag{9}$$

---

[2] In some papers, the symmetric version of the KL divergence can be also defined as the mean between $KL_k^{(1,2)}$ and $KL_k^{(2,1)}$. In others, the minimum value is rather considered.

At this stage, the cross entropy $H_k^{(1,2)}$ can be expressed from the Shannon entropy and the KL as follows:

$$H_k^{(1,2)} = KL_k^{(1,2)} + H_k^{(1)} \tag{10}$$

Given the expression above and as the KL is not negative, the cross entropy is always larger than the entropy. In addition, in many estimation issues, minimizing the cross entropy amounts to minimizing the KL.

Let us now deduce the expressions of the entropy, the KL and the cross entropy for Gaussian processes. For this purpose, let us first recall that the probability density function of the ith real random Gaussian column vector $x_{1:k}$ of size k, mean $\mu_{k,i}$ and covariance matrix $Q_{k,i}$, is defined by:

$$p_i(x_{1:k}) = \frac{1}{(\sqrt{2\pi})^k |Q_{k,i}|^{1/2}} \exp\left(-\frac{1}{2}[x_{1:k} - \mu_{k,i}]^T Q_{k,i}^{-1}[x_{1:k} - \mu_{k,i}]\right) \tag{11}$$

with $i = 1, 2$ and $|Q_{k,i}|$ the determinant of the covariance matrix.

By using (11), the entropy and the KL defined in (7) and (8) become:

$$H_k^{(1)} = \frac{k}{2}\ln(2\pi) + \frac{1}{2}\left(\ln|Q_{k,1}| + \mathrm{Tr}(Q_{k,1}^{-1}Q_{k,1})\right) = \frac{k}{2}(1 + \ln(2\pi)) + \frac{1}{2}\ln|Q_{k,1}| \tag{12}$$

and

$$KL_k^{(1,2)} = \frac{1}{2}\left[\mathrm{Tr}(Q_{k,2}^{-1}Q_{k,1}) - k - \ln\frac{|Q_{k,1}|}{|Q_{k,2}|} + (\mu_{k,2} - \mu_{k,1})^T Q_{k,2}^{-1}(\mu_{k,2} - \mu_{k,1})\right] \tag{13}$$

For real zero-mean processes, (13) reduces to:

$$KL_k^{(1,2)} = \frac{1}{2}\left[\mathrm{Tr}(Q_{k,2}^{-1}Q_{k,1}) - k - \ln\frac{|Q_{k,1}|}{|Q_{k,2}|}\right] \tag{14}$$

Combining (10), (12) and (14) leads to:

$$H_k^{(1,2)} = \frac{1}{2}\left[k\ln(2\pi) + \mathrm{Tr}(Q_{k,2}^{-1}Q_{k,1}) + \ln|Q_{k,2}|\right] \tag{15}$$

3.2   Evolution of the Cross Entropy for Gaussian ARMA Processes

Let us study how the cross entropy evolves when k increases. More particularly, let us analyze the increment $\Delta H_k^{(1,2)} = H_{k+1}^{(1,2)} - H_k^{(1,2)}$. Given (15), one has:

$$\Delta H_k^{(1,2)} = \frac{1}{2}\left(\ln(2\pi) + \mathrm{Tr}(Q_{k+1,2}^{-1}Q_{k+1,1}) - \mathrm{Tr}(Q_{k,2}^{-1}Q_{k,1}) + \ln\frac{|Q_{k+1,2}|}{|Q_{k,2}|}\right) \tag{16}$$

The above expression consists of three terms: the constant $\frac{\ln(2\pi)}{2}$, the second term which is a difference of traces and a last term related to the determinant of the correlation matrices. Let us focus our attention on the last two terms when dealing with Gaussian ARMA processes.

About the Difference of the Traces. In a recent paper [14], while analyzing the properties of the JD for ARMA(p, q) processes as well as ARFIMA(p, d, q) processes, we showed that:

$$\lim_{k \to +\infty} \left(\mathrm{Tr}(Q_{k+1,2}^{-1} Q_{k+1,1}) - \mathrm{Tr}(Q_{k,2}^{-1} Q_{k,1}) \right) = P^{(1,2)} \tag{17}$$

where $P^{(1,2)}$ is the power of the 1st random process filtered by the inverse filter corresponding to the 2nd one. If the 2nd process has a zero on the unit-circle in the z-plane that is not equal to a zero of the 1st process, $P^{(1,2)}$ is infinite. When the parameters of the processes are known, $P^{(1,2)}$ can be obtained from the inverse Fourier of the PSD of the corresponding filtered process. The resulting sequence corresponds to the correlation function and $P^{(1,2)}$ can then be easily deduced. When the parameters of the processes are unknown, the identification of the two ARMA processes from the data can be done, as a preliminary step, by using the prediction error method (PEM) [20].

About the Term Based on the Determinants of the Covariance Matrices. Taking into account the link between the covariance matrices and the normalized covariance matrices, *i.e.* $C_{k+1,2} = \frac{1}{r_{0,2}} Q_{k+1,2}$ where $r_{0,2}$ is the correlation function of the 2nd process computed for a lag equal to 0, one has:

$$\frac{|Q_{k+1,2}|}{|Q_{k,2}|} = r_{0,2} \frac{|C_{k+1,2}|}{|C_{k,2}|} \tag{18}$$

However, as stated in [21], the determinant of the normalized covariance matrices can be expressed from the PACF[3], denoted as $\phi_{\tau,2}$, as follows:

$$|C_{k,2}| = \prod_{\tau=1}^{k-1} (1 - \phi_{\tau,2}^2)^{k-\tau} \tag{19}$$

This means that:

$$\frac{|C_{k+1,2}|}{|C_{k,2}|} = \prod_{\tau=1}^{k} (1 - \phi_{\tau,2}^2) \tag{20}$$

Depending on the random processes under study, the PACF can become equal to 0 or not. For a white noise, they are all null except for $\tau = 0$. For a pth-order AR process, they become null for $\tau > p$. As minimum-phase moving average processes can be seen as AR(∞)-order AR processes, the corresponding PACF are all non-null.

---

[3] After expressing the ith process at times k and $k - \tau$ as linear combinations of the τ values $x_{k-1,i}, .., x_{k-\tau+1,i}$ and their residuals, the PACF $\phi_{\tau,i}$ is defined as the correlation coefficient computed between both residuals. Its modulus is hence necessarily in the interval $[0,1]$. Up to a multiplication by ± 1, the PACF corresponds to the reflexion coefficient.

Let us now show that the limit of the ratio between the determinants of the covariance matrices tends to a constant denoted L_2 when k increases:

$$\lim_{k \to +\infty} \frac{|C_{k+1,2}|}{|C_{k,2}|} = \lim_{k \to +\infty} \prod_{\tau=1}^{k} (1 - \phi_{\tau,2}^2) = L_2 \tag{21}$$

For this purpose, let us recall the way the variance of the driving process is updated with the Durbin-Levinson algorithm [18] in order to express the constant L_2 from the variance of the driving process of the 2nd process. If $\xi_{\tau,2}$ denotes the τth reflection coefficient of the 2nd process, the variance of the corresponding driving process is updated as follows: $\sigma_{u,2}^2(\tau) = (1 - \xi_{\tau,2}^2)\sigma_{u,2}^2(\tau - 1)$ with $\sigma_{u,2}^2(0) = r_{0,2}$. As the square of the reflexion coefficient, $i.e.$ $\xi_{\tau,2}^2$, is equal to $\phi_{\tau,2}^2$, one has by combining (18), (20) and (21):

$$\lim_{k \to +\infty} \frac{|Q_{k+1,2}|}{|Q_{k,2}|} = \lim_{k \to +\infty} r_{0,2} \prod_{\tau=1}^{k} (1 - \phi_{\tau,i}^2) = \sigma_{u,2}^2 \prod_{l=1}^{q_2} K_{l,2} \tag{22}$$

with $K_{l,2} = 1$ when the zero $z_{l,2}$ of the 2nd process is inside the unit circle in the z-plane and $K_{l,2} = |z_{l,2}|^2$ when it is outside the unit-circle. This difference between both cases is due to (4). Therefore, the limit corresponds to the variance of the driving process associated to the 2nd ARMA process whose transfer function is minimum phase.

Conclusion About the Increment of the Cross Entropy. The increment of the cross entropy between wide sense-stationary Gaussian ARMA processes tends to a limit that can be defined from (16), (17) and (22):

$$\Delta H^{(1,2)} = \lim_{k \to +\infty} \Delta H_k^{(1,2)} = \frac{1}{2} \left(\ln(2\pi) + P^{(1,2)} + \ln \left(\sigma_{u,2}^2 \prod_{l=1}^{q_2} K_{l,2} \right) \right) \tag{23}$$

After a transient behavior, the cross entropy between the pdfs of Gaussian ARMA processes can be approximated by an affine function of k whose slope is defined by the asymptotic increment given above.

If the 2nd process has a zero on the unit-circle in the z-plane that is not equal to a zero of the 1st process, then $\Delta H^{(1,2)}$ is infinite.

Remark: the symmetric cross-entropy could be defined as $SH^{(1,2)} = H^{(1,2)} + H^{(2,1)}$. In this case, the corresponding asymptotic increment is defined by:

$$\Delta SH^{(1,2)} = \ln(2\pi) + \frac{1}{2} \left(P^{(1,2)} + P^{(2,1)} + \ln \left(\sigma_{u,2}^2 \sigma_{u,1}^2 \prod_{l=1}^{q_1} K_{l,1} \prod_{l=1}^{q_2} K_{l,2} \right) \right) \tag{24}$$

meanwhile the asymptotic increment of the JD is given by:

$$\Delta JD^{(1,2)} = -1 + \frac{1}{2}(P^{(1,2)} + P^{(2,1)}) \tag{25}$$

In the following, let us give some illustrations.

4   Illustrations

For two white noises processes, one has:

$$\Delta H^{(1,2)} = \frac{1}{2}\left(\ln(2\pi) + \frac{\sigma_{u,1}^2}{\sigma_{u,2}^2} + \ln \sigma_{u,2}^2 \right) \tag{26}$$

and

$$\Delta SH^{(1,2)} = \ln(2\pi) + \frac{1}{2}\left(\frac{\sigma_1^2}{\sigma_2^2} + \frac{\sigma_2^2}{\sigma_1^2} \right) + \frac{1}{2}\ln(\sigma_1^2\sigma_2^2) \tag{27}$$

Let us now illustrate the results with two 1st-order MA processes. The first one is defined by its zero equal to -0.7 whereas the second is defined by its zero that varies from 1.2 to -0.8 with a step equal to -0.5. The variances of the driving processes are first assumed to be equal to 1. Then, $\sigma_{u,2}^2 = 4$. In Table 1, the evolutions of the increments are given as well as the limits obtained with (23).

Table 1. Evolutions of the increments of the cross entropies -MA case-

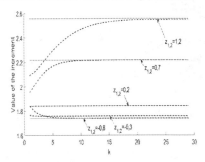

same variances of the driving processes   different variances of the driving processes

Let us now compare two 1st-order AR processes. The first one is defined by its pole equal to 0.99 whereas the second is defined by its pole that varies between 0.8 and 0.95 with a step equal to 0.5. The variances of the driving process are assumed to be equal to 1. In Table 2, the evolutions of the cross-entropies and the increments with respect to k are given. It should be noted that unlike the comparison between 1st-order MA processes, the increment between 1st-order AR processes tends to be rapidly the same and equal to the theoretical value given in (23). They are respectively equal to 2.3260, 1.9114, 1.6225 and 1.4591. They become smaller and smaller as the pole of the second process tends to be closer to the pole of the first process. Note that for the AR cases, $K_{l,i} = 1$ for any l and for $i = 1, 2$.

Table 2. Evolutions of the cross entropies -AR case-

cross entropy

increment

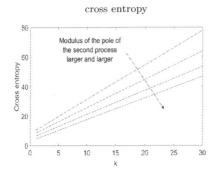
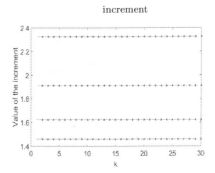

5 Conclusions and Perspectives

When analyzing the cross entropy between the pdfs of k consecutive values of two ARMA processes, the following conclusions can be drawn: firstly, the cross entropy as a function of k may exhibit a transient behavior when k increases. Then, it tends to an affine function of k whose slope is defined by the asymptotic increment. Therefore, unlike the increment of the cross entropy, the asymptotic increment does not change when k is modified. This can be useful to characterize the cross entropy.

References

1. Basseville, M.: Detecting changes in signals and systems – a survey. Automatica **24**(3), 309–326 (1988)
2. Kißlinger, A.-L., Stummer, W.: New model search for nonlinear recursive models, regressions and autoregressions. In: Nielsen, F., Barbaresco, F. (eds.) GSI 2015. LNCS, vol. 9389, pp. 693–701. Springer, Cham (2015). https://doi.org/10.1007/978-3-319-25040-3_74
3. Bhattacharyya, A.: On a measure of divergence between two statistical populations defined by their probability distributions. Bull. Calcutta Math. Soc. **35**, 99–109 (1943)
4. Basseville, M.: Divergence measures for statistical data processing. An annotated bibliography. Signal Process. **93**(4), 621–633 (2013)
5. Van Erven, T., Harremos, P.: Renyi divergence and Kullback-Leibler divergence. IEEE Trans. Inf. Theory **60**(7), 3797–3820 (2014)
6. Diversi, R., Grivel, E., Merchan, F.: ARMA-model identification from noisy observations based on a two-step errors-in-variables approach. In: IFAC Conference (2017)
7. Kullback, S., Leibler, R.A.: On information and sufficiency. Ann. Math. Stat. **22**, 79–86 (1951)
8. Renyi, A.: On measures of entropy and information. Fourth berkeley symposium on mathematical statistics and probability. Bull. Calcutta Math. Soc. **1**, 547–561 (1961)

9. Magnant, C., Giremus, A., Grivel, E.: Jeffrey's divergence between state models: application to target tracking using multiple models. In: Proceedings of EUSIPCO (2013)

10. Magnant, C., Giremus, A., Grivel, E.: On computing Jeffrey's divergence between time-varying autoregressive models. IEEE Signal Process. Lett. **22**(7), 915–919 (2014)

11. Legrand, L., Grivel, E.: Jeffrey's divergence between moving-average models that are real or complex, noise-free or disturbed by additive white noises. Elsevier Signal Process. **131**, 350–363 (2017)

12. Legrand, L., Grivel, E.: Jeffrey's divergence between moving-average and autoregressive models. In: Proceedings of ICASSP, pp. 4291–4295 (2017)

13. Grivel, E., Legrand, L.: Process comparison combining signal power ratio and Jeffrey's divergence between unit-power signals. In: Nielsen, F., Barbaresco, F. (eds.) GSI 2017. LNCS, vol. 10589, pp. 532–540. Springer, Cham (2017). https://doi.org/10.1007/978-3-319-68445-1_62

14. Grivel, E., Saleh, M., Omar, S.-M.: Interpreting the asymptotic increment of Jeffrey's divergence between some random processes. Elsevier Digit. Signal Process. **75**, 120–133 (2018)

15. Legrand, L., Grivel, E.: Jeffrey's divergence between autoregressive processes disturbed by additive white noises. Elsevier Signal Process. **149**, 162–178 (2018)

16. Saleh, M., Grivel, E., Omar, S.-M.: Jeffrey's divergence between ARFIMA processes. Elsevier Igital Signal Process. **82**, 175–186 (2018)

17. Gray, R.M.: Toeplitz and ciruclant matrices: a review. Found. Trends Commun. Inf. Theory **2**(3), 155–239 (2006)

18. Najim, M.: Modeling, Estimation and Optimal filtering in Signal Processing. Wiley, Hoboken (2010)

19. Colwell, P.: Blaschke Products: Bounded Analytical Functions. University of Michigan Press, Ann Arbor (1985)

20. Ljung, L.: System Identification: Theory for the User. Prentice-Hall, Upper Saddle River (1999)

21. Mukherjee, B.N., Maiti, S.S.: On some properties of positive definite toeplitz matrices and their possible applications. Linear Algebr. Appl. **102**, 211–240 (1988)

Statistical Manifold and Hessian Information Geometry

Inequalities for Statistical Submanifolds in Hessian Manifolds of Constant Hessian Curvature

Aliya Naaz Siddiqui[1](✉)(iD), Kamran Ahmad[2](iD), and Cenap Özel[3](iD)

[1] Department of Mathematics, Jamia Millia Islamia, New Delhi 110025, India
aliyanaazsiddiqui9@gmail.com
[2] M.M. Engineering College, Maharishi Markandeshwar (Deemed to be) University,
Mullana, Ambala 133207, India
qskamrankhan@gmail.com
[3] Department of Mathematics, King Abdulaziz University,
Jeddah 21589, Saudi Arabia
cenap.ozel@gmail.com

Abstract. In the present paper, we study Hessian and Einstein-Hessian manifolds with some examples. We establish optimizations of the intrinsic invariant (normalized scalar curvature) for a new extrinsic invariant (generalized normalized Casorati curvatures) on statistical submanifolds in a Hessian manifold of constant Hessian curvature by using algebraic technique. We consider the equality case of the derived inequalities.

Keywords: Hessian manifolds · Einstein-Hessian manifolds ·
Casorati curvatures

1 Introduction

The geometry of Hessian manifolds is a branch of physics, statistics, Kaehlerian and affine differential geometry. This is a new and fruitful field for many researchers. Therefore, many researchers have investigated the similarity between Kaehlerian and Hessian ones. The notion of a Hessian structure is the same that dual affine connections are flat. Hessian metric has many applications. For instance, it arises in the study of optimization, statistical manifolds and string theory (via special Kaehler manifolds). It is also known that there are several smooth families of probability distributions which admit Hessian structures (dually flat affine connections). Hessian sectional curvature and its useful relations with Kaehlerian manifold has been introduced by Shima (see [11,12]).

The Casorati curvature [2] of a submanifold of a Riemannian manifold is an extrinsic invariant defined as the normalized square of the length of the second fundamental form of the submanifold. This curvature, which is of interest in computer vision, was preferred by Casorati over the traditional curvature because it seems to correspond better with the common intuition of curvature.

F. Nielsen and F. Barbaresco (Eds.): GSI 2019, LNCS 11712, pp. 389–397, 2019.
https://doi.org/10.1007/978-3-030-26980-7_40

A family of optimal Casorati inequalities involving the normalized scalar curvature and the Casorati curvatures for any (statistical) submanifold in different ambient space forms have been derived (for example [6], [10], [13]). Therefore, we prove the following:

Theorem 1. *Let N be an $m-$dimensional statistical submanifold in an $n-$dimensional Hessian manifold $\overline{N}(c)$ of constant Hessian curvature c. Then the generalized normalized Casorati curvatures $\delta_C(s; m-1)$ and $\delta_C^*(s; m-1)$ satisfy*

$$\rho \leq \frac{2\delta_C^0(s; m-1)}{m(m-1)} + \frac{C^0}{m-1} + \frac{c}{4} - \frac{m}{2(m-1)}(\|\mathcal{H}\|^2 + \|\mathcal{H}^*\|^2), \qquad (1)$$

where $2C^0 = C + C^*$, $2\delta_C^0(s; m-1) = \delta_C(s; m-1) + \delta_C^*(s; m-1)$ and $s \in \mathbb{R}$ with $0 < s < m(m-1)$.

Theorem 2. *Let N be an $m-$dimensional statistical submanifold in an $n-$dimensional Hessian manifold $\overline{N}(c)$ of constant Hessian curvature c. Then equality holds in the relation (1) if and only if*

$$h_{ij}^k = -h_{ij}^{*k}, \quad \forall\, i, j \in \{1, \ldots, m\},\ i \neq j,$$
$$h_{mm}^{0k} = \frac{m(m-1)}{s} h_{11}^{0k} = \cdots = \frac{m(m-1)}{s} h_{m-1\,m-1}^{0k}, \quad \forall\, k \in \{m+1, \ldots, n\}.$$

2   Hessian Manifolds and Statistical Submanifolds

In 1985, a notion of statistical manifold has been studied by Amari [1]. This concept is of great interest because it connects information geometry, affine differential geometry and Hessian geometry.

Definition 1. *[1] A Riemannian manifold $(\overline{N}, g)$ with a Riemannian metric g is said to be a statistical manifold $(\overline{N}, \overline{\nabla}, g)$ if a pair of torsion-free affine connections on $\overline{N}$ satisfying*

$$Gg(E, F) = g(\overline{\nabla}_G E, F) + g(E, \overline{\nabla}_G^* F), \qquad (2)$$

for any $E, F, G \in \Gamma(T\overline{N})$ and $\overline{\nabla} g$ is symmetric. The connections $\overline{\nabla}$ and $\overline{\nabla}^*$ are called dual connections on $\overline{N}$, which satisfy

$$(\overline{\nabla}^*)^* = \overline{\nabla}. \qquad (3)$$

Remark that if $(\overline{N}, \overline{\nabla}, g)$ is a statistical structure, so is $(\overline{N}, \overline{\nabla}^*, g)$.

Definition 2. *[1] Let $(\overline{N}, \overline{\nabla}, g)$ be a statistical manifold and N be a submanifold of $\overline{N}$. Then (N, ∇, g) is also a statistical manifold with the induced statistical structure (∇, g) on N from $(\overline{\nabla}, g)$ and we call (N, ∇, g) as a statistical submanifold in $\overline{N}$.*

Let N be a statistical submanifold of dimension m in an n−dimensional statistical manifold $\overline{N}$. Let $\overline{\nabla}$ and $\overline{\nabla}^*$ be dual connections on $\overline{N}$, and ∇ and ∇^* be the induced dual connections by $\overline{\nabla}$ and $\overline{\nabla}^*$ on N, respectively. We suppose that $\overline{R}$, R, $\overline{R}^*$ and R^* are the Riemannian curvature tensors with respect to $\overline{\nabla}$, ∇, $\overline{\nabla}^*$ and ∇^* respectively. Then the corresponding Gauss equations are the following:

$$g(\overline{R}(E,F)G,H) = g(R(E,F)G,H) + g(h(E,G),h^*(F,H))$$
$$-g(h^*(E,H),h(F,G)), \tag{4}$$

$$g(\overline{R}^*(E,F)G,H) = g(R^*(E,F)G,H) + g(h^*(E,G),h(F,H))$$
$$-g(h(E,H),h^*(F,G)), \tag{5}$$

for any $E,F,G,H \in \Gamma(TN)$. Here h and h^* are symmetric and bilinear imbedding curvature tensors of N in $\overline{N}$ with respect to $\overline{\nabla}$ and $\overline{\nabla}^*$, respectively.

Definition 3. *[5] A triple $(\overline{N},\overline{\nabla},g)$ is called a Hessian manifold if a pair $(\overline{\nabla},g)$ on $\overline{N}$ satisfies the Codazzi equation*

$$(\overline{\nabla}_E g)(F,G) = (\overline{\nabla}_F g)(E,G), \tag{6}$$

and $\overline{\nabla}$ is flat.

On an n−dimensional Hessian manifold $(\overline{N},\overline{\nabla})$, suppose that $\mathcal{K} = \overline{\nabla} - \overline{\nabla}^0$. The tensor field $\overline{\mathcal{Q}}$ of type $(1,3)$ defined by [11,12]

$$\overline{\mathcal{Q}}(E,F) = [\mathcal{K}_E, \mathcal{K}_F], \tag{7}$$

for any $E,F \in \Gamma(T\overline{N})$ is said to be the Hessian curvature tensor with respect to $\overline{\nabla}$, where the Lie bracket $[,]$ of vector fields is an operator that assigns to any two vector fields on a smooth manifold and give a third vector field. Further, (7) satisfies

$$\overline{R}(E,F) + \overline{R}^*(E,F) = 2\overline{R}^0(E,F) + 2\overline{\mathcal{Q}}(E,F), \tag{8}$$

where $\overline{R}^0$ denotes the Riemannian curvature tensor with respect to $\overline{\nabla}^0$.

Let $\mathcal{L}$ be a plane in $T_\wp\overline{N}$, $\wp \in \overline{N}$. Choose an orthonormal basis $\{E,F\}$ of $\mathcal{L}$ and set

$$K(\mathcal{L}) = g(\overline{\mathcal{Q}}(E,F)F,E). \tag{9}$$

The number $K(\mathcal{L})$ is independent of the choice of an orthonormal basis and is called the Hessian sectional curvature. A Hessian manifold $(\overline{N},\overline{\nabla},g)$ is said to be of constant Hessian sectional curvature c if and only if [11,12]

$$\overline{\mathcal{Q}}(E,F,G,H) = \frac{c}{2}[g(E,F)g(G,H) + g(E,H)g(F,G)], \tag{10}$$

for any $E,F,G,H \in \Gamma(T\overline{N})$.

Generally, one cannot define a sectional curvature with respect to the dual connections (which are not metric) by the standard definitions. However, B. Opozda [8,9] defined a sectional curvature on a statistical manifold.

3   Einstein-Hessian Manifolds and Examples

Recall Theorems 7 and 8 of [11] and Example 2.2 of [12], we construct the following examples of Hessian structure:

Example 1. Let $\mathbb{R}^2$ be the standard affine space with the standard flat connection D and the standard affine coordinate system $\{u^1, u^2\}$. Let Θ be a domain in $\mathbb{R}^2$ equipped with a potential Φ and $g = Dd\Phi$, given by

$$\Theta = \{u \in \mathbb{R}^2 | u^1 > 0\}, \Phi = \frac{(u^2)^2}{4u^1} + log(\frac{1}{u^1}),$$

$$g = \begin{bmatrix} \frac{(u^2)^2}{2(u^1)^3} + 1 & -\frac{u^2}{2(u^1)^2} \\ -\frac{u^2}{2(u^1)^2} & \frac{1}{2u^1} \end{bmatrix}.$$

Then the pair $(D, g = Dd\Phi)$ is a Hessian structure on Θ.

Example 2. Let $\mathbb{R}^n$ be the standard affine space with the standard flat connection D and the standard affine coordinate system $\{u^1, \ldots, u^n\}$. Let Λ be a domain in $\mathbb{R}^n$ equipped with a potential ϕ and $g = Dd\phi$, given by

$$\Lambda = \{u \in \mathbb{R}^n | u^i > 0, i = 1, \ldots, n\}, \phi = \sum_{i=1}^{n} u^i log\ u^i, g = \delta_{ij} \frac{1}{u^i}.$$

Then the pair $(D, g = Dd\phi)$ is a Hessian structure on Λ.

Example 3. We consider a Hessian manifold $(\mathbb{H}^{n+1}, \overline{\nabla}, g)$ of constant Hessian curvature 4 (see [4] for details). We assume that a statistical immersion

$$f_0 : (\mathbb{R}^n, \nabla, g) \rightarrow (\mathbb{H}^{n+1}, \overline{\nabla}, g)$$

of the trivial Hessian manifold $(\mathbb{R}^n, \nabla, g)$ into $(\mathbb{H}^{n+1}, \overline{\nabla}, g)$ is given by

$$f_0(u^1, \ldots, u^n) = (u^1, \ldots, u^n, u^0).$$

It is easy to calculate that $h = 2g$, $h^* = 0$ and $||\mathcal{H}^*|| = 0$. Thus, $\mathbb{R}^n$ is totally umbilical with respect to $\overline{\nabla}$ and totally geodesic with respect to $\overset{*}{\overline{\nabla}}$.

Theorem 3. *Let $(\overline{N}, \overline{\nabla}, g)$ be an $n-$dimensional $(n > 1)$ Hessian manifold of constant Hessian curvature c. Then $(\overline{N}, g)$ is Einstein-Riemannian manifold. Moreover, $(\overline{N}, g)$ is Ricci-flat if $c = 0$.*

Proof. Ricci-flat manifolds are special Riemannian manifolds whose Ricci tensor vanishes. The fact that g is of constant sectional curvature $-\frac{c}{4}$, which directly implies the theorem.

Example 4. We recall Proposition 3.7 of [12]. Following are the Hessian manifolds of constant Hessian curvature zero:

(1) The Euclidean space $(\mathbb{R}^n, D, g = Dd(\frac{1}{2}\sum_{i=1}^{n}(x^i)^2))$,
(2) $(\mathbb{R}^n, D, g = Dd(\sum_{i=1}^{n}(e^{x^i})))$,

where D is a torsion-free connection. Then by Theorem 3, we say that $(\mathbb{R}^n, g)$ is Ricci-flat in both the cases.

4   Optimal Casorati Inequalities

Let $\overline{N}(c)$ be a Hessian manifold of constant Hessian curvature c and N be an $n-$dimensional statistical submanifold of $\overline{N}(c)$. Then $\overline{N}(c)$ is flat with respect to $\overline{\nabla}$ and $\overline{\nabla}^*$. Moreover, $\overline{N}(c)$ is a Riemannian space form of constant sectional curvature $\frac{-c}{4}$ (with respect to the Levi-Civita connection $\overline{\nabla}^0$). For $\wp \in N$, we choose $\{\mathcal{E}_1, \dots, \mathcal{E}_m\}$ and $\{\mathcal{E}_{m+1}, \dots, \mathcal{E}_n\}$ respectively a local orthonormal tangent frame of $T_\wp N$ and a local orthonormal normal frame of $T_\wp^\perp N$ in $\overline{N}$. The scalar curvature $\sigma(\wp)$ of N with respect to the Hessian curvature tensor $\mathcal{Q}$ is given by

$$\sigma(\wp) = \frac{1}{2} \sum_{1 \leq i < j \leq m} \mathcal{Q}(\mathcal{E}_i, \mathcal{E}_j, \mathcal{E}_j, \mathcal{E}_i)$$

$$= \frac{1}{2} \sum_{1 \leq i < j \leq m} [g(R(\mathcal{E}_i, \mathcal{E}_j)\mathcal{E}_j, \mathcal{E}_i) + g(R^*(\mathcal{E}_i, \mathcal{E}_j)\mathcal{E}_j, \mathcal{E}_i)$$

$$- 2g(R^0(\mathcal{E}_i, \mathcal{E}_j)\mathcal{E}_j, \mathcal{E}_i)]. \tag{11}$$

Then the normalized scalar curvature ρ of N is defined as

$$\rho = \frac{2\sigma}{m(m-1)}. \tag{12}$$

The mean curvature vectors $\mathcal{H}$ and $\mathcal{H}^*$ of N are respectively defined by

$$\mathcal{H} = \frac{1}{m} \sum_{i=1}^{m} h(\mathcal{E}_i, \mathcal{E}_i), \quad and \quad \mathcal{H}^* = \frac{1}{m} \sum_{i=1}^{m} h^*(\mathcal{E}_i, \mathcal{E}_i).$$

The squared norm of $\mathcal{H}$ and $\mathcal{H}^*$ of N are respectively given by

$$||\mathcal{H}||^2 = \frac{1}{m^2} \sum_{k=m+1}^{n} \left(\sum_{i=1}^{m} h_{ii}^k\right)^2, \quad and \quad ||\mathcal{H}^*||^2 = \frac{1}{m^2} \sum_{k=m+1}^{n} \left(\sum_{i=1}^{m} h_{ii}^{*k}\right)^2,$$

where

$$h_{ij}^k = g(h(\mathcal{E}_i, \mathcal{E}_j), \mathcal{E}_k), \quad and \quad h_{ij}^{*k} = g(h^*(\mathcal{E}_i, \mathcal{E}_j), \mathcal{E}_k),$$

for $i, j \in \{1, \dots, m\}$ and $k \in \{m+1, \dots, n\}$.

The normalized squared norms of second fundamental forms h and h^* are denoted by $\mathcal{C}$ and $\mathcal{C}^*$, respectively, called the *Casorati curvatures* of N in $\overline{N}$. Therefore, we have

$$\mathcal{C} = \frac{1}{m} \sum_{k=m+1}^{n} \sum_{i,j=1}^{m} (h_{ij}^k)^2, \quad and \quad \mathcal{C}^* = \frac{1}{m} \sum_{k=m+1}^{n} \sum_{i,j=1}^{m} (h_{ij}^{*k})^2.$$

If we consider a r-dimensional subspace $\mathcal{W}$ of TN, $r \geq 2$, and an orthonormal basis $\{\mathcal{E}_1, \ldots, \mathcal{E}_r\}$ of $\mathcal{W}$. Then the Casorati curvatures of the subspace $\mathcal{W}$ are the following:

$$\mathcal{C}(\mathcal{W}) = \frac{1}{r} \sum_{k=m+1}^{n} \sum_{i,j=1}^{r} (h_{ij}^k)^2, \quad and \quad \mathcal{C}^*(\mathcal{W}) = \frac{1}{r} \sum_{k=m+1}^{n} \sum_{i,j=1}^{r} (h_{ij}^{*k})^2.$$

The normalized Casorati curvatures $\delta_\mathcal{C}(m-1)$, $\delta_\mathcal{C}^*(m-1)$, $\widehat{\delta}_\mathcal{C}(m-1)$ and $\widehat{\delta}_\mathcal{C}^*(m-1)$ are defined as

(1) $[\delta_\mathcal{C}(m-1)]_\wp = \frac{1}{2}\mathcal{C}_\wp + (\frac{m+1}{2m})inf\{\mathcal{C}(\Psi)|\Psi : \text{a hyperplane of } T_\wp N\}$,

(2) $[\delta_\mathcal{C}^*(m-1)]_\wp = \frac{1}{2}\mathcal{C}_\wp^* + (\frac{m+1}{2m})inf\{\mathcal{C}^*(\Psi)|\Psi : \text{a hyperplane of } T_\wp N\}$,

(3) $[\widehat{\delta}_\mathcal{C}(m-1)]_\wp = 2\mathcal{C}_\wp - (\frac{2m-1}{2m})sup\{\mathcal{C}(\Psi)|\Psi : \text{a hyperplane of } T_\wp N\}$,

(4) $[\widehat{\delta}_\mathcal{C}^*(m-1)]_\wp = 2\mathcal{C}_\wp^* - (\frac{2m-1}{2m})sup\{\mathcal{C}^*(\Psi)|\Psi : \text{a hyperplane of } T_\wp N\}$.

Further, the generalized normalized Casorati curvatures $\delta_\mathcal{C}(s; m-1)$, $\delta_\mathcal{C}^*(s; m-1)$, $\widehat{\delta}_\mathcal{C}(s; m-1)$ and $\widehat{\delta}_\mathcal{C}^*(s; m-1)$ are given by [3]

(1) For $0 < s < m^2 - m$
$[\delta_\mathcal{C}(s; m-1)]_\wp = s\mathcal{C}_\wp + \zeta(s)inf\{\mathcal{C}(\Psi)|\Psi : \text{a hyperplane of } T_\wp N\}$,
$[\delta_\mathcal{C}^*(s; m-1)]_\wp = s\mathcal{C}_\wp^* + \zeta(s)inf\{\mathcal{C}^*(\Psi)|\Psi : \text{a hyperplane of } T_\wp N\}$.

(2) For $s > m^2 - m$
$[\widehat{\delta}_\mathcal{C}(s; m-1)]_\wp = s\mathcal{C}_\wp + \zeta(s)sup\{\mathcal{C}(\Psi)|\Psi : \text{a hyperplane of } T_\wp N\}$,
$[\widehat{\delta}_\mathcal{C}^*(s; m-1)]_\wp = s\mathcal{C}_\wp^* + \zeta(s)sup\{\mathcal{C}^*(\Psi)|\Psi : \text{a hyperplane of } T_\wp N\}$,

where $\zeta(s) = \frac{1}{sm}(m-1)(m+s)(m^2 - m - s), s \neq m(m-1)$.

Proof of Theorem 1:

We need the following lemma to prove Theorem 1:

Lemma 1. *[13] Let $\vartheta = \{(u_1, u_2, \ldots, u_m) \in \mathbb{R}^m : u_1 + u_2 + \cdots + u_m = l\}$ be a hyperplane of $\mathbb{R}^m$ and $\pi : \mathbb{R}^m \to \mathbb{R}$ a quadratic form given by*

$$\pi(u_1, u_2, \ldots, u_m) = \mu \sum_{i=1}^{m-1} (u_i)^2 + \nu(u_m)^2 - 2 \sum_{1 \leq i < j \leq m} u_i u_j,$$

where $\mu > 0, \nu > 0$.

Then, by the constrained extremum problem, π has a global solution as follows

$$u_1 = u_2 = \cdots = u_{m-1} = \frac{l}{\mu+1},$$

$$u_m = \frac{l}{\nu+1} = \frac{l(m-1)}{(\mu+1)\nu} = (\mu - m + 2)\frac{l}{\mu+1},$$

provided that $\nu = \frac{m-1}{\mu-m+2}$.

Following [7] and (11), we derive

$$2\sigma = m(m-1)\frac{c}{4} + m^2||\mathcal{H}^0||^2 - m\mathcal{C}^0 + \frac{m}{2}(\mathcal{C} + \mathcal{C}^*)$$
$$-\frac{m^2}{2}(||\mathcal{H}||^2 + ||\mathcal{H}^*||^2),$$

where $\mathcal{H}^0$ is the mean curvature vector with respect to $\overline{\nabla}^0$.

Let us define a quadratic polynomial $\mathbb{P}$ in the components of the second fundamental form h^0

$$\mathbb{P} = s\mathcal{C}^0 + \zeta(s)\mathcal{C}^0(\Psi) - 2\sigma + m(m-1)\frac{c}{4}$$
$$-\frac{m^2}{2}(||\mathcal{H}||^2 + ||\mathcal{H}^*||^2) + \frac{m}{2}(\mathcal{C} + \mathcal{C}^*). \tag{13}$$

Without loss of generality, we assume that Ψ is spanned by $\mathcal{E}_1, \ldots, \mathcal{E}_m$, combining (13), it follows that

$$\mathbb{P} = \frac{m+s}{m}\sum_{k=m+1}^{n}\sum_{i,j=1}^{m}(h_{ij}^{0k})^2 + \frac{\zeta(s)}{m-1}\sum_{k=m+1}^{n}\sum_{i,j=1}^{m-1}(h_{ij}^{0k})^2 - \sum_{k=m+1}^{n}(\sum_{i=1}^{m}h_{ii}^{0k})^2$$

$$= \sum_{k=m+1}^{n}\sum_{i=1}^{m-1}[2(\iota+1)(h_{ii}^{0k})^2 + \frac{2(m+s)}{m}(h_{im}^{0k})^2]$$

$$+[\iota\sum_{1\le i\ne j\le m-1}(h_{ij}^{0k})^2 - 2\sum_{1\le i\ne j\le m}(h_{ii}^{0k}h_{jj}^{0k}) + \frac{s}{m}(h_{mm}^{0k})^2]$$

$$\ge \sum_{k=m+1}^{n}[\sum_{i=1}^{m-1}\iota(h_{ii}^{0k})^2 - 2\sum_{1\le i\ne j\le m}h_{ii}^{0k}h_{jj}^{0k} + \frac{s}{m}(h_{mm}^{0})^2], \tag{14}$$

where $\iota = (\frac{s}{m} + \frac{\zeta(s)}{m-1})$.

We consider the quadratic form $\pi^k : \mathbb{R}^m \to \mathbb{R}$, for $k = m+1, \ldots n$, defined by

$$\pi^k(h_{11}^{0k}, \ldots, h_{mm}^{0k}) = \sum_{i=1}^{m-1}\iota(h_{ii}^{0k})^2 - 2\sum_{1\le i\ne j\le m}h_{ii}^{0k}h_{jj}^{0k} + \frac{s}{m}(h_{mm}^{0})^2, \tag{15}$$

and the problem as follows:

$$min\{\pi^k(h_{11}^{0k}, \ldots, h_{mm}^{0k}) : h_{11}^{0k} + \cdots + h_{mm}^{0k} = \gamma^k, \gamma^k \in \mathbb{R}\}.$$

From Lemma 1, we see that the critical point $(h_{11}^{0k}, \ldots, h_{mm}^{0k})$ is given by

$$h_{11}^{0k} = h_{22}^{0k} = \cdots = h_{m-1m-1}^{0k} = \frac{\gamma^k}{\mu+1} = \frac{\gamma^k}{\iota+1}, h_{mm}^{0k} = \frac{\gamma^k}{\nu+1} = \frac{m\gamma^k}{m+s}, \quad (16)$$

and hence it is the global minimum point.

Plugging (16) into (15), we have

$$\pi^k(h_{11}^{0k}, \ldots, h_{mm}^{0k}) = 0. \quad (17)$$

From (14) and (17), we get $\mathbb{P} \geq 0$ and hence we have

$$2\sigma \leq s\mathcal{C}^0 + \zeta(s)\mathcal{C}^0(\Psi) + \frac{m}{2}(\mathcal{C} + \mathcal{C}^*) + +m(m-1)\frac{c}{4}$$
$$- \frac{m^2}{2}(||\mathcal{H}||^2 + ||\mathcal{H}^*||^2).$$

Further, we find that

$$\rho \leq \frac{s}{m(m-1)}\mathcal{C}^0 + \frac{\zeta(s)}{m(m-1)}\mathcal{C}^0(\Psi) + \frac{1}{2(m-1)}(\mathcal{C} + \mathcal{C}^*)$$
$$+ \frac{c}{4} - \frac{m}{2(m-1)}(||\mathcal{H}||^2 + ||\mathcal{H}^*||^2),$$

where we used (12). Thus, we get the required inequality (1).

Remark 1. It is easy to prove that the normalized scalar curvature is bounded above by the generalized normalized Casorati curvatures $\widehat{\delta}_\mathcal{C}(s; m-1)$ and $\widehat{\delta}_\mathcal{C}^*(s; m-1)$ for $s \in \mathbb{R}$, $s > m(m-1)$.

Corollary 1. *Let N be an $m-$dimensional statistical submanifold in an $n-$dimensional Hessian manifold $\overline{N}(c)$ of constant Hessian curvature c. Then the normalized Casorati curvatures $\delta_\mathcal{C}(m-1)$ and $\delta_\mathcal{C}^*(m-1)$ satisfy*

$$\rho \leq 2\delta_\mathcal{C}^0(m-1) + \frac{\mathcal{C}^0}{m-1} + \frac{c}{4} - \frac{m}{2(m-1)}(||\mathcal{H}||^2 + ||\mathcal{H}^*||^2),$$

where $2\mathcal{C}^0 = \mathcal{C} + \mathcal{C}^*$ and $2\delta_\mathcal{C}^0(m-1) = \delta_\mathcal{C}(m-1) + \delta_\mathcal{C}^*(m-1)$.

Remark 2. One can prove Corollary 1 by using Lemma 1 and taking

$$s = \frac{m(m-1)}{2}$$

in $\delta_\mathcal{C}(s; m-1)$ and $\delta_\mathcal{C}^*(s; m-1)$ in Theorem 1, we have

$$\left[\delta_\mathcal{C}\left(\frac{m(m-1)}{2}; m-1\right)\right]_\wp = m(m-1)\left[\delta_\mathcal{C}(m-1)\right]_\wp,$$
$$\left[\delta_\mathcal{C}^*\left(\frac{m(m-1)}{2}; m-1\right)\right]_\wp = m(m-1)\left[\delta_\mathcal{C}^*(m-1)\right]_\wp$$

at any point $\wp \in N$.

Acknowledgment. The authors would like to thank to anonymous referees for their comments to improve the manuscript.

References

1. Amari, S.: Differential-Geometrical Methods in Statistics. Lecture Notes in Statistics, vol. 28. Springer, New York (1985). https://doi.org/10.1007/978-1-4612-5056-2

2. Casorati, F.: Mesure de la courbure des surfaces suivant l'idee commune. Acta Math. **14**(1), 95–110 (1890)

3. Decu, S., Haesen, S., Verstraelen, L.: Optimal inequalities characterising quasi-umbilical submanifolds. J. Inequal. Pure Appl. Math. **9**(3), 1–7 (2008)

4. Furuhata, H.: Hypersurfaces in statistical manifolds. Differ. Geom. Appl. **27**, 420–429 (2009)

5. Furuhata, H.: Statistical hypersurfaces in the space of Hessian curvature zero. Differ. Geom. Appl. **29**, 586–590 (2011)

6. Lee, C.W., Yoon, D.W., Lee, J.W.: A pinching theorem for statistical manifolds with Casorati curvatures. J. Nonlinear Sci. Appl. **10**, 4908–4914 (2017)

7. Mihai, A., Mihai, I.: Curvature invariants for statistical submanifolds of Hessian manifolds of constant Hessian curvature. Mathematics **44**(6) (2018)

8. Opozda, B.: Bochner's technique for statistical structures. Ann. Glob. Anal. Geom. **48**(4), 357–395 (2015)

9. Opozda, B.: A sectional curvature for statistical structures. Linear Algebra Appl. **497**, 134–161 (2016)

10. Siddiqui, A.N. and Shahid, M.H.: Optimizations on statistical hypersurfaces with Casorati curvatures. Accepted for the publication in Kragujevac Journal of Mathematics (2019)

11. Shima, H.: Hessian manifolds of constant Hessian sectional curvature. J. Math. Soc. Jpn. **47**(4), 735–753 (1995)

12. Shima, H.: The Geometry of Hessian Structures. World Scientific, Singapore (2007)

13. Tripathi, M.M.: Inequalities for algebraic Casorati curvatures and their applications. Note Mat. **37**, 161–186 (2017)

B. Y. Chen Inequalities for Statistical Submanifolds in Sasakian Statistical Manifolds

Mohd. Aquib[1]([✉]), Michel Nguiffo Boyom[2], Ali H. Alkhaldi[3], and Mohammad Hasan Shahid[1]

[1] Department of Mathematics, Jamia Millia islamia, New Delhi, India
`aquib80@gmail.com`, `mshahid@jmi.ac.in`
[2] Nguiffo Boyom M. IMAG: Alexander Grothendieck Research Institute, University of Montpellier, Montpellier, France
`nguiffo.boyom@gmail.com`
[3] Department of Mathematics, College of Science, King Khalid University, P.O. Box 9004, Abha 62529, Saudi Arabia
`ahalkhaldi@kku.edu.sa`

Abstract. In this paper, we derive a statistical version of B. Y. Chen inequality for statistical submanifolds in the Sasakian statistical manifolds with constant curvature and discuss the equality case of the inequality. We also give some applications of the inequalities obtained.

Keywords: Chen's inequality · Statistical manifolds · Sasakian statistical manifolds

1 Introduction

In 1989, the notion of statistical submanifolds was introduced and studied by Vos [10]. Though, till the date it has made very little progress due to the hardness to find classical differential geometric approaches for study of statistical submanifolds. Furuhata [6], studied statistical hypersurfaces in the space of Hessian curvature zero and provided some examples as well. In 2017, Furuhata et al. [5] studied Sasakian statistical manifolds and obtained some results. Geometry of statistical submanifolds is still young and efforts are on, so it is growing [1–3,6–9].

In 1993 Chen [4] has obtained a sharp inequality for the sectional curvature of a submanifold in a real space forms in term of the scalar curvature (intrinsic invariant) and squared mean curvature (extrinsic invariant). Afterward, several geometers obtained similar inequality for various submanifolds in various ambient spaces due to its rich geometric importance.

In the present article, we obtain B. Y. Chen inequality for statistical submanifolds in Sasakian statistical manifold with constant curvature and obtain the equality case of the inequality. We also give some applications of the inequalities we derived.

© Springer Nature Switzerland AG 2019
F. Nielsen and F. Barbaresco (Eds.): GSI 2019, LNCS 11712, pp. 398–406, 2019.
https://doi.org/10.1007/978-3-030-26980-7_41

2   Preliminaries

Let $(\overline{N}, g)$ be a Riemannian manifold and $\overline{\nabla}$ and $\overline{\nabla}^*$ be torsion-free affine connections on $\overline{N}$ such that

$$Gg(E, F) = g(\overline{\nabla}_G E, F) + g(E, \overline{\nabla}_G^* F), \tag{1}$$

for $E, F, G \in \Gamma(T\overline{N})$. Then Riemannian manifold $(\overline{N}, g)$ is called a statistical manifold. It is denoted by $(\overline{N}, g, \overline{\nabla}, \overline{\nabla}^*)$. The connections $\overline{\nabla}$ and $\overline{\nabla}^*$ are called dual connections. The pair $(\overline{\nabla}, g)$ is said to be a statistical structure.

If $(\overline{\nabla}, g)$ is a statistical structure on $\overline{N}$, then $(\overline{\nabla}^*, g)$ is also statistical structure on $\overline{N}$.

For the dual connections $\overline{\nabla}$ and $\overline{\nabla}^*$ we have

$$2\overline{\nabla}^{\circ} = \overline{\nabla} + \overline{\nabla}^*, \tag{2}$$

where $\overline{\nabla}^{\circ}$ is Levi-Civita connection for g.

Let $\overline{N}$ be a $(2m + 1)$-dimensional manifold and let N be an n-dimensional submanifolds of $\overline{N}$. Then, the Gauss formulae are [10]

$$\begin{cases} \overline{\nabla}_E F = \nabla_E F + \zeta(E, F), \\ \overline{\nabla}_E^* F = \nabla_E^* F + \zeta^*(E, F), \end{cases} \tag{3}$$

where ζ and ζ^* are symmetric, bilinear, imbedding curvature tensors of N in $\overline{N}$ for $\overline{\nabla}$ and $\overline{\nabla}^*$, respectively.

The $\overline{R}$ and $\overline{R}^*$ be Riemannian curvature tensor fields of $\overline{\nabla}$ and $\overline{\nabla}^*$, respectively. Then [10]

$$\begin{aligned} g(\overline{R}(E, F)G, W) = {} & g(R(E, F)G, W) + g(\zeta(E, G), \zeta^*(F, W)) \\ & - g(\zeta^*(E, W), \zeta(F, G)), \end{aligned} \tag{4}$$

and

$$\begin{aligned} g(\overline{R}^*(E, F)G, W) = {} & g(R^*(E, F)G, W) + g(\zeta^*(E, G), \zeta(F, W)) \\ & - g(\zeta(E, W), \zeta^*(F, G)), \end{aligned} \tag{5}$$

where

$$g(\overline{R}^*(E, F)G, W) = -g(G, \overline{R}(E, F)W). \tag{6}$$

Let us denote the normal bundle of N by $T N^{\perp}$. The linear transformations A_N and A_N^* are defined by

$$\begin{cases} g(A_N E, F) = g(\zeta(E, F), N), \\ g(A_N^* E, F) = g(\zeta^*(E, F), N), \end{cases} \tag{7}$$

for any $N \in \Gamma(T\mathbb{N}^\perp)$ and $\mathrm{E}, \mathrm{F} \in \Gamma(T\mathbb{N})$. The corresponding Weingarten formulas are [10]

$$\begin{cases} \overline{\nabla}_{\mathrm{E}} N = -A_N^* \mathrm{E} + \nabla_{\mathrm{E}}^\perp N, \\ \overline{\nabla}_{\mathrm{E}}^* N = -A_N \mathrm{E} + \nabla_{\mathrm{E}}^{*\perp} N, \end{cases} \tag{8}$$

where $N \in \Gamma(T\mathbb{N}^\perp)$, $\mathrm{E} \in \Gamma(T\mathbb{N})$ and $\nabla_{\mathrm{E}}^\perp$ and $\nabla_{\mathrm{E}}^{*\perp}$ are Riemannian dual connections with respect to the induced metric on $\Gamma(T\mathbb{N}^\perp)$.

Let $\overline{\mathbb{N}}$ be an odd dimensional manifold and ϕ be a tensor of type $(1,1)$, ξ a vector field, and a 1-form η on $\overline{\mathbb{N}}$ satisfying the conditions

$$\eta(\xi) = 1,$$
$$\phi^2 \mathrm{E} = -\mathrm{E} + \eta(\mathrm{E})\xi,$$

for any vector field E on $\overline{\mathbb{N}}$, then $\overline{\mathbb{N}}$ is said to have an almost contact structure (ϕ, ξ, η).

Definition 1. *An almost contact structure (ϕ, ξ, g) on $\overline{\mathbb{N}}$ is said to be a Sasakian structure if*

$$(\overline{\nabla}_{\mathrm{E}}^\circ \phi)\mathrm{F} = \mathrm{g}(\mathrm{F}, \xi)\mathrm{E} - \mathrm{g}(\mathrm{F}, \mathrm{E})\xi,$$

holds for any $\mathrm{E}, \mathrm{F} \in T\overline{\mathbb{N}}$.

Definition 2 ([5]). *A quadruple $(\overline{\nabla}, \mathrm{g}, \phi, \xi)$ is called a Sasakian statistical structure on $\overline{\mathbb{N}}$ if $(\overline{\nabla}, \mathrm{g})$ is a statistical structure, (g, ϕ, ξ) is a Sasakian structure on $\overline{\mathbb{N}}$ and the formula*

$$K_{\mathrm{E}}\phi\mathrm{F} + \phi K_{\mathrm{E}}\mathrm{F} = 0$$

holds for any $\mathrm{E}, \mathrm{F} \in T\overline{\mathbb{N}}$, where $K_{\mathrm{E}}\mathrm{F} = \overline{\nabla}_{\mathrm{E}}\mathrm{F} - \overline{\nabla}_{\mathrm{E}}^\circ \mathrm{F}$.

Definition 3 ([5]). *Let $(\overline{\mathbb{N}}, \overline{\nabla}, \mathrm{g}, \phi, \xi)$ be a Sasakian statistical manifold and $c \in \mathbb{R}$. The Sasakian statistical structure is said to be of constant ϕ-sectional curvature c if the curvature tensor $\overline{S}$ is given by*

$$\overline{S}(\mathrm{E}, \mathrm{F})\mathrm{G} = \frac{c+3}{4}\{\mathrm{g}(\mathrm{F}, \mathrm{G})\mathrm{E} - \mathrm{g}(\mathrm{E}, \mathrm{G})\mathrm{F}\} + \frac{c-1}{4}\{\mathrm{g}(\phi\mathrm{F}, \mathrm{G})\phi\mathrm{E} - \mathrm{g}(\phi\mathrm{E}, \mathrm{G})\phi\mathrm{F}$$
$$- 2\mathrm{g}(\phi\mathrm{E}, \mathrm{F})\phi\mathrm{G} - \mathrm{g}(\mathrm{F}, \xi)\mathrm{g}(\mathrm{G}, \xi)\mathrm{E} + \mathrm{g}(\mathrm{E}, \xi)\mathrm{g}(\mathrm{G}, \xi)\mathrm{F} + \mathrm{g}(\mathrm{F}, \xi)\mathrm{g}(\mathrm{G}, \mathrm{E})\xi$$
$$- \mathrm{g}(\mathrm{E}, \xi)\mathrm{g}(\mathrm{G}, \mathrm{F})\xi\}, \quad where \quad \mathrm{E}, \mathrm{F}, \mathrm{G} \in T\overline{\mathbb{N}} \tag{9}$$

and

$$2\overline{S}(\mathrm{E}, \mathrm{F})\mathrm{G} = \overline{R}(\mathrm{E}, \mathrm{F})\mathrm{G} + \overline{R}^*(\mathrm{E}, \mathrm{F})\mathrm{G}. \tag{10}$$

We denote a Sasakian statistical manifold with constant ϕ-sectional curvature c by $\overline{\mathbb{N}}(c)$.

Let ξ be tangent to the submanifolds N and let $\{e_1, \ldots, e_n = \xi\}$ and $\{e_{n+1}, \ldots, e_{2m+1}\}$ be tangent orthonormal frame and normal orthonormal frame, respectively, on N. Then, the mean curvature vector fields H, H^*, H° are given by

$$H = \frac{1}{n} \sum_{i=1}^{n} \zeta(e_i, e_i), \tag{11}$$

$$H^* = \frac{1}{n} \sum_{i=1}^{n} \zeta^*(e_i, e_i), \tag{12}$$

and

$$H^\circ = \frac{1}{n} \sum_{i=1}^{n} \zeta^\circ(e_i, e_i). \tag{13}$$

We also set

$$\|\zeta\|^2 = \sum_{i,j=1}^{n} g(\zeta(e_i, e_j), \zeta(e_i, e_j)), \tag{14}$$

$$\|\zeta^*\|^2 = \sum_{i,j=1}^{n} g(\zeta^*(e_i, e_j), \zeta^*(e_i, e_j)), \tag{15}$$

and

$$\|\zeta^\circ\|^2 = \sum_{i,j=1}^{n} g(\zeta^\circ(e_i, e_j), \zeta^\circ(e_i, e_j)). \tag{16}$$

The second fundamental form ζ° (resp. ζ, or ζ^*) has several geometric properties due to which we got following different classes of the submanifolds.

– A submanifold is said to be totally geodesic submanifold with respect to $\overline{\nabla}^\circ$ (resp. $\overline{\nabla}$, or $\overline{\nabla}^*$), if the second fundamental form ζ° (resp. ζ, or ζ^*) vanishes identically, that is $\zeta^\circ = 0$ (resp. $\zeta = 0$, or $\zeta^* = 0$).
– A submanifold is said to be minimal submanifold with respect to $\overline{\nabla}^\circ$ (resp. $\overline{\nabla}$, or $\overline{\nabla}^*$), if the mean curvature vector H° (resp. H, or H^*) vanishes identically, that is $H^\circ = 0$ (resp. $H = 0$, or $H^* = 0$).

Let $K(\pi)$ denotes the sectional curvature of a Riemannian manifold N of the plane section $\pi \subset T_pN$ at a point $p \in N$. If $\{e_1, \ldots, e_n\}$ be the orthonormal basis of T_pN and $\{e_{n+1}, \ldots, e_{2m+1}\}$ be the orthonormal basis of $T_p^\perp N$ at any $p \in N$, then

$$\tau(p) = \sum_{1 \leq i < j \leq n} K(e_i \wedge e_j), \tag{17}$$

where τ is the scalar curvature. The normalized scalar curvature ρ is defined as

$$2\tau = n(n-1)\rho. \tag{18}$$

We also put

$$\zeta_{ij}^{\gamma} = g(\zeta(e_i, e_j), e_{\gamma}), \quad \zeta_{ij}^{*\gamma} = g(\zeta^*(e_i, e_j), e_{\gamma}),$$

$i, j \in 1, \ldots, n, \gamma \in \{n+1, \ldots, 2m+1\}$.

3   B. Y. Chen Inequalities

In this section, we obtain statistical version of well known B. Y. Chen inequality for statistical submanifolds of Sasakian statistical manifolds with constant ϕ-sectional curvature.

Theorem 1. *Let* N *be a statistical submanifold in a Sasakian statistical manifold* $\bar{N}(c)$ *with* $\sum_{\alpha} \left[\zeta_{11}^{*\alpha} \zeta_{22}^{\alpha} + \zeta_{11}^{\alpha} \zeta_{22}^{*\alpha} \right] = 2 \sum_{\alpha} \zeta_{12}^{*\alpha} \zeta_{12}^{\alpha}$ *such that the structure vector field* ξ *of* $\bar{N}(c)$ *is tangent to* N. *Then*

$$K(\pi) \leq \tau + \frac{c+3}{4}(1+n-n^2) + \frac{c-1}{4}\{3(\Theta(\pi) - \|P\|^2) - \Phi(\pi) - 2(1-n)\}$$

$$+ \frac{n^2}{2}(\|H\|^2 + \|H^*\|^2) - 2n^2\|H^{\circ}\|^2 + \|\zeta\|\|\zeta^*\|, \tag{19}$$

where $\Theta(\pi) = g^2(\phi e_1, e_2)$, $\Phi(\pi) = \eta^2(e_1) + \eta^2(e_2)$, $\pi = e_1 \wedge e_2$ *and* $\|P\|^2 = g^2(\phi e_i, e_j)\}$. *Moreover, the equality holds if* ζ *and* ζ^* *are parallel. That is*

$$\zeta = k\zeta^*, k \in \mathbb{R}^+. \tag{20}$$

Proof. From (4), (5), (9) and (10), we have

$$g(R(E, F)G, W) + g(R^*(E, F)G, W) = \frac{c+3}{2}\{g(F, G)g(E, W) - g(E, G)g(F, W)$$

$$+ \frac{c-1}{2}\{g(\phi F, G)g(\phi E, W) - g(\phi E, G)g(\phi F, W) - 2g(\phi E, F)g(\phi G, W)$$

$$- g(F, \xi)g(G, \xi)g(E, W) + g(E, \xi)g(G, \xi)g(F, W) + g(F, \xi)g(G, E)g(\xi, W)$$

$$- g(E, \xi)g(G, F)g(\xi, W)\} - g(\zeta(E, G), \zeta^*(F, W)) + g(\zeta^*(E, W), \zeta(F, G))$$

$$- g(\zeta^*(E, G), \zeta(F, W)) + g(\zeta(E, W), \zeta^*(F, G)). \tag{21}$$

Putting $F = W = e_i$ and $E = G = e_j$, in (21), we get

$$g(R(e_i, e_j)e_j, e_i) + g(R^*(e_i, e_j)e_j, e_i) = \frac{c+3}{2}\{g(e_j, e_j)g(e_i, e_i) - g(e_i, e_j)g(e_j, e_i)\}$$

$$+ \frac{c-1}{2}\{g(\phi e_j, e_j)g(\phi e_i, e_i) - g(\phi e_i, e_j)g(\phi e_j, e_i)$$

$$- 2g(\phi e_i, e_j)g(\phi e_j, e_i) - g(e_j, \xi)g(e_j, \xi)g(e_i, e_i)$$

$$+ g(e_i, \xi)g(e_j, \xi)g(e_j, e_i) + g(e_j, \xi)g(e_j, e_i)g(\xi, e_i)$$

$$- g(e_i, \xi)g(e_j, e_j)g(\xi, e_i)\} - g(\zeta(e_i, e_j), \zeta^*(e_j, e_i))$$
$$+ g(\zeta^*(e_i, e_i), \zeta(e_j, e_j)) - g(\zeta^*(e_i, e_j), \zeta(e_j, e_i))$$
$$+ g(\zeta(e_i, e_i), \zeta^*(e_j, e_j)). \tag{22}$$

Applying summation $1 \leq i, j \leq n$ and using (11)–(16) in (22), we obtain

$$\sum_{1 \leq i,j \leq n} [g(R(e_i, e_j)e_j, e_i) + g(R^*(e_i, e_j)e_j, e_i)] = \frac{c+3}{2}n(n-1) + 2n^2 g(H, H^*)$$

$$+ \frac{c-1}{2}\{2(1-n) + 3g^2(\phi e_i, e_j)\} - g(\zeta(e_i, e_j), \zeta^*(e_j, e_i))$$
$$- g(\zeta^*(e_i, e_j), \zeta(e_j, e_i))$$

$$= \frac{c+3}{2}n(n-1) + n^2\{g(H^* + H, H^* + H) - g(H, H) - g(H^*, H^*)\}$$

$$+ \frac{c-1}{2}\{2(1-n) + 3g^2(\phi e_i, e_j)\}$$

$$- \{g(\zeta(e_i, e_j) + \zeta^*(e_j, e_i), \zeta^*(e_i, e_j) + \zeta(e_j, e_i))$$
$$- g(\zeta(e_i, e_j), \zeta(e_i, e_j)) - g(\zeta^*(e_j, e_i), \zeta^*(e_j, e_i))\}. \tag{23}$$

Since from Eq. (2) $2H^\circ = H + H^*$, it follows from the above equation that

$$2\tau = \frac{c+3}{2}n(n-1) + \frac{c-1}{2}\{2(1-n) + 3\|P\|^2\}$$
$$+ 4n^2\|H^\circ\|^2 - n^2(\|H\|^2 + \|H^*\|^2) + 4\|\zeta^\circ\|^2 - (\|\zeta\|^2 + \|\zeta^*\|^2). \tag{24}$$

On the other hand we know that

$$K(\pi) = \frac{1}{2}\big[g(R(e_1, e_2)e_2, e_1) + g(R^*(e_1, e_2)e_2, e_1)\big]$$

$$= \frac{1}{2}\big[g(\overline{R}(e_1, e_2)e_2, e_1) + g(\overline{R}^*(e_1, e_2)e_2, e_1)$$
$$- 2g(\zeta^*(e_1, e_2), \zeta(e_2, e_1)) + 2g(\zeta(e_1, e_1), \zeta^*(e_2, e_2))\big]$$

$$= g(\overline{S}(e_1, e_2)e_2, e_1) + \sum_\alpha \big[\tfrac{1}{2}\zeta^{*\alpha}_{11}\zeta^\alpha_{22} + \tfrac{1}{2}\zeta^\alpha_{11}\zeta^{*\alpha}_{22} - \zeta^{*\alpha}_{12}\zeta^\alpha_{12}\big]. \tag{25}$$

Taking inner product of (9) with W and setting $E = W = e_1$ and $F = G = e_2$, we find

$$g(\overline{S}(e_1, e_2)e_2, e_1) = \frac{c+3}{4}\{g(e_2, e_2)g(e_1, e_1) - g(e_1, e_2)g(e_2, e_1)\}$$

$$+ \frac{c-1}{4}\{g(\phi e_2, e_2)g(\phi e_1, e_1) - g(\phi e_1, e_2)g(\phi e_2, e_1)$$
$$- 2g(\phi e_1, e_2)g(\phi e_2, e_1) - g(e_2, \xi)g(e_2, \xi)g(e_1, e_1)$$
$$+ g(e_1, \xi)g(e_2, \xi)g(e_2, e_1) + g(e_2, \xi)g(e_2, e_1)g(\xi, e_1)$$
$$- g(e_1, \xi)g(e_2, e_2)g(\xi, e_1)\}$$

$$= \frac{c+3}{4} + \frac{c-1}{4}\{g(\phi e_2, e_2)g(\phi e_1, e_1) + 3g^2(\phi e_1, e_2)$$

$$- \mathbf{g}^2(e_2, \xi) - \mathbf{g}^2(e_1, \xi)\}$$
$$= \frac{c+3}{4} + \frac{c-1}{4}\{\mathbf{g}(\phi e_2, e_2)\mathbf{g}(\phi e_1, e_1) + 3\Theta(\pi) - \Phi(\pi)\}. \quad (26)$$

From (25) and (26), we get

$$2K(\pi) = \frac{c+3}{2} + \frac{c-1}{2}\{3\Theta(\pi) - \Phi(\pi)\} + \sum_\alpha \left[\zeta_{11}^{*\alpha}\zeta_{22}^\alpha + \zeta_{11}^\alpha\zeta_{22}^{*\alpha} - 2\zeta_{12}^{*\alpha}\zeta_{12}^\alpha\right]. \quad (27)$$

Taking into account (24) and (27), we have

$$2K(\pi) - 2\tau = \frac{c+3}{2} + \frac{c-1}{2}\{3\Theta(\pi) - \Phi(\pi)\} + \sum_\alpha \left[\zeta_{11}^{*\alpha}\zeta_{22}^\alpha + \zeta_{11}^\alpha\zeta_{22}^{*\alpha} - 2\zeta_{12}^{*\alpha}\zeta_{12}^\alpha\right]$$
$$- \frac{c+3}{2}n(n-1) - \frac{c-1}{2}\{2(1-n) + 3\|P\|^2\} - 4n^2\|\mathbf{H}^\circ\|^2$$
$$+ n^2(\|\mathbf{H}\|^2 + \|\mathbf{H}^*\|^2) + 4n\mathbf{C}^\circ - n(\mathbf{C} + \mathbf{C}^*)$$
$$= \frac{c+3}{2}(1 + n - n^2) + \frac{c-1}{2}\{3\Theta(\pi) - \Phi(\pi) - 2(1-n) - 3\|P\|^2\}$$
$$+ \sum_\alpha \left[\zeta_{11}^{*\alpha}\zeta_{22}^\alpha + \zeta_{11}^\alpha\zeta_{22}^{*\alpha} - 2\zeta_{12}^{*\alpha}\zeta_{12}^\alpha\right] - 4n^2\|\mathbf{H}^\circ\|^2$$
$$+ n^2(\|\mathbf{H}\|^2 + \|\mathbf{H}^*\|^2) + 4\|\zeta^\circ\|^2 - (\|\zeta\|^2 + \|\zeta^*\|^2). \quad (28)$$

On the other hand,

$$\|\zeta + \zeta^*\|^2 = g(\zeta + \zeta^*, \zeta + \zeta^*)$$
$$= \|\zeta\|^2 + g(\zeta, \zeta^*) + g(\zeta^*, \zeta) + \|\zeta^*\|^2$$
$$= \|\zeta\|^2 + 2g(\zeta, \zeta^*) + \|\zeta^*\|^2$$
$$\leq \|\zeta\|^2 + 2\|\zeta\|\|\zeta^*\| + \|\zeta^*\|^2, \quad (29)$$

and the equality holds if

$$\zeta = k\zeta^*, \quad k \in \mathbb{R}^+ \quad (30)$$

Equation (29) implies

$$\|\zeta\|^2 + \|\zeta^*\|^2 \geq \|\zeta + \zeta^*\|^2 - 2\|\zeta\|\|\zeta^*\| \quad (31)$$

Using (31) in (28), we obtain

$$2K(\pi) - 2\tau \leq \frac{c+3}{2}(1 + n - n^2) + \frac{c-1}{2}\{3\Theta(\pi) - \Phi(\pi) - 2(1-n) - 3\|P\|^2\}$$
$$+ \sum_\alpha \left[\zeta_{11}^{*\alpha}\zeta_{22}^\alpha + \zeta_{11}^\alpha\zeta_{22}^{*\alpha} - 2\zeta_{12}^{*\alpha}\zeta_{12}^\alpha\right] - 4n^2\|\mathbf{H}^\circ\|^2 + n^2(\|\mathbf{H}\|^2 + \|\mathbf{H}^*\|^2)$$
$$+ 4\|\zeta^\circ\|^2 - \|\zeta + \zeta^*\|^2 + 2\|\zeta\|\|\zeta^*\|$$
$$= \frac{c+3}{2}(1 + n - n^2) + n^2(\|\mathbf{H}\|^2 + \|\mathbf{H}^*\|^2) + 2\|\zeta\|\|\zeta^*\|$$

$$+ \frac{c-1}{2}\big\{3\Theta(\pi) - \Phi(\pi) - 2(1-n) - 3\|P\|^2\big\}$$

$$+ \sum_\alpha \big[\zeta_{11}^{*\alpha}\zeta_{22}^{\alpha} + \zeta_{11}^{\alpha}\zeta_{22}^{*\alpha} - 2\zeta_{12}^{*\alpha}\zeta_{12}^{\alpha}\big] - 4n^2\|\mathrm{H}^\circ\|^2. \tag{32}$$

Using the hypothesis of the theorem in (32), we have

$$2K(\pi) - 2\tau \leq \frac{c+3}{2}(1+n-n^2) + \frac{c-1}{2}\big\{3\Theta(\pi) - \Phi(\pi) - 2(1-n) - 3\|P\|^2\big\}$$

$$- 4n^2\|\mathrm{H}^\circ\|^2 + n^2(\|\mathrm{H}\|^2 + \|\mathrm{H}^*\|^2) + 2\|\zeta\|\|\zeta^*\|. \tag{33}$$

Moreover, equality holds if and only if it satisfies (30). Hence we have the required result.

The following result is immediate consequence of Theorem 1.

Corollary 1. *Let* N *be a statistical submanifold in a Sasakian statistical manifold* $\overline{\mathrm{N}}(c)$ *with* $\sum_\alpha \big[\zeta_{11}^{*\alpha}\zeta_{22}^{\alpha} + \zeta_{11}^{\alpha}\zeta_{22}^{*\alpha}\big] = 2\sum_\alpha \zeta_{12}^{*\alpha}\zeta_{12}^{\alpha}$ *such that the structure vector field* ξ *of* $\overline{\mathrm{N}}(c)$ *is tangent to* N*. Then*

$$K(\pi) - \tau \leq \frac{c+3}{4}(1+n-n^2) + \frac{c-1}{4}\big\{3(\Theta(\pi) - \|P\|^2) - \Phi(\pi) - 2(1-n)\big\}, \tag{34}$$

if N *is totally geodesic with respect to* $\overline{\nabla}$ *or* N *is totally geodesic with respect to* $\overline{\nabla}^*$.

Further, we state similar result when the structure vector field ξ of $\overline{\mathrm{N}}(c)$ is normal to N.

Theorem 2. *Let* N *be a statistical submanifold in a Sasakian statistical manifold* $\overline{\mathrm{N}}(c)$ *with* $\sum_\alpha \big[\zeta_{11}^{*\alpha}\zeta_{22}^{\alpha} + \zeta_{11}^{\alpha}\zeta_{22}^{*\alpha}\big] = 2\sum_\alpha \zeta_{12}^{*\alpha}\zeta_{12}^{\alpha}$ *such that the structure vector field* ξ *of* $\overline{\mathrm{N}}(c)$ *is Normal to* N*. Then*

$$K(\pi) \leq \tau + \frac{c+3}{4}(1+n-n^2) + \frac{c-1}{4}\big\{3\Theta(\pi) - 3\|P\|^2\big\}$$

$$+ \frac{n^2}{2}(\|\mathrm{H}\|^2 + \|\mathrm{H}^*\|^2) - 2n^2\|\mathrm{H}^\circ\|^2 + \|\zeta\|\|\zeta^*\|. \tag{35}$$

From the above result we deduce the following corollary.

Corollary 2. *Let* N *be a statistical submanifold in a Sasakian statistical manifold* $\overline{\mathrm{N}}(c)$ *with* $\sum_\alpha \big[\zeta_{11}^{*\alpha}\zeta_{22}^{\alpha} + \zeta_{11}^{\alpha}\zeta_{22}^{*\alpha}\big] = 2\sum_\alpha \zeta_{12}^{*\alpha}\zeta_{12}^{\alpha}$ *such that the structure vector field* ξ *of* $\overline{\mathrm{N}}(c)$ *is Normal to* N*. Then*

$$K(\pi) \leq \tau + \frac{c+3}{4}(1+n-n^2) + \frac{c-1}{4}\big\{3\Theta(\pi) - 3\|P\|^2\big\}, \tag{36}$$

if N *is totally geodesic with respect to* $\overline{\nabla}$ *or* N *is totally geodesic with respect to* $\overline{\nabla}^*$.

4 Conclusion and Future work

We obtained the B. Y. Chen inequality for the statistical submanifolds in Sasakian statistical manifolds having constant curvature. In fact, this is the first such attempt for any statistical case. Therefore, I hope it will open the door for the researcher to obtain such inequality, which has the great geometric importance, for different ambient such as **Holomorphic statistical manifolds, Kenmotsu Statistical manifolds, Cosymplectic statistical manifolds, Quaternion Kaehler-like statistical manifolds** etc. with constant curvatures. The forthcoming challenge is to improve the result by weakening the condition.

References

1. Aquib, M., Shahid, M.H.: Generalized normalized δ-Casorati curvature for statistical submanifolds in quaternion Kaehler-like statistical space forms. J. Geom. **109**, 13 (2018)
2. Aydin, M.E., Mihai, A., Mihai, I.: Generalized wintgen inequality for statistical submanifolds in statistical manifolds of constant curvature. Bull. Math. Sci. **7**, 155 (2017)
3. Nguiffo Boyom, M., Aquib, M., Shahid, M.H., Jamali, M.: Generalized wintegen type inequality for lagrangian submanifolds in holomorphic statistical space forms. In: Nielsen, F., Barbaresco, F. (eds.) GSI 2017. LNCS, vol. 10589, pp. 162–169. Springer, Cham (2017). https://doi.org/10.1007/978-3-319-68445-1_19
4. Chen, B.Y.: Some pinching and classification theorems for minimal submanifolds. Arch. Math. **60**, 568–578 (1993)
5. Furuhata, H., Hasegawa, I., Okuyama, Y., Sato, K., Shahid, M.H.: Sasakian statistical manifolds. J. Geom. Phys. **117**, 179–186 (2017)
6. Furuhata, H.: Hypersurfaces in statistical manifolds. Differ. Geom. Appl. **67**, 420–429 (2009)
7. Milijevic, M.: Totally real statistical submanifolds. Interdiscip. Inf. Sci. **21**, 87–96 (2015)
8. Takano, K.: Statistical manifolds with almost contact structures and its statistical submersions. J. Geom. **85**, 171–187 (2006)
9. Vilcu, A., Vilcu, G.E.: Statistical manifolds with almost quaternionic structures and quaternionic Kaehler-like statistical submersions. Entropy **17**, 6213–6228 (2015)
10. Vos, P.: Fundamental equations for statistical submanifolds with applications to the bartlett correction. Ann. Inst. Stat. Math. **41**(3), 429–450 (1989)

Generalized Wintgen Inquality for Legendrian Submanifolds in Sasakian Statistical Manifolds

Michel Nguiffo Boyom[1], Zamrooda Jabeen[2(✉)], Mehraj Ahmad Lone[2(✉)], Mohamd Saleem Lone[3], and Mohammad Hasan Shahid[4]

[1] IMAG, Alexander Grothendieck Research Institute, Université of Montpellier, Montpellier, France
nguiffo.boyom@gmail.com
[2] Department of Mathematics, NIT Srinagar, Srinagar, J&K, India
zjabeen@gmail.com, mehraj.jmi@gmail.com
[3] International Centre for Theoretical Sciences TIFR, Bengaluru, India
saleemraja2008@gmail.com
[4] Department of Mathematics, Jamia Millia Islamia University, New Delhi, India
hasan_jmi@yahoo.com

Abstract. In 1999, Smet et al. conjectured the generalized Wintgen inequality for submanifolds in real space forms. The commonly name used for this conjecture is DDVV conjecture proved independently by Ge and Tang (2008) and Lu (2011). Mihai Proved the Wintgen inequality for lagrangian and Legendrian submanifolds in complex space forms and Sasakian space forms in 2014 and 2017 respectively. In the present paper, we proved the same inequality for Legendrian submanifolds of Sasakian statistical manifolds.

Keywords: DDVV conjucture · Legendrian submanifold · Sasakian statistical manifold · Wintgen inequality

1 Introduction

In 1985, statistical manifolds emerges from statistical distributions due to Amari [1] in terms of information geometry. The geometry of such manifolds uses the notion of dual connections, also referred as conjugate connections. It is closely related to affine geometry and Hessian geometry. Due to lot of applications of statistical manifolds to various fields of science and technology, it attracts the attention of distinguished geometers from last few years. The geometry of submanifolds of statistical manifolds have been studied by different authors and obtain interesting results between intrinsic invariants and extrinsic invariants. Furuhata [3] introduced the notion of Sasakian statistical structure an analogues to Sasakian structure in contact geometry and obtain some beautiful results.

Supported by NIT Srinagar.

F. Nielsen and F. Barbaresco (Eds.): GSI 2019, LNCS 11712, pp. 407–412, 2019.
https://doi.org/10.1007/978-3-030-26980-7_42

Wintgen [7] established an inequality

$$K \leq \|H\|^2 - |K|^\perp$$

where K is Gauss curavture, $K^\perp$ is normal curvature and H is squared mean curvature of M^2 in an Euclidean Space E^4. The equality case holds if and only if the ellipse of the curvature of M^2 in E^4 is circle. The above inequality is known as Wintgen inequality.

Later on Smet et al. [2] extended the inequality and formulated the conjecture on the Wintgen inequality for n-dimensional submanifolds of a real space form $\bar{M}^{n+m}(c)$. The conjucture is also known as DVVV conjecture and is formulated as:

Conjecture 1. *Let $f : M^n \rightarrow \bar{M}^{m+n}(c)$ be an isometric immersion of n-dimensional submanifolds of a real space form $\bar{M}^{n+m}(c)$ of constant sectional curvature c, then*

$$\rho \leq \|H\|^2 - \rho^\perp + c$$

where ρ and $\rho^\perp$ are the normalized scalar curvature and the normalized normal scalar curvature respectively.

The normalized scalar curvature is defined by

$$\rho = \frac{2\tau}{n(n-1)} = \frac{2}{n(n-1)} \sum_{1 \leq i < j \leq n} K(e_i \wedge e_j). \tag{1}$$

where τ is scalar curvature.

The normalized normal scalar curvature are defined as

$$\rho = \frac{2\tau^\perp}{n(n-1)} = \frac{2}{n(n-1)} \sqrt{\sum_{1 \leq i < j \leq n} \sum_{1 \leq \alpha < \beta \leq m} (R^\perp(e_i, e_j, \xi_\alpha, \xi_\beta))}. \tag{2}$$

where $\tau^\perp$ is normal scalar curvature.

The distinguished geometers prove the conjecture for different submanifolds and ambient spaces in complex as well as in contact geometry and also discuss the equality case. Finally Ge and Tang [4] and Lu [5] proved the DDVV conjecture independently for general case.

In the present paper, we obtain the generalized Wintgen inequality for Legendrian submanifolds of Sasakian statistical manifolds with constant ϕ-sectional curvature w.r.t dual connections $\bar{\nabla}$ and $\bar{\nabla}^*$.

2   Preliminaries

Let $\bar{M}, \bar{g}$ be a Riemannian manifold with a pair of torsion free affine connections $\bar{\nabla}$ and $\bar{\nabla}^*$. Then $(\bar{\nabla}, \bar{g})$ is called statistical structure on if

$$(\bar{\nabla}_X \bar{g})(Y, Z) - (\bar{\nabla}_Y \bar{g})(X, Z) = 0 \tag{3}$$

for $X, Y, Z \in T\bar{M}$. If the Riemannian manifold $\bar{M}, \bar{g}$ with statistical structure satisfies

$$X\bar{g}(Y, Z) = \bar{g}(\bar{\nabla}_X Y, Z) + \bar{g}(Y, \nabla_X^* Z), \tag{4}$$

is said to be a statistical manifold and is denoted as $(\bar{M}, \bar{g}, \bar{\nabla}, \bar{\nabla}^*)$. Any torsion-free connection $\bar{\nabla}$ has a dual connection $\bar{\nabla}^*$ and satisfy

$$\bar{\nabla}^\circ = \frac{\bar{\nabla} + \bar{\nabla}^*}{2} \tag{5}$$

where $\bar{\nabla}^\circ$ is the Levi-Civita connection on $\bar{M}$.

Consider $\bar{M}$ be an odd dimensional manifold. Suppose η, ϕ and ξ be the 1-form, a tensor field and structural vector field or Reeb vector field respectively on $\bar{M}$ satisfying

$$\phi\xi = 0, \phi^2 = -I + \eta \otimes \xi$$

then (η, ϕ, ξ) is called almost contact structure on $\bar{M}$. Consider $(\bar{M}, \bar{g})$ Riemannian manifold and (η, ϕ, ξ) is almost contact structure. Suppose ϕ^* be another tensor field of type $(1, 1)$ on $\bar{M}$ satisfying

$$\bar{g}(\phi X, Y) = -\bar{g}(X, \phi^* Y)$$

for $X, Y \in T\bar{M}$. Then $(\bar{M}, \bar{g}, \phi, \xi, \eta)$ is said to be almost contact metric-like manifold ad satisfying

$$\phi^{*2} - I + \eta \otimes \xi$$

and

$$\bar{g}(\phi X, \phi^* Y) = \bar{g}(X, Y) - \eta(X)\eta(Y).$$

If a statistical manifold $(\bar{M}, \bar{g}, \bar{\nabla})$ has an almost contact metric-like structure (ϕ, ξ, η), then $(\bar{M}, \bar{g}, \bar{\nabla}, \phi, \xi, \eta)$ is known as almost contact metric-like statistical manifold.

A quadruple $(\bar{\nabla}, \bar{g}, \phi, \xi)$ be Sasakian statistical structure on $\bar{M}$ and so is $(\bar{\nabla}^*, \bar{g}, \phi^*, \xi)$, if $(\bar{\nabla}, \bar{g}])$ is a statistical structure.

Definition 1. *Let* $(\bar{\nabla}, \bar{g}, \phi, \xi)$ *be a Sasakian statistical structure on* $\bar{M}$, *and* $c \in \mathbf{R}$. *The Sasakian statistical structure is said to be of constant* ϕ-*sectional curvature if*

$$\begin{aligned} S(X, Y)Z = \frac{c+3}{4}\left(\bar{g}(Y, Z)X - \bar{g}(X, Z)Y\right) + \frac{c-1}{4}(\bar{g}(\phi Y, Z)\phi X \\ - \bar{g}(\phi X, Z)\phi Y - 2\bar{g}(\phi X, Y)\phi Z - \bar{g}(Y, \xi)\bar{g}(Z, \xi)X \\ + \bar{g}(X, \xi)\bar{g}(Z, \xi)Y + \bar{g}(Y, \xi)\bar{g}(Z, X)\xi - \bar{g}(X, \xi)\bar{g}(Z, Y)\xi) \end{aligned} \tag{6}$$

for $X, Y, Z \in T\bar{M}$.

A submanifold M which is normal to the structural vector field ξ in a Sasakian statistical manifold is known as c-totally real submanifold. In this particular case, for every $p \in M$, $\phi(T_pM) \subseteq T_p^\perp M$. if $n = m$, the M is known as Legendrian submanifold.

Let (M, g, ∇, ∇^*) be statistical submanifold of $(\bar{M}, \bar{g}, \bar{\nabla}, \bar{\nabla}^*)$. The Gauss and Weingarten formulae are given as

$$\bar{\nabla}_X Y = \nabla_X Y + \sigma X, Y \ , \ \bar{\nabla}_X \xi = -A_\xi X + \nabla_X^\perp \xi \tag{7}$$
$$\bar{\nabla}^*_X Y = \nabla^*_X Y + \sigma^* X, Y \ , \ \bar{\nabla}^*_X \xi = -A^*_\xi X + \nabla_X^{*\perp} \xi \tag{8}$$

for all $X, Y \in TM$ and $\xi \in T^\perp M$ respectively. Moreover, we have the following equations

$$X g(Y, Z) = g(\nabla_X Y, Z) + g(Y, \nabla_X^* Z)$$
$$\bar{g}(\sigma(X, Y), \xi) = g(A^*_\xi X, Y), \bar{g}(\sigma^*(X, Y), \xi) = g(A_\xi X, Y)$$
$$and \quad X \bar{g}(\xi, \eta) = \bar{g}(\nabla_X^\perp \xi, \eta) + \bar{g}(\xi, \nabla_X^{*\perp} \eta).$$

The mean curvature vector fields are defined as

$$H = \frac{1}{n} \sum_{i=1}^{n} \sigma(e_i, e_i) = \frac{1}{n} \sum_{l=1}^{m} \left(\sum_{i=1}^{n} \sigma_{ii}^l \right) \xi_l \tag{9}$$

$$and \quad H^* = \frac{1}{n} \sum_{i=1}^{n} \sigma^*(e_i, e_i) = \frac{1}{n} \sum_{l=1}^{m} \left(\sum_{i=1}^{n} \sigma_{ii}^{*l} \right) \xi_l \tag{10}$$

for $1 \leq i, j \leq n$ and $1 \leq \alpha \leq m$.

Proposition 1. *[6] Let (M, g, ∇, ∇^*) be statistical submanifold of $(\bar{M}, \bar{g}, \bar{\nabla}, \bar{\nabla}^*)$. Let $\bar{R}$ and $\bar{R}^*$ be the Riemannian curvature tensors on $\bar{M}$ for $\bar{\nabla}$ and $\bar{\nabla}^*$ respectively, then the Gauss, Codazzi and Ricci equations are given by the following result.*

$$\bar{g}(\bar{R}(X, Y)Z, W) = g(R(X, Y)Z, W) + \bar{g}(\sigma(X, Z), \sigma^*(Y, W))$$
$$- \bar{g}(\sigma^*(X, W), \sigma(Y, Z))$$
$$\bar{g}(\bar{R}^*(X, Y)Z, W) = g(R^*(X, Y)Z, W) + \bar{g}(\sigma^*(X, Z), \sigma(Y, W))$$
$$- \bar{g}(\sigma(X, W), \sigma^*(Y, Z))$$
$$\bar{g}(R^\perp(X, Y)\xi, \eta) = \bar{g}(\bar{R}(X, Y)\xi, \eta) + g([A^*_\xi, A_\eta]X, Y)$$
$$\bar{g}(R^* \perp (X, Y)\xi, \eta) = \bar{g}(\bar{R}^*(X, Y)\xi, \eta) + g([A_\xi, A^*_\eta]X, Y)$$
$$(\bar{R}(X, Y)Z)^\perp = (\bar{\nabla}_X \sigma)(Y, Z) - (\bar{\nabla}_Y \sigma)(X, Z)$$
$$(\bar{R}^*(X, Y)Z)^\perp = (\bar{\nabla}_X^* \sigma^*)(Y, Z) - (\bar{\nabla}_Y^* \sigma^*)(X, Z)$$

where $[A_\xi, A^*_\eta] = A_\xi A^*_\eta - A^*_\eta A_\xi$ and $[A^*_\xi, A_\eta] = A^*_\xi A_\eta - A_\eta A^*_\xi$, for $X, Y, Z, W \in TM$ and $\xi, \eta \in T^\perp M$.

3   Main Result

In the present section, we prove the main result of the paper.

Theorem 2. *Let N be the Legendrian submanifold of Sasakian statistical manifold $\bar{N}$ of constant ϕ-sectional curvature c. Then*

$$\rho^{\perp \bar{\nabla}, \bar{\nabla}^*} \leq 6\left(\|H\|^2 + \|H^*\|^2 + 16\|H^\circ\|^2 \right)$$
$$- 12\rho - 9 + 3c + 120(\bar{\rho}^\circ - \rho^\circ)$$

Proof. Suppose $\{e_1, e_2, \ldots, e_n\}$ and $\{e_{n+1} = \phi e_1, e_{n+2} = \phi e_2, \ldots, e_{2n} = \phi e_n, e_{2n+1} = \xi\}$ be the orthonormal frame on N and $T^\perp N$ respectively. Using Gauss equation, we have

$$2\tau = 2 \sum_{1 \leq i < j \leq n} g(S(e_i, e_j)e_j, e_i)$$

$$= \sum_{1 \leq i < j \leq n} \{\bar{g}((R(e_i, e_j)e_j, e_i) + \bar{g}((R^*(e_i, e_j)e_j, e_i))\}$$

$$= \sum_{1 \leq i < j \leq n} \{\frac{c+3}{2} + \bar{g}(\sigma(e_i, e_i), \sigma^*(e_j, e_j)) + \bar{g}(\sigma^*(e_i, e_i), \sigma(e_j, e_j))$$
$$- 2\bar{g}(\sigma^*(e_i, e_j), \sigma(e_i, e_j)) \tag{11}$$

Using the definition of normalized scalar curvature and (11), we obtain

$$\rho^{\bar{\nabla}, \bar{\nabla}^*} = \frac{3-c}{2} + \frac{1}{n(n-1)} \sum_{r=1}^{n} \sum_{1 \leq i < j \leq n} \left(2\sigma_{ij}r\sigma_{ij}^{*r} - \sigma_{ii}*r\sigma_{jj}^r - \sigma_{ii}r\sigma_{jj}^{*r} \right) \tag{12}$$

By the definition of normalize scalar curvature, we have

$$\rho^{\perp \bar{\nabla}, \bar{\nabla}^*} = \frac{1}{n(n-1)} \left(\sum_{1 \leq s < t \leq n} \sum_{1 \leq i < j \leq n} \left[g(R^\perp(e_i, e_j)e_r, e_s) + g(R^* \perp (e_i, e_j)e_r, e_s) \right]^2 \right)^{\frac{1}{2}}$$

which implies that

$$\rho^{\perp \bar{\nabla}, \bar{\nabla}^*} = \frac{1}{n(n-1)} \left(\sum_{1 \leq s < t \leq n} \sum_{1 \leq i < j \leq n} \left(\sigma_{ik}^s \sigma_{jk}^{*r} - \sigma_{ik}^{*r} \sigma_{jk}^s + \sigma_{ik}^{*s} \sigma_{jk}^r - \sigma_{ik}^r \sigma_{jk}^{*s} \right]^2 \right)^{\frac{1}{2}}$$

Using the relation $2\sigma_{ik}^{\circ r} = \sigma_{ik}^{\circ r} + \sigma_{ik}^{*r}$ for $1 \leq i < j \leq n$ and $1 \leq r \leq n$ for the submanifold w.r.t. $\bar{\nabla}^\circ$, we have

$$\rho^{\perp \bar{\nabla}, \bar{\nabla}^*} = \frac{1}{n(n-1)} \left[\sum_{1 \leq s < t \leq n} \sum_{1 \leq i < j \leq n} \left(\sum_{k=1}^{n} (4(\sigma_{ik}^{\circ s} \sigma_{jk}^{\circ r} - \sigma_{ik}^{\circ r} \sigma_{jk}^{\circ s}) + (\sigma_{ik}^r \sigma_{jk}^s \right. \right.$$

$$\left. \left. - \sigma_{ik}^s \sigma_{jk}^r) + (\sigma_{ik}^{*r} \sigma_{jk}^{*s} - \sigma_{ik}^{*s} \sigma_{jk}^{*r})) \right)^2 \right]^{\frac{1}{2}}$$

From the inequality $(x + y + z)^2 \leq 3(x^2 + y^2 + z^2)$ for $x, y, z \in \mathbf{R}$ and the equation (18) of [5], we find

$$\rho^{\perp \bar{\nabla}, \bar{\nabla}^*} \leq \frac{3}{2n^2(n-1)} \sum_{r=1}^{m} \sum_{1 \leq i < j \leq n} \left[(\sigma_{ii}^r - \sigma_{jj}^r)^2 + (\sigma_{ii}^{*r} - \sigma_{jj}^{*r})^2 + 16(\sigma_{ii}^{\circ r} - \sigma_{jj}^{\circ r})^2 \right]$$

$$+ \frac{3}{n(n-1)} \sum_{r=1}^{m} \sum_{1 \leq i < j \leq n} \left[(\sigma_{ij}^r)^2 + (\sigma_{ij}^{*r})^2 + 16(\sigma_{ij}^{\circ r})^2 \right]$$

Using simple calculations for mean curvature vector in view of statistical structures, we arrive at

$$\rho^{\perp \bar{\nabla}, \bar{\nabla}^*} \leq \frac{3}{2} \left(\|H\|^2 + \|H^*\|^2 + 16\|H^\circ\|^2 \right) - \frac{3}{n(n-1)} \sum_{r=1}^{m} \sum_{1 \leq i < j \leq n} \left[20\sigma_{ii}^{\circ r} \sigma_{jj}^{\circ r} \right.$$

$$\left. - \sigma_{ii}^{*r} \sigma_{jj}^{\circ r} - \sigma_{ii}^r \sigma_{jj}^{*r} - 20(\sigma_{ij}^{0r})^2 + 2\sigma_{ij}^r \sigma_{ij}^{*r} \right] \tag{13}$$

By Gauss equation for $\bar{\nabla}^\circ$ and substituting the value of (12) in (13), we get

$$\rho^{\perp \bar{\nabla}, \bar{\nabla}^*} \leq \frac{3}{2} \left(\|H\|^2 + \|H^*\|^2 + 16\|H^\circ\|^2 \right)$$

$$- 3\rho^{\bar{\nabla}, \bar{\nabla}^*} - 3\left(\frac{3-c}{4} \right) + 30(\bar{\rho}^\circ - \rho^\circ)$$

which implies that

$$\rho^{\perp \bar{\nabla}, \bar{\nabla}^*} \leq 6 \left(\|H\|^2 + \|H^*\|^2 + 16\|H^\circ\|^2 \right)$$

$$- 12\rho^{\bar{\nabla}, \bar{\nabla}^*} - 9 + 3c + 120(\bar{\rho}^\circ - \rho^\circ)$$

References

1. Amari, S.: Differential-Geometrical Methods in Statistics. Lecture Notes in Statistics, vol. 28. Springer, Berlin (1985). https://doi.org/10.1007/978-1-4612-5056-2
2. De Smet, P.J., Dillen, F., Verstraelen, L., Vranken, L.: A pointwise inequality in submanifold theory. Arch. Math. (Brno) **35**, 221–235 (1990)
3. Furuhata, H., Hasegawa, I., Okuyama, Y., Sato, K., Shahid, M.: Sasakian statistical manifolds. J. Geom. Phys. **117**, 179–186 (2017)
4. Ge, J., Tang, Z.: A proof of the DDVV conjecture and its equality case. Pacific J. Math. **237**, 87–95 (2008)
5. Lu, Z.: Normal scalar curvature conjecture and its applications. J. Funct. Anal. **261**, 1284–1308 (2011)
6. Vos, P.W.: Fumdamental equations for statistical submanifolds with applications to the Bartlett correction. Ann. Inst. Statist. Math. **41**(3), 429–450 (1989)
7. Wintgen, P.: Sur l'inégalité de Chen-Willmore. C. R. Acad. Sci. Paris Sér. A-B **288**, A993–A995 (1979)

Logarithmic Divergences: Geometry and Interpretation of Curvature

Ting-Kam Leonard Wong[1(✉)] and Jiaowen Yang[2]

[1] Department of Statistical Sciences, University of Toronto, Toronto, Canada
tkl.wong@utoronto.ca
[2] Department of Mathematics, University of Southern California, Los Angeles, USA
jiaoweny@usc.edu

Abstract. We study the logarithmic $L^{(\alpha)}$-divergence which extrapolates the Bregman divergence and corresponds to solutions to novel optimal transport problems. We show that this logarithmic divergence is equivalent to a conformal transformation of the Bregman divergence, and, via an explicit affine immersion, is equivalent to Kurose's geometric divergence. In particular, the $L^{(\alpha)}$-divergence is a canonical divergence of a statistical manifold with constant sectional curvature $-\alpha$. For such a manifold, we give a geometric interpretation of its sectional curvature in terms of how the divergence between a pair of primal and dual geodesics differ from the dually flat case. Further results can be found in our follow-up paper [27] which uncovers a novel relation between optimal transport and information geometry.

Keywords: Logarithmic divergence · Bregman divergence · Conformal divergence · Affine immersion · Constant sectional curvature · Optimal transport

1 Introduction

Let $\Omega \subset \mathbf{R}^n$ be an open convex set, $n \geq 2$. For $\alpha > 0$ fixed, we say that a function $\varphi : \Omega \to \mathbf{R}$ is α-exponentially concave if $e^{\alpha\varphi}$ is concave on Ω. All functions in this paper are assumed to be smooth. Given such a function φ, we define its $L^{(\alpha)}$-divergence by

$$\mathbf{L}_\varphi^{(\alpha)}[\xi : \xi'] := \frac{1}{\alpha}\log(1 + \alpha D\varphi(\xi') \cdot (\xi - \xi')) - (\varphi(\xi) - \varphi(\xi')), \quad \xi, \xi' \in \Omega, \quad (1)$$

where $D\varphi$ is the Euclidean gradient and $\cdot$ is the dot product. We always assume the Hessian $D^2 e^{\alpha\varphi}$ is strictly negative definite on Ω. Then $\mathbf{L}_\varphi^{(\alpha)}$ is a divergence on Ω, regarded as a manifold, in the sense of [1, Definition 1.1]. As $\alpha \downarrow 0$, the $L^{(\alpha)}$-divergence (with φ fixed) converges to the Bregman divergence defined by

$$\mathbf{B}_\phi[\xi : \xi'] := (\phi(\xi) - \phi(\xi')) - D\phi(\xi') \cdot (\xi - \xi'), \quad (2)$$

© Springer Nature Switzerland AG 2019
F. Nielsen and F. Barbaresco (Eds.): GSI 2019, LNCS 11712, pp. 413–422, 2019.
https://doi.org/10.1007/978-3-030-26980-7_43

where $\phi = -\varphi$ is convex with $D^2\phi > 0$. Thus the family $\{\mathbf{L}_\varphi^{(\alpha')}\}_{0<\alpha'\leq\alpha}$ of logarithmic divergences extrapolate the Bregman divergence $\mathbf{B}_\phi$.

Originally motivated by applications in stochastic portfolio theory [7], the $L^{(1)}$-divergence (and its extension to the $L^{(\alpha)}$-divergence) was introduced by Pal and the first author in [19,26] and was studied further in [20,24,25]. There are two main results proved in these papers. First, the $L^{(\alpha)}$-divergence corresponds to the solution to an optimal transport problem with a logarithmic cost function; this is formulated using the general framework of c-divergence, see [20,25,27]. Also see [9,18,21] for recent results about the optimal transport problem which have independent mathematical interest. Second, the induced statistical manifold $(\mathcal{M}, g, \nabla, \nabla^*)$ (see [1, Section 6.2] for the definition) is dually projectively flat with constant sectional curvature $-\alpha$. In [26] we also defined an $L^{(-\alpha)}$-divergence corresponding to constant positive sectional curvature α. For expositional simplicity we only consider the $L^{(\alpha)}$-divergence in this paper and [27], but similar results hold for the $L^{(-\alpha)}$-case as well.

In this paper we develop two geometric aspects of the logarithmic divergence. First, we connect the $L^{(\alpha)}$-divergence with classical topics in information geometry, namely conformal transformation and affine differential geometry. In particular, by using an explicit affine immersion, we show that the $L^{(\alpha)}$-divergence is equivalent to the canonical geometric divergence constructed by Kurose [11]. Second, we provide a geometric interpretation of the sectional curvature for a statistical manifold with constant negative sectional curvature. By analyzing a canonical divergence between a pair of primal and dual geodesics, we show that the sectional curvature can be quantified in terms of the deviation from the generalized Pythagorean relation of a dually flat manifold (see Theorem 4 below). This extends the geometric interpretation of sectional curvature in Riemannian geometry. In our follow-up work [27] we proved a more general result (see [27, Theorem 3.13]) that holds for any divergence (though it is not intrinsic in the information geometric sense). This was achieved by a novel relation between information geometry and the pseudo-Riemannian framework of Kim and McCann [10] concerning the Ma-Trudinger-Wang tensor in optimal transport.

2   Conformal Divergence and Its Geometry

We refer the reader to [1] for general background in information geometry. Conformal transformations of divergence have been studied in the literature; see for example [2,12,15,17] and the references therein. An important application is robust clustering [14,23].

Definition 1. *Let $\phi : \Omega \to \mathbb{R}$ be convex (with $D^2\phi > 0$) and let $\kappa : \Omega \to (0,\infty)$. We define the (left-sided) conformal transformation of the Bregman divergence $\mathbf{B}_\phi$ by*

$$\mathbf{D}_{\phi,\kappa}[\xi : \xi'] := \kappa(\xi)\mathbf{B}_\phi[\xi : \xi']. \tag{3}$$

To abbreviate we call $\mathbf{D}_{\phi,\kappa}$ a conformal divergence.

Note that a right-sided conformal transformation can be converted to a left-sided one by considering the convex conjugate of ϕ (see [1, p. 17]).

Our first result is that the $L^{(\alpha)}$-divergence is, up to a monotone transformation, equal to a conformal transformation of a Bregman divergence. This shows that the geometry induced by the $L^{(\alpha)}$-divergence can be studied using results of Bregman divergence and conformal transformation.

Theorem 1. *Consider an $L^{(\alpha)}$-divergence $\mathbf{L}^{(\alpha)}_\varphi$ on Ω as in (1). Let $\phi = -e^{\alpha\varphi}$ which is convex and let $\kappa = -\frac{1}{\alpha\phi} > 0$. Then, with $T(x) = \frac{1}{\alpha}(e^{\alpha x} - 1)$, we have*

$$T(\mathbf{L}^{(\alpha)}_\varphi) \equiv \mathbf{D}_{\phi,\kappa}. \tag{4}$$

In particular, the conformal divergence $\mathbf{D}_{\phi,h}$ induces the same dualistic structure (g, ∇, ∇^*) as that of $\mathbf{D}^{(\alpha)}_\varphi$.

Proof. The identity (4), once conceived, can be verified by a straightforward computation. The second statement is a consequence of the following lemma which can be proved again by a computation. Note that similar reasonings are used in [25, Lemma 3] and [25, Theorem 17]. □

Lemma 1. *Let $\tilde{\mathbf{D}}$ and $\mathbf{D}$ be divergences related by a monotone transformation: $\tilde{\mathbf{D}} = T(\mathbf{D})$, where $T : [0, \infty) \to [0, \infty)$ is strictly increasing with $T(0) = 0$. Let (g, ∇, ∇^*) and $(\tilde{g}, \tilde{\nabla}, \tilde{\nabla}^*)$ be respectively the dualistic structures induced by $\mathbf{D}$ and $\tilde{\mathbf{D}}$. Then, in any local coordinate system, the coefficients of the dualistic structures are related by*

$$\tilde{g}_{ij} = T'(0)g_{ij}, \quad \tilde{\Gamma}_{ijk} = T'(0)\Gamma_{ijk}, \quad \tilde{\Gamma}^*_{ijk} = T'(0)\Gamma^*_{ijk}. \tag{5}$$

In particular, we have $\tilde{\Gamma}_{ij}{}^k = \Gamma_{ij}{}^k$ and $\tilde{\Gamma}^*_{ij}{}^k = \Gamma_{ij}{}^k$, and the primal and dual curvature tensors are the same.

Remark 1. By Lemma 1, we say that two divergences $\mathbf{D}$ and $\tilde{\mathbf{D}}$ are equivalent if there exists T (as in Lemma 1 with $T'(0) = 1$) such that $\tilde{\mathbf{D}} = T(\mathbf{D})$. Clearly this defines an equivalence relation among divergences on a manifold. Theorem 1 thus states that the $L^{(\alpha)}$-divergence is equivalent to a conformal divergence.

Theorem 1 motivates us to study conformal divergences in general. Recall that two torsion-free affine connections ∇ and $\tilde{\nabla}$ are projectively equivalent if there exists a 1-form τ such that

$$\nabla_X Y = \tilde{\nabla}_Y X + \tau(X)Y + \tau(Y)X$$

for any vector fields X and Y. For its geometric interpretation see [16, p. 17]. In particular, ∇ and $\tilde{\nabla}$ have the same geodesics up to time reparameterizations. By definition, ∇ is projectively flat if it is projectively equivalent to a flat connection. When considering the $L^{(\alpha)}$-divergence or a conformal divergence, we think of $\mathcal{M}$ (equal to Ω as a set) as a manifold, and ξ is the primal (global) coordinate system with values in the convex set Ω.

Proposition 1. *Let $(\mathcal{M}, g, \nabla, \nabla^*)$ be the statistical manifold induced by a conformal divergence $\mathbf{D}_{\phi,\kappa}$.*

(i) The primal connection ∇ is projectively flat and the primal geodesics are, up to time reparameterization, straightlines in the ξ-coordinate system. (In fact, using the language of [3, Section 8.4], ∇ is (-1)-conformally flat and ∇^* is 1-conformally flat.)

(ii) ∇ has constant sectional curvature $\lambda \in \mathbb{R}$ with respect to g if and only if

$$\frac{1}{\kappa(\xi)} \equiv \lambda \phi(\xi) + a + \sum_{i=1}^{n} b_i \xi^i \tag{6}$$

for some real constants a and b^i. In this case, the dual sectional curvature is also constant and is equal to λ.

Remark 2. Note that if (6) holds then one may absorb the linear terms in the definition of ϕ. On the other hand, we observe that if $\lambda < 0$ then $\lambda \phi$ is concave. Since on $\mathbb{R}^d$ there are no non-trivial positive concave functions, from (6) we see that if the sectional curvature is constant and negative, the domain Ω must be a proper subset of $\mathbb{R}^d$.

Proof (of Proposition 1). Consider the dualistic structure (g, ∇, ∇^*) induced by the conformal divergence. Consider the Euclidean coordinate ξ on Ω. By a direct computation, the coefficients of g and ∇ are given by

$$g_{ij}(\xi) = \kappa(\xi) \partial_i \partial_j \phi(\xi),$$
$$\Gamma_{ij}{}^k(\xi) = \frac{\partial_i \kappa(\xi)}{\kappa(\xi)} \delta_j^k + \frac{\partial_j \kappa(\xi)}{\kappa(\xi)} \delta_i^k. \tag{7}$$

Since $\kappa > 0$, the 1-form $\tau = d \log \kappa$ is well-defined. From (7), we have that $\nabla_X Y = \tilde{\nabla}_Y X + \tau(X)Y + \tau(Y)X$ we $\tilde{\nabla}$ is the Euclidean flat connection on Ω. Thus ∇ is projectively flat and we have (i). A further computation shows that

$$R_{ijk}{}^\ell(\xi) = \kappa(\xi) \left(\partial_{jk} \frac{1}{\kappa}(\xi) \delta_i^\ell - \partial_{ik} \frac{1}{\kappa}(\xi) \delta_j^\ell \right). \tag{8}$$

Using (8), we see that ∇ has constant sectional curvature $\lambda \in \mathbb{R}$ with respect to g (see [25, Definition 12]) if and only if

$$\kappa(\xi) \partial_{jk} \frac{1}{\kappa}(\xi) = \kappa(\xi) g_{jk}(\xi) = \lambda \kappa(\xi) \partial_{jk} \phi(\xi),$$

which is equivalent to (6) after integration.   □

3   Realization by Affine Immersion

Consider a statistical manifold $(\mathcal{M}, g, \nabla, \nabla^*)$. In [25, Theorem 18] we proved that if both ∇ and ∇^* are dually projectively flat with constant sectional curvature

$-\alpha < 0$, then one can define intrinsically a local divergence of $L^{(\alpha)}$-type which induces the given geometric structure. In this result, a key idea is that the primal and dual coordinates are related by an optimal transport map (this leads to the self-dual representation given by (21) below). In fact, by [3, Theorem 8.3], if a statistical manifold has constant sectional curvature, then we automatically have dual projective flatness. Thus the condition about projective flatness is redundant and we may modify the statement as follows:

Theorem 2. *[25, Theorem 18] The $L^{(\alpha)}$-divergence is a (local) intrinsic divergence for a statistical manifold with constant negative sectional curvature.*

On the other hand, for a (simply connected) statistical manifold with constant sectional curvature, Kurose [11] defined globally a canonical, intrinsic divergence using affine differential geometry and proved that it satisfies a generalized Pythagorean theorem. In this section we show that if $(\Omega, g, \nabla, \nabla^*)$ is induced by an $L^{(\alpha)}$-divergence $\mathbf{L}_\varphi^{(\alpha)}$, then the geometric divergence is the conformal divergence $\mathbf{D}_{\phi,\kappa}$ in (4). While these canonical divergences are equivalent, our approach in [20,25] gives an explicit construction in Kurose's work, covers the Bregman and $L^{(\alpha)}$-divergences under the same framework, and suggests previously unknown connections with optimal transport maps.

To state the main result we recall some concepts of affine differential geometry; for details see [16] and [13]. Let $\mathcal{M}$ be an n-dimensional manifold. An affine immersion of M into $\mathbb{R}^{n+1}$ consists of an immersion $f : M \to \mathbb{R}^{n+1}$ and a transversal vector field $\mathbf{n}$ with values in $\mathbb{R}^{n+1}$ on $M \cong f(M)$. The last statement means that

$$T_{f(p)}\mathbb{R}^{n+1} = f_*(T_p M) \oplus \mathrm{span}(\mathbf{n}(p))$$

for all $p \in M$. Let $\overline{\nabla}$ be the standard (flat) affine connection on $\mathbb{R}^{n+1}$. Then the covariant derivative decomposes as

$$\overline{\nabla}_X f_* Y = f_*(\nabla_X Y) + g(X, Y)\mathbf{n}. \tag{9}$$

We call ∇ and g the induced connection and bilinear form respectively. If the induced connection and bilinear form are equal to the Riemannian metric and primal connection of a dualistic structure (g, ∇, ∇^*), we say that the affine immersion realizes the given structure. By [13, Theorem 5.3], this is possible when the statistical manifold is simply connected and 1-conformally flat. This is true in particular when the statistical manifold has constant sectional curvature.

Let $(\mathbb{R}^{n+1})^*$ be the dual space of $\mathbb{R}^n$, and let $\langle ., . \rangle$ be the dual pairing. Given an affine immersion $(f, \mathbf{n})$, the conormal vector field $\mathbf{n}^* : M \to (\mathbb{R}^{n+1})^*$ is defined by the conditions

$$\langle \mathbf{n}^*(p), \mathbf{n}(p) \rangle = 1, \quad \langle \mathbf{n}^*(p), f_* X \rangle = 0 \quad \forall X \in T_p M. \tag{10}$$

Definition 2 (Kurose's geometric divergence). *For an affine immersion $(f, \mathbf{n})$ with conormal field $\mathbf{n}^*$, the geometric divergence is defined by*

$$\rho(p, q) := \langle f(p) - f(q), \mathbf{n}^*(q) \rangle, \quad p, q \in M. \tag{11}$$

In [11] it was shown that if (g, ∇) is 1-conformally flat, then the geometric divergence does not depend on the choice of the immersion and recovers the given dualistic structure. (The dual connection ∇^* is uniquely determined given g and ∇.) Hence, it can be viewed as a *canonical divergence* (see the next section for more discussion).

The following result connects the $L^{(\alpha)}$-divergence with the geometric divergence. It shows that the geometric divergence, the $L^{(\alpha)}$-divergence and the conformal divergence are all equivalent. In particular, they are all intrinsically defined (at least locally) for the given dualistic structure.

Theorem 3. *Consider a convex domain $\Omega \subset \mathbb{R}^n$ equipped with an $L^{(\alpha)}$-divergence $\mathbf{L}_\varphi^{(\alpha)}$ and its induced geometry (g, ∇, ∇^*). Let $\phi = -e^{\alpha\varphi}$ and $\kappa = -\frac{1}{\alpha\phi}$ as in Theorem 1. Consider the affine immersion defined by*

$$f(\xi) = \kappa(\xi)(\xi^1, \xi^2, ..., \xi^n, 1),$$
$$\mathbf{n}(\xi) = \alpha f(\xi), \tag{12}$$

where ξ is the Euclidean coordinate system on Ω. Then this affine immersion realizes (g, ∇). Moreover, the geometric divergence is given by

$$\rho(\xi, \xi') = \mathbf{D}_{\phi,\kappa}(\xi, \xi') = \frac{1}{\alpha}\left(e^{\mathbf{L}_\varphi^{(\alpha)}[\xi:\xi']} - 1\right). \tag{13}$$

Proof. The choice of our immersion (12) is motivated by the proof of [16, Proposition 2.7]. It is easy to see that f is an immersion and $\mathbf{n}$ is transversal. Let $\tilde{\mathbf{e}}_j := \frac{\partial}{\partial\xi^j} f$ and $\partial_k\tilde{\mathbf{e}}_j := \frac{\partial}{\partial\xi^k}\frac{\partial}{\partial\xi^j} f$. Then, it can be verified by a straightforward computation that

$$\partial_k\tilde{\mathbf{e}}_j \equiv \Gamma_{kj}{}^m\tilde{\mathbf{e}}_m + g_{ij}(\alpha f). \tag{14}$$

We refer the reader to [25, Section 5] for expressions of the coefficients $\Gamma_{ij}{}^k$. Thus the affine immersion $(f, \mathbf{n})$ realizes the given dualistic structure.

Next we construct the conormal vector field. Using the relations in (10), we can show that the conormal field is given by

$$\mathbf{n}^*(p_\xi) = (-\partial_1\phi(\xi), \ldots, -\partial_n\phi(\xi), -\alpha\phi(\xi) + D\phi(\xi) \cdot \xi). \tag{15}$$

We obtain (13) by plugging (15) into (13). $\qquad\qquad\qquad\qquad\qquad\qquad\square$

4   Interpretation of Sectional Curvature

Consider a statistical manifold $(\mathcal{M}, g, \nabla, \nabla^*)$. Given $q \in \mathcal{M}$ and $\mathbf{v}, \mathbf{w} \in T_q\mathcal{M}$ which are linearly independent, we can define the primal sectional curvature $\text{sec}(\mathbf{v}, \mathbf{w})$ by

$$\text{sec}(\mathbf{v}, \mathbf{w}) := \frac{\langle R(\mathbf{w}, \mathbf{v})\mathbf{v}, \mathbf{w}\rangle}{\|\mathbf{v}\|^2\|\mathbf{w}\|^2 - \langle\mathbf{v}, \mathbf{w}\rangle^2}, \tag{16}$$

where $\langle., .\rangle$ is the Riemannian inner product and R is the primal curvature tensor. Similarly, we can define the dual sectional curvature sec^*. What are the geometric interpretations of these sectional curvatures? Interestingly, to the best of our knowledge, this natural question has not been satisfactorily answered in the literature.

For motivations, let us consider a Riemannian manifold $(\mathcal{M}, g)$. In this case, it is well-known that the sectional curvature (given by (16) using the Levi-Civita connection) can be interpreted in terms of the Riemannian distance, defined by

$$d(x, y) := \inf_{\gamma:\gamma(0)=x,\gamma(1)=y} \int_0^1 \|\dot\gamma(t)\| dt, \tag{17}$$

between a pair of geodesics. For $t_1, t_2 > 0$ small, let $r(t_1) = \exp_q(t_1 \mathbf{v})$ and $p(t_2) = \exp_q(t_2 \mathbf{w})$ be geodesics starting at q, where $\exp_q$ is the exponential map. Then, we have

$$\begin{aligned} & d^2(r(t_1), p(t_2)) \\ & = \|\mathbf{v}\|^2 t_1^2 + \|\mathbf{w}\|^2 t_2^2 - 2\langle \mathbf{v}, \mathbf{w}\rangle t_1 t_2 - \frac{1}{3}\langle R(\mathbf{w}, \mathbf{v})\mathbf{v}, \mathbf{w}\rangle t_1^2 t_2^2 + \cdots, \end{aligned} \tag{18}$$

where the higher order terms are omitted (see [22]). This implies that

$$\begin{aligned} & d^2(r(t_1), p(t_2)) - d^2(r(t_1), q) - d^2(q, p(t_2)) \\ & = -2\langle \mathbf{v}, \mathbf{w}\rangle t_1 t_2 - \frac{1}{3}(\|\mathbf{v}\|^2\|\mathbf{w}\|^2 - \langle \mathbf{v}, \mathbf{w}\rangle^2)\sec(\mathbf{v}, \mathbf{w}) t_1^2 t_2^2 + \cdots. \end{aligned} \tag{19}$$

We look for analogous geometric interpretations for a statistical manifold. Given a statistical manifold $(\mathcal{M}, g, \nabla, \nabla^*)$, in order to formulate a statement in the form of (18) or (19), we need to have an intrinsically defined divergence corresponding to the given geometry. This is the problem about constructing a canonical divergence and was studied by several papers including [4–6,8].

Using the $L^{(\alpha)}$-divergence which is explicit, intrinsically defined and has special properties, in this section we study the geometric interpretation for a statistical manifold with constant sectional curvature $-\alpha \le 0$. Let $q \in \mathcal{M}$ and $\mathbf{v}, \mathbf{w} \in T_q\mathcal{M}$. Motivated by the generalized Pythagorean theorem which holds for the Bregman and $L^{(\alpha)}$-divergences, let

$$r(t_1) = \exp_q(t_1 \mathbf{v}) \text{ and } p(t_2) = \exp_q^*(t_2 \mathbf{w}),$$

where $\exp_q$ and $\exp_q^*$ are respectively the exponential maps corresponding respectively to the primal and dual connections ∇ and ∇^*. With $\mathbf{D}$ being an intrinsic local $L^{(\alpha)}$-divergence (see Theorem 2), consider the expression H defined by

$$H(t_1, t_2) := \mathbf{D}[r(t_1) : p(t_2)] - \mathbf{D}[r(t_1) : q] - \mathbf{D}[q : p(t_2)]. \tag{20}$$

By the generalized Pythagorean theorem proved in [20, Theorem 1.2] and [25, Theorem 16], if $\langle \mathbf{v}, \mathbf{w}\rangle = 0$ then $H(t_1, t_2) \equiv 0$. This motivates the definition of H and the comparison with (19). Note that if $\alpha = 0$ then the manifold is dually flat. In this case, there is a canonical divergence $\mathbf{D}$ of Bregman type.

With the Bregman divergence and with H defined by (20), we have the identity $H(t_1, t_2) \equiv -\langle v, w \rangle t_1 t_2$.

Now let $\alpha > 0$ and let $\mathbf{D}$ be the canonical (local) $L^{(\alpha)}$-divergence. By [25, Theorem 18], there exists a local coordinate system ξ and an α-exponentially concave function $\varphi = \varphi(\xi)$ such that $\mathbf{D}[y : x] = \mathbf{D}_\varphi^{(\alpha)}[\xi_y : \xi_x]$. Here ξ_x is the primal coordinate of $x \in \mathcal{M}$. Moreover, letting

$$\eta = \frac{D\varphi(\xi)}{1 - \alpha D\varphi(\xi) \cdot \xi}, \quad \psi(\eta) = \frac{1}{\alpha} \log(1 + \alpha \xi \cdot \eta) - \varphi(\xi),$$

be respectively the dual coordinate and α-conjugate of φ, we have $\mathbf{D}[y : x] = \mathbf{D}_\psi^{(\alpha)}[\eta_x : \eta_y]$ and the self-dual representation

$$\mathbf{D}[y : x] = \frac{1}{\alpha} \log(1 + \alpha \xi_y \cdot \eta_x) - \varphi(\xi_y) - \psi(\eta_x). \tag{21}$$

As $\alpha \downarrow 0$, these identities reduce to well-known properties of the Bregman divergence [1, Chapter 1]. By analyzing carefully the primal and dual geodesics as well as the self-dual representation (21), we have the following result.

Theorem 4. For $t_1, t_2 > 0$ small, we have

$$H(t_1, t_2) = -\langle \mathbf{v}, \mathbf{w} \rangle t_1 t_2 - \alpha \langle \mathbf{v}, \mathbf{w} \rangle \left[\frac{\|\mathbf{v}\|^2}{3} t_1^3 t_2 + \frac{\|\mathbf{w}\|^2}{3} t_1 t_2^3 + \frac{\langle \mathbf{v}, \mathbf{w} \rangle}{2} t_1^2 t_2^2 \right] \tag{22}$$

$$+ \text{ higher order terms.}$$

Proof. By [25, Corollary 2], the primal/dual geodesics of $L^{(\alpha)}$-divergence are straight lines in the primal/dual coordinate systems, up to time changes. Thus we can write $\xi_r(t_1) = \xi_q + s_1(t_1)v$ and $\eta_p(t_2) = \eta_q + s_2(t_2)w$, where v and w are the coordinate representations of $\mathbf{v}$ and $\mathbf{w}$, and s_1 and s_2 are time changes. For notational simplicity we suppress the parameters t_1 and t_2. Using [25, (89)], we have

$$\langle \mathbf{v}, \mathbf{w} \rangle = \left(\frac{v \cdot w}{1 + \alpha(\xi_q \cdot \eta_q)} - \frac{\alpha}{(1 + \alpha(\xi_q \cdot \eta_q))^2} (\eta_q \cdot v)(\xi_q \cdot w) \right). \tag{23}$$

Differentiating (21) and using (23), we expand $H(t_1, t_2)$ in terms of s_1 and s_2:

$$H(t_1, t_2) = -\langle \mathbf{v}, \mathbf{w} \rangle s_1 s_2 + \frac{\alpha \langle \mathbf{v}, \mathbf{w} \rangle}{1 + \alpha(\xi_q \cdot \eta_q)} \cdot \left((\eta_q \cdot v) s_1^2 s_2 + (\xi_q \cdot w) s_1 s_2^2 \right)$$

$$+ (C_3 - \alpha C_1 C_2) \alpha^2 (C_1^2 + C_2^2 + C_1 C_2) - \frac{\alpha}{2} (C_3 - \alpha C_1 C_2)^2 \tag{24}$$

$$+ \text{ higher order terms,}$$

where $C_1 = \frac{\eta_q \cdot v}{1 + \alpha(\xi_q \cdot \eta_q)} s_1$, $C_2 = \frac{\xi_q \cdot w}{1 + \alpha(\xi_q \cdot \eta_q)} s_2$, and $C_3 = \frac{v \cdot w}{1 + \alpha(\xi_q \cdot \eta_q)} s_1 s_2$.

On the other hand, the geodesic equations (see [25, (86)]) give us, after some simplifications, Taylor expansions of s_1 and s_2:

$$s_1(t_1) = t_1 + \alpha(D\varphi(q) \cdot v) t_1^2 + T_1 t_1^3 + O(t_1^4), \tag{25}$$

$$s_2(t_2) = t_2 + \alpha(D\psi(q) \cdot w)t_2^2 + T_2 t_2^3 + O(t_2^4), \tag{26}$$

where $T_1 = \frac{1}{3}(4(\alpha(D\varphi(q) \cdot v))^2 + \alpha(v^\top D^2\varphi(q)v))$ and $T_2 = \frac{1}{3}(4(\alpha(D\psi(q) \cdot w))^2 + \alpha(w^\top D^2\psi(q)w))$. The proof is completed by combining (24), (25) and (26). $\square$

This result gives a geometric interpretation of the negative sectional curvature $-\alpha$ in terms of the canonical local $L^{(\alpha)}$-divergence $\mathbf{D}$. Note that if we use another intrinsic divergence (such as the conformal divergence) we will get a different expression in (22). Analogous results can be derived for the $L^{(-\alpha)}$-divergence.

Note that Theorem 4 implies that $\frac{\partial^2}{\partial t_1^2 \partial t_2^2}\mathbf{D}[r(t_1) : p(t_2)]\Big|_{t_1=t_2=0} = -2\alpha\langle \mathbf{v}, \mathbf{w}\rangle^2$, so the sectional curvature $-\alpha$ may be interpreted in terms of this fourth order mixed derivative. In [27, Theorem 3.13] we extended this result to any divergence. This is formulated using a novel connection between the information geometry of c-divergence (which covers all divergences) and the pseudo-Riemannian framework of Kim and McCann [10]. In particular, for any divergence $\mathbf{D}$, the mixed derivative $\frac{\partial^2}{\partial t_1^2 \partial t_2^2}\mathbf{D}[r(t_1) : p(t_2)]\Big|_{t_1=t_2=0}$ is equal to -2 times an un-normalized cross curvature of the Kim-McCann metric induced by the cost function. The reader is referred to [27] for more details. To conclude this paper, let us remark that for a statistical manifold with non-constant sectional curvature, this cross sectional curvature is not intrinsic as there are different divergences (and hence Kim-McCann metrics) which induce the same dualistic structure. A natural starting point is to analyze the canonical divergence of Ay and Amari constructed in [4]. We leave this as a problem for future research.

References

1. Amari, S.: Information Geometry and Its Applications. Springer, Tokyo (2016). https://doi.org/10.1007/978-4-431-55978-8
2. Amari, S., Cichocki, A.: Information geometry of divergence functions. Bull. Pol. Acad. Sci. Tech. Sci. **58**(1), 183–195 (2010)
3. Amari, S., Nagaoka, H.: Methods of Information Geometry, vol. 191. American Mathematical Society, Providence (2000)
4. Ay, N., Amari, S.: A novel approach to canonical divergences within information geometry. Entropy **17**(12), 8111–8129 (2015)
5. Felice, D., Ay, N.: Dynamical systems induced by canonical divergence in dually flat manifolds. arXiv preprint arXiv:1812.04461 (2018)
6. Felice, D., Ay, N.: Towards a canonical divergence within information geometry. arXiv preprint arXiv:1806.11363 (2018)
7. Fernholz, E.R.: Stochastic portfolio theory. In: Fernholz, E.R. (ed.) Stochastic Portfolio Theory, pp. 1–24. Springer, New York (2002). https://doi.org/10.1007/978-1-4757-3699-1_1
8. Henmi, M., Kobayashi, R.: Hooke's law in statistical manifolds and divergences. Nagoya Math. J. **159**, 1–24 (2000)
9. Khan, G., Zhang, J.: On the Kähler geometry of certain optimal transport problems. arXiv preprint arXiv:1812.00032v2 (2019)

10. Kim, Y.-H., McCann, R.J.: Continuity, curvature, and the general covariance of optimal transportation. J. Eur. Math. Soc. **12**(4), 1009–1040 (2010)
11. Kurose, T.: On the divergences of 1-conformally flat statistical manifolds. Tohoku Math. J. Second. Ser. **46**(3), 427–433 (1994)
12. Matsuzoe, H.: Geometry of contrast functions and conformal geometry. Hiroshima Math. J. **29**(1), 175–191 (1999)
13. Matsuzoe, H.: Statistical manifolds and affine differential geometry. In: Probabilistic Approach to Geometry, pp. 303–321. Mathematical Society of Japan (2010)
14. Nielsen, F., Nock, R.: Total Jensen divergences: definition, properties and clustering. In: 2015 IEEE International Conference on Acoustics, Speech and Signal Processing (ICASSP), pp. 2016–2020. IEEE (2015)
15. Nock, R., Nielsen, F., Amari, S.: On conformal divergences and their population minimizers. IEEE Trans. Inf. Theory **62**(1), 527–538 (2016)
16. Nomizu, K., Sasaki, T.: Affine Differential Geometry: Geometry of Affine Immersions. Cambridge University Press, Cambridge (1994)
17. Okamoto, I., Amari, S.-I., Takeuchi, K.: Asymptotic theory of sequential estimation: differential geometrical approach. Ann. Stat. **19**(2), 961–981 (1991)
18. Pal, S.: On the difference between entropic cost and the optimal transport cost. arXiv preprint arXiv:1905.12206 (2019)
19. Pal, S., Wong, T.-K.L.: The geometry of relative arbitrage. Math. Financ. Econ. **10**(3), 263–293 (2016)
20. Pal, S., Wong, T.-K.L.: Exponentially concave functions and a new information geometry. Ann. Probab. **46**(2), 1070–1113 (2018)
21. Pal, S., Wong, T.-K.L.: Multiplicative Schröodinger problem and the Dirichlet transport. arXiv preprint arXiv:1807.05649 (2018)
22. Sternberg, S.: Curvature in Mathematics and Physics. Dover, Mineola (2012)
23. Vemuri, B.C., Liu, M., Amari, S., Nielsen, F.: Total Bregman divergence and its applications to DTI analysis. IEEE Trans. Med. Imaging **30**(2), 475–483 (2010)
24. Wong, T.-K.L.: Optimization of relative arbitrage. Ann. Financ. **11**(3–4), 345–382 (2015)
25. Wong, T.-K.L.: Logarithmic divergences from optimal transport and Rényi geometry. Inf. Geom. **1**(1), 39–78 (2018)
26. Wong, T.-K.L.: Information geometry in portfolio theory. In: Nielsen, F. (ed.) Geometric Structures of Information. SCT, pp. 105–136. Springer, Cham (2019). https://doi.org/10.1007/978-3-030-02520-5_6
27. Wong, T.-K.L., Yang, J.: Optimal transport and information geometry. arXiv preprint arXiv:1906.00030 (2019)

Hessian Curvature and Optimal Transport

Gabriel Khan[✉] and Jun Zhang

University of Michigan, Ann Arbor, USA
{gabekhan,junz}@umich.edu

Abstract. We consider the problem of optimal transport where the cost function is given by a $\mathcal{D}_{\Psi}^{(\alpha)}$-divergence for some convex function Ψ [21], where $\alpha = \pm 1$ gives the Bregman divergence. For costs of this form, we introduce a new complex geometric interpretation of the optimal transport problem by considering an induced Sasaki metric on the tangent bundle of the domain of Ψ. In this framework, the Ma-Trudinger-Wang (MTW) tensor [12] is proportional to the orthogonal bisectional curvature. This geometric framework for optimal transport is complementary to the pseudo-Riemannian approach of Kim and McCann [10].

1 Introduction

Optimal transport is a classical field of mathematics which dates back to the work of Monge in 1781 [13]. In its original formulation, it considered finding the most efficient way to move piles of rubble from one configuration to another. In the modern framework, we consider X and Y as Borel subsets of two metric spaces, equipped with probability measures μ and ν, respectively, and a lower-semicontinuous cost function $c : X \times Y \to \mathbb{R}$. The optimal transport problem is to find the non-negative measure γ on $X \times Y$ which minimizes

$$W_c(\mu, \nu) = \min_{\gamma \in \Gamma(\mu,\nu)} \int_{X \times Y} c(x, y) d\gamma(x, y).$$

Here, $\Gamma(\mu, \nu)$ is the set of joint probabilities with the same marginal distributions as $\mu \otimes \nu$ and γ is known as the *optimal coupling*.

In this paper, we study the regularity theory of this problem, specializing to the case where the cost function is given by a $\mathcal{D}_{\Psi}^{(\alpha)}$-divergence.

Definition 1 ($\mathcal{D}_{\Psi}^{(\alpha)}$-divergence). *Let $\Psi : M \to \mathbb{R}$ be a convex function on a convex domain M in Euclidean space. For two points $x, y \in M$ and $\alpha \in (-1, 1)$, a $\mathcal{D}_{\Psi}^{(\alpha)}$-divergence is a function of the form*

$$\mathcal{D}_{\Psi}^{(\alpha)}(x, y) = \frac{4}{1-\alpha^2} \left[\frac{1-\alpha}{2} \Psi(x) + \frac{1+\alpha}{2} \Psi(y) - \Psi\left(\frac{1-\alpha}{2} x + \frac{1+\alpha}{2} y \right) \right].$$

© Springer Nature Switzerland AG 2019
F. Nielsen and F. Barbaresco (Eds.): GSI 2019, LNCS 11712, pp. 423–430, 2019.
https://doi.org/10.1007/978-3-030-26980-7_44

These divergences were introduced by the second author [21] and form a one-parameter family of statistical divergences. As α converges to either ± 1, the $\mathcal{D}_\psi^{(\alpha)}$-divergence converges to the Bregman divergence [1]. Bregman divergences play an important role in information geometry as they provide a generalization of distance functions which satisfy the generalized Pythagorean theorem. Due to this interpolation property, $\mathcal{D}_\psi^{(\alpha)}$-divergences form a natural class worthy of investigation.

Our main results provide a new geometric interpretation for the necessary conditions to ensure that the associated optimal transport is smooth. More concretely, we want to understand whether the rubble is moved in a continuous way so that nearby piles remain close after transport. In this paper, we give a summary of our results, omitting proofs and a more complete exposition. We refer the interested reader to [9] for a complete description of our results, accompanied with proofs and extensive exposition.

2   Preliminaries on Optimal Transport

We briefly discuss some background on the regularity theory of optimal transport. For a more complete overview of the subject, we recommend the book by Villani [19] and the survey paper of De Philippis and Figalli [3]. In the following, we use c to refer to the cost function and $c_{I,J}$ to denote $\partial_{x^I}\partial_{y^J}c$ for multi-indices I and J. Furthermore, $c^{i,j}$ is the matrix inverse of the mixed derivative $c_{i,j}$.

In order to state the background results, it is necessary to first define the c-exponential map, which plays a crucial role throughout.

Definition 2 (**c-exponential map**). *For any* $x \in X, y \in Y, p \in \mathbb{R}^n$, *the c-exponential map satisfies the following identity.*

$$c\text{-}\exp_x(p) = y \iff p = -c_x(x,y).$$

Our primary interest is in the case when the optimal coupling γ is supported on the graph of a function. In this case, the optimal transport is said to be *deterministic* and the associated function is known as the *optimal map*. The following theorem, originally proved by Brenier [2] and extended by Gangbo and McCann [8], provides sufficient conditions for optimal transport to be deterministic and shows that the associated optimal maps are induced by solutions to Monge-Ampere type equations.

Theorem 1. *Let* X *and* Y *be two open subsets of* $\mathbb{R}^n$ *and consider a cost function* $c : X \times Y \to \mathbb{R}$. *Suppose that* $d\mu$ *is a smooth probability density supported on* X *and that* $d\nu$ *is a smooth probability density supported on* Y. *Suppose that the following conditions hold:*

1. *The cost function* c *is of class* C^4 *with* $\|c\|_{C^4(X \times Y)} < \infty$
2. *For any* $x \in X$, *the map* $Y \ni y \to c_x(x,y) \in \mathbb{R}^n$ *is injective.*
3. *For any* $y \in Y$, *the map* $X \ni x \to c_y(x,y) \in \mathbb{R}^n$ *is injective.*

4. $\det(c_{x,y})(x,y) \neq 0$ *for all* $(x,y) \in X \times Y$.

Then there exists a c-convex function $u : X \to \mathbb{R}$ such that the map T_u : $X \to Y$ defined by $T_u(x) := c\text{-}\exp_x(\nabla u(x))$ is the unique optimal transport map sending μ onto ν. Furthermore, T_u is injective $d\mu$-a.e.,

$$|\det(\nabla T_u(x))| = \frac{d\mu(x)}{d\nu(T_u(x))} \qquad d\mu - a.e., \tag{1}$$

and its inverse is given by the optimal transport map sending ν onto μ.

The regularity problem for optimal transport studies the smoothness of the potential u. For this question, most of the initial work was done for the squared-distance cost $c(x,y) = \frac{1}{2}\|x-y\|^2$ in Euclidean space, known as the 2-Wasserstein distance. For more general cost functions, the breakthrough work was done by Ma, Trudinger and Wang [12], who gave three conditions that ensure smoothness for the solutions of Monge-Ampere equations. In this paper, we use a modified version of their result, originally proved by Trudinger and Wang [18].

Theorem 2. Suppose that $c : X \times Y \to \mathbb{R}$ satisfies the hypothesis of the previous theorem, and that the smooth densities $d\mu$ and $d\nu$ are bounded away from zero and infinity on their respective supports X and Y. Suppose further that the following holds:

1. X and Y are smooth.
2. The domain X is strictly c-convex relative to the domain Y.
3. The domain Y is strictly c^*-convex relative to the domain X.
4. The following condition (known as MTW(0)) holds:

For all vectors $\xi, \eta \in \mathbb{R}^n$ with $\xi \perp \eta$, the following inequality holds.

$$\mathfrak{S}(\xi,\eta) := \sum_{i,j,k,l,p,q,r,s} (c_{ij,p}c^{p,q}c_{q,rs} - c_{ij,rs})c^{r,k}c^{s,l}\xi^i\xi^j\eta^k\eta^l \geq 0 \tag{2}$$

Then $u \in C^\infty(\overline{X})$ and $T : \overline{X} \to \overline{Y}$ is a smooth diffeomorphism, where $T(x) = c\text{-}\exp_x(\nabla u(x))$.

We will discuss the assumptions of Theorem 2 in a bit more detail. The first condition is self-explanatory, while the second and third define the proper notions of convexity, which are necessary to establish regularity of optimal transport [11].

Definition 3 (c-segment). A c-segment in X with respect to a point y is a solution set $\{x\}$ to $c_y(x,y) = z$ for z on a line segment in $\mathbb{R}^n$. A c^*-segment in Y with respect to a point x is a solution set $\{y\}$ to $c_x(x,y) = z$ for z on a line segment in $\mathbb{R}^n$.

Definition 4 (c-convexity). A set E is c-convex relative to a set E^* if for any two points $x_0, x_1 \in E$ and any $y \in E^*$, the c-segment relative to y connecting x_0 and x_1 lies in E. Similarly we say E^* is c^*-convex relative to E if for any two points $y_0, y_1 \in E^*$ and any $x \in E$, the c^*-segment relative to x connecting y_0 and y_1 lies in E^*.

Finally, we discuss the inequality (2), which is known as the $MTW(0)$ condition, and is a weakened version of the $MTW(\kappa)$ condition.

Definition 5 ($MTW(\kappa)$). *A cost function c satisfies the $MTW(\kappa)$ condition if for any orthogonal vector-covector pair η and ξ, $\mathfrak{S}(\eta,\xi) \geq \kappa|\eta|^2|\xi|^2$ for $\kappa > 0$.*

Ma, Trudinger and Wang's original work used $MTW(\kappa)$, and this stronger assumption is used in many applications. There is another strengthening of the $MTW(0)$ assumption that appears in the literature.

Definition 6 (Non-negative cross-curvature). *A cost function c has non-negative (resp. strictly positive) cross-curvature if, for any vector-covector pair η and ξ,*

$$\mathfrak{S}(\eta,\xi) \geq 0 \; (\text{resp. } \kappa|\eta|^2|\xi|^2).$$

Non-negative cross-curvature is stronger than $MTW(0)$, as the non-negativity must hold for all pairs η and ξ, not simply orthogonal ones. This notion was introduced by Figalli, Kim, and McCann [5] to study a problem in microeconomics and in later work, they showed that stronger regularity for optimal maps can be proven with this assumption [6].

The geometric significance of these notions is a topic of active research. Although it is not immediately clear, $\mathfrak{S}$ is in fact tensorial (coordinate-invariant) and transforms quadratically in η and ξ [10]. On a Riemannian manifold, Loeper [11] gave some insight into the behavior of the MTW tensor by showing that for the 2-Wasserstein distance, the tensor is proportional to the sectional curvature on the diagonal. This was extended by Kim and McCann, who gave a pseudo-Riemannian framework for optimal transport [10], in which the MTW-tensor becomes the curvature of light-like planes.

3   Geometry of TM

In order to associate optimal transport with Kähler geometry, we consider the tangent bundle TM where M is the domain of Ψ. On any Riemannian manifold (M,g) endowed with an affine connection D, it is possible to induce TM with an almost Hermitian structure known as the Sasaki metric [4]. For brevity, we will not review the construction here, but complete details can be found in the paper by Satoh [16].

We are primarily interested in the case where TM is Kähler, which occurs when M is Hessian (also known as *affine-Kähler*). There are two equivalent definitions for such manifolds; with the former definition primarily used in differential geometry and the latter primarily used in information geometry. For details on how these definitions are equivalent, we refer readers to the book by Shima [17].

Definition 7 (Differential geometric). *A Riemannian manifold (M, g) is said to be Hessian if there is an atlas of local coordinates $\{u^i\}_{i=1}^n$ so that for each coordinate chart, there is a convex potential Ψ such that*

$$g_{ij} = \frac{\partial^2 \Psi}{\partial u^i u^j}.$$

Furthermore, the transition maps between these coordinate charts are affine.

Definition 8 (Information geometric). *A Riemannian manifold (M, g) is said to be Hessian if it admits dually flat connections. Namely, it admits two flat (torsion- and curvature-free) connections D and D^* satisfying*

$$\mathcal{X}(g(\mathcal{Y}, \mathcal{Z})) = g(D_{\mathcal{X}}\mathcal{Y}, \mathcal{Z}) + g(\mathcal{Y}, D^*_{\mathcal{X}}\mathcal{Z}) \tag{3}$$

for all vector fields $\mathcal{X}$, $\mathcal{Y}$, and $\mathcal{Z}$.

For our purposes, we are interested in the curvature of this metric, which can be derived using the work of Satoh [16].

Proposition 1. *Let (M, g, D) be a Hessian manifold with metric g and flat connection D. Suppose $\{x^i\}$ are the coordinates where the Christoffel symbols of D vanishes and Ψ is the Hessian potential of g. In the associated holomorphic coordinates $\{z^i\}$ on TM, the holomorphic bisectional curvature of the Sasaki metric satisfies the following identity:*

$$\tilde{R}_{\tilde{g}^D}\left(\partial_{z^i}, \overline{\partial}_{z^j}, \partial_{z^k}, \overline{\partial}_{z^l}\right) = -\frac{1}{2}\Psi_{ijkl} + \frac{1}{2}\Psi^{rs}\Psi_{iks}\Psi_{jlr}.$$

It is worth noting that the holomorphic bisectional curvature is negative to what Shima defined as the *Hessian curvature* [17, p. 38].

4   Our Results

Our main result is to relate the MTW tensor to the bisectional curvature of the Sasaki metric. To do so, we note that if $c : M \times M \to \mathbb{R}$ is a $D_{\Psi}^{(\alpha)}$-divergence, then the MTW tensor takes the following form:

$$\mathfrak{S}_{(x,y)}(\xi, \eta) = \frac{1 - \alpha^2}{4}\left(\Psi_{ijp}\Psi_{rsq}\Psi^{pq} - \Psi_{ijrs}\right)\Psi^{rk}\Psi^{sl}\xi^i\xi^j\eta^k\eta^l. \tag{4}$$

To relate this to a complex metric, we define M to be the domain of Ψ and use Ψ as a potential for a Riemannian metric on M. This immediately implies the main theorem of our current paper.

Theorem 3. *Let X and Y be open sets in $\mathbb{R}^n$ and c be a $D_{\Psi}^{(\alpha)}$-divergence. Then the MTW tensor is proportional to the orthogonal bisectional curvature of the Sasaki metric on (TM, g^D, J^D), after flatting the latter two indices (i.e. treating η as a covector with $\eta(\xi) = 0$). Furthermore, the cross curvature is proportional to the bisectional curvature of (TM, g^D, J^D).*

As an immediate consequence, this shows that the MTW tensor is non-negative if the orthogonal bisectional holomorphic curvature of TM is non-negative. By considering the dually flat structure of a Hessian manifold, we can also give an alternative characterization of relative c-convexity.

Theorem 4. *For a $D_{\Psi}^{(\alpha)}$-divergence, a set Y is c-convex relative to X if and only if, for all $x \in X$, the set $\frac{1+\alpha}{2}x + \frac{1-\alpha}{2}Y \subset M$ is geodesically convex with respect to the dual connection D^*.*

Our results also hold for cost functions of the form $c(x, y) = \Psi(x - y)$ with $\Psi : \mathbb{R}^n \to \mathbb{R}$ a strongly convex function. This includes many of the common examples in the literature, including the p-Wasserstein costs in Euclidean space. These costs were studied by Gangbo and McCann [7] and Ma, Trudinger and Wang [12], derived an expression proportional to Eq. (4) for their MTW tensor.

5 The Regularity of Pseudo-Arbitrages

For brevity, we will focus on a single application and refer readers to the full paper for others [9]. A recent series of papers by Pal and Wong (see, e.g., [14] and [20]) have studied the problem of finding *pseudo-arbitrages*, which are investment strategies which outperform the market portfolio under "mild and realistic assumptions". Their work combines information geometry, optimal transport and mathematical finance to reduce this problem to solving optimal transport problems where the cost function is given by a so-called log-divergence.

A central result in [14] shows that a portfolio map T outperforms the market portfolio almost surely in the long run iff it is a solution to the Monge problem for the cost function $c : \mathbb{R}^{n-1} \times \mathbb{R}^{n-1} \to \mathbb{R}$ given by

$$c(x, y) := \log \left(1 + \sum_{i=1}^{n-1} e^{x^i - y^i} \right) - \log(n) - \frac{1}{n} \sum_{i=1}^{n-1} x^i - y^i. \tag{5}$$

For this problem, we study the Kähler geometry of the Sasaki metric associated to the potential $\Psi(u) = \log \left(1 + \sum_{i=1}^{n-1} e^{u^i} \right)$. Doing so, we find that the Sasaki metric has vanishing orthogonal bisectional curvature and constant positive holomorphic sectional curvature [9]. Applying our main results and Theorem 2, we find the following regularity theorem.

Theorem 5. *Suppose μ and ν are probability measures supported respectively on subsets X and Y of the probability simplex. Suppose further that the following regularity assumptions hold:*

1. *X and Y are smooth, strictly convex and uniformly bounded from the boundary of the probability simplex. More precisely, there exists $\delta > 0$, so that for all $x \in X$, $1 \le i \le n$, $x^i \ge \delta$.*
2. *μ and ν are absolutely continuous with respect to the Lebesgue measure and $d\mu$ and $d\nu$ are smooth functions which are bounded away from zero and infinity on their supports.*

Let $\hat{c}(p, q)$ be the cost function given by

$$\hat{c}(p, q) = \log \left(\frac{1}{n} \sum_{i=1}^{n} \frac{q_i}{p_i} \right) - \frac{1}{n} \sum_{i=1}^{n} \log \frac{q_i}{p_i}.$$

Then the $\hat{c}$-optimal map T_u taking μ to ν is smooth.

This also provides a regularity theorem for the associated displacement interpolation, which was asked in [15].

Corollary 1. *Suppose μ and ν are smooth probability measures satisfying the assumptions of Corollary 5 and that T_u is the $\hat{c}$-optimal map transporting μ to ν. Suppose further that $T(t)\mu$ is the displacement interpolation from μ to ν defined by $T(t) = t \cdot Id + (1 - t)T_u$. Then $T(t)$ is smooth, both as a map for fixed t and also in terms of the t parameter.*

Acknowledgment. The first author would like to thank Fangyang Zheng and Bo Guan for their helpful discussions about this project. The project is supported by DARPA/ARO Grant W911NF-16-1-0383 ("Information Geometry: Geometrization of Science of Information", PI: Zhang).

References

1. Bregman, L.M.: The relaxation method of finding the common point of convex sets and its application to the solution of problems in convex programming. USSR Comput. Math. Math. Phys. **7**(3), 200–217 (1967)
2. Brenier, Y.: Décomposition polaire et réarrangement monotone des champs de vecteurs. C.R. Acad. Sci. Paris Sér. I Math. **305**, 805–808 (1987)
3. De Philippis, G., Figalli, A.: The Monge-Ampère equation and its link to optimal transportation. Bull. Am. Math. Soc. **51**(4), 527–580 (2014)
4. Dombrowski, P.: On the geometry of the tangent bundle. Journal für Mathematik. Bd **210**(1/2), 10 (1962)
5. Figalli, A., Kim, Y.H., McCann, R.J.: When is multidimensional screening a convex program? J. Econ. Theory **146**(2), 454–478 (2011)
6. Figalli, A., Kim, Y.H., McCann, R.J.: Hölder continuity and injectivity of optimal maps. Arch. Rat. Mech. Anal. **209**(3), 747–795 (2013)
7. Gangbo, W., McCann, R.J.: Optimal maps in Monge's mass transport problem. Comptes Rendus de l'Academie des Sciences-Serie I-Mathematique **321**(12), 1653 (1995)
8. Gangbo, W., McCann, R.J.: The geometry of optimal transportation. Acta Math. **177**(2), 113–161 (1996)
9. Khan, G., Zhang, J.: On the Kähler geometry of certain optimal transport problems (2018). arXiv preprint arXiv:1812.00032
10. Kim, Y.H., McCann, R.J.: Continuity, curvature, and the general covariance of optimal transportation (2007). arXiv preprint arXiv:0712.3077
11. Loeper, G.: On the regularity of solutions of optimal transportation problems. Acta Math. **202**(2), 241–283 (2009)
12. Ma, X.N., Trudinger, N.S., Wang, X.J.: Regularity of potential functions of the optimal transportation problem. Arch. Rat. Mech. Anal. **177**(2), 151–183 (2005)

13. Monge, G.: Mémoire sur la théorie des déblais et des remblais. Histoire de l'Académie Royale des Sciences de Paris (1781)
14. Pal, S., Wong, T.K.L.: Exponentially concave functions and a new information geometry. Ann. Probab. **46**(2), 1070–1113 (2018)
15. Pal, S., Wong, T.K.L.: Multiplicative Schrödinger problem and the Dirichlet transport (2018). arXiv preprint arXiv:1807.05649
16. Satoh, H.: Almost Hermitian structures on tangent bundles. In: Proceedings of the Eleventh International Workshop on Differential. Geometry, Kyungpook Nat. Univ., Taegu, vol. 11, pp. 105–118 (2007)
17. Shima, H.: Geometry of Hessian Structures. World Scientific Publishing Co., Singapore (2007)
18. Trudinger, N.S., Wang, X.J.: On the second boundary value problem for Monge-Ampere type equations and optimal transportation. Annali della Scuola Normale Superiore di Pisa-Classe di Scienze-Serie IV **8**(1), 143 (2009)
19. Villani, C.: Optimal Transport: Old and New, vol. 338. Springer, Heidelberg (2008). https://doi.org/10.1007/978-3-540-71050-9
20. Wong, T.K.L.: Logarithmic divergences from optimal transport and Rényi geometry. Inf. Geom. **1**(1), 39–78 (2018)
21. Zhang, J.: Divergence function, duality, and convex analysis. Neural Comput. **16**(1), 159–195 (2004)

Non-parametric Information Geometry

Divergence Functions in Information Geometry

Domenico Felice[1(✉)] and Nihat Ay[1,2,3]

[1] Max Planck Institute for Mathematics in the Sciences,
Inselstrasse 22–04103, Leipzig, Germany
`felice@mis.mpg.de`
[2] Santa Fe Institute, Santa Fe, NM 87501, USA
[3] Faculty of Mathematics and Computer Science, University of Leipzig,
PF 100920, 04009 Leipzig, Germany

Abstract. A recently introduced canonical divergence $\mathcal{D}$ for a dual structure (g, ∇, ∇^*) on a smooth manifold M is discussed in connection to other divergence functions. Finally, general properties of $\mathcal{D}$ are outlined and critically discussed.

Keywords: Classical differential geometry · Riemannian geometries · Information Geometry

1 Introduction

The geometrical structure induced by a *divergence function* (or *contrast function*) on a smooth manifold M provides a unified approach to measurement of notions as information, energy, entropy, playing an important role in mathematical sciences to research random phenomena [1]. In the mathematical formulation, a divergence function $\mathcal{D}(p, q)$ on a smooth manifold M is defined by the first requirement for a distance:

$$\mathcal{D}(p, q) \geq 0, \qquad \mathcal{D}(p, q) = 0 \quad \text{iff} \quad p = q. \tag{1}$$

An important example of a divergence function is given by the Kullback-Leibler divergence $K(p, q)$ in the context that p and q are the vectors of probabilities of disjoint events [2], namely

$$K(p, q) = \sum_{i=1}^{n+1} p_i \, \log\left(\frac{p_i}{q_i}\right) \tag{2}$$

is a function on the n-simplex $\mathcal{S} := \{p = (p_1, \ldots, p_{n+1}) \,|\, p_i \geq 0, \sum_i p_i = 1\}$. Given a smooth n-dimensional manifold M, we assume that $\mathcal{D} : M \times M \to \mathbb{R}^+$ is a C^∞-differentiable function. Working with the local coordinates $\{\boldsymbol{\xi}_p := (\xi_p^1, \ldots, \xi_p^n)\}$

© Springer Nature Switzerland AG 2019
F. Nielsen and F. Barbaresco (Eds.): GSI 2019, LNCS 11712, pp. 433–442, 2019.
https://doi.org/10.1007/978-3-030-26980-7_45

and $\{\boldsymbol{\xi}_q := (\xi_q^1, \ldots, \xi_q^n)\}$ at p and q, respectively, it follows from Eq. (1) that

$$\partial_i \, \mathcal{D}(\boldsymbol{\xi}_p, \boldsymbol{\xi}_q)\big|_{p=q} = 0, \quad \partial_i' \, \mathcal{D}(\boldsymbol{\xi}_p, \boldsymbol{\xi}_q)\big|_{p=q} = 0 \tag{3}$$

$$\partial_j \partial_i \, \mathcal{D}(\boldsymbol{\xi}_p, \boldsymbol{\xi}_q)\big|_{p=q} = - \partial_j' \partial_i \, \mathcal{D}(\boldsymbol{\xi}_p, \boldsymbol{\xi}_q)\big|_{p=q} = \partial_j' \partial_i' \, \mathcal{D}(\boldsymbol{\xi}_p, \boldsymbol{\xi}_q)\big|_{p=q}, \tag{4}$$

where $\partial_i = \frac{\partial}{\partial \xi_p^i}$ and $\partial_i' = \frac{\partial}{\partial \xi_q^i}$. Moreover, under the assumption that

$$g_{ij} := \partial_i \partial_j \mathcal{D}(\boldsymbol{\xi}_p, \boldsymbol{\xi}_q)\big|_{p=q} > 0 \tag{5}$$

we can see that the manifold M is endowed, through the divergence function $\mathcal{D}$, with the Riemannian metric tensor given by $g = g_{ij} \, d\xi^i \otimes d\xi^j$, where the Einstein notation is adopted. The symmetry of g immediately follows from the requirement that $\mathcal{D}$ is a C^∞ function on M × M.

From Eq. (2) we can see that, in general, a divergence function $\mathcal{D}$ is not symmetric. The asymmetry of $\mathcal{D}$ leads to two different affine connections, ∇ and ∇^*, on M such that $1/2(\nabla + \nabla^*)$ is the Levi-Civita connection with respect to the metric tensor $g = g_{ij} d\xi^i \otimes d\xi^j$ defined by Eq. (5). More precisely, working with the local coordinates $\{\boldsymbol{\xi}_p\}$ and $\{\boldsymbol{\xi}_q\}$, we can define the symbols Γ_{ijk} and Γ_{ijk}^* of the connections ∇ and ∇^*, i.e. $\Gamma_{ijk} = g\left(\nabla_{\partial_i}\partial_j, \partial_k\right)$ and $\Gamma_{ijk}^* = g\left(\nabla_{\partial_i}^*\partial_j, \partial_k\right)$, by means of the following relations

$$\Gamma_{ijk}(p) = - \partial_i \partial_j \partial_k' \mathcal{D}(\boldsymbol{\xi}_p, \boldsymbol{\xi}_q)\big|_{p=q}, \quad \Gamma_{ijk}^*(p) = - \partial_i' \partial_j' \partial_k \mathcal{D}(\boldsymbol{\xi}_p, \boldsymbol{\xi}_q)\big|_{p=q}. \tag{6}$$

To sum up, a divergence function $\mathcal{D}$ on a smooth manifold M induces a metric tensor on M by Eq. (5). In addition, the divergence $\mathcal{D}$ yields two linear torsion-free connections, ∇ and ∇^*, on TM which are dual with respect to the metric tensor g [2]:

$$X \, g\left(Y, Z\right) = g\left(\nabla_X Y, Z\right) + g\left(Y, \nabla_X^* Z\right), \ \forall \ X, Y, Z \in \mathcal{T}(\mathrm{M}), \tag{7}$$

where $\mathcal{T}(\mathrm{M})$ denotes the space of vector fields on M. Finally, we refer to the quadruple $(\mathrm{M}, g, \nabla, \nabla^*)$ as a *statistical manifold* [3].

1.1   The Inverse Problem Within Information Geometry

The inverse problem is to find a divergence $\mathcal{D}$ which generates a given geometrical structure $(\mathrm{M}, g, \nabla, \nabla^*)$. For any such statistical manifold there exists a divergence $\mathcal{D}$ such that Eqs. (5) and (6) hold true [4]. However, this divergence is not unique and there are infinitely many divergences generating the same geometrical structure (g, ∇, ∇^*). When this structure is dually flat, namely the curvature tensors of ∇ and ∇^* are null $(\mathrm{R}(\nabla) \equiv 0$ and $\mathrm{R}^*(\nabla^*) \equiv 0)$, Amari and Nagaoka introduced a canonical divergence which is a Bregman divergence [5]. The canonical divergence has nice properties such as the generalized Pythagorean theorem and the geodesic projection theorem and it turns out to be of uppermost importance to define a canonical divergence in the general case. A first attempt to answer this fundamental issue is provided by Ay and Amari in [6] where a canonical divergence for a general statistical manifold (M, g, ∇, ∇^*) is given by using the geodesic integration of the inverse exponential map. This one is understood as a *difference vector* that translates q to p for all $q, p \in M$ sufficiently close to each other.

To be more precise, the inverse exponential map supplies a generalization to M of the concept of difference vector in $\mathbb{R}^n$. In detail, let $p, q \in \mathbb{R}^n$, the difference between p and q is the vector $p - q$ pointing to p (see side (**A**) of Fig. 1). Then, the difference between p and q in M is provided by the inverse exponential map. In particular, given p, q suitably close in M, the difference vector from q to p is defined as (see (**B**) of Fig. 1)

$$X_q(p) := X(q, p) := \exp_q^{-1}(p) = \dot{\tilde{\sigma}}(0), \tag{8}$$

where $\tilde{\sigma}$ is the ∇-geodesic from q to p. Therefore, the divergence D introduced in [6] is defined as the path integral

$$D(p, q) := \int_0^1 \langle X_t(p), \dot{\tilde{\sigma}}(t) \rangle_{\tilde{\sigma}(t)} \, dt, \quad X_t(p) := X(\tilde{\sigma}(t), p) := \exp_{\tilde{\sigma}(t)}^{-1}(p), \tag{9}$$

where $\langle \cdot, \cdot \rangle_{\tilde{\sigma}(t)}$ denotes the inner product induced by g on $\tilde{\sigma}(t)$. After elementary computation Eq. (9) reduces to,

$$D(p, q) = \int_0^1 t \, \|\dot{\sigma}(t)\|^2 \, dt, \tag{10}$$

where $\sigma(t)$ is the ∇-geodesic from p to q [6]. The divergence $D(p, q)$ has nice properties such as the positivity and it reduces to the canonical divergence proposed by Amari and Nagaoka when the manifold M is dually flat. However, if we consider definition (9) for a general path γ, then $D_\gamma(p, q)$ will be depending on γ. On the contrary, if the vector field $X_t(p)$ is integrable, then $D_\gamma(p, q) =: D(p, q)$ turns out to be independent of the path from q to p.

2   The Recent Canonical Divergence

In this article, we discuss about a divergence function recently introduced in [7]. This turns out to be a generalization of the divergence introduced by Ay and Amari. The definition of the recent divergence (see below Eqs. (18), (19)) relies on an extended analysis of the intrinsic structure of the dual geometry of a general statistical manifold (M, g, ∇, ∇^*). In particular, we introduced a vector at q by modifying the definition (8) of the difference vector $X(q, p)$. Consider $p, q \in M$ such that there exist both, a unique ∇-geodesic σ and a unique ∇^*-geodesic σ^*, connecting p with q. Moreover, let $X_p(q) := \exp_p^{-1}(q) = \dot{\sigma}(0) \in T_pM$. Then, we ∇-parallel translate it along σ^* from p to q (see (**C**) of Fig. 1), and obtain

$$\Pi_q(p) := P_{\sigma^*} X_p(q) \in T_qM. \tag{11}$$

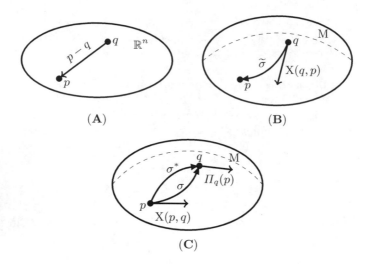

Fig. 1. Illustration (**A**) of the difference vector in $\mathbb{R}^n$ pointing from q to p; and (**B**) the difference vector $X(q,p) = \dot{\tilde{\sigma}}(0)$ as the inverse of the exponential map in q. The novel vector $\Pi_q(p)$ at q is described in (**C**) as the ∇ parallel transport of $X(p,q) = \dot{\sigma}(0)$ along the ∇^*-geodesic σ^* from p to q.

(Note that $\Pi_q(p)$ corresponds to minus a difference vector.) Analogously, we introduce the dual vector of $\Pi_q(p)$ as the ∇^*-parallel transport of $\dot{\sigma}^*(0)$ along the ∇-geodesic σ,

$$\Pi_q^*(p) := P_\sigma^* X_p^*(q), \quad X_p^*(q) := \overset{*}{\exp}{}_p^{-1}(q) = \dot{\sigma}^*(0), \tag{12}$$

where $\overset{*}{\exp}{}_p$ denotes the exponential map of the ∇^*-connection. A fundamental result obtained in [7] is that the sum of $\Pi_q(p)$ and $\Pi_q^*(p)$ is the gradient of a function $r_p(q) = r(p,q)$:

$$\Pi_q(p) + \Pi_q^*(p) = \mathrm{grad}_q r_p, \quad r(p,q) := \langle \exp_p^{-1}(q), \overset{*}{\exp}{}_p^{-1}(q) \rangle_p. \tag{13}$$

From here on, we refer to $r(p,q)$ as the pseudo-squared-distance between p and q. Note that the function $r(p,q)$, in general, is not symmetric in its arguments.

For $p \in M$ fixed and q varying in M, we then obtain two vector fields whenever p and q are connected by a unique ∇-geodesic and a unique ∇^*-geodesic. Then, we can introduce two vector fields on an arbitrary path $\gamma : [0,1] \to M$ connecting p and q in the following way. Let us firstly assume that for each $t \in [0,1]$ there exist a unique ∇-geodesic σ_t and a unique ∇^*-geodesic σ_t^* connecting p with $\gamma(t)$. Then, we define

$$\Pi_t(p) = P_{\sigma_t^*} X_p(t), \quad X_p(t) = \exp_p^{-1}(\gamma(t)) \tag{14}$$

$$\Pi_t^*(p) = P_{\sigma_t}^* X_p^*(t), \quad X_p^*(t) = \overset{*}{\exp}{}_p^{-1}(\gamma(t)). \tag{15}$$

From Eq. (13) we have that the sum

$$\int_0^1 \langle \Pi_t(p), \dot{\gamma}(t) \rangle_{\gamma(t)} \, dt + \int_0^1 \langle \Pi_t^*(p), \dot{\gamma}(t) \rangle_{\gamma(t)} \, dt = r(p, q) \qquad (16)$$

is independent of the particular path from p to q. This potential property of the pseudo-squared-distance $r(p, q)$ has been exploited for introducing a canonical divergence $\mathcal{D}$ as potential function of Π and a dual divergence $\mathcal{D}^*$ as potential function of Π^*, i.e. $\operatorname{grad} \mathcal{D} = \Pi$ and $\operatorname{grad} \mathcal{D}^* = \Pi^*$ [7]. However, this does not turn out in general. Indeed, on a general statistical manifold, the canonical divergence $\mathcal{D}$ and its dual function $\mathcal{D}^*$ are defined by the following unique and orthogonal decompositions,

$$\Pi = \operatorname{grad} \mathcal{D} + V, \qquad \Pi^* = \operatorname{grad} \mathcal{D}^* + V^*, \qquad (17)$$

where V is a vector field orthogonal to ∇-geodesics whereas V^* is orthogonal to ∇^*-geodesics *(we will prove later the orthogonality of these decompositions)*.

When the vector field Π is integrated along the ∇-geodesic $\sigma(t)$ $(0 \leq t \leq 1)$ connecting p with q, the recent canonical divergence assumes the following form

$$\mathcal{D}(p, q) = \int_0^1 \langle \Pi_t(p), \dot{\sigma}(t) \rangle_{\sigma(t)} \, dt, \quad \Pi_t(p) = \mathrm{P}_{\sigma_t^*} \exp_p^{-1}(\sigma(t)), \qquad (18)$$

where $\sigma_t^*(s)$ $(0 \leq s \leq 1)$ denotes the ∇^*-geodesic such that $\sigma_t^*(0) = p$ and $\sigma_t^*(1) = \sigma(t)$. On the contrary, if the vector field Π^* is integrated along the ∇^*-geodesic $\sigma^*(t)$ $(0 \leq t \leq 1)$ connecting p with q, the dual divergence is given by

$$\mathcal{D}^*(p, q) = \int_0^1 \langle \Pi_t^*(p), \dot{\sigma}^*(t) \rangle_{\sigma^*(t)} \, dt, \quad \Pi_t^*(p) = \mathrm{P}_{\sigma_t}^* \exp_p^{*-1}(\sigma^*(t)), \qquad (19)$$

where $\sigma_t(s)$ $(0 \leq s \leq 1)$ is the ∇-geodesic such that $\sigma_t(0) = p$ and $\sigma_t(1) = \sigma(t)$.

In this manuscript we review the relation of the canonical divergence $\mathcal{D}(p, q)$ of (18) to other divergence functions in Sect. 3. Finally, we outline in Sect. 4 the orthogonality and the symmetry properties of $\mathcal{D}$.

3   Comparison with Previous Divergence Functions

Given a general statistical manifold $(\mathrm{M}, \mathrm{g}, \nabla, \nabla^*)$, the basic requirement for a smooth function $\mathcal{D} : \mathrm{M} \times \mathrm{M} \to \mathbb{R}$ to be a divergence on M is its consistency with the dual structure $(\mathrm{g}, \nabla, \nabla^*)$ through Eqs. (5)–(6) and the positivity $\mathcal{D}(p, q) > 0$ for all $p, q \in \mathrm{M}$ sufficiently close to each other such that $p \neq q$. The recent canonical divergence (18) holds these properties [7].

In this section, we will show that the canonical divergence $\mathcal{D}$ can be interpreted as a generalization of the divergence D introduced by Ay and Amari. Indeed, we will see that these two divergences coincide on particular classes of statistical manifolds. In order to achieve this result, we investigate some geometric properties of the vector field $\Pi_t(p)$ given by Eq. (14) aiming to split such a vector field in terms of the difference vector $X_t(p)$ given in Eq. (9). To be more precise, let us refer to Fig. 2 where the ∇-geodesic $\sigma(t)$ $(0 \leq t \leq 1)$ connecting $\sigma(0) = p$ with $\sigma(1) = q$ is drawn. Then, for each $t \in [0,1]$ we can consider the ∇-geodesic $\sigma_t(s)$ $(0 \leq s \leq 1)$ connecting p with $\sigma(t)$ and the ∇^*-geodesic $\sigma_t^*(s)$ $(0 \leq s \leq 1)$ connecting p with $\sigma(t)$. The difference vector $X_t(p) = X(\sigma(t), p)$ at $\sigma(t)$ pointing to p is given in terms of the inverse exponential map by $X_t(p) := \exp_{\sigma(t)}^{-1}(p)$. Therefore, the opposite of $X_t(p)$ can be viewed as the ∇-parallel translation of $X_p(t) = \exp_p^{-1}(\sigma(t))$ along the ∇-geodesic σ_t, namely $-X_t(p) = P_{\sigma_t} X_p(t)$. Consider now the loop Σ_t based at p and given by first traveling from p to $\sigma(t)$ along the ∇^*-geodesic σ_t^* and then back from $\sigma(t)$ to p along the reverse of the ∇-geodesic σ_t. If Σ_t lies in a sufficiently small neighborhood of p, then [7]

$$P_{\Sigma_t} X_p(t) = X_p(t) + \mathcal{R}_{\Sigma_t}\left(X_p^*(t), X_p(t)\right),$$

where

$$\mathcal{R}_{\Sigma_t}\left(X_p^*(t), X_p(t)\right) := \int_{B_t} \frac{P\left[\mathcal{R}\left(X^*(t), X(t)\right) X(t)\right]}{\|X_p^*(t) \wedge X_p(t)\|} \, dA$$

with $X^*(t)$ and $X(t)$ being the ∇-parallel transport of $X_p^*(t)$ $(:=\exp_p^{*-1}(\sigma(t)))$ and $X_p(t)$, respectively, from p to each point of B_t along the unique ∇-geodesic joining them. Here, $\mathcal{R}$ is the curvature tensor of ∇, B_t denotes the disc defined by the curve Σ_t and $X_p^*(t)$, $X_p(t)$ are linearly independent. In addition, P within the integral denotes the ∇-parallel translation from each point in B_t to p along the unique ∇-geodesic segment joining them. Finally, by means of the property of the parallel transport, we obtain the following geometric relation between the vector $\Pi_t(p)$ and the opposite of the difference vector $X_t(p)$ [7],

$$\Pi_t(p) = P_{\sigma_t} X_p(t) + P_{\sigma_t}\left[\mathcal{R}_{\Sigma_t}\left(X_p^*(t), X_p(t)\right)\right]. \tag{20}$$

By noticing that $P_{\sigma_t} X_p(t) = \dot{\sigma}_t(1) = t\,\dot{\sigma}(t)$ and inserting Eq. (20) into the definition (18) of $\mathcal{D}$, we obtain

$$\mathcal{D}(p, q) = D(p, q) + \int_0^1 \langle P_{\sigma_t}\left[\mathcal{R}_{\Sigma_t}\left(X_p^*(t), X_p(t)\right)\right], \dot{\sigma}(t) \rangle_{\sigma(t)} \, dt, \tag{21}$$

where $D(p, q)$ is the divergence introduced in [6] and given by Eq. (10).

It is clear from Eq. (21) that particular conditions on the curvature tensor would lead to the required equivalence between $\mathcal{D}$ and D. Actually, in Information Geometry classes of statistical manifolds are characterized by the conditions on the curvature tensors of ∇ and ∇^* (see for instance Refs. [5], [8], [9]). In the Table 1 we can see the categories of statistical manifolds on which the canonical divergence $\mathcal{D}$ reduces to the divergence D introduced in [6]. A statistical manifold (M, g, ∇, ∇^*) is self-dual when $\nabla = \nabla^*$. Therefore, in this case M becomes a Riemannian manifold endowed with the Levi-Civita connection. Hence, the vectors $X_p(t)$ and $X_p^*(t)$ coincide for all $t \in [0, 1]$. Finally, the skew-symmetry of the

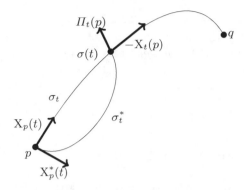

Fig. 2. The vector $\Pi_t(p)$ is obtained by ∇-parallel translating the vector $X_p(t) := \dot{\sigma}_t(0)$ along the ∇^*-geodesic σ_t^*. The opposite of the difference vector $X_t(p)$ at $\sigma(t)$ can be understood as the ∇-parallel translation of the vector $X_p(t)$ along the ∇-geodesic σ_t.

curvature tensor $\mathcal{R}$ yields the property $\mathcal{R}(X, X) = 0$ for any $X \in \mathcal{T}(M)$. When a manifold M is dually flat, it has a mutually dual affine coordinates $\{\boldsymbol{\theta}\}$, $\{\boldsymbol{\eta}\}$ and two potentials $\{\psi, \varphi\}$ such that $D(p, q) = \psi(\boldsymbol{\theta}_p) - \varphi(\boldsymbol{\eta}_q) + \boldsymbol{\theta}_p \cdot \boldsymbol{\eta}_q$ [6]. This claims that the canonical divergence $\mathcal{D}$ coincides with the canonical divergence of Bregman type introduced in [5] by Amari and Nagaoka on dually flat manifolds. The concept of a symmetric statistical manifold, that is the information geometric analogue to a symmetric space in Riemannian geometry, was introduced in [10]. There, the authors employed the following conditions on the curvature tensor, $\nabla\mathcal{R} = 0$ and $R(Y, X, X, X) := \langle \mathcal{R}(Y, X)X, X \rangle = 0$, in order to prove that their divergence function is independent of the particular path connecting any two points p, q sufficiently close to each other. The connection between the canonical divergence $\mathcal{D}$ and the divergence introduced by Henmi and Kobayashi is widely discussed in [7].

Table 1. The column on the left describes the category of statistical manifolds on which the canonical divergence $\mathcal{D}$ reduces to the divergence D. The column on the right shows the properties of the curvature tensor characterizing the corresponding manifolds and supplying the equivalence $\mathcal{D} \equiv D$ through the Eq. (21). Here, X, Y are vector fields on M, i.e. $X, Y \in \mathcal{T}(M)$.

Statistical manifold	Condition on $\mathcal{R}$
Self-dual	$\mathcal{R}(X, X) = 0$
Dually flat	$\mathcal{R} \equiv 0$
Symmetric	$\nabla\mathcal{R} \equiv 0, R(Y, X, X, X) = 0$

To summarize, Table 1 describes, from the top to the bottom, the statistical manifolds ordered from less generality to more generality where the equivalence between $\mathcal{D}$ and D is achieved. In this view, we can consider $\mathcal{D}$ as an extension of the divergence D to the very general statistical manifold (M, g, ∇, ∇^*).

Since Eq. (2) we know that in general a divergence function is not symmetric in its argument. However, the symmetry property owned by the canonical divergence of Bregman type on dually flat manifolds, namely $D(q, p) = D^*(p, q)$, shows the way for the further investigation about symmetry properties of $\mathcal{D}$ in the very general context of Information Geometry.

4   General Properties of $\mathcal{D}$

In this section we prove that the decompositions given in Eq. (17) are orthogonal ones. To this aim, we rely on the *gradient–based approach* to divergence which was introduced in [6] and further developed in [3]. This approach has led to the following decompositions of $\Pi_q(p)$ and $\Pi_q^*(p)$ in terms of the canonical divergence gradient and its dual [7],

$$\Pi_q(p) = \mathrm{grad}_q \mathcal{D}_p(q) + V_q, \quad \langle V_q, \dot{\sigma}(1) \rangle_q = 0 \tag{22}$$

$$\Pi^* q(p) = \mathrm{grad}_q \mathcal{D}_p^*(q) + V_q^*, \quad \langle V_q^*, \dot{\sigma}^*(1) \rangle_{\gamma(t)} = 0, \tag{23}$$

where $\sigma(t), \sigma^*(t), (0 \leq t \leq 1)$ are the ∇ and ∇^* geodesics, respectively, from p to q. In order to prove that these decompositions are orthogonal ones, we exploit the theory of minimum contrast geometry by Eguchi [1]. This allows us to show that $\mathrm{grad}_q \mathcal{D}_p$ is parallel to the tangent vector of the ∇-geodesic starting from p whereas $\mathrm{grad}_q \mathcal{D}_p^*$ is parallel to the tangent vector of the ∇^*-geodesic starting from p. Thus, let us consider the level sets of $\mathcal{D}_p$ and $\mathcal{D}_p^*$:

$$\mathcal{H}(\kappa) = \{q \in M \mid \mathcal{D}_p(q) = \kappa\}, \quad \mathcal{H}^*(\kappa) = \{q \in M \mid \mathcal{D}_p^*(q) = \kappa\}. \tag{24}$$

Then to each $q \in \mathcal{H}(\kappa)$ we can define the minimum contrast leaf of $\mathcal{D}$ at q [2]:

$$L_q := \{p \in M \mid \mathcal{D}(p, q) = \min_{q' \in \mathcal{H}} \mathcal{D}(p, q')\}. \tag{25}$$

Let us now fix q. Since q minimizes the set $\{\mathcal{D}(p, q) \mid p \in L_q, q \in \mathcal{H}\}$ it follows that the derivative of $\mathcal{D}(p, q)$ at q along any direction U tangent to $\mathcal{H}$ vanishes, namely

$$\partial'_U \mathcal{D}(p, q) = 0, \quad \forall U \in T_q \mathcal{H}, \, p \in L_q,$$

where ∂'_U denotes the derivative at q along the direction U. Thus we have that $\langle U, \Xi \rangle_q = 0$ for all $U \in T_q \mathcal{H}$ and $\Xi \in T_q L_q$, or equivalently that the tangent space of L_q coincides with the normal space of $\mathcal{H}$ at q (see Fig. 3 for a cross-reference). In addition, by taking derivatives at q along directions Ξ, Υ normal to $\mathcal{H}$ we have that [1]

$$g\left(II(\Upsilon, \Xi), U\right) := -\partial_\Xi \partial_\Upsilon \partial'_U \mathcal{D}(p, q)|_{p=q} = 0, \quad \forall \Xi, \Upsilon \in T_q L_q, \, U \in T_q \mathcal{H}, \tag{26}$$

where the first relation defines the second fundamental tensor II with respect to the ∇-connection. This implies that the second fundamental tensor with respect to ∇ for L_q vanishes at q. Therefore, according to the well-known Gauss formula [11]

$$\nabla_\Xi \Upsilon = \nabla_\Xi^{L_q} \Upsilon + II(\Xi, \Upsilon)$$

we can see from Eq. (26) that the family of all curves which are orthogonal to the level set $\mathcal{H}$ are all ∇-geodesics ending at q (with a suitable choice of the parameter). Since $\mathrm{grad}_q \mathcal{D}_p$ is orthogonal to $\mathcal{H}$ at q, we finally obtain that $\mathrm{grad}_q \mathcal{D}_p$ is parallel to the ∇-velocity $\dot{\sigma}(1)$. This proves that Eq. (22) provides an orthogonal decomposition.

Analogously, we have that the family of all curves which are orthogonal to the level set $\mathcal{H}^*$ are all ∇^*-geodesics ending at q (with a suitable choice of the parameter). Since $\mathcal{H}^*$ is the hypersurface of constant $\mathcal{D}^*$ it turns out that $\mathrm{grad}_q \mathcal{D}^*$ is orthogonal to $\mathcal{H}^*$ at q. Therefore, we can conclude that the decomposition in Eq. (23) is an orthogonal one.

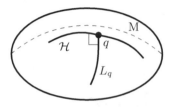

Fig. 3. According to the Eguchi's theory [1], the minimum contrast leaf L_q turns out to be orthogonal at q to the level-set $\mathcal{H}$ generated by the canonical divergence $\mathcal{D}$.

In general, the canonical divergence $\mathcal{D}(p, q)$ is not symmetric in its argument, i.e. $\mathcal{D}(p, q) \neq \mathcal{D}(q, p)$. However, the Kullback-Leibler divergence $\mathrm{K}(p, q)$ and the canonical divergence $\mathrm{D}(p, q)$ of Bregman type suggest a symmetry property that a divergence should satisfy, namely $\mathrm{K}^*(p, q) = \mathrm{K}(q, p)$ and $\mathrm{D}^*(p, q) = \mathrm{D}(q, p)$ [5]. In order to answer what the relation of $\mathcal{D}(p, q)$ and $\mathcal{D}(q, p)$ is, let us set $\widetilde{\mathcal{D}}(p, q) := \mathcal{D}(q, p)$. Consider the hypersurface $\widetilde{\mathcal{H}}$ of constant $\widetilde{\mathcal{D}}$ given by $\widetilde{\mathcal{H}} = \{q \in \mathrm{M} \mid \widetilde{\mathcal{D}}(p, q) = \kappa\}$. Then, by noticing that $-\partial_i \partial_j \partial'_k \widetilde{\mathcal{D}}(p, q) = \Gamma^*_{ijk}$ we have that $\mathrm{grad}_q \widetilde{\mathcal{D}}_p$ is parallel to $\dot{\sigma}^*(1)$ as well as $\mathrm{grad}_q \mathcal{D}_p^*(q)$. Here, $\sigma^*(t)$ is the ∇^*-geodesic from p to q. Recall that, since both functions, $\widetilde{\mathcal{D}}_p(q)$ and $\mathcal{D}_p^*(q)$, are divergences, we have that $\widetilde{\mathcal{D}}(p, q) \geq 0$, $\mathcal{D}^*(p, q) \geq 0$ and $\widetilde{\mathcal{D}}(p, q) = 0$, $\mathcal{D}^*(p, q) = 0$ if and only if $p = q$. Furthermore, the gradient flows of these functions are identical as a family of curves. This implies that, let p fixed, both, $\mathrm{grad}_q \widetilde{\mathcal{D}}_p(q)$ and $\mathrm{grad}_q \mathcal{D}_p^*(q)$, point from q to the opposite direction, along the curve $\sigma^*(t)$, with respect to p. Therefore, we can find a positive constant depending on p and q such that $\mathrm{grad}_q \widetilde{\mathcal{D}}_p = c(p, q) \, \mathrm{grad}_q \mathcal{D}_p^*$. This proves that there exists a function $f : [0, \infty] \to \mathbb{R}^+$ satisfying $f(0) = 0$ and $f'(0) > 0$ such that [7]

$$\mathcal{D}(q, p) = f\left(\mathcal{D}^*(p, q)\right).$$

Though this relation holds for a very general statistical manifold (M, g, ∇, ∇^*), this result is still not satisfactory. However, it allows to face the issue of the symmetry property originally established by the canonical divergence of Bregman type on dually flat manifolds. By analysing the cases where it holds true, we may conjecture through our approach that it is satisfied whenever $r(p, q) = \mathcal{D}(p, q) + \mathcal{D}^*(p, q)$. Or equivalently, whenever the pseudo-squared-distance is symmetric in its argument [7]. Forthcoming investigation will address this fundamental issue of Information Geometry.

References

1. Eguchi, S.: Geometry of minimum contrast. Hiroshima Math. J. **22**, 631–647 (1992)
2. Eguchi, S.: A differential geometric approach to statistical inference on the basis of contrast functions. Hiroshima Math. J. **15**, 341–391 (1985)
3. Ay, N., Jost, J., Van Le, H., Schwachhöfer, L.: Information Geometry, 1st edn. Springer, Cham (2017). https://doi.org/10.1007/978-3-319-56478-4
4. Matumoto, T.: Any statistical manifold has a contrast function-On the C3-functions taking the minimum at the diagonal of the product manifold. Hiroshima Math. J. **23**, 327–332 (1993)
5. Amari, S.-I., Nagaoka, H.: Methods of Information Geometry. Oxford University Press, Oxford (2000)
6. Ay, N., Amari, S.-I.: A novel approach to canonical divergences within information geometry. Entropy **17**, 8111–8129 (2015)
7. Felice, D., Ay, N.: Towards a canonical divergence within information geometry. arXiv:1806.11363 [math.DG] (2018)
8. Lauritzen, S.L.: Differential geometry in statistical inference. Lect. Notes-Monogr. Ser. **10**, 163–218 (1987)
9. Zhang, J.: A note on curvature of α-connections of a statistical manifold. Ann. Inst. Stat. Math. **59**, 161–170 (2007)
10. Henmi, M., Kobayashi, R.: Hooke's law in statistical manifolds and divergences. Nagoya Math. J. **159**, 1–24 (2000)
11. Lee, M.J.: Riemannian Manifolds: An Introduction to Curvature, 1st edn. Springer, New York (1997). https://doi.org/10.1007/b98852

Sobolev Statistical Manifolds
and Exponential Models

Nigel J. Newton[✉]

School of Computer Science and Electronic Engineering, University of Essex,
Wivenhoe Park, Colchester CO4 3SQ, UK
njn@essex.ac.uk

Abstract. This paper develops Sobolev variants of the non-parametric
statistical manifolds appearing in [10] and [11]. The manifolds are modelled on a particular class of weighted, mixed-norm Sobolev spaces,
including a Hilbert-Sobolev space. Densities are expressed in terms of
a *deformed exponential* function having linear growth, which lifts to a
continuous nonlinear superposition (Nemytskii) operator. This property
is used in the construction of finite-dimensional mixture and exponential
submanifolds, on which approximations can be based. The manifolds of
probability measures are developed in their natural setting, as embedded
submanifolds of those of finite measures.

Keywords: Banach manifold · Fisher-Rao metric · Hilbert manifold ·
Information geometry · Log sobolev inequality

1 Introduction

This paper develops non-parametric statistical manifolds modelled on spaces
of Sobolev type. It applies some of the results of [12] to a particular class of
manifolds, and develops smoothly embedded finite-dimensional exponential submanifolds. The non-parametric manifolds are natural refinements of those in [10]
and [11]; they employ charts that are "balanced" between the density function
and its log. The inverses of the charts are expressed in terms of a *deformed
exponential* function having linear growth, a property shared by other deformed
exponentials (notably the Kaniadakis exponential with parameter $\kappa = 1$). (See
[8] and the discussion in [12]). The linear growth property is highly advantageous
in the Sobolev context because the deformed exponential then "lifts" to a nonlinear superposition (Nemytskii) operator that *acts continuously* on particular
classes of model spaces.

For some $d \in \mathbb{N}$, let $\mathcal{X}$ be the σ-algebra of Lebesgue measurable subsets of $\mathbb{R}^d$,
and let μ be a probability measure on $\mathcal{X}$ that is mutually absolutely continuous
with respect to Lebesgue (volume) measure. $\mathcal{X}$ is a very rich collection of subsets,
$A \subset \mathbb{R}^d$, for which the Lebesgue measure $dx(A)$ is well defined. Each $A \in \mathcal{X}$ has a

© Springer Nature Switzerland AG 2019
F. Nielsen and F. Barbaresco (Eds.): GSI 2019, LNCS 11712, pp. 443–452, 2019.
https://doi.org/10.1007/978-3-030-26980-7_46

well-defined probability $\mu(A)$, which can be expressed in terms of the probability density function $r : \mathbb{R}^d \to [0, \infty)$ as follows:

$$\mu(A) = \int_A r(x) \, dx. \tag{1}$$

The simplest example of a statistical manifold over the sample space $\mathbb{R}^d$ is the *finite-dimensional exponential model* [1,3]. This is based on a finite set of linearly independent random variables $\eta_1, \ldots \eta_n$ defined on $(\mathbb{R}^d, \mathcal{X}, \mu)$. Let B be an open subset of $\mathbb{R}^n$ such that, for any $y \in B$, $\mathbf{E}_\mu \exp(\sum_i y_i \eta_i) < \infty$, where $\mathbf{E}_\mu$ is expectation (integration) with respect to μ. Any $y \in B$ represents the probability measure P_y, defined by

$$P_y(A) = \int_A \exp\left(\sum_i y_i \eta_i - c \right) \mu(dx), \tag{2}$$

where $c = \log \mathbf{E}_\mu \exp(\sum_i y_i \eta_i)$. The set $N := \{P_y : y \in B\}$ is a *manifold* of probability measures, with a differentiable structure in terms of which the important statistical divergences of estimation theory are suitably smooth.

The first fully successful infinite-dimensional (non-parametric) statistical manifold was constructed in [14], and further developed in [2,5,13]. This is the natural extension of exponential models such as N to the non-parametric setting. The chart is a centred version of the log of the probability density function $p := dP/d\mu$, and so, as in (2), p is represented in terms of the exponential of the model space variable. The model space used is the *exponential Orlicz space*, which has a stronger topology than the Lebesgue $L^\lambda(\mu)$ spaces for $1 \le \lambda < \infty$.

A central requirement of a chart in a statistical manifold is that it should induce a topology with respect to which statistical divergences, such as the *Kullback-Leibler* (KL)-divergence, are appropriately smooth. The KL-divergence between finite measures P and Q on $\mathcal{X}$ is defined as follows [1,3]:

$$\mathcal{D}(P \,|\, Q) := Q(\mathbb{R}^d) - P(\mathbb{R}^d) + \mathbf{E}_\mu p \log(p/q). \tag{3}$$

It is of class C^∞ on the exponential Orlicz manifold. As (3) shows, the KL-divergence is bilinear in the density p and its log, and so its smoothness properties are closely connected with those of p and $\log p$ considered as elements of dual function spaces. This is why the following *deformed logarithm* $\log_d : (0, \infty) \to \mathbb{R}$ was introduced in the Hilbert setting of [10]:

$$\log_d(y) = y - 1 + \log y. \tag{4}$$

This is composed with probability density functions to realise a chart on a manifold of finite measures that maps into the Lebesgue space $L^\lambda(\mu)$, for any $2 \le \lambda < \infty$ [10,11]. A centred version of this can be used as a chart on the submanifold of probability measures. The inverse of $\log_d$ can be thought of as a *deformed exponential* function. It has linear growth, as a result of which the density, p, and its log both belong to the same space as $\log_d(p)$ (i.e. the model space $L^\lambda(\mu)$). This property is not shared by the exponential Orlicz manifold. Reference [12] shows that it is retained when the sample space is $\mathbb{R}^d$ and the model space $L^\lambda(\mu)$ is replaced by particular spaces of Sobolev type.

The natural domain of statistical divergences, such as the KL-divergence, is a space of measures defined on an abstract measurable space $(\Omega, \mathcal{F})$. Since the primary concern of "raw" information geometry is the smoothness of these divergences, the exponential Orlicz and $L^\lambda(\mu)$ manifolds of [14] and [11], in their general form, make no reference to any other structures that the sample space Ω may possess. However, in the special case that $\Omega = \mathbb{R}^d$, the topology and linear structure of $\mathbb{R}^d$ play important roles in many applications. For example, the Fokker-Planck equation makes direct reference to the linear structure of $\mathbb{R}^d$ through a differential operator. For this reason, it is of interest to develop "hybrid" information manifolds, in which the topology of the sample space is somehow incorporated into the model space. One way of achieving this is to use model spaces of Sobolev type. This approach is taken here, in the context of the $L^\lambda(\mu)$ manifolds of [10, 11]. For the development of Sobolev variants of the exponential Orlicz manifold, the reader is referred to [7].

The paper is structured as follows. Section 2 introduces the spaces on which the manifolds are modelled. Section 3 presents the principal results on the non-parametric manifolds constructed from these spaces; it discusses both manifolds of *finite* measures and submanifolds of *probability* measures. Finally Sect. 4 develops a class of smoothly embedded finite-dimensional exponential manifolds that are of potential use in applications.

2   The Model Spaces

For some $t \in (1, 2]$, let $\theta_t : [0, \infty) \to [0, \infty)$ be a strictly increasing, convex function that is twice continuously differentiable on $(0, \infty)$, such that $-\sqrt{\theta_t}$ is convex, $\lim_{z \downarrow 0} \theta'_t(z) < \infty$, and

$$\theta_t(z) = \begin{Bmatrix} 0 & \text{if } z = 0 \\ c_t + z^t \text{ if } z \geq z_t \end{Bmatrix}, \quad \text{where } z_t \geq 0, \text{ and } c_t \in \mathbb{R}. \tag{5}$$

Examples, including some for which $\mathbb{R} \ni z \mapsto \theta_t(|z|) \in \mathbb{R}$ is of class C^2 are given in [12], which also develops variants for which $t \in (0, 1]$. (The restriction $t \in (1, 2]$, used here, simplifies the presentation in what follows while retaining a useful subset of the manifolds developed in [12]). For each $t \in (1, 2]$, we define a reference probability measure, $\mu = \mu_t$, as follows.

$$\mu_t(dx) = \exp(l_t(x))dx, \quad \text{where } l_t(x) := \sum_i (C_t - \theta_t(|x_i|)), \tag{6}$$

and $C_t \in \mathbb{R}$ is such that $\int \exp(C_t - \theta_t(|z|))dz = 1$. For any $1 \leq \lambda < \infty$, let $L^\lambda(\mu)$ be the Banach space of (equivalence classes of) measurable functions $u : \mathbb{R}^d \to \mathbb{R}$ for which $\|u\|_{L^\lambda(\mu)} := (\int |u|^\lambda d\mu)^{1/\lambda} < \infty$.

For $k \in \mathbb{N}_0$, let $S := \{0, \dots, k\}^d$ be the set of d-tuples of integers in the range $0 \leq s_i \leq k$. For $s \in S$, we define $|s| = \sum_i s_i$, and denote by 0 the d-tuple for which $|s| = 0$. For any $0 \leq i \leq k$, $S_i := \{s \in S : i \leq |s| \leq k\}$ is the set of d-tuples of weight at least i and at most k. Let $\Lambda = (\lambda_0, \lambda_1, \ldots, \lambda_k)$, where $\lambda_i \in [1, \infty)$ for $0 \le i \le k$, and let $W^{k,\Lambda}(\mu)$ be the mixed-norm, weighted Sobolev space comprising functions $a \in L^{\lambda_0}(\mu)$ that have weak partial derivatives $D^s a \in L^{\lambda_{|s|}}(\mu)$, for all $s \in S_1$. For $a \in W^{k,\Lambda}(\mu)$ we define the mixed norm

$$\|a\|_{W^{k,\Lambda}(\mu)} := \left(\sum_{s \in S_0} \|D^s a\|_{L^{\lambda_{|s|}}(\mu)}^{\lambda_0} \right)^{1/\lambda_0} < \infty. \tag{7}$$

$W^{k,\Lambda}(\mu)$ is a Banach space with respect to this norm. (See Theorem 2.1 in [12]).

We shall confine our attention here to the following special class of model spaces, parametrised by $k \in \mathbb{N}$ and $\lambda \in [k, \infty)$:

$$G_m^{k,\lambda} = W^{k,\Lambda}(\mu) \quad \text{with } \lambda_0 = \lambda \text{ and } \lambda_i = \lambda/i \text{ for } 1 \le i \le k. \tag{8}$$

This includes the Hilbert-Sobolev space $G_m^{1,2}$. Let $\psi = \exp_d : \mathbb{R} \to (0, \infty)$ be the inverse of the deformed logarithm of (4). The following is proved as part of Proposition 2 in [12].

Proposition 1. (i) *For any $a \in G_m^{k,\lambda}$, $\psi(a) \in G_m^{k,\lambda}$.*
(ii) *The nonlinear superposition (Nemytskii) operator $\Psi_m^{k,\lambda} : G_m^{k,\lambda} \to G_m^{k,\lambda}$, defined by $\Psi_m^{k,\lambda}(a)(x) = \psi(a(x))$, is continuous.*

This is rare property in the theory of nonlinear maps between Sobolev spaces, and has its origins in the boundedness of the derivatives of ψ. It is useful in the construction of finite-dimensional mixture or exponential submanifolds. Normally the Sobolev space forming the domain of a continuous nonlinear superposition operator would need a stronger topology than that forming its range [15]. This would require it to have larger Lebesgue exponents, or to control a greater number of derivatives.

Remark 1. It is shown in [12] that the continuity of Ψ between identical Sobolev spaces is also true of the fixed-norm space $W^{2,(1,1,1)}(\mu)$, but of no other fixed-norm spaces (except for $G_m^{1,\lambda}$ with $\lambda \in [1, \infty)$).

3   The Nonparametric Manifolds

Let $G = G_m^{k,\lambda}$ be the mixed-norm Sobolev space as defined in (8). Let $\Psi := \Psi_m^{k,\lambda}$ be as defined in Proposition 1. We consider the set M of finite measures on $\mathcal{X}$ satisfying the following:

(M1) P is mutually absolutely continuous with respect to μ;
(M2) $p, \log p \in G$, where $p = dP/d\mu$.

This is equipped with the global chart $\phi : M \to G$, defined by

$$\phi(P) = \log_d p = p - 1 + \log p. \tag{9}$$

In view of Proposition 1, it is not difficult to show that ϕ is a bijection onto G. (See Proposition 1 in [12]). For any $P \in M$, let $\tilde{P}_a \in M$ have density $d\tilde{P}_a/d\mu = \psi^{(1)}(a)$, where $a = \phi(P)$ and $\psi^{(1)} = \psi/(1 + \psi)$ is the first derivative of ψ. We define a tangent vector, U at $P \in M$, to be a *signed* measure on $\mathcal{X}$ of finite total variation such that

(T1) U is mutually absolutely continuous with respect to $\tilde{P}_a$;
(T2) $dU/d\tilde{P}_a \in G$, where $a = \phi(P)$.

The tangent space $T_P M$ is the linear space of all such signed measures, and the tangent bundle is the disjoint union $TM = \cup_{P \in M}(P, T_P M)$. This admits the global chart $\Phi : TM \to G \times G$, defined by

$$\Phi(P, U) = (\phi(P), dU/d\tilde{P}_{\phi(P)}). \tag{10}$$

The derivative of a (Fréchet) differentiable, Banach-space-valued map $f : M \to \mathbb{Y}$ (at P and in the "direction" U) is defined as follows: (clearly $u = U\phi$).

$$Uf = (f \circ \phi^{-1})_a^{(1)} u, \quad \text{where } (a, u) = \Phi(P, U). \tag{11}$$

Let $m_\lambda, e_\lambda : M \to L^\lambda(\mu)$ be the nonlinear superposition operators defined by

$$m_\lambda(P)(x) = p(x) - 1 \quad \text{and} \quad e_\lambda(P)(x) = \log p(x). \tag{12}$$

The map m_λ is the composition of $\Psi_m^{k,\lambda} - 1$ with the inclusion map $\imath : G \to L^\lambda(\mu)$. It is smoother than $\Psi_m^{k,\lambda}$ since its range has a weaker topology. The following is a corollary of Lemma 4 in [12].

Lemma 1. $m_\lambda, e_\lambda \in C^1(M; L^\lambda(\mu))$.

The smoothness properties of the KL-divergence on manifolds modelled on $L^\lambda(\mu)$ is investigated in detail in [11]. Its derivatives can be used to construct the Fisher-Rao metric and Amari-Chentsov tensor on M by the Eguchi method [4]. The Fisher-Rao metric is the covariant 2-tensor field defined, for $\lambda \geq 2$, by

$$\langle U, V \rangle_P = \mathbf{E}_\mu U m_\lambda V e_\lambda = \mathbf{E}_\mu U e_\lambda V m_\lambda = \mathbf{E}_\mu \frac{p}{(1+p)^2} U\phi V\phi. \tag{13}$$

The Amari-Chentsov tensor is the covariant 3-tensor field defined, for $\lambda \geq 3$, by

$$\tau_P(U, V, W) = \mathbf{E}_\mu U m_\lambda V e_\lambda W e_\lambda = \mathbf{E}_\mu \frac{p}{(1+p)^3} U\phi V\phi W\phi \tag{14}$$

Corollary 1. (i) *If $\lambda \geq 2$ then the Fisher-Rao metric is a continuous covariant 2-tensor field on M.*
(ii) *If $\lambda \geq 3$ then the Amari-Chentsov tensor is a continuous covariant 3-tensor field on M.*

Proof. Both parts follows from the first representations in (13) and (14), Lemma 1 and the chain rule of Fréchet derivatives. □

In the raw (non-Sobolev) Hilbert manifold of [10], the composition map $M \ni P \mapsto \langle \mathbf{U}(P), \mathbf{V}(P) \rangle_P \in \mathbb{R}$ is continuous for all continuous vector fields $\mathbf{U}$, $\mathbf{V}$. However, the metric is not continuous in the stronger "operator topology" of Corollary 1(i). The extra regularity here arises from the log-Sobolev embedding theorem, and is not retained if $t \in (0, 1]$. (See Lemma 4 in [12]). Similarly, in the raw Banach manifold of [11] with $\lambda \geq 3$, the composition map $M \ni P \mapsto \tau_P(\mathbf{U}(P), \mathbf{V}(P), \mathbf{W}(P)) \in \mathbb{R}$ is continuous for all continuous vector fields $\mathbf{U}$, $\mathbf{V}$, $\mathbf{W}$, but not continuous in the sense of Corollary 1(ii) unless $\lambda > 3$.

Let $M_0 \subset M$ be the subset of M whose members are *probability* measures. These satisfy the additional hypothesis:

(M3) $\mathbf{E}_\mu p = 1$.

The co-dimension 1 subspace of G whose members, a, satisfy $\mathbf{E}_\mu a = 0$ will be denoted G_0. Let $\phi_0 : M_0 \to G_0$ be defined by

$$\phi_0(P) = \phi(P) - \mathbf{E}_\mu \phi(P) = \log_d p - \mathbf{E}_\mu \log_d p. \tag{15}$$

The following is a special case of parts of Propositions 5 and 6 in [12].

Proposition 2. **(i)** (M_0, G_0, ϕ_0) *is a* $C^{\lfloor \lambda \rfloor}$-*embedded submanifold of* (M, G, ϕ).
(ii) *In terms of the charts* ϕ_0 *and* ϕ, *the inclusion map* $\rho : G_0 \to G$ *has the following form*

$$\rho(a) = a + Z(a), \tag{16}$$

where $Z : G_0 \to \mathbb{R}$ *is an (implicitly defined) normalisation function.*
(iii) *The first and (if* $\lambda \geq 2$) *second derivatives of* ρ *are as follows:*

$$\rho_a^{(1)} u = u - \mathbf{E}_{P_a} u,$$

$$\rho_a^{(2)}(u, v) = -\frac{\mathbf{E}_\mu \psi^{(2)}(\rho(a))(u - \mathbf{E}_{P_a} u)(v - \mathbf{E}_{P_a} v)}{\mathbf{E}_\mu \psi^{(1)}(\rho(a))}, \tag{17}$$

where $P_a := \tilde{P}_a / \tilde{P}_a(\mathbb{R}^d)$, $\tilde{P}_a$ *is the finite measure defined after (9) and* $\psi^{(i)}$ *is the* i*th derivative of* ψ.

For any $P \in M_0$, tangent vectors $U \in T_P M_0$ are distinguished from those only in $T_P M$ by the fact that $U(\mathbb{R}^d) = 0$. The pushforward of the inclusion map $\imath : M_0 \to M$ *splits* $T_P M$ into $T_P M_0$ and the complementary subspace of signed measures $\{y \tilde{P}_a, y \in \mathbb{R}\}$.

Remark 2. The probability measure P_a in (17) is the *escort* probability in the interpretation of M_0 as a *deformed exponential model* [9].

4   Finite-Dimensional Exponential Models

The α-divergences (and their derivatives such as the Fisher-Rao metric) are at least as smooth on the Sobolev manifolds of Sect. 3 as they are on their non-Sobolev counterparts (as developed in [10,11]) because the Sobolev manifolds have stronger topologies. When it comes to embedded submanifolds, however, this benefit is reversed. Theorem 5.1 in [10] shows that any finite dimensional exponential manifold that is contained in the raw Hilbert manifold of [10] is smoothly embedded in that manifold. That this is not so of the Hilbert-Sobolev manifold M modelled on $G_m^{1,2}$ is demonstrated by the following example.

Example 1. Let $d = 1$, let μ be the standard Gaussian measure, $k = 1$ and $\lambda = 2$. For $i = 0, 1$, let $p_i := \exp(\eta_i)$ where $\eta_0 = 3x^2/16$ and $\eta_1 = \sin(\exp(3x^2/16))$. Then P_0 and P_1 are both in M, but the measure with density $\exp((\eta_0 + \eta_1)/2)$ is not, since its derivative is not square integrable.

Nevertheless, the smooth embedding property can be recovered under additional hypotheses. Let $G = G_m^{k,\lambda}$ be the general mixed-norm space of Sect. 2, and let M be the associated manifold of finite measures. Since the Fisher-Rao metric is positive definite on M, it is a (strong) Riemannian metric on any finite-dimensional, smoothly embedded submanifold, N, and the full geometry of dual $\pm\alpha$-covariant derivatives (for $\alpha \in [-1, 1]$) is realised on N.

For some $n \in \mathbb{N}$, let $1, \eta_1, \dots, \eta_n$ be linearly independent members of G, and for any $y \in \mathbb{R}^{n+1}$ let $P(y)$ be the measure on $\mathcal{X}$ with density

$$p(y) := \exp\gamma(y), \quad \text{where} \quad \gamma(y) = \sum_{j=0}^{n} y_j\eta_j \quad \text{and} \quad \eta_0 \equiv 1. \qquad (18)$$

The function $\gamma : \mathbb{R}^{n+1} \to G$ is clearly injective. Let $B \subset \mathbb{R}^n$ be open and such that $P(y) \in M$ for every $y \in \mathbb{R} \times B$, and let $N := \{P(y) : y \in \mathbb{R} \times B\}$. As well as being a subset of M, N is a finite-dimensional exponential model with chart $\theta : N \to \mathbb{R} \times B$, defined by $\theta = \gamma^{-1} \circ e_\lambda \circ \imath$, where e_λ is as defined in (12) and $\imath : N \to M$ is the inclusion map.

Theorem 1. *Suppose that $\lambda \geq \max\{2, k\}$ and that, for every $y \in B$, $1 \leq j \leq n$ and $s \in S_1$,*

$$\mathbf{E}_\mu \, |p(y)D^s\eta_j|^{\lambda/|s|} < \infty; \qquad (19)$$

then N is a C^∞-embedded submanifold of M.

Proof. A *partition* of $s \in S_1$ is a set $\pi = \{\sigma_1, \dots, \sigma_n \in S_1\}$ such that $\sum_i \sigma_i = s$. Let $\Pi(s)$ denote the set of partitions of s. According to the Faá di Bruno formula, for any $s \in S_1$,

$$D_x^s p(y) = p(y) \sum_{\pi \in \Pi(s)} K_\pi \prod_{\sigma \in \pi} D_x^\sigma \gamma(y), \qquad (20)$$

where the $K_\pi < \infty$ are combinatoric constants, and x is made explicit in D_x^s for the sake of clarity.

As in the proof of Theorem 5.1 in [10], we define a local coordinate system around a generic $y \in \mathbb{R} \times B$. Let $\epsilon > 0$ be such that the ball of centre y and radius ϵ is contained in $\mathbb{R} \times B$, and let $B(y, r)$ be the ball of centre y and radius r. For any $\tilde{y} \in B(y, \epsilon/2n)$ let $\zeta \in (1/4, 3/4)^{n+1}$ be defined by $\zeta_j = (1 + (n+1)\epsilon^{-1}(\tilde{y}-y)_j)/2$; then

$$\tilde{y} = \frac{1}{n+1} \sum_{j=0}^{n} \left((1-\zeta_j)(y - \epsilon \mathbf{e}_j) + \zeta_j (y + \epsilon \mathbf{e}_j) \right),$$

where $(\mathbf{e}_j \in \mathbb{R}^{n+1}, 0 \leq j \leq n)$ is the coordinate orthonormal basis. Differentiating p with respect to y, for any $\alpha \in \mathbb{N}_0^{n+1}$,

$$D_y^\alpha p(\tilde{y}) = (2\epsilon)^{-|\alpha|} \prod_{j=0}^{n} \left(p_{j-}^{1-\zeta_j} p_{j+}^{\zeta_j} \log^{(n+1)\alpha_j} (p_{j+}/p_{j-}) \right)^{1/(n+1)}, \qquad (21)$$

where $p_{j\pm} = p(y \pm \epsilon \mathbf{e}_j)$. The product rule now shows that, for any $\tilde{y} \in B(y, \epsilon/2n)$,

$$D_y^\alpha D_x^s p(\tilde{y}) = \sum_{\beta \leq \alpha} K_\beta D_y^{\alpha-\beta} p(\tilde{y}) \sum_{\pi \in \Pi(s)} K_\pi D_y^\beta \prod_{\sigma \in \pi} D_x^\sigma \gamma(\tilde{y}), \qquad (22)$$

where the $K_\beta < \infty$ are combinatoric constants. For any $m \in \mathbb{N}_0$ there is a $K_m < \infty$ such that, for all $q, r \in (0, \infty)$ and all $\delta \in (1/4, 3/4)$,

$$q^{1-\delta} r^\delta |\log(q/r)|^m = \frac{q+r}{(q/r)^\delta + (r/q)^{1-\delta}} |\log(q/r)|^m \leq K_m(q+r).$$

Applying this to (21) and (22), we obtain the bound

$$\left| D_y^\alpha D_x^s p(\tilde{y}) \right| \leq K \sum_{\beta \leq \alpha} \sum_{\pi \in \Pi(s)} \prod_{j=0}^{n} \left| (p_{j-} + p_{j+}) D_y^\beta \prod_{\sigma \in \pi} D_x^\sigma \gamma(\tilde{y}) \right|^{1/(n+1)},$$

for some $K < \infty$. It follows from (19) and Hölder's inequality that the term, whose absolute value is taken on the right-hand side here, belongs to $L^{\lambda/|s|}(\mu)$, and a further application of Hölder's inequality shows that

$$\mathbf{E}_\mu \sup_{\tilde{y} \in B(y, \epsilon/2n)} \left| D_y^\alpha D_x^s p(\tilde{y}) \right|^{\lambda/|s|} < \infty. \qquad (23)$$

A Taylor expansion of $D_y^\alpha D_x^s p_y$ about y, in the direction $\mathbf{e}_j$, yields

$$D_y^\alpha D_x^s p(y + t\mathbf{e}_j) = D_y^\alpha D_x^s p(y) + D_y^{\alpha+\mathbf{e}_j} D_x^s p(y)t + D_y^{\alpha+2\mathbf{e}_j} D_x^s p(y + \delta t \mathbf{e}_j)t^2/2,$$

for some $\delta = \delta(y, t, j, x) \in [0, 1]$. Together with (23) and the dominated convergence theorem, this shows that $(-\epsilon/2n, \epsilon/2n) \ni t \mapsto D_y^\alpha D_x^s p(y + t\mathbf{e}_j) \in L^{\lambda|s|}(\mu)$ is differentiable at $t = 0$, with derivative $D_y^{\alpha+\mathbf{e}_j} D_x^s p(y)$. An inductive argument thus establishes the infinite differentiability of $\mathbb{R} \times B \in y \mapsto D_x^s p(y) \in L^{\lambda|s|}(\mu)$. The same is clearly true of $\mathbb{R} \times B \in y \mapsto p(y) \in L^\lambda(\mu)$, and so we have shown that the inclusion map $\imath$ is of class C^∞.

Expressed in terms of the charts, $\imath$ takes the form $f = \phi \circ e_\lambda^{-1} \circ \gamma$. Let $g : G \to \mathbb{R}^{n+1}$ be defined by $g = \pi \circ e_\lambda \circ \phi^{-1}$, where $\pi : L^\lambda(\mu) \to \mathbb{R}^{n+1}$ is the $L^2(\mu)$-projection onto the subspace spanned by $(1, \eta_1, \ldots, \eta_n)$. It follows from Lemma 1 that g is of class C^1. Now $g \circ f$ is the identity function of $\mathbb{R} \times B$, and so f is a homeomorphism onto its image (endowed with the subspace topology), and its derivative, $f^{(1)}$, is a toplinear isomorphism onto its image. So $f^{(1)}\mathbb{R}^{n+1}$ is a finite-dimensional closed linear subspace of both G and $L^2(\mu)$. Let H^c be its orthogonal complement in $L^2(\mu)$; then $f^{(1)}$ *splits* G into the components $f^{(1)}\mathbb{R}^{n+1}$ and $G \cap H^c$. So f is an *immersion* and an embedding. (See Proposition 2.3 in [6]). This completes the proof. □

Condition (19) is clearly satisfied if $p(y) \in L^\infty(\mu)$ for all $y \in \mathbb{R} \times B$; this is so, for example, if N is the exponential manifold of all non-singular (scaled) Gaussian measures on $\mathcal{X}$, and $t < 2$. However, there are other possibilities, for example that in which $\gamma(y)$, $p(y)$ and their x derivatives have sub-exponential growth in x for all $y \in B$.

Let $N_0 := M_0 \cap N$ be the subset of probability measures. This is, itself, a finite dimensional exponential manifold with chart $\theta_0 : N \to B$ defined by $\theta_0(P) = (\theta(P)_1, \ldots, \theta(P)_n)$.

Corollary 2. *Under the hypotheses of Theorem 1:*

(i) N_0 *is a C^∞-embedded submanifold of N;*
(ii) N_0 *is a C^∞-embedded submanifold of M_0.*

Proof. The map $\mathbb{R} \times B \ni y \mapsto \mathbf{E}_\mu p(y) \in (0, \infty)$ was shown, in the proof of Theorem 1, to be of class C^∞. Its first derivative with respect to y_0 is $\mathbf{E}_\mu p(y)$. Since this is strictly positive on $\mathbb{R} \times B$, the implicit function theorem shows that $f : B \to \mathbb{R}$, defined by $\exp(f(z) + \sum_{j=1}^n z_j \eta_j) = 1$, is of class C^∞. In terms of the charts θ and θ_0, the inclusion map $\imath : N_0 \to N$ takes the form $\varphi(z) = (f(z), z)$. This is clearly a C^∞-embedding, which proves part (i).

Let $\tau : G \to G_0$ be defined by $\tau(a) = a - \mathbf{E}_\mu a$. This is of class C^∞, and so the same is true of the map $g := \tau \circ \phi \circ \imath \circ \theta_0^{-1} : B \to G_0$, where $\imath : N_0 \to M$ is the inclusion map. Now g is injective and, since $\tau^{(1)} \rho^{(1)}$ is the identity map of G_0, the same is true of its first derivative, at all points in B. The latter clearly splits G_0, and so the inclusion map $\imath : N_0 \to M_0$ (which is expressed in charts by g) is an embedding. This completes the proof of part (ii). □

References

1. Amari, S.-I., Nagaoka, H.: Methods of Information Geometry. American Mathematical Society, Providence (2000)
2. Cena, A., Pistone, G.: Exponential statistical manifold. Ann. Inst. Stat. Math. **59**, 27–56 (2007)
3. Chentsov, N.N.: Optimal Decision Rules and Optimal Inference. American Mathematical Society, Providence (1982)

4. Eguchi, S.: Second order efficiency of minimum contrast estimators in a curved exponential family. Ann. Stat. **11**, 793–803 (1983)
5. Gibilisco, P., Pistone, G.: Connections on non-parametric statistical manifolds by Orlicz space geometry. Infin. Dimens. Anal. Quantum Probab. Relat. Top. **1**, 325–347 (1998)
6. Lang, S.: Fundamentals of Differential Geometry. Springer, New York (2001)
7. Lods, B., Pistone, G.: Information geometry formalism for the spatially homogeneous Boltzmann equation. Entropy **17**, 4323–4363 (2015)
8. Montrucchio, L., Pistone, G.: A class of non-parametric deformed exponential statistical models, arXiv:1709.01430 (2018)
9. Naudts, J.: Generalised Thermostatistics. Springer, London (2011). https://doi.org/10.1007/978-0-85729-355-8
10. Newton, N.J.: An infinite-dimensional statistical manifold modelled on Hilbert space. J. Funct. Anal. **263**, 1661–1681 (2012)
11. Newton, N.J.: Infinite-dimensional statistical manifolds based on a balanced chart. Bernoulli **22**, 711–731 (2016)
12. Newton, N.J.: A class of non-parametric statistical manifolds modelled on Sobolev space, arXiv:1808.06451 (2018)
13. Pistone, G., Rogantin, M.P.: The exponential statistical manifold: mean parameters, orthogonality and space transformaions. Bernoulli **5**, 721–760 (1999)
14. Pistone, G., Sempi, C.: An infinite-dimensional geometric structure on the space of all the probability measures equivalent to a given one. Ann. Stat. **23**, 1543–1561 (1995)
15. Runst, T., Sickel, W.: Sobolev Spaces of Fractional Order, Nemytskij Operators, and Nonlinear Partial Differential Equations. de Gruyter, Berlin (2011)

Minimization of the Kullback-Leibler Divergence over a Log-Normal Exponential Arc

Paola Siri$^{(\boxtimes)}$ and Barbara Trivellato

Dipartimento di Scienze Matematiche "G.L. Lagrange",
Dipartimento di Eccellenza 2018–2022, Politecnico di Torino,
Corso Duca degli Abruzzi 24, 10129 Torino, Italy
{paola.siri,barbara.trivellato}@polito.it

Abstract. The Kullback-Leibler divergence of a given log-normal density from a log-normal exponential arc is minimized, obtaining an optimal value which is robust with respect to homogeneous transformations leaving the correlation of the random variables involved unchanged.

Keywords: Exponential models · Orlicz spaces ·
Kullback-Leibler divergence

1 Introduction

The geometry of nonparametric exponential models and its analytical properties in the topology of Orlicz spaces started with the paper of Pistone and Sempi [5]. In their framework, the starting point is the notion of *maximal exponential model* centered at a given positive density p, which is defined using the Orlicz space associated to an exponentially growing Young function. One of the main result in the subsequent work by Cena and Pistone [1] states that any density belonging to the maximal exponential model centered at p is connected by an *open* exponential arc to p and viceversa (by *open*, we essentially mean that the two densities are not the extremal points of the arc). Further upgrades have been proved in Imparato and Trivellato [2] and, more recently, in Santacroce, Siri and Trivellato [6–9]. In particular, in [6], the equivalence between the equality of the maximal exponential models centered at two (connected) densities p and q and the equality of the Orlicz spaces referred to the same densities is proved. Recently, different authors have generalized the exponential setting by replacing the exponential function with a new class of functions, called deformed exponentials (see, e.g., Vigelis and Cavalcante [10]).

Applications of statistical exponential models built on Orlicz spaces can be found in several fields, such as differential geometry, algebraic statistics, information theory. In mathematical finance, applications to convex duality have been recently given by [8] and [9], while in physics by, e.g., [4] and [3].

© Springer Nature Switzerland AG 2019
F. Nielsen and F. Barbaresco (Eds.): GSI 2019, LNCS 11712, pp. 453–461, 2019.
https://doi.org/10.1007/978-3-030-26980-7_47

In this paper, we briefly recall some results on exponential connections by arc and their relation to Orlicz spaces. We then deal with the minimization of the Kullback-Leibler divergence of a given density from an open exponential arc. The optimal density is explicitly computed in the log-normal case, and the corresponding minimal divergence turns out to be robust with respect to homogeneous transformations which leave the correlation of the random variables involved unchanged. Finally, we specify our results in the Merton's financial model.

2   Orlicz Spaces and Exponential Models

In this section we recall the notions of Orlicz space, exponential arc and maximal exponential model, as well as the corresponding main results.

Let $(\mathcal{X}, \mathcal{F}, \mu)$ be a fixed probability space and denote with $\mathcal{P}$ the set of all densities which are positive μ-a.s. and with $\mathbb{E}_p$ the expectation with respect to $p \cdot \mu$, for each fixed $p \in \mathcal{P}$.

Let us consider the Young function $\Phi_1(x) = \cosh(x) - 1$, equivalent to the more commonly used $\Phi_2(x) = e^{|x|} - |x| - 1$.

Its conjugate function is $\Psi_1(y) = \int_0^y \sinh^{-1}(t)dt$, which, in its turn, is equivalent to $\Psi_2(y) = (1 + |y|)\log(1 + |y|) - |y|$.

Given $p \in \mathcal{P}$, the Orlicz space associated to Φ_1 is defined by

$$L^{\Phi_1}(p) = \{u \text{ measurable } : \ \exists \ \alpha > 0 \ s.t. \ \mathbb{E}_p(\Phi_1(\alpha u)) < +\infty\} \qquad (1)$$

and it is a Banach space when endowed with the *Luxembourg norm*

$$\|u\|_{\Phi_1, p} = \inf\left\{k > 0 : \ \mathbb{E}_p\left(\Phi_1\left(\frac{u}{k}\right)\right) \le 1\right\}. \qquad (2)$$

Moreover, it is worth to note the following chain of inclusions:

$$L^{\infty}(p) \subseteq L^{\Phi_1}(p) \subseteq L^a(p) \subseteq L^{\psi_1}(p) \subseteq L^1(p), \ a > 1.$$

Definition 1. *$p, q \in \mathcal{P}$ are connected by an open exponential arc if there exists an open interval $I \supset [0, 1]$ such that one of the following equivalent relations is satisfied:*

1. *$p(\theta) \propto p^{(1-\theta)}q^{\theta} \in \mathcal{P}, \ \forall \theta \in I$;*
2. *$p(\theta) \propto e^{\theta u}p \in \mathcal{P}, \ \forall \theta \in I$, where $u \in L^{\Phi_1}(p)$ and $p(0) = p, \ p(1) = q$.*

Observe that connection by open exponential arcs is an equivalence relation.

Let us consider the cumulant generating functional map defined on $L_0^{\Phi_1}(p) = \{u \in L^{\Phi_1}(p) : \mathbb{E}_p(u) = 0\}$, by the relation $K_p(u) = \log \mathbb{E}_p(e^u)$. We recall from Pistone and Sempi [5] that K_p is a positive convex and lower semicontinuous function, vanishing at zero, and that the interior of its proper domain, denoted here by $\overset{\circ}{\text{dom}} K_p$, is a non empty convex set.

For every density $p \in \mathcal{P}$, we can then define the maximal exponential model at p as

$$\mathcal{E}(p) = \left\{ q = e^{u - K_p(u)} p : u \in \overset{\circ}{\mathrm{dom}}\, K_p \right\} \subseteq \mathcal{P}. \tag{3}$$

The following theorem, which is one of the central results of [1,6,9], gives equivalent conditions to open exponential connection by arcs. We state it in a complete version, containing all the recent improvements.

Theorem 1. *(Portmanteau Theorem)*
Let $p, q \in \mathcal{P}$. The following statements are equivalent.

(i) $q \in \mathcal{E}(p)$;
(ii) q is connected to p by an open exponential arc;
(iii) $\mathcal{E}(p) = \mathcal{E}(q)$;
(iv) $\log \frac{q}{p} \in L^{\Phi_1}(p) \cap L^{\Phi_1}(q)$;
(v) $L^{\Phi_1}(p) = L^{\Phi_1}(q)$;
(vi) $\frac{q}{p} \in L^{1+\varepsilon}(p)$ *and* $\frac{p}{q} \in L^{1+\varepsilon}(q)$, *for some* $\varepsilon > 0$;
(vii) $^m\mathbb{U}_p^q : L^{\Phi_1}(p) \longrightarrow L^{\Psi_1}(q)$ *s.t.* $^m\mathbb{U}_p^q(v) = \frac{p}{q}v$ *is an isomorphism of Banach spaces.*

3   Minimization of the Kullback-Lebleir Divergence

Let

$$D(q\|p) = \mathbb{E}_p\left(\frac{q}{p} \log \frac{q}{p} \right) = \mathbb{E}_q\left(\log \frac{q}{p} \right)$$

be the Kullback-Leibler divergence of $q \cdot \mu$ with respect to $p \cdot \mu$. We will simply refer to it as the divergence of q from p.

The following result relates the divergence finiteness to Orlicz spaces and the proof can be find in Cena and Pistone [1] for the first statement and in Santacroce, Siri and Trivellato [6] for the second one.

Proposition 1. *Let* $p, q \in \mathcal{P}$. *Then*

(i) $D(q\|p) < \infty \Leftrightarrow \log \frac{q}{p} \in L^1(q) \Leftrightarrow \frac{q}{p} \in L^{\Psi_1}(p)$;
(ii) $D(q\|p) < \infty \Rightarrow L^{\Phi_1}(p) \subseteq L^1(q)$ *(not only* $L^{\Phi_1}(p) \subseteq L^1(p)$*)*.

Moreover, when working with the maximal exponential model, a stronger result holds, implied by the Portmanteau Theorem:

Proposition 2. *If* $q \in \mathcal{E}(p)$, *then the Kullback-Leibler divergences* $D(q\|p)$ *and* $D(p\|q)$ *are both finite.*

Nevertheless the converse is not true, as the counterexamples in Santacroce, Siri and Trivellato [6,9] show.

For a fixed $p \in \mathcal{P}$, let us now consider $q, r \in \mathcal{E}(p)$.

Denote with $r(\theta)$ a generic element of the open arc between r and p, i.e. for every $\theta \in I = (-\varepsilon, 1 + \varepsilon), \ \varepsilon > 0$

$$r(\theta) = \frac{p^{(1-\theta)} r^\theta}{c(\theta)} \in \mathcal{P}, \quad \text{where} \ \ c(\theta) = \mathbb{E}_\mu \left(p^{(1-\theta)} r^\theta \right) = \mathbb{E}_p \left(\left(\frac{r}{p} \right)^\theta \right),$$

and suppose that q does not belong to the arc.

We are interested in minimizing the divergence of q from $r(\theta)$ over the exponential arc, i.e. in finding

$$\min_{\theta \in I} D(q \| r(\theta)). \tag{4}$$

Explicit computations give

$$D(q \| r(\theta)) = \mathbb{E}_q \left(\log \frac{q}{r(\theta)} \right) = \mathbb{E}_q (\log q) - \mathbb{E}_q \left(\log \frac{p^{(1-\theta)} r^\theta}{c(\theta)} \right)$$

$$= \mathbb{E}_q (\log q) - (1 - \theta) \mathbb{E}_q (\log p) - \theta \mathbb{E}_q (\log r) + \log \mathbb{E}_p \left(\left(\frac{r}{p} \right)^\theta \right)$$

$$= \mathbb{E}_q \left(\log \frac{q}{p} \right) - \theta \mathbb{E}_q \left(\log \frac{r}{p} \right) + \log \mathbb{E}_p \left(\left(\frac{r}{p} \right)^\theta \right).$$

The corresponding stationary condition is

$$\frac{d}{d\theta} D(q \| r(\theta)) = -\mathbb{E}_q \left(\log \frac{r}{p} \right) + \frac{\mathbb{E}_p \left(\left(\frac{r}{p} \right)^\theta \log \frac{r}{p} \right)}{\mathbb{E}_p \left(\left(\frac{r}{p} \right)^\theta \right)} = 0,$$

i.e.

$$\mathbb{E}_p \left(\left(\frac{r}{p} \right)^\theta \log \frac{r}{p} \right) = \mathbb{E}_p \left(\left(\frac{r}{p} \right)^\theta \right) \mathbb{E}_q \left(\log \frac{r}{p} \right).$$

Since $q, r \in \mathcal{E}(p)$, then $q = e^{u - K_p(u)} p$ and $r = e^{v - K_p(v)} p$, with $u, v \in \overset{\circ}{\text{dom}} K_p$, and we can rewrite the divergence as

$$D(q \| r(\theta)) = D(q \| p) - \theta \mathbb{E}_q (v - K_p(v)) + \log \mathbb{E}_p \left(e^{\theta(v - K_p(v))} \right)$$

$$= D(q \| p) - \theta \mathbb{E}_q (v) + \log \mathbb{E}_p \left(e^{\theta v} \right).$$

The stationary condition then becomes

$$\mathbb{E}_p \left(v e^{\theta v} \right) = \mathbb{E}_p \left(e^{\theta v} \right) \mathbb{E}_q (v).$$

Let us now introduce $M_v(\theta) = \mathbb{E}_p(e^{\theta v}) = e^{K_p(\theta v)}$, the moment generating function of v with respect to $p \cdot \mu$, which is well defined on $I = (-\varepsilon, 1 + \varepsilon)$ since $v \in \overset{\circ}{\text{dom}} K_p$.

In terms of the moment generating function, we then get

$$D(q\|r(\theta)) = D(q\|p) - \theta\mathbb{E}_q(v) + \log M_v(\theta), \qquad (5)$$

with the stationary condition

$$M_v'(\theta) = M_v(\theta)\mathbb{E}_q(v). \qquad (6)$$

3.1   The Log-Normal Case

Let us now suppose that (u, v) is a non-degenerate Gaussian vector with p-mean (m_u, m_v) and p-covariance matrix $\begin{pmatrix} \sigma_u^2 & c \\ c & \sigma_v^2 \end{pmatrix}$, with parameters $m_u, m_v \in \mathbb{R}$, $\sigma_u, \sigma_v > 0$ and $c \in \mathbb{R}$ such that $-\sigma_u\sigma_v < c < \sigma_u\sigma_v$.

Since, $\forall(\xi, \theta) \in \mathbb{R}^2$, the moment generating function of the Gaussian vector is

$$M_{(u,v)}(\xi, \theta) = \mathbb{E}_p\left(e^{\xi u + \theta v}\right) = \exp\left(m_u\xi + m_v\theta + \frac{1}{2}\sigma_u^2\xi^2 + \frac{1}{2}\sigma_v^2\theta^2 + c\xi\theta\right)$$

and

$$\frac{\partial}{\partial\theta}M_{(u,v)}(\xi, \theta) = \exp\left(m_u\xi + m_v\theta + \frac{1}{2}\sigma_u^2\xi^2 + \frac{1}{2}\sigma_v^2\theta^2 + c\xi\theta\right) \cdot (m_v + \sigma_v^2\theta + c\xi),$$

we can derive

$$\mathbb{E}_p(ve^u) = \frac{\partial}{\partial\theta}M_{(u,v)}(1, 0) = \exp\left(m_u + \frac{1}{2}\sigma_u^2\right) \cdot (m_v + c)$$
$$= M_u(1) \cdot (m_v + c). \qquad (7)$$

Specifically, in our case $u, v \in \overset{\circ}{\operatorname{dom}} K_p$ so that $(m_u, m_v) = (0, 0)$ and thus $\mathbb{E}_p(ve^u) = cM_u(1) = ce^{K_p(u)}$.

We can use this equality in order to compute the term $\mathbb{E}_q(v)$ in the divergence expression (5):

$$\mathbb{E}_q(v) = \mathbb{E}_p\left(v\frac{q}{p}\right) = \mathbb{E}_p\left(ve^{u - K_p(u)}\right) = e^{-K_p(u)}\mathbb{E}_p(ve^u) = c.$$

As a byproduct of our computations, let us observe that, for p-densities q/p and r/p, with a log-normal distribution, the following meaningful equality holds:

$$\mathbb{E}_q(v) = \mathbb{E}_r(u) = Cov_p(u, v) = \mathbb{E}_p(uv). \qquad (8)$$

Moreover the divergence of q from the arc $r(\theta)$, with $\theta \in I = \mathbb{R}$, becomes a simple parabola

$$D(q\|r(\theta)) = D(q\|p) - \theta Cov_p(u, v) + \log M_v(\theta)$$
$$= D(q\|p) - \theta c + \frac{1}{2}\sigma_v^2\theta^2.$$

So the minimum is reached in

$$\bar{\theta} = \frac{c}{\sigma_v^2} = \frac{Cov_p(u,v)}{Var_p(v)} \tag{9}$$

and its value is

$$\min_{\theta \in \mathbb{R}} D(q\|r(\theta)) = D(q\|r(\bar{\theta})) = D(q\|p) - \bar{\theta}c + \frac{1}{2}\sigma_v^2\bar{\theta}^2 = D(q\|p) - \frac{1}{2}\frac{c^2}{\sigma_v^2}$$

$$= D(q\|p) - \frac{1}{2}\frac{Cov_p^2(u,v)}{Var_p(v)}. \tag{10}$$

The representation in $\mathcal{E}(p)$ of the density which minimizes the divergence is $r(\bar{\theta}) = e^{\bar{v} - K_p(\bar{v})}p$, where

$$\bar{v} = \frac{c}{\sigma_v^2}v = \frac{Cov_p(u,v)}{Var_p(v)}v, \qquad K_p(\bar{v}) = \frac{1}{2}\frac{c^2}{\sigma_v^2} = \frac{1}{2}\frac{Cov_p^2(u,v)}{Var_p(v)}. \tag{11}$$

Let us observe that, if we explicit our results in terms of the correlation $\varrho = \varrho_p(u,v) \in (-1,1)$ we get

$$\bar{\theta} = \varrho\frac{\sigma_u}{\sigma_v}, \qquad \bar{v} = \varrho\frac{\sigma_u}{\sigma_v}v, \qquad K_p(\bar{v}) = \frac{1}{2}\varrho^2\sigma_u^2, \tag{12}$$

and the minimal divergence is

$$\min_{\theta \in \mathbb{R}} D(q\|r(\theta)) = D(q\|p) - \frac{1}{2}\varrho^2\sigma_u^2. \tag{13}$$

Remark 1. By (13), let us note that the minimum of the divergence of q from the arc connecting p and r depends on the choice of r only through the correlation ϱ between u and v.

On the other hand, let us consider another density $s = e^{w - K_p(w)}p$ and suppose that (u,w) is a Gaussian vector with the same correlation as (u,v). Then, there exists $a \in I = \mathbb{R}$ such that $w = av$, which in turn means that s belongs to the exponential arc $r(\theta)$, connecting r and p, with $\theta = a$.

Remark 2. If q and r are independent, which means that u and v are independent, then we immediately get $Cov_p(u,v) = c = 0$, so that the density which minimizes the divergence turns out to be $r(0) = p$.

Let us show that it is in fact a general result, not only true for log-normal densities.

In case of independence between u and v, since $v \in L_0^{\Phi_1}(p)$, we have

$$\mathbb{E}_q(v) = \mathbb{E}_p\left(ve^{u - K_p(u)}\right) = e^{-K_p(u)}\mathbb{E}_p(ve^u) = e^{-K_p(u)}\mathbb{E}_p(v)\mathbb{E}_p(e^u) = 0.$$

As a consequence

$$D(q\|r(\theta)) = D(q\|p) + \log M_v(\theta)$$

and the stationary condition immediately becomes

$$M_v'(\theta) = 0.$$

Since M_v is a strictly positive, continuous and convex function on its proper domain, which contains $(-\varepsilon, 1 + \varepsilon)$, and $M_v'(0) = \mathbb{E}_p(v) = 0$, we deduce that $\theta = 0$ is the unique minimum point for M_v. As a consequence, the density which minimizes the divergence is $p = r(0)$.

Example 1. Merton's model

Let us now consider an application of the previous result to a financial model in an uncertainty framework (see [9]).

Let $\mathbb{F}$ be the augmented filtration generated by a Brownian motion $(B_t)_{0 \le t \le T < \infty}$. Assume that the price process X of a financial market follows the Black and Scholes dynamics

$$dX_t = X_t \left(\sigma dB_t + m dt \right), \quad \forall\, 0 \le t \le T,$$

with $\sigma > 0$ and $m \in \mathbb{R}$.

Since the market is complete, the unique martingale measure obviously coincides with the minimal entropy martingale measure, which has the form

$$q* = e^{-\frac{m}{\sigma} B_T - \frac{1}{2} \frac{m^2}{\sigma^2} T}.$$

It can be proved, by condition (vi) of the Portmanteau Theorem, that $q*$ belongs to the maximal exponential model $\mathcal{E}(p)$ with $p = 1$, i.e. $q* = e^{u* - K_1(u*)}$, where

$$u* = -\frac{m}{\sigma} B_T, \qquad K_1(u*) = D(1\|q*) = \frac{1}{2} \frac{m^2}{\sigma^2} T. \qquad (14)$$

In fact, we immediately get

$$\mathbb{E}((q*)^{1+\varepsilon}) = \mathbb{E}\left(e^{-\frac{m}{\sigma}(1+\varepsilon)B_T - \frac{1}{2}\frac{m^2}{\sigma^2}(1+\varepsilon)T} \right) < +\infty,$$

$$\mathbb{E}((q*)^{-\varepsilon}) = \mathbb{E}\left(e^{\frac{m}{\sigma}\varepsilon B_T + \frac{1}{2}\frac{m^2}{\sigma^2}\varepsilon T} \right) < +\infty.$$

From (iii) of the Portmanteau Theorem, if we choose a density r connected to $p = 1$ by an open arc, then $q* \in \mathcal{E}(r) = \mathcal{E}(1)$ and trivially, since the set of the equivalent martingale measures is a singleton, it turns out to be the minimal entropy martingale measure with respect to r.

Let us now consider the generic element $r(\theta)$ of the open arc connecting r and p. The minimization of the entropy over the arc

$$\min_{\theta \in I} D(q*\|r(\theta)) \qquad (15)$$

is the dual problem of a classical exponential utility maximization problem in which an uncertainty exists on the choice of the reference probability measure.

As we have seen before, modeling this uncertainty by varying the reference measure over an exponential arc $r(\theta) \propto e^{\theta v}$ ensures that, in the Gaussian case, the optimal value of the divergence is robust with respect to homogeneous transformations of v which leave its correlation with $u*$ unchanged.

Indeed, a particular choice of r is made by selecting v such that $(u*, v)$ is a non-degenerate Gaussian vector with covariance matrix $\begin{pmatrix} \frac{m^2}{\sigma^2}T & c \\ c & \gamma^2 \end{pmatrix}$, picking arbitrarily the parameters γ^2 and c.

Using (10) we obtain the solution of the dual problem (15) as

$$D\left(q*\|r(\bar{\theta})\right) = D(q*\|1) - \frac{1}{2}\frac{c^2}{\gamma^2} = \frac{1}{2}\frac{m^2}{\sigma^2}\left(T - \frac{Cov(B_T, v)^2}{Var(v)}\right). \tag{16}$$

The representation in $\mathcal{E}(1)$ of the corresponding optimal density is then given by $r(\bar{\theta}) = e^{\bar{v} - K_1(\bar{v})}$, where

$$\bar{v} = \frac{c}{\gamma^2}v = -\frac{m}{\sigma}\frac{Cov(B_T, v)}{Var(v)}v, \qquad K_1(\bar{v}) = \frac{1}{2}\frac{c^2}{\gamma^2} = \frac{1}{2}\frac{m^2}{\sigma^2}\frac{Cov(B_T, v)^2}{Var(v)}. \tag{17}$$

4   Conclusions

In the paper, we have dealt with the minimization of the Kullback-Leibler divergence over a log-normal exponential arc. Explicit expressions have been done for both the optimal density and the minimal divergence, showing their robustness with respect to homogeneous transformations which leave the correlation of the random variables involved unchanged. The results obtained have been then applied to the complete Merton's market model.

References

1. Cena, A., Pistone, G.: Exponential statistical manifold. AISM **59**, 27–56 (2007)
2. Imparato, D., Trivellato, B.: Geometry of extendend exponential models. In: Algebraic and Geometric Methods in Statistics pp. 307–326 (2009)
3. Lods, B., Pistone, G.: Information geometry formalism for the spatially homogeneous Boltzmann equation. Entropy **17**, 4323–4363 (2015)
4. Pistone, G.: Examples of the application of nonparametric information geometry to statistical physics. Entropy **15**, 4042–4065 (2013)
5. Pistone, G., Sempi, C.: An infinite-dimensional geometric structure on the space of all the probability measures equivalent to a given one. Ann. Stat. **23**(5), 1543–1561 (1995)
6. Santacroce, M., Siri, P., Trivellato, B.: New results on mixture and exponential models by Orlicz spaces. Bernoulli **22**(3), 1431–1447 (2016)
7. Santacroce, M., Siri, P., Trivellato, B.: On mixture and exponential connection by open arcs. In: Nielsen, F., Barbaresco, F. (eds.) GSI 2017. LNCS, vol. 10589, pp. 577–584. Springer, Cham (2017). https://doi.org/10.1007/978-3-319-68445-1_67

8. Santacroce, M., Siri, P., Trivellato, B.: An application of maximal exponential models to duality theory. Entropy **20**(495), 1–9 (2018)
9. Santacroce, M., Siri, P., Trivellato, B.: Exponential models by Orlicz spaces and applications. J. Appl. Probab. **55**, 682–700 (2018)
10. Vigelis, R.F., Cavalcante, C.C.: On φ-families of probability distributions. J. Theoret. Probab. **26**(3), 870–884 (2013)

Riemannian Distance and Diameter of the Space of Probability Measures and the Parametrix

Mitsuhiro Itoh[1(⊠)] and Hiroyasu Satoh[2]

[1] University of Tsukuba, Ibaraki 305–8571, Japan
itohm@math.tsukuba.ac.jp
[2] Liberal Arts and Sciences, Nippon Institute of Technology, Saitama, Japan
hiroyasu@nit.ac.jp

Abstract. Information geometry of the space $\mathcal{P}(M)$ of probability measures defined on a compact smooth Riemannian manifold M and equipped with Fisher metric G, is investigated from the viewpoint of Riemannian geometry. The function $\ell : \mathcal{P}(M) \times \mathcal{P}(M) \to [0, \pi)$ associated to the geometric mean of two probability measures is introduced. From the formulae of Levi-Civita geodesics the Riemannian distance $d(\cdot, \cdot)$ of $(\mathcal{P}(M), G)$ is exactly given by $\ell(\cdot, \cdot)$. By applying the parametrix $H(x, x_0; t)$ of the heat kernel of M it is shown that the diameter D satisfies $D = \pi$.

Keywords: Riemannian distance · Geodesic · Diameter · Parametrix

1 Geodesics and Riemannian Distance Function

Information geometry of statistical models has been developed in a parametric case. Duality of α-connections together with Fisher information metric and also α-divergences are of great importance for a general framework in a parametric model. For their development refer to [17], [1], [2] and the references therein. The Fisher information metric G is well defined on the space of all probability measures on a smooth manifold M, similarly to the parametric case, as shown by T. Friedrich [7] and many interesting aspects and properties of the Fisher metric G have been studied from information geometry. The α-connections and the α-divergences can be provided associated to the Fisher metric G on the space of all probability measures on M, as a non-parametric model. Refer to [10] for these and the dual flatness of (± 1)-connections and the subjects related to them.

In this paper we study, from a viewpoint of Riemannian geometry, information geometry of a non-parametric model, namely of the space $\mathcal{P}(M)$ of all probability measures on a manifold M, equipped with the Fisher metric G and hence with the Levi-Civita connection ∇, i.e., the α-connection of $\alpha = 0$. We focus on geometric aspects associated to the Levi-Civita connection and give a formula describing geodesics and geodesic segments and then exhibit an exact

© Springer Nature Switzerland AG 2019
F. Nielsen and F. Barbaresco (Eds.): GSI 2019, LNCS 11712, pp. 462–471, 2019.
https://doi.org/10.1007/978-3-030-26980-7_48

form of the Riemannian distance function for the space $\mathcal{P}(M)$ with respect to the Fisher metric G. The diameter, the supremum of the distance of two measures of $\mathcal{P}(M)$ is discussed in Sect. 2 by applying the notion of parametrix for the heat kernel of M, namely a certain approximation of the fundamental solution of the heat equation on M and then the diameter of $(\mathcal{P}(M), G)$ turns out to equal π (see Theorem 7).

To be precise, let (M, g) be a connected, compact smooth n-dimensional Riemannian manifold with the Riemannian volume measure $\lambda = \sqrt{|\det g_{ij}|}\, dx^1 \cdots dx^n$ of $\int_M d\lambda = 1$ and let $\mathcal{P}(M)$ be the space of probability measures on M defined as

$$\mathcal{P}(M) = \left\{ \mu \text{ is a measure on } M \,\middle|\, \int_M d\mu = 1, \mu \ll \lambda, \frac{d\mu}{d\lambda} \in C^0_+(M) \right\}. \tag{1}$$

Here $\mu \ll \lambda$ means that μ is absolutely continuous with respect to λ so that μ is represented by $\mu = p\lambda$ with density function $p = p(x)$, $x \in M$. The symbol $d\mu/d\lambda$ denotes the Radon-Nikodym derivative of μ with respect to λ, given by the density function p and $C^0_+(M)$ denotes the space of positive continuous functions on M.

Notice that there is a natural embedding $\rho : \mathcal{P}(M) \to L_2(M, \lambda)$, $\mu = p\lambda \mapsto \rho(\mu) = 2\sqrt{p}$. The embedding ρ induces naturally a topology on $\mathcal{P}(M)$ with an ε-open ball $\{\mu_1 = p_1\lambda \in \mathcal{P}(M) \,|\, 2|\sqrt{p_1} - \sqrt{p}|_{L_2} < \varepsilon\}$ around $\mu = p\lambda$.

Definition 1. *Let $\varphi : \mathcal{P}(M) \times \mathcal{P}(M) \to \mathcal{P}(M)$ be a map, called the normalized geometric mean;*

$$\varphi(\mu_1, \mu_2) = \left(\int_{x \in M} \sqrt{\frac{d\mu_2}{d\mu_1}(x)}\, d\mu_1(x) \right)^{-1} \sqrt{\frac{d\mu_2}{d\mu_1}}\, \mu_1 \tag{2}$$

and let $\ell : \mathcal{P}(M) \times \mathcal{P}(M) \to [0, \pi)$ be a function defined by

$$\ell(\mu_1, \mu_2) = 2 \arccos \left(\int_{x \in M} \sqrt{\frac{d\mu_2}{d\mu_1}(x)}\, d\mu_1(x) \right). \tag{3}$$

The map φ and the function ℓ are continuous with respect to the induced topology [11]. The function ℓ provides the Riemannian distance function with respect to the Fisher metric G, as stated in Theorem 5. In [7] the metric G is defined at each $\mu \in \mathcal{P}(M)$ as the inner product of tangent vectors $\tau_1, \tau_2 \in T_\mu \mathcal{P}(M)$

$$G_\mu(\tau_1, \tau_2) = \int_M \frac{d\tau_1}{d\mu} \frac{d\tau_2}{d\mu}\, d\mu = \int_{x \in M} \frac{q_1(x)}{p(x)} \frac{q_2(x)}{p(x)} p(x)\, d\lambda(x). \tag{4}$$

Here the tangent space at μ is defined by the space of signed measures on M

$$T_\mu \mathcal{P}(M) = \left\{ \tau = q\lambda \,\middle|\, q \in C^0(M) \cap L_2(M, \lambda), \int_M q(x)\, d\lambda(x) = 0 \right\}. \tag{5}$$

Note that the right hand side of (5) is an infinite dimensional vector space which does not depend on a choice of μ, so we denote it $\mathcal{V}(M)$. The symbol $d\tau_i/d\mu$, $i = 1, 2$ means the Radon-Nikodym derivative of τ_i with respect to μ, given by $(d\tau_i/d\lambda)/(d\mu/d\lambda) = q_i/p$. If $\tau \in \mathcal{V}(M)$, then for any $\mu \in \mathcal{P}(M)$ there exists $\varepsilon > 0$ such that $\mu + t\tau \in \mathcal{P}(M)$, $|t| < \varepsilon$, since M is compact and the density functions are continuous.

The Fisher metric G is a natural generalization of the classical Fisher information for statistical models in parametric statistics and information theory [2]. The uniqueness of the metric G for probability measures with smooth density function in the sense of diffeomorphism-invariance is shown in [3].

The Levi-Civita connection ∇ of the metric G is given explicitly in [7, p.276] by applying an idea of *constant vector fields*:

$$(\nabla_\tau \tau_1)_\mu = -\frac{1}{2} \left(\frac{d\tau}{d\mu} \frac{d\tau_1}{d\mu} - G_\mu(\tau, \tau_1) \right)_\mu, \quad \tau, \tau_1 \in \mathcal{V}(M). \tag{6}$$

Here the Levi-Civita connection is an affine connection on $\mathcal{P}(M)$ which is torsion free and leaves invariant the metric G. It is shown that the Riemannian curvature tensor of the manifold $(\mathcal{P}(M), G)$ is given by

$$R_\mu(\tau_1, \tau_2)\tau_3 = \frac{1}{4} \left(G_\mu(\tau_2, \tau_3)\tau_1 - G_\mu(\tau_1, \tau_3)\tau_2 \right) \tag{7}$$

so the sectional curvature is constant $1/4$. Refer to [7].

Remark 1. The α-connections ∇^α, $\alpha \in \mathbb{R}$ associated to the metric G and the Riemannian curvature tensor R^α of ∇^α admit the following form, respectively; $\nabla^\alpha_\tau \tau_1 = (1+\alpha)\nabla_\tau \tau_1$ and $R^\alpha_\mu(\cdot, \cdot) = (1 - \alpha^2)R_\mu(\cdot, \cdot)$ so that $\{\nabla^\alpha, \nabla^{-\alpha}\}$ is a dual pair and the pair $\{\nabla^{+1}, \nabla^{-1}\}$ is dually flat. For the details refer to [10].

We obtain the explicit formula for a geodesic associated with the Levi-Civita connection ∇. In fact, let $\gamma : I \to \mathcal{P}(M)(I$ is an open interval, $0 \in I)$ be a geodesic, parametrized by arc-length satisfying $\gamma(0) = p_0\lambda$, $\dot{\gamma}(0) = \dot{p}_0\lambda$ of $|\dot{\gamma}(0)|_\mu (:= G_\mu(\dot{\gamma}(0), \dot{\gamma}(0))^{1/2}) = 1$.

Theorem 1. (T. Friedrich [7]) *The density function $p_t = p_t(x)$ of $\gamma(t)$ with respect to λ has the form*

$$p_t(x) = \frac{1}{1 + \tan^2(\frac{t}{2})} \left\{ p_0(x) + 2\tan\left(\frac{t}{2}\right)\dot{p}_0(x) + \tan^2\left(\frac{t}{2}\right)\frac{\dot{p}_0^2(x)}{p_0(x)} \right\}. \tag{8}$$

From (8) each geodesic of unit speed is periodic and of period π. However every geodesic is not complete, since $\gamma(\pi) \notin \mathcal{P}(M)$ at $t = \pi$. One is able to obtain a density free description for a geodesic.

Theorem 2. ([9–11]) *Let $\gamma(t)$ be a geodesic with $\gamma(0) = \mu$ and $\dot{\gamma}(0) = \tau \in T_\mu\mathcal{P}(M)$. If τ is of unit norm, i.e., $|\tau|_\mu = 1$ with respect to G, then $\gamma(t)$ is represented by*

$$\gamma(t) = \left(\cos\frac{t}{2} + \frac{d\tau}{d\mu} \cdot \sin\frac{t}{2} \right)^2 \mu. \tag{9}$$

Next we consider the problem whether any two measures of $\mathcal{P}(M)$ can be joined by a geodesic segment. For given probability measures $\mu, \mu_1 \in \mathcal{P}(M)$ we give affirmatively an explicit formula representing a geodesic segment joining μ and μ_1, by using above theorem.

Theorem 3. ([10,11]) *Let μ, μ_1 be arbitrary probability measures of $\mathcal{P}(M)$ with $\mu \neq \mu_1$. Then there exists a unique geodesic segment $\gamma(t)$, $t \in [0,l]$ such that $\gamma(0) = \mu$, $\gamma(l) = \mu_1$. Here the length l of γ is given by $\ell(\mu, \mu_1)$. Moreover, $\gamma(t)$ is represented by*

$$\gamma(t) = \left(\cos \frac{t}{2} + \frac{d\tau}{d\mu} \cdot \sin \frac{t}{2} \right)^2 \mu$$

with initial velocity vector of unit norm

$$\tau = \frac{1}{\tan(l/2)} \left(\varphi(\mu, \mu_1) - \mu \right).$$

This theorem indicates that any measures μ, μ_1 can be joined by a unique geodesic segment whose length is $\ell(\mu, \mu_1)$.

Proof. Assume that μ and μ_1 are joined by (9). Then there exists a number $l > 0$ such that

$$\left(\cos \frac{l}{2} + \frac{d\tau}{d\mu} \cdot \sin \frac{l}{2} \right)^2 \mu = \mu_1. \tag{10}$$

Rewrite this as

$$\left(\cos \frac{l}{2} + \frac{d\tau}{d\mu} \cdot \sin \frac{l}{2} \right)^2 = \frac{d\mu_1}{d\mu}$$

and solve this with respect to $d\tau/d\mu$, by an analogous argument in [9, p.1830, Assertion 3], in fact, actually from the connectedness of the manifold M as

$$\cos \frac{l}{2} + \frac{d\tau}{d\mu} \cdot \sin \frac{l}{2} = \sqrt{\frac{d\mu_1}{d\mu}}. \tag{11}$$

We find then that the initial velocity τ of unit norm is uniquely determined by

$$\tau = \frac{1}{\sin(l/2)} \left(\sqrt{\frac{d\mu_1}{d\mu}} - \cos \frac{l}{2} \right) \mu \tag{12}$$

immediate from (11). By integrating (11) with respect to μ we have

$$\int_M \sqrt{\frac{d\mu_1}{d\mu}} \, d\mu = \cos \frac{l}{2} \tag{13}$$

from the tangent vector condition; $\int_M d\tau = 0$. We can write (12) and (13), respectively as

$$\tau = \frac{1}{\tan(l/2)} \left(\varphi(\mu, \mu_1) - \mu \right) \tag{14}$$

and

$$l = 2 \arccos \left(\int_M \sqrt{\frac{d\mu_1}{d\mu}} \, d\mu \right) = \ell(\mu, \mu_1). \tag{15}$$

The following is a direct consequence of above theorem.

Theorem 4. ([10,11]) *Let $\mu, \mu_1 \in \mathcal{P}(M)$ with $\mu \neq \mu_1$. As is stated in above Theorem, there exists a unique geodesic $\gamma(t)$ with respect to G parametrized by arc-length, joining μ and μ_1. This geodesic segment can be expressed in the form*

$$\gamma(t) = c_1(t)\, \mu + c_2(t)\, \mu_1 + c_3(t)\, \varphi(\mu, \mu_1), \qquad t \in [0, l]. \tag{16}$$

Here $\varphi(\mu, \mu_1)$ is the normalized geometric mean of measures μ, μ_1, as introduced in Definition 1. Moreover, $\gamma(0) = \mu$, $\gamma(l) = \mu_1$, $l = \ell(\mu, \mu_1)$ and $c_i(t)$, $i = 1, 2, 3$ are the non-negative functions of t satisfying

$$c_1(t) + c_2(t) + c_3(t) = 1$$

which are written respectively by

$$c_1(t) = \frac{\sin^2 (l-t)/2}{\sin^2 (l/2)}, \quad c_2(t) = \frac{\sin^2 (t/2)}{\sin^2 (l/2)},$$
$$c_3(t) = \frac{2 \cos(l/2) \cdot \sin(t/2) \cdot \sin(l-t)/2}{\sin^2 (l/2)}.$$

As a consequence of Theorem 4 one has the following theorem.

Theorem 5. ([10,11]) *Let $\gamma = \gamma(t)$, $t \in [0, l]$ $(l = \ell(\mu, \mu_1))$ be a geodesic segment joining $\mu, \mu_1 \in \mathcal{P}(M)$, $\mu \neq \mu_1$ satisfying $\gamma(0) = \mu$, $\gamma(l) = \mu_1$. Then,*

1. *$\gamma(t) \in \mathcal{P}(M)$ at every $t \in [0, l]$, i.e., the density function of $\gamma(t)$ belongs to $C_+^0(M)$ for every $t \in [0, l]$,*
2. *the geodesic segment $\gamma : [0, l] \to \mathcal{P}(M)$ is a curve lying on the 2-simplex spanned by μ, μ_1 and $\varphi(\mu, \mu_1)$,*
3. *Two tangent lines defined at the endpoints of the geodesic segment always intersect each other at $\varphi(\mu, \mu_1)$,*
4. *the midpoint of the geodesic segment $\gamma(t)$, $t \in [0, l]$ is represented by*

$$\gamma(l/2) = \frac{1}{4 \cos^2 (l/2)} \left(1 + \sqrt{\frac{d\mu_1}{d\mu}} \right)^2 \mu \tag{17}$$

which is the $(1/2)$–power mean of μ, μ_1. Refer to [11], [14].

One of our main theorems is the following.

Theorem 6. ([11]) *The Riemannian distance d between μ and μ_1 with respect to the Fisher metric G is given by the function $\ell(\mu, \mu_1)$;*

$$d(\mu, \mu_1) = \ell(\mu, \mu_1),$$

where

$$d(\mu, \mu_1) := \inf \left\{ L(c) := \int_0^1 \sqrt{G\left(\frac{dc}{dt}, \frac{dc}{dt}\right)} \, dt \, : \, c \in \mathcal{C}(\mu, \mu_1) \right\}.$$

Here $\mathcal{C}(\mu, \mu_1)$ denotes the space of piecewise C^1-curves $c : [0,1] \to \mathcal{P}(M)$ joining μ and μ_1.

Notice $0 \le d(\mu, \mu_1) = \ell(\mu, \mu_1) < \pi$ for any $\mu, \mu_1 \in \mathcal{P}(M)$. The function $\ell(\cdot, \cdot)$ is related to certain informations called the Hellinger integral and the Hellinger distance [13].

This theorem is verified as follows. First we restrict our argument to the space $\mathcal{P}^\infty(M)$ of measures in $\mathcal{P}(M)$ having *smooth* density function and then define the exponential map over $\mathcal{P}^\infty(M)$ with the Fisher metric G and prove basic facts, familiar in ordinary Riemannian geometry, Gauss lemma, the existence of totally normal neighborhood and the minimizing length properties of geodesics, cf. [6, Chap. 3],[8],[12]. We then show that the function ℓ exactly gives the Riemannian distance for the space $(\mathcal{P}^\infty(M), G)$. We prove secondly that $\mathcal{P}^\infty(M)$ is dense in $\mathcal{P}(M)$ with respect to the C^0-norm and show that the Riemannian distance of $\mu, \mu_1 \in \mathcal{P}^\infty(M)$ in $\mathcal{P}(M)$ is actually given by $\ell(\mu, \mu_1)$ by using the argument of mollifiers. We verify finally from the continuity that $\ell(\mu, \mu_1)$ is properly the Riemannian distance of μ, μ_1 of $\mathcal{P}(M)$. For the details refer to [11].

2   Parametrix and Diameter

The diameter of a Riemannian manifold is defined by the supremum of the distance over two points.

Theorem 7. *The diameter D of the space $(\mathcal{P}(M), G)$ satisfies $D = \pi$.*

Note that since $\ell(\mu, \mu_1) < \pi$ for μ, μ_1, $D \le \pi$. We will give a proof by taking a sequence $\{\mu_i\}$ in $\mathcal{P}(M)$ satisfying

$$\int_{x \in M} \sqrt{\frac{d\mu_i}{d\lambda}}(x) \, d\lambda(x) \to 0, \quad i \to \infty. \tag{18}$$

Then, $\ell(\mu_i, \lambda) \to \pi$. This is because

$$\cos \frac{\ell_i}{2} = \int_M \sqrt{\frac{d\mu_i}{d\lambda}}(x) \, d\lambda(x), \quad \ell_i = \ell(\mu_i, \lambda) \in [0, \pi). \tag{19}$$

We will now construct such a sequence by using a parametrix of the heat kernel, the fundamental solution of the heat equation $(\Delta + \frac{\partial}{\partial t})u = 0$ over a compact smooth manifold M. Here $\Delta = -\sum g^{ij} \nabla_i \nabla_j$ is the Laplacian of (M, g). See [4, Cap. III, E], [18, VI], [20, III, §2] for the heat kernel and its parametrix. We also refer to [22] for the heat kernel. Recall that on the Euclidean space $(\mathbb{R}^n, g_o)$ the heat kernel is $E(x, x'; t) = (4\pi t)^{-n/2} \exp\{-r^2/4t\}$, $r = d(x, x')$. Refer to [21].

Take a point $x_o \in M$ as a reference point. Let $B(x_o, \varepsilon) = \{x \in M \,; d_g(x, x_o) < \varepsilon\}$ be an ε-open ball centered at x_o, endowed with a geodesic normal coordinates $\{x^1, \cdots, x^n\}$. Let $H_k(x, x_o; t) = S_k(x, x_o; t)\eta(x)$, $k > \dim M/2$, be a parametrix of the heat kernel, where $S_k(x, x_o; t)$ is the function defined on $B(x_o, \varepsilon) \times \mathbb{R}_+^*$ by

$$S_k(x, x_o; t) = (4\pi t)^{-n/2} \exp\left\{-\frac{r^2(x)}{4t}\right\} \left(u_0(x) + t u_1(x) + \cdots + t^k u_k(x)\right),$$
(20)

where $u_j(x) = u_j(x, x_o)$, $0 \le j \le k$ are smooth functions inductively defined on $B(x_o, \varepsilon)$. Notice that $u_0(x, x_o) = \sqrt{\det g_{ij}}$ on the normal coordinates around x_o in $B(x_o, \varepsilon)$ and $u_0(x_o, x_o) = 1$. The principal part of (20) is just the heat kernel of the Euclidean space. Refer to [4] for these. Moreover $\eta = \eta(r)$, $r = r(x) := d_g(x, x_o)$ is a bump function; $\eta = \eta(r)$; $0 \le \eta \le 1$ satisfying $\eta(r) \equiv 1$, $r \le \varepsilon/4$, $\eta(r) \equiv 0$, $r \ge \varepsilon/2$.

Definition 2. *([4]) A function H satisfying the following is called a parametrix for the heat kernel;*

1. $H(\cdot, \cdot; \cdot) \in C^\infty(M \times M \times \mathbb{R}_+^*)$,
2. $\left(\Delta_1 + \frac{\partial}{\partial t}\right) H(\cdot, \cdot; \cdot)$ *extends to a function in* $C^0(M \times M \times \mathbb{R}_+)$,
3. $\lim_{t \to 0+} H(\cdot, x_o; t) = \delta_{x_o}$,

where $\mathbb{R}_+ = \{x \in \mathbb{R} \,|\, x \ge 0\}$ and $\mathbb{R}_+^* = \{x \in \mathbb{R} \,|\, x > 0\}$ and the symbol δ_{x_o} means the delta function centered at x_o. In condition 2 Δ_1 is the Laplacian operating on the first variable.

It is shown in [4, Lemma E.III.3] that the above $H_k(x, x_o; t)$ gives a parametrix, for $x, x_o \in M$. Define a probability measure $\mu(t)$ on M parametrized in $t > 0$ as

$$\mu(t) := \frac{1}{V(t)}\{S_k(x, x_o; t)\eta(r(x)) + (1 - \eta(r(x)))t\}\,\lambda,$$
(21)

where $V(t)$ is given by

$$V(t) = \int_{B(x_o, \varepsilon)} S_k(x, x_o; t)\eta(x)\,d\lambda + t\int_M (1 - \eta)\,d\lambda,$$
(22)

the total integral of the measure $\{S_k(x, x_o; t)\eta(r(x)) + (1 - \eta(r(x)))t\}\,\lambda$, whose first integral tends to 1, when $t \to 0$ so we may assume that the first integral is greater than $1/2$ for any t, $|t| < \delta$, for a sufficiently small δ. The second term is estimated as

$$t\int_M (1 - \eta)\,d\lambda \ge t\,\mathrm{Vol}\,(M \setminus B(x_o, \varepsilon/2)).$$
(23)

Then, $\mu(t) \in \mathcal{P}(M)$ for any $0 < t < \delta$.

Consider next

$$\sqrt{\frac{d\mu(t)}{d\lambda}}(x)\lambda = \frac{1}{\sqrt{V(t)}}\{S_k(x, x_o; t)\,\eta(r(x)) + (1 - \eta(x))t\}^{1/2}\lambda$$
(24)

and its integral

$$\int_M \sqrt{\frac{d\mu(t)}{d\lambda}}(x)\, d\lambda = \frac{1}{\sqrt{V(t)}} \int_M \{S_k(x, x_o; t)\eta(r(x)) + (1 - \eta(x))t\}^{1/2}\, d\lambda. \quad (25)$$

Since $\sqrt{a+b} \le \sqrt{a} + \sqrt{b}$ for $a, b \ge 0$, the above is estimated from above as

$$\int_M \sqrt{\frac{d\mu(t)}{d\lambda}}(x)\, d\lambda \le \frac{1}{\sqrt{V(t)}} \int_M \{\{S_k(x, x_o; t)\,\eta(r(x))\}^{1/2}$$

$$+ \{(1 - \eta(x))t\}^{1/2}\}\, d\lambda$$

$$\le \frac{1}{\sqrt{V(t)}} \int_{B_{\varepsilon/4}} \{S_k(x, x_o; t)\eta(r(x))\}^{1/2}\, d\lambda$$

$$+ \frac{\sqrt{t}}{\sqrt{V(t)}} \mathrm{Vol}(M \setminus B_{\varepsilon/2}), \quad B_{\varepsilon/2} := B(x_o, \varepsilon/2).$$

From the above estimation of $V(t)$ we have $V(t) \ge \frac{1}{2} + t\,\mathrm{Vol}(M \setminus B_{\varepsilon/2})$ and consequently

$$\int_M \sqrt{\frac{d\mu(t)}{d\lambda}}(x) d\lambda \le \frac{1}{\sqrt{\frac{1}{2} + t\,\mathrm{Vol}(M \setminus B_{\varepsilon/2})}} \int_{B_{\varepsilon/2}} \{S_k(x, x_o; t)\eta(r(x))\}^{1/2} d\lambda$$

$$+ \frac{\sqrt{t}}{\sqrt{\frac{1}{2} + t\,\mathrm{Vol}(M \setminus B_{\varepsilon/2})}} \mathrm{Vol}(M \setminus B_{\varepsilon/2}).$$

$$(26)$$

We estimate then the integrand of the first term, $\sqrt{S_k(x, x_o; t)}\sqrt{\eta(r(x))}$. From (20), we have

$$\sqrt{S_k(x, x_o; t)} = (8\pi t)^{n/2}(4\pi t)^{-n/4}(4\pi \times 2t)^{-n/2} \exp\left\{-\frac{r^2(x)}{4(2t)}\right\} \sqrt{\sum_{i=0}^k t^i u_i(x)}.$$

Since $t^{n/2} \cdot t^{-n/4} = t^{n/4}$,

$$\int_{B_{\varepsilon/2}} \sqrt{S_k(x, x_o; t)}\,\sqrt{\eta(r(x))}\, d\lambda$$

$$= \frac{(8\pi)^{n/2}}{(4\pi)^{n/4}} \cdot t^{n/4} \int_{B_{\varepsilon/2}} (4\pi(2t))^{-n/2} \exp\left\{-\frac{r^2(x)}{4 \cdot (2t)}\right\} \sqrt{\sum t^i u_i(x)}\,\sqrt{\eta(r(x))}\, d\lambda. \quad (27)$$

Here $\sqrt{\eta(r(x))}$ is a bump function and the square root term is estimated on $B_{\varepsilon/2}$ by

$$\sqrt{\sum_{i=0}^k t^i u_i(x)} = \left(\sqrt{\sum_{i=0}^k t^i u_i(x)}\right)^{-1} \left(\sum_{i=0}^k t^i u_i(x)\right) \le C \sum_{i=0}^k (2t)^i u_i(x), \quad C > 0$$

for sufficiently small $t > 0$, since $u_0(x) = \sqrt{\det g_{ij}(x)} > 0$ on $B_{\varepsilon/2}$ and $u_0(x_o) = 1$.

Therefore the left hand of (27) must tend to 0, as $t \to 0$, since the right hand integral $\int_{B_{\varepsilon/2}} \cdots d\lambda$ takes a finite value from the form of $S_k(x, x_o; 2t)$, or rather of $H_k(x, x_o; 2t)$ (in fact, from Condition 3 of Definition 2). Therefore, $\lim_{t\to 0} \ell(\mu(t), \lambda) = \pi$ and hence $D = \pi$.

3   Final Remarks

Our theorems, thus exhibited above, are derived from the argument of constant vector fields employed in [7]. For a non-compact manifold case, one might need a hypothesis of connectedness by an open mixture arc for probability measures [11]. Probability measures μ and $\mu_1 \in \mathcal{P}(M)$ are said to be connected by an open mixture arc (denoted $\mu \sim_m \mu_1$) if there exists an open interval $I \supset [0, 1]$ such that $t\mu + (1 - t)\mu_1$ belongs to $\mathcal{P}(M)$ for $t \in I$. For this connectedness refer to [11],[5] and [19]. Let $\mathcal{P}_m(M) = \{\mu \in \mathcal{P}(M) \,|\, \mu \sim_m \lambda\}$. Define an infinite dimensional vector space

$$\mathcal{V}_m(M) = \left\{\nu = q\lambda \,\middle|\, q \in C^0(M), \int_M d\nu = 0, \lambda + t\nu \in \mathcal{P}_m(M) \,\forall t, |t| < \varepsilon\right\} \quad (28)$$

which coincides with $\mathcal{V}(M)$ (see (5)), when M is compact. Every $\tau \in \mathcal{V}_m(M)$ induces a vector field at each $\mu \in \mathcal{P}_m(M)$. Therefore, the argument of constant vector fields can be applied to a non-compact manifold M.

Example 1. For simplicity we let $M = \mathbb{R}$ and $\lambda = (4\pi)^{-1/2} \exp(-x^2/4)dx$ be the Gaussian measure on $\mathbb{R}$. Let $\mathcal{P}_m(\mathbb{R}) = \{\mu \in \mathcal{P}(\mathbb{R}) \,|\, \mu \sim_m \lambda\}$. Then the theorems above are all valid for $\mathcal{P}_m(\mathbb{R})$. The diameter is π, since the heat kernel $E(x, x'; t) = (4\pi t)^{-1/2} \exp\{-|x' - x|^2/4t\}$, $t > 0$ on $\mathbb{R}$ itself plays a role of a parametrix.

We remark that a Gaussian measure $\mu_{(m,\sigma)}$ of mean value m and variance $\sigma > 0$ is connected with a Gaussian measure $\mu_{(m',\sigma')}$ by an open mixture arc if and only if $(m, \sigma) = (m', \sigma')$. Thus, a space of probability measures on $\mathbb{R}$ including all Gaussian measures does not admit globally defined constant vector fields. To overcome this difficulty we might employ other methods, for example, the method of Orlicz spaces given by Pistone et al. [16], [15] and [5], and formulate then information geometry for non-parametric model over the space $\widehat{\mathcal{P}}(M)$ of probability measures having L_1-integrable positive density function which is endowed with an affine smooth structure.

For a study of parametric models from our non-parametric information geometry we take an embedding of a parametric model over M into our space $(\mathcal{P}(M), G)$ as a submanifold. Then the Riemannian submanifold theory will be applied.

References

1. Amari, S.: Information Geometry and its Applications. Applied Mathematical Sciences, vol. 194. Springer, Cham (2016). https://doi.org/10.1007/978-4-431-55978-8
2. Amari, S., Nagaoka, H.: Methods of Information Geometry, Translations of Mathematical Monographs, vol. 191. AMS, Oxford (2000)
3. Bauer, M., Bruveris, M., Michor, P.W.: Uniqueness of the Fisher-Rao metric on the space of smooth densities. Bull. London Math. Soc. **48**, 499–506 (2016)
4. Berger, M., Gauduchon, P., Mazet, E.: Le Spectre d'une Variété Riemannienne. Lecture Notes, vol. 194. Springer-Verlag, Berlin (1971). https://doi.org/10.1007/BFb0064643
5. Cena, A., Pistone, G.: Exponential statistical manifold. Ann. Inst. Stat. Math. **59**, 27–56 (2007)
6. do Carmo, M.P.: Riemannian Geometry. Birkhäuser, Boston (1992)
7. Friedrich, T.: Die Fisher-Information und symplektische Strukturen. Math. Nachr. **153**, 273–296 (1991)
8. Gallot, S., Hulin, D., Lafontaine, J.: Riemannian Geometry, 2nd edn. Springer-Verlag, Berlin (1993). https://doi.org/10.1007/978-3-642-97242-3
9. Itoh, M., Satoh, H.: Geometry of Fisher information metric and the barycenter map. Entropy **17**, 1814–1849 (2015)
10. Itoh, M., Satoh, H.: Information geometry of the space of probability measures and barycenter maps. Sugaku **69**, 387–406 (2017). (in Japanese. English version is to appear in Sugaku Expositions, AMS)
11. Itoh, M., Satoh, H.: Geometric mean of probability measures and geodesics of Fisher information metric. arXiv:1708.07211, submitted
12. Milnor, J.: Morse Theory. Princeton University Press, Princeton (1963)
13. Nikulin, M.S.: Hellinger distance, in Encyclopedia of Mathematics, Springer. http://www.encyclopediaofmath.org/index.php?title=Hellinger_distance&oldid=16453
14. Ohara, A.: Geodesics for dual connections and means on symmetric cones. Integr. Equ. Oper. Theor. **50**, 537–548 (2004)
15. Pistone, Giovanni: Nonparametric information geometry. In: Nielsen, Frank, Barbaresco, Frédéric (eds.) GSI 2013. LNCS, vol. 8085, pp. 5–36. Springer, Heidelberg (2013). https://doi.org/10.1007/978-3-642-40020-9_3
16. Pistone, G., Sempi, C.: An infinite-dimensional geometric structure on the space of all the probability measures equivalent to a given one. Ann. Stat. **23**, 1543–1561 (1995)
17. Rao, C.R.: Information and the accuracy attainable in the estimation of statistical parameters. Bull. Calcutta Math. Soc. **37**, 81–91 (1945)
18. Sakai, T.: Riemannian Geometry, Translations of Mathematical Monographs, vol. 149. A.M.S. Providence, Oxford (1996)
19. Santacroce, M., Siri, P., Trivellato, B.: New results on mixture and exponential models by Orlicz spaces. Bernoulli **22**, 1431–1447 (2016)
20. Schoen, R., Yau, S.-T.: Lectures on Differential Geometry. Intern. Press, Boston (1994)
21. Schwartz, L.: Séminaire SCHWARTZ, 2ème année, 1954/1955. Faculté des Sciences de Paris, Paris (1955)
22. Szabó, Z.: The Lichnerowicz conjecture on harmonic manifolds. J. Differential Geom. **31**, 1–28 (1990)

Statistics on Non-linear Data

A Unified Formulation for the Bures-Wasserstein and Log-Euclidean/Log-Hilbert-Schmidt Distances between Positive Definite Operators

Hà Quang Minh$^{(\boxtimes)}$ [ID]

RIKEN Center for Advanced Intelligence Project, Tokyo, Japan
minh.haquang@riken.jp

Abstract. This work presents a parametrized family of distances, namely the Alpha Procrustes distances, on the set of symmetric, positive definite (SPD) matrices. The Alpha Procrustes distances provide a unified formulation encompassing both the Bures-Wasserstein and Log-Euclidean distances between SPD matrices. This formulation is then generalized to the set of positive definite Hilbert-Schmidt operators on a Hilbert space, unifying the infinite-dimensional Bures-Wasserstein and Log-Hilbert-Schmidt distances. The presented formulations are new both in the finite and infinite-dimensional settings.

Keywords: Procrustes distance · Bures-Wasserstein distance · Log-Euclidean distance · Log-Hilbert-Schmidt distance · Positive definite matrices · Positive definite operators

1 Introduction and Motivations

The purpose of the current work is to provide a unified formulation linking two important distances on the set of symmetric, positive definite (SPD) matrices, namely the Bures-Wasserstein and Log-Euclidean distances, along with their infinite-dimensional generalizations on the set of positive definite Hilbert-Schmidt operators on an infinite-dimensional Hilbert space.

Let $\text{Sym}^+(n)$ denote the set of $n \times n$ real, symmetric, positive semi-definite matrices and $\text{Sym}^{++}(n) \subset \text{Sym}^+(n)$ denote the set of symmetric positive definite (SPD) matrices. Let $\mathbb{U}(n)$ denote the set of $n \times n$ unitary matrices. In the context of optimal transport theory [15], the Bures-Wasserstein distance on $\text{Sym}^+(n)$ arises as follows. Let $\mu_X \sim \mathcal{N}(m_1, A)$ and $\mu_Y \sim \mathcal{N}(m_2, B)$ be two Gaussian probability distributions on $\mathbb{R}^n$. Let $\Gamma(\mu_X, \mu_Y)$ be the set of all probability distributions on $\mathbb{R}^n \times \mathbb{R}^n$ whose marginal distributions are μ_X and μ_Y.

F. Nielsen and F. Barbaresco (Eds.): GSI 2019, LNCS 11712, pp. 475–483, 2019.
https://doi.org/10.1007/978-3-030-26980-7_49

It was proved [3,5,6,13] that the following is a squared distance, the so-called $\mathcal{L}^2$-Wasserstein distance, between μ_X and μ_Y

$$d_W^2(\mu_X, \mu_Y) = \inf_{\mu \in \Gamma(\mu_X, \mu_Y)} \int_{\mathbb{R}^n \times \mathbb{R}^n} ||x - y||^2 d\mu(x, y)$$
$$= ||m_1 - m_2||^2 + \text{tr}[A + B - 2(A^{1/2}BA^{1/2})^{1/2}]. \qquad (1)$$

For $m_1 = m_2 = 0$, we obtain the Bures-Wasserstein distance on $\text{Sym}^+(n)$

$$d_{\text{BW}}(A, B) = \left(\text{tr}[A + B - 2(A^{1/2}BA^{1/2})^{1/2}]\right)^{1/2}. \qquad (2)$$

From the viewpoint of Procrustes distances [2,9], d_{BW} is obtained via the following optimization problem

$$d_{\text{BW}}(A, B) = \min_{U,V \in \mathbb{U}(n)} ||A^{1/2}U - B^{1/2}V||_F = \min_{U \in \mathbb{U}(n)} ||A^{1/2} - B^{1/2}U||_F, \qquad (3)$$

where $|| \ ||_F$ is the Frobenius norm. Both the optimal transport and Procrustes distance formulations remain valid in the infinite-dimensional settings, where μ_X, μ_Y are two Gaussian measures on a Hilbert space $\mathcal{H}$, A, B are two covariance operators on $\mathcal{H}$ [5,9], with $|| \ ||_F$ replaced by the Hilbert-Schmidt norm $|| \ ||_{\text{HS}}$.

The Log-Euclidean distance, on the other hand, is the Riemannian distance associated with the bi-invariant Riemannian metric on $\text{Sym}^{++}(n)$ [1], considered as a Lie group under the commutative multiplication $A \odot B = \exp(\log(A) + \log(B))$, where log denotes the principal matrix logarithm. It is given by

$$d_{\text{logE}}(A, B) = ||\log(A) - \log(B)||_F. \qquad (4)$$

Contributions of this work. While the two distances given in Eqs.(2) and (4) appear quite different and unrelated, we show that

1. By generalizing the Procrustes distance optimization problem in Eq.(3), we obtain a parametrized family of distances on $\text{Sym}^{++}(n)$ that includes both the Bures-Wasserstein and Log-Euclidean distances as special cases. We call this family *Alpha Procrustes distances*.
2. The Alpha Procrustes distances are then generalized from $\text{Sym}^{++}(n)$ to the set of positive definite unitized Hilbert-Schmidt operators on an infinite-dimensional Hilbert space $\mathcal{H}$. This setting is more general than the setting of covariance operators on $\mathcal{H}$. In particular, we recover the infinite-dimensional Bures-Wasserstein and Log-Hilbert-Schmidt distances [10] as special cases.

2   Finite-Dimensional Setting

We start with the sets $\text{Sym}^+(n)$ and $\text{Sym}^{++}(n)$. The Procrustes distance formulation in Eq.(3) can be generalized as follows.

Definition 1 (Alpha Procrustes distance - finite-dimensional version).
Let $\alpha \in \mathbb{R}, \alpha \neq 0$ be fixed. The α-Procrustes distance between two matrices $A, B \in \mathrm{Sym}^{++}(n)$ is defined to be

$$d^{\alpha}_{\mathrm{proE}}(A, B) = \min_{U, V \in \mathbb{U}(n)} \left\| \frac{A^{\alpha}U - B^{\alpha}V}{\alpha} \right\|_F = \min_{U \in \mathbb{U}(n)} \left\| \frac{A^{\alpha} - B^{\alpha}U}{\alpha} \right\|_F. \tag{5}$$

For $\alpha > 0$, we define this distance over the larger set $\mathrm{Sym}^+(n)$.

Theorem 1 (Explicit expression). Let either (i) $A, B \in \mathrm{Sym}^{++}(n)$, $\alpha \in \mathbb{R}, \alpha \neq 0$, or (ii) $A, B \in \mathrm{Sym}^+(n), \alpha \in \mathbb{R}, \alpha > 0$. Then

$$d^{\alpha}_{\mathrm{proE}}(A, B) = \left(\frac{1}{\alpha^2} \mathrm{tr}(A^{2\alpha} + B^{2\alpha} - 2\mathrm{tr}(A^{\alpha}B^{2\alpha}A^{\alpha})^{1/2}) \right)^{1/2}. \tag{6}$$

Special case: Bures-Wasserstein-Fréchet distance. For $\alpha = 1/2$, $A, B \in \mathrm{Sym}^+(n)$, we obtain

$$d^{1/2}_{\mathrm{proE}}(A, B) = 2(\mathrm{tr}[A + B - (A^{1/2}BA^{1/2})^{1/2}])^{1/2} = 2d_{\mathrm{BW}}(A, B). \tag{7}$$

This is precisely twice the Bures-Wasserstein-Fréchet distance [2,3,8,13] between $A, B \in \mathrm{Sym}^+(n)$.

Special case: A, B commute. In this case, Eq.(6) becomes

$$d^{\alpha}_{\mathrm{proE}}(A, B) = \left\| \frac{A^{\alpha} - B^{\alpha}}{\alpha} \right\|_F. \tag{8}$$

This is precisely the power Euclidean distance [4]. For $A, B \in \mathrm{Sym}^{++}(n)$,

$$\lim_{\alpha \to 0} \left\| \frac{A^{\alpha} - B^{\alpha}}{\alpha} \right\|_F = \| \log(A) - \log(B) \|_F = d_{\mathrm{logE}}(A, B). \tag{9}$$

The following shows that this limit also holds for the Alpha Procrustes distance.

Theorem 2 (Limiting case - Log-Euclidean distance). Let $A, B \in \mathrm{Sym}^{++}(n)$ be fixed. Then

$$\lim_{\alpha \to 0} \frac{1}{\alpha^2} [\mathrm{tr}(A^{2\alpha} + B^{2\alpha} - 2(A^{\alpha}B^{2\alpha}A^{\alpha})^{1/2}] = \| \log(A) - \log(B) \|_F^2. \tag{10}$$

We have then the following result.

Theorem 3. The function $d^{\alpha}_{\mathrm{proE}}$, as defined in Eq.(6), is a metric on the set $\mathrm{Sym}^{++}(n)$ for all $\alpha \in \mathbb{R}$, with twice the Bures-Wasserstein-Fréchet distance corresponding to $\alpha = 1/2$ and the Log-Euclidean distance corresponding to $\alpha = 0$. Furthermore, $d^{\alpha}_{\mathrm{proE}}$ is a metric on the set $\mathrm{Sym}^+(n)$ for all $\alpha > 0$.

3   Infinite-Dimensional Setting

We now generalize the results in Sect. 2 to the infinite-dimensional setting. Throughout the following, let $\mathcal{H}$ denote a real, separable, infinite-dimensional Hilbert space, unless explicitly stated otherwise. Let $\mathcal{L}(\mathcal{H})$ denote the set of bounded linear operators on $\mathcal{H}$, $\mathrm{Sym}(\mathcal{H}) \subset \mathcal{L}(\mathcal{H})$ the set of bounded, self-adjoint operators, $\mathrm{Sym}^+ \subset \mathrm{Sym}(\mathcal{H})$ and $\mathrm{Sym}^{++}(\mathcal{H}) \subset \mathrm{Sym}^+(\mathcal{H})$ denote, respectively, the sets of positive and strictly positive operators on $\mathcal{H}$. Let $\mathbb{U}(\mathcal{H})$ denote the set of unitary operators on $\mathcal{H}$ and $\mathrm{Tr}(\mathcal{H})$ and $\mathrm{HS}(\mathcal{H})$ denote the sets of trace class and Hilbert-Schmidt operators on $\mathcal{H}$, respectively.

In the case A, B are two positive trace class operators on $\mathcal{H}$, we have

Theorem 4 ([5,9]). *Let* $A, B \in \mathrm{Sym}^+(\mathcal{H}) \cap \mathrm{Tr}(\mathcal{H})$ *be fixed. Then*

$$\min_{U \in \mathbb{U}(\mathcal{H})} ||A^{1/2} - B^{1/2}U||_{\mathrm{HS}}^2 = \mathrm{tr}[A + B - 2(A^{1/2}BA^{1/2})^{1/2}] \qquad (11)$$

$$= d_{\mathrm{W}}^2(\mathcal{N}(0, A), \mathcal{N}(0, B)). \qquad (12)$$

Corollary 1. *Let* $\alpha \in \mathbb{R}, \alpha > 0$ *be fixed. Let* $A, B \in \mathrm{Sym}^+(\mathcal{H})$ *be fixed, such that* $A^\alpha, B^\alpha \in \mathrm{HS}(\mathcal{H})$. *Then*

$$\min_{U \in \mathbb{U}(\mathcal{H})} ||A^\alpha - B^\alpha U||_{\mathrm{HS}}^2 = \mathrm{tr}[A^{2\alpha} + B^{2\alpha} - 2(A^\alpha B^{2\alpha} A^\alpha)^{1/2}]. \qquad (13)$$

While Eq.(13) is valid for any pair $A, B \in \mathrm{Sym}^+(\mathcal{H})$ such that $A^\alpha, B^\alpha \in \mathrm{HS}(\mathcal{H})$, in general the limit $\lim_{\alpha \to 0} \frac{1}{\alpha} \min_{U \in \mathbb{U}(\mathcal{H})} ||A^\alpha - B^\alpha U||_{\mathrm{HS}}$ does not exist in a form similar to Eq.(10), since $\log(A)$ is unbounded even when A is strictly positive. To obtain the infinite-dimensional generalization of Theorem 2, we consider the setting of positive definite unitized Hilbert-Schmidt operators. Let us denote by $\mathbb{P}(\mathcal{H})$ the set of self-adjoint, positive definite operators on $\mathcal{H}$ $\mathbb{P}(\mathcal{H}) = \{A \in \mathcal{L}(\mathcal{H}), A^* = A, \exists M_A > 0 \text{ } s.t. \langle x, Ax \rangle \geq M_A ||x||^2 \text{ } \forall x \in \mathcal{H}\}$. We write $A > 0 \iff A \in \mathbb{P}(\mathcal{H})$. We recall that in [7], the author defined the set of *extended (or unitized) Hilbert-Schmidt operators* on $\mathcal{H}$ to be

$$\mathrm{HS}_X(\mathcal{H}) = \{A + \gamma I : A \in \mathrm{HS}(\mathcal{H}), \gamma \in \mathbb{R}\}, \qquad (14)$$

which becomes a Hilbert space under the *extended Hilbert-Schmidt inner product*

$$\langle (A + \gamma I), (B + \nu I) \rangle_{\mathrm{HS}_X} = \langle A, B \rangle_{\mathrm{HS}} + \gamma\nu = \mathrm{tr}(A^*B) + \gamma\nu. \qquad (15)$$

The set of positive definite unitized (or extended) Hilbert-Schmidt operators, which is an infinite-dimensional generalization of $\mathrm{Sym}^{++}(n)$, is defined to be

$$\mathscr{P}\mathscr{C}_2(\mathcal{H}) = \mathbb{P}(\mathcal{H}) \cap \mathrm{HS}_X(\mathcal{H}) = \{A + \gamma I > 0 : A \in \mathrm{HS}(\mathcal{H}), \gamma \in \mathbb{R}\}. \qquad (16)$$

The set $\mathscr{P}\mathscr{C}_2(\mathcal{H})$ is then a Hilbert manifold on which one can define the infinite-dimensional generalizations of the affine-invariant Riemannian metric [7] and the

Log-Determinant divergences [11,12]. In [10], we introduced the Log-Hilbert-Schmidt distance on $\mathscr{PC}_2(\mathcal{H})$, which generalizes the Log-Euclidean distance on $\mathrm{Sym}^{++}(n)$. For $(A + \gamma I), (B + \nu I) \in \mathscr{PC}_2(\mathcal{H})$, this distance is defined by

$$d_{\log\mathrm{HS}}[(A + \gamma I), (B + \nu I)] = \|\log(A + \gamma I) - \log(B + \nu I)\|_{\mathrm{HS}_X}. \quad (17)$$

We next define a family of distances that includes both the infinite-dimensional Bures-Wasserstein and Log-Hilbert-Schmidt distances as special cases.

3.1   The Case $\gamma = \nu = 1$

For a fixed $\gamma \in \mathbb{R}, \gamma > 0$, we consider the following subset of $\mathscr{PC}_2(\mathcal{H})$

$$\mathscr{PC}_2(\mathcal{H})(\gamma) = \{A + \gamma I > 0 : A \in \mathrm{HS}\} \subset \mathscr{PC}_2(\mathcal{H}). \quad (18)$$

We first generalize the Alpha Procrustes distance in Definition 1 to the set

$$\mathscr{PC}_2(\mathcal{H})(1) = \{I + A > 0 : A \in \mathrm{HS}\} \subset \mathscr{PC}_2(\mathcal{H}). \quad (19)$$

We first note that for any $(I + A) \in \mathscr{PC}_2(\mathcal{H})(1)$, $(I + A)^\alpha = \exp[\alpha \log(I + A)] = I + \sum_{k=1}^\infty \frac{\alpha^k}{k!} [\log(I + A)]^k = I + C$ where $C \in \mathrm{Sym}(\mathcal{H}) \cap \mathrm{HS}(\mathcal{H})$, since $\|C\|_{\mathrm{HS}} \leq \sum_{k=1}^\infty \frac{|\alpha|^k}{k!} \|\log(I + A)\|_{\mathrm{HS}}^k = \exp(\|\log(I + A)\|_{\mathrm{HS}}) - 1 < \infty$.

Proposition 1. *Let* $(I + A), (I + B) \in \mathscr{PC}_2(\mathcal{H})(1)$ *and* $\alpha \in \mathbb{R}$ *be fixed. Then*

$$\min_{(I+U),(I+V)\in\mathbb{U}(\mathcal{H})\cap\mathrm{HS}_X(\mathcal{H})} \|(I + A)^\alpha(I + U) - (I + B)^\alpha(I + V)\|_{\mathrm{HS}} \quad (20)$$

$$= \min_{(I+V)\in\mathbb{U}(\mathcal{H})\cap\mathrm{HS}_X(\mathcal{H})} \|(I + A)^\alpha - (I + B)^\alpha(I + V)\|_{\mathrm{HS}}. \quad (21)$$

The operators of the form $(I + U) \in \mathbb{U}(\mathcal{H}) \cap \mathrm{HS}_X(\mathcal{H})$ in Proposition 1 are motivated by the following polar decomposition.

Lemma 1. *Let* $(A + \gamma I) \in \mathrm{HS}_X(\mathcal{H})$ *be invertible. Then its polar decomposition has the form*

$$A + \gamma I = S|A + \gamma I| = (I + R)|A + \gamma I|, \quad (22)$$

where $S = I + R \in \mathbb{U}(\mathcal{H}) \cap \mathrm{HS}_X(\mathcal{H})$.

We note also that if $U \in \mathbb{U}(\mathcal{H})$ and $(I + A) \in \mathrm{HS}_X(\mathcal{H})$, then generally $(I + A)U = U + AU \notin \mathrm{HS}_X(\mathcal{H})$. With operators of the form $(I + U) \in \mathbb{U}(\mathcal{H}) \cap \mathrm{HS}_X(\mathcal{H})$, we have $(I + A)(I + U) = I + A + U + AU \in \mathrm{HS}_X(\mathcal{H})$. Furthermore, for $(I + A), (I + B), (I + U), (I + V) \in \mathrm{HS}_X(\mathcal{H})$, we have $(I + A)(I + U) - (I + B)(I + V) \in \mathrm{HS}(\mathcal{H})$, so that the expressions in Eqs.(20) and (21) are both well-defined and finite.

Motivated by Proposition 1, the following is our definition for the Alpha Procrustes distance between operators of the form $(I + A) > 0$, $A \in \mathrm{HS}(\mathcal{H})$.

Definition 2 (Alpha Procrustes distance between positive definite Hilbert-Schmidt operators - special case). *Let* $\alpha \in \mathbb{R}, \alpha \neq 0$ *be fixed. The* α-*Procrustes distance between* $(I + A), (I + B) \in \mathscr{PC}_2(\mathcal{H})(1)$ *is defined to be*

$$d^\alpha_{\text{proHS}}[(I + A), (I + B)] = \min_{(I+U)\in\mathbb{U}(\mathcal{H})\cap\text{HS}_X(\mathcal{H})} \left\| \frac{(I + A)^\alpha - (I + B)^\alpha(I + U)}{\alpha} \right\|_{\text{HS}}. \tag{23}$$

To state the explicit expressions for $d^\alpha_{\text{proHS}}[(I + A), (I + B)]$, as defined in Eq.(23), we first need the following result.

Proposition 2. *Let* $(I + A), (I + B) \in \mathscr{PC}_2(\mathcal{H})$. *Let* $\alpha \in \mathbb{R}$ *be fixed. Then*

$$(I + A)^{2\alpha} + (I + B)^{2\alpha} - 2[(I + A)^\alpha(I + B)^{2\alpha}(I + A)^\alpha]^{1/2} \in \text{Tr}(\mathcal{H}), \tag{24}$$

$$[(I + A)^\alpha(I + B)^{2\alpha}(I + A)^\alpha]^{1/2} - (I + A)^\alpha - (I + B)^\alpha + I \in \text{Tr}(\mathcal{H}). \tag{25}$$

With Proposition 2, we are ready to state the following.

Theorem 5. *Let* $(I + A), (I + B) \in \mathscr{PC}_2(\mathcal{H})$ *and* $\alpha \in \mathbb{R}, \alpha \neq 0$ *be fixed. Then*

$$(d^\alpha_{\text{proHS}}[(I + A), (I + B)])^2 = \frac{1}{\alpha^2} \min_{(I+U)\in\mathbb{U}(\mathcal{H})\cap\text{HS}_X(\mathcal{H})} \|(I + A)^\alpha - (I + B)^\alpha(I + U)\|^2_{\text{HS}}$$

$$= \frac{1}{\alpha^2} \left(\text{tr}[(I + A)^{2\alpha} + (I + B)^{2\alpha} - 2[(I + A)^\alpha(I + B)^{2\alpha}(I + A)^\alpha]^{1/2}] \right) \tag{26}$$

$$= \frac{1}{\alpha^2} \left(\|(I + A)^\alpha\|^2_{\text{HS}_X} + \|(I + B)^\alpha\|^2_{\text{HS}_X} - 2 \right)$$

$$- \frac{2}{\alpha^2} \left(\text{tr}[[(I + A)^\alpha(I + B)^{2\alpha}(I + A)^\alpha]^{1/2} - (I + A)^\alpha - (I + B)^\alpha + I] \right). \tag{27}$$

3.2   The Case $\gamma = \nu > 0$

The case $\gamma = \nu = 1$ generalizes to the case $\gamma = \nu > 0$ as follows.

Definition 3 (Alpha Procrustes distance between positive definite Hilbert-Schmidt operators). *Let* $\gamma > 0$, $\alpha \in \mathbb{R}, \alpha \neq 0$ *be fixed. The Alpha Procrustes distance between two operators* $(A + \gamma I), (B + \gamma I) \in \mathscr{PC}_2(\mathcal{H})$ *is defined to be*

$$d^\alpha_{\text{proHS}}[(A + \gamma I), (B + \gamma I)]$$

$$= \min_{(I+U)\in\mathbb{U}(\mathcal{H})\cap\text{HS}_X(\mathcal{H})} \left\| \frac{(A + \gamma I)^\alpha - (B + \gamma I)^\alpha(I + U)}{\alpha} \right\|_{\text{HS}_X}. \tag{28}$$

Theorem 6 (Explicit expression). *Let* $(A + \gamma I), (B + \gamma I) \in \mathscr{PC}_2(\mathcal{H})$ *be fixed. Let* $\alpha \in \mathbb{R}, \alpha \neq 0$ *be fixed. Then*

$$(d^\alpha_{\text{proHS}}[(A + \gamma I), (B + \gamma I)])^2 \tag{29}$$

$$= \frac{1}{\alpha^2} \text{tr}[(A + \gamma I)^{2\alpha} + (B + \gamma I)^{2\alpha} - 2[(A + \gamma I)^\alpha(B + \gamma I)^{2\alpha}(A + \gamma I)^\alpha]^{1/2}]$$

$$= \frac{\|(A + \gamma I)^\alpha\|^2_{\text{HS}_X} - \gamma^{2\alpha}}{\alpha^2} + \frac{\|(B + \gamma I)^\alpha\|^2_{\text{HS}_X} - \gamma^{2\alpha}}{\alpha^2} \tag{30}$$

$$- \frac{2}{\alpha^2} \text{tr}[[(A + \gamma I)^\alpha(B + \gamma I)^{2\alpha}(A + \gamma I)^\alpha]^{1/2} - \gamma^\alpha(A + \gamma I)^\alpha - \gamma^\alpha(B + \gamma I)^\alpha + \gamma^{2\alpha} I].$$

Special case: Finite-dimensional setting. For $A, B \in \mathrm{Sym}^{++}(n)$ ($\alpha \neq 0$), or $A, B \in \mathrm{Sym}^{+}(n)$ ($\alpha > 0$), setting $\gamma = 0$ in Eq.(29) gives

$$(d^{\alpha}_{\mathrm{proHS}}[A, B])^2 = \frac{1}{\alpha^2}\mathrm{tr}[A^{2\alpha} + B^{2\alpha} - 2(A^{\alpha}B^{2\alpha}A^{\alpha})^{1/2}] = (d^{\alpha}_{\mathrm{proE}}[A, B])^2. \quad (31)$$

Corollary 2 (Special case - Bures-Wasserstein distance). *Let $A, B \in \mathrm{Sym}^{+}(\mathcal{H}) \cap \mathrm{Tr}(\mathcal{H})$. Then*

$$\lim_{\gamma \to 0} d^{1/2}_{\mathrm{proHS}}[(A + \gamma I), (B + \gamma I)] = 2(\mathrm{tr}[A + B - 2(A^{1/2}BA^{1/2})^{1/2}])^{1/2}. \quad (32)$$

Theorem 7 (Limiting case - Log-Hilbert-Schmidt distance). *Let $(A + \gamma I), (B + \gamma I) \in \mathscr{PC}_2(\mathcal{H})$ be fixed. Then*

$$\lim_{\alpha \to 0} d^{\alpha}_{\mathrm{proHS}}[(A + \gamma I), (B + \gamma I)] = ||\log(A + \gamma I) - \log(B + \gamma I)||_{\mathrm{HS_X}}. \quad (33)$$

Theorem 8. *Let $\gamma > 0$ be fixed. The function $d^{\alpha}_{\mathrm{proHS}}$, as defined in Eq.(28), is a metric on the set $\mathscr{PC}_2(\mathcal{H})(\gamma) = \{A + \gamma I > 0 : A \in \mathrm{HS}(\mathcal{H})\}$ for all $\alpha \in \mathbb{R}$.*

3.3   The RKHS Setting

We now present explicit expressions for the Alpha Procrustes distances between RKHS covariance operators. Let $\mathcal{X}$ be a separable topological space and K be a continuous positive definite kernel on $\mathcal{X} \times \mathcal{X}$. Then the reproducing kernel Hilbert space (RKHS) $\mathcal{H}_K$ induced by K is separable ([14], Lemma 4.33). Let $\Phi : \mathcal{X} \to \mathcal{H}_K$ be the corresponding feature map, so that $K(x, y) = \langle \Phi(x), \Phi(y) \rangle_{\mathcal{H}_K}$ $\forall (x, y) \in \mathcal{X} \times \mathcal{X}$. Let $\mathbf{X} = [x_1, \ldots, x_m], m \in \mathbb{N}$, be a data matrix randomly sampled from $\mathcal{X}$ according to a Borel probability distribution ρ, where $m \in \mathbb{N}$ is the number of observations. The feature map Φ on $\mathbf{X}$ defines the bounded linear operator $\Phi(\mathbf{X}) : \mathbb{R}^m \to \mathcal{H}_K, \Phi(\mathbf{X})\mathbf{b} = \sum_{j=1}^{m} b_j \Phi(x_j), \mathbf{b} \in \mathbb{R}^m$. The corresponding empirical covariance operator for $\Phi(\mathbf{X})$ is defined to be

$$C_{\Phi(\mathbf{X})} = \frac{1}{m}\Phi(\mathbf{X})J_m\Phi(\mathbf{X})^* : \mathcal{H}_K \to \mathcal{H}_K, \quad (34)$$

where $J_m = I_m - \frac{1}{m}\mathbf{1}_m\mathbf{1}_m^T$ is the centering matrix, with $\mathbf{1}_m = (1, \ldots, 1)^T \in \mathbb{R}^m$.

Let $\mathbf{X} = [x_i]_{i=1}^{m}$, $\mathbf{Y} = [y_i]_{i=1}^{m}$, be two random data matrices sampled from $\mathcal{X}$ according to two Borel probability distributions and $C_{\Phi(\mathbf{X})}, C_{\Phi(\mathbf{Y})}$ be the corresponding covariance operators induced by K. Define the $m \times m$ Gram matrices

$$K[\mathbf{X}] = \Phi(\mathbf{X})^*\Phi(\mathbf{X}), \; K[\mathbf{Y}] = \Phi(\mathbf{Y})^*\Phi(\mathbf{Y}), K[\mathbf{X}, \mathbf{Y}] = \Phi(\mathbf{X})^*\Phi(\mathbf{Y}). \quad (35)$$

Define $A = \frac{1}{\sqrt{m}}\Phi(\mathbf{X})J_m : \mathbb{R}^m \to \mathcal{H}_K$, $B = \frac{1}{\sqrt{m}}\Phi(\mathbf{Y})J_m : \mathbb{R}^m \to \mathcal{H}_K$, so that

$$A^*A = \frac{1}{m}J_mK[\mathbf{X}]J_m, \; B^*B = \frac{1}{m}J_mK[\mathbf{Y}]J_m, \; A^*B = \frac{1}{m}J_mK[\mathbf{X}, \mathbf{Y}]J_m. \quad (36)$$

To state our next result, let $E : \mathcal{H}_1 \to \mathcal{H}_1$ be a self-adjoint, positive, compact operator on a separable Hilbert space $\mathcal{H}_1$, with nonzero eigenvalues $\{\lambda_k(E)\}_{k=1}^{N_E}$,

$1 \leq N_E \leq \infty$, and corresponding orthonormal eigenvectors $\{\phi_k(E)\}_{k=1}^{N_E}$. Consider the following operator

$$h_\alpha(E) = \sum_{k=1}^{N_E} \frac{(1 + \lambda_k(E))^\alpha - 1}{\lambda_k(E)} \phi_k(E) \otimes \phi_k(E). \tag{37}$$

When (Lemma 10, [12]) $\dim(\mathcal{H}_1) < \infty$, let $E = U_E \Sigma_E U_E^T$ be the reduced singular value decomposition of E, where U_E is a unitary matrix of size $\dim(\mathcal{H}_1) \times N_E$. Then

$$h_\alpha(E) = U_E[(\Sigma_E + I_{N_E})^\alpha - I_{N_E}]\Sigma_E^{-1}U_E^T. \tag{38}$$

Theorem 9 (Alpha Procrustes distance between RKHS covariance operators). *Let A^*A, A^*B, B^*B be as defined in Eq.(36) and h_α be as defined in Eq.(38), with $\mathcal{H}_1 = \mathbb{R}^m$. Then*

$$\alpha^2(d_{\text{proHS}}^\alpha[(C_{\Phi(\mathbf{X})} + \gamma I_{\mathcal{H}_K}), (C_{\Phi(\mathbf{Y})} + \gamma I_{\mathcal{H}_K})])^2 \tag{39}$$
$$= \text{tr}[(A^*A + \gamma I_m)^{2\alpha} - \gamma^{2\alpha}I_m] + \text{tr}[(B^*B + \gamma I_m)^{2\alpha} - \gamma^{2\alpha}I_m]$$

$$- 2\gamma^{2\alpha}\text{tr}\left[\left[I_{3m} + \begin{pmatrix} C_{11} & C_{12} & C_{13} \\ C_{21} & C_{22} & C_{23} \\ C_{21} & C_{22} & C_{23} \end{pmatrix}\right]^{1/2} - I_{3m}\right], \tag{40}$$

where the $m \times m$ matrices C_{ij}, $i = 1, 2$, $j = 1, 2, 3$, are given by

$$C_{11} = [(I_m + A^*A/\gamma)^{2\alpha} - I_m], \quad C_{12} = \frac{1}{\gamma}A^*Bh_{2\alpha}(B^*B/\gamma), \tag{41}$$

$$C_{13} = \frac{1}{\gamma}[(I_m + A^*A/\gamma)^{2\alpha} - I_m]A^*Bh_{2\alpha}(B^*B/\gamma), \tag{42}$$

$$C_{21} = \frac{1}{\gamma}B^*Ah_{2\alpha}(A^*A/\gamma), \quad C_{22} = [(I_m + B^*B/\gamma)^{2\alpha} - I_m], \tag{43}$$

$$C_{23} = \frac{1}{\gamma^2}B^*Ah_{2\alpha}(A^*A/\gamma)A^*Bh_{2\alpha}(B^*B/\gamma). \tag{44}$$

Remark. We will present in the full version of the current paper the associated geometrical structures, the more technically involved case $\gamma \neq \nu$ in the infinite-dimensional setting, along with all the proofs.

References

1. Arsigny, V., Fillard, P., Pennec, X., Ayache, N.: Geometric means in a novel vector space structure on symmetric positive-definite matrices. SIAM J. Matrix Anal. Appl. **29**(1), 328–347 (2007)
2. Bhatia, R., Jain, T., Lim, Y.: On the Bures-Wasserstein distance between positive definite matrices. Expositiones Mathematicae (2018)
3. Dowson, D., Landau, B.: The Fréchet distance between multivariate normal distributions. J. Multivar. Anal. **12**(3), 450–455 (1982)

4. Dryden, I., Koloydenko, A., Zhou, D.: Non-euclidean statistics for covariance matrices, with applications to diffusion tensor imaging. Ann. Appl. Stat. **3**, 1102–1123 (2009)
5. Gelbrich, M.: On a formula for the L2 Wasserstein metric between measures on Euclidean and Hilbert spaces. Mathematische Nachrichten **147**(1), 185–203 (1990)
6. Givens, C.R., Shortt, R.M.: A class of Wasserstein metrics for probability distributions. Michigan Math. J. **31**(2), 231–240 (1984)
7. Larotonda, G.: Nonpositive curvature: a geometrical approach to Hilbert-Schmidt operators. Differ. Geom. Appl. **25**, 679–700 (2007)
8. Malagò, L., Montrucchio, L., Pistone, G.: Wasserstein Riemannian geometry of Gaussian densities. Inf. Geom. **1**(2), 137–179 (2018)
9. Masarotto, V., Panaretos, V., Zemel, Y.: Procrustes metrics on covariance operators and optimal transportation of Gaussian processes. Sankhya A **80**, 1–42 (2018)
10. Minh, H.Q., Biagio, M.S., Murino, V.: Log-Hilbert-Schmidt metric between positive definite operators on Hilbert spaces. Adv. Neural Inf. Process. Syst. **27**, 388–396 (2014). (NIPS 2014)
11. Minh, H.: Infinite-dimensional log-determinant divergences between positive definite trace class operators. Linear Algebra Appl. **528**, 331–383 (2017)
12. Minh, H.: Infinite-dimensional Log-Determinant divergences III: Log-Euclidean and Log-Hilbert–Schmidt divergences. In: Ay, N., Gibilisco, P., Matus, F. (eds.) Inf. Geom. Appl. Springer Proceedings in Mathematics & Statistics, vol. 252. Springer, Cham (2016). https://doi.org/10.1007/978-3-319-97798-0_8
13. Olkin, I., Pukelsheim, F.: The distance between two random vectors with given dispersion matrices. Linear Algebra Appl. **48**, 257–263 (1982)
14. Steinwart, I., Christmann, A.: Support Vector Machines. Springer Science & Business Media, Berlin (2008)
15. Villani, C.: Optimal Transport: Old and New, vol. 338. Springer Science & Business Media, Berlin (2008)

Exploration of Balanced Metrics on Symmetric Positive Definite Matrices

Yann Thanwerdas$^{(\boxtimes)}$ and Xavier Pennec

Université Côte d'Azur, Inria, Epione, France
yann.thanwerdas@inria.fr

Abstract. Symmetric Positive Definite (SPD) matrices have been used in many fields of medical data analysis. Many Riemannian metrics have been defined on this manifold but the choice of the Riemannian structure lacks a set of principles that could lead one to choose properly the metric. This drives us to introduce the principle of balanced metrics that relate the affine-invariant metric with the Euclidean and inverse-Euclidean metric, or the Bogoliubov-Kubo-Mori metric with the Euclidean and log-Euclidean metrics. We introduce two new families of balanced metrics, the mixed-power-Euclidean and the mixed-power-affine metrics and we discuss the relation between this new principle of balanced metrics and the concept of dual connections in information geometry.

1 Introduction

Symmetric Positive Definite (SPD) matrices are used in many applications: for example, they represent covariance matrices in signal or image processing [1–3] and they are diffusion tensors in diffusion tensor imaging [4–6]. Many Riemannian structures have been introduced on the manifold of SPD matrices depending on the problem and showing significantly different results from one another on statistical procedures such as the computation of barycenters or the principal component analysis. Non exhaustively, we can cite Euclidean metrics, power-Euclidean metrics [7], log-Euclidean metrics [8], which are flat; affine-invariant metrics [5,6,9] which are negatively curved; the Bogoliubov-Kubo-Mori metric [10] whose curvature has a quite complex expression.

Are there some relations between them? This question has practical interests. First, understanding the links between these metrics could lead to interesting formulas and allow to perform more efficient algorithms. Second, finding families of metrics that comprise these isolated metrics could allow to perform optimization on the parameters of these families to find a better adapted metric. Some relations already exist. For example, the power-Euclidean metrics [7] (resp. power-affine metrics [11]) comprise the Euclidean metric (resp. affine-invariant metric) and tend to the log-Euclidean metric when the power tends to zero.

We propose the principle of *balanced metrics* after observing two facts. The affine-invariant metric $g_\Sigma^A(X, Y) = \mathrm{tr}((\Sigma^{-1} X \Sigma^{-1}) Y)$ on SPD matrices may be seen as a balanced hybridization of the Euclidean metric $g_\Sigma^E(X, Y) = \mathrm{tr}(XY)$

© Springer Nature Switzerland AG 2019
F. Nielsen and F. Barbaresco (Eds.): GSI 2019, LNCS 11712, pp. 484–493, 2019.
https://doi.org/10.1007/978-3-030-26980-7_50

on one vector and of the inverse-Euclidean metric (the Euclidean metric on precision matrices) $g_\Sigma^I(X,Y) = \mathrm{tr}((\Sigma^{-1}X\Sigma^{-1})(\Sigma^{-1}Y\Sigma^{-1}))$ on the other vector. Moreover, the definition of the Bogoliubov-Kubo-Mori metric can be rewritten as $g_\Sigma^{BKM}(X,Y) = \mathrm{tr}(\partial_X \log(\Sigma)\, Y)$ where it appears as a balance of the Euclidean metric and the log-Euclidean metric $g_\Sigma^{LE}(X,Y) = \mathrm{tr}(\partial_X \log(\Sigma)\, \partial_Y \log(\Sigma))$. These observations raise a few questions. Given two metrics, is it possible to define a balanced bilinear form in general? If yes, is it clear that this bilinear form is symmetric and positive definite? If it is a metric, are the Levi-Civita connections of the two initial metrics dual in the sense of information geometry?

In this work, we explore this principle through the affine-invariant metric, the Bogoliubov-Kubo-Mori metric and we define two new families of balanced metrics, the mixed-power-Euclidean and the mixed-power-affine metrics. In Sect. 2, we show that if a balanced metric comes from two flat metrics, the three of them define a dually flat structure. In particular, we show that the balanced structure defined by the Euclidean and the inverse-Euclidean metrics corresponds to the dually flat structure given by the ±1-connections of Fisher information geometry. In Sect. 3, we enlighten the balanced structure of the BKM metric and we generalize it by defining the family of mixed-power-Euclidean metrics. In Sect. 4, we define the family of mixed-power-affine metrics and we discuss the relation between the concepts of balanced metric and dual connections when the two initial metrics are not flat.

2   Affine-Invariant Metric as a Balance of Euclidean and Inverse-Euclidean Metrics

Because the vocabulary may vary from one community to another, we shall first introduce properly the main geometric tools that we use in the article (Sect. 2.1). Then we examine in Sect. 2.2 the principle of balanced metric in the particular case of the pair Euclidean/inverse-Euclidean metrics and we formalize it in the general case of two flat metrics. In Sect. 2.3, we show that the ±1-connections of the centered multivariate normal model are exactly the Levi-Civita connections of the Euclidean and inverse-Euclidean metrics.

2.1   Reminder on Metrics, Connections and Parallel Transport

On a manifold $\mathcal{M}$, we denote $\mathcal{C}^\infty(\mathcal{M})$ the ring of smooth real functions and $\mathfrak{X}(\mathcal{M})$ the $\mathcal{C}^\infty(\mathcal{M})$-module of vector fields.

Connection. A *connection* is an $\mathbb{R}$-bilinear map $\nabla : \mathfrak{X}(\mathcal{M}) \times \mathfrak{X}(\mathcal{M}) \longrightarrow \mathfrak{X}(\mathcal{M})$ that is $\mathcal{C}^\infty(\mathcal{M})$-linear in the first variable and satisfies the Leibniz rule in the second variable. It gives notions of parallelism, parallel transport and geodesics. A vector field V is *parallel* to the curve γ if $\nabla_{\dot\gamma} V = 0$. The *parallel transport* of a vector v along a curve γ is the unique vector field $V_{\gamma(t)} = \Pi_\gamma^{0 \to t} v$ that extends v and that is parallel to γ. Thus, the connection is an infinitesimal parallel transport, that is $\Pi_\gamma^{t \to 0} V_{\gamma(t)} = V_{\gamma(0)} + t\nabla_{\dot\gamma} V + o(t)$. The *geodesics* are autoparallel curves, that is curves γ satisfying $\nabla_{\dot\gamma} \dot\gamma = 0$.

Levi-Civita Connection. Given a metric g on a manifold $\mathcal{M}$, the *Levi-Civita connection* is the unique torsion-free connection ∇^g compatible with the metric g, that is $\nabla^g g = 0$ or more explicitly $X(g(Y,Z)) = g(\nabla^g_X Y, Z) + g(Y, \nabla^g_X Z)$ for all vector fields $X, Y, Z \in \mathfrak{X}(\mathcal{M})$. Thus a metric inherits notions of parallel transport and geodesics. Note that geodesics coincide with distance-minimizing curves with constant speed.

Dual Connections. Given a metric g and a connection ∇, the *dual connection* of ∇ with respect to g is the unique connection ∇^* satisfying the following equality $X(g(Y,Z)) = g(\nabla_X Y, Z) + g(Y, \nabla^*_X Z)$ for all vector fields $X, Y, Z \in \mathfrak{X}(\mathcal{M})$. It is characterized by Lemma 1 below. In this sense, the Levi-Civita connection ∇^g is the unique torsion-free self-dual connection with respect to g. We say that $(\mathcal{M}, g, \nabla, \nabla^*)$ is a *dually-flat manifold* when ∇, ∇^* are dual with respect to g and ∇ is flat (then ∇^* is automatically flat [12]).

Lemma 1 (Characterization of dual connections). *Two connections ∇, ∇' with parallel transports Π, Π' are dual with respect to a metric g if and only if the dual parallel transport preserves the metric, i.e. for all vector fields $X, Y \in \mathfrak{X}(\mathcal{M})$ and all curve γ, $g_{\gamma(t)}(X_{\gamma(t)}, Y_{\gamma(t)}) = g_{\gamma(0)}(\Pi^{t \to 0}_\gamma X_{\gamma(t)}, (\Pi')^{t \to 0}_\gamma Y_{\gamma(t)})$.*

Proof The direct sense is proved in [12]. Let us assume that the dual parallel transport preserves the metric and let $X, Y, Z \in \mathfrak{X}(\mathcal{M})$ be vector fields. Let $x \in \mathcal{M}$ and let γ be a curve such that $\gamma(0) = x$ and $\dot{\gamma}(0) = X_x$. Using the first order approximation of the parallel transport, our assumption leads to:

$$g_{\gamma(t)}(Y_{\gamma(t)}, Z_{\gamma(t)}) = g_x(\Pi^{t \to 0}_\gamma Y_{\gamma(t)}, (\Pi')^{t \to 0}_\gamma Z_{\gamma(t)})$$
$$= g_x(Y_x + t\nabla_{\dot{\gamma}} Y + o(t), Z_x + t\nabla'_{\dot{\gamma}} Z + o(t))$$
$$= g_x(Y_x, Z_x) + t[g_x(\nabla_{\dot{\gamma}} Y, Z) + g_x(Y, \nabla'_{\dot{\gamma}} Z)] + o(t).$$

So $X_x(g(Y,Z)) = g_x(\nabla_{\dot{\gamma}} Y, Z) + g_x(Y, \nabla'_{\dot{\gamma}} Z)$ and ∇, ∇' are dual w.r.t. g. $\qquad\square$

2.2   Principle of Balanced Metrics

Observation. We denote $\mathcal{M} = SPD_n$ the manifold of SPD matrices and $N = \dim \mathcal{M} = \frac{n(n+1)}{2}$. The (A)ffine-invariant metric g^A on SPD matrices [5,6,9], i.e. satisfying $g^A_{M\Sigma M^\top}(MXM^\top, MYM^\top) = g^A_\Sigma(X,Y)$ for $M \in GL_n$, is defined by:

$$g^A_\Sigma(X,Y) = \mathrm{tr}((\Sigma^{-1} X \Sigma^{-1})Y) = \mathrm{tr}(X(\Sigma^{-1} Y \Sigma^{-1})). \tag{1}$$

The (E)uclidean metric g^E on SPD matrices is the pullback metric by the embedding $id : \mathcal{M} \hookrightarrow (\mathrm{Sym}_n, \langle \cdot | \cdot \rangle_{\mathrm{Frob}})$:

$$g^E_\Sigma(X,Y) = \mathrm{tr}(XY). \tag{2}$$

The (I)nverse-Euclidean metric g^I on SPD matrices belongs to the family of power-Euclidean metrics [7] with power -1. If SPD matrices are seen as covariance matrices Σ, the inverse-Euclidean metric is the Euclidean metric on precision matrices Σ^{-1}:

$$g^I_\Sigma(X,Y) = \mathrm{tr}(\Sigma^{-2} X \Sigma^{-2} Y) = \mathrm{tr}((\Sigma^{-1} X \Sigma^{-1})(\Sigma^{-1} Y \Sigma^{-1})). \tag{3}$$

Observing these definitions, the affine-invariant metric (1) appears as a *balance* of the Euclidean metric (2) and the inverse-Euclidean metric (3). We formalize this idea thanks to parallel transport.

Formalization. The diffeomorphism inv $: (\mathcal{M}, g^I) \longrightarrow (\mathcal{M}, g^E)$ is an isometry. Since these two metrics are flat, the parallel transports do not depend on the curve. On the one hand, the Euclidean parallel transport from Σ to I_n is the identity map $\Pi^E : X \in T_\Sigma \mathcal{M} \longmapsto X \in T_{I_n} \mathcal{M}$ since all tangent spaces are identified to the vector space of symmetric matrices Sym_n by the differential of the embedding $id : \mathcal{M} \hookrightarrow \mathrm{Sym}_n$. On the other hand, the isometry inv gives the inverse-Euclidean parallel transport from Σ to I_n, $\Pi^I : X \in T_\Sigma \mathcal{M} \longmapsto \Sigma^{-1} X \Sigma^{-1} \in T_{I_n} \mathcal{M}$. We generalize this situation in Definition 1. Given Lemma 1, it automatically leads to Theorem 1.

Definition 1 (Balanced bilinear form). *Let g, g^* be two flat metrics on SPD_n and Π, Π^* the associated parallel transports that do not depend on the curve. We define the balanced bilinear form $g^0_\Sigma(X, Y) = \mathrm{tr}((\Pi_{\Sigma \to I_n} X)(\Pi^*_{\Sigma \to I_n} Y))$.*

Theorem 1 (A balanced metric defines a dually flat manifold). *Let g, g^* be two flat metrics and let ∇, ∇^* be their Levi-Civita connections. If the balanced bilinear form g^0 of g, g^* is a metric, then $(\mathcal{M}, g^0, \nabla, \nabla^*)$ is a dually flat manifold.*

If two connections ∇ and ∇^* are dual connections with respect to a metric g^0, there is no reason for them to be Levi-Civita connections of some metrics. Therefore, the main advantage of the principle of balanced metrics on the concept of dual connections seems to be the metric nature of the dual connections.

Corollary 1 (Euclidean and inverse-Euclidean are dual with respect to affine-invariant). *We denote ∇^E and ∇^I the Levi-Civita connections of the Euclidean metric g^E and the inverse-Euclidean metric g^I. Then g^A is the balanced metric of g^I, g^E and $(SPD_n, g^A, \nabla^I, \nabla^E)$ is a dually flat manifold.*

2.3 Relation with Fisher Information Geometry

We know from [13] that the affine-invariant metric is the Fisher metric of the centered multivariate normal model. Information geometry provides a natural one-parameter family of dual connections, called α-connections [12]. In the following table, we recall the main quantities characterizing the *centered multivariate normal model* $\mathcal{P} = \{p_\Sigma : \mathbb{R}^n \longrightarrow \mathbb{R}^*_+, \Sigma \in \mathcal{M}\}$, where $\mathcal{M} = SPD_n$.

Densities	$p_\Sigma(x) = \frac{1}{\sqrt{2\pi}^n} \frac{1}{\sqrt{\det \Sigma}} \exp\left(\frac{1}{2} x^\top \Sigma^{-1} x\right)$
Log likelihood	$l_\Sigma(x) = \log p_\Sigma(x) = \frac{1}{2}\left(-n \log(2\pi) - \log \det \Sigma + x^\top \Sigma^{-1} x\right)$
Differential	$d_\Sigma l(V)(x) = -\frac{1}{2}\left[\mathrm{tr}(\Sigma^{-1}V) + x^\top \Sigma^{-1} V \Sigma^{-1} x\right]$
Fisher metric [13]	$g_\Sigma(V, W) = \frac{1}{2}\mathrm{tr}(\Sigma^{-1} V \Sigma^{-1} W)$

We recall that the α-*connections* $\nabla^{(\alpha)}$ [12] of a family of densities $\mathcal{P}$ are defined by their Christoffel symbols $\Gamma_{ijk} = g_{lk}\Gamma_{ij}^l$ in the local basis $(\partial_i)_{1 \leqslant i \leqslant N}$ at $\Sigma \in \mathcal{M}$:

$$\left(\Gamma_{ijk}^{(\alpha)}\right)_{\Sigma} = \mathbb{E}_{\Sigma}\left[\left(\partial_i\partial_j l + \frac{1-\alpha}{2}\partial_i l \partial_j l\right)\partial_k l\right]. \tag{4}$$

We give in Theorem 2 the expression of the α-connections of the centered multivariate normal model and we notice that the Euclidean and inverse-Euclidean Levi-Civita connections belong to this family.

Theorem 2 (α-*connections of the centered multivariate normal model*). *In the global basis of* $\mathcal{M} = SPD_n$ *given by the inclusion* $\mathcal{M} \hookrightarrow \mathrm{Sym}(n) \simeq \mathbb{R}^N$, *writing* $\partial_X Y = X^i(\partial_i Y^j)\partial_j$, *the α-connections of the multivariate centered normal model are given by the following formula:*

$$\nabla_X^{(\alpha)}Y = \partial_X Y - \frac{1+\alpha}{2}(X\Sigma^{-1}Y + Y\Sigma^{-1}X). \tag{5}$$

The mixture m-connection (α = −1) is the Levi-Civita connection of the Euclidean metric $g_\Sigma^E(X,Y) = \mathrm{tr}(XY)$. *The exponential e-connection (α = 1) is the Levi-Civita connection of the inverse-Euclidean metric, i.e. the pullback of the Euclidean metric by matrix inversion,* $g_\Sigma^I(X,Y) = \mathrm{tr}(\Sigma^{-2}X\Sigma^{-2}Y)$.

The formula (5) can be proved thanks to Lemma 2 which gives the results of expressions of type $\int_{\mathbb{R}^n} x^\top\Sigma^{-1}X\Sigma^{-1}Y\Sigma^{-1}Z\Sigma^{-1}x \exp\left(-\frac{1}{2}x^\top\Sigma^{-1}x\right)dx$, with $y = \Sigma^{-1/2}x$, $A = \Sigma^{-1/2}X\Sigma^{-1/2}$, $B = \Sigma^{-1/2}Y\Sigma^{-1/2}$ and $C = \Sigma^{-1/2}Z\Sigma^{-1/2}$. If one wants to avoid using the third formula of Lemma 2, one can rely on the formula (5) in the case $\alpha = 0$ which is already known from [13].

Lemma 2. *For* $A, B, C \in \mathrm{Sym}_n$:

$$\mathbb{E}_{I_n}(y \longmapsto y^\top Ay) = \mathrm{tr}(A),$$

$$\mathbb{E}_{I_n}(y \longmapsto y^\top Ayy^\top By) = \mathrm{tr}(A)\mathrm{tr}(B) + 2\mathrm{tr}(AB),$$

$$\mathbb{E}_{I_n}(y \longmapsto y^\top Ayy^\top Byy^\top Cy) = \mathrm{tr}(A)\mathrm{tr}(B)\mathrm{tr}(C) + 8\mathrm{tr}(ABC)$$
$$+ 2(\mathrm{tr}(AB)\mathrm{tr}(C) + \mathrm{tr}(BC)\mathrm{tr}(A) + \mathrm{tr}(CA)\mathrm{tr}(B)).$$

Proof (Theorem 2). Given Lemma 2, the computation of the Christoffel symbols $\left(\Gamma_{ijk}^{(\alpha)}\right)_\Sigma$ *leads to* $\left(\Gamma_{ijk}^{(\alpha)}\right)_\Sigma X^i Y^j Z^k = -\frac{1+\alpha}{4}\mathrm{tr}(\Sigma^{-1}[X\Sigma^{-1}Y + Y\Sigma^{-1}X]\Sigma^{-1}Z)$. *On the other hand, the relation* $\Gamma_{ijk} = g_{lk}\Gamma_{ij}^l$ *between Christoffel symbols gives* $\left(\Gamma_{ijk}^{(\alpha)}\right)_\Sigma X^i Y^j Z^k = \frac{1}{2}\mathrm{tr}\left(\Sigma^{-1}\left[(\Gamma_{ij}^l)_\Sigma^{(\alpha)} X^i Y^j \partial_l\right]\Sigma^{-1}Z\right)$. *So we get:*

$$\nabla_X^{(\alpha)}Y = \partial_X Y + \left[(\Gamma_{ij}^l)_\Sigma^{(\alpha)} X^i Y^j \partial_l\right] = \partial_X Y - \frac{1+\alpha}{2}(X\Sigma^{-1}Y + Y\Sigma^{-1}X). \tag{6}$$

It is clear that the mixture connection (α = −1) is the Euclidean connection. The inverse-Euclidean connection can be computed thanks to the Koszul formula. This calculus drives exactly to the exponential connection (α = 1). $\qquad\square$

In the next section, we apply the principle of balanced metrics to the pairs Euclidean/log-Euclidean (Bogoliubov-Kubo-Mori metric) and power-Euclidean/power-Euclidean (mixed-power-Euclidean metrics).

3   The Family of Mixed-power-Euclidean Metrics

3.1   Bogoliubov-Kubo-Mori Metric

The Bogoliubov-Kubo-Mori metric g^{BKM} is a metric on symmetric positive definite matrices used in quantum physics. It was introduced as $g_\Sigma^{BKM}(X,Y) = \int_0^\infty \mathrm{tr}((\Sigma + tI_n)^{-1}X(\Sigma + tI_n)^{-1}Y)dt$ and can be rewritten [10] thanks to the differential of the symmetric matrix logarithm $\log : \mathcal{M} = SPD_n \longrightarrow \mathrm{Sym}_n$ as:

$$g_\Sigma^{BKM}(X,Y) = \mathrm{tr}(\partial_X \log(\Sigma)\, Y) = \mathrm{tr}(X\, \partial_Y \log(\Sigma)). \tag{7}$$

The log-Euclidean metric g^{LE} [8] is the pullback metric of the Euclidean metric by the symmetric matrix logarithm $\log : (\mathcal{M}, g^{LE}) \longrightarrow (\mathrm{Sym}_n, g^E)$:

$$g_\Sigma^{LE}(X,Y) = \mathrm{tr}(\partial_X \log(\Sigma)\, \partial_Y \log(\Sigma)). \tag{8}$$

Therefore, the BKM metric (7) appears as the balanced metric of the Euclidean metric (2) and the log-Euclidean metric (8). As the Euclidean and log-Euclidean metrics are flat, the parallel transport does not depend on the curve and Theorem 1 ensures that they form a dually flat manifold.

Corollary 2 (Euclidean and log-Euclidean are dual with respect to BKM). *We denote ∇^E and ∇^{LE} the Levi-Civita connections of the Euclidean metric g^E and the log-Euclidean metric g^{LE}. Then g^{BKM} is the balanced metric of g^{LE}, g^E and $(SPD_n, g^{BKM}, \nabla^{LE}, \nabla^E)$ is a dually flat manifold.*

3.2   Mixed-power-Euclidean

Up to now, we observed that existing metrics (affine-invariant and BKM) were the balanced metrics of pairs of flat metrics (Euclidean/inverse-Euclidean and Euclidean/log-Euclidean). Thus, the symmetry and the positivity of the balanced bilinear forms were obvious. From now on, we build new bilinear forms thanks to the principle of balanced metrics. Therefore, it is not as obvious as before that these bilinear forms are metrics.

The family of power-Euclidean metrics $g^{E,\theta}$ [7] indexed by the power $\theta \neq 0$ is defined by pullback of the Euclidean metric by the power function $\mathrm{pow}_\theta = \exp \circ\, \theta \log : (\mathcal{M}, \theta^2 g^{E,\theta}) \longrightarrow (\mathcal{M}, g^E)$:

$$g_\Sigma^{E,\theta}(X,Y) = \frac{1}{\theta^2}\mathrm{tr}(\partial_X \mathrm{pow}_\theta(\Sigma)\, \partial_Y \mathrm{pow}_\theta(\Sigma)). \tag{9}$$

This family comprise the Euclidean metric for $\theta = 1$ and tends to the log-Euclidean metric when the power θ goes to 0. Therefore, we abusively consider that the log-Euclidean metric belongs to the family and we denote it $g^{E,0} := g^{LE}$.

We define the mixed-power-Euclidean metrics g^{E,θ_1,θ_2} as the balanced bilinear form of the power-Euclidean metrics g^{E,θ_1} and g^{E,θ_2}, where $\theta_1, \theta_2 \in \mathbb{R}$:

$$g_\Sigma^{E,\theta_1,\theta_2}(X,Y) = \frac{1}{\theta_1\theta_2}\mathrm{tr}(\partial_X \mathrm{pow}_{\theta_1}(\Sigma)\, \partial_Y \mathrm{pow}_{\theta_2}(\Sigma)). \tag{10}$$

Note that the family of mixed-power-Euclidean metrics contains the BKM metric for $(\theta_1, \theta_2) = (1, 0)$ and the θ-power-Euclidean metric for $(\theta_1, \theta_2) = (\theta, \theta)$, including the Euclidean metric for $\theta = 1$ and the log-Euclidean metric for $\theta = 0$.

At this stage, we do not know that the bilinear form g^{E,θ_1,θ_2} is a metric. This is stated by Theorem 3. As the power-Euclidean metrics are flat, Theorem 1 combined with Theorem 3 ensure that the Levi-Civita connections ∇^{E,θ_1} and ∇^{E,θ_2} of the metrics g^{E,θ_1} and g^{E,θ_2} are dual with respect to the (θ_1, θ_2)-mixed-power-Euclidean metric. This is stated by Corollary 3.

Theorem 3. *The bilinear form g^{E,θ_1,θ_2} is symmetric and positive definite so it is a metric on SPD_n. Moreover, the symmetry ensures that $g^{E,\theta_1,\theta_2} = g^{E,\theta_2,\theta_1}$.*

Corollary 3 (θ_1 and θ_2-power-Euclidean are dual with respect to (θ_1, θ_2)-mixed-power-Euclidean). *For $\theta_1, \theta_2 \in \mathbb{R}$, we denote ∇^{E,θ_1} and ∇^{E,θ_2} the Levi-Civita connections of the power-Euclidean metrics g^{E,θ_1} and g^{E,θ_2}. Then g^{E,θ_1,θ_2} is the balanced metric of $g^{E,\theta_1}, g^{E,\theta_2}$ and $(SPD_n, g^{E,\theta_1,\theta_2}, \nabla^{E,\theta_1}, \nabla^{E,\theta_2})$ is a dually flat manifold.*

To prove Theorem 3, we show that for all spectral decomposition $\Sigma = PDP^\top$ of an SPD matrix, there exists a matrix A with positive coefficients $A(i,j) > 0$ such that $g_\Sigma^{E,\theta_1,\theta_2}(X,Y) = \operatorname{tr}((A \bullet P^\top XP)(A \bullet P^\top YP))$, where $\bullet$ is the Hadamard product, i.e. $(A \bullet B)(i,j) = A(i,j)B(i,j)$, which is associative, commutative, distributive w.r.t. matrix addition and satisfies $\operatorname{tr}((A \bullet B)C) = \operatorname{tr}(B(A \bullet C))$ for symmetric matrices $A, B, C \in \operatorname{Sym}_n$. The existence of A relies on Lemma 3.

Lemma 3. *Let $\Sigma = PDP^\top$ be a spectral decomposition of $\Sigma \in \mathcal{M}$, with $P \in O(n)$ and D diagonal. For $f \in \{\exp, \log, \operatorname{pow}_\theta\}$, $\partial_V f(\Sigma) = P(\delta(f,D) \bullet P^\top VP)P^\top$ where $\delta(f,D)(i,j) = \frac{f(d_i) - f(d_j)}{d_i - d_j}$. Note that $\frac{1}{\theta}\delta(\operatorname{pow}_\theta, D)(i,j) > 0$ for all $\theta \in \mathbb{R}^*$ and $\delta(\log, D)(i,j) > 0$.*

Proof (Lemma 3). Once shown for $f = \exp$, it is easy to get for $f = \log$ by inversion and for $f = \operatorname{pow}_\theta = \exp \circ \theta \log$ by composition. But the case $f = \exp$ itself reduces to the case $f = \operatorname{pow}_k$ with $k \in \mathbb{N}$ by linearity, so we focus on this last case. As $\partial_V \operatorname{pow}_k(\Sigma) = \sum_{l=0}^{k-1} \Sigma^l V \Sigma^{k-1-l} = P\partial_{P^\top VP}\operatorname{pow}_k(D)P^\top$ and $\partial_{P^\top VP}\operatorname{pow}_k(D)(i,j) = \sum_{l=0}^{k-1} D^l P^\top VPD^{k-1-l}(i,j) = \frac{d_i^k - d_j^k}{d_i - d_j}P^\top VP(i,j)$, we get $\partial_V \operatorname{pow}_k(\Sigma) = P(\delta(\operatorname{pow}_k, D) \bullet P^\top VP)P^\top$.

Proof (Theorem 3). Let $\theta_1, \theta_2 \in \mathbb{R}^*$. For a spectral decomposition $\Sigma = PDP^\top$, the matrix A defined by $A(i,j) = \sqrt{\frac{1}{\theta_1}\delta(\operatorname{pow}_{\theta_1}, D)(i,j)\frac{1}{\theta_2}\delta(\operatorname{pow}_{\theta_2}, D)(i,j)} > 0$ satisfies $g_\Sigma^{E,\theta_1,\theta_2}(X,Y) = \operatorname{tr}((A \bullet P^\top XP)(A \bullet P^\top YP))$. Symmetry and non-negativity are clear since they come from the Frobenius scalar product. Finally, if $g_\Sigma^{E,\theta_1,\theta_2}(X,X) = 0$, then $A \bullet P^\top XP = 0$ so $X = 0$. So g^{E,θ_1,θ_2} is a metric. If $\theta_1 = 0$, the matrix A defined by $A(i,j) = \sqrt{\delta(\log, D)(i,j)\frac{1}{\theta_2}\delta(\operatorname{pow}_{\theta_2}, D)(i,j)} > 0$ satisfies the same property and $g^{E,0,\theta_2}$ is a metric. $\square$

4   The Family of Mixed-power-affine Metrics

In previous sections, we defined our balanced metric from a pair of two flat metrics and we showed that it corresponded to the duality of (Levi-Civita) connections in information geometry. In this section, we investigate the balanced metric of two non-flat metrics and we observe that the corresponding Levi-Civita connections cannot be dual with respect to this balanced metric.

The family of power-affine metrics $g^{A,\theta}$ [11] indexed by the power $\theta \neq 0$ are defined by pullback of the affine-invariant metric by the power function $\mathrm{pow}_\theta : (\mathcal{M}, \theta^2 g^{A,\theta}) \longrightarrow (\mathcal{M}, g^A)$:

$$g_\Sigma^{A,\theta}(X,Y) = \frac{1}{\theta^2} \mathrm{tr}(\Sigma^{-\theta}\, \partial_X \mathrm{pow}_\theta(\Sigma)\, \Sigma^{-\theta}\, \partial_Y \mathrm{pow}_\theta(\Sigma)) \tag{11}$$

This family comprise the affine-invariant metric for $\theta = 1$ and tends to the log-Euclidean metric when the power θ goes to 0. We consider that the log-Euclidean metric belongs to the family and we denote $g^{A,0} := g^{LE}$.

As these metrics have no cut locus because they endow the manifold with a negatively curved Riemannian symmetric structure, there exists a unique geodesic between two given points. Therefore, a canonical parallel transport can be defined along geodesics. This allows to define the balanced bilinear form of two metrics without cut locus.

Definition 2 (Balanced bilinear form). *Let g, g^* be two metrics without cut locus on SPD_n and Π, Π^* the associated geodesic parallel transports. We define the balanced bilinear form $g_\Sigma^0(X,Y) = \mathrm{tr}((\Pi_{\Sigma \to I_n} X)(\Pi_{\Sigma \to I_n}^* Y))$.*

Given that the geodesic parallel transport on the manifold $(\mathcal{M}, g^{A,\theta})$ is $\Pi_{\Sigma \to I_n} : X \in T_\Sigma \mathcal{M} \longmapsto \frac{1}{\theta}\Sigma^{-\theta/2}\partial_X \mathrm{pow}_\theta(\Sigma)\Sigma^{-\theta/2} \in T_{I_n}\mathcal{M}$, we define the mixed-power-affine metrics g^{A,θ_1,θ_2} as the balanced metric of the power-affine metrics g^{A,θ_1} and g^{A,θ_2}, where $\theta_1, \theta_2 \in \mathbb{R}$ and $\theta = (\theta_1 + \theta_2)/2$:

$$g_\Sigma^{A,\theta_1,\theta_2}(X,Y) = \frac{1}{\theta_1\theta_2} \mathrm{tr}(\Sigma^{-\theta}\, \partial_X \mathrm{pow}_{\theta_1}(\Sigma)\, \Sigma^{-\theta}\, \partial_Y \mathrm{pow}_{\theta_2}(\Sigma)). \tag{12}$$

Note that the family of mixed-power-affine metrics contains the θ-power-affine metric for $(\theta_1, \theta_2) = (\theta, \theta)$, including the affine-invariant metric for $\theta = 1$ and the log-Euclidean metric for $\theta = 0$. This family has two symmetries since $g^{A,\theta_1,\theta_2} = g^{A,\pm\theta_1,\pm\theta_2}$, they come from the inverse-consistency of the affine-invariant metric. This family has a non-empty intersection with the family of mixed-power-Euclidean metrics since $g^{A,\theta_1,-\theta_1} = g^{E,\theta_1,-\theta_1} = g^{A,\theta_1}$ for all $\theta_1 \in \mathbb{R}$.

The fact that g^{A,θ_1,θ_2} is a metric can be shown exactly the same way as in the mixed-power-Euclidean case thanks to the equality $\Sigma^{-\theta/2}\partial_V \mathrm{pow}_\theta(\Sigma)\Sigma^{-\theta/2} = P(\varepsilon(\mathrm{pow}_\theta, D) \bullet P^\top V P)P^\top$ where $\varepsilon(\mathrm{pow}_\theta, D) = (d_i d_j)^{-\theta/2}\delta(\mathrm{pow}_\theta, D)$ and where $\delta(\mathrm{pow}_\theta, D)$ has been defined in Lemma 3. This is stated in Theorem 4.

Theorem 4. *The bilinear form g^{A,θ_1,θ_2} is symmetric and positive definite. Hence it is a metric on SPD_n and $g^{A,\theta_1,\theta_2} = g^{A,\theta_2,\theta_1}$.*

Power-affine metrics being non-flat, $(\mathcal{M}, g^{A,\theta_1,\theta_2}, \nabla^{A,\theta_1}, \nabla^{A,\theta_2})$, where ∇^{A,θ_1} and ∇^{A,θ_2} are Levi-Civita connections of g^{A,θ_1} and g^{A,θ_2}, cannot be a dually-flat manifold. Actually, the two connections are even not dual. It can be understood by comparison with previous sections since the duality was a consequence of the independence of the parallel transport with respect to the chosen curve, which was a consequence of the flatness of the two connections. Moreover, in the Definition 2, the vectors are parallel transported along two different curves (the geodesics relative to each connection) so it may exists a better definition for the balanced bilinear form of two metrics without cut locus or even of two general metrics.

5   Conclusion

The principle of balanced bilinear form is a procedure on SPD matrices that takes a pair of flat metrics or metrics without cut locus and builds a new metric based on the parallel transport of the initial metrics. When the two initial metrics are flat, we showed that the two Levi-Civita connections are dual with respect to the balanced metric. When the two initial metrics are not flat, the two Levi-Civita connections seem not to be dual, so the principle of balanced metrics does not reduce to the concept of dual Levi-Civita connections. A challenging objective for future works is to define properly this principle for other general pairs of metrics and to find conditions under which the balanced bilinear form is a metric.

Acknowledgements. This project has received funding from the European Research Council (ERC) under the European Union's Horizon 2020 research and innovation programme (grant G-Statistics agreement No 786854). This work has been supported by the French government, through the UCAJEDI Investments in the Future project managed by the National Research Agency (ANR) with the reference number ANR-15-IDEX-01.

References

1. Barachant, A., Bonnet, S., Congedo, M., Jutten, C.: Classification of covariance matrices using a Riemannian-based kernel for BCI applications. Elsevier Neurocomput. **112**, 172–178 (2013)
2. Deligianni, F., et al.: A probabilistic framework to infer brain functional connectivity from anatomical connections. In: IPMI Conference, pp. 296–307 (2011)
3. Cheng, G., Vemuri, B.: A novel dynamic system in the space of SPD matrices with applications to appearance tracking. SIAM J. Imag. Sci. **6**(16), 592–615 (2013)
4. Lenglet, C., Rousson, M., Deriche, R., Faugeras, O.: Statistics on the manifold of multivariate normal distributions: theory and application to diffusion tensor MRI processing. J. Math. Imag. Vis. **25**(3), 423–444 (2006)
5. Pennec, X., Fillard, P., Ayache, N.: A Riemannian framework for tensor computing. Int. J. Comput. Vis. **66**(1), 41–66 (2006)
6. Fletcher, T., Joshi, S.: Riemannian geometry for the statistical analysis of diffusion tensor data. Signal Process. **87**, 250–262 (2007)

7. Dryden, I., Pennec, X., Peyrat, J.-M.: Power euclidean metrics for covariance matrices with application to diffusion tensor imaging (2010)
8. Arsigny, V., Fillard, P., Pennec, X., Ayache, N.: Log-Euclidean metrics for fast and simple calculus on diffusion tensors. Magn. Reson. Med. **56**, 411–421 (2006)
9. Lenglet, C., Rousson, M., Deriche, R.: DTI segmentation by statistical surface evolution. IEEE Trans. Med. Imag. **25**, 685–700 (2006)
10. Michor, P., Petz, D., Andai, A.: The curvature of the Bogoliubov-Kubo-Mori scalar product on matrices. Infinite Dimensional Anal., Quant. Prob. Relat. Top. **3**, 1–14 (2000)
11. Thanwerdas, Y., Pennec, X.: Is affine-invariance well defined on SPD matrices?. A principled continuum of metrics. In: Proceedings Geometric Science of Information (2019)
12. Amari, S., Nagaoka, H.: Methods of Information Geometry (2000)
13. Skovgaard, L.: A Riemannian geometry of the multivariate normal model. Scand. J. Stat. **11**, 211–223 (1984)

Affine-Invariant Midrange Statistics

Cyrus Mostajeran$^{(\boxtimes)}$, Christian Grussler, and Rodolphe Sepulchre

Department of Engineering, University of Cambridge, Cambridge CB2 1PZ, UK
csm54@cam.ac.uk

Abstract. We formulate and discuss the affine-invariant matrix midrange problem on the cone of $n \times n$ positive definite Hermitian matrices $\mathbb{P}(n)$, which is based on the Thompson metric. A particular computationally efficient midpoint of this metric is investigated as a highly scalable candidate for an average of two positive definite matrices within this context, before studying the N-point problem in the vector and matrix settings.

Keywords: Positive definite matrices · Statistics · Optimization

1 Introduction

In this paper, we develop a framework for affine-invariant midrange statistics on the cone of positive definite Hermitian matrices $\mathbb{P}(n)$ of dimension n. In Subsection 1.1, we briefly note the basic elements of Finsler geometry relevant to the problem. In Subsection 1.2, we define the affine-invariant midrange problem for a collection of N points. In Sect. 2, we study a particular midrange of two positive definite matrices arising as the midpoint of a suitable geodesic curve. In Sect. 3, we briefly review the scalar and vector affine-invariant midrange problem before returning to the N-point problem on $\mathbb{P}(n)$ in Sect. 4.

1.1 Affine-Invariant Finsler Metrics on $\mathbb{P}(n)$

Consider the family of affine-invariant metric distances d_Φ on $\mathbb{P}(n)$ defined as

$$d_\Phi(A, B) = \| \log A^{-1/2} B A^{-1/2} \|_\Phi, \tag{1}$$

where $\| \cdot \|_\Phi$ is any unitarily invariant norm on the space of Hermitian matrices of dimension n defined by $\|X\|_\Phi := \Phi(\lambda_1(X), \ldots, \lambda_n(X))$, with $\lambda_{\min}(X) = \lambda_n(X) \leq \ldots \leq \lambda_1(X) = \lambda_{\max}(X)$ denoting the n real eigenvalues of X and Φ

This work has received support from the European Research Council under the Advanced ERC Grant Agreement Switchlet n.670645. Cyrus Mostajeran is supported by the Cambridge Philosophical Society.

F. Nielsen and F. Barbaresco (Eds.): GSI 2019, LNCS 11712, pp. 494–501, 2019.
https://doi.org/10.1007/978-3-030-26980-7_51

a symmetric gauge norm on $\mathbb{R}^n$ (i.e. a norm that is invariant under permutations and sign changes of the coordinates) [1]. For any such Φ, X is said to be a d_Φ-midpoint of A and B if

$$d_\Phi(A, X) = d_\Phi(X, B) = \frac{1}{2} d_\Phi(A, B). \tag{2}$$

The curve $\gamma_\mathcal{G} : [0, 1] \to \mathbb{P}(n)$ defined by

$$\gamma_\mathcal{G}(t) = A^{1/2} \left(A^{-1/2} B A^{-1/2} \right)^t A^{1/2} \tag{3}$$

is geometrically significant as a minimal geodesic for any of the affine-invariant metrics d_Φ [1]. The midpoint of $\gamma_\mathcal{G}$ is the matrix geometric mean $A \# B$, which is a metric midpoint in the sense of $d_\Phi(A, A \# B) = d_\Phi(A \# B, B) = \frac{1}{2} d_\Phi(A, B)$ for any Φ. For the choice of $\Phi(x_1, \ldots, x_n) = (\sum_i x_i^2)^{1/2}$, $d_2 := d_\Phi$ corresponds to the metric distance generated by the standard affine-invariant Riemannian structure given by $\langle X, Y \rangle_\Sigma = \operatorname{tr}(\Sigma^{-1} X \Sigma^{-1} Y)$ for $\Sigma \in \mathbb{P}(n)$ and Hermitian matrices $X, Y \in T_\Sigma \mathbb{P}(n)$. For the choice of $\Phi(x_1, \ldots, x_n) = \max_i |x_i|$ on the other hand, $d_\infty := d_\Phi$ yields the distance function that coincides with the Thompson metric [7] on the cone $\mathbb{P}(n)$

$$d_\infty(A, B) = \| \log A^{-1/2} B A^{-1/2} \|_\infty = \max\{\log \lambda_{\max}(BA^{-1}), \log \lambda_{\max}(AB^{-1})\}. \tag{4}$$

While the minimal geodesic $\gamma_\mathcal{G}$ in (3) and the midpoint is unique for the Riemannian distance function d_2, it is not unique with respect to the d_∞ metric which generally admits infinitely many minimal geodesics and midpoints between a given pair of matrices $A, B \in \mathbb{P}(n)$ [6], as expected from the analogous picture concerning the associated norms in $\mathbb{R}^n$. As we shall see in Sect. 2, some of these minimal geodesics are much more readily constructible than others from a computational standpoint and yield midpoints that satisfy many of the properties expected of an affine-invariant matrix mean. Specifically, we will use the midpoint of a particular minimal geodesic of d_∞ from a construction by Nussbaum as a scalable relaxation of the matrix geometric mean that is much cheaper to construct than $A \# B$.

1.2 The Affine-Invariant Midrange Problem

Given a collection of N points Y_i in $\mathbb{P}(n)$, the midrange problem can be formulated as the following optimization problem

$$\min_{X \succeq 0} \max_i d_\infty(X, Y_i). \tag{5}$$

We call a solution $X^\star$ to the above problem a midrange of $\{Y_i\}$.

Proposition 1. *Let $X^\star$ be a solution to (5) with optimal cost $t^\star = f(X^\star) = \max_i d_\infty(X^\star, Y_i)$. We have $l \leq t^\star \leq u$, where the lower and upper bounds are given by*

$$l = \frac{1}{2} \operatorname{diam}_\infty(\{Y_i\}) := \frac{1}{2} \max_{i,j} d_\infty(Y_i, Y_j), \quad u = \min_i \max_j d_\infty(Y_i, Y_j) \leq 2l. \tag{6}$$

Proof. By the triangle inequality, we have for any $i, j = 1, \ldots, N$,

$$d_\infty(Y_i, Y_j) \leq d_\infty(Y_i, X^\star) + d_\infty(X^\star, Y_j) \leq t^\star + t^\star = 2t^\star, \tag{7}$$

since $t^\star = \max_i d_\infty(X^\star, Y_i)$. Taking the maximum of the left-hand side of (7) over i, j, we arrive at $l = \frac{1}{2} \max_{i,j} d_\infty(Y_i, Y_j) \leq t^\star$. For the upper bound, note that taking $X = Y_i$ for each i, we obtain a cost $f(Y_i) = \max_j d_\infty(Y_i, Y_j)$. Since the minimum of these N cost evaluations will still yield an upper bound on the optimum cost $t^\star$, we have $t^\star \leq u = \min_i \max_j d_\infty(Y_i, Y_j)$. $\qquad\square$

2   The 2-point Midrange Problem in $\mathbb{P}(n)$

2.1   Scalable d_∞ geodesic midpoints

As $\gamma_\mathcal{G}$ is a minimal geodesic for the d_∞ metric, $A \# B$ lies in the set of midpoints of this metric. In particular,

$$A \# B \in \operatorname{argmin}_{X \in \mathbb{P}(n)} \left(\max_{Y \in \{A,B\}} d_\infty(X, Y) \right). \tag{8}$$

Recall that the metric distance d_∞ coincides with the Thompson metric d_T on the cone $\mathbb{P}(n)$. If C is a closed, solid, pointed, convex cone in a vector space V, then C induces a partial order on V given by $x \leq y$ if and only if $y - x \in C$. The Thompson metric on C is defined as $d_T(x, y) = \log \max\{M(x/y; C), M(y/x; C)\}$, where $M(y/x; C) := \inf\{\lambda \in \mathbb{R} : y \leq \lambda x\}$ for $x \in C \setminus \{0\}$ and $y \in V$. For $A, B \in \mathbb{P}(n)$, we have $M(A/B) = \lambda_{\max}(AB^{-1})$, so that

$$d_T(A, B) = \log \max\{\lambda_{\max}(AB^{-1}), \lambda_{\max}(BA^{-1})\}, \tag{9}$$

which indeed coincides with d_∞ in (4).

It is known that the Thompson metric does not admit unique minimal geodesics. Indeed, a remarkable construction by Nussbaum describes a family of geodesics that generally consists of an infinite number of curves connecting a pair of points in a cone C. In particular, setting $\alpha := 1/M(x/y; C)$ and $\beta := M(y/x; C)$, the curve $\phi : [0, 1] \to C$ given by

$$\phi(t; x, y) := \begin{cases} \left(\frac{\beta^t - \alpha^t}{\beta - \alpha} \right) y + \left(\frac{\beta\alpha^t - \alpha\beta^t}{\beta - \alpha} \right) x & \text{if } \alpha \neq \beta, \\ \alpha^t x & \text{if } \alpha = \beta, \end{cases} \tag{10}$$

is always a minimal geodesic from x to y with respect to the Thompson metric. The curve ϕ defines a projective straight line in the cone [6]. If we take C to be the cone of positive semidefinite matrices with interior $\operatorname{int} C = \mathbb{P}(n)$, then for a pair of points $A, B \in \mathbb{P}(n)$, we have $\beta = M(B/A; C) = \lambda_{\max}(BA^{-1})$ and $\alpha = 1/M(A/B; C) = \lambda_{\min}(BA^{-1})$. Thus, the minimal geodesic described by (10) takes the form

$$\phi(t) := \begin{cases} \left(\frac{\lambda_{\max}^t - \lambda_{\min}^t}{\lambda_{\max} - \lambda_{\min}} \right) B + \left(\frac{\lambda_{\max}\lambda_{\min}^t - \lambda_{\min}\lambda_{\max}^t}{\lambda_{\max} - \lambda_{\min}} \right) A & \text{if } \lambda_{\min} \neq \lambda_{\max}, \\ \lambda_{\min}^t A & \text{if } \lambda_{\min} = \lambda_{\max}, \end{cases} \tag{11}$$

where $\lambda_{\max}$ and $\lambda_{\min}$ denote the largest and smallest eigenvalues of BA^{-1}, respectively. Taking the midpoint $t = 1/2$ of this geodesic, we arrive at a computationally convenient d_∞-midpoint of A and B, which we will denote by $A * B$.

Proposition 2. *For $A, B \in \mathbb{P}(n)$, we have*

$$A * B = \frac{1}{\sqrt{\lambda_{\min}} + \sqrt{\lambda_{\max}}} \left(B + \sqrt{\lambda_{\min}\lambda_{\max}} A \right). \qquad (12)$$

The result follows from elementary algebraic simplification upon setting $A * B = \phi(1/2; A, B)$ in the case $\lambda_{\min} \neq \lambda_{\max}$. If $\lambda_{\min} = \lambda_{\max}$, then $\phi(1/2; A, B) = \sqrt{\lambda_{\min}} A$ also agrees with the formula in (12). The geometry of the set of d_∞-midpoints of a given pair of points in $\mathbb{P}(n)$ is studied in detail in [5], where it is shown that there is a unique d_∞ minimal geodesic from A to B if and only if the spectrum of BA^{-1} lies in a set $\{\lambda, \lambda^{-1}\}$ for some $\lambda > 0$. Moreover, it is shown that otherwise there are infinitely many d_∞ minimal geodesics from A to B, and that the set of d_∞-midpoints of A and B is compact and convex in both Riemannian and Euclidean senses [5].

We now consider the merits of $A * B$ as a mean of A and B. The following are a number of properties that should be satisfied by a sensible notion of a mean $\mu : \mathbb{P}(n) \times \mathbb{P}(n) \to \mathbb{P}(n)$ of a pair of positive definite matrices [2,4].

(i) Continuity: μ is a continuous map.
(ii) Symmetry: $\mu(A, B) = \mu(B, A)$ for all $A, B \in \mathbb{P}(n)$.
(iii) Affine-invariance: $\mu(XAX^*, XBX^*) = X\mu(A, B)X^*$, for all $X \in GL(n)$.
(iv) Order property: $A \preceq B \implies A \preceq \mu(A, B) \preceq B$.
(v) Monotonicity: $\mu(A, B)$ is monotone in its arguments.

Note that X^* in (iii) denotes the conjugate transpose of X. It is relatively straightforward to show that the map $\mu(A, B) := A * B$ satisfies properties (i)–(iii) listed above. In the remainder of this section, we will turn our attention to the order and monotonicity properties (iv) and (v).

2.2 Order and Monotonicity Properties of $\mu(A, B) = A * B$

Condition (iv) is a generalization of the property of means of positive numbers whereby a mean of a pair of points is expected to lie between the two points on the number line. For Hermitian matrices, a standard partial order $\preceq$ exists according to which $A \preceq B$ if and only if $B - A$ is positive semidefinite. This partial order is known as the Löwner order and the monotonicity of condition (v) is also with reference to this order. It is well-known that the Löwner order is affine-invariant in the sense that for all $A, B \in \mathbb{P}(n)$, $X \in GL(n)$, $A \preceq B$ implies that $XAX^* \preceq XBX^*$. In particular, $A \preceq B$ if and only if $I \preceq A^{-1/2}BA^{-1/2}$. Thus, by affine-invariance of μ, it suffices to prove (iv) in the case where $A = I$ since $A \preceq \mu(A, B) \preceq B$ if and only if $I \preceq \mu(I, A^{-1/2}BA^{-1/2}) \preceq A^{-1/2}BA^{-1/2}$. To establish condition (iv) for $\mu(A, B) = A*B$, we shall make use of the following lemma whose proof is elementary.

Lemma 1. *If $c_1, c_2 \in \mathbb{R}$ and M is an $n \times n$ matrix with eigenvalues $\lambda_i(M)$, then $c_1 M + c_2 I$ has eigenvalues $c_1 \lambda_i(M) + c_2$.*

Let $\Sigma \in \mathbb{P}(n)$ be such that $I \preceq \Sigma$ and note that this is equivalent to $\lambda_i(\Sigma) \geq 1$ for $i = 1, \ldots, n$. Writing $\lambda_{\min} = \lambda_{\min}(\Sigma)$ and $\lambda_{\max} = \lambda_{\max}(\Sigma)$, we have by Lemma 1 that

$$\lambda_i(I * \Sigma) - 1 = \lambda_i\left(\frac{1}{\sqrt{\lambda_{\min}} + \sqrt{\lambda_{\max}}} \left(\Sigma + \sqrt{\lambda_{\min}\lambda_{\max}} I \right) \right) - 1 \qquad (13)$$

$$= \frac{\lambda_i(\Sigma) + \sqrt{\lambda_{\min}\lambda_{\max}} - \sqrt{\lambda_{\min}} - \sqrt{\lambda_{\max}}}{\sqrt{\lambda_{\min}} + \sqrt{\lambda_{\max}}} \qquad (14)$$

$$\geq \frac{\left(\sqrt{\lambda_{\min}} + \sqrt{\lambda_{\max}} \right) \left(\sqrt{\lambda_{\min}} - 1 \right)}{\sqrt{\lambda_{\min}} + \sqrt{\lambda_{\max}}} \geq 0, \qquad (15)$$

since $\lambda_i(\Sigma) \geq \lambda_{\min} \geq 1$. Thus, we have shown that $I \preceq \Sigma$ implies $I \preceq I * \Sigma$. To prove the other inequality, note that

$$\lambda_i(\Sigma - I * \Sigma) = \lambda_i\left(\left(\frac{\sqrt{\lambda_{\min}} + \sqrt{\lambda_{\max}} - 1}{\sqrt{\lambda_{\min}} + \sqrt{\lambda_{\max}}} \right) \Sigma - \frac{\sqrt{\lambda_{\min}\lambda_{\max}}}{\sqrt{\lambda_{\min}} + \sqrt{\lambda_{\max}}} I \right) \qquad (16)$$

$$= \left(\frac{\sqrt{\lambda_{\min}} + \sqrt{\lambda_{\max}} - 1}{\sqrt{\lambda_{\min}} + \sqrt{\lambda_{\max}}} \right) \lambda_i(\Sigma) - \frac{\sqrt{\lambda_{\min}\lambda_{\max}}}{\sqrt{\lambda_{\min}} + \sqrt{\lambda_{\max}}} \qquad (17)$$

$$\geq \frac{(\sqrt{\lambda_{\min}} + \sqrt{\lambda_{\max}} - 1)\lambda_{\min} - \sqrt{\lambda_{\min}\lambda_{\max}}}{\sqrt{\lambda_{\min}} + \sqrt{\lambda_{\max}}} \qquad (18)$$

$$= \sqrt{\lambda_{\min}} \left(\sqrt{\lambda_{\min}} - 1 \right) \geq 0. \qquad (19)$$

Therefore, we have also shown that $\Sigma - I * \Sigma \succeq 0$. That is,

$$I \preceq \Sigma \implies I \preceq I * \Sigma \preceq \Sigma, \qquad (20)$$

for all $\Sigma \in \mathbb{P}(n)$. In particular, upon substituting $\Sigma = A^{-1/2} B A^{-1/2}$ in (20) and using the affine-invariance properties of both the Löwner order and the mean $\mu(A, B) = A * B$, we establish the following important property.

Proposition 3. *For $A, B \in \mathbb{P}(n)$, $A \preceq B$ implies that $A \preceq A * B \preceq B$.*

We now consider the monotonicity of μ in its arguments. First recall that a map $F : \mathbb{P}(n) \to \mathbb{P}(n)$ is said to be monotone if $\Sigma_1 \preceq \Sigma_2$ implies that $F(\Sigma_1) \preceq F(\Sigma_2)$. By symmetry and affine-invariance, it is sufficient to consider monotonicity of $\mu(I, \Sigma)$ with respect to Σ. That is, monotonicity is established by showing that

$$\Sigma_1 \preceq \Sigma_2 \implies I * \Sigma_1 \preceq I * \Sigma_2. \qquad (21)$$

Unfortunately, it turns out that $F(\Sigma) := I * \Sigma$ is not monotone with respect to Σ as we shall demonstrate below. Nonetheless, F is seen to enjoy certain weaker monotonicity properties, which is interesting and insightful. Considering the eigenvalues of $I * \Sigma$, we find that

$$\lambda_i(I * \Sigma) = \frac{\lambda_i(\Sigma) + \sqrt{\lambda_{\min}\lambda_{\max}}}{\sqrt{\lambda_{\min}} + \sqrt{\lambda_{\max}}}, \qquad (22)$$

where $\lambda_{\min}$ and $\lambda_{\max}$ refer to the smallest and largest eigenvalues of Σ.

Proposition 4. *The maximum and minimum eigenvalues of $F(\Sigma) = I * \Sigma$ are monotone with respect to Σ.*

Proof. Considering the cases $i = 1$ and $i = n$, we find that (22) yields

$$\lambda_{\min}(I * \Sigma) = \sqrt{\lambda_{\min}(\Sigma)} \quad \text{and} \quad \lambda_{\max}(I * \Sigma) = \sqrt{\lambda_{\max}(\Sigma)}, \tag{23}$$

both of which are seen to be monotone functions of Σ.   □

It is in the sense of the above that $\mu(A, B) = A * B$ inherits a weak monotonicity property. The monotonic dependence of the extremal eigenvalues of $I * \Sigma$ on Σ ensures that if $\Sigma_1 \preceq \Sigma_2$, then we can at least rule out the possibility that $I * \Sigma_1 \succ I * \Sigma_2$, where $\succ 0$ here denotes positive definiteness. To prove that monotonicity is generally not satisfied in the full sense of condition (v), consider a diagonal matrix $\Sigma = \mathrm{diag}(a, b, x) \in \mathbb{P}(3)$, where $\lambda_{\min}(\Sigma) = a < b \leq x = \lambda_{\max}(\Sigma)$ and x is thought of as a variable. We have $I * \Sigma = \mathrm{diag}\left(\sqrt{a}, f(x), \sqrt{x}\right)$, where

$$\lambda_2(I * \Sigma) = f(x) := \frac{b + \sqrt{ax}}{\sqrt{a} + \sqrt{x}}. \tag{24}$$

Taking the derivative of f with respect to x, we find that

$$f'(x) = \frac{a - b}{2\sqrt{x}(\sqrt{a} + \sqrt{x})^2} < 0, \quad \forall x \geq b, \tag{25}$$

which shows that the second eigenvalue of $I * \Sigma$ decreases as x increases. Thus, we see that $I * \Sigma$ cannot depend monotonically on Σ in this example.

2.3   A Geometric Scaling Property

Before completing this section on the midrange $\mu(A, B) = A * B$ of a pair of positive definite matrices, we note a key scaling property satisfied by μ which suggests that it is a plausible candidate as a scalable substitute for the standard matrix geometric mean $A \# B$.

Proposition 5. *For any real scalars $a, b > 0$ and matrices $A, B \in \mathbb{P}(n)$, we have*

$$(aA) * (bB) = \sqrt{ab}(A * B). \tag{26}$$

Proof. The result follows upon substituting $\lambda_i\left((bB)(aA)^{-1}\right) = \frac{b}{a}\lambda_i(BA^{-1})$ into the formula (12).   □

The scaling in (26) of course does not generally hold for a mean of two matrices. Indeed, it does not generally hold for means arising as d_∞-midpoints either. For instance, [5] identifies

$$A \diamond B = \begin{cases} \frac{\sqrt{\lambda_{\max}}}{1 + \lambda_{\max}}(A + B) & \text{if} \quad \lambda_{\min}\lambda_{\max} \geq 1, \\ \frac{\sqrt{\lambda_{\min}}}{1 + \lambda_{\min}}(A + B) & \text{if} \quad \lambda_{\min}\lambda_{\max} \leq 1, \end{cases} \tag{27}$$

as another d_∞-midpoint of A and B corresponding to the midpoint of a different d_∞ minimal geodesic to the one considered in this paper. It is clear that (27) does not scale geometrically as in (26). As a summary, we collect the key results of this section in the following theorem.

Theorem 1. *The mean $\mu(A, B) = A * B$ defined in (12) yields a d_∞-midpoint of $A, B \in \mathbb{P}(n)$ that is continuous, symmetric, affine-invariant, and scales geometrically as in (26). Moreover, if $A \preceq B$, then $A \preceq \mu(A, B) \preceq B$, and the extremal eigenvalues of $\mu(I, \Sigma)$ depend monotonically on $\Sigma \in \mathbb{P}(n)$.*

3   The Midrange Problem in $\mathbb{R}_+$ and $\mathbb{R}_+^n$

It is instructive to consider the N-point affine-invariant midrange problem in the scalar and vector cases, corresponding to the cones $\mathbb{R}_+$ and $\mathbb{R}_+^n$, respectively. In the scalar case, we are given N positive numbers $y_i > 0$ that can be ordered such that $\min_i y_i \le y_k \le \max_i y_i$ for each $k = 1, \ldots, N$. By the monotonicity of the log function, we have $\log\left(\min_i y_i\right) \le \log y_k \le \log\left(\max_i y_i\right)$. The midrange is uniquely given by

$$x = \exp\left(\frac{1}{2}\left[\log\left(\min_i y_i\right) + \log\left(\max_i y_i\right)\right]\right) = \left(\min_i y_i \cdot \max_i y_i\right)^{1/2}. \tag{28}$$

Note that (28) is the unique solution of the optimization problem

$$\min_{x>0}\ \max_i\ |\log x - \log y_i|. \tag{29}$$

In the vector case, the midrange problem in $\mathbb{R}_+^n$ takes the form

$$\min_{\boldsymbol{x}>0}\ \max_i\ \|\log \boldsymbol{x} - \log \boldsymbol{y}_i\|_\infty := \min_{\boldsymbol{x}>0}\ \max_i\ \max_a\ |\log x^a - \log y_i^a|, \tag{30}$$

where $\boldsymbol{x} > 0$ means that $\boldsymbol{x} = (x^a)$ satsifies $x^a > 0$ for $a = 1, \ldots, n$ and $\boldsymbol{y}_i$ are a collection of N given points in $\mathbb{R}_+^n$. As in the matrix case, the optimum cost $t^\star$ has a lower bound

$$l = \frac{1}{2} \max_{i,j} \|\log \boldsymbol{y}_i - \log \boldsymbol{y}_j\|_\infty. \tag{31}$$

Proposition 6. *The lower bound (31) is attained by $\boldsymbol{x} = (x^a) \in \mathbb{R}_+^n$ given by*

$$x^a = \left(\min_i y_i^a \cdot \max_i y_i^a\right)^{1/2}. \tag{32}$$

4   The N-point Midrange Problem in $\mathbb{P}(n)$

In the matrix setting, the midrange problem (5) takes the form

$$\min_{X \succeq 0}\ \max_i\ \|\log(Y_i^{-1/2} X Y_i^{-1/2})\|_\infty, \tag{33}$$

which can be rewritten as

$$\begin{cases} \min_{X \succeq 0, t} \; t \\ e^{-t} Y_i \preceq X \preceq e^t Y_i \end{cases} \tag{34}$$

While this problem is not convex due to the presence of the log function, the feasibility condition $e^{-t} Y_i \preceq X \preceq e^t Y_i$ is convex for *fixed* t and can be solved using standard convex optimization packages such as CVX [3]. Given a t that is greater than or equal to the optimum value $t^\star = \min_X \max_i d_\infty(X, Y_i)$, we can solve (34) by successively solving the feasibility condition as we decrease t. In the bisection method it is very desirable to have a good estimate for the initial t as the successive reductions in t can be quite slow. In particular, if the lower bound $l = \frac{1}{2} \operatorname{diam}(\{Y_i\})$ is attained as in the vector case, then we can solve (34) in one step by taking $t = l$ and solving the feasibility condition once. Unfortunately, and rather remarkably, numerical examples show that unlike the scalar and vector cases, the lower bound l is not always attained in the affine-invariant matrix midrange problem.

Proposition 7. *The lower bound $l = \frac{1}{2} \operatorname{diam}_\infty(\{Y_i\})$ is not necessarily attained in (33).*

The above result suggests that the N-point matrix midrange problem is more challenging than the vector case in fundamental ways. While the bisection method offers a solution to this problem, we expect that significantly more efficient solutions to the problem can be found. In particular, we expect to find algorithms that rely principally on the computation of dominant generalized eigenpairs that would be considerably more efficient and scalable than existing algorithms for computing matrix geometric means, as in the $N = 2$ case.

References

1. Bhatia, R.: On the exponential metric increasing property. Linear Algebra Appl. **375**, 211–220 (2003)
2. Bhatia, R.: Positive Definite Matrices. Princeton University Press, New Jersey (2007)
3. Grant, M., Boyd, S.: CVX: Matlab software for disciplined convex programming, version 2.1. http://cvxr.com/cvx March 2014
4. Kubo, F., Ando, T.: Means of positive linear operators. Math. Ann. **246**(3), 205–224 (1980)
5. Lim, Y.: Geometry of midpoint sets for Thompson's metric. Linear Algebra Appl. **439**(1), 211–227 (2013)
6. Nussbaum, R.D.: Finsler structures for the part metric and Hilbert's projective metric and applications to ordinary differential equations. Differ. Integr. Equ. **7**(5–6), 1649–1707 (1994)
7. Thompson, A.C.: On certain contraction mappings in a partially ordered vector space. Proc. Am. Math. Soc. **14**(3), 438–443 (1963)

Is Affine-Invariance Well Defined on SPD Matrices? A Principled Continuum of Metrics

Yann Thanwerdas$^{(\boxtimes)}$ and Xavier Pennec

Université Côte d'Azur, Inria, Epione, France
yann.thanwerdas@inria.fr

Abstract. Symmetric Positive Definite (SPD) matrices have been widely used in medical data analysis and a number of different Riemannian metrics were proposed to compute with them. However, there are very few methodological principles guiding the choice of one particular metric for a given application. Invariance under the action of the affine transformations was suggested as a principle. Another concept is based on symmetries. However, the affine-invariant metric and the recently proposed polar-affine metric are both invariant and symmetric. Comparing these two cousin metrics leads us to introduce much wider families: power-affine and deformed-affine metrics. Within this continuum, we investigate other principles to restrict the family size.

Keywords: SPD matrices · Riemannian symmetric space

1 Introduction

Symmetric positive definite (SPD) matrices have been used in many different contexts. In diffusion tensor imaging for instance, a diffusion tensor is a 3-dimensional SPD matrix [1–3]; in brain-computer interfaces (BCI) [4], in functional MRI [5] or in computer vision [6], an SPD matrix can represent a covariance matrix of a feature vector, for example a spatial covariance of electrodes or a temporal covariance of signals in BCI. In order to make statistical operations on SPD matrices like interpolations, computing the mean or performing a principal component analysis, it has been proposed to consider the set of SPD matrices as a manifold and to provide it with some geometric structures like a *Riemannian metric*, a transitive *group action* or some *symmetries*. These structures can be more or less natural depending on the context of the applications, and they can provide closed-form formulas and consistent algorithms [2,7].

Many Riemannian structures have been introduced over the manifold of SPD matrices [7]: Euclidean, log-Euclidean, affine-invariant, Cholesky, square root, power-Euclidean, Procrustes... Each of them has different mathematical properties that can fit the data in some problems but can be inappropriate in some other contexts: for example the curvature can be null, positive, negative, constant, not constant, covariantly constant... These properties on the curvature

© Springer Nature Switzerland AG 2019
F. Nielsen and F. Barbaresco (Eds.): GSI 2019, LNCS 11712, pp. 502–510, 2019.
https://doi.org/10.1007/978-3-030-26980-7_52

have some important consequences on the way we interpolate two points, on the consistence of algorithms, and more generally on every statistical operation one could want to do with SPD matrices. Therefore, a natural question one can ask is: given the practical context of an application, how should one choose the metric on SPD matrices? Are there some relations between the mathematical properties of the geometric structure and the intrinsic properties of the data?

In this context, the affine-invariant metric [2,3,8] was introduced to give an invariant computing framework under affine transformations of the variables. This metric endows the manifold of SPD matrices with a structure of a Riemannian symmetric space. Such spaces have a covariantly constant curvature, thus they share some convenient properties with constant curvature spaces but with less constraints. It was actually shown that there exists not only one but a one-parameter family that is invariant under these affine transformations [9]. More recently, [10–12] introduced another Riemannian symmetric structure that does not belong to the previous one-parameter family: the polar-affine metric.

In this work, we unify these two frameworks by showing that the polar-affine metric is a square deformation of the affine-invariant metric (Sect. 2). We generalize in Sect. 3.1 this construction to a family of power-affine metrics that comprises the two previous metrics, and in Sect. 3.2 to the wider family of deformed-affine metrics. Finally, we propose in Sect. 4 a theoretical approach in the choice of subfamilies of the deformed-affine metrics with relevant properties.

2   Affine-Invariant Versus Polar-Affine

The affine-invariant metric [2,3,8] and the polar-affine metric [12] are different but they both provide a Riemannian symmetric structure to the manifold of SPD matrices. Moreover, both claim to be very naturally introduced. The former uses only the action of the real general linear group GL_n on covariance matrices. The latter uses the canonical left action of GL_n on the left coset space GL_n/O_n and the polar decomposition $GL_n \simeq SPD_n \times O_n$, where O_n is the orthogonal group. Furthermore, the affine-invariant framework is exhaustive in the sense that it provides *all* the metrics invariant under the chosen action [9] whereas the polar-affine framework only provides *one* invariant metric.

In this work, we show that the two frameworks coincide on the same quotient manifold GL_n/O_n but differ because of the choice of the diffeomorphism between this quotient and the manifold of SPD matrices. In particular, we show that there exists a one-parameter family of polar-affine metrics and that any polar-affine metric is a square deformation of an affine-invariant metric.

In 2.1 and 2.2, we build the affine-invariant metrics g^1 and the polar-affine metric g^2 in a unified way, using indexes $i \in \{1, 2\}$ to differentiate them. First, we give explicitly the action $\eta^i : GL_n \times SPD_n \longrightarrow SPD_n$ and the quotient diffeomorphism $\tau^i : GL_n/O_n \longrightarrow SPD_n$; then, we explain the construction of the orthogonal-invariant scalar product $g^i_{I_n}$ that characterizes the metric g^i; finally, we give the expression of the metrics g^1 and g^2. In 2.3, we summarize the results and we focus on the Riemannian symmetric structures of SPD_n.

2.1   The One-Parameter Family of Affine-Invariant Metrics

Affine Action and Quotient Diffeomorphism. In many applications, one would like the analysis of covariance matrices to be invariant under affine transformations $X \longmapsto AX + B$ of the random vector $X \in \mathbb{R}^n$, where $A \in GL_n$ and $B \in \mathbb{R}^n$. Then the covariance matrix $\Sigma = \mathrm{Cov}(X)$, is modified under the transformation $\Sigma \longmapsto A\Sigma A^\top$. This transformation can be thought as a transitive Lie group action η^1 of the general linear group on the manifold of SPD matrices:

$$\eta^1 : \begin{cases} GL_n \times SPD_n \longrightarrow & SPD_n \\ (A, \Sigma) & \longmapsto \eta_A^1(\Sigma) = A\Sigma A^\top \end{cases}. \tag{1}$$

This transitive action induces a diffeomorphism between the manifold SPD_n and the quotient of the acting group GL_n by the stabilizing group $\mathrm{Stab}^1(\Sigma) = \{A \in GL_n, \eta^1(A, \Sigma) = \Sigma\}$ at any point Σ. It reduces to the orthogonal group at $\Sigma = I_n$ so we get the quotient diffeomorphism τ^1:

$$\tau^1 : \begin{cases} GL_n/O_n \longrightarrow & SPD_n \\ [A] = A.O_n \longmapsto \eta^1(A, I_n) = AA^\top \end{cases}. \tag{2}$$

Orthogonal-Invariant Scalar Product. We want to endow the manifold $\mathcal{M} = SPD_n$ with a metric g^1 invariant under the affine action η^1, i.e. an affine-invariant metric. As the action is transitive, the metric at any point Σ is characterized by the metric at one given point I_n. As the metric is affine-invariant, this scalar product g_{I_n} has to be invariant under the stabilizing group of I_n. As a consequence, the metric g^1 is characterized by a scalar product $g_{I_n}^1$ on the tangent space $T_{I_n}\mathcal{M}$ that is invariant under the action of the orthogonal group.

The tangent space $T_{I_n}\mathcal{M}$ is canonically identified with the vector space Sym_n of symmetric matrices by the differential of the canonical embedding $\mathcal{M} \hookrightarrow \mathrm{Sym}_n$. Thus we are now looking for all the scalar products on symmetric matrices that are invariant under the orthogonal group. Such scalar products are given by the following formula [9], where $\alpha > 0$ and $\beta > -\frac{\alpha}{n}$: for all tangent vectors $V_1, W_1 \in T_{I_n}\mathcal{M}$, $g_{I_n}^1(V_1, W_1) = \alpha \, \mathrm{tr}(V_1 W_1) + \beta \, \mathrm{tr}(V_1)\mathrm{tr}(W_1)$.

Affine-Invariant Metrics. To give the expression of the metric, we need a linear isomorphism between the tangent space $T_\Sigma \mathcal{M}$ at any point Σ and the tangent space $T_{I_n}\mathcal{M}$. Since the action $\eta_{\Sigma^{-1/2}}^1$ sends Σ to I_n, its differential given by $T_\Sigma \eta_{\Sigma^{-1/2}}^1 : V \in T_\Sigma \mathcal{M} \longmapsto V_1 = \Sigma^{-1/2} V \Sigma^{-1/2} \in T_{I_n}\mathcal{M}$ is such a linear isomorphism. Combining this transformation with the expression of the metric at I_n and reordering the terms in the trace, we get the general expression of the affine-invariant metric: for all tangent vectors $V, W \in T_\Sigma \mathcal{M}$,

$$g_\Sigma^1(V, W) = \alpha \, \mathrm{tr}(\Sigma^{-1}V\Sigma^{-1}W) + \beta \, \mathrm{tr}(\Sigma^{-1}V)\mathrm{tr}(\Sigma^{-1}W). \tag{3}$$

As the geometry of the manifold is not much affected by a scalar multiplication of the metric, we often drop the parameter α, as if it were equal to 1, and we consider that this is a one-parameter family indexed by $\beta > -\frac{1}{n}$.

2.2   The Polar-Affine Metric

Quotient Diffeomorphism and Affine Action. In [12], instead of defining a metric directly on the manifold of SPD matrices, a metric is defined on the left coset space $GL_n/O_n = \{[A] = A.O_n, A \in GL_n\}$, on which the general linear group GL_n naturally acts by the left action $\eta^0 : (A, [A']) \longmapsto [AA']$. Then this metric is pushed forward on the manifold SPD_n into the polar-affine metric g^2 thanks to the polar decomposition $\mathrm{pol} : A \in GL_n \longmapsto (\sqrt{AA^\top}, \sqrt{AA^\top}^{-1} A) \in SPD_n \times O_n$ or more precisely by the quotient diffeomorphism τ^2:

$$\tau^2 : \begin{cases} GL_n/O_n \longrightarrow SPD_n \\ A.O_n \longmapsto \sqrt{AA^\top} \end{cases}. \tag{4}$$

This quotient diffeomorphism induces an action of the general linear group GL_n on the manifold SPD_n, under which the polar-affine metric will be invariant:

$$\eta^2 : \begin{cases} GL_n \times SPD_n \longrightarrow SPD_n \\ (A, \Sigma) \longmapsto \eta^2_A(\Sigma) = \sqrt{A\Sigma^2 A^\top} \end{cases}. \tag{5}$$

It is characterized by $\eta^2(A, \tau^2(A'.O_n)) = \tau^2(\eta^0(A, A'.O_n))$ for $A, A' \in GL_n$.

Orthogonal-Invariant Scalar Product. The polar-affine metric g^2 is characterized by the scalar product $g^2_{I_n}$ on the tangent space $T_{I_n}\mathcal{M}$. This scalar product is obtained by pushforward of a scalar product $g^0_{[I_n]}$ on the tangent space $T_{[I_n]}(GL_n/O_n)$. It is itself induced by the Frobenius scalar product on $\mathfrak{gl}_n = T_{I_n}GL_n$, defined by $\langle v|w \rangle_{\mathrm{Frob}} = \mathrm{tr}(vw^\top)$, which is orthogonal-invariant. This is summarized on the following diagram.

$$
\begin{array}{ccccc}
GL_n & \xrightarrow{\ s\ } & GL_n/O_n & \xrightarrow{\ \tau^2\ } & \mathcal{M} = SPD_n \\
A & \longmapsto & A.O_n & \longmapsto & \sqrt{AA^\top} \\
\langle \cdot | \cdot \rangle_{\mathrm{Frob}} & & g^0_{[I_n]} & & g^2_{I_n}
\end{array}
$$

Finally, we get the scalar product $g^2_{I_n}(V_2, W_2) = \mathrm{tr}(V_2 W_2)$ for $V_2, W_2 \in T_{I_n}\mathcal{M}$.

Polar-Affine Metric. Since the action $\eta^2_{\Sigma^{-1}}$ sends Σ to I_n, a linear isomorphism between tangent spaces is given by the differential of the action $T_\Sigma \eta^2_{\Sigma^{-1}} : V \in T_\Sigma M \longrightarrow V_2 = \Sigma^{-1} T_\Sigma \mathrm{pow}_2(V) \Sigma^{-1} \in T_{I_n}\mathcal{M}$. Combined with the above expression of the scalar product at I_n, we get the following expression for the polar affine metric: for all tangent vectors $V, W \in T_\Sigma \mathcal{M}$,

$$g^2_\Sigma(V, W) = \mathrm{tr}(\Sigma^{-2} T_\Sigma \mathrm{pow}_2(V)\, \Sigma^{-2} T_\Sigma \mathrm{pow}_2(W)). \tag{6}$$

2.3   The Underlying Riemannian Symmetric Manifold

In the affine-invariant framework, we started from defining the affine action η^1 (on covariance matrices) and we inferred the quotient diffeomorphism

$\tau^1 : (GL_n/O_n, \eta^0) \longrightarrow (SPD_n, \eta^1)$. In the polar-affine framework, we started from defining the quotient diffeomorphism $\tau^2 : GL_n/O_n \longrightarrow SPD_n$ (corresponding to the polar decomposition) and we inferred the affine action η^2. The two actually correspond to the same underlying affine action η^0 on the quotient GL_n/O_n. Then there is also a one-parameter family of affine-invariant metrics on the quotient GL_n/O_n and a one-parameter family of polar-affine metrics on the manifold SPD_n. This is stated in the following theorems.

Theorem 1 (Polar-affine is a square deformation of affine-invariant).

1. *There exists a one-parameter family of affine-invariant metrics on the quotient GL_n/O_n.*
2. *This family is in bijection with the one-parameter family of affine-invariant metrics on the manifold of SPD matrices thanks to the diffeomorphism τ^1 : $A.O_n \longmapsto AA^\top$. The corresponding action is $\eta^1 : (A, \Sigma) \longmapsto A\Sigma A^\top$.*
3. *This family is also in bijection with a one-parameter family of polar-affine metrics on the manifold of SPD matrices thanks to the diffeomorphism τ^2 : $A.O_n \longmapsto \sqrt{AA^\top}$. The corresponding action is $\eta^2 : (A, \Sigma) \longmapsto \sqrt{A\Sigma^2 A^\top}$.*
4. *The diffeomorphism* $\mathrm{pow}_2 : \begin{cases} (SPD_n, 4g^2) \longrightarrow (SPD_n, g^1) \\ \quad\quad \Sigma \quad\quad \longmapsto \quad\quad \Sigma^2 \end{cases}$ *is an isometry between polar-affine metrics g^2 and affine-invariant metrics g^1.*

In other words, performing statistical analyses (e.g. a principal component analysis) with the polar-affine metric on covariance matrices is equivalent to performing these statistical analyses with the classical affine-invariant metric on the *square* of our covariance matrix dataset.

All the metrics mentioned in Theorem 1 endow their respective space with a structure of a Riemannian symmetric manifold. We recall the definition of that geometric structure and we give the formal statement.

Definition 1 (Symmetric manifold, Riemannian symmetric manifold).

A manifold $\mathcal{M}$ is symmetric if it is endowed with a family of involutions $(s_x)_{x\in\mathcal{M}}$ called symmetries such that $s_x \circ s_y \circ s_x = s_{s_x(y)}$ and x is an isolated fixed point of s_x. It implies that $T_x s_x = -\mathrm{Id}_{T_x\mathcal{M}}$. A Riemannian manifold $(\mathcal{M}, g)$ is symmetric if it is endowed with a family of symmetries that are isometries of $\mathcal{M}$, i.e. that preserve the metric: $g_{s_x(y)}(T_y s_x(v), T_y s_x(w)) = g_y(v, w)$ for $v, w \in T_y\mathcal{M}$.

Theorem 2 (Riemannian symmetric structure on SPD_n).

The Riemannian manifold (SPD_n, g^1), where g^1 is an affine-invariant metric, is a Riemannian symmetric space with symmetry $s_\Sigma : \Lambda \longmapsto \Sigma\Lambda^{-1}\Sigma$. The Riemannian manifold (SPD_n, g^2), where g^2 is a polar-affine metric, is also a Riemannian symmetric space whose symmetry is $s_\Sigma : \Lambda \longmapsto \sqrt{\Sigma^2\Lambda^{-2}\Sigma^2}$.

This square deformation of affine-invariant metrics can be generalized into a power deformation to build a family of affine-invariant metrics that we call power-affine metrics. It can even be generalized into any diffeomorphic deformation of SPD matrices. We now develop these families of affine-invariant metrics.

3   Families of Affine-Invariant Metrics

There is a theoretical interest in building families comprising some of the known metrics on SPD matrices to understand how one can be deformed into another. For example, power-Euclidean metrics [13] comprise the Euclidean metric and tends to the log-Euclidean metric [14] when the power tends to 0. We recall that the log-Euclidean metric is the pullback of the Euclidean metric on symmetric matrices by the symmetric matrix logarithm $\log : SPD_n \longrightarrow \mathrm{Sym}_n$. There is also a practical interest in defining families of metrics: for example, it is possible to optimize the power to better fit the data with a certain distribution [13].

First, we generalize the square deformation by deforming the affine-invariant metrics with a power function $\mathrm{pow}_\theta : \Sigma \in SPD_n \longmapsto \Sigma^\theta = \exp(\theta \log \Sigma)$ to define the power-affine metrics. Then we deform the affine-invariant metrics by any diffeomorphism $f : SPD_n \longrightarrow SPD_n$ to define the deformed-affine metrics.

3.1   The Two-Parameter Family of Power-Affine Metrics

We recall that $\mathcal{M} = SPD_n$ is the manifold of SPD matrices. For a power $\theta \neq 0$, we define the θ-power-affine metric g^θ as the pullback by the diffeomorphism $\mathrm{pow}_\theta : \Sigma \longmapsto \Sigma^\theta$ of the affine-invariant metric, scaled by a factor $1/\theta^2$.

Equivalently, the θ-power-affine metric is the metric invariant under the θ-affine action $\eta^\theta : (A, \Sigma) \longmapsto (A\Sigma^\theta A^\top)^{1/\theta}$ whose scalar product at I_n coincides with the scalar product $g^1_{I_n} : (V, W) \longmapsto \alpha \operatorname{tr}(VW) + \beta \operatorname{tr}(V)\operatorname{tr}(W)$. The θ-affine action induces an isomorphism $V \in T_\Sigma M \longmapsto V_\theta = \frac{1}{\theta} \Sigma^{-\theta/2} \partial_V \mathrm{pow}_\theta(\Sigma) \Sigma^{-\theta/2} \in T_{I_n} \mathcal{M}$ between tangent spaces. The θ-power-affine metric is given by:

$$g^\theta_\Sigma(V, W) = \alpha \operatorname{tr}(V_\theta W_\theta) + \beta \operatorname{tr}(V_\theta)\operatorname{tr}(W_\theta). \tag{7}$$

Because a scaling factor is of low importance, we can set $\alpha = 1$ and consider that this family is a two-parameter family indexed by $\beta > -1/n$ and $\theta \neq 0$.

We have chosen to define the metric g^θ so that the power function $\mathrm{pow}_\theta : (\mathcal{M}, \theta^2 g^\theta) \longrightarrow (\mathcal{M}, g^1)$ is an isometry. Why this factor θ^2? The first reason is for consistence with previous works: the analogous power-Euclidean metrics have been defined with that scaling [13]. The second reason is for continuity: when the power tends to 0, the power-affine metric tends to the log-Euclidean metric.

Theorem 3 (Power-affine tends to log-Euclidean for $\theta \to 0$). *Let $\Sigma \in \mathcal{M}$ and $V, W \in T_\Sigma \mathcal{M}$. Then $\lim_{\theta \to 0} g^\theta_\Sigma(V, W) = g^{LE}_\Sigma(V, W)$ where the log-Euclidean metric is $g^{LE}_\Sigma(V, W) = \alpha \operatorname{tr}(\partial_V \log(\Sigma) \, \partial_W \log(\Sigma)) + \beta \operatorname{tr}(\partial_V \log(\Sigma))\operatorname{tr}(\partial_W \log(\Sigma))$.*

3.2   The Continuum of Deformed-Affine Metrics

In the following, we call a diffeomorphism $f : SPD_n \longrightarrow SPD_n$ a deformation. We define the f-deformed-affine metric g^f as the pullback by the diffeomorphism f of the affine-invariant metric, so that $f : (\mathcal{M}, g^f) \longrightarrow (\mathcal{M}, g^1)$ is an isometry. (Regarding the discussion before the Theorem 3, $g^{\mathrm{pow}_\theta} = \theta^2 g^\theta$.)

The f-deformed-affine metric is invariant under the f-affine action η^f : $(A, \Sigma) \longmapsto f^{-1}(Af(\Sigma)A^\top)$. It is given by $g_\Sigma^f(V, W) = \alpha\mathrm{tr}(V_f W_f) + \beta\mathrm{tr}(V_f)\mathrm{tr}(W_f)$ where $V_f = f(\Sigma)^{-1/2}\,\partial_V f(\Sigma)\,f(\Sigma)^{-1/2}$. The basic Riemannian operations are obtained by pulling back the affine-invariant operations.

Theorem 4 (Basic Riemannian operations). *For SPD matrices $\Sigma, \Lambda \in \mathcal{M}$ and a tangent vector $V \in T_\Sigma\mathcal{M}$, we have at all time $t \in \mathbb{R}$:*

Geodesics	$\gamma_{(\Sigma,V)}^f(t) = f^{-1}(f(\Sigma)^{1/2}\exp(tf(\Sigma)^{-1/2}T_\Sigma f(V)f(\Sigma)^{-1/2})f(\Sigma)^{1/2})$
Logarithm	$\mathrm{Log}_\Sigma^f(\Lambda) = (T_\Sigma f)^{-1}(f(\Sigma)^{1/2}\log(f(\Sigma)^{-1/2}f(\Lambda)f(\Sigma)^{-1/2})f(\Sigma)^{1/2})$
Distance	$d_f(\Sigma, \Lambda) = d_1(f(\Sigma), f(\Lambda)) = \sum_{k=1}^n (\log\lambda_k)^2$

where $\lambda_1, ..., \lambda_n$ are the eigenvalues of the symmetric matrix $f(\Sigma)^{-1/2} f(\Lambda)f(\Sigma)^{-1/2}$.

All tensors are modified thanks to the pushforward f_* and pullback f^* operators, e.g. the Riemann tensor of the f-deformed metric is $R^f(X, Y)Z = f^*(R(f_*X, f_*Y)(f_*Z))$. As a consequence, the deformation f does not affect the values taken by the sectional curvature and these metrics are negatively curved.

From a computational point of view, it is very interesting to notice that the identification $\mathcal{L}_\Sigma' : V \in T_\Sigma\mathcal{M} \longmapsto V' = T_\Sigma f(V) \in T_{f(\Sigma)}\mathcal{M}$ simplifies the above expressions by removing the differential $T_\Sigma f$. This change of basis can prevent from numerical approximations of the differential but one must keep in mind that $V \neq V'$ in general. This identification was already used for the polar-affine metric ($f = \mathrm{pow}_2$) in [12] without explicitly mentioning.

4   Interesting Subfamilies of Deformed-Affine Metrics

Some deformations have already been used in applications. For example, the family $A_r : \mathrm{diag}(\lambda_1, \lambda_2, \lambda_3) \longmapsto \mathrm{diag}(a_1(r)\lambda_1, a_2(r)\lambda_2, a_3(r)\lambda_3)$ where $\lambda_1 \geqslant \lambda_2 \geqslant \lambda_3 > 0$ was proposed to map the anisotropy of water measured by diffusion tensors to the one of the diffusion of tumor cells in tumor growth modeling [15]. The inverse function $\mathrm{inv} = \mathrm{pow}_{-1} : \Sigma \longmapsto \Sigma^{-1}$ or the adjugate function $\mathrm{adj} : \Sigma \longmapsto \det(\Sigma)\Sigma^{-1}$ were also proposed in the context of DTI [16,17]. Let us find some properties satisfied by some of these examples. We define the following subsets of the set $\mathcal{F} = \mathrm{Diff}(SPD_n)$ of diffeomorphisms of SPD_n.

(Spectral) $\mathcal{S} = \{f \in \mathcal{F} | \forall U \in O_n, \forall D \in \mathrm{Diag}_n^{++}, f(UDU^\top) = Uf(D)U^\top\}$. Spectral deformations are characterized by their values on sorted diagonal matrices so the deformations described above are spectral: $A_r, \mathrm{adj}, \mathrm{pow}_\theta \in \mathcal{S}$.

For a spectral deformation $f \in \mathcal{S}$, $f(\mathbb{R}_+^* I_n) = \mathbb{R}_+^* I_n$ so we can unically define a smooth diffeomorphism $f_0 : \mathbb{R}_+^* \longrightarrow \mathbb{R}_+^*$ by $f(\lambda I_n) = f_0(\lambda)I_n$.

(Univariate) $\mathcal{U} = \{f \in \mathcal{S} | f(\mathrm{diag}(\lambda_1, ..., \lambda_n)) = \mathrm{diag}(f_0(\lambda_1), ..., f_0(\lambda_n))\}$. The power functions are univariate. Any polynomial $P = \lambda X \prod_{k=1}^p (X - a_i)$ null at 0, with non-positive roots $a_i \leqslant 0$ and positive coefficient $\lambda > 0$, also gives rise to a univariate deformation.

(Diagonally-stable) $\mathcal{D} = \{f \in \mathcal{F} | f(\text{Diag}_n^{++}) \subset \text{Diag}_n^{++}\}$. The deformations described above $A_r, \text{adj}, \text{pow}_\theta$ and the univariate deformations are clearly diagonally-stable: $A_r, \text{adj}, \text{pow}_\theta \in \mathcal{D}$ and $\mathcal{U} \subset \mathcal{D} \cap \mathcal{S}$.

(Log-linear) $\mathcal{L} = \{f \in \mathcal{F} | \log_* f = \log \circ f \circ \exp \text{ is linear}\}$. The adjugate function and the power functions are log-linear deformations. More generally, the functions $f_{\lambda,\mu} : \Sigma \longmapsto (\det \Sigma)^{\frac{\lambda-\mu}{n}} \Sigma^\mu$ for $\lambda, \mu \neq 0$, are log-linear deformations. We can notice that the $f_{\lambda,\mu}$-deformed-affine metric belongs to the one-parameter family of μ-power-affine metrics with $\beta = \frac{\lambda^2-\mu^2}{n\mu^2} > -\frac{1}{n}$.

The deformations $f_{\lambda,\mu}$ just introduced are also spectral and the following result states that they are the only spectral log-linear deformations.

Theorem 5 (Characterization of the power-affine metrics). *If $f \in \mathcal{S} \cap \mathcal{L}$ is a spectral log-linear diffeomorphism, then there exist real numbers $\lambda, \mu \in \mathbb{R}^*$ such that $f = f_{\lambda,\mu}$ and the f-deformed-affine metric is a μ-power-affine metric.*

The interest of this theorem comes from the fact that the group of spectral deformations and the vector space of log-linear deformations have large dimensions while their intersection is reduced to a two-parameter family. This strong result is a consequence of the theory of Lie group representations because the combination of the spectral property and the linearity makes $\log_* f$ a homomorphism of O_n-modules (see the sketch of proof below).

Sketch of the proof. Thanks to Lie group representation theory, the linear map $F = \log_* f : Sym_n \longrightarrow Sym_n$ appears as a homomorphism of O_n-modules for the representation $\rho : P \in O_n \longmapsto (V \longmapsto PVP^\top) \in GL(Sym_n)$. Once shown that $Sym_n = \text{span}(I_n) \oplus \ker \text{tr}$ is a ρ-irreducible decomposition of Sym_n and that each one is stable by F, then according to Schur's lemma, F is homothetic on each subspace, i.e. there exist $\lambda, \mu \in \mathbb{R}^*$ such that for $V \in Sym_n$,
$$F(V) = \lambda \frac{\text{tr}(V)}{n} I_n + \mu \left(V - \frac{\text{tr}(V)}{n} I_n \right) = \log_* f_{\lambda,\mu}(V), \text{ so } f = f_{\lambda,\mu}.$$

5 Conclusion

We have shown that the polar-affine metric is a square deformation of the affine-invariant metric and this process can be generalized to any power function or any diffeomorphism on SPD matrices. It results that the invariance principle of symmetry is not sufficient to distinguish all these metrics, so we should find other principles to limit the scope of acceptable metrics in statistical computing. We have proposed a few characteristics (spectral, diagonally-stable, univariate, log-linear) that include some functions on tensors previously introduced. Future work will focus on studying the effect of such deformations on real data and on extending this family of metrics to positive semi-definite matrices. Finding families that comprise two non-cousin metrics could also help understand the differences between them and bring principles to make choices in applications.

Acknowledgements. This project has received funding from the European Research Council (ERC) under the European Union's Horizon 2020 research and innovation program (grant G-Statistics agreement No 786854). This work has been supported by the French government, through the UCAJEDI Investments in the Future project managed by the National Research Agency (ANR) with the reference number ANR-15-IDEX-01.

References

1. Lenglet, C., Rousson, M., Deriche, R., Faugeras, O.: Statistics on the Manifold of multivariate normal distributions: theory and application to diffusion tensor MRI processing. J. Math. Imag. Vis. **25**(3), 423–444 (2006)
2. Pennec, X., Fillard, P., Ayache, N.: A Riemannian framework for tensor computing. Int. J. Comput. Vis. **66**(1), 41–66 (2006)
3. Fletcher, T., Joshi, S.: Riemannian geometry for the statistical analysis of diffusion tensor data. Signal Process. **87**, 250–262 (2007)
4. Barachant, A., Bonnet, S., Congedo, M., Jutten, C.: Classification of covariance matrices using a Riemannian-based kernel for BCI applications. Elsevier Neurocomput. **112**, 172–178 (2013)
5. Deligianni, F., et al.: A probabilistic framework to infer brain functional connectivity from anatomical connections. In: IPMI Conference, pp. 296–307 (2011)
6. Cheng, G., Vemuri, B.: A novel dynamic system in the space of SPD matrices with applications to appearance tracking. SIIMS **6**(16), 592–615 (2013)
7. Dryden, I., Koloydenko, A., Zhou, D.: Non-Euclidean statistics for covariance matrices with applications to diffusion tensor imaging. Ann. Appl. Stat. **3**, 1102–1123 (2009)
8. Lenglet, C., Rousson, M., Deriche, R.: DTI segmentation by statistical surface evolution. IEEE Trans. Med. Imag. **25**, 685–700 (2006)
9. Pennec, X.: Statistical computing on manifolds: from Riemannian geometry to computational anatomy. Emerg. Trends Vis. Comp. **5416**, 347–386 (2008)
10. Su, J., Dryden, I., Klassen, E., Le, H., Srivastava, A.: Fitting optimal curves to time-indexed, noisy observations on non-linear manifolds. Image Vis. Comput. **30**, 428–442 (2018)
11. Su, J., Kurtek, S., Klassen, E., Srivastava, A.: Statistical analysis of trajectories on Riemannian manifolds: bird migration, hurricane tracking and video surveillance. Ann. Appl. Stat. **8**, 530–552 (2014)
12. Zhang, Z., Su, J., Klassen, E., Le, H., Srivastava, A.: Rate-invariant analysis of covariance trajectories. J. Math. Imag. Vis. **60**, 1306–1323 (2018)
13. Dryden, I., Pennec, X., Peyrat, J.-M.: Power Euclidean metrics for covariance matrices with application to diffusion tensor imaging (2010)
14. Fillard, P., Arsigny, V., Pennec, X., Ayache, N.: Clinical DT-MRI estimation, smoothing and fiber tracking with Log-Euclidean metrics. IEEE Trans. Med. Imag. **26**, 1472–1482 (2007)
15. Jbabdi, S., Mandonnet, E., Duffau, H., Capelle, L., Swanson, K., Pélégrini-Issac, M., Guillevin, R., Benali, H.: Simulation of anisotropic growth of low-grade gliomas using diffusion tensor imaging. Magn. Reson. Med. **54**, 616–624 (2005)
16. Lenglet, C., Deriche, R., Faugeras, O.: Inferring white matter geometry from diffusion tensor MRI: application to connectivity mapping. In: Pajdla, T., Matas, J. (eds.) ECCV 2004. LNCS, vol. 3024, pp. 127–140. Springer, Heidelberg (2004). https://doi.org/10.1007/978-3-540-24673-2_11
17. Fuster, A., Dela Haije, T., Tristán-Vega, A., Plantinga, B., Westin, C.-F., Florack, L.: Adjugate diffusion tensors for geodesic tractography in white matter. J. Math. Imag. Vis. **54**, 1–14 (2016)

Shape Part Transfer via Semantic Latent Space Factorization

Raphaël Groscot[1]([✉]), Laurent Cohen[1], and Leonidas Guibas[2]

[1] University Paris Dauphine, PSL Research University CEREMADE, CNRS,
UMR 7534, 75016 Paris, France
groscot@ceremade.dauphine.fr
[2] Stanford University, Stanford, USA

Abstract. We present a latent space factorization that controls a generative neural network for shapes in a semantic way. Our method uses the segmentation data present in a collection of shapes to explicitly factorize the encoder of a pointcloud autoencoder network, replacing it by several sub-encoders. This allows to learn a semantically-structured latent space in which we can uncover statistical modes corresponding to semantically similar shapes, as well as mixing parts from several objects to create hybrids and quickly explore design ideas through varying shape combinations. Our work differs from existing methods in two ways: first, it proves the usefulness of neural networks to achieve shape combinations and second, adapts the whole geometry of the object to accommodate for its different parts.

Keywords: Autoencoder · Pointcloud · Latent space

1 Introduction

Design ideas exploration is a necessary step for creative modeling. Building tools that help quickly prototyping ideas can significantly improve designers' workflow. Given the tremendous size of 3D shape repositories, scanning all previously existing models can be cumbersome. This is why we propose, in this work, a first step to building such a tool: a shape composer that allows to combine parts coming from different objects into a single and coherent new object. Unlike other works that extract and snap different parts into new positions, we explore the possibility of holistic composition with the use of generative neural networks.

This paper presents a semantically-rich way of controlling generative networks for 3D shapes, without limiting the user to predefined labels. On the contrary, our approach is essentially data-driven in two ways. First, because we rely on a large collection of shapes to train our generative model; second, because the dataset itself is used by the user to tweak the output. More specifically, the

Raphaël Groscot—This work was initiated during a long visit in the Geometric Computation group at Stanford University.

F. Nielsen and F. Barbaresco (Eds.): GSI 2019, LNCS 11712, pp. 511–519, 2019.
https://doi.org/10.1007/978-3-030-26980-7_53

dataset contains various shapes along with their segmentations into meaningful object parts. Our generative network is then trained to produce shapes in a way that is compatible with the segmentation. This is achieved by factorizing the latent space of the generative model according to the different possible shape parts. Thanks to this, a user can edit any given shape and decide to only change part of it, by picking the desired geometry within the dataset. Moreover, the network automatically adapts the final shape in a holistic way to make sure the new part fits naturally. This leads to an inherent ambiguity in our task. On the one hand, if we want to change the wings of a plane (for instance), we do not want *too much* change in the other parts of the plane. On the other hand, we still want the rest of the plane to adapt for the change. This is why we propose an asymmetry in the design of our model.

As a matter of fact, our method relies on the Variational Auto Encoder (VAE) framework (Sect. 3.1), but where the encoder is subdivided in several partial encoders, one for each semantic part, that are mixed into a global code which is then given to the decoder (Sect. 3.3). The decoder is structure-agnostic: it only knows to transform a general code into a plausible shape, so that when codes are manipulated and changed, the reconstruction should still look like a plausible shape.

2   Related Work

Our method is related to different research efforts in 3D shapes analysis and generation. We separate our review in three categories: generative modeling, shapes neural networks, and data-driven shapes editing.

Generative Neural Networks. Generative models suchs as Generative Adversarial Networks (GANs) [6] and Variational Auto Encoders (VAEs) [10] both offer ways to sample from a distribution that matches a given dataset. VAEs rely on an *autoencoder* scheme, where a network is asked to project data samples to a subspace of much lower dimensionality (*encode*), while being able to reconstruct the original data (*decode*). Adding a variational constraint that imposes a prior (e.g. gaussian) on the latent distribution makes sure that the model generalizes well. Their compression-like behavior can then be used for several tasks among which unsupervised learning, sampling, interpolation and denoising [4]. One drawback is that the output is typically blurred, because their loss does not account for a perceptual term. On the contrary, GANs aim at *mimicking* a given distribution by generating samples that are indistinguishable from the original dataset; they can hence generate much sharper results, to the cost of harder training and difficulty to control for mode collapse [12]. Conditioning on the likelihood [3,16] allows to have a finer control on their outputs. Our work aims at the same property by means of imposing a specific factorization on an autoencoder latent space.

Shape Neural Networks. As opposed to images, 3D shapes do not naturally fit in a neural network framework. The main issue is to represent them in a fixed-size Euclidian domain. The most direct way to do so is to use voxel grids and directly transpose Convolutional Neural Networks in 3D [7]. However, even if this approach can yield good results, generated shapes quality is limited by the grid discretization and the $O(n^3)$ complexity. To overcome these limits, Pointnet [18] introduced a neural network architecture based on pointclouds and permutation-invariant operators, which characterizes well an *unordered* set such as a pointcloud. It has successfully shown its usefulness for tasks such as classification and segmentation, and even has an extension that exploits hierarchical analysis [19]. This architecture can also be used to generate pointclouds from photographs [5]. Lastly, [2] has replaced the permutation invariance constraint by imposing a lexicographic order on the pointset, leading to pointcloud GANs with high reconstruction accuracy. Using shapes segmented into semantical parts, [17] learns the joint probability for structure and geometry – for instance, the presence or absence of engines on a plane will constraint the profile of the wings. While producing good quality results, their method does not allow to exchange parts between shapes. Our method relies on a variation of such a shape neural network, tailored at being used for shape combinations.

Data-Driven Shape Editing. Many existing methods give automated tools for shapes editing and design exploration. Existing works range from shape correspondences [7] to style similarity and transfer [13,14]. Others focus on generating diversity, by extracting and snapping parts together [8], or by hierchical shape analysis and synthesis [11]. While [8] creates a combinatorial diversity, our method focuses on the geometric prior for the whole shape, as contained in the computed latent space. We also share a common usage as [14], but while they use an example to guide the overall style of the reconstruction, we use multiple examples, each one guiding a specific part of the shape.

3 Method

3.1 Autoencoder Foundation

Our goal is to create new object shapes, by generating variations within their different parts, in a data-driven process. The first step is to be able to recreate objects from the dataset. A natural choice is to use a generative model, we chose autoencoders. Formally, the goal is to learn the two functions E (encoder) and D (decoder) such that, for all X in the dataset:

$$X = D(E(X)) \tag{1}$$

These two functions are implemented as neural networks that operate on pointclouds, either taken as a source (for E) or as a target (for D). The key specificity of our method is our factorization of E based on the available segmentation data.

The architecture of our foundational autoencoder is the following ($N = 1024$ points):

Input: a minibatch of 32 pointclouds, each represented as a $N \times 3$ matrix, accompanied by their segmentation data (see part 3.2);
Encoder: based on Pointnet [18] but in a much simpler version, with successive layers of per-point filters followed by ReLU layers;
Code mixer: the latent space factorization step, as explained in part 3.3;
Decoder: three fully connected layers with biases, except on the last layer;
Output: the last layer is ultimately reshaped to a $N \times 3$ matrix.

3.2 Consistent Segmentation data

To demonstrate our method, we use the *airplane* category from ShapenetCore and its segmentation obtained from [20], comprising of the four following parts: body, wings, engine, tail. We restricted our analysis only to models containing the four parts, but these parts need not have the same number of points. Since all models are aligned in a consistent manner (the plane body is aligned with the Z axis), our neural networks does not need any rotational invariance, and can leverage from the strong spatial relations of the models' parts for both the encoder and decoder. Note that the value of K depends on the given dataset: for the *airplane* category, $K = 4$.

3.3 Semantic Latent Space factorization

We use pointclouds to represent surfaces, a choice that leads to the following remark: any subset of a pointcloud is a pointcloud. Although this may seem trivial, note that this is not a property that usually holds in a machine learning setting: for instance, a segmented region in an image is not typically rectangular. This allows us to replace the encoder by K encoders, each for a part, which yields the following factorization:

$$E = E_1 * E_2 * ... * E_K \qquad (2)$$

$$E(X) = C = [c_1, c_2, ..., c_n], c_i = E_i(X) \qquad (3)$$

where each E_i represents a *partial encoder* for part i, and evaluates what we call a *subcode*. The above product corresponds to vector concatenation. In this form, the factorization of the latent space simply corresponds to assigning parts to dedicated coordinates. Figure 1 shows a diagram of the corresponding pipeline. Note that for a given part i, the corresponding partial encoder E_i will take as input pointclouds of different sizes, since one should not assume equal parts sizes across the dataset. This limitation can be lifted thanks to the Pointnet [18] max-pooling operation. Since part i is included in the whole shape of size $N = 1024$, we know its size has to be smaller than 1024, so we can add zeros until we fill a 1024×3 matrix. Then, one just has to make sure these padded zeros remain through all the layers of the network, until the final max-pooling discards them.

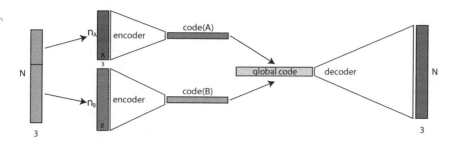

Fig. 1. Structure of our *hybrid encoder* model, illustrated for simplicity with $K = 2$ parts

3.4   Loss and Training

When it comes to pointclouds, two reconstruction losses can be considered: Earth Mover's Distance (EMD) and Chamfer Distance (CD). The former solves the optimal transport problem of transporting S_1 (where each point is seen as a Dirac delta function) onto S_2, and computing the optimal bijection ϕ:

$$d_{EMD}(S_1, S_2) = \min_{\phi: S_1 \to S_2} \sum_{x \in S_1} \|x - \phi(x)\|_2 \qquad (4)$$

Relaxing the global optimality of the assignment, Chamfer Distance computes the squared distance between each point of one pointcloud to its nearest neighbor in the other pointcloud:

$$d_{CD}(S_1, S_2) = \sum_{x \in S_1} \min_{y \in S_2} \|x - y\|_2^2 + \sum_{y \in S_2} \min_{x \in S_1} \|x - y\|_2^2 \qquad (5)$$

Chamfer distance is easier to implementat, has a shorter computation time, and produces acceptable results for our usecase, so we chose to use it over EMD. The interested reader will find a comparison of generation results for both losses in [5].

This reconstruction loss becomes the objective function that is to be minimized. The reconstruction itself depends on the partial encoders and decoder networks, which are simply non-linear parametric functions. So, the learning task ultimately consists of finding the values for these parameters that minimize the objective function. As typically in machine learning, this is done by a stochastic gradient descent.

4   Experiments

We implemented our architecture using Tensorflow [1] and ran it on an Nvidia Gti1080 GPU. We trained over 40 epochs using Adam optimizer [9] with learning rate of 0.9 and a batchsize of 32.

4.1 Basic Autoencoder Mode

Since our network is based on an autoencoder, we first demonstrate its ability to reconstruct objects from the training set. Figure 2 shows examples of reconstructions, chosen to be representative of the type of objects present in our dataset. We can notice that the reconstruction quality highly depends on the sub category (not available) of the object: the typical plane present in the dataset is similar to the second column, so this is where the autoencoder concentrated most of its capacity.

Fig. 2. Example of some reconstructions. Top row: original. Bottom row: reconstructed. The pairs of colored arrows point at errors in the predictions (Color figure online)

Clustering. The latent codes computed by E can be explored using standard dimensionality reduction techniques, such as PCA and tSNE [15]. Figure 3 shows the tSNE projection of our latent space over 2 dimensions, and snapshots of certain blobs with their corresponding shapes. Note how similar shapes live in the same blob. As with any tSNE projection, we remind the reader that distances between blobs are not significant.

Continuous Part Transfer. Thanks to the factorization of E, by simply interpolating on a given E_i, we can easily transfer a part of an object to another one while keeping the rest of the object unchanged. Let S be a source object, T the target and i the index of the part we wish to transfer from S to T. This is done with $E(T) = E_1(T) * E_2(T) * ... * E_K(T)$, replacing $E_i(T)$ by $E_i(S)$. A linear interpolation between $E_i(T)$ and $E_i(S)$ effectively realizes the continuous morphing of the part. Figure 4 shows the results of selectively transfering parts of a plane onto another one.

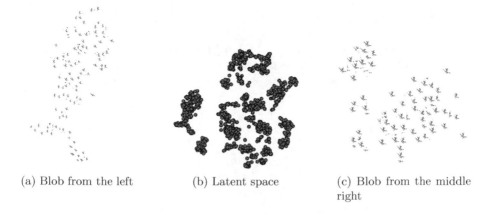

(a) Blob from the left   (b) Latent space   (c) Blob from the middle right

Fig. 3. tSNE projection of the encoder latent space, with close-ups of two blobs

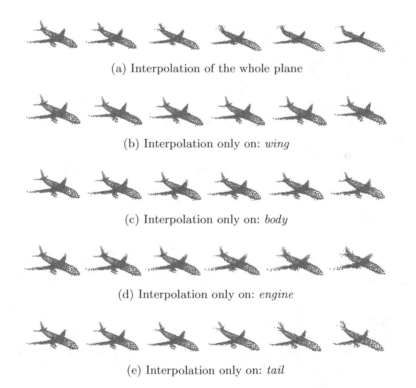

(a) Interpolation of the whole plane

(b) Interpolation only on: *wing*

(c) Interpolation only on: *body*

(d) Interpolation only on: *engine*

(e) Interpolation only on: *tail*

Fig. 4. Selective part transfer, compared to the global interpolation from a source to a target plane

5   Limitations and Discussion

An inherent limitation of our model is that of the autoencoder it is based on. Indeed, it suffers from a problem slightly similar to mode collapse, as shown in Fig. 2: it focuses all its reconstruction capacity towards the most frequent shapes from the dataset. In these reconstructions – running in simple autoencoder mode – see how the secondary engines (blue arrows) are just partially recovered: points that should be dedicated to them stayed on the wings. This is because such planes belong to a rare class. During training, the decoder converged to a state that favors the majority of wings, to the detriment of a minority of engines. It also means that our model cannot be suited for part transfer when one part belongs to an atypical object. Another consequence of lost small details is when they belong to a discriminative part. Let us consider once again the example of the plane with four engines: overall, the engines only have a mild contribution to the reconstruction loss. Adopting a part-specific loss could be a way of circumventing this problem.

As for part transfer, the ambiguity of our holistic design choice yields results which are sometimes hard to predict. Since we want the whole model to adapt for the new shape, we do not want to limit the geometry changes to the region of the transferred part. We are still able to swap parts but this process does not work for all parts. For several shapes, only one part – say the body of the plane – determines the geometry of the other parts, at least in the learnt latent space. So, swapping the body of such a plane with another one might lead to undesired changes in the other parts of the plane.

Moreover, our autoencoder foundation suffers the same unbalanced latent activation as reported in [17]: only a fraction of the latent dimensions have a significant contribution to the reconstruction. All these aforementioned effects, combined together, limit the current predictability of our part transfers. Further investigations are required to improve this point.

Acknowledgement. This work was made possible by an exchange program between PSL–ITI and Stanford University. The authors gratefully acknowledge the financial support coming from PSL, FJF and Fondation Armand et Janet Sibony.

References

1. Abadi, M., et al.: TensorFlow: large-scale machine learning on heterogeneous systems (2015). http://tensorflow.org/, software available from tensorflow.org
2. Achlioptas, P., Diamanti, O., Mitliagkas, I., Guibas, L.J.: Learning Representations and Generative Models for 3D Point Clouds (2017)
3. Chen, X., Duan, Y., Houthooft, R., Schulman, J., Sutskever, I., Abbeel, P.: Infogan: Interpretable representation learning by information maximizing generative adversarial nets (2016)
4. Doersch, C.: Tutorial on Variational Autoencoders. arXiv e-prints arXiv:1606.05908 (2016)

5. Fan, H., Su, H., Guibas, L.J.: A point set generation network for 3D object reconstruction from a single image. CoRR abs/1612.00603 (2016). http://arxiv.org/abs/1612.00603
6. Goodfellow, I.J., et al.: Generative adversarial networks (2014)
7. van Kaick, O., Zhang, H., Hamarneh, G., Cohen-Or, D.: A survey on shape correspondence. Comput. Graph. Forum **30**(6), 1681–1707 (2011). https://doi.org/10.1111/j.1467-8659.2011.01884.x. https://onlinelibrary.wiley.com/doi/abs/10.1111/j.1467-8659.2011.01884.x
8. Kalogerakis, E., Chaudhuri, S., Koller, D., Koltun, V.: A probabilistic model for component-based shape synthesis. ACM Trans. Graph. **31**(4), 55:1–55:11 (2012). https://doi.org/10.1145/2185520.2185551. http://doi.acm.org/10.1145/2185520.2185551
9. Kingma, D.P., Ba, J.: Adam: A Method for Stochastic Optimization. arXiv e-prints arXiv:1412.6980 (2014)
10. Kingma, D.P., Welling, M.: Auto-Encoding Variational Bayes. arXiv e-prints arXiv:1312.6114 (2013)
11. Li, J., Xu, K., Chaudhuri, S., Yumer, E., Zhang, H., Guibas, L.J.: GRASS: generative recursive autoencoders for shape structures. In: ACM Transactions on Graphics (Proceedings of SIGGRAPH 2017) (2017)
12. Locatello, F., Vincent, D., Tolstikhin, I.O., Rätsch, G., Gelly, S., Schölkopf, B.: Clustering meets implicit generative models. CoRR abs/1804.11130 (2018). http://arxiv.org/abs/1804.11130
13. Lun, Z., Kalogerakis, E., Sheffer, A.: Elements of style: learning perceptual shape style similarity. ACM Trans. Graph. **34**(4), 84 (2015)
14. Lun, Z., Kalogerakis, E., Wang, R., Sheffer, A.: Functionality preserving shape style transfer. ACM Trans. Graph. **35**(6), 209:1–209:14 (2016). https://doi.org/10.1145/2980179.2980237. http://doi.acm.org/10.1145/2980179.2980237
15. van der Maaten, L., Hinton, G.E.: Visualizing high-dimensional data using t-SNE. J. Mach. Learn. Res. **9**, 2579–2605 (2008)
16. Mirza, M., Osindero, S.: Conditional Generative Adversarial Nets. arXiv e-prints arXiv:1411.1784 (2014)
17. Nash, C., Williams, C.K.I.: The shape variational autoencoder: a deep generative model of part-segmented 3d objects. Comput. Graph. Forum **36**(5), 1 12 (2017). https://doi.org/10.1111/cgf.13240. https://doi.org/10.1111/cgf.13240
18. Qi, C.R., Su, H., Mo, K., Guibas, L.J.: Pointnet: deep learning on point sets for 3D classification and segmentation (2016)
19. Qi, C.R., Yi, L., Su, H., Guibas, L.J.: Pointnet++: deep hierarchical feature learning on point sets in a metric space. CoRR abs/1706.02413 (2017). http://arxiv.org/abs/1706.02413
20. Yi, L., et al.: A scalable active framework for region annotation in 3D shape collections. ACM Trans. Graph. **35**(6), 2101–21012 (2016). https://doi.org/10.1145/2980179.2980238. http://doi.acm.org/10.1145/2980179.2980238

Geometric and Structure Preserving Discretizations

Variational Discretization Framework for Geophysical Flow Models

Werner Bauer[1,2(✉)] [iD] and François Gay-Balmaz[3] [iD]

[1] Inria Rennes, Campus de Beaulieu, 35042 Rennes Cedex, France
`werner.bauer.email@gmail.com`
[2] École Normale Supérieure, Laboratoire de Météorologie Dynamique, Paris, France
[3] CNRS/École Normale Supérieure, Laboratoire de Météorologie Dynamique,
Paris, France
`francois.gay-balmaz@lmd.ens.fr`

Abstract. We introduce a geometric variational discretization framework for geophysical flow models. The numerical scheme is obtained by discretizing, in a structure-preserving way, the Lie group formulation of fluid dynamics on diffeomorphism groups and the associated variational principles. Being based on a discrete version of the Euler-Poincaré variational method, this discretization approach is widely applicable. We present an overview of structure-preserving variational discretizations of various equations of geophysical fluid dynamics, such as the Boussinesq, anelastic, pseudo-incompressible, and rotating shallow-water equations. We verify the structure-preserving nature of the resulting variational integrators for test cases of geophysical relevance. Our framework applies to irregular mesh discretizations in 2D and 3D in planar and spherical geometry and produces schemes that preserve invariants of the equations such as mass and potential vorticity. Descending from variational principles, the discussed variational schemes exhibit a discrete version of Kelvin circulation theorem and show excellent long term energy behavior.

Keywords: Anelastic and pseudo-incompressible equations ·
Rotating shallow-water equations ·
Soundproof and compressible fluids · Variational principle ·
Euler-Poincaré equations · Structure-preserving discretizations

1 Introduction

Variational methods are a powerful tool to derive consistent models from Hamilton's principle of least action. The equations of motion follow by computing the critical curve of the action functional associated to the Lagrangian of the system. When derived from a discrete version of variational principles, the resulting discretizations preserve important geometric properties of their underlying continuous equations, such as long term stability, consistency in statistical properties and conservation of stationary solutions, see e.g. [9,12,14].

Supported by ANR (ANR-14-CE23-0002-01) and Horizon 2020 (No 657016).

F. Nielsen and F. Barbaresco (Eds.): GSI 2019, LNCS 11712, pp. 523–531, 2019.
https://doi.org/10.1007/978-3-030-26980-7_54

Given the generality of the approach, variational methods are applied in various fields of interest. While most of the literature covers variational integrators for ordinary differential equations (ODEs), in recent years they have been developed also for partial differential equations (PDEs), in particular for fluid and geophysical fluid dynamics (GFD), see e.g. [13] and [1–3,6]. The numerical schemes descend from discretizing the Lie group formulation of fluid dynamics on diffeomorphism groups and the associated variational principles.

In the field of GFD, variational integrators are of particular interest given their conservation of invariants which is a crucial property for long time integrations to guarantee accurate representation of the statistical properties of these models [4,7]. In this context, the rotating shallow-water (RSW) equations, both in the plane and on the sphere, have received considerable attention because they allow us to study the essential features of the full 3D equations in an idealized setting. Another interesting approximation of the Euler equations is to filter out sound waves as they are assumed to be negligible for atmospheric models. In this context, mostly anelastic and pseudo-incompressible approximations are studied, cf. [8,11].

Here, we present a unified variational discretization framework that covers both the soundproof approximations of the Euler equations [2,3] and the compressible rotating shallow-water case [1,5]. As shown in Sect. 2, this framework will allow us to stress differences in the description of compressible and incompressible flows. Naturally, the Lie groups approximations describing the different flow models differ from each other, but the derivation of the corresponding Lie algebras and fluid's vector fields follows for all cases the same procedure. In contrast, the treatment of advected quantities differs between soundproof and compressible models. For some of these models, we present in Sect. 3 some numerical results of the schemes from Sect. 2.4 focusing again on the similarities and differences between them. Finally, in Sect. 4 we draw some conclusions.

2   Variational Discretization Framework

The discretization procedure mimics the continuous variational principle step by step. In Table 1 the corresponding continuous definitions are given that have to be suitably approximated.

Recall that in the Lagrangian representation, the variational principle is the Hamilton principle $\delta \int L(\varphi, \dot{\varphi}) \mathrm{d}t = 0$ written on the appropriate diffeomorphism group of the fluid domain $\mathcal{M}$. For instance for the RSW equations the group $\mathrm{Diff}(\mathcal{M})$ of all diffeomorphisms is used, whereas for soundproof models, one chooses the group $\mathrm{Diff}_{\bar{\sigma}\mu}(\mathcal{M})$ that preserve a weighted volume $\bar{\sigma} d\mathbf{x}$, with $\bar{\sigma} = 1$, $\bar{\sigma} = \bar{\rho}$, or $\bar{\sigma} = \bar{\rho}\bar{\theta}$, for the Boussinesq, anelastic, or pseudo-incompressible model, in which $\bar{\rho}(z)$ and $\bar{\theta}(z)$ characterize vertically varying reference states for density and potential temperature, respectively, in hydrostatic balance [2].

The variational principle inherited from the Hamilton principle in Eulerian (spatial) representation is the Euler-Poincaré principle ([10]). The spatial Lagrangian is expressed in terms of the Eulerian velocity $\mathbf{u}$, the fluid depth

h, and/or the potential temperature θ, and the fluid equations follow from $\delta \int_0^T \ell dt = 0$ with respect to constrained variations $\delta\mathbf{u}$, δh or $\delta\theta$, see Table 1.

2.1 Discrete Diffeomorphism Groups

The discretization procedure starts with the choice of a discrete version of the diffeomorphism group, [1, 13], obtained by first discretizing the space of functions $\mathcal{F}(\mathcal{M})$ on which the group acts by composition on the right, and then identifying a finite dimensional group acting by matrix multiplication on the finite dimensional space of discrete functions, while preserving some properties of the action by diffeomorphisms (constant functions are preserved). Given a mesh $\mathbb{M}$ of $\mathcal{M}$ and choosing as discrete functions the space $\mathbb{R}^N$ of piecewise constant functions on $\mathbb{M}$, this results in the following matrix groups:

- for compressible flow: $\mathsf{D}(\mathbb{M}) = \{q \in \mathrm{GL}(N)^+ \mid q \cdot \mathbf{1} = \mathbf{1}\}$, with $\mathbf{1} = (1, ..., 1)^\mathsf{T}$, in which the condition $q \cdot \mathbf{1} = \mathbf{1}$ encodes, at the discrete level, the fact that constant functions are preserved under composition by a diffeomorphism;
- for soundproof flow: $\mathsf{D}_{\bar{\sigma}}(\mathbb{M}) = \{q \in \mathrm{GL}(N)^+ \mid q \cdot \mathbf{1} = \mathbf{1} \text{ and } q^T \Omega^{\bar{\sigma}} q = \Omega^{\bar{\sigma}}\}$, with $\Omega_i^{\bar{\sigma}} := \int_{C_i} \bar{\sigma}(z)d\mathbf{x}$, for a cell C_i, and where the additional constraint imposes the preservation of the weighted volume at the discrete level.

The action of the groups $\mathsf{D}(\mathbb{M})$ and $\mathsf{D}_{\bar{\sigma}}(\mathbb{M})$ by matrix multiplication on discrete functions $F \in \mathbb{R}^N$ is denoted as

$$F \in \mathbb{R}^N \mapsto qF = F \circ q^{-1} \in \mathbb{R}^N, \quad q \in \mathsf{D}(\mathbb{M}), \tag{1}$$

where the suggestive notation $F \circ q^{-1}$ for the multiplication of the vector F by the matrix q is introduced to indicate that this action is understood as a discrete version of the action of $\mathrm{Diff}(\mathcal{M})$ and $\mathrm{Diff}_{\bar{\sigma}}(\mathcal{M})$ by composition on $\mathcal{F}(\mathcal{M})$. The situation is formally illustrated by the diagram

$$\begin{array}{ccc} f \in \mathcal{F}(\mathcal{M}) & \xrightarrow{\mathrm{Diff}(\mathcal{M}) \text{ or } \mathrm{Diff}_{\bar{\sigma}}(\mathcal{M})} & f \circ \varphi^{-1} \in \mathcal{F}(\mathcal{M}) \\ \downarrow \text{Discretization} & & \downarrow \text{Discretization} \\ F \in \mathbb{R}^N & \xrightarrow{\mathsf{D}(\mathbb{M}) \text{ or} \mathsf{D}_{\bar{\sigma}}(\mathbb{M})} & q \cdot F \in \mathbb{R}^N \end{array}$$

2.2 Discrete Lie Algebra and Discrete Vector Fields

By taking the derivative of continuous and discrete actions at the identity, we get $\frac{d}{dt}\big|_{t=0} f \circ \varphi_t^{-1} = -\mathbf{d}f \cdot \mathbf{u}$ and $\frac{d}{dt}\big|_{t=0} F \circ q_t^{-1} = AF$, where $\frac{d}{dt}\big|_{t=0} \varphi_t = \mathbf{u}$ and $\frac{d}{dt}\big|_{t=0} q_t = A$. Hence AF, with A an element of the Lie algebra of $\mathsf{D}(\mathbb{M})$ or $\mathsf{D}_{\bar{\sigma}}(\mathcal{S})$ is a discretization of (minus) the derivative of f in the direction $\mathbf{u}$. The Lie algebras of $\mathsf{D}(\mathbb{M})$ and $\mathsf{D}_{\bar{\sigma}}(\mathcal{S})$ are:

- for compressible flow: $\mathfrak{d}(\mathbb{M}) = \{A \in \mathrm{Mat}(N) \mid A \cdot \mathbf{1} = 0\}$,
- for soundproof flow: $\mathfrak{d}_{\bar{\sigma}}(\mathbb{M}) = \{A \in \mathrm{Mat}(N) \mid A \cdot \mathbf{1} = 0, \ A^\mathsf{T} \Omega^{\bar{\sigma}} + \Omega^{\bar{\sigma}} A = 0\}$.

However, not all $A \in \mathfrak{d}(\mathbb{M})$ or $\mathfrak{d}_{\bar{\sigma}}(\mathbb{M})$ can be interpreted as discrete vector fields. This induces *nonholonomic constraints* on the Lie algebras which have to be appropriately taken into account in the variational principle.

Nonholonomic Constaints. For both *soundproof* and *compressible* fluids it is required that fluxes are nonzero only between neighboring cells, hence we have the linear constraint $\mathcal{S} = \{A \in \mathfrak{d}(\mathbb{M})/\mathfrak{d}_{\bar{\sigma}}(\mathbb{M}) \mid A_{ij} = 0, \ \forall \ j \notin N(i)\}$ where $N(i)$ is the set of indices of those cells adjacent to cell i.

For the *compressible* case, we have the additional constraint, $\Omega_{ii} A_{ij} = -\Omega_{jj} A_{ji}$, for all $j \neq i$, where $\Omega = \Omega^{\bar{\sigma}}$ with $\bar{\sigma} = 1$. This gives the additional linear constraint $\mathcal{R} = \{A \in \mathfrak{d}(\mathbb{M}) \mid A^{\mathsf{T}} \Omega + \Omega A \text{ is diagonal}\}$. These nonholonomic constraints are taken into account by using the *Euler–Poincaré–d'Alembert principle*, which is the nonholonomic version of the Euler–Poincaré principle.

Discrete Vector Fields. Taking into account these non-holonomic constraints, it can be shown that if a matrix A approximates a vector field $\mathbf{u}$, then,

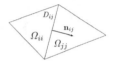

Fig. 1. Flux associated to A_{ij}.

– for compressible flow: matrix elements of $A \in \mathcal{S} \cap \mathcal{R}$ satisfy $A_{ij} \simeq -\frac{1}{2\Omega_{ii}} \int_{D_{ij}} (\mathbf{u} \cdot \mathbf{n}_{ij}) dS$, $A_{ii} \simeq \frac{1}{2\Omega_{ii}} \int_{C_i} (\operatorname{div} \mathbf{u}) d\mathbf{x}$,
– for soundproof flow: matrix elements of $A \in \mathcal{S}$ satisfy $A_{ij} \simeq -\frac{1}{2\Omega_{ii}^{\bar{\sigma}}} \int_{D_{ij}} (\bar{\sigma} \mathbf{u} \cdot \mathbf{n}_{ij}) dS$,

for all $j \in N(i)$, $j \neq i$, with D_{ij} the hyperface common to cells C_i and C_j and $\mathbf{n}_{ij}$ is the normal vector on D_{ij} pointing from C_i to C_j, cf. Fig. 1.

Discrete Advected Quantities. To formulate the discrete Euler–Poincaré–d'Alembert principle, we need to define appropriate actions of $\mathsf{D}(\mathbb{M})$ on discrete fluid depth D for RSW and of $\mathsf{D}_{\bar{\sigma}}(\mathbb{M})$ on discrete potential temperature Θ for SP. In both cases, the action results from the definition in (1), namely

– for compressible flow: D is a discrete density so the action, $D \mapsto D \bullet q$, is dual to the action on discrete functions: $\langle D \bullet q, F \rangle = \langle D, F \circ q^{-1} \rangle$ for all $F \in \mathbb{R}^N$, with respect to the discrete L^2 pairing. It results in $D \bullet q = \Omega^{-1} q^{\mathsf{T}} \Omega D$.
– for soundproof flow: Θ is a discrete function so the action is $q\Theta = \Theta \circ q^{-1}$ as in (1). Then, $\Theta(t) = q(t)\Theta_0$.

2.3   Euler–Poincaré–d'Alembert (EPA) Variational Principle

Consider the spatial discrete Lagrangian $\ell_d = \ell_d(A, Q) : \mathfrak{d}(\mathbb{M})/\mathfrak{d}_{\bar{\sigma}}(\mathbb{M}) \times \mathbb{R}^N \to \mathbb{R}$ with $Q \in \mathbb{R}^N$ an advected quantity, $(D \text{ or } \Theta)$. The *discrete EPA principle* reads: $\delta \int_0^T \ell_d(A, Q) dt = 0$ for variations $\delta A = \partial_t B + [B, A]$, $B(0) = B(T) = 0$, and

– for compressible flow: $\delta Q = -Q \bullet B$, with $A, B \in \mathcal{S} \cap \mathcal{R}$,
– for soundproof flow: $\delta Q = BQ$, with $A, B \in \mathcal{S}$.

Case 1: soundproof model. The discrete EPA principle, with $Q = \Theta$, yields the following semidiscrete equations for $(A(t), \Theta(t)) \in \mathfrak{d}_{\bar{\sigma}}(\mathbb{M}) \times \mathbb{R}^N$, [2]:

$$\left(\frac{d}{dt} \frac{\delta \ell_d}{\delta A} + \left[\frac{\delta \ell_d}{\delta A} \Omega^{\bar{\sigma}}, A \right] (\Omega^{\bar{\sigma}})^{-1} + \left(\Theta \frac{\delta \ell_d}{\delta \Theta} \right)^{\top (A)} + dP \right)_{ij} = 0, \quad \text{for all } i \in N(j) \quad (2)$$

for some discrete function P (the discrete pressure). Here $(dP)_{ij} = P_j - P_i$ and $()^{(A)}$ denotes the skew-symmetric part of a matrix.

Case 2: compressible model. The discrete EPA principle, with $Q = D$, yields the following semidiscrete equations for $(A(t), D(t)) \in \mathfrak{d}(\mathbb{M}) \times \mathbb{R}^N$, [1]:

$$\mathbf{P} \left(\frac{d}{dt} \frac{\delta \ell}{\delta A} + \Omega^{-1} \left[A^{\top}, \Omega \frac{\delta \ell}{\delta A} \right] + D \frac{\delta \ell}{\delta D}^{\top} \right)_{ij} = 0, \quad \text{for all } i \in N(j), \quad (3)$$

where $\mathbf{P}$ is the projection associated to the nonholonomic constraint, [1]. These equations are accompanied with the discrete continuity equation $\frac{d}{dt} D + D \bullet A = 0$.

We provide in Table 1 a summary that enlightens the correspondence between the continuous and discrete objects. Note that in both cases, the resulting equations of motion for soundproof and compressible flows are valid on any reasonable mesh (e.g. not degenerated cells [1]). To result in implementable code, we have to choose a mesh and a suitable discrete flat operators such as in [13].

Table 1. Continuous and discrete objects for soundproof (SP) and compressible (CP) discretizations. The divergence is denoted by div and the Jacobian by J.

Continuous diffeomorphisms		Discrete diffeomorphisms	
$\text{Diff}(\mathcal{M}) \ni \varphi$		$\text{D}(\mathbb{M}) \ni q$	
Group action on functions		Group action on discrete functions	
$f \mapsto f \circ \varphi^{-1}$		$F \mapsto F \circ q^{-1} =: qF$	
Group action on densities		Group action on discrete densities	
CP: $\quad h \mapsto h \bullet \varphi = (h \circ \varphi) J\varphi$		CP: $\quad D \mapsto D \bullet q = \Omega^{-1} q^{\top} \Omega D$	
Eulerian velocity and advected quantity		Disc. Eulerian veloc. and advec. quantity	
$\mathbf{u} = \dot{\varphi} \circ \varphi^{-1}, \begin{cases} \text{SP:} & \theta = \Theta_0 \circ \varphi^{-1} \\ \text{CP:} & h = (h_0 \circ \varphi^{-1}) J\varphi^{-1} \end{cases}$		$A = \dot{q} q^{-1}, \begin{cases} \text{SP:} & \Theta - q\Theta_0 \\ \text{CP:} & D = \Omega^{-1} q^{-\top} \Omega D_0 \end{cases}$	
Euler-Poincaré principle		Euler-Poincaré-d'Alembert principle	
$\delta \int_0^T \ell(\mathbf{u}, h/\theta) dt = 0, \delta \mathbf{u} = \partial_t \mathbf{v} + [\mathbf{v}, \mathbf{u}]$		$\delta \int_0^T \ell(A, D/\Theta) dt = 0, \delta A = \partial_t B + [B, A]$	
$\begin{cases} \text{SP:} & \delta\theta = -\mathbf{d}\theta \cdot \mathbf{v} \\ \text{CP:} & \delta h = -\text{div}(h\mathbf{v}) \end{cases}$		$\begin{cases} \text{SP:} & \delta D = B\Theta, \qquad A, B \in \mathcal{S} \\ \text{CP:} & \delta D = -\Omega^{-1} B^{\top} \Omega D, \quad A, B \in \mathcal{S} \cap \mathcal{R} \end{cases}$	

2.4   Numerical Schemes on Irregular Simplicial Meshes

The numerical schemes are obtained from (2) and (3), by specializing them to the chosen mesh and the chosen discrete Lagrangian, which requires the construction of a discrete "flat" operator $A \in \mathcal{S} \cap \mathcal{R} \mapsto A^{\flat}$ associated to the given mesh, see [13]. For instance, for the RSW case the discrete Lagrangian is

$$\ell_d(A, D) = \frac{1}{2} \sum_{i,j=1}^N D_i A_{ij}^{\flat} A_{ij} \Omega_{ii} + \sum_{i,j=1}^N D_i R_{ij}^{\flat} A_{ij} \Omega_{ii} - \frac{1}{2} \sum_{i=1}^N g(D_i + B_i)^2 \Omega_{ii}. \quad (4)$$

The discrete flat operator is defined from the two conditions $A^\flat_{ij} = 2\Omega_{ii}\frac{h_{ij}}{f_{ij}}A_{ij}$, and $A^\flat_{ij} + A^\flat_{jk} + A^\flat_{ki} = K^e_j \langle \omega(A^\flat), \zeta_e \rangle$, for $i, k \in N(j)$, $k \notin N(i)$, with e the node common to cells C_i, C_j, C_k, where $K^e_k := \frac{|\zeta_e \cap C_k|}{|\zeta_e|}$, $\langle \omega(A^\flat), \zeta_e \rangle := \sum_{h_{mn} \in \partial \zeta_e} A^\flat_{mn}$, and where $|\zeta_e \cap C_k|$ is the area of the intersection of C_k with the dual cell ζ_e, f_{ij} is the length of the triangle edge between C_i and C_j, and h_{ij} is the length of the dual edge connecting the circumcenters of C_i and C_j.

The numerical scheme is then obtained by applying a variational discretization in time, [6].

3   Numerical Results

On a small selection of test cases of fluid and geophysical fluid dynamics on an f-plane or the sphere, we show the performance of the variational integrators developed in [1,2]. We focus here on illustrating similarities and differences between the models and their variational discretizations.

Consider first the RSW scheme on both an f-plane and the sphere. We study the scheme's capability to conserve steady state solutions, cf. [1,5], and invariants such as mass, energy, potential vorticity and enstrophy. Figure 2 shows that for long term integrations, the total energy (kinetic + potential energy) is well preserved for simulations on a regular (left) and irregular (middle) f-plane mesh, but also rather well on the sphere (right).

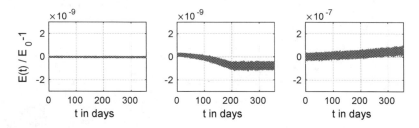

Fig. 2. Relative errors in total energy for the RSW scheme over 1 year. Left and middle: isolated vortex solutions [1] on uniform, resp. non-uniform meshes (64^2 triangles) on an f-plane; right: TC 2 solutions [5] on the sphere (10242 Voronoi cells).

Note that mass and potential vorticity are preserved at machine precision for all cases. Although not by construction, potential enstrophy is well preserved too on both f-plane and the sphere: for 50 days run of TC 2 of [15] at 10^{-7} and for TC 5 at 10^{-3}. In general we notice that the solutions on the f-plane and on the sphere behave very similarly.

Consider further the convergence plots for the RSW scheme on a regular and irregular f-plane mesh (left) and on the sphere (right) of Fig. 3. The plots show the convergence of the numerical results after 1 day, resp. 12 days of simulations against the corresponding steady state solutions. On the f-plane, our

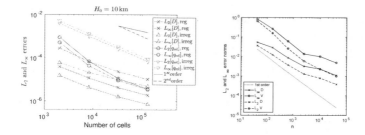

Fig. 3. Convergence of RSW scheme. Left: on the sphere for steady state solution (WTC2) after 12 days. Right: on an f-plane for steady state solutions after 1 day.

scheme shows at least 1st order convergence rates, while on the sphere, given the additional curvature, it reduces to the order of about 0.5.

Figure 4 shows the fluid depth after a simulation of 14 days for TC 2 (left) and a Rossby wave [5]. A comparison to literature confirms that these solutions are accurately represented by our RSW integrator.

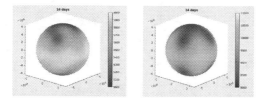

Fig. 4. RSW scheme on the sphere. Left: Williamson test case 5 (flow over a mountain) after 14 days. Right: Rossby wave test case. Colorbars indicate fluid depth in [m]. (Color figure online)

Finally, Fig. 5 shows solutions of the cold bubble test case [3], i.e. a falling cold bubble in a warm environment. The left panel presents the solution of the anelastic model with a linearized buoyancy term, the right one the corresponding

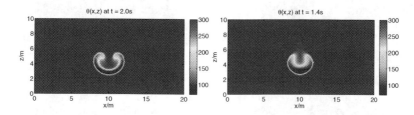

Fig. 5. Potential temperature θ on regular meshes: comparison of results of the anelastic (left) and pseudo-incompressible schemes (right) for the falling cold air bubble with $\theta_{min}/\theta_{max} = 90\,\text{K}/300\,\text{K}$. Colorbar indicates $[\theta]$ in K.

solution for the pseudo-incompressible (PI) scheme applying a nonlinear buoyancy approximation. Our PI captures well the physical meaningful nonlinear effect that prevents the bubble from stretching [3,11], in contrast to the anelastic scheme. Matching well the results from literature, this confirms the accuracy of the variational schemes for the soundproof models.

4   Conclusions

We presented a variational discretization framework for geophysical flow models. This framework unifies the integrators for soundproof and compressible flow models developed in [1,2]. In particular, we could illustrate that the methodology of deriving discrete velocity fields as elements of discrete Lie algebras of the fluid models has many steps in common, while the discrete Lie groups that approximate the configuration space and the advection of either buoyancy or the fluid density of soundproof or compressible fluids, respectively, naturally differ. We illustrated on some selected numerical results simularities and differences between these variational integrators while confirming their excellent conservation properties.

References

1. Bauer, W., Gay-Balmaz, F.: Towards a geometric variational discretization of compressible fluids: the rotating shallow water equations, J. Comp. Dyn., 6(1) (2019). https://doi.org/10.3934/jcd.2019001
2. Bauer, W., Gay-Balmaz, F.: Variational integrators for anelastic and pseudo-incompressible flows, accepted with minor revisions at J. Geom. Mech., Preprint: https://arxiv.org/abs/1701.06448 (2019)
3. Bauer, W., Gay-Balmaz, F.: Variational integrators for soundproof models on arbitrary triangular C-grids. Preprint: https://hal.inria.fr/view/index/docid/1970335 (2019)
4. Bendall, T.M., Cotter, C.J.: Statistical properties of an enstrophy conserving discretisation for the stochastic quasi-geostrophic equation. Geophysical and Astrophysical Fluid Dynamics, ISSN: 0309–1929 (2018)
5. Brecht, R., Bauer, W., Bihlo, A., Gay-Balmaz, F., MacLachlan, S.: Variational integrator for the rotating shallow-water equations on the sphere. Q. J. Meteorol. Soc. 145, 1070–1088 (2018)
6. Desbrun, M., Gawlik, E.S., Gay-Balmaz, F., Zeitlin, V.: Variational discretization for rotating stratified fluids. Discr. Cont. Dyn. Syst. - A 34, 479–511 (2014)
7. Dubinkina, S., Frank, J.: Statistical mechanics of Arakawa's discretizations. J. Computat. Phys. 227, 1286–1305 (2007)
8. Durran, D.R.: Improving the anelastic approximation. J. Atmos. Sci. 46, 1453–1461 (1989)
9. Hairer, E., Lubich, C., Wanner, G.: Geometric Numerical Integration: Structure-Preserving Algorithms for Ordinary Differential Equations. Springer, Berlin (2006). https://doi.org/10.1007/3-540-30666-8
10. Holm, D.D., Marsden, J.E., Ratiu, T.S.: The Euler-Poincaré equations and semidirect products with applications to continuum theories. Adv. Math. 137, 1–81 (1998)

11. Klein, R.: Asymptotics, structure, and integration of sound-proof atmospheric flow equations. Theor. Comput. Fluid Dyn. **23**, 161–195 (2009)
12. Leimkuhler, B., Reich, S.: Simulating Hamiltonian Dynamics. Cambridge University Press, Cambridge (2004)
13. Pavlov, D., Mullen, P., Tong, Y., Kanso, E., Marsden, J., Desbrun, M.: Structure-preserving discretization of incompressible fluids. Phys. D: Nonlinear Phenom. **240**(6), 443–458 (2011)
14. Wan, A.T.S., Nave, J.C.: On the arbitrarily long-term stability of conservative methods. SIAM J. Numer. Anal. **56**(5), 2751–2775 (2018)
15. Williamson, D.L., Drake, J.B., Hack, J.J., Jakob, R., Swarztrauber, P.N.: A standard test set for numerical approximations to the shallow water equations in spherical geometry. J. Comput. Phys. **102**, 211–224 (1992)

Finite Element Methods for Geometric Evolution Equations

Evan S. Gawlik[✉]

University of Hawaii at Manoa, Honolulu, HI 96822, USA
egawlik@hawaii.edu

Abstract. We study finite element methods for the solution of evolution equations in Riemannian geometry. Our focus is on Ricci flow and Ricci-DeTurck flow in two dimensions, where one of the main challenges from a numerical standpoint is to discretize the scalar curvature of a time-dependent Riemannian metric with finite elements. We propose a method for doing this which leverages Regge finite elements – piecewise polynomial symmetric $(0,2)$-tensors possessing continuous tangential-tangential components across element interfaces. In the lowest order setting, the finite element method we develop for two-dimensional Ricci flow is closely connected with a popular discretization of Ricci flow in which the scalar curvature is approximated with the so-called angle defect: 2π minus the sum of the angles between edges emanating from a common vertex. We present some results from our ongoing work on the analysis of the method, and we conclude with numerical examples.

Keywords: Finite element · Ricci flow · Scalar curvature · Angle defect

1 Introduction

Partial differential equations governing the evolution of time-dependent Riemannian metrics are ubiquitous in geometric analysis. In this work, we study finite element discretizations of such problems.

The model problem we consider consists of finding a Riemannian metric $g(t)$ on a smooth manifold Ω satisfying

$$\frac{\partial}{\partial t} g = \sigma, \quad g(0) = g_0, \tag{1}$$

where g_0 is given and σ is a symmetric $(0,2)$-tensor field depending on g and/or t. We are particularly interested in two special cases: (i) two-dimensional normalized Ricci flow, in which case $\sigma = (\bar{R} - R)g$, R is the scalar curvature of g, and $\bar{R}$ is the average of R over Ω (or some other prescribed scalar function); and (ii)

Supported by the NSF under grant DMS-1703719.

F. Nielsen and F. Barbaresco (Eds.): GSI 2019, LNCS 11712, pp. 532–540, 2019.
https://doi.org/10.1007/978-3-030-26980-7_55

two-dimensional Ricci-DeTurck flow, in which case $\sigma = -Rg + \mathcal{L}_w g$ and w is a certain vector field depending on g.

In both Ricci flow and Ricci-DeTurck flow, the problem can be recast as a coupled system of differential equations by treating the (densitized) scalar curvature R and the metric g as independent variables. As we show below, the system reads

$$\frac{\partial}{\partial t}(R\mu) = (\operatorname{div}_g \operatorname{div}_g S_g \sigma)\mu, \qquad\qquad R(0) = R_0, \qquad (2)$$

$$\frac{\partial}{\partial t} g = \sigma, \qquad\qquad g(0) = g_0, \qquad (3)$$

where R_0 is the scalar curvature of g_0, div_g is the covariant divergence operator, $\mu = \mu(g)$ is the volume form on Ω determined by g, and $(S_g \sigma)_{ij} = \sigma_{ij} - g_{ij} g^{k\ell} \sigma_{k\ell}$. An advantage of this formulation is that it eliminates the need to discretize the scalar curvature operator (the nonlinear second-order differential operator sending g to R). The scalar curvature R is instead initialized at $t = 0$ and evolved forward in time by solving the differential Eq. (2). The latter equation involves a differential operator $\operatorname{div}_g \operatorname{div}_g$ which is somewhat easier to discretize.

To fix ideas, let us consider the setting in which Ω is a 2-torus. Let $\mathcal{T}_h$ be a triangulation of Ω with maximum element diameter h. Assume that $\mathcal{T}_h$ belongs to a shape-regular, quasi-uniform family of triangulations parametrized by h. Let $\mathcal{E}_h$ denote the set of edges of $\mathcal{T}_h$. Let $q \in \mathbb{N}$ and $r \in \mathbb{N}_0$. Define finite element spaces

$$V_h = \{v \in H^1(\Omega) \mid v|_K \in \mathcal{P}_q(K), \forall K \in \mathcal{T}_h\},$$
$$\Sigma_h = \{\sigma \in L^2(\Omega) \otimes \mathbb{S} \mid \sigma|_K \in \mathcal{P}_r(K) \otimes \mathbb{S}, \forall K \in \mathcal{T}_h, \text{ and } [\![\tau^T \sigma \tau]\!] = 0, \forall e \in \mathcal{E}_h\},$$

where $\mathcal{P}_r(K)$ denotes the space of polynomials of degree $\leq r$ on K, $[\![\tau^T \sigma \tau]\!]$ denotes the jump in the tangential-tangential component of σ across an edge $e \in \mathcal{E}_h$, and $\mathbb{S} = \{\sigma \in \mathbb{R}^{2 \times 2} \mid \sigma = \sigma^T\}$. The space Σ_h is the space of *Regge finite elements* of degree r [4,13].

For scalar fields u and v on Ω, denote $\langle u, v \rangle_g = \int_\Omega uv\, \mu(g)$. For symmetric $(0,2)$-tensor fields σ and ρ defined on $K \in \mathcal{T}_h$, let $\langle \sigma, \rho \rangle_{g,K} = \int_K g^{ij} \sigma_{jk} g^{k\ell} \rho_{\ell i}\, \mu(g)$. For $e \in \mathcal{E}_h$, denote $\langle u, v \rangle_{g,e} = \int_e uv \sqrt{\tau^T g \tau}\, d\ell$, where τ is the unit vector tangent to e relative to the Euclidean metric δ, and $d\ell$ is the Euclidean line element along e. With $J = \left(\begin{smallmatrix} 0 & 1 \\ -1 & 0 \end{smallmatrix}\right)$, let $\tau_g = \tau / \sqrt{\tau^T g \tau}$, $n_g = Jg\tau / \sqrt{\tau^T g \tau \det g}$, and $\frac{\partial v}{\partial n_g} = n_g^T g \nabla_g v$. Let $\operatorname{Hess}_g v$ denote the Riemannian Hessian of v.

To discretize the operator $\operatorname{div}_g \operatorname{div}_g S_g$ appearing in (2), we make use of the metric-dependent bilinear form

$$b_h(g; \sigma, v) = \sum_{K \in \mathcal{T}_h} \langle S_g \sigma, \operatorname{Hess}_g v \rangle_{g,K} + \sum_{e \in \mathcal{E}_h} \left\langle \tau_g^T \sigma \tau_g, \left[\!\!\left[\frac{\partial v}{\partial n_g} \right]\!\!\right] \right\rangle_{g,e}.$$

This bilinear form is a non-Euclidean generalization of the bilinear form used in the classical Hellan-Herrmann-Johnson mixed discretization of the biharmonic equation [9, p. 237]. Using integration by parts, it can be shown that for smooth g, σ, and v, we have $b_h(g; \sigma, v) = \int_\Omega (\text{div}_g \, \text{div}_g \, S_g \sigma) v \, \mu(g)$.

To discretize (2), (3), we choose approximations $R_{h0} \in V_h$ and $g_{h0} \in \Sigma_h$ of R_0 and g_0, respectively. We then seek $R_h(t) \in V_h$ and $g_h(t) \in \Sigma_h$ such that $R_h(0) = R_{h0}$, $g_h(0) = g_{h0}$, and

$$\frac{\partial}{\partial t} \langle R_h, v_h \rangle_{g_h} = b_h(g_h; \sigma_h, v_h), \qquad\qquad \forall v_h \in V_h, \qquad (4)$$

$$\frac{\partial}{\partial t} g_h = \sigma_h, \qquad\qquad\qquad\qquad (5)$$

where $\sigma_h = \sigma_h(g_h, R_h, t)$ is a discretization of σ. For the moment, we postpone discussing our choice of σ_h; this will be addressed in the next sections. We assume throughout what follows that (1) and (4), (5) preserve the signature of g and g_h, in the sense that the eigenvalues of g and g_h are bounded from below by a positive constant independent of h, x, and t.

1.1   Connection with the Angle Defect

An important feature of (4) is its connection with the widely studied *angle defect* from discrete differential geometry [2,5,14]. Recall that the angle defect Θ_i at the ith vertex $y^{(i)} \in \Omega$ of the triangulation $\mathcal{T}_h$ measures the failure of the angles incident at $y^{(i)}$ to sum up to 2π:

$$\Theta_i = 2\pi - \sum_{K \in \omega_i} \theta_{iK}. \qquad (6)$$

Here, ω_i denotes the set of triangles in $\mathcal{T}_h$ having $y^{(i)}$ as a vertex, and θ_{iK} denotes the interior angle of K at $y^{(i)}$. The following proposition shows that in the lowest order setting ($r = 0$ and $q = 1$), the differential Eq. (4) reproduces the angle defect if R_{h0} is chosen appropriately.

Proposition 1. *Let $r = 0$ and $q = 1$. Let $\{\phi_i\}_i$ be the basis for V_h satisfying $\phi_i(y^{(j)}) = \delta_i^j$, and let Θ_{i0} be the angle defect at vertex $y^{(i)}$ as measured by g_{h0}. If*

$$\langle R_{h0}, \phi_i \rangle_{g_{h0}} = 2\Theta_{i0}, \qquad (7)$$

then the solution of (4), (5) satisfies

$$\langle R_h(t), \phi_i \rangle_{g_h(t)} = 2\Theta_i(t)$$

for every t, where $\Theta_i(t)$ is the angle defect at vertex $y^{(i)}$ as measured by $g_h(t)$.

Proof. It is shown in [8, Lemma 3.3] that

$$\frac{\partial}{\partial t}(2\Theta_i(t)) = b_h\left(g_h(t); \frac{\partial}{\partial t} g_h(t), \phi_i\right), \qquad (8)$$

so $2\Theta_i(t)$ and $\langle R_h(t), \phi_i \rangle_{g_h(t)}$ obey the same ordinary differential equation.

The relation (8) is a discrete analogue of the following relation which holds in the smooth setting.

Proposition 2. *Let $g(t)$ be a smooth Riemannian metric on Ω depending smoothly on t. Then, for every smooth scalar field v,*

$$\frac{\partial}{\partial t}\langle R(g(t)), v\rangle_{g(t)} = \left\langle \operatorname{div}_{g(t)} \operatorname{div}_{g(t)} S_{g(t)}\frac{\partial g}{\partial t}, v\right\rangle_{g(t)}.$$

Remark 1. The relation above is not valid in dimensions greater than 2.

Proof. We have

$$\frac{\partial}{\partial t}\langle R(g(t)), v\rangle_{g(t)} = \int_\Omega (DR(g(t))\cdot\sigma(t))\, v\, \mu(g(t)) + \int_\Omega R(g(t))v\,(D\mu(g(t))\cdot\sigma(t)),$$

where $\sigma(t) = \frac{\partial}{\partial t}g(t)$. The linearizations of R and μ are given by [6, Lemma 2]

$$DR(g)\cdot\sigma = \operatorname{div}_g \operatorname{div}_g \sigma - \Delta_g(g^{ij}\sigma_{ij}) - g^{ij}\sigma_{jk}g^{k\ell}\operatorname{Ric}_{\ell i},$$

$$D\mu(g)\cdot\sigma = \frac{1}{2}g^{ij}\sigma_{ij}\mu(g).$$

Since $\operatorname{Ric} = \frac{1}{2}Rg$ in two dimensions and $\Delta_g u = \operatorname{div}_g \operatorname{div}_g(gu)$ for any scalar field u, the first expression simplifies to

$$DR(g)\cdot\sigma = \operatorname{div}_g \operatorname{div}_g S_g\sigma - \frac{1}{2}Rg^{ij}\sigma_{ij}.$$

Combining these gives

$$\frac{\partial}{\partial t}\langle R(g(t)), v\rangle_{g(t)} = \int_\Omega (\operatorname{div}_g \operatorname{div}_g S_g\sigma)\, v\, \mu.$$

2   Ricci Flow

Let us now focus on two-dimensional normalized Ricci flow, which corresponds to the choice $\sigma = (\bar{R} - R)g$ in (1). As before, R is the scalar curvature of g and $\bar{R}$ is the average of R over Ω (or some other prescribed scalar function).

Several simplifications can be made in this setting. Since σ is proportional to g, we have $\operatorname{div}_g \operatorname{div}_g S_g\sigma = \Delta_g(\bar{R} - R) - 2\Delta_g(\bar{R} - R) = \Delta_g(R - \bar{R})$, so that (2) reduces to

$$\frac{\partial}{\partial t}(R\mu) = (\Delta_g(R - \bar{R}))\mu.$$

This offers us some flexibility in our choice of discretization. One option is to use (4), (5) as it is written, choosing σ_h equal to

$$\sigma_h = P_h((\bar{R}_h - R_h)g_h), \tag{9}$$

where P_h is any projector onto Σ_h whose domain contains $\{v_h\rho_h \mid v_h \in V_h, \rho_h \in \Sigma_h\}$, and $\bar{R}_h \in V_h$ is equal to $\bar{R}$ or an approximation thereof. Another option is to use the discretization

$$\frac{\partial}{\partial t}\langle R_h, v_h\rangle_{g_h} = \langle \nabla_{g_h}(\bar{R}_h - R_h), \nabla_{g_h} v_h\rangle_{g_h}, \qquad \forall v_h \in V_h, \qquad (10)$$

$$\frac{\partial}{\partial t} g_h = \sigma_h, \qquad (11)$$

again with σ_h given by (9).

The next proposition gives an example of a setting in which (4), (5) and (10), (11) are equivalent. In it, we denote by $z^{(e)} \in \Omega$ the midpoint of an edge $e \in \mathcal{E}_h$. Note that when $r = 0$, the linear functionals

$$\rho \mapsto \tau^T \rho(z^{(e)})\tau, \quad e \in \mathcal{E}_h$$

form a basis for the dual of Σ_h. We denote by $\{\psi_e\}_{e \in \mathcal{E}_h} \subset \Sigma_h$ the basis for Σ_h satisfying

$$\tau^T \psi_e(z^{(e')})\tau = \begin{cases} 1, & \text{if } e = e', \\ 0, & \text{otherwise.} \end{cases}$$

Proposition 3. *Let $r = 0$ and $q = 1$. Let P_h be given by*

$$P_h\rho = \sum_{e \in \mathcal{E}_h}(\tau^T\rho(z^{(e)})\tau)\psi_e,$$

and let σ_h be given by (9). Choose R_{h0} equal to the unique element of V_h satisfying (7) for every i. Then, with initial conditions $R_h(0) = R_{h0}$ and $g_h(0) = g_{h0}$, problems (4), (5) and (10), (11) are equivalent. Furthermore, the solution $g_h(t)$ satisfies

$$g_h(t) = P_h(e^{u_h(t)}g_{h0}), \qquad (12)$$

where $u_h(t) \in V_h$ obeys the differential equation

$$\frac{\partial}{\partial t}u_h = \bar{R}_h - R_h, \quad u_h(0) = 0, \qquad (13)$$

and the solution $R_h(t)$ satisfies

$$\langle R_h(t), \phi_i\rangle_{g_h(t)} = 2\Theta_i(t) \qquad (14)$$

for every t and every i, where $\Theta_i(t)$ is the angle defect at vertex $y^{(i)}$ as measured by $g_h(t)$.

Proof. Using the fact that functions in V_h are piecewise linear when $q = 1$, one verifies through integration by parts that

$$b_h(g_h; P_h((\bar{R}_h - R_h)g_h), v_h) = b_h(g_h; (\bar{R}_h - R_h)g_h, v_h)$$
$$= \langle \nabla_{g_h}(\bar{R}_h - R_h), \nabla_{g_h} v_h\rangle_{g_h}$$

for every $v_h \in V_h$. This demonstrates the equivalence of (4), (5) and (10), (11). To deduce (12), (13), observe that differentiating (12) and invoking (13) gives

$$\frac{\partial}{\partial t} g_h = P_h \left((\bar{R}_h - R_h) e^{u_h} g_{h0} \right)$$
$$= P_h \left((\bar{R}_h - R_h) P_h (e^{u_h} g_{h0}) \right)$$
$$= P_h \left((\bar{R}_h - R_h) g_h \right)$$
$$= \sigma_h$$

where the second line above follows from our choice of P_h. The relation (14) between $R_h(t)$ and the angle defect follows from Proposition 1.

2.1   Connection with Other Discretizations of Ricci Flow

Proposition 3 reveals a close connection between the lowest-order version of our finite element discretization of Ricci flow and another popular finite difference scheme for Ricci flow [3,11]. In this popular method, (Ω, g) is discretized with a triangulation having time-dependent edge lengths ℓ_{ij} between adjacent vertices i and j. The scalar curvature $R(g)$ (which is twice the Gaussian curvature) is then approximated by (two times) the angle defect. The method stores a time-dependent scalar u_i at each vertex i which evolves according to

$$\frac{\partial}{\partial t} u_i = 2(\bar{\Theta}_i - \Theta_i). \tag{15}$$

where $\bar{\Theta}_i$ is prescribed. (Note that in [3], (15) is expressed in terms of $r_i := e^{u_i/2}$ rather than u_i.) This collection of scalars determines the lengths ℓ_{ij} of all edges at time t in terms of their lengths at $t = 0$ via a relation which is analogous to (12) but is motivated by circle packing theory [12] rather than finite element theory. (Other choices are also possible; see [10, Section 5] and [15] for a discussion of alternatives.)

The connection with our finite element discretization is now transparent. In the lowest order instance of our finite element discretization ($r = 0$ and $q = 1$), the degrees of freedom for u_h and g_h are the values of u_h at each vertex and the squared length of each edge as measured by g_h. According to Eq. (13) and (12), these degrees of freedom evolve in nearly the same way that u_i and ℓ_{ij} evolve in [3,11].

There is one important discrepancy, however: Equation (15) is not a consistent discretization of normalized Ricci flow. This is because the angle defect (6) approximates the *integral* of the Gaussian curvature over a cell which is dual to vertex i, not its average over the cell. See [1, Remark B.2.4] for more insight. In many applications, this is not a serious concern, since very often the goal is not to accurately approximate Ricci flow, but rather to construct a discrete conformal mapping from a given triangulation to one with prescribed discrete curvature.

Putting this discrepancy aside, the similarities noted above suggest that our finite element method (with $r \geq 0$ and $q \geq 1$) can be loosely regarded as a high-order generalization of the scheme studied in [3,11]. A link like this does not appear to hold for some other finite element discretizations of Ricci flow such as the one studied in [7]. In particular, [7] relies on the existence of an embedding of (Ω, g) into $\mathbb{R}^3$.

3 Error Analysis

We now discuss some of our ongoing work on the analysis of the accuracy of the discretization (4), (5). One setting which is particularly easy to analyze is that in which σ and σ_h are prescribed functions of t. Then estimates for $g_h - g$ are immediate, and it remains to estimate $R_h - R$. The following proposition gives estimates for $R_h - R$ in the metric-dependent negative-order Sobolev-norm (recall that Ω has no boundary)

$$\|v\|_{H^{-1}(\Omega,g)} = \sup_{u \in H^1(\Omega)} \frac{\langle v, u \rangle_g}{\|u\|_{H^1(\Omega)}}. \tag{16}$$

In what follows, we take $\bar{R} = \bar{R}_h$ to be constant, and we assume $r > 0$.

Proposition 4. *If σ and σ_h depend only on t, and if g and R are sufficiently regular, then for $T > 0$ small enough, the solutions of (2), (3) and (4), (5) satisfy*

$$\|g_h(T) - g(T)\|_{L^2(\Omega)} \leq \|g_{h0} - g_0\|_{L^2(\Omega)} + \int_0^T \|\sigma_h(t) - \sigma(t)\|_{L^2(\Omega)} \, dt,$$

$$\|R_h(T) - R(T)\|_{H^{-1}(\Omega,g(T))}$$

$$\leq C \Bigg(\int_0^T \left(h^{-1} \|\sigma_h(t) - \sigma(t)\|_{L^2(\Omega)} + |\sigma_h(t) - \sigma(t)|_{H^1_h(\Omega)} \right) dt$$

$$+ \inf_{u_h \in V_h} \|R(T) - u_h\|_{H^{-1}(\Omega,g(T))} + \|R_{h0} - R_0\|_{H^{-1}(\Omega,g(T))} \Bigg).$$

Proof. The estimate for $g_h(T) - g(T)$ is immediate, and the estimate for $R_h(T) - R(T)$ can be obtained by extending the analysis in [8], which studies the case in which $g(t) = \frac{T-t}{T}\delta + \frac{t}{T}g(T)$, $g_h(t) = \frac{T-t}{T}\delta + \frac{t}{T}g_h(T)$, and $R_{h0} = R_0 = 0$.

Choosing g_{h0}, R_{h0}, and $\sigma_h(t)$ equal to suitable interpolants of g_0, R_0, and $\sigma(t)$, one obtains from Proposition 4 estimates of the form

$$\|g_h(T) - g(T)\|_{L^2(\Omega)} \leq Ch^{r+1},$$

$$\|R_h(T) - R(T)\|_{H^{-1}(\Omega,g(T))} \leq C(h^r + h^{q+2})$$

for sufficiently regular solutions.

4 Numerical Examples

Figure 1 plots a numerical simulation of normalized Ricci flow obtained using the finite element method (4), (5) with the parameter choices described in Proposition 3. Here, $\mathcal{T}_h$ was taken equal to a triangulation of a 2-sphere rather than a 2-torus. At each instant $t \geq 0$, we visualized $g_h(t)$ by numerically determining an embedding of the vertices of $\mathcal{T}_h$ into $\mathbb{R}^3$ with the property that the distances between adjacent vertices agree with the edge lengths determined by $g_h(t)$.

Fig. 1. Numerical solution at $t = 0$, $t = 0.05$, and $t = 0.75$.

References

1. Bobenko, A.I., Pinkall, U., Springborn, B.A.: Discrete conformal maps and ideal hyperbolic polyhedra. Geom. Topol. **19**(4), 2155–2215 (2015)
2. Cheeger, J., Müller, W., Schrader, R.: On the curvature of piecewise flat spaces. Commun. Math. Phys. **92**(3), 405–454 (1984)
3. Chow, B., Luo, F.: Combinatorial Ricci flows on surfaces. J. Differ. Geom. **63**(1), 97–129 (2003)
4. Christiansen, S.H.: On the linearization of Regge calculus. Numer. Math. **119**(4), 613–640 (2011)
5. Crane, K., Desbrun, M., Schröder, P.: Trivial connections on discrete surfaces. Computer Graphics Forum, vol. 29, pp. 1525–1533. Wiley Online Library, New Jersey (2010)
6. Fischer, A.E., Marsden, J.E.: Deformations of the scalar curvature. Duke Math. J. **42**(3), 519–547 (1975)
7. Fritz, H.: Numerical Ricci-DeTurck flow. Numer. Math. **131**(2), 241–271 (2015)
8. Gawlik, E.S.: High order approximation of Gaussian curvature with Regge finite elements. arXiv preprint arXiv:1905.07004 (2019)
9. Girault, V., Raviart, P.A.: Finite Element Methods for Navier-Stokes Equations: Theory and Algorithms. Springer-Verlag, Berlin (1986). https://doi.org/10.1007/978-3-642-61623-5
10. Glickenstein, D.: Discrete conformal variations and scalar curvature on piecewise flat two-and three-dimensional manifolds. J. Differ. Geom. **87**(2), 201–238 (2011)
11. Jin, M., Kim, J., Luo, F., Gu, X.: Discrete surface Ricci flow. IEEE Trans. Vis. Comput. Graph. **14**(5), 1030–1043 (2008)

12. Kharevych, L., Springborn, B., Schröder, P.: Discrete conformal mappings via circle patterns. ACM Trans. Graph. (TOG) **25**(2), 412–438 (2006)
13. Li, L.: Regge Finite Elements with Applications in Solid Mechanics and Relativity. Ph.D. thesis, University of Minnesota (2018)
14. Regge, T.: General relativity without coordinates. Il Nuovo Cimento (1955–1965) **19**(3), 558–571 (1961)
15. Springborn, B., Schröder, P., Pinkall, U.: Conformal equivalence of triangle meshes. ACM Trans. Graph. (TOG) **27**(3), 77 (2008)

Local Truncation Error of Low-Order Fractional Variational Integrators

Fernando Jiménez$^{(\boxtimes)}$ and Sina Ober-Blöbaum

Department of Engineering Science, University of Oxford,
Parks Road, Oxford OXI 3PJ, UK
{fernando.jimenez,sina.ober-blobaum}@eng.ox.ac.uk

Abstract. We study the local truncation error of the so-called fractional variational integrators, recently developed in [1,2] based on previous work by Riewe and Cresson [3,4]. These integrators are obtained through two main elements: the enlarging of the usual mechanical Lagrangian state space by the introduction of the fractional derivatives of the dynamical curves; and a discrete restricted variational principle, in the spirit of discrete mechanics and variational integrators [5]. The fractional variational integrators are designed for modelling fractional dissipative systems, which, in particular cases, reduce to mechanical systems with linear damping. All these elements are introduced in the paper. In addition, as original result, we prove (Sect. 3, Theorem 2) the order of local truncation error of the fractional variational integrators with respect to the dynamics of mechanical systems with linear damping.

Keywords: Fractional variational integrators · Dissipative systems · Local truncation error

1 Preliminaries

1.1 Local Truncation Error

Let $z : [a, b] \to \mathbb{R}^d$ and $f : \mathbb{R}^d \to \mathbb{R}^d$ a smooth curve and a smooth vector field, respectively, for $d \in \mathbb{N}$ and $[a, b] \subset \mathbb{R}$. Using the usual dot notation as time derivative we can define the initial value problem:

$$\dot{z} = f(z), \quad z(a) = z_0, \tag{1}$$

$z_0 \in \mathbb{R}^d$, with smooth solution $z(t) \subset \mathbb{R}^d$. On the other hand, we define an implicit one-step numerical method:

$$z_{k+1} = z_k + h\, f_h(z_k, z_{k+1}, h), \tag{2}$$

This work has been funded by the EPSRC project: "Fractional Variational Integration and Optimal Control"; ref: EP/P020402/1.

F. Nielsen and F. Barbaresco (Eds.): GSI 2019, LNCS 11712, pp. 541–548, 2019.
https://doi.org/10.1007/978-3-030-26980-7_56

where $h \in \mathbb{R}_+$ is the time step, $f_h : \mathbb{R}^d \times \mathbb{R}^d \times \mathbb{R}_+ \to \mathbb{R}^d$ is smooth, and z_k is considered an approximation of $z(t_k)$ for the time grid $t_k = \{a + hk \mid k = 0, \cdots, N\}$, with $N = (b - a)/h$. We say that the **local truncation error order** of the method (2) with respect to (1) **is** p if

$$||z(t_{k+1}) - z_{k+1}|| = O(h^{p+1}), \tag{3}$$

when $h \to 0$ and where $|| \cdot ||$ is the Euclidean norm in $\mathbb{R}^d$ [6].

1.2   Conservative Mechanical Systems

The dynamics of conservative simple mechanical systems, subject to a potential force [7], is described by the second-order differential equation:

$$m\ddot{x} = -\nabla U(x), \quad x(a) = x_0, \quad \dot{x}(a) = v_0, \tag{4}$$

where $x_0, v_0 \in \mathbb{R}$, $m \in \mathbb{R}_+$ is the mass of the system (for simplicity, we will set $m = 1$), $x : [a, b] \to \mathbb{R}^1$ is the dynamical curve and the potential energy $U : \mathbb{R} \to \mathbb{R}$ is a smooth function. This equation can be transformed into a first-order differential equation:

$$\begin{aligned} \dot{x} &= v, \\ \dot{v} &= -\nabla U(x), \quad x(a) = x_0, \quad v(a) = v_0. \end{aligned} \tag{5}$$

The dynamical equation (4) can be obtained as a critical condition for extremals from the Hamilton's principle [8], given the action integral

$$S(x) = \int_a^b L(x(t), \dot{x}(t)) \, dt \tag{6}$$

for a Lagrangian function $L : T\mathbb{R} \to \mathbb{R}$ (we shall consider the tangent bundle $T\mathbb{R}$, i.e. the state space, as the space locally isomorphic to $\mathbb{R} \times \mathbb{R}$, with coordinates $(x, \dot{x})$) defined by

$$L(x, \dot{x}) = \frac{1}{2}\dot{x}^2 - U(x). \tag{7}$$

Remarkable geometric properties of the flow generated by (4) (equivalently (5)) are its symplecticity (it preserves the symplectic form $\Omega_L := dx \wedge d\dot{x} = dx \wedge dv \in \bigwedge^2(T\mathbb{R})$) and the preservation of symmetries (Noether's theorem) [8].

Remark 1. Observe that we are choosing a "Lagrangian version" (5) of Hamilton equations for simple mechanical systems. In the picked setup, i.e. the configuration manifold is the real space and the particular Lagrangian function (7), both Lagrangian and Hamiltonian pictures are equivalent. Therefore, the theorems about the local truncation error order of variational integrators in [5,9], apply for (5).

---

[1] We will restrict to the real space $\mathbb{R}$ for sake of simplicity, but all results in these paper are straightforwardly extended to $\mathbb{R}^d$.

1.3   Variational Integrators

A natural way of obtaining integrators preserving the symplectic form Ω_L and the symmetries of the system is to construct **variational integrators** [5]. For that, we replace the continuous curves $x(t)$ by discrete ones $x_d = \{x_k\}_{0:N} := \{x_0, x_1, \cdots, x_N\} \in \mathbb{R}^{N+1}$. Moreover, we define the discrete Lagrangian $L_d :$ $\mathbb{R} \times \mathbb{R} \to \mathbb{R}$ as an approximation in one time step of the action integral (6), say

$$L_d(x_k, x_{k+1}, h) \simeq \int_{t_k}^{t_k+h} L(x(t), \dot{x}(t))\, dt, \tag{8}$$

where we shall omit the h dependence of the discrete Lagrangian unless needed. Given this, we define the discrete action sum $S_d(x_d) = \sum_{k=0}^{N-1} L_d(x_k, x_{k+1})$; the discrete Hamilton's principle applied upon this action sum provides the so-called discrete Euler-Lagrange equations:

$$D_1 L_d(x_k, x_{k+1}) + D_2 L_d(x_{k-1}, x_k) = 0, \quad k = 1, ..., N-1, \tag{9}$$

which, under the condition $[D_{12}L_d]$ is regular, define a discrete flow $F_{L_d} : \mathbb{R} \times \mathbb{R} \to \mathbb{R} \times \mathbb{R}$; $(x_k, x_{k+1}) \mapsto (x_{k+1}, x_{k+2})$, that we call **variational integrator**. Alternatively, the transformation[2]:

$$\begin{aligned} v_k^- &= -D_1 L_d(x_k, x_{k+1}), \\ v_{k+1}^+ &= D_2 L_d(x_k, x_{k+1}), \end{aligned} \tag{10}$$

defines an alternate discrete flow $\tilde{F}_{L_d} : \mathbb{R} \times \mathbb{R} \to \mathbb{R} \times \mathbb{R}$; $(x_k, v_k) \mapsto (x_{k+1}, v_{k+1})$, which, when we pick the Lagrangian (7), will be a variational integrator for (5) (observe that, in the general case, the "velocity matching" condition $v_k^- = v_k^+$ reproduces the discrete Euler-Lagrange equations (9)). As mentioned above, $\tilde{F}_{L_d}$ is symplectic and momentum preserving. Moreover, the symplecticity ensures a bounded energy behaviour in the long-term, which is explained by Backward Error Analysis [6]. Another advantage of the variational approach is that the local truncation error order of the integrators can be determined from the approximation in (8). In particular, we can establish the following result, which is a direct application of the order theorems in [5] and [9]:

Theorem 1. *Given the Lagrangian $L(x(t), \dot{x}(t))$ (7) and $L_d(x_k, x_{k+1})$ an order p approximation of the action integral (8), then the local truncation error of the variational integrator $\tilde{F}_{L_d}$ determined by (10) with respect to (5) is of order p.*

Low-order integrators (up to 2)[3] can be obtained through a first order quadrature and the following linear interpolation between the points $[x_k, x_{k+1}]$: $\dot{x}(t_k) \simeq (x_{k+1} - x_k)/h$ and $x(t_k) \simeq \gamma x_k + (1 - \gamma)x_{k+1}$, where $\gamma \in [0, 1] \subset \mathbb{R}$. Namely:

$$L_d(x_k, x_{k+1}) = \frac{1}{2h}(x_{k+1} - x_k)^2 - hU(\gamma x_k + (1 - \gamma)x_{k+1}).$$

---

[2] Naturally, this transformation is nothing but the discrete Legendre transform [5], which is shown here in a Lagrangian version.

[3] High-order variational integrators can be obtained via the use of inner discrete nodes and more involved interpolations, see [10].

From this discrete Lagrangian, the discrete Euler-Lagrange equations read:

$$\frac{x_{k+1} - 2x_k + x_{k-1}}{h^2} = -\gamma \nabla U(\gamma x_k + (1-\gamma)x_{k+1}) - (1-\gamma)\nabla U(\gamma x_{k-1} + (1-\gamma)x_k),$$

which are a discretization in finite differences of (4); whereas the flow $\tilde{F}_{L_d}$ defined by (10) reads:

$$\begin{aligned} x_{k+1} &= x_k + hv_k - h^2\gamma\nabla U(\gamma x_k + (1-\gamma)x_{k+1}), \\ v_{k+1} &= v_k - h\nabla U(\gamma x_k + (1-\gamma)x_{k+1}). \end{aligned} \tag{11}$$

Using the Taylor expansion and the definition in Sect. 1.1, it is easy to see that the order 2 of this integrator w.r.t. (5) is achieved when $\gamma = 1/2$, i.e. for the midpoint rule, circumstance which is consistent with Theorem 1.

1.4 Linearly Damped Mechanical Systems

The dynamical equations of a mechanical system subject to linear damping are:

$$\ddot{x} = -\nabla U(x) - \rho \dot{x}, \quad x(a) = x_0, \quad \dot{x}(a) = v_0, \tag{12}$$

with $\rho \in \mathbb{R}_+$. In the first-order version:

$$\begin{aligned} \dot{x} &= v, \\ \dot{v} &= -\nabla U(x) - \rho v, \quad x(a) = x_0, \quad v(a) = v_0. \end{aligned} \tag{13}$$

There is no Lagrangian function such that (12) are its Euler-Lagrange equations [11]. With our fractional approach [1,2], explained in Sect. 2, we have designed a restricted variational principle surpassing this issue.

2   Fractional Variational Integrators

2.1   Continuous and Discrete Fractional Derivatives

Given a smooth function $g : [a, b] \to \mathbb{R}$, the α-fractional derivatives (Riemann-Liouville version), with $\alpha \in [0, 1]$ are:

$$D_-^\alpha g(t) = \frac{1}{\Gamma(1-\alpha)}\frac{d}{dt}\int_a^t (t-\tau)^{-\alpha}g(\tau)d\tau, \quad D_+^\alpha g(t) = -\frac{1}{\Gamma(1-\alpha)}\frac{d}{dt}\int_t^b (\tau-t)^{-\alpha}g(\tau)d\tau,$$

where $\Gamma(z)$ is the Gamma function [12]. Relevant properties in our approach are

$$\int_a^b h(t)D_\sigma^\alpha g(t)dt = \int_a^b \left(D_{-\sigma}^\alpha h(t)\right)g(t)dt, \quad D_-^{1/2}D_-^{1/2}g(t) = \dot{g}(t), \quad D_+^{1/2}D_+^{1/2}g(t) = -\dot{g}(t), \tag{14}$$

with $\sigma = \{-, +\}$. On the other hand, for a discrete curve $\{x_k\}_{0:N}$ and the time step $h \in \mathbb{R}_+$, we can define the following discrete α-fractional derivatives [4]:

$$\Delta_-^\alpha x_k := \frac{1}{h^\alpha}\sum_{n=0}^k \alpha_n x_{k-n}, \qquad \Delta_+^\alpha x_k := \frac{1}{h^\alpha}\sum_{n=0}^{N-k} \alpha_n x_{k+n}, \tag{15}$$

where $\alpha_n := -\alpha(1-\alpha)(2-\alpha)\cdots(n-1-\alpha)/n!$ and $\alpha_0 := 1$. It is proven in [13] (Theorem 2.4) that $\Delta_\pm^\alpha x_k$ is an order 0 approximation (i.e. consistent) of $D_\pm^\alpha x(t)$.

2.2   Continuous Restricted Variational Principle

In [1,2], the fractional state space $T\mathbb{R}$ is defined, which is a vector bundle over $\mathbb{R} \times \mathbb{R}$ with coordinates $(x, y, \dot{x}, \dot{y}, D^\alpha_- x, D^\alpha_+ y)$ over the point (x, y). This is an extension of the usual tangent bundle, including the fractional derivatives after doubling the space of curves (note that we are considering an extra curve $y(t)$). The necessity of this doubling comes out of the assymetric integration by parts rule in (14). Given this fractional phase space, we define a Lagrangian function $\mathcal{L} : T\mathbb{R} \to \mathbb{R}$ and the action integral:

$$S((x, y)) = \int_a^b \mathcal{L}(x(t), y(t), \dot{x}(t), \dot{y}(t), D^\alpha_- x(t), D^\alpha_+ y(t))\, dt. \qquad (16)$$

Using a particular set of varied curves $(x(t), y(t))_\epsilon := (x(t), y(t)) + \epsilon(\delta x(t), \delta x(t))$ (observe that we are "restricting" the variations of both curves to be equal), where $\epsilon \in \mathbb{R}_+$ and $\delta x : [a, b] \to \mathbb{R}$ is smooth and defined such that $\delta x(a) = \delta x(b) = 0$, and considering the extremal condition of the action as $d/d\epsilon\big|_{\epsilon=0} S((x, y)_\epsilon) = 0$, we obtain the next result.

Proposition 1. *Given the Lagrangian function*

$$\mathcal{L}(x, y, \dot{x}, \dot{y}, D^\alpha_- x, D^\alpha_+ y) = \left(\frac{1}{2}\dot{x}^2 - U(x)\right) + \left(\frac{1}{2}\dot{y}^2 - U(y)\right) - \rho D^\alpha_- x D^\alpha_+ y, \quad (17)$$

then, a sufficient condition for the extremals of (16) *subject to restricted variations* $(x, y)_\epsilon$ *are the equations:*

$$\begin{aligned}
\ddot{x} &= -\nabla U(x) - \rho\, D^\alpha_- D^\alpha_- x \to (\alpha = 1/2) \to \ddot{x} = -\nabla U(x) - \rho\,\dot{x}, \\
\ddot{y} &= -\nabla U(y) - \rho\, D^\alpha_+ D^\alpha_+ y \to (\alpha = 1/2) \to \ddot{y} = -\nabla U(y) + \rho\,\dot{y}.
\end{aligned} \qquad (18)$$

The previous equations are the so-called **restricted fractional Euler-Lagrange equations** in [1,2] (see these references for the proof) for the particular Lagrangian (17). It can be rigorously proven that the y-system is nothing but the x-system in reversed time (even for more general Lagrangians), and therefore these equations do not imply extra physics. For a general α we obtain the equations of a mechanical system subject to fractional damping. When $\alpha \to 1/2$, according to (14), we recover the dynamics of systems with linear damping (12).

2.3   Discrete Restricted Variational Principle

Given discrete sequences $x_d = \{x_k\}_{0:N}$ and $y_d = \{y_k\}_{0:N}$ and defining $\dot{x}_k := (x_{k+1} - x_k)/h$; equiv. y_k; $x_{k+1/2} := (x_{k+1} + x_k)/2$ and $x_{k-1/2} := (x_k + x_{k-1})/2$; equiv. $y_{k\pm1/2}$; (we pick the midpoint rule because it provides the maximum order of (11) w.r.t. (5)), the discrete action sum for the fractional problem is

$$S_d((x_d, y_d)) = \sum_{k=0}^{N-1} h\mathcal{L}(x_{k+1/2}, y_{k+1/2}, \dot{x}_k, \dot{y}_k, \Delta^\alpha_- x_k, \Delta^\alpha_+ y_k), \qquad (19)$$

where the discrete fractional derivatives are defined in (15). As in the continuous case, we pick a particular set of restricted discrete variations, namely $(x_d, y_d)_\epsilon := (\{x_k\}_{0:N}, \{y_k\}_{0:N}) + \epsilon(\{\delta x_k\}_{0:N}, \{\delta x_k\}_{0:N})$, where $\{\delta x_k\}_{0:N}$ is such that $\delta x_0 = \delta x_N = 0$. Considering the extremal condition of the discrete action as $d/d\epsilon\big|_{\epsilon=0} S_d((x_d, y_d)_\epsilon) = 0$, we get the next result:

Proposition 2. *Given the Lagrangian $\mathcal{L}$ (17), a sufficient condition for the extremals of (19) subject to restricted variations $(x_d, y_d)_\epsilon$ is*

$$
\begin{aligned}
\frac{x_{k+1} - 2x_k + x_k}{h^2} &= -\frac{1}{2}\nabla U(x_{k+1/2}) - \frac{1}{2}\nabla U(x_{k-1/2}) - \rho\, \Delta_-^\alpha \Delta_-^\alpha x_k, \\
\frac{y_{k+1} - 2y_k + y_k}{h^2} &= -\frac{1}{2}\nabla U(y_{k+1/2}) - \frac{1}{2}\nabla U(y_{k-1/2}) - \rho\, \Delta_+^\alpha \Delta_+^\alpha y_k,
\end{aligned}
\tag{20}
$$

for $k = 1, \cdots, N - 1$.

The previous equations are the so-called **discrete restricted fractional Euler-Lagrange equations** in [1,2] for the particular Lagrangian (17). In (20) we recognize a discretization in finite differences of (18) for a general α. Moreover, it can be also rigorously proven that the discrete y-system is the discrete x-system in reversed (discrete) time.

3   Order Result

As original result, we explore the local truncation error order of (20) with respect to (13). With that aim, we need to establish an equivalent to (10) in the fractional case. Based on [1,2], we pick (restricting to the x-system):

$$
v_k^- = \frac{x_{k+1} - x_k}{h} + \frac{h}{2}\nabla U(x_{k+1/2}) + h\rho\, \Delta_-^\alpha \Delta_-^\alpha x_k,
\tag{21a}
$$

$$
v_{k+1}^+ = \frac{x_{k+1} - x_k}{h} - \frac{h}{2}\nabla U(x_{k+1/2}).
\tag{21b}
$$

Note that the first two terms in the right hand side of both equations corresponds to (10) for $L_d(x_k, x_{k+1}) = (x_{k+1} - x_k)/2h - hU(x_{k+1/2})$. In addition, observe that the "velocity matching" condition $v_k^- = v_k^+$ reproduces the discrete dynamics (20). Finally, according to (15) it can be proven [1,2] that

$$
\Delta_-^{1/2} \Delta_-^{1/2} x_k = (x_k - x_{k-1})/h, \quad k = 1, \cdots, N.
\tag{22}
$$

With these elements, we can establish the following order result

Theorem 2. *The local truncation order of the fractional variational integrators for simple mechanical systems (21) when $\alpha = 1/2$, with respect to the continuous dynamics (13), is one.*

Proof. First, using Taylor expansions and setting the notation $x(t_k) := x_k$, $\dot{x}(t_k) := v_k$, $v(t_k) := v_k$, we deliver expressions for $x(t_{k+1})$ and $v(t_{k+1})$ in terms of the dynamics (13), namely:

$$x(t_k + h) = x_k + hv_k - \frac{h^2}{2}\nabla U(x_k) - \frac{h^2}{2}\rho v_k + O(h^3), \tag{23a}$$

$$v(t_k + h) = v_k - h\nabla U(x_k) - h\rho v_k - \frac{h^2}{2}\Delta U(x_k)v_k \tag{23b}$$

$$+ \frac{h^2}{2}\rho\nabla U(x_k) + \frac{h^2}{2}\rho^2 v_k + O(h^3).$$

On the other hand, from (21) we get the integrator:

$$x_{k+1} = x_k + hv_{k+1} + \frac{h^2}{2}\nabla U(x_{k+1/2}), \tag{24a}$$

$$v_{k+1} = v_k - h\nabla U(x_{k+1/2}) - h\rho\, \Delta_-^{1/2}\Delta_-^{1/2}x_k. \tag{24b}$$

Replacing (24b) into (24a) we get

$$x_{k+1} = x_k + hv_k - \frac{h^2}{2}\nabla U(x_{k+1/2}) - h^2\rho\Delta_-^{1/2}\Delta_-^{1/2}x_k$$

$$=^1 x_k + hv_k - \frac{h^2}{2}\nabla U(x_{k+1/2}) - h^2\rho\left(\frac{x_k - x_{k-1}}{h}\right)$$

$$=^2 x_k + hv_k - \frac{h^2}{2}\nabla U(x_k) - h^2\rho v_k + O(h^3),$$

where in $=^1$ we have used (22) and in $=^2$ we have used $x_{k+1/2} = x_k + hv_k/2 + O(h^2)$ and $(x_k - x_{k-1})/h = v_k + h\nabla U(x_{k-1/2})/2$, according to (21b). Thus, from the last expression and (23a), it follows that $\|x(t_{k+1}) - x_{k+1}\| = O(h^2)$. From (24b) we get

$$v_{k+1} = v_k - h\nabla U(x_{k+1/2}) - h\rho\, \Delta_-^{1/2}\Delta_-^{1/2}x_k = v_k - h\nabla U(x_{k+1/2}) - h\rho\left(\frac{x_k - x_{k-1}}{h}\right)$$

$$= v_k - h\nabla U(x_k) - h\rho v_k - \frac{h^2}{2}\Delta U(x_k)v_k - \frac{h^2}{2}\rho\nabla U(x_{k-1/2}) + O(h^3)$$

$$= v_k - h\nabla U(x_k) - h\rho v_k - \frac{h^2}{2}\Delta U(x_k)v_k - \frac{h^2}{2}\rho\nabla U(x_k) + O(h^3),$$

where we have taken into account that $x_{k-1/2} = x_k + O(h)$ and used the same expressions as above for $x_{k+1/2}$ and $(x_k - x_{k-1})/h$. From this last expression and (23b), we obtain that $\|v(t_{k+1}) - v_{k+1}\| = O(h^2)$, and the result follows from the definition of local truncation error in Sect. 1.1. □

Remark 2. The alternate integrator:

$$v_k^- = \frac{x_{k+1} - x_k}{h} + \frac{h}{2}\nabla U(x_{k+1/2}), \quad v_{k+1}^+ = \frac{x_{k+1} - x_k}{h} - \frac{h}{2}\nabla U(x_{k+1/2}) - h\rho\, \Delta_-^{\alpha}\Delta_-^{\alpha}x_{k+1},$$

which also reproduces (20) via velocity matching $v_k^- = v_k^+$, delivers the same result.

4   Conclusions

We prove that the local truncation error order of the fractional variational integrators (21), with respect to the dynamics of linearly damped mechanical systems (13), is one. These integrators are designed in the spirit of variational integrators [5], i.e. by means of the discretization of variational principles, in our case the Hamilton's principle with restricted variations. Thus, we expect similar behaviour in terms of Theorem 1, i.e. the order of the approximation of the action is equal to the order of the integrator. Our result is not coherent in the fractional case. On the one hand, we pick the midpoint rule approximation $x(t_k) \simeq (x_k + x_{k+1})/2$, which is the case where the maximum order (2) is achieved for the usual variational integrators and conservative mechanical systems. On the other, the approximation of the fractional derivative that we use, $\Delta_-^\alpha x_k$, is only consistent (order 0) w.r.t. $D_-^\alpha x(t)$ ([13], Theorem **2.4**). Thus, the approximation of the action (16) is limited to $O(h)$, whereas the integrator is $O(h^2)$. This represents an improvement from the expected result, and its numerical demonstration can be found in ([2],§5) through several simulations.

References

1. Jiménez, F., Ober-Blöbaum, S.: A fractional variational approach for modelling dissipative mechanical systems: continuous and discrete settings. IFAC-PapersOnLine **51**(3), 50–55 (2018). 6th IFAC LHMNC-2018 Proceedings
2. Jiménez, F., Ober-Blöbaum, S.: Fractional damping through restricted calculus of variations. Submitted arXiv:1905.05608 (2019)
3. Riewe, F.: Nonconservative Lagrangian and Hamiltonian mechanics. Phys. Rev. E **53**(2), 1890–1899 (1996)
4. Cresson, J., Inizan, P.: Variational formulations of differential equations and asymmetric fractional embedding. J. Math. Anal. Appl. **385**(2), 975–997 (2012)
5. Marsden, J.E., West, M.: Discrete mechanics and variational integrators. Acta Numerica **10**, 357–514 (2001)
6. Hairer, E., Lubich, C., Wanner, G.: Geometric Numerical Integration: Structure-Preserving Algorithms for Ordinary Differential Equations. Springer Series in Computational Mathematics, vol. 31. Springer, Heidelberg (2002). https://doi.org/10.1007/978-3-662-05018-7
7. Goldstein, H., Poole, C., Safko, J.: Classical Mechanics, 3rd edn. Addison Wesley, Reading (2001)
8. Abraham, R., Marsden, J.E.: Foundations of Mechanics. Benjamin-Cummings Publ. Co., San Francisco (1978)
9. Patrick, C.W., Cuell, C.: Error analysis of variational integrators of unconstrained Lagrangian systems. Numer. Math. **113**(2), 243–264 (2009)
10. Ober-Blöbaum, S., Saake, N.: Construction and analysis of higher order Galerkin variational integrators. Adv. Comput. Math. **41**(6), 955–986 (2015)
11. Bauer, P.S.: Dissipative dynamical systems. Proc. Nat. Acad. Sci. **17**, 311–314 (1931)
12. Samko, S., Kilbas, A., Marichev, O.: Fractional Integrals and Derivatives: Theory and Applications. Gordon and Breach, Yverdon (1993)
13. Meerschaert, M.M., Tadjeran, C.: Finite difference approximations for fractional advection-dispersion flow equations. J. Comput. Appl. Math. **172**, 65–77 (2004)

A Partitioned Finite Element Method for the Structure-Preserving Discretization of Damped Infinite-Dimensional Port-Hamiltonian Systems with Boundary Control

Anass Serhani, Denis Matignon, and Ghislain Haine$^{(\boxtimes)}$

ISAE-SUPAERO, Université de Toulouse, Toulouse, France
{anass.serhani,denis.matignon,ghislain.haine}@isae.fr

Abstract. Many boundary controlled and observed Partial Differential Equations can be represented as port-Hamiltonian systems with dissipation, involving a Stokes-Dirac geometrical structure together with constitutive relations. The Partitioned Finite Element Method, introduced in Cardoso-Ribeiro et al. (2018), is a structure preserving numerical method which defines an underlying Dirac structure, and constitutive relations in weak form, leading to finite-dimensional port-Hamiltonian Differential Algebraic systems (pHDAE). Different types of dissipation are examined: internal damping, boundary damping and also diffusion models.

Keywords: Port-Hamiltonian systems · Dissipation · Structure preserving method · Partitioned Finite Element Method

1 Introduction

In this work, we are interested in infinite-dimensional dynamical systems representing *open* physical systems, *i.e.* with control v_∂ and observation y_∂ located at the boundary $\partial\Omega$ of the geometrical domain $\Omega \subset \mathbb{R}^d$. When the corresponding closed physical system proves conservative w.r.t. a given Hamiltonian functional H, the open system is said to be *lossless*. When it proves dissipative, the open system is said to be *lossy*. Here we use the port-Hamiltonian formalism, introduced a few decades ago, see e.g. [9,17,21,22]. Note that very different multiphysics applications can be described through it, e.g. plasmas in tokamaks [27], or fluid structure interaction [6]. The underlying geometry of the dynamical systems relies on a so-called Stokes-Dirac structure, see [8]; for the system to be

This work has been performed in the frame of the Collaborative Research DFG and ANR project INFIDHEM, entitled *Interconnected Infinite-Dimensional systems for Heterogeneous Media*, n° ANR-16-CE92-0028. Further information is available at https://websites.isae-supaero.fr/infidhem/the-project/.

© Springer Nature Switzerland AG 2019
F. Nielsen and F. Barbaresco (Eds.): GSI 2019, LNCS 11712, pp. 549–558, 2019.
https://doi.org/10.1007/978-3-030-26980-7_57

well-defined, some constitutive equations have to be added to the geometrical structure.

Our main concern is to provide a numerical method that preserves, at the discrete level, the geometrical structure of the original controlled PDE; for short, we look for a structure-preserving method which automatically transforms the Stokes-Dirac structure into a finite-dimensional Dirac structure: in the last decade, quite a number of ways have been explored, see e.g. [10,13,19,20,26]. Recently in [4], a method based on the weak formulation of the Partial Differential Equation and the use of the celebrated Finite Element Method has emerged. One of its many advantages is the preservation of the geometrical structure. It has successfully been applied to 1-D and also n-D systems, linear and nonlinear systems, with uniform or space-varying coefficients; it enables to deal with scalar-valued fields, vector-valued fields and also tensor-valued fields. Wave equations are tackled in [4,25], Mindlin's or Kirchhoff's plate equations are considered in [2,3], the treatment of the shallow water equations together with a general presentation of the Partitioned Finite Element Method (PFEM) is to be found in [5]. However, only lossless open systems have been addressed up to now: thus, the present paper intends to enlarge the scope of PFEM to lossy open systems, based on dissipative closed systems. These can be nicely accounted for in the port-Hamiltonian framework by introducing specific interaction ports: resistive ports.

The paper is organized as follows: in Sect. 2 the structure preserving discretization procedure is presented on a damped wave equation (with both internal and boundary damping) by introduction of resistive ports, in Sect. 3 the extension is proposed to a diffusion model as another class of dissipative PDE. Conclusions are drawn and a few perspectives are given in Sect. 4.

2   A General Result of Structure-Preserving Discretization for Damped pHs

To fix ideas and notations, a simple 1-D PDE model borrowed from [28] is first recalled: the lossy transmission line, on domain $\Omega = (0, \ell)$.

Example 1: The Lossless Transmission Line. Let us choose as energy variables or state variables $q(z, t)$ the linear charge density, and $\varphi(z, t)$ the magnetic flux density. With uniform or space-varying coefficients $C(z)$ the distributed capacitance, and $L(z)$ the distributed inductance, let us define the Hamiltonian density $\mathcal{H}(q, \varphi) := \frac{1}{2}(\frac{q^2}{C} + \frac{\varphi^2}{L})$, and the Hamiltonian $H(q, \varphi) := \int_0^\ell \mathcal{H}(q, \varphi) \, dz$. With a slight abuse of notation, $H(q(t), \varphi(t))$ will be denoted $H(t)$ in the sequel. The co-energy variables are defined as the variational derivatives of the Hamiltonian w.r.t. the energy variables: $u_C := \delta_q H = \frac{q}{C}$ is the voltage, and $i_L := \delta_\varphi H = \frac{\varphi}{L}$ is the current. The conservation laws for the lossless transmission line read $\partial_t q = -\partial_z i_L$ and $\partial_t \varphi = -\partial_z u_C$. This can be rewritten in vector form $\partial_t \vec{X} = \mathcal{J}\, \delta_{\vec{X}} H$, or in a more compact and abstract form:

$$\vec{f} = \mathcal{J}\vec{e}, \quad \text{with } \mathcal{J} = \begin{pmatrix} 0 & -\partial_z \\ -\partial_z & 0 \end{pmatrix},$$

$$\text{and } \vec{X} = \begin{pmatrix} q \\ \varphi \end{pmatrix}, \quad \vec{e} := \delta_{\vec{X}} H = \begin{pmatrix} e_q \\ e_\varphi \end{pmatrix} = \begin{pmatrix} u_C \\ i_L \end{pmatrix}, \quad \vec{f} = \partial_t \vec{X} = \begin{pmatrix} \partial_t q \\ \partial_t \varphi \end{pmatrix}.$$

$\vec{e}$ are the effort, or co-energy variables, $\vec{f}$ are the flows, and $\mathcal{J}$ the interconnection matrix. It is easy to prove that $\mathcal{J}$ is a formally skew-symmetric differential operator on $L^2(0, \ell; \mathbb{R}^2)$. This results in a conservative closed system: indeed $d_t H = \int_0^\ell \partial_t \vec{X} \cdot \delta_{\vec{X}} H = \int_0^\ell \vec{X} \cdot \mathcal{J}\vec{X} = (\vec{f}, \vec{e}) = 0$, with the usual scalar product in $L^2(0, \ell; \mathbb{R}^2)$.

For the study of the open system, we need to introduce boundary ports at the boundary $\partial\Omega = \{0\} \times \{\ell\}$, such as $e_\partial := \big(e_q(0), e_q(\ell)\big)^\top$, $f_\partial := \big(e_\varphi(0), -e_\varphi(\ell)\big)^\top$. With $(\vec{e}, e_\partial) \in \mathcal{E}$ the effort space and $(\vec{f}, f_\partial) \in \mathcal{F}$ the flow space, we define the bond space $\mathcal{B} := \mathcal{E} \times \mathcal{F}$, and introduce a bilinear product on $\mathcal{B}$, namely:

$$<(\vec{e}, e_\partial), (\vec{f}, f_\partial)> := \int_0^\ell \vec{e} \cdot \vec{f} \ dz + (e_\partial, f_\partial)_{\mathbb{R}^2}.$$

A Dirac structure is a subset $\mathcal{D} \subset \mathcal{B}$ which is maximally isotropic w.r.t. the symmetrized product on $\mathcal{B} \times \mathcal{B}$, $<<(e_1, f_1), (e_2, f_2)>> := <e_1, f_2> + <e_2, f_1>$.

Proposition 1. *The subspace:*

$$\mathcal{D} := \{(e, f) \in \mathcal{B} \ | \ \vec{f} = \mathcal{J}\vec{e}, \ e_\partial := \big(e_q(0), e_q(\ell)\big)^\top, \ f_\partial := \big(e_\varphi(0), -e_\varphi(\ell)\big)^\top\},$$

is indeed a Stokes-Dirac structure.

As a consequence, the former conservative property of the closed system now generalizes into the following losslessness property for the open system:

$$d_t H(t) = -(e_\partial(t), f_\partial(t))_{\mathbb{R}^2}.$$

Example 2: The Lossy Transmission Line. Taking into account some losses with $R(z)$ the distributed resistance coefficient, leads to a new balance equation: $\partial_t q = -\partial_z i_L$ and $\partial_t \varphi = -\partial_z u_C - R i_L$. This can be first seen as a pHs with dissipation: $\partial_t \vec{X} = (\mathcal{J} - \mathcal{R}) \delta_{\vec{X}} H$ with some positive symmetric bounded operator $\mathcal{R}$, implying the dissipativity of the closed system: indeed, $d_t H = \int_0^\ell \partial_t \vec{X} \cdot \delta_{\vec{X}} H = \int_0^\ell \delta_{\vec{X}} H \cdot (\mathcal{J} - \mathcal{R}) \delta_{\vec{X}} H = -(\vec{e}, \mathcal{R}\vec{e}) \leq 0$.

But the construction of a Stokes-Dirac structure associated to it requires the definition of extra resistive ports (e_R, f_R) which will now be related by an extra *constitutive relation* $e_R = R f_R$. Let us consider the natural extension $\vec{f}_e = \mathcal{J}_e \vec{e}_e$, where $\vec{e}_e = (\vec{e}, e_R)$, $\vec{f}_e = (\vec{f}, f_R)$ and the extended interconnection operator:

$$\mathcal{J}_e = \begin{pmatrix} 0 & -\partial_z & 0 \\ -\partial_z & 0 & -1 \\ 0 & 1 & 0 \end{pmatrix}.$$

With the extended bilinear product:

$$<(\overrightarrow{e}, e_\partial, e_R), (\overrightarrow{f}, f_\partial, f_R)> := \int_0^\ell (\overrightarrow{e} \cdot \overrightarrow{f} + e_R f_R) \ \mathrm{d}z + (e_\partial, f_\partial)_{\mathbb{R}^2},$$

a new Stokes-Dirac structure can be defined. As a consequence, thanks to $e_R = R f_R$, the former dissipative property of the closed system now generalizes into the following lossy property for the open system:

$$d_t H(t) = -\int_0^\ell R f_R^2(t) - (e_\partial(t), f_\partial(t))_{\mathbb{R}^2} \leq -(e_\partial(t), f_\partial(t))_{\mathbb{R}^2}.$$

Note that for the dissipative system to be correctly defined, one actually needs an extra constitutive relation to close the system. In fact, we have:

$$\begin{pmatrix} \partial_t q \\ \partial_t \varphi \\ f_R \end{pmatrix} = \begin{pmatrix} 0 & -\partial_z & 0 \\ -\partial_z & 0 & -1 \\ 0 & 1 & 0 \end{pmatrix} \begin{pmatrix} e_q \\ e_\varphi \\ e_R \end{pmatrix}.$$

The first two lines are dynamical equations (once the link between the efforts $\overrightarrow{e}$ and the state variables $\overrightarrow{X}$, called a *constitutive relation*, has been made explicit: in the present case it is a diagonal linear transform) which must be complemented by initial data, while the third line is an algebraic equation, to which a *closure equation* must be added, namely $e_R = R f_R$.

PFEM consists of two steps: the definition of the Dirac structure from the original Stokes-Dirac structure in Sect. 2.1, and the definition of the constitutive relations at the discrete level in Sect. 2.2. Both steps are now detailed on the n-D case of a wave equation with internal damping, see e.g. [18].

Let us consider the damped wave equation of the form:

$$\rho(\mathbf{x}) \, \partial_{tt}^2 w(t, \mathbf{x}) + \epsilon(\mathbf{x}) \, \partial_t w(t, \mathbf{x}) = \mathrm{div} \left(\overline{\overline{T}}(\mathbf{x}) \cdot \overrightarrow{\mathbf{grad}} \, w(t, \mathbf{x}) \right), \quad \mathbf{x} \in \Omega,$$

with $\epsilon \geq 0$. Define as energy variables the strain $\alpha_q(t, \mathbf{x}) := \overrightarrow{\mathbf{grad}} \, w(t, \mathbf{x})$, and the linear momentum $\alpha_p(t, \mathbf{x}) := \rho(\mathbf{x}) \partial_t w(t, \mathbf{x})$. Taking the mechanical energy as Hamiltonian $H(t) := \frac{1}{2} \int_\Omega \alpha_q(t, \mathbf{x})^\top \cdot \overline{\overline{T}}(\mathbf{x}) \cdot \alpha_q(t, \mathbf{x}) + \frac{1}{\rho(\mathbf{x})} \alpha_p(t, \mathbf{x})^2 \ \mathrm{d}\mathbf{x}$, the corresponding co-energy variables are the stress $e_q := \delta_{\alpha_q} H = \overline{\overline{T}} \cdot \alpha_q$, and the velocity $e_p := \delta_{\alpha_p} H = \frac{1}{\rho} \alpha_p$. Introducing damping ports, the PDE can be written:

$$\begin{bmatrix} \partial_t \alpha_q \\ \partial_t \alpha_p \\ f_r \end{bmatrix} = \begin{bmatrix} 0 & \overrightarrow{\mathbf{grad}} & 0 \\ \mathrm{div} & 0 & -1 \\ 0 & 1 & 0 \end{bmatrix} \begin{bmatrix} e_q \\ e_p \\ e_r \end{bmatrix}, \tag{1}$$

together with the closure relation $e_r = \epsilon f_r$.

As seen in Example 1, boundary ports can be taken as traces of the efforts. Let us then denote $\overrightarrow{\mathbf{n}}$ the outward normal to Ω, and define the boundary ports:

$$u_\partial := e_p \big|_{\partial \Omega}, \qquad y_\partial := \overrightarrow{e}_q \cdot \overrightarrow{\mathbf{n}}.$$

This also gives rise to a Stokes-Dirac structure (thanks to Green's formula), and taking $e_\partial := u_\partial$ and $f_\partial := -y_\partial$, one immediately has the following lossy property:

$$d_t H(t) = -\int_0^\ell \epsilon f_r^2(t) - (e_\partial(t), f_\partial(t))_{\partial\Omega} \leq (u_\partial(t), y_\partial(t))_{\partial\Omega}. \qquad (2)$$

Impedance Boundary Conditions (IBC) can easily be taken into account within this formalism: for $\mathbf{x} \in \partial\Omega$, let $Z(\mathbf{x}) \geq 0$ be the impedance, and take $u_\partial = -Zy_\partial + v_\partial$ as control, where v_∂ is an extra boundary control. Indeed, it means that the IBC $\partial_t w + Z\left(\overline{\overline{T}} \cdot \overrightarrow{\mathbf{grad}}(w)\right) \cdot \overrightarrow{\mathbf{n}} = v_\partial$ is imposed to the original system. The previous power balance then reads:

$$d_t H(t) = -\int_0^\ell \epsilon f_r^2(t) - (Zy_\partial(t), y_\partial(t))_{\partial\Omega} + (v_\partial(t), y_\partial(t))_{\partial\Omega} \leq (v_\partial(t), y_\partial(t))_{\partial\Omega}.$$

Note that, as it has been said in Example 2, and as it has been done above with the introduction of the resistive ports f_r and e_r, the construction of a Stokes-Dirac structure for the wave equation with IBC requires another extension of the interconnection operator, *i.e.* boundary resistive ports have to be added. However, this latter task is not that straightforward, since it involves unbounded trace operators.

2.1   Stokes-Dirac Structure Translates into a Dirac Structure

Let us write a weak form of (1) with v_q and v_p as test functions, and apply Green's formula to the first line only, to make the boundary control term appear, $u_\partial = e_p|_{\partial\Omega}$. Thus, we get $(\partial_t \alpha_q, v_q)_\Omega = -(e_p, \mathrm{div}\, v_q)_\Omega + (u_\partial, v_q \cdot \overrightarrow{\mathbf{n}})_{\partial\Omega}$ and $(\partial_t \alpha_p, v_p)_\Omega = (\mathrm{div}\, e_q, v_p)_\Omega - (e_r, v_p)_\Omega$. Let us choose finite-element bases: $\alpha_q^d(t, \mathbf{x}) := \Sigma_{i=1}^{N_q} \alpha_q^i(t)\varphi_q^i(\mathbf{x}) = \Phi_q^\top \cdot \underline{\alpha}_q(t)$ and similarly for e_q^d for the q vector-valued variables in the basis φ_q; $\alpha_p^d(t, \mathbf{x}) := \Sigma_{k=1}^{N_p} \alpha_p^k(t)\varphi_p^k(\mathbf{x}) = \Phi_p^\top \cdot \underline{\alpha}_p(t)$ and similarly for e_p^d for the p scalar-valued variables in the basis φ_p; $f_r^d(t, \mathbf{x}) := \Sigma_{k=1}^{N_r} f_r^k(t)\varphi_p^k(\mathbf{x}) = \Phi_r^\top \cdot \underline{f}_r(t)$ and similarly for e_r^d for the r scalar-valued variables in the basis φ_r; and $u_\partial^d(t, \mathbf{x}) := \Sigma_{m=1}^{N_\partial} u_\partial^k(t)\psi_\partial^m(\mathbf{x}) = \Psi_\partial^\top \cdot \underline{u}_\partial(t)$ and similarly for y_∂^d for the boundary variables in the basis ψ_∂. Plugging the finite-dimensional approximations into the above weak form leads to the following pHs:

$$\begin{cases} M_q \, d_t \underline{\alpha}_q = D\,\underline{e}_p + B\underline{u}_\partial, \\ M_p \, d_t \underline{\alpha}_p = -D^\top \, \underline{e}_q + G\,\underline{e}_r, \\ M_r \, \underline{f}_r = -G^\top \, \underline{e}_p, \\ M_\partial \, \underline{y}_\partial = B^\top \underline{e}_q, \end{cases} \qquad (3)$$

with mass matrices $M_q = \int_\Omega \Phi_q \cdot \Phi_q^\top \in \mathbb{R}^{N_q \times N_q}$, $M_p = \int_\Omega \Phi_p \cdot \Phi_p^\top \in \mathbb{R}^{N_p \times N_p}$ and $M_\partial = \int_{\partial\Omega} \Psi_\partial \cdot \Psi_\partial^\top \in \mathbb{R}^{N_\partial \times N_\partial}$, a control matrix $B := \int_{\partial\Omega} \Phi_q \cdot \overrightarrow{\mathbf{n}} \cdot \Psi_\partial^\top \in \mathbb{R}^{N_q \times N_\partial}$, and a structure matrix J composed of $D := -\int_\Omega \operatorname{div} \Phi_q \cdot \Phi_p^\top \in \mathbb{R}^{N_q \times N_p}$ and $G := \int_\Omega \Phi_p \cdot \Phi_r^\top \in \mathbb{R}^{N_p \times N_r}$. It is then straightforward to define a bilinear product on $\mathcal{B}^d := \mathbb{R}^{N_q+N_p+N_r+N_\partial} \times \mathbb{R}^{N_q+N_p+N_r+N_\partial}$ as $<(\underline{e}_q, \underline{e}_p, \underline{e}_r, \underline{e}_\partial), (\underline{f}_q, \underline{f}_p, \underline{f}_r, \underline{f}_\partial)> := \underline{e}_q^\top M_q \underline{f}_q + \underline{e}_p^\top M_p \underline{f}_p + \underline{e}_r^\top M_r \underline{f}_r + \underline{e}_\partial^\top M_\partial \underline{f}_\partial$.

Proposition 2. *The subspace:*

$$\mathcal{D}^d := \{(e, f) \in \mathcal{B}^d \mid (\underline{f}_q, \underline{f}_p, \underline{f}_r)^\top = J(\underline{e}_q, \underline{e}_p, \underline{e}_r)^\top, \, e_\partial := \underline{u}_\partial, \, f_\partial := -\underline{y}_\partial\},$$

is a Dirac structure.

Remark 1. Moreover, contrarily to other structure-preserving methods relying on Stokes-Dirac structure, like [13,20], there is no need here to project, reduce, some non square matrices in order to recover a full rank system at the discrete level, which is, at least from the numerical point of view, a severe limitation indeed.

2.2 Constitutive Relation Are Approximated in Weak Form

The idea is fairly simple: the constitutive equation of the resistive port $e_r = \epsilon f_r$ is written in weak form, and using the previously defined approximation f_r^d and e_r^d, one gets:

$$M_r \, \underline{e}_r(t) = <R> \underline{f}_r(t),$$

involving two symmetric $N_r \times N_r$ matrices, the mass matrix $M_r := \int_\Omega \Phi_r^\top \Phi_r \, \mathrm{dx}$ which is positive definite, and $<R> := \int_\Omega \Phi_r^\top \epsilon(\mathbf{x}) \Phi_r \, \mathrm{dx}$, the averaged *resistive* matrix which is positive.

Finally, once these two steps have been carried out, we can prove the following

Proposition 3. *Defining the discrete Hamiltonian as:*

$$H_d(t) = H_d(\underline{\alpha}_q(t), \underline{\alpha}_p(t)) := H(\boldsymbol{\alpha}_q^d(t, \mathbf{x}), \alpha_p^d(t, \mathbf{x})),$$

the discrete counterpart of the continuous lossy property (2) holds for the finite-dimensional system (3) obtained with PFEM: $d_t H_d \leq \langle \underline{u}_\partial, \underline{y}_\partial \rangle_\partial := \underline{y}_\partial^\top M_\partial \underline{u}_\partial$.

Indeed, thanks to the Dirac structure and the constitutive relations, we have:
$d_t H_d = \underline{e}_q^\top M_q \, d_t \underline{\alpha}_q + \underline{e}_p^\top M_p \, d_t \underline{\alpha}_p = -\underline{f}_r^\top <R> \underline{f}_r + \underline{y}_\partial^\top M_\partial \underline{u}_\partial \leq \langle \underline{u}_\partial, \underline{y}_\partial \rangle_\partial$.

Now, at the boundary, the IBC is discretized in the same manner above: let $<Z> := \int_{\partial\Omega} \Psi_\partial^\top Z(\mathbf{x}) \Psi_\partial \, \mathrm{dx} \in \mathbb{R}^{N_\partial \times N_\partial}$ be the averaged resistive matrix taking Z into account on the boundary only. Then, define $v_\partial^d(t, \mathbf{x}) := \Sigma_{m=1}^{N_\partial} v_\partial^k(t) \psi_\partial^m(\mathbf{x}) = \Psi_\partial^\top \cdot v_\partial(t)$ the approximation of the extra control v_∂, and add to system (3) the following algebraic equation: $M_\partial \underline{u}_\partial(t) = -<Z> \underline{y}_\partial(t) + M_\partial \underline{v}_\partial(t)$, which mimics $u_\partial = -Z y_\partial + v_\partial$ by finite element discretization on the boundary. We finally get:

$$d_t H_d = -\underline{f}_r^\top <R> \underline{f}_r - \underline{y}_\partial^\top <Z> \underline{y}_\partial + \underline{y}_\partial^\top M_\partial \underline{v}_\partial \leq \langle \underline{v}_\partial, \underline{y}_\partial \rangle_\partial.$$

Remark 2. At the continuous level, we have seen that the extension of the inter-connection operator which gives rise to the Stokes-Dirac structure could be a difficult task, since it would involve unbounded operators (typically trace opera-tors). However, once PFEM has been applied, it proves straightforward to define the resistive ports, both internal and at the boundary. Indeed we can write, with obvious notations:

$$
\begin{pmatrix} M_q & 0 & 0 & 0 & 0 \\ 0 & M_p & 0 & 0 & 0 \\ 0 & 0 & M_r & 0 & 0 \\ 0 & 0 & 0 & M_i & 0 \\ 0 & 0 & 0 & 0 & M_\partial \end{pmatrix} \begin{pmatrix} d_t\underline{\alpha_q} \\ d_t\underline{\alpha_p} \\ \underline{f_r} \\ \underline{f_i} \\ \underline{f_\partial} \end{pmatrix} = \begin{pmatrix} 0 & D & 0 & -B & B \\ -D^\top & 0 & -1 & 0 & 0 \\ 0 & 1 & 0 & 0 & 0 \\ B^\top & 0 & 0 & 0 & 0 \\ -B^\top & 0 & 0 & 0 & 0 \end{pmatrix} \begin{pmatrix} \underline{e_q} \\ \underline{e_p} \\ \underline{e_r} \\ \underline{e_i} \\ \underline{e_\partial} \end{pmatrix},
$$

together with the two constitutive relations:

$$
M_r\underline{e_r} = <R>\underline{f_r}, \qquad M_i\underline{e_i} = <Z>\underline{f_i},
$$

and the definitions $\underline{f_\partial} = -\underline{y_\partial}$ and $\underline{e_\partial} = \underline{v_\partial}$ that are now usual in our approach. All together, the desired lossy property of the system is ensured at the discrete level.

Remark 3. Let us point out that the mass matrices on the left-hand side are required in order to preserve the underlying geometry. To some extent, they do discretize the bilinear form used to define the Stokes-Dirac structure, w.r.t. the chosen finite element families, as seen before in Proposition 2.

3   Diffusion Model in Dissipative Formulation

The heat or diffusion PDE is most often considered as a dissipative infinite-dimension system from a mathematical point of view, and examplifies the cat-egory of parabolic PDEs: this approach is being recalled here with the choice of a quadratic potential as Hamiltonian function, though its thermodynamical meaning is far from clear, see [23,24] for details and the choice of either energy or entropy as thermodynamically meaningful Hamiltonian.

Port-Hamiltonian System Model. Let $H(t) := \frac{1}{2} \int_\Omega \rho(\mathbf{x}) \frac{(u(t,\mathbf{x}))^2}{C_V(t,\mathbf{x})} \, d\mathbf{x}$ be the Hamiltonian with u the energy variable: ρ is the mass density, C_V is the isochoric heat capacity and u the internal energy density. The co-energy variable is $\delta_u\mathcal{H} = \frac{u}{C_V} = T$ (the temperature), assuming $u(t,\mathbf{x}) = C_V(\mathbf{x})\,T(t,\mathbf{x})$. Let us define $f_u := \partial_t u$, $e_u := T$ and $\overrightarrow{e}_Q := \overrightarrow{\mathbf{J}}_Q$ (the heat flux). Then, with $\overrightarrow{f}_Q := -\overrightarrow{\mathbf{grad}}(T)$, we get:

$$
\begin{pmatrix} \rho f_u \\ \overrightarrow{f}_Q \end{pmatrix} = \begin{pmatrix} 0 & -\operatorname{div} \\ -\overrightarrow{\mathbf{grad}} & 0 \end{pmatrix} \begin{pmatrix} e_u \\ \overrightarrow{e}_Q \end{pmatrix}.
$$

The system must be completed by e.g. Fourier's law as constitutive relation:

$$\vec{\mathbf{J}}_Q(t,\mathbf{x}) = -\overline{\overline{\lambda}}(\mathbf{x}) \cdot \overrightarrow{\mathbf{grad}}\,(T(t,\mathbf{x})), \quad \forall t \geq 0, \mathbf{x} \in \Omega, \tag{4}$$

where $\overline{\overline{\lambda}}$ is a tensor representing the thermal conductivity; it is a positive symmetric tensor thanks to Onsager's reciprocal relations.

For the boundary ports, one possible choice is $\mathcal{B}\,\vec{e} := e_{u|\partial\Omega}$ the boundary temperature, and $\mathcal{C}\,\vec{e} := -(\vec{e}_Q \cdot \mathbf{n})_{|\partial\Omega}$ the incoming boundary flux. Hence, the power balance for this lossy open system is:

$$d_t H(t) = -\int_\Omega \vec{f}_Q(t,\mathbf{x}) \cdot \overline{\overline{\lambda}}(\mathbf{x}) \cdot \vec{f}_Q(t,\mathbf{x})\,\mathrm{d}\mathbf{x} + \int_{\partial\Omega} v_\partial(t,\gamma)y_\partial(t,\gamma)\,\mathrm{d}\gamma. \tag{5}$$

Partitioned Finite Element Method. Following the procedure explained in Sect. 2 and with obvious notations, we get:

$$\begin{cases} M_\rho\,\underline{f_u}(t) = D\,\underline{e_Q}(t) + B\,\underline{v_\partial}(t), \\ \vec{M}\,\underline{f_Q}(t) = -D^\top\,\underline{e_u}(t), \\ M_\partial\,\underline{y_\partial}(t) = C\,\underline{e_u}(t), \end{cases}$$

where for example $D := \int_\Omega \overrightarrow{\mathbf{grad}}\,(\Phi) \cdot \vec{\Phi}^\top\,\mathrm{d}\mathbf{x} \in \mathbb{R}^{N \times \vec{N}}$. The weak version of the constitutive law (4) reads:

$$\vec{M}\,\underline{e_Q}(t) = \vec{\Lambda}\,\underline{f_Q}(t), \quad \text{where}\,\vec{\Lambda} := \int_\Omega \vec{\Phi} \cdot \overline{\overline{\lambda}} \cdot \vec{\Phi}^\top\,\mathrm{d}\mathbf{x} \in \mathbb{R}^{\vec{N} \times \vec{N}},$$

and thus the coupled system is now a pHDAE, see e.g. [1]; the energy balance (5) becomes at the discrete level:

$$d_t \mathcal{H}_d(t) = -\underline{f_Q}^\top(t)\vec{\Lambda}\,\underline{f_Q}(t) + \underline{v_\partial}^\top(t)M_\partial\underline{y_\partial}(t).$$

4   Conclusion and Perspectives

In this paper, a structure-preserving numerical method has been presented for lossy port-Hamiltonian systems: the so-called Partitioned Finite Element Method (PFEM). It is based on the weak formulation of PDE, the application of a Stokes formula (reduced to Green formula in our examples) to get the useful boundary control explicitly, and the application of the classical finite element method with the choice of conforming elements for the different ports. Boundary damping, such as impedance boundary condition, studied theoretically in [15] as a pHs, becomes particularly straightforward with PFEM, since it results in a sparse damping matrix R at the discrete level (see [25] for more details). The following perspectives seem relevant and promising:

– the choice of the finite element family remains quite open so far, but indeed, from first numerical experiments, some optimal choices can be observed in practice: the careful numerical analysis must still be investigated,

– structure-preserving model reduction can be carried out using methods presented in [11],
– for the time-domain discretization as last step procedure for numerical simulation, specific approaches should be followed, see e.g. [7],
– following [28], the introduction of entropy ports enables to transform dissipative systems into conservative systems, taking into account some thermodynamical laws; see [23,24] for the application to the heat equation.
– an alternative computational solution consists in making use of the transformation of the lossy system into a lossless one, and only then apply classical symplectic numerical schemes, see e.g. [12,16]; the difficulty lies in the fact that the obtained finite-dimensional systems are necessarily differential algebraic equations that should be treated with some specific care, see e.g. [14].

References

1. Beattie, C., Mehrmann, V., Xu, H., Zwart, H.: Linear port-Hamiltonian descriptor systems. Math. Control Signals Syst. **30**(4), 17 (2018)
2. Brugnoli, A., Alazard, D., Pommier-Budinger, V., Matignon, D.: Port-Hamiltonian formulation and symplectic discretization of plate models Part I: Mindlin model for thick plates. Appl. Math. Model. (2019, in press). https://doi.org/10.1016/j.apm.2019.04.035
3. Brugnoli, A., Alazard, D., Pommier-Budinger, V., Matignon, D.: Port-Hamiltonian formulation and symplectic discretization of plate models Part II: Kirchhoff model for thin plates. Applied Mathematical Modelling (2019, in press). https://doi.org/10.1016/j.apm.2019.04.036
4. Cardoso-Ribeiro, F.L., Matignon, D., Lefèvre, L.: A structure-preserving Partitioned Finite Element Method for the 2D wave equation. IFAC-PapersOnLine **51**(3), 119–124 (2018). 6th IFAC Workshop on Lagrangian and Hamiltonian Methods for Nonlinear Control LHMNC 2018
5. Cardoso-Ribeiro, F.L., Matignon, D., Lefèvre, L.: A Partitioned Finite-Element Method (PFEM) for power-preserving discretization of open systems of conservation laws (2019). arXiv:1906.05965
6. Cardoso-Ribeiro, F.L., Matignon, D., Pommier-Budinger, V.: A port-Hamiltonian model of liquid sloshing in moving containers and application to a fluid-structure system. J. Fluids Struct. **69**, 402–427 (2017)
7. Celledoni, E., Høiseth, E.H.: Energy-preserving and passivity-consistent numerical discretization of port-Hamiltonian systems (2017). arXiv:1706.08621
8. Courant, T.J.: Dirac manifolds. Trans. Am. Math. Soc. **319**(2), 631–661 (1990)
9. Duindam, V., Macchelli, A., Stramigioli, S., Bruyninckx, H.: Modeling and Control of Complex Physical Systems: The Port-Hamiltonian Approach. Springer, Heidelberg (2009). https://doi.org/10.1007/978-3-642-03196-0
10. Egger, H.: Structure preserving approximation of dissipative evolution problems. Numerische Mathematik (2019, in press). https://doi.org/10.1007/s00211-019-01050-w
11. Egger, H., Kugler, T., Liljegren-Sailer, B., Marheineke, N., Mehrmann, V.: On structure-preserving model reduction for damped wave propagation in transport networks. SIAM J. Sci. Comput. **40**(1), A331–A365 (2018)

12. Hairer, E., Lubich, C., Wanner, G.: Geometric Numerical Integration: Structure-preserving Algorithms for Ordinary Differential Equations. Springer Series in Computational Mathematics, vol. 31. Springer, Heidelberg (2006). https://doi.org/10.1007/3-540-30666-8

13. Kotyczka, P., Maschke, B., Lefèvre, L.: Weak form of Stokes-Dirac structures and geometric discretization of port-Hamiltonian systems. J. Comput. Phys. **361**, 442–476 (2018)

14. Kunkel, P., Mehrmann, V.: Differential-Algebraic Equations: Analysis and Numerical Solution. EMS Textbooks in Mathematics. European Mathematical Society, Zurich (2006)

15. Kurula, M., Zwart, H.: Linear wave systems on n-D spatial domains. Int. J. Control. **88**(5), 1063–1077 (2015)

16. Leimkuhler, B., Reich, S.: Simulating Hamiltonian Dynamics. Cambridge Monographs on Applied and Computational Mathematics. Cambridge University Press, Cambridge (2004)

17. Macchelli, A., van der Schaft, A., Melchiorri, C.: Port Hamiltonian formulation of infinite dimensional systems I. Modeling. In: 2004 43rd IEEE Conference on Decision and Control (CDC), vol. 4, pp. 3762–3767 (2004)

18. Matignon, D., Hélie, T.: A class of damping models preserving eigenspaces for linear conservative port-Hamiltonian systems. Eur. J. Control **19**(6), 486–494 (2013)

19. Mehrmann, V., Morandin, R.: Structure-preserving discretization for port-Hamiltonian descriptor systems (2019). arXiv:1903.10451

20. Moulla, R., Lefèvre, L., Maschke, B.: Pseudo-spectral methods for the spatial symplectic reduction of open systems of conservation laws. J. Comput. Phys. **231**(4), 1272–1292 (2012)

21. van der Schaft, A.J., Jeltsema, D.: Port-Hamiltonian systems theory: an introductory overview. Now Found. Trends **1**, 173–378 (2014)

22. van der Schaft, A.J., Maschke, B.: Hamiltonian formulation of distributed-parameter systems with boundary energy flow. J. Geom. Phys. **42**(1–2), 166–194 (2002)

23. Serhani, A., Matignon, D., Haine, G.: Anisotropic heterogeneous n-D heat equation with boundary control and observation: I. Modeling as port-Hamiltonian system. In: 2019 3rd IFAC Workshop on Thermodynamical Foundation of Mathematical Systems Theory (TFMST) (2019). Accepted for publication in IFAC-PapersOnLine

24. Serhani, A., Matignon, D., Haine, G.: Anisotropic heterogeneous n-D heat equation with boundary control and observation: II. Structure-preserving discretization. In: 2019 3rd IFAC Workshop on Thermodynamical Foundation of Mathematical Systems Theory (TFMST) (2019). Accepted for publication in IFAC-PapersOnLine

25. Serhani, A., Matignon, D., Haine, G.: Partitioned Finite Element Method for port-Hamiltonian systems with Boundary Damping: Anisotropic Heterogeneous 2-D wave equations. In: 2019 3rd IFAC workshop on Control of Systems Governed by Partial Differential Equations (CPDE) (2019). To appear in IFAC-PapersOnLine

26. Trenchant, V., Ramirez, H., Le Gorrec, Y., Kotyczka, P.: Finite differences on staggered grids preserving the port-Hamiltonian structure with application to an acoustic duct. J. Comput. Phys. **373**, 673–697 (2018)

27. Vu, N.M.T., Lefèvre, L., Maschke, B.: A structured control model for the thermo-magneto-hydrodynamics of plasmas in tokamaks. Math. Comput. Model. Dyn. Syst. **22**(3), 181–206 (2016)

28. Zhou, W., Hamroun, B., Couenne, F., Le Gorrec, Y.: Distributed port-Hamiltonian modelling for irreversible processes. Math. Comput. Model. Dyn. Syst. **23**(1), 3–22 (2017)

Geometry, Energy, and Entropy Compatible (GEEC) Variational Approaches to Various Numerical Schemes for Fluid Dynamics

Antoine Llor[(⊠)] and Thibaud Vazquez-Gonzalez

CEA, DAM/DIF, 91297 Arpajon Cedex, France
`antoine.llor@cea.fr`

Abstract. Since WW2, computer fluid dynamics has seen a staggering expansion of methods and applications fueled by the development of computer power. Now, of the main numerical approaches that have been explored over these years, only a few have become mainstream and make the vast majority of theoretical investigations in academia and practical usage in applications. These mostly hinge on concepts of finite volume discretization, monotonicity preservation, flux upwinding, and the analysis of the associated numerical dissipation processes—common tools here are the Riemann problem at cell interfaces and the Godunov scheme, more or less adapted from their original versions.

However, application to what looks as "niche" problems shows that these dominant approaches may not be as effective as generally accepted and have unduly benefited from a "winner-takes-it-all" effect. One of these problems is the simulation of isentropic flows which is actually "not-so-niche" as it is of high practical interest, especially in multi-fluid systems which involve complex energy transfers.

The present contribution aims at providing some perspective on CFD numerical schemes recently designed in order to better capture isentropic flows. The basic principle is that isentropic flow is *geometric*, i.e. potential (or internal) energy only depends on fluid density which in turn is defined by fluid element trajectories. A numerical scheme can thus be obtained by a variational, least action principle. Corrections must be further added to enforce other properties such as energy conservation and positive dissipation. This Geometry, Energy, and Entropy Compatible approach (GEEC) is illustrated here on the historical von Neumann–Richtmyer Lagrangian scheme and on our recently developed multi-fluid Arbitrary Lagrangian–Eulerian scheme.

Keywords: Variational schemes · Computer fluid dynamics · Isentropic flow · Energy conservation · Entropy production

© Springer Nature Switzerland AG 2019
F. Nielsen and F. Barbaresco (Eds.): GSI 2019, LNCS 11712, pp. 559–567, 2019.
https://doi.org/10.1007/978-3-030-26980-7_58

1   Motivations

1.1   A Short Historical Perspective

The very first CFD scheme was designed by von Neumann and Richtmyer (VNR) at Los Alamos in 1944—but published somewhat later [12]. It has the following properties:

- 1D Lagrangian discretization of the (compressible) fluid,
- positions and internal energies defined at cells centers x_c and integer time steps t^n,
- velocities (and momenta) defined at nodes x_p and half-integer time steps $t^{n+1/2}$—the so-called staggered discretization,
- propagation cycle involving three discrete increment equations, for node velocities, node positions, and cell energies,
- dissipation for shock capture explicitly added as an "artificial" numerical viscosity,
- approximate conservation of total energy (to the scheme's order).

The scheme was later extended to 2D geometry with elasto-plastic material behavior [13] and many other features where added over the years [1].

Concurrently, a more general approach applicable to hyperbolic coupled equations was designed by Godunov in 1954—but published somewhat later [3]—from which many other methods were derived. The basic building block here consists in considering the discontinuities of field values at cell interfaces as perturbations whose propagation is calculated and remapped on the original cells. The calculation can take many different forms and levels of accuracy producing a wealth of different schemes—Godunov's original being the solution of the Riemann problem at cell interfaces. However, this *always* introduces some form of *dissipation* which has the advantage of removing almost all over- and undershoots, oscillations, and other artifacts, at the expense of poor behavior in isentropic conditions. The only options for recovering proper isentropic behavior are then an artificial correction of over-dissipation or an expensive crank up of the schemes' order—both options being potentially very fragile for complex non-linear systems. Intrinsic to these methods is the upwinding along the propagated discontinuities which was critically reviewed by Roe [7,8].

Over the last decades, because exact conservation of mass, momentum, and energy are required in order to properly capture Rankine-Hugoniot jump conditions, the VNR schemes have eventually been deprecated by most numericists while being preserved and used by practitioners for many specific applications involving isentropic phases. However, in a recent co-publication by one of the authors [4], a simple correction to the VNR scheme was designed so as to recover exact conservation while preserving its isentropic behavior to its order of accuracy. For this purpose the scheme was first reinterpreted as deriving from a variational least-action principle which made it practically identical to the very popular Størmer–Verlet symplectic scheme used for astrophysics and molecular dynamics.

1.2   GEEC: A General Framework for Hydro-Scheme Design

The approach to generate the conservative VNR scheme—here designated as CSTS for Conservative Space- and Time-Staggered—appeared to be applicable in many other numerical settings. We here provide an overview of two such schemes: the original CSTS compressible single-fluid Lagrangian [4], possibly with variable node masses [5], and a compressible multi-fluid ALE [10] (Arbitrary Lagrangian-Eulerian were the mesh moves and distorts according to user's free prescriptions, with unlimited number of fluids) and its single-fluid ALE reduction [11]. The designing approach, designated as "GEEC," consists of three steps:

G: *Geometry* – Energy in isentropic flow only depends on the *geometry* of the system, whose dynamics is thus defined by a Lagrangian and its action integral. Mimicking the derivation of the continuous evolution equations [9], the minimization of the discretized action integral provides a numerical scheme which in principle is *symplectic*;

E: *Energy* – According to a theorem of Ge and Marsden [2], a numerical scheme cannot simultaneously preserve momentum, energy, and symplecticity but only two out of the three. Hence, the scheme from step "G" is modified by adding a conservative energy equation obtained from the energy balance analysis and the thermodynamic consistency. It is then symplectic or entropic to the scheme's order only;

E: *Entropy* – For non-isentropic behavior, dissipation terms are added, positivity of entropy production being easily enforced by appropriate algebraic closures such as positive quadratic forms. Artificial viscosity [6] is merely a special case of dissipation designed for shock capture.

The critical ingredient in the GEEC approach is the action integral discretization as it entirely defines the numerical scheme except for some residual terms of higher than the scheme's order. Thus in all the following, the discrete action integrals will always be provided, but for the sake of legibility, some energy equations will be skipped. The reader is referred to the original publications for full derivations and descriptions of the numerical schemes [4,5,10,11].

1.3   Common Notations

In all the following, the discretization will be carried out in any dimension over an unstructured time-dependent constant-connectivity mesh of polytope-shaped cells labeled c. The cells are defined by their boundary nodes x_p^n labeled p at each time t^n. All fields (density ρ, velocity u, internal energy e, pressure p, etc.) can be defined at cells or nodes, and integer or half-integer labeled times.

2   Simple Example: The GEEC Conservative Space-and-Time Staggered (CSTS) Lagrangian Scheme

2.1   G: Geometry Step

In the Lagrangian setting, the mass of fluid in each cell m_c is constant and all fields except velocity are discretized at cell centers and integer-labeled times. Velocity describes the evolution of the grid nodes and is thus best defined at nodes and half-integer labeled times (space-and-time staggering) $u_p^{n+1/2} = (x_p^{n+1} - x_p^n)/\Delta t^{n+1/2}$ where $\Delta t^{n+1/2} = t^{n+1} - t^n$. A simple second-order accurate action integral is then built as

$$\mathcal{A}[\{x_p^n\}] = \sum_n \left[\sum_p \tfrac{1}{2} m_p^{n+1/2} (u_p^{n+1/2})^2 \Delta t^{n+1/2} - \sum_c m_c e[v_c^n] \Delta t^n \right], \qquad (1)$$

which is to be minimized with respect to the discrete trajectories, that is the sets of positions $\{x_p^n\}$.

The internal energies are functions of the per-mass volume v_c^n of the fluid in cell c, and node masses $m_p^{n+1/2}$ are defined by redistributing the cell masses m_c in a conservative way

$$m_p^{n+1/2} = \sum_c m_{cp}^{n+1/2}[\{x_q^{n+1}, x_q^n\}], \quad \text{with} \quad m_c = \sum_p m_{cp}^{n+1/2}[\{x_q^{n+1}, x_q^n\}],$$

where $m_{cp}^{n+1/2}$ are functions of the cell shapes defined by the set of cell nodes $\{x_q^{n+1}, x_q^n\}$. The most basic choice for quadrilaterals is Wilkins' time independent $m_{cp}^{n+1/2} = m_c/4$ [13].

For constant node masses the least action principle yields the discrete momentum equation

$$m_p(u_p^{n+1/2} - u_p^{n-1/2}) = \sum_c p_c^n \frac{\partial V_c}{\partial x_p}\Big|^n \Delta t^n, \qquad (2)$$

—notice here that pressure is denoted $p_c^n = -\partial e_c^n/\partial v_c^n$ (do not confuse with subscript p). This scheme is very similar to the well known Størmer–Verlet symplectic scheme. On the right-hand side the pressure gradient discretization involves the cells' corner vectors $\partial V_c/\partial x_p$. For variable node masses, the Euler–Lagrange equations produce supplementary terms in (2) involving vectors $\partial m_{cp}/\partial x_q$ [5].

2.2   E: Energy Step

The kinetic energy equation, as defined from the action integral (1), is obtained by rearranging the product (2) $\cdot \tfrac{1}{2}(u_p^{n+1/2} + u_p^{n-1/2})$ into

$$\tfrac{1}{2} m_p \left[(u_p^{n+1/2})^2 - (u_p^{n-1/2})^2 \right] = \sum_c p^n \frac{\partial V_c}{\partial x_p}\Big|^n \cdot \tfrac{1}{2}(u_p^{n+1/2} + u_p^{n-1/2}) \Delta t^n.$$

Energy conservation then requires that the internal energy equation involves pressure work terms that exactly balance with those of the kinetic energy. Further imposing thermodynamic consistency (whereby p_c^n can only appear between t^n and $t^{n\pm 1}$), it is then found

$$m_c(e_c^{n+1} - e_c^n) = \sum_c \Big[-\tfrac{1}{2}\big(p^{n+1}\tfrac{\partial V_c}{\partial x_p}\big|^{n+1} + p_c^n \tfrac{\partial V_c}{\partial x_p}\big|^n\big) \cdot \boldsymbol{u}_p^{n+1/2} \Delta t^{n+1/2}$$
$$+ \tfrac{1}{4} p_c^n \tfrac{\partial V_c}{\partial x_p}\big|^n \cdot \big(\boldsymbol{u}_p^{n+1/2} - \boldsymbol{u}_p^{n-1/2}\big)\big(\Delta t^{n+1/2} - \Delta t^{n-1/2}\big)\Big].$$

The first term on the right hand side is a consistent second order discretization of the pressure work "$-p\,\mathrm{d}\,v$" and the second is a correction for energy conservation which is of third order but may be non negligible at singularities. This term acts as a consistent "flux-in-time" term which ensures conservation despite the different time centering of kinetic and internal energies [4, § 2.3].

2.3   E: Entropy Step

Physical and numerical dissipation terms can now be added to the momentum and energy equations. In particular, for numerical shock capture and mesh regularization it is usual to respectively introduce "artificial" stress q and force $\boldsymbol{f}$ terms as

$$m_p(\boldsymbol{u}_p^{n+1/2} - \boldsymbol{u}_p^{n-1/2}) = \sum_c \big((p_c^n + q_c^n)\tfrac{\partial V_c}{\partial x_p}\big|^n + \boldsymbol{f}_{cp}^n\big)\Delta t^n,$$

Positive entropy production can then be ensured *to no better than the scheme order* if q and $\boldsymbol{f}$ fulfill the condition

$$\sum_p -\tfrac{1}{2}\Big[q_c^n\tfrac{\partial V_c}{\partial x_p}\big|^n + \boldsymbol{f}_{cp}^n\Big]\cdot(\boldsymbol{u}_p^{n+1/2} + \boldsymbol{u}_p^{n-1/2}) \geq \pm\mathcal{O}\big[(\Delta t)^3\big].$$

This is achieved by usual algebraic closures and by introducing a prediction–correction procedure on the momentum equation [4, § 2.5].

The CSTS scheme was extensively tested in one and two dimensions and fulfilled expected properties [4, § 3]: second order behavior on isentropic flows (Kidder test), exact jump conditions and shock propagation (Sod, Noh and Sedov tests), and high maximum operating CFL limit (increased by over 60% thanks to the prediction–correction step). Variable node masses preserved these features and further allowed a very significant reduction of mesh distortions by appropriately choosing the mass functions [5].

3   A GEEC Conservative Direct-ALE Multi-fluid Scheme

3.1   G: Geometry Step

For Eulerian or ALE schemes the evolution equations to be solved do not involve *explicit fluid element coordinates* as in the Lagrangian setting of Sect. 2. However, the various fields (density, velocity, internal energy, etc. which are functions of time and space coordinates) must be constrained to conserve mass and Lagrangian coordinates or they would unduly appear as independent. This is achieved by adding the constraints to the action integral with Lagrange multipliers.

For a system of Φ fluids labeled φ (which describes a conditional ensemble average of individual multi-fluid flow realizations) the action integral involves the per-fluid kinetic and internal energy contributions, the per-fluid mass conservation and Lagrangian coordinate preservation (Lin) constraints, and a volume filling constraint

$$
\mathcal{A} = \sum_{n,c,\varphi} \Big(\Delta t^{n+1/2} V_c^n \tfrac{1}{2} [\alpha \rho]_c^{\varphi n} (u_c^{\varphi n+1/2})^2 - \Delta t^n V_c^n [\alpha \rho]_c^{\varphi n} e^\varphi \big(\rho_c^{\varphi n}, s^\varphi(\xi_c^{\varphi n}) \big)
$$
$$
+ \phi_c^{\varphi n+1} \, \mathrm{D}_{\Delta t}^\varphi [\alpha \rho]_c^{\varphi n} + \psi_c^{\varphi n+1} \, \mathrm{D}_{\Delta t}^\varphi [\alpha \rho \xi]_c^{\varphi n} \Big) - \sum_{n,c} \Pi_c^n \Big(\sum_\varphi \alpha_c^{\varphi n} - 1 \Big), \quad (3)
$$

where α^φ, ρ^φ, u^φ, v^φ, e^φ, s^φ, ξ^φ, ϕ^φ, ψ^φ, Π are respectively the fluid φ's volume fraction, density, absolute velocity, relative-to-the-grid velocity, per-mass internal energy, per-mass entropy, Lagrangian coordinate, and mass, Lin, and volume filling multipliers. In the case of a single-fluid $\Phi = 1$, the action integral simplifies as $\alpha^1 = 1$ and multiplier Φ becomes irrelevant. Relative-to-grid transport is embedded in the evolution equations and is not split as in Lagrange + Remap schemes: this is a direct ALE scheme.

The constraints are expressed with a common transport operator

$$
\mathrm{D}_{\Delta t}^\varphi a_c^{\varphi n} = V_c^{n+1} a_c^{\varphi n+1} - V_c^n a_c^{\varphi n} + \Delta t^{n+1/2} \sum_d \big(\mathring{V}_{cd}^{\varphi n+1/2} a_c^{\varphi n} - \mathring{V}_{dc}^{\varphi n+1/2} a_d^{\varphi n} \big),
$$

where V_c^n is the cell volume at time t^n and $\mathring{V}_{cd}^{\varphi n+1/2}$ are the volume transfer rates of fluid φ from cell c to neighboring cell d during time step $t^{n+1/2}$ given by a first order upwind closure

$$
\mathring{V}_{cd}^{\varphi n+1/2} = \sigma_{cd}^{\varphi n+1/2} s_{cd}^{n+1/2} \cdot v_c^{\varphi n+1/2}, \quad \text{and} \quad \sigma_{cd}^{\varphi n+1/2} = H\big(s_{cd}^{n+1/2} \cdot v_c^{\varphi n+1/2} \big),
$$

$s_{cd}^{n+1/2}$ being the surface vector normal to the edge connecting cells c to d and H being the Heaviside function. Velocities u_c^φ and v_c^φ are related by the grid velocity, $w_c^\varphi = u_c^\varphi - v_c^\varphi$ interpolated from its natural definition at nodes w_p^φ. The transport operator is holonomic to its order of accuracy, which introduces some spurious dissipation.

The momentum equation is obtained from the action integral minimization with respect to the independent fields α^φ, ρ^φ, v^φ, ξ^φ, ϕ^φ, ψ^φ, Π and yields after some lengthy calculations [10]

$$
V_c^n [\alpha \rho]_c^{\varphi n} u_c^{\varphi n+1/2} - V_c^{n-1} [\alpha \rho]_c^{\varphi n-1} u_c^{\varphi n-1/2}
$$
$$
+ \Delta t^{n-1/2} \sum_d \big(\mathring{V}_{cd}^{\varphi n-1/2} [\alpha \rho]_c^{\varphi n-1} u_c^{\varphi n-1/2} - \mathring{V}_{dc}^{\varphi n-1/2} [\alpha \rho]_d^{\varphi n-1} u_d^{\varphi n-1/2} \big)
$$
$$
= -\Delta t^n \sum_d \tfrac{1}{2} \big(\tfrac{1}{2} (\alpha_c^{\varphi n} + \alpha_d^{\varphi n}) + \alpha_c^{\varphi n} \sigma_{cd}^{\varphi n-1/2} - \alpha_d^{\varphi n} \sigma_{dc}^{\varphi n-1/2} \big)
$$
$$
\times s_{cd}^{n-1/2} (P_d^n - P_c^n + Q_d^n - Q_c^n), \quad (4)
$$

all fluids at any time in a given cell being at equal pressures.

The pressure gradient in the momentum equation displays the surprising property of being *downwind*. This is a general property that derives from the *upwind* transport due to a discrete integration by parts in the calculation of the action variation.

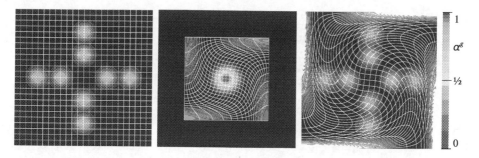

Fig. 1. Volume fraction maps at times $t = 0$, 10^{-3}, and 2×10^{-3} s (left, center, and right) for the nine-fluids crossing test on a shrink-then-stretch swirling grid. One in twenty grid lines represented in both dimensions (white lines). (Color figure online)

3.2   E and E: Energy and Entropy Steps

The per-fluid internal energy equation will not be given here due to the length constraints on the present publication but they are obtained [10, § 3.7] by following an approach identical to that of Sect. 2.2: conservation and thermodynamic consistency. Suffice to say at this stage that they involve numerous terms due to the full coupling by pressure forces of the 2Φ kinetic and internal energy reservoirs. With fluids of highly contrasted volume fractions, densities, and polytropic coefficients such as air and water the coupling terms can become particularly stiff. The final scheme is completely explicit and was found to behave stably and consistently for a wide range of compositions, EOS stiffness, and Mach numbers.

3.3   A Strenuous Nine-Fluid Test: Supersonic Crossing of Eight Gaussian Clouds on Shrink-then-Stretch Swirling ALE Grid

This 2D test involves eight packets or clouds of water in a background of still air. The clouds and the background are each represented by separate fluids thus defining a nine-fluid system whose evolution can thus be captured by the multi-fluid scheme. Though described by separate equations, the eight water clouds have the same stiffened-gas equation of state. Apart of pressure terms, no interactions between the clouds or between clouds and air are added—collisions, drag, thermal transfers, etc. *Clouds can thus cross each other freely.* The computation domain is $[-3; 3] \times [-3; 3]$ m. The initial condition at $t = 0$ consists in cloud volume fractions $\alpha^\varphi(0, \boldsymbol{x})$, all with Gaussian profiles of amplitudes 0.15 and root mean square radii 0.2 m, but centered at different positions $\boldsymbol{x}_0^\varphi = (\pm1, 0)$ m, $(0, \pm1)$ m, $(\pm2, 0)$ m, and $(0, \pm2)$ m and set in respective uniform motions $\boldsymbol{u}^\varphi(0, \boldsymbol{x}) = \boldsymbol{u}_0^\varphi = (\pm1000, 0)$ m/s, $(0, \pm1000)$ m/s, $(\pm2000, 0)$ m/s, and $(0, \pm2000)$ m/s. The velocity field of the air background is initially set so that the mean volume weighted velocity cancels, $\overline{\boldsymbol{u}} = \sum_\varphi \alpha^\varphi \boldsymbol{u}^\varphi = \boldsymbol{0}$. Boundary conditions are perfect zero flux walls. The computation is carried out up to time $t = 2 \times 10^{-3}$ s on an initially-Cartesian shrink-then-stretch swirling ALE grid

[10, eq. 70]: the grid is shrunk to about 60% at $t = 10^{-3}$ s and swirled at the center by about a half turn at final time.

Figure 1 displays the gas volume fraction maps produced at $t = 0$, 10^{-3}, and 2×10^{-3} s (respectively left, center, and right) by the present GEEC multi-fluid scheme on a 480×480 mesh at CFL $= 0.7$. At $t = 10^{-3}$ s, all the clouds cross (without merging) at the origin where the air volume fraction drops to $\alpha^{\text{Air}} \approx 3.8\%$. At final time $t = 2 \times 10^{-3}$ s, the clouds occupy opposite positions with respect to the initial configuration, their trajectory being marginally affected by the crossing and by the air motion despite the large volume fraction variations and the severe mesh distortion. The only visible distortion of the clouds is the expected smearing due to numerical diffusion.

References

1. Benson, D.: Computational methods in Lagrangian and Eulerian hydrocodes. Comput. Methods Appl. Mech. Eng. **99**(2–3), 235–394 (1992). https://doi.org/10.1016/0045-7825(92)90042-I

2. Ge, Z., Marsden, J.: Lie-Poisson Hamilton-Jacobi theory and Lie-Poisson integrators. Phys. Lett. A **133**(3), 134–139 (1988). https://doi.org/10.1016/0375-9601(88)90773-6

3. Godunov, S.: A difference method for numerical calculation of discontinuous solutions of the equations of hydrodynamics. Mat. Sb. (NS) **47(89)**(3), 271–306 (1959). https://mi.mathnet.ru/eng/msb4873

4. Llor, A., Claisse, A., Fochesato, C.: Energy preservation and entropy in Lagrangian space- and time-staggered hydrodynamic schemes. J. Comput. Phys. **309**, 324–349 (2016). https://doi.org/10.1016/j.jcp.2015.12.044

5. Marbœuf, A.: Schémas ALE multi-matériaux totalement conservatifs pour l'hydrodynamique. Ph.D. thesis, Université Paris-Saclay, France (2018). 2018SACLX015, https://www.theses.fr/2018SACLX015

6. Mattsson, A., Rider, W.: Artificial viscosity: back to the basics. Int. J. Numer. Methods Fluids **77**(7), 400–417 (2015). https://doi.org/10.1002/fld.3981

7. Roe, P.: Multidimensional upwinding. In: Abgrall, R., Shu, C. (eds.) Handbook of Numerical Methods for Hyperbolic Problems, Chap. 3. Handbook of Numerical Analysis, vol. 18, pp. 53–80. Elsevier, Amsterdam (2017). https://doi.org/10.1016/bs.hna.2016.10.009

8. Roe, P.: Is discontinuous reconstruction really a good idea? J. Sci. Comput. **73**(2), 1094–1114 (2017). https://doi.org/10.1007/s10915-017-0555-z

9. Tonti, E.: Why starting from differential equations for computational physics? J. Comput. Phys. **257**(B), 1260–1290 (2014). https://doi.org/10.1016/j.jcp.2013.08.016

10. Vazquez-Gonzalez, T., Llor, A., Fochesato, C.: A mimetic numerical scheme for multi-fluid flows with thermodynamic and geometric compatibility on an arbitrarily moving grid. Comput. Fluids (submitted)

11. Vazquez-Gonzalez, T., Llor, A., Fochesato, C.: A novel GEEC (Geometry, Energy, and Entropy Compatible) procedure applied to a staggered direct-ALE scheme for hydrodynamics. Eur. J. Mech. B Fluids **65**, 494–514 (2017). https://doi.org/10.1016/j.euromechflu.2017.05.003

12. von Neumann, J., Richtmyer, R.: A method for the numerical calculation of hydro-dynamic shocks. J. Appl Phys. **21**(3), 232–237 (1950). https://doi.org/10.1063/1.1699639

13. Wilkins, M.: Calculation of elastic-plastic flow. In: Alder, B., Fernbach, S., Rotenberg, M. (eds.) Fundamental Methods in Hydrodynamics, Methods in Computational Physics: Advances in Research and Applications, vol. 3, pp. 211–262. Academic Press, London (1964)

Optimization on Manifold

Canonical Moments for Optimal Uncertainty Quantification on a Variety

Jérôme Stenger[1,2]([⊠]), Fabrice Gamboa[1], Merlin Keller[2], and Bertrand Iooss[1,2]

[1] Institut Mathématiques de Toulouse, 118 route de Narbonne, Toulouse, France
`fabrice.gamboa@math.univ-toulouse.fr`
[2] EDF R&D, 6 quai Watier, Chatou, France
`{jerome.stenger,merlin.keller,bertrand.iooss}@edf.fr`

Abstract. The purpose of this work is to optimize an affine functional over positive measures. More precisely, we deal with a probability of failure (P.O.F). The optimization is realized over a set of distributions satisfying moment constraints, called moment set. The optimum is to be found on an extreme point of this moment set. Winkler's classification of those extreme points states they are finite discrete measures. The set of the support points of all discrete measures in the moment set is a manifold over which the P.O.F is optimized. We characterize the manifold's structure by proving it is an algebraic variety. It is the zero locus of polynomials defined thanks to the canonical moments. This reduces a highly constrained optimization over the moment set onto a constraint free manifold.

Keywords: Canonical moments · Optimal uncertainty quantification · Robustness

1 Introduction

1.1 Probability of Failure Inference

Computer codes are increasingly used to measure safety margins, especially in nuclear accident management analysis. In this context, it is essential to evaluate the accuracy of the numerical model results, whose uncertainties come mainly from the lack of knowledge of the underlying physic and the model input parameters. Methods were developed in safety analyses to quantify those uncertainties [6]. Their common principle relies mainly on a probabilistic modeling of the model input uncertainties, on Monte Carlo sampling for running the computer code on sets of input, and on the application of statistical tools to infer probabilities of failure (P.O.F) of the scalar output variables of interest [13].

This takes place in a more general setting, known as Uncertainty Quantification (UQ) methods [10]. Quantitative assessment of the uncertainties tainting

---

The authors are grateful to the anonymous reviewers for their helpful comments.

F. Nielsen and F. Barbaresco (Eds.): GSI 2019, LNCS 11712, pp. 571–578, 2019.
https://doi.org/10.1007/978-3-030-26980-7_59

the results of computer simulations is a major topic of interest in both industrial and scientific communities.

P.O.F inference is tainted by the uncertainty of the input modeling. More specifically, the inputs probability densities are usually chosen in parametric families (uniform, normal, log-normal, etc.), and their parameters are estimated using available datas and/or the opinion of an expert. However, they may differ from the reality. This uncertainty on the input probability densities is propagated to the P.O.F. As a consequence, different choices of distributions will lead to different P.O.F values, thus different safety margins.

1.2   Optimal Uncertainty Quantification

In this work, we propose to gain robustness on the quantification of this measure of risk. We aim to account for the uncertainty on the input distributions by evaluating the minimal P.O.F over a class of probability measures $\mathcal{A}$. In this optimization problem, the set $\mathcal{A}$ must be large enough to effectively represent our uncertainty on the inputs, but not too large in order to keep the estimation of the quantile representative of the physical phenomena. For example, the minimal P.O.F over the very large class $\mathcal{A} = \{$all distributions$\}$, proposed in [5], will certainly be too conservative to remain physically meaningful. Several articles which discuss possible choices of classes of distributions can be found in the literature of Bayesian robustness (see [11]). Deroberts et al. [2] consider a class of measures specified by a type of upper and lower envelope on their density. Sivaganesan et al. [12] study the class of unimodal distributions. In more recent work, Owhadi et al. [9] propose to optimize the measure of risk over a class of distributions specified by constraints on their *generalized* moments. They call their work Optimal Uncertainty Quantification (OUQ). However, in practical engineering cases, the available information on an input distribution is often reduced to the knowledge of its mean and/or variance. This is why in this paper, we are interested in a specific case of the framework introduced in [9]. We consider the class of measures known by some of their *classical* moments, we refer to it as the moment class:

$$\mathcal{A} = \left\{ \mu = \otimes \mu_i \in \bigotimes_{i=1}^{d} \mathcal{M}_1([l_i, u_i]) \mid \mathbb{E}_{\mu_i}[x^j] = c_j^{(i)}, \right. \tag{1}$$

$$\left. c_j^{(i)} \in \mathbb{R}, \text{ for } 1 \leq j \leq N_i \text{ and } 1 \leq i \leq d \right\},$$

where $\mathcal{M}_1([l_i, u_i])$ denotes the set of scalar probability measures on the interval $[l_i, u_i]$. The tensorial product of measure set traduces the mutual independence of the d-components of the input vector μ.

1.3   Reduction Theorem

The solution of our optimization problem is numerically computed thanks to the OUQ reduction theorem [9,14]. This theorem states that the measure corresponding to the minimal P.O.F is located on the extreme points of the distribution set. In the context of the moment class, the extreme distributions are located on the d-fold product of finite convex combinations of Dirac masses:

$$\mathcal{A}_\Delta = \left\{ \mu \in \mathcal{A} \mid \mu_i = \sum_{k=1}^{N_i+1} w_k^{(i)} \delta_{x_k^{(i)}} \text{ for } 1 \leq i \leq d \right\}, \qquad (2)$$

To be more specific it holds that when n pieces of information are available on the moments of a scalar measure μ, it is enough to pretend that the measure is supported on at most $n + 1$ points. This powerful theorem gives the basis for practical optimization of our optimal quantity of interest. In this matter, Semi-Definite-Programming [4] has been already explored in [1] and [7], but the deterministic solver used rapidly reaches its limitation as the dimension of the problem increases. One can also find in the literature a Python toolbox developed by McKerns [8] called Mystic framework that fully integrates the OUQ framework. However, it was built as a generic tool for generalized moment problems and the enforcement of the moment constraints is not optimal.

By restricting the work to our moment class, we propose an original and practical approach based on the theory of canonical moments [3]. Canonical moments of a measure can be seen as the relative position of its moment sequence in the moment space. They are inherent to the measure and therefore present many interesting properties. Our main contribution is in the proof that the optimization set $\mathcal{A}_\Delta$ is an algebraic manifold, more specifically it is the zero locus of polynomials whose coefficients are function of canonical moments. This geometric approach replaces the optimization on the constrained space in Eq. (2) into a constraint free optimization.

This paper proceeds as follows. In Sect. 2 we develop the reduction theorem and the parameterization of the optimization space, we present the manifold over which the optimization takes place. Section 3 is dedicated to the canonical moments and the construction of the polynomials of interest. We show that the zero locus of those polynomials constitute the optimization space. Section 4 gives some conclusions and perspectives.

2 Problem Reduction

2.1 OUQ Theorem

In this work, we consider a P.O.F on the output of a computer code $G : \mathbb{R}^d \to \mathbb{R}$, seen as a black box function. In order to gain robustness on our safety margin choice, our goal is to find the minimal P.O.F over the moment set $\mathcal{A}$ described in Eq. (1). For a given threshold h, it reads:

$$\inf_{\mu \in \mathcal{A}} P_\mu(G(X) \leq h) \qquad (3)$$

The OUQ reduction theorem applies (Theorem 1). It states that the optimal solution of the optimization problem (3) is a product of discrete measures. A general form of the theorem reads as follows:

Theorem 1 (OUQ reduction [9]**).** *Suppose that* $\mathcal{X} := \mathcal{X}_1 \times \cdots \times \mathcal{X}_d$ *is a product of Radon spaces. Let*

$$
\mathcal{A} := \left\{ (G,\mu) \left| \begin{array}{l} G : \mathcal{X} \to \mathcal{Y}, \text{ is a real valued measurable function,} \\ \mu = \mu_1 \otimes \cdots \otimes \mu_d \in \bigotimes_{i=1}^{d} \mathcal{M}_1(\mathcal{X}_i), \\ \text{for some integers } N_0, \ldots, N_d, \text{ and measurable functions} \\ \varphi_l : \mathcal{X} \to \mathbb{R} \text{ and } \varphi_j^{(i)} : \mathcal{X}_i \to \mathbb{R}, \\ \quad \bullet \ \mathbb{E}_\mu[\varphi_l] \leq 0 \text{ for } l = 1, \ldots, N_0, \\ \quad \bullet \ \mathbb{E}_{\mu_i}[\varphi_j^{(i)}] \leq 0 \text{ for } j = 1, \ldots, N_i \text{ and } i = 1, \ldots, d \end{array} \right. \right\}
$$

Let $\Delta_n(\mathcal{X})$ *be the set of all discrete measures supported on at most* $n+1$ *points of* $\mathcal{X}$*, and*

$$
\mathcal{A}_\Delta := \{(G,\mu) \in \mathcal{A} \mid \mu_i \in \Delta_{N_0 + N_i}(\mathcal{X}_i)\}.
$$

Let q *be a measurable real function on* $\mathcal{X} \times \mathcal{Y}$*. Then*

$$
\sup_{(G,\mu) \in \mathcal{A}} \mathbb{E}_\mu[q(X, G(X))] = \sup_{(G,\mu) \in \mathcal{A}_\Delta} \mathbb{E}_\mu[q(X, G(X))].
$$

This theorem derives from the work of Winkler [14], who has shown that the extreme measures of moment class $\{\mu \in \mathcal{M}_1(\mathcal{X}) \mid \mathbb{E}_\mu[\varphi_1] \leq 0, \ldots, \mathbb{E}_\mu[\varphi_n] \leq 0\}$ are the discrete measures that are supported on at most $n+1$ points. The strength of Theorem 1 is that it extends the result to a tensorial product of moment sets. The proof relies on a recursive argument using Winkler's classification on every set $\mathcal{X}_i$. A remarkable fact is that, as long as the quantity to be optimized is an affine function of the underlying measure μ, this theorem remains true whatever the function G and the quantity of interest q are.

Now, by taking $\varphi_j^{(i)}(x) = x^j$ for $1 \leq i \leq N_i$, we enforced N_i moment constraints to μ_i, as in Eq. (1). Applying Theorem 1 to the function $q(X, G(X)) = -\mathbb{1}_{\{G(X) \leq h\}}$, the P.O.F reaches its optimum on the reduced set $\mathcal{A}_\Delta$, such that for a fixed threshold h we have:

$$
\begin{aligned}
\inf_{\mu \in \mathcal{A}} F_\mu(h) &= \inf_{\mu \in \mathcal{A}_\Delta} \mathbb{E}_\mu[\mathbb{1}_{\{G(X) \leq h\}}], \\
&= \inf_{\mu \in \mathcal{A}_\Delta} P_\mu(G(X) \leq h), \\
&= \inf_{\mu \in \mathcal{A}_\Delta} \sum_{i_1=1}^{N_1+1} \cdots \sum_{i_d=1}^{N_d+1} \omega_{i_1}^{(1)} \ldots \omega_{i_d}^{(d)} \ \mathbb{1}_{\{G(x_{i_1}^{(1)}, \ldots, x_{i_d}^{(d)}) \leq h\}},
\end{aligned} \tag{4}
$$

2.2   Parameterization Simplification

The optimization problem in Eq. (4) shows that the weights and the positions of the input distributions provide a natural parameterization for the computation of the P.O.F. However, we now highlight the fact that the knowledge of the support points of a discrete measure (Eq. (5)) fully determines the corresponding weights. Hence, the support points are sufficient to compute the P.O.F (Eq. (4)). Indeed, we recall that in the optimization set $\mathcal{A}_\Delta$ (Eq. (2)), N_i constraints are enforced on the moment of the ith input. The measure μ_i is therefore supported on at most $N_i + 1$ points, which reads:

$$\mu_i = \sum_{i=1}^{N_i} w_j^{(i)} \delta_{x_j^{(i)}} \tag{5}$$

The $N_i + 1$ Vandermonde system holds

$$\begin{cases} \omega_1^{(i)} & + \ldots + \omega_{N_i+1}^{(i)} & = 1 \\ \omega_1^{(i)} x_1^{(i)} & + \ldots + \omega_{N_i+1}^{(i)} x_{N_i+1}^{(i)} & = c_1^{(i)} \\ \vdots & \vdots & \vdots \\ \omega_1^{(i)} {x_1^{(i)}}^{N_i} & + \ldots + \omega_{N_i+1}^{(i)} {x_{N_i+1}^{(i)}}^{N_i} & = c_{N_i}^{(i)} \end{cases} \tag{6}$$

where the N_i last equations derive from the constraints and the first one is the expression of the measure mass equals to one. Because every support points $(x_j^{(i)})_j$ are distinct, when they are set, the corresponding weights are uniquely determined.

The optimization problem in Eq. (4) is therefore parameterized only with the position of the support points of every input, so that the optimization takes place on the following manifold:

$$\mathcal{V} = \prod_{i=1}^{d} \mathcal{V}_i,$$

$$= \prod_{i=1}^{d} \left\{ \mathbf{x}_i = \left(x_1^{(i)}, \ldots, x_{N_i+1}^{(i)} \right) \in \mathbb{R}^{N_i+1}, \text{ s.t } \mu_i = \sum_{j=1}^{N_i+1} \omega_j^{(i)} \delta_{x_j^{(i)}} \in \mathcal{A}_\Delta^{(i)} \right\}, \tag{7}$$

where $\mathcal{A}_\Delta^{(i)}$ is such that $\mathcal{A}_\Delta = \bigotimes_{i=1}^{d} \mathcal{A}_\Delta^{(i)}$, this reads

$$\mathcal{A}_\Delta^{(i)} = \left\{ \mu_i = \sum_{k=1}^{N_i+1} \omega_k^{(i)} \delta_{x_k^{(i)}} \mid \mathbb{E}_{\mu_i}[x^j] = c_j^{(i)}, \text{ for } 1 \leq j \leq N_i \right\}. \tag{8}$$

$\mathcal{V}_i$ is simply the set of support points of all measures in $\mathcal{A}_\Delta^{(i)}$ respecting the constraints. Our main contribution in this work is to show that $\mathcal{V}_i$ is an algebraic manifold, meaning it is the zero locus of some well defined polynomials.

3 Optimization Space Seen as a Variety

3.1 Canonical Moments

We define the moment space $M := M(a, b) = \{\mathbf{c}(\mu) \mid \mu \in \mathcal{M}_1([a, b])\}$ where $\mathbf{c}(\mu)$ denotes the sequence of all moments of some measure μ. The nth moment space M_n is defined by projecting M onto its first n coordinates, $M_n = \{\mathbf{c}_n(\mu) = (c_1, \ldots, c_n) \mid \mu \in \mathcal{M}_1([a, b])\}$. We now define the extreme values,

$$c_{n+1}^+ = \max\{c \in \mathbb{R} : (c_1, \ldots, c_n, c) \in M_{n+1}\},$$
$$c_{n+1}^- = \min\{c \in \mathbb{R} : (c_1, \ldots, c_n, c) \in M_{n+1}\},$$

which represent the maximum and minimum values of the $(n + 1)$th moment that a measure can have, when its moments up to order n are fixed. The nth canonical moment is then defined recursively as

$$p_n = p_n(\mathbf{c}) = \frac{c_n - c_n^-}{c_n^+ - c_n^-}. \tag{9}$$

Note that the canonical moments are defined up to the degree $N = N(\mathbf{c}) = \min\{n \in \mathbb{N} \mid \mathbf{c}_n \in \partial M_n\}$, and p_N is either 0 or 1. Indeed, we know from [3, Theorem 1.2.5] that $\mathbf{c}_n \in \partial M_n$ implies that the underlying μ is uniquely determined, so that, $c_n^+ = c_n^-$. We also introduce the quantity $\zeta_n = (1 - p_{n-1})p_n$ that will be of some importance in the following. The very nice properties of canonical moments are that, by construction, they belong to $[0, 1]$ and are invariant under linear transformation of the measures, $y = a + (b - a)x$. Hence, we may restrict ourselves to the case $a = 0$, $b = 1$.

3.2   Support Points and Canonical Moments

From a given sequence of canonical moments, one wishes to reconstruct the support of a discrete measure. This link arises through the following theorem

Theorem 2 ([3, Theorem 3.6.1]). *Let μ denote a measure on the interval $[a, b]$ supported on $n + 1$ points with canonical moments $p_1, p_2, \ldots$. Then the support of μ consists of the zeros of $P_{n+1}^*(x)$ where*

$$P_{k+1}^*(x) = (x - a - (b - a)(\zeta_{2k} + \zeta_{2k+1}))P_k^*(x) - (b - a)^2 \zeta_{2k-1}\zeta_{2k}P_{k-1}^*, \tag{10}$$

with $P_{-1}^*(x) = 0$, $P_0^*(x) = 1$ and $\zeta_k = (1 - p_{k-1})p_k$

The polynomial P_{n+1}^* is defined with the sequence of canonical moments up to order $2n + 1$. In the following, we consider a fixed sequence of moments $\mathbf{c}_n = (c_1, \ldots, c_n) \in M_n$, let μ be a measure supported on at most $n + 1$ points, with classical moments $\mathbf{c}_n$. Hence, μ has canonical moments equal to $\mathbf{p}_n = (p_1, \ldots, p_n)$ the corresponding sequence of canonical moments related to $\mathbf{c}_n$, as described in Sect. 3.1. We define the set $\Theta_{n+1} = \{\mathbf{x} \in [0, 1]^{n+1} \mid x_i \in \{0, 1\} \Rightarrow x_k = 0, k > i\}$ and the functional:

$$\phi_{\mathbf{p}_n} : \qquad \Theta_{n+1} \qquad \to \mathbb{R}[X]$$
$$(p_{n+1}, \ldots, p_{2n+1}) \mapsto P_{n+1}^*, \tag{11}$$

The function ϕ computes, from a sequence of canonical moments $(p_1, \ldots, p_{2n+1})$, a polynomial P_{n+1}^* in regards of Theorem 2. Therefore, the roots of P_{n+1}^* correspond to the support of a measure with moments $\mathbf{c}_n$. We derive the following Theorem, it is the geometric version of Theorem 2.

Theorem 3. *The set $\mathcal{V}_i$ of (N_i+1)-tuples corresponding to the support points of a discrete measure with prescribed first N_i moments $(c_1^{(i)}, \ldots, c_{N_i}^{(i)})$ is an algebraic manifold of $\mathbb{R}^{N_i+1}$. It is the zeros locus of the set of polynomials:*

$$\mathcal{S}_i = \left\{ P_{N_i+1}^* = \phi_{\mathbf{p}_{N_i}}(p_{N_i+1}, \ldots, p_{2N_i+1}), \ (p_{N_i+1}, \ldots, p_{2N_i+1}) \in \Theta_{N_i+1} \right\} \quad (12)$$

In order to optimize our quantity of interest in Eq. (3), one need to explore the space of admissible measures $\mathcal{A}_\Delta$. More precisely, the P.O.F in Eq. (4) is computed over the space $\mathcal{V}$. This space corresponds to the support points of all discrete measures respecting the constraints in $\mathcal{A}_\Delta$. What is interesting is that Θ_{n+1} provides a very simple parameterization of $\mathcal{V}$ through the computation of roots of some well defined polynomial.

An optimization over the highly constrained space $\mathcal{A}_\Delta$ is therefore simplified into a constraint free optimization program over the space Θ_{n+1}.

4   Conclusion

This work aims to evaluate the maximum quantile over a class of distributions constrained by some of their moments. We used the theory of canonical moments into an improved methodology for solving OUQ problems. The set of optimization corresponds to the support points of the discrete measures in the moment set. We have successfully shown it is the zero locus of a set of polynomials defined with canonical moments. The knowledge of the shape of this manifold allows a computational constraint free optimization program, instead of a highly constrained optimization over the moment set.

References

1. Betrò, B., Guglielmi, A.: Methods for global prior robustness under generalized moment conditions. In: Insua, D.R., Ruggeri, F. (eds.) Robust Bayesian Analysis. Lecture Notes in Statistics. Springer, New York (2000). https://doi.org/10.1007/978-1-4612-1306-2_15. www.springer.com/la/book/9780387988665
2. DeRoberts, L., Hartigan, J.A.: Bayesian inference using intervals of measures. Ann. Stat. **9**(2), 235–244 (1981). https://doi.org/10.1214/aos/1176345391. https://projecteuclid.org/euclid.aos/1176345391
3. Dette, H., Studden, W.J.: The Theory of Canonical Moments with Applications in Statistics, Probability, and Analysis. Wiley-Blackwell, New York (1997)
4. Henrion, D., Lasserre, J.B., Löfberg, J.: GloptiPoly 3: moments, optimization and semidefinite programming. Optim. Methods Softw. **24**(4–5), 761–779 (2009). https://doi.org/10.1080/10556780802699201. http://www.tandfonline.com/doi/abs/10.1080/10556780802699201
5. Huber, P.J.: The use of Choquet capacities in statistics. Bull. Int. Stat. Inst. **45**(4), 181–188 (1973)

6. Iooss, B., Marrel, A.: Advanced methodology for uncertainty propagation in computer experiments with large number of inputs. Nuclear Technol. 1–19 (2019). https://doi.org/10.1080/00295450.2019.1573617
7. Lasserre, J.B.: Moments, Positive Polynomials and Their Applications. Imperial College Press Optimization Series, vol. 1. Imperial College Press, London (2010). Distributed by World Scientific Publishing Co., Singapore. oCLC: ocn503631126
8. McKerns, M., Owhadi, H., Scovel, C., Sullivan, T.J., Ortiz, M.: The optimal uncertainty algorithm in the mystic framework. CoRR abs/1202.1055 (2012). http://arxiv.org/abs/1202.1055
9. Owhadi, H., Scovel, C., Sullivan, T.J., McKerns, M., Ortiz, M.: Optimal uncertainty quantification. SIAM Rev. **55**(2), 271–345 (2013). https://doi.org/10.1137/10080782X. arXiv:1009.0679
10. de Rocquigny, E., et al.: Uncertainty in industrial practice: a guide to quantitative uncertainty management (2008). https://doi.org/10.1002/9780470770733
11. Ruggeri, F., Rios-Insua, D., Martin, J.: Robust Bayesian analysis. In: Dey, D.K., Rao, C.R. (eds.) Handbook of Statistics, Bayesian Thinking, vol. 25, pp. 623–667. Elsevier, Amsterdam (2005). https://doi.org/10.1016/S0169-7161(05)25021-6. http://www.sciencedirect.com/science/article/pii/S0169716105250216
12. Sivaganesan, S., Berger, J.O.: Ranges of posterior measures for priors with unimodal contaminations. Ann. Stat. **17**(2), 868–889 (1989). https://doi.org/10.1214/aos/1176347148. https://projecteuclid.org/euclid.aos/1176347148
13. Wallis, G.: Uncertainties and probabilities in nuclear reactor regulation. Nuclear Eng. Des. **237**, 1586–1592 (2004)
14. Winkler, G.: Extreme points of moment sets. Math. Oper. Res. **13**(4), 581–587 (1988). https://doi.org/10.1287/moor.13.4.581

Computational Investigations of an Obstacle-Type Shape Optimization Problem in the Space of Smooth Shapes

Daniel Luft[1] and Kathrin Welker[2(✉)]

[1] Trier University, Universitätsring 15, 54296 Trier, Germany
luft@uni-trier.de
[2] Helmut-Schmidt-University/University of the Federal Armed Forces Hamburg,
Holstenhofweg 85, 22043 Hamburg, Germany
welker@hsu-hh.de

Abstract. We investigate and computationally solve a shape optimization problem constrained by a variational inequality of the first kind, a so-called obstacle-type problem, with a gradient descent and a BFGS algorithm in the space of smooth shapes. In order to circumvent the numerical problems related to the non-linearity of the shape derivative, we consider a regularization strategy leading to novel possibilities to numerically exploit structures, as well as possible treatment of the regularized variational inequality constrained shape optimization in the context of optimization on infinite dimensional Riemannian manifolds.

Keywords: Variational inequality · Obstacle problem ·
Shape manifold · Gradient method · BFGS method

1 Introduction

Shape optimization is a classical topic in mathematics which is of high importance in a wide range of applications, e.g., acoustics [23], aerodynamics [19] and electrostatics [6]. Qualitative properties of optimal shapes such as minimum surfaces are investigated in classical shape optimization. In select cases, an analytical solution can be derived. In contrast, modern and application-oriented questions in shape optimization are concerned with specific calculations of shapes which are optimal with respect to a process which is mostly described by partial differential equations (PDE) or variational inequalities (VI). Consequently, the area of shape optimization builds a bridge between pure and applied mathematics. Recently, shape optimization gained new interest due to novel developments such as the usage of volumetric/weak formulations of shape derivatives. This paper, which focuses on VI constrained shape optimization problems, is based on recent results in the field of PDE constrained shape optimization and carries the achieved methodology over to shape optimization problems with constraints in the form of VIs. Thus, this paper can be seen as an extension of the Riemannian shape optimization framework for PDEs formulated in [24] to VI. Note that

© Springer Nature Switzerland AG 2019
F. Nielsen and F. Barbaresco (Eds.): GSI 2019, LNCS 11712, pp. 579–588, 2019.
https://doi.org/10.1007/978-3-030-26980-7_60

VI constrained shape optimization problems are very challenging because of the two main reasons: One needs to operate in inherently non-linear, non-convex and infinite-dimensional shape spaces and—in contrast to PDEs—one cannot expect the existence of shape derivatives for an arbitrary shape functional depending on solutions to variational inequalities.

So far, there are only very few approaches in the literature to the problem class of VI constrained shape optimization problems. In [12], shape optimization of 2D elasto-plastic bodies is studied, where the shape is simplified to a graph such that one dimension can be written as a function of the other. In [22, Chap. 4], shape derivatives of elliptic VI problems are presented in the form of solutions to again VIs. In [18], shape optimization for 2D graph-like domains are investigated. Also [14] presents existence results for shape optimization problems which can be reformulated as optimal control problems, whereas [4,7] show existence of solutions in a more general set-up. In [18], level-set methods are proposed and applied to graph-like two-dimensional problems. Moreover, [8] presents a regularization approach to the computation of shape and topological derivatives in the context of elliptic VIs and, thus, circumventing the numerical problems in [22, Chap. 4]. However, all these mentioned problems have in common that one cannot expect for an arbitrary shape functional depending on solutions to VIs to obtain the shape derivative as a linear mapping (cf. [22, Example in Chap. 1]). E.g., in general, the shape derivative for the obstacle problems fails to be linear with respect to the normal component of the vector field defined on the boundary of the domain under consideration. In order to circumvent the numerical problems related to the non-linearity of the shape derivative (cf., e.g., [22, Chap. 4]) and in particular the non-existence of the shape derivative of a VI constrained shape optimization problem, [8] presents a regularization approach to the computation of shape and topological derivatives in the context of elliptic VIs. In this paper, we consider this regularization strategy, leading to novel possibilities to numerically exploit structures, as well as possible treatment of the regularized VI constrained shape optimization in the context of optimization on infinite dimensional manifolds.

This paper is structured as follows. In Sect. 2, we give a brief overview of the VI constrained shape optimization model class and regularization techniques on which we focus in this paper. Section 3 presents a way to solve the VI constrained shape model problem in the space of smooth shapes based on gradient representations via Steklov-Poincaré metrics. Finally, numerical results of the gradient descent and a Broyden-Fletcher-Goldfarb-Shanno (BFGS) algorithm are presented in Sect. 4.

2   VI Constrained Model Problem

Let $\Omega \subset \mathbb{R}^2$ be a bounded domain equipped with a sufficiently smooth boundary $\partial\Omega$, which we will specify in more detail after stating the model problem. This domain is assumed to be partitioned in a subdomain $\Omega_{\text{out}} \subset \Omega$ and an interior domain $\Omega_{\text{int}} \subset \Omega$ with boundary $\Gamma_{\text{int}} := \partial\Omega_{\text{int}}$ such that $\Omega_{\text{out}} \sqcup \Omega_{\text{int}} \sqcup \Gamma_{\text{int}}$

$= \Omega$, where $\sqcup$ denotes the disjoint union. We consider Ω depending on Γ_{int}, i.e., $\Omega = \Omega(\Gamma_{\text{int}})$. In the following, the boundary Γ_{int} of the interior domain is called the interface. In contrast to the outer boundary $\partial\Omega$, which is fixed, the inner boundary Γ_{int} is variable. The interface is an element of an appropriate shape space. In this paper, we focus on the space of one-dimensional smooth shapes (cf. [16]) characterized by $B_e := B_e(S^1, \mathbb{R}^2) := \text{Emb}(S^1, \mathbb{R}^2)/\text{Diff}(S^1)$, i.e., the orbit space of $\text{Emb}(S^1, \mathbb{R}^2)$ under the action by composition from the right by the Lie group $\text{Diff}(S^1)$. Here, $\text{Emb}(S^1, \mathbb{R}^2)$ denotes the set of all embeddings from the unit circle S^1 into $\mathbb{R}^2$, which contains all simple closed smooth curves in $\mathbb{R}^2$. Note that we can think of smooth shapes as the images of simple closed smooth curves in the plane of the unit circle because the boundary of a shape already characterizes the shape. The set $\text{Diff}(S^1)$ is the set of all diffeomorphisms from S^1 into itself, which characterize all smooth reparametrizations. These equivalence classes are considered because we are only interested in the shape itself and images are not changed by reparametrizations. More precisely, shapes that have been translated represent the same shape. In contrast, shapes with different scaling are not equivalent in this shape space. In [13], it is proven that the shape space $B_e(S^1, \mathbb{R}^2)$ is a smooth manifold. For the sake of completeness it should be mentioned that the shape space $B_e(S^1, \mathbb{R}^2)$ together with appropriate inner products is even a Riemannian manifold. In [17], a survey of various suitable inner products is given. In the following, we assume $\Gamma_{\text{int}} \in B_e$.

Let $\nu > 0$ be an arbitrary constant, $\bar{y} \in L^2(\Omega)$ and y solving the VI formulated in (3). For the objective function

$$J(y, \Omega) := \mathcal{J}(y, \Omega) + \mathcal{J}_{\text{reg}}(\Omega) := \frac{1}{2} \int_{\Omega(\Gamma_{\text{int}})} |y - \bar{y}|^2 \, dx + \nu \int_{\Gamma_{\text{int}}} 1 \, ds \quad (1)$$

we consider the following VI constrained shape optimization problem:

$$\min_{\Gamma_{\text{int}} \in B_e} J(y, \Omega) \quad (2)$$

with y solving the following obstacle type variational inequality:

$$a(y, v - y) \geq \langle f, v - y \rangle \quad \forall v \in K := \{\theta \in H_0^1(\Omega) : \theta(x) \leq \psi(x) \text{ in } \Omega\}, \quad (3)$$

where $f \in L^2(\Omega)$ dependents on the shape, $\langle \cdot, \cdot \rangle$ denotes the duality pairing and $a(\cdot, \cdot)$ is a general bilinearform $a \colon H_0^1(\Omega) \times H_0^1(\Omega) \to \mathbb{R}$, $(y, v) \mapsto \sum_{ij} a_{ij} y_{x_i} v_{x_j} + byv$ defined by coefficient functions $a_{ij}, b \in L^\infty(\Omega)$, $b \geq 0$.

With the tracking-type objective $\mathcal{J}$ the model is fitted to data measurements $\bar{y} \in L^2(\Omega)$. The second term $\mathcal{J}_{\text{reg}}$ in the objective function J is a perimeter regularization, which is frequently used to overcome ill-posedness of shape optimization problems. In (3), ψ denotes an obstacle which needs to be an element of $L^1_{\text{loc}}(\Omega)$ such that the set of admissible functions K is non-empty (cf. [22]). Then smoothness of the boundary $\partial\Omega$, where $C^{1,1}$ regularity or polyhedricity is sufficient, and $\psi \in H^2(\Omega)$ ensure that the solution to (3) satisfies $y \in H_0^1(\Omega)$, see, e.g., [9, Remark 2.3]. Further, (3) can be equivalently expressed as a PDE with complementary constraints (cf. [11]):

$$a(y,v) + (\lambda,v)_{L^2(\Omega)} = (f,v)_{L^2(\Omega)} \quad \forall v \in H_0^1(\Omega) \tag{4}$$

$$\lambda \geq 0, \ y \leq \psi, \ \lambda(y-\psi) = 0 \quad \text{in } \Omega \tag{5}$$

The existence of solutions of any shape optimization problem is a non-trivial question. Shape optimization problems constrained by VIs are especially challenging because, in general, the shape derivative of VI constrained shape optimization problems is not linear (cf. [8,22]). This potential non-linearity of the shape derivative complicates its use in algorithms. In order to circumvent the problems related to the non-linearity, we consider a regularized version of (2) constrained by (4)–(5). For convenience, we focus on a special bilinearform: We assume the bilinearform $a(\cdot,\cdot)$ of the state equation to correspond to the Laplacian $-\Delta$. In this setting, a regularized version is given by:

$$\min_{\Gamma_{\text{int}} \in B_e} J(y_{\gamma,c}, \Omega) \tag{6}$$

$$\text{s.t.} \ -\Delta y_{\gamma,c} + \lambda_{\gamma,c} = f \quad \text{in } \Omega \tag{7}$$

$$y_{\gamma,c} = 0 \quad \text{on } \partial\Omega \tag{8}$$

with $\lambda_{\gamma,c} = \max_\gamma(0, \overline{\lambda} + c(y_{\gamma,c} - \psi))$, where $\gamma, c > 0$, $0 \leq \overline{\lambda} \in L^2(\Omega)$ fixed and

$$\max{}_\gamma(x) := \begin{cases} \max(0,x) & \text{for } x \in \mathbb{R}\backslash[-\frac{1}{\gamma}, \frac{1}{\gamma}] \\ \frac{\gamma}{4}x^2 + \frac{1}{2}x + \frac{1}{4\gamma} & \text{else} \end{cases} \tag{9}$$

being a smoothed max-function. The convergence $y_{\gamma,c} \to y$ in $H_0^1(\Omega)$ of the regularized solution $y_{\gamma,c}$ to the unregularized solution y of (4) is guaranteed by a result in [15, Proposition 1]. Furthermore, the smoothness of the regularized PDE (7) guarantees the existence of adjoints, which in turn gives possibility to characterize a corresponding shape derivative of (6). In [8], it is mentioned that for a large parameter c the associated solution of the regularized state equation (7)–(8) using the unsmoothed max-function is an excellent approximation of the solution to the unregularized VI. Moreover, it is shown in [8] that the shape derivative for the regularized problem converges to the solution of a linear problem which depends linearly on a perturbation vector field. Numerical tests in [8] show the efficiency of the approach to introduce a regularization of the VI, which allows to apply the usual theory for obtaining shape derivatives. We refer to [15] for the shape derivative of (6)–(8) and the adjoint equation to (6)–(8), as well as the limiting objects and equations for $\gamma, c \to \infty$. However, we want to point out that a proof of convergence of the optimal shapes generated by the steepest descent or BFGS method using the regularization parameters $\gamma, c > 0$ for $\gamma, c \to \infty$ is yet to be done.

3   Algorithmic Details

This section presents a way to solve (6)–(8) computationally in the Riemannian manifold of smooth shapes. If we want to optimize on a Riemannian shape manifold, we have to find a representation of the shape derivative with respect to the Riemannian metric under consideration, called the *Riemannian shape gradi-ent*, which is required to formulate optimization methods on a shape manifold. In [21], the authors present a metric based on the Steklov-Poincaré operator, which allows for the computation of the Riemannian shape gradient as a repre-sentative of the shape derivative in volume form. Besides saving analytical effort during the calculation process of the shape derivative, this technique is compu-tationally more efficient than using an approach which needs the surface shape derivative form. For example, the volume form allows us to optimize directly over the hold-all domain Ω containing one or more elements $\Gamma_{\text{int}} \in B_e$, whereas the surface formulation would give us descent directions (in normal directions) for the boundary Γ_i only, which would not help us to move mesh elements around the shape. Additionally, when we are working with a surface shape derivative, we need to solve another PDE in order to get a mesh deformation in the hold-all domain Ω as outlined for example in [24].

We denote the shape derivative of J in direction of a vector field U which can be given in volume or surface form by $DJ(\cdot)[U]$. In order to distinguish between surface and volume formulation, we use the notation $DJ^{\text{surf}}(\cdot)[U]$, $DJ^{\text{vol}}(\cdot)[U]$. Following the ideas presented in [21], we choose the Steklov-Poincaré metric

$$G^S : H^{1/2}(\Gamma_{\text{int}}) \times H^{1/2}(\Gamma_{\text{int}}) \to \mathbb{R}, (\alpha, \beta) \mapsto \int_{\Gamma_{\text{int}}} \alpha(s) \cdot [(S^{pr})^{-1}\beta](s) \, ds,$$

where $S^{pr} : H^{-1/2}(\Gamma_{\text{int}}) \to H^{1/2}(\Gamma_{\text{int}})$, $\alpha \mapsto (\gamma_0 V)^T n$ denotes the projected Poincaré-Steklov operator with $\text{tr} : H_0^1(\Omega, \mathbb{R}^2) \to H^{1/2}(\Gamma_{\text{int}}, \mathbb{R}^2)$ denoting the trace operator on Sobolev spaces for vector-valued functions and $V \in H_0^1(X, \mathbb{R}^2)$ solving the Neumann problem

$$a^{\text{deform}}(U, V) = \int_{\Gamma_{\text{int}}} \alpha \cdot (\text{tr}(U))^T n \, ds \quad \forall U \in H_0^1(\Omega, \mathbb{R}^2), \tag{10}$$

where $a^{\text{deform}} : H_0^1(\Omega, \mathbb{R}^2) \times H_0^1(\Omega, \mathbb{R}^2) \to \mathbb{R}$ is a symmetric and coercive bilinear form. If $r \in L^1(\Gamma_{\text{int}})$ denotes the L^2-shape gradient given by the surface formula-tion of the shape derivative $DJ^{\text{surf}}(y_{\gamma,c}, \Omega)[V] = \int_{\Gamma_{\text{int}}} r \langle V, n \rangle \, ds$ with n denoting the normal vector field and $y_{\gamma,c}$ denoting the solution of the regularized state equation (7)–(8), then a representation $h \in T_{\Gamma_{\text{int}}} B_e \cong \mathcal{C}^\infty(\Gamma_{\text{int}})$ of the shape gra-dient in terms of G^S is determined by $G^S(\phi, h) = (r, \phi)_{L^2(\Gamma_{\text{int}})} \ \forall \phi \in \mathcal{C}^\infty(\Gamma_{\text{int}})$. From this we get that the mesh deformation vector $V \in H_0^1(\Omega, \mathbb{R}^2)$ can be viewed as an extension of a Riemannian shape gradient to the hold-all domain Ω because of the identities

$$G^S(v, u) = DJ(y_{\gamma,c}, \Omega)[U] = a^{\text{deform}}(V, U) \quad \forall U \in H_0^1(\Omega, \mathbb{R}^2), \tag{11}$$

where $v = (\text{tr}(V))^T n, u = (\text{tr}(U))^T n \in T_{\Gamma_{\text{int}}} B_e$ with $T_{\Gamma_{\text{int}}} B_e \cong \mathcal{C}^\infty(\Gamma_{\text{int}})$.

One option to choose the operator $a^{\text{deform}}(\cdot, \cdot)$ is the bilinear form associated with the linear elasticity problem, i.e., $a^{\text{elas}}(V, U) := \int_\Omega (\lambda \text{tr}(\epsilon(V)) \text{id} + 2\mu\epsilon(V)) : \epsilon(U) \, dx$, where $\epsilon(U) := \frac{1}{2}(\nabla U + \nabla U^T)$, $A : B$ denotes the Frobenius inner product for two matrices A, B and λ, μ denote the so-called Lamé parameters. To summarize, we need to solve the following *deformation equation*: find $V \in H_0^1(\Omega, \mathbb{R}^2)$ s.t.

$$a^{\text{elas}}(V, U) = DJ(y_{\gamma, c}, \Omega)[U] \quad \forall U \in H_0^1(\Omega, \mathbb{R}^2). \tag{12}$$

In this equation, we need the solution $y_{\gamma, c}$ of the regularized state equation (7)–(8), and the solution $p_{\gamma, c}$ of the corresponding adjoint equation (cf. [15, Chapter 3]) in order to construct $DJ(y_{\gamma, c}, \Omega)[\cdot]$. An alternative strategy to the regularization outlined is the *linearized modified primal-dual active set (lmP-DAS) algorithm* formulated in [5, Algorithm 2]. The lmPDAS algorithm is based on the primal-dual active set (PDAS) algorithm given in [10] and on a linearization technique inspired by the concept of internal numerical differentiation [3].

The Riemannian shape gradient is required to formulate optimization methods in the shape space B_e. In the setting of constrained shape optimization problems, a Lagrange-Newton method is obtained by applying a Newton method to find stationary points of the Lagrangian of the optimization problem. In contrast to this method, which requires the Hessian in each iteration, quasi-Newton methods only need an approximation of the Hessian. Such an approximation is realized, e.g., by a limited memory Broyden-Fletcher-Goldfarb-Shanno (BFGS) update. In the Steklov-Poincaré setting, such an update can be computed with the representation of the shape gradient with respect to G^S and a suitable vector transport (cf. [21]). The limited memory BFGS method (l-BFGS) for iteration j is summarized in Algorithm 1, where $l \in \{2, 3, \dots\}$ is the memory-length, $V_i \in H_0^1(\Omega, \mathbb{R}^2)$ are the volume representations of the gradients in $T_{\Gamma_{\text{int}_i}} B_e$ as by (12), $S_i \in H_0^1(\Omega)$ are the BFGS deformations generated in iteration i, $\mathcal{T}_{S_{j-1}}$ denotes the vector transport associated to the update $\Omega_j = \exp_{\Omega_{j-1}}(\text{tr}(S_{j-1})^T n)$ and $Y_i := V_{i+1} - \mathcal{T}_{S_i} V_i \in H_0^1(\Omega, \mathbb{R}^2)$.

Remark 1. In general, we need the concept of the exponential map and vector transports in order to formulate optimization methods on a shape manifold. The calculations of optimization methods have to be performed in tangent spaces because manifolds are not necessarily linear spaces. This means points from a tangent space have to be mapped to the manifold in order to get a new shape-iterate, which can be realized with the help of the exponential map as used in Algorithm 1. However, the computation of the exponential map is prohibitively expensive in the most applications because a calculus of variations problem must be solved or the Christoffel symbols need be known. It is much easier and much faster to use a first-order approximation of the exponential map. In [1], it is shown that a so-called *retraction* is such a first-order approximation and sufficient in most applications. We refer to [20], where a suitable retraction and vector transport on B_e are given.

4   Numerical Results

We focus on a numerical experiment which is selected in order to demonstrate challenges arising for VI constrained shape optimization problems. To be more precise, we move and magnify a circle in the domain $\Omega = (0,1)^2$.

Algorithm 1. Inverse limited memory BFGS update in terms of the metric G^S.

$q \leftarrow V_j$
$Y_{j-1} \leftarrow V_j - \mathcal{T}_{S_{j-1}} V_{j-1}$
for $i = j-2, \ldots, j-l$ **do**
$\quad Y_i \leftarrow \mathcal{T}_{S_{j-1}} Y_i$
end for
for $i = j-1, \ldots, j-l$ **do**
$\quad S_i \leftarrow \mathcal{T}_{S_{j-1}} S_i$
$\quad \rho_i \leftarrow G^S(\mathrm{tr}(Y_i)^T n, \mathrm{tr}(S_i)^T n)^{-1} = a^{\mathrm{deform}}(Y_j, S_j)^{-1}$
$\quad \alpha_i \leftarrow \rho_i G^S(\mathrm{tr}(S_i)^T n, \mathrm{tr}(q)^T n) = \rho_i a^{\mathrm{deform}}(S_i, q)$
$\quad q \leftarrow q - \alpha_i Y_i$
end for
$q \leftarrow \dfrac{G^S(\mathrm{tr}(Y_{j-1})^T n, \mathrm{tr}(S_{j-1})^T n)}{G^S(\mathrm{tr}(Y_{j-1})^T n, \mathrm{tr}(Y_{j-1})^T n)} q = \dfrac{a^{\mathrm{deform}}(Y_{j-1}, S_{j-1})}{a^{\mathrm{deform}}(Y_{j-1}, Y_{j-1})} q$
for $i = j-l, \ldots, j-1$ **do**
$\quad z \leftarrow U_j$
$\quad q \leftarrow \dfrac{G^S((\gamma_0 Y_{j-1})^T n, (\gamma_0 S_{j-1})^T n)}{G^S((\gamma_0 Y_{j-1})^T n, (\gamma_0 Y_{j-1})^T n)} U_j = \dfrac{a^{\mathrm{deform}}(Y_{j-1}, S_{j-1})}{a^{\mathrm{deform}}(Y_{j-1}, Y_{j-1})} U_j$
end for
for $i = j-l, \ldots, j-1$ **do**
$\quad \beta_i \leftarrow \rho_i G^S(\mathrm{tr}(Y_i)^T n, \mathrm{tr}(q)^T n) = \rho_i a^{\mathrm{deform}}(Y_i, q)$
$\quad q \leftarrow q + (\alpha_i - \beta_i) S_i$
end for
$S_j \leftarrow q$

The right-hand side of (7), $f \in L^2(\Omega)$, is chosen as a shape dependent piecewise constant function $f(x) = 100$ for $x \in \Omega_{\mathrm{int}}$ and $f(x) = -10$ for $x \in \Omega \setminus \bar{\Omega}_{\mathrm{int}}$. Further, the perimeter regularization in Eq. (1) is weighted by $\nu = 0.00001$. The constants $\gamma, c > 0$ in the regularized state equation are set to $\gamma = 100, c = 25$. The obstacle is given by

$$\psi \colon (0,1)^2 \to \mathbb{R}, \quad (x,y) \mapsto \begin{cases} 0.25 & \text{if } (x,y) \in (0.75,1) \times (0,1) \\ 100 & \text{if } (x,y) \in (0,0.75] \times (0,1) \end{cases}. \tag{13}$$

For our numerical test, we build artificial data $\bar{y}$ by solving the state equation without obstacle for the setting that $\Gamma_{\mathrm{int}} := \{(x,y) \in (0,1)^2 \colon (x-0.6)^2 + (y-0.5)^2 = 0.2^2\}$. Then, we add noise to the measurements $\bar{y}$, which is i.i.d. with $\mathcal{N}(0.0, 0.5)$ for each mesh node. The Lamé parameter are chosen by $\lambda = 0$ and μ as the solution of the following Laplace equation:

$$-\Delta \mu = 0 \text{ in } \Omega \quad \text{with } \mu = 20 \text{ on } \Gamma_{\mathrm{int}} \text{ and } \mu = 5 \text{ on } \partial\Omega$$

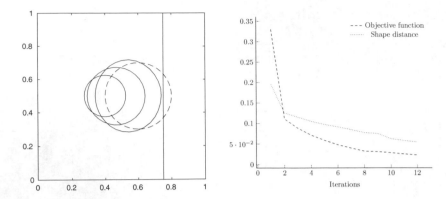

Fig. 1. *Left:* Ω with the initial (small circle) and expected shape (dashed circle) and the shape iterates of the gradient descent method. *Right:* Values of the objective function and the shape distance in each iteration of the gradient descent method.

In order to solve our model problem formulated in Sect. 2, we focus on the strategy described in Sect. 3. This means after solving the state and adjoint equation, we compute a mesh deformation vector field by solving the deformation equation. The regularized state and adjoint equations are solved using the following discretizition. We use a Finite Element Method (FEM) with continuous Galerkin ansatz functions of first order and perform computations on unstructured meshes with up to approx. 2 300 vertices and 4 300 cells. All linear systems are solved with the preconditioned conjugate gradient solver of the software PETSc, which is used as a backend to the open source Finite Element Software FEniCS, see [2].

First, we focus on a steepest descent strategy. This means we add the mesh deformation vector field, which we get by solving the deformation equation, to all nodes in the finite element mesh. We implemented also a full BFGS strategy as described in Algorithm 1. Figures 1 and 2 present the results of the gradient and the BFGS method. The left pictures show the domain Ω together with the initial shape (small circle), the expected shape (dashed circle) and the shape iterates. One can see that the expected shape is only achieved with the BFGS method an not with the gradient method. This is due to some loss of shape information in $(0.75, 1) \times (0, 1)$. This could be explained by the structure of the limiting object $p \in H_0^1(\Omega)$ of the adjoints to the regularized problem, which is given by

$$-\Delta p = -(y - \bar{y}) \text{ in } \Omega \backslash A \quad \text{with } p = 0 \text{ in } A \text{ and } p = 0 \text{ on } \partial\Omega, \qquad (14)$$

where $A := \{x \in \Omega : y(x) \geq \psi(x)\}$ denotes the active set of the variational inequality (4) (cf. [15, Theorem 1]). Due to the low obstacle ψ in $(0.75, 1) \times (0, 1)$, see (13), and the target $\bar{y}$ being above ψ in $(0.75, 1) \times (0, 1)$, we have $(0.75, 1) \times (0, 1) \subset A$. Hence we have $p_{|(0.75,1)\times(0,1)} = 0$ for the sensitivities, leaving no information in this area, resulting in shape derivatives which are 0 for directions $V \in H_0^1(\Omega, \mathbb{R}^2)$ with supp$(V) \subset (0.75, 1) \times (0, 1)$. In this sense,

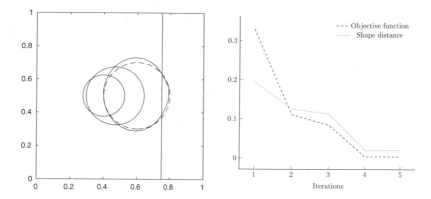

Fig. 2. *Left:* Ω with the initial (small circle) and expected shape (dashed circle) and the shape iterates of the BFGS method. *Right:* Values of the objective function and the shape distance in each iteration of the BFGS method.

only target shapes Ω that correspond to solutions $\bar{y} \in H_0^1(\Omega)$ which are below $0.25 = \psi_{|(0.75,1) \times (0,1)}$ can be reached. This creates "blind areas" in the space of shapes B_e for the shapes not fulfilling this correspondence, which, due to the Laplace equation regarded, is mostly the case for shapes that partly lie in $(0.75, 1) \times (0, 1)$. The BFGS method only manages to reach the globally optimal shape since it generates a large step while the current shape iterate is still outside the blind area $(0.75, 1) \times (0, 1) \subset A$.

The right pictures of Figs. 1 and 2 show the decrease of the objective function and the mesh distance. In both methods, the shape distance between two shapes $\Gamma_{\text{int}}^1, \Gamma_{\text{int}}^2$ is approximated by the integral $\int_{x \in \Gamma_{\text{int}}^1} \max_{y \in \Gamma_{\text{int}}^2} \|x - y\|_2 \, dx$, where $\|\cdot\|_2$ denotes the euclidean norm. One sees that the BFGS method is superior to the gradient method: 5 iterations (BFGS) vs. 12 iterations (gradient method).

References

1. Absil, P.A., Mahony, R., Sepulchre, R.: Optimization Algorithms on Matrix Manifolds. Princeton University Press, Princeton (2008)
2. Alnæs, M.S., et al.: The FEniCS project version 1.5. Arch. Numer. Softw. **3**(100), 9–23 (2015)
3. Bock, H.: Numerical treatment of inverse problems in chemical reaction kinetics. In: Ebert, K., Deuflhard, P., Jäger, W. (eds.) Modelling of Chemical Reaction Systems. Springer Series Chemical Physics, vol. 18, pp. 102–125. Springer, Heidelberg (1981). https://doi.org/10.1007/978-3-642-68220-9_8
4. Denkowski, Z., Migorski, S.: Optimal shape design for hemivariational inequalities. Universitatis Iagellonicae Acta Matematica **36**, 81–88 (1998)
5. Führ, B., Schulz, V., Welker, K.: Shape optimization for interface identification with obstacle problems. Vietnam. J. Math. (2018). https://doi.org/10.1007/s10013-018-0312-0
6. Gangl, P., Laurain, A., Meftahi, H., Sturm, K.: Shape optimization of an electric motor subject to nonlinear magnetostatics. SIAM J. Sci. Comput. **37**(6), B1002–B1025 (2015)

7. Gasiński, L.: Mapping method in optimal shape design problems governed by hemi-variational inequalities. In: Cagnol, J., Polis, M., Zolésio, J.-P. (eds.) Shape Optimization and Optimal Design. Lecture Notes in Pure and Applied Mathematics, vol. 216, pp. 277–288. Marcel Dekker, New York (2001)
8. Hintermüller, M., Laurain, A.: Optimal shape design subject to elliptic variational inequalities. SIAM J. Control. Optim. 49(3), 1015–1047 (2011)
9. Ito, K., Kunisch, K.: Semi-smooth Newton methods for variational inequalities and their applications. M2AN Math. Model. Numer. Anal. 37, 41–62 (2003)
10. Ito, K., Kunisch, K.: Semi-smooth Newton methods for variational inequalities of the first kind. ESAIM Math. Model. Numer. Anal. 37(1), 41–62 (2003)
11. Kinderlehrer, D., Stampacchia, G.: An Introduction to Variational Inequalities and Their Applications, vol. 31. SIAM, Philadelphia (1980)
12. Kocvara, M., Outrata, J.: Shape optimization of elasto-plastic bodies governed by variational inequalities. In: Zolésio, J.-P. (ed.) Boundary Control and Variation. Lecture Notes in Pure and Applied Mathematics, vol. 163, pp. 261–271. Marcel Dekker, New York (1994)
13. Kriegl, A., Michor, P.: The Convient Setting of Global Analysis. Mathematical Surveys and Monographs, vol. 53. American Mathematical Society, Providence (1997)
14. Liu, W., Rubio, J.: Optimal shape design for systems governed by variational inequalities, Part 1: existence theory for the elliptic case. J. Optim. Theory Appl. 69(2), 351–371 (1991)
15. Luft, D., Schulz, V.H., Welker, K.: Efficient techniques for shape optimization with variational inequalities using adjoints. arXiv:1904.08650 (2019)
16. Michor, P., Mumford, D.: Riemannian geometries on spaces of plane curves. J. Eur. Math. Soc. 8, 1–48 (2006)
17. Michor, P., Mumford, D.: An overview of the Riemannian metrics on spaces of curves using the Hamiltonian approach. Appl. Comput. Harmon. Anal. 23, 74–113 (2007)
18. Myśliński, A.: Level set method for shape and topology optimization of contact problems. In: Korytowski, A., Malanowski, K., Mitkowski, W., Szymkat, M. (eds.) CSMO 2007. IAICT, vol. 312, pp. 397–410. Springer, Heidelberg (2009). https://doi.org/10.1007/978-3-642-04802-9_23
19. Schmidt, S., Ilic, C., Schulz, V., Gauger, N.: Three dimensional large scale aerodynamic shape optimization based on the shape calculus. AIAA J. 51(11), 2615–2627 (2013)
20. Schulz, V., Welker, K.: On optimization transfer operators in shape spaces. In: Schulz, V., Seck, D. (eds.) Shape Optimization. Homogenization and Optimal Control, pp. 259–275. Springer, Cham (2018). https://doi.org/10.1007/978-3-319-90469-6_13
21. Schulz, V.H., Siebenborn, M., Welker, K.: Efficient PDE constrained shape optimization based on Steklov-Poincaré type metrics. SIAM J. Optim. 26(4), 2800–2819 (2016)
22. Sokolowski, J., Zolésio, J.-P.: An Introduction to Shape Optimization. Springer, Heidelberg (1992). https://doi.org/10.1007/978-3-642-58106-9
23. Udawalpola, R., Berggren, M.: Optimization of an acoustic horn with respect to efficiency and directivity. Int. J. Numer. Methods Eng. 73(11), 1571–1606 (2007)
24. Welker, K.: Optimization in the space of smooth shapes. In: Nielsen, F., Barbaresco, F. (eds.) GSI 2017. LNCS, vol. 10589, pp. 65–72. Springer, Cham (2017). https://doi.org/10.1007/978-3-319-68445-1_8

Bézier Curves and C^2 Interpolation in Riemannian Symmetric Spaces

Chafik Samir[1,3]($\boxtimes$) and Ines Adouani[2]

[1] CNRS LIMOS (UMR 6158),
University of Clermont Auvergne, Clermont-Ferrand, France
chafik.samir@uca.fr
[2] Higher Institute of Applied Sciences and Technology of Sousse,
University of Sousse, Sousse, Tunisia
[3] Institut de Mathématiques de Toulouse, Université Paul Sabatier, Toulouse, France

Abstract. We consider the problem of interpolating a finite set of observations at given time instant. In this paper, we introduce a new method to compute the optimal intermediate control points that define a C^2 interpolating Bézier curve. We prove this concept for interpolating data points belonging to a Riemannian symmetric spaces. The main property of the proposed method is that the control points minimize the mean square acceleration. Moreover, potential applications of fitting smooth paths on Riemannian manifold include applications in robotics, animations, graphics, and medical studies.

Keywords: Riemannian Bézier curves ·
Regression on Riemannian manifolds · Curve fitting ·
Mean square acceleration · Special orthogonal group

1 Introduction

The problem of constructing smooth interpolating curves in non-linear spaces, or manifolds plays an important role in a wide variety of applications. For instance, interpolation in the rotation group $SO(3)$ has immediate application not only in computer graphics and animation of 3D objects [1–3], but also in applications ranging from robot motion planning to machine vision [4–6]. Such applications encourage us to further search for some efficient methods to generate smooth interpolating curves on non-linear spaces.

Motivated by potential applications in engineering science and technology, our goal is to develop a new framework for generating C^2 Bézier curves on Riemannian manifolds that interpolate a given ordered set of points at specified time instants. While quite general, we will focus on a special class of Riemannian symmetric spaces. The task of constructing interpolating curve on $SO(n)$ has attracted the attention of several authors. One of the most widely cited approaches is the work of Shoemake [7] on $SO(3)$, who adopts a reparametrization of the rotation matrices based on unit quaternion representation. Shoemake's approach can essentially be viewed as a generalization of the de Casteljau's algorithm for Bézier curves to $SU(2)$ in which two elements

© Springer Nature Switzerland AG 2019
F. Nielsen and F. Barbaresco (Eds.): GSI 2019, LNCS 11712, pp. 589–598, 2019.
https://doi.org/10.1007/978-3-030-26980-7_61

of $SO(3)$ are interpolated by the geodesic that joins them. Although this algorithm seems computationally efficient, unfortunately the resulting curve depends on the choice of local system coordinates. A few years later, taking into account the Shoemake algorithm, a more careful geometric analysis of unit-quaternion-based method was introduced by Barr et al. [1], Hart et al. [8], Ge and Ravani [9], and Nielson et al. [10]. Despite the fact of producing an intrinsic curves, these approaches does not generalize to higher-dimensional manifold.

In this paper, we present a novel framework to treat the interpolation problem in the setting of Riemannian geometry and Bézier curve approach. We show that it makes sense to define a C^2 interpolating Bézier curve on Riemannian symmetric spaces as the result of a least squares minimization and a recursive algorithm. In particular, we will focus on a special class of Riemannian symmetric spaces: the special orthogonal group $SO(n)$. Indeed, working in such Riemannian manifold allows nice properties to solve the issues above. The key point to give explicit solution for the interpolation problem and ensures the C^2 differentiability condition at joint points is the use of global symmetries in these last points. In fact, we will first derive equations for control points of a C^2 Bézier curve on the Euclidean space $\mathbb{R}^m$. Then, building upon prior works [6,11], we use these equations to find the control points of a C^1 interpolating Bézier curve on Riemannian manifolds as a generalization of the Bézier based fitting in the Euclidean space and by means of methods of Riemannian geometry. These results are sufficient to give explicit formula for control points of the C^2 interpolating Bézier curve on $SO(n)$. The proposed method will be shown to enjoy a number of nice properties and the solution is unique in many common situations.

The rest of the paper is organized as follows. In Sect. 2, we present our new algorithm to construct a C^2 Bézier curve on the Euclidean space. This will help with the visualization of its main features and motivate its generalization on $SO(n)$. In Sect. 3, the generalization of our approach on the Lie group $SO(n)$ is prescribed. We conclude the paper with numerical examples and a conclusion.

2 C^2 Interpolating Bézier Curves on $\mathbb{R}^m$

In this section, we first describe our approach on the Euclidean space $\mathbb{R}^m$. For simplicity we will assume that the time instants are $t_i = i$. In this work, we only use Bézier curves of degree 2 and 3 such that the segment joining p_0 and p_1, as well as the segment joining p_{N-1} and p_N are Bézier curves of order two, while all the other segments are Bézier curves of order three. Explicitly, the Bézier curve β_k of degree $k \in \{2, 3\}$ are expressed in $\mathbb{R}^m$ with a number of control points b_i, represented as their coefficients in the Bernstein basis polynomials by:

$$\beta_2(t; b_0, b_1, b_2) = b_0(1-t)^2 + 2b_1(1-t)t + b_2t^2,$$
$$\beta_3(t; b_0, b_1, b_2, b_3) = b_0(1-t)^3 + 3b_1t(1-t)^2 + 3b_2t^2(1-t) + b_3t^3.$$

Moreover, we assume that there exists two artificial control points $(\widehat{b_i^-}, \widehat{b_i^+})$ on the left and on the right hand side of the interpolation point p_i for $i = 1, ..., (N-1)$. Consequently, the Bézier curve β on $\mathbb{R}^m$ is given by:

$$\beta(t) = \begin{cases} \beta_2(t; p_0, \widehat{b_1^-}, p_1), & \text{if } t \in [0, 1] \\ \beta_3(t - (i-1); p_{i-1}, \widehat{b_{i-1}^+}, \widehat{b_i^-}, p_i), & \text{if } t \in [i-1, i], i = 2, ..., N-1 \\ \beta_2(t - (N-1); p_{N-1}, \widehat{b_{N-1}^+}, p_N), & \text{if } t \in [N-1, N] \end{cases}$$

Then β is C^∞ on $[t_i, t_{i+1}]$, for $i = 0, ..N-1$. To ensure that β is C^1 at knots p_i, for $i = 1, ..N-1$, we shall make the following assumption:

$$\dot{\beta}_{k^i}(b_0^i, ..., b_{k^i}^i; t - i + 1)|_{t=i} = \dot{\beta}_{k^{i+1}}(b_0^{i+1}, ..., b_{k^{i+1}}^{i+1}; t - i)|_{t=i} \qquad i = 0, ..., N-2. \tag{1}$$

This differentiability condition allows us to express $\widehat{b_i^+}$ in terms of $\widehat{b_i^-}$ as:

$$\widehat{b_1^+} = \frac{5}{3}p_1 - \frac{2}{3}\widehat{b_1^-}, \tag{2}$$

$$\widehat{b_i^+} = 2p_i - \widehat{b_i^-}, i = 2, ..., N-2 \tag{3}$$

$$\widehat{b_{N-1}^+} = \frac{5}{2}p_{N-1} - \frac{3}{2}\widehat{b_{N-1}^-}, \tag{4}$$

We are left with the task of computing the control points $\widehat{b_i^-}$, for $i = 1, ..., N-1$, that generate the C^1 Bézier curve β. In [11], we have shown that solutions of the problem of minimization of the mean square acceleration of the Bézier curve β are exactly the control points of the curve:

$$\min_{\widehat{b_1^-}, ..., \widehat{b_{N-1}^-}} E(\widehat{b_1^-}, ..., \widehat{b_{N-1}^-}) := \min_{\widehat{b_1^-}, ..., \widehat{b_{N-1}^-}} \int_0^1 \|\ddot{\beta}_2^0(t; p_0, \widehat{b_1^-}, p_1)\|^2$$

$$+ \sum_{i=1}^{N-2} \int_0^1 \|\ddot{\beta}_3^i(t; p_i, \widehat{b_i^-}, \widehat{b_{i+1}^-}, p_{i+1})\|^2 + \int_0^1 \|\ddot{\beta}_2^{N-1}(t; p_{N-1}, \widehat{b_{N-1}^-}, p_N)\|^2 \tag{5}$$

It turns out that the optimal solution $Y = [\widehat{b_1^-}, ..., \widehat{b_{N-1}^-}]^T \in \mathbb{R}^{(N-1) \times m}$ of (5) is the unique solution of a tridiagonal linear system

$$Y = A^{-1}CP = DP \text{ with } \sum_{j=0}^{j=N+1} d_{ij} = 1. \tag{6}$$

where A is a tridiagonal sparse square matrix of size $(N-1) \times (N-1)$ with a dominant diagonal, C a matrix of size $(N-1) \times (N+1)$ and P the matrix of p_i's of size $(N+1) \times m$ given by:

$$A_{(1,1:2)} = [16\ 6] \qquad (7)$$
$$A_{(2,1:3)} = [6\ 36\ 9] \qquad (8)$$
$$A_{(i,i-1:i+1)} = [9\ 36\ 9], \qquad (9)$$
$$A_{(n-1,n-2:n-1)} = [9\ 36] \qquad (10)$$

$$C_{(1,1:2)} = [16\ 6] \qquad (11)$$
$$C_{(2,2:3)} = [6\ 36\ 9] \qquad (12)$$
$$C_{(i,i:i+1)} = [9\ 36\ 9], i = 3, ..., n-2 \qquad (13)$$
$$C_{(n-1,n-1:n+1)} = [9\ 36] \qquad (14)$$

Now, let us assume that β is C^1, so that (1) is met and the solution Y given by (6) is obtained. The additional C^2 condition for a C^1 curve is the equality of the second derivative at the joint point p_i, for $i = 1, ..., N-1$:

$$\ddot{\beta}_{k^i}(b_0^i, ..., b_{k^i}^i; t - i + 1)|_{t=i} = \ddot{\beta}_{k^{i+1}}(b_0^{i+1}, ..., b_{k^{i+1}}^{i+1}; t - i)|_{t=i} \qquad i = 0, ..., N-2.$$

It is obvious that with this C^2 condition the position of the control points $\widehat{b}_i^-$ and $\widehat{b}_i^+$ that generate the curve β will be modified. Therefore, it is more convenient to use another notation. Let us denote by b_i^- and b_i^+ the new control points on the left and on the right hand side of the interpolation point p_i, for $i = 1, ..., N-1$. Computing the acceleration of β on respective intervals and taking into account that β is C^1, we shall replace b_1^+ by (2), b_i^+ by (3), and b_{N-1}^+ by (4). We deduce that:

$$b_2^- = \frac{1}{3}p_0 - \frac{1}{2}b_1^- + \frac{8}{3}p_1, \tag{15a}$$

$$b_{i+1}^- = b_{i-1}^+ + 4p_i - 4b_i^-, i = 2, ..., N-2 \tag{15b}$$

$$p_N = 2p_{N-1} + 2b_{N-1}^+ - 6b_{N-1}^- + 3b_{N-2}^+, \tag{15c}$$

We see at once that points that will be modified by the additional C^2 condition are $\widehat{b}_i^-$ and hence $\widehat{b}_i^+$, for $i = 2, ..., N-1$. The point $\widehat{b}_1^-$ remains invariant and consequently it will be the case for $\widehat{b}_1^+$. We thus get $b_1^- = \widehat{b}_1^-$, with $\widehat{b}_1^-$ is the first row of the matrix Y obtained as a solution of the optimization problem (5). However, the endpoint p_N is affected as we can deduce from Eq. (15c). Nevertheless, it follows that giving the control point b_1^- allows us to find all the other control points including b_2^- with Eq. (15a) and hence b_2^+ with (3), then b_{i+1}^- for $i = 2, ..., N-2$ with (15b) and therefore b_i^+, for $i = 3, ..., N-2$ with (3) and b_{N-1}^+ with (4).

3 C^2 Interpolating Bézier Curves On $SO(n)$

Our objective in this section is to work out concretely the extension of our approach used to find control points that define a C^2 Bézier curve in the Euclidean space to the Riemannian manifold $SO(n)$. In other words, given $R_0, ..., R_N$ a set of $(N+1)$ distinct points in $SO(n)$ and $0 = t_0 < t_1 < ... < t_N = N$ an increasing sequence of time instants, we present a conceptually simple framework to construct a C^2 Bézier curve $\gamma : [0, N] \to SO(n)$ such that $\gamma(t_k) = R_k$, $k = 0, ..., N$. For the most part of Riemannian manifolds, the generalization of our approach is not straightforward. For the case treated here, of the Lie group $SO(n)$, since it is a symmetric space and all the important geometric functions have nice, closed-form expressions, the problem of finding a C^2 Bézier curve that interpolates a given set of points in such space can be completely solved.

Let us start by briefly sketch the differential structure of $SO(n)$. We illustrate this with the geometric toolbox described in Table 1. For more details concerning the differential geometry of $SO(n)$, see [12,13].

Table 1. Geometric toolbox for the Riemannian manifold $SO(n)$

Set	$SO(n) = \{R \in \mathbb{R}^{n \times n} \mid R^T R = I_n \text{ and } \det(R) = 1\}$
Tangent spaces	$T_R SO(n) = \{H \in \mathbb{R}^{n \times n} \mid RH^T + HR^T = 0\}$
Inner product	$<H_1, H_2>_R = \text{trace}(H_1^T H_2)$
Exponential	$\text{Exp}_R(H) = \text{Exp}_I(R^T H) = R \exp(R^T H)$
Logarithm	$\text{Log}_{R_1}(R_2) = R_1 \log(R_1^T R_2)$

The shortest geodesic arc joining R_1 to R_2 in $SO(n)$ can be parameterized explicitly by:

$$\alpha(t, R_1, R_2) = R_1 \exp(t \log(R_1^T R_2)), \ t \in [0, 1]. \tag{16}$$

and we write:

$$\dot{\alpha}(t, R_1, R_2) := \frac{\partial}{\partial u}\big|_{u=t} \, \alpha(t, R_1, R_2).$$

Furthermore, for each $R_1 \in SO(n)$, there exists a symmetry

$$\varphi_{R_1} : SO(n) \longrightarrow SO(n), \ R_2 \longrightarrow R_1 R_2^T R_1$$

that reverses geodesics through R_1. It is easy to check that φ_{R_1} is an isometry and thus $SO(n)$ turns into a Riemannian symmetric space. For $R_1, R_2 \in SO(n)$, let us denote by $(d\text{Exp}_{R_1})_H$ the derivative of Exp_{R_1} at $H \in T_{R_1} SO(n)$ and by $(d\varphi_{R_1})_{R_2}$ the derivative of the geodesic symmetry φ_{R_1} at R_2. Then, the following result can be easily proved and will be very important for the derivation of the results presented along this section.

Lemma 1. *Let $R_1 \in SO(n)$.*

(i) $(d\varphi_{R_1})_{R_2}^{-1} = (d\varphi_{R_1})_{\varphi_{R_1}(R_2)}$, *for all $R_2 \in SO(n)$*
(ii) $(d\text{Exp}_{R_1})_H^{-1} = -(d\text{Exp}_{R_1})_{-H} \circ (d\varphi_{R_1})_{\text{Exp}_{R_1}(H)}$, *for all $H \in T_{R_1} SO(n)$*

Let us now denote by $\gamma_k(t, V_0, ..., V_k)$ the Bézier curve of order $k \in \{2, 3\}$ on $SO(n)$ with a number of control points V_i for $i = 0, ..., k$. Furthermore, similar to the Euclidean case, we will suppose that there exists two artificial control points $(\widehat{Z}_i^-, \widehat{Z}_i^+)$ on the left and on the right hand side of the interpolation point R_i for $i = 1, ..., (N-1)$. Hence, the Bézier Curve $\gamma : [0, N] \longrightarrow SO(n)$ is defined by:

$$\gamma(t) = \begin{cases} \gamma_2(t; R_0, \widehat{Z}_1^-, R_1), & \text{if } t \in [0, 1] \\ \gamma_3(t - (i - 1); R_{i-1}, \widehat{Z}_{i-1}^+, \widehat{Z}_i^-, R_i), & \text{if } t \in [i - 1, i], i = 2, ..., N - 1 \\ \gamma_2(t - (N - 1); R_{N-1}, \widehat{Z}_{N-1}^+, R_N), & \text{if } t \in [N - 1, N] \end{cases}$$

In order to obtain equations that govern the control points of the C^2 Bézier curve on $SO(n)$, one should begin to compute $(\widehat{Z}_i^-, \widehat{Z}_i^+)$, for $i = 1, ..., N - 1$, control

---

Algorithm 1. Construction of the C^1 interpolating Bézier curve on $SO(n)$.

Input: $N \geq 3$, $R = [R_0, ..., R_N]^T$ a matrix of size $n(N+1) \times n$ containing the $(N+1)$ interpolation points on $SO(n)$.

Output: $\widehat{Z}$ and $\tilde{R}$.

1: **for** $i = 1 : N - 1$ **do**
2:     Calculate $Q = [Q_0^i, ..., Q_N^i]^T$ a matrix of size $n(N+1) \times n$ containing the $(N+1)$ interpolation points on $T_{R_i}SO(n)$:
3:     **for** $k = 0 : N$ **do**
4:         $Q_k^i = \text{Log}_{R_i}(R_k) = R_i \log(R_i^T R_k)$
5:         Calculate $X_i = [(B_1^i)^-, ..., (B_{N-1}^i)^-]^T$ a matrix of size $n(N-1) \times n$ containing the $(N-1)$ control points of the C^2 Bézier curve β_i on $T_{R_i}SO(n)$, and $\tilde{Q} = [\tilde{Q}_0^i, ..., \tilde{Q}_N^i]^T$ a matrix of size $n(N+1) \times n$ containing the new interpolation points on $T_{R_i}SO(n)$ using the prescribed method on section 2.
6:         Calculate control point $\widehat{Z}_i^-$ with $\widehat{Z}_i^- = \text{Exp}_{R_i}((B_i^i)^-)$
7:         Calculate the new interpolation points $\tilde{R}_k = \text{Exp}_{R_i}(\tilde{Q}_k^i)$.
8:     **end for**
9: **end for**
10: **return** $\widehat{Z}$ and $\tilde{R}$,

---

points of the Bézier curve γ that ensure the C^1 differentiability condition of γ at knots R_i on $SO(n)$. To do this, our main idea is to treat the fitting problem on the tangent space $T_{R_i}SO(n)$ at a point $R_i \in SO(n)$ as for the Euclidean case. Consequently, for each $i = 1, ..., N - 1$, we would like to transfer the data $R_0, ..., R_N$ in each tangent space $T_{R_i}SO(n)$ using Riemannian logarithmic map. The mapped data are then given by $Q = (Q_0^i, ..., Q_N^i)$ with $Q_k^i = \text{Log}_{R_i}(R_k)$ for $k = 0, ..., N$. Applying our approach used to define a C^2 Bézier curve on the Euclidean space $\mathbb{R}^m$ in each tangent space $T_{R_i}SO(n)$, for $i = 1, ..., N - 1$, provides a natural and intrinsic method to compute control points $(\widehat{Z}_i^-, \widehat{Z}_i^+)$ of the desired C^1 Bézier curve γ on $SO(n)$.

Theorem 1. *Let $R_0, ..., R_N$ be a finite sequence of distinct points in the special orthogonal group $SO(n)$ with $R_i^T R_k$, $i \neq k$, sufficiently close to I_n. For each $i = 1, ..., N - 1$, $Q = (Q_0^i, ..., Q_N^i)$ are the corresponding mapped data in the tangent space $T_{R_i}SO(n)$ at R_i defined by $Q_k^i = \text{Log}_{R_i}(R_k)$ for $k = 0, ..., N$. Set $t_0 = 0 < ... < t_N = N$ a sequence of time instants. Then, there exists a unique matrix $X_i = [(B_1^i)^-, ..., (B_{N-1}^i)^-]^T \in \mathbb{R}^{n(N-1) \times n}$ containing the $(N-1)$ control points that generate the C^2 Bézier curve β_i, in each tangent space $T_{R_i}SO(n)$ and a matrix $\tilde{Q} = [\tilde{Q}_0^i, ..., \tilde{Q}_N^i]^T$ of size $n(N+1) \times n$ containing the new $(N+1)$ interpolation points in each tangent space $T_{R_i}SO(n)$.*

Proposition 1. *Under the same hypotheses of Theorem 1, there exists a unique matrix $Z = [\widehat{Z}_1^-, ..., \widehat{Z}_{N-1}^-]^T \in \mathbb{R}^{n(N-1) \times n}$, containing the $(N-1)$ control points that generate the Bézier curve γ interpolating the points R_i at t_i on $SO(n)$, for $i = 0, ..., N$. The rows of $\widehat{Z}$ are given by:*

$$\widehat{Z}_i^- = \text{Exp}_{R_i}(\tilde{x}_i), \ i = 1, ..., N - 1. \tag{17}$$

where $\tilde{x}_i$, represent the row i of X_i in $T_{R_i} SO(n)$, for $i = 1, .., N-1$. Moreover, the new $(N+1)$ interpolation points in $SO(n)$ are given by:

$$\tilde{R}_k = Exp_{R_i}(\tilde{Q}_k^i), \quad k = 0, ..., N; \quad i = 1, ..., N-1. \tag{18}$$

Algorithm 1 provides a detailed exposition of the steps of the proof of Theorem 1 and Proposition 1.

Corollary 1. *The Bézier path* $\gamma : [0,1] \to SO(n)$ *is* C^1 *on* $SO(n)$.

Proof. The following result may be proved in much the same way as Corollary 3.3. in [11].

We are now in a position to formulate the main theorem of this section, which contains the counterpart of the equations derived in the last section that generate control points of a C^2 Bézier curve on $\mathbb{R}^m$. Let us assume that γ is C^1, so that the solution $\hat{Z}$ is obtained. Let us denote by Z_i^- and Z_i^+ the new control points on the left and on the right side of the interpolation point $\tilde{R}_i$ that generate the C^2 Bézier curve γ on $SO(n)$. The key point to find the control points Z_i^-, for $i = 1, ..., N-1$ is similar to the Euclidean case. That is, we might know Z_1^- (and therefore Z_1^+ by the C^1 differentiability condition on $SO(n)$) and wish to define iteratively Z_i^- for $i = 2, ..., N-1$ (and obviously Z_i^+ in much the same way as Z_1^+).

Algorithm 2. Construction of the C^2 interpolating Bézier curve on $SO(n)$.

Input: $N \geq 3$, $\tilde{R} = [\tilde{R}_0, ..., \tilde{R}_N]^T$ a matrix of size $n(N+1) \times n$ containing the $(N+1)$ interpolation points on $SO(n)$.

Output: Z.

1: Calculate $\hat{Z} = [\hat{Z}_1^-, ..., \hat{Z}_{N-1}^-]^T$ using Algorithm 1.

2: Set $Z_1^- = \hat{Z}_1^-$.

3: Calculate control point Z_1^+:

4:    $Z_1^+ = Exp_{\tilde{R}_1}(-\frac{2}{3} Exp_{\tilde{R}_1}^{-1}(Z_1^-))$

5: Calculate control point Z_2^-:

6:    $Z_2^- = Exp_{Z_1^+}\left(\frac{1}{3}\left((d\varphi_{\tilde{R}_1})_{Z_1^-}\left(\dot{\alpha}(1, \tilde{R}_0, Z_1^-)\right) - 4\dot{\alpha}(0, Z_1^-, \tilde{R}_1)\right)\right)$

7: **for** $i = 2 : N-2$ **do** do

8:      $Z_i^+ = Exp_{\tilde{R}_i}(-Exp_{\tilde{R}_i}^{-1}(Z_i^-))$

9:      $Z_{i+1}^- = Exp_{Z_i^+}\left(\left((d\varphi_{\tilde{R}_i})_{Z_i^-}\left(\dot{\alpha}(1, Z_{i-1}^+, Z_i^-)\right) - 2\dot{\alpha}(0, Z_i^-, \tilde{R}_i)\right)\right)$

10: **end for**

11: Calculate control point Z_{N-1}^+:

12:    $Z_{N-1}^+ = Exp_{\tilde{R}_{N-1}}(-\frac{2}{3} Exp_{\tilde{R}_{N-1}}^{-1}(Z_{N-1}^-))$

13: **return** Z,

Theorem 2. *Let $\tilde{R}_0, ..., \tilde{R}_N$ be a set of distinct points in the special orthogonal group $SO(n)$ given by Eq. (18) and $\alpha(t)$ the shortest geodesic arc joining control points of the curve γ on $SO(n)$ given by Eq. (16). Let $X_1 = [(B_1^1)^-, ..., (B_{N-1}^1)^-]^T$ be the matrix of size $n(N-1) \times n$ containing the control points of the C^2 Bézier curve β_1 in $T_{R_1}SO(n)$. Then, there exists a unique matrix $Z = [Z_1^-, ..., Z_{N-1}^-]^T \in \mathbb{R}^{n(N-1) \times n}$, containing the $(N-1)$ control points that generate the C^2 Bézier curve γ interpolating the points $\tilde{R}_i$ at t_i on $SO(n)$, for $i = 0, ..., N$. The rows of Z are given by:*

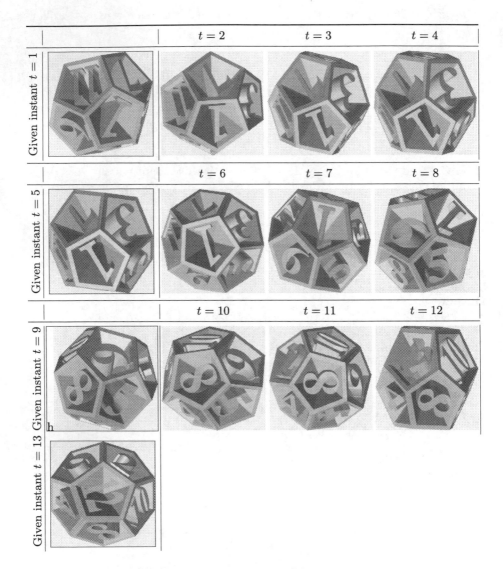

Fig. 1. Example of an interpolating path on $SO(3)$ applied to rotate a 12 sided dice at given time instants $(1, 5, 9, 13)$.

(i) $Z_1^- = Exp_{R_1}((B_1^1)^-).$

(ii) $Z_2^- = Exp_{Z_1^+} \left(\frac{1}{3} \left((d\varphi_{\tilde{R}_1})_{Z_1^-} \left(\dot{\alpha}(1, \tilde{R}_0, Z_1^-) \right) - 4\dot{\alpha}(0, Z_1^-, \tilde{R}_1) \right) \right).$

(iii) $Z_{i+1}^- = Exp_{Z_i^+} \left(\left((d\varphi_{\tilde{R}_i})_{Z_i^-} \left(\dot{\alpha}(1, Z_{i-1}^+, Z_i^-) \right) - 2\dot{\alpha}(0, Z_i^-, \tilde{R}_i) \right) \right),$
     $i = 2, ..., N - 2.$

We illustrate the proposed method to construct a smooth interpolating path on $SO(3)$ from four rotation matrices R_1, R_2, R_3, and R_4. We display the result in Fig. 1 where rotations are applied to rotate a 12 sided dice and the given time instants are displayed in a box. We can easily check that the resulting curve path is smooth including at the interpolation points.

4   Conclusion

In this paper, we have introduced a new framework and algorithms to study the fitting problem of C^2 Bézier curves to a finite set of time-indexed data points on the special orthogonal group $SO(n)$. The proposed method takes into account the global symmetries defined in the joint points. Therefore, the presented approach is valid on any locally symmetric space and other Riemannian symmetric spaces. In the future, we intend to extend the theory and then apply it to more general nonlinear manifolds.

Acknowledgement. This research was partially supported by The National Center for Scientific Research as CNRS PRIME Grant and the I-Site Clermont Auvergne project.

References

1. Barr, A., Currin, B., Gabriel, S., Hughes, J.: Smooth interpolation of orientations with angular velocity constraints using quaternions. ACM SIGGRAPH Comput. Graph. **26**, 313–320 (1992)
2. Popiel, T., Noakes, L.: Bézier curves and C2 interpolation in Riemannian manifolds. J. Approx. Theory **148**(2), 111–127 (2007)
3. Fang, Y., Hsieh, C., Kim, M., Chang, J., Woo, T.: Real time motion fairing with unit quaternions. Comput. Aided Des. **30**, 191–198 (1998)
4. Zefran, M., Kumar, V., Croke, C.: On the generation of smooth three-dimensional rigid body motions. IEEE Trans. Robot. Autom. **14**, 576–589 (1998)
5. Camarinha, M., Silva Leite, F., Crouch, P.: Splines of class C^k on non-Euclidean spaces. IMA J. Math. Control Inf. **12**(4), 399–410 (1995)
6. Arnould, A., Gousenbourger, P.-Y., Samir, C., Absil, P.-A., Canis, M.: Fitting smooth paths on Riemannian manifolds: endometrial surface reconstruction and preoperative MRI-based navigation. In: Nielsen, F., Barbaresco, F. (eds.) GSI 2015. LNCS, vol. 9389, pp. 491–498. Springer, Cham (2015). https://doi.org/10.1007/978-3-319-25040-3_53
7. Shoemake, K.: Animating rotation with quaternion curves. ACM SIGGRAPH **85**(19), 245–254 (1985)

8. Hart, J., Francis, G., Kaufman, L.: Visualizing quaternion rotation. ACM Trans. Graph. **13**(3), 256–276 (1994)
9. Ge, Q., Ravani, B.: Geometric construction of Bézier motions. ASME J. Mech. Des. **116**, 749–755 (1994)
10. Nielson, G., Heiland, R.: Animated rotationsusing quaternions and splines on a 4D sphere. Program. Comput. Softw. **18**(4), 145–154 (1992)
11. Samir, C., Adouani, I.: C^1 interpolating Bézier path on Riemannian manifolds, with applications to 3D shape space. Appl. Math. Comput. **348**, 371–384 (2019)
12. Helgason, S.: Differential Geometry, Lie Groups, and Symmetric Spaces (1978)
13. Adams, F.: Lectures on Lie Groups (1982)

A Formalization of the Natural Gradient Method for General Similarity Measures

Anton Mallasto$^{(\boxtimes)}$, Tom Dela Haije, and Aasa Feragen

University of Copenhagen, Copenhagen, Denmark
{mallasto,haije,aasa}@di.ku.dk

Abstract. In optimization, the natural gradient method is well-known for likelihood maximization. The method uses the Kullback–Leibler (KL) divergence, corresponding infinitesimally to the Fisher–Rao metric, which is pulled back to the parameter space of a family of probability distributions. This way, gradients with respect to the parameters respect the Fisher–Rao geometry of the space of distributions, which might differ vastly from the standard Euclidean geometry of the parameter space, often leading to faster convergence. The concept of natural gradient has in most discussions been restricted to the KL-divergence/Fisher–Rao case, although in information geometry the local C^2 structure of a general divergence has been used for deriving a closely related Riemannian metric analogous to the KL-divergence case. In this work, we wish to cast natural gradients into this more general context and provide example computations, notably in the case of a Finsler metric and the p-Wasserstein metric. We additionally discuss connections between the natural gradient method and multiple other optimization techniques in the literature.

Keywords: Optimization · Natural gradient · Statistical manifolds

1   Introduction

The natural gradient method [2] in optimization originates from *information geometry* [4], which utilizes the Riemannian geometry of statistical manifolds (the parameter spaces of model families) endowed with the *Fisher–Rao metric*. The natural gradient is used for minimizing the *Kullback–Leibler* (KL) divergence, a *similarity measure* between a model distribution and a target distribution, that can be shown to be equivalent to maximizing model likelihood of given data. The success of natural gradient in optimization stems from accelerating likelihood maximization and providing infinitesimal invariance to reparametrizations of the model, providing robustness towards arbitrary parametrization choices.

In the modern formulation of the natural gradient, a *Riemannian metric* on the statistical manifold is chosen, with respect to which the gradient of the given similarity is computed [4, Sec. 12]. The choice of the Riemannian metric should, however, relate closely to the similarity measure being minimized.

F. Nielsen and F. Barbaresco (Eds.): GSI 2019, LNCS 11712, pp. 599–607, 2019.
https://doi.org/10.1007/978-3-030-26980-7_62

We have illustrated this in Fig. 1, where model selection for Gaussian process regression is carried out by maximizing the prior-likelihood of the data with natural gradients stemming form different metrics. Clearly, the Fisher–Rao metric— which infinitesimally corresponds to the KL-divergence—achieves the fastest convergence.

An example of an approach to choose a related Riemannian metric is the classical Newton's method that derives a metric from the Hessian of a convex objective function, or its absolute value in the non-convex case [7]. Unfortunately, evaluating the Hessian is not feasible in some cases. Instead, we can compute a *local Hessian*, which corresponds to a local second order expansion of the similarity measure [3]. This approach generalizes the natural gradient from the KL-divergence case to general similarity measures, and to avoid confusion with the well-known KL-divergence setting, we refer to this approach as the *formal natural gradient*. We furthermore discuss the similarities between the trust region, proximal, and natural gradient methods in Sect. 3 and provide example computations in Sect. 4.

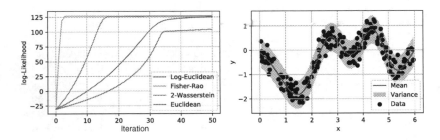

Fig. 1. Maximizing prior likelihood for Gaussian process regression using natural gradients under different metrics on Gaussian distributions. Convergence plots on left. Data and model fit, with optimal exponentiated quadratic kernel parameters, on right.

2   Useful Metrics via Formalizing the Natural Gradient

The natural gradient is computed with respect to a chosen metric on the statistical manifold, which often results from pulling back a metric between distributions. This way, the gradient takes into account how the metric on distributions penalizes movement into different directions. We will now review how the natural gradient is computed given a Riemannian metric. Then, we introduce the formal natural gradient, which derives this metric from the similarity measure.

Statistical Manifold. Let $AC(X)$ denote the set of absolutely continuous probability distributions on some manifold X. A *statistical manifold* is defined by a triple (X, Θ, ρ), where X is called the *sample space* and $\Theta \subseteq \mathbb{R}^n$ the *parameter space*. Then, $\rho \colon \Theta \to AC(X)$ maps a parameter to a density, given by $\rho \colon \theta \mapsto \rho_\theta(\cdot)$, for any $\theta \in \Theta$. Abusing terminology, we also call Θ the statistical manifold.

Cost Function. Let a *similarity measure* $c^* \colon \mathrm{AC}(X) \times \mathrm{AC}(X) \to \mathbb{R}_{\geq 0}$ (e.g. a metric or an information divergence) be defined on $\mathrm{AC}(X)$ satisfying $c^*(\rho, \rho') = 0$ if and only if $\rho = \rho'$. Assume c^* to be strictly convex in ρ. Given a target distribution $\rho \in \mathrm{AC}(X)$ and a statistical manifold (X, Θ, ρ), we wish to minimize the *cost function* $c \to \Theta \times \mathrm{AC}(X) \to \mathbb{R}_{\geq 0}$ given by

$$c(\theta, \rho) = c^*(\rho_\theta, \rho). \tag{2.1}$$

If $\rho = \rho_{\theta'}$ for some $\theta' \in \Theta$, then by abuse of notation we write $c(\theta, \theta')$. We finally assume that $\theta \mapsto c(\theta, \theta')$ is C^2 whenever $\theta \neq \theta'$.

Natural Gradient. Assume a Riemannian structure (Θ, g^Θ) on the statistical manifold. The *Riemannian metric* g^Θ induces a *metric tensor* G^Θ, given by $g^\Theta_\theta(u, v) = u^T G^\Theta_\theta v$ and a *distance function* which we denote by d_Θ. The vectors u, v belong to the *tangent space* $T_\theta\Theta$ at θ. It is common intuition that the negative gradient $v = -\nabla_\theta c(\theta, \rho)$ gives the direction of maximal descent for c. However, this is only true on a Euclidean manifold. Consider

$$\hat{v} = \underset{v \in T_\theta\Theta : d_\Theta(\theta, \theta+v) = \Delta}{\arg\min} c(\theta + v, \rho), \tag{2.2}$$

where $\theta + v$ is to be understood in a chart of Θ, and $\Delta > 0$ defines the radius of the trust region. Linearly approximating the objective and quadratically approximating the constraint, this is solved using Lagrangian multipliers, giving the *natural gradient*

$$\hat{v} = -\frac{1}{\lambda} \left[G^\Theta_\theta \right]^{-1} \nabla_\theta c(\theta, \rho), \tag{2.3}$$

for some Lagrangrian multiplier $\lambda > 0$, which we refer to as the *learning rate*. Below, a similar derivation is carried out in more detail.

Formal Natural Gradient. Traditionally, the natural gradient uses the Fisher–Rao metric when the similarity measure used is the KL-divergence. We will now show, how a trust region formulation with respect to the chosen similarity measure can be used to derive a natural metric under which the natural gradient can be computed, resulting in the *formal natural gradient*. Thus, consider the minimization task

$$\hat{v} := \underset{v \in T_\theta\Theta, \ c(\theta+v, \theta) = \Delta}{\arg\min} c(\theta + v, \rho). \tag{2.4}$$

We approximate the constraint by the second degree Taylor expansion

$$c(\theta + v, \theta) \approx \frac{1}{2} v^T \left(\nabla^2_{\eta \to \theta} c(\eta, \theta) \right) v, \tag{2.5}$$

where the 0th and 1st degree terms disappear as $c(\theta + v, \theta)$ has a minimum 0 at $v = 0$. We call the symmetric positive definite matrix $H^c_\theta := \nabla^2_{\eta \to \theta} c(\eta, \theta)$ the *local Hessian*. Then, we further approximate the objective function

$$c(\theta + v, \rho) \approx c(\theta, \rho) + \nabla_\theta c(\theta, \rho)^T v. \tag{2.6}$$

Writing the approximate Langrangian $\mathcal{L}(v)$ of (2.4) with a multiplier $\lambda > 0$, we get

$$\mathcal{L}(v) \approx c(\theta, \rho) + \nabla_\theta c(\theta, \rho)^T v + \frac{\lambda}{2} v^T \left(\nabla^2_{\eta \to \theta} c(\eta, \theta) \right) v. \qquad (2.7)$$

Thus by the method of Langrangian multipliers, (2.4) is solved as

$$\hat{v} = -\frac{1}{\lambda} [H^c_\theta]^{-1} \nabla_\theta c(\theta, \rho). \qquad (2.8)$$

We refer to $\hat{v}$ as the *formal natural gradient* with respect to c.

Remark 1. We could have just substituted $\eta = \theta$ in the local Hessian if $\nabla^2_\eta c(\eta, \theta)$ was continuous at η. However, when studying Finsler metrics later in this work, the expression has a discontinuity at $\eta = \theta$. Therefore, a direction for a limit has to be chosen, and as a straight-forward candidate we compute the limit from the direction of the gradient.

Metric Interpretation. The local Hessian G^c_θ can be seen as a metric tensor at any $\theta \in \Theta$, inducing an inner product $g^c_\theta \colon T_\theta\Theta \times T_\theta\Theta \to \mathbb{R}$ given by $g^c_\theta(v, u) = v^T H^c_\theta u$. This imposes a *pseudo-Riemannian* structure on Θ, forming the pseudo-Riemannian manifold (Θ, g^c). Therefore, G^c_x provides us a natural metric under which to compute the natural gradient for a general c^*. If ρ has a full rank Jacobian everywhere, then a Riemannian metric is retrieved. Also, there is an obvious *pullback* structure at play. Recall, that the cost is defined by $c(\theta, \theta') = c^*(\rho_\theta, \rho_{\theta'})$. Then, computing the local Hessian yields

$$H^c_\theta = J^T_\theta H^{c^*}_{\rho_\theta} J_\theta, \qquad (2.9)$$

where $H^{c^*}_{\rho_\theta} = \nabla^2_{\rho \to \rho_\theta} c^*(\rho, \rho_\theta)$. Thus, H^c results from pulling back the c^* induced metric tensor H^{c^*} on $\mathrm{AC}(X)$ to the statistical manifold Θ. In information geometry, this Riemannian metric is said to be induced by the corresponding divergence (similarity measure) [3]. Therefore, the formal natural gradient is just the Riemannian gradient under the aforementioned induced metric.

Asymptotically Newton's Method. We provide a straightforward result, stating that the local Hessian approaches the actual Hessian in the limit, thus the formal natural gradient method approaches Newton's method. This is well known in the Fisher–Rao case, but for completeness we provide the result for the formal natural gradient.

Proposition 1. *Assume $c(\theta, \rho) = c(\theta, \theta')$ for some $\theta' \in \Theta$, and that c is C^2 in θ. Then, the natural gradient yields asymptotically Newton's method.*

Proof. The Hessian at θ is given by $\nabla^2_\theta c(\theta, \theta')$. Then, as c is C^2 in the first argument, passing the limit $\theta \to \theta'$ yields

$$H^c_\theta = \nabla^2_{\eta \to \theta} c(\eta, \theta) \overset{\theta \to \theta'}{\to} \nabla^2_{\eta \to \theta'} c(\eta, \theta') = \nabla^2_{\eta = \theta'} c(\eta, \theta'), \qquad (2.10)$$

where the last expression is the Hessian at θ'.

3   Loved Child has Many Names – Related Methods

In this section, we discuss connections between seemingly different optimization methods. Some of these connections have already been reported in the literature, some are likely to be known to some extent in the community. However, the authors are unaware of previous work drawing out these connections in their full extent. We provide such a discussion, and then present other related connections.

As discussed in [14], *proximal methods* and *trust region methods* are equivalent up to learning rate. Trust region methods employ an l^2-metric constraint

$$x_{t+1} = \operatorname*{arg\,min}_{x:\|x-x_t\|_2 \le \Delta} f(x), \ \Delta > 0, \tag{3.1}$$

whereas proximal methods include a l^2-metric penalization term

$$x_{t+1} = \operatorname*{arg\,min}_x \left\{ f(x) + \frac{1}{2\lambda}\|x - x_t\|_2^2 \right\}, \ \lambda > 0, \tag{3.2}$$

The two can be shown to be equivalent up to learning rate via Lagrangian duality.

Instead of the l^2 metric penalization, *mirror gradient descent* [13] employs a more general *proximity function* $\Psi \colon \mathbb{R}^n \times \mathbb{R}^n \to \mathbb{R}_{>0}$, that is strictly convex in the first argument. Then, the mirror descent step is given by

$$x_{t+1} = \operatorname*{arg\,min}_x \left\{ \langle x - x_t, \nabla f(x_t)\rangle + \frac{1}{\lambda}\Psi(x, x_t) \right\}. \tag{3.3}$$

Commonly, Ψ is chosen to be a *Bregman divergence* D_g, defined by choosing a strictly convex C^2 function g and writing

$$D_g(x, x') = g(x) - g(x') - \langle \nabla g(x'), x - x' \rangle. \tag{3.4}$$

To explain how these methods are related to the natural gradient, assume that we are minimizing a general similarity measure $c(x, y)$ with respect to x, as in Sect. 2. Recall, that we first defined the natural gradient as a *trust region step*. In order to derive an analytical expression for the iteration, we approximated the objective function with the first order Taylor polynomial and the constraints by the local Hessian and then used Lagrangian duality to yield a *proximal expression*, which yields the formal natural gradient when solved. In Sect. 4, we will show how this workflow indeed corresponds to known examples of the natural gradient.

Further Connections. Raskutti and Mukherjee [16] showed, that Bregman divergence proximal mirror gradient descent is equivalent to the natural gradient method on the *dual manifold* of the Bregman divergence. Khan et al. [8], consider a KL divergence proximal algorithm for learning *conditionally conjugate exponential families*, which they show to correspond to a natural gradient step. For exponential families, the KL-divergence corresponds to a Bregman divergence, and so the natural gradient step is on the *primal manifold* of the Bregman divergence. Thus the result seems to conflict with the resut in [16]. However, this can be explained, as the gradient is taken with respect to a different argument of the divergence, i.e., they consider $\nabla_x D_g(x', x)$ and not $\nabla_x D_g(x, x')$. It is intriguing how two different geometries are involved in this choice.

Pascanu and Bengio [15] remarked on the connections between the natural gradient method and Hessian-free optimization [11], Krylov Subspace Descent [17], and TONGA [9]. The main connection between Hessian-free optimization and Krylov subspace descent is the use of *extended Gauss–Newton approximation of the Hessian* [18], which gives a similar square form involving the Jacobian as the *pullback* Fisher–Rao metric on a statistical manifold. The connection was further studied by Martens [12], where an equivalence criterion between the Fisher–Rao natural gradient and extended Gauss–Newton was given.

4   Example Computations

We will now provide example computations for the local Hessian H^c of different similarity measures c, as it is the essential object in computing the natural gradient given in (2.8). We first show that in the cases of KL-divergence and a Riemannian metric, the definition of the formal natural gradient matches the classical definition, as expected. Furthermore, we contribute local Hessians for general f-divergences and Finsler metrics, specifically for the p-Wasserstein metrics.

Natural Gradient of f-Divergences. Let $\rho, \rho' \in AC(X)$ and $f : \mathbb{R}_{>0} \to \mathbb{R}_{\geq 0}$ be a convex function satisfying $f(1) = 0$. Then, the f-divergence from ρ' to ρ is

$$D_f(\rho||\rho') = \int_X \rho(x) f\left(\frac{\rho'(x)}{\rho(x)}\right) dx. \tag{4.1}$$

Now, consider the statistical manifold $(\mathbb{R}^d, \Theta, \rho)$, and compute the local Hessian

$$\left[H_\theta^{D_f}\right]_{ij} = \nabla^2 f(1) \int_X \frac{\partial \log \rho_\theta(x)}{\partial \theta_i} \frac{\partial \log \rho_\theta(x)}{\partial \theta_j} \rho_\theta(x) dx. \tag{4.2}$$

Substituting $f = -\log$ in (4.1) results in the KL-divergence, denoted by $D_{KL}(\rho||\rho')$. Noticing that $\nabla^2 f(1) = 1$ with this substitution, we can write (4.2) as $H_\theta^{D_f} = \nabla^2 f(1) H_\theta^{D_{KL}}$, where the local Hessian $H_\theta^{D_{KL}}$ is also the Fisher–Rao metric tensor at θ, and thus the natural gradient of Amari [2] is retrieved.

Natural Gradient of Riemannian Distance. Let (M, g) be a Riemannian manifold with the induced distance function d_g and the metric tensor at $\rho \in M$ denoted by G_ρ^M. Finally, denote by ρ_θ a submanifold of M parametrized by $\theta \in \Theta$. Then, when $c = \frac{1}{2}d^2$, we compute $G_\theta^{\frac{1}{2}d_g}$ as follows

$$\begin{aligned}
\left[H_\theta^{\frac{1}{2}d^2}\right]_{ij} =&\frac{1}{2}\left(\frac{\partial}{\partial\theta_j}\rho_\theta\right)^T \left[\nabla^2_{\rho_\eta \to \rho_\theta} d^2(\rho_\eta, \rho_\theta)\right]\left(\frac{\partial}{\partial\theta_i}\rho_\theta\right) \\
&+\frac{1}{2}\left[\frac{\partial^2}{\partial\theta_j\partial\theta_i}\rho_\theta\right]\left[\nabla_{\rho_\eta \to \rho_\theta} d^2(\rho_\eta, \rho_\theta)\right],
\end{aligned} \tag{4.3}$$

as $\theta' \to \theta$, the second term vanishes. Finally, $\nabla^2_{\rho_\eta \to \rho_\theta} d^2(\rho_\eta, \rho_\theta) = 2G^M_{\rho_\theta}$, thus

$$H^{\frac{1}{2}d_g}_\theta = J^T_\theta G^M_{x_\theta} J_\theta, \tag{4.4}$$

where $J_\theta = \frac{\partial}{\partial \theta} \rho_\theta$ denotes the Jacobian. Therefore, the formal natural gradient corresponds to the traditional coordinate-free definition of a gradient on a Riemannian manifold, when the metric is given by the pullback.

Natural Gradient of Finsler Distance. Let (M, F) denote a Finsler manifold, where $F_\rho \colon T_\rho M \to \mathbb{R}_{\geq 0}$, for any $\rho \in M$, is a *Finsler metric*, satisfying the properties of strong convexity, positive 1-homogeneity and positive definiteness. Then, a distance d_F is induced on M by

$$d_F(\rho, \rho') = \inf_\gamma \int_0^1 F_{\gamma(t)}(\dot\gamma(t))dt, \;\; \rho, \rho' \in M \tag{4.5}$$

where γ is any continuous, unit-parametrized curve with $\gamma(0) = \rho$ and $\gamma(1) = \rho'$.

The *fundamental tensor* G^F of F at (ρ, v) is defined as $G^F_\rho(v) = \frac{1}{2}\nabla^2_\rho F^2_\rho(v)$. Then, G^F_ρ is 0-homogeneous as the second differential of a 2-homogeneous function. Therefore, $G^F_\rho(\lambda v) = G^F_\rho(v)$ for any $\lambda > 0$. Furthermore, $G^F_\rho(v)$ is positive-definite when $v \neq 0$. Now, let $u = -J_\theta \nabla_\theta d^2_F(\rho_\theta, \rho')$, and as we can locally write $d^2_F(\rho, \rho') = F^2_{\rho\theta}(v)$ for a suitable v, then

$$H^{\frac{1}{2}d^2_F}_\theta = \frac{1}{2}\nabla^2_{\eta \to \theta} d^2_F(\rho_\eta, \rho_\theta) = \frac{1}{2}\lim_{\lambda \to 0}\nabla^2_{v = \lambda u} F^2_{\rho_\theta}(v) = J^T_\theta G^F_{\rho_\theta}(u) J_\theta. \tag{4.6}$$

Coordinate-free gradient descent on Finsler manifolds has been studied by Bercu [5]. The formal natural gradient differs slightly from this, as we use $v = -J_\theta \nabla_\theta d^2_F(\rho_\theta, \rho')$ in the preconditioning matrix $G^F_{(\rho_\theta, v)}$ (see Remark 1), where as in [5], v is chosen to maximize the descent. Thus the natural gradient descent in the Finsler case approximates the geometry in the direction of the gradient quadratically to improve the descent, but fails to take the entire local geometry into account.

p-Wasserstein Metric. Let $X = \mathbb{R}^n$ and $\rho \in \mathcal{P}_p(X)$ if

$$\int_X d^p_2(x_0, x)\rho(x)dx, \text{ for some } x_0 \in X, \tag{4.7}$$

where d_2 is the Euclidean distance. Then, the p-Wasserstein distance W_p between $\rho, \rho' \in \mathcal{P}_p(X)$ is given by

$$W_p(\rho, \rho') = \left(\inf_{\gamma \in \mathrm{ADM}(\rho, \rho')} \int_{X \times X} d^p_2(x, x')d\gamma(x, x') \right)^{\frac{1}{p}}, \tag{4.8}$$

where $\mathrm{ADM}(\rho, \rho')$ is the set of joint measures with marginal densities ρ and ρ'. The p-Wasserstein distance is induced by a Finsler metric [1], given by

$$F_\rho(v) = \left(\int_X \|\nabla \Phi_v\|^p_2 d\rho \right)^{\frac{1}{p}}, \tag{4.9}$$

where $v \in T_\rho \mathcal{P}_p(X)$ and Φ_v satisfies $v(x) = -\nabla \cdot (\rho(x)\nabla_x \Phi_v(x))$ for any $x \in X$, where $\nabla \cdot$ is the divergence operator. Now, choose $v = -J_\theta \nabla_\theta W_p^2(\rho_\theta, \rho)$. Then, through a cumbersome computation, we compute how the local Hessian acts on two tangent vectors $d\theta_1, d\theta_2 \in T_\theta \Theta$

$$
\begin{aligned}
H_\theta^{\frac{1}{2}W_p^2}&(d\theta_1, d\theta_2) \\
&= (2-p)F_{\rho_\theta}^{2(1-p)}(v)\left(\int_X \|\nabla\Phi_v\|_2^{p-2}\langle\nabla\Phi_{d\theta_1}, \nabla\Phi_v\rangle d\rho_\theta\right) \\
&\quad \times \left(\int_X \|\nabla\Phi_v\|_2^{p-2}\langle\nabla\Phi_{d\theta_2}, \nabla\Phi_v\rangle d\rho_\theta\right) \\
&\quad + F_{\rho_\theta}^{2-p}(v)\int_X \|\nabla\Phi_v\|_2^{p-2}\langle\nabla\Phi_{d\theta_1}, \nabla\Phi_{d\theta_2}\rangle d\rho_\theta \\
&\quad + (p-2)F_{\rho_\theta}^{2-p}(v)\int_X \|\nabla\Phi_v\|_2^{p-4}\langle\nabla\Phi_{d\theta_1}, \nabla\Phi_v\rangle\langle\nabla\Phi_{d\theta_2}, \nabla\Phi_v\rangle d\rho_\theta,
\end{aligned}
\tag{4.10}
$$

where $J_\theta d\theta_i = -\nabla \cdot (\rho_\theta \nabla\Phi_{d\theta_i})$ for $i = 1, 2$. The case $p = 2$ is special, as the 2-Wasserstein metric is induced by a Riemannian metric, whose pullback can be recovered by substituting $p = 2$ in (4.10), yielding

$$
H_\theta^{\frac{1}{2}W_2^2}(d\theta_1, d\theta_2) = \int_X \langle\nabla\Phi_{d\theta_1}, \nabla\Phi_{d\theta_2}\rangle d\rho_\theta.
\tag{4.11}
$$

This yields the natural gradient of W_2^2 as introduced in [6,10].

Acknowledgements. The authors were supported by Centre for Stochastic Geometry and Advanced Bioimaging, and a block stipendium, both funded by a grant from the Villum Foundation. We furthermore wish to thank the anonymous reviewers for their very useful comments.

References

1. Agueh, M.: Finsler structure in the p-Wasserstein space and gradient flows. Comptes Rendus Mathematique **350**(1–2), 35–40 (2012)
2. Amari, S.I.: Natural gradient works efficiently in learning. Neural Comput. **10**(2), 251–276 (1998)
3. Amari, S.i.: Divergence function, information monotonicity and information geometry. In: Workshop on Information Theoretic Methods in Science and Engineering (WITMSE). Citeseer (2009)
4. Amari, S.I.: Information Geometry and Its Applications. Springer, Tokyo (2016). https://doi.org/10.1007/978-4-431-55978-8
5. Bercu, G.: Gradient methods on Finsler manifolds. In: Proceedings of the Workshop on Global Analysis, Differential Geometry and Lie Algebras, pp. 230–233 (2000)
6. Chen, Y., Li, W.: Natural gradient in Wasserstein statistical manifold. arXiv preprint arXiv:1805.08380 (2018)
7. Dauphin, Y.N., Pascanu, R., Gulcehre, C., Cho, K., Ganguli, S., Bengio, Y.: Identifying and attacking the saddle point problem in high-dimensional non-convex optimization. In: Advances in Neural Information Processing Systems, pp. 2933–2941 (2014)

8. Khan, M.E., Baqué, P., Fleuret, F., Fua, P.: Kullback-Leibler proximal variational inference. In: Advances in Neural Information Processing Systems, pp. 3402–3410 (2015)
9. Le Roux, N., Manzagol, P.A., Bengio, Y.: Topmoumoute online natural gradient algorithm. In: Advances in Neural Information Processing Systems, pp. 849–856 (2008)
10. Li, W., Montúfar, G.: Natural gradient via optimal transport. Inf. Geom. **1**(2), 181–214 (2018)
11. Martens, J.: Deep learning via Hessian-free optimization. In: ICML, vol. 27, pp. 735–742 (2010)
12. Martens, J.: New insights and perspectives on the natural gradient method. arXiv preprint arXiv:1412.1193 (2014)
13. Nemirovsky, A.S., Yudin, D.B.: Problem complexity and method efficiency in optimization (1983)
14. Parikh, N., Boyd, S., et al.: Proximal algorithms. Found. Trends Optim. **1**(3), 127–239 (2014)
15. Pascanu, R., Bengio, Y.: Revisiting natural gradient for deep networks. arXiv preprint arXiv:1301.3584 (2013)
16. Raskutti, G., Mukherjee, S.: The information geometry of mirror descent. IEEE Trans. Inf. Theory **61**(3), 1451–1457 (2015)
17. Saad, Y.: Krylov subspace methods for solving large unsymmetric linear systems. Math. Comput. **37**(155), 105–126 (1981)
18. Schraudolph, N.N.: Fast curvature matrix-vector products for second-order gradient descent. Neural Comput. **14**(7), 1723–1738 (2002)

The Frenet-Serret Framework
for Aligning Geometric Curves

Nicolas J.-B. Brunel[1(✉)] and Juhyun Park[2]

[1] ENSIIE, 91025 Evry, France
nicolas.brunel@ensiie.fr
[2] Lancaster University, Lancaster LA1 4YF, UK
juhyun.park@lancaster.ac.uk

Abstract. Variations of the curves and trajectories in 1D can be analysed efficiently with functional data analysis tools. The main sources of variations in 1D curves have been identified as amplitude and phase variations. Dealing with the latter gives rise to the problem of curve alignment and registration problems. It has been recognised that it is important to incorporate geometric features of the curves in developing statistical approaches to address such problems. Extending these techniques to multidimensional curves is not obvious, as the notion of multidimensional amplitude can be defined in multiple ways. We propose a framework to deal with the curve alignment in multidimensional curves as 3D objects. In particular, we propose a new distance between the curves that utilises the geometric information of the curves through the Frenet-Serret representation of the curves. This can be viewed as a generalisation of the elastic shape analysis based on the square root velocity framework. We develop an efficient computational algorithm to find an optimal alignment based on the proposed distance using dynamic programming.

Keywords: Curve registration · Functional data analysis · Frenet-Serret frames

1 Introduction

We consider the general problem of aligning multidimensional curves as 3D objects. The curve alignment and registration problems are well studied for scalar curves (1D) under functional data analysis framework [7,10]. The richness of the registration problem comes from the variety of the criterion for comparing and measuring the similarity between the curves, which may also depend on the context. Nevertheless, in practice, good registration techniques aim to align significant features of the curves, called *landmarks*, such as peaks and valleys, and more generally geometric patterns of the curves. Many statistical approaches have been developed to automate this process, without the need of manually identifying the landmarks. As the geometric information is contained

© Springer Nature Switzerland AG 2019
F. Nielsen and F. Barbaresco (Eds.): GSI 2019, LNCS 11712, pp. 608–617, 2019.
https://doi.org/10.1007/978-3-030-26980-7_63

in the derivatives, it is often better to align the curves based on the derivatives. A related problem is to identify the sources of variations, in particular, decoupling amplitude and phase variations has been the main framework to study the variations of 1D curves [6].

While the notion of amplitude is univocally defined for scalar curves, the generalisation to curves in Euclidean space $\mathbb{R}^d$ can be done in multiple ways [5]. Among possible approaches, the use of geometric features is shown to be effective for registering curves. This idea is formalised within the framework of elastic shape analysis [9], by considering significant landmarks and looking for invariant properties through group actions such as isometries or invariance by re-parametrisation. The basis of shape analysis is provided by the definition of proper spaces for representing objects, and the definition of an adapted distance. One of the successful applications of shape analysis for curves in general Euclidean spaces or more exotic ones is found with the use of the square root velocity transform (SRVT, [9]). The geometric feature is embedded in the first order derivative of the curves, the tangent of the curves.

In this article, we generalise the methodology based on the SRVT for the registration of two curves. Instead of using only the tangent information, we use an exhaustive description of the geometry of curves given by the so-called Frenet frame, which corresponds to the higher order information. This moving frame gives an explicit link to the complete geometric characterisation of a curve (curvature and torsion) through the Frenet-Serret formula. We propose a new distance between the curves based on the Frenet frame and demonstrates that the registration of the curves based on the Frenet frames is equivalent to stretching the curvatures and torsions. We show how to find an optimal solution using dynamic programming.

The article is organised as follows. In Sect. 2, we introduce the Frenet framework and review the square root velocity framework. Section 3 present our proposed methodology of curve alignment under the Frenet framework. Section 4 develops a computational algorithm.

2   Preliminaries

2.1   Frenet-Serret Framework

We consider regular curves x, i.e, functions such that the derivatives $x^{(k)}(\cdot)$, $k = 0, \ldots 3$ exist, are continuous, and for all t in $[0, T]$, we have $\dot{x}(t) = x^{(1)}(t) \neq 0$ and $\det\left(x^{(1)}(t), x^{(2)}(t), x^{(3)}(t)\right) \neq 0$. Consequently, we can write $x(t) = X\left(s(t)\right)$ where $s \mapsto X(s)$ is the arclength parametrised curve and $t \mapsto s(t)$ is the curvilinear speed $\dot{s}(t) = \|\dot{x}(t)\|$ and $s(t) = \int_0^t \|\dot{x}(u)\| \, du$. The length of the curve is $L = s(T)$. For clarity, we write $\frac{d}{dt}x = \dot{x}(t)$ for differentiation with respect to time and $\frac{d}{ds}X = X'(s)$ for differentiation with respect to the curvilinear abscissa s. We denote the space of warping functions as

$$\mathcal{W}_{T,L}^+ = \left\{ h : [0, T] \to [0, L] \mid h, h^{-1} \in C^1, \dot{h} > 0 \right\}$$

which corresponds to the space of increasing diffeormorphisms.

The parametrised curve $\{s \mapsto X(s), \ s \in [0, L]\}$ is the geometric curve associated with x. For each $s \in [0, L]$, the tangent vector $\mathbf{T}(s) = X'(s)$ is normalised, and we can define additional normalised vectors $\mathbf{N}$ and $\mathbf{B}$ such that $\mathbf{N}(s) \propto \mathbf{T}'(s)$ and $\mathbf{B}(s) \propto \mathbf{T}(s) \times \mathbf{N}(s)$. Then, the matrix $\mathbf{Q}(s) = [\mathbf{T}(s)|\mathbf{N}(s)|\mathbf{B}(s)]$ is an orthonormal frame, which can be obtained by Gram-Schmidt orthonormalisation of the frame $[X'(s)|X''(s)|X'''(s)]$. Quite remarkably, the Frenet frames are shown to be the solution of the following ODE:

$$\begin{cases} \mathbf{T}'(s) = \kappa(s)\mathbf{N}(s), \\ \mathbf{N}'(s) = -\kappa(s)\mathbf{T}(s) + \tau(s)\mathbf{B}(s), \\ \mathbf{B}'(s) = -\tau(s)\mathbf{N}(s), \end{cases} \tag{1}$$

where the functions $s \mapsto \kappa(s), \tau(s)$ are respectively the curvature and torsion (with a positivity condition for the curvature κ). In the rest of the paper, we will denote $\boldsymbol{\theta} : s \mapsto (\kappa(s), \tau(s))$ the corresponding $\mathbb{R}^2$-valued function, and $\boldsymbol{\theta}$ will be called the generalised curvature. An alternative interpretation of this Frenet-Serret formula is that it defines an ODE in the Lie group $SO(3)$ as:

$$\dot{\mathbf{Q}}(s) = \mathbf{Q}(s)A_{\boldsymbol{\theta}}(s) \tag{2}$$

where the matrix

$$A_{\boldsymbol{\theta}}(s) \triangleq \begin{bmatrix} 0 & -\kappa(s) & 0 \\ \kappa(s) & 0 & -\tau(s) \\ 0 & \tau(s) & 0 \end{bmatrix}, \tag{3}$$

is in the Lie algebra of skew-symmetric matrices, with the generalised curvature $\boldsymbol{\theta}$ and the initial condition $\mathbf{Q}(0) = Q_0$.

The fundamental theorem of Differential Geometry of curves [2] is based on the Frenet-Serret Eq. (2) and claims that two curves x_0, x_1 with the same generalised curvature $\boldsymbol{\theta}$ (hence $L_0 = L_1$) differ only by a rigid (Euclidean) transformation and a re-parametrisation: there exists a unique $(a, O) \in \mathbb{R}^3 \times SO(3)$ and $h \in \mathcal{W}_{T,L}^+$ such that

$$x_1(t) = a + Ox_0 \circ h(t).$$

Obviously this means that the Frenet frames $\mathbf{Q}_0$ and $\mathbf{Q}_1$ satisfy $\mathbf{Q}_1(s) = O\mathbf{Q}_0(\gamma(s))$ for all $s \in [0, L]$, and an appropriate diffeomorphism γ. It is clear then that $\mathbf{Q}_i$ or $\boldsymbol{\theta}_i$ represent the shape of the curves x_i, for $i = 0, 1$.

2.2   Elastic Shape Analysis

The shapes X_0, X_1 are what is left invariant under the actions of the rigid group and the group of re-parametrisations. We now focus on the development of an elastic shape analysis framework for the comparison of two different shapes through the action of specific groups of local (and nonlinear) deformations. This is motivated by finding the most appropriate warping $h : [0, T] \longrightarrow [0, T] \in \mathcal{W}_{T,T}^+$ such that the two curves $x_1(h(t)) = X_1(s_1(h(t)))$ and $x_0(t) = X_0(s_0(t))$ looks similar. This is the standard alignment or registration problem, which has been studied in various ways, in particular based on geometric features through the geodesic distance between curves, [1,4,7,11]. We focus here on the geometric features of the curves described by Frenet frames that can be seen as an extension of the Square Root Velocity Transform (SRVT) [9]. For each curve x, the square root velocity function is defined as

$$q_x(t) = \frac{\dot{x}(t)}{\sqrt{\|\dot{x}(t)\|}} = \sqrt{\dot{s}(t)}\mathbf{T}(s(t)) ,$$

which can be viewed as a representation of the shape of the curve. The distance between two curves is then defined as the L^2 distance between q_x and is parametrisation-independent.

The SRVF transformation $F : x \mapsto \dot{x}(t)/\sqrt{\|\dot{x}(t)\|}$ helps defining a pre-shape space that is used for characterising the underlying shape of a given function. In order to align the curves x_0, x_1 with SRVF, we solve the following minimisation problem that defines at the same time a geodesic distance:

$$d_{srvf}(x_0, x_1) = \inf_{O \in SO(3), h \in \mathcal{W}_{T,T}^+} \int_0^T \left\| q_0(t) - O\sqrt{\dot{h}(t)}q_1(h(t)) \right\|^2 dt. \quad (4)$$

As the distance d_{srvf} is invariant under re-parametrisation (and translation and rotation), it can be conveniently re-written by using the increasing diffeomorphisms $\gamma : [0, L_0] \to [0, L_1] \in \mathcal{W}_{L_0,L_1}^+$ such that $s_1 \circ h = \gamma \circ s_0$ and by defining the discrepancy

$$\mathcal{R}(O, \gamma) = \int_0^{L_0} \left\| \mathbf{T}_0(s) - \sqrt{\dot{\gamma}(s)}O\mathbf{T}_1(\gamma(s)) \right\|_2^2 ds . \quad (5)$$

The distance d_{srvf} and the corresponding optimal registration is obtained by solving the following program $(O^*, \gamma^*) = \min_{\gamma, O} \mathcal{R}(O, \gamma)$ for $\gamma \in \mathcal{W}_{L_0,L_1}^+$. The optimal registration $h^* \in \mathcal{W}_{T,T}^+$ is then decomposed as $h^* = s_1^{-1} \circ \gamma^* \circ s_0$. While the warping functions $s_i, i = 0, 1$ are related to curvilinear speeds along the shapes $X_i, i = 0, 1$, the diffeomorphism γ^* is a non-linear curve stretching that induces a deformation of the shape X_1 towards the shape X_0. We elaborate on this analysis for defining a distance and the corresponding alignment problem that fully exploit the geometry of the curve.

3   Elastic Shape Analysis and Curvature Stretching

3.1   Geometry Stretching

The previous section shows the alignment of x_0, x_1 is not done by changing the curvilinear speed but by changing the geometry of the curves. Indeed, the elastic distance induces a specific family of transformations on the geometry of the curves. We have seen that the warping γ^* permits to transform a curve of length L_0 to a curve of length L_1 by a non-linear (diffeomorphic) stretching.

The two Frenet paths $\mathbf{Q}_0 : [0, L_0] \longrightarrow SO(3)$ and $\mathbf{Q}_1 : [0, L_1] \longrightarrow SO(3)$ are stretched with $\gamma \in \mathcal{W}_{L_0, L_1}^+$. From Sect. 2.1, we know that the Frenet path $s \mapsto \tilde{\mathbf{Q}}_1(s) = \mathbf{Q}_1(\gamma(s))$ is the solution to a new Frenet-Serret ODE:

$$
\begin{aligned}
\frac{d}{ds} \tilde{\mathbf{Q}}_1(s) &= \mathbf{Q}_1'(\gamma(s))\gamma'(s), \\
&= \mathbf{Q}_1(\gamma(s)) A_{\boldsymbol{\theta}}(\gamma(s))\gamma'(s), \\
&= \tilde{\mathbf{Q}}_1(s) A_{\tilde{\boldsymbol{\theta}}}(s),
\end{aligned}
$$

This means that the shape X_1 is stretched to a new shape $\tilde{X}_1$ that possesses a generalised curvature $\tilde{\boldsymbol{\theta}}(s) \triangleq \boldsymbol{\theta}(\gamma(s))\gamma'(s)$. In the case of equal length curves $(L_0 = L_1 = L)$, the standard group of non-linear stretching $\mathcal{W}_{L,L}^+$ defines a family of deformations that corresponds to a group action on the set of generalised curvatures $\boldsymbol{\Theta} = \{\boldsymbol{\theta} = (\kappa, \tau), \kappa > 0\}$: for any $\gamma \in \mathcal{W}_{L,L}^+$, we have $\boldsymbol{\theta} \mapsto \gamma \cdot \boldsymbol{\theta} = \gamma'\boldsymbol{\theta} \circ \gamma$. Indeed, we can check that for $\gamma_1, \gamma_2 \in \mathcal{W}_{L,L}^+$, we have

$$
(\gamma_2 \circ \gamma_1) \cdot \boldsymbol{\theta} = \gamma_2 \cdot (\gamma_1 \cdot \boldsymbol{\theta}).
$$

In general, the problem of finding a proper stretching γ between $\boldsymbol{\theta}_0$ and $\boldsymbol{\theta}_1$ can be expressed as solving the boundary value problem of finding γ such that $\boldsymbol{\theta}_1(\gamma(s))\gamma'(s) = \boldsymbol{\theta}_0(s)$ for all $s \in [0, L_0]$ with the constraint $\gamma(0) = 0$, $\gamma(L_0) = L_1$ and $\gamma' > 0$. Nevertheless, it is easy to see that there is no solution in general for stretching any geometry into another: if $\boldsymbol{\theta}_0$ and $\boldsymbol{\theta}_1$ are two generalised curvatures such that the torsions are $\tau_0 > 0$ and $\tau_1 < 0$, then we cannot find γ such that $\gamma'\tau_1(\gamma) = \tau_0$.

A relaxation of that problem can be turned into the standard registration problem considered by SRVF, where we introduce a distance d defined on $SO(3)$ and we aim at solving the problem of calculus of variations

$$
\begin{cases}
\min_\gamma \int_0^{L_0} d\left(\mathbf{Q}_0(s), \mathbf{Q}_1(\gamma(s))\right) \sqrt{\gamma'(s)} ds \\
\gamma(0) = 0, \ \gamma(L_0) = L_1 \\
\gamma' > 0
\end{cases},
$$

This problem might be solved with the corresponding Euler-Lagrange equation, but we will focus instead on a dynamic programming algorithm that solves efficiently a discretised version of the problem, adapted to sampled curves.

3.2   Registration with Frenet-Serret Frames

We define in this section precisely our Frenet-Serret framework for the registration of two curves. We aim at finding a warping $h : [0, T] \longrightarrow [0, T]$ that minimises the discrepancy between the moving frames $\mathbf{Q}_1(s_1(h(t)))$ to $\mathbf{Q}_0(s_0(t))$. Similarly to the elastic distance, we propose the following distance between the curves

$$
D(x_0, x_1) = \int_0^T d\left(\mathbf{Q}_0(s_0(t)), \mathbf{Q}_1(s_1(t))\right) \sqrt{\dot{s}_0(t)\dot{s}_1(t)} dt, \tag{6}
$$

where $d(\mathbf{Q}_0, \mathbf{Q}_1)$ is a distance between the frames in $SO(3)$. Standard choices are the Frobenius distance $\|\mathbf{Q}_0 - \mathbf{Q}_1\|_F^2$ or the geodesic distance $\left\|\log \mathbf{Q}_1^\top \mathbf{Q}_0\right\|_F^2$, where log is the matrix logarithm. More generally, we can consider a distance based on the weighted norms such as $\|\mathbf{Q}\|_{W,F}^2 = Trace(\mathbf{Q}^\top W \mathbf{Q})$, indicating preferred directions in the frame.

If we introduce the non-linear stretching diffeomorphism $s_1 \circ s_0^{-1} = \gamma \in \mathcal{W}_{L_0,L_1}^+$, this leads to

$$D(x_0, x_1) = \int_0^{L_0} d\left(\mathbf{Q}_0\left(s\right), \mathbf{Q}_1\left(\gamma\left(s\right)\right)\right) \sqrt{\gamma'(s)} ds \,.$$

The distance between curves can be seen as a weighted distance between the Frenet path $\mathcal{D}\left(\mathbf{Q}_0, \mathbf{Q}_1; \gamma\right) = \int_0^{L_0} d\left(\mathbf{Q}_0\left(s\right), \mathbf{Q}_1\left(\gamma\left(s\right)\right)\right) \sqrt{\gamma'(s)} ds$. A direct extension of the distance d_{srvf} is then the elastic Frenet-Serret distance

$$D_{FS}(x_0, x_1) = \min_{h \in \mathcal{W}_{T,T}^+, O \in SO(3)} D(x_0, Ox_1 \circ h). \tag{7}$$

We can also consider a distance that does not respect rotation invariance, but only reparametrisation, defined by

$$D_{FS}^0(x_0, x_1) = \min_{h \in \mathcal{W}_{T,T}^+} D(x_0, x_1 \circ h). \tag{8}$$

As with elastic distance based on the SRVF, the registration problem is the computation of the distance function:

$$\gamma^* = \arg \min_{\gamma \in \mathcal{W}_{L_0,L_1}^+} \int_0^{L_0} d\left(\mathbf{Q}_0\left(s\right), \mathbf{Q}_1\left(\gamma\left(s\right)\right)\right) \sqrt{\gamma'(s)} ds \,. \tag{9}$$

The optimal warping $h \in \mathcal{W}_{T,T}^+$ for aligning $x_1(h(t))$ to $x_0(t)$ is given by $h^* = s_1^{-1} \circ \gamma^* \circ s_0$, where γ^* is the optimal non-linear stretching. Similarly, we can find the best reparametrisation and rotation (γ^*, O^*) that solves the optimisation problem (7), and the curve $O^* x_1(h^*(t))$ is aligned to $x_0(t)$ with $h^* = s_1^{-1} \circ \gamma^* \circ s_0$.

Remark 1. D_{FS}^0 *is a direct generalisation of the standard elastic distance. If* $d\left(\mathbf{Q}_0, \mathbf{Q}_1\right) = \|\mathbf{Q}_0 - \mathbf{Q}_1\|_F^2 = \left(\|\mathbf{Q}_0\|_F^2 + \|\mathbf{Q}_1\|_F^2 - 2\,Trace\left(\mathbf{Q}_0^\top \mathbf{Q}_1\right)\right)$*, the minimisation of* $\int_0^{L_0} d\left(\mathbf{Q}_0\left(s\right), \mathbf{Q}_1\left(\gamma\left(s\right)\right)\right) \sqrt{\gamma'(s)} ds$ *is then equivalent to the maximisation of* $\int_0^{L_0} Trace\left(\mathbf{Q}_0^\top(s) \mathbf{Q}_1\left(\gamma\left(s\right)\right)\right) \sqrt{\gamma'(s)} ds$. *In the same way, the minimisation of* $\int_0^{L_0} \left\|T_0(s) - \sqrt{\gamma'(s)} OT_1(\gamma(s))\right\|_2^2 ds$ *is equivalent to the maximisation of* $\int_0^{L_0} T_0^\top(s) OT_1\left(\gamma\left(s\right)\right) \sqrt{\gamma'(s)} ds$. *This demonstrates that warping Frenet-Serret frames requires a higher degree of agreement between the geometries of* x_0 *and* x_1.

Remark 2. *We put emphasis on the fact that registering two curves* x_0 *and* x_1 *with an elastic distance (based on SRVF or Frenet-Serret) is not equivalent to aligning the curvatures and torsions.*

4   Algorithm for Pairwise Alignment

Our objective is to obtain an algorithm that computes the optimal stretching and rotation by minimising $\int_0^{L_0} d\left(\mathbf{Q}_0\left(s\right), O\mathbf{Q}_1\left(\gamma\left(s\right)\right)\right)\sqrt{\gamma'(s)}ds$ in O, γ. We derive an iterative algorithm that works with discretely sampled data from $\mathbf{Q}_0$ and $\mathbf{Q}_1$. It starts by finding the best rotation $O^{[0]}$ that minimises the distance between the normalised Frenet paths $\int_0^{L_0} d\left(\mathbf{Q}_0\left(s\right), O\mathbf{Q}_1\left(s\right)\right)ds$ (with same length L_0). Then it implements an alternate optimisation based on a discretised criterion denoted by $\mathcal{D}_{N_0}\left(\mathbf{Q}_0, O\mathbf{Q}_1; \gamma\right)$: Repeat steps 1 and 2 until convergence for $m \geq 0$

1. $\gamma^{[m]} = \arg\min_\gamma \mathcal{D}_{N_0}\left(\mathbf{Q}_0, O^{[m]}\mathbf{Q}_1; \gamma\right)$ computed by dynamic programming.
2. $O^{[m+1]} = \arg\min_O \mathcal{D}_{N_0}\left(\mathbf{Q}_0, O\mathbf{Q}_1; \gamma^{[m]}\right)$ computed by weighted averaging of rotations.

4.1   Discretisation and Dynamic Programming

We consider two regular grids G_i, $i = 0, 1$ defined on $[0, L_i], i = 0, 1$, with stepsize $h_i = \frac{L_i}{N_i}, i = 0, 1$ and N_0, N_1 are the number of points used. The points of the grid G_0 are denoted $s_k = kh_0, k \leq N_0$, and the point of the grid G_1 are denoted $x_j = jh_1, j \leq N_1$. We use a quadrature formula for approximating the integrals $\int_{s_k}^{s_{k+1}} d\left(\mathbf{Q}_0(s), \mathbf{Q}_1\left(\gamma(s)\right)\right)\sqrt{\gamma'(s)}ds$ and we use a piece-wise linear approximation of the function γ on the same grid G_0, with values in G_1. The function values are denoted $x_k = \gamma(s_k)$, and the derivatives are piece-wise constant on $]s_k, s_{k+1}[$ and are denoted $u_k = \gamma'(s_k)$. Consequently, we have $x_{k+1} = x_k + h_0 u_k$, for $k = 0, \ldots, N_0 - 1$, and we must have $x_{N_0} = \gamma(s_{N_0}) = L_1$. As the boundary conditions are fixed and known, the computation of γ is equivalent to optimising with respect to $\boldsymbol{u} = (u_k)_{k=0,\ldots,N_0-1}$. Moreover, the state dynamics is $x_{k+1} = x_k + h_0 u_k$, which means that there exists $i_k \in 1, \ldots, N_1$, such that $i_k h_1 = u_k h_0$, i.e. the possible values of the derivatives u_k are multiple of h_1/h_0.

On each segment $[s_k, s_{k+1}]$, our approximation of the integral is $g_k(x_k, u_k)$ and is obtained by using the trapezoidal rule

$$g_k(x_k, u_k) = h_0 \frac{d\left(\mathbf{Q}_0(s_k), \mathbf{Q}_1\left(x_k\right)\right) + d\left(\mathbf{Q}_0(s_{k+1}), \mathbf{Q}_1\left(x_{k+1}\right)\right)}{2} \times \sqrt{u_k}.$$

where we have made use of the fact that $\gamma' = u_k$ is constant on the segment. Finally, our approximation of the integral criterion is the following sum

$$\mathcal{D}_{N_0}\left(\mathbf{Q}_0, \mathbf{Q}_1; \gamma\right) \triangleq \mathcal{D}_{N_0}\left[\boldsymbol{u}\right] = \sum_{k=0}^{N_0-1} g_k(x_k, u_k) + g_{N_0}(x_{N_0}).$$

The terminal cost $g_N(x_N)$ is such that $g_{N_0}(x_{N_0}) = +\infty$ if $x_{N_0} \neq L_1$ and $g_{N_0}(x_{N_0}) = 0$ if $x_{N_0} = L_1$. We need to solve the following program with constraints on the state and the control variables:

$$\begin{cases} \min_{\boldsymbol{u}} \sum_{k=0}^{N_0-1} g_k(x_k, u_k) + g_{N_0}(x_{N_0}) \\ s.t. \quad x_{k+1} = x_k + h_0 u_k \\ \quad x_0 = 0, x_{N_0} = L_1, 0 < x_k < L_1, \forall k \in [1, \ldots, N_0 - 1] \\ \quad \forall k \leq N_0 - 1, u_k > 0 \end{cases}$$

We use Dynamic Programming for computing the criterion and optimising in $\boldsymbol{u}$. This is based on the backward computation of the value function $J_k(x)$ ($k = 0, \ldots, N_0$) defined on the state space, i.e the grid G_1. For every initial state x, the optimal cost is given by

$$\begin{cases} J_{N_0}(x) = g_{N_0}(x), \forall x \in G_1 \\ J_k(x_k) = \min_{u \in U_k(x_k)} g_k(x_k, u) + J_{k+1}(x_k + h_0 u), \ k = 0, 1, \ldots, N_0 - 1, \end{cases}$$

where $U_k(x_k)$ is the set of admissible control at time k and state x_k. For computation speed, we may impose a more stringent constraint on the control such as $h_0 u_k \leq \Delta_{Max}$. The optimal control found at each time k is denoted by u_k^*, and is computed during the forward pass. The optimal alignment function γ^* is sampled on the grid G_0, such that $x_k^* = \gamma^*(s_k)$ is obtained by starting at $x_0^* = 0$ and by using the optimal decision $u_k^*, k = 0, \ldots, N_0 - 1$.

Remark 3. *In practice we use this Dynamic Programming algorithm for computing warping functions defined on $[0, 1]$. A straightforward change of variable gives*

$$\mathcal{D}(\boldsymbol{Q}_0, \boldsymbol{Q}_1; \gamma) = \int_0^1 d\left(\tilde{\boldsymbol{Q}}_0(u), \tilde{\boldsymbol{Q}}_1(\tilde{\gamma}(s))\right) \sqrt{\tilde{\gamma}'(s)} \sqrt{L_0 L_1} ds \qquad (10)$$

where $\tilde{\gamma} : [0, 1] \longrightarrow [0, 1]$ is the warping function defined for all $u \in [0, 1]$ such as $\tilde{\gamma}(u) = L_1 \gamma(u L_0)$. The Frenet path $\tilde{\boldsymbol{Q}}_0, \tilde{\boldsymbol{Q}}_1$ have been defined by normalizing the paths by their length, i.e $\tilde{\boldsymbol{Q}}_i(u) = \boldsymbol{Q}_i(u L_i), i = 0, 1.$

Remark 4. *We deal with two grids defined on $[0, 1]$: $G_0 = \{i \frac{1}{N_0}, i = 0, \ldots, N_0\}$ and $G_1 = \{j \frac{1}{N_1}, j = 0, \ldots, N_1\}$. It is important to resample the data and to interpolate the data of the Frenet Path $\tilde{\boldsymbol{Q}}_1$ such that $N_1 = 2 \times N_0$. This is needed because the warping function is piece-wise linear, and we control only the slope u_k on the interval $[s_k, s_{k+1}]$. We impose $u_k > 0$, but the slope is quantified and is proportional to $\frac{h_1}{h_0}$, consequently if we want a good approximation of the exact warping function γ we need a fine grid G_1. In practice, doubling and interpolating the number of points in $\tilde{\boldsymbol{Q}}_1$ is sufficient. On the other side, the trapezoid approximation might be significantly biased, hence refining the grid G_0 by interpolating the data $\tilde{\boldsymbol{Q}}_0(s_k)$ is sometimes needed.*

Remark 5. *We use linear interpolation in the Lie Algebra: let $A, B \in SO(3)$, we define the smooth path $\varphi : [0, 1] \longrightarrow SO(3)$ such that $\varphi(0) = A$ and $\varphi(1) = B$, by $\varphi(s) = \exp\left(s \log(BA^\top)\right) A.$*

4.2 Optimal Rotation

The computation of the minimum depends on the type of the distance function used. For the standard Frobenius distance, the solution is found by solving

$$\min_O \sum_{k=0}^{N_0-1} \frac{h_0 \sqrt{u_k^*}}{2} \text{Trace}\left(\left(\mathbf{Q}_1(x_{k_1}^*)\mathbf{Q}_0^\top(s_{k+1}) + \mathbf{Q}_1(x_k^*)\mathbf{Q}_0^\top(s_k)\right) O\right). \qquad (11)$$

The solution is the polar part of the weighted mean $\sum_{k=0}^{N_0-1} \frac{h_0 \sqrt{u_k^*}}{2} \mathbf{Q}_1(x_{k_1}^*)$ $\mathbf{Q}_0^\top(s_{k+1}) + \mathbf{Q}_1(x_k^*)\mathbf{Q}_0^\top(s_k)$. If we use the geodesic distance, the problem is equivalent to computing a weighted geodesic, which can be computed by gradient descent in $SO(3)$, see [3,8].

5   Examples and Simulations

We show that our DP algorithm can estimate properly the warping function. In our simulations, we define a reference generalised curvature $\boldsymbol{\theta_0}$ and the associated

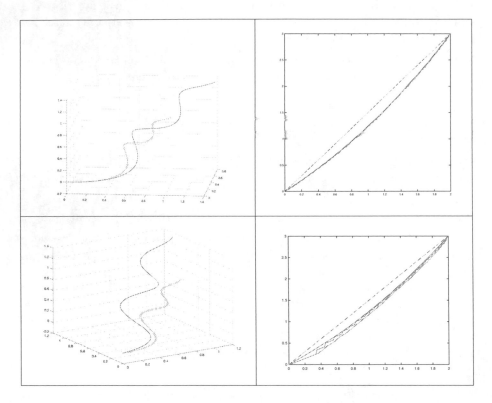

Fig. 1. Alignement of two shapes X_0 and X_1, when $\boldsymbol{\theta_1} = w^* \cdot \boldsymbol{\theta_0}$ and estimation of the warping function γ^*.
First Row: No rotation. Top Left: The two curves X_0 (in blue) and X_1 (in red). Top Right: the estimates of γ^*: the truth γ^* (dark blue), $\hat{\gamma}_{fs}$ (red), and $\hat{\gamma}_{srvf}$ (magenta). The red and magenta are superimposed.
2nd Row: With a random rotation. Bottom Left: The two curves X_0 (in blue) and X_1 (in red). X_1 is a stretched and rotated version of X_0. Bottom Right: the estimates of γ^*: the truth γ^* (dark blue), $\hat{\gamma}_{fs}$ (red), and $\hat{\gamma}_{srvf}$ (magenta). The estimate $\hat{\gamma}_{fs}$ is closer to the truth blue line than $\hat{\gamma}_{srvf}$. The rotations are both estimated by SRVF and FS, and we obtain also more precise results with Frenet-Serret, which might be natural because we use more information. Finally, we should notice that the convergence is much faster for the Frenet-Serret version than for the SRVF (less than 5 iterations against more than 50 iterations) (color figure online).

Frenet path $\mathbf{Q_0}$ and curve X_0 of length L_0. The curve X_1 is obtained from X_0 by applying the elastic deformation given by w^*, i.e $\boldsymbol{\theta_1} = w^* \cdot \boldsymbol{\theta_0}$, and the Frenet paths are such that $\forall s \in [0, L_1], \mathbf{Q_1}(s) = \mathbf{Q_0}\big(w^*(s)\big)$. Our objective is then to estimate $w^{*(-1)} = \gamma^*$ by minimizing our criteria $\mathcal{D}_{N_0}(\mathbf{Q_0}, \mathbf{Q_1}; \gamma)$. We denote $\hat{\gamma}_{fs}$ the estimate obtained by Frenet-Serret frames and $\hat{\gamma}_{srvf}$ obtained by standard SRVF. We consider a warping functions defined in $[0, 1]$ (see remark 3) $w(s) = \frac{\log(s+1)}{\log(2)}$ with $\gamma(s) = \exp\big(\log(2)s\big) - 1$. We consider that we observe directly the Frenet paths $\mathbf{Q_0}, \mathbf{Q_1}$ on grids G_0, G_1, with $L_0 = 2, N_0 = 100$ and $L_1 = 3, N_0 = 150$. The curves have a shape defined by a curvature $\kappa_0(s) = \exp(\theta \sin(s))$ and a torsion $\tau_0(s) = \eta s - 0.5$. We consider a simple registration case without any rotation O, and we consider also the case of a additional random rotation i.e $\mathbf{Q_1} = O\mathbf{Q_0} \circ w^*$. The results are available in Fig. 1.

References

1. Dryden, I., Mardia, K.: Statistical Shape Analysis, vol. 4. Wiley, Chichester (1998)
2. Kühnel, W.: Differential Geometry, vol. 77. American Mathematical Society, Boston (2015)
3. Le, H.: Estimation of Riemannian barycentres. LMS J. Comput. Math. **7**, 193–200 (2004)
4. Le Brigant, A., Arnaudon, M., Barbaresco, F.: Optimal matching between curves in a manifold. In: Nielsen, F., Barbaresco, F. (eds.) GSI 2017. LNCS, vol. 10589, pp. 57–64. Springer, Cham (2017). https://doi.org/10.1007/978-3-319-68445-1_7
5. Marron, J., Alonso, A.: Overview of object oriented data analysis. Biometrical J. **56**(5), 732–753 (2014)
6. Marron, J., Ramsay, J., Sangalli, L., Srivastava, A.: Functional data analysis of amplitude and phase variation. Stat. Sci. **30**, 468–484 (2015)
7. Ramsay, J., Silverman, B.: Functional Data Analysis. Springer Series in Statistics, 2nd edn. Springer, New York (2005). https://doi.org/10.1007/b98888
8. Rentmeesters, Q., Absil, P.A.: Algorithm comparison for karcher mean computations of rotation matrices and diffusion tensors. In: 19th European Signal Processing Conference (EUSPICO 2011) (2011)
9. Srivastava, A., Klassen, E.P.: Functional and Shape Data Analysis. SSS. Springer, New York (2016). https://doi.org/10.1007/978-1-4939-4020-2
10. Wang, J.L., Chiou, J.M., Müller, H.G.: Functional data analysis. Annu. Rev. Stat. Appl. **3**, 257–295 (2016)
11. Younes, L.: Shapes and Diffeomorphisms. AMS, vol. 171. Springer, Heidelberg (2019). https://doi.org/10.1007/978-3-662-58496-5

Geometry of Quantum States

When Geometry Meets Psycho-Physics and Quantum Mechanics: Modern Perspectives on the Space of Perceived Colors

Michel Berthier[1] and Edoardo Provenzi[2(✉)]

[1] Laboratoire MIA, Pôle Sciences et Technologie, Université de La Rochelle, 23, Avenue A. Einstein, BP 33060, 17031 La Rochelle, France
`michel.berthier@univ-lr.fr`
[2] Université de Bordeaux, CNRS, Bordeaux INP, IMB, UMR 5251, 351 Cours de la Libération, 33400 Talence, France
`edoardo.provenzi@math.u-bordeaux.fr`

Abstract. We discuss some modern perspectives about the mathematical formalization of colorimetry, motivated by the analysis of a groundbreaking, yet poorly known, model of the color space proposed by H.L. Resnikoff and based on differential geometry. In particular, we will underline two facts: the first is the need of novel, carefully implemented, psycho-physical experiments and the second is the role that Jordan algebras may have in the development of a more rigorously founded colorimetry.

Keywords: Color space · Resnikoff's model · Jordan algebras

1 Introduction

In 1974, H.L. Resnikoff published a revolutionary paper about the geometry of the space of perceived colors $\mathcal{P}$ [12]. Starting from the axiomatic set for colorimetry provided by Schrödinger [15], he added a new axiom, the homogeneity of $\mathcal{P}$ with respect to a suitable group of transformations, and proved that only two geometrical structures were coherent with the new set of axioms: the first is isomorphic to the well-known tristimulus flat space $\mathbb{R}^+ \times \mathbb{R}^+ \times \mathbb{R}^+ \equiv \mathcal{P}_1$, while the second, totally new, is isomorphic to $\mathbb{R}^+ \times SL(2,\mathbb{R})/SO(2) \equiv \mathcal{P}_2$, thus confirming the interest about hyperbolic geometry in colorimetry, already pointed out by Yilmaz in [21].

Resnikoff was also able to single out a unique Riemannian metric on the two geometrical structures by requiring it to be invariant with respect to the group transformations: the resulting metric on $\mathcal{P}_1$ coincides with the well-known Helmholtz-Stiles flat metric [20], while that on $\mathcal{P}_2$ is constant negative curvature Rao-Siegel metric, which is analogous to the Fisher metric in the geometric theory of information [1].

F. Nielsen and F. Barbaresco (Eds.): GSI 2019, LNCS 11712, pp. 621–630, 2019.
https://doi.org/10.1007/978-3-030-26980-7_64

Finally, Resnikoff provided an elegant framework to treat the two cases $\mathcal{P}_1$ and $\mathcal{P}_2$ as special instances of a unique theory based on Jordan algebras. In spite of its elegance and innovative character, Resnikoff's and Yilmaz's paper remained practically ignored until today, receiving only a few quotation since their publication.

With this contribution, we would like to share our ideas about the influence that these pioneers might have for the development of a modern, geometry-based, colorimetry. The aim is both to overcome the lack of mathematical rigor that affects the foundation of this discipline and to create a theory more suited for color image processing applications.

2   Description of Resnikoff's Model of Color Space

As we said in the introduction, in the paper [12], Resnikoff analyzed the geometrical properties of the space of perceived colors $\mathcal{P}$ with a high level mathematical rigor. He started from Schrödinger's axioms [15] for $\mathcal{P}$:

Axiom 1 (Newton 1704): if $x \in \mathcal{P}$ and $\alpha \in \mathbb{R}^+$, then $\alpha x \in \mathcal{P}$.

Axiom 2: if $x \in \mathcal{P}$ then it does not exist any $y \in \mathcal{P}$ such that $x + y = 0$.

Axiom 3 (Grassmann 1853, Helmholtz 1866): for every $x, y \in \mathcal{P}$ and for every $\alpha \in [0, 1]$, $\alpha x + (1 - \alpha)y \in \mathcal{P}$.

Axiom 4 (Grassmann 1853): every collection of more than three perceived colors is a linear dependent family in the vector space V spanned by the elements of $\mathcal{P}$.

The axioms imply that $\mathcal{P}$ *is a convex cone* embedded in a vector space V of dimension 3, for standard observers non affected by color blindness, as it will be implicitly assumed in the following part of the paper.

Resnikoff added another axiom, that of *local homogeneity* of $\mathcal{P}$ with respect to changes of background. To correctly introduce this axiom, it is worthwhile showing the observational arrangement that he considered, which is depicted in Fig. 1: a standard observer is watching a simple color stimulus embedded in a uniform background.

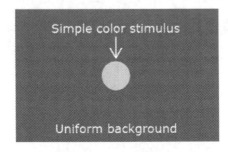

Fig. 1. The observational arrangement considered by Resnikoff.

When the background of the color stimulus is modified, our perception of the stimulus changes. Resnikoff identified the change of background transformations B with the following group:

$$GL_+(\mathcal{P}) := \{B \in GL(V) \ : \ \det(B) > 0, \text{ and } B(x) \in \mathcal{P} \ \forall x \in \mathcal{P}\},$$

where $GL(V)$ is the group of invertible linear operators on V, the requirements $\det(B) > 0$ and $B(x) \in \mathcal{P}$ guarantee that these transformations preserve the orientation of the cone and that $\mathcal{P}$ is stable under their action.

The previous observation about how a change of background modifies the perception of the colors stimulus can thus be formalized by saying that $\mathcal{P}$ is locally homogeneous with respect to $GL_+(\mathcal{P})$. However, thanks to the convex nature of $\mathcal{P}$, it is clear that local homogeneity implies global homogeneity.

For this reason, Resnikoff postulates a fifth axiom on the structure of the color space:

Axiom 5 (Resnikoff 1974): $\mathcal{P}$ is globally homogeneous with respect to the group of background transformations $GL_+(\mathcal{P})$.

Starting from the set of axioms 1–5 and by using standard results from the theory of Lie groups and algebras, Resnikoff managed to show that the only two geometrical structures compatible with these axioms are:

$$\mathcal{P}_1 \simeq \mathbb{R}^+ \times \mathbb{R}^+ \times \mathbb{R}^+, \tag{1}$$

or

$$\mathcal{P}_2 \simeq \mathbb{R}^+ \times SL(2, \mathbb{R})/SO(2), \tag{2}$$

where $SL(2, \mathbb{R})$ is the group of 2×2 matrices with real entries and determinant $+1$ and $SO(2)$ is the group of matrices that perform rotations in the plane $\mathbb{R}^2$.

The first geometrical structure, $\mathcal{P}_1$ agrees with the usual trichromatic space, such as RGB, XYZ, and so on. The second one, $\mathcal{P}_2$, on the contrary, is a totally new geometrical structure for the space $\mathcal{P}$.

If a Riemannian metric on $\mathcal{P}$ existed, then the difference between the perceived colors $x, y \in \mathcal{P}$ would be calculated with the integral

$$d(x, y) = \int_\gamma ds, \qquad \gamma(0) = x, \ \gamma(1) = y, \tag{3}$$

where γ is the unique geodesic arc between x and y.

Resnikoff proved that the only Riemannian $GL_+(\mathcal{P})$-invariant metric on $\mathcal{P}_1$ is precisely the Helmholtz-Stiles metric (obtained with a totally different method), i.e.,

$$ds^2 = \alpha_1 \left(\frac{dx_1}{x_1}\right)^2 + \alpha_2 \left(\frac{dx_2}{x_2}\right)^2 + \alpha_3 \left(\frac{dx_3}{x_3}\right)^2, \tag{4}$$

where $x_j \in \mathbb{R}^+$ and α_j are positive real constants for $j = 1, 2, 3$.

Turning his attention to $\mathcal{P}_2$, he showed that the only Riemannian $GL_+(\mathcal{P})$-invariant metric on it can be written like this:

$$ds^2 = \mathrm{tr}(x^{-1}dxx^{-1}dx), \tag{5}$$

which is equivalent to the Rao-Siegel metric [4,17].

Resnikoff concluded his paper by showing that the models $\mathcal{P}_1$ and $\mathcal{P}_2$ are particular instances of a unified framework based on the use of Jordan algebras. We recall that a Jordan algebra $\mathcal{A}$ is an algebra over a field whose multiplication $\circ$ is commutative but non-associative and it satisfies the so-called Jordan's identity:

$$(x \circ y) \circ x^2 = x \circ (y \circ x^2), \tag{6}$$

for all x and y in $\mathcal{A}$. Such an algebra is power-associative in the sense that the sub-algebra generated by any of its element is associative.

In the Sect. 3 we will underline some problems that remained opened since the appearance of Resnikoff's paper, while in Sect. 4 we will discuss how Resnikoff's use of Jordan algebras was ahead of his time and it can be rescued from oblivion and used to define the colorimetric attributes of a color.

3   Missing Pieces in Resnikoff's Model from the Viewpoint of Modern Colorimetry

Resnikoff's paper remains, after more than 40 years since its publication, an example of elegance, originality and independent research. However, in the light of nowadays knowledge about color perception, there are three issues that must be discussed carefully.

The first, and more delicate, one is the hypothesis of linearity for the background transformations $B \in GL_+(\mathcal{P})$. Resnikoff himself, in a subsequent paper [13] recognized that this hypothesis is a very strong one with the sentence: '*the least verified aspect of Axiom 5 is its assertion of the linearity of transitive group of changes of background*'.

Without linearity, the whole mathematical structure built by Resnikoff to arrive to the identification of $\mathcal{P}_1$ and $\mathcal{P}_2$ as the only two possible geometrical representations of $\mathcal{P}$ loses its foundation. Thus, a carefully developed psycho-physical experiment based on color matching [6] is needed to check the linearity of background transformations $B \in GL_+(\mathcal{P})$.

An original experiment to check *additivity*, i.e. the fact that $B(x + y)$ color matches $B(x) + B(y)$, would be the following: on one side, we superpose the lights $\mathbf{x}$ and $\mathbf{y}$ which generate the perceived colors x and y, respectively, with respect to the same background b, then we perform a change of background with the transformation B and we call b' the new background. Finally, we color match what we obtain, this will give us the perceived color $B(x + y) \in \mathcal{P}$ which represents $\mathbf{x} + \mathbf{y}$ in the context b'.

On the other side, we separately perform the change of context B on x and y and we color match the results, obtaining $B(x) \in \mathcal{P}$ and $B(y) \in \mathcal{P}$, respectively.

We then match $B(x)$ with the physical light $\mathbf{x}'$ and $B(y)$ with the physical light $\mathbf{y}'$. If the color sensations produced by $x + y$ in the background b' matches that of $\mathbf{x}' + \mathbf{y}'$ in the same background, then the change of context is additive.

To test *homogeneity*, i.e. the fact that $B(\alpha x)$ color matches $\alpha B(x)$, we must use a similar procedure for at least a sufficiently large range of coefficients $\alpha \in \mathbb{R}$.

The question about how large this range of coefficients must be leads us directly to the second issue, which is shared by any model of visual perception. We are referring to the fact that Axiom 1, i.e. the fact that $\mathcal{P}$ is an *infinite cone*, is only an idealization: for any $x \in \mathcal{P}$ and very large α, αx will cease to be perceived, and this it will not belong to $\mathcal{P}$ anymore, because the retinal photoreceptors will be firstly saturated and then permanently damaged [16]. Similarly, for $\alpha \neq 0$, but $\alpha \simeq 0$, αx will firstly switch the human visual system to mesopic and then scotopic vision via the Purkinje effect [8], and then it will fall below the threshold limit to be perceived. Thus, more than an infinite cone, $\mathcal{P}$ has the structure of a *truncated cone*. In classical colorimetry, one bypasses this last observation by working far from these limits, however, in order to build a modern theory of colorimetry we must start taking into account more seriously the lower and upper perceptual bound.

The third issue about Resnikoff's model is the lack of locality, i.e. the fact the observational configuration considered, that of Fig. 1 is a over-simplified version of a real-world visual condition. Everyday vision deals with what is commonly called 'color in context', i.e. the fact that a non-uniform background strongly influences color perception, this phenomenon is referred to as color induction [18]. Actually, many papers has been emphasized the role of context for color vision to the point that a standalone definition of the color of a surface, without the specification of the context in which the surface is embedded does not make sense anymore: color *is* color in context [7,10,11,14].

This observation implies that the Resnikoff model, and any other color perception model based on the observational configuration of Fig. 1, can only be viewed as a first step towards a local theory of color, mathematically similar to a field theory. A more thorough understanding of the color induction phenomenon and properties as color constancy, i.e. the robustness of color perception with respect to changes of illumination [10] or the invariance of saturation perception for monochromatic light stimuli [21], are likely to play a fundamental role in the construction of this kind of color field theory.

4 On the Role of Jordan Algebras in Colorimetry

The second part of [12] devoted to Jordan algebras may suggest that Resnikoff had already a quantum interpretation of his new hyperbolic model. Jordan algebras are non-associative commutative algebras that have been classified by Jordan, Von Neumann and Wigner [9] under the assumptions that they are of finite dimension and formally real. They are considered as a fitting alternative to the usual associative non-commutative framework for the geometrization of quantum mechanics. One of the main motivation to introduce Jordan algebras in our context is Koecher-Vinberg theorem which states that every open convex regular homogeneous and self-dual cone is the interior of the positive domain of a Jordan algebra [5]. From a quantum viewpoint, this means that such a cone is the set of positive observables of a quantum system.

Contrary to Resnikoff, one may postulate at first that $\mathcal{P}$ can be described from the state space of a quantum system characterized by a formally real Jordan algebra $\mathcal{A}$ of real dimension 3, according to the dimension of $\mathcal{P}$. Such an algebra $\mathcal{A}$ is necessarily isomorphic to one of the following two: either $\mathbb{R} \oplus \mathbb{R} \oplus \mathbb{R}$ or $\mathcal{H}(2, \mathbb{R})$, which is the algebra of real symmetric 2×2 matrices, with Jordan product given by

$$x \circ y = \frac{1}{2}(xy + yx). \tag{7}$$

The classification by Resnikoff can be simply recovered by taking the symmetric cone of the positive elements of $\mathcal{A}$ [5].

Let us, in particular, concentrate on the geometrical structure $\mathcal{P}_2$ of the color space: the algebra $\mathcal{H}(2, \mathbb{R})$ is isomorphic to the so-called spin factor $\mathbb{R} \oplus \mathbb{R}^2$ via the transformation defined by

$$(\alpha + v) \longmapsto \begin{pmatrix} \alpha + v_1 & v_2 \\ v_2 & \alpha - v_1 \end{pmatrix}, \tag{8}$$

with $\alpha \in \mathbb{R}$ and $v = (v_1, v_2) \in \mathbb{R}^2$, where v_1 and v_2 are the components of v with respect to the canonical basis of $\mathbb{R}^2$. One may consider the spin factor $\mathbb{R} \oplus \mathbb{R}^2$ as a 3-dimensional Minkowski space-time equipped with the metric

$$(\alpha + v) \cdot (\beta + w) = \alpha\beta - \langle v, w \rangle, \tag{9}$$

where α and β are reals and v and w are vectors of $\mathbb{R}^2$. Let us also recall that the light-cone $\mathcal{C}$ of $\mathbb{R} \oplus \mathbb{R}^2$ is the set of elements $x = (\alpha + v)$ that satisfy

$$x \cdot x = 0, \tag{10}$$

and that a light ray is a 1-dimensional subspace of $\mathbb{R} \oplus \mathbb{R}^2$ spanned by an element of $\mathcal{C}$. It is clear that every light ray is spanned by a unique element of the form $(1 + v)$ with v a unit vector of $\mathbb{R}^2$ and, therefore, that the space of light rays coincides with the projective real space $\mathbb{P}_1(\mathbb{R})$. In other words, we have the following result.

Proposition 1. *There is a one to one correspondence between the light rays of the spin factor $\mathbb{R} \oplus \mathbb{R}^2$ and the rank 1 projections of the Jordan algebra $\mathcal{H}(2, \mathbb{R})$.*

The correspondance is given by

$$(1 + v) \longmapsto \frac{1}{2} \begin{pmatrix} 1 + v_1 & v_2 \\ v_2 & 1 - v_1 \end{pmatrix}. \tag{11}$$

This correspondence has a meaningful interpretation: the light rays of the spin factor $\mathbb{R} \oplus \mathbb{R}^2$, as a Minkowski space-time of dimension 3, are precisely the pure states of the algebra $\mathcal{H}(2, \mathbb{R})$, as a quantum system over $\mathbb{R}^2$.

A state of $\mathcal{A}$ is a linear functional $\langle \cdot \rangle : \mathcal{A} \longrightarrow \mathbb{R}$ that is nonnegative and normalized, i.e. $\langle 1 \rangle = 1$. It can be shown that the states of $\mathcal{A}$ are given by density matrices, namely by the elements of $\mathcal{H}(2, \mathbb{R})$ that are non-negative and have trace 1 [2]. They correspond precisely to the elements $x = (1 + v)/2$ of the spin factor with $\|v\| \leq 1$. The pure states, i.e those which can be characterized as projections, form the boundary of this disk since, for pure states, it holds that $\|v\| = 1$. It is clear that this boundary can be identified with $\mathbb{P}_1(\mathbb{R})$. Contrary to the usual context of quantum mechanics, the system that we consider is real, i.e. the algebra $\mathcal{H}(2, \mathbb{C})$ is replaced by $\mathcal{H}(2, \mathbb{R})$.

This system is a so-called rebit, a real qubit [2], that has no classic physical interpretation because there is no space with a rotation group of dimension two. As explained in the sequel, it appears that this kind of system is relevant to explain color perception. We refer also to [19] for information on real-vector-space quantum theory and its consistency regarding optimal information transfer.

An element ρ of $\mathcal{H}(2, \mathbb{R})$ is a state density matrix if and only if it can be written as:

$$\rho(v_1, v_2) = \frac{1}{2}(Id_2 + v \cdot \sigma), \tag{12}$$

where $\sigma = (\sigma_1, \sigma_2)$ with:

$$\sigma_1 = \begin{pmatrix} 1 & 0 \\ 0 & -1 \end{pmatrix} \quad \sigma_2 = \begin{pmatrix} 0 & 1 \\ 1 & 0 \end{pmatrix}, \tag{13}$$

and $v = v_1 e_1 + v_2 e_2$ is a vector of $\mathbb{R}^2$ with $\|v\| \leq 1$. The matrices σ_1 and σ_2 are Pauli-like matrices. In the usual framework of quantum mechanics, the Bloch body [2], is the unit Bloch ball in $\mathbb{R}^3$. It represents the states of the two-level quantum system of a spin-$\frac{1}{2}$ particle, also called a qubit. In the present context, the Bloch body is the unit disk of $\mathbb{R}^2$ associated to a rebit.

More precisely, let us consider the four state vectors:

$$|u_1\rangle = \begin{pmatrix} 1 \\ 0 \end{pmatrix}, \ |d_1\rangle = \begin{pmatrix} 0 \\ 1 \end{pmatrix}, \ |u_2\rangle = \frac{1}{\sqrt{2}} \begin{pmatrix} 1 \\ 1 \end{pmatrix}, \ |d_2\rangle = \frac{1}{\sqrt{2}} \begin{pmatrix} -1 \\ 1 \end{pmatrix}. \tag{14}$$

We have:

$$\sigma_1 = |u_1\rangle\langle u_1| - |d_1\rangle\langle d_1|, \ \sigma_2 = |u_2\rangle\langle u_2| - |d_2\rangle\langle d_2|. \tag{15}$$

The state vectors $|u_1\rangle$ and $|d_1\rangle$, resp. $|u_2\rangle$ and $|d_2\rangle$, are eigenstates of σ_1, resp. σ_2, with eigenvalues 1 and -1. Using polar coordinates $v_1 = r \cos\theta$, $v_2 = r \sin\theta$, we can write $\rho(v_1, v_2)$ as:

$$\begin{aligned} \rho(r, \theta) &= \frac{1}{2} \begin{pmatrix} 1 + r \cos\theta & r \sin\theta \\ r \sin\theta & 1 - r \cos\theta \end{pmatrix} \\ &= \frac{1}{2} \{(1 + r \cos\theta)|u_1\rangle\langle u_1| + (1 - r \cos\theta)|d_1\rangle\langle d_1| \\ &\quad + (r \sin\theta)|u_2\rangle\langle u_2| - (r \sin\theta)|d_2\rangle\langle d_2|\}. \end{aligned} \tag{16}$$

In particular, every pure state density matrix can be written as:

$$\rho(1, \theta) = |(1, \theta)\rangle\langle(1, \theta)|, \tag{17}$$

with:

$$|(1, \theta)\rangle = \cos(\theta/2)|u_1\rangle + \sin(\theta/2)|d_1\rangle. \qquad (18)$$

This means that we can identify the pure state density matrices $\rho(1, \theta)$ with the state vectors $|(1, \theta)\rangle$ and also with the points of the unit disk boundary of coordinate θ. More generally, every state density matrix can be written as a mixture:

$$\rho(r, \theta) = \rho_0 + \frac{r\cos\theta}{2}\left(\rho(1, 0) - \rho(1, \pi)\right) + \frac{r\sin\theta}{2}\left(\rho(1, \pi/2) - \rho(1, 3\pi/2)\right), \quad (19)$$

with:

$$\rho_0 = \frac{1}{2}\begin{pmatrix} 1 & 0 \\ 0 & 1 \end{pmatrix}. \qquad (20)$$

Such a mixture is given by the point of the unit disk of polar coordinates (r, θ).

It is important to notice that the four state density matrices $\rho(1, 0)$, $\rho(1, \pi)$, $\rho(1, \pi/2)$ and $\rho(1, 3\pi/2)$ correspond to two pairs of state vectors $(|u_1\rangle, |d_1\rangle)$, $(|u_2\rangle, |d_2\rangle)$, the state vectors $|u_i\rangle$ and $|d_i\rangle$, for $i = 1, 2$, being linked by the "up and down" Pauli-like matrix σ_i. It can be shown that this Bloch disk coincides with Hering's disk given by the color opponency mechanism. Details will appear elsewhere [3].

Among all the states, the normalized identity $\rho_0 = (1 + 0)/2 = 1/2 I_2$ plays a significant role: it is the state of maximal von Neumann entropy. It is characterized by:

$$\rho_0 = \underset{\rho}{\mathrm{argmax}}(-\mathrm{Trace}(\rho \log \rho)). \qquad (21)$$

Actually, $-\mathrm{Trace}(\rho_0 \log \rho_0) = \log 2$ and $-\mathrm{Trace}(\rho \log \rho) = 0$ for a pure state ρ.

This quantum interpretation allows us to define the three main attributes of a color, without any reference to physical colors or even to an observer. In fact, we can define a perceived color as a non-negative normalized element $(\alpha + v)/2$ of the spin factor $\mathbb{R} \oplus \mathbb{R}^2$. Nonnegativity is equivalent to $\alpha^2 \geq \|v\|^2$, so that a perceived color can be identified with a time-like element of the 3-dimensional Minkowski space-time.

The real value α is naturally interpreted as the 'luminance' of the perceived color, so that $(\alpha + v)/2\alpha$ is a 'chromatic' state. Pure 'chromatic' states are primary, monochromatic, colors and form the 'hue' circle $\mathbb{P}_1(\mathbb{R})$ equipped with the projective metric. Finally, since $0 \leq -\mathrm{Trace}(\rho \log \rho) \leq \log 2$, for all states ρ, the entropy measure $-\mathrm{Trace}(x \log x)$ provides the description of the 'saturation', the state of maximal entropy ρ_0 being perceived as 'achromatic'.

5   Conclusions

A critical analysis of the mathematically elegant and theoretically avant-garde model of Resnikoff for the space of perceived colors led us to propose a psychophysical experiment to verify one of the fundamental hypothesis on which the model is based.

We have also underlined that the finite threshold and saturation limit of retinal photoreceptors should be taken into account in a moder rigorous description of the color space geometry.

Moreover, we have motivated through the very important (and often undervalued) phenomenon of color induction, why a color theory should be constructed with the building blocks of local field theories.

Finally, we have sketched our ideas about how Jordan algebras can be used to define the colorimetric attributes by pointing out the similarities to the formalism of quantum mechanics. The quantum description that we propose creates a deep connection between the pioneering works of Yilmaz and Resnikoff. A more detailed study shows that the above mentioned rebit makes it possible to recover Hering's disk and thus to obtain a mathematical justification of the coherence between trichromatic and color opponency theories [3].

References

1. Amari, S.: Differential-Geometrical Methods in Statistics, vol. 28. Springer, New York (2012)
2. Bengtsson, I., Zyczkowski, K.: Geometry of Quantum States: An Introduction to Quantum Entanglement. Cambridge University Press, Cambridge (2017)
3. Berthier, M.: Perceived colors from real quantum states: Hering's rebit. Preprint (2019)
4. Calvo, M., Oller, J.: A distance between multivariate normal distributions based in an embedding into the Siegel group. J. Multivar. Anal. **35**(2), 223–242 (1990)
5. Faraut, J., Koranyi, A.: Analysis on Symmetric Cones. Clarendon Press, Oxford (1994)
6. Goldstein, B.: Sensation and Perception, 9th edn. Cengage Learning, Boston (2013)
7. Gronchi, G., Provenzi, E.: A variational model for context-driven effects in perception and cognition. J. Math. Psychol. **77**, 124–141 (2017)
8. Hubel, D.: Eye, Brain, and Vision. Scientific American Library, New York (1995)
9. Jordan, P., Von Neumann, J., Wigner, E.: On an algebraic generalization of the quantum mechanical formalism. Ann. Math. **35**, 29–64 (1934)
10. Land, E., McCann, J.: Lightness and Retinex theory. J. Opt. Soc. Am. **61**(1), 1–11 (1971)
11. Palma-Amestoy, R., Provenzi, E., Bertalmío, M., Caselles, V.: A perceptually inspired variational framework for color enhancement. IEEE Trans. Pattern Anal. Mach. Intell. **31**(3), 458–474 (2009)
12. Resnikoff, H.: Differential geometry and color perception. J. Math. Biol. **1**, 97–131 (1974)
13. Resnikoff, H.: On the geometry of color perception, vol. 7, pp. 217–232 (1974)
14. Rudd, M., Zemach, I.: Quantitive properties of achromatic color induction: An edge integration analysis. Vis. Res. **44**, 971–981 (2004)
15. Schrödinger, E.: Grundlinien einer theorie der farbenmetrik im tagessehen (Outline of a theory of colour measurement for daylight vision). In: Macadam, D.L. (ed.) Available in English in Sources of Colour Science, pp. 134–182. The MIT Press (1970). Annalen der Physik **63**(4), 397–456; 481–520 (1920)
16. Shapley, R., Enroth-Cugell, C.: Visual adaptation and retinal gain controls. In: Progress in Retinal Research, Chap. 9, vol. 3, pp. 263–346 (1984)

17. Siegel, C.L.: Symplectic Geometry. Elsevier, Amsterdam (2014)
18. Wallach, H.: Brightness constancy and the nature of achromatic colors. J. Exp. Psychol. **38**(3), 310–324 (1948)
19. Wootters, W.K.: Optimal information transfer and real-vector-space quantum theory. In: Chiribella, G., Spekkens, R.W. (eds.) Quantum Theory: Informational Foundations and Foils. Fundamental Theories of Physics, vol. 181, pp. 21–43. Springer, Dordrecht (2016). https://doi.org/10.1007/978-94-017-7303-4_2
20. Wyszecky, G., Stiles, W.S.: Color Science: Concepts and Methods, Quantitative Data and Formulas. Wiley, New York (1982)
21. Yilmaz, H.: Color vision and a new approach to general perception. In: Bernard, E.E., Kare, M.R. (eds.) Biological Prototypes and Synthetic Systems, pp. 126–141. Springer, Boston (1962). https://doi.org/10.1007/978-1-4684-1716-6_22

Quantum Statistical Manifolds: The Finite-Dimensional Case

Jan Naudts$^{(\boxtimes)}$ ⓘ

Departement Fysica, Universiteit Antwerpen,
Universiteitsplein 1, 2610 Wilrijk, Antwerpen, Belgium
jan.naudts@uantwerpen.be

Abstract. Quantum information geometry studies families of quantum states by means of differential geometry. A new approach is followed. The emphasis is shifted from a manifold of strictly positive density matrices to a manifold $\mathbb{M}$ of faithful quantum states on a von Neumann algebra of bounded linear operators working on a Hilbert space. In order to avoid technicalities the theory is developed for the algebra of n-by-n matrices. A chart is introduced which is centered at a given faithful state ω_ρ. It maps the manifold $\mathbb{M}$ onto a real Banach space of self-adjoint operators belonging to the commutant algebra. The operator labeling any state ω_σ of $\mathbb{M}$ also determines a tangent vector in the point ω_ρ along the exponential geodesic in the direction of ω_σ. A link with the theory of the modular automorphism group is worked out. Explicit expressions for the chart can be derived in terms of the modular conjugation and the relative modular operators.

Keywords: Information geometry · Quantum statistical manifolds · Quantum exponential family · Modular automorphism group · Relative modular operator

1 Introduction

In quantum information theory [1] the state of the system is described either by a wave function, which is a normalized element of a Hilbert space, or, more generally, by a density matrix. Density matrices are also used in quantum statistical physics. The equilibrium state of a quantum system at the inverse temperature β is given by the density matrix

$$\rho_\beta = \frac{1}{Z(\beta)} \exp(-\beta H).$$

Here, H is a Hermitian matrix, called the Hamiltonian, and $Z(\beta) = \mathrm{Tr} \exp(-\beta H)$ is the normalizing factor and is called the partition sum. The expression is the quantum analogue of a Boltzmann-Gibbs distribution. It is the prototype of a model belonging to the quantum exponential family.

© Springer Nature Switzerland AG 2019
F. Nielsen and F. Barbaresco (Eds.): GSI 2019, LNCS 11712, pp. 631–637, 2019.
https://doi.org/10.1007/978-3-030-26980-7_65

Information geometry deals with the application of differential geometry to statistical models. A quantum version of Amari's dually flat geometry [2] was studied by Hasegawa [3,4], Jenčová [5] and others. The approach of Pistone and Sempi [6] was transferred to the quantum setting by Streater [7,8].

In [9], the present author proposes to shift the emphasis from manifolds of density matrices to manifolds of states on a von Neumann algebra. The intention of this move is to get rid of tracial states and, by doing so, to facilitate further generalizations. However, both [9] and the present work are limited to the case of a finite-dimensional Hilbert space in order to avoid technicalities. A first attempt to advance with the general case is found in [10].

The next three sections review the mathematical formalism, the notion of a manifold of quantum states and some elements of the theory of the modular automorphism group. Section 5 derives an explicit expression for the positive operators wich belong to the commutant algebra and characterize the states of the manifold. In Sect. 6 the vectors tangent to an exponential geodesic are characterized. The paper finishes with a short discussion in Sect. 7.

2   The GNS Representation

The space of n-by-n matrices forms a Hilbert space $\mathcal{H}^{\mathrm{HS}}$ for the Hilbert-Schmidt inner product

$$\langle A, B \rangle^{\mathrm{HS}} = \mathrm{Tr}\, B^* A.$$

Here, B^* is the adjoint matrix, i.e. the Hermitian conjugate of the matrix B. Operators on $\mathcal{H}^{\mathrm{HS}}$ are sometimes called *superoperators* because the matrices are themselves already operators on the Hilbert space $\mathbb{C}^n$. An alternative view is offered by the Gelfand-Naimark-Segal (GNS) representation. It is more powerful and very general. The starting point is the remark that the Hilbert-Schmidt inner product can be written as

$$\langle A, B \rangle^{\mathrm{HS}} = n \, \mathrm{Tr}\, \rho_0 B^* A,$$

where the *density matrix* ρ_0 is the identity matrix $\mathbb{I}$ divided by n.

By definition, a density matrix ρ is positive and has trace equal to one: $\mathrm{Tr}\, \rho = 1$. By the GNS theorem there exists a Hilbert space $\mathcal{H}$, a vector Ω_ρ in $\mathcal{H}$ and a *-representation of the algebra $\mathcal{A}$ of n-by-n matrices as operators on $\mathcal{H}$ such that

$$\mathrm{Tr}\, \rho A = (A\Omega_\rho, \Omega_\rho) \quad \text{for all } A \in \mathcal{A}.$$

The map $A \mapsto \omega_\rho(A) = (A\Omega_\rho, \Omega_\rho)$ belongs to the dual $\mathcal{A}^*$ of $\mathcal{A}$ and is called a *state*. Its defining properties are that $\omega_\rho(A^*A) \geq 0$ for all A (positivity) and $\omega_\rho(\mathbb{I}) = 1$ (normalization). The state is said to be *faithful* if $\omega_\rho(A^*A) = 0$ implies $A = 0$. This is the case if the density matrix ρ is non-degenerate.

In the case of a non-degenerate density matrix ρ the GNS representation can be made explicit as follows. The Hilbert space $\mathcal{H}$ is the tensor product $\mathbb{C}^n \otimes \mathbb{C}^n$. Each matrix A is replaced by the matrix $A \otimes \mathbb{I}$, which has dimension n^2-by-n^2.

Choose an orthonormal basis ψ_j, $j = 1, 2, \cdots, n$ of eigenvectors of ρ. One has $\rho\psi_j = p_j\psi_j$ with eigenvalues $p_j > 0$. Let

$$\Omega_\rho = \sum_j \sqrt{p_j}\psi_j \otimes \psi_j.$$

A short calculation then shows that for any n-by-n matrix A one has $\omega_\rho(A) = \text{Tr}\,\rho A = (A\Omega_\rho, \Omega_\rho)$.

The main advantage of this representation of the state ω is that the commutant $\mathcal{A}'$ of the algebra $\mathcal{A}$ is explicitly present. It consists of all matrices of the form $\mathbb{I} \otimes A$. They clearly commute with all matrices of the form $A \otimes \mathbb{I}$. In particular, with any density matrix σ there corresponds a unique positive operator X_σ in the commutant $\mathcal{A}'$ such that

$$\omega_\sigma(A) \equiv \text{Tr}\,\sigma A = (AX_\sigma^{1/2}\Omega_\rho, X_\sigma^{1/2}\Omega_\rho) \quad \text{for all } A \in \mathcal{A}. \tag{1}$$

3   The Manifold

The manifold $\mathbb{M}$ consists of all states ω_σ on the von Neumann algebra $\mathcal{A}$, where σ is any non-degenerate density matrix of dimension n-by-n. Guided by recent works of Pistone et al. [11,12] a chart χ_ρ is introduced which is centered at a state ω_ρ, corresponding with an arbitrary chosen but fixed non-degenerate density matrix ρ. For any state ω_σ in $\mathbb{M}$ the chart defines an element $\chi_\rho(\omega_\sigma)$ of the commutant $\mathcal{A}'$. Its construction follows later on in Sect. 6.

If $t \mapsto \omega_t \in \mathbb{M}$ is a smooth curve then tangent vectors f_t are defined by

$$f_t(A) = \frac{\mathrm{d}}{\mathrm{d}t}\omega_t(A), \quad A \in \mathcal{A}. \tag{2}$$

They belong to the dual $\mathcal{A}^*$ of the algebra $\mathcal{A}$ and satisfy $f_t(\mathbb{I}) = 0$ and $f_t(A^*) = \overline{f_t(A)}$. The tangent plane at the point ω_ρ is denoted $T_\rho\mathbb{M}$ and consists of all linear functionals f_K of the form

$$f_K(A) = (A\Omega_\rho, K\Omega_\rho), \quad A \in \mathcal{A},$$

where K belongs to the Banach space $\mathcal{B}_x$ of all self-adjoint elements of the commutant algebra $\mathcal{A}'$ and satisfies $(\Omega_\rho, K\Omega_\rho) = 0$.

The metric chosen on the tangent plane is that of Bogoliubov (see for instance [1,2,13,14]). It can be derived from Umegaki's relative entropy [15]

$$D(\sigma||\tau) = \text{Tr}\,\sigma\,[\log\sigma - \log\tau].$$

by taking twice a derivative. The metric is not discussed in the present paper. Details can be found in [9].

A geodesic corresponding to the exponential connection and connecting the state $\omega_\sigma \in \mathbb{M}$ to the state ω_ρ at the center is of the form $t \mapsto \omega_{\sigma,t}$, where

$$\omega_{\sigma,t}(A) = \text{Tr}\,\sigma_t A, \quad A \in \mathcal{A}, \quad \text{with}$$
$$\log\sigma_t = \log\rho + tH_\sigma - \zeta(tH_\sigma)$$
$$H_\sigma = \log\sigma - \log\rho,$$
$$\zeta(tH_\sigma) = \log\,\text{Tr}\,\exp(\log\rho + tH_\sigma). \tag{3}$$

The chart χ_ρ is constructed in such a way that the tangent in the point ω_ρ equals the linear functional f_K with $K = \chi_\rho(\omega_\sigma)$.

4   Relative Modular Operators

In [9] the chart χ_ρ is constructed in an indirect manner. A more direct construction is given below in Sect. 6. It is based on Araki's notion of *relative modular operators* [16, 17].

Let S denote the *modular conjugation operator* [18] determined by the vector Ω_ρ. It is the anti-linear operator defined by $SA\Omega_\rho = A^*\Omega_\rho$ for all A in $\mathcal{A}$. The *modular operator* Δ equals S^*S and does not depend on the specific choice of the vector Ω_ρ representing the state ω_ρ. The polar decomposition of S reads $S = J\Delta^{1/2}$. The operator J satisfies $J^* = J$, $J^2 = \mathbb{I}$. An important result of Tomita-Takesaki theory, needed further on, states that JAJ belongs to the commutant algebra $\mathcal{A}'$ if and only if A belongs to $\mathcal{A}$.

Given any vector Ξ in the Hilbert space $\mathcal{H}$ the *relative modular conjugation operator* S_{Ξ,Ω_ρ} is defined by

$$S_{\Xi,\Omega_\rho}A\Omega_\rho = A^*\Xi, \qquad A \in \mathcal{A}.$$

For convenience, let us introduce the notation $S_\sigma \equiv S_{\Xi,\Omega_\rho}$ when $\Xi = X_\sigma^{1/2}\Omega_\rho$. The *relative modular operator* Δ_σ is then given by

$$\Delta_\sigma = S_\sigma^*S_\sigma = S^*X_\sigma S = \Delta\rho^{-1}\sigma.$$

Consider now the geodesic $t \mapsto \omega_{\sigma,t}$ given by (3). Because $\Delta\rho^{-1}$ commutes with σ_t it follows from $\Delta_{\sigma,t} = \Delta\rho^{-1}\sigma_t$ that

$$\begin{aligned}
\log\Delta_{\sigma,t} &= \log\Delta\rho^{-1} + \log\rho + tH_\sigma - \zeta(tH_\sigma) \\
&= \log\Delta + tH_\sigma - \zeta(tH_\sigma).
\end{aligned} \tag{4}$$

One concludes that the operator H_σ which generates the exponential geodesic also describes the relative modular operator $\Delta_{\sigma,t}$ for all states $\omega_{\sigma,t}$ along the geodesic.

5   Explicit Expressions

In [9] the operator X_σ, which characterizes the state ω_σ via (1) is defined in an indirect manner by requiring that

$$X_\sigma\Omega_\rho = \sigma\rho^{-1}\Omega_\rho.$$

An explicit expression is given by the following proposition.

Proposition 1. *The operator X_σ satisfies $X_\sigma = S\rho^{-1}\sigma S$. The relative modular operator Δ_σ satisfies*

$$\Delta_\sigma = S^*X_\sigma S \quad \text{and} \quad \Delta_\sigma^{1/2} = (\rho^{-1}\Delta)^{1/2}\sigma^{1/2}.$$

Proof. One has for all A in $\mathcal{A}$

$$X_\sigma A\Omega_\rho = AX_\sigma \Omega_\rho = A\sigma\rho^{-1}\Omega_\rho = S\rho^{-1}\sigma SA\Omega_\rho$$

Because Ω_ρ is cyclic for $\mathcal{A}$ this implies that $X_\sigma = S\rho^{-1}\sigma S$. Next use that $\rho^{-1}\Delta = \Delta\rho^{-1}$ belongs to the commutant $\mathcal{A}'$ to obtain

$$X_\sigma = J\Delta^{1/2}\rho^{-1}\sigma S = J\Delta^{-1/2}[\Delta\rho^{-1}]\sigma S = S^* \left([\Delta\rho^{-1}]^{1/2}\sigma^{1/2}\right)^2 S.$$

Finally, one has for all A in $\mathcal{A}$

$$S_\sigma A\Omega_\rho = A^* X_\sigma^{1/2}\Omega_\rho = X_\sigma^{1/2} A^*\Omega_\rho = X_\sigma^{1/2} SA\Omega_\rho.$$

This implies $S_\sigma = X_\sigma^{1/2} S$ and hence

$$\Delta_\sigma = S_\sigma^* S_\sigma = S^* X_\sigma S = \left([\Delta\rho^{-1}]^{1/2}\sigma^{1/2}\right)^2.$$

$\square$

6   The Chart

In [9] the operator $\chi_\rho(\omega_\sigma)$ belonging to the commutant $\mathcal{A}'$ is defined by the relation

$$\chi_\rho(\omega_\sigma)\Omega_\rho = \int_0^1 \mathrm{d}u \, \rho^u \left[\log\sigma - \log\rho + D(\rho||\sigma)\right] \rho^{-u}\Omega_\rho. \tag{5}$$

Its main property is that, given an exponential geodesic $t \mapsto \omega_{\sigma,t}$ of the form (3), the tangent vector at $t = 0$ satisfies

$$\left.\frac{\mathrm{d}}{\mathrm{d}t}\right|_{t=0} \omega_{\sigma,t}(A) = (A\Omega_\rho, \chi_\rho(\omega_\sigma)\Omega_\rho), \qquad A \in \mathcal{A}. \tag{6}$$

There is a one-to-one correspondence between the tangent vectors of $T_\rho\mathbb{M}$ and the elements of the Banach space $\mathcal{B}_x$.

Alternative proof of (6)

Starting point is the following relation between any state ω_σ in the manifold $\mathbb{M}$ and the corresponding relative modular operator Δ_σ

$$\omega_\sigma(A) = (A\Omega_\rho, X_\sigma\Omega_\rho) = (A\Omega_\rho, S^*\Delta_\sigma S\Omega_\rho), \qquad A \in \mathcal{A}. \tag{7}$$

Consider a geodesic $t \mapsto \omega_{\sigma,t}$ of the form (3). The relative modular operator $\Delta_{\sigma,t}$ satisfies

$$\frac{\mathrm{d}}{\mathrm{d}t}\Delta_{\sigma,t} = \frac{\mathrm{d}}{\mathrm{d}t}\exp\left(\log\Delta + tH_\sigma - \zeta(tH_\sigma)\right).$$

Use the identity

$$\frac{\mathrm{d}}{\mathrm{d}t}\bigg|_{t=0} e^{AH+tH} = \int_0^1 \mathrm{d}u\, e^{uA} H e^{(1-u)A}$$

to obtain

$$\begin{aligned}
\frac{\mathrm{d}}{\mathrm{d}t}\bigg|_{t=0} \Delta_{\sigma,t} &= \int_0^1 \mathrm{d}u\, \Delta^u \left(H_\sigma - \frac{\mathrm{d}}{\mathrm{d}t}\zeta(tH_\sigma)\bigg|_{t=0} \right) \Delta^{1-u} \\
&= \left[\int_0^1 \mathrm{d}u\, \rho^u H_\sigma \rho^{-u} - \frac{\mathrm{d}}{\mathrm{d}t}\zeta(tH_\sigma)\bigg|_{t=0} \right] \Delta.
\end{aligned} \tag{8}$$

From (5) one obtains

$$\chi_\rho(\omega_\sigma) = S \left(\int_0^1 \mathrm{d}u\, \rho^u \left[H + D(\rho||\sigma) \right] \rho^{-u} \right)^* S.$$

Combine this with

$$\frac{\mathrm{d}}{\mathrm{d}t}\bigg|_{t=0} \zeta(tH_\sigma) = -D(\rho||\sigma).$$

and (8) to obtain

$$\frac{\mathrm{d}}{\mathrm{d}t}\bigg|_{t=0} \Delta_{\sigma,t} = S^* \chi_\rho(\omega_\sigma) S. \tag{9}$$

Putting the pieces together one obtains (6) from (7) and (9)

7   Discussion

The manifold $\mathbb{M}$ of faithful states on the von Neumann algebra $\mathcal{A}$ of n-by-n matrices is studied. An arbitrary state ω_ρ in $\mathbb{M}$ is selected as the reference state. Other states in $\mathbb{M}$ are labeled with operators in the commutant of the G.N.S.-representation of the selected state. Tangent vectors are linear functionals on the von Neumann algebra. They are also labeled with operators belonging to the commutant algebra. In particular, the vectors tangent to an exponential geodesic are characterized. By the use of the theory of the modular automorphism group [18] and the relative modular operators [16,17] explicit expressions are obtained for operators which were introduced already in a previous paper [9]. The explicit expressions should help in generalizing the present approach to manifolds of states on arbitrary σ-finite von Neumann algebras.

The expression (4) for the logarithm of the relative modular operator is affine along the exponential geodesic, up to a scalar function which is due to normalization. It resembles the similar expression (3) for the density matrix. Both use the operator $H_\sigma = \log \sigma - \log \rho$ as the generator of the geodesic connecting the state ω_σ to the state ω_ρ. The map $\omega_\sigma \mapsto H_\sigma$ labels the states of the manifold with operators belonging to the algebra $\mathcal{A}$. On the other hand, the chart χ_ρ, defined in Sect. 6, uses operators belonging to the commutant algebra $\mathcal{A}'$. This chart determines in a direct manner the vectors tangent to the exponential geodesics at the state ω_ρ.

References

1. Petz, D.: Quantum Information Theory and Quantum Statistics. Springer, Berlin (2008). https://doi.org/10.1007/978-3-540-74636-2
2. Amari, S., Nagaoka, H.: Methods of Information Geometry. Oxford University Press, Oxford (2000)
3. Hasegawa, H.: α-divergence of the non-commutative information geometry. Rep. Math. Phys. **33**, 87–93 (1993)
4. Hasegawa, H.: Exponential and mixture families in quantum statistics: dual structure and unbiased parameter estimation. Rep. Math. Phys. **39**, 49–68 (1997)
5. Jenčová, A.: Geometry of quantum states: dual connections and divergence functions. Rep. Math. Phys. **47**, 121–138 (2001)
6. Pistone, G., Sempi, C.: An infinite-dimensional structure on the space of all the probability measures equivalent to a given one. Ann. Stat. **23**, 1543–1561 (1995)
7. Streater, R.F.: Duality in quantum information geometry. Open Syst. Inf. Dyn. **11**, 71–77 (2004)
8. Streater, R.F.: Quantum Orlicz spaces in information geometry. Open Syst. Inf. Dyn. **11**, 359–375 (2004)
9. Naudts, J.: Quantum statistical manifolds. Entropy **20**, 472 (2018). https://doi.org/10.3390/e20060472. Correction **20**, 796 (2018)
10. Naudts, J.: Quantum statistical manifolds: the linear growth case. Rep. Math. Phys., in print; arXiv:1801.07642v2 (2019)
11. Pistone, G.: Nonparametric information geometry. In: Nielsen, F., Barbaresco, F. (eds.) GSI 2013. LNCS, vol. 8085, pp. 5–36. Springer, Heidelberg (2013). https://doi.org/10.1007/978-3-642-40020-9_3
12. Montrucchio, L., Pistone, G.: Deformed exponential bundle: the linear growth case. In: Nielsen, F., Barbaresco, F. (eds.) GSI 2017. LNCS, vol. 10589, pp. 239–246. Springer, Cham (2017). https://doi.org/10.1007/978-3-319-68445-1_28
13. Naudts, J., Verbeure, A., Weder, R.: Linear response theory and the KMS condition. Commun. math. Phys. **44**, 87–99 (1975)
14. Petz, D., Toth, G.: The Bogoliubov inner product in quantum statistics. Lett. Math. Phys. **27**, 205–216 (1993)
15. Umegaki, H.: Conditional expectation in an operator algebra. IV. entropy and information. Kodai Math. Sem. Rep. **14**, 59–85 (1962)
16. Araki, H.: Relative entropy of states of von neumann algebras. Publ. RIMS Kyoto Univ. **11**, 809–833 (1976)
17. Araki, H.: Relative entropy for states of von neumann algebras II. Publ. RIMS Kyoto Univ. **13**, 173–192 (1977)
18. Takesaki, M.: Tomita's Theory of Modular Hilbert Algebras and its Applications. LNM, vol. 129. Springer, Heidelberg (1970). https://doi.org/10.1007/BFb0065832

Generalized Gibbs Ensembles in Discrete Quantum Gravity

Goffredo Chirco[1,2]([⊠]) [iD] and Isha Kotecha[1,2]([⊠]) [iD]

[1] Max Planck Institute for Gravitational Physics (Albert Einstein Institute),
Am Mühlenberg 1, 14476 Potsdam-Golm, Germany
{goffredo.chirco,isha.kotecha}@aei.mpg.de
[2] Institute for Physics, Humboldt-Universität zu Berlin,
Newtonstraße 15, 12489 Berlin, Germany

Abstract. Maximum entropy principle and Souriau's symplectic generalization of Gibbs states have provided crucial insights leading to extensions of standard equilibrium statistical mechanics and thermodynamics. In this brief contribution, we show how such extensions are instrumental in the setting of discrete quantum gravity, towards providing a covariant statistical framework for the emergence of continuum spacetime. We discuss the significant role played by information-theoretic characterizations of equilibrium. We present the Gibbs state description of the geometry of a tetrahedron and its quantization, thereby providing a statistical description of the characterizing quanta of space in quantum gravity. We use field coherent states for a generalized Gibbs state to write an effective statistical field theory that perturbatively generates 2-complexes, which are discrete spacetime histories in several quantum gravity approaches.

Keywords: Maximum entropy principle · Constrained systems ·
Quantum gravity and quantum geometry · Gibbs states

1 Discrete Quantum Spacetime

From the existence of singularities in classical gravitational theory to the discovery of horizon entropies in semiclassical settings, many studies have hinted at a discrete quantum microstructure of spacetime. Precisely what these quanta of spacetime are, and how they give rise to a continuum gravitational field is the holy grail of non-perturbative discrete quantum gravity. It is a complex open issue, being tackled from various sides. Despite many conceptual and technical differences between the different formalisms, they admit an interesting commonality: modelling of spacetime quanta as geometric polyhedra.

In particular, tetrahedra are the candidates of choice in 4d models for quantum excitations of geometry in several approaches, such as loop quantum gravity, spin foams, group field theory, dynamical triangulations and simplicial gravity. Collective dynamics of such degrees of freedom is then expected to give rise to an emergent spacetime. Statistical mechanical and field theoretic techniques are

© Springer Nature Switzerland AG 2019
F. Nielsen and F. Barbaresco (Eds.): GSI 2019, LNCS 11712, pp. 638–646, 2019.
https://doi.org/10.1007/978-3-030-26980-7_66

thus crucial from the point of view of an emergent spacetime, not only for providing tools to extract effective spacetime 'macroscopic' physics from the quantum gravitational 'microscopics', but also as probes to investigate non-perturbative features. A natural way for such explorations is to consider quantum spacetime as a many-body system [1], which is complementary with the view of classical continuum spacetime as an effective thermodynamic system.

The procedure of maximizing information entropy subject to a given set of constraints as presented by Jaynes [2,3] is uniquely positioned to be utilised in background independent systems for defining an equilibrium Gibbs state. The primary reasons for this are that this method does not rely on the existence of any 1-parameter automorphism of the system (such as physical time evolution), unlike the customary Kubo-Martin-Schwinger condition of non-relativistic statistical mechanics. It also allows for considering observables other than energy, such as geometric volume, which may not necessarily be naturally understood as symmetry generators. These two technical features of not requiring a 1-parameter group of symmetries a priori, and an inclusion of other observables of interest, makes this procedure particularly valuable in background independent quantum gravity settings. Moreover, in an almost unassuming way, it points toward a fundamental status of information entropy in quantum gravity, which has been a recurring theme across various avenues in modern theoretical physics.

2   Generalized Gibbs States

A macrostate of a system with many underlying degrees of freedom is given in terms of a finite number of observable averages. Jaynes [2,3] argued that the least biased statistical distribution over the microscopics of the system, compatible with our limited knowledge of its macroscopics in terms of these averages, is that which maximizes the information entropy. The resultant distribution is Gibbs, which faithfully encodes our partial knowledge of the system. Maximizing the uncertainty in this manner ensures that we are using exactly only the information that we have access to, not less or more.

Let $\{\mathcal{O}_a\}_{a=1,2,\dots}$ be a finite set of smooth, real-valued functions on a finite-dimensional symplectic phase space Γ_{ex}. It is the unconstrained, extended state space with respect to all constraints. Let ρ be a statistical density (real-valued and positive function, normalised with respect to Liouville measure) on Γ_{ex}. The functions $\mathcal{O}_a$ are such that their statistical averages in ρ are well-defined and constant, that is

$$\langle \mathcal{O}_a \rangle_\rho \equiv \int_{\Gamma_{\mathrm{ex}}} d\lambda \, \mathcal{O}_a \, \rho \;=\; U_a. \tag{1}$$

Shannon entropy of ρ is,

$$S[\rho] = -\langle \ln \rho \rangle_\rho \tag{2}$$

and its normalization is $\langle 1 \rangle_\rho = 1$. Now consider maximization of $S[\rho]$ under the constraints of state normalization and Eq. (1) [2]. This optimization problem can be phrased in the language of Lagrange multipliers and imposing stationarity of an auxiliary functional,

$$L[\rho, \beta_a, \kappa] = S[\rho] - \sum_a \beta_a(\langle \mathcal{O}_a \rangle_\rho - U_a) - \kappa(\langle 1 \rangle_\rho - 1) \tag{3}$$

with multipliers $\beta_a, \kappa \in \mathbb{R}$. Stationarity with respect to variations in ρ then results in a generalized Gibbs state of the form,

$$\rho_{\{\beta_a\}} = \frac{1}{Z_{\{\beta_a\}}} e^{-\sum_a \beta_a \mathcal{O}_a} \tag{4}$$

with the partition function,

$$Z_{\{\beta_a\}} \equiv \int_{\Gamma_{\text{ex}}} d\lambda \, e^{-\sum_a \beta_a \mathcal{O}_a} = e^{1+\kappa} \tag{5}$$

where $\{\beta_a\}$ and $\mathcal{O}_a$ are such that the above integral converges.

The above procedure can be carried out analogously for finite quantum systems [3], given that the operators under consideration have well-defined trace averages on a kinematic, unconstrained Hilbert space. Here statistical states are density operators (self-adjoint, positive and trace-class), and the ensemble averages for (self-adjoint) operators $\hat{\mathcal{O}}$ are,

$$\langle \hat{\mathcal{O}}_a \rangle_\rho \equiv \text{Tr}(\hat{\rho} \hat{\mathcal{O}}_a) = U_a. \tag{6}$$

Then Jaynes' method gives a generalized Gibbs density operator,

$$\hat{\rho}_{\{\beta_a\}} = \frac{1}{Z_{\{\beta_a\}}} e^{-\sum_a \beta_a \hat{\mathcal{O}}_a}. \tag{7}$$

Averages U_a are generalized energies, $\beta \equiv \{\beta_a\}$ is a generalized vector-valued (inverse) temperature, and $dQ_a \equiv dU_a - \langle d\mathcal{O}_a \rangle$ are generalized heat differentials.

This information-theoretic manner of defining equilibrium statistical mechanics is to elevate the status of entropy as being more fundamental than energy. This perspective can prove instrumental in background independent settings [4]. As long as the system is equipped with a well-defined state space and an observable algebra, and is described macroscopically with a few observables $\{\mathcal{O}_a\}$ in terms of its averages $\{U_a\}$, the maximum entropy principle can be applied to characterize a notion of generalized statistical equilibrium.

3   Statistically Constrained Tetrahedra

Jaynes' characterization of equilibrium also allows for a natural group-theoretic generalization of thermodynamics, whenever the constraint is associated to some (dynamical) symmetry of the system. In this case, the momentum map associated to the Hamiltonian action of the symmetry group on the covariant (extended) phase space of the system plays the role of a generalized energy function, comprising the full set of conserved quantities. Moreover, its convexity properties allow for a generalization of standard equilibrium thermodynamics [5].

This approach is useful also in the simplicial geometric context of non-perturbative quantum gravity [6–8]. We will use the generalized Gibbs states to define along these lines a statistical characterization of tetrahedral geometry in terms of its closure, starting from the extended phase space of a single open tetrahedron. The closure constraint is what allows to interpret geometrically a set of 3d vectors as the normal vectors to the faces of a polyhedron, and thus to fully capture its intrinsic geometry in terms of them. Subsequently, we will consider a system of many closed tetrahedra (or polyhedra in general) and demonstrate its relation to the group field theory approach to quantum gravity.

3.1   Classical Closure Fluctuations

The symplectic space $\Gamma_{\{A_I\}} = \{(X_I) \in \mathfrak{su}(2)^{*4} \cong \mathbb{R}^{3\times 4} \mid ||X_I|| = A_I\} \cong S^2_{A_1} \times \ldots \times S^2_{A_4}$, is the space of intrinsic geometries of an open tetrahedron. Each $S^2_{A_I}$ is a 2-sphere with radius A_I, and $I \in \{1, 2, 3, 4\}$. When the four vectors X_I are constrained to sum to zero, the orthogonal surfaces associated to them close, giving a tetrahedron in Euclidean $\mathbb{R}^3$ with face areas $\{A_I\}^1$. In this subsection we take $\Gamma_{\mathrm{ex}} = \Gamma_{\{A_I\}}$.

The diagonal action of the Lie group $SU(2)$ (rotations) on Γ_{ex} has an associated momentum map $J : \Gamma_{\mathrm{ex}} \to \mathfrak{su}(2)^*$ defined by,

$$J = \sum_{I=1}^{4} X_I \tag{8}$$

where $||X_I|| = A_I$. Symplectically reducing Γ_{ex} with respect to $J = 0$ level set gives the Kapovich-Millson phase space [9] $\mathcal{S}_4 = \Gamma_{\mathrm{ex}}//SU(2) = J^{-1}(0)/SU(2)$ of a closed tetrahedron with the given face areas, where notation $//$ means a symplectic reduction. It imposes closure of the four faces, with space $J^{-1}(0)$ being the constraint submanifold. But what we are interested in here is to define a Gibbs probability distribution on Γ_{ex} by imposing closure only on average, using the method of Sect. 2.

From a statistical perspective, the exact (or strong) fulfilment of closure can be understood as defining a microcanonical statistical state on Γ_{ex} with respect to this constraint. On the other hand, a weak fulfilment of the same constraint can be thought of as being implemented by a generalized Gibbs state. Their respective partition functions on the extended state space are then formally related by a Laplace transform.

---

[1] Analogous arguments hold for the case of an open d-polyhedron and its associated closure condition.

To define a Gibbs state with respect to closure for an open tetrahedron, we maximize the Shannon entropy (2) under the constraints of state normalization and the following three,

$$\langle J_i \rangle_\rho \equiv \int_{\Gamma_{\text{ex}}} d\lambda \, \rho \, J_i = 0 \quad (i = 1, 2, 3). \tag{9}$$

Here ρ is a statistical state defined on Γ_{ex}, and J_i are components of J in a basis of $\mathfrak{su}(2)^*$. Clearly the above equation (for each i) is a weaker condition than imposing closure exactly by $J_i = 0$. Functions J_i are smooth and real-valued on Γ_{ex}, taking on the role of quantities $\mathcal{O}_a$ used in Eq. (1). Then optimizing the auxiliary functional of Eq. (3) gives a Gibbs state on Γ_{ex} of the form,

$$\rho_\beta = \frac{1}{Z(\beta)} e^{-\beta \cdot J} \tag{10}$$

where $\beta \in \mathfrak{su}(2)$ is a vector-valued temperature, with components β_i. Moreover, the function $\beta \cdot J = \sum_{i=1}^{3} \beta_i J_i$ is the corresponding co-momentum map on Γ_{ex} (equivalently, the modular Hamiltonian).

Evidently, the state ρ_β is an example of a generalization of Souriau's Gibbs states [5,10], to the case of Lie group actions of gauge symmetries generated by first class constraints, in a fully background independent setting.

3.2   Quantum Statistical Mechanics

In a quantum setting, each tetrahedron face I is prescribed an $SU(2)$ representation label j_I and Hilbert space $\mathcal{H}_{j_I}$. The tetrahedron itself is assigned an invariant tensor (intertwiner) of the four incident representation spaces. The full space of 4-valent intertwiners is $\bigoplus_{j_I} \text{Inv} \otimes_{I=1}^{4} \mathcal{H}_{j_I}$, where $\text{Inv}\otimes_{I=1}^{4}\mathcal{H}_{j_I}$ is the space of 4-valent intertwiners with given fixed spins $\{j_I\}$ (given fixed face areas), corresponding to a quantization of $\mathcal{S}_4$. A collection of neighbouring quantum tetrahedra has been associated to a 4-valent spin network [11], with the labelled nodes and links of the latter being dual to labelled tetrahedra and their shared faces respectively of the former. Then taking the viewpoint of tetrahedra as being extended 'particles', the single particle Hilbert space of interest here is

$$\mathcal{H} = \bigoplus_{j_I} \text{Inv} \otimes_{I=1}^{4} \mathcal{H}_{j_I} \tag{11}$$

and, quantum states of a system of N such tetrahedra are elements of $\mathcal{H}_N = \mathcal{H}^{\otimes N}$. We can equivalently work with the holonomy representation of the same quantum system in terms of $SU(2)$ group data, which is also the state space of a single gauge-invariant quantum of a group field theory defined on an $SU(2)^4$ base manifold [6,12],

$$\mathcal{H} = L^2(SU(2)^4/SU(2)). \tag{12}$$

Mechanical models of N quantum tetrahedra can be defined by a set of gluing operators defined on $\mathcal{H}_N$. Thus a quantum mechanical model of a system of N

tetrahedra consists of the unconstrained Hilbert space $\mathcal{H}_N$, an operator algebra defined over it and a set of gluing operators specifying the model.

Now for a quantum multi-particle system, a Fock space is a suitable home for configurations with varying particle numbers. For bosonic (indistinguishable) quanta, each N-particle sector is the symmetric projection of the full N-particle Hilbert space, so that the Fock space is,

$$\mathcal{H}_F = \bigoplus_{N \geq 0} \mathrm{sym}\, \mathcal{H}_N. \tag{13}$$

Fock vacuum $|0\rangle$ is the cyclic state with no tetrahedron degrees of freedom. Then, a system of an arbitrarily large number of quantum tetrahedra is described by the state space $\mathcal{H}_F$, an algebra of operators over it with a special subset of them identified as gluing constraints. Quantum statistical states of tetrahedra are density operators (self-adjoint, positive and trace-class) on $\mathcal{H}_F$ [4].

As before, a generalized Gibbs state in a Fock system of quantum tetrahedra with a constraint operator $\hat{\mathbb{C}}$ is of the form,

$$\hat{\rho}_\beta = \frac{1}{Z_\beta} e^{-\beta \hat{\mathbb{C}}} \tag{14}$$

where β is associated with the condition $\langle \hat{\mathbb{C}} \rangle = 0$. In particular, a density operator with a contribution from a grand-canonical weight of the form $e^{\mu \hat{N}}$, will correspond to a situation with varying particle number, where $\hat{N}$ is the number operator associated with the Fock vacuum. The corresponding partition function is

$$Z(\mu, \beta) = \mathrm{Tr}_{\mathcal{H}_F} \left[e^{-\beta \hat{\mathbb{C}} + \mu \hat{N}} \right]. \tag{15}$$

If $\hat{\mathbb{C}}$ is a dynamical constraint of the system, which in general could include number- and graph-changing interactions, then one obtains a grand-canonical state of the type above with respect to $\hat{\mathbb{C}}$.

3.3   Field Theory of Quantum Tetrahedra

Hilbert space $\mathcal{H}_F$ is generated by ladder operators acting on the vacuum $|0\rangle$, and satisfying the algebra,

$$[\hat{\varphi}(\boldsymbol{g}), \hat{\varphi}^*(\boldsymbol{g}')] = \delta(\boldsymbol{g}, \boldsymbol{g}') \tag{16}$$

where δ is a delta distribution on the space of smooth, complex-valued L^2 functions on $SU(2)^4$, and $\boldsymbol{g} \equiv (g_1, ..., g_4)$.

For a state $e^{-\beta \hat{\mathbb{C}}}$, the traces in the partition function and other observable averages can be evaluated using an overcomplete basis of coherent states,

$$|\psi\rangle = e^{-\frac{||\psi||^2}{2}} e^{\int d\boldsymbol{g}\, \psi(\boldsymbol{g})\hat{\varphi}^*(\boldsymbol{g})} |0\rangle. \tag{17}$$

These states are labelled by $\psi \in \mathcal{H}$ and $||.||$ is the L^2 norm in the single particle Hilbert space $\mathcal{H}$. This gives,

$$\text{Tr}(e^{-\beta \hat{C}} \hat{O}) = \int [D\mu(\psi, \bar{\psi})] \langle \psi | e^{-\beta \hat{C}} \hat{O} | \psi \rangle, \quad \text{with } Z = \text{Tr}(e^{-\beta \hat{C}} \mathbb{I}). \quad (18)$$

Resolution of identity is $\mathbb{I} = \int [D\mu(\psi, \bar{\psi})] |\psi\rangle \langle \psi |$, and the standard coherent state measure is $D\mu(\psi, \bar{\psi}) = \prod_{k=1}^{\infty} d\,\text{Re}\psi_k \, d\,\text{Im}\psi_k / \pi$. The set of all such observable averages formally defines the total statistical system. In the following, we show how the quantum statistical partition function can be reinterpreted as the partition function for a field theory (of complex-valued L^2 fields) of the underlying quantum tetrahedra.

For generic operators $\hat{C}(\hat{\varphi}, \hat{\varphi}^*)$ and $\hat{O}(\hat{\varphi}, \hat{\varphi}^*)$ as polynomials in the algebra generators, and an arbitrary choice of operator ordering defining the exponential operator, the integrand of the statistical averages can be treated as follows.

$$\langle \psi | e^{-\beta \hat{C}} \hat{O} | \psi \rangle = \langle \psi | \sum_{k=0}^{\infty} \frac{(-\beta)^k}{k!} \hat{C}^k \hat{O} | \psi \rangle \quad (19)$$

$$= \langle \psi | : e^{-\beta \hat{C}} \hat{O} : | \psi \rangle + \langle \psi | : \text{po}_{C,O}(\hat{\varphi}, \hat{\varphi}^*, \beta) : | \psi \rangle \quad (20)$$

where the second equality is gotten by using the commutation relations (16) on each $\hat{C}^k \hat{O}$ and collecting all normal ordered terms $: \hat{C}^k \hat{O} :$ to get the normal ordered $: e^{-\beta \hat{C}} \hat{O} :$. The second term in the last line of (19) is a collection of the remaining terms arising as a result of exchanging $\hat{\varphi}$'s and $\hat{\varphi}^*$'s. In general, it will be a normal ordered series in powers of $\hat{\varphi}$ and $\hat{\varphi}^*$, with coefficient functions of β. The exact form of this series will depend on both $\hat{C}$ and $\hat{O}$.

Further recalling that coherent states are eigenstates of the annihilation operator, $\hat{\varphi}(g) | \psi \rangle = \psi(g) | \psi \rangle$, we have

$$\langle \psi | : e^{-\beta \hat{C}} \hat{O} : | \psi \rangle = e^{-\beta C[\bar{\psi}, \psi]} O[\bar{\psi}, \psi] \quad (21)$$

where $C[\bar{\psi}, \psi] = \langle \psi | \hat{C} | \psi \rangle$ and $O[\bar{\psi}, \psi] = \langle \psi | \hat{O} | \psi \rangle$. Defining operators $\hat{A}_{C,O} \equiv \text{po}_{C,O}(\hat{\varphi}, \hat{\varphi}^*, \beta)$, we have

$$\langle \psi | : \hat{A}_{C,O}(\hat{\varphi}, \hat{\varphi}^*, \beta) : | \psi \rangle = A_{C,O}[\bar{\psi}, \psi, \beta], \quad (22)$$

encoding all higher order corrections. Averages (18) can thus be written as

$$\text{Tr}(e^{-\beta \hat{C}} \hat{O}) = \int [D\mu(\psi, \bar{\psi})] \left(e^{-\beta C[\bar{\psi}, \psi]} O[\bar{\psi}, \psi] + A_{C,O}[\bar{\psi}, \psi, \beta] \right). \quad (23)$$

In particular, the quantum statistical partition function for a dynamical system of complex-valued L^2 fields ψ defined on the base manifold $SU(2)^4$ is

$$Z = \int [D\mu(\psi, \bar{\psi})] \left(e^{-\beta C[\bar{\psi}, \psi]} + A_{C,\mathbb{I}}[\bar{\psi}, \psi, \beta] \right) \equiv Z_0 + Z_{\mathcal{O}(\hbar)}. \quad (24)$$

Here, by notation $\mathcal{O}(\hbar)$ we mean simply that this sector of the full theory includes all higher orders in quantum corrections relative to Z_0. This full set of observable averages (or correlation functions) (23), including the above partition function, defines thus a statistical field theory of quantum tetrahedra (or in general, polyhedra with a fixed number of boundary faces), characterized by a combinatorially non-local statistical weight. In other words, a group field theory. This statistical foundation of group field theories was first suggested in [12].

Whenever it is possible to reformulate $A_{C,\mathcal{O}}$ such that

$$A_{C,\mathcal{O}} = A_{C,\mathbb{I}}[\bar{\psi}, \psi, \beta]\, \mathcal{O}[\bar{\psi}, \psi] \tag{25}$$

then (24) defines a statistical field theory for the algebra of observables $\mathcal{O}[\bar{\psi}, \psi]$. Further by rewriting Z in terms of a simple exponential measure (under some approximations), we would get

$$Z_{\text{eff}} = \int [D\mu(\psi, \bar{\psi})]\, e^{-C_{\text{eff}}[\bar{\psi}, \psi, \beta, C, A]}, \tag{26}$$

making the correspondence with a standard field theory manifest. A detailed discussion of the relation of the resulting statistical field theory to existing group field theory models, for topological BF theories, is given in [12,13].

4 Conclusion

We generalized Jaynes' information-theoretic approach of statistical equilibrium to a background independent system of many geometric tetrahedra. Using the symplectic description of classical tetrahedron geometry, we presented a natural generalization of Souriau's Gibbs states to a constrained system. Using a Fock space description, a quantum canonical partition function was put in relation with the generating function of labelled 2-complexes of discrete quantum gravity.

References

1. Oriti, D.: Spacetime as a quantum many-body system. arXiv:1710.02807 (2017)
2. Jaynes, E.T.: Information theory and statistical mechanics. Phys. Rev. **106**, 620–630 (1957)
3. Jaynes, E.T.: Information Theory and Statistical Mechanics. II. Phys. Rev. **108**, 171–190 (1957)
4. Kotecha, I., Oriti, D.: Statistical equilibrium in quantum gravity: Gibbs states in group field theory. New J. Phys. **20**(7), 073009 (2018)
5. Souriau, J.-M.: Structure des Systemes Dynamiques. Dunod (1969)
6. Oriti, D.: The Group field theory approach to quantum gravity. In: Oriti, D. (ed.) Approaches to Quantum Gravity [gr-qc/0607032]
7. Rovelli, C.: Quantum Gravity. Cambridge University Press, Cambridge
8. Baez, J.C.: Spinfoam models. Class. Quantum Gravity **15**, 1827–1858 (1998)
9. Kapovich, M., Millson, J.: The symplectic geometry of polygons in Euclidean space. J. Differ. Geom. **44**(3), 479–513 (1996)

10. Marle, C.-M.: From tools in symplectic and poisson geometry to J.-M. Souriau's theories of statistical mechanics and thermodynamics. Entropy **18**(10), 370 (2016)
11. Bianchi, E., Dona, P., Speziale, S.: Polyhedra in loop quantum gravity. Phys. Rev. D **83**, 044035 (2011)
12. Oriti, D.: Group field theory as the 2nd quantization of Loop Quantum Gravity. Class. Quant. Grav. **33**(8), 085005 (2016)
13. Chirco, G., Kotecha, I., Oriti, D.: Statistical equilibrium of tetrahedra from maximum entropy principle. Phys. Rev. D **99**(8), 086011 (2019)

On the Notion of Composite System

Florio Maria Ciaglia[1]([✉]) [ID], Alberto Ibort[2] [ID], and Giuseppe Marmo[3,4] [ID]

[1] Max Planck Institute for Mathematics in the Sciences, Leipzig, Germany
ciaglia@mis.mpg.de
[2] ICMAT, Instituto de Ciencias Matemáticas (CSIC-UAM-UC3M-UCM)
and Depto. de Matemáticas, Univ. Carlos III de Madrid, Leganés, Madrid, Spain
albertoi@math.uc3m.es
[3] INFN-Sezione di Napoli, Napoli, Italy
marmo@na.infn.it
[4] Dipartimento di Fisica, Università Federico II, Napoli, Italy

Abstract. The notion of composite system made up of distinguishable parties is investigated in the context of arbitrary convex spaces.

Keywords: Composite-systems · Tensor-product · Entanglement

1 Introduction

When dealing with composite systems, one of the most striking features of quantum theories is undoubtedly the existence of non-classical correlations between subsystems of the given system, a phenomenon known under the name of 'entanglement'.

In the context of standard quantum mechanics [10,24,29], where a physical system is described by means of a Hilbert space $\mathcal{H}$ and the physical states are density operators on $\mathcal{H}$, entanglement is associated with the fact that the Hilbert space of a composite system is not the Cartesian product of the Hilbert spaces of the subsystems as it happens for the phase space of a classical composite system, but, rather, it is taken to be the tensor product of the Hilbert spaces of the subsystems.

A more refined formalism for quantum theories is the algebraic formulation in terms of C^*-algebras [1,12,13,18,26,27]. In this context, a physical system is described in terms of the C^*-algebra $\mathscr{A}$ of (bounded) observables and the physical states are the mathematical states on $\mathscr{A}$, that is, the positive linear functionals on $\mathscr{A}$ normalized to 1.

The reformulation of quantum theories in terms of C^*-algebras also helps to clearly see the link between quantum theories and classical probability theory. Indeed, the space of quantum states and the space of probability distributions on

The authors acknowledge the support of the QUITEMAD project. G.M. acknowledges the support provided by the the Santander/UC3M Excellence Chair Programme 2019/2020.

© Springer Nature Switzerland AG 2019
F. Nielsen and F. Barbaresco (Eds.): GSI 2019, LNCS 11712, pp. 647–654, 2019.
https://doi.org/10.1007/978-3-030-26980-7_67

a topological/measure space may be described by means of the "same object", namely the space of mathematical states on a C^*-algebra. When this algebra is Abelian (commutative) we obtain the case of classical probability theory, while when the algebra is non-Abelian, we enter in the quantum realm. Analogously to what happens in the Hilbert-space formalism of quantum theories, the entanglement content of the theory comes from the fact that the C^*-algebra of a composite system is taken to be a suitable tensor product of the C^*-algebras of some subsystems.

In this contribution, we want to understand the mathematical requirements we should impose on the description of the notion of composite system in a given theoretical framework in order for the tensor product of suitable objects to necessarily come out. What we have in mind is a rather elementary discussion on the mathematical features characterizing the relation between composite systems and tensor products. Accordingly, in order to avoid as much as we can to rely on the specific traits of some given theoretical framework, we will not focus much on the technical and interpretational details.

Essentially, we will model the space of states of a physical system by means of a real, convex space $\mathscr{S}$. This is a very broad theoretical framework of which the spaces of states of both classical probability theory and quantum theories are a particular instance. From the operational point of view, this perspective is motivated by the idea that the states of a physical system are associated with equivalence classes of preparation procedures yielding the same measurement statistics, and that inequivalent preparation procedures may be "mixed in arbitrary proportions" resulting in operations that may be considered as admissible preparation procedures (see [9,11,14–16,19–23]). Mathematically speaking, this instance is then translated in the possibility of taking arbitrary convex combinations of elements in $\mathscr{S}$ describing physical states.

From the purely mathematical point of view, the fact that $\mathscr{S}$ is a convex set implies the existence of the vector space $\mathscr{S}^*$ of real-valued, affine linear functionals on $\mathscr{S}$, and this space will be the only ingredient, beside $\mathscr{S}$, we will need in our discussion. Note that $\mathscr{S}^*$ coincides with the dual space $\mathscr{V}^*$ of the vector space $\mathscr{V}$ generated by formal linear combinations of elements in $\mathscr{S}$. For the sake of linguistic simplicity, we **define** $\mathscr{S}^*$ to be the **dual space** of the convex set $\mathscr{S}$ with an evident abuse of nomenclature.

We want to stress that, by focusing only on the convex structure of $\mathscr{S}$ and its dual space $\mathscr{S}^*$, our analysis clearly applies to both the space of quantum states and the space of classical probability distributions on a topological/measure space, while maintaining open the possibility of considering different types of theories like those considered in the so-called **generalized probabilistic theories** (see [2,3,5,17]).

2   Composite Systems and Tensor Products

When describing composite systems from a theoretical point of view, there are, essentially, two possible perspectives: either we start from the total system and then proceed in determining a suitable notion of subsystem, or we start from the subsystems and then proceed in determining a suitable notion of composite system. Here, we will investigate the latter case in the context of composite systems made of distinguishable parties (we refer to [4] for a modern approach to the former case).

For the purpose of this contribution, we represent a composite system by means of the family $\{\mathscr{S}_j\}_{j \in [1,...n]}$ of spaces of states of the n subsystems together with the space $\mathscr{S}$ of states of the total system. As said before, we consider the spaces of states of the subsystems as given, and we want to characterize the admissible candidates for the convex set of the total system on the basis of additional constraints associated with the notion of composite system. We shall not deal with indistinguishable "particles", i.e., neither Bosons, Fermions or other parastatistics. These additional aspects would only add complications without helping in addressing the core problem. If needed, we can include other types of "statistics" at later time.

First of all, we want to implement a notion of "independence" among the states (preparation procedures) of the subsystems. Roughly speaking, we want to formalize the idea according to which there are no constraints among the preparation procedures of the subsystems, that is, each party is free to prepare its associated subsystem in any of the possible states independently from the preparations of the other parties. Mathematically speaking, we implement this idea by assuming the existence of an injective map

$$\mathrm{I} \colon \mathscr{S}_1 \times \cdots \times \mathscr{S}_n \longrightarrow \mathscr{S} \tag{1}$$

so that for every n-tuple $(\rho_1, ..., \rho_n) \in \mathscr{S}_1 \times \cdots \times \mathscr{S}_n$ of states there is a corresponding $\rho \in \mathscr{S}$ representing the n-tuple of independent states (preparation procedures) as a state of the total system. The notion of independence among the states of the subsystems (see Eq. (1)) appears also in the context of algebraic quantum field theory. For instance, in [25], this condition, together with a commutativity assumption, is used to prove that the algebra $\mathfrak{A}$ generated by the algebras $\mathfrak{A}_1$ and $\mathfrak{A}_2$ of observables associated with two space-like separated spacetime regions is isomorphic with the algebraic tensor product $\mathfrak{A}_1 \otimes \mathfrak{A}_2$. Here, we will obtain a similar result in the framework of convex spaces (of which the spaces of states of C^*-algebras typical of algebraic quantum field theory form a subfamily) by replacing the commutativity assumption with an interdependence condition among the dual spaces of the subsystems (see below).

Before proceeding further, we note that the choice $\mathscr{S} = \mathscr{S}_1 \times \cdots \times \mathscr{S}_n$, where $\mathscr{S}$ is endowed with the convex sum obtained by the component-wise application of the convex sums of the $\mathscr{S}_j$'s, is clearly the minimal choice compatible with the assumption of independence among the states of the subsystems. In this case, denoting by $\mathscr{V}_j$ the vector space canonically generated by $\mathscr{S}_j$ by means of formal linear combinations of elements in $\mathscr{S}_j$, it is clear that $\mathscr{S} = \mathscr{S}_1 \times \cdots \times \mathscr{S}_n$ is a subset of the vector space $\mathscr{V} = \oplus_{j=1}^n \mathscr{V}_j$. Consequently, the dual space $\mathscr{S}^*$ of $\mathscr{S} = \mathscr{S}_1 \times \cdots \times \mathscr{S}_n$ is just the dual space of $\mathscr{V}$, that is, $\mathscr{V}^* = \oplus_{j=1}^n \mathscr{V}_j^*$. In particular, this means that, for every n-tuple $(f_{a_1}, \cdots, f_{a_n})$ with $f_{a_j} \in \mathscr{S}_j^*$ for

$j \in [1, ..., n]$, there is an element $f_{a_1,...,a_n} \in \mathscr{S}^*$ such that

$$f_{a_1,...,a_n}(\rho_1, \cdots, \rho_n) = \sum_{j=1}^{n} f_{a_j}(\rho_j). \tag{2}$$

Clearly, $f_{a_1,...,a_n}$ vanishes on the product space $\mathscr{S}_1 \times \cdots \times \mathscr{S}_n$ representing independent equivalence classes of preparation procedures if and only if $f_{a_j} = 0$ for all $j \in [1, ..., n]$. Intuitively speaking, we may say that the dual spaces of the subsystems do not "compose" with each other. This means that the system described by $\mathscr{S} = \mathscr{S}_1 \times \cdots \times \mathscr{S}_n$ endowed with the component-wise convex sum should be interpreted more as a juxtaposition rather than a composition of systems, and, in general, we want to avoid the possibility of this convex set as an admissible space of states.

At this point we may say that this is exactly what happens in the groupoid interpretation of Quantum Mechanics [6–8]. There are two natural operations with groupoids: disjoint union and direct product. The first corresponds to juxtaposition (the corresponding algebra and space of states are direct sums) and the second is the proper composition (tensor product).

A possible way to overcome this instance and force the subsystems to "compose" is to implement a notion of interdependence for the dual spaces of the subsystems. Before introducing this notion of interdependence, we want to point out that there is no clear and unambiguous physical interpretation for it at this moment because the theoretical framework of arbitrary convex spaces does not allow a clear and unambiguous physical interpretation for the dual spaces. Having cleared this point, we proceed by introducing the interdependence condition among the dual spaces of the subsystems. First of all, we consider the injective map we introduced in Eq. (1) implementing the notion of independence among the states of the subsystems. Then, we should implement the idea that, while a dual space possesses a "linearity" property, our "composite" objects are "multilinear". Accordingly, we assume the existence of an injective map

$$I^* : \mathscr{S}_1^* \times \cdots \times \mathscr{S}_n^* \longrightarrow \mathscr{S}^* \tag{3}$$

such that, introducing the notation

$$f_{a_1,...,a_n} := I^*(f_{a_1}, \cdots, f_{a_n}), \tag{4}$$

we have

$$f_{a_1,...,a_n}(\rho) = \prod_{j=1}^{n} f_{a_j}(\rho_j), \tag{5}$$

for every $\rho = I(\rho_1, \cdots, \rho_n) \in I(\mathscr{S}_1 \times \cdots \times \mathscr{S}_n) \subset \mathscr{S}$. We define elements of this type in $\mathscr{S}^*$ to be **simple**. The simple element $f_{a_1,...,a_n}$ defined by Eq. (5) vanishes on (the injective image of) $\mathscr{S}_1 \times \cdots \times \mathscr{S}_n$ whenever there is at least one $j \in [1, ..., n]$ for which $f_{a_j} = 0$. This is in sharp contrast with what happens in the case $\mathscr{S} = \mathscr{S}_1 \times \cdots \times \mathscr{S}_n$ (see Eq. (2)) where we need $f_{a_j} = 0$ to be true for all $j \in [1, ..., n]$ in order for the associated element in $\mathscr{S}^*$ to vanish on $\mathscr{S}_1 \times \cdots \times \mathscr{S}_n$. It is in this sense that we interpret Eq. (5) as an interdependence relation among the dual spaces of the subsystems. There is a fully mature theory of non-commutative measure spaces called "free probability theory" (essentially, C^*-algebras with a tracial state), where it is introduced the notion of independence in a genuine non-commutative way and, what is more important, the notion of freeness (see [28]). We believe that there is a connection between the notion of independence and freeness as defined in the context of free probability theory and the notions of independence and interdependence introduced above, however, we will analyse this connection elsewhere.

Now, we note that the existence of simple elements allows us to introduce the notion of **separable** and **entangled** states as follows. First of all, consider the set $\mathscr{S}_{fs}$ composed by all those $\rho \in \mathscr{S}$ such that, for every simple element $f_{a_1,...,a_n} \in \mathscr{S}^*$, there is a finite N, there are n-tuples $(\rho_1^j, ..., \rho_n^j)$ with $j = 1, ..., N$ and ρ_k^j in $\mathscr{S}_k$ for every $k \in [1, ..., n]$, and there is a probability vector $\boldsymbol{p} = (p_1, ..., p_N)$ such that

$$(f_{a_1,...,a_n})(\rho) = \sum_{j=1}^{N} p_j \prod_{k=1}^{n} f_{a_k}(\rho_k^j). \tag{6}$$

Elements in $\mathscr{S}_{fs}$ are called **finitely-separable**. Then, the space of separable elements $\mathscr{S}_s$ is given by the closure of $\mathscr{S}_{fs}$ in $\mathscr{S}$ with respect to a suitable topology that, in general, will depend on the specific situation considered. For instance, if $\mathscr{S}$ is the space of states of a C^* algebra $\mathscr{A}$ (i.e., the space of positive, normalized linear functionals on $\mathscr{A}$), the closure of $\mathscr{S}_{fs}$ in $\mathscr{S}$ is taken with respect to the weak* topology on $\mathscr{S}$ induced by $\mathscr{A}$ when thought of as a subset of its double dual $\mathscr{A}^{**}$. It is not hard to see that the space of separable elements is a convex cone in $\mathscr{S}$. An element in $\mathscr{S}$ which is not separable will be called **entangled**, and, in general, composite systems admit entangled states. In finite dimensions, classical probability theory is the only case in which there are no entangled states.

Below, we will show that the product convex set $\mathscr{S} = \mathscr{S}_1 \times \cdots \times \mathscr{S}_n$ considered above is ruled out as a valid candidate because the interdependence condition among the dual spaces of the subsystems forces $\mathscr{S}^*$ to "contain" a copy of the tensor product $\otimes_{j=1}^{n} \mathscr{S}_j^*$ of the dual spaces of the single subsystems. For this purpose, we define $W \subseteq \mathscr{S}^*$ to be the vector space obtained taking arbitrary but finite linear combinations of simple elements in $\mathscr{S}^*$, and we prove that W is isomorphic, as a vector space, with the (algebraic) tensor product $\otimes_{j=1}^{n} \mathscr{S}_j^*$ by exploiting the universal property of the (algebraic) tensor product. Essentially, we will see that, given any vector space X, and any multilinear map

$$\phi : \mathscr{S}_1^* \times \cdots \times \mathscr{S}_n^* \longrightarrow X, \tag{7}$$

there is a unique linear map

$$\Phi : W \longrightarrow X \tag{8}$$

such that

$$\phi = \Phi \circ \mathrm{I}^*. \tag{9}$$

Recall that the range of I^* is in W because it coincides with the set of simple elements generating W. We start defining Φ on the simple elements in W by setting

$$\Phi(f_{a_1,\ldots,a_n}) := \phi(f_{a_1}^1, \cdots, f_{a_n}^n). \tag{10}$$

Since the set of simple elements is a generating set for W, we can extend Φ to the whole W by linearity so that, by construction, we have that Eq. (9) holds. Furthermore, again because the set of simple elements is a generating set for W, the map Φ is unique by construction. Consequently, the universal property of the algebraic tensor product implies the existence of a vector space isomorphism between W and $\otimes_{j=1}^n \mathscr{S}_j^*$. It is important to note that, in general, W is only a proper subspace of $\mathscr{S}^*$.

Now, it is not hard to see that a convex set $\mathscr{S}$ generating a vector space $\mathscr{V}$ which is isomorphic with the tensor product $\otimes_{j=1}^n \mathscr{V}_j$ of the vector spaces generated by the single $\mathscr{S}_j$'s may always be interpreted as the convex set of a composite system implementing the independence condition among states of the subsystems (see Eq. (1)) and with the interdependence condition among the dual spaces of the subsystems (see Eq. (5)). Indeed, we can define the map $I \colon \mathscr{S}_1 \times \cdots \times \mathscr{S}_n \longrightarrow \mathscr{S}$ given by

$$(\rho_1, \cdots, \rho_n) \mapsto I(\rho_1, \cdots, \rho_n) = \rho_1 \otimes \cdots \otimes \rho_n, \tag{11}$$

and a general result from linear algebra assures us that $\otimes_{j=1}^n \mathscr{S}_j^*$ is always a subset of $\mathscr{S}^*$ (recall that we defined the dual space of a convex set to be the dual space of the vector space generated by the convex set itself). Furthermore, in the finite-dimensional case where $\dim(\mathscr{V}_j) < \infty$ for all $j \in [1, \ldots, n]$, we have that

$$W \cong \otimes_{j=1}^n \mathscr{S}_j^* \cong \left(\otimes_{j=1}^n \mathscr{V}_j \right)^*, \tag{12}$$

where $\mathscr{V}_j$ is the vector space generated by $\mathscr{S}_j$. Consequently, choosing the vector space $\mathscr{V}$ generated by $\mathscr{S}$ to be the tensor product $\otimes_{j=1}^n \mathscr{V}_j$ is equivalent to impose the minimality condition $\mathscr{S}^* = W$ for the dual space of $\mathscr{S}$. Note that this is no longer true in the infinite-dimensional case because the dual space of a tensor product need not be the tensor product of the dual spaces. However, it is reasonable to say that the subleties associated with infinite dimensions requires more structures to be handled, and the framework of arbitrary convex spaces is too broad to provide these structures.

As a final comment, let us point out that, even if we consider the finite-dimensional case with the minimality condition $\mathscr{S}^* = W$, there is no way to single out unambiguously an explicit candidate for $\mathscr{S}$ without introducing further assumptions. Again, this should not come as a surprise because the theoretical framework of arbitrary convex spaces considered here is too broad.

3   Conclusions

We investigated the notion of composite system made of distinguishable parties in the context of physical theories for which the admissible spaces of states are real, convex spaces. Essentially, we modelled a composite system by means of the family $\{\mathscr{S}_j\}_{j\in[1,...n]}$ of spaces of states of the n subsystems together with the space $\mathscr{S}$ of states of the total system "endowed" with two mathematical constraints. First of all, we imposed an independence relation among the states of the subsystems in terms of an injective linear map I: $\mathscr{S}_1 \times \cdots \times \mathscr{S}_n \longrightarrow \mathscr{S}$, where $\mathscr{S}$ is the space of states of the total system, and $\mathscr{S}_1 \times \cdots \times \mathscr{S}_n$ is the Cartesian product of the spaces of states of the subsystems. From the operational point of view, the existence of the map I should be thought of as guaranteeing that each party of the system is free to prepare its associated subsystem in any of the possible states independently from the preparations of the other parties. Then, we introduced an interdependence condition among the dual spaces of the subsystems (see Eq. (5)). We saw that these two mathematical conditions are enough to introduce the notion of separable and entangled states in the context of arbitrary convex spaces, and to prove that the dual space $\mathscr{S}^*$ of a composite system must contain a copy of the tensor product $\otimes_{j=1}^n \mathscr{S}_j^*$ of the dual spaces of the single subsystems. Furthermore, in the finite-dimensional case, $\mathscr{S}$ generates a vector space $\mathscr{V}$ which is isomorphic with the tensor product $\otimes_{j=1}^n \mathscr{V}_j$ of the vector spaces generated by the single subsystems if and only if $\mathscr{S}^*$ satisfies a minimality condition.

We must stress that the interdependence condition expressed by Eq. (5) has not yet a clear physical interpretation, but its mathematical expression points toward a connection with the notions of independence and freeness as defined in the context of free probability theory (see [28]) which will be explored elsewhere.

References

1. Araki, H.: Mathematical Theory of Quantum Fields. Oxford University Press, New York (1999)
2. Barnum, H., Wilce, A.: Post-classical probability theory. In: Chiribella, G., Spekkens, R.W. (eds.) Quantum Theory: Informational Foundations and Foils. FTP, vol. 181, pp. 367–420. Springer, Dordrecht (2016). https://doi.org/10.1007/978-94-017-7303-4_11
3. Barrett, J.: Information processing in generalized probabilistic theories. Phys. Rev. A **75**, 032304 (2007)
4. Chiribella, G.: Agents, subsystems, and the conservation of information. Entropy **20**, 358 (2018)
5. Chiribella, G., D'Ariano, G.M., Perinotti, P.: Probabilistic theories with purification. Phys. Rev. A **81**, 062348 (2010)
6. Ciaglia, F.M., Ibort, A., Marmo, G.: A gentle introduction to Schwinger's formulation of quantum mechanics: the groupoid picture. Mod. Phys. Lett. A **33**(20), 1850122 (2018)
7. Ciaglia, F.M., Ibort, A., Marmo, G.: Schwinger's picture of quantum mechanics: groupoids. Int. J. Geom. Methods Mod. Phys. (2019)
8. Ciaglia, F.M., Ibort, A., Marmo, G.: Schwinger's picture of quantum mechanics II: algebras and observables. Int. J. Geom. Methods Mod. Phys. (2019)
9. Cornette, W.M., Gudder, S.P.: The mixture of quantum states. J. Math. Phys. **15**(6), 842–850 (1974)

10. Dirac, P.A.M.: Principles of Quantum Mechanics. Oxford university Press, Oxford (1958)
11. Gudder, S.P.: Convex structures and operational quantum mechanics. Commun. Math. Phys. **29**, 249–264 (1973)
12. Haag, R.: Local Quantum Physics: Fields, Particles, Algebras. Springer, Berlin (1996). https://doi.org/10.1007/978-3-642-61458-3
13. Haag, R., Kastler, D.: An algebraic approach to quantum field theory. J. Math. Phys. **5**(7), 848–861 (1964)
14. Hellwig, K.-E., Kraus, K.: Pure operations and measurements. Commun. Math. Phys. **11**, 214–220 (1969)
15. Holevo, A.S.: Statistical Structure of Quantum Theory. Springer, Berlin (2001). https://doi.org/10.1007/3-540-44998-1
16. Holevo, A.S.: Probabilistic and Statistical Aspects of Quantum Theory. Edizioni della Normale, Basel (2011)
17. Janotta, P., Hinrichsen, H.: Generalized probability theories: what determines the structure of quantum theory? J. Phys. A Math. Theor. **47**(32), 323001 (2014)
18. Jordan, P., von Neumann, J., Wigner, E.P.: On an algebraic generalization of the quantum mechanical formalism. Ann. Math. **35**, 29–64 (1934)
19. Kraus, K.: States, Effects, and Operations. Springer, Berlin (1983). https://doi.org/10.1007/3-540-12732-1
20. Ludwig, G.: Foundations of Quantum Mechanics I. Springer, Berlin (1983). https://doi.org/10.1007/978-3-642-86751-4
21. Mielnik, B.: Geometry of quantum states. Commun. Math. Phys. **9**, 55–80 (1968)
22. Mielnik, B.: Theory of Filters. Commun. Math. Phys. **15**, 1–46 (1969)
23. Mielnik, B.: Generalized quantum mechanics. Commun. Math. Phys. **37**, 221–256 (1974)
24. Pauli, W.: General Principles of Quantum Mechanics. Springer, Berlin (1980). https://doi.org/10.1007/978-3-642-61840-6
25. Roos, H.: Independece of local algebras in quantum field theory. Commun. Math. Phys. **16**, 238–246 (1970)
26. Segal, I.E.: Postulates for general quantum mechanics. Ann. Math. **48**(4), 930–948 (1947)
27. Segal, I.E.: C^*-algebras and quantization. Contemp. Math. **167**, 55–65 (1994)
28. Voicolescu, D.V., Dykema, K.J., Nica, A.: Free Random Variables. American Mathematical Society, Providence (1992)
29. von Neumann, J.: Mathematical Foundations of Quantum Mechanics. Princeton University Press, Princeton (1955)

Probability on Riemannian Manifolds

The Riemannian Barycentre as a Proxy for Global Optimisation

Salem Said[1(✉)] and Jonathan H. Manton[2]

[1] Laboratoire IMS (CNRS 5218), Université de Bordeaux, Bordeaux, France
salem.said@u-bordeaux.fr
[2] Department of Electrical and Electronic Engineering, The University of Melbourne, Melbourne, Australia
j.manton@ieee.org

Abstract. Let M be a simply-connected compact Riemannian symmetric space, and U a twice-differentiable function on M, with unique global minimum at $x^* \in M$. The idea of the present work is to replace the problem of searching for the global minimum of U, by the problem of finding the Riemannian barycentre of the Gibbs distribution $P_T \propto \exp(-U/T)$. In other words, instead of minimising the function U itself, to minimise $\mathcal{E}_T(x) = \frac{1}{2} \int d^2(x, z) P_T(dz)$, where $d(\cdot, \cdot)$ denotes Riemannian distance. The following original result is proved: if U is invariant by geodesic symmetry about x^*, then for each $\delta < \frac{1}{2} r_{cx}$ (r_{cx} the convexity radius of M), there exists T_δ such that $T \leq T_\delta$ implies $\mathcal{E}_T$ is strongly convex on the geodesic ball $B(x^*, \delta)$, and x^* is the unique global minimum of $\mathcal{E}_T$. Moreover, this T_δ can be computed explicitly. This result gives rise to a general algorithm for black-box optimisation, which is briefly described, and will be further explored in future work.

Keywords: Riemannian barycentre · Black-box optimisation · Symmetric space

It is common knowledge that the Riemannian barycentre $\bar{x}$, of a probability distribution P defined on a Riemannian manifold M, may fail to be unique. However, if P is supported inside a geodesic ball $B(x^*, \delta)$ with radius $\delta < \frac{1}{2} r_{cx}$ (r_{cx} the convexity radius of M), then $\bar{x}$ is unique and also belongs to $B(x^*, \delta)$. In fact, Afsari has shown this to be true, even when $\delta < r_{cx}$ (see [1,2]).

Does this statement continue to hold, if P is not supported inside $B(x^*, \delta)$, but merely concentrated on this ball? The answer to this question is positive, assuming that M is a simply-connected compact Riemannian symmetric space, and $P = P_T \propto \exp(-U/T)$, where the function U has unique global minimum at $x^* \in M$. This is given by Proposition 2, in Sect. 2 below.

Proposition 2 motivates the main idea of the present work: the Riemannian barycentre $\bar{x}_T$ of P_T can be used as a proxy for the global minimum x^* of U. In general, $\bar{x}_T$ only provides an approximation of x^*, but the two are equal if U is invariant by geodesic symmetry about x^*, as stated in Proposition 3, in Sect. 4 below.

© Springer Nature Switzerland AG 2019
F. Nielsen and F. Barbaresco (Eds.): GSI 2019, LNCS 11712, pp. 657–664, 2019.
https://doi.org/10.1007/978-3-030-26980-7_68

The following Sect. 1 introduces Proposition 2, which estimates the Riemannian distance between $\bar{x}_T$ and x^*, as a function of T.

1   Concentration of the Barycentre

Let P be a probability distribution on a complete Riemannian manifold M. A (Riemannian) barycentre of P is any global minimiser $\bar{x} \in M$ of the function

$$\mathcal{E}(x) = \frac{1}{2} \int_M d^2(x,z) P(dz) \quad \text{for } x \in M \tag{1}$$

The following statement is due to Karcher, and was improved upon by Afsari [1,2]: *if P is supported inside a geodesic ball $B(x^*, \delta)$, where $x^* \in M$ and $\delta < \frac{1}{2} r_{cx}$ (r_{cx} the convexity radius of M), then $\mathcal{E}$ is strongly convex on $B(x^*, \delta)$, and P has a unique barycentre $\bar{x} \in B(x^*, \delta)$.*

On the other hand, the present work considers a setting where P is not supported inside $B(x^*, \delta)$, but merely concentrated on this ball. Precisely, assume P is equal to the Gibbs distribution

$$P_T(dz) = (Z(T))^{-1} \exp\left[-\frac{U(z)}{T} \right] \text{vol}(dz); \; T > 0 \tag{2}$$

where $Z(T)$ is a normalising constant, U is a C^2 function with unique global minimum at x^*, and vol is the Riemannian volume of M. Then, let $\mathcal{E}_T$ denote the function $\mathcal{E}$ in (1), and let $\bar{x}_T$ denote any barycentre of P_T.

In this new setting, it is not clear whether $\mathcal{E}_T$ is differentiable or not. Therefore, statements about convexity of $\mathcal{E}_T$ and uniqueness of $\bar{x}_T$ are postponed to the following Sect. 2. For now, it is possible to state the following Proposition 1. In this proposition, $d(\cdot, \cdot)$ denotes Riemannian distance, and $W(\cdot, \cdot)$ denotes the Kantorovich (L^1-Wasserstein) distance [3,4]. Moreover, $(\mu_{\min}, \mu_{\max})$ is any open interval which contains the spectrum of the Hessian $\nabla^2 U(x^*)$, considered as a linear mapping of the tangent space $T_{x^*} M$.

Proposition 1. *Assume M is an n-dimensional compact Riemannian manifold with non-negative sectional curvature. Denote δ_{x^*} the Dirac distribution at x^*. The following hold,*
(i) for any $\eta > 0$,

$$W(P_T, \delta_{x^*}) < \frac{\eta^2}{(4 \operatorname{diam} M)} \implies d(\bar{x}_T, x^*) < \eta \tag{3}$$

(ii) for $T \leq T_o$ (which can be computed explicitly)

$$W(P_T, \delta_{x^*}) \leq \sqrt{2} \, (\pi/2)^{n-1} \, B_n^{-1} \, (\mu_{\max}/\mu_{\min})^{n/2} \, (T/\mu_{\min})^{1/2} \tag{4}$$

where $B_n = B(1/2, n/2)$ in terms of the Beta function.

Proposition 1 is motivated by the idea of using $\bar{x}_T$ as an approximation of x^*. Intuitively, this requires choosing T so small that P_T is sufficiently close to δ_{x^*}. Just how small a T may be required is indicated by the inequality in (4). This inequality is optimal and explicit, in the following sense.

It is optimal because the dependence on $T^{1/2}$ in its right-hand side cannot be improved. Indeed, by the multi-dimensional Laplace approximation (see [5], for example), the left-hand side is equivalent to $\mathrm{L} \cdot T^{1/2}$ (in the limit $T \to 0$). While this constant L is not tractable, the constants appearing in Inequality (4) depend explicitly on the manifold M and the function U. In fact, this inequality does not follows from the multi-dimensional Laplace approximation, but rather from volume comparison theorems of Riemannian geometry [6].

In spite of these nice properties, Inequality (4) does not escape the curse of dimensionality. Indeed, for fixed T, its right-hand side increases exponentially with the dimension n (note that B_n decreases like $n^{-1/2}$). On the other hand, although T_o also depends on n, it is typically much less affected by dimensionality, and decreases slower that n^{-1} as n increases.

2 Convexity and Uniqueness

Assume now that M is a simply-connected, compact Riemannian symmetric space. In this case, for any T, the function $\mathcal{E}_T$ turns out to be C^2 throughout M. This results from the following lemma.

Lemma 1. *Let M be a simply-connected compact Riemannian symmetric space. Let $\gamma : I \to M$ be a geodesic defined on a compact interval I. Denote $\mathrm{Cut}(\gamma)$ the union of all cut loci $\mathrm{Cut}(\gamma(t))$ for $t \in I$. Then, the topological dimension of $\mathrm{Cut}(\gamma)$ is strictly less than $n = \dim M$. In particular, $\mathrm{Cut}(\gamma)$ is a set with volume equal to zero.*

Remark: *The assumption that M is simply-connected cannot be removed, as the conclusion does not hold if M is a real projective space.*

The proof of Lemma 1 uses the structure of Riemannian symmetric spaces, as well as some results from topological dimension theory [7] (Chapter VII). The notion of topological dimension arises because it is possible $\mathrm{Cut}(\gamma)$ is not a manifold. The lemma immediately implies, for all t,

$$\mathcal{E}_T(\gamma(t)) = \frac{1}{2} \int_M d^2(\gamma(t), z) P_T(dz) = \frac{1}{2} \int_{M - \mathrm{Cut}(\gamma)} d^2(\gamma(t), z) P_T(dz)$$

Then, since the domain of integration avoids the cut loci of all the $\gamma(t)$, it becomes possible to differentiate under the integral. This is used in obtaining the following (the assumptions are the same as in Lemma 1).

Corollary 1. *For $x \in M$, let $G_x(z) = \nabla f_z(x)$ and $H_x(z) = \nabla^2 f_z(x)$, where f_z is the function $x \mapsto \frac{1}{2} d^2(x, z)$. The following integrals converge for any T*

$$G_x = \int_{M - \mathrm{Cut}(x)} G_x(z) \, P_T(dz); \quad H_x = \int_{M - \mathrm{Cut}(x)} H_x(z) \, P_T(dz)$$

and both depend continuously on x. Moreover,

$$\nabla \mathcal{E}_T(x) = G_x \ \text{and} \ \nabla^2 \mathcal{E}_T(x) = H_x \tag{5}$$

so that $\mathcal{E}_T$ is C^2 throughout M.

With Corollary 1 at hand, it is possible to obtain Proposition 2, which is concerned with the convexity of $\mathcal{E}_T$ and uniqueness of $\bar{x}_T$. In this proposition, the following notation is used

$$f(T) = (4/\pi) \, (\pi/8)^{n/2} \, (\mu_{\max}/T)^{n/2} \exp\left(-U_\delta/T\right) \tag{6}$$

where $U_\delta = \inf\{U(x) - U(x^*) \, ; \, x \notin B(x^*, \delta)\}$ for positive δ. The reader may wish to note the fact that $f(T)$ decreases to 0 as T decreases to 0.

Proposition 2. *Let M be a simply-connected compact Riemannian symmetric space. Let κ^2 be the maximum sectional curvature of M, and $r_{cx} = \kappa^{-1}\frac{\pi}{2}$ its convexity radius. If $T \leq T_o$ (see (ii) of Proposition 1), then the following hold for any $\delta < \frac{1}{2}r_{cx}$.*
(i) for all x in the geodesic ball $B(x^*, \delta)$,

$$\nabla^2 \mathcal{E}_T(x) \geq \mathrm{Ct}(2\delta) \, (1 - \mathrm{vol}(M) f(T)) - \pi A_M f(T) \tag{7}$$

where $\mathrm{Ct}(2\delta) = 2\kappa\delta \cot(2\kappa\delta) > 0$ and $A_M > 0$ is a constant given by the structure of the symmetric space M.
(ii) there exists T_δ (which can be computed explicitly), such that $T \leq T_\delta$ implies $\mathcal{E}_T$ is strongly convex on $B(x^*, \delta)$, and has a unique global minimum $\bar{x}_T \in B(x^*, \delta)$. In particular, this means $\bar{x}_T$ is the unique barycentre of P_T.

Note that *(ii)* of Proposition 2 generalises the statement due to Karcher [1], which was recalled in Sect. 1.

3   Finding T_o and T_δ

Propositions 1 and 2 claim that T_o and T_δ can be computed explicitly. This means that, with some knowledge of the Riemannian manifold M and the function U, T_o and T_δ can be found by solving scalar equations. The current section gives the definitions of T_o and T_δ.

In the notation of Proposition 1, let $\rho > 0$ be small enough, so that,

$$\mu_{\min} d^2(x, x^*) \leq 2 \, (U(x) - U(x^*)) \leq \mu_{\max} d^2(x, x^*)$$

whenever $d(x, x^*) \leq \rho$, and consider the quantity

$$f(T, m, \rho) = (2/\pi)^{1/2} \, (\mu_{\max}/T)^{m/2} \exp\left(-U_\rho/T\right)$$

where U_ρ is defined as in (6). Note that $f(T, m, \rho)$ decreases to 0 as T decreases to 0, for fixed m and ρ. Now, it is possible to define T_o as

$$T_o = \min\left\{T_o^1, T_o^2\right\} \quad \text{where} \tag{8}$$

$$T_o^1 = \inf \left\{ T > 0 \,:\, f(T, n-2, \rho) > \rho^{2-n} A_{n-1} \right\}$$
$$T_o^2 = \inf \left\{ T > 0 \,:\, f(T, n+1, \rho) > (\mu_{\max}/\mu_{\min})^{n/2} C_n \right\}$$

Here, $A_n = E|X|^n$ for $X \sim N(0,1)$, and $C_n = \omega_n A_n/(\operatorname{diam} M \times \operatorname{vol} M)$, where ω_n is the surface area of a unit sphere S^{n-1}.

With regard to Proposition 2, define T_δ as follows,

$$T_\delta = \min \left\{ T_\delta^1, T_\delta^2 \right\} - \varepsilon \tag{9}$$

for some arbitrary $\varepsilon > 0$. Here, in the notation of (4), (6) and (7),

$$T_\delta^1 = \inf \left\{ T \le T_o \,:\, \sqrt{2\pi}\,(T/\mu_{\min})^{1/2} > \delta^2 \,(\mu_{\min}/\mu_{\max})^{n/2} D_n \right\}$$
$$T_o^2 = \inf \left\{ T \le T_o \,:\, f(T) > \operatorname{Ct}(2\delta)\,(\operatorname{Ct}(2\delta)\operatorname{vol} M + \pi A_M)^{-1} \right\}$$

where $D_n = (2/\pi)^{n-1} B_n/(4\operatorname{diam} M)$.

4   Black-Box Optimisation

Consider the problem of searching for the unique global minimum x^* of U. In black-box optimisation, it is only possible to evaluate $U(x)$ for given $x \in M$, and the cost of this evaluation precludes numerical approximation of derivatives. Then, the problem is to find x^* using successive evaluations of $U(x)$ (hopefully, as few of these evaluations as possible).

Here, a new algorithm for solving this problem is described. The idea of this algorithm is to find $\bar{x}_T$ using successive evaluations of $U(x)$, in the hope that $\bar{x}_T$ will provide a good approximation of x^*. While the quality of this approximation is controlled by Inequalities (3) and (4) of Proposition 1, in some cases of interest, $\bar{x}_T$ is exactly equal to x^*, for correctly chosen T, as in the following proposition 3.

To state this proposition, let s_{x^*} denote geodesic symmetry about x^* (see [7]). This is the transformation of M, which leaves x^* fixed, and reverses the direction of geodesics passing through x^*.

Proposition 3. *Assume that U is invariant by geodesic symmetry about x^*, in the sense that $U \circ s_{x^*} = U$. If $T \le T_\delta$ (see (ii) of Proposition 2), then $\bar{x}_T = x^*$ is the unique barycentre of P_T.*

Proposition 3 follows rather directly from Proposition 2. Precisely, by *(ii)* of Proposition 2, the condition $T \le T_\delta$ implies $\mathcal{E}_T$ is strongly convex on $B(x^*, \delta)$, and $\bar{x}_T \in B(x^*, \delta)$. Thus, $\bar{x}_T$ is the unique stationary point of $\mathcal{E}_T$ in $B(x^*, \delta)$. But, using the fact that U is invariant by geodesic symmetry about x^*, it is possible to prove that x^* is a stationary point of $\mathcal{E}_T$, and this implies $\bar{x}_T = x^*$. The two following examples verify the conditions of Proposition 3.

Example 1. Assume $M = \mathrm{Gr}(k, \mathbb{C}^n)$ is a complex Grassmann manifold. In particular, M is a simply-connected, compact Riemannian symmetric space. Identify M with the set of Hermitian projectors $x : \mathbb{C}^n \to \mathbb{C}^n$ such that $\mathrm{tr}(x) = k$, where tr denotes the trace. Then, define $U(x) = -\mathrm{tr}(C\,x)$ for $x \in \mathrm{Gr}(k, \mathbb{C}^n)$, where C is a Hermitian positive-definite matrix with distinct eigenvalues. Now, the unique global minimum of U occurs at x^*, the projector onto the principal k-subspace of C. Also, the geodesic symmetry s_{x^*} is given by $s_{x^*} \cdot x = r_{x^*} x\, r_{x^*}$, where $r_{x^*} : \mathbb{C}^n \to \mathbb{C}^n$ denotes reflection through the image space of x^*. It is elementary to verify that U is invariant by this geodesic symmetry.

Example 2. Let M be a simply-connected, compact Riemannian symmetric space, and U_o a function on M with unique global minimum at $o \in M$. Assume moreover that U_o is invariant by geodesic symmetry about o. For each $x^* \in M$, there exists an isometry g of M, such that $x^* = g \cdot o$. Then, $U(x) = U_o(g^{-1} \cdot x)$ has unique global minimum at x^*, and is invariant by geodesic symmetry about x^*.

Example 1 describes the standard problem of finding the principal subspace of the covariance matrix C. In Example 2, the function U_o is a known template, which undergoes an unknown transformation g, leading to the observed pattern U. This is a typical situation in pattern recognition problems.

Of course, from a mathematical point of view, Example 2 is not really an example, since it describes the completely general setting where the conditions of Proposition 3 are verified. In this setting, consider the following algorithm.

Description of the algorithm:

– input : $T \leq T_\delta$ % to find such T, see Section 3

      $Q(x, dz) = q(x, z)\mathrm{vol}(dz)$ % symmetric Markov kernel

      $\hat{x}_0 = z_0 \in M$ % initial guess for x^*

– iterate : for $n = 1, 2, \ldots$

      (1) sample $z_n \sim q(z_{n-1}, z)$

      (2) compute $r_n = 1 - \min\{1, \exp[(U(z_{n-1}) - U(z_n))/T]\}$

      (3) reject z_n with probability r_n % then, $z_n = z_{n-1}$

      (4) $\hat{x}_n = \hat{x}_{n-1} \#_{\frac{1}{n}} z_n$ % see definition (10) below

– until : $\hat{x}_n$ does not change sensibly

– output : $\hat{x}_n$ % approximation of x^*

The above algorithm recursively computes the Riemannian barycentre $\hat{x}_n$ of the samples z_n generated by a symmetric Metropolis-Hastings algorithm (see [8]). Here, The Metropolis-Hastings algorithm is implemented in lines (1)--(3). On the other hand, line (4) takes care of the Riemannian barycentre. Precisely, if $\gamma : [0, 1] \to M$ is a length-minimising geodesic connecting $\hat{x}_{n-1}$ to z_n, let

$$\hat{x}_{n-1} \#_{\frac{1}{n}} z_n = \gamma(1/n) \tag{10}$$

This geodesic γ need not be unique.

The point of using the Metropolis-Hastings algorithm is that the generated z_n eventually sample from the Gibbs distribution P_T. The convergence of the distribution P_n of z_n to P_T takes place exponentially fast. Indeed, it may be inferred from [8] (see Theorem 8, Page 36)

$$\|P_n - P_T\|_{TV} \leq (1 - p_T)^n \tag{11}$$

where $\|\cdot\|_{TV}$ is the total variation norm, and $p_T \in (0,1)$ verifies

$$p_T \leq (\text{vol}(M)) \inf_{x,z} q(x,z) \exp(-\sup_x U(x)/T)$$

so the rate of convergence is degraded when T is small.

Accordingly, the intuitive justification of the above algorithm is the following. Since the z_n eventually sample from the Gibbs distribution P_T, and the desired global minimum x^* of U is equal to the barycentre $\bar{x}_T$ of P_T (by Proposition 3), then the barycentre $\hat{x}_n$ of the z_n is expected to converge to x^*.

It should be emphasised that, in the present state of the literature, there is no rigorous result which confirms this convergence $z_n \to x^*$. It is therefore an open problem, to be confronted in future work.

For a basic computer experiment, consider $M = S^2 \subset \mathbb{R}^3$, and let

$$U(x) = -P_9(x^3) \quad \text{for } x = (x^1, x^2, x^3) \in S^2 \tag{12}$$

where P_9 is the Legendre polynomial of degree 9 [9]. The unique global minimiser of U is $x^* = (0,0,1)$, and the conditions of Proposition 3 are verified, since U is invariant by reflection in the x^3 axis, which is geodesic symmetry about x^*.

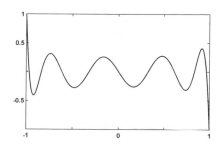

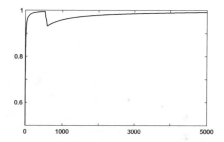

Fig. 1. graph of $-P_9(x^3)$ **Fig. 2.** $\hat{x}_n^3$ versus n

Figure 1 shows the dependence of $U(x)$ on x^3, displaying multiple local minima and maxima. Figure 2 shows the algorithm overcoming these local minima and maxima, and converging to the global minimum $x^* = (0,0,1)$, within $n = 5000$ iterations. The experiment was conducted with $T = 0.2$, and the Markov kernel Q obtained from the von Mises-Fisher distribution (see [10]). The initial guess $\hat{x}_0 = (0,0,-1)$ is not shown in Fig. 2.

In comparison, a standard simulated annealing method offered less robust performance, which varied considerably with the choice of annealing schedule.

Proofs

The proofs of all results stated in this work are detailed in the extended version, available online: https://arxiv.org/abs/1902.03885

References

1. Karcher, H.: Riemannian centre of mass and mollifier smoothing. Commun. Pure. Appl. Math. **30**(5), 509–541 (1977)
2. Afsari, B.: Riemannian L^p center of mass: existence, uniqueness, and convexity. Proc. Am. Math. Soc. **139**(2), 655–673 (2010)
3. Kantorovich, L.V., Akilov, G.P.: Functional Analysis, 2nd edn. Pergamon Press, Oxford (1982)
4. Villani, C.: Optimal Transport, Old and New, 2nd edn. Springer, Heidelberg (2009). https://doi.org/10.1007/978-3-540-71050-9
5. Wong, R.: Asymptotic Approximations of Integrals. Society for Industrial and Applied Mathematics, Philadelphia (2001)
6. Chavel, I.: Riemannian Geometry: A Modern Introduction. Cambridge University Press, Cambridge (2006)
7. Helgason, S.: Differential Geometry, Lie Groups, and Symmetric Spaces. American Mathematical Society, Providence (1978)
8. Roberts, G.O., Rosenthal, J.S.: General state space Markov chains and MCMC algorithms. Probab. Surv. **1**, 20–71 (2004)
9. Beals, R., Wong, R.: Special Functions: A Graduate Text. Cambridge University Press, Cambridge (2010)
10. Mardia, K.V., Jupp, P.E.: Directional Statistics. Academic Press Inc., London (1972)

Hamiltonian Monte Carlo on Lie Groups and Constrained Mechanics on Homogeneous Manifolds

Alessandro Barp[1,2]([✉])

[1] Imperial College London, Kensington, London SW7 2AZ, UK
a.barp16@imperial.ac.uk
[2] The Alan Turing Institute, 96 Euston Rd, Kings Cross, London NW1 2DB, UK

Abstract. In this paper we show that the Hamiltonian Monte Carlo method for compact Lie groups constructed in [20] using a symplectic structure can be recovered from canonical geometric mechanics with a bi-invariant metric. Hence we obtain the correspondence between the various formulations of Hamiltonian mechanics on Lie groups, and their induced HMC algorithms. Working on $\mathcal{G} \times \mathfrak{g}$ we recover the Euler-Arnold formulation of geodesic motion, and construct explicit HMC schemes that extend [20,21] to non-compact Lie groups by choosing metrics with appropriate invariances. Finally we explain how mechanics on homogeneous spaces can be formulated as a constrained system over their associated Lie groups. In some important cases the constraints can be naturally handled by the symmetries of the Hamiltonian.

Keywords: Hamiltonian Monte Carlo · Lie groups ·
Homogeneous manifolds · MCMC · Symmetric spaces · Sampling ·
Symmetries · Symplectic integrators

1 Introduction

The HMC algorithm [16] is a method that generate samples from a probability distribution which has a density known up to constant factor with respect to a reference measure [5,7,8]. The method was originally extended to compact semisimple Lie groups in [20], while Poisson brackets, shadow Hamiltonians and higher order integrators were discussed in [21]. The algorithm was designed by constructing a symplectic structure (and thus a mechanics) on $\mathcal{G} \times \mathfrak{g}^*$, but the relation with the standard mechanics induced by the Liouville form on $T^*\mathcal{G}$ was never explained. One of the difficulties that arise when sampling distributions on manifolds using HMC is the requirement to compute the geodesic flow accurately in order to maintain the motion on the manifold [6,12,17].

Electronic supplementary material The online version of this chapter (https://doi.org/10.1007/978-3-030-26980-7_69) contains supplementary material, which is available to authorized users.

F. Nielsen and F. Barbaresco (Eds.): GSI 2019, LNCS 11712, pp. 665–675, 2019.
https://doi.org/10.1007/978-3-030-26980-7_69

In the context of Lie groups, the geodesic motion was handled in [4, 14, 20–22] by assuming the Killing form defined a positive-definite Riemannian metric, in which case geodesics are given by 1-parameter subgroups. This holds for compact semisimple Lie groups (for example $SO(n)$ and $SU(n)$) but is not case for most other Lie groups, such as GL(n) or $SL(n)$.[1]

In this article we formalise the HMC schemes given in [20, 21] and connect them to standard geometric mechanics on Lie groups, which allows us to remove the compactness assumption on the Lie algebra (or equivalently, the assumption that the Killing form is negative definite).

In Sect. 2.2 we show the symplectic structure used in [20, 21] is symplectomorphic to the canonical one on $T^*\mathcal{G}$. It follows that many results from geometric mechanics that assume standard mechanics, such as symplectic reduction on cotangent bundles, can be used in HMC [1]. Since HMC is usually implemented on the tangent bundle, we derive the corresponding symplectic structure on $\mathcal{G} \times \mathfrak{g}$ induced by a metric. In [20, 21] the kinetic flow was simply given by left-invariant vector fields as a result of the bi-invariance of the Killing-form. This corresponds to solving the reconstruction equation for a constant curve on $\mathfrak{g}$, but, as we show in Sect. 2.3, the geodesic flow of more general left-invariant metrics must solve both the Euler-Arnold and reconstruction equations. Hence in Sect. 3 we consider the Euler-Arnold equation and explain how it can be simplified and solved by choosing inner products with appropriate symmetries. In particular we derive explicit HMC schemes for GL(n) and semi-simple Lie groups. In Sect. 2.4 we provide the explicit relation between the leapfrog scheme on Lie groups as used in [6] and the general one on manifolds; while Sect. 2.5 discusses more efficient integrators. In Sect. 4 we explain how more generally the mechanical system on homogeneous manifolds can be derived from a constrained mechanics defined on Lie groups. For reductive homogeneous manifolds, the constraints can be naturally handled by choosing a Hamiltonian with sufficient symmetries. Hence we recover the HMC scheme proposed in [6] to sample distributions on naturally reductive homogeneous spaces using Lie group mechanics and symplectic reduction. We will see that this is still true for the discretised motion defined by the leapfrog method or force-gradient integrators (i.e., these preserve the symmetries).

2   Hamiltonian Monte Carlo on Lie Groups

2.1   Lie Groups

Let $L_g : \mathcal{G} \to \mathcal{G}$ denote the left translation $L_g h = gh$ on a Lie group $\mathcal{G}$, and $\mathfrak{g}$ (and $\mathfrak{g}_L$) denote the Lie algebra (of left-invariant vector fields) (see Sect. 6.10 for more details). The Maurer-Cartan form $\theta \in \Gamma(T^*\mathcal{G} \otimes \mathfrak{g})^2$ is defined by $\theta : g \mapsto \partial_g L_{g^{-1}}$

---

[1] In [21] the equations of motion are given for semisimple lie algebras, but the derivation uses the Killing form which is not Riemannian in general and as a result cannot be used for HMC (due to the momentum refreshment step).

[2] $\Gamma(\mathcal{M})$ denotes the set of smooth sections of a bundle $\mathcal{M}$. If the base space $\mathcal{B}$ is ambiguous we write $\Gamma(\mathcal{B}, \mathcal{M})$.

where ∂ denotes the tangent map. Let ξ_i be a basis of $\mathfrak{g}$, $e_i \in \mathfrak{g}_L$ be the induced left-invariant vector fields (so $e_i(1) = \xi_i$), and θ^i the dual 1-forms, $\theta^i(e_j) = \delta^i_j$. We will adopt the Einstein notation and sum over repeated indices.

2.2   Symplectic Structures on Lie Groups

In this section we derive the symplectic structures on $\mathcal{G} \times \mathfrak{g}^*$, $T\mathcal{G}$ and $\mathcal{G} \times \mathfrak{g}$ induced by the canonical 1-form on $T^*\mathcal{G}$.

The canonical Liouville 1-form can be expressed in terms of the momentum $p \equiv (\theta^{-1})^* : T^*\mathcal{G} \to \mathfrak{g}^*$, as $p(\pi^*\theta)$ [2]. We can expand θ and p using a basis e_i of $\mathfrak{g}_L$, which yields $\theta = e_i(1) \otimes \theta^i$ and $p = \theta^i(1) \otimes e_i$ where we use the canonical isomorphism $T_g\mathcal{G} \cong T_g^{**}\mathcal{G}$ to view e_i as a function $T^*\mathcal{G} \to \mathbb{R}$, i.e., $e_i : (g, \alpha_g) \mapsto \alpha_g(e_i(g))$. The Liouville 1-form is then $\eta \equiv e_i\pi^*\theta^i \in \Gamma(T^*\mathcal{G}, T^*T^*\mathcal{G})$ on $T^*\mathcal{G}$ (here $\pi : T^*\mathcal{G} \to \mathcal{G}$ is the usual projection), and the canonical symplectic structure is its exterior derivative[3]

$$\omega \equiv -\mathrm{d}\eta = -\mathrm{d}(e_i\pi^*\theta^i) = \pi^*\theta^i \wedge \mathrm{d}e_i + \tfrac{1}{2}e_ic^i_{jk}\pi^*\theta^j \wedge \pi^*\theta^k. \tag{1}$$

The symplectic volume form is then given by the product of the Haar measure $\theta^1 \wedge \cdots \wedge \theta^n$ on $\mathcal{G}$ with the Lebesgue measure on the fibres

$$\omega^n \equiv \bigwedge_{i=1}^{n} \omega \propto \mathrm{d}e_1 \wedge \cdots \wedge \mathrm{d}e_n \wedge \pi^*\theta^1 \wedge \cdots \wedge \pi^*\theta^n.$$

When $\mathcal{G} = \mathbb{R}^n$, then by the Maurer-Cartan equation and Poincaré lemma the 1-forms θ^i can be written as $\theta^i = \mathrm{d}x^i$, and we recover the usual volume form.

Originally HMC on compact semisimple matrix groups was constructed on the trivial bundle $\mathcal{G} \times \mathfrak{g}^*$, and although the momentum coordinates were left undefined,[4] we can recover the original mechanical system by defining $\tilde{p}_i : \mathcal{G} \times \mathfrak{g}^* \to \mathbb{R}$ s.t. $\tilde{p}_i : (g, \alpha_1) \to \alpha_1(\xi_i)$, from which we can form the symplectic structure $\tilde{\omega} = -\mathrm{d}(\tilde{p}_i\pi^*\theta^i)$. Clearly ω and $\tilde{\omega}$ are symplectomorphic with respect to the canonical isomorphism $T^*\mathcal{G} \cong \mathcal{G} \times \mathfrak{g}^*$, $(g, \alpha_g) \mapsto (g, L_g^*\alpha_g)$ (see Sect. 6.1 of the Supplementary material).

In practice however we usually work on the tangent bundle. Consider a Riemannian metric $\langle \cdot, \cdot \rangle$ on $T\mathcal{G}$. The associated musical isomorphism, $\mathbb{FL} : T\mathcal{G} \to T^*\mathcal{G}$, $v \mapsto \langle v, \cdot \rangle$, enable us to pull-back ω to a symplectic structure $\omega_{\mathcal{L}}$ on $T\mathcal{G}$. The latter is the exterior derivative of the 1-form $-v^j\pi^*(\langle e_i, e_j \rangle\theta^i) \in \Gamma(T\mathcal{G}, T^*T\mathcal{G})$, where $v^j : T\mathcal{G} \to \mathbb{R}$ satisfies $(g, u_g) \mapsto \theta^j_g(u_g)$, and now $\pi : T\mathcal{G} \to \mathcal{G}$ is the tangent bundle projection (see Sect. 6.2 of the Supplementary material). To take advantage of the symmetries of $\mathcal{G}$ we will restrict to left-invariant Riemannian

---

[3] Here we use the fact that θ is a flat gauge-field, and thus satisfy Maurer-Cartan equation $0 = \mathrm{d}\theta + \tfrac{1}{2}[\theta \wedge \theta] = e_k(1) \otimes (\mathrm{d}\theta^k + \tfrac{1}{2}c^k_{ij}\theta^i \wedge \theta^j)$.

[4] In [21] the symplectic structure is defined as $\mathrm{d}p$ for $p \in T_g^*\mathcal{G}$ which does not define a differential form over $T^*\mathcal{G}$ since the components p_i are then constant.

metric, $\langle \cdot, \cdot \rangle \equiv g_{ij}\theta^i \otimes \theta^j$ where g_{ij} is symmetric positive definite, so the metric is determined by its value at the identity, $g_{ij} = \langle \xi_i, \xi_j \rangle_1$. Then

$$\omega_{\mathcal{L}} \equiv \mathbb{F}\mathcal{L}^*\omega = -\mathrm{d}(g_{ij}v^j\pi^*\theta^i) = g_{ij}\pi^*\theta^i \wedge \mathrm{d}v^j + \tfrac{1}{2}g_{ir}v^r c^i_{jk}\pi^*\theta^j \wedge \pi^*\theta^k. \quad (2)$$

As above, $\omega_{\mathcal{L}}$ can be pulled-back to define an equivalent mechanics on $\mathcal{G} \times \mathfrak{g}$ (the coordinate functions v^j are now the maps $(g, X) \mapsto \theta^j_1(X)$). In particular, if the metric is left-invariant the Hamiltonian $H \equiv V \circ \pi + \tfrac{1}{2}\langle \cdot, \cdot \rangle$ on $T\mathcal{G}$ defines an equivalent Hamiltonian $V \circ \pi + \tfrac{1}{2}\langle \cdot, \cdot \rangle_1$ on $\mathcal{G} \times \mathfrak{g}$ (see Sect. 6.3 of the Supplementary material).

2.3   Hamiltonian Fields and Euler-Arnold Equation

We now focus on the mechanical system on $\mathcal{G} \times \mathfrak{g}$ induced by a left-invariant metric $T = v^i v^j g_{ij} : \mathfrak{g} \to \mathbb{R}$. For any function $f : \mathcal{G} \times \mathfrak{g} \to \mathbb{R}$, its Hamiltonian vector field $\hat{f} \in \Gamma(T\mathcal{G} \oplus T\mathfrak{g})$ is defined by $\omega_{\mathcal{L}}(\hat{f}, \cdot) = \mathrm{d}f$. We have (see Sect. 6.4 of the Supplementary material)

$$\hat{f} = g^{jk}\xi_k(f)e_j + \left(g^{ik}g^{lj}v^r c_{rji}\xi_l(f) - g^{jk}e_j(f) \right)\xi_k. \quad (3)$$

For the potential energy and kinetic energy induced by a left-invariant metric

$$\hat{T} = v^k e_k + g^{ik}v^r v^s c_{rsi}\xi_k \qquad \hat{V} = -g^{jk}e_j(V)\xi_k$$

The Poisson brackets associated to $\omega_{\mathcal{L}}$ are defined by $\{r, h\} \equiv \omega_{\mathcal{L}}(\hat{r}, \hat{h})$ for any functions $r, h : \mathcal{G} \times \mathfrak{g} \to \mathbb{R}$, and describe the rate of change of r along the flow of h since $\{r, h\} = \omega_{\mathcal{L}}(\hat{r}, \hat{h}) = \mathrm{d}r(\hat{h}) = \hat{h}[r]$. This is often written $\dot{r} = \{r, h\}$ where $\dot{r} : \mathcal{G} \times \mathfrak{g} \to \mathbb{R}$ is the function giving the rate of change of r in the direction $\hat{h}$, $\dot{r}(u) \equiv \frac{\mathrm{d}r(\Phi^{\hat{h}}_u(t))}{\mathrm{d}t}|_{t=0}$ (here $\Phi^{\hat{h}} : \mathbb{R} \times (\mathcal{G} \times \mathfrak{g}) \to \mathcal{G} \times \mathfrak{g}$ denotes the flow of $\hat{h}$). It follows immediately that a function does not vary along its Hamiltonian flow: $\dot{h} = \{h, h\} = 0$, and in fact neither does the symplectic structure $\Phi^{\hat{h}}(t)^*\omega_{\mathcal{L}} = \omega_{\mathcal{L}}$, which enables the construction of symplectic integrators (see Sect. 6.9 of the Supplementary material). It is also common to write the Poisson bracket of a curve $\gamma : [a, b] \to T\mathcal{G}$ with a function r by setting $\dot{\gamma} = \{\gamma, h\} \equiv \hat{h}(\gamma)$. This is a differential equation stating that $\dot{\gamma}$ is the integral curve of $\hat{h}$. In particular, a curve satisfying $\dot{\gamma} = \{\gamma, H\}$ with initial conditions $(g_0, v_0) = \gamma(0)$ is known as a solution of Hamilton's equations.

An integral curve γ_V of $\hat{V}$ satisfies $\dot{\gamma}_V = \hat{V}(\gamma)$. Let $\gamma_V(t) = (q(t), v(t))$. The equation then reads $(\dot{q}, \dot{v}) = (0, -g^{jk}e_j(V)\xi_k)|_{(q(t),v(t))}$, which implies

$$q(t) = q(0) \qquad v(t) = v(0) - tg^{jk}e_j(V)|_{q(0)}\xi_k. \quad (4)$$

For the integral curves of $\hat{T}$, we have $(\dot{q}, \dot{v}) = (v^k e_k, g^{ik}v^r v^s c_{rsi}\xi_k)|_{\gamma_T(t)}$. The mysterious term $g^{ik}v^r v^s c_{rsi}\xi_k$ disappeared in both [20,21] (and didn't appear in [22]) since the structure constants of the Killing form are totally anti-symmetric (see Sect. 3). It is in fact nothing else than the $\langle \cdot, \cdot \rangle$-adjoint of the adjoint representation. Recall that the adjoint representation $\mathrm{ad} : \mathfrak{g} \to \mathrm{End}(\mathfrak{g})$ satisfies $\mathrm{ad}_\xi \equiv \mathrm{ad}(\xi) = [\xi, \cdot]$. For each $\xi \in \mathfrak{g}$ we can define its adjoint $\mathrm{ad}_\xi^\top$ with respect to the inner product $\langle \mathrm{ad}_\xi^\top X, Y \rangle = \langle X, \mathrm{ad}_\xi Y \rangle$. It follows that $\mathrm{ad}^\top$ is bilinear so

$$\langle \mathrm{ad}_v^\top v, \xi_k \rangle = v^i v^j \langle \mathrm{ad}_{\xi_i}^\top \xi_j, \xi_k \rangle = v^i v^j \langle \xi_j, \mathrm{ad}_{\xi_i} \xi_k \rangle = v^i v^j \langle \xi_j, [\xi_i, \xi_k] \rangle$$
$$= v^i v^j \langle \xi_j, c_{ik}^r \xi_r \rangle = c_{ik}^r v^i v^j g_{jr} = v^i v^j c_{jik}.$$

On the other hand $\langle \mathrm{ad}_v^\top v, \xi_k \rangle = \langle \left(\mathrm{ad}_v^\top v \right)^s \xi_s, \xi_k \rangle = \left(\mathrm{ad}_v^\top v \right)^s g_{sk}$. Combining these two equations gives $\mathrm{ad}_v^\top v = g^{kl} v^i v^j c_{jik} \xi_l$. Hence the integral curves of $\hat{T}$ satisfy

$$\left(\dot{q}, \dot{v} \right) = \left(v^k e_k, \mathrm{ad}_v^\top v \right)|_{\gamma_T(t)}.$$

The equation $\dot{v} = \mathrm{ad}_v^\top v$ is a first order differential equation on the Lie algebra $\mathfrak{g}$ known as the Euler-Arnold equation. In order to derive the integral curves of $\hat{T}$ we can proceed in two steps. We first solve the Euler-Arnold equation to give a solution $s : [0, T] \to \mathfrak{g}$. The equation for $\dot{q}$ then states that $\dot{q} = v^k e_k|_{(q(t), s(t))} = s^k(t) e_k(q(t)) = \partial_1 L_{q(t)} s(t)$, which is a first order ODE for q called the reconstruction equation.

In summary we have shown that the Hamilton's equation $\dot{\gamma} = \hat{H}(\gamma)$ defined by the symplectic structure on $\mathcal{G} \times \mathfrak{g}$ reads

$$\dot{q} = v^j e_j(q) = \partial_1 L_q v \qquad \dot{v} = \mathrm{ad}_v^\top v - g^{jk} e_j(V)(q) \xi_k,$$

which can be recognised as the equations arising from the Hamilton-Pontryagin variational principle.

We finally note that the left-invariance of the metric enabled us to obtain two first order differential equations for the geodesic flow, rather than the usual second-order differential equation on a general Riemannian manifold. This is known as the Euler-Poincaré reduction.

2.4   Leapfrog Integrator

On an arbitrary Riemannian manifold $\mathcal{M}$, we can define the leapfrog integrator for the Hamiltonian system $H = V + T$ as a composition of the kinetic $\hat{T}$ and potential $\hat{V}$ fields defined by the canonical symplectic structure on $T\mathcal{M}$ [24]. The potential flow leads to the velocity update

$$q(t) = q(0) \qquad v(t) = v(0) - t \nabla_{q(0)} V, \tag{5}$$

(here ∇ is the Riemannian gradient) and the kinetic (or geodesic) flow leads to the update

$$\left(q(t), v(t) \right) = \left(\exp_{q(0)}(v(0)), \mathbb{P}_{q(t)} v(0) \right),$$

where $\exp_q v$ denotes the Riemann exponential map and $\mathbb{P}_q v$ the parallel transport along the geodesic. Let us see the explicit relation between these updates and the ones derived in Sect. 2.3. Suppose that $\mathcal{M} = \mathcal{G}$ is a Lie group and T

is defined by a left-invariant metric $\langle \cdot, \cdot \rangle$. The components of the curvature are determined by the structure constants (see appendix B [15])

$$\Gamma_{ijk} = \tfrac{1}{2}(c_{ikj} - c_{jki} - c_{kji}), \qquad R^i_{jkl} = -\tfrac{1}{4}c^i_{mj}c^m_{kl}.$$

From Sect. 2.2 we know that, under the symplectomorphism $T\mathcal{G} \cong \mathcal{G} \times \mathfrak{g}$, the Hamiltonian vector-field induced by the inner product $\langle \cdot, \cdot \rangle_1 : \mathfrak{g} \to \mathbb{R}$ push-forwards to the geodesic flow induced by $T : T\mathcal{G} \to \mathbb{R}$. More precisely if $(q(t), v(t)) \in \mathcal{G} \times \mathfrak{g}$ is the integral curve of the Hamiltonian vector field of $\langle \cdot, \cdot \rangle_1$, then the geodesic flow of the Riemannian metric $T = \langle \cdot, \cdot \rangle$ on $T\mathcal{G}$ is given by $(q(t), \partial_1 L_{q(t)} v(t)) = (q(t), \dot{q}(t))$.

For the potential term, observe that[5]

$$\nabla_g V = g^{jk}(g) \mathrm{d}_g V(e_k) e_j(g) = g^{jk}(1) e_k(V)(g) \partial_1 L_g \xi_j$$

(by left-invariance g^{ij} is constant), and $v(t) = \theta^i_g(v(t)) e_i(g) = v^i(t) \partial_1 L_g \xi_i$. Thus under the symplectomorphism $\mathcal{G} \times \mathfrak{g} \to T\mathcal{G}$, we have $(\xi_i \to \partial_1 L_g \xi_i)$ and (5) push-forwards to the velocity update (4).

Finally let us see how the velocity update (4) is computed in practice when $\mathcal{G}$ is a matrix group (or we have an injective representation $\mathcal{G} \to \mathrm{GL}(n)$ and identify $\mathcal{G}$ with its image). In that case we need an extension W of V defined on an open subset of $\mathrm{GL}(n)$. Then

$$e_i(V)(g) = \mathrm{d}(W \circ \iota)(g \cdot \xi_i) = \frac{\partial W}{\partial x^{jk}}\left(\partial(\iota \circ L_g)\xi_i\right)_{jk} = \mathrm{tr}\left(\frac{\partial W}{\partial x}^\top \cdot g \cdot \xi_i\right),$$

where $\cdot$ is matrix multiplication, $\iota : \mathcal{G} \to \mathrm{GL}(n)$ is the inclusion (or a more general injective representation), and in the last line we have identified $g \sim \iota(g), \xi_i \sim \partial\iota(\xi_i)$. We arrive at the general velocity update formula for a matrix group

$$v^i(t)\xi_i = v^i(0)\xi_i - tg^{ij}(1)\,\mathrm{tr}\left(\frac{\partial W}{\partial x}^\top \cdot g \cdot \xi_j\right)\xi_i.$$

In lattice gauge theory we typically have $W(x) = \mathrm{Re}\,\mathrm{tr}\,(Ux)$ for a constant complex matrix U. Given $H : \mathcal{G} \times \mathfrak{g} \to \mathbb{R}$, the following procedure produces a sample from $e^{-H}\omega^n_{\mathcal{L}}$; to obtain an HMC sample from the marginal $e^{-V}\theta^1 \wedge \cdots \wedge \theta^n$ we simply ignore the velocity component. Given current state $(q(0), v(0)) \in \mathcal{G} \times \mathfrak{g}$

2.5   Efficient Integrators

While the leapfrog integrator is the most common choice of integrator, it is not always the optimal choice. For example [28,29] consider a symplectic integrator of the form[6] $\exp\left(\lambda \delta t \hat{T}\right) \exp\left(\tfrac{1}{2}\delta t \hat{V}\right) \exp\left((1-2\lambda)\delta t \hat{T}\right) \exp\left(\tfrac{1}{2}\delta t \hat{V}\right) \exp\left(\lambda \delta t \hat{T}\right)$ which,

---

[5] Note that if $\alpha \in \Gamma(T^*\mathcal{G})$, then $\alpha = e_i(\alpha)\theta^i = \alpha(e_i)\theta^i = \alpha_i \theta^i$. Moreover $u = g^{ij}\alpha_i e_j$ is the vector field associated to α by the metric since $\alpha(v) = \alpha_i \theta^i(v^j e_j) = \alpha_i v^i = \alpha_k \delta^k_i v^i = \alpha_k g^{kr} g_{ri} v^i = u^r g_{ri} v^i = \langle u, v \rangle$.

[6] Using the notation $\exp(t\hat{H}) \equiv \Phi^{\hat{H}}(t)$ for the Hamiltonian flow.

Algorithm 1. Algorithm to generate a sample from $e^{-H}\omega_{\mathcal{L}}^n$.

1: sample $v \sim e^{-\frac{1}{2}g_{ij}v^j v^i}\mathrm{d}v = \mathcal{N}(0, g_{ij}^{-1})$
2: $v \leftarrow v - \frac{1}{2}tg^{jk}\,\mathrm{tr}\left(\frac{\partial W}{\partial x}^\top \cdot q(0)\cdot\xi_j\right)\xi_k$
3: Solve the Euler-Arnold and Reconstruction equations $(q, v) \leftarrow (q(0), v)$
4: $v \leftarrow v - \frac{1}{2}tg^{jk}\,\mathrm{tr}\left(\frac{\partial W}{\partial x}^\top \cdot q\cdot\xi_j\right)\xi_k$
5: $\Delta H \leftarrow H(q, v) - H(q(0), v(0))$
6: $U \sim \mathrm{Uniform}[0, 1]$
7: If $U < \exp(-\Delta H)$, then return (q, v); else return $(q(0), v(0))$

---

despite being computationally more expensive, they show is 50% more efficient than leapfrog. In [21] a "force-gradient" integrator is examined which involves the Hamiltonian field $\{V, \widehat{\{V, T\}}\}$, and thus contains second-order information (derivatives) about V. The question of tuning HMC using Poisson brackets to minimise the cost was considered in [14]; see also [13] for a discussion on Lie group integrators, and [10] for integrators which outperform leapfrog on $\mathbb{R}^n$.

3 Geodesics

We now analyse how Euler-Arnold equation $\dot{v} = \mathrm{ad}_v^\top v$ simplifies when the inner product on $\mathfrak{g}$ is $\mathrm{ad}_{\mathfrak{k}}$-invariant for a subalgebra $\mathfrak{k} \subset \mathfrak{g}$. When $\mathcal{G}$ is connected, $\mathrm{ad}_{\mathfrak{k}}$-invariant inner products correspond to left $\mathcal{G}$-invariant and right $\mathcal{K}$-invariant Riemannian metrics on $\mathcal{G}$.[7] When $\mathfrak{k} = \mathfrak{g}$ these correspond to bi-invariant Riemannian metric on $\mathcal{G}$, which always exists when $\mathcal{G}$ is compact [27].

Ad $\mathcal{G}$-Invariant Inner Products. The simplest case arises when $\mathrm{ad}_v^\top v = 0$. This situation occurs for example when the inner product on $\mathfrak{g}$ is ad-invariant, $\langle \mathrm{ad}_u v, w\rangle + \langle v, \mathrm{ad}_u w\rangle = 0$, which implies $\mathrm{ad}^\top = -\,\mathrm{ad}$ and $\mathrm{ad}_v^\top v = -[v, v] = 0$. We can find ad-invariant inner products on $\mathfrak{g}$ when $\mathcal{G}$ has an $\mathrm{Ad}\,\mathcal{G}$-invariant inner product, or equivalently if it has a bi-invariant metric. A particular $\mathrm{Ad}\,\mathcal{G}$-invariant inner product is available on Lie groups whose lie algebra is compact, which means the Killing-form $B(u, v) \equiv \mathrm{tr}\left(\mathrm{ad}(u)\,\mathrm{ad}(v)\right)$ is negative definite, and so $-B$ defines a positive definite ad-invariant inner product.

When $\mathrm{ad}_v^\top v = 0$, then $v(t) = v_0$, so the reconstruction equation becomes $\dot{q} = v^k e_k|_{(q(t), v_0)} = v_0^k e_k(q(t)) = \partial_1 L_{q(t)}v_0 = \frac{\mathrm{d}}{\mathrm{d}t}L_{q(t)}e^{tv_0}|_{t=0}$, which implies $q(t) = q(0)e^{tv_0}$ (here we used the fact that e^{tv_0} is tangent to v_0 at $t = 0$). Hence we obtain the kinetic flow as in [21]

$$\gamma_T(t) = \left(q(0)e^{tv_0}, v_0\right).$$

---

[7] i.e., the left and right actions L_g and R_k are isometries for $g \in \mathcal{G}, k \in \mathcal{K}$ (here $\mathcal{K}$ is the Lie group associated to $\mathfrak{k}$). When $\mathcal{G}$ is not connected an analogous correspondence exists with $\mathrm{Ad}_{\mathcal{K}}$-invariant inner products [6], i.e., inner products s.t. $\langle \mathrm{Ad}_k u, \mathrm{Ad}_k v\rangle_1 = \langle u, v\rangle_1$ for $u, v \in \mathfrak{g}$, which are automatically $\mathrm{ad}_{\mathfrak{k}}$-invariant.

Reductive Decomposition. Sometimes it is not possible to find a bi-invariant metric on $\mathcal{G}$. The natural next step is to consider inner products on $\mathfrak{g}$ which are $\mathrm{ad}_{\mathfrak{k}}$-invariant, where $\mathfrak{k}$ is a subalgebra of $\mathfrak{g}$. This means that $\langle \mathrm{ad}_u v, w \rangle + \langle v, \mathrm{ad}_u w \rangle = 0$ for all $u \in \mathfrak{k}$, and it follows that $\mathrm{ad}_u^\top = -\mathrm{ad}_u$. Then $\mathcal{K}$ is totally geodesic in $\mathcal{G}$ and its geodesics are given by 1-parameter subgroups (see Sect. 6.5 of the Supplementary material). Let $\mathfrak{p} \equiv \mathfrak{k}^\perp$, so $\mathfrak{g} = \mathfrak{k} \oplus \mathfrak{p}$ and $v = v_{\mathfrak{k}} + v_{\mathfrak{p}}$. The Euler-Arnold equation reads $\dot{v} = \dot{v}_{\mathfrak{k}} + \dot{v}_{\mathfrak{p}} = \mathrm{ad}_v^\top v = \mathrm{ad}_{v_{\mathfrak{k}}}^\top v_{\mathfrak{k}} + \mathrm{ad}_{v_{\mathfrak{k}}}^\top v_{\mathfrak{p}} + \mathrm{ad}_{v_{\mathfrak{p}}}^\top v_{\mathfrak{k}} + \mathrm{ad}_{v_{\mathfrak{p}}}^\top v_{\mathfrak{p}} = -[v_{\mathfrak{k}}, v_{\mathfrak{p}}] + \mathrm{ad}_{v_{\mathfrak{p}}}^\top v_{\mathfrak{k}} + \mathrm{ad}_{v_{\mathfrak{p}}}^\top v_{\mathfrak{p}}$. We can simplify this further if the decomposition $\mathfrak{k} \oplus \mathfrak{p}$ is naturally reductive, i.e., $[\mathfrak{k}, \mathfrak{p}] \subset \mathfrak{p}$ and for all $u, v, w \in \mathfrak{p}$, $\langle [u, v]_{\mathfrak{p}}, w \rangle = \langle u, [v, w]_{\mathfrak{p}} \rangle$ [3]. This implies $\mathrm{ad}_{v_{\mathfrak{p}}}^\top v_{\mathfrak{p}} = 0$ (see Sect. 6.6 of the Supplementary material), and we are left with

$$\dot{v} = -[v_{\mathfrak{k}}, v_{\mathfrak{p}}] + \mathrm{ad}_{v_{\mathfrak{p}}}^\top v_{\mathfrak{k}}.$$

Notably the constant curve $v = v_0$ with either $v_0 \in \mathfrak{k}$ or $v_0 \in \mathfrak{p}$ is a solution.

Matrix Groups. Consider a subgroup $\mathcal{G} \subset \mathrm{GL}(n)$ and set $\mathfrak{g} = \mathfrak{p} \oplus \mathfrak{k}$ where $\mathfrak{p}$ is a subspace of the space of symmetric matrices, while $\mathfrak{k} = \mathfrak{so}(n)$. We have in mind the cases $\mathrm{GL}(n), SL(n)$ and $SO(n)$ with $\mathfrak{p}$ the space of symmetric matrices, the space of traceless symmetric matrices, and $\{0\}$ respectively. We equip $\mathfrak{g}$ with the standard inner product $\langle A, B \rangle \equiv \mathrm{tr}(A^\top B)$ which is $\mathrm{Ad}\,\mathcal{K}$-invariant. Since the product of a symmetric and an antisymmetric matrix is traceless, we have $\langle A, B \rangle = \mathrm{tr}(A_{\mathfrak{p}} B_{\mathfrak{p}}) - \mathrm{tr}(A_{\mathfrak{k}} B_{\mathfrak{k}})$.

Note that for $\mathcal{G} = SO(n)$, this is just the negative of the Killing form (up to a positive constant); while on $\mathrm{GL}(n)$ and $SL(n)$, the Killing form, respectively given by $2n\,\mathrm{tr}(AB) - 2\,\mathrm{tr}(A)\,\mathrm{tr}(B)$ and $2n\,\mathrm{tr}(AB)$, doesn't define an inner product (it is degenerate and pseudo-Riemannian respectively [25]).

For our inner product we have $\mathrm{ad}_S^\top A = [S, A]$ for any $S \in \mathfrak{p}, A \in \mathfrak{k}$, hence Euler-Arnold equation simplifies to $\dot{v} = -2[v_{\mathfrak{k}}, v_{\mathfrak{p}}]$ with solution (see Sect. 6.7 of the Supplementary material)

$$\big(q(t), v(t)\big) = \Big(q(0)e^{\left(v_{\mathfrak{p}}(0) - v_{\mathfrak{k}}(0)\right)t}e^{2v_{\mathfrak{k}}(0)t}, v_{\mathfrak{k}}(0) + \mathrm{Ad}\,e^{-2v_{\mathfrak{k}}(0)t}(v_{\mathfrak{p}}(0))\Big). \quad (6)$$

Semi-simple Symmetric Spaces. The equation for the geodesics above also holds for a semi-simple Lie group $\mathcal{G}$ using an inner product based on the Killing form B (note $\mathrm{GL}(n)$ is not semi-simple). Let us consider a subgroup $\mathcal{K}$ for which $\mathcal{G}/\mathcal{K}$ is a symmetric space. In that case we can find a Cartan decomposition $\mathfrak{g} = \mathfrak{k} \oplus \mathfrak{p}$. By corollary 5.4.3 [19] the Killing form is negative-definite on $\mathcal{K}$. We say $\mathcal{G}/\mathcal{K}$ is of compact (non-compact) type if B is negative (positive) definite on $\mathfrak{p}$. If $\mathcal{G}/\mathcal{K}$ is of compact type the inner product $\langle v, u \rangle \equiv B(v_{\mathfrak{k}}, u_{\mathfrak{k}}) + B(v_{\mathfrak{p}}, u_{\mathfrak{p}})$ is $\mathrm{Ad}\,\mathcal{G}$-invariant, while if it is of non-compact type $\langle v, u \rangle \equiv -B(v_{\mathfrak{k}}, u_{\mathfrak{k}}) + B(v_{\mathfrak{p}}, u_{\mathfrak{p}})$ is $\mathrm{Ad}\,\mathcal{K}$-invariant. Then $\mathrm{ad}_{v_{\mathfrak{p}}}^\top v_{\mathfrak{k}} = [v_{\mathfrak{p}}, v_{\mathfrak{k}}]$ (see Lemma 5.5.4 [19]) and we can find the integral curves of $\hat{T}$ as in Sect. 3.

Approximating the Matrix Exponential. In practice there are several ways to approximate the matrix exponential, for example by combining a Padé approximant with a projection from $\mathrm{GL}(n)$ to $\mathcal{G}$ [9]. In particular the Cayley transform $z \mapsto (1+\frac{1}{2}z)/(1-\frac{1}{2}z) = (1-\frac{1}{2}z)^{-1}(1+\frac{1}{2}z)$, which is the Padé-$(1,1)$ approximant, maps exactly onto $\mathcal{G}$ for quadratic groups[8] (in fact diagonal Padé approximants are product of Cayley transforms) [18].

4   Constrained Mechanics on Homogeneous Manifolds

It was shown in [6] that HMC on naturally reductive manifolds can be implemented using mechanics on Lie groups. Consider a Homogeneous manifold $\mathcal{G}/\mathcal{K}$ (where $\mathcal{K}$ is a closed subgroup of $\mathcal{G}$) and a Hamiltonian $\tilde{H} = \tilde{V} + \tilde{T}$ on $T(\mathcal{G}/\mathcal{K})$. Let $\pi : \mathcal{G} \to \mathcal{G}/\mathcal{K}$ be the quotient projection. We can lift $\tilde{H}$ to $T\mathcal{G}$ by defining $H \equiv (\partial \pi)^* \tilde{H}$. The lift $T \equiv (\partial \pi)^* \tilde{T}$ defines a degenerate kinetic energy on $T\mathcal{G}$, since it maps all vectors in the vertical space $\mathrm{ver}_g \equiv \ker(\partial_g \pi)$ to zero. The vertical space is described by the action fields $\mathrm{ver}_g = \{\xi_{\mathcal{G}}(g) : \xi \in \mathfrak{k}\}$ where $\xi_{\mathcal{G}}(g) = \frac{\mathrm{d}}{\mathrm{d}t}e^{t\xi} \cdot g|_{t=0} = \partial_1 L_g(\xi)$.

In order to circumvent the issues related to the degeneracy of T, we need to make a choice of horizontal space complementary to ver_g, or equivalently choose a connection on the principal bundle $\mathcal{G} \to \mathcal{G}/\mathcal{K}$. This defines a unique way to lift a curve on $\mathcal{G}/\mathcal{K}$ to a curve on $\mathcal{G}$, and a complementary subspace $\mathfrak{p}$ to $\mathfrak{k}$, i.e., $\mathfrak{g} = \mathfrak{k} \oplus \mathfrak{p}$ (see Sect. 6.11 of the Supplementary material). Then, if (q, v) satisfies the constrained equation

$$\dot{q} = v^j e_j(q) \qquad \frac{\mathrm{d}}{\mathrm{d}t}\left(\frac{\partial T(q,v)}{\partial v}\right) = \mathrm{ad}_v^* \frac{\partial T(q,v)}{\partial v} - L_g^* \mathrm{d}V \qquad v(t) \in \mathfrak{p},$$

the projected curve $\pi(q(t))$ solves Euler-Lagrange equations on $\mathcal{G}/\mathcal{K}$ [23]. We note the constraint $v(t) \in \mathfrak{p}$ can be rewritten in the form $\mathrm{ver}(v(t)) = 0$ where $\mathrm{ver} : \mathfrak{g} \to \mathfrak{k}$ is the vertical projection.

If $\mathcal{G}/\mathcal{K}$ is a reductive homogeneous manifold, there is a one-to-one correspondence between $\mathrm{Ad}_{\mathcal{G}}\mathcal{K}$-invariant inner products on $\mathfrak{p}$ and $\mathcal{G}$-invariant metric on $\mathcal{G}/\mathcal{K}$. In particular if we extend the inner product on $\mathfrak{p}$ to a non-degenerate quadratic form on $\mathfrak{g}$ s.t. $\mathfrak{p} = \mathfrak{k}^{\perp}$, it defines a pseudo-Riemannian metric on $\mathcal{G}$ that is left $\mathcal{G}$-invariant and right $\mathcal{K}$-invariant, and a kinetic energy T with $T = (\partial \pi)^* \tilde{T}$ on $\mathfrak{p}$. The resulting Hamiltonian $H = V + T$ is right $\mathcal{K}$-invariant so by Noether's theorem, if $v(0) \in \mathfrak{p}$, then $v(t) \in \mathfrak{p}$ for all t. This means the constraint $v(t) \in \mathfrak{p}$ is naturally handled by the symmetries of H (see Sect. 6.8 of the Supplementary material).

If $\mathcal{G}/\mathcal{K}$ is further naturally reductive, since $\mathrm{ad}_v^{\top} v = 0$ for $v \in \mathfrak{p}$ the above system becomes

$$\dot{q} = v^j e_j(q) \qquad \dot{v} = -g^{jk} e_j(V)(q)\xi_k \qquad v(0) \in \mathfrak{p},$$

---

[8] By a quadratic group (sometimes called J-orthogonal) we mean a subgroup $\mathcal{G} = \{M \in \mathrm{GL}(n) : M^T J M = J\}$ for some $J \in \mathrm{GL}(n)$.

and we recover the HMC algorithm on naturally reductive homogeneous spaces proposed in [6].

It is clear that integrators that are built by alternating between the flows of V and T are automatically horizontal (the trajectory stays in $\mathcal{G} \times \mathfrak{p}$) since both V and T are $\mathcal{K}$-invariant. This is still true for the force-gradient integrators Sect. 2.5. Indeed, from the $\mathrm{ad}_{\mathfrak{k}}$-invariance of the inner product on $\mathfrak{g}$, we have $c_{rij} = -c_{jir}$ for any $i \in \mathfrak{k}$ (the structure constants always satisfy $c^i_{jk} = -c^i_{kj}$). Hence if V is invariant under the right action of $\mathcal{K}$ on $\mathcal{G}$, then $\{V, \widehat{\{V, T\}}\} = -2g^{jk}g^{ls}e_l(V)e_je_s(V)\xi_k \in \mathfrak{p}$ (see Sect. 6.12 of the Supplementary material).

Acknowledgements. We thank Prof. Anthony Kennedy and Prof. Mark Girolami for useful discussions. AB was supported by a Roth Scholarship funded by the Department of Mathematics, Imperial College London, by EPSRC Fellowship (EP/J016934/3) and by The Alan Turing Institute under the EPSRC grant [EP/N510129/1].

References

1. Abraham, R., Marsden, J.E., Marsden, J.E.: Foundations of Mechanics, vol. 36. Benjamin/Cummings Publishing Company Reading, Massachusetts (1978)
2. Alekseevsky, D., Grabowski, J., Marmo, G., Michor, P.W.: Poisson structures on the cotangent bundle of a Lie group or a principle bundle and their reductions. J. Math. Phys. **35**(9), 4909–4927 (1994)
3. Alekseevsky, D., Arvanitoyeorgos, A.: Riemannian flag manifolds with homogeneous geodesics. Trans. Am. Math. Soc. **359**(8), 3769–3789 (2007)
4. Arnaudon, A., Barp, A., Takao, S.: Irreversible Langevin MCMC on Lie groups. arXiv preprint arXiv:1903.08939 (2019)
5. Barp, A., Briol, F.X., Kennedy, A.D., Girolami, M.: Geometry and dynamics for Markov chain Monte Carlo. Annu. Rev. Stat. Appl. (2018). https://doi.org/10.1146/annurev-statistics-031017-100141
6. Barp, A., Kennedy, A.D., Girolami, M.: Hamiltonian Monte Carlo on symmetric and homogeneous spaces via symplectic reduction. arXiv preprint arXiv:1903.02699 (2019)
7. Betancourt, M.: A conceptual introduction to Hamiltonian Monte Carlo (2017), preprint arXiv:1701.02434
8. Betancourt, M., Byrne, S., Livingstone, S., Girolami, M., et al.: The geometric foundations of Hamiltonian Monte Carlo. Bernoulli **23**(4A), 2257–2298 (2017). https://doi.org/10.3150/16-BEJ810
9. Bou-Rabee, N., Marsden, J.E.: Hamilton–pontryagin integrators on lie groups Part I: introduction and structure-preserving properties. Found. Comput. Math. **9**(2), 197–219 (2009)
10. Bou-Rabee, N., Sanz-Serna, J.M.: Geometric integrators and the Hamiltonian Monte Carlo method. Acta Numer. **27**, 113–206 (2018)
11. Bryant, R.L.: An introduction to Lie groups and symplectic geometry. Geom. Quantum Field Theory **1**, 321–347 (1995)
12. Byrne, S., Girolami, M.: Geodesic Monte Carlo on embedded manifolds. Scand. J. Stat. **40**(4), 825–845 (2013)
13. Celledoni, E., Marthinsen, H., Owren, B.: An introduction to Lie group integrators– basics, new developments and applications. J. Comput. Phys. **257**, 1040–1061 (2014)

14. Clark, M.A., Kennedy, A.D., Silva, P.: Tuning HMC using poisson brackets. arXiv preprint arXiv:0810.1315 (2008)
15. Del Castillo, G.F.T.: Differentiable Manifolds: A Theoretical Physics Approach. Springer, Boston (2011). https://doi.org/10.1007/978-0-8176-8271-2
16. Duane, S., Kennedy, A.D., Pendleton, B.J., Roweth, D.: Hybrid Monte Carlo. Phys. Lett. B **195**(2), 216–222 (1987)
17. Holbrook, A., Lan, S., Vandenberg-Rodes, A., Shahbaba, B.: Geodesic Lagrangian Monte Carlo over the space of positive definite matrices: with application to bayesian spectral density estimation. J. Stat. Comput. Simul. **88**(5), 982–1002 (2018)
18. Iserles, A.: On cayley-transform methods for the discretization of lie-group equations. Found. Comput. Math. **1**(2), 129–160 (2001)
19. Jost, J.: Riemannian Geometry and Geometric Analysis, vol. 42005. Springer, Heidelberg (2008). https://doi.org/10.1007/978-3-540-77341-2
20. Kennedy, A.D., Rossi, P.: Classical mechanics on group manifolds. Nucl. Phys. B **327**, 782–790 (1989). https://doi.org/16/0550-3213(89)90315-5
21. Kennedy, A.D., Silva, P.J., Clark, M.A.: Shadow Hamiltonians, poisson brackets, and gauge theories. Phys. Rev. D **87**(3), 034511 (2013). https://doi.org/10.1103/PhysRevD.87.034511
22. Knechtli, F., Günther, M., Peardon, M.: Lattice Quantum Chromodynamics: Practical Essentials. Springer, Dordrecht (2017). https://doi.org/10.1007/978-94-024-0999-4
23. Lee, T., Leok, M., McClamroch, N.H.: Global Formulations of Lagrangian and Hamiltonian Dynamics on Manifolds. Springer, Cham (2017). https://doi.org/10.1007/978-3-319-56953-6
24. Leimkuhler, B., Patrick, G.W.: A symplectic integrator for Riemannian manifolds. J. Nonlinear Sci. (1996). https://doi.org/10.1007/BF02433475
25. Miolane, N., Pennec, X.: Computing bi-invariant pseudo-metrics on lie groups for consistent statistics. Entropy **17**(4), 1850–1881 (2015)
26. Modin, K., Perlmutter, M., Marsland, S., McLachlan, R.: Geodesics on Lie groups: Euler equations and totally geodesic subgroup (2010)
27. Nielsen, F., Bhatia, R.: Matrix Information Geometry. Springer, Heidelberg (2012). https://doi.org/10.1007/978-3-642-30232-9
28. Omelyan, I., Mryglod, I., Folk, R.: Algorithm for molecular dynamics simulations of spin liquids. Phys. Rev. Lett. **86**(5), 898 (2001)
29. Takaishi, T., De Forcrand, P.: Testing and tuning symplectic integrators for the hybrid Monte Carlo algorithm in lattice QCD. Phys. Rev. E **73**(3), 036706 (2006)

On the Fisher-Rao Information Metric in the Space of Normal Distributions

Julianna Pinele[1]([⊠]) [ID], Sueli I. R. Costa[2] [ID], and João E. Strapasson[3] [ID]

[1] CETEC, Federal University of Recôncavo of Bahia, Cruz das Almas, Brazil
`julianna.pinele@ufrb.edu.br`
[2] Institute of Mathematics, University of Campinas, Campinas, Brazil
`sueli@ime.unicamp.br`
[3] School of Applied Sciences, University of Campinas, Campinas, Brazil
`joao.strapasson@fca.unicamp.br`

Abstract. The Fisher-Rao distance between two probability distribution functions, as well as other divergence measures, is related to entropy and is in the core of the research area of information geometry. It can provide a framework and enlarge the perspective of analysis for a wide variety of domains, such as statistical inference, image processing (texture classification and inpainting), clustering processes and morphological classification. We present here a compact summary of results regarding the Fisher-Rao distance in the space of multivariate normal distributions including some historical background, closed forms in special cases, bounds, numerical approaches and references to recent applications.

Keywords: Fisher-Rao distance · Information geometry · Multivariate normal distributions

1 Introduction

The Fisher-Rao distance is a measure of dissimilarity between two probability distributions and, as well as other divergence measures, has had applications in several areas. Let us start with the classical example of the univariate normal distributions with media $\mu \in \mathbb{R}$ and standard deviation $\sigma \in (0, +\infty)$,

$$p(x; \mu, \sigma) = \frac{1}{\sqrt{2\pi}\sigma} \exp\left(-\frac{1}{2}\left(\frac{x-\mu}{\sigma}\right)^2\right), \quad x \in \mathbb{R}. \tag{1}$$

Figure 1 illustrates a comparison between four univariate normal distributions represented by their (μ, σ) pairs. We can see on the left that the dissimilarity between probability distributions associated with C and D is smaller than the one between the distributions associated with A and B. This means that a proper distance between points in the parameter space given by the media-standard deviation plane (right) representing those normal distributions cannot be Euclidean [1].

Supported by FAPESP (13/25997-7) and CNPq (313326/2017-7) foundations.

F. Nielsen and F. Barbaresco (Eds.): GSI 2019, LNCS 11712, pp. 676–684, 2019.
https://doi.org/10.1007/978-3-030-26980-7_70

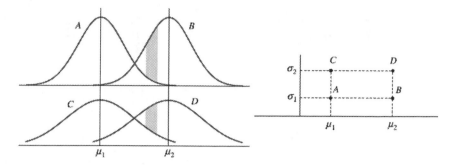

Fig. 1. Univariate normal distributions and their representations in the (μ, σ) plane.

After some previous papers (Mahalanobis [2], Bhattacharyya [3], Hotelling [4]) connecting geometry and statistics, Rao [5] in the search for an adequate measure to determine the distance between two populations, introduced differential geometry methods in the study of a space composed by parameters of probability distributions (statistical models). In this landmark paper he consider the statistical models with the Riemannian metric induced by the information matrix defined by Fisher in 1921 [6]. The geodesic distance in this model is usually called Fisher distance or, as in this paper, the Fisher-Rao distance.

Given a statistical model $S = \{p_\theta = p(x; \boldsymbol{\theta}); \boldsymbol{\theta} = (\theta_1, \theta_2, \ldots, \theta_m) \in \Theta\}$, a natural Riemannian structure [17] can be provided by the Fisher information matrix:

$$g_{ij}(\boldsymbol{\theta}) = \int \frac{\partial}{\partial \theta_i} \log p(x; \boldsymbol{\theta}) \frac{\partial}{\partial \theta_j} \log p(x; \boldsymbol{\theta}) p(x; \boldsymbol{\theta}) dx, \tag{2}$$

This matrix can also be viewed as the Hessian matrix of the Shannon Entropy

$$H(p) = -\int p(x; \boldsymbol{\theta}) \log p(x; \boldsymbol{\theta}) dx \tag{3}$$

and is used to establish connections between inequalities in information theory and geometrical inequalities.

The Fisher-Rao distance, d_F, between two distributions $\boldsymbol{\theta}_1$ and $\boldsymbol{\theta}_2$ is given by the shortest length of a curve $\gamma(t)$ in the parameter space Θ connecting $\boldsymbol{\theta}_1$ and $\boldsymbol{\theta}_2$. Such a curve is called a geodesic and is given by the solutions of the Euler-Lagrange equations

$$\frac{d^2\theta_k}{dt^2} + \sum_{i,j} \Gamma_{ij}^k \frac{d\theta_i}{dt} \frac{d\theta_j}{dt} = 0, \quad k = 1, \cdots, m, \tag{4}$$

where Γ_{ij}^k are the Christoffel symbols attached to this metric.

In the Fisher-Rao model of univariate normal distributions parametrized by (μ, σ), the Fisher matrix is

$$\begin{pmatrix} \frac{1}{\sigma^2} & 0 \\ 0 & \frac{2}{\sigma^2} \end{pmatrix} \tag{5}$$

and a closed form for the Fisher-Rao distance, d_F, in this case is known via an association with the classical model of the hyperbolic plane [1,5,7,8].

An analytic expression for d_F is

$$d_F((\mu_1, \sigma_1), (\mu_2, \sigma_2)) = \sqrt{2} \cosh^{-1} \left(1 + \frac{\left| \left(\frac{\mu_1}{\sqrt{2}}, \sigma_1\right) - \left(\frac{\mu_2}{\sqrt{2}}, \sigma_2\right) \right|^2}{2\sigma_1 \sigma_2} \right) \qquad (6)$$

where $|.|$ is the standard Euclidean norm in $\mathbb{R}^2$. The geodesics in this half-plane model are half-vertical lines and half-ellipses centered at $\sigma = 0$, with eccentricity $\frac{1}{\sqrt{2}}$, [1]. Connections between the Fisher-Rao distance and the well-know Kullback-Leibler divergence of univariate normals can also be devised [1].

The Fisher-Rao distance in the space of univariate normal distributions was applied to quantization of hyperspectral images [9] and to the space of projected lines in the paracatadioptric image [10]. In [11], this Fisher-Rao model was used to simplify Gaussian mixtures through the k-means method and in [12] to a hierarchical clustering technique.

The approach of Rao [5] for distances between univariate normal distributions encourage several authors to use geometric tools in the study of statistical models and the calculus of the Fisher-Rao distance between other probability distributions as well as stimulated the approach of other dissimilarity measures [7,8,13]. In 1975, Efron [14] introduced the curvature concept in statistical models and was followed by important contributions of Dawid [15] and Amari-Nakaoga [16,17] who unified the information geometry theory by introducing other concepts of connection in statistical models. In an independent work, Chentsov [18] proved that the Fisher-Rao information metric is, up to a scaling factor, the unique yielding statistical invariance under Markov morphisms [19]. This metric has been considered in several statistical models aiming applications to different fields [19,20].

In this paper we gathered results and applications of the Fisher-Rao distance in the statistical model composed by multivariate normal distributions which has also been the focus of our previous papers [1,21–23]. In this space a closed form for the Fisher-Rao distance is not yet known. Section 2 is devoted to main concepts and important results to approach this distance for general normal distributions. In Sect. 3 we collect closed forms for special submanifolds and related applications. Bounds for the Fisher-Rao distance regarding general normal distributions are described in Sect. 4.

2   The Fisher-Rao Distance on Multivariated Normal Distribution Space

We consider here multivariate normal distributions given by

$$p(\boldsymbol{x}; \boldsymbol{\mu}, \Sigma) = \frac{(2\pi)^{-\left(\frac{n}{2}\right)}}{\sqrt{Det(\Sigma)}} \exp \left(-\frac{(\boldsymbol{x} - \boldsymbol{\mu})^t \Sigma^{-1} (\boldsymbol{x} - \boldsymbol{\mu})}{2} \right), \qquad (7)$$

where $\boldsymbol{x}^t = (x_1, \cdots, x_n) \in \mathbb{R}^n$ is the variable vector, $\boldsymbol{\mu}^t = (\mu_1, \cdots, \mu_n) \in \mathbb{R}^n$ is the mean vector and $\Sigma \in P_n(\mathbb{R})$ is the covariance matrix ($P_n(\mathbb{R})$ is the space of order n positive definite symmetric matrices). $\mathcal{M} = \{p_\theta; \boldsymbol{\theta} \in \mathbb{R}^n \times P_n(\mathbb{R})\}$ is the statistical $\left(n + \frac{n(n+1)}{2}\right)$-dimensional model composed by these distributions.

We observe that, for any $(\boldsymbol{c}, Q) \in \mathbb{R}^n \times GL_n(\mathbb{R})$, where $GL_n(\mathbb{R})$ is the space of non-singular order n matrices, the mapping $(\boldsymbol{\mu}, \Sigma) \mapsto (Q\boldsymbol{\mu} + \boldsymbol{c}, Q\Sigma Q^t)$ is an isometry in $\mathcal{M}$ [7]. Consequently, the Fisher-Rao distance between $\boldsymbol{\theta}_1 = (\boldsymbol{\mu}_1, \Sigma_1)$ and $\boldsymbol{\theta}_2 = (\boldsymbol{\mu}_2, \Sigma_2)$ in $\mathcal{M}$ satisfies

$$d_F(\boldsymbol{\theta}_1, \boldsymbol{\theta}_2) = d_F((Q\boldsymbol{\mu}_1 + \boldsymbol{c}, Q\Sigma_1 Q^t), (Q\boldsymbol{\mu}_2 + \boldsymbol{c}, Q\Sigma_2 Q^t)), \tag{8}$$

for any $(\boldsymbol{c}, Q) \in \mathbb{R}^n \times GL_n(\mathbb{R})$. In particular, taking $Q = \Sigma_1^{-(1/2)}$ and $\boldsymbol{c} = -\Sigma_1^{(-1/2)}\boldsymbol{\mu}_1$, $\boldsymbol{\theta} = (\boldsymbol{\mu}, \Sigma) = (\Sigma_1^{-(1/2)}(\boldsymbol{\mu}_2 - \boldsymbol{\mu}_1), \Sigma_1^{-(1/2)}\Sigma_2\Sigma_1^{-(1/2)})$ the Fisher-Rao distance admits the form

$$d_F(\boldsymbol{\theta}_1, \boldsymbol{\theta}_2) = d_F(\boldsymbol{\theta}_0, \boldsymbol{\theta}), \tag{9}$$

where $\boldsymbol{\theta}_0 = (\boldsymbol{0}, I_n)$, I_n is the n-order identity matrix and $\boldsymbol{0} \in \mathbb{R}^n$ is the null vector.

The geodesic equations in $\mathcal{M}$ can be expressed as [13]

$$\begin{cases} \frac{d^2\mu}{dt^2} - \left(\frac{d\Sigma}{dt}\right)\Sigma^{-1}\left(\frac{d\mu}{dt}\right) = 0 \\ \frac{d^2\Sigma}{dt^2} + \left(\frac{d\mu}{dt}\right)\left(\frac{d\mu}{dt}\right)^t - \left(\frac{d\Sigma}{dt}\right)\Sigma^{-1}\left(\frac{d\Sigma}{dt}\right) = 0. \end{cases} \tag{10}$$

These equations could be partially integrated and, under certain initial conditions, Eriksen [24] and Calvo and Oller [25] solved them, independently.

Han and Park in [26] proposed a numerically shooting method for computing the minimum geodesic distance between two normal distributions, through parallel transport of a vector field defined along of a geodesic. They applied their method for the segmentation of diffusion tensor magnetic resonance images.

In general, a closed form for the Fisher-Rao distance between normal distributions is known only for the univariate case and for some submanifolds as described next.

3   Closed Forms for Special Submanifolds and Related Applications

We collect closed forms which are known for special submanifolds of $\mathcal{M}$. One important aspect to be discussed here is when the considered submanifold is totally geodesic, that is, when the distance restricted to the submanifold is the same that the distance in $\mathcal{M}$.

3.1 The Submanifold $\mathcal{M}_\Sigma$ Where Σ Is Constant

In the n-dimensional manifold composed by multivariate normal distributions with common covariance matrix Σ, $\mathcal{M}_\Sigma = \{p_\theta; \theta = (\mu, \Sigma), \; \Sigma = \Sigma_0 \in P_n(\mathbb{R})$ constant$\}$, the Fisher-Rao distance is equal to the Mahalanobis distance [2], d_Σ, which and for $\theta_1 = (\mu_1, \Sigma_0)$ and $\theta_2 = (\mu_2, \Sigma_0)$ is defined as

$$d_\Sigma(\theta_1, \theta_2) = \sqrt{(\mu_1 - \mu_2)^t \Sigma_0^{-1} (\mu_1 - \mu_2)}. \tag{11}$$

This distance was one of the first dissimilarity measure between data sets with some correlation.

We remark that this submanifold is not totally geodesic. This can be seen even in the space of univariate normal distributions described in Sect. 1. For $\sigma = \sigma_0$, $d_F((\mu_1, \sigma_0), (\mu_2, \sigma_0)) < d_\sigma((\mu_1, \sigma_0), (\mu_2, \sigma_0)) = |\mu_1 - \mu_2|/\sigma_0$ [1].

3.2 The Submanifold $\mathcal{M}_\mu$ Where μ Is Constant

In 1976, S. T. Jensen deduced a closed expression for the Fisher-Rao distance in the $\frac{n(n+1)}{2}$-dimensional totally geodesic submanifold $\mathcal{M}_\mu = \{p_\theta; \theta = (\mu, \Sigma), \; \mu = \mu_0 \in \mathbb{R}^n$ constant$\}$, composed by distributions which have the same mean vector μ [8]. We denote $d_F(,)$ as d_μ,

$$d_\mu((\mu, \Sigma_1), (\mu, \Sigma_2)) = \sqrt{\frac{1}{2} \sum_{i=1}^n [\log(\lambda_i)]^2}, \tag{12}$$

where $0 < \lambda_1 \leq \lambda_2 \leq \cdots \leq \lambda_n$ are the eigenvalues of $\Sigma_1^{-1/2} \Sigma_2 \Sigma_1^{-1/2}$.

This submanifold was considered in the analysis and processing of diffusion tensor images (DTI) [27,28]. In [29], the authors used this submanifold to analyze color texture discrimination in several classification experiments.

3.3 The Submanifold $\mathcal{M}_D$ Where Σ Is Diagonal

Let $\mathcal{M}_D = \{p_\theta; \; \theta = (\mu, \Sigma), \; \Sigma = diag(\sigma_1^2, \sigma_2^2, \cdots, \sigma_n^2), \; \sigma_i > 0, \; i = 1, \cdots, n\}$, the submanifold of $\mathcal{M}$ composed by distributions with diagonal covariance matrix. If we consider the parameter $\theta = (\mu_1, \sigma_1, \mu_2, \sigma_2, \cdots, \mu_n, \sigma_n)$, it can be shown [1] that the metric in the parametric space of $\mathcal{M}_D$ is the product metric of the univariate Gaussian distributions:

$$d_D(\theta_1, \theta_2) = \sqrt{\sum_{i=1}^n d_F^2((\mu_{1i}, \sigma_{1i}), (\mu_{2i}, \sigma_{2i}))}, \tag{13}$$

where d_F is the distance given in (6). In this space, a curve $\alpha(t) = (\alpha_1(t), \cdots, \alpha_n(t))$ is a geodesic if and only if, $\alpha_i(t), \forall i$, is a geodesic in $\mathbb{H}_F^2$.

In [30] the authors described shapes representing each landmark by a Gaussian model and considering the Fisher-Rao distance in $\mathcal{M}_D$ to quantifying the difference between two shapes and in [13] Skovgaard applied this model in statistical inference. The Fisher-Rao distance in $\mathcal{M}_D$ also was used in [23] to simplify of Gaussian mixtures models similarly of what was done in [31] with the Bregman divergence. A comparison between these two mixture simplifications can be done by using the recent results of [32] which provides a closed formula for the Minkowski distance for Gaussian mixture models.

It is important to note that $\mathcal{M}_D \subset \mathcal{M}$ is not totally geodesic. The submanifold of $\mathcal{M}_D$ composed only by normal distributions with covariance matrices which are multiples of the identity (round normals) is totally geodesic. In fact, this is a special case of the submanifold described next.

3.4   The Submanifold $\mathcal{M}_{D\mu}$ Where Σ Is Diagonal and μ Is an Eigenvector of Σ

Let the $n+1$-dimensional submanifold composed by distributions with the mean vector $\boldsymbol{\mu} = \mu_1 \mathbf{e}_i$ and diagonal covariance matrix Σ. Without loss of generality we shall assume that $e_i = e_1$.

An analytic expression for the distance in $\mathcal{M}_{D\mu}$ is

$$d_{D\mu}^2(\boldsymbol{\theta}_1, \boldsymbol{\theta}_2) = d_F^2\left((\mu_{11}, \sigma_{11}), (\mu_{21}, \sigma_{21})\right) + \sum_{i=2}^{n} d_F^2\left((0, \sigma_{1i}), (0, \sigma_{2i})\right). \quad (14)$$

The distance in this submanifold was derived in [21] and it was proved that $\mathcal{M}_{D\mu}$ is a totally geodesic submanifold of $\mathcal{M}$. By considering this submanifold it is possible to calculate the distance in $\mathcal{M}$ between two distributions with the same covariance matrix.

4   Bounds for the Fisher-Rao Distance

Some bounds for the Fisher-Rao distance between two general normal distributions were derived in [21, 22, 25, 33].

A Lower Bound. In 1990 Calvo and Oller [33] derived a lower bound for the Fisher-Rao distance through an isometric embedding of the parametric space $\mathcal{M}$ into the manifold of the positive definite matrices. Given $\boldsymbol{\theta}_1 = (\boldsymbol{\mu}_1, \Sigma_1)$ and $\boldsymbol{\theta}_2 = (\boldsymbol{\mu}_2, \Sigma_2)$, let

$$S_i = \begin{pmatrix} \Sigma_i + \boldsymbol{\mu}_i^t \boldsymbol{\mu}_i & \boldsymbol{\mu}_i^t \\ \boldsymbol{\mu}_i & 1 \end{pmatrix}, \quad (15)$$

$i = 1, 2$. A lower bound for the distance between $\boldsymbol{\theta}_1$ and $\boldsymbol{\theta}_2$ is

$$LB(\boldsymbol{\theta}_1, \boldsymbol{\theta}_2) = \sqrt{\frac{1}{2} \sum_{i=1}^{n} [log(\lambda_i)]^2}, \quad (16)$$

where λ_k, $1 \leq k \leq n+1$, are the eigenvalues of $S_1^{-1/2} S_2 S_1^{-1/2}$.

Upper Bounds. In [21] we have proposed an upper bound based on an isometry in the manifold $\mathcal{M}$ and on the distance in the submanifold $\mathcal{M}_D$, as follows: The Fisher-Rao distance between two multivariate normal distributions $\boldsymbol{\theta}_1 = (\boldsymbol{\mu}_1, \Sigma_1)$ and $\boldsymbol{\theta}_2 = (\boldsymbol{\mu}_2, \Sigma_2)$ is upper bounded by,

$$U1(\boldsymbol{\theta}_1, \boldsymbol{\theta}_2) = \sqrt{\sum_{i=1}^{n} d_F^2((0,1),(\mu_i, \lambda_i))}, \tag{17}$$

where, λ_i are the diagonal terms of the matrix Λ given by the eigenvalues of $A = \Sigma_1^{-(1/2)} \Sigma_2 \Sigma_1^{-(1/2)} = Q\Lambda Q^t$ and μ_i are the coordinates of $\boldsymbol{\mu} = Q^t \Sigma_1^{-(1/2)}(\boldsymbol{\mu}_2 - \boldsymbol{\mu}_1)$, where Q is the orthogonal matrix whose columns are the respective eigenvectors of A.

Another upper bound was derived in [22]. Given $\boldsymbol{\theta}_1 = (\boldsymbol{\mu}_1, \Sigma_1)$ and $\boldsymbol{\theta}_2 = (\boldsymbol{\mu}_2, \Sigma_2)$, by Eq. (9) we can consider the distance between $\boldsymbol{\theta}_0 = (\mathbf{0}, I_n)$ and $\boldsymbol{\theta} = (\boldsymbol{\mu}, \Sigma)$. Let P be an orthogonal matrix such that $P\boldsymbol{\mu} = (|\boldsymbol{\mu}|, 0, \cdots, 0)$ and $D = \text{diag}(\sqrt{(|\boldsymbol{\mu}| + 2)/2}, 1, \cdots, 1)$ a diagonal matrix. Taking $\boldsymbol{\theta}_P = (P\boldsymbol{\mu}, D)$ and $\bar{\boldsymbol{\theta}} = (\boldsymbol{\mu}, \bar{\Sigma})$, with $\bar{\Sigma} = P^{-1}DP^{-t}$, the bound is given by

$$U2(\boldsymbol{\theta}_1, \boldsymbol{\theta}_2) = d_{D\boldsymbol{\mu}}(\boldsymbol{\theta}_0, \boldsymbol{\theta}_P) + d_{\boldsymbol{\mu}}(\bar{\boldsymbol{\theta}}, \boldsymbol{\theta}). \tag{18}$$

From some comparisons in [22] it is possible to note that in some cases these bounds are very tight. In [34] the upper bound U1 was used to tracking quality monitoring.

5 Concluding Remarks

In this paper we approach the statistical model of multivariate normal distributions with the Fisher-Rao information distance, summarizing results and applications with the aim of widening the range of possible interpretations and the use of this Riemannian manifold for data analysis in different contexts.

The Fisher-Rao distance between probability distributions has important characteristics such as its strong connection with the Shannon entropy and its uniqueness regarding statistical invariance under Markov morphisms. On the other hand, as described here, even for normal multivariate distributions closed forms for this distance is only known in special cases, what requires the use of bounds and may demand somehow high computational cost. As pointed out in [32], selecting a proper divergence measure for a specific context is usually a difficult task involving the classical trade-off between accuracy and computational complexity.

References

1. Costa, S.I.R., Santos, S.A., Strapasson, J.E.: Fisher information distance: a geometrical reading. Discret. Appl. Math. **197**, 59–69 (2015)

2. Mahalanobis, P.C.: On the generalized distance in statistics. Proc. Natl. Inst. Sci. **2**, 49–55 (1936)
3. Bhattacharyya, A.: On a measure of divergence between two statistical populations defined by their probability distributions. Bull. Calcutta Math. Soc. **35**, 99–110 (1943)
4. Hotelling, H.: Spaces of statistical parameters. Bull. Am. Math. Soc. (AMS) **36**, 191 (1930)
5. Rao, C.R.: Information and the accuracy attainable in the estimation of statistical parameters. Bull. Calcutta Math. Soc. **37**, 81–91 (1945)
6. Fisher, R.A.: On the mathematical foundations of theoretical statistics. Philos. Trans. R. Soc. Lond. **222**, 309–368 (1921)
7. Burbea, J.: Informative geometry of probability spaces. Expositiones Mathematica **4**, 347–378 (1986)
8. Atkinson, C., Mitchell, A.F.S.: Rao's distance measure. Samkhyā Indian J. Stat. **43**, 345–365 (1981)
9. Angulo, J., Velasco-Forero, S.: Morphological processing of univariate Gaussian distribution-valued images based on Poincaré upper-half plane representation. In: Nielsen, F. (ed.) Geometric Theory of Information, pp. 331–366. Springer, Cham (2014). https://doi.org/10.1007/978-3-319-05317-2_12
10. Maybank, S.J., Ieng, S., Benosman, R.: A Fisher-Rao metric for paracatadioptric images of lines. Int. J. Comput. Vis. **99**(2), 147–165 (2012)
11. Schwander, O., Nielsen, F.: Model centroids for the simplification of kernel density estimators. In: IEEE - Acoustics, Speech and Signal Processing (ICASSP) (2012)
12. Taylor, S.: Clustering financial return distributions using the Fisher information metric. Entropy **21**(2), 110 (2019)
13. Skovgaard, L.T.: A Riemannian geometry of the multivariate normal model. Scand. J. Stat. **11**, 211–223 (1984)
14. Efron, B.: Defining the curvature of a statistical problem (with applications to second order efficiency). Ann. Stat. **3**, 1189–1242 (1975)
15. Dawid, A.P.: Discussions to Efron's paper. Ann. Stat. **3**, 1231–1234 (1975)
16. Amari, S., Nagaoka, H.: Differential Geometrical Methods in Statistics. Lecture Notes in Statistics, vol. 28. Springer, New York Heidelberg (1986). https://doi.org/10.1007/978-1-4612-5056-2
17. Amari, S., Nagaoka, H.: Methods of Information Geometry. Translations of Mathematical Monographs, vol. 191. American Mathematical Society, New York (2000)
18. Chentsov, N.N.: Statistical Decision Rules and Optimal Inference, vol. 53. AMS Bookstore, New York (1982)
19. Nielsen, F.: An elementary introduction to information geometry. arXiv preprint arXiv:1808.08271 (2018)
20. Amari, S.: Information Geometry and Its Applications, vol. 194. Springer, Tokyo (2016). https://doi.org/10.1007/978-4-431-55978-8
21. Strapasson, J.E., Porto, J., Costa, S.I.R.: On bounds for the Fisher-Rao distance between multivariate normal distributions. In: Bayesian Inference and Maximum Entropy Methods in Science and Engineering (MAXENT 2014). AIP, vol. 1641 (2015)
22. Strapasson, J.E., Pinele, J., Costa, S.I.R.: A totally geodesic submanifold of the multivariate normal distributions and bounds for the Fisher-Rao distance. In: IEEE Information Theory Workshop (ITW) (2016)
23. Strapasson, J.E., Pinele, J., Costa, S.I.R.: Clustering using the Fisher-Rao distance. In: Sensor Array and Multichannel Signal Processing Workshop. IEEE (2016)

24. Eriksen, P.S.: Geodesics connected with the fischer metric on the multivariate normal manifold. Aalborg University Centre, Institute of Electronic Systems (1986)
25. Calvo, M., Oller, J.M.: An explicit solution of information geodesic equations for the multivariate normal model. Stat. Decis. **9**, 119–138 (1991)
26. Han, M., Park, F.C.: DTI segmentation and fiber tracking using metrics on multivariate normal distributions. J. Math. Imaging Vis. **49**(2), 317–334 (2014)
27. Lenglet, C., Rousson, M., Deriche, R., Faugeras, O.: Statistics on the manifold of multivariate normal distributions: theory and application to diffusion tensor MRI processing. J. Math. Imaging Vis. **25**(3), 423–444 (2006)
28. Moakher, M., Mourad, Z.: The Riemannian geometry of the space of positive-definite matrices and its application to the regularization of positive-definite matrix-valued data. J. Math. Imaging Vis. **40**(2), 171–187 (2011)
29. Verdoolaege, G., Scheunders, P.: Geodesics on the manifold of multivariate generalized Gaussian distributions with an application to multicomponent texture discrimination. Int. J. Comput. Vis. **95**(3), 265 (2011)
30. Gattone, S., et al.: On the geodesic distance in shapes K-means clustering. Entropy **20**(9), 647 (2018)
31. Garcia, V., Nielsen, F.: Simplification and hierarchical representations of mixtures of exponential families. Signal Process. **90**(12), 3197–3212 (2010)
32. Nielsen, F., The statistical Minkowski distances: closed-form formula for Gaussian mixture models. arXiv preprint arXiv:1901.03732 (2019)
33. Calvo, M., Oller, J.M.: A distance between multivariate normal distributions based in an embedding into the Siegel group. J. Multivar. Anal. **35**(2), 223–242 (1990)
34. Pilté, M., Barbaresco, F.: Tracking quality monitoring based on information geometry and geodesic shooting. In: Radar Symposium (IRS). IEEE (2016)

Simulation of Conditioned Diffusions on the Flat Torus

Mathias Højgaard Jensen$^{(\boxtimes)}$, Anton Mallasto, and Stefan Sommer

Department of Computer Science, Copenhagen University, Copenhagen, Denmark
{matje,mallasto,sommer}@di.ku.dk

Abstract. Diffusion processes are fundamental in modelling stochastic dynamics in natural sciences. Recently, simulating such processes on complicated geometries has found applications for example in biology, where toroidal data arises naturally when studying the backbone of protein sequences, creating a demand for efficient sampling methods. In this paper, we propose a method for simulating diffusions on the flat torus, conditioned on hitting a terminal point after a fixed time, by considering a diffusion process in $\mathbb{R}^2$ which we project onto the torus. We contribute a convergence result for this diffusion process, translating into convergence of the projected process to the terminal point on the torus. We also show that under a suitable change of measure, the Euclidean diffusion is locally a Brownian motion.

Keywords: Simulation · Conditioned diffusion · Manifold diffusion · Flat Torus

1 Introduction

Stochastic differential equations are ubiquitous in models describing evolution of dynamical systems with, e.g. in modelling the evolution of DNA or protein structure, in pricing financial derivatives, or for modelling changes in landmark configurations which are essential in shape analysis and computational anatomy. In settings where the beginning and end values are known on some fixed time interval, the use of Brownian bridges becomes natural to evaluate the uncertainty on the intermediate time interval.

When the data elements are elements of non-linear spaces, here differentiable manifolds, methodology for simulating bridge processes is lacking. In particular, in cases where the transition probability densities are intractable, it is of interest to use simulation schemes that can numerically approximate the true densities. In this paper we propose a method for simulating diffusion bridges on the flat torus, $\mathbb{T}^2 = \mathbb{R}^2/\mathbb{Z}^2$, i.e. we propose a process that can easily be simulated and satisfies that the distribution of the true bridge of interest is absolutely continuous with respect to the distribution of this proposal process. This specific

---

MHJ, AM, and SS are supported by the CSGB Centre for Stochastic Geometry and Advanced Bioimaging funded by a grant from the Villum Foundation.

© Springer Nature Switzerland AG 2019
F. Nielsen and F. Barbaresco (Eds.): GSI 2019, LNCS 11712, pp. 685–694, 2019.
https://doi.org/10.1007/978-3-030-26980-7_71

case will serve as an example of the more general setting of simulating diffusion bridge processes on Riemannian manifolds. Because of the non-trivial topology of the torus $\mathbb{T}^2$, the conditioned process will be equivalent to a process in $\mathbb{R}^2$ that is conditioned on ending up in a set of points. Therefore, we will address the question of conditioning a process on infinitely many points. Secondly, we will handle the case when the process crosses the cut locus of the target point, i.e. the set of points with no unique distance minimizing geodesic.

It is a basic consequence of Doob's h-transform that the distribution of a conditioned diffusion process is the same as another diffusion process with the drift depending on the transition density. However, as mentioned in [1], using this transform directly is undesirable for simulation purposes as the transition density is often intractable. Instead, the authors introduce a diffusion process which can easily be simulated and with the property that the distribution of the true conditioned diffusion is absolutely continuous wrt. the diffusion used for simulation. We here use this approach that in [1] covers the Euclidean case as the starting point for developing a simulation scheme on the torus.

Recent papers have considered diffusion processes on the torus, for example, Langevin diffusions on the torus were studied in [3] and [4], in the latter to describe protein evolution. In this paper, we introduce a diffusion process in $\mathbb{R}^2$ which can easily be simulated and projects onto a bridge process on the torus. More generally, Brownian bridges on manifolds have been studied for example in the context of landmark manifolds [9] and used for approximating the transition density of the Brownian motion. The present paper uses bridges on the flat torus to exemplify how some of the challenges of bridge simulation on Riemannian manifolds can be addressed, here in particular non-trivial topology of the manifold.

We begin in Sect. 2 with a short introduction to Brownian bridge processes in the standard Eucliden case and how it relates to the definition of a Brownian bridge process on the flat torus. At the end we introduce the stochastic differential equation (SDE) which will be used for simulating the bridge process. In Sect. 3 we argue that a strong solution of our proposed SDE exist. We show results about convergence and absolute continuity in Sect. 4. Numerical examples are presented in Sect. 5.

2 Theoretical Setup

This section will briefly review some Brownian bridges theory and discuss the torus case. A more general theory of diffusion bridges can be found in [1], constituting the main reference for this work. At the end, we introduce our proposal process.

Consider a Brownian motion $W = (W_t)_{t\geq 0}$ in $\mathbb{R}^n$. By conditioning, it can be shown that W will end up at a given point at a given time. For example, the process given by $B_t = W_t - \frac{t}{T}W_T$ defines a Brownian bridge conditioned to return to 0 at time T. It can be shown that the diffusion process given by

$$dX_t = \frac{b - X_t}{T - t}dt + dW_t; \quad 0 \leq t < T \quad \text{and} \quad X_0 = a, \tag{1}$$

for given $a, b \in \mathbb{R}^d$ and W a d-dimensional standard Brownian motion, is a d-dimensional Brownian bridge from a to b on $[0, T]$ (see e.g. [6, sec. 5.6]). More generally, diffusion bridges can be defined through Doob's h-transform, that is, the distribution of a diffusion

$$dX_t = b(t, X_t)dt + \sigma(t, X_t)dW_t, \qquad X_0 = a,$$

conditioned on $X_T = b$ is the same as that of

$$dY_t = \tilde{b}(t, Y_t)dt + \sigma(t, Y_t)dW_t,$$
$$\tilde{b}(t, x) = b(t, x) + \sigma(t, x)\sigma^T(t, x)\nabla_x \log(p(t, x; T, b)),$$

where $p(t, x; T, b)$ denotes the transition density of the process X. In the usual setting where p is the transition density of a Brownian motion it has the form

$$p(s, x; t, y) = \frac{1}{\sqrt{2\pi(t-s)}} \exp\left(-\frac{||x - y||^2}{2(t-s)}\right), \quad s < t,$$

which yields (1).

We propose a method similar to the Euclidean scheme [1] for simulating Brownian bridges on the flat torus, which is of the form

$$dX_t = b(t, X_t)dt + \sigma dW_t; \quad 0 \leq t < T \quad \text{and} \quad X_0 = a \quad \text{a.s.,} \qquad (2)$$

where $\sigma > 0$, $a \in \mathbb{T}^2$ is given, and W is a two-dimensional standard Brownian motion. The exact form of $b(t, x)$ will become apparent below. It is important here to note that in the particular case of the flat torus the transition density for the Brownian motion is known and therefore it is possible to simulate from the distribution of the true Brownian bridge on $\mathbb{T}^2$, however, it requires the calculation of the distance to infinitely many points which the proposed model does not. In Fig. 4 is shown paths of the proposed model and the corresponding paths of the true bridge process.

Let $\pi: \mathbb{R}^2 \to \mathbb{T}^2 = \mathbb{R}^2/\mathbb{Z}^2$ denote the canonical projection onto the torus. The standard two-dimensional Brownian motion $W = (W^1, W^2)$, for two independent one-dimensional Brownian motions W^1 and W^2, is mapped to a Brownian motion $B = (B_t)_{t \geq 0}$ on the flat torus $\mathbb{T}^2$ by the projection map π. Indeed, we can identify the torus $\mathbb{T}^2$ with the unit cube $Q = \{x \in \mathbb{R}^2 : -\frac{1}{2} \leq x_k < \frac{1}{2}, k = 1, 2\}$. Then for $g \in C^\infty(\mathbb{T}^2)$ the Laplace-Beltrami operator, $\Delta_{\mathbb{T}^2}$, on $\mathbb{T}^2$ corresponds to the restriction to Q of the usual Euclidean Laplacian, $\Delta_{\mathbb{R}^2}\tilde{g}$, where $\tilde{g}$ denotes the periodic extension of g, i.e. $\tilde{g} = g \circ \pi$ (see [8, Sec. 3.5]). Since W is a Brownian motion in $\mathbb{R}^2$ if and only if it satisfies the diffusion equation

$$h(W_t) \overset{m}{=} h(W_0) - \frac{1}{2} \int_0^t \Delta_{\mathbb{R}^2} h(W_s)ds,$$

for all smooth functions h, where $X \overset{m}{=} Y$ means that the difference $X - Y$ is a local martingale (see e.g. [2, Sec. 1.5]), it follows that, for $h = \tilde{g}$,

$$\tilde{g}(W_t) \overset{m}{=} \tilde{g}(W_0) - \frac{1}{2} \int_0^t \Delta_{\mathbb{R}^2} \tilde{g}(W_s)ds = g(B_0) - \frac{1}{2} \int_0^t \Delta_{\mathbb{T}^2} g(B_s)ds \overset{m}{=} g(B_t).$$

As this holds for all smooth functions g on $\mathbb{T}^2$, we get that B is a Brownian motion on $\mathbb{T}^2$ in agreement with the definition of a manifold-valued Brownian motion given in [5, Sec. 3.2].

By conditioning B on $\mathbb{T}^2$ to hit a given point $a \in \mathbb{T}^2$, at some fixed time $0 \leq T < \infty$, it is seen that

$$\{\omega \in \Omega : B_T(\omega) = a\} = \{\omega \in \Omega : W_T(\omega) \in \pi^{-1}(a)\},$$

and so simulating a Brownian bridge on the flat torus $\mathbb{T}^2$ is equivalent to simulating a two-dimensional standard Brownian motion conditioned to end up in the set $\pi^{-1}(a)$ at time T. The diffusion given by (1) will not suffice as it is constructed to hit exactly one point. It will, however, provide one subset of sample paths of the Brownian bridge on $\mathbb{T}^2$, corresponding to subset of paths that will "unwrap" the same number of times that it "wraps" around the cut locus. This is illustrated in Fig. 1. To give a precise meaning to this statement we consider the h-transform

$$h(t, z) = \sum_{y \in \pi^{-1}(a)} \frac{p(t, z; T, y)}{p(0, z_0; T, y)},$$

with p denoting the transition density of the two-dimensional Brownian motion, which by Doob's h-transform implies that the distribution of W conditioned on $W_T \in \pi^{-1}(a)$ is the same as the distribution of the diffusion

$$dZ_t = \sigma^2 \nabla_z \log \left(\sum_{y \in \pi^{-1}(a)} p(t, z; T, y) \right) \Bigg|_{x = Z_t} dt + \sigma dW_t$$

$$= \sum_{y \in \pi^{-1}(a)} g_y(t, Z_t) \frac{y - Z_t}{T - t} dt + \sigma dW_t, \quad Z_0 = z_0, \tag{3}$$

where $g_y(t, x) = \dfrac{\exp\left(-\frac{\|y - z\|^2}{2\sigma^2(T-t)}\right)}{\sum_{y \in \pi^{-1}(a)} \exp\left(-\frac{\|y - z\|^2}{2\sigma^2(T-t)}\right)}.$

Instead, we propose to consider the diffusion process on $[0, T)$, for some fixed positive T, defined by

$$dX_t = 1_{G^c}(X_t) \frac{\alpha(X_t) - X_t}{T - t} dt + \sigma dW_t, \quad X_0 = x_0 \tag{4}$$

where $\sigma > 0$ and α is defined by

$$\alpha(X_t) = \arg \min_{y \in \pi^{-1}(a)} \|y - X_t\|,$$

with $a \in \mathbb{T}^2$, and where G is the set of "straigt lines" of the form $\mathbb{R} \times \{x\}$ (resp. $\{x\} \times \mathbb{R}$) in $\mathbb{R}^2$ where $\alpha(X_t)$ is not unique (see Fig. 1). The indicator function removes the drift when the process does not have a natural attraction point.

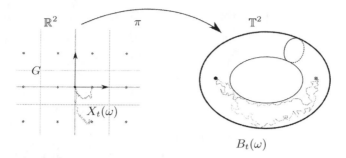

Fig. 1. The figure illustrates the possibility of the diffusion path going an arbitrary number of times around the torus, starting at the black dot and ending in the red. This is illustrated by the red path. The conditioning on single point in $\mathbb{T}^2$ therefore leads to conditioning on multiple points in $\mathbb{R}^2$. Left: Two paths from the same two-dimensional process with multiple endpoints. Right: The projection of the two paths onto the torus. (Color figure online)

3   Existence of Strong Solution

The drift term in Eq. (4) is discontinuous. However, we below show that it posses certain regularity conditions and use this to show that a strong solution to the SDE exist.

In order to ensure the existence of a solution to the diffusion in (4), we need some regularity of the drift term. The drift coefficient is given by

$$
1_{G^c}(X_t)\frac{\alpha(X_t) - X_t}{T - t} = \begin{cases} \frac{\alpha(X_t) - X_t}{T-t}, & \text{if } X_t \in G^c \\ 0, & \text{otherwise,} \end{cases} \tag{5}
$$

for every $0 \le t < T$, where the superscript c denotes the complement. It is a discontinuous process with the set of discontinuities being the set G consisting of the set of straight lines in $\mathbb{R}^2$ where the argmin process is non-unique. It is not even clear that the drift term is suitably measurable as the argmin map in general is not.

Lemma 1. *Let $b\colon [0,T) \times \mathbb{R}^2 \to \mathbb{R}^2$ be the map given by (5). Then b is $\mathcal{B}([0,T)) \otimes \mathcal{B}(\mathbb{R}^2) - \mathcal{B}(\mathbb{R}^2)$ measurable. Furthermore, the map $(s, \omega) \mapsto b(s, X_s(\omega))$ is $\mathcal{B}([0,t]) \otimes \mathcal{F}_t^0$ measurable, for every $0 \le t < T$, where $(\mathcal{F}_t^0)$ denotes the natural filtration generated by X. This is called progressive measurability.*

Proof. First note that G^c is a Borel measurable set as we can write it as a countable union of open sets, i.e., for $y = (y_1, y_2)$ we have

$$
G^c = \bigcup_{y \in \pi^{-1}(a)} \left(y_1 - \frac{1}{2}, y_1 + \frac{1}{2} \right) \times \left(y_2 - \frac{1}{2}, y_2 + \frac{1}{2} \right) =: \bigcup_{y \in \pi^{-1}(a)} V_y.
$$

Now, we need to show that for all $A \in \mathcal{B}(\mathbb{R}^2)$, the set $b^{-1}(A)$ is an element of $\mathcal{B}([0,T)) \otimes \mathcal{B}(\mathbb{R}^2)$. It is enough to consider all open subsets $U \subseteq \mathbb{R}^2$ as these sets generate the Borel algebra on $\mathbb{R}^2$. So let U be an arbitrary open subset, then we have that

$$b^{-1}(U) = b^{-1}(U) \cap \big([0,T) \times G^c\big) \cup b^{-1}(U) \cap \big([0,T) \times G\big).$$

As b is continuous on each of the sets $[0,T) \times V_y$ we have that $b^{-1}(U) \cap \big([0,T) \times G^c\big)$ is a countable union of open sets and therefore an element of $\mathcal{B}([0,T)) \otimes \mathcal{B}(\mathbb{R}^2)$. For the second part we see that

$$b^{-1}(U) \cap ([0,T) \times G) = \begin{cases} [0,T) \times G, & \text{if } (0,0) \in U \\ \emptyset, & \text{otherwise}, \end{cases}$$

where both are elements of $\mathcal{B}([0,T)) \otimes \mathcal{B}(\mathbb{R}^2)$. This shows that b is Borel measurable.

Progressive measurability follows by a very similar argument. $\square$

Usually, global or local Lipschitz conditions are imposed on the drift and diffusion coefficients in order to secure global (resp. local) strong solutions to an SDE. This is a too strong condition for the drift term in this case, however, it is bounded in the following sense.

Lemma 2. *The drift coefficient in (5) is uniformly bounded in x and in t on $[0,S]$, for any $0 \leq S < T$.*

Proof. The first assertion is clear. Let $S \in [0,T)$ be arbitrary and $0 \leq t \leq S$. For every $x \in G^c$ there exist a $y \in \pi^{-1}(a)$ such that we have

$$\left\| 1_{G^c}(x) \frac{\alpha(x) - x}{T - t} \right\|^2 = \left\| \frac{y - x}{T - t} \right\|^2 \leq \frac{C}{(T - S)^2} = C_S,$$

for some positive constants $C > 0$. $\square$

We now come to the main result of this section.

Proposition 1. *There exist a strong solution of (4) on $[0,T)$, which is strongly unique.*

Proof. The drift term is Borel measurable and bounded on $[0,S]$ by Lemmas 1 and 2. As indicated in [10, Thm. 2] and [11, Thm. 1] (4) has a strong solution which is strongly unique. $\square$

Remark 1. The assumption in [10, Thm. 2] can be verified by using smooth bump functions.

4   Convergence and Absolute Continuity

The considerations above make the solution of (4) into a continuous semimartingale. If a semimartingale X takes its values in an open set U of $\mathbb{R}^2$ then Itô's formula holds true for any $C^{1,2}([0,T) \times U)$ functions as well.

Proposition 2. *Let X be a solution to (4) on the filtered probability space $(\Omega, \mathcal{F}, (\mathcal{F}_t), P)$. For every $\omega \in \Omega$ for which there exist an $S < T$ such that $X_t(\omega)$ stays in G^c on $[S,T)$, then X converges pointwise almost surely to $\pi^{-1}(a)$.*

Proof. Assume that for some $\omega \in \Omega$ there exist some $S < T$ such that on $[S,T)$ the process $X_t(\omega)$ takes its values in G^c. By continuity of the process it will take it its values in some open neighborhood V_y of the point $y \in \pi^{-1}$. The proof is then identical to the proof in [1, Lemma 4].   □

Remark 2. It is of course of interest to show that for almost every path the process will converge. This can be obtained by showing that the process will not intersect G infinitely many times close to T.

Consider the stochastic process $\mathcal{E}$ on $0 \le t \le S$ defined by

$$\mathcal{E}(L)_t = \exp\left(-\int_0^t b(s, X_s) dW_s - \frac{1}{2} \int_0^t \|b(s, X_s)\|^2 \, ds \right), \tag{6}$$

where L is the local martingale in the exponential. This is known as the Doléans-Dade exponential. From Lemma 2 it follows that, for all $t \le S$,

$$\mathbb{E}\left[\exp\left(\int_0^t \|b(s, X_s)\|^2 \, ds \right) \right] \le \exp\left(tC_S \right) < \infty$$

The above is known as the Novikov condition (cf. [7]) which ensures that (6) is a martingale on $[0,T)$. Girsanov's theorem ([6, Thm. 5.1 Chap. 3]) then provides that the process defined by

$$\widetilde{W}_t = W_t + \int_0^t b(s, X_s) ds$$

is a Brownian motion under the new measure Q introduced below.

Theorem 1. *Let X defined on $(\Omega, \mathcal{F}, (\mathcal{F}_t), P)$ be a solution of (4) on $[0,S]$ for $S < T$. The process in (6) defined on $0 \le t \le S$ $(S < T)$ is a true martingale and so there exists a measure Q which is absolutely continuous wrt. P such that X is Q-Brownian motion.*

Proof. The martingale property of (6) on $[0,S]$ is a consequence of the Novikov condition. Then Girsanov's theorem gives us that X is a Q-Brownian motion on $[0,S]$.   □

(a) Paths visualized on an embedded torus.

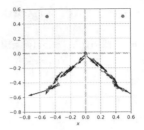

(b) The two Euclidean paths that
are mapped onto the torus.

Fig. 2. Two different paths visualized both on the torus and in Eucliden space. The blue dot represents the starting point and the red represents the end point. (Color figure online)

From the (perhaps obvious) fact that the distribution of the true Brownian bridge is locally equivalent to the distribution of the Brownian motion up to time $t < T$, it follows that the distribution of the Brownian bridge is absolutely continuous wrt. The proposed process up to time $t < T$.

Remark 3. A bit of extra work is needed to obtain the correction term as in [1]. There are indications that it is possible to simulate from the true distribution of the Brownian bridge on the torus, however, Theorem 1 shows that (4) can approximate it.

5   Numerical Experiments

For the numerical implementation of the proposed SDE in Eq. (4) we implemented the Euler-Maruyama scheme, i.e. taking n equidistant discretization points of the time interval $t_1, ..., t_n$, with $t_{i+1} - t_i = \Delta t$, the numerical equation becomes

$$x_{t_{i+1}} = x_{t_i} + \frac{\arg\min_{y \in \pi^{-1}(a)} (\|y - x_{t_i}\|) - x_{t_i}}{T - t_i} \Delta t + \sigma \Delta W_{t_i},$$

where $\Delta W_{t_{i+1}} = W_{t_{i+1}} - W_{t_i}$ is equal in distribution to a normal random variable with mean zero and variance Δt.

Figure 2a shows the implementation of the numerical scheme on an embedded torus and Fig. 2b its Euclidean counterpart. Figure 3a shows the behaviour of the drift term along a given path, illustrating that the attraction becomes stronger as time approaches the terminal time. The vector fields in Fig. 3b shows the constant attraction to the center of the open subsets.

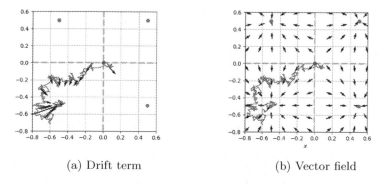

(a) Drift term   (b) Vector field

Fig. 3. Figure a depicts the evolution of the drift term. It shows how the pull from the drift becomes stronger near the end. Figure b shows the underlying vector field.

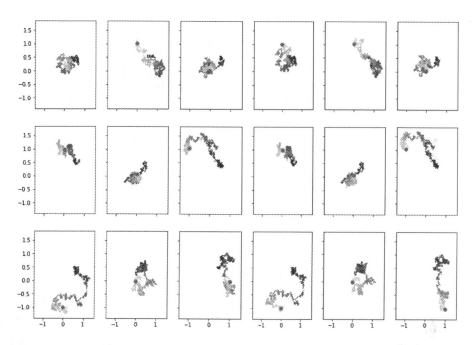

Fig. 4. Figure shows 9 paths from the proposed model (4) on the left and the corresponding paths from the true bridge (3) on the right. It is seen that the first and the last paths disagree on the limiting point, whereas the rest looks fairly similar. The picture agree with the fact that roughly four in five have the same limiting point. Here $\sigma = 0.8$ and the conditioning points being the integers in $[-2, 2] \times [-2, 2]$.

Acknowledgements. We acknowledge F. van der Meulen for discussions and insights on conditioned diffusions.

References

1. Delyon, B., Hu, Y.: Simulation of conditioned diffusion and application to parameter estimation. Stochast. Processes Appl. **116**(11), 1660–1675 (2006)
2. Emery, M.: Stochastic Calculus in Manifolds. Springer, Heidelberg (1989). https://doi.org/10.1007/978-3-642-75051-9
3. García-Portugués, E., Sørensen, M., Mardia, K.V., Hamelryck, T.: Langevin diffusions on the torus: estimation and applications. Stat. Comput. **29**, 1–22 (2017)
4. Golden, M., García-Portugués, E., Sørensen, M., Mardia, K.V., Hamelryck, T., Hein, J.: A generative angular model of protein structure evolution. Mol. Biol. Evol. **34**(8), 2085–2100 (2017)
5. Hsu, E.P.: Stochastic Analysis on Manifolds, vol. 38. American Mathematical Soc., Providence (2002)
6. Karatzas, I., Shreve, S.: Brownian Motion and Stochastic Calculus, 2nd edn. Springer, New York (1991)
7. Novikov, A.A.: On an identity for stochastic integrals. Theory Probab. Appl. **17**(4), 717–720 (1973)
8. Sogge, C.D.: Hangzhou Lectures on Eigenfunctions of the Laplacian (AM-188). Princeton University Press, Princeton (2014)
9. Sommer, S., Arnaudon, A., Kuhnel, L., Joshi, S.: Bridge simulation and metric estimation on landmark manifolds. In: Cardoso, M.J., et al. (eds.) GRAIL/MFCA/MICGen -2017. LNCS, vol. 10551, pp. 79–91. Springer, Cham (2017). https://doi.org/10.1007/978-3-319-67675-3_8
10. Veretennikov, A.Y.: On the strong solutions of stochastic differential equations. Theory Probab. Appl. **24**(2), 354–366 (1980)
11. Veretennikov, A.J.: On strong solutions and explicit formulas for solutions of stochastic integral equations. Sbornik: Mathematics **39**, 387–403 (1981)

Towards Parametric Bi-Invariant Density Estimation on $SE(2)$

Emmanuel Chevallier$^{(\boxtimes)}$

Aix Marseille Univ, CNRS, Centrale Marseille, Institut Fresnel, Marseille, France
emmanuel.chevallier@fresnel.com

Abstract. This papers aims at describing a novel framework for bi-invariant density estimation on the group of planar rigid motion $SE(2)$. Probability distributions on the group are constructed from distributions on tangent spaces pushed to the group by the exponential map. The exponential mapping on Lie groups presents two key particularities: it is compatible with left and right multiplications and its Jacobian can be computed explicitly. These two properties enable to define probability densities with tractable expressions and bi-invariant procedures to estimate them from a set of samples. Sampling from these distributions is easy since it is sufficient to draw samples in Euclidean tangent spaces. This paper is a preliminary work and the convergences of these estimators are not studied.

Keywords: Density estimation · Lie groups · Rigid motions

1 Introduction

Probability density estimation problems generally fall in one of the two categories: estimating a density on a Euclidean vector space and estimating a density on a non-Euclidean manifold. In turn, estimation problems on non-Euclidean manifolds can be divided in different categories depending on the nature of the manifold. The two main classes of non-Euclidean manifold encountered in statistics are Riemannian manifolds and Lie groups. On Riemannian manifolds, the objects studied in statistics should be consistent with the Riemannian distance. For instance, means of distributions are defined as points minimizing the average square Riemannian distances. On a Lie group, the objects should be consistent with the group law. Products of compact Lie groups and vector spaces belong to both of these categories, they admit a Riemannian metric invariant by left and right multiplications. On the other hand, other Lie groups do not admit such nice metrics, hence the need for statistical tools based solely on group law and not on the Riemannian distance. This problem has been properly addressed for the first time by Pennec and Arsigny in [1] where authors define bi-invariant mean on arbitrary Lie groups. Once the bi-invariant mean has been defined, higher order

Supported by organization x.

F. Nielsen and F. Barbaresco (Eds.): GSI 2019, LNCS 11712, pp. 695–702, 2019.
https://doi.org/10.1007/978-3-030-26980-7_72

bi-invariant centered moments can be defined in the tangent space at the mean. We address the problem of estimating densities on $SE(2)$, the group of direct isometries of the Euclidean plane. To do so, we introduce a parametric model on $SE(2)$ similar to the probability kernels defined in [4,7–9] on Riemannian manifolds. Harmonic analysis is another well known approach to density estimation, see [12] for $SE(2)$ and [5,6,11] on other manifolds. Beside the technicalities and numerical difficulties introduced by harmonic analysis on non abelian and non compact groups, the main interest of our approach over harmonic analysis techniques, is that it enables to define parametric models.

This work is based on two facts. First, the exponential map can be transported from the identity element to any point of the group regardless of the choice of left or right multiplication. This property was already of primary importance in the construction of the bi-invariant mean [1] and enables to define bi-invariant estimation procedures. The second important fact is that the Jacobian of the exponential map on $SE(2)$ has a simple expression. This Jacobian provides an easy way to define probability densities with explicit expressions on the group by pushing densities from tangent spaces using the exponential map.

The paper is organized as follow. Section 2 describes the group of direct isometries of the Euclidean plane. Section 3 reviews the relevant properties of the exponential mapping. Section 4 recalls the definitions of the first and second centered moments on a Lie group. A statistical model on Lie groups together with an estimation procedure is introduced in Sect. 5. The estimation in a mixture model is also discussed. Section 6 concludes the paper.

2   The Group $SE(2)$

$SE(2)$ is the set of all direct isometries on the Euclidean space $\mathbb{R}^2$. The composition law of functions makes $SE(2)$ a group. For each element g of $SE(2)$ there are a unique rotation R and a unique vector t such that

$$g(u) = Ru + t$$

A convenient way to represent elements of $SE(2)$ is to identify the isometry g with the matrix

$$\begin{pmatrix} R & t \\ 0 & 1 \end{pmatrix} \in Gl_3(\mathbb{R}).$$

It is easy to check that the composition of isometries corresponds to the matrix multiplication. $SE(2)$ is thus seen as a subgroup of $Gl_3(\mathbb{R})$. The tangent space at the identity element, noted $T_e SE(2)$, is spanned by the matrices

$$v_1 = \begin{pmatrix} 0 & -1 & 0 \\ 1 & 0 & 0 \\ 0 & 0 & 0 \end{pmatrix}, \quad v_2 = \begin{pmatrix} 0 & 0 & 1 \\ 0 & 0 & 0 \\ 0 & 0 & 0 \end{pmatrix}, \text{ and } v_3 = \begin{pmatrix} 0 & 0 & 0 \\ 0 & 0 & 1 \\ 0 & 0 & 0 \end{pmatrix}$$

and tangent matrix X is identified with its three coordinates $(\theta, a, b) \in \mathbb{R}^3$. Since $SE(2)$ is identified with a subgroup of $Gl_3(\mathbb{R})$, its group exponential is simply the matrix exponential. Classical calculations on matrix exponential lead to

$$\exp\left(\theta v_1 + a v_2 + b v_3\right) = \begin{pmatrix} \cos(\theta) & -\sin(\theta) & a\frac{\sin(\theta)}{\theta} - b\frac{1-\cos(\theta)}{\theta} \\ \sin(\theta) & \cos(\theta) & a\frac{1-\cos(\theta)}{\theta} + b\frac{\sin(\theta)}{\theta} \\ 0 & 0 & 1 \end{pmatrix}.$$

Let

$$S_{\theta[2\pi]} = \begin{pmatrix} \frac{\sin(\theta)}{\theta} & -\frac{1-\cos(\theta)}{\theta} \\ \frac{1-\cos(\theta)}{\theta} & \frac{\sin(\theta)}{\theta} \end{pmatrix}.$$

Given any matrix

$$M = \begin{pmatrix} \cos(\theta) & -\sin(\theta) & x \\ \sin(\theta) & \cos(\theta) & y \\ 0 & 0 & 1 \end{pmatrix},$$

the solutions of $\exp\left(\theta' u_1 + a v_2 + b v_3\right) = M$ are given by

$$\theta' = \theta[2\pi], \quad \begin{pmatrix} a \\ b \end{pmatrix} = S_{\theta[2\pi]}^{-1} \begin{pmatrix} x \\ y \end{pmatrix},$$

hence the exponential map is a bijection between $U = [-\pi, \pi[\times\mathbb{R}^2$ and $SE(2)$. The logarithm is defined as the inverse of the exponential on $[-\pi, \pi[\times\mathbb{R}^2$.

Even though our density modelling framework is intrinsic, it is useful to define a reference basis in each tangent space. Let $\mathcal{B}_e$ be the basis (v_1, v_2, v_3) of $T_e SE(2)$ and $dL_g(v_1, v_2, v_3)$ the basis $\mathcal{B}_g$ of $T_g SE(2)$, where dL_g is the differential of the left multiplication by g. In the remainder of the paper, coordinates of vectors and matrices are expressed in these reference basis. The Lebesgue measure on the tangent space $T_g SE(2)$ induced by $\mathcal{B}_g$ is noted λ_g.

In order to define densities on $SE(2)$, it is necessary to define a reference measure. Let ω be the left invariant volume form with $\omega_e(v_1, v_2, v_3) = 1$ and choose the orientation according to ω. The corresponding volume measure μ_G is defined by

$$\mu_G(A) = \int_A \omega.$$

(μ_G is a Haar measure on $SE(2)$). On $SE(2)$ it can be checked that any left (right) invariant volume form is also right (left) invariant hence μ_G is a bi-invariant measure. In the rest of the paper, densities on $SE(2)$ are expressed with respect to μ_G.

3   Bi-Invariant Local Linearizations

Densities are modelled using local linearizations. This section describes natural maps between the tangent spaces at each point of the group, and the group.

3.1   The Exponential at Point g

Since the exponential maps the lines of the tangent space at e to the one parameter subgroups of G, it is a natural candidate to linearize the group near the identity. Let Ad be the group adjoint representation. Recall that on a Lie group,

$$g\exp(v)g^{-1} = \exp(Ad_g(v)) = \exp(dL_g(dR_{g^{-1}}(v))) = \exp(dR_{g^{-1}}(dL_g(v))),$$

where dL_g and dR_g are the differentials of the left and right multiplication. This property enables the transport of the exponential application to any element of the group without ambiguity on the choice of left or right multiplication,

$$\exp_g : T_g SE(2) \to SE(2)$$
$$v \mapsto \exp_g(v) = g.\exp\left(dL_{g^{-1}}v\right) = \exp\left(dR_{g^{-1}}v\right)g.$$

Note $U_g \subset T_g SE(2) = dL_g(U)$ the injectivity domain of $\exp_g$. The logarithm $\log_g : SE(2) \to U_g$ becomes

$$\log_{g_1}(g_2) = dL_{g_1}\left(\log\left(g_1^{-1}g_2\right)\right).$$

We now have a natural linearization of the group around an arbitrary $g \in SE(2)$ and the two commutative diagrams,

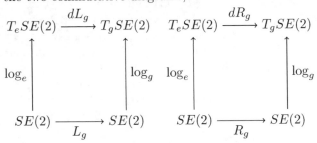

hence the name bi-invariant linearization.

Independence from the choice of left or right multiplication in the definition of the exponential at an arbitrary point was the key ingredient of the definition of the bi-invariant mean in [1]. It is again a key property in our framework for bi-invariant density modelling.

The strength of the exponential map is that it turns some general algebra problems into linear algebra. Once the space has been lifted to a tangent space, the problem of left and right invariances is reduced to the study of the commutation with the differentials of left and right multiplications. Since the tangent spaces do not have a canonical basis or scalar product, the manipulations we perform such as computing a mean, a covariance or estimating a density should not depend on a particular coordinates choice. Hence if these manipulations commute with all the linear invertible transformations, in particular with the left and right differentials, they induce bi-invariant operations.

3.2 Jacobian Determinant of the Exponential

A measure ν on $T_g SE(2)$ can be pushed forward to the group using the exponential at g. This push forward measure is noted $\exp_{g*}(\nu)$. Since $\exp_g$ commutes with the right and left actions, so does the push forward of measures. In order to obtain expressions of the densities on the group, it is necessary to compute the Jacobian determinant of the exponential.

Let $J_e(v)$ be the determinant of the differential of the exponential $d\exp_{e,v}$ at the vector v in the basis $\mathcal{B}_e$ and $\mathcal{B}_{\exp_e(v)}$. Calculations show that on $SE(2)$

$$J_e(v) = 2\frac{1 - \cos(\theta)}{\theta^2}, \tag{1}$$

for $\theta \neq 0$, and $J(v) = 1$ when $\theta = 0$. Since $\exp_g(u) = g . \exp_e\left(dL_{g^{-1}}u\right)$,

$$d\exp_{g,u} = dL_g \circ d\exp_{e, dL_{g^{-1}}(u)} \circ dL_{g^{-1}}.$$

Furthermore,

$$dL_{g^{-1}}\left(\mathcal{B}_g\right) = \mathcal{B}_e \text{ and } \mathcal{B}_{\exp_g(u)} = dL_g\left(\mathcal{B}_{\exp_e\left(dL_{g^{-1}}(u)\right)}\right).$$

Hence expressed in the basis $\mathcal{B}_g$ and $\mathcal{B}_{\exp_g(u)}$, the determinant of $d\exp_{g,u}$ is given by

$$J_g(u) = J_e\left(dL_{g^{-1}}(u)\right).$$

When all tangent vectors are expressed in the left invariant basis, it is possible to drop the subscripts and write

$$J(v) = J(\theta, a, b) = 2\frac{1 - \cos(\theta)}{\theta^2}. \tag{2}$$

Assume ν has a density f with respect to a Lebesgue measure of $T_g SE(2)$ and that its support is contained in an injectivity domain U_g of $\exp_g$. The density $f_{SE(2)}$ of the measure pushed on the group is given by

$$f_{SE(2)}(\exp_g v) = \frac{d\exp_{g*}(\nu)}{d\mu_G}(\exp_g(v)) = \frac{f(v)}{J(v)}.$$

Since $SE(2)$ is unimodular, i.e. μ_G is bi-invariant, the density of the pushed forward measure also commutes with the left and right actions of $SE(2)$.

4 First and Second Moments of a Distribution on a Group

4.1 Bi-Invariant Means

Bi-invariant means on Lie groups have been introduced by Pennec and Arsigny, see [1]. An element $\bar{g} \in G$ is said to be a bi-invariant mean of $g_1, \ldots, g_k \in G$ or of probability distribution ν on G, if

$$\sum_i \log_{\bar{g}}(g_i) = 0 \text{ or } \int_G \log_{\bar{g}}(g) d\nu(g) = 0.$$

Observe that $\bar{g}$ is not necessarily unique, see [1–3] for more details.

4.2   Covariance of a Distribution on $SE(2)$

Let ν be a distribution on $SE(2)$ such that its bi-invariant mean $\bar{\nu}$ is uniquely defined. The covariance matrix of ν in the basis $\mathcal{B}_{\bar{\nu}}$ is defined by

$$\Sigma = \int_{SE(2)} \log_{\bar{g}}(g) \log_{\bar{g}}(g)^t d\nu(g).$$

In the principal geodesic analysis, the matrix Σ is sometimes referred to as a linearized quantity in contrast to the 'exact' principal geodesic analysis, see [10].

5   Statistical Models for Bi-Invariant Density Estimation

5.1   Density Modelling from a Single Tangent Space

Let $K : \mathbb{R}_+ \to \mathbb{R}_+$ be such that

$$\int_{\mathbb{R}_+} x^2 K(x) dx = \frac{1}{4\pi} (K \text{ defines a probability density on } \mathbb{R}^3),$$

$$\int_{\mathbb{R}_+} x^4 K(x) dx = \frac{1}{2\pi} \text{ (the identity matrix is the covariance matrix)},$$

$$K(x > 1) = 0.$$

Let C_g be the set of covariance matrices compatible with the injectivity domain U_g,

$$C_g = \left\{ \Sigma | v \notin U_g, v\Sigma^{-1}v^t > 1 \right\}.$$

When $\Sigma \in C_g$, the support of the probability distribution ν on $T_g SE(2)$,

$$\frac{d\nu}{d\lambda_g} = \frac{1}{\sqrt{\det(\Sigma)}} K\left(\sqrt{v\Sigma^{-1}v^t}\right)$$

is contained in U_g. The density of the push forward of ν is then

$$f(\exp_g(v)) = \frac{1}{J(v)\sqrt{\det(\Sigma)}} K\left(\sqrt{v\Sigma^{-1}v^t}\right),$$

where J is given in Eq. 2. The set of such probability densities when g and Σ vary form a natural parametric model for bi-invariant density estimation:

$$\mathcal{M} = \{ f_{g,\Sigma} : g \in SE(2) \text{ and } \Sigma \in C_g \}.$$

The commutation diagrams 3.1 imply that $\mathcal{M}$ is closed under left and right action. The fact that g and Σ are the moments of $f_{g,\Sigma}\mu_G$ plays a major role in the relevance of the model $\mathcal{M}$. This fact holds when Σ is small enough. A more precise result will follow in a future work.

Let $g_1, .., g_k$ be points in $SE(2)$ with a unique bi-invariant mean $\bar{g}$ and such that the empirical covariance

$$\Sigma = \sum_i \log_{\bar{g}}(g) \log_{\bar{g}}(g)^t$$

is contained in $C_{\bar{g}}$. On the hand, finding the maximum likelihood estimation when $g_1, \ldots, g_k$ are i.i.d. requires an optimization procedure. On the other hand, matching moments is straightforward, provided that the moments of $f_{g, \Sigma \mu_G}$ are (g, Σ). In most cases, this moment matching estimator is expected to have reasonable convergence properties. It can be checked that both the maximum likelihood and the moment matching estimators are bi-invariant.

5.2 Mixture Models from Multiple Tangent Spaces

On a general Lie group, there are two reasons why it is useful, and sometimes necessary, to model a distribution using lifts in more than one tangent space. The first one is when there is no point g such that all the points g_i are contained in the domain of definition of the logarithm g. Note that on $SE(2)$ this never happens since the domain of definition of $\log_g$ is $SE(2)$. The second is that linearizing the space far from its origin distorts its structure.

We provide now the outlines of a mixture estimator. Let $g_1, \ldots, g_k$ be sample points in $SE(2)$. If the mean of these points is not defined or not uniquely defined, we partition this set of sample points in subsets $E_1, \ldots, E_p$ such that

- the mean l_j of each E_j is defined and unique,
- the covariance Σ_j is in the admissible set C_{l_j}.

For each j there is a density $f_j = f_{l_j, \Sigma_j}$ in the parametric model $\mathcal{M}$. The mixed estimator is then the density

$$T(g_1, \ldots, g_k) = f = \frac{1}{k} \sum_j \mathrm{card} E_j f_j.$$

If the construction of the partition $E_1, \ldots, E_p$ from $g_1, \ldots, g_k$ is bi-invariant then the estimator T is bi-invariant.

6 Conclusion and Perspectives

In this paper, we proposed a new statistical model $\mathcal{M}$ of densities on $SE(2)$. The strength of this model is that densities have explicit expressions and that the moment matching estimator is straightforward to compute. Further works should focus on analyzing the performances of the moment matching estimator and on proposing detailed algorithms to estimate densities in a mixture model.

References

1. Pennec, X., Arsigny, V.: Exponential barycenters of the canonical Cartan connection and invariant means on Lie groups. In: Nielsen, F., Bhatia, R. (eds.) Matrix Information Geometry, pp. 123–166. Springer, Heidelberg (2013). https://doi.org/10.1007/978-3-642-30232-9_7

2. Pennec, X.: Bi-invariant means on lie groups with cartan-schouten connections. In: Nielsen, F., Barbaresco, F. (eds.) GSI 2013. LNCS, vol. 8085, pp. 59–67. Springer, Heidelberg (2013). https://doi.org/10.1007/978-3-642-40020-9_5

3. Arsigny, V., Pennec, X., Nicholas Ayache, N., Bi-invariant Means in Lie Groups: Application to Left-invariant Polyaffine Transformations. Research report RR-5885, INRIA Sophia-Antipolis, April 2006

4. Pelletier, B.: Kernel density estimation on Riemannian manifolds. Stat. Probab. Lett. **73**, 297–304 (2005)

5. Kim, P., Richards, D.: Deconvolution density estimation on the space of positive definite symmetric matrices. In: Nonparametric Statistics and Mixture Models: A Festschrift in Honor of Thomas P. Hettmansperger, pp. 147–168. World Scientific Publishing, Singapore (2008)

6. Huckemann, S., Kim, P., Koo, J., Munk, A.: Mobius deconvolution on the hyperbolic plane with application to impedance density estimation. Ann. Stat. **38**, 2465–2498 (2010)

7. Chevallier, E.: A family of anisotropic distributions on the hyperbolic plane. In: Nielsen, F., Barbaresco, F. (eds.) GSI 2017. LNCS, vol. 10589, pp. 717–724. Springer, Cham (2017). https://doi.org/10.1007/978-3-319-68445-1_83

8. Chevallier, E., Forget, T., Barbaresco, F., Angulo, J.: Kernel density estimation on the siegel space with an application to radar processing. Entropy **18**(11), 396 (2016)

9. Chevallier, E., Kalunga, E., Angulo, J.: Kernel density estimation on spaces of Gaussian distributions and symmetric positive definite matrices. SIAM J. Imaging Sci. **10**(1), 191–215 (2017)

10. Sommer, S., Lauze, F., Hauberg, S., Nielsen, M.: Manifold valued statistics, exact principal geodesic analysis and the effect of linear approximations. In: Daniilidis, K., Maragos, P., Paragios, N. (eds.) ECCV 2010. LNCS, vol. 6316, pp. 43–56. Springer, Heidelberg (2010). https://doi.org/10.1007/978-3-642-15567-3_4

11. Hendriks, H.: Nonparametric estimation of a probability density on a Riemannian manifold using Fourier expansions. Ann. Stat. **18**, 832–849 (1990)

12. Lesosky, M., Kim, P.T., Kribs, D.W.: Regularized deconvolution on the 2D-Euclidean motion group. Inverse Prob. **24**(5), 055017 (2008)

13. Rossmann, W.: Lie Groups: An Introduction Through Linear Groups, vol. 5. Oxford University Press on Demand, Oxford (2002)

Wasserstein Information Geometry/Optimal Transport

Affine Natural Proximal Learning

Wuchen Li[1], Alex Tong Lin[1], and Guido Montúfar[1,2,3]($\boxtimes$)

[1] Department of Mathematics, UCLA, Los Angeles, CA 90095, USA
[2] Department of Statistics, UCLA, Los Angeles, CA 90095, USA
[3] Max Planck Institute for Mathematics in the Sciences, 04103 Leipzig, Germany
`montufar@mis.mpg.de`

Abstract. We revisit the natural gradient method for learning in statistical manifolds. We consider the proximal formulation and obtain a closed form approximation of the proximity term over an affine subspace of functions in the Legendre dual formulation. We consider two important types of statistical metrics, namely the Wasserstein and Fisher-Rao metrics, and introduce numerical methods for high dimensional parameter spaces.

Keywords: Optimal transport · Information geometry · Proximal operator

1 Introduction

Learning algorithms often proceed by minimizing a loss function that measures the discrepancy between a data distribution and a model distribution. Given a parametric model and a metric in probability space, the loss can be minimized by the Riemannian gradient descent method, also known as the natural gradient method. An important metric in this context is the Fisher-Rao information metric [4,18], which induces the Fisher-Rao natural gradient [1]. Another important metric is the Wasserstein metric [15,20], which induces the Wasserstein natural gradient [8,9,12,14]. Natural gradient methods have numerous applications in learning; see, e.g., [2,3,10,13,16,17].

In spite of having numerous theoretical advantages, applying natural gradient methods is often challenging. In particular, machine learning models usually have many parameters, making the direct computation of the parameter updates too costly. Each update requires to compute the Jacobi matrix of the model and the inverse of the metric tensor in parameter space. An alternative, implicit, way to formulate the update is via a proximal operator. Recently [11] proposed proximal methods as an approach to natural gradients and demonstrated their viability in state of the art generative modeling. The idea is to compute the proximity penalty in closed form over an approximation space. This results in a tractable iterative regularization for the parameter updates.

We develop this idea to obtain a general natural proximal method, and provide explicit formulas for the Fisher-Rao and the Wasserstein metrics. These

© Springer Nature Switzerland AG 2019
F. Nielsen and F. Barbaresco (Eds.): GSI 2019, LNCS 11712, pp. 705–714, 2019.
https://doi.org/10.1007/978-3-030-26980-7_73

serve three purposes: (i) The proximal operator and its approximation can enable efficient and effective expressions for the time discretized parameter updates of the natural gradient flow. (ii) The proximal method, as an implicit method, naturally regularizes the objective function, and can be used to optimize non-smooth objective functions. (iii) The metric regularization is expressed in terms of statistics, such as mean and variance, and can be estimated from samples.

2   Natural Proximal Gradient

We review the natural gradient flow in a statistical manifold with Wasserstein and Fisher-Rao metrics, present the natural proximal operators, and introduce a systematic approximation which is suitable for estimation from samples.

2.1   Natural Gradients Flows

Learning problems are often formulated as the minimization of a loss function, as $\min_{\theta \in \Theta} F(\theta)$, where $\Theta \in \mathbb{R}^d$ is the parameter of the hypothesis class, and $F \colon \Theta \to \mathbb{R}$ is the loss function. As the hypothesis class, we consider a parametrized probability model $\rho \colon \Theta \to \mathcal{P}(\Omega)$, where Ω is the sample space, which is a discrete or continuous set on which the distributions are supported. The loss is usually a divergence (sometimes distance) function between the empirical data distribution $\hat{\rho}_{\text{data}}$ and the model distribution ρ_θ.

To find a minimizer, the gradient flow approach is often considered. This flow follows the steepest descent direction of the loss function with respect to a given Riemannian metric. In general, this is defined by

$$\dot{\theta}(t) = -G(\theta(t))^{-1} \nabla_\theta F(\theta(t)), \tag{1}$$

where $G(\theta) \in \mathbb{R}^{d \times d}$ is the matrix representation of the Riemannian metric tensor (for our choice of coordinates), and $\nabla_\theta = (\frac{\partial}{\partial \theta_1}, \ldots, \frac{\partial}{\partial \theta_d})^\top$ is the standard (Euclidean) gradient operator. In the context of probability distributions, the metric $G(\theta)$ is pulled back from a natural metric structure on probability space. This implies that for any choice of the parametrization, (1) defines the same flow of probability distributions. Hence it is said to be parametrization invariant.

We will focus on two important statistical metrics on probability space: the Wasserstein metric and the Fisher-Rao metric. These metrics induce the following metric tensors in parameter space. We write $(\cdot, \cdot)$ for the Euclidean or L^2 inner product on the sample space Ω (which might be continuous or discrete).

Definition 1 (Statistical metric tensor on parameter space). *Consider the probability space $(\mathcal{P}(\Omega), g)$ with metric tensor g, and a smoothly parametrized probability model ρ_θ with parameter $\theta \in \Theta$. Then the pull-back G of g is given by*

$$G(\theta) = \Big(\nabla_\theta \rho_\theta, g(\rho_\theta) \nabla_\theta \rho_\theta \Big).$$

(i) If $g_\theta = -(\Delta_{\rho_\theta})^{-1}$, with $\Delta_{\rho_\theta} = \nabla \cdot (\rho_\theta \nabla)$ being the weighted elliptic operator [6, 7, 15], then $G(\theta)$ is the Wasserstein metric tensor, given by

$$G_W(\theta)_{ij} = \left(\nabla_{\theta_i} \rho_\theta, (-\Delta_{\rho_\theta})^{-1} \nabla_{\theta_j} \rho_\theta \right),$$

(ii) If $g_\theta = \frac{1}{\rho_\theta}$, then $G(\theta)$ is the Fisher-Rao metric tensor, given by

$$G_{FR}(\theta)_{ij} = \left(\nabla_{\theta_i} \rho_\theta, \frac{1}{\rho_\theta} \nabla_{\theta_j} \rho_\theta \right).$$

Given a metric tensor on parameter space, the standard approach for numerical computation of the gradient flow (1) is the forward Euler method, i.e.,

$$\theta^{k+1} = \theta^k - h G(\theta^k)^{-1} \nabla_\theta F(\theta^k),$$

where $h > 0$ is a step-size. This is known as the natural gradient descent method [2]. In practice, we need to compute the matrix $G(\theta)$ and its inverse at each parameter update, which is difficult in high dimensional parameter spaces.

2.2 Natural Proximal Operators

We next present another way to approximate the gradient flow, known as the backward Euler or proximal operator method. The proximal operator refers to

$$\theta^{k+1} = \text{Prox}_{hF}(\theta^k) = \arg\min_\theta \ F(\theta) + \frac{D(\theta, \theta^k)}{2h}, \tag{2}$$

where D is a proximity term that penalizes the distance from the current point, and h adjusts the strength. When h is infinity, the proximal operator returns the global minimizer of F. The proximity term is given by the metric function:

$$\begin{aligned}
D(\theta, \theta^k) &= \inf_{\theta(t)} \left\{ \int_0^1 \dot{\theta}(t)^\top G(\theta(t)) \dot{\theta}(t) dt : \theta_0 = \theta, \ \theta_1 = \theta^k \right\} \\
&= \inf_{\theta(t)} \left\{ \int_0^1 (\partial_t \rho_{\theta(t)}, g(\rho_{\theta(t)}) \partial_t \rho_{\theta(t)}) dt : \theta_0 = \theta, \ \theta_1 = \theta^k \right\}.
\end{aligned} \tag{3}$$

In rare cases, the proximal operator (2) can be written explicitly.

We shall approximate D in a way that allows for a more friendly computation of the proximal operator. Consider the iterative proximal update

$$\theta^{k+1} = \arg\min_\theta \ F(\theta) + \frac{1}{2h} \left(\rho_\theta - \rho_{\theta^k}, g(\rho_{\tilde{\theta}})(\rho_\theta - \rho_{\theta^k}) \right), \tag{4}$$

where $\tilde{\theta} = \frac{\theta + \theta^k}{2}$. Here the D term in (2) is replaced by a mid-point expression, which is exact up to the order $o(\|\theta - \theta^k\|^2)$. This new proximal operator corresponds to a numerical method known as the semi-backward Euler method. Both (2) and (4) are time discretizations of (1) with first order accuracy. We shall focus on (4), and derive a tractable approximation of the regularization term.

3   Affine Space Approximation of the Metric

Consider the proximity term (similar to a squared Mahalanobis distance)

$$\tilde{D}(\theta, \theta^k) = \left(\rho_\theta - \rho_{\theta^k}, g(\rho_{\tilde{\theta}})(\rho_\theta - \rho_{\theta^k}) \right). \tag{5}$$

In the following we derive an explicit and computer friendly approximation. To this end, we first consider the variational formulation

$$\frac{1}{2}\tilde{D}(\theta, \theta^k) = \sup_{\Phi:\, \Omega \to \mathbb{R}} (\Phi, \rho_\theta - \rho_{\theta^k}) - \frac{1}{2}\left(\Phi, g(\rho_{\tilde{\theta}})^\dagger \Phi \right), \tag{6}$$

where $\dagger$ is the pseudo-inverse operator and the maximizer $\Phi = g(\rho_{\tilde{\theta}})(\rho_\theta - \rho_{\theta^k})$ recovers the previous formula. This corresponds to a expressing (5) in terms of its Legendre dual between tangent space and cotangent space in probability space; for a discussion see [7].

Now we restrict the optimization domain (i.e., the set of functions $\Phi\colon \Omega \to \mathbb{R}$) to an affine space of functions of the form

$$\mathcal{F}_\Psi = \left\{ \Phi(x) = \sum_{j=1}^{n} \xi_j \psi_j(x) = \xi^\top \Psi(x) : \xi \in \mathbb{R}^n \right\},$$

where $\xi = (\xi_j)_{j=1}^n$ is a parameter vector and $\Psi = (\psi_j)_{j=1}^n$ collects a choice of basis functions $\psi_j\colon \Omega \to \mathbb{R}$. This results in following optimization problems:

(i) For the Wasserstein metric, we have

$$\frac{1}{2}\tilde{D}_\Psi^W(\theta, \theta^k) = \sup_{\Phi = \xi^\top \Psi} \mathbb{E}_\theta[\Phi] - \mathbb{E}_{\theta^k}[\Phi] - \frac{1}{2}\mathbb{E}_{\tilde{\theta}}[\|\nabla \Phi\|^2];$$

(ii) For the Fisher-Rao metric, we have

$$\frac{1}{2}\tilde{D}_\Psi^{FR}(\theta, \theta^k) = \sup_{\Phi = \xi^\top \Psi} \mathbb{E}_\theta[\Phi] - \mathbb{E}_{\theta^k}[\Phi] - \frac{1}{2}\mathbb{E}_{\tilde{\theta}}\left[(\Phi - \mathbb{E}_{\tilde{\theta}}[\Phi])^2 \right].$$

These are quadratic semi-definite programs in ξ. In practice, if using small sample estimates for the expectations, one can add a regularization $-\lambda\|\xi\|^2$, with a small $\lambda > 0$, to ensure strict definiteness and existence of a solution. We proceed to solve these problems. We write $\mathbb{E}_\theta[\psi] = \mathbb{E}_{x \sim \rho_\theta}[\psi(x)]$ and $\partial_l = \frac{\partial}{\partial x_l}$ for the partial derivative w.r.t. the lth sample space variable.[1]

Theorem 1 (Affine space approximation). *Given a basis Ψ, the proximity term $\tilde{D}$ within the affine function space $\mathcal{F}_\Psi = \{\xi^\top \Psi : \xi \in \mathbb{R}^n\}$ is given by*

$$\tilde{D}_\Psi(\theta, \theta^k) = (\mathbb{E}_\theta[\Psi] - \mathbb{E}_{\theta^k}[\Psi])^\top \left(\Psi, g(\rho_\theta)^\dagger \Psi \right)^\dagger (\mathbb{E}_\theta[\Psi] - \mathbb{E}_{\theta^k}[\Psi]).$$

---

[1] If the sample space is discrete, we use the discrete differential operator. For an edge weighted graph $G = (V, E, \omega)$, the gradient of $\Phi \in \mathbb{R}^{|V|}$ is $\nabla \Phi = (\sqrt{\omega_{ij}}\Phi_i - \Phi_j)_{(i,j) \in E} \in \mathbb{R}^{|E|}$, and $\mathbb{E}_\theta[\|\nabla \Phi\|^2] = \frac{1}{2}\sum_{i \in V} p_i(\theta) \sum_{j \in V} \omega_{ij}(\Phi_i - \Phi_j)^2$. For details see [8].

(i) *For the Wasserstein metric, we have*

$$\tilde{D}^W_\Psi(\theta, \theta^k) = (\mathbb{E}_\theta[\Psi] - \mathbb{E}_{\theta^k}[\Psi])^\top \left(\mathfrak{C}^W(\tilde\theta)\right)^{-1} (\mathbb{E}_\theta[\Psi] - \mathbb{E}_{\theta^k}[\Psi]),$$

where $\mathfrak{C}^W(\tilde\theta) = \mathbb{E}_{\tilde\theta}[\sum_l \left(\partial_l\Psi\right)\left(\partial_l\Psi\right)^\top].$

(ii) *For the Fisher-Rao metric, we have*

$$\tilde{D}^{FR}_\Psi(\theta, \theta^k) = (\mathbb{E}_\theta[\Psi] - \mathbb{E}_{\theta^k}[\Psi])^\top \left(\mathfrak{C}^{FR}(\tilde\theta)\right)^{-1} (\mathbb{E}_\theta[\Psi] - \mathbb{E}_{\theta^k}[\Psi]),$$

where $\mathfrak{C}^{FR}(\tilde\theta) = \mathbb{E}_{\tilde\theta}[\left(\Psi(x) - \mathbb{E}_{\tilde\theta}[\Psi]\right)\left(\Psi(x) - \mathbb{E}_{\tilde\theta}[\Psi]\right)^\top].$

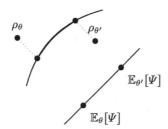

Fig. 1. Illustration of the proximity term over an affine space. Intuitively, the metric between two distributions is measured along a chosen set of statistics.

Remark 1. The matrix $\mathfrak{C}$ has size $n \times n$, corresponding to the dimension of Ψ. For the Fisher-Rao metric, it is the covariance of the basis functions Ψ w.r.t. $\rho_{\tilde\theta}$. This corresponds to the Fisher-Rao matrix when the basis is a sufficient statistics of the model. See Fig. 1. The resulting metric bears a similarity with the Relative Fisher Information Metric approach proposed in [19]. Similar observations apply for the Wasserstein metric.

Remark 2. In the case of implicit generative models (used in GANs), where ρ_θ is expressed as the push-forward measure of a latent variable z by a parametrized family of functions $\mathfrak{g}_\theta$, we obtain

$$\tilde{D}(\theta, \theta^k) = (\mathbb{E}_z[\Psi(\mathfrak{g}_\theta(z))] - \mathbb{E}_z[\Psi(\mathfrak{g}_{\theta^k}(z))])^\top \mathbb{E}_z[C(\mathfrak{g}_{\tilde\theta}(z))]^{-1}(\mathbb{E}_z[\Psi(\mathfrak{g}_\theta(z))] - \mathbb{E}_z[\Psi(\mathfrak{g}_{\theta^k}(z))]),$$

where C is the corresponding term inside the expectation in Theorem 1.

Proof. (i) For the constrained Wasserstein metric, the gradient of Φ w.r.t. the sample space variable x is $\nabla\Phi(x) = (\sum_{i=1}^n \xi_i \partial_l \psi_i(x))_l$. The squared norm is then

$$\|\nabla\Phi(x)\|^2 = \sum_l (\sum_i \xi_i \partial_l \psi_i(x))^2 = \sum_l \sum_i \xi_i \partial_l \psi_i(x) \sum_j \xi_j \partial_l \psi_j(x) = \xi^\top C^W(x)\xi,$$

where $C_{ij}^W(x) = \sum_l \partial_l \psi_i(x) \partial_l \psi_j(x)$. Now we consider the distance

$$\frac{1}{2}\tilde{D}_\Psi^W(\theta, \theta^k) = \sup_{\Phi = \xi^\top \Psi} \left(\Phi, \rho_\theta - \rho_{\theta^k}\right) - \frac{1}{2}\left((\nabla\Phi)^2, \rho_{\tilde\theta}\right)$$

$$= \sup_\xi \xi^\top (\mathbb{E}_\theta[\Psi] - \mathbb{E}_{\theta^k}[\Psi]) - \frac{1}{2}\xi^\top \mathbb{E}_{\tilde\theta}[C^W]\xi.$$

In turn, by first order optimality conditions, at the maximizer we have

$$\xi^* = (\mathbb{E}_{\tilde\theta}[C^W])^{-1}(\mathbb{E}_\theta[\Psi] - \mathbb{E}_{\theta^k}[\Psi]).$$

Thus $\tilde{D}_\Psi^W(\theta, \theta^k) = (\mathbb{E}_\theta[\Psi] - \mathbb{E}_{\theta^k}[\Psi])(\mathbb{E}_{\tilde\theta}[C^W])^{-1}(\mathbb{E}_\theta[\Psi] - \mathbb{E}_{\theta^k}[\Psi])$.

(ii) For the Fisher-Rao metric, the term $\|\Phi(z) - \mathbb{E}_{\tilde\theta}[\Phi]\|^2$ equals

$$\|\xi^\top \Psi(z) - \xi^\top \mathbb{E}_{\tilde\theta}[\Psi]\|^2 = \xi^\top (\Psi(z) - \mathbb{E}_{\tilde\theta}[\Psi])(\Psi(z) - \mathbb{E}_{\tilde\theta}[\Psi])^\top \xi = \xi^\top C^{FR}(z)\xi,$$

where $C^{FR}(z) = (\Psi(z) - \mathbb{E}_{\tilde\theta}[\Psi])(\Psi(z) - \mathbb{E}_{\tilde\theta}[\Psi])^\top$.   □

Example 1 (**Order 1 approximation**). For the metric approximation with the space of linear functions, $\mathcal{F}_1 = \left\{\Phi(x) = a^\top x + b : a \in \mathbb{R}^m, \ b \in \mathbb{R}\right\}$, we have:

(i)
$$\tilde{D}_1^W(\theta, \theta^k) = (\mathbb{E}_\theta[x] - \mathbb{E}_{\theta^k}[x])^\top (\mathbb{E}_\theta[x] - \mathbb{E}_{\theta^k}[x]).$$

(ii)
$$\tilde{D}_1^{FR}(\theta, \theta^k) = (\mathbb{E}_\theta[x] - \mathbb{E}_{\theta^k}[x])^\top \left(\mathbb{E}_{\tilde\theta}\left[(x - \mathbb{E}_{\tilde\theta}x)(x - \mathbb{E}_{\tilde\theta}x)^\top\right]\right)^{-1}(\mathbb{E}_\theta[x] - \mathbb{E}_{\theta^k}[x]).$$

Example 2 (**Order 2 approximation**). For the space of quadratic functions, $\mathcal{F}_2 = \left\{\Phi(x) = \frac{1}{2}x^\top Q x + a^\top x + b : Q \in \mathbb{R}^{m \times m}, \ a \in \mathbb{R}^m, \ b \in \mathbb{R}\right\}$, we have:

(i)
$$\tilde{D}_2^W(\theta, \theta^k) = \left(\mathbb{E}_\theta\begin{bmatrix} x \\ \frac{x \otimes x}{2}\end{bmatrix} - \mathbb{E}_{\theta^k}\begin{bmatrix} x \\ \frac{x \otimes x}{2}\end{bmatrix}\right)^\top \mathbb{E}_{\tilde\theta}\begin{bmatrix} I_m & x^\top \otimes I_m \\ x \otimes I_m & I_m \otimes xx^\top \end{bmatrix}^{-1} \left(\mathbb{E}_\theta\begin{bmatrix} x \\ \frac{x \otimes x}{2}\end{bmatrix} - \mathbb{E}_{\theta^k}\begin{bmatrix} x \\ \frac{x \otimes x}{2}\end{bmatrix}\right).$$

(ii)
$$\tilde{D}_2^{FR}(\theta, \theta^k) = \left(\mathbb{E}_\theta\begin{bmatrix} x \\ \frac{x \otimes x}{2}\end{bmatrix} - \mathbb{E}_{\theta^k}\begin{bmatrix} x \\ \frac{x \otimes x}{2}\end{bmatrix}\right)^\top \left(\mathfrak{C}^{FR}(\tilde\theta)\right)^{-1}\left(\mathbb{E}_\theta\begin{bmatrix} x \\ \frac{x \otimes x}{2}\end{bmatrix} - \mathbb{E}_{\theta^k}\begin{bmatrix} x \\ \frac{x \otimes x}{2}\end{bmatrix}\right),$$

where $\otimes$ is the Kronecker product (e.g., $x \otimes x$ is an $m^2 \times 1$ vector), and

$$\mathfrak{C}^{FR} = \mathbb{E}_{\tilde\theta}\left[\left(\begin{bmatrix} x \\ \frac{x \otimes x}{2}\end{bmatrix} - \mathbb{E}_{\tilde\theta}\begin{bmatrix} x \\ \frac{x \otimes x}{2}\end{bmatrix}\right)\left(\begin{bmatrix} x \\ \frac{x \otimes x}{2}\end{bmatrix} - \mathbb{E}_{\tilde\theta}\begin{bmatrix} x \\ \frac{x \otimes x}{2}\end{bmatrix}\right)^\top\right].$$

4   Numerical Examples

The optimization loop can be implemented as shown in Algorithm 1. Here the proximal operator is computed by a short gradient iteration. In practice we can replace the expectations by sample averages, $\mathbb{E}_\theta[f] \approx \frac{1}{N}\sum_{i=1}^N f(x^{(i)})$, with $x^{(i)}$ i.i.d. from ρ_θ. For the basis Ψ we can choose low order polynomials, as in Examples 1 and 2, but even random functions worked well in our experiments. The optimal choice will balance low dimension and relevant statistics for the model under consideration. Orthogonality tends to be beneficial.

Algorithm 1. Natural gradient with affine space proximal approximation.

Require: Loss F, basis of affine space Ψ, proximal step-size h, step-size α

  for $t = 0$ **to** max outer iterations **do**

    $\mathfrak{C}(\theta) = \mathrm{cov}_\theta[\Psi]^{-1}$ (Fisher-Rao); $\mathfrak{C}(\theta) = \mathbb{E}_\theta[\sum_l (\partial_l \Psi)(\partial_l \Psi)^\top]^{-1}$ (Wasserstein)

    for $t' = 0$ **to** max inner iterations **do**

      $\nabla_{\theta'} D(\theta, \theta') \leftarrow \frac{1}{2}\nabla_{\theta'}\mathbb{E}_{\theta'}[\Psi^\top]\mathfrak{C}(\theta)(\mathbb{E}_{\theta'}[\Psi] - \mathbb{E}_\theta[\Psi])$

      $\theta' \leftarrow \theta' - \alpha(\nabla_{\theta'}F(\theta') + \frac{1}{2h}\nabla D_{\theta'}(\theta, \theta'))$

    $\theta \leftarrow \theta'$

4.1 Maximum Likelihood Estimation for Hierarchical Models

We consider binary k-interaction models, which are exponential families $\rho_\theta(x) = \exp(\theta^\top A(x))/Z(\theta)$, $x \in \{0,1\}^m$, with sufficient statistics $A_\lambda(x) = \prod_{i \in \lambda}(-1)^{x_i}$, for $\lambda \subseteq \{1, \ldots, m\}$, $|\lambda| \leqslant k$. We use $\Psi_j(x) = (-1)^{x_j}$, $j \in \{1, \ldots, m\}$, which are sufficient statistics for the 1-interaction model (independence model). We draw target distributions uniformly from the simplex and compute the MLEs. We compare Euclidean, Fisher-Rao, Wasserstein, and proximals. For each problem and method we run grid search over the step size α and proximal strength h, which are kept fixed during optimization. The results are shown in Fig. 2.

4.2 Classification on CIFAR-10

Here we present an image classification task on the CIFAR-10 dataset [5] using the Wasserstein proximal method. We use a simple CNN with two convolutional layers followed by two fully-connected layers, with ReLU activations. In this experiment F is the categorical cross-entropy loss and $D = \tilde{D}_\Psi^W$ is the Order 1 or Order 2 Wasserstein approximation. The specific details of our experiments can be found in Appendix A. Figure 2 provides the results, where we give curves

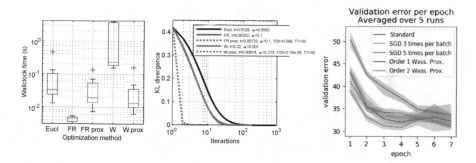

Fig. 2. Left: MLE wall-clock computation times until the KL-divergence is within 10^{-9} of optimal, for 4 binary variables and Ψ the independence model, and typical optimization curves. Right: The learning curves for the image classification task on CIFAR-10. Each experiment was averaged over 5 runs. The bold lines represent the average, and the envelopes are the minimum and maximum achieved.

for the validation error per epoch. As a baseline, we also give results when performing SGD many times per epoch, but without regularization. We see that the best result comes from the Order 2 Wasserstein distance approximation.

5   Discussion

We studied sampling–friendly implementations of the natural gradient based on the proximal operator. We approximate the proximity penalty by an affine space restriction in the Legendre dual formulation. This gives rise to a lower dimensional metric, expressed in expectation parameters, which can be estimated from samples. We cover both Fisher-Rao and Wasserstein metrics. Especially for the Wasserstein proximal, our method offers significant savings in computation time and provide improvement in validation error (in CIFAR-10 classification).

Acknowledgement. This project has received funding from AFOSR MURI FA9550-18-1-0502 and the European Research Council (ERC) under the European Union's Horizon 2020 research and innovation programme (grant agreement n° 757983).

A   Appendix for image classification on CIFAR-10

Here is the detailed version of our experiments, for image classification on CIFAR-10. We use a simple CNN with two convolutional layers (each with 32 filters, with a kernel size of 3×3, a stride of 1, and zero padding), followed by two fully-connected layers each having 512 nodes. For the optimizer, we use standard stochastic gradient descent (SGD) with momentum value 0.95 and learning rate of 0.001.

For the Wasserstein distance, if we denote the (deterministic) output of our neural network as $f(x, \theta)$ (the log probability vector), the loss function as $L(y, f(x, \theta))$ where x is the image and y the label, and the dataset as $\mathcal{D}$, then the Order 1 and Order 2 approximations for the Wasserstein distance on image classification on CIFAR-10 are: Order 1 approximation:

$$\tilde{D}_1^W(\theta, \theta^k) = \|\mathbb{E}_{x \sim \mathcal{D}}[f(x, \theta)] - \mathbb{E}_{x \sim \mathcal{D}}[f(x, \theta^k)]\|^2, \tag{7}$$

and the Order 2 approximation:

$$\tilde{D}_2^W(\theta, \theta^k) = \|\mathbb{E}_{x \sim \mathcal{D}}[f(x, \theta)] - \mathbb{E}_{x \sim \mathcal{D}}[f(x, \theta^k)]\|^2$$
$$+ \mathrm{tr}\left(\mathrm{var}_{x \sim \mathcal{D}}[f(x, \theta)] + \mathrm{var}_{x \sim \mathcal{D}}[f(x, \theta^k)] \right.$$
$$\left. - 2 \left(\mathrm{var}_{x \sim \mathcal{D}}[f(x, \theta^k)]^{1/2} \, \mathrm{var}_{x \sim \mathcal{D}}[f(x, \theta)] \, \mathrm{var}_{x \sim \mathcal{D}}[f(x, \theta^k)]^{1/2} \right) \right).$$
$$\tag{8}$$

We present our experiments on 5 different settings: (1) Standard learning with no regularization, (2) performing SGD 3 times per batch, (3) performing SGD 5 times per batch, (4) using the Order 1 Wasserstein Proximal (with $m = 3$

and $h = 2$), (5) and using the Order 2 Wasserstein proximal (with $m = 5$ and $h = 1$). From Fig. 2, we see that using the Order 2 Wasserstein proximal provides the best results. We note that performing SGD a number of times per batch is presented as a baseline, as we experimentally found that they also provided improvements in validation error per epoch (but they are not the best as we can see from Fig. 2).

Algorithm 2. Wasserstein Proximal Natural Gradient for Neural Networks

---

Require: Loss function L, neural network $f(x, \theta)$, Order 1 or 2 Wasserstein distance approximation D, and data-label pairs $\{(x, y)\}$ from dataset $\mathcal{D}$.
Require: m number of gradient descent steps, and h strength of the proximal term
  while stopping criteria not met **do**
    Sample a mini-batch of image-label pairs $\{(x_b, y_b)\}_{b=1}^{B} \in \mathcal{D}$
    Approximately solve (by performing SGD m times)

$$\theta^{k+1} \leftarrow \operatorname{argmin}_{\theta} \left\{ \frac{1}{B} \sum_{b=1}^{B} L(y, f(x, \theta)) + \frac{1}{2h} D(\theta, \theta^k) \right\}$$

References

1. Amari, S.: Differential-Geometrical Methods in Statistics. Lecture Notes in Statistics. Springer-Verlag, New York (1985)
2. Amari, S.: Natural gradient works efficiently in learning. Neural Comput. **10**(2), 251–276 (1998)
3. Desjardins, G., Simonyan, K., Pascanu, R., Kavukcuoglu, K.: Natural neural networks. In: Cortes, C., Lawrence, N.D., Lee, D.D., Sugiyama, M., Garnett, R. (eds.) NIPS 28, pp. 2071–2079. Curran Associates Inc. (2015)
4. Fisher, R.A.: On the mathematical foundations of theoretical statistics. Philos. Trans. Roy. Soc. London Ser. A **222**, 309–368 (1922)
5. Krizhevsky, A.: Learning multiple layers of features from tiny images. Technical report, University of Toronto (2009)
6. Lafferty, J.D.: The density manifold and configuration space quantization. Trans. Am. Math. Soc. **305**(2), 699–741 (1988)
7. Li, W.: Geometry of probability simplex via optimal transport. arXiv:1803.06360 [math] (2018)
8. Li, W., Montúfar, G.: Natural gradient via optimal transport. Inf. Geom. **1**(2), 181–214 (2018)
9. Li, W., Montúfar, G.: Ricci curvature for parametric statistics via optimal transport. arXiv:1807.07095 [cs, math, stat] (2018)
10. Liang, T., Poggio, T.A., Rakhlin, A., Stokes, J.: Fisher-Rao metric, geometry, and complexity of neural networks. CoRR, abs/1711.01530 (2017)
11. Lin, A., Li, W., Osher, S., Montúfar, G.: Wasserstein proximal of GANs. In: CAM reports (2018)
12. Malagò, L., Montrucchio, L., Pistone, G.: Wasserstein Riemannian geometry of Gaussian densities. Inf. Geom. **1**(2), 137–179 (2018)

13. Martens, J., Grosse, R.: Optimizing neural networks with Kronecker-factored approximate curvature. In: ICML 32. PMLR, vol. 37, pp. 2408–2417 (2015)
14. Modin, K.: Geometry of matrix decompositions seen through optimal transport and Information Geometry. J. Geom. Mech. **9**(3), 335–390 (2017)
15. Otto, F.: The geometry of dissipative evolution equations the porous medium equation. Commun. Partial Differ. Eqn. **26**(1–2), 101–174 (2001)
16. Park, H., Amari, S., Fukumizu, K.: Adaptive natural gradient learning algorithms for various stochastic models. Neural Networks **13**(7), 755–764 (2000)
17. Peters, J., Schaal, S.: Natural actor-critic. Neurocomputing **71**(7), 1180–1190 (2008). Progress in Modeling, Theory, and Application of Computational Intelligenc
18. Rao, C.R.: Information and the accuracy attainable in the estimation of statistical parameters. Bull. Calcutta Math. Soc. **37**, 81–89 (1945)
19. Sun, K., Nielsen, F.: Relative Fisher information and natural gradient for learning large modular models. In: ICML 34. PMLR, vol. 70, pp. 3289–3298 (2017)
20. Villani, C.: Optimal Transport: Old and New. Grundlehren der mathematischen Wissenschaften, vol. 338. Springer, Heidelberg (2009). https://doi.org/10.1007/978-3-540-71050-9

Parametric Fokker-Planck Equation

Wuchen Li[1], Shu Liu[2]($\boxtimes$), Hongyuan Zha[2], and Haomin Zhou[2]

[1] University of California, Los Angeles, USA
[2] Georgia Institute of Technology, Atlanta, USA
Sliu459@gatech.edu

Abstract. We derive the Fokker-Planck equation on the parametric space. It is the Wasserstein gradient flow of relative entropy on the statistical manifold. We pull back the PDE to a finite dimensional ODE on parameter space. Some analytical examples and numerical examples are presented.

Keywords: Optimal transport · Information geometry ·
Statistical manifold · Fokker-Planck equation · Gradient flow

1 Introduction

Fokker-Planck equation, a linear evolution partial differential equation (PDE), plays a crucial role in stochastic calculus, statistical physics and modeling [13,16,18]. Recently, people also discover its importance in statistics and machine learning [11,15,17]. Fokker-Planck equation describes the evolution of density functions of the stochastic process driven by a stochastic differential equation (SDE).

There is another viewpoint of Fokker-Planck equation based on optimal transport theory. It treats the equation as the gradient flow of relative entropy on probability manifold equipped with Wasserstein metric [5,14]. Recently, the studies have been extended to information geometry [1–3], creating a new area known as Wasserstein information geometry [7,9,10]. Inspired by those studies, in this paper, we derive the metric tensor on parameter space by pulling back the Wasserstein metric via the parameterized pushforward map. Then we compute the Wasserstein gradient flow (an ODE system) of relative entropy defined on parameter space. This leads to a statistical manifold version of Fokker Planck equation, which can be viewed as an approximation of the original PDE.

Our work is motivated by two purposes, (1) reducing the evolution PDE to a finite dimensional ODE system on parameter space; (2) applying parameterized pushforward map to obtain an efficient sampling method to generate samples from SDE. This is different from Markov Chain Monte Carlo (MCMC) methods [12] or momentum methods [16]. In this brief presentation, we sketch the theoretical framework with illustrations on several examples. The complete results will be reported in an extended version [8].

© Springer Nature Switzerland AG 2019
F. Nielsen and F. Barbaresco (Eds.): GSI 2019, LNCS 11712, pp. 715–724, 2019.
https://doi.org/10.1007/978-3-030-26980-7_74

2   Parametric Fokker-Planck Equation

In this section, we briefly review the fact that Fokker-Planck equation is a Wasserstein gradient flow of relative entropy. We then introduce a Wasserstein statistical manifold generated by parameterized mapping function. Based on it, we derive the parametric Fokker-Planck equation as the gradient flow of parameterized relative entropy.

2.1   Fokker-Planck Equation

Consider the Fokker-Planck equation:

$$\frac{\partial \rho(t,x)}{\partial t} = \nabla \cdot (\rho(t,x)\nabla V(x)) + \beta \Delta \rho(t,x), \quad \rho(0,x) = \rho_0(x). \tag{1}$$

Here $\nabla \cdot$, ∇ is the divergence and gradient operator in $\mathbb{R}^d$, ∇V is the drift function and $\beta > 0$ is a diffusion constant. There are several understandings for the Eq. (1).

On the one hand, consider the stochastic differential equation:

$$d\boldsymbol{X}_t = -\nabla V(\boldsymbol{X}_t) + \sqrt{2\beta}d\boldsymbol{B}_t, \quad \boldsymbol{X}_0 \sim \rho_0. \tag{2}$$

Here $\{\boldsymbol{B}_t\}_{t\geq0}$ is the standard Brownian motion. It is well known that the density function $\rho(t,x)$ of stochastic process $\boldsymbol{X}_t$, i.e. $\boldsymbol{X}_t \sim \rho(t,x)$, satisfies the Fokker-Planck equation (1).

On the other hand, Eq. (1) is the Wasserstein gradient flow of relative entropy. Denote the probability space supported on $\mathbb{R}^d$:

$$\mathcal{P} = \left\{\rho\colon \int \rho(x)dx = 1, \ \rho(x) \geq 0, \ \int |x|^2\rho(x)\ dx < \infty\right\}$$

Equipped with the Wasserstein metric [6,14], $\mathcal{P}$ is an infinite dimensional Riemannian manifold. Denote

$$T_\rho\mathcal{P} = \left\{\dot{\rho}\colon \int \dot{\rho}(x)dx = 0\right\}.$$

Consider a specific $\rho \in \mathcal{P}$ and $\dot{\rho}_i \in T_\rho\mathcal{P}$, $i = 1,2$. The Wasserstein metric tensor g^W is defined as:

$$g^W(\rho)(\dot{\rho}_1, \dot{\rho}_2) = \int \nabla \psi_1(x) \cdot \nabla \psi_2(x)\rho(x)\ dx,$$

where $\dot{\rho}_i = -\nabla \cdot (\rho_i \nabla \psi_i)$ for $i = 1,2$. Here g^W is a metric tensor, which is a positive definite bilinear form defined on tangent bundle $T\mathcal{P} = \{(\rho, \dot{\rho})\colon \rho \in \mathcal{P}, \ \dot{\rho} \in T_\rho\mathcal{P}\}$.

The Riemannian gradient in $(\mathcal{P}, g^W)$ is given as follows. Consider a smooth functional $\mathcal{F} \colon \mathcal{P} \to \mathbb{R}$, then

$$
\begin{aligned}
\operatorname{grad}_W \mathcal{F}(\rho) &= g^W(\rho)^{-1}\left(\frac{\delta \mathcal{F}}{\delta \rho}\right)(x) \\
&= -\nabla \cdot (\rho(x)\nabla \frac{\delta}{\delta \rho(x)}\mathcal{F}(\rho)),
\end{aligned}
\tag{3}
$$

where $\frac{\delta}{\delta \rho(x)}$ is the L^2 first variation at variable $x \in \mathbb{R}^d$. In particular, consider the relative entropy

$$
\mathcal{F}(\rho) = \beta \int \rho(x) \log \frac{\rho(x)}{\frac{1}{Z}e^{-\frac{V(x)}{\beta}}}\ dx = \int V(x)\rho(x)dx + \beta \int \rho(x)\log \rho(x)dx + \beta \log(Z).
\tag{4}
$$

Here $Z = \int e^{\frac{V(x)}{\beta}}\ dx$ is the normalizing constant for $e^{\frac{V(x)}{\beta}}$.
Then $\nabla\left(\frac{\delta \mathcal{F}}{\delta \rho}\right) = \nabla V + \beta \nabla \log \rho$, and (3) forms

$$
\frac{\partial \rho}{\partial t} = -\operatorname{grad}_W \mathcal{F}(\rho) = \nabla \cdot (\rho \nabla V) + \beta \nabla \cdot (\rho \nabla \log \rho)).
$$

Notice $\nabla \log \rho = \frac{\nabla \rho}{\rho}$, then $\nabla \cdot (\rho \nabla \log \rho) = \nabla \cdot (\nabla \rho) = \Delta \rho$. The above equation is exactly Fokker-Planck equation (1).

From now on, we apply the above geometric gradient flow formulation and derive the Fokker-Planck equation (1) on parameter space.

2.2 Parameter Space Equipped with Wasserstein Metric

We consider a parameter space Θ as an open set in $\mathbb{R}^m$. Denote the sample space $M = \mathbb{R}^d$. Suppose T_θ is a pushforward map from M to M, which is parametrized by θ. For example, we can set $T_\theta(x) = Ux + b$, with $\theta = (U, b), U \in GL_d(\mathbb{R})$, $b \in \mathbb{R}^d$; we can also let T_θ be a neural network with parameter θ. We further assume that T_θ is invertible and smooth with respect to parameter θ and variable x.

Denote $p \in \mathcal{P}$ as a reference probability measure with positive density defined on M. For example, we can choose p as the standard Gaussian. We denote ρ_θ as the density of $T_{\theta\#}p$.[1] We further require: $\int |T_\theta(x)|^2\ dp(x) < \infty$ holds for all $\theta \in \Theta$. Then $\rho_\theta \in \mathcal{P}$ for each $\theta \in \Theta$. Denote $\mathcal{P}_\Theta = \{\rho_\theta = \rho(\theta, x)|\theta \in \Theta\}$, then $\mathcal{P}_\Theta \subset \mathcal{P}$.

Now the connection between $\mathcal{P}$ and Θ is the pushforward operation $T_\# \colon \Theta \to \mathcal{P}_\Theta \subset \mathcal{P}, \theta \mapsto \rho_\theta$. In order to introduce the Wasserstein metric to parameter space Θ, we assume that $T_\#$ is an isometric immersion from Θ to $\mathcal{P}$. Under this assumption, the pullback $(T_\#)^* g^W$ of the Wasserstein metric g^W by $T_\#$ is the metric tensor on Θ. Let us denote $G = (T_\#)^* g^W$. Then for each θ, $G(\theta)$ is a

---

[1] Let X, Y be two measurable spaces, λ is a probability measure defined on X; let $T \colon X \to Y$ be a measurable map, then $T_\#\lambda$ is defined as: $T_\#\lambda(E) = \lambda(T^{-1}(E))$ for all measurable $E \subset Y$. We call $T_\# p$ the pushforward of measure p by map T.

bilinear form on $T_\theta \Theta \simeq \mathbb{R}^m$, thus $G(\theta)$ can be treated as an $m \times m$ matrix. Computation of $G(\theta)$ is illustrated in the following theorem:

Theorem 1. *Suppose $T_\# : \Theta \to \mathcal{P}$ is isometric immersion from Θ to $\mathcal{P}$. Then the metric tensor $G(\theta)$ at $\theta \in \Theta$ is $m \times m$ non-negative definite symmetric matrix and can be computed as:*

$$G(\theta) = \int \nabla \boldsymbol{\Psi}(T_\theta(x)) \nabla \boldsymbol{\Psi}(T_\theta(x))^T \, dp(x), \tag{5}$$

Or in entry-wised form:

$$G_{ij}(\theta) = \int \nabla \psi_i(T_\theta(x)) \cdot \nabla \psi_j(T_\theta(x)) \, dp(x), \quad 1 \le i,j \le m.$$

Here $\boldsymbol{\Psi} = (\psi_1, ... \psi_m)^T$ and $\nabla \boldsymbol{\Psi}$ is $m \times d$ Jacobian matrix of $\boldsymbol{\Psi}$. For each $k = 1, 2, ..., m$, ψ_k solves the following equation:

$$\nabla \cdot (\rho_\theta \nabla \psi_k(x)) = \nabla \cdot (\rho_\theta \, \partial_{\theta_k} T_\theta(T_\theta^{-1}(x))). \tag{6}$$

Proof. Suppose $\xi \in T\Theta$ is a vector field on Θ, for a fixed $\theta \in \Theta$, we first compute the pushforward $(T_\# |_\theta)_* \xi(\theta)$ of ξ at point θ: We choose any differentiable curve $\{\theta_t\}_{t \ge 0}$ on Θ with $\theta_0 = \theta$ and $\dot{\theta}_0 = \xi(\theta)$. If we denote $\rho_{\theta_t} = T_{\theta_t} \# p$, then we have $(T_\#)_* \xi(\theta) = \frac{\partial \rho_{\theta_t}}{\partial t} \big|_{t=0}$. To compute $\frac{\partial \rho_{\theta_t}}{\partial t} \big|_{t=0}$, we consider for any $\phi \in C_0^\infty(M)$:

$$\int \phi(y) \frac{\partial \rho_{\theta_t}}{\partial t}(y) dy = \frac{\partial}{\partial t} \left(\int \phi(T_{\theta_t}(x)) dp \right) = \int \dot{\theta}_t^T \partial_\theta T_{\theta_t}(x) \nabla \phi(T_{\theta_t}(x)) dp$$

$$= \int \dot{\theta}_t^T \partial_\theta T_{\theta_t}(T_{\theta_t}^{-1}(x)) \nabla \phi(x) \, \rho_{\theta_t}(x) \, dx$$

$$= \int \phi(x) \left(-\nabla \cdot (\rho_{\theta_t} \partial_\theta T_{\theta_t}(T_{\theta_t}^{-1}(x))^T \, \dot{\theta}_t) \right) \, dx$$

This weak formulation reveals that

$$(T_\# |_\theta)_* \xi(\theta) = \frac{\partial \rho_{\theta_t}}{\partial t} \bigg|_{t=0} = -\nabla \cdot (\rho_\theta \, \partial_\theta T_\theta(T_\theta^{-1}(x))^T \, \xi(\theta)) \tag{7}$$

Now let us compute the metric tensor G. Since $T_\#$ is isometric immersion from Θ to $\mathcal{P}$, the pullback of g^W by $T_\#$ gives G, i.e. $(T_\#)^* g^W = G$. By definition of pullback map, for any $\xi \in T\Theta$ and for any $\theta \in \Theta$, we have:

$$G(\theta)(\xi(\theta), \xi(\theta)) = g^W(\rho_\theta)((T_\# |_\theta)_* \xi(\theta), (T_\# |_\theta)_* \xi(\theta)) \tag{8}$$

To compute the right hand side of (8), recall (3), we need to solve for φ from:

$$\frac{\partial \rho_{\theta_t}}{\partial t} \bigg|_{t=0} = -\nabla \cdot (\rho_\theta \nabla \varphi(x)) \tag{9}$$

By (7), (9) is:

$$\nabla \cdot (\rho_\theta \nabla \varphi(x)) = \nabla \cdot (\rho_\theta \partial_\theta T_\theta(T_\theta^{-1}(\cdot))^T \, \xi(\theta)) \tag{10}$$

We can straightforwardly check that $\varphi(x) = \boldsymbol{\Psi}^T(x)\xi(\theta)$ is the solution of (10). Then $G(\theta)$ is computed as:

$$G(\theta)(\xi,\xi) = \int |\nabla\varphi(y)|^2 \, \rho_\theta(y) \, dy = \int |\nabla\varphi(T_\theta(x))|^2 \, dp(x)$$

$$= \int |\nabla\boldsymbol{\Psi}(T_\theta(x))^T\xi|^2 dp(x) = \xi^T \left(\int \nabla\boldsymbol{\Psi}(T_\theta(x))\nabla\boldsymbol{\Psi}(T_\theta(x))^T dp(x) \right) \xi$$

Thus we can verify that:

$$G(\theta) = \int \nabla\boldsymbol{\Psi}(T_\theta(x))\nabla\boldsymbol{\Psi}(T_\theta(x))^T \, dp(x)$$

Generally speaking, the metric tensor G doesn't have an explicit form when $d \geq 2$; but for $d = 1$, G has an explicit form and can be computed directly.

Corollary 1. *When dimension d of M equals 1. And we further assume that: $\rho_\theta > 0$ on M and $\lim_{x\to\pm\infty}\rho_\theta(x) = 0$. Then $G(\theta)$ has an explicit form:*

$$G(\theta) = \int \partial_\theta T_\theta(x)^T \partial_\theta T_\theta(x) \, dp(x). \tag{11}$$

The following theorem ensures the positive definiteness of the metric tensor G:

Theorem 2. *We follow the notations and conditions in Sects. 2.2 and 2.3. Then G is Riemmanian metric on $T\Theta$ iff For each $\theta \in \Theta$, for any $\xi \in T_\theta\Theta$ ($\xi \neq 0$), we can find $x \in M$ such that $\nabla \cdot (\rho_\theta \, \partial_\theta T_\theta(T_\theta^{-1}(x)\xi) \neq 0$.*

To keep our discussion concise, in the following sections, we will always assume that G is positive definite on $T\Theta$.
From now on, following [9,10], we call (Θ, G) Wasserstein statistical manifold.

2.3 Fokker-Planck Equation on Statistical Manifold

Recall the relative entropy functional $\mathcal{F}$ defined in (4), we consider $F = \mathcal{F} \circ T_\# : \Theta \to \mathbb{R}$. Then:

$$F(\theta) = \mathcal{F}(\rho_\theta) = \int V(x)\rho_\theta(x) \, dx + \beta \int \rho_\theta(x) \log \rho_\theta(x) \, dx. \tag{12}$$

As in [1], the gradient flow of F on Wasserstein statistical manifold (Θ, G) satisfies

$$\dot{\theta} = -G(\theta)^{-1}\nabla_\theta F(\theta). \tag{13}$$

We call (13) *parametric Fokker-Planck equation*. The ODE (13) as the Wasserstein gradient flow on parameter space (Θ, G) is closely related to Fokker-Planck equation on probability submanifold $\mathcal{P}_\Theta$. We have the following theorem, which is a natural result derived from submanifold geometry:

Theorem 3. *Suppose $\{\theta_t\}_{t\geq 0}$ solves (13). Then $\{\rho_{\theta_t}\}$ is the gradient flow of $\mathcal{F}$ on probability submanifold $\mathcal{P}_\Theta$.*

3   Example on Fokker-Planck Equations with Quadratic Potential

The solution of Fokker-Planck equation on statistical manifold (13) can serve as an approximation to the solution of the original Eq. (1). However, in some special cases, ρ_{θ_t} exactly solves (1). In this section, we demonstrate such examples.

Let us consider Fokker-Planck equations with quadratic potentials whose initial conditions are Gaussian, i.e.

$$V(x) = \frac{1}{2}(x - \mu)^T \Sigma^{-1}(x - \mu) \quad \text{and} \quad \rho_0 \sim \mathcal{N}(\mu_0, \Sigma_0). \tag{14}$$

Consider parameter space $\Theta = (\Gamma, b) \subset \mathbb{R}^m$ $(m = d(d+1))$, where Γ is a $d \times d$ invertible matrix with $\det(\Gamma) > 0$ and $b \in \mathbb{R}^d$. We define the parametric map as $T_\theta(x) = \Gamma x + b$. We choose the reference measure $p = \mathcal{N}(0, I)$. Here is the lemma we have to use:

Lemma 1. *Let $\mathcal{F}$ be the relative entropy defined in (4) and F defined in (12). For $\theta \in \Theta$, If the vector function $\nabla\left(\frac{\delta\mathcal{F}}{\delta\rho}\right) \circ T_\theta$ can be written as the linear combination of $\{\frac{\partial T_\theta}{\partial\theta_1}, ..., \frac{\partial T_\theta}{\partial\theta_m}\}$, i.e. there exists $\zeta \in \mathbb{R}^m$, such that $\nabla\left(\frac{\delta\mathcal{F}}{\delta\rho}\right) \circ T_\theta(x) = \partial_\theta T_\theta(x)\zeta$. Then:*

(1) $\zeta = G(\theta)^{-1}\nabla_\theta F(\theta)$, which is the Wasserstein gradient of F at θ.
(2) If we denote the gradient of $\mathcal{F}$ on $\mathcal{P}$ as $\mathrm{grad}\mathcal{F}(\rho_\theta)$ and the gradient of $\mathcal{F}$ on the submanifold $\mathcal{P}_\Theta$ as $\mathrm{grad}\mathcal{F}(\rho_\theta)|_{\mathcal{P}_\Theta}$, then $\mathrm{grad}\mathcal{F}(\rho_\theta)|_{\mathcal{P}_\Theta} = \mathrm{grad}\mathcal{F}(\rho_\theta)$.

Proof. The detailed proof is provided in [8]. Here is an intuitive explanation: $\nabla\left(\frac{\delta\mathcal{F}}{\delta\rho}\right) = \nabla V + \beta\nabla\log\rho_\theta$ is the real vector field that moves the particles in Fokker-Planck equation; and $\partial_\theta T_\theta(T_\theta^{-1}(\cdot))\dot\theta$ is the approximate vector field induced by the pushforward map T_θ. If such approximate is perfect with zero error, i.e. exits ζ such that $\nabla\left(\frac{\delta\mathcal{F}}{\delta\rho}\right) \circ T_\theta(x) = \partial_\theta T_\theta(x)\zeta$, then $\zeta = \dot\theta = G(\theta)^{-1}\nabla_\theta F(\theta)$ and the submanifold gradient agrees with entire manifold gradient.

Now, let us come back to our example, we can compute

$$\rho_\theta(x) = T_{\theta\#}p(x) = \frac{f(T_\theta^{-1}(x))}{|\det(\Gamma)|} = \frac{f(\Gamma^{-1}(x-b))}{|\det(\Gamma)|}, \quad f(x) = \frac{\exp(-\frac{1}{2}|x|^2)}{(2p)^{\frac{d}{2}}}.$$

Then we have:

$$\nabla\left(\frac{\delta\mathcal{F}(\rho_\theta)}{\delta\rho}\right) \circ T_\theta(x) = \nabla(V + \beta\log\rho_\theta) \circ T_\theta(x) = \Sigma^{-1}(\Gamma x + b - \mu) - \beta\Gamma^{-T}x$$

is affine w.r.t. x.

Notice that $\partial_{\Gamma_{ij}} T_\theta(x) = (..0.. \ \underset{i-\text{th}}{x_j} \ ..0..)^T$ and $\partial_{b_i} T_\theta = (..0.. \ \underset{i-\text{th}}{1} \ ..0..)^T$. We can

verify that $\zeta = (\Sigma^{-1}\Gamma - \beta\Gamma^{-T}, \Sigma^{-1}(b-\mu))$ solves $\nabla\left(\frac{\delta \mathcal{F}(\rho_\theta)}{\delta\rho}\right) \circ T_\theta(x) = \partial_\theta T_\theta(x)\zeta$.
By (1) of Lemma 1, $\zeta = G(\theta)^{-1}\nabla_\theta F(\theta)$. Thus ODE (13) for our example is:

$$\dot{\Gamma} = -\Sigma^{-1}\Gamma + \beta\Gamma^{-T} \quad \Gamma_0 = \sqrt{\Sigma_0} \tag{15}$$

$$\dot{b} = \Sigma^{-1}(\mu - b) \quad b_0 = \mu_0 \tag{16}$$

By (2) of Lemma 1, we know $\text{grad}\mathcal{F}(\rho_\theta)|_{\mathcal{P}_\Theta} = \text{grad}\mathcal{F}(\rho_\theta)$ for all $\theta \in \Theta$. This indicates that there is no local error for our approximation, one can verify that the solution to the parametric Fokker-Planck equation also solves the original equation.

In addition to previous results, we have the following corollary:

Corollary 2. *The solution of Fokker-Planck equation (1) with condition (14) is Gaussian distribution for all $t > 0$.*

Proof. If we denote $\{\Gamma_t, b_t\}$ as the solutions to (15), (16), set $\theta_t = (\Gamma_t, b_t)$, then $\rho_t = T_{\theta_t \#}p$ solves the Fokker Planck Equation (1) with conditions (14). Since the pushforward of Gaussian distribution p by an affine transform T_θ is still a Gaussian, we conclude that for any $t > 0$, the solution $\rho_t = T_{\theta_t \#}p$ is always Gaussian distribution. This is already a well known result about Fokker-Planck equation. We reprove it under our framework.

4 Numerical Examples for 1D Fokker-Planck Equation

Since the Wasserstein metric tensor G has an explicit solution when dimension $d = 1$, it is convenient to numerically compute ODE (13).

For example, we can choose a series of basis functions $\{\varphi_k\}_{k=1}^n$. Each φ_k can be chosen as a sinusoidal function or a piece-wise linear function defined on a certain interval $[-l, l]$. It is also beneficial to choose orthogonal or near-orthogonal basis functions because they will keep the metric tensor G far away from ill-posedness. We set $T_\theta(x) = \sum_{k=1}^m \theta_k \varphi_k(x)^2$. Then according to (11), we can compute G as

$$G_{ij}(\theta) = \mathbb{E}_{\mathbf{X} \sim p}\left[\varphi_i(\mathbf{X})\varphi_j(\mathbf{X})\right] \quad 1 \leq i, j \leq m$$

Recall that $F(\theta) = \int V(x)\rho_\theta(x)dx + \beta \int \rho_\theta(x)\log\rho_\theta(x)dx$. The second part of F is the entropy of ρ_θ, which can be computed by solving the following optimization problem [4]:

$$\int \rho_\theta(x)\log\rho_\theta(x) \ dx = \sup_h\left\{\int h(x)\rho_\theta(x) \ dx - \int e^{h(x)}dx\right\} + 1 \tag{17}$$

---

[2] In application, carefully choosing T_θ which is not necessarily invertibile or smooth can still provide valid results.

We can solve (17) by parametrizing h. Suppose the optimal solution is h^*. Then by envelope theorem, we know $\nabla_\theta F(\theta)$ can be computed as

$$\nabla_\theta F(\theta) = \partial_\theta \left(\int V(x)\rho_\theta(x) \, dx + \beta \int h^*(x)\rho_\theta(x) \, dx \right)$$
$$= \mathbb{E}_{\mathbf{x} \sim p} \left[\partial_\theta T_\theta(\mathbf{X})^T \nabla_y (V(y) + \beta h^*(y))|_{y=T_\theta(\mathbf{X})} \right] \quad (18)$$

Notice that both the metric tensor G and $\nabla_\theta F(\theta)$ are written in forms of expectations, thus we can compute them by Monte Carlo simulations. And finally, (13) can be computed by forward Euler method.

Our numerical results are always demonstrated by sample points: For each time node t, we sample points $\{\mathbf{X}_1, ..., \mathbf{X}_N\}$ from p, then $\{T_{\theta_t}(\mathbf{X}_1), ..., T_{\theta_t}(\mathbf{X}_N)\}$ are our numerical samples from distribution ρ_t which solves the Fokker-Planck equation.

Here are several numerical results based on our method. We exhibit them in the form of histograms. Consider the potential $V(x) = (x+1)^2(x-1)^2$. Suppose the initial distribution is $\rho_0 = \mathcal{N}(0, I)$. Figure 1 contains histograms of ρ_t which solves $\frac{\partial \rho}{\partial t} = \nabla \cdot (\rho \nabla V)$ at different time nodes; we know ρ_t converges to $\frac{\delta_{-1} + \delta_{+1}}{2}$

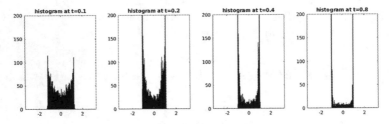

Fig. 1. Histograms of ρ_t solving $\frac{\partial \rho}{\partial t} = \nabla \cdot (\rho \nabla V)$

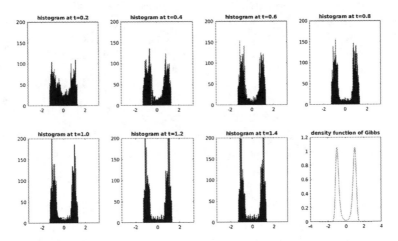

Fig. 2. Histograms of ρ_t solving $\frac{\partial \rho}{\partial t} = \nabla \cdot (\rho \nabla V) + \frac{1}{4}\Delta \rho$

as $t \to \infty$. Here δ_a is the Dirac distribution concentrated on point a. Figure 2 contains histograms of ρ_t which solves $\frac{\partial \rho}{\partial t} = \nabla \cdot (\rho \nabla V) + \frac{1}{4}\Delta \rho$ at different time nodes, we know ρ_t will converge to Gibbs distribution $\rho_* = \frac{1}{Z}\exp(-4(x+1)^2(x-1)^2)$, with Z being a normalizing constant, as $t \to \infty$. The density function of ρ_* is exhibited in Fig. 2.

5   Discussion

We presented a new approach for approximating Fokker-Planck equations by parameterized push-forward mapping functions. Compared to the classical moment method and MCMC method, we propose a systematic way for obtaining a finite dimensional ODE on parameter space. The ODE represents the evolution of statistical information conveyed in the original Fokker-Planck equation. In the future, we will study its geometric and statistical properties, and derive practical numerical methods for applications in scientific computing and machine learning. To be specific, in scientific computing, our techniques can be used to provide numerical solutions (samples) to those evolution PDEs that can be treated as Wasserstein gradient flows of certain functions defined on probability manifold; in area of machine learning, we wish to create efficient sampling methods based on our computational tools designed for Wasserstein gradients.

Acknowledgement. This project has received funding from AFOSR MURI FA9550-18-1-0502 and NSF Awards DMS–1419027, DMS-1620345, and ONR Award N000141310408.

References

1. Amari, S.: Natural gradient works efficiently in learning. Neural Comput. **10**(2), 251–276 (1998)
2. Amari, S.: Information Geometry and Its Applications. AMS, vol. 194. Springer, Tokyo (2016). https://doi.org/10.1007/978-4-431-55978-8
3. Ay, N., Jost, J., Lê, H.V., Schwachhöfer, L.: Information Geometry. EMG-FASMSM, vol. 64. Springer, Cham (2017). https://doi.org/10.1007/978-3-319-56478-4
4. Essid, M., Laefer, D., Tabak, E.G.: Adaptive Optimal Transport. arXiv:1807.00393 [math] (2018)
5. Jordan, R., Kinderlehrer, D., Otto, F.: The variational formulation of the fokker-planck equation. SIAM J. Math. Anal. **29**(1), 1–17 (1998)
6. Lafferty, J.D.: The density manifold and configuration space quantization. Trans. Am. Math. Soc. **305**(2), 699–741 (1988)
7. Li, W.: Geometry of probability simplex via optimal transport. arXiv:1803.06360 [math] (2018)
8. Li, W., Liu, S., Zha, H., Zhou, H.: Scientific computing via parametric fokker-planck equations. In preparation (2019)
9. Li, W., Montufar, G.: Natural gradient via optimal transport. arXiv:1803.07033 [cs, math] (2018)

10. Li, W., Montufar, G.: Ricci curvature for parametric statistics via optimal transport (2018)
11. Liu, Q., Wang, D.: Stein variational gradient descent: a general purpose bayesian inference algorithm. arXiv:1608.04471 [cs, stat] (2016)
12. Liu, Q., Wang, D.: Stein variational gradient descent as moment matching. arXiv:1810.11693 [cs, stat] (2018)
13. Nelson, E.: Quantum Fluctuations. Princeton Series in Physics. Princeton University Press, Princeton (1985)
14. Otto, F.: The geometry of dissipative evolution equations: the porous medium equation. Commun. Partial Differ. Eqn. **26**(1–2), 101–174 (2001)
15. Pavon, M., Tabak, E.G., Trigilal, G.: The data-driven Schroedinger bridge. arXiv:1806.01364 [math] (2018)
16. Qi, D., Majda, A.J.: Low-dimensional reduced-order models for statistical response and uncertainty quantification: barotropic turbulence with topography. Phys. D **343**, 7–27 (2017)
17. Rezende, D.J., Mohamed, S.: Variational inference with normalizing flows. arXiv:1505.05770 [cs, stat] (2015)
18. Risken, H.: The Fokker-Planck Equation. Springer Series in Synergetics, vol. 18. Springer, Heidelberg (1989)

Multi-marginal Schrödinger Bridges

Yongxin Chen[1][iD], Giovanni Conforti[2], Tryphon T. Georgiou[3(✉)][iD],
and Luigia Ripani[3]

[1] Georgia Institute of Technology, Atlanta, GA 30332, USA
[2] École Polytechnique, Route de Saclay, 91128 Palaiseau Cedex, France
[3] University of California, Irvine, CA 92697, USA
tryphon@uci.edu

Abstract. We consider the problem to identify the most likely flow
in phase space, of (inertial) particles under stochastic forcing, that is
in agreement with spatial (marginal) distributions that are specified
at a set of points in time. The question raised generalizes the classical Schrödinger Bridge Problem (SBP) which seeks to interpolate two
specified end-point marginal distributions of overdamped particles driven
by stochastic excitation. While we restrict our analysis to second-order
dynamics for the particles, the data represents *partial* (i.e., only positional) information on the flow at *multiple* time-points. The solution
sought, as in SBP, represents a probability law on the space of paths that
is closest to a uniform prior while consistent with the given marginals.
We approach this problem as an optimal control problem to minimize an
action integral *a la* Benamou-Brenier, and derive a time-symmetric formulation that includes a Fisher information term on the velocity field. We
underscore the relation of our problem to recent measure-valued splines
in Wasserstein space, which is akin to that between SBP and Optimal
Mass Transport (OMT). The connection between the two provides a
Sinkhorn-like approach to computing measure-valued splines. We envision that interpolation between measures as sought herein will have a
wide range of applications in signal/images processing as well as in data
science in cases where data have a temporal dimension.

Keywords: Schrödinger bridge · Optimal mass transport ·
Optimal control · Multi-marginal

1 Introduction

In 1931/32, in an attempt to gain insights into the stochastic nature of quantum
mechanics, Schrödinger [22,23] raised the following question regarding a system
of a large number of classical independent identically distributed (i.i.d.) Brownian particles. He hypothesized that this "cloud" of particles is observed to have

---

Partial support was provided by NSF under grants 1665031, 1807664, 1839441 and
1901599, and by AFOSR under grant FA9550-17-1-0435.

F. Nielsen and F. Barbaresco (Eds.): GSI 2019, LNCS 11712, pp. 725–732, 2019.
https://doi.org/10.1007/978-3-030-26980-7_75

(empirical) distributions $\rho_0(x_0)$ and $\rho_1(x_1)$ at two points in time $t_0 = 0$ and $t_1 = 1$, respectively, and further that $\rho_1(x_1)$ differs from what is dictated by the law of large numbers, i.e., that

$$\rho_1(x_1) \neq \int_{\mathbb{R}^n} q(t_0, x_0, t_1, x_1) \rho_0(x_0) dx_0,$$

where

$$q(s, x, t, y) = (2\pi)^{-n/2} (t - s)^{-n/2} \exp\left(-\frac{1}{2} \frac{\|x - y\|^2}{(t - s)}\right)$$

denotes the Brownian transition probability kernel. Schrödinger then sought to find the "most likely" evolution for the cloud of particles to have transitioned from ρ_0 to ρ_1. In the language or large deviation theory (which was not in place at the time), Schrödinger's question amounts to seeking a probability law on the path space that is in agreement with the two marginals while being the closest to the Brownian prior in the sense of relative entropy [12]. The solution is known as the Schrödinger bridge since the law "bridges" the two given end-point marginal distributions.

Renewed interest in the Schrödinger Bridge Problem (SBP) has been fueled by its connections to the Monge-Kantorovic Optimal Mass Transport (OMT) and a wide range aplications in image analysis, stochastic control, and physics [4,6,9,13,18,19,21]. More specifically, SBP, seen as a suitable regularization of OMT, provides a natural model for uncertainty in the transport of distributions as well as a valuable computational tool for interpolating distributional data.

In this work we consider a natural generalization of the Schrödinger bridge theory to address the situation where the data consist of possible partial marginal distributions at various points in time. Thus, we postulate a similar experiment with stochastic particles. However, in contrast to the standard SB theory, we conceive these particles to obey second order stochastic differential equations with Brownian stochastic forcing that accounts for random acceleration along trajectories in phase space (see Sect. 2). This new setting connects with recent results on measure-valued splines [1,3,8], which are general notions of splines on the Wasserstein space of measures. In this short paper, we provide a summary of the theory. A more detailed account will appear in a forthcoming publication (in preparation).

2   Multi-marginal Schrödinger Bridges for Inertial Particles

Suppose we are given a large number of independent inertial particles driven by white noise, that is, they follow the dynamics

$$dx = vdt, \tag{1a}$$

$$dv = dw, \tag{1b}$$

where dw denotes standard Brownian motion and (x, v) is the phase space (x denoting position and v velocity). The flow of probability densities $\mu_t(x, v)$ in phase space obeys the Fokker-Planck equation

$$\frac{\partial \mu}{\partial t} + v \cdot \nabla_x \mu - \frac{1}{2} \Delta_v \mu = 0, \tag{2}$$

with initial condition μ_0 at time $t = 0$. The law of large numbers dictates that, when the number of particles is large enough, the distributions of the particles will be closed to μ_t. The data of the problem we consider will, as in the standard SB problem, be inconsistent with the Fokker-Planck equation, and viewed as a "atypical/rare" event. In standard SB setting, where only two end-point marginals are specified, the "most likely" evolution amounts to an adjustment of the Fokker-Planck equation by adding a suitable drift term to match the two marginals. For inertial particles driven by white noise, the generator is hypoelliptic [16] and the SB theory carries over to matching marginals in phase space.

Throughout, we suppose that we have only access to position x and that empirical marginals $\rho_0 = \mathrm{Proj}_x\, \mu_0, \rho_1 = \mathrm{Proj}_x\, \mu_1$ represent projections, accordingly. The extra degree of freedom, since μ's are partially specified, make the corresponding multi-marginal problem nontrivial. Specifically, we seek the most likely paths the particles have taken that match positional distributions $\rho_0, \rho_1, \cdots, \rho_N$ at times $0 = t_0 < t_1 < \cdots < t_N = 1$. To this end, we let $\mathcal{Q}$ be the law of (1) (on path space) and we let $\mathcal{P}$ be any other law. We seek the minimizer of

$$H(\mathcal{P}, \mathcal{Q}) = \int d\mathcal{P} \log \frac{d\mathcal{P}}{d\mathcal{Q}} \tag{3}$$

over all laws $\mathcal{P}$ that are consistent with the marginals $\rho_0, \rho_1, \cdots, \rho_N$.

To guarantee the boundedness of the relative entropy H between $\mathcal{P}$ and the prior process $\mathcal{Q}$, $\mathcal{P}$ has to be of the form

$$dx = v dt, \tag{4a}$$

$$dv = a dt + dw, \tag{4b}$$

where a is a suitable drift that may depend on the current and past values of the process state. Invoking Girsanov's theorem [7,11,14,17], we obtain that

$$H(\mathcal{P}, \mathcal{Q}) = \mathbb{E}_{\mathcal{P}} \left\{ \int_0^1 \|a(t)\|^2 dt \right\} = \int_0^1 \int \|a\|^2 \mu dx dv dt.$$

Thus, we arrive at the optimal control formulation

$$\min\ \mathbb{E} \left\{ \int_0^1 \|a(t)\|^2 dt \right\}, \tag{5a}$$

$$dx = v dt, \tag{5b}$$

$$dv = a dt + dw, \tag{5c}$$

$$x(t_i) \sim \rho_i, \quad i = 0, 1, \ldots, N. \tag{5d}$$

The difference to the standard Schrödinger bridge problem lies in the constraint (5d); *multiple* marginals are *partially* specified. The existence and uniqueness of the solution follow from the fact that it is a strongly convex optimization problem on path space measures. The argument is similar to the standard argument in SB theory [5, 12, 18, 19] and will be presented in an extended version of the paper. By utilizing the Fokker-Planck equation, we can rewrite the above as

$$\min \int_0^1 \int \|a\|^2 \mu dx dv dt, \tag{6a}$$

$$\frac{\partial \mu}{\partial t} + v \cdot \nabla_x \mu + \nabla_v \cdot (a\mu) - \frac{1}{2}\Delta_v \mu = 0, \tag{6b}$$

$$\int \mu_{t_i}(x, v)dv = \rho_i(x), \quad i = 0, 1, \ldots, N. \tag{6c}$$

Here $\int \mu_{t_i}(x, v)dv = \rho_i(x)$ since $\text{Proj}_x(\mu_{t_i}) = \rho_i$.

Now let $\hat{a} = a - \frac{1}{2}\nabla_v \log \mu$, then the diffusion term in (6b) can be absorbed into the convection terms, and then the cost becomes

$$\int_0^1 \int \|a\|^2 \mu dx dv dt = \int_0^1 \int \|\hat{a} + \frac{1}{2}\nabla_v \log \mu\|^2 \mu dx dv dt$$

$$= \int_0^1 \int \left\{ \|\hat{a}\|^2 \mu + \frac{1}{4}\|\nabla_v \log \mu\|^2 \mu \right\} dx dv dt$$

$$+ \int_0^1 \int \langle \hat{a}, \nabla_v \log \mu \rangle \mu dx dv dt.$$

Direct calculation yields

$$\int_0^1 \int \langle \hat{a}, \nabla_v \log \mu \rangle \mu dx dv dt = \int \{\mu_1 \log \mu_1 - \mu_0 \log \mu_0\} dx dv, \tag{7}$$

which only depends on the two end distributions. Thus, we need to consider

$$\min \int_0^1 \int \left\{ \|\hat{a}\|^2 \mu + \frac{1}{4}\|\nabla_v \log \mu\|^2 \mu \right\} dx dv dt + \int \{\mu_1 \log \mu_1 - \mu_0 \log \mu_0\} dx dv, \tag{8a}$$

$$\frac{\partial \mu}{\partial t} + v \cdot \nabla_x \mu + \nabla_v \cdot (\hat{a}\mu) = 0, \tag{8b}$$

$$\int \mu_{t_i}(x, v)dv = \rho_i(x), \quad i = 0, 1, \ldots, N. \tag{8c}$$

A similar formulation with a Fisher information term has been studied in [6, 13, 20, 25] for standard Schrödinger bridge problems. However, in the standard setting, the term $\int \{\mu_1 \log \mu_1 - \mu_0 \log \mu_0\} dx dv$ can be dropped since the full-state marginal distributions are already specified. It important to note that, compared to (6), (8) is time symmetric.

3   Connections to Measure-Valued Splines

A variational formulation [15] of splines going through $\{x_1, x_2, \ldots, x_N\}$ in Euclidean space is given by

$$\min \quad \int_0^1 \|\ddot{x}\|^2 dt$$
$$x(t_i) = x_i, \quad i = 0, 1, \ldots, N,$$

where the minimization is taken over all twice-differentiable trajectories that satisfy the constraints. This formation has been generalized to the Wasserstein space of measures [1,3]. In particular, the fluid dynamic formulation for measure-valued splines in [3] with marginals $\rho_0, \rho_1, \cdots, \rho_N$ at $0 = t_0 < t_1 < \cdots < t_N = 1$ reads

$$\min \quad \int_0^1 \int \|a\|^2 \mu dx dv dt, \tag{9a}$$

$$\frac{\partial \mu}{\partial t} + v \cdot \nabla_x \mu + \nabla_v \cdot (a\mu) = 0, \tag{9b}$$

$$\int \mu_{t_i}(x, v) dv = \rho_i(x), \quad i = 0, 1, \ldots, N. \tag{9c}$$

We note that the above formulation is almost the same as (6) except for a missing diffusion term in the constraint. This resembles the relation between standard Schrödinger bridges and optimal mass transport [5,18,19].

Indeed, a zero-noise limit argument follows. If we replace the dynamics (1) of the inertial particles by

$$dx = vdt, \tag{10a}$$
$$dv = \sqrt{\epsilon}dw, \tag{10b}$$

then the multi-marginal SB problem becomes

$$\min \quad \int_0^1 \int \|a\|^2 \mu dx dv dt, \tag{11a}$$

$$\frac{\partial \mu}{\partial t} + v \cdot \nabla_x \mu + \nabla_v \cdot (a\mu) - \frac{\epsilon}{2} \Delta_v \mu = 0, \tag{11b}$$

$$\int \mu(t_i, x, v) dv = \rho_i(x), \quad i = 0, 1, \ldots, N. \tag{11c}$$

The "slowed down" formulation (11) reduces to (6) when we take the limit $\epsilon \to 0$. Therefore, we establish the measure-valued spline as a zero-noise limit of a multi-marginal SB, and SB as a regularized version of measure-valued spline. Rigorous proof of these conclusions will be presented in a forthcoming paper.

4   Algorithms

The Sinkhorn [10,24] algorithm is a natural iterative scheme in SB problems. Due to its efficiency and simplicity, it has became a workhorse for data science applications of optimal transport [10]. In this section, we develop a Sinkhorn-type algorithm for multi-marginal Schrödinger bridge problems. The relation established in Sect. 3 implies that the same algorithm can also be used to approximate measure-valued splines.

Using a measure decomposition argument, the Schrödinger problem (11) can be rewritten as

$$\min J(\pi) := \sum_{i=0}^{N-1} KL(\pi_{i,i+1} \mid e^{-C_{i,i+1}/\epsilon}) \tag{12a}$$

$$\int \pi_{i,i+1} dx_{i+1} dv_{i+1} = \mu_i, \quad i = 0, \dots, N-1 \tag{12b}$$

$$\int \pi_{i,i+1} dx_i dv_i = \mu_{i+1}, \quad i = 0, \dots, N-1 \tag{12c}$$

$$\int \mu_i dv_i = \rho_i, \quad i = 0, \dots, N \tag{12d}$$

where $KL(\alpha|\beta) = \int \alpha \log \frac{\alpha}{\beta} - \alpha + \beta$ and

$$C_{i,i+1}(x_i, v_i, x_{i+1}, v_{j+1}) = (12\|x_{i+1} - x_i - v_i\|^2 - 12\langle x_{i+1} - x_i - v_i, v_{i+1} - v_i\rangle \\ + 4\|v_{i+1} - v_i\|^2)/(t_{i+1} - t_i).$$

The optimization variables are joint distributions on the consecutive time points over the phase space. Since the cost is a summation of relative entropies and the constraints are convex, a natural algorithm is that of Bregman projections [2].

Define convex constraint sets

$$K_0 = \{\int \pi_{01} dx_1 dv_1 = \mu_0, \ \int \mu_0 dv_0 = \rho_0\},$$

$$K_N = \{\int \pi_{N-1,N} dx_{N-1} dv_{N-1} = \mu_N, \ \int \mu_N dv_N = \rho_N\}, \text{ and}$$

$$K_i = \{\int \pi_{i,i+1} dx_{i+1} dv_{i+1} = \mu_i, \ \int \pi_{i-1,i} dx_{i-1} dv_{i-1} = \mu_i, \ \int \mu_i dv_i = \rho_i\},$$

for $i = 1, \dots, N-1$, then the Bregman iterative projection becomes

$$\pi^n = P_{K_{i_n}}^{KL}(\pi^{n-1}), \quad n = 1, 2, 3, \dots$$

where i_n enumerates $\{0, 1, \dots, N\}$ repeatedly. The initial condition is

$$\pi^0 = e^{-C_{i,i+1}/\epsilon}, \quad i = 0, \dots, N-1. \tag{13}$$

The projection operator is

$$P_K^{KL}(\bar{\pi}) := \mathrm{argmin}_{\pi \in K} KL(\pi \mid \bar{\pi}). \tag{14}$$

These projections can be derived via Lagrangian method. The projections to K_0, K_N are easier and the projections to $K_1, \ldots, K_{N-1}$ are more involved. Specifically,

$$P_{K_0}: \quad \pi_{01} = \frac{\rho_0 \bar{\pi}_{01}}{\int \bar{\pi}_{01} dv_0 dx_1 dv_1}$$

$$P_{K_N}: \quad \pi_{N-1,N} = \frac{\rho_N \bar{\pi}_{N-1,N}}{\int \bar{\pi}_{N-1,N} dv_N dx_{N-1} dv_{N-1}}$$

whereas for P_{K_i}, $i = 1, \ldots, N-1$:

$$\pi_{i-1,i} = \frac{\rho_i (\int \bar{\pi}_{i-1,i} dx_{i-1} dv_{i-1} \int \bar{\pi}_{i,i+1} dx_{i+1} dv_{i+1})^{1/2}}{\int \bar{\pi}_{i-1,i} dx_{i-1} dv_{i-1} \int (\int \bar{\pi}_{i-1,i} dx_{i-1} dv_{i-1} \int \bar{\pi}_{i,i+1} dx_{i+1} dv_{i+1})^{1/2} dv_i} \bar{\pi}_{i-1,i}$$

$$\pi_{i,i+1} = \frac{\rho_i (\int \bar{\pi}_{i-1,i} dx_{i-1} dv_{i-1} \int \bar{\pi}_{i,i+1} dx_{i+1} dv_{i+1})^{1/2}}{\int \bar{\pi}_{i,i+1} dx_{i+1} dv_{i+1} \int (\int \bar{\pi}_{i-1,i} dx_{i-1} dv_{i-1} \int \bar{\pi}_{i,i+1} dx_{i+1} dv_{i+1})^{1/2} dv_i} \bar{\pi}_{i,i+1}$$

In real implementation, we need to discretize the phase space over a grid. After discretization, the algorithm only involves matrix multiplication, pointwise-division, multiplication, square root, and therefore can be parallelized easily. The linear convergence rate is guarantee by the property of Bregman projections [2]. Our algorithm should be compared to that developed in [1]. A major difference is that our algorithm doesn't require discretization over the time domain.

5 Conclusion

We considered a natural extension of the Schrödinger bridge problems to multi-marginal partially observable setting. We focused on inertial particles, but more general dynamics can be examined similarly. We discussed the physical meaning, stochastic control formulation and several other aspects of the problems. Just like in the standard SB problem, it has a natural relation to the measure-valued spline theory. An efficient algorithm was also developed, which is available for possible applications. We envision that this line of research will spark interest in optimal transport theory for applications that involve constraints on multiple time-points along transport paths.

References

1. Benamou, J.D., Gallouët, T., Vialard, F.X.: Second order models for optimal transport and cubic splines on the Wasserstein space. arXiv:1801.04144 (2018)
2. Benamou, J.D., Carlier, G., Cuturi, M., Nenna, L., Peyré, G.: Iterative bregman projections for regularized transportation problems. SIAM J. Sci. Comput. **37**(2), A1111–A1138 (2015)
3. Chen, Y., Conforti, G., Georgiou, T.: Measure-valued spline curves: an optimal transport viewpoint. SIAM J. Math. Anal. **50**(6), 5947–5968 (2018)

4. Chen, Y., Georgiou, T., Pavon, M.: Optimal steering of a linear stochastic system to a final probability distribution, Part I. IEEE Trans. Autom. Control **61**(5), 1158–1169 (2016)
5. Chen, Y., Georgiou, T., Pavon, M.: Entropic and displacement interpolation: a computational approach using the Hilbert metric. SIAM J. Appl. Math. **76**(6), 2375–2396 (2016)
6. Chen, Y., Georgiou, T.T., Pavon, M.: On the relation between optimal transport and Schrödinger bridges: A stochastic control viewpoint. J. Optim. Theory Appl. **169**(2), 671–691 (2016)
7. Chen, Y., Georgiou, T.T., Pavon, M.: Optimal transport over a linear dynamical system. IEEE Trans. Autom. Control **62**(5), 2137–2152 (2017)
8. Chen, Y., Karlsson, J.: State tracking of linear ensembles via optimal mass transport. IEEE Control Syst. Lett. **2**(2), 260–265 (2018)
9. Conforti, G.: A second order equation for schrödinger bridges with applications to the hot gas experiment and entropic transportation cost. Probab. Theory Relat. Fields **17**, 1–47 (2018)
10. Cuturi, M.: Sinkhorn distances: lightspeed computation of optimal transport. In: Advances in Neural Information Processing Systems, pp. 2292–2300 (2013)
11. Dai Pra, P.: A stochastic control approach to reciprocal diffusion processes. Appl. Math. Optim. **23**(1), 313–329 (1991)
12. Föllmer, H.: Random fields and diffusion processes. In: Hennequin, P.-L. (ed.) École d'Été de Probabilités de Saint-Flour XV–XVII, 1985–87. LNM, vol. 1362, pp. 101–203. Springer, Heidelberg (1988). https://doi.org/10.1007/BFb0086180
13. Gentil, I., Léonard, C., Ripani, L.: About the analogy between optimal transport and minimal entropy. Annales de la Faculté des Sciences de Toulouse. Mathématiques. Série 6 **3**, 569–600 (2017)
14. Girsanov, I.V.: On transforming a certain class of stochastic processes by absolutely continuous substitution of measures. Theory Probab. Appl. **5**(3), 285–301 (1960)
15. Holladay, J.: A smoothest curve approximation. Math. Tables Aids Comput. **11**(60), 233–243 (1957)
16. Hörmander, L.: Hypoelliptic second order differential equations. Acta Math. **119**(1), 147–171 (1967)
17. Ikeda, N., Watanabe, S.: Stochastic differential equations and diffusion processes. Elsevier (2014)
18. Léonard, C.: From the Schrödinger problem to the Monge-Kantorovich problem. J. Funct. Anal. **262**(4), 1879–1920 (2012)
19. Léonard, C.: A survey of the Schrödinger problem and some of its connections with optimal transport. Dicrete Contin. Dyn. Syst. A **34**(4), 1533–1574 (2014)
20. Li, W., Yin, P., Osher, S.: Computations of optimal transport distance with fisher information regularization. J. Sci. Comput. **75**(3), 1581–1595 (2018)
21. Peyré, G., Cuturi, M.: Computational optimal transport. arXiv:1803.00567 (2018)
22. Schrödinger, E.: Über die Umkehrung der Naturgesetze. Sitzungsberichte der Preuss Akad. Wissen. Phys. Math. Klasse, Sonderausgabe **IX**, 144–153 (1931)
23. Schrödinger, E.: Sur la théorie relativiste de l'électron et l'interprétation de la mécanique quantique. In: Annales de l'institut Henri Poincaré. vol. 2, NO. (4), pp. 269–310. Presses universitaires de France (1932)
24. Sinkhorn, R.: A relationship between arbitrary positive matrices and doubly stochastic matrices. Ann. Math. Stat. **35**(2), 876–879 (1964)
25. Yasue, K.: Stochastic calculus of variations. J. Funct. Anal. **41**(3), 327–340 (1981)

Hopf–Cole Transformation
and Schrödinger Problems

Flavien Léger[⊠] and Wuchen Li

Department of Mathematics, UCLA, Los Angeles, USA
flavien.leger@nyu.edu

Abstract. We study generalized Hopf–Cole transformations motivated by the Schrödinger bridge problem. We present two examples of canonical transformations, including a Schrödinger problem associated with a quadratic Rényi entropy.

Keywords: Hopf–Cole transformation · Schrödinger bridge problem

1 Schrödinger Bridge Problem

The Schrödinger problem, also named Schrödinger bridge problem (SBP), was first introduced by Schrödinger in 1931 [4,6–8,16,18–21,24]. It is closely related to, but different from the famous Schrödinger equation. The SBP searches for the minimal kinetic energy density path for drift-diffusion processes with fixed initial and final distributions.

Here we study a general family of SBPs. This family of problems consists of controlled gradient flows of general potential energies on the Wasserstein density manifold [13,22]. We study a generalized Hopf–Cole transformation for general potential energies and show that the Hopf–Cole change of variables is a symplectic embedding in the symplectic geometry of density manifold [15]. In the last section we present two examples for which this Hopf–Cole transformation is canonical.

1.1 The Classical Schrödinger Problem

Let M be a finite-dimensional manifold, which for simplicity we assume to be compact and without boundary. We denote by ∇ and div the gradient and divergence operators on M.

Fixing two probability measures μ and ν on M, the Schrödinger bridge problem (SBP) is defined by

$$\inf_{\rho,b} \int_0^1 \int_M \frac{1}{2} |b_t(x)|^2 \rho_t(x)\, dx\, dt, \tag{1}$$

where the infimum runs over all probability densities ρ and vector fields b satisfying

$$\partial_t \rho_t + \mathrm{div}(\rho_t b_t) = \gamma \Delta \rho_t, \quad \rho_0 = \mu, \quad \rho_1 = \nu.$$

© Springer Nature Switzerland AG 2019
F. Nielsen and F. Barbaresco (Eds.): GSI 2019, LNCS 11712, pp. 733–738, 2019.
https://doi.org/10.1007/978-3-030-26980-7_76

The constant $\gamma \in \mathbb{R}$ is a diffusion parameter which can be positive or negative; note that $\gamma = 0$ corresponds to the Wasserstein distance with quadratic cost between μ and ν via the Benamou–Brenier formula [2]. The solution to problem (1) is of the form $b_t = \nabla \phi_t$ for some scalar field ϕ, and the optimal pair (ρ, ϕ) satisfies a Fokker–Planck equation and a Hamilton–Jacobi–Bellman equation:

$$\begin{cases} \partial_t \rho_t + \operatorname{div}(\rho_t \nabla \phi_t) = \gamma \Delta \rho_t \\ \partial_t \phi_t + \dfrac{1}{2}|\nabla \phi_t|^2 = -\gamma \Delta \phi_t, \end{cases} \tag{2}$$

for all $t \in (0, 1)$.

1.2 Hopf–Cole Transformation

Rather formidably, the optimality conditions (2) can be rewritten into a simpler and more symmetric way thanks to a *Hopf–Cole transformation*. To this effect define

$$\eta_t(x) = e^{\phi_t(x)/(2\gamma)} \quad \text{and} \quad \eta_t^*(x) = \rho_t(x)e^{-\phi_t(x)/(2\gamma)}. \tag{3}$$

If (ρ, ϕ) satisfies the system (2) then (η, η^*) solves the backward-forward heat system

$$\begin{cases} \partial_t \eta_t = -\gamma \Delta \eta_t \\ \partial_t \eta_t^* = \gamma \Delta \eta_t^*. \end{cases} \tag{4}$$

Integrating the above system in time with the appropriate boundary conditions on ρ leads to the so-called Schrödinger system, see [3,9,11].

1.3 Generalized Schrödinger Problem

Let $\mathcal{P}(M)$ be the space of probability measures on M and consider a regular functional $\mathcal{F} \colon \mathcal{P}(M) \to \mathbb{R}$. In [14] was introduced the generalized Schrödinger problem (GSP)

$$\inf_{\rho, b} \int_0^1 \int_M \frac{1}{2}|b_t(x)|^2 \rho_t(x) \, dx \, dt, \tag{5}$$

subject to the constraint

$$\partial_t \rho_t + \operatorname{div}\big(\rho_t(b_t - \nabla \delta \mathcal{F}(\rho_t))\big) = 0, \tag{6}$$

as well as the boundary conditions $\rho_0 = \mu$, $\rho_1 = \nu$. Here μ and ν are fixed probability measures on M and $\delta \mathcal{F}$ denotes the first variation (i.e. the L^2 gradient) of $\mathcal{F}$. This model was later considered in [10] where the authors prove various convexity inequalities.

In the entropic case $\mathcal{F}(\rho) = \gamma \int \rho \log \rho \, dx$ we recover the classical Schrödinger bridge problem (1), since

$$\operatorname{div}(\rho \nabla \delta \mathcal{F}(\rho)) = \operatorname{div}(\rho \nabla(\gamma \log \rho + \gamma)) = \gamma \Delta \rho.$$

This is remindful of Otto's interpretation of the heat equation as gradient flow of the entropy in the Wasserstein metric; indeed the GSP can be seen as a controlled gradient flow problem with respect to the Wasserstein metric.

As can be expected from control theory, solutions to the GSP can be written as Hamiltonian flows. For a GSP with potential $\mathcal{F}$ the Hamiltonian is given by

$$\mathcal{H}(\rho, \phi) = \int \frac{1}{2} |\nabla \phi|^2 \rho - \nabla \phi \cdot \nabla \delta \mathcal{F}(\rho) \, \rho \, dx,$$

where the dual variable ϕ is related to the control b by $b = \nabla \phi$. Then, a solution $(\rho, \nabla \phi)$ to the GSP satisfies the Hamiltonian flow

$$\begin{cases} \partial_t \rho_t = \delta_\phi \mathcal{H}(\rho_t, \phi_t) \\ \partial_t \phi_t = -\delta_\rho \mathcal{H}(\rho_t, \phi_t). \end{cases} \tag{7}$$

This is an example of a Hamiltonian flow in the Wasserstein space [1,5,17].

2   Canonical Transformations

Let $\mathcal{F}$ be a regular functional defined over the space of probability measures and consider (5) the GSP with potential $\mathcal{F}$. In this context, the work [14] introduced a type of Hopf–Cole transformation.

Definition 1. *The generalized Hopf–Cole transformation* $(\eta, \eta^*) \rightarrow (\rho, \phi)$ *is defined by*

$$\begin{cases} \delta \mathcal{F}(\rho) = \delta \mathcal{F}(\eta) + \delta \mathcal{F}(\eta^*) \\ \phi = 2 \, \delta \mathcal{F}(\eta). \end{cases} \tag{8}$$

Here $\delta \mathcal{F}$ denotes the first variation (i.e. the L^2 gradient) of $\mathcal{F}$ and we assume it is invertible.

It is a simple matter to check that in the entropic case $\mathcal{F}(\rho) = \gamma \int \rho \log \rho \, dx$ we recover the classical Hopf–Cole transformation (3).

The first result of this paper describes the optimality conditions written in the new variables η and η^*.

Theorem 1. *Given a solution (ρ, ϕ) to the Hamiltonian flow (7), the new variables (η, η^*) satisfy*

$$\begin{cases} \partial_t \eta_t = \sigma(\eta_t, \eta_t^*) \, \delta_{\eta^*} \mathcal{K}(\eta_t, \eta_t^*) \\ \partial_t \eta_t^* = -\sigma(\eta_t^*, \eta_t) \, \delta_\eta \mathcal{K}(\eta_t, \eta_t^*). \end{cases} \tag{9}$$

Here $\mathcal{K}$ is the Hamiltonian in the new variables: $\mathcal{K}(\eta, \eta^*) = \mathcal{H}(\rho, \phi)$. *Moreover σ is defined by*

$$\sigma(\eta, \eta^*)(x, w) = -\frac{1}{2} \iint \left[\delta^2 \mathcal{F}(\eta) \right]^{-1} (x, y) \, \delta^2 \mathcal{F}(\rho)(y, z) \left[\delta^2 \mathcal{F}(\eta^*) \right]^{-1} (z, w) \, dy \, dz,$$

therefore $\sigma(\eta, \eta^*)$ can be regarded as a linear map (which has the above kernel). Here $\delta^2 \mathcal{F}$ denotes the second variation (i.e. L^2 Hessian) of $\mathcal{F}$.

Theorem 1 shows that the optimal (η, η^*) satisfies a system of Hamiltonian-type equations. This is to be expected from symplectic theory since operating a change of variables $(\eta, \eta^*) \to (\rho, \phi)$ leads to Hamiltonian flows involving pulled-back symplectic forms. We discuss this fact in more details in the general version of this paper [15]. In the next section we focus on two potentials $\mathcal{F}$, the Boltzmann–Shannon entropy and the quadratic Rényi entropy, for which our Hopf–Cole transformation is *canonical*, i.e. the form of Hamilton's equations are exactly preserved.

3   Examples

3.1   Classical Schrödinger Problem

In this first section we show that the classical Hopf–Cole transformation does more than simplifying the optimality conditions of the SBP; from a symplectic point of view it is in fact *canonical*, i.e. it preserves the form of the Hamiltonian flows.

Proposition 1. *Let (ρ, ϕ) be a solution to the SBP* (1) *and let (η, η^*) be the new variables given by the Hopf–Cole transformation* (3). *Denoting by $\mathcal{K}$ the Hamiltonian in the new variables, we have*

$$\begin{cases} 2\gamma \, \partial_t \eta_t = \delta_{\eta^*} \mathcal{K}(\eta_t, \eta_t^*) \\ 2\gamma \, \partial_t \eta_t^* = -\delta_\eta \mathcal{K}(\eta_t, \eta_t^*). \end{cases}$$

Note that here the Hamiltonian is $\mathcal{H}(\rho, \phi) = \int \frac{1}{2} |\nabla \phi|^2 \rho - \gamma \nabla \log \rho \cdot \nabla \phi \, \rho \, dx$ and therefore $\mathcal{K}$ takes the simple form

$$\mathcal{K}(\eta, \eta^*) = -2\gamma^2 \int \nabla \eta \cdot \nabla \eta^* \, dx.$$

We note that in the above Hamiltonian flow, as well as in the next example, one could get rid of the constant 2γ by a simple rescaling of time.

3.2   Slow Schrödinger Problem

Let $\gamma > 0$ and consider the quadratic Rényi entropy

$$\mathcal{F}(\rho) = \frac{\gamma}{2} \int \big(\rho(x) \big)^2 \, dx.$$

Solutions to the corresponding GSP (5) satisfy the system

$$\begin{cases} \partial_t \rho_t + \operatorname{div}(\rho_t \nabla \phi_t) = \frac{\gamma}{2} \Delta(\rho_t^2) \\ \partial_t \phi_t + \frac{1}{2} |\nabla \phi_t|^2 = -\gamma \, \rho_t \, \Delta \, \phi_t. \end{cases}$$

This is a "slow diffusion" type of Schrödinger problem which exhibits a nonlinear diffusion term $\Delta(\rho^2)$. In this case the generalized Hopf–Cole transformation (8) is given by the linear formula

$$\rho = \eta + \eta^*, \quad \phi = 2\gamma\,\eta.$$

The next result shows that this transformation preserves the form of the Hamiltonian flow.

Proposition 2. *The optimality conditions for the slow Schrödinger problem can be written in the new variables as*

$$\begin{cases} 2\gamma\,\partial_t\eta_t = \delta_{\eta^*}\mathcal{K}(\eta_t,\eta_t^*) \\ 2\gamma\,\partial_t\eta_t^* = -\,\delta_\eta\mathcal{K}(\eta_t,\eta_t^*), \end{cases}$$

where the Hamiltonian $\mathcal{K}$ takes the form

$$\mathcal{K}(\eta,\eta^*) = \frac{\gamma}{2}\int \big(\eta(x) + \eta^*(x)\big)\,\nabla\eta(x)\cdot\nabla\eta^*(x)\,dx.$$

More specifically, this is a system of PDEs

$$\begin{cases} \gamma^{-1}\,\partial_t\eta_t + \tfrac{1}{2}|\nabla\eta_t|^2 + (\eta_t + \eta_t^*)\,\Delta\,\eta_t = 0 \\ -\gamma^{-1}\,\partial_t\eta_t^* + \tfrac{1}{2}|\nabla\eta_t^*|^2 + (\eta_t + \eta_t^*)\,\Delta\,\eta_t^* = 0. \end{cases}$$

3.3 Schrödinger Equation

To conclude this list of examples we would like to mention the case of the Schrödinger equation, which is closely connected to the GSP. Formally speaking, the Schrödinger equation can be associated to an imaginary entropy $\mathcal{F}(\rho) = \gamma \int \rho\log\rho\,dx$ where $\gamma = i$ is the imaginary unit; then the generalized Hopf–Cole transformation (8) forms exactly the so-called *Madelung transformation*. Studies of the Madelung transformation as a symplectic change of variables include [5,12,13,23]. With this perspective, we also propose a generalized Madelung transformation in [15].

Acknowledgement. This project has received funding from AFOSR MURI FA9550-18-1-0502.

References

1. Ambrosio, L., Gangbo, W.: Hamiltonian ODEs in the Wasserstein space of probability measures. Commun. Pure and Appl. Math.: A J. Issued Courant Inst. Math. Sci. **61**(1), 18–53 (2008)
2. Benamou, J.-D., Brenier, Y.: A computational fluid mechanics solution to the Monge-Kantorovich mass transfer problem. Numer. Math. **84**(3), 375–393 (2000)
3. Beurling, A.: An automorphism of product measures. Ann. Math. **72**(1), 189–200 (1960)

4. Carlen, E.A.: Stochastic Mechanics: A Look Back and a Look Ahead. Princeton University Press, Berlin (2014)
5. Chow, S.-N., Li, W., Zhou, H.: A discrete Schrödinger equation via optimal transport on graphs. J. Funct. Anal. **276**(8), 2440–2469 (2019)
6. Conforti, G.: A second order equation for Schrödinger bridges withapplications to the hot gas experiment and entropic transportation cost. Probab. Theor. Relat. Fields **174**, 1–47 (2018)
7. Conforti, G., Pavon, M.: Extremal flows in Wasserstein space. J. Math. Phys. **59**(6), 063502 (2018)
8. Föllmer, H.: Random fields and diffusion processes. In: Hennequin, P.-L. (ed.) École d'Été de Probabilités de Saint-Flour XV–XVII, 1985–87. LNM, vol. 1362, pp. 101–203. Springer, Heidelberg (1988). https://doi.org/10.1007/BFb0086180
9. Fortet, R.: Résolution d'un système d'équations de M. Schrödinger. J. Math. Pures Appl. **19**, 83–105 (1940)
10. Gentil, I., Léonard, C., Ripani, L.: Dynamical aspects of generalized Schrödinger problem via Otto calculus – a heuristic point of view. arXiv:1806.01553 (2018)
11. Jamison, B.: The Markov processes of Schrödinger. Zeitschrift für Wahrscheinlichkeitstheorie und Verwandte Gebiete **32**(4), 323–331 (1975)
12. Khesin, B., Misiolek, G., Modin, K.: Geometric hydrodynamics via Madelung transform. Proc. Nat. Acad. Sci. **115**(24), 6165–6170 (2018)
13. Lafferty, J.D.: The density manifold and configuration space quantization. Trans. Am. Math. Soc. **305**(2), 699–699 (1988)
14. Léger, F.: A geometric perspective on regularized optimal transport. J. Dyn. Differ. Equ. **2018**, 1–15 (2018)
15. Léger, F., Li, W.: Hopf-Cole transformation via generalized Schrödinger bridge problem. arXiv:1901.09051 (2019)
16. Léonard, C.: A survey of the Schrödinger problem and some of its connections with optimal transport. Discrete and Continuous Dyn. Syst. **34**(4), 1533–1574 (2013)
17. Li, W.: Geometry of probability simplex via optimal transport. arXiv:1803.06360 (2018)
18. Nelson, E.: Derivation of the Schrödinger equation from Newtonian mechanics. Phys. Rev. **150**(4), 1079–1085 (1966)
19. Nelson, E.: Quantum Fluctuations. Princeton Series in Physics. Princeton University Press, Princeton (1985)
20. Schrödinger, E.: Über die Umkehrung der Naturgesetze. Sitzungsber. Preuß. Akad. Wiss. Phys. Math.- Kl **9**, 144–153 (1931)
21. Schrödinger, E.: Sur la théorie relativiste de l'électron et l'interprétation de la mécanique quantique. Ann. Inst. Henri Poincaré **2**(4), 269–310 (1932)
22. Villani, C.: Topics in Optimal Transportation, Graduate studies in mathematics, vol. 58. American Mathematical Society, Providence (2003)
23. von Renesse, M.-K.: An optimal transport view of Schrödinger's equation. Canad. Math. Bull. **55**(4), 858–869 (2012)
24. Yasue, K.: Stochastic calculus of variations. J. Funct. Anal. **41**(3), 327–340 (1981)

Curvature of the Manifold of Fixed-Rank Positive-Semidefinite Matrices Endowed with the Bures–Wasserstein Metric

Estelle Massart[1]([✉]), Julien M. Hendrickx[1,2], and P.-A. Absil[1]

[1] ICTEAM, UCLouvain, Louvain-la-Neuve, Belgium
estelle.massart@uclouvain.be
[2] CISE Resident Scholar at Boston University, Boston, USA

Abstract. We consider the manifold of rank-p positive-semidefinite matrices of size n, seen as a quotient of the set of full-rank n-by-p matrices by the orthogonal group in dimension p. The resulting distance coincides with the Wasserstein distance between centered degenerate Gaussian distributions. We obtain expressions for the Riemannian curvature tensor and the sectional curvature of the manifold. We also provide tangent vectors spanning planes associated with the extreme values of the sectional curvature.

1 Introduction

Positive-semidefinite (PSD) matrices appear, e.g., as covariance matrices in statistics, kernels in machine learning, and variables in semidefinite optimization; see, e.g., [MA18] for pointers to the literature.

The set of PSD matrices of size $n \times n$ is a stratified space [Tak11, Thm. C], in which the strata are the manifolds

$$\mathbb{S}_+(p,n) = \{S \in \mathbb{R}^{n \times n} | S \succeq 0,\ \text{rank}(S) = p\},$$

of PSD matrices of rank p, for $p = 0, \ldots, n$. In many practical applications, the rank of all the datapoints can be truncated to a common value, so that algorithms can be restricted to handle datapoints lying on the same stratum (see [MA18] and references within). This is for example the case when the data points are low-rank approximations of large PSD matrices. Each stratum $\mathbb{S}_+(p,n)$, with $p \geq 1$, can be given a Riemannian structure.

Classical algorithms on Riemannian manifolds can thus be used for processing data on $\mathbb{S}_+(p,n)$. For example, optimization on $\mathbb{S}_+(p,n)$ has been used in [MBS11, MMS11, MHB+16] for distance learning, distance matrix completion,

This work was supported by (i) the Fonds de la Recherche Scientifique – FNRS and the Fonds Wetenschappelijk Onderzoek – Vlaanderen under EOS Project no 30468160, (ii) "Communauté française de Belgique - Actions de Recherche Concertées" (contract ARC 14/19-060), (iii) the WBI-World Excellence Fellowship.

F. Nielsen and F. Barbaresco (Eds.): GSI 2019, LNCS 11712, pp. 739–748, 2019.
https://doi.org/10.1007/978-3-030-26980-7_77

and role model extraction. The works [LB14, GMM+17, KDB+18, MGS+19] run interpolation algorithms on $\mathbb{S}_+(p,n)$ for generating protein conformation transitions, modeling wind field, video classification and parametric model order reduction.

In the full-rank case, i.e., when $p = n$, the manifold $\mathbb{S}_+(n,n)$ is classically identified to the reductive homogeneous space $\mathbb{S}_+(n,n) \simeq \mathrm{GL}_n/\mathcal{O}_n$, where GL_n is the general linear group. Therefore, there exists a GL_n-invariant metric on $\mathrm{GL}_n/\mathcal{O}_n$ which leads (up to a scaling factor) to the natural, affine-invariant metric, or Fisher-Rao metric on $\mathbb{S}_+(n,n)$, see [Smi05]. When $p \neq n$, the set $\mathbb{S}_+(p,n)$ can be identified to a homogeneous space (see [VAV13]), but this homogeneous space is shown to be nonreductive, and there is no metric invariant under the group action. There is thus no wide agreement on a preferred metric on $\mathbb{S}_+(p,n)$.

In this work, we consider the identification $\mathbb{S}_+(p,n) \simeq \mathbb{R}_*^{n \times p}/\mathcal{O}_p$, with $\mathbb{R}_*^{n \times p}$ the set of full-rank n-by-p matrices. The quotient manifold $\mathbb{R}_*^{n \times p}/\mathcal{O}_p$ is endowed with the metric induced from the Euclidean metric in $\mathbb{R}_*^{n \times p}$. This geometry was already proposed in [JBAS10] (which contains, e.g., expressions for the Riemannian exponential and for the projector on the horizontal space) and more recently described in [MA18]. In this last paper, we obtained expressions for the Riemannian logarithm, the injectivity radius and the cut locus. We mention that several other geometries have been proposed on $\mathbb{S}_+(p,n)$: [VAV09] represents $\mathbb{S}_+(p,n)$ as an embedded submanifold of $\mathbb{R}^{n \times n}$, [BS09] identifies it to the quotient manifold $(\mathrm{St}(p,n) \times \mathbb{S}_+(p,p))/\mathcal{O}_p$, and, as already mentioned, [VAV13] identifies $\mathbb{S}_+(p,n)$ to a homogeneous space endowed with a right-invariant metric.

Even though the metric resulting from the identification $\mathbb{S}_+(p,n) \simeq \mathbb{R}_*^{n \times p}/\mathcal{O}_p$ does not lead to a complete metric space, there are two main motivations to consider it. The first one is the low computation cost associated with the most common operations on the manifold. Indeed, the operations are directly performed on the representatives in $\mathbb{R}_*^{n \times p}$ of the matrices, which are smaller than the initial $n \times n$ matrices. As shown in [JBAS10, MA18], the Riemannian exponential and logarithm have a computational cost that evolves linearly with n. Among all the geometries proposed for $\mathbb{S}_+(p,n)$, this is to our knowledge the only one that leads to expressions for both the logarithm and the exponential maps that are cheap to evaluate.

The second motivation to consider this quotient geometry is its interpretation with respect to optimal transport theory. Indeed, there exists a bijection between the set of $n \times n$ PSD matrices and the set of (possibly degenerate) centered Gaussian distributions on $\mathbb{R}^n$. Let $C_1, C_2 \in \mathbb{S}_+(n,n)$, two nonsingular covariance matrices, and let $\mathrm{W}_2(\mu_1, \mu_2)$ be the 2-Wasserstein distance between the nondegenerate centered Gaussian distributions $\mu_1 := \mathcal{N}(0, C_1)$ and $\mu_2 := \mathcal{N}(0, C_2)$. It is well-known that $\mathrm{W}_2(\mu_1, \mu_2)$ coincides with the Riemannian distance between C_1 and C_2, for the metric inherited from the quotient representation $\mathbb{S}_+(n,n) \simeq \mathrm{GL}(n)/\mathcal{O}_n$ (see, e.g., [Tak11, BJL18]). When $C_1, C_2 \in \mathbb{S}_+(p,n)$, for $p < n$, the same conclusion holds: $\mathrm{W}_2(\mu_1, \mu_2)$ is equal to the Riemannian distance between the low-rank covariance matrices C_1 and C_2, for the metric induced by the quotient $\mathbb{S}_+(p,n) \simeq \mathbb{R}_*^{n \times p}/\mathcal{O}_p$ [Gel90, Corollary 2.5]. Specifically, the distance is given by (see [MA18, Sect. 2.10]):

$$d(C_1, C_2) = \left[\operatorname{tr}(C_1) + \operatorname{tr}(C_2) - 2\operatorname{tr}\left(\left(C_1^{1/2} C_2 C_1^{1/2} \right)^{1/2} \right) \right]^{1/2}.$$

The Wasserstein metric is also known as the Bures metric in quantum theory (see [BJL18] and references therein).

Geometric properties of the manifold $\mathbb{S}_+(n,n) \simeq \mathrm{GL}(n)/\mathcal{O}_n$ have been widely studied, see, e.g., [Tak11, BJL18, MMP18]. In particular, its sectional curvature has been computed in [Tak11]. The contribution of this paper is to compute the Riemannian curvature tensor and the sectional curvature of the manifold $\mathbb{S}_+(p,n) \simeq \mathbb{R}_*^{n \times p}/\mathcal{O}_p$. We also provide tangent vectors spanning tangent planes associated with the maximal and minimal sectional curvatures. Bounds on the curvature of the manifold appear, e.g., in some optimization algorithms and associated convergence results on manifolds [ATV13, Bon13], and in guarantees for the continuity of the result of some curve fitting algorithms [AGSW16]. The Riemannian curvature tensor is, e.g., used in [SASK12] for curve fitting on manifolds. We show that the sectional curvature is non-negative, and may become infinitely large when approaching the boundary of the manifold (specifically, if two singular values go simultaneously to zero). A consequence is that some of the above-mentioned results involving bounds on the sectional curvature (in optimization or curve fitting) do not directly apply on this manifold. Our conclusions agree with the work [Dit95], which computes the curvature of the manifold of density matrices ($n \times n$ positive-definite complex matrices of unit trace), endowed with the Bures metric, and observes a similar unboundedness of the sectional curvature as the rank of the matrix goes to $n - 2$.

The structure of this paper is as follows. Section 2 presents a brief summary of the geometry of $\mathbb{R}_*^{n \times p}/\mathcal{O}_p$. In Sect. 3, we derive expressions for the Riemannian curvature tensor and the sectional curvature. Finally, we compute in Sect. 4 the extreme values of the sectional curvature.

2 Geometry of the Manifold $\mathbb{S}_+(p,n) \simeq \mathbb{R}_*^{n \times p}/\mathcal{O}_p$

This quotient geometry, described in [MA18], relies on the characterization $\mathbb{S}_+(p,n) = \{YY^\top | Y \in \mathbb{R}_*^{n \times p}\}$. The quotient representation comes from the fact that the set of points $Y\mathcal{O}_p := \{YQ | Q \in \mathcal{O}_p\}$ is a fiber under the map $Y \mapsto YY^\top$. The tangent space $T_Y \mathbb{R}_*^{n \times p} \simeq \mathbb{R}^{n \times p}$ is the direct sum of two orthogonal subspaces: the vertical space (the tangent space of the fiber $Y\mathcal{O}_p$), and the horizontal space (its orthogonal complement, with respect here to the Euclidean metric). The vertical space at Y is given by $\mathcal{V}_Y = \{Y\Omega | \Omega = -\Omega^\top \in \mathbb{R}^{p \times p}\}$, while the horizontal space is $\mathcal{H}_Y = \{\bar{\eta}_Y = Y(Y^\top Y)^{-1}S + Y_\perp K | S \in \mathbb{R}^{p \times p}, S = S^\top, K \in \mathbb{R}^{(n-p) \times p}\}$. Let $\pi : \mathbb{R}_*^{n \times p} \to \mathbb{R}_*^{n \times p}/\mathcal{O}_p$ be the quotient map, mapping points from $\mathbb{R}_*^{n \times p}$ to their fibers. For any $Y \in \mathbb{R}_*^{n \times p}$, any tangent vector $\xi_{\pi(Y)} \in T_{\pi(Y)} \mathbb{R}_*^{n \times p}/\mathcal{O}_p$ is associated to a unique horizontal lift

$\bar{\xi}_Y \in \mathcal{H}_Y$, such that $\xi_{\pi(Y)} = D\pi(Y)[\bar{\xi}_Y]$. The metric in $\mathbb{R}_*^{n \times p}/\mathcal{O}_p$ is defined as $g_{\pi(Y)}\left(\xi_{\pi(Y)}, \eta_{\pi(Y)}\right) := \operatorname{tr}\left(\bar{\xi}_Y^\top \bar{\eta}_Y\right)$, which turns the quotient map π into a Riemannian submersion. Finally, given two horizontal vector fields $\bar{\xi}, \bar{\eta}$, the projection on the vertical space of the bracket $[\bar{\xi}, \bar{\eta}]$ is:

$$\mathrm{P^v}_Y[\bar{\xi}, \bar{\eta}] = Y\mathbf{T}_{Y^\top Y}^{-1}\left(2\left(\bar{\eta}_Y^\top \bar{\xi}_Y - \bar{\xi}_Y^\top \bar{\eta}_Y\right)\right), \tag{1}$$

with $\mathbf{T}_{Y^\top Y}^{-1}(\Omega)$ the unique solution X to the Sylvester equation $Y^\top Y X + XY^\top Y = \Omega$, see [MA18, Proposition 2.37].

3 Curvature of the Manifold $\mathbb{R}_*^{n \times p}/\mathcal{O}_p$

In this section, we obtain expressions for the Riemannian curvature tensor and the sectional curvature of the manifold $\mathbb{R}_*^{n \times p}/\mathcal{O}_p$. We rely on the fact that the operator $\pi : \mathbb{R}_*^{n \times p} \to \mathbb{R}_*^{n \times p}/\mathcal{O}_p$ is a Riemannian submersion.

Theorem 1. *Let ξ, η, α and β be vector fields on $\mathbb{R}_*^{n \times p}/\mathcal{O}_p$, and let $\bar{\xi}, \bar{\eta}, \bar{\alpha}$ and $\bar{\beta}$ be their horizontal lifts. The Riemannian curvature tensor at $\pi(Y)$ satisfies:*

$$\langle R_{\mathbb{R}_*^{n \times p}/\mathcal{O}_p}(\xi_{\pi(Y)}, \eta_{\pi(Y)})\alpha_{\pi(Y)}, \beta_{\pi(Y)}\rangle = \frac{1}{2}\langle \mathrm{P^v}_Y[\bar{\xi}, \bar{\eta}], \mathrm{P^v}_Y[\bar{\alpha}, \bar{\beta}]\rangle$$
$$-\frac{1}{4}\left(\langle \mathrm{P^v}_Y[\bar{\eta}, \bar{\alpha}], \mathrm{P^v}_Y[\bar{\xi}, \bar{\beta}]\rangle - \langle \mathrm{P^v}_Y[\bar{\xi}, \bar{\alpha}], \mathrm{P^v}_Y[\bar{\eta}, \bar{\beta}]\rangle\right),$$

with $\mathrm{P^v}_Y[\bar{\xi}, \bar{\eta}]$ given by (1).

Proof. According to [O'N66, Thm 2], there holds:

$$\langle R_{\mathbb{R}_*^{n \times p}/\mathcal{O}_p}(\xi_{\pi(Y)}, \eta_{\pi(Y)})\alpha_{\pi(Y)}, \beta_{\pi(Y)}\rangle = \langle R_{\mathbb{R}_*^{n \times p}}(\xi_Y, \eta_Y)\alpha_Y, \beta_Y\rangle$$
$$+\frac{1}{2}\langle \mathrm{P^v}_Y[\bar{\xi}, \bar{\eta}], \mathrm{P^v}_Y[\bar{\alpha}, \bar{\beta}]\rangle - \frac{1}{4}\langle \mathrm{P^v}_Y[\bar{\eta}, \bar{\alpha}], \mathrm{P^v}_Y[\bar{\xi}, \bar{\beta}]\rangle - \frac{1}{4}\langle \mathrm{P^v}_Y[\bar{\alpha}, \bar{\xi}], \mathrm{P^v}_Y[\bar{\eta}, \bar{\beta}]\rangle.$$

Since $\mathbb{R}_*^{n \times p}$ is an open subset of $\mathbb{R}^{n \times p}$, its Riemannian curvature tensor is zero [O'N83, p. 79], hence the first term of the previous expression vanishes. □

The sectional curvature is then obtained as a corollary, see [O'N66, Corollary. 1, Eq. 3]. In the case $n = p$, these results are already given in [Tak11].

Corollary 1. *Let $\xi_{\pi(Y)}, \eta_{\pi(Y)}$ be (independent) tangent vectors on $\mathbb{R}_*^{n \times p}/\mathcal{O}_p$, with horizontal lifts $\bar{\xi}_Y, \bar{\eta}_Y$. The sectional curvature at $\pi(Y)$ in $\mathbb{R}_*^{n \times p}/\mathcal{O}_p$ is*

$$K_{\mathbb{R}_*^{n \times p}/\mathcal{O}_p}(\xi_{\pi(Y)}, \eta_{\pi(Y)}) = \frac{3\left\|Y\mathbf{T}_{Y^\top Y}^{-1}\left(\bar{\eta}_Y^\top \bar{\xi}_Y - \bar{\xi}_Y^\top \bar{\eta}_Y\right)\right\|_\mathrm{F}^2}{\langle \bar{\xi}_Y, \bar{\xi}_Y\rangle\langle \bar{\eta}_Y, \bar{\eta}_Y\rangle - \langle \bar{\xi}_Y, \bar{\eta}_Y\rangle^2}. \tag{2}$$

The rest of the paper aims at computing the maximal and minimal sectional curvatures at an arbitrary $\pi(Y) \in \mathbb{R}_*^{n \times p}/\mathcal{O}_p$.

4   Extreme Values of the Sectional Curvature

We first introduce two lemmas. The first one solves for X a Sylvester equation of the form $Y^\top Y X + X Y^\top Y = \Omega$, a step required to evaluate (2).

Lemma 1. *Let* $Y \in \mathbb{R}_*^{n \times p}$, *with* $Y =: U\Sigma V^\top$ *a singular value decomposition, with singular values* $\sigma_1 \geq \cdots \geq \sigma_p > 0$, *and let* $\Omega \in \mathbb{R}^{p \times p}$. *The solution* X *to the Sylvester equation* $Y^\top Y X + X Y^\top Y = \Omega$ *is*

$$X = V\tilde{X}V^\top, \ \text{with} \ \tilde{X} \in \mathbb{R}^{p \times p}, \ \tilde{X}_{ij} := \frac{\tilde{\Omega}_{ij}}{(\sigma_i^2 + \sigma_j^2)}, \ \tilde{\Omega} := V^\top \Omega V. \tag{3}$$

Moreover, if the matrix Ω *is skew-symmetric, then so are* $\tilde{X}$ *and* X.

Proof. We sketch the proof, presented in [BR97, Sect. 10], for the reader's convenience. Since $Y^\top Y = V\Sigma^2 V^\top$, the Sylvester equation becomes: $V\Sigma^2 V^\top X + XV\Sigma^2 V^\top = \Omega$. Applying a similarity associated with V to both sides of the equation yields: $\Sigma^2 V^\top X V + V^\top X V\Sigma^2 = V^\top \Omega V$. Now, defining $\tilde{X} := V^\top X V$ and $\tilde{\Omega} := V^\top \Omega V$, the equation becomes: $\Sigma^2 \tilde{X} + \tilde{X}\Sigma^2 = \tilde{\Omega}$, which implies that $(\sigma_i^2 + \sigma_j^2)\tilde{X}_{ij} = \tilde{\Omega}_{ij}$. □

The second lemma provides an upper bound on the Frobenius norm of the skew part of the product of two matrices with unit norm. We will need this result when computing the maximal sectional curvature of $\mathbb{R}_*^{n \times p}/\mathcal{O}_p$ at some point $\pi(Y) \in \mathbb{R}_*^{n \times p}/\mathcal{O}_p$.

Lemma 2. *Let* $A, B \in \mathbb{R}^{n \times p}$, *such that* $\|A\|_F = \|B\|_F = 1$. *Then,*

$$\left\|A^\top B - B^\top A\right\|_F^2 \leq 2.$$

Proof. Let us consider the optimization problem:

$$\max_{\|A\|_F = \|B\|_F = 1} \left\|A^\top B - B^\top A\right\|_F^2.$$

Observe that, by symmetry of the problem, the Lagrange multipliers associated with the constraints $\|A\|_F = 1$ and $\|B\|_F = 1$ are equal, and that the linear independence constraint qualification (LICQ) condition holds. Hence the KKT first-order necessary optimality conditions are:

$$\begin{cases} 2B(B^\top A - A^\top B) - \lambda A = 0 & \text{(a)} \\ -2A(B^\top A - A^\top B) - \lambda B = 0 & \text{(b)} \\ \|A\|_F = \|B\|_F = 1. & \text{(c)} \end{cases} \tag{4}$$

Premultiplying 4a by $A^\top$, 4b by $B^\top$, and taking the sum of the two yields:

$$\lambda(A^\top A + B^\top B) = 2(A^\top B - B^\top A)(B^\top A - A^\top B).$$

Taking the trace of both sides of the equation, we obtain:

$$\lambda = \operatorname{tr}\left((B^\top A - A^\top B)^\top(B^\top A - A^\top B)\right) = \left\|A^\top B - B^\top A\right\|_{\mathrm{F}}^2.$$

We will show that $\lambda \le 2$, which will conclude the proof. If $B = 0$, then the claim obviously holds, hence we assume from now on that $B \ne 0$. Let $B = U\Sigma V^\top$ be a compact singular value decomposition, where $U \in \mathbb{R}^{n \times r}$, $\Sigma \in \mathbb{R}^{r \times r}$ and $V \in \mathbb{R}^{p \times r}$, with r the rank of B and $U^\top U = V^\top V = I_r$. Equation (4) becomes:

$$2U\Sigma^2 U^\top A - 2U\Sigma V^\top A^\top U\Sigma V^\top = \lambda A.$$

Left- and right-multiplying this equation by respectively $U^\top$ and V yields:

$$2\Sigma^2 U^\top AV - 2\Sigma V^\top A^\top U\Sigma = \lambda U^\top AV.$$

Now, defining $\tilde{A} := \Sigma U^\top AV$, we get:

$$2\Sigma\tilde{A} - 2\Sigma\tilde{A}^\top = \lambda\Sigma^{-1}\tilde{A},$$

which can be written as:

$$2\Sigma^2\tilde{A} - 2\Sigma^2\tilde{A}^\top = \lambda\tilde{A}. \tag{5}$$

Assume first that $\tilde{A} \ne 0$. Then, if $r = 1$, $\lambda = 0$. If $r \ge 2$, the coefficients $\tilde{A}_{ij}$, $i, j = 1, \ldots, r$ of the matrix $\tilde{A}$ satisfy the equation:

$$\lambda(\tilde{A}_{ij} - \tilde{A}_{ji}) = 2(\sigma_i^2 + \sigma_j^2)(\tilde{A}_{ij} - \tilde{A}_{ji}).$$

If for some $i, j \in \{1, \ldots, r\}$, $\tilde{A}_{ij} \ne \tilde{A}_{ji}$ there holds $\lambda = 2(\sigma_i^2 + \sigma_j^2) \le 2\|B\|_{\mathrm{F}}^2 = 2$. Otherwise (i.e., $\tilde{A} \ne 0$ is symmetric), $\lambda = 0$ by (5).

There remains to check the value of λ when $\tilde{A} = 0$. It can be readily checked that the matrix $[V, V_\perp]^\top B^\top A[V, V_\perp]$ is of the form:

$$[V, V_\perp]^\top B^\top A[V, V_\perp] = \begin{bmatrix} \tilde{A} & \Sigma U^\top AV_\perp \\ 0_{p-r \times r} & 0_{p-r \times p-r} \end{bmatrix}.$$

Since $\tilde{A} = 0$, the matrix is strictly upper triangular. There holds

$$\left\|B^\top A - A^\top B\right\|_{\mathrm{F}}^2 = \left\|[V, V_\perp]^\top(B^\top A - A^\top B)[V, V_\perp]\right\|_{\mathrm{F}}^2 = 2\left\|\Sigma U^\top AV_\perp\right\|_{\mathrm{F}}^2 \le 2,$$

which concludes the proof. $\square$

We are now able to compute the minimum and maximum values of the sectional curvature of $\mathbb{R}_*^{n \times p}/\mathcal{O}_p$ at some point $\pi(Y)$. Observe that, since the sectional curvature is associated to a tangent plane, it does not depend on the choice of the vectors $\xi_{\pi(Y)}$, $\eta_{\pi(Y)}$ that span this tangent plane. As a result, we make the assumption in the rest of the document that the horizontal lifts $\bar{\xi}_Y$ and $\bar{\eta}_Y$ are orthonormal vectors, i.e., $\langle\bar{\xi}_Y, \bar{\xi}_Y\rangle = \langle\bar{\eta}_Y, \bar{\eta}_Y\rangle = 1$ and $\langle\bar{\xi}_Y, \bar{\eta}_Y\rangle = 0$. This makes the denominator of (2) equal to one.

Proposition 1. *The minimum of the sectional curvature at $\pi(Y)$ of the quotient manifold $\mathbb{R}_*^{n \times p} / \mathcal{O}_p$ is always zero. If $p = 1$, the sectional curvature is equal to zero.*

Proof. By (2), the sectional curvature associated with a pair of orthonormal tangent vectors $\xi_{\pi(Y)}, \eta_{\pi(Y)}$ is defined as:

$$K_{\mathbb{R}_*^{n \times p} / \mathcal{O}_p}(\xi_{\pi(Y)}, \eta_{\pi(Y)}) = 3 \left\| Y \mathbf{T}_{Y^\top Y}^{-1} (\bar{\eta}_Y^\top \bar{\xi}_Y - \bar{\xi}_Y^\top \bar{\eta}_Y) \right\|_{\mathrm{F}}^2.$$

Using Lemma 1, with $Y = U \Sigma V^\top$ a singular value decomposition and $\tilde{\Omega} := V^\top (\bar{\eta}_Y^\top \bar{\xi}_Y - \bar{\xi}_Y^\top \bar{\eta}_Y) V$, there holds:

$$K_{\mathbb{R}_*^{n \times p} / \mathcal{O}_p}(\xi_{\pi(Y)}, \eta_{\pi(Y)}) = 3 \left\| (U \Sigma V^\top)(V \tilde{X} V^\top) \right\|_{\mathrm{F}}^2, \quad \tilde{X}_{ij} = \frac{\tilde{\Omega}_{ij}}{(\sigma_i^2 + \sigma_j^2)}.$$

Due to the unitarily invariance of the Frobenius norm, there holds:

$$K_{\mathbb{R}_*^{n \times p} / \mathcal{O}_p}(\xi_{\pi(Y)}, \eta_{\pi(Y)}) = 3 \left\| \Sigma \tilde{X} \right\|_{\mathrm{F}}^2 = 3 \sum_{i,j=1}^{p} \frac{\sigma_i^2 \tilde{\Omega}_{ij}^2}{(\sigma_i^2 + \sigma_j^2)^2}. \tag{6}$$

This is zero if and only if $\tilde{\Omega}$ is zero. If $p = 1$, the sectional curvature is always zero since $\bar{\eta}_Y^\top \bar{\xi}_Y \in \mathbb{R}$. If $p \geq 2$, take for example $\bar{\xi}_Y = Y(Y^\top Y)^{-1} S$ with $S = S^\top$, and $\bar{\eta}_Y = Y \|Y\|_{\mathrm{F}}^{-1}$. Then, $\tilde{\Omega} = 0$, and if the matrix S is chosen such that $\|\bar{\xi}_Y\|_{\mathrm{F}} = 1$ and $\mathrm{Diag}(S) = 0$, the two vectors $\bar{\xi}_Y$ and $\bar{\eta}_Y$ are orthonormal.   □

The following result characterizes the maximum of the sectional curvature of $\mathbb{R}_*^{n \times p} / \mathcal{O}_p$ at some point $\pi(Y)$.

Proposition 2. *Let $Y \in \mathbb{R}_*^{n \times p}$ and $Y = U \Sigma V^\top$ a singular value decomposition, with singular values $\sigma_1 \geq \sigma_2 \geq \cdots \geq \sigma_p > 0$. If $p = 1$, the sectional curvature is always zero. If $p \geq 2$, the maximum of the sectional curvature at $\pi(Y)$ of the quotient $\mathbb{R}_*^{n \times p} / \mathcal{O}_p$ is:*

$$K_{\mathbb{R}_*^{n \times p} / \mathcal{O}_p}(\xi_{\pi(Y)}^*, \eta_{\pi(Y)}^*) = \frac{3}{\sigma_{p-1}^2 + \sigma_p^2}. \tag{7}$$

This value is reached for, e.g., $\xi_{\pi(Y)}^* = D\pi(Y)[\bar{\xi}_Y^*]$ and $\eta_{\pi(Y)}^* = D\pi(Y)[\bar{\eta}_Y^*]$, with $\bar{\xi}_Y^* = Y(Y^\top Y)^{-1} S_\xi$ and $\bar{\eta}_Y^* = Y(Y^\top Y)^{-1} S_\eta$, where

$$S_\xi := \frac{V(E_{p-1,p-1} - E_{p,p})V^\top}{\sqrt{\sigma_{p-1}^{-2} + \sigma_p^{-2}}} \qquad S_\eta := \frac{V(E_{p-1,p} + E_{p,p-1})V^\top}{\sqrt{\sigma_{p-1}^{-2} + \sigma_p^{-2}}},$$

with E_{ij} the matrix whose elements are zero excepted $E(i,j) = 1$.

Proof. Similarly as in the proof of Proposition 1, let us write:

$$K_{\mathbb{R}^{n \times p}_*/\mathcal{O}_p}(\xi_{\pi(Y)}, \eta_{\pi(Y)}) = 3 \sum_{i,j=1}^{p} \frac{\sigma_i^2 \tilde{\Omega}_{ij}^2}{(\sigma_i^2 + \sigma_j^2)^2} = 3 \sum_{i>j} \frac{\tilde{\Omega}_{ij}^2}{(\sigma_i^2 + \sigma_j^2)}, \qquad (8)$$

where the last inequality comes from the fact that $\tilde{\Omega} := V^\top(\bar{\eta}_Y^\top \bar{\xi}_Y - \bar{\xi}_Y^\top \bar{\eta}_Y)V$ is skew-symmetric. According to Lemma 2, the squared Frobenius norm of $\tilde{\Omega}$ is upper bounded by 2:

$$\left\| \tilde{\Omega} \right\|_F^2 = \left\| V^\top(\bar{\eta}_Y^\top \bar{\xi}_Y - \bar{\xi}_Y^\top \bar{\eta}_Y)V \right\|_F^2 = \left\| \bar{\eta}_Y^\top \bar{\xi}_Y - \bar{\xi}_Y^\top \bar{\eta}_Y \right\|_F^2 \leq 2.$$

Therefore:

$$K_{\mathbb{R}^{n \times p}_*/\mathcal{O}_p}(\xi_{\pi(Y)}, \eta_{\pi(Y)}) \leq \frac{3\sum_{i>j} \tilde{\Omega}_{ij}^2}{(\sigma_{p-1}^2 + \sigma_p^2)} \leq \frac{3\|\tilde{\Omega}\|_F^2}{2(\sigma_{p-1}^2 + \sigma_p^2)} \leq \frac{3}{\sigma_{p-1}^2 + \sigma_p^2}.$$

To finish the proof, we show that this bound is reached for the vectors $\bar{\xi}_Y^*$ and $\bar{\eta}_Y^*$ given in the proposition. It can be readily checked that $\bar{\xi}_Y^*$ and $\bar{\eta}_Y^*$ are orthogonal and have unit norm. There remains to compute $\tilde{\Omega}^*$:

$$\bar{\eta}_Y^{*\top} \bar{\xi}_Y^* = S_\eta (Y^\top Y)^{-1} S_\xi = \frac{V(E_{p-1,p} + E_{p,p-1})\Sigma^{-2}(E_{p-1,p-1} - E_{p,p})V^\top}{\sigma_{p-1}^{-2} + \sigma_p^{-2}},$$

which simply becomes

$$\bar{\eta}_Y^{*\top} \bar{\xi}_Y^* = \frac{V(\sigma_{p-1}^{-2} E_{p,p-1} - \sigma_p^{-2} E_{p-1,p})V^\top}{\sigma_{p-1}^{-2} + \sigma_p^{-2}}.$$

Therefore, $\tilde{\Omega}^*$ is:

$$\tilde{\Omega}^* = \frac{(\sigma_{p-1}^{-2} + \sigma_p^{-2})E_{p,p-1} - (\sigma_{p-1}^{-2} + \sigma_p^{-2})E_{p-1,p}}{\sigma_{p-1}^{-2} + \sigma_p^{-2}} = (E_{p,p-1} - E_{p-1,p}),$$

such that

$$K_{\mathbb{R}^{n \times p}_*/\mathcal{O}_p}(\xi_{\pi(Y)}^*, \eta_{\pi(Y)}^*) = \frac{3}{\sigma_{p-1}^2 + \sigma_p^2}. \qquad \square$$

5   Conclusion

We have computed the curvature of the manifold $\mathbb{S}_+(p,n)$ endowed with the Bures–Wasserstein metric. We have provided expressions for the Riemannian curvature tensor and the sectional curvature of the manifold. We have shown that in the case $p = 1$ the sectional curvature is always zero. If $p \geq 2$, the minimum over the tangent planes of the sectional curvature is zero, while the maximum goes to infinity as the pth and $p-1^{\text{th}}$ eigenvalues of the PSD matrix go simultaneously to zero. Further works might aim at computing the curvature of $\mathbb{S}_+(p,n)$ endowed with the other metrics proposed in the literature (see [VAV09, BS09, VAV13]), which to our knowledge are still unknown.

References

[AGSW16] Absil, P.A., Gousenbourger, P.-Y., Striewski, P., Wirth, B.: Differentiable piecewise-Bézier surfaces on Riemannian manifolds. SIAM J. Imaging Sci. **9**(4), 1788–1828 (2016). http://dx.doi.org/10.1137/16M1057978

[ATV13] Afsari, B., Tron, R., Vidal, R.: On the convergence of gradient descent for finding the Riemannian center of mass. SIAM J. Control Optim. **51**(3), 2230–2260 (2013). https://doi.org/10.1137/12086282X

[BJL18] Bhatia, R., Jain, T., Lim, Y.: On the Bures-Wasserstein distance between positive definite matrices. Expositiones Mathematicae (2018). https://doi.org/10.1016/j.exmath.2018.01.002

[Bon13] Bonnabel, S.: Stochastic gradient descent on Riemannian manifolds. IEEE Trans. Autom. Control **58**(9), 2217–2229 (2013). https://doi.org/10.1109/TAC.2013.2254619

[BR97] Bhatia, R., Rosenthal, P.: How and why to solve the operator equation AX- XB= Y. Bull. Lond. Math. Soc. **29**(1), 1–21 (1997)

[BS09] Bonnabel, S., Sepulchre, R.: Riemannian metric and geometric mean for positive semidefinite matrices of fixed rank. SIAM J. Matrix Anal. Appl. **31**(3), 1055–1070 (2009). https://doi.org/10.1137/080731347

[Dit95] Dittmann, J.: On the Riemannian metric on the space of density matrices. Rep. Math. Phys. **36**(2–3), 309–315 (1995). https://doi.org/10.1016/0034-4877(96)83627-5

[Gel90] Gelbrich, G.: On a formula for the L^2 Wasserstein metric between measures on Euclidean and Hilbert spaces. Math. Nachr. **147**(1), 185–203 (1990). https://doi.org/10.1002/mana.19901470121

[GMM+17] Gousenbourger, P.-Y., et al.: Piecewise-Bézier C^1 smoothing on manifolds with application to wind field estimation. In: Proceedings of the 25th European Symposium on Artificial Neural Networks, Computational Intelligence and Machine Learning (ESANN), pp. 305–310 (2017)

[JBAS10] Journée, M., Bach, F., Absil, P.-A., Sepulchre, R.: Low-rank optimization on the cone of positive semidefinite matrices. SIAM J. Opti. **20**(5), 2327–2351 (2010). https://doi.org/10.1137/080731359

[KDB+18] Kacem, A., Daoudi, M., Amor, B.B, Berretti, S., Alvarez-Paiva, J.C.: A Novel geometric framework on gram matrix trajectories for human behavior understanding. IEEE Trans. Pattern Anal. Mach. Intell. (T-PAMI) (2018). https://doi.org/10.1109/tpami.2018.2872564

[LB14] Li, X.-B., Burkowski, F.J.: Conformational transitions and principal geodesic analysis on the positive semidefinite matrix manifold. In: Basu, M., Pan, Y., Wang, J. (eds.) ISBRA 2014. LNCS, vol. 8492, pp. 334–345. Springer, Cham (2014). https://doi.org/10.1007/978-3-319-08171-7_30

[MA18] Massart, E., Absil, P.-A.: Quotient geometry of the manifold of fixed-rank positive-semidefinite matrices. Technical report UCL-INMA-2018.06, UCLouvain, November 2018, Preprint. http://sites.uclouvain.be/absil/2018.06

[MBS11] Meyer, G., Bonnabel, S., Sepulchre, R.: Regression on fixed-rank positive semidefinite matrices: a Riemannian approach. J. Mach. Learn. Res. **12**, 593–625 (2011)

[MGS+19] Massart, E., Gousenbourger, P.-Y., Son, N.T., Stykel, T., Absil, P.-A.: Interpolation on the manifold of fixed-rank positive-semidefinite matrices for parametric model order reduction: preliminary results. In: Proceedings of the 27th European Symposium on Artifical Neural Networks, Computational Intelligence and Machine Learning (ESANN2019), pp. 281–286 (2019)

[MHB+16] Marchand, M., Huang, W., Browet, A., Van Dooren, P., Gallivan, K.A.: A Riemannian optimization approach for role model extraction. In: Proceedings of the 22nd International Symposium on Mathematical Theory of Networks and Systems, pp. 58–64 (2016)

[MMP18] Malagò, L., Montrucchio, L., Pistone, G.: Wasserstein Riemannian geometry of Gaussian densities. Inf. Geom. **1**(2), 137–179 (2018). https://doi.org/10.1007/s41884-018-0014-4

[MMS11] Mishra, B., Meyer, G., Sepulchre, R.: Low-rank optimization for distance matrix completion. In: Proceedings of the 50th IEEE Conference on Decision and Control and European Control Conference (CDC-ECC), pp. 4455–4460 (2011). https://doi.org/10.1109/CDC.2011.6160810

[O'N66] O'Neill, B.: The fundamental equations of a submersion. Mich. Math. J. **13**(4), 459–469 (1966). https://doi.org/10.1307/mmj/1028999604

[O'N83] O'Neill, B.: Semi-Riemannian geometry. In: Pure and Applied Mathematics, vol. 103. Academic Press Inc., Harcourt Brace Jovanovich Publishers, New York (1983)

[SASK12] Samir, C., Absil, P.-A., Srivastava, A., Klassen, E.: A gradient-descent method for curve fitting on Riemannian manifolds. Found. Comput. Math. **12**(1), 49–73 (2012)

[Smi05] Smith, S.T.: Covariance, subspace, and intrinsic Cramér-Rao bounds. IEEE Trans. Signal Process **53**(5), 1610–1630 (2005)

[Tak11] Takatsu, A.: Wasserstein geometry of Gaussian measures. Osaka J. Math. **48**(4), 1005–1026 (2011)

[VAV09] Vandereycken, B., Absil, P.-A., Vandewalle, S.: Embedded geometry of the set of symmetric positive semidefinite matrices of fixed rank. In: IEEE/SP 15th Workshop on Statistical Signal Processing, pp. 389–392 (2009). https://doi.org/10.1109/SSP.2009.5278558

[VAV13] Vandereycken, B., Absil, P.-A., Vandewalle, S.: A Riemannian geometry with complete geodesics for the set of positive semidefinite matrices of fixed rank. IMA J. Numer. Anal. **33**(2), 481–514 (2013). https://doi.org/10.1093/imanum/drs006

Geometric Science of Information
Libraries

Second-Order Networks in PyTorch

Daniel Brooks[1,2(✉)], Olivier Schwander[2], Frédéric Barbaresco[1],
Jean-Yves Schneider[1], and Matthieu Cord[2]

[1] Thales Land and Air Systems, Advanced Radar Concepts, Limours, France
daniel.brooks@lip6.fr
[2] Sorbonne Université, CNRS, LIP6 - Laboratoire d'Informatique de Paris 6,
75005 Paris, France

Abstract. Classification of Symmetric Positive Definite (SPD) matrices
is gaining momentum in a variety machine learning application fields. In
this work we propose a Python library which implements neural networks
on SPD matrices, based on the popular deep learning framework Pytorch.

Keywords: SPD matrix · Covariance · Second-order neural network ·
Riemannian machine learning

1 Introduction

Information geometry-based machine learning has recently been rapidly emerging in a broad spectrum of learning scenarios, and deep learning has been no
exception. Notably, works such as [13–15] introduce neural networks respectively
operating on Lie groups, Grassmann spaces, and SPD matrices. The natural representation of any temporally or spatially structured signal as a Gaussian process
allows for a near universal possible interpretation of the signal as its temporal
or spatial covariance, which is an SPD matrix, i.e. which belongs to the SPD
Riemannian manifold, which we note $\mathcal{S}_*^+$. Previous works make use of the SPD
representation in other contexts than deep learning: for instance, Riemannian
metric learning on $\mathcal{S}_*^+$ is developed in [24], while [23] review kernel methods
on $\mathcal{S}_*^+$, with a primary applicative focus on electro-encephalogram/cardiogram
(EEG/ECG) classification. In a similar vein, [2,4] extend barycenter-based classification methods to the SPD Riemannian framework. On the other hand, [9]
propose the usage of SPD matrices as a region descriptor in images, with applications in image segmentation. The work in [17] pushed the idea further by allowing
the region covariance descriptor to be appended to a deep neural representation
of an image, and by doing so introduced the first hints of automatic backpropagation in a Riemannian setting. Finally, the older theoretical developments in [7]
notably allowed the extension of optimization methods to manifold-valued neural networks as later utilized in [8,10,13]. Even more recent works, namely [1,25]
have appended SPD neural networks to classical, Euclidean ones, by considering the second-order moments of the learnt feature representations as a suitable
representation for the data.

© Springer Nature Switzerland AG 2019
F. Nielsen and F. Barbaresco (Eds.): GSI 2019, LNCS 11712, pp. 751–758, 2019.
https://doi.org/10.1007/978-3-030-26980-7_78

In this environment of popularization of deep learning on SPD matrices, we propose *torchspdnet*, a Python library featuring many relevant modules necessary to build a neural network operating on SPD matrices. We do so in the popular PyTorch framework [21]. While other libraries were proposed for general learning on manifolds (Geomstats [20]), deep learning on manifolds (McTorch [19]), optimization on manifolds (Manopt [5]) and SPD matrix manipulation (PyRiemann [3]), ours focusses exclusively on deep learning architectures for SPD matrices, providing seamless integration with any PyTorch development framework. In the following section we describe the core components of a SPD neural network, which we may call SPDNet. The third section deals with the optimization of a manifold-valued network. Finally, we show some use cases.

2   Second Order Networks

Here we describe the architecture of an SPDNet. We begin with the core building blocks, then show how to build a network using these blocks in various scenarios. Following the logic of most modern deep learning frameworks including PyTorch, the core building blocks, or layers of the network, are implemented as individual modules.

2.1   SPD Layers

Similarly to a classical neural network, an SPDNet aims at building a hierarchical sequence of more compact and discriminative manifolds as illustrated in Fig. 1. Three main layers are introduced in [13], described below.

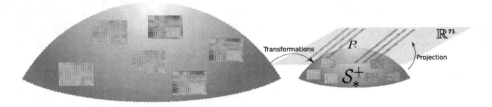

Fig. 1. Illustration of a generic SPD neural network. Successive bilinear layers followed by activations build a feature SPD manifold, which is then transformed to a Euclidean space to allow for classification.

BiMap. The bilinear mapping (BiMap) layer transforms an input matrix $X^{(l-1)}$ of size $n^{(l-1)}$ at layer $(l-1)$ into an SPD matrix $X^{(l)}$ of size $n^{(l)}$ at layer (l) using a basis change matrix $W^{(l)}$, required to be full-rank, which in turn constrains $n^{(l)} \leq n^{(l-1)}$. In practice $W^{(l)}$ is in fact constrained to be semi-orthogonal:

$$X^{(l)} = W^{(l)^T} X^{(l-1)} W^{(l)} \text{ with } W^{(l)} \in \mathcal{O}(n^{(l-1)}, n^{(l)}) \quad (1)$$

In the equation above, $\mathcal{O}(n^{(l-1)}, n^{(l)})$ is the manifold of semi-orthogonal rectangular matrices, also called Stiefel manifold, and $X^{(l-1)} = U^{(l-1)} \Sigma^{(l-1)} U^{(l-1)^T}$ designates the eigenvalue decomposition of $X^{(l-1)}$

ReEig. The transformation layer is followed by an activation, in this case a rectified eigenvalues (ReEig) layer:

$$X^{(l)} = U^{(l-1)} \max(\Sigma^{(l-1)}, \epsilon I_{n^{(l-1)}}) U^{(l-1)^T} \text{ with } P^{(l-1)} = U^{(l-1)} \Sigma^{(l-1)} U^{(l-1)^T}$$

(2)

The ReEig layer also makes use of an eigenvalue decomposition as it operates directly on the eigenvalues, with ϵ being a fixed threshold set to a default value of $1e - 4$.

LogEig. After a succession of transformations and activations, the final feature manifold is then transformed via a logarithmic mapping to a Euclidean space (LogEig layer) to perform the actual classification:

$$X^{(l)} = vec(\ U^{(l)} \log(\Sigma^{(l)}) U^{(l)^T})\ , \text{ with } P^{(l)} = U^{(l)} \Sigma^{(l)} U^{(l)^T}$$

(3)

The LogEig layer is justified in the Log-Euclidian Metric (LEM) framework, independently introduced in [11,22], which shows a correspondence from the manifold $\mathcal{S}^+_*$ to the Euclidean space $\mathcal{S}^+$ of symmetric matrices through the matrix logarithm. The *vec* operator denotes matrix vectorization.

3   Training

The main difficulties of learning an SPDNet lie both in the backpropagation through structured Riemannian functions [6,16], and in the manifold-constrained optimization [7].

3.1   Structured Derivatives

Manifold-valued functions, such as the LogEig and ReEig layers, require a generalization of the chain rule, key to the backpropagation algorithm. Both these layers can be represented in a unified fashion as a non-linear function f acting directly on the eigenvalues of the input matrix $X^{(l-1)} = U^{(l-1)} \Sigma^{(l-1)} U^{(l-1)^T}$. Then, the backpropagation goes as follows: given the succeeding gradient $\frac{\partial L^{(l)}}{\partial X^{(l)}}$, the output gradient $\frac{\partial L^{(l-1)}}{\partial X^{(l-1)}}$ is:

$$\frac{\partial L^{(l-1)}}{\partial X^{(l-1)}} = U \left(L \odot (U^T(\frac{\partial L^{(l)}}{\partial X^{(l)}})U) \right) U^T$$

(4)

In the previous equation, the Loewner matrix of finite differences L is defined as:

$$L_{ij} = \begin{cases} \frac{f(\sigma_i) - f(\sigma_j)}{\sigma_i - \sigma_j} & \text{if } \sigma_i \neq \sigma_j \\ f'(\sigma_i) & \text{otherwise} \end{cases}$$

(5)

3.2   Constrained Optimization

In the specific case of the BiMap layer, the transformation matrix W is constrained to the Stiefel manifold. The Euclidean gradient $\frac{\partial \mathcal{L}}{\partial G}$ of the loss function $\mathcal{L}$ does not respect the geometry of the manifold: as such the gradient descent is ill-defined. $\frac{\partial \mathcal{L}}{\partial G}$. The correct Riemannian gradient is obtained by tangent projection $\Pi \mathcal{T}_W$ on the manifold at W. The update is then obtained by computing the geodesic on the manifold from W towards the Riemannian gradient, also called exponential mapping $Exp_W(X)$. We illustrate this process in Fig. 2. Both the tangent projection and geodsic are known on the Stiefel manifold [7]: ction Exp_W have a closed form [7]:

$$\Pi \mathcal{T}_W(X) = X - WW^T X$$
$$Exp_W(X) = Orth(W + X) \tag{6}$$

The operator $Orth$ represents the orthonormalization of a free family of vectors, i.e. the Q matrix in the QR decomposition.

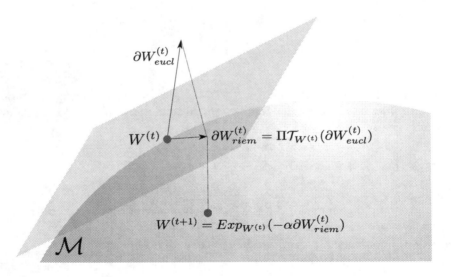

Fig. 2. Illustration of manifold-constrained gradient update. The Euclidean gradient is projected to the tangent space, then mapped to the manifold.

3.3   Summary

The library we propose seamlessly integrates orthogonally-constrained optimization on $\mathcal{S}_*^+$: the code for setting up the learning of a model in PyTorch is only modified in the usage of the *MixOptimizer* class, which mixes a conventional optimizer with the Riemannian ones:

```
import torch.nn as nn
from mixoptimizer import MixOptimizer
...
model=... #define the model
...
l=nn.CrossEntropyLoss()
opt=MixOptimizer(model.parameters(),lr=lr,momentum=0.9,
weight_decay=5e-4) #define the loss function and mixed
optimizer
...
l.backward()
opt.step() #in the training loop, compute gradients and
update weights as usually done
```

4   Use Cases

Here we show how to use the library in practice. Following the PyTorch logic, elementary functions are defined in *torchspdnet.functional* and high-level modules in *torchspdnet.nn*.

4.1   Basic SPDNet Model

Here we give the most basic use case scenario: given input covariance data of size 20×20, we build an SPDNet which reduces its size to 15 then 10 through two BiMaps and a ReEig activation, followed by the LogEig and vectorization. Finally, a standard fully-connected layer allows for classification over the 3 classes

```
import torch.nn as nn
import torchspdnet.nn as nn_spd

model=nn.Sequential(
        nn_spd.BiMap(1,1,20,15),
        nn_spd.ReEig(),
        nn_spd.BiMap(1,1,15,10),
        nn_spd.LogEig(),
        nn_spd.Vectorize(),
        nn.Linear(10**2,3)
    )
```

Note that our implementation of the BiMap module supports an arbitrary number of channels, represented by the additional parameters all set to 1 in this example.

4.2   First-Order and Second-Order Combined

In a more complex example, an SPDNet acts upon the features maps of a convolutional network. For an image recognition task, these features may come from a pre-trained deep network but nothing keeps from training the whole network in an end-to-end fashion or to fine-tune the parameters. Here we describe the combination of a pre-trained ResNet-18 [12] on the CIFAR10 [18] challenge and of SPDNet layers. We call such a model a second-order neural network (SOCNN).

```python
import torch.nn as nn
import torchspdnet.nn as nn_spd
from resnet import ResNet18

class SOCNN(nn.Module):
    def __init__(self):
        super(__class__, self).__init__()

        self.model_fo=ResNet18() #first-order model
        self.model_fo.load_state_dict(th.load
          ('pretrained/ResNet18.pth')['state_dict'])

        self.connection=nn.Conv2d(512,256,kernel_size
          =(1,1)) #convolutional connection

        self.model_so=nn.Sequential( #second-order model
            nn_spd.BiMap(1,1,256,128),
            nn_spd.ReEig(),
            nn_spd.BiMap(1,1,128,64),
        ).to(self.device_so)

        self.dense=nn.Sequential(
            nn.Linear(64**2,1024),
            nn.Linear(1024,10)
        )

    def forward(self,x):
        x_fo=self.model_fo(x)
        x_co=self.connection(x_fo)
        x_sym=nn_spd.CovPool()(x_co.view
          (x_co.shape[0],x_co.shape[1],-1))
        x_so=self.model_so(x_sym)
        x_vec=nn_spd.LogEig()(x_so).view
          (x_so.shape[0],x_so.shape[-1]**2)
        y=self.dense(x_vec)
        return y
```

5   Conclusion

We have proposed a PyTorch library for deep learning on SPD matrices. We hope its versatility and natural integration in any PyTorch workflow will allow future projects to more readily make use of the potential of exploiting covariance structure in data at any level.

References

1. Acharya, D., Huang, Z., Paudel, D.P., Gool, L.V.: Covariance Pooling for Facial Expression Recognition, p. 8
2. Barachant, A., Bonnet, S., Congedo, M., Jutten, C.: Multiclass brain-computer interface classification by riemannian geometry. IEEE Trans. Biomed. Eng. **59**(4), 920–928 (2012). https://doi.org/10.1109/TBME.2011.2172210. http://ieeexplore.ieee.org/document/6046114/
3. Barachant, A.: Python package for covariance matrices manipulation and Biosignal classification with application in Brain Computer interface: alexandrebarachant/pyRiemann, February 2019. https://github.com/alexandrebarachant/pyRiemann, original-date: 2015–04-19T16:01:44Z
4. Barachant, A., Bonnet, S., Congedo, M., Jutten, C.: Classification of covariance matrices using a Riemannian-based kernel for BCI applications. Neurocomputing **112**, 172–178 (2013). https://doi.org/10.1016/j.neucom.2012.12.039. https://hal.archives-ouvertes.fr/hal-00820475
5. Boumal, N., Mishra, B., Absil, P.A., Sepulchre, R.: Manopt, a matlab toolbox for optimization on manifolds. J. Mach. Learn. Res. **15**, 1455–1459 (2014). http://jmlr.org/papers/v15/boumal14a.html
6. Brodskiĭ, M., et al.: Thirteen Papers on Functional Analysis and Partial Differential Equations, American Mathematical Society Translations: Series 2, vol. 47. American Mathematical Society, December 1965. https://doi.org/10.1090/trans2/047. http://www.ams.org/home/page/
7. Edelman, A., Arias, T., Smith, S.: The geometry of algorithms with orthogonality constraints. SIAM J. Matrix Anal. Appl. **20**(2), 303–353 (1998). https://doi.org/10.1137/S0895479895290954. https://epubs.siam.org/doi/abs/10.1137/S0895479895290954
8. Engin, M., Wang, L., Zhou, L., Liu, X.: DeepKSPD: Learning Kernel-matrix-based SPD Representation for Fine-grained Image Recognition. arXiv:1711.04047 [cs], November 2017
9. Faulkner, H., Shehu, E., Szpak, Z.L., Chojnacki, W., Tapamo, J.R., Dick, A., Hengel, A.V.D.: A study of the region covariance descriptor: impact of feature selection and image transformations. In: 2015 International Conference on Digital Image Computing: Techniques and Applications (DICTA), pp. 1–8, November 2015. https://doi.org/10.1109/DICTA.2015.7371222
10. Gao, Z., Wu, Y., Bu, X., Jia, Y.: Learning a Robust Representation via a Deep Network on Symmetric Positive Definite Manifolds. arXiv:1711.06540 [cs], November 2017
11. Harris, W.F.: The average eye. Ophthalmic Physiol. Opt. **24**(6), 580–585 (2004). https://doi.org/10.1111/j.1475-1313.2004.00239.x. https://onlinelibrary.wiley.com/doi/abs/10.1111/j.1475-1313.2004.00239.x

12. He, K., Zhang, X., Ren, S., Sun, J.: Deep residual learning for image recognition. In: 2016 IEEE Conference on Computer Vision and Pattern Recognition (CVPR), pp. 770–778. IEEE, Las Vegas, June 2016. https://doi.org/10.1109/CVPR.2016. 90. http://ieeexplore.ieee.org/document/7780459/
13. Huang, Z., Van Gool, L.J.: A Riemannian network for SPD matrix learning. In: AAAI, vol. 1, p. 3 (2017)
14. Huang, Z., Wan, C., Probst, T., Van Gool, L.: Deep Learning on Lie Groups for Skeleton-based Action Recognition. arXiv:1612.05877 [cs], December 2016
15. Huang, Z., Wu, J., Van Gool, L.: Building Deep Networks on Grassmann Manifolds. arXiv:1611.05742 [cs], November 2016
16. Ionescu, C., Vantzos, O., Sminchisescu, C.: Matrix backpropagation for deep networks with structured layers. In: 2015 IEEE International Conference on Computer Vision (ICCV), pp. 2965–2973. IEEE, Santiago, December 2015. https://doi.org/ 10.1109/ICCV.2015.339. http://ieeexplore.ieee.org/document/7410696/
17. Ionescu, C., Vantzos, O., Sminchisescu, C.: Training Deep Networks with Structured Layers by Matrix Backpropagation. arXiv:1509.07838 [cs], September 2015
18. Krizhevsky, A.: Learning Multiple Layers of Features from Tiny Images, p. 60
19. Meghwanshi, M., Jawanpuria, P., Kunchukuttan, A., Kasai, H., Mishra, B.: McTorch, a manifold optimization library for deep learning. arXiv:1810.01811 [cs, stat], October 2018
20. Miolane, N., Mathe, J., Donnat, C., Jorda, M., Pennec, X.: geomstats: a Python Package for Riemannian Geometry in Machine Learning. arXiv:1805.08308 [cs, stat], May 2018
21. Paszke, A., et al.: Automatic differentiation in PyTorch, October 2017. https:// openreview.net/forum?id=BJJsrmfCZ
22. Pennec, X., Fillard, P., Ayache, N.: A Riemannian framework for tensor computing. Int. J. Comput. Vis. **66**(1), 41–66 (2006). https://doi.org/10.1007/s11263-005-3222-z. http://link.springer.com/10.1007/s11263-005-3222-z
23. Yger, F.: A review of kernels on covariance matrices for BCI applications. In: 2013 IEEE International Workshop on Machine Learning for Signal Processing (MLSP), pp. 1–6, September 2013. https://doi.org/10.1109/MLSP.2013.6661972
24. Yger, F., Sugiyama, M.: Supervised LogEuclidean Metric Learning for Symmetric Positive Definite Matrices. arXiv:1502.03505 [cs], Febraury 2015
25. Yu, K., Salzmann, M.: Second-order Convolutional Neural Networks. arXiv:1703.06817 [cs], March 2017

Symmetric Algorithmic Components
for Shape Analysis with Diffeomorphisms

Nicolas Guigui[(✉)], Shuman Jia, Maxime Sermesant, and Xavier Pennec

Université Côte d'Azur, Inria, Epione Project-Team, Valbonne, France
nicolas.guigui@inria.fr

Abstract. In computational anatomy, the statistical analysis of tempo-ral deformations and inter-subject variability relies on shape registration. However, the numerical integration and optimization required in diffeo-morphic registration often lead to important numerical errors. In many cases, it is well known that the error can be drastically reduced in the presence of a symmetry. In this work, the leading idea is to approxi-mate the space of deformations and images with a possibly non-metric symmetric space structure using an involution, with the aim to perform parallel transport. Through basic properties of symmetries, we investi-gate how the implementations of a midpoint and the involution compare with the ones of the Riemannian exponential and logarithm on diffeo-morphisms and propose a modification of these maps using registration errors. This leads us to identify transvections, the composition of two symmetries, as a mean to measure how far from symmetric the underly-ing structure is. We test our method on a set of 138 cardiac shapes and demonstrate improved numerical consistency in the Pole Ladder scheme.

Keywords: Shape registration · Parallel transport · Symmetric spaces

1 Introduction

Computational anatomy aims at modeling the temporal evolution and cross-sectional variability of anatomical shapes. The deformations between shapes are obtained by applying non-rigid registration algorithms that seek the *smallest* trans-formation - in a sense that will be defined precisely - of the ambient space to match two shapes. In the diffeomorphic registration setting, the deformations are modeled by diffeomorphisms, that provide invertible and folding-free transformations.

In the Large Deformation Diffeomorphic Metric Mapping (LDDMM) frame-work, the space of diffeomorphisms is endowed with a right invariant metric and deformations of interest are obtained by geodesic flows from the identity transformation. They are parameterized by their initial velocity fields, which are tangent vectors at the identity deformation, defined by initial control points and dual momenta. A transformation is then computed by integration of differential equations. Registration is performed by solving an optimization problem on the initial momenta and control points with a gradient descent. These numerical

© Springer Nature Switzerland AG 2019
F. Nielsen and F. Barbaresco (Eds.): GSI 2019, LNCS 11712, pp. 759–768, 2019.
https://doi.org/10.1007/978-3-030-26980-7_79

schemes efficiently implement an exponential and logarithm map on a subspace of diffeomorphisms.

However in practice, the optimization problem is relaxed to enforce smooth deformations and results in inexact matching. Therefore we can think of the exponential map as not going "far enough". In this work we introduce a modified exponential map that accounts for a residual error due to registration, indifferent to the choice of the regularization parameter.

The choice of the metric and regularization parameter affects the geometric structure of the space of deformations under consideration. Many convergence results depend on the curvature of this space, e.g. [9], and especially on its covariant derivative. This gradient being null in locally symmetric spaces, they form a very convenient setting to perform statistics on shapes. In order to assess how far from symmetric our structure is, it would therefore be valuable to develop a procedure to measure this gradient. In this paper we build on a specific parallel transport scheme.

In the statistical analysis of temporal deformations, parallel transport along geodesics is commonly used to perform inter-subject normalization, that is the vector transport of velocity fields from each subject's space to a common atlas' space. An approximation based on Jacobi fields was proposed in [7,11]. A numerical implementation named Pole Ladder (PL) was proposed in [5] and relies only on the computation of exponential and logarithm maps. Following the Shild's ladder, it consists in the construction of geodesic parallelograms.

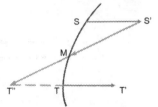

Fig. 1. Pole Ladder

The progression between two shapes S and S' is transported to T by:

- first computing a "midpoint" M on the geodesic between S and T. It is seen as the diagonal of the geodesic parallelogram;
- then extending the geodesic from S' to M by the same length to obtain T'';
- similarly extending the geodesic from T'' to T to obtain the parallel deformation of the template T;

See Fig. 1 for a schematic representation.

For large deformations that typically arise in inter-subject registration, this procedure is usually iterated. We therefore expect numerical errors to grow linearly in the number of steps, and lose crucial numerical accuracy. The accuracy of this scheme was analyzed in [9] and shown to be a third order scheme in general, and exact in symmetric spaces where it is equivalent to a single transvection, that is, a composition of two symmetries, in our case symmetries with respect to M and T.

In many cases, it is well known that numerical errors can be drastically reduced in the presence of a symmetry. This is for instance the case to diagonalize a symmetric matrix versus an arbitrary one. Using the Stationary Velocity Fields (SVF) framework for registration, a symmetric variant of Pole Ladder built on a Lie Group intrinsic symmetric structure was proposed in [3]. This procedure is recalled in Sect. 2.2. In this work, we build on this idea, but rely on LDDMM to implement a more general involution that accounts for the registration residual. This elementary algorithmic component constructs the symmetric shape of an original shape with respect to another one. It is presented in Sect. 2.3. In Sect. 2.4, we introduce the basic properties that symmetries must verify in an affine symmetric space and discuss whether these are fulfilled by our implementation. From a theoretical perspective, deviations from these properties are due to a non zero covariant derivative of the curvature tensor with the LDDMM metric. Conversely, we may interpret transvection errors as estimates of the numerical curvature gradient that encompass all the approximations due to the implementation. The numerical experiments of Sect. 3 show that there is an optimal regularisation parameter for which the space of deformations can best be approximated by a symmetric space.

The paper is organised as follows: in Sect. 2.1, we recall the LDDMM framework following [2] and in Sect. 2.2 the Pole Ladder procedure. We then introduce in Sects. 2.3 and 2.4 the main contributions of the paper, namely accounting for residuals, defining symmetries and their properties: centrality, involutivity and transvectivity. In Sect. 3 we present the numerical experiments and comment on the results.

2   Background and Method

2.1   The LDDMM Framework

In this work we consider shapes represented by 3D meshes. However, the methodology seamlessly applies to images. In order to define a practical finite dimensional parameterization of a subspace of diffeomorphisms G acting on the ambient space $\mathbb{R}^d$, we consider time-varying velocity fields $v_t(x) = \sum_{k=1}^{N_c} K(x, c_k^{(t)}) \mu_k^{(t)}$ obtained by convolution of a Gaussian kernel $K(x, y) = \exp(-\frac{\|x-y\|^2}{\sigma^2})$ over N_c control points $c^{(t)} = [c_k^{(t)}]_k$ and momenta $\mu^{(t)} = [\mu_k^{(t)}]_k$.

The set of such fields forms a pre-Hilbert space with scalar product between $v = \sum_k K(\cdot, c_k) \mu_k$ and $v' = \sum_k K(\cdot, c_k') \mu_k'$ defined by

$$<v, v'>_H = \sum_i \sum_j K(c_i, c_j') \mu_i^T \mu_j'. \tag{1}$$

Diffeomorphisms are then defined as flows of velocity fields from $\phi_0 = Id$. This amounts to integrating the ordinary differential equation (ODE) $\partial \phi_t(\cdot) = v_t[\phi_t(\cdot)]$ between 0 and 1. The scalar product on velocity fields induces a right-invariant Riemannian metric on the obtained subspace of diffeomorphisms, and geodesics of this metric are parameterized by control points and momenta that satisfy the following Hamiltonian equations:

$$\begin{cases} \dot{c}_k^{(t)} = \sum_j K(c_k^{(t)}, c_j^{(t)}) \mu_j^{(t)} \\ \dot{\mu}^{(t)} = -\sum_j \nabla_1 K(c_k^{(t)}, c_j^{(t)}) \mu_k^{(t)^T} \mu_j^{(t)} \end{cases} \tag{2}$$

ϕ_t and v_t are thus uniquely determined by their initial control points μ, this dependence will be explicitly written $\phi_t^{c,\mu}$ and $v_t^{c,\mu}$. In practice the interval $[0, 1]$ is discretized with n time steps and the ODE is solved with an iterative Euler forward or Runge-Kutta 2 method. This implements an exponential map at identity. The registration problem between a template shape T and a target mesh S optimizes the following criterion over initial control points and momenta c, μ:

$$C(c, \mu) = \|S - \phi_1^{c,\mu}(T)\|_2^2 + \alpha^2 \cdot \|v_0^{c,\mu}\|_H^2. \tag{3}$$

For simplicity, we measure the distance between shapes by the L_2 distance between nodes of the meshes. $\| \cdot \|_H$ is the norm defined by the scalar product of Eq. 1, which is actually the metric on G. The resulting v_0 is a logarithm at identity of ϕ_1. In fact the metric is scaled by a factor α, and this impacts the geometry of the underlying space as will be demonstrated. It allows to smoothly interpolate between solutions that belong to two paradigms:

- $\alpha \to 0$: exact matching between shapes;
- $\alpha \to \infty$: point distribution models (PDM), no deformation.

2.2 Symmetric Pole Ladder for Parallel Transport

In the context of computational anatomy, the aforementioned registration framework is used to represent a subject-specific temporal deformation between shapes S and S', and to transport this deformation to a common atlas or template T along the geodesic segment $[S, T]$. The anatomical shapes are modeled as points in a manifold $\mathcal{V}$ under the action of the space of diffeomorphisms G described above. We suppose here that this manifold is equipped with an affine connection, which defines parallel transport and the Exp map. Locally it further defines the Log map.

Algorithm 1 presents a symmetric variant of Pole Ladder [5] introduced in [3] to approximately perform parallel transport in $\mathcal{V}$.

Algorithm 1. Mid-point symmetric Pole Ladder transport of the geodesic segment $[S, S']$ along the geodesic $[S, T]$

- Compute the midpoint $M = \mathrm{Exp}_T(\frac{1}{2}Log_T(S))$ on the inter-subject geodesic;
- Compute the symmetric point $T'' = \mathrm{Exp}_M(-\mathrm{Log}_M(S'))$ of S' with respect to M;
- Compute the symmetric point $T' = \mathrm{Exp}_T(-\mathrm{Log}_T(T''))$ of T'' with respect to T, and return the geodesic segment $[T, T']$.

A Taylor expansion at the midpoint M of the error between the vector transported by Pole Ladder and exact parallel transport of the vector $u_S = \mathrm{Log}_S(S')$ is derived in [9]. We denote by $u = \Pi_S^M u_S$ the exact transport to M, and $u' = \Pi_T^M \mathrm{Log}_T(T')$ where T' is obtained by Pole Ladder. Let also $v = \mathrm{Log}_M(T)$. Then

$$u' - u = \frac{1}{12}\left((\nabla_v R)(u, v)(5u - 2v) + (\nabla_u R)(u, v)(v - 2u)\right) + O(\|v + u\|^5). \tag{4}$$

In fact this scheme is exact in an affine locally symmetric space, where $\nabla R = 0$. In this case, using the local symmetries s_X at point X, we have $T' = s_T \circ s_M(S')$ meaning that Pole Ladder is equivalent to a transvection. As local symmetries are affine mappings, the following diagrams commute:

$$
\begin{array}{ccc}
T_S\mathcal{V} & \xrightarrow{(ds_M)_S} & T_T\mathcal{V} \\
\downarrow{\scriptstyle \Pi_S^M} & & \downarrow{\scriptstyle \Pi_T^M} \\
T_M\mathcal{V} & \xrightarrow{-Id} & T_M\mathcal{V}
\end{array}
\qquad
\begin{array}{ccc}
T_S\mathcal{V} & \xrightarrow{(ds_M)_S} & T_T\mathcal{V} \\
\downarrow{\scriptstyle \mathrm{Exp}_S} & & \downarrow{\scriptstyle \mathrm{Exp}_T} \\
\mathcal{V} & \xrightarrow{s_M} & \mathcal{V}
\end{array}
$$

Thus, $\Pi_T^M \circ (ds_M)_S = -\Pi_S^M$ [10, Prop. 4.3]. So with the previous notations, $(ds_M)_S u_S = -\Pi_M^T u$ and $(ds_M)_S u_S = Log_T(T") = -\Pi_M^T u'$ yielding $u' = u$.

2.3   Accounting for Residuals to Improve Centrality and Symmetry

In practice the registration is never exact due to the regularization. Thus, we propose to decompose the space of shapes into
a deformation part encoding the orbit of the template, and a
Euclidean space of residual displacement fields

$$S = \phi_1(T) + \delta, \tag{5}$$

where δ is a displacement field between corresponding points of
the two meshes. Of course different possibilities exist to compute the residual, to transport it, and to apply it to different
shapes, but our experiments suggest that this very simple formulation may be sufficient, so that we will not detail other approaches in this
paper. With this decomposition, a midpoint between T and S is defined by
(Fig. 2):

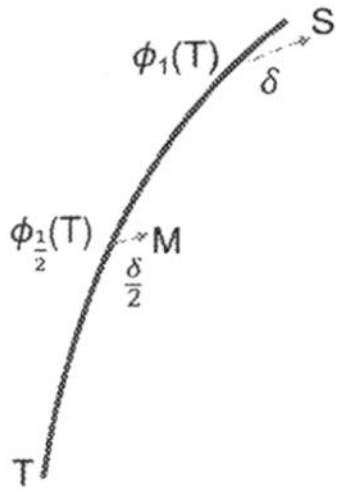

Fig. 2.   Midpoint with residuals

$$M = \phi_{\frac{1}{2}}(T) + \frac{1}{2}\delta = \mathrm{Exp}_T\left(\frac{1}{2}\mathrm{Log}_T(S)\right) + \frac{1}{2}\delta. \tag{6}$$

Unfortunately this formulation is not symmetric in T and S, and registering S
on T and shooting from S results in a different midpoint in general. We will
see however that using residuals decreases the distance between the midpoints
obtained with the two initial points.

Similarly, a symmetry is defined by inverting the
geodesics and the residuals (Fig. 3):

$$s_T(S) = \mathrm{Exp}_T\left(-\mathrm{Log}_T(S)\right) - \delta. \tag{7}$$

The first desired consistency property which we refer to
as **centrality** is the compatibility of the midpoint with
the symmetry, namely $s_M(S) = T$. Note that this construction of a central midpoint and a local geodesic symmetry is possible in arbitrary affine connection manifolds
for close enough points. However, these symmetries are
affine mappings if and only if the space is locally symmetric [10, Prop. 4.2].

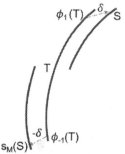

Fig. 3.   Symmetry with residuals

2.4   Involutivity and Transvectivity to Measure Curvature Gradient

By construction our symmetry verifies $s_T(T) = T$ for all $T \in \mathcal{V}$, and if the
log and exp maps are exact, $s_T \circ s_T = Id$. We will evaluate the exactitude of
this property, called **involutivity**. Note that in a Lie Group of transformations,
natural symmetries may be defined at ϕ by $\psi \mapsto \phi \circ \psi^{-1} \circ \phi$, which, at $\phi = Id$ is
in fact the inversion. Involutivity in this case reduces to **inverse consistency**:

$(\psi^{-1})^{-1} = \psi \iff \psi \circ \psi^{-1} = Id$. However, in our framework, both types of errors are different because the metric exponential differs from the one defined by the canonical Cartan-Shouten connection, which is the only one compatible with the group operations.

Finally, in an affine globally symmetric space, symmetries must verify the following property that we will call **transvectivity** [10, Prop. 5.3]:

$$\forall M, S \in \mathcal{V} \quad s_M \circ s_S = s_T \circ s_M, \quad \text{where } T = s_M(S). \tag{8}$$

We want to evaluate the exactitude of this property, and use the deviation to this ideal case as a proxy to measure the gradient of the curvature of the space in the directions of interest. At this point the role of α becomes clearer, it allows to form a continuum of decompositions of the space of diffeomorphisms under consideration:

- $\alpha \to 0$, $\delta \to 0$: the Riemannian space of deformations where the metric is not compatible with the Cartan-Shouten connection. This space is not symmetric, which shows in the transvectivity error.
- $\alpha \to \infty$, $v_0 \to 0$: the shapes are considered in the ambient space with Euclidean norm, this space is of course symmetric. This is the PDM framework.

We saw in Sect. 2.2 that Pole Ladder was doing the transvection $s_T \circ s_M$. With the same notations and using a Taylor expansion from [9], the deviation to parallel transport when applying the opposite transvection $s_M \circ s_S$ is:

$$u'' - u = -\frac{1}{12} \left((\nabla_v R)(u,v)(5u+v) + (\nabla_u R)(u,v)(v+2u) \right) + O(\|v+u\|^5). \tag{9}$$

Thus when measuring the transvectivity error $\|s_M \circ s_S(S') - s_T \circ s_M(S')\|_2$, we in fact measure $\|\text{Exp}_T(\Pi_M^T u'') - \text{Exp}_T(\Pi_M^T u')\|_2$ where

$$u' - u'' = \frac{1}{12} \left((\nabla_v R)(u,v)(10u-v) - 4(\nabla_u R)(u,v)u \right) + O(\|v+u\|^5). \tag{10}$$

The transvectivity error thus provides a practical way to measure the gradient of the curvature of the space even in the absence of any closed-form expression. This is noticeable in regards to the complexity of the curvature tensor itself in Mario's formula [8] and it may lead in the future to new ways of estimating the curvature and its gradient.

3   Experiments and Application to Cardiac Shapes

In this section we assess the consistency of our numerical implementation of the symmetry compared to its theoretical properties. We compare the symmetry with residuals to standard symmetry without residuals ($\delta = 0$) for different values of the parameter α. We used Deformetrica [1] for all our experiments. We also compare the parallel transport obtained by Pole Ladder with both types of symmetry to the one implemented in Deformetrica [6] using the fanning scheme.

We use a database of cardiac shapes from 138 subjects [3] for which the shapes at two time-points are available: at end-diastole (S) and at end-systole (S'). We use a population atlas as template T. Four types of errors are first measured. The first is the distance between midpoints when shooting from T or from S. The three others are, where M is the midpoint obtained by shooting from T:

- $\|s_M(T) - S\|_2$: the centrality error (Fig. 4a)
- $\|s_M \circ s_M(S') - S'\|_2$: the involutivity error (Fig. 4b)
- $\|s_T \circ s_M(S') - s_M \circ s_S(S')\|_2$: the transvectivity error (Fig. 4c).

Mean results for extreme values of α are given in Table 1. The average registration error of T on each subject's S, as well as the norm of the deformation and inverse consistency (by registering S on T) for each regularisation parameter are also given for reference.

Table 1. Mean errors measured on cardiac shapes, in millimeters

Error type	$\alpha^2 = 0.01$		$\alpha^2 = 1$		$\alpha^2 = 1089$	
	Residual	No residual	Residual	No residual	Residual	No residual
Centrality	0.36	0.43	0.10	0.50	<0.01	5.76
Involutivity	1.42	1.55	0.33	0.80	<0.01	9.39
Transvectivity	1.98	2.16	0.58	0.62	<0.01	0.16
Reg. error	0.23		0.41		5.61	
Reg. norm	42		30		1	
Inverse cons.	0.13		0.14		0.10	

These results illustrate the two contributions of this paper. Firstly, using residuals in the symmetry considerably improves the numerical accuracy of the computation of a midpoint. As we can see on Fig. 5a, the distance between midpoints computed by shooting from T or from S is reduced when using the residuals. This error compares well with the inverse consistency error in general and is even significantly lower for $\alpha \geq 1$.

Moreover, this increase in numerical accuracy is also visible on the centrality and involutivity errors. Indeed, for centrality (Fig. 5b), the error when using residuals is consistently smaller than without residuals, and this gain becomes

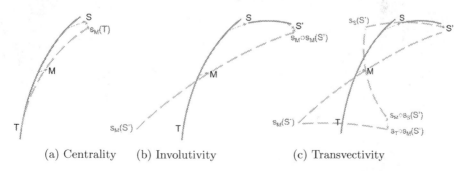

(a) Centrality   (b) Involutivity   (c) Transvectivity

Fig. 4. The three types of errors measured

larger as α grows. It is also remarkable that this error is significantly lower than the registration error for $\alpha \geq 1$. The same behavior is observed for the involutivity on Fig. 5c. This means that for reasonable values of α we obtain reliable implementations of the midpoint and symmetry.

Secondly, the LDDMM space endowed with the right invariant metric is not symmetric, and the transvectivity error reflects the covariant derivative of the curvature. Figure 5d gives further details: pushing registration with a small α generates larger errors, and there is an optimal $\alpha \in [1; 2]$ for which the space is closest to being symmetric. Using residuals, this error decreases as the space flattens to a Euclidean space and deformations tend to the identity.

Finally, in order to evaluate the result of symmetric Pole Ladder and compare it to the fanning scheme, we compute the local area strain (LAS) between end-diastole and end-systole at every landmark of the mesh. For two corresponding points m_i, m'_i that belong to k_i triangular cells, we compute the mean of the difference of area of each of these cells between S and S'.

$$LAS_i^{S-S'} = \frac{1}{k_i} \sum_{j=1}^{k_i} \frac{(a_j - a'_j)}{a_j} \tag{11}$$

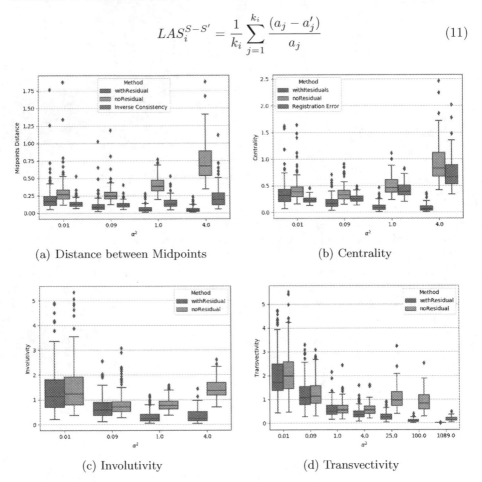

(a) Distance between Midpoints

(b) Centrality

(c) Involutivity

(d) Transvectivity

Fig. 5. The four errors for different values of α^2, with and without residuals in the computation of the midpoint and the symmetries.

This feature is commonly used by clinicians to characterise the cardiac motion [4]. Here we use it to test the isometric property of the parallel transport scheme: we measure the area strain between the original subject's meshes S and S', and compare it to the one measured between the atlas T and the deformed atlas T' obtained by the parallel transport algorithm. We report in Fig. 6a the square root of the sum of squared differences over all landmarks:

$$ASE^2 = \sum_i (LAS_i^{S-S'} - LAS_i^{T-T'})^2 \tag{12}$$

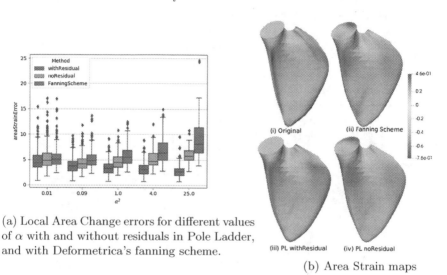

(a) Local Area Change errors for different values of α with and without residuals in Pole Ladder, and with Deformetrica's fanning scheme.

(b) Area Strain maps

Fig. 6. (b) Area strain maps for one patient computed between S and S' and represented on S (top left), and computed between T and T' and represented on T, where T' is obtained with the different methods and $\alpha = 1$.

As α grows, both the temporal and subject-to-atlas deformations decrease, thus generating less area changes, which explains the growing errors. Furthermore, we can see on Fig. 6b that the area strain is dominated by a bending artifact of the valve sections at the borders. Using residuals reduces this effect in this example. Although more suitable methods may exist to directly map scalar functions from one shape to another, these results emphasize the contribution of this paper: using residuals improves the symmetry and parallel transport with Pole Ladder.

4   Conclusion

We introduced residuals from registration errors to compute midpoints and symmetries between shapes. This results in improved numerical consistency for the centrality and involutivity properties. Furthermore the transvectivity error reflects the curvature of the underlying Riemannian space of deformations, and allows to estimate how far from symmetric this space is, depending on the regularisation parameter. Performing the same experiments in the framework of SVF with a similar implementation would yield very interesting comparison as the space is naturally symmetric, and accounting for residuals would thus provide a more consistent method for parallel transport. This gain could be even more interesting when considering registration between images. Indeed images are not in the template's orbit in practice and residuals would encode intensity bias, which is a key source of error in image registration.

Acknowledgements. This project has received funding from the European Research Council (ERC) under the European Union's Horizon 2020 research and innovation program (grant agreement G-Statistics No. 786854).

References

1. Bône, A., Louis, M., Martin, B., Durrleman, S.: Deformetrica 4: an open-source software for statistical shape analysis. In: Reuter, M., Wachinger, C., Lombaert, H., Paniagua, B., Lüthi, M., Egger, B. (eds.) ShapeMI 2018. LNCS, vol. 11167, pp. 3–13. Springer, Cham (2018). https://doi.org/10.1007/978-3-030-04747-4_1
2. Durrleman, S., et al.: Morphometry of anatomical shape complexes with dense deformations and sparse parameters. NeuroImage **101**, 35–49 (2014)
3. Jia, S., Duchateau, N., Moceri, P., Sermesant, M., Pennec, X.: Parallel transport of surface deformations from pole ladder to symmetrical extension. In: Reuter, M., Wachinger, C., Lombaert, H., Paniagua, B., Lüthi, M., Egger, B. (eds.) ShapeMI 2018. LNCS, vol. 11167, pp. 116–124. Springer, Cham (2018). https://doi.org/10.1007/978-3-030-04747-4_11
4. Kleijn, S.A., Aly, M.F.A., Terwee, C.B., van Rossum, A.C., Kamp, O.: Three-dimensional speckle tracking echocardiography for automatic assessment of global and regional left ventricular function based on area strain. J. Am. Soc. Echocardiogr. **24**(3), 314–321 (2011)
5. Lorenzi, M., Pennec, X.: Efficient parallel transport of deformations in time series of images: from Schild's to pole ladder. JMIV **50**(1–2), 5–17 (2013)
6. Louis, M., Bône, A., Charlier, B., Durrleman, S.: Parallel transport in shape analysis: a scalable numerical scheme. In: Nielsen, F., Barbaresco, F. (eds.) GSI 2017. LNCS, vol. 10589, pp. 29–37. Springer, Cham (2017). https://doi.org/10.1007/978-3-319-68445-1_4
7. Louis, M., Charlier, B., Jusselin, P., Pal, S., Durrleman, S.: A fanning scheme for the parallel transport along geodesics on Riemannian manifolds. SIAM J. Numer. Anal. **56**(4), 2563–2584 (2018)
8. Micheli, M., Michor, P.W., Mumford, D.: Sobolev metrics on diffeomorphism groups and the derived geometry of spaces of submanifolds. Izvestiya: Math. **77**, 541–570 (2013)
9. Pennec, X.: Parallel transport with pole ladder: a third order scheme in affine connection spaces which is exact in affine symmetric spaces (2018)
10. Postnikov, M.M.: Geometry VI: Riemannian Geometry. Encyclopaedia of Mathematical Sciences, Geometry. Springer, Heidelberg (2001). https://doi.org/10.1007/978-3-662-04433-9
11. Younes, L.: Jacobi fields in groups of diffeomorphisms and applications. Q. Appl. Math. **65**(1), 113–134 (2007)

Author Index